# Holmes Principles of Physical Geology

**Arthur Holmes** DSc LLD FRS
Sometime Professor of Geology and Mineralogy, University of Edinburgh
Fellow of Imperial College, University of London
Vetlesen Prizewinner, 1964

Third Edition, revised by

**Doris L. Holmes** DSc FRSE FGS



Parícutin, Mexico, erupting early in April 1943, six weeks after the first appearance of the volcano. See pp. 207 ff.
*(Three Lions Inc., New York)*

NELSON

Thomas Nelson and Sons Ltd
Lincoln Way Windmill Road
Sunbury-on-Thames
Middlesex TW16 7HP
P.O. Box 73146 Nairobi Kenya
P.O. Box 943 95 Church Street
Kingston Jamaica

Thomas Nelson (Australia) Ltd
19–39 Jeffcott Street
West Melbourne Victoria 3003

Thomas Nelson and Sons (Canada) Ltd
81 Curlew Drive
Don Mills Ontario

Thomas Nelson (Nigeria) Ltd
8 Ilupeju Bypass
PMB 1303 Ikeja Lagos

First edition published in
Great Britain 1944
*Reprinted 18 times*

Second edition published in
Great Britain 1965
*Reprinted 6 times*

Third edition published in
Great Britain 1978



ISBN
0 17 771299 6 (paperback)
0 17 761298 3 (cased)



Cover/jacket photograph by courtesy of NASA

Filmset and printed in Great Britain by
BAS Printers Limited, Over Wallop, Hampshire

# Preface

It would be difficult to find a more attractive and rewarding introduction to the basic concepts of science than physical geology. The activities of our planet may be compared to the combined operations of the four Elements of the ancient Greek philosophers: Fire, Earth, Air and Water, to which we should now add Life. The ever-changing interplay of these operations is responsible for a fascinating variety of natural phenomena, ranging from landscape forms and scenery to the catastrophes brought about by earthquakes, volcanic eruptions, floods and hurricanes, all of which are of daily interest and concern to a high proportion of the world's inhabitants. The effects of these activities, and—so far as current knowledge permits—their causes, are the chief topics of our subject.

In the preface to the first edition I expressed the hope that the book would appeal not only to university students and the senior classes in schools and their teachers, but also to the wide range of general readers whose wonder and curiosity are excited by the behaviour of this mysterious world of ours. My hopes have been surprisingly surpassed. Partly, I expect, this has been because the subject was presented with a minimum of jargon, with constant reference to observational evidence and with copious illustrations. Since I have found this method of treatment to be consistently successful in arousing and developing the interest of students, including those who had no preliminary acquaintance with the elementary principles of science, I have endeavoured to follow it again in this new edition. Partly, too, the timing of the original edition was auspicious, for the book appeared just as the services of geologists and geophysicists began to be required in ever-increasing numbers all over the world.

Writing in 1952 of *The Next Million Years*, Sir Charles Darwin pointed out that 'we are living in the middle of an entirely exceptional period, the age of the scientific revolution'. Throughout this century the rate of scientific publication has been doubling every few years and in many sciences, including the geological group, the rate of increase is itself increasing, like compound interest. It is not generally realised that, of all the geologists who have ever lived, about ninety per cent are now alive and actively at work. For geophysicists the percentage is naturally higher. The curves representing these remarkable increases have been rising much more steeply than those of the world 'population explosion'. Obviously, since the contrast applies to scientists in general, the upward-sloping curves must gradually turn and flatten out, probably with a few fluctuations dependent on future developments in China and Africa. Otherwise there would sooner or later be as many scientists as people. But despite the enormous wealth of geological data still to be collected and the certainty that some major surprises will continue to emerge (e.g. when the various Mohole projects are completed), it is doubtful whether there will ever again be such a profusion of unexpected discoveries concentrated into so short an interval of time as there has been during the last twenty years.

The above considerations serve to explain why the revision on which I started full-time work seven years ago has taken so long; why its completion now instead of a few years ago is more timely; and why it has resulted in what is practically a new book, almost entirely rewritten and of necessity greatly enlarged. Old chapters have been extended and new ones added, to deal with the more significant results of the explorations and researches that have revolutionised geology and geophysics during the last two decades. These include the radiometric dating of minerals and rocks, and the consequent possibility of estimating the rates of some of the earth's long-term activities; the concept of rheidity and its applications to the flow of glaciers, the formation of salt domes and the intrusion of granitic and associated rocks; fluidisation as the key to many aspects of volcanic and igneous geology; the host of complex issues raised by the occurrence of ice ages—some involving the destiny of mankind in the relatively near future and others the totally different distribution of climates in the far-off past; the extraordinary surprises, contradicting all expectations, that have rewarded the explorers of the ocean floors, particularly the astonishing thinness of the oceanic crust and its youthful veneer of deep-sea sediments, paradoxically combined in most places with a far higher rate of heat flow than anyone could have foreseen; the growing evidence that the earth is expanding rather than shrinking; the possible energy sources required to keep the earth's internal 'machinery' going; the maintenance of continents above sea-level and the uplift of plateaux; the contorted structures of mountain ranges and the recognition of gravity gliding as a major factor in the folding of rocks; and finally, the magnetism preserved in ancient rocks as a compass-guide to the wandering of continents.

In terms of observational facts alone the behaviour of the inner earth now appears even more fantastic than when we knew much less about it. In this rejuvenated book I have tried to present a balanced view of the new fields of knowledge which have been so explosively opened up. Every chapter contains exciting stories of man's achievements and speculations, tempered by the sobering reflection that while we may often be wrong, Nature cannot be. It may be noticed that inconsistencies have arisen here and there, when certain problems are approached along different lines. Such difficulties are not glossed over. By bringing them into the open, instead of sweeping them under the carpet of outworn doctrines and traditional assumptions, the reader is helped to see where further research is needed and may even be stimulated to take part in it. Primarily, however, the book has been designed to meet the needs of the same sort of audience as before: this time enlarged, I hope, by those of my former readers who may be finding it increasingly time-consuming to dig out the essentials of recent progress from the avalanche of publications in which they are recorded.

It is a pleasure to acknowledge the considerable help I have myself received in this way from the publications and pre-prints kindly sent to me by friends and correspondents in many parts of the world. By discussing various problems and reading parts of the book, generous assistance has been given by Professor Tom F. W. Barth, Dr Lucien Cahen, Dr Lauge Koch, Professor L. Egyed, Professor Maurice Ewing (and some of his colleagues at the Lamont Geological Observatory), Dr R. W. Girdler, Professor W. Nieuwenkamp and Professor C. E. Wegmann. To all of these I owe my grateful thanks for help, suggestions and constructive criticisms. To my wife and fellow geologist, Dr Doris L. Reynolds, I am, as always, more deeply indebted than can be adequately expressed. Not only has her unselfish devotion ensured the completion of this book, but her professional influence and her flair for the utmost rigour of scientific method have been never-failing sources of inspiration and encouragement. It is, however, only fair to add that I remain entirely responsible for any defects of fact, treatment, judgment or style.

As befits the subject, special care has again been taken to illustrate the book as fully and effectively as possible. Nearly all the line diagrams have been specially drawn, many of them being based on diagrams already published by others, with only slight modifications or translations of the lettering. Particular sources are credited in the captions, and a general expression of my indebtedness and thanks is given here to the many authors whose diagrams have been so freely called upon. In addition to the photographs procured from professional photographers and press agencies (all duly acknowledged in the captions) many striking and instructive subjects have been contributed by friends and official organisations.

It will be noticed that the illustrations are not systematically listed in these preliminary pages. When such lists become unduly long, as would here have been the case, they tend to defeat their purpose. As a practical alternative, which it is hoped will facilitate easy reference, all illustrations are indexed, with their page numbers in italics, under each relevant subject or key word.

Finally, I wish to express my thanks to the staff of my publishers for their enthusiastic and helpful co-operation throughout the production of this book, and particularly to Miss Nina T. Yule and her colleagues for their editorial advice and guidance.

ARTHUR HOLMES

London
*October* 1964

# Preface to Third Edition

IN 1944, when the first edition of this book was published, it concluded with a chapter on continental drift. At that time this was very daring. Apart from masters of drift, like A. L. du Toit, who had proved its geological reality, even the few geologists who were prepared to discuss continental drift in the evening were inclined to dismiss it as too fantastic in the sober light of morning. Physicists, moreover, found it unacceptable.

By the time the second edition was published in 1965, continental drift was a confirmed reality. From remanent magnetism in rocks, geophysicists had determined the positions of the poles for past periods, and proved that the continents have moved both relative to the poles and to one another since the Cretaceous. This then formed the highlight of the second edition together with such evidence as then existed—the migration of volcanic islands for example—that the ocean floors are moving away from the mid-oceanic ridges.

During the following years geophysical discoveries continued to gain impetus, and there were many exciting and even fantastic revelations; movement of the ocean floors was firmly established from palaeomagnetic evidence, and their varying rates of movement determined. The culminating triumph was the evolution of the hypothesis of Plate Tectonics, which indeed might now be described as a theory, because it explains and correlates so many major features of the earth.

These fundamental additions to knowledge have made it necessary to reorganize and rewrite the final chapters of the book and, in conformity, to adapt parts of the early chapters. The chapter on orogenesis has been rewritten in order to present some important but overlooked discoveries that have been made by Emile Argand's successors. Argand's hypothesis, of more than 60 years ago, that collision between Africa and Europe was the cause of the uprise of the Alps is becoming regarded as fundamental to understanding the *cause* of orogenesis. Discoveries by his successors, however, cast doubt on the supposedly predominant role played by compression during orogenesis.

The sections on batholiths, granite, granitization and migmatites, although brief, have been rewritten so as to be of more general application, and 'ring-dykes' have been revised.

I am very grateful to many earth scientists for allowing me to reproduce their illustrations, and to the Officers of the Ulster Museum for generously waiving their royalties on some photographs from the R. Welch collection, of which they hold the copyright. I am particularly indebted to Dr P. S. Doughty, Keeper of Geology at the Ulster Museum, for the trouble he took examining the cliffs for many miles along the Antrim coast road in order to provide me with a photograph of faults showing black basalt downthrown against white chalk, here reproduced as Figure 17.14. To Professor P. McL. D. Duff I owe a debt of gratitude for much help concerning recent work on the production and reserves of petroleum and coal. I should hasten to add, however, that I am entirely responsible for any defects or errors that may have been introduced in presenting this material.

Finally I wish to thank the staff of Thomas Nelson and Sons for their very helpful co-operation, and particularly Dr Dominic Recaldin and Mr. John E. Padfield on the editorial staff.

DORIS L. HOLMES, 1978

# Contents

# Chapter 1

# Science and the World We Live in

In the first place, there can be no living science unless there is a widespread instinctive conviction in the existence of an *Order of Things*, and, in particular, of an *Order of Nature.*

*Alfred North Whitehead* 1927

### Interpretations of Nature: Ancient and Modern

The world we live in presents an endless variety of fascinating problems which excite our wonder and curiosity. The scientific worker, like a detective, attempts to formulate these problems in accurate terms and, so far as is humanly possible, to solve them in the light of all the relevant facts that can be collected by observation and experiment. Such questions as What? How? Where? and When? challenge him to find clues that may suggest possible answers. Confronted by the many problems presented by, let us say, an active volcano (Frontispiece), we may ask: What are the lavas made of? How does the volcano work and how is the heat generated? Where do the lavas and gases come from? When did the volcano first begin to erupt and when is it likely to erupt again?

Here and in all such queries the question What? commonly refers to the stuff things are made of, and an answer can be given in terms of chemical compounds and elements. Not in terms of the four 'elements' of the Greek philosopher Empedocles (about 490–430 B.C.), who considered the ultimate ingredients of things to be Fire, Water, Earth and Air, but chemical elements such as hydrogen, oxygen, silicon, iron and aluminium. The question What? also refers to the names of things, and particularly to their forms and structures: things ranging in size from the elementary particles of atoms to the galaxies of stars that make up the Universe, with our own Earth falling manageably between these extremes of the inconceivably large and small.

The question How? refers to natural processes and events—the way things originate or happen or change—and in human activities to methods and techniques. This question leads to the very root of most natural problems, and satisfactory answers, although amongst the hardest to find, are to the scientist the most rewarding.

Where? refers to everything connected with space, and particularly to the relative positions and distributions of things. The location and distribution of oil and uranium are topical examples of current interest.

The question When? raises all the problems connected with the history of things and events. To unfold the history of the Earth and its inhabitants is the most ambitious aim of geological endeavour.

The scientific worker of today differs in one very important respect from the detective with whom we have compared him. Except when he is concerned with human nature, as the detective invariably is, and perhaps with some types of animal behaviour, the scientific worker nevers asks the question Why? in its strict sense of implying motive or purpose. It is quite useless, for example, to ask why a volcano erupts, or why the sky is blue, or why there are earthquakes, because there is no possible means of finding answers to such questions. Of course we all commonly ask Why? in the loose sense of meaning 'How does it happen or come about . . . ?' and so long as it is clearly understood that this usage is just scientific slang to avoid the appearance of pedantry, no great harm may be done.

Why? always implies the further question Who? Thus the old legends of Ireland tell us that giants were responsible for many natural phenomena.

**Figure 1.1** Giant's Causeway, Co. Antrim, Northern Ireland. View of the Grand Causeway, showing vertical columns. Steeply inclined columns are seen in the cliff to the right of the house. (*Northern Ireland Tourist Board*)

They were stone throwers or builders. One of them is reputed to have flung the Isle of Man into the Irish Sea, Lough Neagh being the place it was taken from. The Giant's Causeway, which is a terrace carved by the weather and the sea from a lava of columnar basalt (Figure 1.1), was 'explained' as the work of the giant Fionn MacComhal (or Finn MacCoul). This defiant stonemason began to construct a causeway across the sea, the better to attack his hated rival Fingal, who had built himself a stronghold of similar construction in the Isle of Staffa (Figure 1.2). Complacent satisfaction with such 'explanations' obviously stifles the spirit of eager enquiry which is essential for the birth and development of science.

It is only during the last century or so that the futility of reading motives and purposes into events has been at all widely realized. The ancients in particular, including Sumerians, Hindus, Babylonians and Greeks, all of whom made remarkable discoveries in mathematics and astronomy, went sadly astray as a result of inventing answers to the dangerous question Why? They regarded the phenomena of Nature as manifestations of power by mythical deities who were thought to behave like irresponsible men of highly uncertain temper, but who might nevertheless be propitiated by suitable sacrifices.

In the Mediterranean region, for example, many of the Olympian gods made familiar by the epics ascribed to Homer (about 850 B.C.) were personifications of various aspects of Nature, including quite a number of geological processes which have been named after them. Poseidon (later identified with the Roman water-god Neptune) was the ruler of the seas and underground waters. As the waters confined below the surface struggled to escape, Poseidon assisted them by shaking the earth and fissuring the ground. He thus became the god of earthquakes. Typhon, the source of destructive tempests, was a 'many-headed monster of malignant ferocity' who was eventually vanquished by the thunderbolts of Zeus, the sky and weather god, who imprisoned him in the earth. Hades, corresponding to Pluto of the Romans, was the deity presiding over the nether regions which share his name. He was not, however, the god of

**Figure 1.2** Fingal's Cave, Island of Staffa, west of Mull, Strathclyde (*Paul Popper Limited*)

subterranean fire. That responsibility was given to Hephaestus, a son of Zeus, who himself controlled fire in the form of lightning. When the Greeks settled in Sicily, Hephaestus was identified with the local volcano-god Vulcan or Volcanus. The eruptions of lava and volcanic bombs from Etna and Stromboli were feared both for their danger and as expressions of the fire-god's wrath.

The epic poems of Homer had biblical authority for the ancient Greeks, and therefore all the more honour is due to Thales of Miletus (about 624–565 B.C.) for his courage in making a clean break with all these traditional beliefs. He regarded the activities of Nature not as indications of supernatural intervention but as natural and orderly events which could be investigated in the light of observation and reason. He thus became the first Greek prophet of science. Having noticed that deposition of silt from the waters of the Nile led to the outward growth of the delta, he developed the hypothesis that Water was the source of earth and everything else. No doubt crude, but probably little more so than a present-day hypothesis that hydrogen is the primordial element. The important point is that the natural and observable activities of water took the place of the imaginary and therefore inscrutable activities of Poseidon. Science had replaced superstition.

Later Heraclitus (about 500 B.C.), celebrated for his philosophy that 'all is perpetual flux and nothing abides', picked on Fire, the most active agent of change, as the fundamental principle behind phenomena. The observable manifestations of fire took the place of the fire-gods. The next step was taken by Empedocles, who saw that both Fire and Water were necessary, for when natural water is heated it evaporates into Air (in this case steam) and leaves a residue of Earth (the material originally in solution). Thus he established the four 'elements' which roughly correspond to our concepts of energy, and the solid, liquid and gaseous states of matter.

Despite this promising start, science languished for two thousand years. The Ionian philosophers who began the scientific quest were well aware that another 'element' was necessary to account for Man. This they called Consciousness or Soul, and some of them regarded it as divine. In concentrating on this aspect of existence, Plato (427–347 B.C.)

and Aristotle (384–322 B.C.) revived the idea that Earth, Sun and Planets were all deities. So, unwittingly, one effect of their unrivalled authority was to lend support to the old Babylonian cult of astrology, as against the sterner discipline of science. Now what has this to do with geology? Two examples will suffice. Herodotus (484–426 B.C.) was a great traveller who made many significant geological observations. He speculated about the effect of earthquakes on landscapes, but nevertheless he thought it quite reasonable to ascribe the earthquakes themselves to Poseidon. Much later Pliny (A.D. 23–79), who, like Empedocles, lost his life while investigating a volcano too closely, 'explained' earthquakes as an expression of the Earth's resentment against those who mutilated and plundered her skin by mining for gold and silver and iron.

With the rise of Christianity there was no longer any incentive to study the ways of Nature. Because men had a complete theory as to *why* things happen, they were not interested in *how* they happen. Moreover, it was widely believed that the Earth had been created ready-made only a few thousand years before, and that it would soon come to an end. Two factors, amongst others, slowly brought this period of stagnation to a close. The Earth inconsiderately failed to come to an end, until at last it began to seem worth while to collect facts about the world instead of merely arguing about ideas. Another factor was a social one. Among the Greeks it had been bad form for a philosopher to become proficient in the craftsmanship and special skills demanded by any form of technology. For all such practical activities slaves were available. However, as the practice of alchemy came increasingly into vogue during the Middle Ages, a philosopher's study was quite likely to become his laboratory. Philosophy and technology—or more generally theory and practice—gradually ceased to be divorced, and with the coming of the Renaissance the occasional and increasing alliance of the two fostered afresh the spirit of science. By his unsurpassed and versatile genius Leonardo da Vinci (1452–1519) firmly established the new pattern. Were he not more renowned as painter and engineer, he would still be famous as a pioneer in many fields of natural science. He recognized that landscapes are sculptured and worn away by erosion; that the fossil shells found in the limestones of the Apennines are the remains of marine organisms that lived on the floor of a long-vanished sea that once extended over Italy; and that it could not have been the Noachian Deluge that swept them into the rocks of which they now form so important a part. But Leonardo was far in advance of his time, and three more centuries had to elapse before the Deluge ceased to obstruct the progress of geology.

Today we think of natural processes as manifestations of energy acting on or through matter. We no longer accept events as the results of arbitrary—and therefore unpredictable—interference by mythological deities. Typhoons and hurricanes are no longer interpreted as the destructive breath of an angry wind-god: they arise from the heating of the air over sun-scorched lands. The source of the energy is heat from the sun. Volcanic eruptions and earthquakes no longer reflect the revengeful behaviour of the gods of underground fire and water: they arise from the stresses and strains of the earth's unstable interior, and from the action of escaping gases and heat on the outer and crustal parts of the globe. The source of the energy lies mainly in the material of the inner earth.

In many directions, of course, and particularly where great catastrophes are concerned, our knowledge is still woefully incomplete. The point is not that we now pretend to understand everything—if we did, the task of science would be over—but that we have faith in the orderliness of natural processes. The steadily accelerating researches of the last two or three centuries have unfailingly justified our belief that Nature is understandable: understandable in the sense that if we ask her questions by way of appropriate observations and experiments, she will answer truly and reward us with discoveries that endure. But it must never be forgotten that Nature is the most perfect of expert witnesses. It is often far from easy to find the right questions and to ask them in the right way. For this reason, to say nothing of human fallibility, the discipline of science does not preclude the making of errors. 'What it does preclude,' as Norbert Wiener has expressed it, 'is the retention of an error which has clearly and distinctly betrayed its wrongness.'

### The Major Fields of Scientific Study

The questions we ask when faced with a volcano in eruption are typical of the kinds suggested by all natural phenomena. They indicate that—in general terms—scientific investigation is concerned

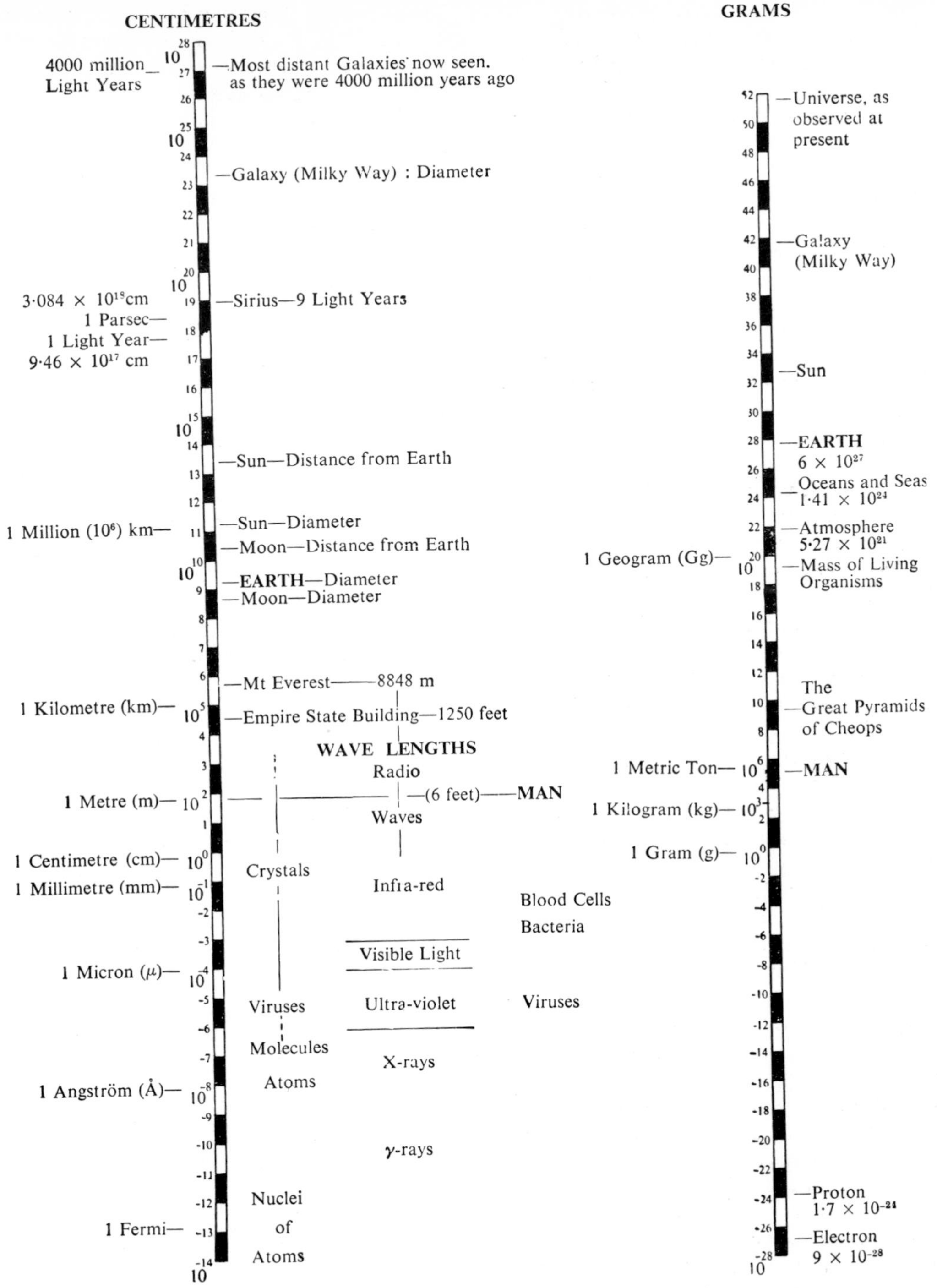


**Figure 1.3** The relative place of the Earth and Man in the Universe expressed in powers of ten centimetres and grams

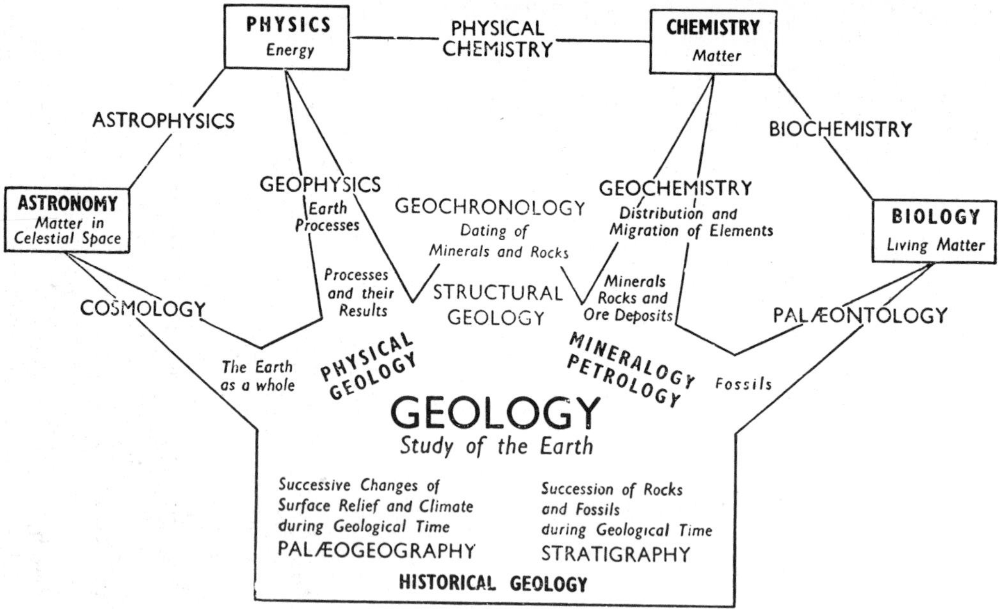


**Figure 1.4** Subdivisions of the science of geology and their relation to other sciences

with the manifestations and transformations of matter and energy in space and time. Put more briefly and philosophically, science is concerned with relationships between events.

Of all the sciences *physics* is the most fundamental, for it deals with all the manifestations of energy and with the nature and properties of matter in their most general aspects, and on all scales from sub-atomic particles to the observable limits of the expanding universe (Figure 1.3). It overlaps to some extent with *chemistry*, which is particularly concerned with the composition and interactions of substances of every kind in terms of atoms and molecules, elements and compounds. *Biology* is the science of living matter. The nature of life still remains an elusive mystery. As examples of a vast range of phenomena that are quite inexplicable in terms of what we know of matter and energy, despite the vaunted powers of electronic robots, it is sufficient to recall our faculties for transforming physical waves into the conscious miracle of colours and sounds. Nevertheless there are innumerable aspects of organisms and their fossil remains that can be investigated by scientific methods with astonishing success.

All the other sciences have their own fields of interest, but these 'fields' overlap in all directions and trespassing is regarded as welcome co-operation. *Astronomy*, with the unfathomed universe of stars and nebulae as its field of study, deals essentially with the distribution and movements of matter in space on a celestial scale. Its interest in the earth is limited to the purely planetary aspects of our globe. From an astronomical point of view the earth may seem to be only an insignificant speck in the immensity of space. But space has no downward limit, and the earth can equally well be seen as a galaxy of atoms, as Figure 1.3, designed to restore our sense of proportion, clearly shows. As the home of mankind and the mother of all the life we known, we naturally regard the earth as a supremely important field for special investigation. So we come to the science devoted to the earth, appropriately known as *geology* (from the Greek *gaia* or *ge*, the ancestral Earth-goddess of Greek mythology; and *logy*, a suffix denoting 'knowledge of').

From the earliest days of exploration *geography* has been recognized as the study of the 'home of mankind'. Modern geography focuses attention on man's physical, biological and cultural environ-

**Figure 1.5** Folded strata of Jurassic age (see p. 98). The Stair Hole, west of Lulworth Cove, Dorset (*F. N. Ashcroft*)

ment and on the relationships between man and his environment. The study of the physical environment by itself is physical geography, which includes consideration of the surface relief of the continents and ocean floors (geomorphology), of the seas and oceans (oceanography), and of the air (meteorology and climatology), all of which have developed into important members of the commonwealth of sciences on their own account. As a major science, however, geology deals not merely with the land-forms and other surface features of the earth's rocky crust, but with the structure and behaviour of every part of the earth, with special reference to the rocks and structures of the visible crust and all that can be learned from them.

## The Scope and Subdivisions of Geology

Modern geology has for its aim the deciphering of the whole evolution of the earth and its inhabitants from the time of the earliest records that can be recognized in the rocks right down to the present day. So ambitious a programme requires much subdivision of effort, and in practice it is convenient to divide the subject into a number of branches, as shown in Figure 1.4, which also indicates the chief relationships between geology and the other major sciences. The key words of the main branches are the *materials* of the earth's rocky framework (mineralogy and petrology), and their *dispositions*, i.e. their forms, structures and interrelationships (structural geology); the geological *processes* or machinery of the earth, by means of which changes of all kinds are brought about (physical geology); and finally the succession of these changes in time, or the *history* of the earth (historical geology).

The earth is made up of a great variety of materials, such as air, water, ice and living organisms, as well as minerals and rocks and the useful deposits of metallic ores and fuels which are associated with them. The relative movements of these materials (wind, rain, rivers, waves, currents and glaciers; the growth and movements of plants and animals; and the movements of hot materials inside the earth, as witnessed by volcanic activity) all bring about changes in the earth's crust and on its surface. The changes involve the development of new rocks from old; new structures in the crust; and new distributions of land and sea, mountains

**Figure 1.6** A long sequence of horizontal strata exposed by river erosion; looking up Bright Angel Canyon, a tributary of the Grand Canyon of the Colorado River (*U.S. Geological Survey*)

and plains, and even of climate and weather. The scenery of today is only the latest stage of an ever-changing kaleidoscopic series of widely varied landscapes—and seascapes. *Physical geology* is concerned with all the terrestrial agents and processes of change and with the effects brought about by them. This branch of geology is by no means restricted to geomorphology, as we have seen. Its main interest is in the machinery of the earth, and in the results, past and present, of the various processes concerned, all of which are still actively in operation at or near the earth's surface or out of sight in the depths. Of these results the changing positions of continents and oceans, of fold mountains, rift valleys and ocean troughs, are important examples. Others are the rock structures—such as folds, Figure 1.5—that have resulted from movements and deformation of the earth's crust. *Tectonics*, the study of these structures, is an important part of *structural geology*, which is also concerned with the forms and structures that characterize rocks when they are first formed.

Changes of all kinds have been going on continuously throughout the lifetime of the earth—that is, for something like 4600 million years. To the geologist a rock is more than an aggregate of minerals; it is a page of the earth's autobiography with a story to unfold, if only he can read the language in which the record is written. Placed in their proper order from first to last (*stratigraphy*, Figure 1.6), and dated where possible by determining the ages of radioactive minerals and rocks (*geochronology*), these pages embody the history of the earth. Moreover, it is familiar knowledge that many beds of rock contain the remains or impressions of shells or bones or leaves. These objects are called fossils, a term that was first applied by Agricola (1494–1555) to anything of interest dug out of the ground, including minerals. Since the end of the eighteenth century, however, the term has been used only for the relics of animals and plants that inhabited the earth in former times. *Palaeontology* is the study of the remains of these ancestral forms of life, some of which, resembling certain types of seaweed, can be traced back for at least 3000 million years. Thus we see that *historical geology* deals not only with the nature and sequence of events brought about by the operation of the physical processes, but also with the history of the long procession of life through the ages.

Geology is by no means without practical importance in relation to the needs and industries of mankind. Indeed, the search for fuels has always been a powerful stimulant to geological progress. During the eighteenth century it became increasingly recognized that the work done by horses and windmills and waterwheels could be enormously increased by utilizing the energy stored in coal. Prospecting for coal, and later for the ores of iron and other useful metals, gave geology its first effective start nearly two hundred years ago. During the present century the search for oil, uranium and other sources of 'atomic fuel' have led to renewed geological efforts. With increasing demand for fuel, and depletion of reserves of oil on land areas, the co-operation of geophysicists has resulted in discoveries of gas and oil in the rocks of continental shelves. Texas and Louisiana have long been supplied with oil from this source, and Britain already receives oil and gas from beneath the North Sea. According to estimates made in early 1975 the recoverable reserves of oil in the British sector of the North Sea shelf area are of the order of 12,000 to 14,400 million barrels, and the recoverable reserves of gas from that source about $1158 \times 10^9$ cubic metres. Geologists are also directly concerned with the vital subject of water supply. Many engineering projects, such as tunnels, reservoirs, hydroelectric schemes and docks, call for geological advice in the selection of sites and materials. In these and many other ways geology is applied to the service of mankind.

Although geology has its own laboratory methods for studying minerals, rocks and fossils, it is essentially an open-air science. It attracts its followers to crags and waterfalls, glaciers and volcanoes, beaches and coral reefs, ever farther and farther afield in the search for information about the earth and her often puzzling behaviour. Wherever rocks are to be seen, for example, in cliffs and quarries, their arrangement and sequence can be observed and their story deciphered. The hidden parts of the earth beneath the oceans, however, are the province of recent geophysical investigations. Since 1960, geophysical discoveries about the changed positions of the magnetic poles relative to the continents, and the discovery of magnetic anomalies displayed by the basaltic crust of the ocean floors, combined with increasing accuracy in radiometric dating of rocks, have resulted in a new understanding of the earth, and a veritable revolution in geological thought.

### Selected References


ADAMS, F. D., 1938, *The Birth and Development of the Geological Sciences*, Dover Publications, London.

BEVERIDGE, W. I. B., 1950, *The Art of Scientific Investigation*, Heinemann, London.

FARRINGTON, B., 1953, *Greek Science*, Pelican Books, London.

GAMOW, G., 1963, *A Planet Called Earth*, Viking Press, New York.

GEIKIE, A., 1905, *The Founders of Geology*, Macmillan, London.

GILLISPIE, C. C., 1951, *Genesis and Geology*, Harvard University Press, Cambridge, U.S.A.

MERRILL, G. P., 1924, *The First Hundred Years of American Geology*, Yale University Press.

PEGRUM, R. M., REES, C., and NAYLOR, D., 1975, *Geology of the North-West European Continental Shelf*, vol. 2, The North Sea, Graham, Trotman, Dudley, London.

TOMKEIEFF, S. I., 1950, 'Geology in historical perspective', *Advancement of Science*, vol. 7, pp. 63–7.

WHITEHEAD, A. N., 1927, *Science and the Modern World*, Cambridge University Press, London.

# Chapter 2
# The Dynamic Earth

Let the great world spin for ever down
the ringing grooves of change.

*A. Tennyson* (1809–1892)

## The Outer Zones of the Earth

As it presents itself to direct experience, the earth can be physically described as a ball of rock (the crust), partly covered by water (the hydrosphere) and wrapped in an envelope of air (the atmosphere). To these three physical zones it is convenient to add a biological zone (the biosphere). The system, crust, atmosphere and hydrosphere is usually regarded as being a closed system, i.e. to be in a steady state. This means that losses from any members of the system are balanced by additions to the others. Only hydrogen and helium are light enough to escape from the system.

The *atmosphere* is the layer of gases and vapour which envelops the earth. It is essentially a mixture of nitrogen and oxygen with smaller quantities of water vapour, carbon dioxide, and inert gases such as argon. Geologically it is important as the medium of climate and weather, of wind, cloud, rain and snow.

The *hydrosphere* includes all the natural waters of the outer earth. Oceans, seas, lakes and rivers cover about three-quarters of the surface. But this is not all. Underground, for a few hundred metres in some places, the pore spaces and fissures of the rocks are also filled with water. This groundwater, as it is called, is tapped in springs and wells, and is sometimes encountered in disastrous quantities in mines. Thus there is a somewhat irregular but nearly continuous mantle of water around the earth, saturating the rocks, and over the enormous depressions of the ocean floors completely submerging them. If it were uniformly distributed over the earth's surface it would form an ocean nearly 2750 m deep.

The *biosphere*, the sphere of life, is probably a less familiar conception. But think of the great forests and prairies with their countless swarms of animals and insects. Think of the tangles of seaweed, of the widespread banks of molluscs, of reefs of coral and shoals of fishes, Add to these the inconceivable numbers of bacteria and other microscopic plants and animals. Myriads of these minute organisms are present in every cubic centimetre of air and water and soil. Taken altogether, the diverse forms of life constitute an intricate and ever-changing network, clothing the surface with a tapestry that is nearly continuous. Even high snows and desert sands fail to interrupt it completely, and lava fields fresh from the craters of volcanoes are quickly invaded by the pressure of life outside. Such is the sphere of life, and both geologically and geographically it is of no less importance than the physical zones. Amongst its many products are coal, oil, natural gas, most of the oxygen of the air we breathe, and limestones in great abundance (Figure 2.1).

The *crust* (Figure 2.2) is the outer shell of the solid earth. It is made up of rocks in great variety. On the lands its uppermost layer is commonly a blanket of soil or other loose deposits, such as desert sands. The idea that the earth has a crust can be traced back to Descartes (1596–1650), who thought of it as a shell of heavy rock, covered with lighter sands and clays and resting on a metallic interior. Leibnitz (1646–1716) suggested that the earth had cooled from an incandescent state and that the outside rocky part—the first to cool and consolidate—had formed a crust which covered a still molten interior. This view, that the crust is a relatively thin layer of solid rock on a liquid

**Figure 2.1** Two celebrated products of the Biosphere. Many millions of years ago the coin-shaped shells of innumerable generations of nummulites accumulated on the floor of a vanished sea to form the thick and widespread deposits of Nummulitic Limestone that is now the bedrock of much of the Egyptian and Libyan Desert. From a conspicuous outcrop of this rock the Sphinx was carved, probably during the reign of Chephren, about 2900 B.C., whose Pyramid is seen close by, built of gigantic blocks of the same Nummulitic Limestone quarried from the hills on the other (east) side of the Nile. Stone and builders alike were once part of the Biosphere. (*Syndication International*)

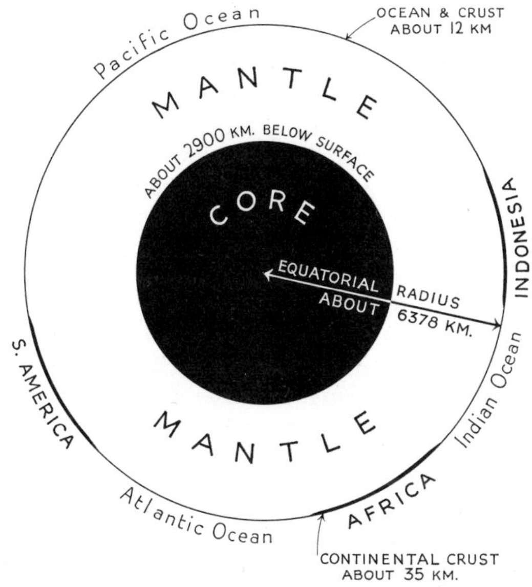


**Figure 2.2** Equatorial section through the Earth, showing crust (continental and oceanic), mantle and core

interior, became widely prevalent until about a century ago, when it was shown to be unsound.

At that time it began to be suspected that if the bulk of the interior were liquid the oceanic tides would rise and fall through a smaller range than they actually do. Tides demonstrate a very conspicuous movement of sea water relative to land. Clearly, if land and sea yielded equally to the attractive forces exerted by the moon and sun—as they would do if the interior were liquid—then there would be no relative movement and no advance and retreat of the sea. Tides do occur in the earth's crust, but the rise and fall of the land is imperceptible to its inhabitants and can be detected and measured only by special techniques. This elastic response of the earth to the moon and sun turns out to be just what it would be if the earth consisted throughout of solid steel. This remarkable conclusion, first reached by Kelvin in 1862, undermined the term 'crust' by depriving it of its original significance. The term was never abandoned, however, and by another route it has now acquired a precise meaning which is generally accepted.

## The Crust and Inner Zones of the Earth

In the light of modern knowledge the structure of the earth's interior has turned out to be not unlike the model conceived by Descartes. The deep interior is called the *core*. The surrounding zone of

**Figure 2.3** Basaltic lava-flows of the Antrim plateau (see Fig. 12.8) looking eastward from the Giant's Causeway. Above the cliff path three tiers of columns can be seen, corresponding to three successive lava-flows, the lowest of which is that of the Causeway itself. The older lavas below the path are not columnar. (*J. Allan Cash, Northern Ireland Tourist Board*)

'heavy rock' is known as the *mantle* and it extends up to a boundary surface above which the rocks have physical properties that are very different from those of the mantle. This outermost envelope is the *crust*. The core has metallic properties and a very high density. It was at one time thought to be composed of material having a chemical composition similar to that of the mantle, but changed into a physically different condition by intense compression. Geophysical evidence now suggests that it is composed largely of iron. It is likely, however, that there is some admixture of nickel, or possibly magnesium oxide in a metallic phase, whilst an admixture of silica is favoured by some geophysicists.

By 'X-raying' the earth with its own earthquake waves, or with similar waves specially generated for the purpose by controlled explosions, it is possible to estimate with fair accuracy the depth at which the mantle material begins in different parts of the world. The boundary surface or *discontinuity* between the mantle and the crust was discovered in 1909 by A. Mohorovičić. Since then it has become familiarly known as the Mohorovičić discontinuity, or as the M-discontinuity or *Moho*. In passing through the rocks immediately above this surface, earthquake waves reach a velocity of about 7·2 km/s, whereas in the rocks below the M-discontinuity the velocity suddenly jumps to about 8·1 km/s. Obviously the crust can only be defined with precision as consisting of the varied assemblage of rocks overlying the M-discontinuity and thus forming an envelope surrounding the mantle.

The dominant rocks occurring in the crust fall into two contrasted groups:

(*a*) A group of light rocks, including granite and related types, and sediments such as sandstones and shales, forming an assemblage having an average specific gravity or density of about 2·7. Chemically these rocks, again on average, are very rich in *si*lica (65–75 per cent), while *al*umina is the most abundant of the remaining constituents. Since it is often desirable to refer to them as a whole, these crustal rocks are collectively known by the mnemonic term *sial*.

(*b*) A group of dark and heavy rocks, consisting

mainly of basalt and related types (having a density of about 2·8–3·0), collectively known as *basic* rocks (with about 50 per cent of silica), but also including certain still heavier rocks (with a density up to about 3·4), which are distinguished as *ultrabasic* rocks (with about 40–45 per cent of silica). In these rocks *si*lica is still the most abundant single constituent, but *i*ron oxides and *ma*gnesia, singly or together, generally take second place, and the whole group is conveniently known as *sima*.

As proposed by Suess (1831–1914), these group terms were intended to denote the dominant materials of the crust (sial) and the mantle (sima). However, further discoveries have shown that the real structure is not so simple. The most notable of these has been the discovery that practically no trace of sial can be detected in the crust underlying the deep ocean floors.

Apart from a thin carpet of loose sediments and organic oozes, the sub-oceanic crust seems to be made almost entirely of basic rocks, most of them probably submarine flows of basalt. This basaltic layer is only a few miles thick (5–6 km); then comes the M-discontinuity. Samples of ultrabasic rocks thought to represent the hidden sima of the mantle are brought to the surface by the action of many oceanic volcanoes, such as those of Hawaii. But since the basaltic lavas of these volcanoes must themselves have arisen from the heated depths of the mantle, the term *sima* appropriately reminds us that under the Pacific and the other oceans, both crust and mantle are closely related.

In the early days of crustal exploration by earthquake waves it was thought that the basaltic layer of the ocean floor continued beneath the continents: that there was, in fact, a world-wide basaltic layer, surmounted here and there by slabs of sial which were the continents. This interpretation has also turned out to be too simple, in the sense that sialic and basaltic rocks are so inextricably mixed in the continental crust, both mechanically and chemically, that it would now be misleading to picture the continents as having everywhere a basaltic layer as their foundation. Sialic rocks are undoubtedly the dominant materials of the continental crust down to a depth of many kilometres; but the continents also have basaltic volcanoes, and in some places, such as the Antrim plateau, the sial is covered by thick accumulations of basaltic lava-flows (Figure 2.3). Apart from these local volcanic accidents, however, it is safe to say that in most of the regions so far tested basaltic and other basic rocks of the sima group become increasingly abundant in the lower part of the continental crust.

## Continents and Ocean Floors

The surface of the crust reaches very different levels in different places. The areas of land and sea floor between successive levels have been estimated, and the results can be graphically represented as shown in Figure 2.4. From this diagram

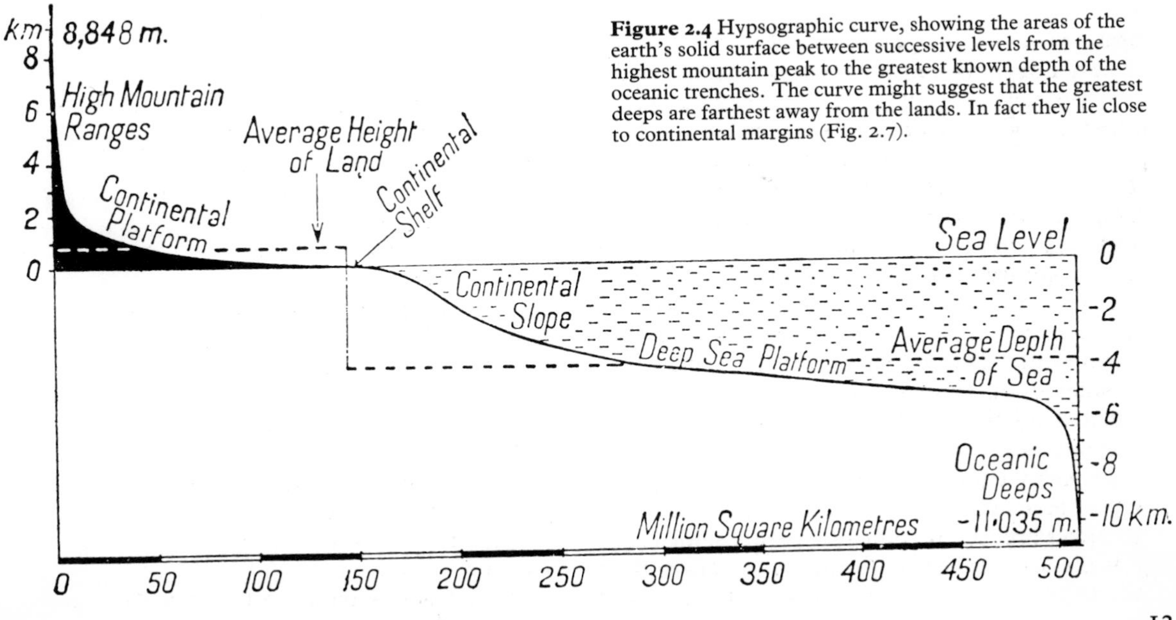


**Figure 2.4** Hypsographic curve, showing the areas of the earth's solid surface between successive levels from the highest mountain peak to the greatest known depth of the oceanic trenches. The curve might suggest that the greatest deeps are farthest away from the lands. In fact they lie close to continental margins (Fig. 2.7).

**Figure 2.5** View of the Everest group of peaks, seen from a height of about 6000 m. From left to right: Everest, 8848 m; Lhotse, 8501 m; Nuptse, 7827 m. The valley between Everest and Nuptse is occupied by the Khumbu Glacier. (*Royal Geographical Society*)

it is clear that there are two dominant levels: the continental platform and the oceanic or deep-sea platform. The slope connecting them, which is actually quite gentle, is called the continental slope.

The continental platform includes a submerged outer border known as the *continental shelf*, which ranges in width up to about 1500 km, but may be absent along mountainous coasts. The older rocks which form the basements of the shelves are overlain by sediments with a thickness up to about 2 km. At one time it was thought that the sediments of the shelves became finer and finer in grain size with increasing distance from the shore line. As a result of investigations begun during the Second World War it has been discovered that it is only where the grain-size is sorted by wave-action, down to a depth of about 20 m, that the sediments become finer grained away from the shore. The shelves are mostly overlain by coarse sands and the shells of molluscs.

Structurally, the real ocean basins commence, not at the visible shore line, but at the edge of the shelf. The basins, however, are more than full, and the overflow of sea water inundates nearly 26 million square kilometres of continental shelf. The North Sea, the Baltic and Hudson Bay are examples of shallow seas (epicontinental seas) which lie on the shelf. It is of interest to notice that during the Ice Age, when enormous quantities of water were abstracted from the oceans to form the great ice sheets that then lay over Europe and North America, much of the continental shelf must have been land. Conversely, if the ice now covering Antarctica and Greenland were to melt away, the sea-level would rise and the continents would be still further submerged.

Economically, the continental shelves are of prime importance; the epicontinental seas provide fertilizers, and add considerably to the world's food resources, whilst the shelves themselves are a source of oil and gas. At present about one fifth of the world's oil and gas comes from continental shelves, and progressively more will be obtained from them in the future as supplies are used up.

The continents themselves have a varied relief of plains, plateaus and mountain ranges, the latter rising to a maximum height of 8848 m in Mount Everest (Figure 2.5). The ocean floors, once thought to be monotonous slopes and plains, are characterized by submarine basaltic mountain

# Some Numerical Facts about the Earth

## Land

| *Greatest known height:* | Feet | Metres |
|---|---|---|
| Mount Everest | 29,028 | 8,848 |
| Average height | 2,757 | 840 |

## Oceans and Seas

| *Greatest known depth:* | Feet | Metres |
|---|---|---|
| Marianas Trench | 36,204 | 11,035 |
| Average depth | 12,460 | 3,808 |

## Size and Shape

| | Miles | Km |
|---|---|---|
| Equatorial semi-axis, $a$ | 3,963·2 | 6,378·2 |
| Polar semi-axis, $b$ | 3,950·0 | 6,356·8 |
| Mean radius | 3,956·4 | 6,371·0 |
| Equatorial circumference | 24,902 | 40,076 |
| Polar (meridian) circumference | 24,857 | 40,009 |
| Ellipticity, $(a+b)/a$ | 1/298 | |

## Area

| | Millions Sq. Miles | Millions Sq. Km |
|---|---|---|
| Land (29·22 per cent) | 57·5 | 149 |
| Ice sheets and glaciers | 6 | 15·6 |
| Oceans & seas (70·78 per cent) | 139·4 | 361 |
| Land plus continental shelf | 68·5 | 177·4 |
| Oceans & seas minus continental shelf | 128·4 | 332·6 |
| Total area of the earth | 196·9 | 510·0 |

## Volume, Density and Mass

| | Average thickness or radius (km) | Volume ($\times 10^6$ km³) | Mean Density (g/cm³) | Mass ($\times 10^{24}$g) |
|---|---|---|---|---|
| Atmosphere | — | — | — | 0·005 |
| Oceans and seas | 3·8 | 1,370 | 1·03 | 1·41 |
| Ice sheets and glaciers | 1·6 | 25 | 0·90 | 0·023 |
| Continental crust* | 35 | 6,210 | 2·8 | 17·39 |
| Oceanic crust† | 8 | 2,660 | 2·9 | 7·71 |
| Mantle | 2,881 | 898,000 | 4·53 | 4,068 |
| Core | 3,473 | 175,500 | 10·72 | 1,881 |
| Whole Earth | 6,371 | 1,083,230 | 5·517 | 5,976 |

*Including continental shelves

†Excluding continental shelves

## Conversion factors

| | |
|---|---|
| 1 miles = 1·609 km | 1 km = 0·621 mile |
| 1 foot = 0·3048 metre (m) | 1 m = 3·281 feet |
| 1 sq. mile = 2·59 km² | 1 km² = 0·386 sq. mile |
| 1 cubic mile = 4·17 km³ | 1 km³ = 0·24 cubic mile |

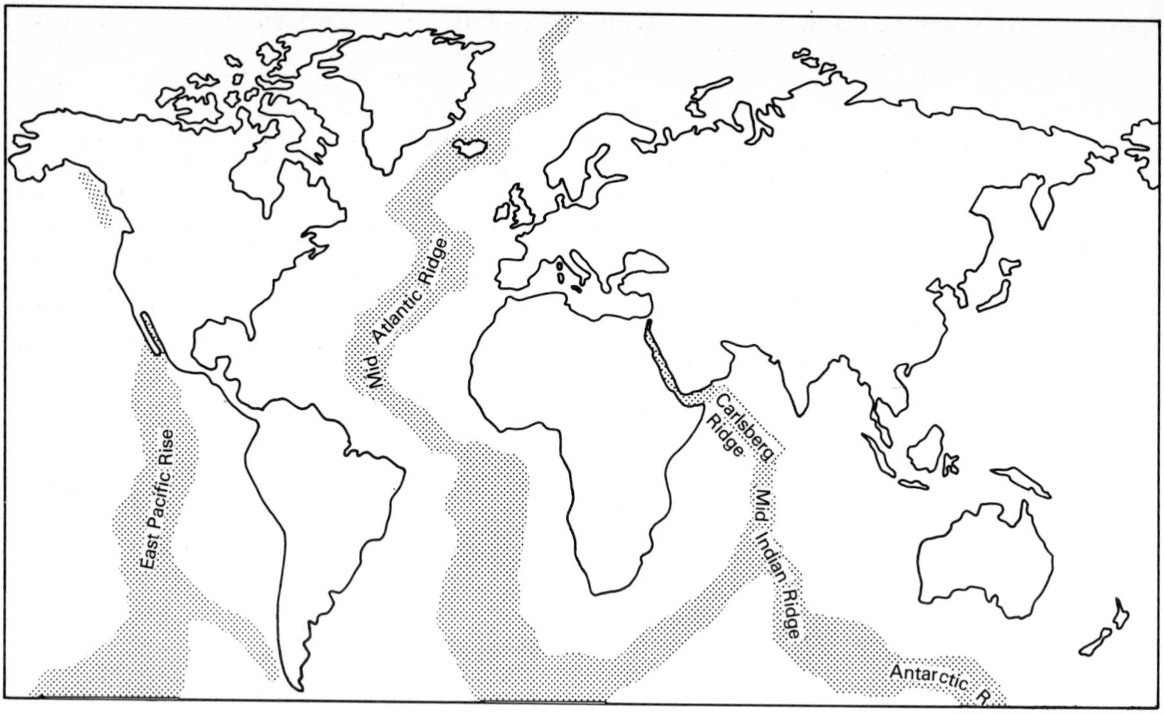


**Figure 2.6** Submarine mountain chains extending around the earth as mid-oceanic ridges that rise to heights of three or four thousand metres above the ocean plains. Unlike continental mountains, submarine mountains are formed of basalt.

ranges which encircle the earth for more than 40,000 km (Figure 2.6). This underwater mountain chain extends along the middle of the oceans and is known as the mid-oceanic ridge or rise.

Other features of the ocean floors are great numbers of sea-mounts, representing the deeply denuded ruins of ancient volcanic islands (Figures 24.32 and 26.10). Submarine canyons, comparable to the Colorado canyon, are of common occurrence, and deep ocean trenches, the subject of many recent investigations, carry the ocean floor down to more than twice its average depth, the greatest depth so far determined being 11,033 m in the Nero deep in the Mariana Trench (Figure 2.7).

From the figures given above it is clear that the total vertical range of the surface of the crust (continental and oceanic) is well over 19,881 m. To grasp the true relationship between the surface relief and the earth itself, draw a circle with a radius of 5 cm, representing 6,371 km, then the thickness of the outline of the circle represents 32 km. On this scale the relief is all contained well within the thickness of a pencil line (cf. Figure 2.2).

## The Shape of the Earth

The first voyage around the world, begun at Seville by Magellan in 1519 and completed at Seville by del Cano in 1522, established beyond dispute that the earth is a globe. Today it is possible to girdle the earth, like *Puck*, 'in forty minutes' and to photograph its surface at heights from which the curvature of the globe is plainly to be seen (Figure 2.8). Pythagoras (about 530 B.C.) was probably the first to consider the possibility that the earth might be a sphere. By observing the approach of ships from beyond the horizon—first the masts and sails and then the hull—he realized that the surface of the sea is not flat, but curved. Three centuries later, when it was already known that the distance of the sun was so great that the direction of the sun's rays at any moment could be regarded as parallel, Eratosthenes (276–196 B.C.), the Chief Librarian at Alexandria, devised a simple and elegant method for estimating the size of the globe. He had heard that at Syene on the Nile (the Aswan of today) the sun shines vertically at noon on Midsummer's Day, so that a vertical stick or plumb line then throws no shadow. He observed, however, that at Alexandria, roughly 800 km to the north of Syene, there were very perceptible shadows at that time. Figure 2.9 illustrates the conditions, but with greatly exaggerated angles and lengths. At Alexandria a plumb line having a length AB would throw a shadow of length AC. These two lengths determine the angle ABC which, on the simplifying assumptions made, equals the angle SOA. Eratosthenes made the necessary measurements and found the angle ABC to be just over 7°, or almost exactly one fiftieth of 360°. The approximate length of the whole circumference would therefore be fifty times the

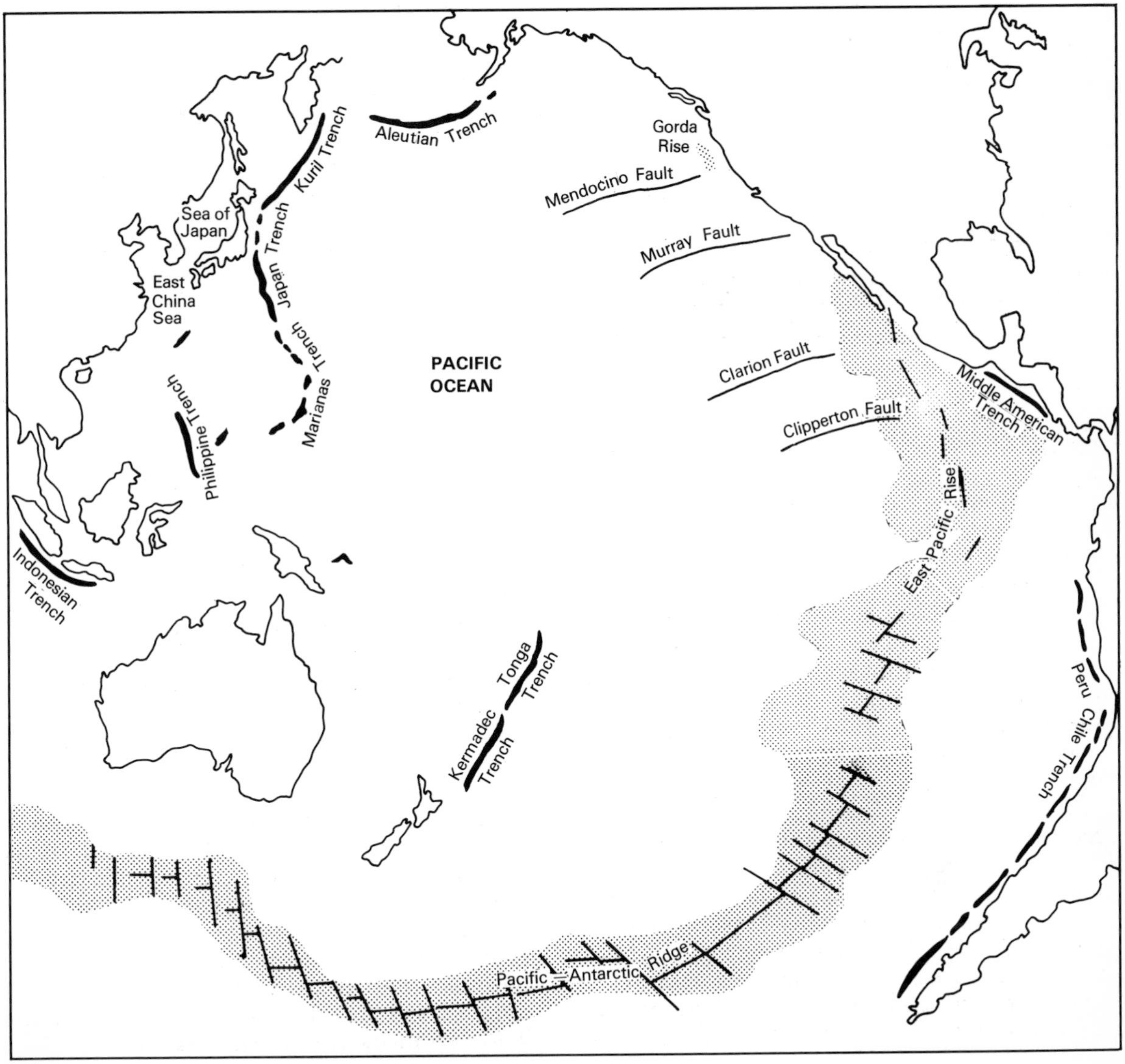


**Figure 2.7** Oceanic trenches (black) encircling the Pacific Ocean. The line along the middle of the East Pacific and Pacific–Antarctic ridges represents the rift along which basaltic magma rises and moves away on either side as new ocean floor. The lines intersecting the rift are a diagrammatic illustration of transform faults (pp. 620–2).

**Figure 2.8** Mosaic of photographs, covering the area within the semicircle of the index map, taken from a rocket near the summit of its flight, 16 km above the earth, on 5 October 1954. Apart from its spectacular, but now familiar, demonstration of the curvature of the earth, the outstanding feature of the mosaic photograph is the portrait it provides of an unexpected tropical storm. The cloud spirals (upper left) represent a hurricane which invaded Texas and adjoining States from the Gulf of Mexico. At the time of the rocket flight the storm, already 1600 km across, was entirely a high-altitude disturbance with no surface winds from which its existence could have been recognized at weather stations on the ground. (*Photograph obtained by the United States Naval Research Laboratory*)

distance between Syene and Alexandria—that is, $50 \times 800 = 40{,}000$ km. Eratosthenes measured the distance in *stades*, his result for the circumference being 252,000 *stades*. It must, however, be remarked that the result is less accurate than it appears to be, because Alexandria lies well to the west of the meridian through Syene, while Syene itself is several kilometres north of the Tropic of Cancer, where the midsummer sun shines vertically.

The reason for the spherical shape of the earth became clear when Newton formulated his law of gravitation. All the particles of the earth are pulled towards the centre of gravity and the spherical shape is the natural response to the maximum possible concentration. Even if a body the size of the earth were stronger than steel, it could not maintain a shape such as, let us say, that of a cube. The pressure exerted by the weight of the edges and corners would squeeze out material in depth. Equilibrium would be reached only when the faces had bulged out, and the edges and corners had sunk in, until every part of the surface was equidistant from the centre.

The earth is not exactly spherical, however. Again it was Newton who first showed that,

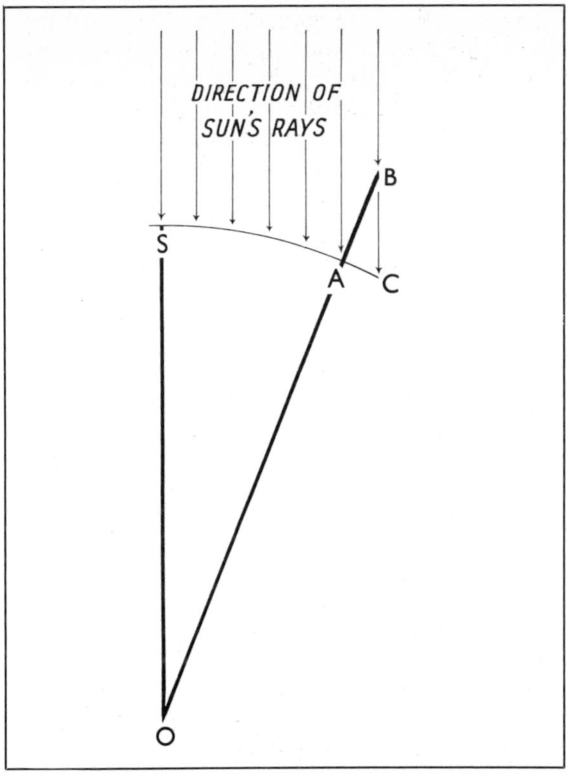


**Figure 2.9** Illustrating the method devised by Eratosthenes for measuring the circumference of the earth. S represents Syene, and A Alexandria. It is assumed that the arc SA lies on a meridian and that O is the centre of the earth. (Not to scale)

because of the earth's daily rotation, its matter is affected not only by inward gravitation, but also by an outward centrifugal force, which reaches its maximum at the equator. He inferred that there should be an equatorial bulge where the apparent value of gravity was reduced, and a complementary polar flattening, where the centrifugal forces becomes vanishingly small. Clearly, if this were so, the length of a degree of latitude along a meridian would be greatest across the poles, where the curvature is flattened, and least across the equator, where the curvature bulges out. It is of interest to notice that Newton's inference was at variance with the few crude measurements that had been made. According to these the earth was shaped, not like an orange with a short polar axis, but like a lemon with a long polar axis. To settle the matter the French Academy in 1735 dispatched a surveying expedition to the neighbourhood of Chimborazo in the Andes of what is now Ecuador, and followed it up in 1736 by another to Lapland. The results showed that Newton was right. It is, moreover, highly significant that before these expeditions returned, the celebrated mathematician, Clairaut, had calculated what the shape of the earth would be, assuming the earth to be a fluid and subject only to the effects of its own rotation and gravitational attraction. The ellipsoid of rotation now internationally adopted for surveying purposes as most closely representing the real shape of the earth corresponds almost exactly to that calculated by Clairaut.

To sum up: if the surface of the earth were everywhere at sea-level its shape—the *geoid* or figure of the earth—would closely approximate to that of an ellipsoid of rotation (i.e. an oblate spheroid) with the equatorial axis 42·8 km longer than the polar axis, as estimated from data derived from satellites. From such data it is now known that the polar axis is slightly longer from the centre of the earth to the north pole, than from the centre to the south pole, and the earth is now sometimes described as pear-shaped (see p. 711). The deviation in its form from that of an oblate spheroid, however, is very small.

How is it, then, that the earth is not exactly a spheroid? The reason is that the crustal rocks are not everywhere of the same density. Since the equatorial bulge is a consequence of the relatively low value of gravity around the equatorial zone, it follows that there should be bulges in other places where gravity is relatively low—that is to say, wherever light, sialic rocks make up an appreciable part of the crust. Such places are the continents. On the other hand, wherever the crust is composed of heavy rocks gravity is relatively high and the surface should be correspondingly depressed. Such regions are the ocean basins.

The earth continuously tends towards a state of gravitational equilibrium. If there were no rotation and no lateral differences in the density of the rocks, the earth would be a sphere. As a result of rotation it becomes a spheroid. As a further result of differences of density and thickness in the crustal rocks and in the underlying mantle, continents, mountain ranges and oceanic basins occur as irregularities superimposed upon the surface of the spheroid.

### Isostasy

For the ideal condition of gravitational equilibrium that controls the heights of continents and ocean floors, in accordance with the densities of their underlying rocks, the term *isostasy* (Gr.

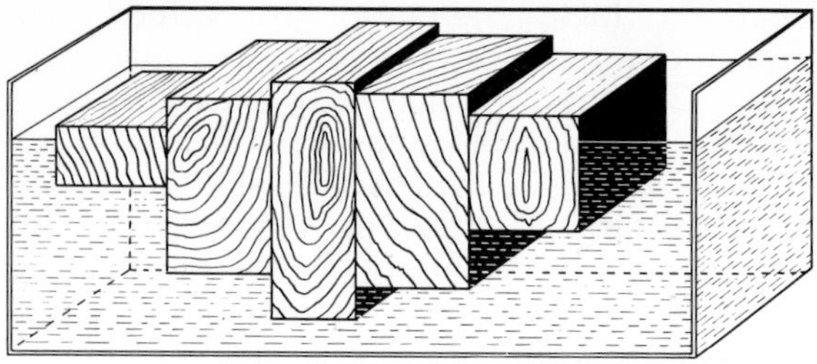

**Figure 2.10** Wooden blocks of different heights floating in water (shown in front as a section through the tank), to illustrate the concept of isostatic balance between adjacent columns of the earth's crust

*isostasios*, 'in equipoise') was proposed by Dutton, an American geologist, in 1889. The idea may be grasped by thinking of a series of wooden blocks of different heights floating in water (Figure 2.10). The blocks emerge by amounts which are proportional to their respective heights; they are said to be in a state of hydrostatic balance. Isostasy is the corresponding state of balance between extensive blocks of the earth's crust which rise to different levels and appear at the surface as mountain ranges, plateaus, plains or ocean floors. The idea implies that there is a certain minimum depth below sea-level where the pressure due to the weight of the overlying material in each unit column is everywhere the same. In Fig. 2.10 this level of uniform pressure is that of the base of the highest block. The earth's major relief is said to be *compensated* by the underlying differences of density, and the level where the compensation is estimated to be complete—i.e. the level of uniform pressure—is often referred to as the *level of compensation*. Naturally, individual peaks and valleys are not separately balanced or compensated in this way; the minor relief features of the surface are easily maintained by the strength of the crustal rocks. As indicated on pp. 22, 38 and 39, perfect isostasy is rarely attained, because of the restlessness of our globe; however, in regions which have remained free from geological disturbances for long periods of time there is generally a remarkably close approach to a state of equilibrium.

If a mountain range were simply a protuberance of rock resting on the continental platform and wholly supported by the strength of the foundation, then a plumb line—such as is used for levelling surveying instruments—would be deflected from the true vertical by an amount proportional to the gravitational attraction of the mass of the mountain range. The first hint that mountains are not merely masses of rock stuck on an unyielding crust was provided by the Andes expedition of 1735. Pierre Bouguer, the leader of the expedition, made observations both north and south of Chimborazo, and found to his surprise that the deflection of the plumb line towards this towering volcanic peak was very much less than he had estimated. He recorded his suspicion that the gravitational attraction of the Andes 'is much smaller than that to be expected from the mass represented by these mountains!'

Similar discrepancies were met with during the survey of the Indo-Gangetic plain, south of the Himalayas, carried out by Sir George Everest, Surveyor-General of India, over a century ago. The difference in latitude between Kalianpur and Kaliana (603 km due north) was determined astronomically, and also by direct triangulation on the ground. The two results differed by 5·23 seconds of arc, corresponding to a distance on the ground of 168 m. The discrepancy was ascribed to the attraction exerted by the enormous mass of the Himalayas and Tibet (Figure 2.11) on the bob of the plumb-line used for levelling the astronomical instruments. The error so introduced does not arise in the triangulation method. A few years later (1855) Archdeacon Pratt made a minimum es-

**Figure 2.11** Meridional section (about Long. 78°E) through northern India and Tibet, indicating the enormous mass tending to deflect the plumb-line of surveying instruments used on the plains to the south

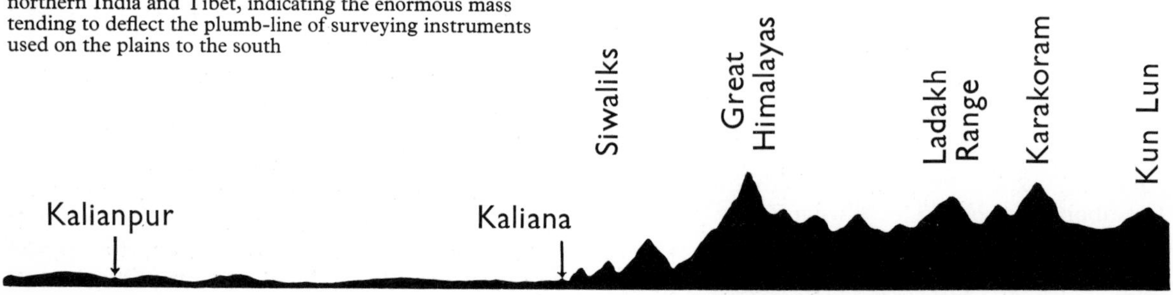

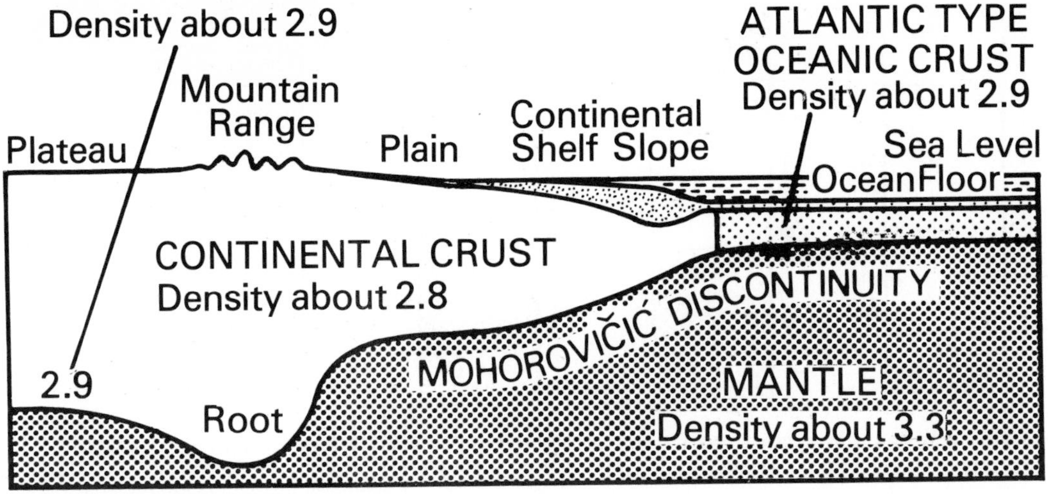


**Figure 2.12** Diagrammatic section through the earth's crust and the upper part of the mantle to illustrate the relationship between surface features and crustal structure. Based on gravity determinations and exploration of the distribution of sial, crustal sima and mantle sima by earthquake waves.

timate of the mass of the mountains and calculated what the corresponding gravitational effects would be at the two places on the plains to the south. His estimates for the deflections of the plumb line towards the mountains were

27·853″ at Kaliana, and
11·968″ at Kalianpur.

The difference, 15·885″, was more than three times the observed deflection, 5·23″. Bouguer's suspicion that the mountains were apparently not pulling their weight was now a demonstrated fact. Even more spectacular evidence was provided by French surveyors, who found that in some of the coastal regions of south-western France the plumb-line was deflected, not towards the mountains, but away from them, towards the Bay of Biscay.

Clearly the mountains seem to behave almost as if they were hollow, and this apparent gravitational anomaly has since been amply confirmed by the results of innumerable measurements of gravity on and near mountains and high plateaus. The observed results are found to be much lower than those to be expected. Only one physical explanation of these discrepancies or anomalies is available. Since the mountains are *not* hollow, there must be a compensating deficiency of mass in the columns underlying the visible mountain ranges. In simpler language, the density of the rocks must be relatively low down to considerable depths. The possible combinations of density distributions and depths are, of course, theoretically infinite, but unfortunately we know something about the crustal rocks and their densities in most regions, so that in practice the probable combinations are limited. Moreover, as we have seen, exploration of the crust by earthquake waves confirms the inference that mountain ranges have roots, largely composed of sialic rocks, going down to depths of as much as 50 or 60 km. Under plains near sea-level the thickness of the sial and other crustal rocks is only about 30 km, sometimes less. Beneath all the deeper parts of the oceans sial cannot be detected at all. The suggestion that the crust is supported by underlying denser material and that the weight of mountains is balanced by light materials extending as roots into the denser—just as icebergs are balanced in water—was first made in 1855 by Sir George Airy, who was then the Astronomer Royal. Figure 2.12 illustrates these relationships between surface relief and crustal structure in a general way.

Figure 2.13 shows characteristic examples of crustal columns, each of which has the same area and extends downwards to the same depth below sea-level, the depth at which the weight of each column exerts approximately the same pressure on the underlying material, irrespective of its surface elevation. For the columns selected, this depth—50 km—would be the depth of isostatic com-

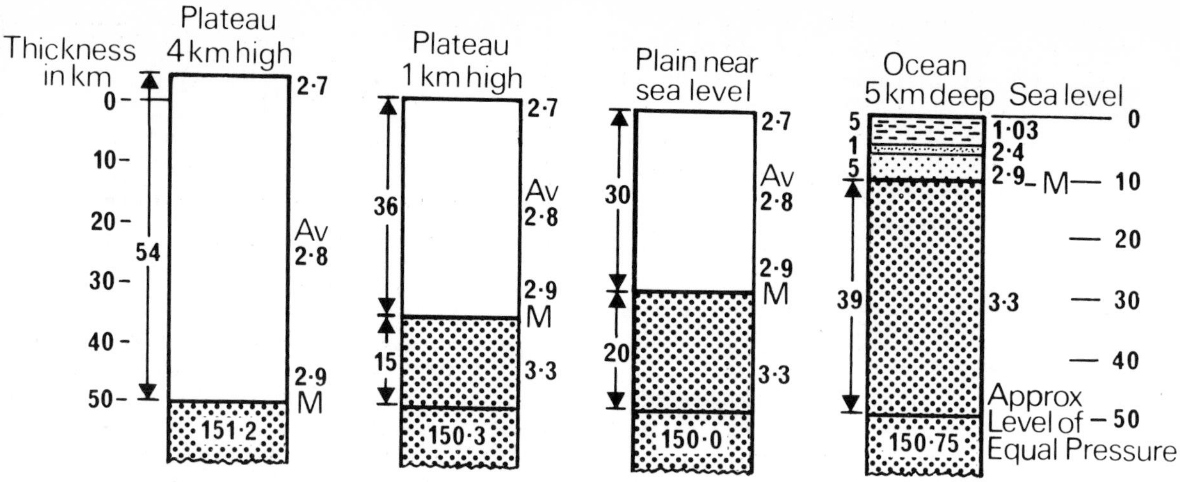


**Figure 2.13** Columns of equal cross-section through characteristic parts of the continents and ocean floor. Shading as in Fig. 2.12. The figures to the right of each column indicate approximate densities. These, multiplied by the corresponding thickness, down to a depth of 50 km, give a total figure proportional to the pressure at that depth; e.g. for the oceanic column: $5 \times 1{\cdot}03$ (sea water) $+ 1 \times 2{\cdot}4$ (sediments) $+ 5 \times 2{\cdot}9$ (crustal sima, probably basaltic rock) $+ 39 \times 3{\cdot}3$ (mantle sima) $= 150{\cdot}75$. M is the Mohorovičić discontinuity (the 'Moho').

pensation if the regions concerned had long remained undisturbed by the activities of our restless earth. When the gravitational effects of the appropriate underlying densities are taken into account in such regions the anomalies referred to above (known as *Bouguer anomalies*) disappear on average to the extent of about 85 per cent. Any residual discrepancy that still remains is called the *isostatic anomaly*.

In some regions, however, as we shall see, the crust has recently been very actively disturbed, and is indeed still undergoing geologically rapid change. As a result certain elevations and depressions cannot be explained as simply as Figure 2.13 would suggest. Well-known examples are the Colorado Plateau and the deep oceanic trenches. Gravity measurements over such regions reveal abnormal isostatic anomalies, and provide the geological detective with some of the clues he needs towards understanding the nature and behaviour of the underlying materials.

## The Moving Lithosphere

The material of the continents (sial) might be regarded as a kind of light slag that accumulated at the surface of the earth at an early stage in its history. We might have expected it to have accumulated uniformly and have formed a continuous layer. Indeed some geologists, notably Carey (1958), think that this was so, and that the sialic crust has been pulled apart as a result of gradual expansion of the earth's interior (Figure 2.14). If, on the other hand, sialic crust was from the start concentrated in continental rafts, a clue as to how such concentration might be brought about is provided by the behaviour of scum on the surface of gently boiling jam (Figure 2.15). A hot current ascends near the middle, and, turning outwards at the surface, it sweeps the scum to the edges, where the current descends. The scum is too light to be carried down, so it accumulates and can be skimmed off. Actual examples of convection occur in the celebrated Pitch Lake of Trinidad, the surface effects of which are illustrated in Figure 16.16. There are reasons, as we shall see later, for suspecting that similar circulations may be going on inside the 'solid' earth, though on a much larger scale and at a much slower rate. If at one time such convection currents had been sufficiently vigorous, the horizontal currents spreading out from each ascending column might well have swept some regions clear of sial. These regions would become ocean basins. Where the horizontal current of one convection system met those of a neighbouring system they would be obliged to turn downwards, and the sial, too light to be carried down, would be left behind. Thus regions overlying descending currents would develop into continents.

Since early in the century, a minority of geologists have followed Wegener (1912) in thinking that about 300 million years ago the continents were joined together as one supercontinent,

*Pangaea* (Gr., all earth). The earth as we see it today, largely from palaeoclimatic evidence, was thought by Wegener to have gradually evolved by the break-up of Pangaea, and the drifting apart of the separate pieces. This process is known as continental drift. It is only since about 1960, however, that new geophysical discoveries have made this conclusion inescapable except to one or two diehards, like Beloussov in Russia.

The first surprise was the discovery that sediments overlying the basaltic oceanic crust are very thin, and surprise turned into astonishment when it was found that these sediments are nowhere more than about 150 million years old, and rarely older than 80 million years, whereas the oldest continental rocks have for long been known to be more than 3000 million years old. The next step forward was the discovery that the earth's magnetic field has reversed its direction several times during the last four million years. This has been determined from the magnetism of lava-flows from widely spaced regions. By determining the radiometric ages of the same lava-flows a time scale of reversals has been constructed (Figure 27.10). Further back in time than five million years, errors in the method of radiometric dating become too great, relative to the lengths of magnetic periods, for it to be practicable to establish a time-scale of magnetic reversals for older periods.

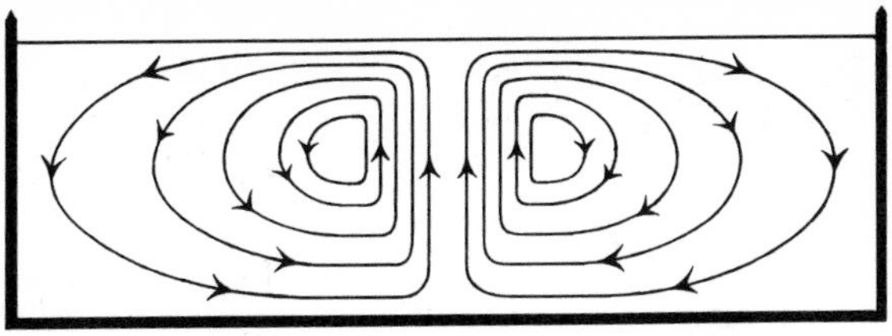

**Figure 2.15** Section through a convection 'cell' showing the directions of convection currents in a layer of liquid uniformly heated from below

In the meantime it had been discovered that the basalt crust of the ocean floors is characterized by 'zebra-like' stripes of magnetic anomalies, which in the environs of the mid-oceanic ridges correspond to the reversals of the magnetic time-scale. The anomalies of the oceanic crust are parallel to the mid-oceanic ridges and bilaterally symmetrical on either side of them (Figure 27.15). Moreover they become older away from the ridges. It thus became apparent that new oceanic crust, formed by eruption of basaltic lava at the ridges, slowly moves away on either side. Magnetic anomalies have been detected over much of the oceanic crust, and by calculating the rate of movement from the positions of magnetic anomalies of known age close to the ridges, and assuming that these rates can be extrapolated back through older times, it has been possible to estimate the ages of the older anomalies (Figure 27.16). In this way magnetic anomalies away from the mid-oceanic ridges have been found to be as old as 70 or 80 million years. These conclusions

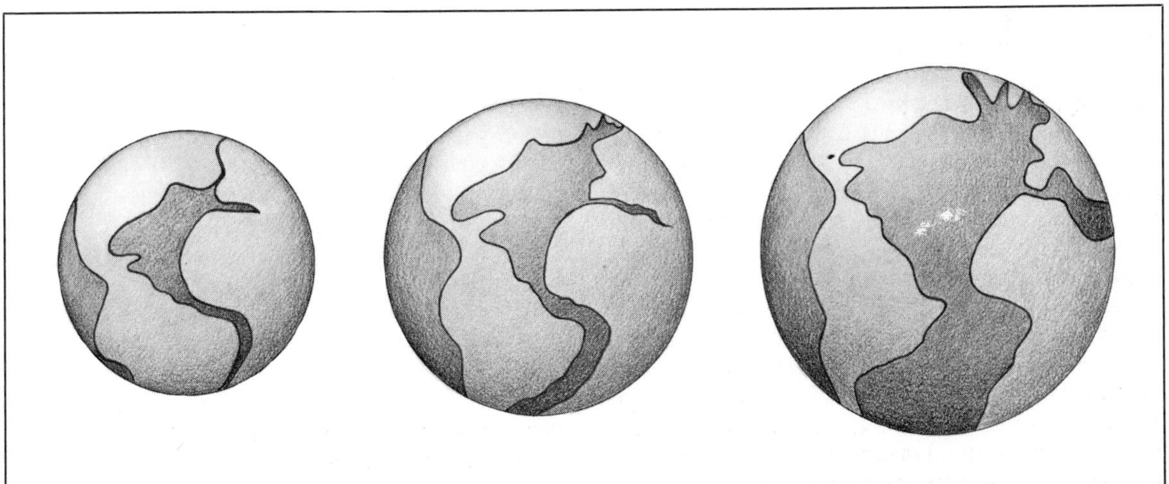

**Figure 2.14** Schematic illustration to indicate the effects of expansion of the earth's interior on a sialic crust that may originally have covered the entire globe. Primordial continents (white) first outlined by stretching and fragmentation, are gradually separated as they move outwards in consequence of the earth's increasing volume, Heavy material from the mantle flows into the intervening gaps and forms the floors of the growing ocean basins (black). (*After O. C. Hilgenberg*, 1933)

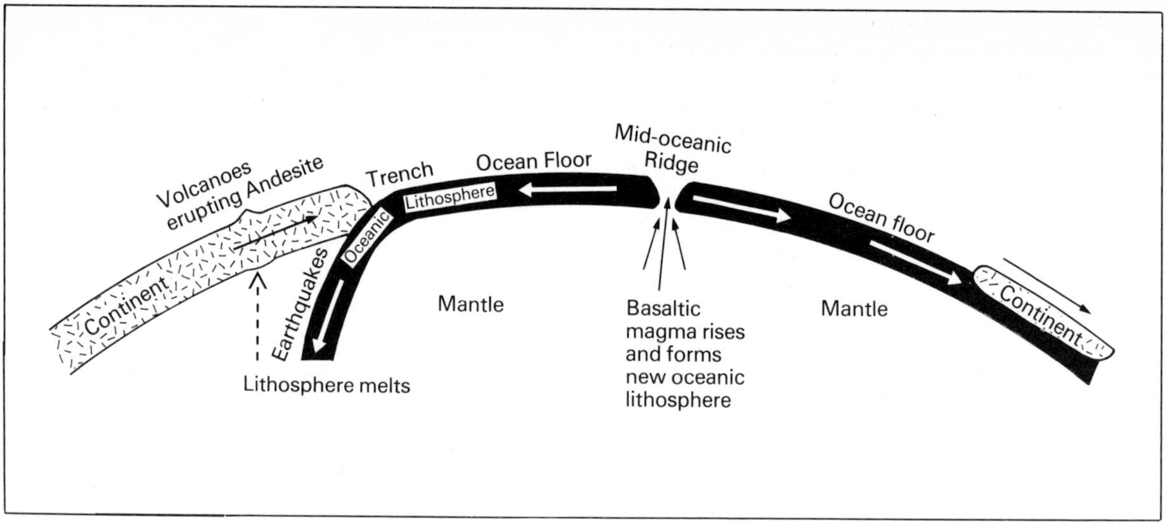


**Figure 2.16** Diagrammatic representation of new oceanic crust (black) forming along a mid-oceanic ridge and, together with the upper part of the mantle moving away as oceanic lithosphere on either side of the ridge. The deep ocean trenches represent the sites of subduction zones, and on their landward side the foci of medium and deep earthquakes outline the surface of oceanic lithosphere descending into the mantle. Movement of oceanic lithosphere away from mid-oceanic ridges is variously attributed to push by basaltic magma emplaced at mid-oceanic ridges, to gravity-gliding away from mid-oceanic ridges, or to downpull exerted on descending oceanic lithospere.

from geophysical discoveries have been confirmed by lifting cores of sediment from the ocean floors, and dating the sediment, from each core, that immediately overlay the basalt. From core lifted from the Atlantic it has been found that the lowermost sediment overlying the basalt increases in age with distance away from the mid-oceanic ridge.

The question at once arises as to what has happened to the crust older than 80 to 150 million years, and here lies an appreciation, if not a complete understanding, of the deep ocean trenches which surprisingly lie close to continental margins (Figure 2.7). The ocean trenches which reach depths below sea-level greater than the heights attained by the highest mountains on land, are characterized by deep earthquakes on their landward sides. Indeed, the points of origin (foci) of earthquakes increase in depth, from shallow to deep, along planes dipping at about 45° towards the adjacent continents. Earthquakes are the passage of vibrations from a focus where, for example, rocks subjected to stress suddenly break. The foci of earthquakes, on the continental sides of ocean troughs, are thought to lie on the surface of oceanic lithosphere moving down into the mantle (Figure 2.16).

Out of these discoveries revealing ocean floor spreading, a hypothesis known as *plate tectonics*, has evolved. The earth is envisaged as being covered by six large rigid carapace-like plates, and several smaller ones. The plates, about 100 km thick, consist of crust plus the uppermost part of the mantle, which are jointly known as the *lithosphere*. The part of the mantle immediately underlying the lithosphere is distinguished as the *asthenosphere*, and over the asthenosphere the lithosphere moves as new crust is created at the mid-oceanic ridges, and disappears into the depth at ocean trenches (Figure 2.16). The rate of movement varies from 1 cm a year in the North Atlantic to about 6 cm a year in the South Pacific. The upper part of plates may consist entirely of oceanic crust, or of both oceanic and continental crust. Where the leading edge of a plate is formed of continental crust, this is too light to sink into the mantle, and fold mountains, thought to result from crumpling of the sialic crust on impact, occur along such continental margins.

The plates are the relatively inert parts of the earth's surface, and they are separated from one another by mobile belts characterized by earthquakes, volcanic activity and fold mountains. The plates move slowly over the earth's surface, and since they are interlocked, movement of one plate must affect all the others.

In the later chapters of this book, details of the discoveries outlined above are described. Clearly these discoveries must influence our understanding of some of the earth's surface features, such as fold mountain ranges, ocean troughs, earthquake zones, and belts of igneous activity, volcanic and intrusive. The earth's surface features, however, arise in two different, but not completely separable ways (pp. 37, 38). They may arise from processes within the earth itself, or from external processes, such as the action of the sun, the atmosphere, rivers, ocean currents, waves, etc. It is, of course, our understanding of surface phenomena that arise from the earth's internal processes that is affected by the new discoveries. Explanations of the earth's surface features arising from external processes are not affected, except through the immensity of geological time. The earth's climatic zones, for example, as they exist now, are explained as previously, but it is now no longer a hypothesis but an accepted fact that the climatic zones, whilst maintaining their relative positions to one another, have in the past occupied different positions on the earth's surface; the continents have moved relative to the poles, and to one another.

Selected References

BULLARD, E., 1969, *The Origin of the Oceans*, Scientific American Offprint No. 880, Freeman, San Francisco.
CALDER, N., 1972, *Restless Earth*, B.B.C., London.
EMERY, K. O., 1969, *The Continental Shelves*, Scientific American Offprint No. 826, Freeman, San Francisco.
FISHER, R. L. and REVELLE, R., 1955, *The Trenches of the Pacific*, Scientific American Offprint No. 814, Freeman, San Francisco.
HALLAM, A., 1973, 'A Revolution in the Earth Sciences', from *Continental Drift to Plate Tectonics*, Clarendon Press, Oxford.
KING, L. C., 1967, *The Morphology of the Earth* (2nd edn.). Oliver and Boyd, Edinburgh.
TARLING, D. H., and M. P., 1971, *Continental Drift: A Study of the Earth's Moving Surface*, Penguin, Harmondsworth.
GEOLOGICAL MUSEUM, 1972, *The Story of the Earth*, H.M.S.O. for the Institute of Geological Sciences.

# Chapter 3

# The Changing Continental Surfaces

The same regions do not remain always sea or always land, but all change their condition in the course of time.

*Aristotle* (384–322 B.C.)

## Weathering, Erosion and Denudation

The circulations of matter that are continually going on in the zones of air and water, and even of life, constitute a very complicated mechanism which is maintained, essentially, by the heat from the sun. A familiar example of such a circulation is that of the winds. Another, more complex, is the circulation of water. Heat from the sun lifts water vapour from the surface of oceans, seas, lakes and rivers; and wind distributes the vapour far and wide through the lower levels of the atmosphere. Clouds are formed, rain and snow are precipitated, and on the land these gather into rivers and glaciers. Finally, most of the water is returned to the oceanic and other reservoirs from which it came (Figure 3.1). These circulations are responsible for an important group of geological processes, for the agents involved—wind (moving air), rain and rivers (moving water) and glaciers (moving ice)—act on the land by breaking up the rocks and so producing rock-waste which is gradually carried away.

Part of every shower of rain sinks into the soil and promotes the work of decay by solution and by loosening the particles. Every frost shatters the rocks with its expanding wedges of freezing water. Water expands on freezing, and through repeated alternations of frost and thaw in water-filled pores and cracks the rocks are relentlessly broken to bits. Fallen fragments dislodged from cliffs and precipitous crags accumulate at the base as screes or talus (Figure 3.2). Life also co-operates in the work of destruction. The roots of trees grown down into cracks, and assist in splitting up the rocks (Figure 3.3). Worms and other burrowing animals bring up the finer particles of soil to the surface, where they fall a ready prey to wind and rain. The soil is a phase through which much of the rock-waste of the lands must pass before it is ultimately removed. The production of rock-waste by these various agents, partly by mechanical breaking and partly by solution and chemical decay, is described as *weathering*. A familiar example is the fact that inscriptions on exposed

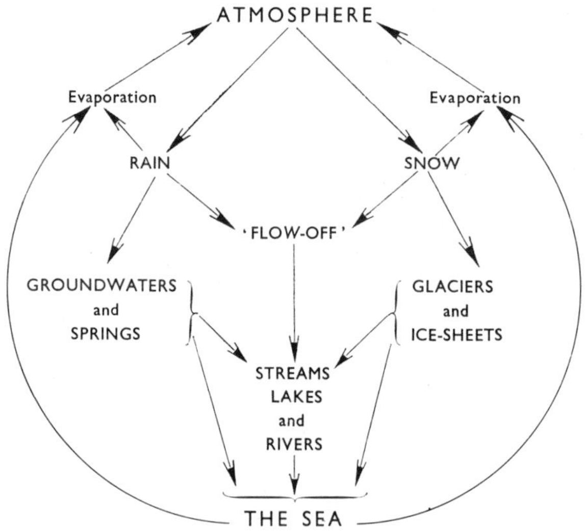


**Figure 3.1** The circulation of meteoric water. In addition to the main evaporations here indicated it should be noted that evaporation takes place from all exposed surfaces of water and ice (e.g. lakes, rivers, glaciers and ice sheets) and also from the soil and from plants and animals. Part of the water that ascends from the depths by way of volcanoes reaches the surface for the first time; such water is called *juvenile water* to distinguish it from the *meteoric water* already present in the hydrosphere and other outer zones of the earth.

**Figure 3.2** Screes of Doe Crag (Borrowdale Volcanic Series), Old Man of Coniston, English Lake District (*G. P. Abraham Limited, Keswick*)

**Figure 3.3** Ice-transported boulder, with crack enlarged by the growth of tree roots, Trefarthen, Anglesey (*Institute of Geological Sciences*)

slabs of marble—and other rocks—rarely survive the ravages of rain, frost and wind for more than two or three centuries.

Sooner or later the products of weathering are removed from their place of formation, Blowing over the lands, the wind becomes armed with dust and sand, carrying them far and wide and often becoming a powerful sand-blasting agent as it sweeps across areas of exposed rock. Glaciers, similarly armed with morainic and other debris, grind down the rocks over which they pass during their slow descent from ice-fields and high mountain valleys. Rainwash, sliding screes and landslips feed the rivers with fragments, large and small, and these are not only carried away, but are used by the rivers as tools to excavate their floors and sides. And in addition to their visible burden of mud and sand, the river waters carry an invisible load of dissolved material, extracted from rocks and soils by the solvent action of rain and soil water, and by that of the river water itself. Winds, rivers and glaciers, the agents that carry away the products of rock-waste, are known as transporting agents. All the destructive processes due to the effects of the transporting agents are described as *erosion* (L. *erodere*, to gnaw; *erosus*, eaten away).

It is convenient to regard weathering as rock decay by agents involving little or no transport of the resulting products except by gravity, and erosion as land destruction by agents which simultaneously remove the debris. Both sets of processes co-operate in wearing away the land surface, and their combined effects are described by the term *denudation* (L. *denudare*, to strip bare).

## Deposition of Sediment

The sediment carried away by the transporting agents is sooner or later deposited again. Sand blown by the wind collects into sand dunes in the desert or bordering the seashore (Figure 3.4). Where glaciers melt away, the debris gathered up during their journey is dumped down unsorted

**Figure 3.4** Sand dunes in the Colorado Desert, Imperial County, California (*W. C. Mendenhall, United States Geological Survey*)

(Figure 3.5), to be dealt with later by rivers or the sea. When a stream enters a lake the current is checked and the load of sand and mud gradually settles to the bottom. Downsteam in the open valley sand and mud are spread over the alluvial flats during floods, while the main stream continues, by way of estuary or delta, to sweep the bulk of the material into the sea. Storm waves thundering against rocky coasts provide still more rock-waste, and the whole supply is sorted out and widely distributed by waves and currents. Smooth and rounded water-worn boulders collect beneath the cliffs. Sandy beaches accumulate in quiet bays. Out on the sea floor the finer particles are deposited as broad fringes of sediment, the finest material of all being swept far across the continental shelves, and even over the edge towards the deeper ocean floor, before it finally comes to rest. All these deposits are examples of sedimentary rocks in the making.

We have still to trace what happens to the invisible load of dissolved mineral matter that is removed from the land by rivers. Some rivers flow into lakes that have no outlet save by evaporation into the air above them. The waters of such lakes rapidly become salt because, as the famous astronomer Edmund Halley realized in 1715, 'the saline particles brought in by the rivers remain behind, while the fresh evaporate'. Gradually the lake waters become saturated and rock-salt and other saline deposits, like those on the shores and floor of the Dead Sea, are precipitated. Most rivers, however, reach the sea and pour into it the greater part of the material dissolved from the land. So, as Halley pointed out, 'the ocean itself is become salt from the same cause'. But while, on balance, the salinity of the sea may be slowly increasing, much of the mineral matter contributed to the sea is taken out again by living organisms. Cockles and mussels, sea-urchins and corals, and many other sea creatures, make shells for themselves out of calcium carbonate abstracted from the water in which they live. When the creatures die, most of their soft parts are eaten and the rest decays. But their hard parts remain, and these accumulate as the shell banks of shallow seas (Figure 3.6), the coral reefs of tropical coasts and islands, and the grey globigerina ooze of the deep-sea floor. All of these are limestones in the making. Life, as a builder of organic sediments, is a geological agent of the first importance.

## The Importance of Time

It will now be realized that while the higher regions of the earth's crust are constantly wasting away, the lower levels are just as steadily being built up. Evidently denudation and deposition are great levelling processes. In the course of a single lifetime their effects may not be everywhere perceptible. Nevertheless they are not too slow to be measured. An average thickness of about half an inch has already been worn from the outer surface of the Portland stone with which St Paul's Cathedral was built over 250 years ago. Portland

**Figure 3.5** Öraefajökull Glacier, Iceland, showing its source in an upland ice-cap. Its snout is just off the bottom-right of the photograph, and there it melts away amidst the moraines of rock-waste which it transported and deposited before shrinking to its present size. See Fig. 12.9 for locality. (*Thorvardur R. Jonsson*)

**Figure 3.6** Limestone in the making: a deposit of shell-gravel (*S. H. Reynolds*)

stone has justified Wren's confidence in its suitability to withstand the London atmosphere, for the land of Britain as a whole is being worn down rather faster—at an average rate of about 30 cm in three or four thousand years. At this rate a few million years would suffice to reduce the varied landscapes of our country to a monotonous plain. Evidently slowly-acting causes are competent to produce enormous changes if only they continue to operate through sufficiently long periods.

Now, geologically speaking, a million years is a comparatively short time, just as a million miles is a very short distance from an astronomical point of view. One of the modern triumphs of geology and physics is the demonstration that the age of the earth cannot be less than 4500 million years (Ch. 13). Geological processes act slowly, but geological time is inconceivably long. The effects of slow processes acting for long periods have been fully adequate to account for all the successive transformations of landscape that the earth has witnessed.

It was James Hutton (1726–97), the founder of modern geology, who first clearly grasped the full significance and immensity of geological time. In his epoch-making *Theory of the Earth*, communicated to the Royal Society of Edinburgh in 1785, he presented an irrefutable body of evidence to prove that the hills and mountains of the present day are far from being everlasting, but have themselves been sculptured by slow processes of erosion such as those now in operation. He showed that the alluvial sediment continually being removed from the land by rivers is eventually deposited as sand and mud on the sea floor. He observed that the sedimentary rocks of the earth's crust bear all the hall-marks of having accumulated exactly like those now being deposited. He realized, as no one had done before, that the vast thicknesses of these older sedimentary rocks implied the operation of erosion and sedimentation throughout a period of time that could only be described as inconceivably long.

Hutton was the first to apply to geological problems as a whole that faith in the orderliness of nature which is the basic principle on which all science is founded. The doctrine known as *catastrophism*—the myth of successive destructions of the face of the earth by violent and supernatural cataclysms, of which the Noachian Flood was the classic example—was still widely prevalent in Hutton's day. With such extravagances Hutton would have nothing to do. He realized with the clarity of sheer genius that 'the

**Figure 3.7** Late Glacial raised beach, west of Rhuvaal Lighthouse, Islay, Inner Hebrides, Scotland (*Institute of Geological Sciences*)

past history of our globe must be explained by what can be seen to be happening now'. This principle, which gradually replaced the catastrophic conception of earth history, was eventually called *uniformitarianism* by Sir Charles Lyell—an unhappy word, liable to be taken too literally. Hutton did not exclude temporary and local crises: such natural catastrophes, for example, as the great Lisbon earthquake of 1755 or the eruption of Vesuvius which overwhelmed Pompeii and Herculaneum in A.D. 79. But Lyell's term seemed to do so, since it inevitably suggests uniformity of *rate*, whereas what is meant is uniformity of *natural law*. Hutton made this quite clear when he wrote: 'No powers are to be employed that are not natural to the globe, no action to be admitted except those of which we know the principle.' He insisted that the ways and means of nature could be discovered only by observation. On the Continent the term *uniformitarianism* never won favour, but was gradually replaced by *actualism*, which conveys much more appropriately the real meaning of Hutton's inspired appeal to 'actual causes': the principle that the same processes and natural laws prevailed in the past as those we can now observe or infer from observations.

Hutton's genius received scant recognition in his lifetime, mainly because of the prevalent belief that the world had been created in the year 4004 B.C. Far from being welcomed, Hutton's discoveries were generally regarded with righteous horror. Had he lived in India, however, Hutton would have found a ready-made system of world chronology fully adequate for the needs of geology at that time. According to the Hindu calendar, as recorded in the ancient books of Vedic philosophy, the year A.D. 1977 corresponds to 1,972,949,078 years since the present world came into existence. This astonishing concept of the earth's duration has at least the merit of being of the right order.

## Earth Movements

It follows that there has been ample time, since land and sea came into existence, for Britain, and indeed for the highest land areas, to have been worn down to sea-level over and over again. How then does it happen that every continent still has its highlands and mountain peaks? The special creation theory, immortalized in James Thomson's words:

**Figure 3.8** Post Glacial raised beach, with sea cave in the quartzite cliff behind, Loch Tarbert, Jura (*Institute of Geological Sciences*)

When Britain first at Heaven's command
Arose from out of the azure main. . . .

is not very helpful, yet it does suggest a possible answer. The lands, together with adjoining parts of the sea floor, may have been uplifted from time to time. Alternatively, as suggested long ago by Xenophanes (*c.* 500 B.C.), who noticed that fossil shells in the limestone hills of Malta were like those still being washed up on the beaches down below, the level of the sea may have fallen, leaving the land relatively upraised. In either case there would again be land above the sea, and on its surface the agents of denudation would begin afresh their work of sculpturing the land into hills and valleys. An additional factor is the building of new land—like the volcanic islands that rise from oceanic depths—by the accumulated products of volcanic eruptions. Each of these processes of land renewal has been in continual operation during the course of the earth's long history.

Relative movements between land and sea are convincingly proved by the presence in Islay and Jura, and in many other places, of typical sea beaches, now raised far above the reach of the waves (Figure 3.7). Behind these raised beaches the corresponding cliffs, often tunnelled with sea caves, are still preserved (Figure 3.8). In Scandinavia and Peru old strand lines can be traced which rise from near sea-level in the south to heights of hundreds of feet in the north. Such tilting of the shores shows that the movements involved actual upheaval of the crust and not merely a change of sea-level. The former uplift of old sea floors can be recognized in the Pennines, where the grey limestones contain fossil shells and corals that bear silent witness to the fact that the rocks forming the hills of northern England once lay under the sea. Uplift on a far greater scale is demonstrated by the occurrence of marine fossils in the shales that cap the granitic rocks of Mt McKinley in Alaska (Figure 30.1)) and so form the highest point of North America. And even more impressive, being the present world record, is the summit of Mt Everest (Figure 2.5), carved out of sediments that were originally deposited on the sea floor of a former age.

Crustal movements have not always and everywhere resulted in emergence of the land. Relative movements between land and sea may also bring about the submergence of the land. Recent submergence of parts of the land surface is proved

**Figure 3.9** Submerged Forest, Leasowe foreshore, Cheshire (*C. A. Defieux*)

by the local occurrence around our shores of *submerged forests* (Figure 3.9). These are groups of tree-stumps still preserved with their roots in the original position of growth, but now uncovered only at the lowest tides. In Mount's Bay, near Penzance, what is probably the same submergence is indicated by the occurrence of a Stone Age axe-factory, dating from about 1800–1500 B.C., which now lies under the sea. The memory of this and other submerged lands of south-west Britain may well be enshrined in the legends of the lost land of Lyonesse.

When earth movements take place suddenly they are recognized by the passage of earthquake waves. In certain restless belts of the crust, for example in Japan, there may be several shocks every day, occasionally with terribly disastrous consequences. Exceptionally, as in Yakatut Bay, Alaska, in 1899, an upward jerk of as much as 14 m has been measured, but usually these sudden movements are on a much smaller scale (Figure 3.10).

If crustal movements were all uniformly vertical, then beds of sediment uplifted from their original positions on the sea floor would generally be found lying in nearly horizontal positions. So indeed they often are (Figure 1.6), but in many places they have been corrugated and buckled into folds (Figure 3.11), like a wrinkled-up tablecloth which has been pushed along the table. But from a mechanical point of view this is not a good analogy. A familiar small-scale domestic model that in many ways more closely corresponds to the natural conditions of large-scale folding is provided by the skin that forms on hot milk. If the containing vessel is tilted a little the skin slumps into a series of folds, some of which are likely to overlap their down-slope neighbours. Many an Alpine precipice displays great sheets of rock which have been 'overfolded' in much the same way, so that parts of them now lie upside down (Figure 3.11). In other places layers of rock have been folded tightly together, like the pleats of a closed concertina.

Such amazing structures as these suggest that certain parts of the earth's crust have yielded to pressures of irresistible intensity. All the great mountain ranges of the world have been carved out of rocks that have been folded and crumpled and overthrust. Long belts of the crust appear to have been so compressed and thickened that eventually they had no alternative but to rise to mountainous heights.

Although there must have been long periods when much of the land lay under the sea, it is probably thousands of millions of years since all the land was submerged at once. Earth movements and volcanic additions to the surface have evid-

**Figure 3.10** Displacement of a road and production of a fault scarp by a sudden movement on the White Creek Fault (see Fig. 9.39) which was responsible for the Murchison earthquake of 1929, South Island, New Zealand. The uplifted block rose about 5 metres relative to the foreground. (*New Zealand Geological Survey*)

**Figure 3.11** Recumbent folds in the High Calcareous Alps: the Axenstrasse, near Flüelen, at the south end of Lake Lucerne, Switzerland (*F. N Ashcroft*)

ently been fully competent to restore the balance of land and sea whenever that balance has been threatened by the levelling processes of denudation. Most of the sediments originally deposited on the shallow sea floor, sometimes hardened and cemented into firm and durable rocks, sometimes bent and twisted into intricate folds, sometimes accompanied by lavas and volcanic ashes, have sooner or later emerged to form new lands, either by upheaval or by the withdrawal of the sea.

### Volcanic and Igneous Activity

Earth movements are not the only manifestations of the earth's internal activities. Volcanic eruptions provide a most spectacular proof that the earth's interior is so hot that locally even the crustal rocks pass into a molten state. A volcano is essentially a fissure or vent, communicating with the interior, from which flows of lava, fountains of incandescent spray, or explosive bursts of gases and volcanic 'ashes' are erupted at the surface. The fragmental materials produced during volcanic eruptions are collectively known as *pyroclasts* (Gr. *pyros*, fire; *klastos*, broken in pieces).

It is convenient to have a general term for the parental materials of these hot volcanic products, as they occur in the depths before eruption at the surface. The term commonly used for this purpose is *magma*, a Greek word originally meaning any kneaded mixture, such as dough or ointment. The only connection with lava seems to be mobility or capacity to flow. In its geological application the property of high temperature is also implied. Magma, then, means any *hot* mobile material within the earth which is capable of penetrating into or through the crustal rocks. So far as we can judge from their products, magmas may consist of hot liquids, gases and solids in all possible proportions and combinations. The most important point to keep in mind, however, is that magma is not merely molten rock. When confined under pressure it is commonly associated with gases and vapours, sometimes in immense quantities. As a highly gas-charged magma ascends towards the surface and the overhead pressure gradually falls, the gases begin to be liberated. Sooner or later the growing gas pressure overcomes the resistance—the soda-water bottle bursts, so to speak, or blows off its seal-cap—and an explosive eruption breaks out (see Frontispiece).

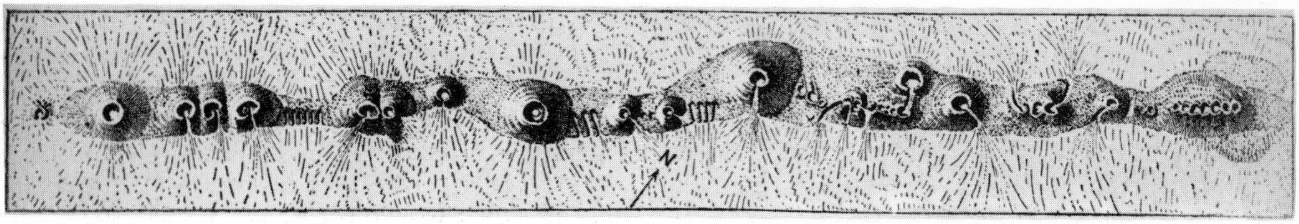


**Figure 3.12** Pictorial representation of a 16 km stretch of the Laki fissure, Iceland, showing conelets formed towards the end of the 1783 eruption (cf. Fig. 3.14) (*After Helland*)

In the more familiar volcanoes the magma ascends through a central pipe, around which the lavas and ashes commonly accumulate to form a more or less conical volcanic mountain. Magma may also reach the surface through long fissures from which lava-flows spread over the surrounding country, filling up the valleys and forming widespread volcanic plains or plateaus. In such cases the lava is generally basaltic, like that of the Giant's Causeway.

The greatest basalt flood of modern times broke out at Laki, in Iceland, during the summer of 1783. From a fissure 32 km long torrents of gleaming lava, amounting in all to over 12 cubic kilometres, overwhelmed 565 square kilometres of country, extending long fiery arms down the valleys, 64 km to the west and 45 to the east. Fertile farmlands were deeply buried beneath a desert of lava (Figure 3.13). As the activity diminished in intensity, obstructions choked the long rent. Gases that had been effervescing freely then began to accumulate below the surface until they raised sufficient pressure to overcome the resistance. At the hundreds of points along the fissure where the pent-up gases eventually broke through, miniature cones-with-craters, ranging in height from a metre or so to thirty metres or more, were built up and great quantities of pyroclasts were erupted (Figures 3.12 and 3.14).

Despite the obvious dangers to life and human welfare, of which the Laki eruption is an impressive example, volcanic activity is geologically to be regarded as a constructive process in so far as new materials are brought to the surface and new topographic forms built up.

As we have seen, volcanic activity is only the

**Figure 3.13** Eastern part of the Laki basalt floods of 1783, as seen in 1957; looking towards the ice-covered Öraefajökull (2119 metres), an active volcano and the highest mountain in Iceland (cf. Fig. 3.5) (*Sigurdur Thorarinsson*)

surface manifestation of the movement through the earth's crust of magma generated in the mantle or in exceptionally heated regions of the crust itself. Naturally not all the magma reaches the surface, and the new rocks formed in the crust by the consolidation of such magma are the chief examples of what are called *intrusive rocks*, to distinguish them from lavas, which are called *extrusive rocks*. In many places such intrusive rocks are well exposed to observation, as a result of the removal of the original cover by denudation. Thus the feeders of fissure eruptions, like that of Laki, appear long afterwards as dykes (Figure 3.15). Moreover, there are innumerable dykes that never succeeded in reaching the surface. In favourable circumstances the magma may open up a passageway along a bedding plane, making room for itself by uplifting the overlying rocks. The resulting tabular sheet of rock is called a sill (Figure 3.16). Exceptionally large intrusions, consisting of granite and similar crystalline rocks, are particularly characteristic of the hearts of mountain ranges, including those of former ages as well as those of the present day (Figure 3.17).

It has been usual to describe the rocks of intrusions or volcanic extrusions as *igneous rock* (from the Latin *ignis*, fire; implying that 'sub-terranean fire' or the earth's internal heat was concerned in their origin). But here we must not generalize too broadly. In recent years it has become evident that by no means all intrusions are of igneous origin in the accepted meaning of the term. Rock-salt, for example, makes surprisingly large intrusions, known as *salt domes*, some of which are comparable in size with the smaller granite intrusions (e.g. the Shap Fell or Dartmoor granite masses). Where salt domes are still actively rising, as in south-west Iran, the salt occasionally reaches the surface, breaking through the summit of the dome to continue its flow downhill like a glacier or a lava-flow. Most salt domes, however, fail to reach the surface and have been discovered only in the course of geophysical prospecting and boring for oil. Now the salt of such intrusive masses has certainly been 'kneaded' during its upward intrusion, and some geologists might regard it as the extreme example of a magma consisting of deformable 'plastic' or 'pseudo-viscous' solid. But no one would regard rock-salt as an igneous rock. It thus becomes evident that

**Figure 3.14** Looking south-west along the Laki fissure, showing some of the conelets and pyroclastic deposits formed during the closing stages of the 1783 eruption (*G. Kjartansson*, 1956)

**Figure 3.15** Dyke of dolerite (a basaltic rock) cutting the Chalk and the overlying Tertiary plateau basalts, Cave Hill, Belfast, Northern Ireland (*R. Welch Collection, Copyright Ulster Museum*)

the term igneous can be properly applied to intrusions only when it can be inferred that they were emplaced at a relatively high temperature—several hundreds of degrees Centigrade or more.

The basaltic dyke illustrated in Figure 3.15 cuts through the white chalk over which the earliest lava-flows of Antrim were poured. Instead of being soft and friable, like the chalk of southern England, the Antrim chalk has become hardened by heat from the overlying basalts. And on both sides of the wall-like dyke the chalk has obviously been heated still more, having been recrystallized here and there into a fine-grained marble. The process of 'baking' and recrystallization by which the chalk has been transformed is an example of what is called *contact* or *thermal* metamorphism. Such evidence, if found in the rocks bordering an intrusion, indicates that the latter was capable of heating up its surroundings when it was emplaced, and that it may therefore be properly described as an igneous intrusion.

## Metamorphism of Rocks

Besides the contact metamorphism referred to above, rocks are subject to many other kinds of transformation to which the term *metamorphism* is applied, and examples of these will be considered in a later chapter. Here the term is being introduced to draw attention to the fact that rocks respond to the earth's internal activities not only by folding but also by recrystallization. When crustal rocks are buried and come under the influence of any combination of the following—

(*a*) the intense pressure or stress differences set up by gravity in association with other processes responsible for earth movements;

(*b*) the increased temperature associated with nearby igneous activity, or caused by internal friction or the through-passage of hot gases;

**Figure 3.16** Dolerite sills intrusive into horizontally bedded sandstones of the Beacon Series, exposed in the cliffs on the southern side of the Taylor Glacier, Southern Victoria Land, Antarctica. The sills range in thickness from 3 to 180 metres. At the lower left an offshoot from a sill can be seen cutting across the bedding in places, with corresponding displacement of the overlying sandstone. (*B. C. McKelvey and P. N. Webb*)

(*c*) the chemical changes stimulated by the through-passage of chemically active hot gases and liquids—

they respond by changes in structure and mineral composition and so become transformed into new types of rocks.

It must be carefully noticed that metamorphism is the very antithesis of weathering. Both processes bring about great changes in pre-existing rocks, but weathering is destructive while metamorphism is constructive. Instead of reducing a pre-existing rock to a decaying mass of rock-waste and soil, metamorphism brings about its transformation, often from a dull and uninteresting-looking stone, into a crystalline rock of bright and shining minerals and attractive appearance.

**Figure 3.17** Granite peaks of the Bergeller Massif across the Swiss-Italian border, north-east of Lake Como. Village of Soglio in the foreground. Between, on each side of the valley, are the metamorphic rocks of the Southern Alps in which the Bergeller granite intrusion is emplaced. (*F. N. Ashcroft*)

## Summary of the Processes of Land Destruction and Renewal

It will now be clear from our rapid survey of the leading geological processes that they fall into two contrasted groups. The first group—denudation and deposition—includes the processes which act on the crust at or very near its surface, as a result of the movements and chemical activities of air, water, ice and living organisms. Such processes are essentially of external origin. The second group—earth movements, igneous activity and metamorphism—includes the processes which act within or through the crust, as a result of the physical and chemical activities of the materials of both crust and mantle. Such processes are essentially of internal origin.

Both groups of processes operate under the control of gravitation (including attractions due to the sun and moon), co-operating with all the earth's bodily movements—of which the chief are rotation about its axis and revolution around the sun. But if these were all, the earth's surface would soon reach a state of approximate equilibrium from which no further changes of geological significance could develop. Each group of processes, to be kept going, requires some additional source of energy. The processes of external origin are specifically maintained by the radiation of heat from the sun. Those of internal origin are similarly maintained by the liberation of heat from the stores of energy locked within the earth.

Throughout the ages the face of the earth has been changing its expression. At times its features have been flat and monotonous. At others—as today—they have been bold and vigorous. But in what appears to be a continual conflict between the sun-born forces of land destruction and the earth-born forces of land renewal neither has yet permanently gained the mastery. The major groups of processes outlined in the adjoining classification do not, in fact, stand so much in contrast as the scheme adopted may suggest. It is only the practical necessity of dealing with one aspect at a time that justifies their separation. The classification should, indeed, be regarded only as a help towards grasping the leading features of a vast and complex subject. In reality many of the processes concerned are intimately geared together; and this can be effectively demonstrated by considering the balancing effects quietly achieved by gravitation.

## Classification of Geological Processes

PROCESSES OF EXTERNAL ORIGIN

1 *Denudation* (Weathering, Erosion, Transport)
Sculpturing of the land surface and removal of the products of rock decay mechanically and in solution.

2 *Deposition*
(*a*) of the debris transported mechanically (e.g. sand and mud)
(*b*) of the materials transported in solution:
  (i) by evaporation and chemical precipitation (e.g. rock-salt)
  (ii) by the intervention of living organisms (e.g. coral limestone)
(*c*) of organic matter, largely the remains of vegetation (e.g. peat).

PROCESSES OF INTERNAL ORIGIN

1 *Earth Movements*
Moving lithosphere; uplift and depression of land areas and sea floors; fold-mountain building; earthquakes.

2 *Metamorphism*
Transformation of pre-existing rocks into new types by the action of heat, pressure, stress and hot, chemically active, migrating fluids.

3 *Igneous Activity*
Emplacement of intrusions; emission of lavas and gases and other volcanic products.

## Isostasy and Geological Processes

It can now be realized that geological processes bring about changes that must inevitably upset the ideal state of isostatic balance which gravitation tends to establish. When a mountain range is carved into peaks and valleys and gradually worn down by the agents of denudation, the load on the underlying column of the crust is reduced by the weight of the rock-waste that has been carried away. At the same time a neighbouring column, underlying a region of delta and sea floor where the rock-waste is being deposited, receives a corresponding increase of load. Unless a compensating transfer of material occurs in depth, the two columns cannot remain in isostatic equilibrium. At the base of the crust the pressure exerted by the loaded column is increased, while that exerted by the unloaded column is decreased. In response to this pressure difference in the mantle a slow migration of material is set going, as illustrated in Figure 3.18. The loaded column sinks and the unloaded one rises. This process, whereby isostasy is restored, is called *isostatic readjustment*.

It may happen that certain processes disturb the pre-existing isostatic balance much more rapidly than it can be restored by deep-seated rock-flowage in the mantle. For example, when the last of the thick European and North American ice-sheets melted away between, say, 11,000 and 8,000 years ago, these regions were quickly relieved of an immense load of ice. The resulting uplifts which then began are still actively in progress. Far above the shores of Finland and Scandinavia there are raised beaches which show that an uplift of about 250 m has already occurred (Figure 3.19), and every twenty-eight years another 30 cm are added to the total all around the northern end of the Gulf of Bothnia. The region is still out of isostatic balance, and it can be estimated that it has still to rise another 200 m or so before equilibrium can be reached.

Similarly around the northern shores of Hudson Bay new rocky islands have appeared within the memory of the older Eskimos, and the land is known to have risen at least 9 m since the Thule Eskimos first established themselves there, indicating an average uprise of about a metre per century.

## The Paradox of Solids that Flow

Careful measurements of the rates at which the region around the Gulf of Bothnia has been uplifted during the last few thousand years (see p. 40) show that the rate has been gradually slowing down in a systematic way. The process corresponds to a sluggish inflow of deep-seated material that behaves in this respect like an extremely viscous fluid. We have now come face to face with an apparent contradiction that has seriously befogged geological thinking in the past, and still leads to much confusion of thought in many important parts of our subject.

Kelvin proved from his study of the tides that the earth behaves like a solid body which is at least

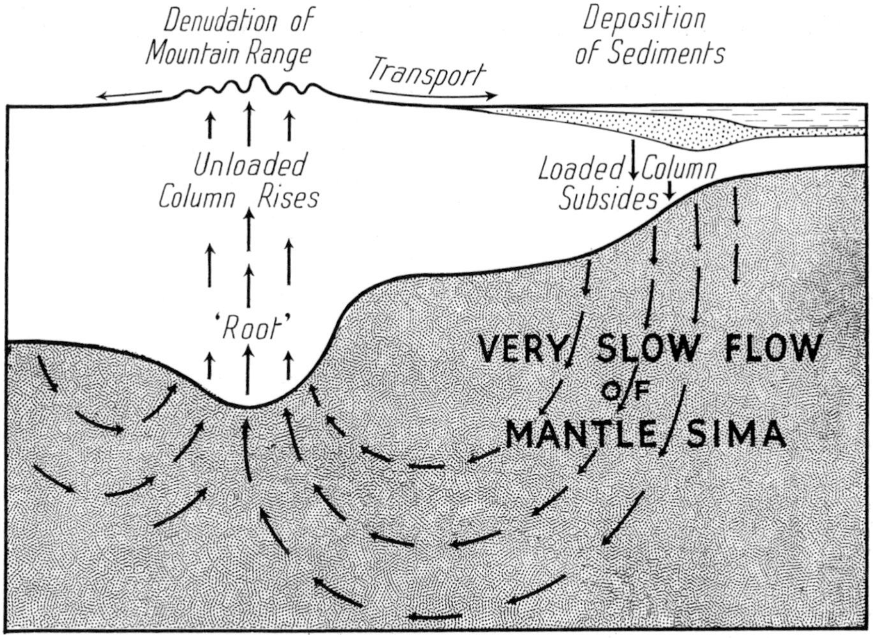


**Figure 3.18** Schematic section to illustrate isostatic readjustment, by slow sub-crustal flow in the mantle, in response to unloading by denudation and loading by deposition. Vertical scale greatly exaggerated.

as rigid as steel. Moreover, the crust and the mantle both transmit earthquake vibrations of a kind that cannot be sheared. Now *rigidity*—elastic resistance to change of shape—is the property we most characteristically associate with solids, and judged by this test the mantle is undoubtedly solid. How then can it also behave like a fluid?

Some authors still assert that the earth retains her shape because of the high rigidity which is demonstrated by tides and earthquakes. Yet we have seen (p. 19) that although Clairaut long ago calculated the shape or 'figure' of the earth and got the right result, his success was based on the assumption that the earth was in a molten state, apart from a thin solid crust. The earth's tidal response to the attraction of the moon is that of a solid body, but the earth's shape is that of a fluid body.

The equatorial bulge is responsible for another phenomenon—called the precession of the equinoxes—which also indicates fluidity. The attractions of the sun and moon on the bulge make the earth's axis wobble, so that the earth behaves like a spinning top that has been given a sideways push. At the present rate of precession the time for a complete wobble is nearly 26,000 years, just what it is calculated to be if the earth behaves as a fluid.

It will be noticed that the evidence for fluidity all comes from phenomena of long duration (from ten thousand years upwards), whereas the evidence for rigidity depends on earthquake waves (each lasting a second or so) and tides (half a day) and a few other short-lived phenomena. Since the mantle may be simultaneously transmitting earthquake waves and undergoing tidal strain, while it is also imperceptibly flowing to restore isostatic equilibrium, it is clearly both rigid and viscous simultaneously. It has a double personality, so to speak, being at the same time an elastic solid and a viscous fluid. It is not, however, a viscous liquid. Proof of rigidity is also a proof that the material concerned is not a liquid. A solid can become liquid only by

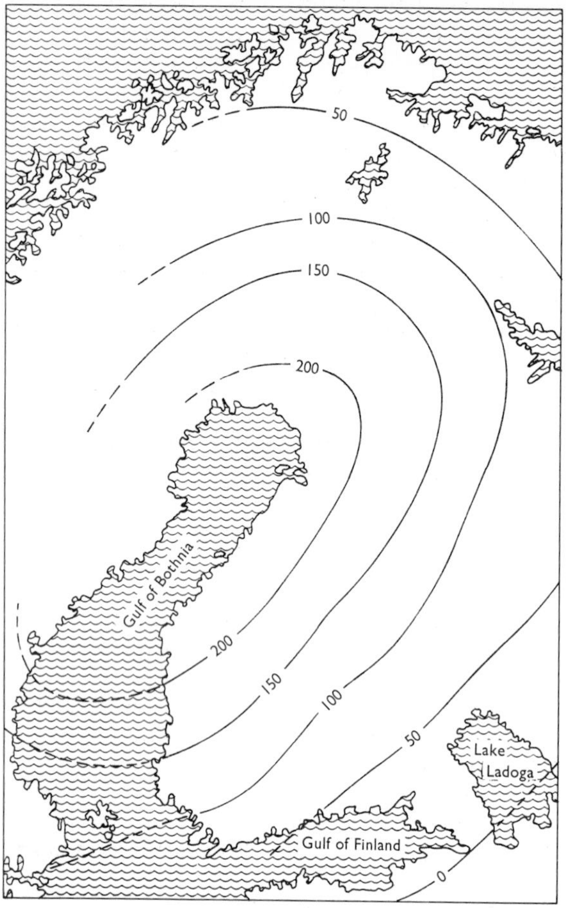


**Figure 3.19** Post-glacial uplift of Fenno-scandia (Finland and Scandinavia). The curves are lines of equal uplift, in metres, from 6800 B.C. to the present day. Around the northern shores of the Gulf of Bothnia the present rate of uplift is 1 cm per year. (*After Niskanen*)

melting or by dissolving in a medium which is already a liquid. But this does not mean that a solid is forever barred from flowing—that is from behaving as a fluid—even while it is still incontestably a solid.

The term *fluid*, as used today, has a much wider connotation than formerly. In its industrial applications it includes anything that can flow: e.g. gas, vapour, liquid, smoke, spray, emulsion, sludge, slurry, powder, and various fluidized systems in which solid particles are transported by pumping high-pressure steam or other gases through them. And geologically the term also includes solids under such conditions of stress and temperature that they can flow.

If a small force (pressure or stress) acts for a long time on a substance, the latter is said to be *viscous* if it flows continuously—however slowly—so long as the force is maintained. Time is, indeed, the clue for understanding the paradox of rocks that flow. Fortunately there are several familiar rocks that display their double personalities in periods of time that are well within the grasp of human experience. Pitch is perhaps the best known. A lump of it is easily broken into angular fragments by a tap from a hammer, but the same lump, left to itself at ordinary temperatures, flattens out under its own weight into a thin sheet. To a sharply applied force it behaves as a brittle solid, while in response to the long continued but much smaller stress of its own weight it flows as a viscous fluid. Figure 16.16, showing a convection cell in the Trinidad Pitch Lake, admirably illustrates this dual behaviour. In the middle the ascending current of asphalt makes a half-turn and then spreads radially outwards on a gradually lengthening circular front. For a time the rate of flow is such that the resulting tension stretches the asphalt crust beyond its elastic limit, so that it cracks and pulls apart into open fissures. Towards the outside, however, where the circumference reaches its greatest length and the flowing asphalt meets another convection cell and begins to turn downwards, the rate of flow is enormously reduced and the crust is wrinkled into concentric folds. The whole outspreading movement from the middle to the outside—about 60 cm—takes several months.

Ice is another material that demonstrably flows—in glaciers and ice-sheets—but, unlike pitch, ice in bulk is a crystalline rock. As might be expected its viscosity is very much higher than that of pitch: more than ten thousand times higher; while pitch, in turn, has a viscosity of about a million million times that of water. To grasp the significance of these differences in terms of time, suppose a small steel ball (such as those used in bicycle ball-bearings) takes a second to sink through the water filling a tall jar of just the right length. If the jar were filled with glycerine or castor oil the time taken to fall through would be about a quarter of an hour. If filled with cobbler's wax it would take the steel ball about a hundred years to reach the bottom. With pitch, the time required would be about 10,000 years; while if the jar were filled with ice, and kept well below freezing-point, the journey from top to bottom would probably take several million years. So sluggish is the flow of ice that it is no wonder we have an intuitive difficulty in understanding how it

**Figure 3.20** The contortions of ice and morainic debris here seen result from the competition for space between glaciers advancing into the same valley at different but comparable rates of flow, Alaska. (*Bradford Washburn*)

manages to flow at all. But flow it does, as Figure 3.20 demonstrates in no uncertain fashion.

Going farther up the viscosity scale we come next to rock-salt. This is also a crystalline rock, and it flows into such extraordinary forms of intrusion and extrusion that we shall have to consider them in detail later on (p. 153). But what of the really strong rocks, like granite? As we see them at the surface, where they are quarried, there is no present indication of any behaviour that could be called 'viscous'. For all practical purposes granite and similar rocks are strong enough to stand up as precipitous crags and mountains without detectable flow at the bottom. But this does not necessarily mean that there is *no* flow. It only means that weathering and erosion destroy the crags and mountains more rapidly than any flowage could do, if flowage there be. In the hotter depths, however, protected from the ravages of rain and rivers, these rocks have undoubtedly been able to flow as easily as ice or rock-salt. Indeed, were it not for such flowage and for the slow but relentless activities of the mantle, we should not be here to discuss the matter. The continents would long ago have disappeared beneath the sea, washed away by rain and rivers, and worn down still further by waves and currents. Only a few volcanic islands and coral reefs might diversify the otherwise monotonous surface. That the earth has avoided this uninteresting fate is due to processes at work in the depths. The changing face of the earth is primarily dependent on the fact that the materials of the core and mantle, and at least some of those of the crust, can flow.

## Selected References


CAREY, S. WARREN, 1954, 'The rheid concept in geotectonics', *Journal of the Geological Society of Australia*, vol. 1, pp. 67–117.

BULLARD, F. M., 1962, *Volcanoes: In History, In Theory, In Eruption*, University of Texas Press, Austin, Texas.

DURY, G. H., 1966, *The Face of the Earth*, Penguin Books, Harmondsworth.

GAMOW, G., 1963, *A Planet Called Earth*, Viking Press, New York.

HUBBERT, M. KING, 1945, 'Strength of the Earth', *Bulletin of the American Association of Petroleum Geologists*, vol. 29, pp. 1630–53.

LEES, G. M. and FALCON, N. L., 1952, 'The geographical history of the Mesopotamian Plains', *Geographical Journal*, vol. 118, pp. 24–39.

NORTH, F. J., *Sunken Cities: Some Legends of the Coast and Lakes of Wales*, University of Wales Press.

REINER, MARCUS, 1959, 'The flow of matter', *Scientific American*, vol. 201, pp. 122–38.

TAZIEFF, H., 1952, *Craters of Fire*, Hamish Hamilton, London.

# Chapter 4
# Materials of the Earth's Crust: Atoms and Minerals

Go, my sons, buy stout shoes, climb the mountains, search the valleys, the deserts, the seas shores, and the deep recesses of the earth. Mark well the various kinds of minerals, note their properties and their mode of origin.

*Petrus Severinus* (1571)

## Elements: Atoms and Isotopes

The vast majority of rocks are aggregates of minerals. Of the remainder some, like obsidian and pumice, are made of volcanic glass, while others, like coal, are composed of the residual products of organic decay. All these materials in turn are made of atoms of the chemical elements, of which 103 are known at the time of writing. Of these only 87 occur naturally in detectable amounts, the others having been synthesized by the modern alchemy of nuclear reactions. With the discovery of radioactivity about sixty years ago the atom lost its age-long status as the fundamental particle of an element that could not be further subdivided. For centuries the old alchemists tried in vain to change one element into another. They failed because they used only chemical reactions, and chemical reactions affect only the arrangements and associations of atoms; the elements themselves are not changed. Atomic or nuclear reactions, however, change the actual identities of the atoms concerned. Moreover, as we all know far too well, enormous amounts of energy are released in such reactions. Less alarmingly—indeed quite otherwise—radioactivity and nuclear physics have geological applications to many problems that otherwise could hardly be effectively tackled at all. It is therefore desirable to learn a little about the atom.

An atom cannot really be pictured in any intelligible way, since it behaves as if it were made up of 'particles' which at the same time are also systems of 'waves'. And it would be quite useless—at the moment—to ask what the particles are made of, or what it is that is waving. However, it is convenient for most purposes to visualize the atom as a constellation of *electrons* surrounding an inconceivably small central nucleus. The orbits in which the electrons revolve and spin occupy a relatively vast space—at least a million times as great as that occupied by the nucleus.

It is well known that an electron has (or perhaps is) an electric charge that is regarded as *negative*. This use of 'negative' can be very misleading unless it is clearly realized that the term is only a verbal convention and does not imply that anything is missing. How it came about was purely by chance. *Electron* is the Greek word for amber. When amber is rubbed with wool or fur it acquires the property of attracting little bits of paper; it is then said to be electrified or to have an electric charge. It was recognized that there must be two kinds of electric charges when it was found that certain other materials, e.g. glass rubbed with silk, develop a charge that is in one sense the 'opposite' of that acquired by amber. Two 'amber' charges repel each other, but an 'amber' charge and a 'glass' charge attract each other. Long ago the glass kind of charge came to be distinguished as positive, and the amber kind as negative. If a positively charged point is connected by a wire to a negatively charged point, a momentary current of electricity flows along the wire. Since flow is naturally from high to low, the direction of the current was assumed to be from positive to negative. We now know that this assumption was wrong. The current is a swift flow of electrons along the wire, but the

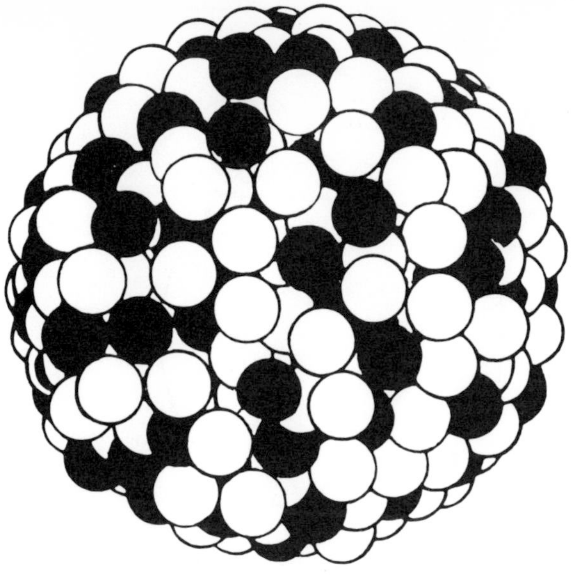

**Figure 4.1** A purely diagrammatic representation of the nucleus of an atom of lead (isotope 206); a cluster of 82 protons (black) and 124 neutrons (white). Magnification about ten million million diameters. It is essential to remember that it is merely for convenience that the protons and neutrons are depicted *as if* they were spherical objects; their real nature is as mysterious as that of electrons and all other 'elementary particles'.

'high' end, where there is an excess of electrons to being with, is the negative end. The excess electrons flash along the wire to the 'low' or positive end, where there is a deficiency of electrons. The charge on an electron has therefore to be thought of as 'negative'. This means, of course, that the charge on the nucleus of an atom must be regarded as 'positive', since an atom as whole is electrically neutral. However, atoms readily become electrically charged by gaining or losing electrons. Such electrified atoms are called *ions*. According to the convention, an atom with more than its normal ration of electrons is a negative ion, whereas one with less than is normal ration is positive.

The simplest atom is that of hydrogen. This has a single electron revolving around a nucleus consisting of a single particle, called a *proton*, the mass of which is 1836 times the mass of an electron. Each particle 'carries' a unit electric charge ($e$): negative ($e^-$) in the electron, positive ($e^+$) in the proton. The nuclei of other elements, however, contain *neutrons* as well as protons. These are particles with nearly the same mass as protons, but without any electric charge. In general the nucleus may be described as a cluster of tightly packed protons and neutrons: a cluster in which nearly the whole mass of the atom is concentrated (Figure 4.1). The fact that the protons do not fly apart as a result of their electric repulsions proves that they must be held together by inconceivably powerful forces of attraction. These nuclear attractions—by far the most powerful forces known to exist—are at least a million times stronger than the electric attractions that hold electrons and the nucleus together. In turn the electric forces are stronger than the chemical forces that hold atoms together in molecules and crystals, while gravitational forces, capable as they are of raising mountains, are utterly negligible on the atomic scale.

An element is now defined as a substance in which all the atoms have the same nuclear charge, i.e. the same number of protons. The number of protons in the nucleus of a given element is called the *atomic number* ($Z$), and the elements are numbered accordingly: starting with hydrogen, 1; helium, 2; and so on, through the periodic table of the elements, up to uranium, 92; and now up to the latest of the synthesized elements, lawrencium, 103. The elements were originally arranged in the periodic table according to their atomic weights, and chemists were puzzled by the fact that some elements have atomic weights that depart considerably from whole numbers. Chlorine, for example, has an atomic weight of 35·46. The reason for this and other apparent anomalies became clear when it was discovered that the number of neutrons ($N$) in the nucleus of a given element may vary, unlike the number of protons, which is invariable. The sum ($A$) of the number of protons ($Z$) and neutrons ($N$) in a nucleus is called the *mass number* ($A = Z+N$). It follows that an element does not necessarily consist of only one kind of atom. It happens that some of the elements with odd atomic numbers have all their atoms alike, e.g. sodium ($Z = 11$; $N = 12$); and gold ($Z = 79$; $N = 118$); but the rest, including all the elements with even atomic numbers, are mixtures of nuclear varieties with different numbers of neutrons in their nuclei and therefore with different mass numbers. These varieties are called *isotopes* (Gr. *isos*, equal; *topos*, place) of the given element, because they are identical in their chemical properties and so occupy the same place in the periodic table of the elements. Chlorine ($Z = 17$) is an example of an odd-numbered element with two isotopes: one with 18 neutrons (making $A = 35$), the other with 20 neutrons

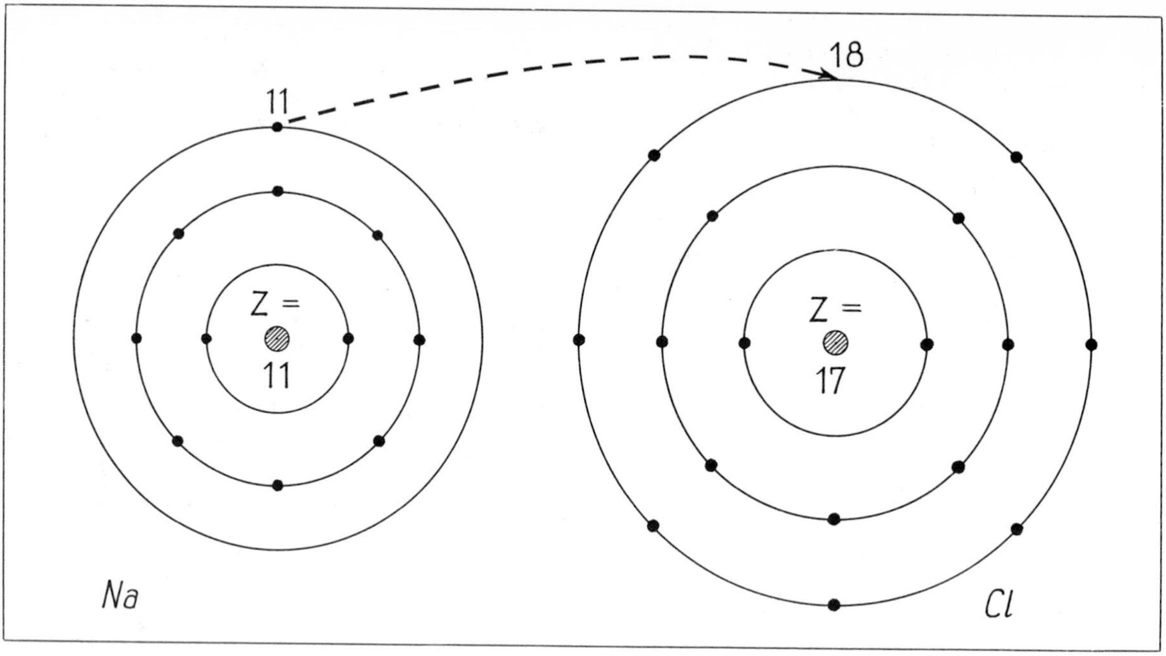


**Figure 4.2** The solitary electron No. 11 from a sodium atom jumps into the unoccupied orbit No. 18 of a chlorine atom, leaving the sodium as an ion $Na^+$ and transforming the chlorine to an ion $Cl^-$. $Na^+$ and $Cl^-$ then unite to form NaCl in which the electric charges are balanced.

(making $A = 37$). Approximately three-quarters of the atoms of natural chlorine are the isotope chlorine-35 (or $^{35}Cl$) and one-quarter the isotope chlorine-37 (or $^{37}Cl$); and accordingly the atomic weight cannot be far from 35·5, thus clearing up the old puzzle. It is of interest that chlorine extracted from meteorites has the same atomic weight and the same mixture of isotopes as terrestrial chlorine, whether the latter is prepared from natural rock-salt (NaCl), from sea-salt or from volcanic gases.

## Elements: Electrons and Ions

In a neutral atom—electrically uncharged—the number of electrons swinging in their orbits around the nucleus is the same as the number ($Z$) of the protons in the nucleus. These electrons do not revolve in a haphazard swarm, but have orbits concentrated in a series of concentric 'shells'. The innermost shell never contains more than two electrons; the second can accommodate up to eight, but no more; the third takes up to eight for some elements, but up to eighteen for others—and so on, The chemical properties of an element depend on the number of electrons in its atoms, and particularly on those in the outermost shell. This can be simply illustrated by considering how sodium and chlorine combine to form sodium chloride (NaCl), familiarly known as common salt or rock-salt (Figure 4.2). The two inner shells of the chlorine atom have their full complement of electrons, but the outer one has only seven of the eight it can take. Now as far as possible, electrons revolve in pairs, one spinning in a clockwise direction as it swings round its orbit, while the other spins in the opposite or anticlockwise direction. So in chlorine, one of the electrons in the outer shell is left without a dancing partner, so to speak: there is a vacancy to be filled. Sodium is a willing candidate. In sodium the inner shells are also fully occupied, but the outer shell has only one electron and this is so loosely held that, if a chlorine atom comes sufficiently near, it can jump into the vacant place. The chlorine atom (Cl), having gained an electron, has now acquired a negative electric charge and has become a negative ion ($Cl^-$), while the sodium atom (Na), which has been deserted by its outer electron, has acquired a positive charge and has become a positive ion ($Na^+$).

The sodium ion, $Na^+$, and the chlorine ion, $Cl^-$, are now mutually attracted and combine to form the uncharged 'molecule', NaCl, of common salt. In practice, if a bit of sodium is introduced into the poisonous green gas, chlorine, it burns to a white vapour from which tiny crystals of salt settle out. Millions of millions of atoms have turned into ions, all competing for partners. And since there are partners for all, they settle down in an orderly fashion which achieves electric balance and perfect uniformity of composition. Figure 4.3 illustrates the three-dimensional pattern of the ions in a crystal of common salt. Each crystal might be regarded as a sort of giant molecule, but more accurately it is a continuous ionic structure rather than a molecule.

The outer-shell electrons which are responsible for binding together various atoms into the molecules or crystals of a chemical compound are called the valency-electrons. *Valency* is the combining power of an atom (or group of atoms), expressed numerically as the number of electrons which each atom (or group of atoms) has gained, lost, or shared. Hydrogen and sodium ions each lose one electron and their valency is 1+. Some atoms, however, like those of oxygen need two electrons to complete their outer shells, Thus an oxygen atom can combine with two atoms of hydrogen and the result is water, $H_2O$. But it could also combine with one atom of an element with two valency-electrons in its outer shell, e.g. magnesium, to make magnesia, MgO; or calcium, to make lime CaO. In these cases the valency is 2+. Where the valency is 3+, as in aluminium, the combination with oxygen requires the proportions represented by $Al_2O_3$; this oxide is called alumina, and as a gemstone it is familiar as ruby and sapphire. Iron is a more complicated case, as its valency can be either 2+ or 3+. Thus it makes two series of oxides—and of other compounds—which are distinguished as ferrous ($Fe^{2+}$) and ferric ($Fe^{3+}$): e.g. ferrous oxide, FeO, and ferric oxide, $Fe_2O_3$.

In the mineral world the most important element with a valency of 4+ is silicon. Its oxide, silica, represented by $SiO_2$, is well known as quartz, agate and flint. Moreover, as indicated below, silica combines with the other common oxides to form a group of crystalline compounds known as *silicates* which, together with quartz, constitute a very large majority of the minerals occurring in rocks. Carbon is another element with a valency of 4, vitally important because the latter may be either 4− or 4+. Carbon with a negative valency (as in the carbon of methane, $CH_4$, and of related hydrocarbons in petrol and petroleum) is the essential element in organic compounds and in all living matter. When the valency is positive, however, as in the carbon of carbon dioxide, $CO_2$, the latter combines with other oxides to form another important group of rock-forming minerals, the *carbonates*, which predominate in limestones and marbles. With this introduction we may now turn to the rocks and minerals themselves.

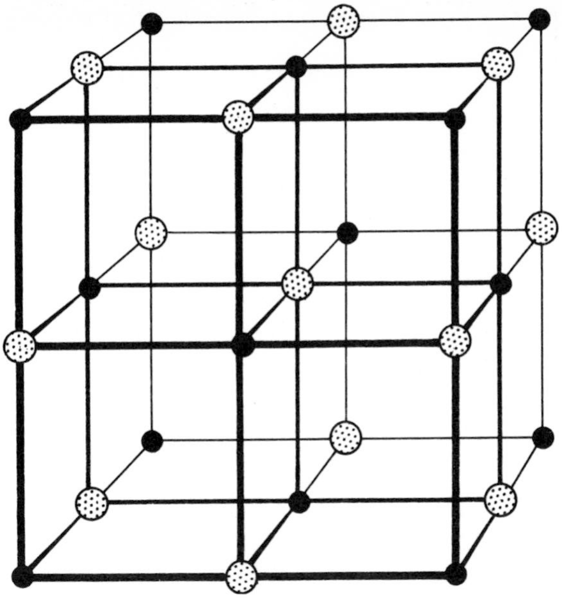

**Figure 4.3.** Lattice structure of a crystal of common salt, NaCl. Relatively small $Na^+$ ions (black) and larger $Cl^-$ ions (dotted) are arranged alternately at the corners of a set of cubes (here drawn in to show the relationship). Each $Na^+$ ion is surrounded by six $Cl^-$ ions, and each $Cl^-$ ion by six $Na^+$. On the scale of the diagram the ionic 'spheres' are shown—for clarity—as having only about one-third of the radii conventionally ascribed to them. See Fig. 4.8.

## Chemical Composition of Crustal Rocks

Although 87 elements occur naturally in minerals, eight of these are so abundant that they make up nearly 99 per cent by weight of all the many thousands of rocks that have been chemically analysed. Many of the others, such as gold, tin, copper and uranium, though present only in traces in ordinary rocks—for which reason they are referred to as *trace elements*—are locally concentrated in mineral veins and other ore deposits sufficiently to make their extraction profitable. Elements 43, 61, 87 and 89 have not yet been detected at all in minerals, but they have been made artificially by nuclear reactions.

# Average Composition of Crustal Rocks

(After V. M. Goldschmidt and Bryan Mason)

In Terms of Elements

| *Name* | *Symbol and Valency* | *Per cent* |
|---|---|---|
| Oxygen | $O^{2-}$ | 46·60 |
| Silicon | $Si^{4+}$ | 27·72 |
| Aluminium | $Al^{3+}$ | 8·13 |
| Iron | $Fe^{3+}$, $Fe^{2+}$ | 5·00 |
| Calcium | $Ca^{2+}$ | 3·63 |
| Sodium | $Na^{+}$ | 2·83 |
| Potassium | $K^{+}$ | 2·59 |
| Magnesium | $Mg^{2+}$ | 2·09 |
| Titanium | $Ti^{4+}$ | 0·44 |
| Hydrogen | $H^{+}$ | 0·14 |
| Phosphorus | $P^{5+}$ | 0·12 |
| | | 99·29 |

In Terms of Oxides

| *Name* | *Formula* | *Per cent* |
|---|---|---|
| Silica | $SiO_2$ | 59·26 |
| Alumina | $Al_2O_3$ | 15·35 |
| Iron oxides: Ferric | $Fe_2O_3$ | 3·14 |
| Iron oxides: Ferrous | $FeO$ | 3·74 |
| Lime | $CaO$ | 5·08 |
| Soda | $Na_2O$ | 3·81 |
| Potash | $K_2O$ | 3·12 |
| Magnesia | $MgO$ | 3·46 |
| Titania | $TiO_2$ | 0·73 |
| Water | $H_2O$ | 1·26 |
| Phosphorus pentoxide | $P_2O_5$ | 0·28 |
| | | 99·23 |

Continuing the above table of abundances, the elements immediately following are *manganese*, Mn, 0·10; *flourine*, F, 0.08; *sulphur*, S, 0·05; *chlorine*, Cl, 0·04; and *carbon*, C, 0·03 per cent. The abundances of the rarer or trace elements are more conveniently expressed in parts per million (p.p.m.), which is the same as grams per metric ton (tonne, $10^6$g=2205 lb.). Gold and platinum, though well known to everyone as precious metals, are amongst the rarest of elements in ordinary rocks, their average abundances being only about 0·005 p.p.m.

## Minerals and Crystals

Some of the elements, e.g. gold, copper, sulphur and carbon (as diamond and graphite), make minerals by themselves, but most minerals are compounds of two or more elements. Oxygen is by far the most abundant element in rocks. In combination with other elements it forms compounds called oxides, some of which occur as minerals. As silicon is the most abundant element after oxygen, it is not surprising that silica, $SiO_2$, should be the most abundant of all oxides. Silica is familiar as **quartz**, a common mineral which is specially characteristic of granites, sandstones and quartz veins. The formula, $SiO_2$, is a simple way of expressing the fact that for every atom of silicon in quartz there are two atoms of oxygen. Perfectly pure quartz has, therefore, a definite composition. The formulae for other oxides and compounds may be similarly interpreted.

In the cavities of mineral veins, quartz can be found as clear transparent prisms, each with six sides and each terminated by a pyramid with six faces. Thinking of the mineral as a variety of ice that had been permanently frozen, the ancient Greeks gave the name *krystallos* (clear ice) to these beautiful forms, and to this day water-clear quartz is still known as rock crystal (Figure 4.4). Most other minerals and a great variety of chemically prepared substances can also develop into symmetrical forms bounded by plane faces, and all of these are now called *crystals*.

In recent years the study of crystals by means of X-rays has revealed the fact that their symmetrical forms are simply the outward expression of a perfectly organized internal structure. As already indicated in the case of rock-salt (Figure 4.3), the electrically charged atoms or ions of which a crystal is composed are arranged in an orderly fashion, the different kinds of atoms being built into a definite pattern which is repeated over and over again, as in the design of a wallpaper. In crystals, however, the design is in three dimensions and for this reason it is referred to as a *space lattice*. Perhaps the nearest approach to 'seeing' atoms is illustrated by Figure 4.6, which shows with remarkable clarity the distribution of atoms in one layer of the crystal lattice of pyrite, $FeS_2$ (Figure 4.7).

It will have been noticed that diamond and graphite are both crystalline forms of carbon. Correspondingly to their strongly contrasted physical properties—one being hard and brilliant, the other soft, opaque and flaky—the crystals of diamond and graphite have very different lattice structures. This contrast in turn reflects the widely different physical conditions under which the two minerals crystallized. Diamond requires a combination of high temperature with such extremely high pressure that it has only recently become possible to make diamonds—though not of gem quality—artificially. For graphite, however, quite moderate conditions of temperature and pressure suffice. This capacity of certain substances to occur as two or more quite different species of crystals, i.e. to crystallize with lattice structures appropriate to the physical conditions at the time of formation, is a phenomenon described as *polymorphism* (Gr. *polys*, many; *morphe*, shape or form). Other well-known examples are $FeS_2$, which occurs not only as pyrite, but also as marcasite; and calcium carbonate, $CaCO_3$, which crystallizes mainly as calcite, but under special conditions as aragonite (e.g. in the shells secreted by certain molluscs and other marine organisms).

Few minerals have the exact chemical com-

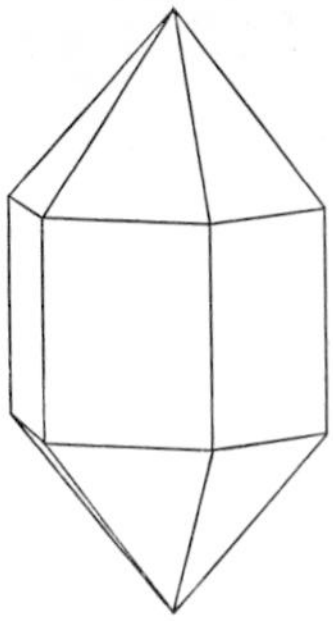

**Figure 4.4** Doubly terminated crystal of quartz. Such crystals grow only where they can develop freely at both ends. More usually quartz crystals grow from the walls of a cavity and are singly terminated, as in Fig. 4.5.

**Figure 4.5** Group of quartz crystals from Minas Gerais, Brazil (*Trustees of the British Museum, Natural History*)

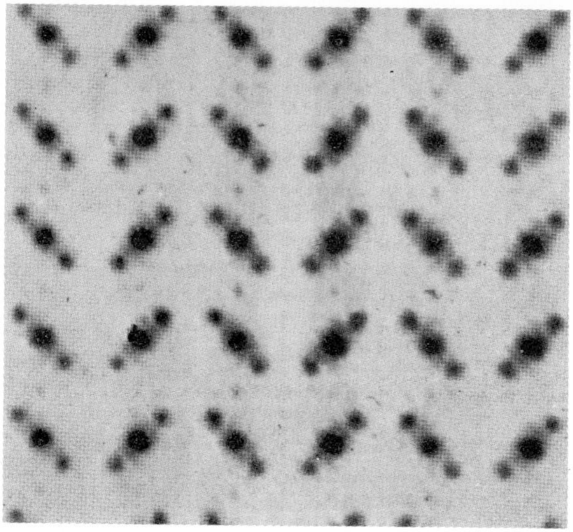

**Figure 4.6** Image of the lattice structure of pyrite, magnified 2·00 million diameters. The large spots represent ions of Fe; the smaller ones, of which there are twice as many, represent paired ions of S. The X-ray diffraction pattern produced by a pyrite crystal is made into a grating and diffracted again, thus giving an enormous magnification. For details see M. J. Buerger, 'The photography of atoms in crystals.' *Proceedings of the National Academy of Sciences*, Washington, vol. 36, pp. 330–5, 1950. (*M. J. Buerger*)

position corresponding to their ideal formulae. The reason is that any ion that happens to be on the spot at the time of crystallization can act as a substitute or proxy for another without seriously disturbing the crystal lattice, provided that the ion of the proxy has nearly the same 'size' (Figure 4.8) as that of the ion whose place it takes in the growing crystal. In just the same way bricks of standard size but of different colours could be built into a wall without altering the structure of the wall or its outward shape. In building the crystal edifice such substitution is especially favoured when both ions have the same electric charge or valency. A good example is provided by the green mineral, olivine. The formula is generally written $(Mg,Fe)_2SiO_4$ to express the fact that it is part of a continuous series ranging from $Mg_2SiO_4$ at one end, through $(Mg,Fe)_2SiO_4$ and $(Fe,Mg)_2SiO_4$ to $Fe_2SiO_4$ at the other end, all the members having essentially the same lattice structures and crystal forms. The intermediate members of the olivine series are not to be thought of as mixtures or combinations of the two 'isomorphous' compounds or molecules represented by the formulae of the end-members. It is not the compounds that are mixed, but the ions in the crystal lattice: in this case $Mg^{2+}$ and $Fe^{2+}$, which are very nearly the same size (see Figure 4.8). Crystals belonging to such series are commonly referred to as solid solutions or, less happily, as 'mixed crystals'.

Equality of valency is, however, by no means essential for effective substitution. All that is necessary, provided that the ionic sizes are not too different, is that electric neutrality is maintained, and this can be achieved by a balancing interchange between the ions of two or more other elements in the lattice. Thus, in the plagioclase series of the feldspar group (see p. 54) there is a continuous range of composition from albite, $NaAlSi_3O_8$, at one end to anorthite, $CaAl_2Si_2O_8$ at the other. Here $Al^{3+}$ replaces one of the $Si^{4+}$ ions, while at the same time, to balance the resulting deficiency in charge, $Ca^{2+}$ replaces $Na^+$.

A glance at Figure 4.8 indicates that $Na^+$ and $K^+$ are too different in size to be readily interchangeable. This corresponds to their actual associations in minerals. On the other hand, the relatively rare element, rubidium, $Rb^+$, has nearly the same size as $K^+$ and consequently growing potash minerals provide an equally good home for rubidium. The natural concentrations of rubidium occur almost entirely in potash feldspars and micas. This is a fortunate circumstance from a geological point of view because both elements have radioactive isotopes, and potash minerals containing rubidium in sufficient quantity can be dated in two independent ways, as we shall see in a

**Figure 4.7** Crystals of pyrite or iron pyrites, $FeS_2$ (brassy-looking striated cubes) from Leadville, Colorado, U.S.A. *Pyrites* is an old Greek name for minerals that give off fiery sparks when struck. (*Ward's Natural Science Establishment, Inc., Rochester 9, N.Y., U.S.A.*)

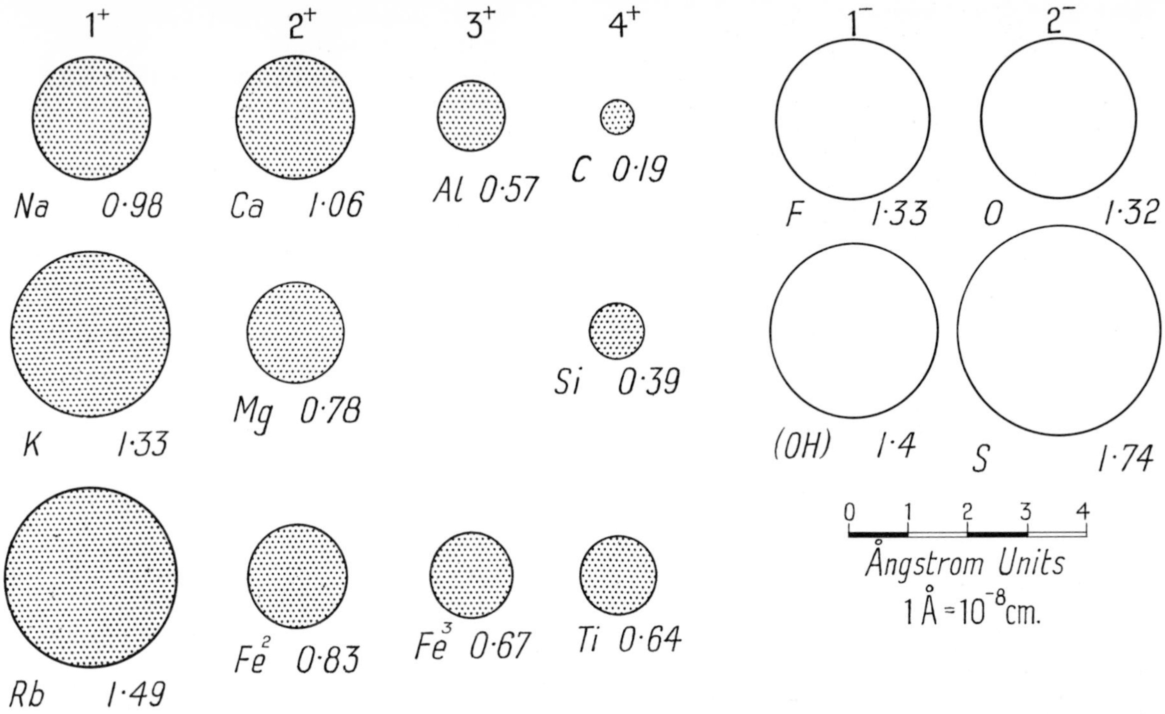


**Figure 4.8** Relative sizes of some positive ions (*cations*) and negative ions (*anions*) in crystals. The figure at the head of each column refers to the ionic charge or valency. The effective ionic radius of the field of influence of each ion within a crystal, conventionally regarded as spherical, is given in ångstroms.

later chapter (p. 235).

A few minerals are non-crystalline in the sense that they never develop crystal forms, and for this reason they are said to be *amorphous*. Examples are *opal*, $SiO_2.nH_2O$, which has been described as 'an incompletely dried-out jelly', and *limonite*, approximately $Fe_2O_3.H_2O$, one of the iron hydroxides familiar as iron rust and as the material responsible for the rusty-brown appearance of many weathered rock surfaces. In such materials very tiny particles are arranged haphazardly, like the bricks in a tumbled heap, but investigations with the electron microscope show that within each of the particles the atoms have a recognizable lattice arrangement.

Glass is also a typically amorphous substance. Most glasses are mixtures of silicates that have had insufficient time for the atoms to arrange themselves into the regular patterns of crystals, either because of rapid cooling from the molten state, or because the original melt was extremely viscous from the start. X-ray study reveals, however, that twisted and misshapen lattices are present, indicating that the first steps towards crystallization have been taken. Moreover, this tendency to bring crystalline order out of amorphous chaos is eventually successful. Despite the fact that glass has the mechanical properties of a solid with very high viscosity, it slowly *devitrifies* and turns into an aggregate of minute crystals. The process of devitrification may take a few years or many hundreds in the case of man-made glass at ordinary temperatures. Natural glass, like the highly viscous volcanic lava that solidifies as pumice or obsidian (Figure 5.11), may require millions of years before it shows visible signs of crystallization. The transformation is speeded up, however, if hot volcanic gases find a way through the glassy material.

Apart from the amorphous materials, minerals are naturally-occurring inorganic crystalline substances, of which each 'species' has its own specific variety of crystal structure. Allowing for the inevitable presence of 'impurities' and trace elements, the chemical composition may be constant (as in quartz) or it may vary (as in the feldspars) within limits that depend on the degree to which the ions of certain elements can substitute for those of other elements without changing the specific pattern of the crystal lattice.

## Rock-forming Minerals

Although about 2000 named minerals are known, most common rocks can be adequately described in terms of a dozen series of minerals, as the following table indicates. It is, therefore, well worth while to become familiar with these essential rock-forming minerals and with a few others of special interest, and especially to learn something of their chemical compositions. An attempt is made here to present this minimum equipment of chemical knowledge as briefly as possible. The student should refer to textbooks for additional information, and above all he should handle typical specimens of minerals and rocks and examine actual rock exposures out of doors whenever opportunity affords.

By far the most abundant of the rock-forming minerals are silicates, but before dealing with these a few other important minerals—oxides, carbonates, etc.—may conveniently be passed in review.

*Quartz* has already been referred to as an oxide mineral. On p. 53 it is classified as a mineral with a silicate structure; chemically it could be regarded as either $SiO_2$ or $Si(SiO_4)$.

Alumina, $Al_2O_3$, occurs naturally as *corundum*, the hardest natural abrasive after diamond, and in its rarer transparent form familiar as the gemstones ruby and sapphire.

In their large-scale occurrences the oxides of iron are, with the carbonate, $FeCO_3$, the chief sources of iron ore; as accessories they are notable constituents in a great variety of common rocks. *Haematite*, $Fe_2O_3$, takes its name from the Greek word for 'blood' in reference to its colour. *Magnetite*, $Fe_3O_4$, is black and strongly magnetic. *Ilmenite*, $FeO.TiO_2$, is often associated with magnetite, especially in basalts and rocks of similar composition. *Limonite*, averaging about $FeO.H_2O$, is the rusty alteration product of other iron minerals.

*Ice*, crystalline water, $H_2O$, is not commonly thought of as a rock-forming mineral, but glaciers and ice-sheets are rocks on the grand scale composed of granules of ice, and they are none the less rocks because they flow, melt and evaporate before our eyes.

After water, the next oxide in order of abundance (p. 46) is that of phosphorus, an element of critical importance in agriculture, and indeed for life generally. Phosphates occur in ordinary rocks as the mineral *apatite*, $Ca_5F(PO_4)_3$, and as a related compound of organic derivation (e.g. from fish bones and teeth and the excreta of birds) which has a similar composition, but with (OH) instead of F. If water charged with traces of fluorine passes or soaks through such *phosphorite*, the latter gradually gives up its (OH) in exchange for F, so approaching apatite in composition and becoming more stable. The same process tends to happen

**Figure 4.9** Models to illustrate the $(SiO_4)^{4-}$ tetrahedron, the fundamental atomic structural unit of all silicate crystals. A small ion of silicon is enclosed by four much later ions of oxygen. On this scale the ions are magnified about 65 million times.

## Average Mineral Composition of some Common Rocks

| *Minerals* | *Granite* | *Basalt* | *Sandstone* | *Shale* | *Limestone* |
|---|---|---|---|---|---|
| Quartz | 31·3 | — | 69·8 | 31·9 | 3·7 |
| Feldspars | 52·3 | 46·2 | 8·4 | 17·6 | 2·2 |
| Micas | 11·5 | — | 1·2 | 18·4 | — |
| Clay minerals | — | — | 6·9 | 10·0 | 1·0 |
| Chlorite | — | — | 1·1 | 6·4 | — |
| Amphiboles (mainly hornblende) | 2·4 | — | — | — | — |
| Pyroxenes (mainly augite) | rare | 36·9 | — | — | — |
| Olivine | — | 7·6 | — | — | — |
| Calcite and dolomite | — | — | 10·6 | 7·9 | 92·8 |
| Iron ores | 2·0 | 6·5 | 1·7 | 5·4 | 0·1 |
| Other minerals | 0·5 | 2·8 | 0·3 | 2·4 | 0·3 |

during the growth of children's teeth, and perhaps later, and this explains how it is that the addition of traces of fluoride salts (1–3 p.p.m.) to drinking water helps to make teeth more resistant to decay in areas where the water supply is naturally deficient in fluorine. The chief fluoride mineral is *fluorspar*, $CaF_2$, which characteristically occurs in association with lead and zinc ores; it appears in a variety of attractive colours and is mainly used as a flux in making iron and steel.

Among the sulphur minerals, sulphur itself and pyrite have already been mentioned. Most of the important ores of lead, zinc, copper and nickel are sulphides. As rock-forming minerals, however, two sulphates are outstanding: *anhydrite*, $CaSO_4$, and *gypsum*, $CaSO_4.2H_2O$. Anhydrite, with or without gypsum according to circumstances, occurs mainly in salt deposits (evaporites), such as are left behind when salt lakes dry up, or when enclosed bodies of sea water are strongly evaporated. When the brine becomes sufficiently concentrated, *rock-salt* or *halite*, the chief chloride mineral, begins to be precipitated together with anhydrite.

Evaporation of sea water might be expected to begin with precipitation of carbonates, and so it does, but only on a trifling scale. Most of the carbonate rocks have different modes of origin, as we shall see later (pp. 82–3). The chief minerals of these rocks are:

*Calcite*, $CaCO_3$, the predominant mineral of limestones

*Dolomite*, $CaMg(CO_3)_2$, which occurs intermixed with calcite in magnesian limestones and, essentially by itself, as the predominant mineral of a carbonate rock that is also called dolomite

*Siderite* or *chalybite*, $FeCO_3$, an important ore of iron.

Carbonates of lead, zinc and copper are locally prominent minerals in many of the ore deposits of these metals.

### Crystal Structures of the Silicate Minerals

As a result of the exploration of the crystal lattices of the silicates by means of X-rays, initiated by Sir Lawrence Bragg and largely carried out by him and his co-workers, these minerals can now be classified in a most elegant way. The fundamental unit of silicate structures consists of a tetrahedral arrangement of four ions of oxygen—one at each corner of the tetrahedron—with an ion of silicon tucked into the interstitial space in the middle (Figure 4.9). Picture three tennis balls (6 cm in diameter) placed at the corners of a triangle and just touching; a marble about 1·6 cm in diameter then just fits into the central dimple (as in Figure 4.9, right); a fourth tennis ball placed on top (as in Figure 4.9, left) completes the structure. In this model the tiny silicon ion and the relatively large oxygen ions are magnified about 230 million times.

The four positive charges of $Si^{4+}$ are balanced by four negative charges, one from each of the four oxygen ions, $O^{2-}$, thus leaving each tetrahedron with four negative charges. By themselves these tetrahedral building units would fly apart in

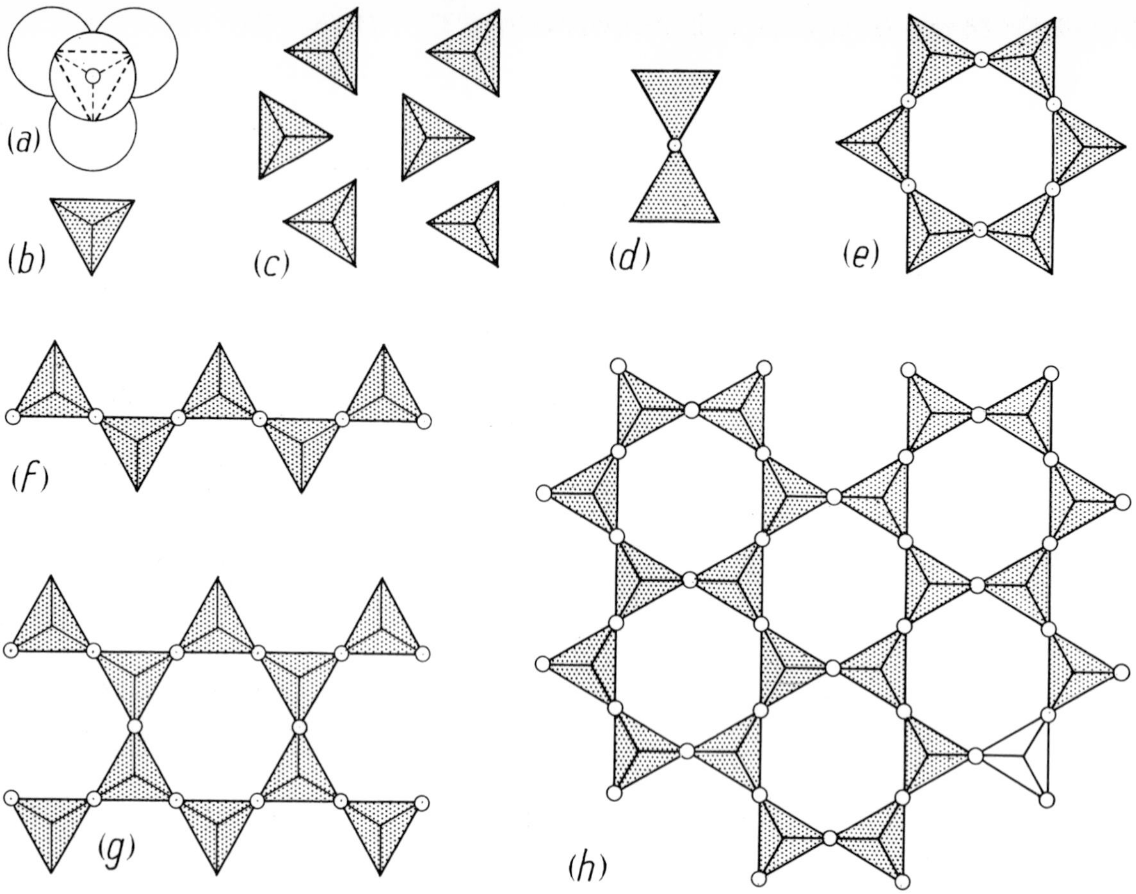


**Figure 4.10** Some of the chief structural arrangements of the $SiO_4$ tetrahedron in crystals:

(*a*) The $SiO_4$ tetrahedron, with the ions approximately to scale

(*b*) Conventional representations of the $SiO_4$ tetrahedron, shown by broken lines in (*a*). In (*d*) to (*h*) shared oxygens are indicated by open circles.

(*c*) No oxygens shared (e.g. olivine)

(*d*) A pair of tetrahedra sharing one oxygen (e.g. melilite)

(*e*) A ring of six tetrahedra, each sharing two oxygens (e.g. beryl)

(*f*) A single chain of tetrahedra, each sharing two oxygens (e.g. pyroxenes)

(*g*) A double chain of tetrahedra; the outward-pointing tetrahedra share two oxygens, as in (*f*), while those pointing inwards share three oxygens and so produce a succession of hexagonal 'holes' large enough to accommodate ions of hydroxyl (OH), or fluorine, F (e.g. amphiboles)

(*h*) A sheet of tetrahedra, each sharing three oxygens and forming a continuous network with hexagonal 'holes' as in (*g*) (e.g. mica)

consequence of the resulting electrical repulsion. To bind them strongly together, as in crystals, they must be cemented or linked so that the charges, are neutralized. The fascinating architecture of crystals is entirely controlled by the various ways in which this is accomplished. The tetrahedra may be held together by the proper proportion of metal ions (e.g. $Mg^{2+}$ and $Fe^{2+}$), as in olivine; or they may be linked together by sharing (and thereby neutralizing) all four oxygen ions with their immediate neighbours, as in quartz; or they may share only one, two or three of their oxygens, leaving the remaining charges to be neutralized by metal ions of appropriate kinds. The linkages result in the building up of structures such as separate pairs and rings, single and double chains, and sheets (as illustrated conventionally in Figure 4.10) and finally as the more complex three-dimensional frameworks of quartz and feldspars—patterns spreading out in all directions—so that they cannot be clearly represented except in three

# Classification of Silicate Structures

| *Structural relationship of Silicon-oxygen tetrahedra* | Si:O *ratio* | *Characteristic examples* |
|---|---|---|
| NESOSILICATES (Gr. *nesos*, an island)<br>Separate $SiO_4$-tetrahedra, held together by ions such as $Mg^{2+}$. No oxygen shared (Fig 4.10c) | $\mathbf{Si_4O_6}$ | olivine $\begin{cases} Mg_2SiO_4 \\ Fe_2SiO_4 \end{cases}$<br>*Garnets*; *Zircon*; *Topaz* |
| SOROSILICATES (Gr. *soros*, a group)<br>Separate pairs of tetrahedra, formed by sharing one oxygen (Fig. 4.10*d*) | $\mathbf{Si_4O_{14}}$<br>$\mathbf{(Al_2Si_2O)_{14}}$ | *Melilite* $\begin{cases} Ca_2MgSi_2O_7 \\ Ca_2Al(AlSi)O_7 \end{cases}$ |
| CYCLOSILICATES (Gr. *kyklos*: L. *cyclus*, a circle or ring)<br>Separate closed rings of 3, 4 or 6 tetrahedra, formed by sharing two oxygens (Fig. 4.10e) | $\mathbf{Si_4O_{12}}$ | beryl (*Emerald*) $Al_2Be_3Si_6O_{18}$<br>*Tourmaline* |
| INOSILICATES (Gr. *inos*, a thread or, fibre)<br>Continuous single chains of tetrahedra, formed by sharing two oxygens (Fig 4.10*f*) | $\mathbf{Si_4O_{12}}$ | pyroxenes, e.g.<br>*Hypersthene* $(Mg,Fe)SiO_3$<br>*Diopside* $Ca(Mg,Fe)Si_2O_6$<br>*Augite* (complex) |
| Continuous double chains, formed by the lateral coalescence of two single chains. Alternately two and three oxygens are shared (Fig. 4.10g). The resulting hexagonal spaces accommodate ions of $(OH)^-$ or $F^-$ | $\mathbf{Si_4O_{11}}$ | amphiboles, e.g.<br>*Tremolite* $Ca_2Mg_5\ Si_8O_{22}(OH)_2$<br>*Hornblende* (complex) |
| PHYLLOSILICATES (Gr. *phyllon*, a leaf)<br>Continuous plane sheets of hexagonal networks, like wire-netting, formed by sharing three oxygens (Fig. 4.10*h*) | $\mathbf{Si_4O_{10}}$ | *Talc* $Mg_3Si_4O_{10}(OH)_2$<br>*Serpentine* $Mg_3Si_2O_5(OH)_4$<br>clay minerals, e.g. $Al_2Si_2O_5(OH)_4$ |
| | $\mathbf{AlSi_3O_{10}}$ | micas, e.g.<br>*Muscovite* $KAl_2(AlSi_3O_{10})\ (OH,F)_2$ |
| TECTOSILICATES (Gr. *tekton*, a builder or frame-maker)<br>Continuous framework of tetrahedra in three dimensions; all four oxygens shared | $\mathbf{Si_4O_8}$ | quartz $SiO_2$ |
| | $\mathbf{AlSi_3O_8}$ | feldspars, e.g.<br>*Orthoclase* $K(AlSi_3)O_8$<br>*Albite* $Na(AlSi_3)O_8$<br>*Anorthite* $Ca(Al_2Si_2)O_8$ |
| | $\mathbf{Al_2Si_2O_8}$ | feldspathoids, e.g.<br>*Nepheline* $Na(AlSi)O_4$ |

dimensions.

The resulting classification is summarized in the table on page 53. As the number of shared oxygens increases, the proportion of O to Si necessarily decreases; this is emphasized in the table by adopting $Si_4$ as a unit of easy comparison throughout. In explanation of some of the mineral formulae it should be mentioned that Al plays a double role in crystals. It may replace Si inside some of the tetrahedra, as in the feldspars. $Si_4$ then becomes ($AlSi_3$) or ($Al_2Si_2$), in which cases corresponding additions of suitable ions are necessary to balance the altered charges. A further complication arises in minerals that contain ions like hydroxyl, $(OH)^-$, or fluorine, $F^-$; these may occupy relatively large 'hexagonal' spaces in the crystal lattice, as in mica (Figures 4.10(*h*) and 4.14), or they may substitute for oxygen. It should be noticed that the somewhat mysterious ion hydroxyl, $(OH)^-$, does not imply that the 'water' introduced into the crystal lattice has lost hydrogen; what happens is that the place of an oxygen ion is taken by two hydroxyl ions: $H_2O + O^{2-} = 2(OH)^-$.

It used to be said that the beauty of a crystal depends on the planeness of its faces. But now it has been realized that the beauty of a crystal is far from being only skin deep, as that old academic quip might suggest. The well-developed faces that are still commonly thought of as the most distinctive features of crystals are only the outward expressions of a pattern of ions cemented by their own electric charges. The internal architecture of the crystal edifice displays a natural beauty that is even more astonishing than the external façade, and one, moreover, that is equally present throughout the most minute fragment, however worn or broken it may be.

**Figure 4.11** Subdivision of the plagioclase series of feldspars by percentages of *Ab* and *An*

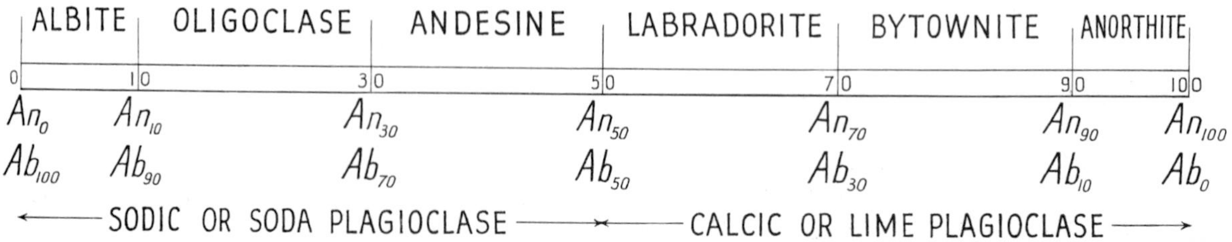


## Rock-forming Silicate Minerals

**Feldspars** are the most abundant minerals of the earth's crust, and, as we have seen, they consist of frameworks of $SiO_4$— and $AlO_4$— tetrahedra, with ions of potassium, sodium or calcium occupying the appropriate places in the structure. These minerals may therefore be considered as solid solutions of three ideal compounds which, as indicated below, are commonly distinguished from the minerals themselves by the use of the convenient symbols *Or*, *Ab* and *An*.

| | | | | |
|---|---|---|---|---|
| *Alkali Feldspars* | Orthoclase | *Or* | $KAlSi_3O_8$ | |
| *Alkali Feldspars* | Albite | *Ab* | $NaAlSi_3O_8$ | *Plagioclase* |
| | Anorthite | *An* | $CaAl_2Si_2O_8$ | *Plagioclase* |

In the alkali feldspars sodium and potassium are interchangeable to only a limited degree. Thus the potash feldspars *orthoclase* or *microcline* generally contain a small amount of *Ab* in solid solution with *Or*, while the soda feldspar *albite* generally contains a little *Or* with the predominant *Ab*. Solid solution or, more accurately, ion substitution is limited at both ends. *Ab* and *An*, however, form a continuous series of minerals, known collectively as *plagioclase*. Because of their abundance and importance in classifying rocks, the series is conventionally subdivided as indicated in Figure 4.11. *Labradorite*, named after Labrador, where a celebrated iridescent variety occurs, is the characteristic feldspar of basalts. *Andesine* shares its name with that of the volcanic rocks in which it is most abundant: rocks called andesite because of their common occurrence as lavas erupted from the volcanoes of the Andes.

Orthoclase can be easily recognized as the pinkish or cream-coloured mineral in granite. In some granites, like those of Cornwall and Shap Fell (Figure 4.12), large slab-like crystals of orthoclase, 2·5 cm or more in length, are sprinkled through the rock. When the crystals are broken

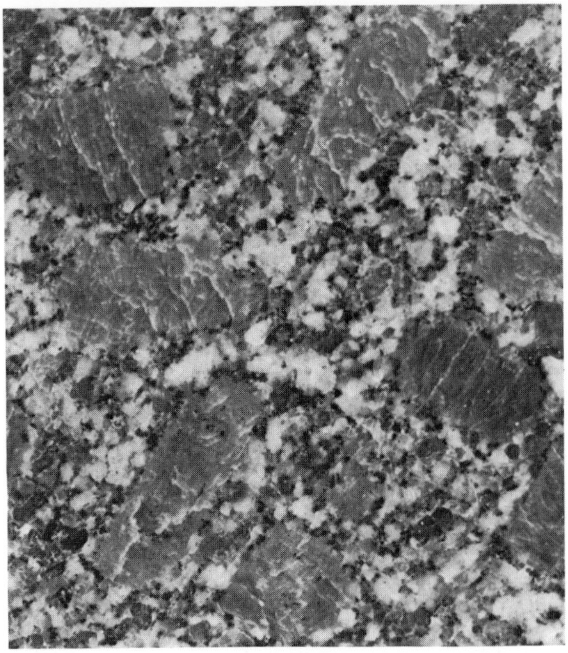

**Figure 4.12** A polished surface of Shap granite (Cumbria) showing large crystals of orthoclase (about 20%) embedded in a groundmass of finer grain. The resulting pattern is described as *porphyritic texture*, p. 67. In approximate percentages the groundmass consists of quartz (24), oligoclase (34), orthoclase (16) and biotite (6). (*The Shap Granite Company Limited*)

across, the surfaces are smooth and glistening. Orthoclase does not break anyhow; it 'cleaves' along parallel planes in the crystal structure across which cohesion is comparatively weak. Just as in many wallpapers the repetition of the unit pattern gives rise to a parallel series of 'open' lanes or bands, so in the atomic pattern of a crystal there may be similar 'open' planes, and it is along these that the crystal splits most readily. Orthoclase has two such sets of cleavage planes, and the mineral takes its name from the fact that they are exactly at right angles (Gr. *orthos*, normal or right; *klastos*, broken). Microcline is another common variety of potash feldspar and its name, meaning 'little slope', refers to the fact that the angle between its cleavages departs by half a degree from a right angle. The plagioclase feldspars also have two cleavages, but here the difference from a right angle is about 4°, varying with the composition from slightly less than 4° at the albite end of the series to slightly more at the anorthite end. Hence the name (*plagios*, oblique).

Orthoclase and other varieties of alkali feldspar are extensively used in making glass and glazes, but workable occurrences of these minerals are found only in certain rocks which are called *pegmatites* (Gr. *pegma*, thick, coarse) because in them the constituent minerals have generally crystallized to an unusually large size; moreover, both the grain size and the distribution of the minerals may be extremely variable. Most pegmatites occur in the form of irregular dykes, or sheet-like veins, or lenticular bodies, consisting largely of quartz and alkali feldspars, commonly accompanied by mica (Figure 4.13). In bulk, therefore, their composition is granitic. But whereas the minerals are more or less uniformly distributed in granite, there are many pegmatites in which they occur as gigantic crystals or massive aggregates, sometimes in sufficient concentration for profitable quarrying or mining. Single crystals of orthoclase or microcline as big as a house, though exceptional, are not unknown (e.g. Norway and Urals). Crystals of mica like giant 'books' of hexagonal shape have been found up to 3 and 5 m across (Transvaal and Ontario). Some of these commercially valuable pegmatites are also noteworthy in being veritable museums of beautifully developed rare minerals. Beryls as long as telegraph poles, and much thicker, have been quarried from some of the North American and Indian pegmatites. Sometimes open cavities are encountered in the larger pegmatites during quarrying, and springing from the walls of these some of

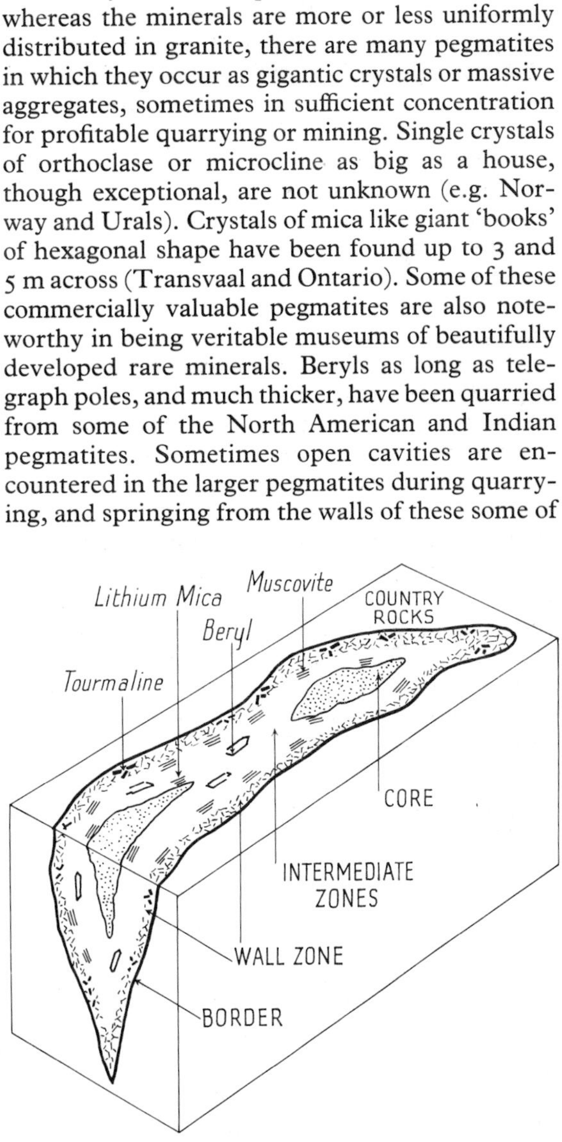


**Figure 4.13** Characteristic structure of a lenticular body of zoned pegmatite

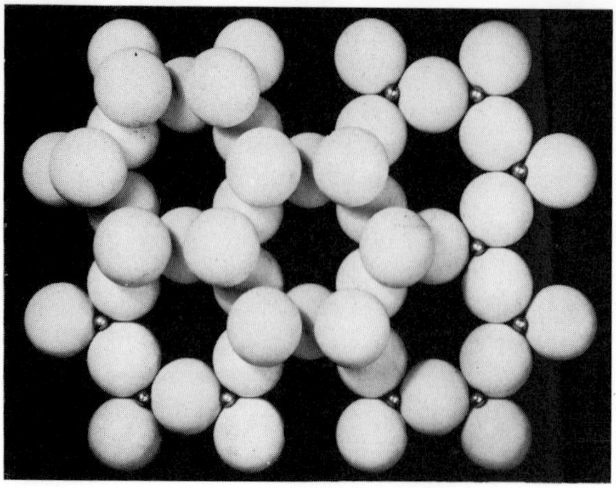

**Figure 4.14** Model of part of a continuous sheet of $SiO_4$ tetrahedra (cf. Fig. 4.10(*h*)). Some of the 'top' oxygen ions have been removed to show the silicon ions. (*Trustees of the British Museum, Natural History*)

the finest crystals of quartz, topaz, tourmaline and other showy minerals are found.

When feldspars are decomposed in the course of weathering, or by other processes involving the solvent action of water, practically none of the aluminium is lost by solution, and consequently the residual products become increasingly aluminous. The usual residues are extremely minute flakes consisting of (*a*) very fine-grained micaceous material called *sericite* or *hydromica*; or (*b*), when all the potassium (or sodium or calcium) has been lost, a *clay mineral*, of which there are several varieties. Most of the clay minerals are essentially hydrous silicates of aluminium, with formulae such as $Al_2Si_2O_5(OH)_4$, as in the kaolin group. The well-known china clay of Cornwall and Devon consists of kaolin formed by the decomposition of feldspar in granite. Some of the clay minerals, however, like those in fuller's earth and bentonite, are of more complex composition, having a small proportion of their $Al^{3+}$ ions replaced by $Mg^{2+}$, plus $K^+$ or $Na^+$ to balance the valency.

Under appropriate conditions in tropical climates all the silica may be removed from feldspars by weathering. The residue then left is *bauxite*, a mixture of two aluminous minerals with the compositions $Al_2O_3.H_2O$ or $AlO(OH)$, and $Al_2O_3.3H_2O$ or $Al(OH)_3$. Bauxite is of great value as the only workable ore of aluminium.

Third in abundance amongst the minerals of granite, and of the sialic rocks in general, are the **micas**, of which there are two leading varieties, one white, silvery and glistening, the other dark and often bronze-like. A third variety of amber tint is found in association with magnesium-rich rocks. All are hydrous alumino-silicates of potassium, as the following formulae indicate:

White mica or *muscovite* $KAl_2(AlSi_3O_{10})(OH,F)_2$

Amber mica or *phlogopite* $KMg_3(AlSi_3O_{10})(OH,F)_2$

Dark mica or *biotite* $K(Mg,Fe)_3(AlSi_3O_{10})(OH,F)_2$

The expression (OH,F) means that fluorine may replace hydroxyl (OH) to a limited extent in the crystal lattice. In the formula for biotite (Mg,Fe) means that Mg and $Fe^{2+}$ are similarly interchangeable; moreover, $Fe^{3+}$ may take the place of some of the Al in biotite, a mineral that can obviously accommodate a wide range of composition within its crystal structure.

Micas all have a perfect cleavage, because their tetrahedral sheets and binding atoms are all arranged in parallel layers. Two layers having the structure represented in Figures 4.10*h* and 4.14 are tightly bound together like a sandwich, with an infilling of Al ions (inmuscovite), or Mg ions (in phlogopite), or Mg and $Fe^{2+}$ ions (in biotite). These 'sandwiches' are held together in turn, though rather more loosely, by layers of K ions, and it is along these 'weak' layers between successive 'sandwiches' that mica cleaves with such remarkable facility, The 'books' of mica extracted from certain pegmatites are so called because when the crystals are seen from the side they generally have the appearance of a thick pile of uncut pages. The cleavage is, indeed, so perfect that each crystal can be separated into sheets very much thinner—if so required—than the pages of this book. The cleavage flakes or sheets are both flexible and elastic. Large transparent sheets of muscovite have long been used for lamp chimneys and furnace windows. But the outstanding property that at one time gave mica, and especially muscovite, a high place amongst industrial minerals is its quality as an insulator.

Biotite serves to introduce the silicate minerals that are characterized by an abundance of magnesium and iron, and are therefore commonly known as ferromagnesian or mafic minerals. The leading groups are the *pyroxenes*, the *amphiboles*, and the *olivine* series.

As indicated on p. 48, **olivine** (or chrysolite) is part of a continuous solid-solution series which ranges in composition from $Mg_2SiO_4$ (forsterite) to $Fe_2SiO_4$ (fayalite). The common rock-forming variety contains more magnesium than iron, as

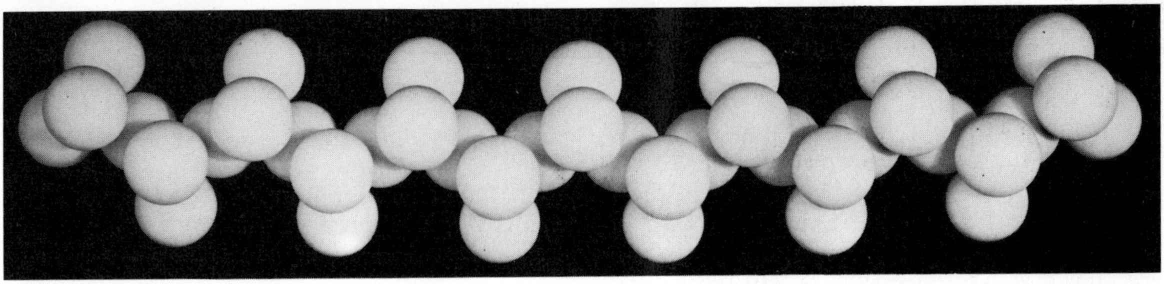

**Figure 4.15** Model of part of a continuous single chain of $SiO_4$ tetrahedra (cf. Fig 4.10(*f*)) (*Trustees of the British Museum, Natural History*)

expressed by the formula $(Mg,Fe)SiO_4$, and is called olivine in reference to its usual olive-green colour; but that there are also golden-yellow varieties is recalled by the older name *chrysolite* (Gr. *chrysos*, gold). The mineral is familiar as the transparent green crystals which are cut as gemstones under the name *peridot*.

Rocks in which olivine is the most abundant mineral (generally in association with pyroxenes) are called *peridotite*. The chief source of olivine of gem quality is the peridotite of a small island in the Red Sea; equally limpid crystals have been found in meteorites and hailed, therefore, like diamonds, as celestial gems. But whereas diamond is extraordinarily rare in meteorites, olivine is one of their most characteristic minerals. It is partly for this reason (meteorites being regarded as fragments of a shattered planet or planets), and partly because olivine-rich peridotite would transmit earthquake waves at the observed speeds, that olivine, or its high-pressure equivalent at great depths, is thought by many geophysicists to be the most abundant material in the earth's mantle. It is curious that Shakespeare, in a flash of poetic imagery, should have selected the same mineral when, through the mouth of Othello, he speaks of heaven making

> . . . another world
> Of one entire and perfect chrysolite.

The name 'chrysolite' is now falling into disuse because it is so readily confused with *chrysotile*, the name of a variety of serpentine which occurs in the form of flexible fibres and is commercially valuable as a heat-resisting material—asbestos—that can readily be woven into fabrics.

*Serpentine*, $Mg_3SiO_5(OH)_4$, is formed from olivine and other magnesium-rich silicates by a process of alteration involving addition of water. Large bodies of peridotite may be partly or wholly replaced by serpentine, which is thus a rock as well as a mineral: a rock composed of a tough but rather slippery felted mesh or network of minute crystals. Massive serpentine is generally mottled or variegated in shades of green and brown, like a serpent's skin, and some occurrences (e.g. at The Lizard in Cornwall) find a limited use as ornamental stones.

**Pyroxenes** are minerals of widespread occurrence in a great variety of rocks. In basalts (including olivine-basalts) and related rocks they are the most abundant constituents after plagioclase. In peridotites they are the most abundant minerals after olivine, and associated masses of rock in which pyroxenes are most abundant are distinguished as *pyroxenites*.

The simplest pyroxenes are *enstatite*, $MgSiO_3$ and *hypersthene*, $(Mg,Fe)SiO_3$. Chemically these are like members of the olivine series, but with more silica. Structurally, however, they are very different. All the pyroxenes are built of innumerable single chains of $SiO_4$- tetrahedra (Figures 4.10*f* and 4.15) with some substitution of $AlO_4$-tetrahedra in the more complex varieties. The chains are parallel and bound together by ions such as $Mg^{2+}$.

The composition of diopside, with little or no alumina, is $Ca(Mg,Fe)Si_2O_6$. The chief aluminous pyroxene, and the commonest member of the whole group, is *augite*, which may be regarded as a solid solution of diopside with a little hypersthene and varied proportions of 'alumina' and 'ferric oxide'. It must be noticed that aluminium plays a double role. Where an $AlO_4$-tetrahedron occurs in the chain an extra valency has to be satisfied. To make up the deficiency and keep the crystal neutral, $Al^{3+}$ or $Fe^{3+}$ could then serve as binding ions between the chains instead of $Mg^{2+}$. $Al^{3+}$ can also serve to bind a chain of $SiO_4$-tetrahedra, provided it is associated with a monovalent ion like $Na^+$. Thus, instead of a diopside, $CaMgSi_2O_6$, one can have $NaAlSi_2O_6$.

Though very rare, this mineral actually occurs, and because it is beautiful as well as rare it is also familiar. It is, in fact, *jadeite*, the mineral of which jade is composed, i.e. the real jade of Upper Burma which is so greatly prized by the Chinese; there is also an amphibole, *nephrite*, which is commercially called 'jade'. Another soda pyroxene, known as *acmite* or *aegirine*, $NaFe^{3+}Si_2O_6$, makes a series with augite, the intermediate members of which are known as *aegirine-augite*. It will be realized from the scheme at the foot of this page that the pyroxenes, augite in particular, include some very complex minerals, as well as a few relatively simple 'end members' of the various series.

Because of the importance of lithium in connection with atomic energy, another alkali-pyroxene has become extremely valuable in recent years. This is *spodumene*, $LiAlSi_2O_6$, now the chief source of lithium. It occurs in certain pegmatites and is one of those minerals which, though rare, is remarkable for the enormous size (e.g. up to 12 m long) attained by some of its crystals.

The **amphibole** group, represented mainly by hornblende and its many varieties, is more abundant and widespread than the table on p. 51 suggests. Many rocks closely related to granite contain more hornblende than biotite; and amphiboles are especially abundant in the metamorphic rocks known as *amphibolite*. In composition the amphiboles are not unlike the pyroxenes, the essential difference being that the amphiboles all contain ions of hydroxyl, $(OH)^-$, which may be proxied by fluorine, $F^-$, as in mica. This is made possible by the fact that they are built of long double chains of tetrahedra, as indicated on p. 52 and illustrated by Figures 4.10*g* and 4.16, which provide 'holes' (again as in mica) into which these large ions can fit. The resulting complexities of composition are well shown by the following comparisons:

| PYROXENES | | AMPHIBOLES | |
|---|---|---|---|
| *Enstatite* | $MgSiO_3$ | *Anthophyllite* | $Mg_7Si_8O_{22}(OH)_2$ |
| *Diopside* | $CaMgSi_2O_6$ | *Tremolite* | $Ca_2Mg_5Si_8O_{22}(OH)_2$ |
| *Acmite* | $NaFe^3Si_2O_6$ | *Riebeckite* | $Na_2Fe^2{}_3Fe^3{}_2Si_8O_{22}(OH)_2$ |

These three amphiboles and some of their varieties occur locally as workable masses of asbestos, known commercially as 'amphibole asbestos' to distinguish it from 'chrysotile asbestos', which is generally of higher quality.

A scheme for the amphiboles, with hornblende in the middle, like that presented below for the pyroxenes, would be too complicated to serve any useful purpose. It will suffice to say that a characteristic composition for hornblende is approximately represented by the formula

$$Na\,Al\,Ca_2\,(Mg,Fe)_4\,(Al_2Si_6)O_{22}\,(OH)_2,$$

keeping in mind that substitutions such as F for (OH), $Fe^3$ for Al, and Ti for Si or $2Fe^2$ are general, and that there are numerous 'end members', like the three listed above, with which limited amounts

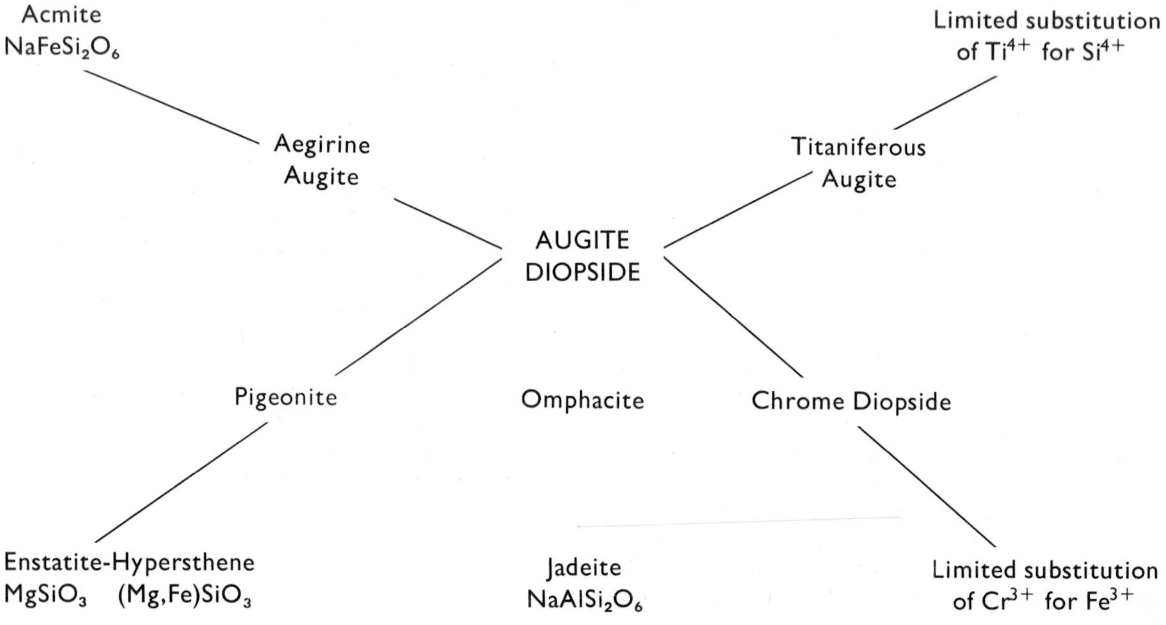

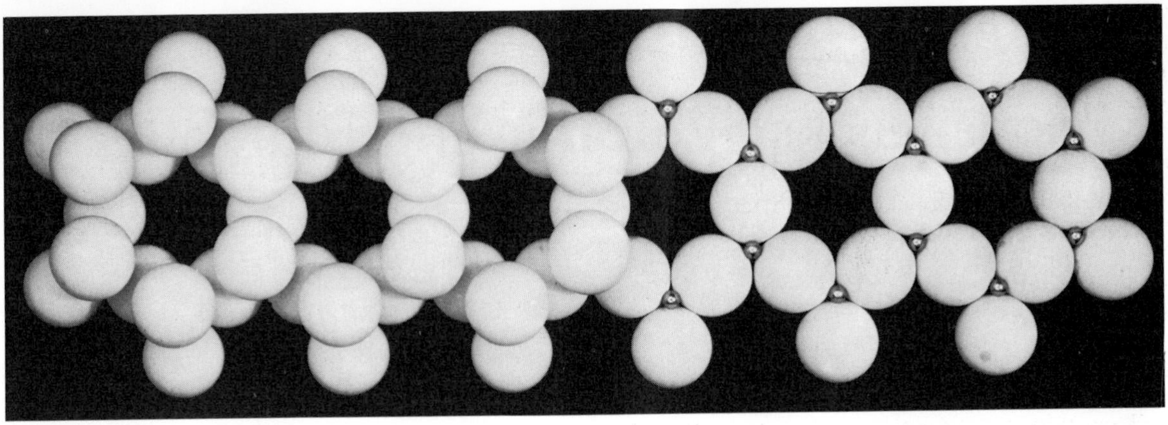

**Figure 4.16** Model of part of a continuous double chain of $SiO_4$ tetrahedra (cf. Fig. 4.10g) (*Trustees of the British Museum, Natural History*)

of solid solution are possible. It is a remarkable fact that hornblende contains notable amounts of all the common elements of the earth's crust except potassium.

Augite and hornblende are readily distinguished by their crystal forms and cleavages, both of which features depend on the internal architecture of the crystals. As in the pyroxenes, the amphibole chains are bound together in long parallel bundles. These bundles can be readily split along two particular planes—the cleavage planes—parallel to the length. Thus it happens that both minerals have two well-developed cleavages; but in augite the angle of intersection between them is 87° or 93°, appropriate to the single-chain structure, while in hornblende the angle is 56° or 124°, corresponding to the double-chain structure. Since these angles are nearly 90° and 120° (or 60°) they can be recognized and distinguished at a glance.

Just as olivine and enstatite are altered to sepentine by processes involving addition of water, so the Al- and Fe-bearing ferromagnesian minerals are altered to a group of greenish minerals collectively called **chlorite** (Gr. *chloros*, green, as in chlorine and chlorophyll). These have a double-sheet or 'sandwich' structure like the micas, but the 'sandwiches' are held together not by K, as in the micas, but by (OH), Mg or $Fe^2$, and Al or $Fe^3$. The particular composition depends mainly on that of the parent mineral. Because of its structure chlorite has an excellent cleavage, but the resulting flakes lack the flexibility of mica. Moreover, the individual crystals are usually very small. The chlorites bear much the same relation to ferromagnesian minerals as the clay minerals do to feldspars, alkalis and lime being lost in both cases, while the water that is added appears as $(OH)^-$ in the crystal lattice of the alteration product.

## Selected References


BRAGG, L., 1968, 'X-Ray crystallography', *Scientific American*, vol. 219, pp. 58–70.

BRAGG, W. L., CLARINGBULL, G. F., and TAYLOR, W. H., 1965, *Crystal Structure of Minerals*, Cornell University Press, New York.

FYFE, W. S., 1964, *Geochemistry of Solids: An Introduction*, McGraw Hill, New York.

GOLDSCHMIDT, V. M., 1954, *Geochemistry*, Clarendon Press, Oxford.

HURLBUT, C., Jr., 18th edn, 1971, *Dana's Manual of Mineralogy*, Wiley-Interscience, New York.

JONES, W. R. and WILLIAMS, DAVID, 1948, *Minerals and Mineral Deposits*, Home University Library, Oxford University Press.

MASON, BRIAN, 3rd edn, 1968, *Principles of Geochemistry*, Wiley-Interscience, Chichester and New York.

MOREY, G. W., 1954, 'Silica and inorganic silicates', *Encyclopedia of Chemical Technology*, vol. 12, pp. 268–303.

POUGH, F. H., 1970, *A Field Guide to Rocks and Minerals*, Constable, London.

READ, H. H., 1963/70, *Rutley's Mineralogy*, Murby (Allen and Unwin), London.

# Chapter 5

# Igneous Rocks: Volcanic and Plutonic

Rocks, like everything else, are subject to change and so also are our views on them.

*F. Y. Loewinson-Lessing*, 1936

## Difficulties of Classification

The few references to rocks that have already been made suffice to show that three major groups have been recognized—sedimentary, metamorphic and igneous—according to the processes that were concerned in their origins. Rocks formed by processes of external origin (exogenetic) are broadly grouped as *sedimentary*, whether they are fragmental, like sandstone; chemical or biochemical precipitates from the sea or other natural waters, like rock salt or limestone; or accumulations of organic matter, like coal. Most of the rocks formed by processes of internal origin (endogenic) belong to the *metamorphic* or *igneous* groups. But it must never be forgotten that there are, inevitably, many rocks that are transitional between these groups. Loose sediments are progressively changed into firm, indurated sedimentary rocks by consolidation and cementation. These changes are described as *lithification* or *diagenesis*, and it is not always clear just where diagenesis ends and metamorphism begins. However, in practice this leads to no serious difficulties.

Much more troublesome is the fact that there are all gradations, not only from igneous rocks to their metamorphosed equivalents, but also from metamorphic rocks to those that everyone would agree in calling igneous. This is because the processes of metamorphism may become so intense, especially when they include the passage of hot gaseous emanations, that the characteristic structures by which metamorphic rocks are commonly recognized are themselves gradually lost. As a result of this *ultra-metamorphism*, as it has been called, a rock may begin to 'look' igneous; that is, to have its minerals more or less uniformly arranged as interlocking crystals (Figures 4.12 and 5.10) of easily visible sizes. If the mineral or chemical composition of the rock corresponds approximately to that of a known volcanic rock, it is regarded as the plutonic equivalent of the latter. The term *plutonic*, (after Pluto, the Roman god of the netherworld) has been generally applied to granite and the other crystalline rocks of major intrusions (*plutons*) to distinguish them from the finer-grained volcanic and associated intrusive rocks, such as those of lavas, dykes and sills. However, as originally used by Lyell, the term specifically included metamorphic rocks. This reminds us that to recognize a rock (e.g. granite) as plutonic is very far from implying that the rock has necessarily crystallized from a molten state. Some plutonic rocks may have done so, but there is convincing evidence, as we shall see, that some of them have not. A rock can lose its metamorphic characteristics and become a plutonic igneous rock without having been completely melted.

Since it is current practice to classify plutonic intrusive rocks, such as granite, as igneous, the best way to avoid confusion is to draw the line between metamorphic and igneous so that granite is on the igneous side. Igneous rocks are then of two main kinds: *volcanic rocks*, including those of the associated minor intrusions; and *plutonic rocks* of similar composition occurring as major intrusions or as associated veins and rock-bodies such as pegmatites.

## Neptunists and Plutonists

Meanwhile it is both interesting and instructive to go back to the early days of geology when both granite and basalt were traditionally considered to be 'sedimentary' rocks which, like rock-salt, had crystallized from oceanic water. Only those lavas and volcanic ashes that were *known* (i.e. by observation or reliable human testimony) to have been erupted from volcanoes were ascribed to the operation of 'subterraneous fires' and so admitted to be, as we now say, of igneous origin. Apart from these, all rocks were thought to be of aqueous origin. Those who held this archaic view, which dates from the time of Thales (p. 3), came to be called *Neptunists*.

Even as late as the last quarter of the eighteenth century the Neptunists were self-deluded by several fatally wrong notions:

(*a*) They believed that crystals could form only from solution in water. Now it was recognized that granite and basalt were crystalline rocks—indeed some thought that the polygonal columns of columnar basalt were gigantic crystals (e.g. Collini, 1776)— and it logically followed, according to their belief, that granite and basalt could have crystallized only from aqueous solution.

(*b*) They believed that molten lavas consolidated only as glass and were incapable of crystallizing. Granting this, it followed that basalt could never have been a molten lava.

(*c*) They believed that volcanic eruptions were due to the burning of coal seams beneath the volcanic vents, an explanation first proposed by Agricola (1494–1555). Yet many occurrences of basalt were known in places where, judging from all experience, there had never been any coal to catch fire, and where, still more convincingly, no sign of a volcano was to be seen (Figure 5.1). Surely it followed that basalt could not be a volcanic rock?

Neptunist doctrines culminated in the eloquent teaching of Abraham Gottlob Werner (1749–1817), who attracted students from all parts of the western world to the modest Mining Academy of Freiberg in Saxony, where he inspired them with his own contagious enthusiasm for everything to do with minerals and rocks. Before exposing Werner's mistakes it is only fair to remember that he was a great pioneer mineralogist and that he was unsparing of himself in giving his students by far the best practical training then available. They responded by having such faith in their master's speculations that in later years it seemed disloyal to have to admit—as many of them were obliged to do from the evidence of the rocks

**Figure 5.1** The south-western bastions of the basaltic plateau of the Coiron, viewed from the road about 16 km west of Montélimar on the Rhône (From *The Geology of the Extinct Volcanoes of Central France*, London, 1858, Plate XII, by G. Poulett Scrope)

**Figure 5.2** Columnar basalt resting on 'alluvial' deposits of gravel and sand in the quarry at Scheibenberg in Saxony (Photograph taken in 1938 by Werner's successor, Professor K. Pietzsch, at the Bergakademie of Freiberg)

themselves—that Werner could have been wrong.

Being of a logical mind, and starting from the above beliefs, plus the further assumption that the rocks of his native Saxony were a fair sample of those to be found elsewhere, Werner concluded that all the known materials of the earth's crust must originally have been dissolved in a primeval ocean that 'in the beginning' covered every part of the globe. To explain the succession of the rocks seen in Saxony he postulated that the first materials to separate formed a thick layer of granite that moulded itself on the highly irregular surface of the earth's nucleus': a layer which was then followed by sheets of other crystalline rocks, of the kind we now call metamorphic. These, with granite as the oldest rock of all, were the *Primitive* rocks. The ocean level now sank a little and exposed the higher protruberances to denudation. The resulting sediments and the accompanying precipitates made an assemblage called the *Transition* rocks, consisting of massive limestones, graywackes and slates. These were often folded and steeply inclined because of slumping down the slopes on which they were deposited. The ocean next withdrew to a level that was lower than all the steeper slopes, which thus became exposed as hills and mountains. Further erosion and deposition provided the *Flötz*, bedded rocks consisting of more or less fossiliferous strata, like shale, coal, sandstone and chalk, occasional layers of basalt, and local occurrences of volcanic rocks. Finally came the *Alluvial* deposits: gravel, sand and clay, representing the recent waste products of the older rocks. In some places these were followed by basalt, as at Scheibenberg (Figure 5.2). This locality was visited by Werner in 1787 and became celebrated in the continuing dispute with the Vulcanists, who had long been convinced that basalt was a lava. Werner, however, drew a sharp distinction between basalt and lava; and damaging though it was to his 'theory', he insisted that the ocean had risen again, this time to cap the hills with 'chemical precipitates of basalt', before retreating to its present confines.

The date 1787 is significant. The terrible Icelandic eruption of white-hot basaltic lava that poured from the Laki fissure in 1783 was widely discussed throughout western Europe and Werner must surely have known about it. One would at least have expected his faith in the adequacy of coal as a volcanic fuel to have been shattered. But no! He continued to deny that any lava could be basalt.

Other dates in this fantastic story are even more dramatically interesting. So far as basalt was concerned the battle for Neptunism was already lost before it had seriously begun—and yet Werner did not know. The scene changes to Auvergne and the year is 1751. Jean Guettard (1715–86), accompanied by a friend, made a journey from Paris to the now celebrated Puys (Figure 5.3). Here, with growing excitement, Guettard recognized that the entire landscape, with its cratered cones and flows of rugged lava overlying burnt and reddened soils, had been the scene of volcanic activity not long ago. He was also able to identify some of the lavas as basalt, a rock with which he was already familiar. The following year he published an account of his discoveries, but his paper was coldly received, not so much because the Vulcanist theory was unorthodox, as because the idea that volcanoes had once erupted 'in the very sunlit heart of France' was felt to be an outrage. It was altogether too reminiscent of Sodom and Gomorrah. However, sentiment apart, the Vulcanist position was still incomplete, because Guettard specifically excluded columnar basalts from his volcanic rocks, having failed to find any field evidence that they, too, might be lavas.

The conclusive discovery was made by Nicolas Desmarest (1725–1815). While beginning his exploration of the Auvergne cones and craters in 1763, he found in a wayside quarry a sheet of black columnar basalt, underlaid by cindery lava and burnt soil. He traced the basalt across country until it finally led him up to the crater of one of the Puys. No further proof was needed that what was now columnar basalt had been erupted from a volcano as a lava flow. This time the French scientists were convinced and Desmarest was commissioned to make a geological map of the volcanic country of Auvergne. He saw that there was every gradation between obvious lava flows of columnar basalt and extensive sheets of the same kind of rock (Figure 5.1) which had been dissected—and sometimes isolated—by valleys cut by running water. The riddle of the Giant's Causeway was now solved. Desmarest had correctly inferred the volcanic origin of the Antrim basalts; William Hamilton, one of the founders of

**Figure 5.3** 'The graveyard of Neptunistic opinions'—the Puys of Auvergne, looking southwards. The most recent flow of lava from the Puy de la Vache has been dated by radiocarbon (p. 244) at 7650 ±350 B.P. (before the present). (*J. Richard, Paris*)

the Royal Irish Academy, clinched the matter in 1784 by demonstrating that the chalk and flints over which the lavas had flowed (Figure 3.15) had been baked and reddened by their heat. Although the Vulcanist position was now unassailable, there were still so many possibilities of local mistakes (see Figure 8.5) and misrepresentations that controversy continued to rage for many years.

Desmarest took no part in this dismal squabble. To his opponents his invariable reply was 'Go and see.' Werner, however, was only poorly paid and could not afford the expense of long journeys. What he did go and see, the Scheibenberg (Figure 5.2), seemed to him to confirm his own views. Two other factors helped to sustain the Neptunists. The French Revolution and the Napoleonic wars made communication difficult. And throughout this turbulent period the Neptunists had another fight on their hands—over the origin of granite. In this controversy the opposing faction came to be known as *Plutonists*. While recognizing the part played by water in sedimentation, the Plutonists insisted that granite, as well as basalt, had crystallized from a molten state.

(a)

(b)

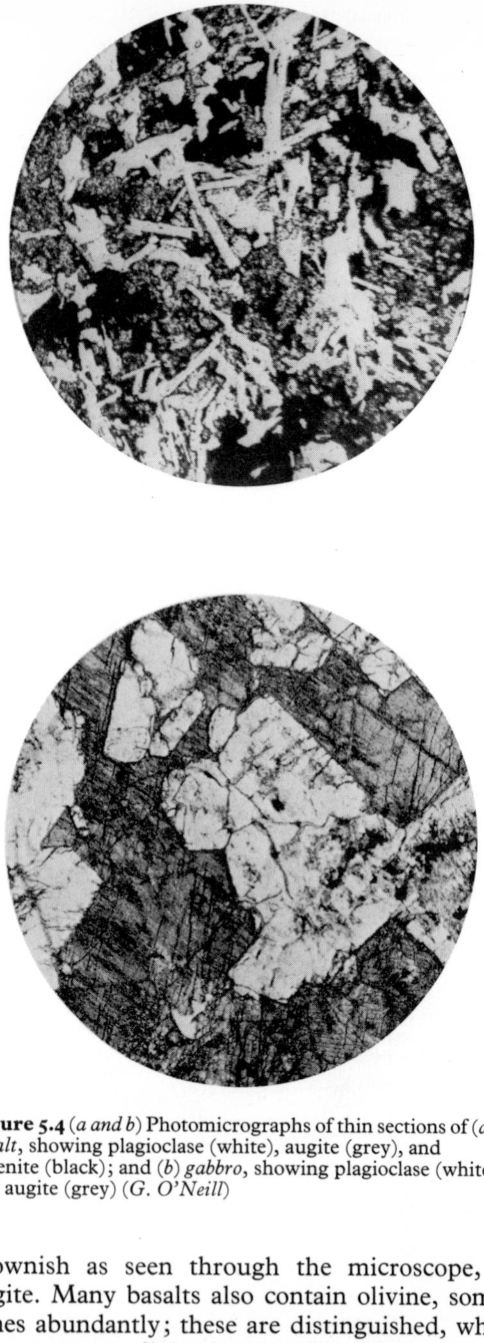

**Figure 5.4** (*a and b*) Photomicrographs of thin sections of (*a*) *basalt*, showing plagioclase (white), augite (grey), and ilmenite (black); and (*b*) *gabbro*, showing plagioclase (white) and augite (grey) (*G. O'Neill*)

## Basalt

Basalt is a dark-coloured, very fine-grained rock, which is of widespread occurrence as lava flows of all geological ages, and which is still the most abundant type of lava erupted from the volcanoes of today. Its igneous origin is, of course, no longer in doubt. Much of the early controversy already passed in review was due to the fact that the pioneer geologists found it extremely difficult to investigate fine-grained rocks like basalt, for with the limited means at their disposal they were rarely able to identify the tiny crystals in such compact materials. In 1851 this difficulty was successfully overcome by Henry Clifton Sorby (1826–1908) of Sheffield, the founder of the microscopal study of rocks, who showed how a slice of rock could be ground down to a film so thin (about 0·0025 mm) that it becomes more or less transparent, thus making it possible, after mounting the film on a glass slide, to view the rock through a microscope and to examine the magnified minerals with ease.

A thin section of basalt prepared in this way has the appearance illustrated in Figure 5.4(a). Lath-like crystals of a clear colourless mineral, which is calcic plagioclase (generally labradorite), form an irregular open network that extends throughout the rock. The grey mineral, which is greenish or brownish as seen through the microscope, is augite. Many basalts also contain olivine, sometimes abundantly; these are distinguished, when necessary, as *olivine-basalts*. The black opaque mineral is magnetite or ilmenite. It is the high proportion of iron in basalt which is responsible for the dark colour of the rock, and for the rusty-looking material (limonite) that encrusts its surface when it has been exposed to the weather.

Basalt is not always wholly made up of crystals.

Varieties that solidified very rapidly, as a result of sudden chilling, had no time to crystallize completely. In consequence, the part that remained uncrystallized had no alternative but to solidify into black volcanic glass. Crystals may already have grown in the magma before its eruption as lava. In this case the resulting basalt is a *porphyritic* variety (see p. 67), with relatively large crystals in a very fine-grained or glassy groundmass.

At the top of a basalt flow the lava may be blown into a cinder-like froth by the expansion of escaping gases. Even in the more compact basalts, gas-blown cavities of various sizes may occur. These may be empty; or lined with crystals, often beautifully developed; or even completely filled with minerals. The filled 'bubbles' sometimes look like almonds, and so the name *amygdale* is given to them (Gr. *amygdalos*, almond). Basalts that are studded with numerous amygdales are called *amygdaloidal* basalts. One of the commonest minerals found in amygdales is *agate*, banded in concentric layers of different tints. Agate is a variety of chalcedony, a crypto-crystalline form of silica; and sometimes, inside a lining of agate, crystals of quartz or of amethyst (purple quartz) may be found projecting into the hollow space within. The occurrence of these minerals in a basalt suggests that there was some free silica left over after the crystallization of the rock. One would not expect this to happen in olivine-basalts, because the presence of olivine implies a deficiency of silica. For this reason olivine-basalts generally have minerals such as zeolites and calcite in their amygdales. *Zeolites* are like the feldspars in composition, but with a good deal of water rather loosely held in channels within the crystal framework. This water is easily driven out by heating, often with an appearance of boiling; whence the name, from the Greek *zeo*, I boil.

## Columnar Jointing

The origin of the columnar structure of certain compact basalts and other fine-grained or glassy igneous rocks is best approached by considering the sets of polygonal cracks that are often to be seen in the dried-up mud of a marsh or river-flat that has been exposed to the sun (Figure 5.5). These mud cracks result from shrinkage due to loss of water by evaporation from the surface layers. Similarly, the polygonal cracks or *joints* of basaltic sheets result from contraction during cooling.

**Figure 5.5** Mud cracks on the dried-out floor of Loughaveema, the 'vanishing lake' between Cushendun and Ballycastle, Co. Antrim. The lake drains underground through the Chalk and rises and falls with the rainfall, vanishing after long dry spells. (*R. Welch Collection, Copyright Ulster Museum*)

When a hot homogeneous rock cools uniformly against a plane surface, the contraction is equally developed in all directions throughout the surface. This condition is mechanically the same as if the contraction acted towards each of a series of equally spaced centres. Such centres (e.g. C, 1, 2, 3, etc. in Figure 5.6(*a*)) form the corners of equilateral triangles, and theoretically this is the only possible arrangement. At the moment of rupture the distance between any given centre C and those nearest to it (e.g. 1–6) is such that the contraction along lines such as C—1 is just sufficient to overcome the tensile strength of the rock. A tension crack then forms half-way between C and 1 and at right angles to the line C—1. As each centre is surrounded by six others (1 to 6 in Figure 5.6(*a*)), the resultant system of cracks is hexagonal. Once a crack occurs somewhere in the cooling layer the centres are definitely localized, and a repeated pattern of hexagonal cracks spreads almost simultaneously throughout the layer (Figure 5.6(*b*)). As cooling proceeds into the sheet of rock the cracks grow inwards at right angles to

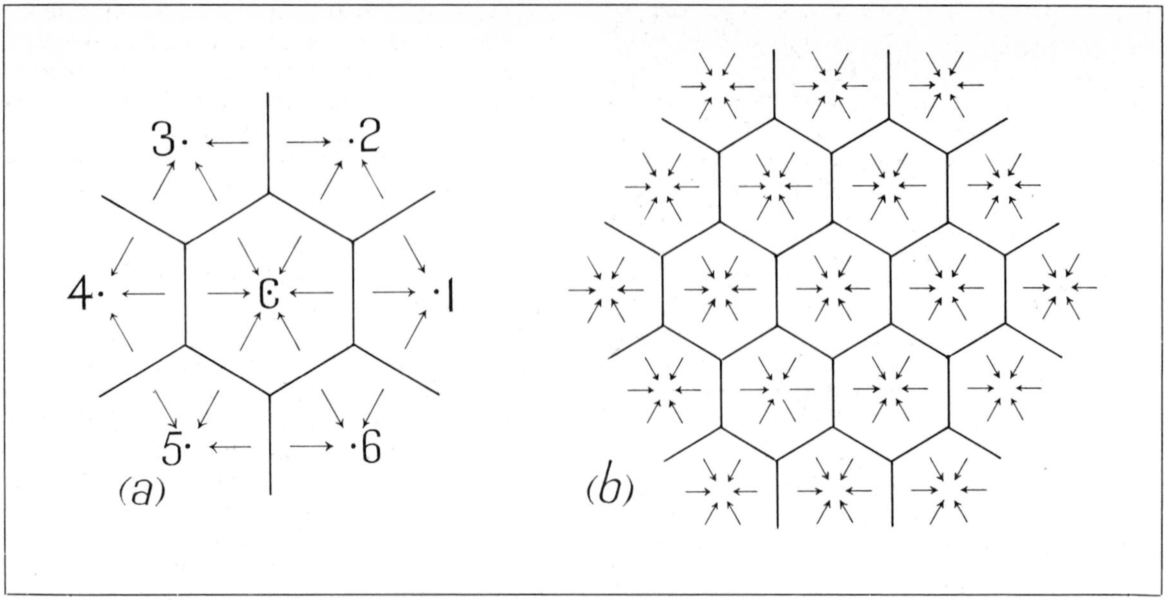


**Figure 5.6** The formation of an ideal hexagonal pattern of joints by uniform contraction towards evenly spaced centres

the cooling surface, and so divide the sheet into a system of hexagonal columns.

Neither the physical conditions nor the rocks concerned are usually sufficiently uniform to ensure perfect symmetry, and the actual result is a set of columns with from three to eight sides, six, however, being by far the commonest number. Vertical contraction is relieved by cross joints, which may be either concave or convex (Figure 5.7), and the columns are thus divided into short lengths. The resulting appearance of well-trimmed masonry is often remarkably impressive, as in the tessellated pavement of the Giant's Causeway, and the amazing architecture of Fingal's Cave in Staffa (Figure 1.2). Comparable, but usually cruder and less regular columnar jointing develops during the cooling of sills and dykes. In a sill the cooling surfaces are the floor and roof. A dyke cools against vertical walls and the resulting columns are therefore horizontal.

A lava flow, once it has come to rest, cools uniformly only through its floor; at the top, surface irregularities and escaping gases introduce complications. Thus it happens that in vertical section, as seen in cliffs, columnar lavas of the Causeway type have a conspicuous twofold arrangement. A lower zone of well-developed columns, appropriately described by Tomkeieff as the *colonnade*, is surmounted, as in classical buildings, by the *entablature*, consisting of a sort of freize of thinner irregularly curving or radiating columns and a cornice of stumpy polygonal blocks, with a slaggy and vesicular zone on top.

## Granite

As a result of early geological explorations in the Alps (Figure 3.17), the Urals, the Andes and the mountains of Norway, it soon came to be realized that granite is the characteristic rock of many of the high peaks of mountain ranges. But the mountains of former geological ages have been worn down by denudation, and so it also happens that granite may be more comfortably examined in the tors and valleys of less elevated regions (Figure 5.8), or in the cliffs of headlands where, for a time, granite resists the attack of the sea (Figure 5.9). In most towns it can be seen as hewn blocks or decorative slabs and columns, and it provides one of the best materials for piers and lighthouses.

Granite is a medium- to coarse-grained rock composed essentially of quartz, feldspar and mica. In some examples (e.g. from Aberdeen, Figure 5.10) the interlocking minerals are uniformly distributed, and all are about the same size. Feldspar, mostly orthoclase, is the most abundant mineral. Gleaming plates of mica (black or bronze-like biotite, accompanied in some varieties by silvery white muscovite) can easily be recognized. Between the feldspars and micas the remaining

**Figure 5.7** Columnar joining in the basalt of the Giant's Causeway, Co. Antrim, showing concave and convex cross joints (*J. Allan Cash*)

spaces are occupied by translucent, glassy-looking quartz.

Instead of being uniformly granular, certain granites (Figure 4.12) have a distinctive pattern or *texture*, clearly seen on polished slabs, due to the development of orthoclase as conspicuous, islated crystals which are much larger than those of the granular groundmass in which they are embedded. This tecture is technically described as *porphyritic*, a term derived from an old Greek word meaning 'purple'. The Romans, prospecting for decorative stones in Egypt nearly 2000 years ago, came upon a chocolate-red to deep-purple rock—which they called *Lapis porphyrites*—of such attractive appearance that they actively quarried it for columns, vases and slabs. In the course of time the same name came to be applied to other rocks which contain large crystals embedded in a finer groundmass, even though they lack the purple hue of the original porphyritic rock.

As already indicated, the problem whether granite crystallized from an aqueous solution or from a hot molten state led to a fierce controversy between the Neptunists, whose most influential advocate was Werner, and the Plutonists, whose founder was James Hutton. According to Werner granite was the oldest of rocks, having been the first material to precipitate from his universal ocean. Thinking that granite might be the cause of the uplift of mountain ranges, Hutton searched to confirm or disprove this hypothesis. In 1789, he examined the granite that outcrops to the north of Glen Tilt in the heart of the Grampian Highlands. Here he discovered that the country rocks were uparched over the granite, and that in Glen Tilt itself they were displaced and veined against the granite. Here was proof that the granite was younger, *not* older than the country rocks, as

**Figure 5.8** Coombestone Tor, Dartmoor, Devon, part of a large granite intrusion. Three sets of joints (see p. 140) have controlled the erosion forms: two roughly vertical, and essentially at right angles to one another, and the third not far from horizontal. (*Fox photos*)

Werner supposed. Moreover, from the highly crystalline character of the invaded rocks, Hutton inferred that they had been raised to a high temperature, and that the granite must have been hotter still. He concluded that sedimentary rocks had melted, in the depth of the earth, and as a result of the melting had expanded upwards, uplifting the overlying rocks to form mountains, within the cores of which the molten rock had consolidated as granite.

By the end of the last century the orthodox view was that granite magma was derived from a supposedly molten layer or chambers within the earth. Bunsen (1851) and Loewingson-Lessing (1911) postulated two primary magmas—granitic and basaltic. Subsequently, mainly as a result of the brilliant exposition of Bowen (1928), based on experimental investigations of dry melts at the Geophysical Laboratory, Washington, it came to be the orthodox view that granite magma was derived from basaltic magma by crystallization differentiation. According to this hypothesis, the final liquid residue remaining after much of the basaltic magma had crystallized, was granite magma. This hypothesis claimed so many adherents that it acquired the status of a dogma. It became useless, in criticism, to point out that if crystallization differentiation of basaltic magma were operative then no more than 5 to 10 per cent of the initial basaltic magma could form granite magma (Grout, 1926; Holmes, 1936), whereas the granite batholiths to be explained are giants in comparison with intrusions of more basic rocks. Nor did the fact that granite batholiths characteristically occur within fold mountains where the sialic crust is thick, and are absent from oceanic islands, where sial is absent, make any impact. World-wide measurements, made during gravity surveys have now revealed low values of gravity over granitic intrusions, indicating that the average density of the underlying rocks is less than that of the surrounding country rocks down to depths of several miles. They thus provide no evidence for the existence of enormous volumes of basic igneous rocks beneath granites, such as the crystalline differentiation hypothesis requires.

On the other hand, since early in the last century there have always been some geologists, particularly in France, who were firmly convinced that granite is formed from rocks of sedimentary origin by a process of granitization involving change of chemical composition (see pp. 115–16). There has, however, been disagreement amongst them as to whether or not the transformation processes culminate in fusion.

Crystallization differentiation, however, re-

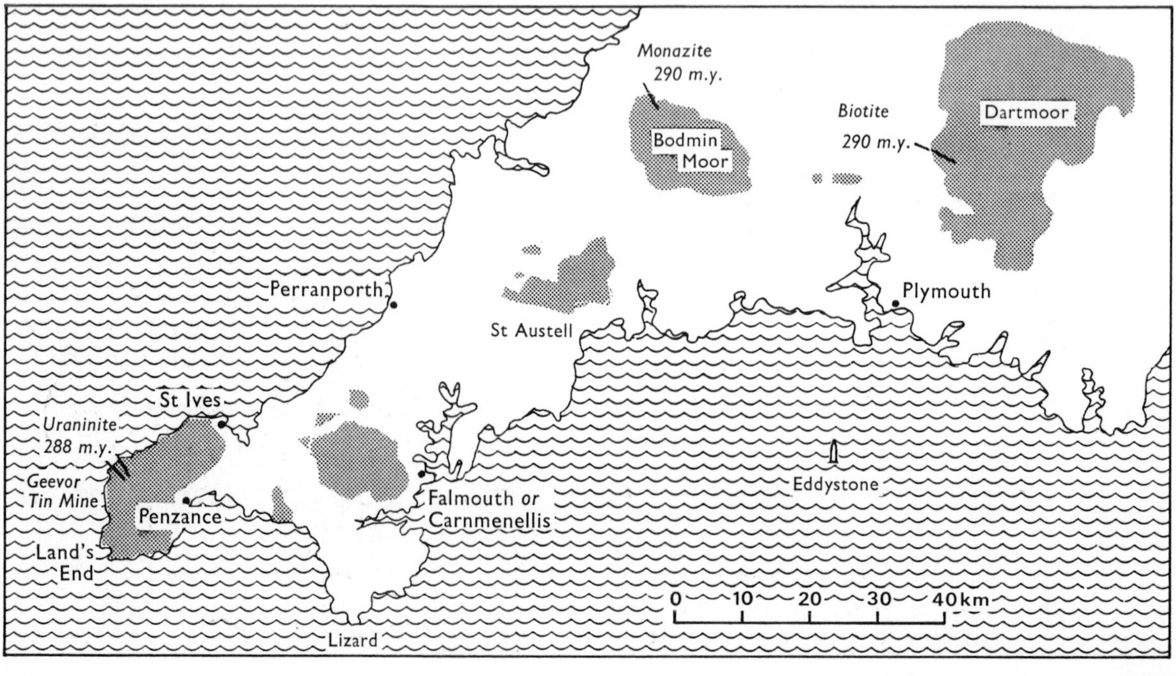


**Figure 5.9** Granite masses of Cornwall and Devon. The ages of some radioactive minerals that have been dated are indicated in millions of years (m.y.).

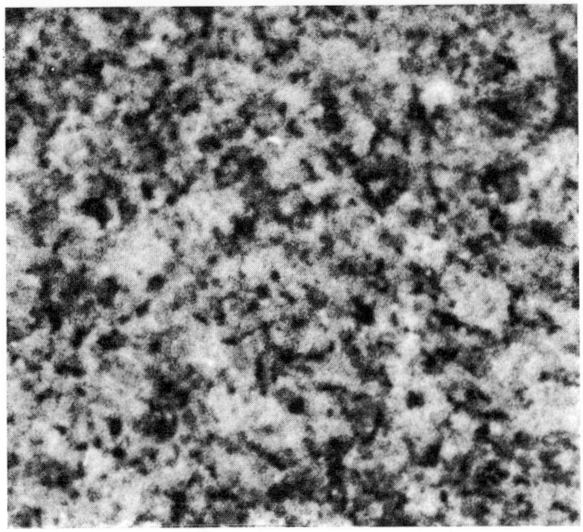

**Figure 5.10** Polished surface of bluish grey granite, Rubislaw, Grampian Region, Scotland. Natural size (*Granite Supply Association, Aberdeen*)

mained the orthodox view until about 1950 when the results of melting experiments, in the presence of water, began to suggest that granite magma might arise as a first formed melt rather than as a residuum. Particularly apposite were the early experimental meltings of sedimentary rocks in Germany (Winkler, 1957), and France (Wyart and Sabatier, 1959). These experiments showed that at water vapour pressures of from 1800 to 2000 atmospheres, after the growth of minerals characteristic of metamorphic rocks, from about 45 to 60 per cent of initial clay melted at temperatures of about 700° to 800°C. The melts consolidated as glass corresponding chemically to mixtures of quartz, and feldspars; the minerals of granite, commonly with an excess of alumina. By adding sodium chloride to the initial clay, to represent the pore solution of sediments, Winkler *et al.* found the resulting melt to correspond chemically with granodiorite, the most common granitic rock.

The difficulty has been to crystallize what any geologist would call granite from the glass made by melting either sedimentary rocks or granite; the results are always so fine-grained that they would be more correctly described as felsite. Tuttle and Bowen (1958, pp. 12 and 14), for example, define the usual fine-grain of their synthetic 'granites' as having a grain-size less than 0·001 mm across. They use the expression coarse-grained to mean less than 1 mm across. Wyart (1955), who completely crystallized a prism of natural glass (obsidian Figure 5.11) in the presence of a 2·5 per cent solution of potassium carbonate ($K_2CO_3$), with water vapour pressure of 1300 atmospheres, and at a temperature of 600°C, was most successful, but he himself compares the crystalline product with felsite (Figure 5.12).

Melting experiments all differ from natural earth processes in that they relate to closed systems. Within the earth's crust, water would escape with rising temperature, and there would be loss of silica, alkalies and alumina by vapour transport, as was found in experimental investigations when the containers were not sufficiently tightly sealed.

## Textures and Modes of Occurrence

Granite and basalt, and types closely related to them are by far the most abundant of the igneous rocks. Granite is the typical example of relatively coarse-grained plutonic rocks that crystallized slowly in large masses (e.g. plutons and batholiths, Figure 11.25) within the crust. Basalt, on the other hand, is the leading example of the fine-grained or even glassy volcanic rocks that cooled quickly from lava flows erupted at the surface. Terms like 'coarse grained' and 'fine grained' refer to the usual sizes of the chief minerals in the above-mentioned rocks (Figures 5.4(a) and 5.10), or in their groundmasses, if the rocks happen to be porphyritic (Figures 4.12 and 5.13). Size of crystals is one of the chief factors that control what is called the

**Figure 5.11** Obsidian, from a lava tongue of volcanic glass, Lake Co., Oregon, U.S.A. The shell-like (conchoidal) fracture is a characteristic feature of glass. (*Ward's Natural Science Establishment, Inc.*)

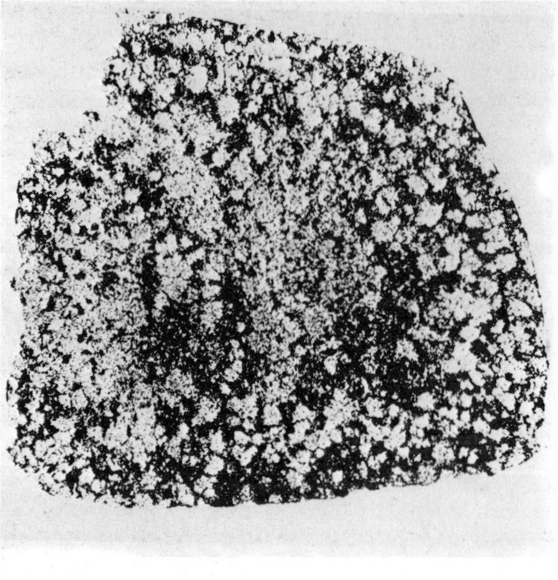

**Figure 5.12** Photomicrograph showing the complete crystallization of obsidian to feldspar, mica and quartz, in the presence of a solution of $K_2CO_3$ at a temperature of 600°C and a pressure of 1000 to 1500 bars. Ordinary light ×12 (*J. Wyart, 1955*)

*texture* of a rock; others are the proportion of crystals to glass, and any characteristic pattern seen on a fresh surface due to, say, contrast in crystal sizes (e.g. porphyritic texture), or intergrowths of different minerals (e.g. graphic texture, Figure 5.14).

Between the two extremes of coarse- and fine-grained rocks there are many of intermediate grain, rocks that probably cooled and crystallized at intermediate rates, generally in minor intrusions such as sills and dykes. For this reason the rocks of minor intrusions have been grouped together by many petrologists as *hypabyssal* rocks (implying that they crystallized at 'intermediate depths'), leaving 'volcanic, for the glassy and fine-grained rocks of lava flows and 'plutonic' for the coarse-grained rocks of major intrusions. Taking basaltic rocks as an example, this scheme gives us four types:

*Tachylyte*[1], a basaltic material in a glassy state;
*Basalt*, the finely crystalline or porphyritic rock forming the bulk of basic lava flows;
*Dolerite* or *Diabase*, the common rock of most basic dykes and sills; and
*Gabbro*, the coarse-grained equivalent, occurring in the larger basic intrusions.

In practice, however, the introduction of a hypabyssal class involves much unsatisfactory overlapping. In addition to rocks of medium grain it would have to include rocks like pitchstone, which are volcanic in their associations, and at the other extreme granite pegmatites, which are plutonic in their associations and the most coarsely crystalline of all rocks. Even in a single dyke, consisting mainly of dolerite, there may be a glassy selvage of tachylyte along each contact with the wall rocks, probably because the magma became too viscous to crystallize: generally as a result of rapid loss of heat or gases, or both. Between the tachylyte skin and the internal dolerite there is likely to be a transitional zone having a typically basaltic grain. Moreover, a thick dolerite sill may become progressively coarser towards the middle until the rock is a gabbro (Figure 5.4(*b*)), and locally there may be patches sufficiently coarse to be described as gabbro pegmatite. It is, in fact, much more convenient to extend the scope of the volcanic class to include the rocks of the associated minor intrusions, and similarly with the plutonic class. Adding the fragmental (pyroclastic) volcanic rocks for completeness, this leads to the scheme on p. 72. Some of the chief departures from the typical textures are indicated between brackets. It should be noted that fragmental deposits have a

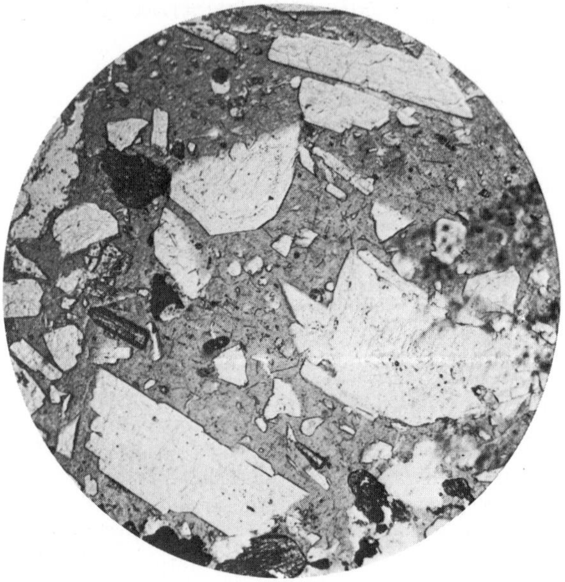

**Figure 5.13** Photomicrograph of hypersthene-andesite, 'spine' of Mt Pelée, Martinique, West Indies. Specimen collected from a fallen block by Dr C. T. Trechmann. Ordinary light ×32 (*Jean Tarrant*)

[1]The term comes from the Greek *tachys*, swift, and *lytos*, soluble, fusible, and should not be written 'tachylite'.

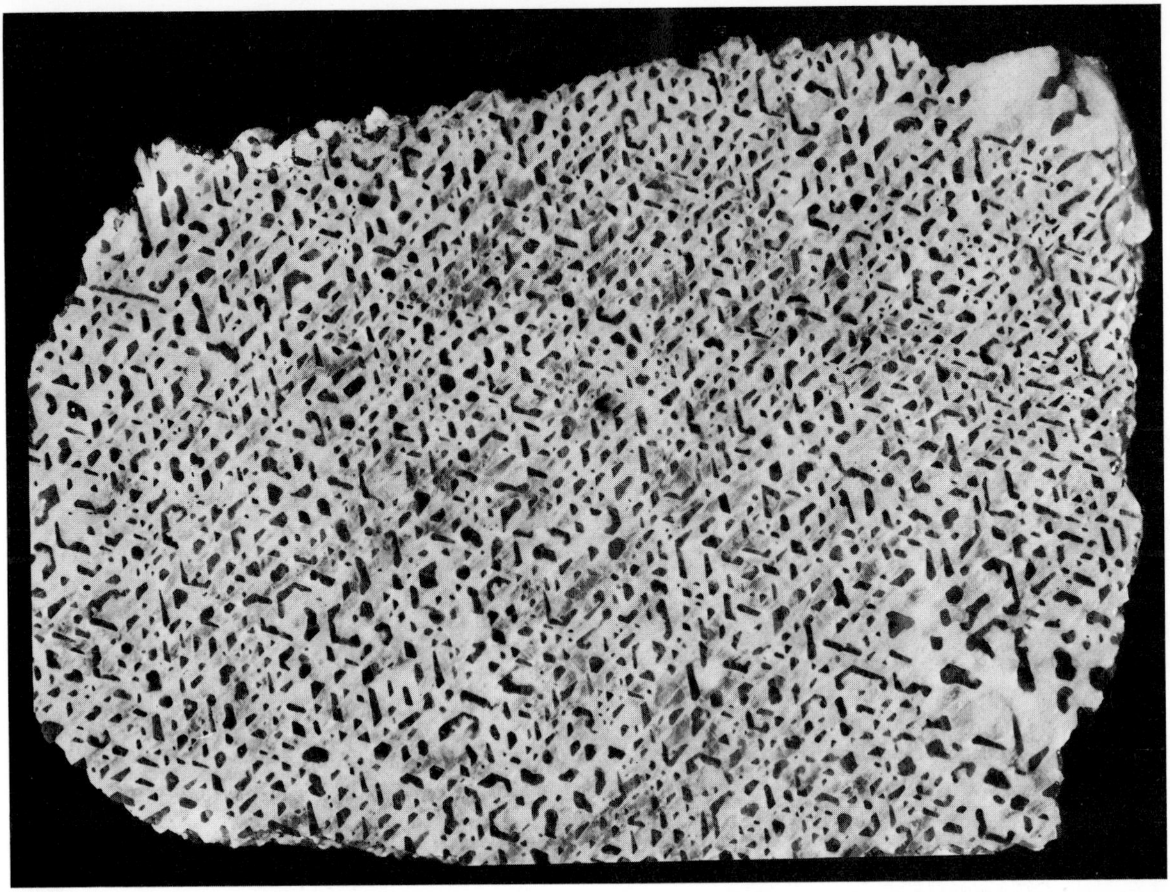

**Figure 5.14** Polished surface of a graphic granite, showing intergrowth of a crystal of quartz (dark) with a crystal of potash feldspar (light) on a large scale. The specimen includes only a small part of each crystal. About half natural size. From a pegmatite, Hale's Quarry, near Glastonbury, Connecticut, U.S.A. (*B. M. Shaub*)

structure rather than a texture. Structure refers to the whole assemblage; texture to the individual parts which make up the assemblage.

As the above scheme implies, the rocks of any given composition may display a great variety of different textures—and structures—according to their origin and mode of occurrence. This diversity is exceptionally well illustrated by rocks of granitic composition, as the following table clearly shows.

*Pyroclasts* are fragmental materials produced by the erosive and explosive action of hot, high-pressure gases. In any given eruption they may consist of materials from three different sources:

(*a*) 'live' lava that at the time of formation of the pyroclasts was molten or partly consolidated, the fragments ranging in size from fine particles discharged as more or less incandescent spray to masses of pumice and scoriae (volcanic 'cinders'), and volcanic bombs that may exceptionally be as big as a small cottage;

(*b*) 'dead' lavas and pyroclasts torn from the walls of the feeding pipe or conduit of an already existing volcanic cone, or from obstructions in the vent;

(*c*) pre-existing crustal rocks from beneath the volcanic cone, carried up and blown out when the volcanic pipe first reached the surface, and also during its subsequent enlargement by blasts of uprushing gases.

The ingredients of types (*b*) and (*c*) range in size from fine dust to large blocks. It should be noted that the term *volcanic bomb* is used for representatives of the live lava (*a*) of the eruption during which they were blown out. The term *ejected block* is used for pre-existing rocks, whether these are materials previously erupted by the volcano (*b*) or samples of the crustal foundations (*c*). Pyroclasts

| Mode of Occurrence | Texture | |
|---|---|---|
| VOLCANIC | | |
| **Pyroclastic deposits** | **Fragmental**: ashes, tuffs, ignimbrites, etc. | |
| **Lava flows** | **Glassy to fine grained** | |
| **Minor intrusions** | **Medium grained**<br>(fragmental in some volcanic pipes, e.g. tuffsite)<br>(glassy, e.g. pitchstone)<br>(pegmatitic patches in thick sills) | and *porphyritic* varieties of some of these |
| PLUTONIC | | |
| **Major intrusions** | **Coarse grained** | |
| **Minor intrusions** | **Medium grained** e.g. aplite<br>**'Giant' grained** e.g. pegmatite | |

## Rocks of Granitic Composition

| | *Pyroclasts* | *Lava flows* | *Minor intrusions* | *Major intrusions* |
|---|---|---|---|---|
| **Fragmental** | Ignimbrite | | Ignimbrite in dykes and volcanic pipes | |
| **Glassy** | Pumice | Pumice<br>Obsidian | Pitchstone | |
| **Fine-grained** | | Rhyolite | Felsite<br>Quartz-porphyry | |
| **Medium-grained** | | | Microgranite<br>Granophyre<br>Aplite | |
| **Coarse-grained** | | | | Granite |
| **Very coarse-grained** | | | Pegmatite<br>Graphic granite | |

of the latter type provide invaluable information about the nature of the crust perforated by the volcanic pipe.

The finer materials that shower down from volcanic 'clouds' are generally described as *volcanic ash*. Deposits of this kind that have become more or less indurated are known as *volcanic tuff*. The coarser pyroclasts form deposits of *agglomerate* (mainly volcanic fragments and bombs) or *volcanic breccia* (with many angular fragments of foundation rocks).

## Rhyolite and Ignimbrite

The name *tachylyte* for basalt-glass, as indicated in the left-hand column of page 70, refers to the rapidity with which tachylyte (or any other basaltic rock) melts to a 'thin', dangerously mobile liquid when heated to the appropriate temperature, say about 1000°C. Granitic rocks, in contrast, when heated to the same temperature or immersed in molten basalt, melt sluggishly and incompletely to what is hardly distinguishable from a highly viscous glass. Even when the fusion is carried out under high pressure with as much water as can be dissolved by the resulting melt, the latter is still extremely viscous. Corresponding to these experimental facts, freshly erupted basaltic lava is able to flow freely for long distances, whereas known lavas of rhyolite or obsidian are so stiff that they quickly congeal to thick glassy or stony tongues, even on steep volcanic slopes. Because of this striking contrast in mobility between freshly erupted basic and acid lavas, it was for long an extremely puzzling fact that enormous areas were known to have been buried beneath what appeared to be floods of rhyolite lavas that must have flowed at least as readily as the most mobile floods of plateau basalts. Indeed, the very word *rhyolite* comes from the Greek *rheo*, I flow. The reason for the apparent contradiction was made clear by P. Marshall in 1932, when he showed that the very thick sheets of what were thought to be lava-flows of rhyolite, covering about 25,900 km² of the Rotorua volcanic region of New Zealand, are really deposits of acid pyroclasts. He recognized that these had been erupted not as coherent lava flows but as vast, swiftly moving clouds of effervescent incandescent spray, possibly discharged in much the same way as the disastrous *nuée ardente* from Mt Pelée that destroyed St Pierre in 1902 (see Figure 12.17 and pp. 203–4). For these widespread sheet-like tuff flows, and also for their constituent rocks, whether tuff-like or lava-like, Marshall suggested the general term *ignimbrite* (L. *ignis*, fire; *nimbus*, cloud), meaning 'fiery cloud rock' or, perhaps better, 'fiery spray rock', since spray involves the additional idea of rapid movement.

The characteristic microscopic structures seen in a typical sheet of ignimbrite are well illustrated by Figure 5.15. Except towards the top (and locally at the bottom) the viscous magmatic particles of the spray must have been sufficiently hot to adhere tightly together as they accumulated, and to flatten out and bend under the weight of later additions, so forming a compact 'welded' assemblage of drawn-out glassy shreds, some of which moulded themselves around crystals and other fragments (Figure 5.15, B–B). Similar rocks in Yellowstone Park and elsewhere have been called *welded tuffs*. The term *ignimbrite*, however, has a wider meaning. Besides the welded varieties it includes the rocks of the associated upper (and sometimes lower) layers. These contain glass shards and bits of pumice that were already solidified as they came to rest, so that they show little or no sign of compression other than fracture (Figure 5.15 A and C). Indeed, the last material to accumulate in a particular eruption might remain more or less incoherent, like loose sand or ashes. The deposits from the *nuées ardentes* of Mt Pelée and the so-called 'sand-flow' of 1912 in what came to be called the 'Valley of the Ten Thousand Smokes' of Alaska are of this kind. Catastrophic though these eruptions were by human standards, the volume of their products fades into insignificance compared with the colossal bulk of the ignimbrite sheets of New Zealand and western America. The corresponding eruptions must have been on a scale that is terrifying to contemplate.

Some layers of the welded varieties are indistinguishable from obsidian except under the microscope, when tell-tale relics of flattened shards or bits of pumice reveal that they originated as hot pyroclasts and not as coherent lava flows. Other layers look like rhyolite, especially when devitrification has set in. This is simply the delayed crystallization that sooner or later takes place in every kind of glass. The usual presence of quartz crystals, cracked, broken or corroded, together with a few of feldspar, embedded in what has become a minutely crystalline groundmass, heightens the resemblance to rhyolite, or even to quartz-porphyry. The term *ignimbrite* is now in

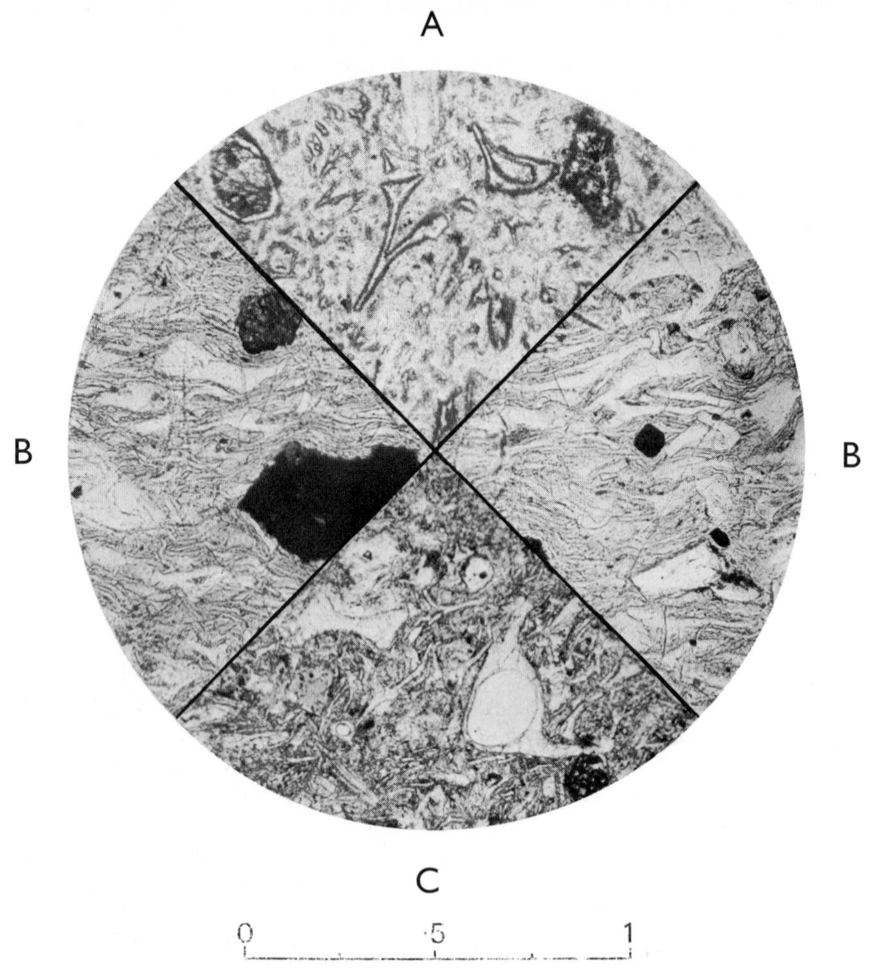


**Figure 5.15** Photomicrograph of ignimbrite from Maraetai Dam, near Mangakino, New Zealand. The scale printed beneath indicates 0·5 and 1 mm. (*R. C. Martin*)

A Shards of broken glass (vitroclastic structure) from the upper tuffaceous zone of the ignimbrite sheet
B Flattened and welded glass shards from the middle part of the sheet
C Shards of broken glass in the tuff forming the basal zone of the ignimbrite

general use for all these varieties, whether recent or geologically very old, whether loose or welded, glassy or devitrified, extrusive of intrusive, provided only that the rocks retain some evidence of having been formed from acid melt in the form of spray.

Spray must be carefully distinguished from foam or froth. *Spray* consists of liquid or solid particles which are being rapidly transported by a stream of gas. Having a viscosity only a little higher than that of the gas concerned, spray can travel with the velocity of a hurricane, far more rapidly than any lava flow. On the other hand, *foam* consists of gas bubbles dispersed through a liquid medium. Observations of the foam produced by breakers on dirty beaches, to say nothing of the domestic uses of the varieties of foam or froth called 'lather', suffice to show that it is much more viscous than the liquid alone would be. The contrast is about as extreme as one could imagine.

Foam is sluggish and cannot make its way into cracks. Spray has almost unlimited powers of penetration, as those who live on a wind-swept sea front or have had the misfortune to be caught in a dust storm know to their cost. These contrasted properties have important geological applications.

The acid lava corresponding to foam is *pumice*. When relief of pressure allows the liberation and rapid expansion of gases through the upper part of an ascending column of obsidian lava in a volcanic pipe, the lava swells into a froth (with minute bubbles) or foam (with larger ones). If the foam solidifies without rupture, the result is pumice, with a pearly-white or silvery-gray lustre. Unruptured masses of pumice are generally quite small, except perhaps where the top of a flow of obsidian has been distended by expanding gas bubbles. More commonly the gases continue to expand during the actual eruption until the tenuous walls confining the bubbles can no longer withstand the strain. Sluggish foam is then rapidly transformed into swiftly moving spray as the released gases respond to the sudden relief of pressure. Bits of white-hot viscous pumice of all shapes and sizes, surrounded by incandescent clouds of the produces of fragmentation (drawn-out shreds, lunate bows and Y-shaped shards), together with any crystals that may have been present in the original magma, may then be carried far and wide over the surrounding country, smothering it beneath a deadly pall of the 'ashes' that eventually become ignimbrite. Marshall showed that the New Zealand ignimbrites must have had a temperature above 1000°C, more probably 1200°C, at the time of eruption. How such temperatures were reached remains an unsolved problem. Marshall himself hazarded the guess that gas reactions facilitated by relief of pressure might generate much of the heat required for these appalling eruptions, He may well have been right.

**Figure 5.16** Columnar jointing in a sheet of ignimbrite, near Waipapa dam, Waikato River, North Island, New Zealand (*R. H. Clark*)

# Mineralogical Classification of Common Igneous Rocks

Fine-grained types in *italics* (mostly Volcanic)
Medium-grained types in ordinary type
Coarse-grained types in **bold** type (mostly Plutonic)

| FELDSPARS<br>———<br>OTHER MINERALS | ORTHOCLASE > SODIC PLAGIOCLASE | SODIC PLAGIOCLASE > ORTHOCLASE | SODIC PLAGIOCLASE (ANDESINE) PREDOMINANT | |
|---|---|---|---|---|
| QUARTZ ESSENTIAL<br>*Ferro-magnesian minerals:*<br>BIOTITE OR HORNBLENDE OR BOTH | *Rhyolite*<br>Quartz porphyry<br>**Granite** | *Rhyodacite*<br>Granodiorite porphyry<br>**Granodiorite** | *Dacite*<br>Quartz porphyrite<br>**Quartz-diorite** | |

| FELDSPARS<br>———<br>OTHER MINERALS | OTHOCLASE PREDOMINANT | ORTHOCLASE AND PLAGIOCLASE ROUGHLY EQUAL | SODIC PLAGIOCLASE (ANDESINE) PREDOMINANT | NO FELDSPAR |
|---|---|---|---|---|
| LITTLE OR NO QUARTZ<br>*Ferro-magnesian minerals:*<br>HORNBLENDE AND/OR BIOTITE AND/OR AUGITE | *Trachyte*<br>Porphyry<br>**Syenite** | *Trachyandesite*<br>Monzonite-porphyry<br>**Monzonite** | *Andesite*<br>Porphyrite<br>**Diorite** | **Hornblende** |

| Feldspars<br>———<br>OTHER MINERALS | | | CALCIC PLAGIOCLASE PREDOMINANT | NO FELDSPAR |
|---|---|---|---|---|
| LITTLE OR NO QUARTZ<br>*Ferro-magnesian minerals:*<br>AUGITE AND IRON ORES | | | *Basalt*<br>Dolerite or Diabase<br>**Gabbro** | **Pyroxenite** |
| NO QUARTZ<br>*Ferro-magnesian minerals:*<br>AUGITE, OLIVINE AND IRON ORES | | | *Olivine-basalt*<br>Olivine-dolerite or Olivine-diabase<br>**Olivine-gabbro** | **Peridotite** |

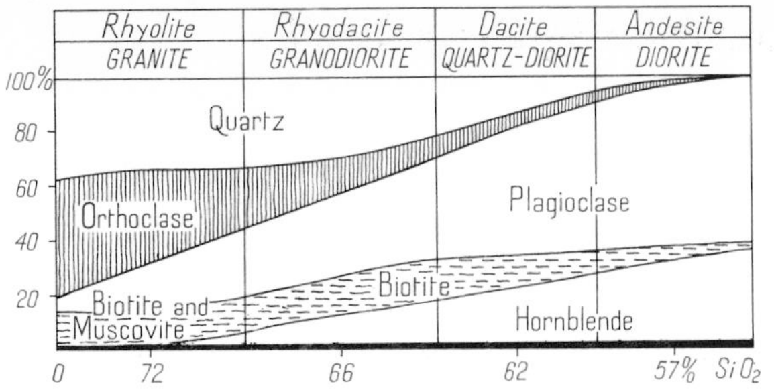


**Figure 5.17** Variation in mineral composition and silica content of the rock series granite-diorite. Accessory minerals (e.g. zircon, apatite, iron ore minerals) in black at the base of the diagram. In some varieties augite or hypersthene may be present.

## Classification of Common Igneous Rocks

We have already seen how many structural and textural varieties can be recognized of rocks having much the same chemical composition as granite. Igneous rocks (apart from the glassy varieties) can also be classified according to the kinds and proportions of their constituent minerals. And here it becomes necessary to say that the term 'granite' has so far been used in this book with the very broad meaning usually extended to it by field geologists. Actually, many of the rocks commonly referred to as 'granite' have more plagioclase than orthoclase and are therefore not 'true' granites; it is convenient to distinguish them by the name granodiorite, because they are intermediate in mineral composition between granite and quartz-diorite. Diorite, with little or no quartz, is made up essentially of plagioclase and hornblende. Many large 'granite' masses, consisting mainly of granite and granidiorite, pass marginally into quartz-diorite and diorite. The continuous mineral variation, and the conventional dividing lines between the types, are shown graphically in Figure 5.17. The names of the corresponding volcanic rocks are added for convenience at the top of the diagram. In the nature of the case it may not be easy to distinguish between rhyodacite and rhyolite, and the latter name is commonly used for both. Similarly, the term ignimbrite may be applied to materials ranging from rhyolite to dacite in composition.

As this example indicates, igneous rocks that are completely crystalline can be classified mineralogically, and therefore identified, by means of criteria such as:

(*a*) the presence or absence of quartz;

(*b*) the kinds of feldspar and the proportions between them (and in some cases by the absence of feldspar); and

(*c*) the kinds of ferromagnesian minerals.

Combining these criteria with the main textural variations, a workable classification such as that on the previous page is arrived at.

## Selected References


BOWEN, N., 1928, *The Evolution of the Igneous Rocks*, Princeton University Press, Princeton, N.J.

MARMO, V., 1971, *Granite Petrology and the Granite Problem*, Elsevier, Amsterdam.

MARSHALL, P., 1935, 'Acid rocks [ignimbrites] of the Rotorua volcanic district', *Transactions of the Royal Society of New Zealand*, vol. 64, pp. 1–44.

MARTIN, R. C., 1959, 'Some field and petrographic features of American and New Zealand ignimbrites', *New Zealand Journal of Geology and Geophysics*, vol. 2, pp. 394–411.

RAGUIN, E., 1965, *Geology of Granite*, Wiley, New York.

TUTTLE, O. F., and BOWEN, N., 1958, '*Origin of granite in the light of experimental studies in the system* $NaAlSi_3O_8—KAlSi_3O_8—SiO_2—H_2O$', Geological Society of America, Memoir 74.

WINKLER, H. G. F. and VON PLATEN, H., 1957–62, 'Experimentelle Gesteinmetamorphose I–IV. *Geochim. Cosmochim. Acta*, vol. 13, pp. 42–69; vol. 15, pp. 91–112; vol. 18, pp. 294–316; vol. 24, pp. 48–69, 250–9.

WYART, J., 1955, 'Cristallisation, par voie hydrothermale, d'un verre naturel et origine du granite', *Colloques Internationaux du Centre National de la Recherche Scientifique*, Nancy, vol. 68, pp. 177–188.

WYART, J. M. and SABATIER, G., 1959, 'Transformation des sédiments pélitiques à 800°C sous une pression d'eau de 1800 bars et granitisation', *Bull. Soc. France. Mineral. Crist.*, vol. 82, pp. 201–10.

# Chapter 6
# Sedimentary Rocks

Sufficient for us is the testimony of things produced in the salt waters and now found again in the high mountains, sometimes far from the sea.

*Leonardo da Vinci* (1452–1519)

## Sandstones

Sandstone is perhaps the most familiar of all rocks, for it is easily quarried, and it has been used more than any other kind of natural stone for building purposes. Examined closely, using a lens if necessary, a piece of sandstone is seen to consist of grains of sand identical in appearance with those that are churned up by the waves breaking on a beach. Most of the grains consist of more or less rounded grains of quartz, but there are others of cloudy, weathered-looking feldspar, and generally a few shining spangles of mica can be seen (Figure 6.1).

Clearly, sandstone is made of second-hand materials, of worn fragments derived from the disintegration of some older rock, such as granite, which contained the same minerals. It differs from deposits of modern sands only in being coherent instead of loose. Calcite is a common cementing material. Brown sandstones are cemented by limonite and red varieties of haematite. In white, extremely hard sandstones, the cement is silica, which has crystallized as quartz. These cementing materials were deposited between the grains by ground-waters which percolated through the sand when it was buried under later sheets of sand or other formations.

White siliceous sandstone in which most of the grains, as well as the cement, consist of quartz are often referred to as *quartzite*. However, when a granitic area is being eroded there may not be time—or the climate may be unsuitable—for all the feldspar to be weathered into clay minerals before the disintegrated debris is deposited again, and grains of feldspar, commonly pinkish orthoclase or microcline, will then be abundant in the resulting sands. The term *arkose* is applied to sandstones rich in feldspar. Another important type of sandstone is known as *graywacke*, a name meaning 'gray grit', first given to it by miners in the Harz Mountains, and later adopted by Werner. Graywacke differs from ordinary sandstone (*a*) in being less well sorted, i.e. its grains are of widely varied sizes (Figure 6.7); and (*b*) in having a wider variety of constituents as grains, matrix and cement. Its grains may include not only quartz, feldspars and micas, but bits of incompletely

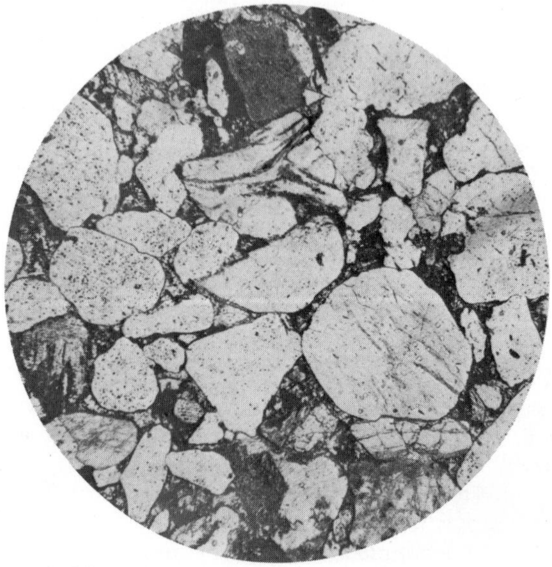

**Figure 6.1** Photomicrograph of sandstone, Torridon Sandstone, North-West Highlands of Scotland (×30) (*G. S. Sweeting*)

## Fragmental Sedimentary Rocks

| | Coarse<br>*Rudaceous*<br>(L. *rudus*, debris) | Medium<br>*Arenaceous*<br>(L. *arena*, sand) | Fine<br>*Argillaceous*<br>(L. *argilla*, clay) |
|---|---|---|---|
| **Loose** | Boulders and pebbles<br>Gravel, shingle and rubble | Sands | Mud<br>Clay |
| **Indurated** | *Conglomerate*<br>(rounded fragments)<br>*Breccia*<br>(angular fragments) | *Sandstones*<br>(including<br>quartzite, arkose<br>and graywacke) | *Mudstone*<br>(compact)<br>*Shale*<br>(laminated) |

weathered ferromagnesian minerals and fragments of rocks from the region that was undergoing erosion, including volcanic types in some areas. The matrix is fine grained and largely composed of the flaky alteration products of weathering, such as the clay, micaceous and chloritic minerals. Evidently graywackes are muddy grits of a kind that can have accumulated only as a result of rapid transport from a region of varied rocks and vigorous erosion.

### Other Fragmental Sedimentary Rocks

Along the beach, and especially near the cliffs, boulders and pebbles are heaped up by storm waves. Then come the sands, and beyond them on the sea floor lie still finer deposits of mud, made up of minute flaky shreds of clay minerals and their micaceous and chloritic associates, together with finely comminuted grains of quartz and altered feldspar. There is, of course, every gradation from the coarsest boulder beds down to the finest muds, and all these deposits are essentially made of bits of pre-existing rocks or their alteration products. The Latin word for rock-waste, *detritus*, is used for such ingredients, as opposed to the cement, which is introduced later from solution. Detrital sediments are said to have a *clastic* structure (Gr. *klastos*, broken in pieces), as opposed to the crystalline structure of chemically deposited sediments, such as rock-salt.

All the varieties of the three main groups of loose sediment recognized above have their compacted or cemented equivalents amongst the sedimentary strata of all geological periods. Traced across the country a sheet of sandstone is often found to thin out and pass laterally into clay or shale. Traced in the other direction, it may become coarser in grain and pass into a boulder bed or conglomerate (Figure 6.2). The term *conglomerate* is applied to cemented fragmental rocks containing rounded fragments such as pebbles and boulders; if the fragments are angular or subangular, the rock is called *breccia*.

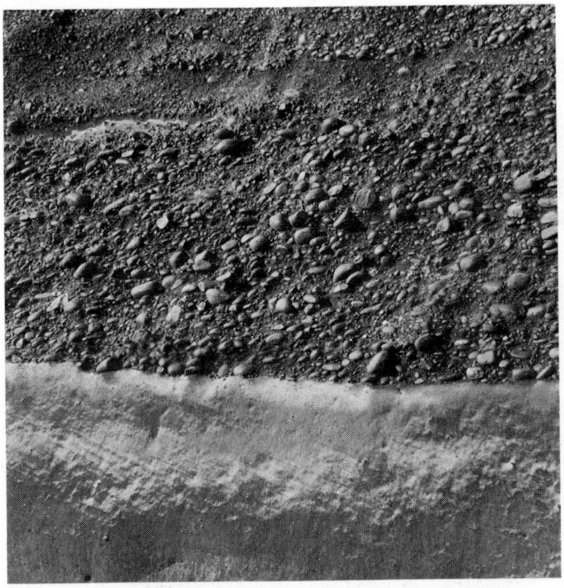

**Figure 6.2** Conglomerate overlying cross-bedded sandstone (Triassic System), The Cliff, Budleigh Salterton, Devon (*F. T. Blackburn*)

When first deposited, wet mud or clay is a very weak sediment, because of the large amount of water filling the spaces between the mineral particles. As more sediment is deposited on top, the growing overburden compresses the buried clay, which gradually consolidates as water is squeezed out. Eventually, with varying degrees of cementation it passes into mudstone or shale. Mudstone is compact, but shale can easily be divided into laminae, sometimes as thin as paper. This structure, which gives shale its characteristic property of fissility, is called *lamination*. The clay and other flaky materials occur as minute films which tend to lie with their flat surfaces parallel to the plane of lamination. The development of this regular arrangement probably began as the flakes were deposited and in the course of time it would naturally be enhanced by the weight of overlying sediments. Lamination is only feebly developed, if at all, in mudstone: perhaps because the minute particles coalesced into flocculent aggregates as they sank, so coming to rest in random positions.

**Figure 6.3** Bedding and jointing in Carboniferous sandstone, Muckross Head, Co. Donegal, Ireland (*R. Welch Collection, Copyright Ulster Museum*)

## Varieties of Bedding

In the steep face of a sandstone quarry or cliff, successive beds or layers can be seen, differing from one another by variations in colour or coarseness of grain (Figure 6.3). At intervals there may be strongly marked bedding planes, along which the sandstone is easily split, due perhaps to the presence of a thin layer packed with flat-lying flakes or mica, or to the intervention of a thin band of clay or shale. Evidently the beds or strata have been formed by the deposition of successive sheets of sediment, The resulting bedding, or *stratification*, is a primary structure of sedimentary rocks. The structural unit of stratified rocks is a bed or layer of uniform or characteristically varied composition, the thickness of which may be anything from a few millimetres to a metre or so (Figure 6.4). The lamination of shale is parallel to the bedding and resembles a sort of microscopic stratification. But it should be remembered that lamination refers to a parallel arrangement of minerals within a bed or stratum, whereas stratification refers to a succession of beds separated by bedding planes.

Most sediments were originally deposited on nearly flat or very gently inclined surfaces. But when sand is deposited from currents in shallow water, shoals and sandbanks are often built up, the

**Figure 6.4** Well bedded limestones alternating with shales and mudstones, cliffs west of Harness, Castlemartin, Dyfed, Wales (*Institute of Geological Sciences*)

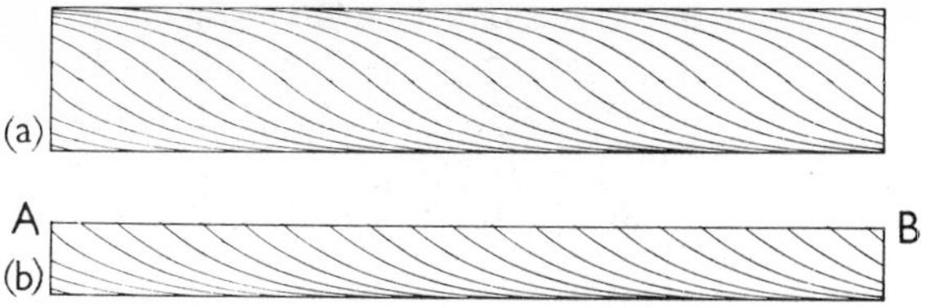


**Figure 6.5** Sections to illustrate current bedding. In (*a*) the structure is complete. In (*b*) the upper part has been eroded down to AB and the current bedding pattern is truncated.

front of deposition advancing in the direction of the current, just as a railway embankment is built forward during its construction. The bedding of a growing sandbank follows the gently curved slopes on which the sand is dropped, giving a pattern in cross-section which, under ideal conditions, resembles that shown in Figure 6.5(*a*). With changing conditions, possibly during a storm, the upper part of the sandbank is swept away, and the bedding planes are sharply truncated by an erosion surface such as AB.

Later on, another sandbank, or possibly a different kind of sediment, may be deposited on the surface thus provided. So, in quarry or cliff exposures of sedimentary beds, we may find that within certain bands the bedding is oblique and variously inclined to the general 'lie' of the formation as a whole (Figure 6.6). This structure, which is original, and not due to tilting or folding, is called *cross bedding* or *current bedding*. Sand dunes, accumulating from wind-blown sands, also exhibit cross bedding, the pattern of which reproduces, wholly or in part, the characteristic outlines of dunes.

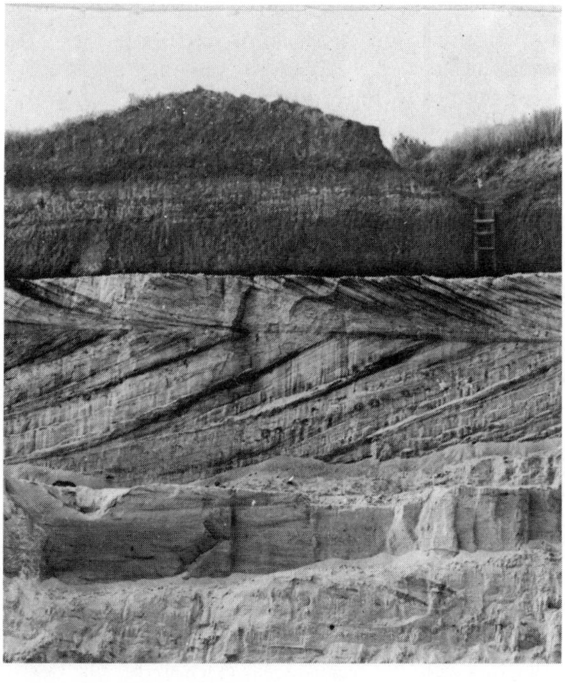

**Figure 6.6** Current-bedded Lower Greensand overlain by Gault, NE of Leighton Linslade, Bedfordshire (*Institute of Geological Sciences*)

If a mixture of loose sediment, ranging in grade size from little pebbles through sand to mud, is dropped into a column of water in a long jar, the large particles reach the bottom first and are followed in turn by successively smaller ones. This arrangement, which is often seen in beds of graywacke, is called *graded* bedding (Figure 6.7). The bottom of the bed generally lies on shale and may consist of a coarse grit from which where is an upward transition towards finer material, commonly shale, at the top, where again there is a well-marked bedding plane. Such beds are generally of fairly uniform thickness over great areas. It is only in recent years that the peculiarities of graywackes and their characteristic graded bedding have been understood. Most graywackes appear to be formed from what are called turbidity currents on the sea floor. A sort of submarine avalanche of mixed sediment is started on a relatively steep slope by an earthquake. The resulting heavy cloud of sediment suspended in sea water travels swiftly down the slope and may spread out far over the sea plains beyond before re-deposition is completed (Figure 24.18).

## Limestones

Limestones of suitable quality are widely used as building stones, because of the ease with which they can be worked, and some varieties, the aristocrats of a very mixed group, have become famous through their lavish use in great public buildings. Portland stone, for example, has been a favourite choice for many of London's greatest buildings, ever since Wren selected it for the rebuilding of St Paul's Cathedral after the Great Fire of 1666. The towers and steeples of the London churches, the government offices in Whitehall, the front of Buckingham Palace and the headquarters of London University all display its use in various styles of architecture.

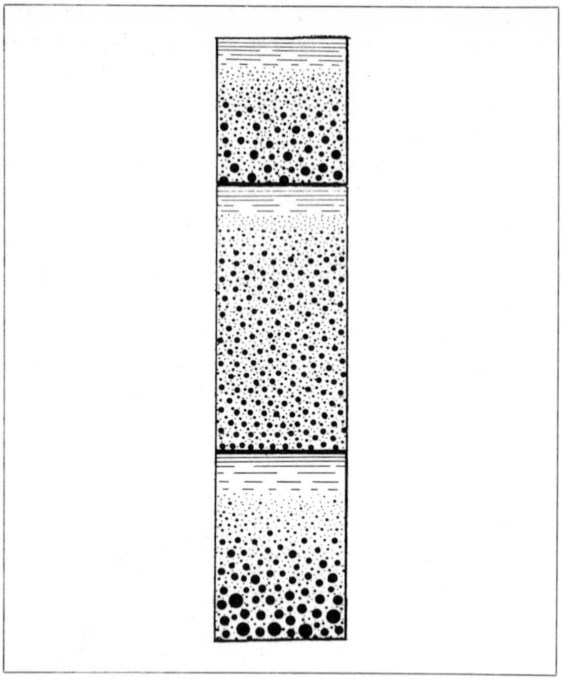

**Figure 6.7** Graded bedding

**Figure 6.8** The Great Scar Limestone (Carboniferous) above Malham Cove, North Yorkshire (*Institute of Geological Sciences*)

The natural architecture of limestones can be studied in the quarries of Portland and the Cotswolds, in the grey scars of the Pennines (Figure 6.8) and in the white gashes cut by lime-makers in the green slopes of the Chalk Downs. Some of the limestones of the Pennines are packed with the remains of corals and marine shells, and the bead-like relics of the stems of sea-lilies (animals like starfishes on long stalks—Figures 6.9 and 6.10). The yellow Cotswold limestones, more open and porous, are often crowded with fossil shells, while belemnites, looking like thick blunt pencils, and the coiled forms of ammonites add further interest and variety. The fine-grained and generally friable limestone known as chalk[1] was shown long ago by Sorby to be composed almost entirely of organic remains. Smooth white fossil shells of molluscs and sea urchins, and much smaller coiled or globular shells of foraminifera (Figure 24.9), visible only under the microscope, are embedded in an extremely fine-grained matrix. This consists largely of highly comminuted shell-debris. The nodules and bands of flint, which are a conspicuous peculiarity of the Chalk, are referred to on pp. 86–7.

Evidently many limestones are accumulations of organic remains, vast cemeteries continually being added to by the teeming life of the sea. The well-known limestones of Portland and Bath are by no means devoid of fossil shells, but they are mainly composed of rounded grains that look like insects' eggs. For this reason they are called *oolites* or oolitic limestones (Gr. *oion*, an egg). Under the microscope each granule or *oolith* is seen to be made of concentric layers of $CaCO_3$, often with a bit of shell at the centre (Figure 6.11).

Ooliths are being formed at the present time in the shallow marine waters of the tidal channels and lagoons of Florida and the Bahamas. Warming and evaporation of sea water increases the concentration to a point at which $CaCO_3$ can be precipitated. Where the water is agitated by waves and currents, tiny grains of $CaCO_3$ serve as nuclei for precipitation and become coated with successive concentric layers. Rounded forms result from the continual rolling on the sea floor. Organisms also contribute to the formation of

[1]It should be noted that the word *chalk* with a small 'c' refers to the rock when it is regarded only as a material. The *Chalk*, with a capital 'C', refers to the strata of chalk which were deposited over an enormous area during the period of geological time known as the Cretaceous Period (see p. 102).

ooliths. Lime-fixing bacteria and calcareous algae encrust nuclei, and even their own cells, with film after film of $CaCO_3$ abstracted from sea water or from the saline waters of lakes such as Great Salt Lake in Utah (Figure 6.12).

The grain size of the ooliths in a bed of oolithic limestone is generally fairly uniform, however they may have originated, because of the sorting action of waves and currents. The rate of growth is very slow. Age measurements by the radiocarbon method (see p. 244) show that the inner rings of an oolith from the Bahamas were precipitated more than 1000 years ago. Occasionally the grains have grown as big as peas and the resulting limestone is then called *pisolite* (Gr. *pison*, a pea).

**Figure 6.9** Weathered surface of crinoidal limestone (Carboniferous Limestone) (*W. W. Watts*)

**Figure 6.10** Crinoids. Slab of calcerous shale, River Liddel, south of Penton Bridge. Scottish/English border, showing two specimens of *Woodocrinus* (*J. Wright*)

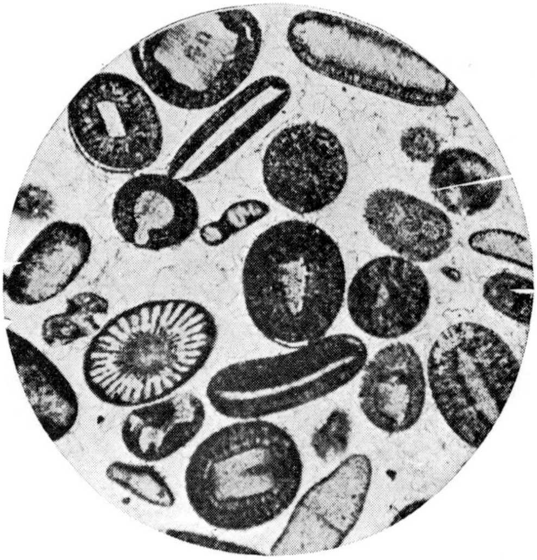

**Figure 6.11** Photomicrograph of oolitic limestone, Farley, Bath ($\times 30$) (*G. S. Sweeting*)

Limestones are thus seen to be deposits formed from dissolved material: generally, but not always, from sea water. They usually accumulate outside the stretches of sand and mud that in most places border the land; but where the sea is uncontaminated by muddy sediment, and especially where the cliffs themselves happen to be made of older limestones, they may form close up to the land. Beaches may locally be composed of sand made up, not of quartz grains but of shell debris (Figure 3.6). Similarly, the sands associated with coral reefs and atolls consist largely of coral debris, ground down by the waves.

Naturally there are many places where muds and calcareous materials are deposited together.

The resulting sediment may be argillaceous limestone or calcareous shale, but two names in common use for such intermediate products are *marl*, when the sediment is soft, and *cementstone*, when it is indurated. The latter term refers to the fact that the $SiO_2$ and $Al_2O_3$ of clay and the CaO of limestone may be present in suitable proportions for the manufacture of Portland cement. Limestone and shale (or mudstone or marl) may also be deposited in sharply alternating beds, as illustrated in Figure 6.4.

### Magnesian Limestone and Dolomite

From the warm seas of the tropics many organisms abstract not only calcium but also a certain amount of magnesium to construct their carbonate shells, and for this reason some of the resulting deposits are *magnesian limestones*. But amongst the sedimentary carbonate rocks every gradation can be found from limestone, through magnesian limestone to dolomite. Although magnesium is about three times as abundant as calcium in sea water, little dolomite, if any, appears to be deposited on the sea floor of the present time, and then probably only with the aid of bacteria and algae and their decay products, which include ammonia. However, in past ages the evaporation of enclosed seas to the right degree of concentration provided conditions favourable to dolomite precipitation. More commonly, dolomite is formed indirectly by the action of ions of magnesium ($Mg^{2+}$) on $CaCO_3$ already deposited as coral reefs or otherwise on the shallower sea floors of tropical regions. The exchange that occurs can be represented by the following equation:

$$\underset{\text{Calcite}}{2CaCO_3} + \underset{\substack{\text{Ion from}\\\text{Sea water}}}{Mg^{2+}} = \underset{\text{Dolomite}}{CaMg(CO_3)_2} + \underset{\substack{\text{Ion into}\\\text{Sea water}}}{Ca^{2+}}$$

But this is not the whole story of what happens. In the equation, as in all such chemical reactions, the masses of the reacting substances on one side are equal to the masses of the products on the other. What commonly happens under natural conditions, however, is that a shell or coral is replaced by dolomite without any change of volume, so that every detail of the original organic structure is still perfectly preserved. Such equal-volume replacements—which cannot be completely expressed by chemical equations—are highly charac-

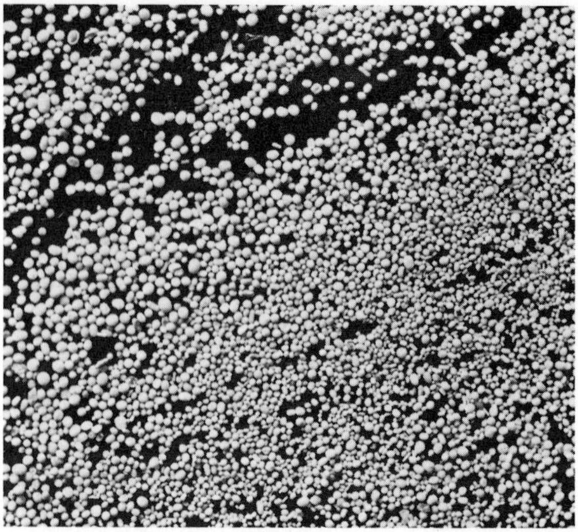

**Figure 6.12** Oolitic grains now forming in the shallows of Great Salt Lake, Utah (*R. Welch Collection, Copyright, Ulster Museum*)

teristic of many mineral and rock transformations. They are described as *metasomatic* replacements (Gr. *meta*, an affix connoting subsequent change; *soma*, body-substance), i.e. replacements involving a change of substance without a change of form.

One of the most celebrated formations of dolomite limestones extends from the south of Northumberland and the Durham coast to Nottingham; it is known as the Magnesian Limestone and is extensively quarried for various purposes in the chemical and steel industries, as well as for building stone. Dolomite from Derbyshire was selected as the building stone for the Houses of Parliament. Unfortunately the stone was subjected for many years to the attack of sulphurous fumes from the potteries that used to be situated on the other side of the Thames. Sulphuric acid and rain corroded the delicate carvings of the buildings turning the magnesia of the stone into easily soluble Epsom salts, so that very expensive repairs, especially to the riverside façade, had eventually to be carried out. But where dolomite is not exposed to such abnormal weathering conditions it makes an excellent building stone.

The Magnesian Limestone of Durham has become celebrated for the extraordinary variety of its concretionary structures. There are two main kinds: spheroidal (or 'cannon-ball') and reticulate. The spheroidal structures include well-shaped spheres ranging in size from a marble to a football, and clusters of interfering spheres, embedded in a powdery or granular matrix (Figure 6.13). The reticulate structures include a diversity of lace-like patterns, such as networks of irregular bands parallel to the bedding crossed by radial groups of rods and spindles. The matrix is the same as in the 'cannon-ball' group, and chemical analysis shows that it consists of dolomite containing only a little calcium carbonate in solid solution. The balls and networks, however, consist of calcite with only a little dolomite in solid solution. The concretionary structures are obviously of later origin than the bedding, relics of which can often be seen, like 'ghosts', still passing through the calcite concretions. Under the influence of percolating waters the original fine-grained and intimately mixed deposit of $CaCO_3$ and $CaMg(CO_3)_2$ has been reconstructed into large-scale spheres and networks of calcite embedded in dolomite.

These remarkable structures illustrate a method of segregation that is widely exemplified in rocks. The principle involved is realistically expressed in the familiar text: 'Unto every one that hath shall be given . . . but from him that hath not shall be taken away even that which he hath.' Minute particles of a given substance are more soluble than large ones. Consequently, solutions that are just saturated for intermediate sizes will be able to dissolve small particles while they are precipitating the same material around bigger ones. A nucleus, such as a shell fragment, thus provides an initial collecting centre and a concretion or nodule then gradually develops at the expense of the surrounding material of finer grain.

## Sedimentary Ironstones and Iron Ores

At one time two thirds—about 20 million tonnes—of all the iron ore annually used in the United Kingdom was produced from a belt of 'ironstones' that is strongly developed from Lincolnshire to Northamptonshire, with important extensions to the south-east in Oxfordshire and to the north in the Cleveland Hills of North Yorkshire. Radical changes have taken place, however, and during the years 1972, 1973 and 1974, the whole British iron ore output was only 9, 7, and 3·6 million tonnes respectively, imports for the year 1974 being 20 million tonnes.

The term *ironstone* refers to deposits of the mineral siderite, $FeCO_3$, and of the commonly associated iron silicate minerals. Deposits of iron oxides and hydroxides are called *iron ores*. Many of the English ironstones are oolitic and some consist of siderite that has metasomatically replaced the ooliths and fossils of what were originally oolitic limestones. But most of the ironstone ooliths appear to have been formed on the sea floor by precipitation of a green iron silicate, called *chamosite*, in concentric layers around suitable nuclei. The matrix in which these ooliths are embedded is a sort of chamosite-mud, and the latter also occurs as individual beds alternating with the oolitic bands. Replacement of chamosite

**Figure 6.13** 'Cannon-ball limestone' formerly exposed at the base of the Magnesian Limestone cliffs of Roker, Sunderland, Tyne and Wear. The powdery dolomite that originally occupied the interstices has been washed out by the sea. (*T. M. Finlay*)

by siderite is locally quite extensive. In the Cleveland district of North Yorkshire there is a remarkable bed in which many of the chamosite ooliths have been metasomatically replaced by pyrite.

When iron is first liberated by weathering from its parent rocks on the land it goes into solution as ferrous bicarbonate, but it can only remain in this condition or pass into it again where there is a lack of oxygen and an excess of carbon dioxide, e.g. in a stagnant waterlogged swamp on land, or in depressions on the sea floor where organic decay is more active than growth. Seasonal lowering of the relative proportion of $CO_2$ then tends to bring about precipitation of the iron as siderite. Clay-ironstones, which are spheroidal concretions and bands of impure muddy siderite, appear to have been formed in this way, since they are particularly characteristic of the argillaceous sediments underlying coal seams (see p. 286).

Far more commonly, however, iron is in the presence of oxygen from the start and is immediately precipitated as one or other of the natural pigments that give weathered rocks their characteristic tints of yellow, brown and red. The actual minerals form a series ranging in composition from $Fe(OH)_3$ to $Fe_2O_3$. Eventually, of course, the ferruginous material is carried away by rain and river, either as a fine suspension which sooner or later settles down and finds a home with any clay or mud that is being deposited, or as submicroscopic particles of $Fe(OH)_3$ which form a *colloidal* system with water. In this peculiar condition of 'colloidal solution' the particles are unable to sink and consequently they may be transported for long distances by rivers and currents. The inability to sink is a result of two curious facts. Colloidal particles of $Fe(OH)_3$ are positively charged and therefore repel one another. At the same time they are attracted to the negative 'poles' of molecules of water. Each molecule of $H_2O$ acts like a tiny magnet because the two very small atoms of hydrogen are practically embedded inside the very large oxygen atom, not one on each side as might have been expected, but both on the same side. The hydrogen side of the molecule has therefore a positive charge, while the other has a negative charge, and this polarity helps to keep $Fe(OH)_3$—and indeed all other colloidal particles—in a state of uniform dispersion through the water. In due course, however, precipitation becomes inevitable as a result of increased concentration or chemical changes in the water. In many of the lakes of Sweden and Finland substantial deposits of this kind, called 'bog iron ore', have accumulated since the withdrawal of the last of the great ice sheets—and the process of accumulation is still going on.

## Siliceous Deposits: Flint and Chert

Although quartz is practically insoluble in ordinary water and mostly turns up again in sandstones after the journey from its weathered source to its place of deposition, it must not be overlooked that a great deal of soluble silica is liberated during the chemical weathering of most silicate minerals. This can be illustrated by considering the alteration of orthoclase to clay:

$$\underset{\textit{Orthoclase}}{2KAlSi_3O_8} + \underset{\textit{Water}}{11H_2O} = \underset{\textit{Clay (kaolin)}}{Al_2Si_2O_5(OH)_4} + \underset{\textit{Removed in solution}}{2KOH + 4Si(OH)_4}$$

Similar equations can be written for albite and anorthite, and it should be emphasized that feldspars of the plagioclase series, especially the calcic varieties, weather in this way more readily than orthoclase. The soda liberated from albite helps to hold the hydrous silica in solution, mainly as $Si(OH)_4$. Still more silica is released by the weathering of pyroxenes and amphiboles.

Practically all this dissolved silica must soon be precipitated again, since many river waters contain five to ten times as much soluble silica as sea water. Probably most of it contributes to the finer constituents of alluvium and offshore muds. A little is fixed in iron silicate minerals and some is used for cementing sandstones. The balance is used by micro-organisms. In freshwater lakes these are mainly diatoms, the remains of which locally accumulate as diatomaceous earth. From the sea both diatoms and radiolarians (pp. 544–7) abstract silica to form their tiny opaline shells, while sponges support or protect their ungainly forms with networks of opaline rods or loose frameworks of needle-like spicules, also made of opal. Much of the silica used in this way is soon restored to circulation by being re-dissolved, but the part that survives contributes to the oozes and other deposits that are slowly accumulating on the ocean floor.

Sponges appear to have contributed most of the silica that now appears in the Chalk as *flint*. The relatively soluble opaline silica originally distributed through the Chalk has since been dissolved by percolating waters and redeposited in the insoluble form of flint. We have here another striking example of segregation of the kind already illustrated by the Magnesian Limestone (pp. 84–5). The original siliceous chalk has become segregated into flint that is nearly pure $SiO_2$, and chalk that is practically all $CaCO_3$. Flint, like chalcedony, is a sort of felt or mosaic of extremely minute but rather imperfect crystals of quartz—imperfect in the sense that there are unfilled gaps here and there in the crystal lattice. For this reason the specific gravity of flint is a trifle lower than that of quartz. Flint occurs as scattered nodules of knobbly and fantastic shapes, commonly concentrated in layers parallel to the bedding planes (Figure 6.14) and in places passing into tabular sheets; occasionally vertical stringers and veins of flint have developed along joint planes.

Remains of sponges are abundant in many parts of the Chalk, but now preserved as calcite, which replaced the original opaline silica. In turn the liberated silica then replaced the finely divided and more vulnerable parts of the Chalk, especially around suitable nuclei such as concentrations of sponge debris, and along bedding planes and joints that served as passageways for migrating solutions. Once deposited as flint, the silica was not again taken into solution.

Such silicification of limestone is not, of course, confined to the formation known as the Chalk, but the related bands and concretions in other limestones (and sometimes in calcareous shales and sandstones) are generally referred to as *chert*. There are also radiolarian cherts: these are ancient organic deposits of radiolarians which were cemented by silica into hard, tough or splintery rocks composed essentially of chalcedony or, if coloured

**Figure 6.14** Chalk with characteristic bands of flint nodules, Beachy Head, East Sussex (*P. Ekin-Wood*)

**Figure 6.15** Wood opal from a silicified log, Petrified Forest National Park, Arizona (*Jean Tarrant*)

by ferruginous impurities, of jasper. Indeed both flint and jasper might be regarded as particular varieties of chert.

Flint, however, has a special interest because one of the most important discoveries made by early man was that flint, of all the natural materials available to him, was unsurpassed for fashioning weapons and tools. Flint, like obsidian (Figure 5.11), breaks with a perfect conchoidal fracture and sharp cutting edges are easily obtained with a few deft blows. In other parts of the world chert and obsidian were similarly utilized, but wherever the Chalk still remains, as it does over extensive areas of Britain and Europe, multitudes of flint implements have been found in the gravels of river terraces and in cave deposits. Flint, for which the earliest known mines were dug, was the raw material that 'enabled the great creative inventors and the skilful and industrious artisans of the Stone Age to lay the material foundations of modern civilizations' (V. M. Goldschmidt, *Geochemistry*, 1954, p. 370).

One of the most remarkable of all replacement phenomena is the transformation of wood into opal, and even into chalcedony or jasper. In *wood opal* the replacement of immense logs is often so perfectly metasomatic that the structures of the wood, down to many details of the original cells, are preserved with astonishing fidelity (Figure 6.15). The process of petrification occurs when waterlogged tree trunks are buried in feldspathic sandy muds, perhaps in the shoals of a river bed. Percolating water becomes charged with additional alkalis and silica as a result of the continuing decay of the feldspar grains. As the solutions soak through the buried wood, some still undiscovered process of natural alchemy slowly operates. The silica so brought in is fixed as opal, taking the place and the form of the wood *without disturbing the structure*. The wood (cellulose) presumably escapes as marsh gas (methane, $CH_4$) and carbon dioxide. If chalcedony or jasper should be formed instead of opal there is usually some internal distortion and loss of detail, but the all-over replacement still remains so perfect that a petrified log may be mistaken for a fallen and weathered telegraph pole until the hammer is applied.

The most celebrated display of silicified logs occurs in the Painted Desert of Arizona, where the number of bark-stripped trunks of tall forest trees, some of them 30 m long, is so great that the region is known as the Petrified Forest (Figure 6.16). But it is more like the remains of a log-jam than a forest, for few roots or branches remain, and the logs had evidently travelled a long way before being entombed in the sandy muds where they were silicified. At some later time the sands and muds were consolidated into brilliantly coloured sandstones and shales by the deposition of a variety of cementing materials, mainly ferruginous, but also including compounds of manganese, copper and uranium. In the bright sunshine of this semi-arid region the name 'Painted Desert' does no more than justice to a landscape of gorgeously coloured rocks.

**Figure 6.16** Silicified logs, Petrified Forest National Park, Arizona. The log forming a bridge (known as Onyx Bridge) is 15 m long. It has resisted erosion while the soft sandy shale in which it was embedded has been washed away. These fossil trees flourished during the Triassic Period, about 200 million years ago. (*Dick Carter*)

## Salt Deposits

The natural history of the potash dissolved from rocks during weathering is very different from that of soda. In average crustal rocks these two constituents are about equally abundant, but while the liberated soda is carried to the oceans where it accumulates for a time, much of the potash remains in the soil, where it serves as an essential nutrient for plants. Eventually, when the soil itself is removed by erosion, the potash goes with it as a constituent of the potash-bearing clay minerals. Fixation of potash by clay continues on the sea floor, together with further fixation in the green mineral glauconite, described on p. 235. Of the small balance that then remains, much is abstracted by seaweeds, so that the amount left dissolved in the oceans (as shown in the adjoining table) is far less than might have been expected.

The average amount of mineral matter dissolved in sea water is about 3·5 per cent. For convenience this is usually expressed in parts per thousand (‰) and the *salinity* is then said to be about 35. When sea water is evaporated (at, say, 30°C) precipitation begins with quite negligible amounts of carbonates and the salinity has to rise to well over 100 before gypsum, $CaSO_4.2H_2O$, begins to be deposited. When the brine is concentrated to about a tenth of the original volume, rock-salt

| *The chief constituents of average sea water* | | *Average salinities of some natural waters* | |
|---|---|---|---|
| Na | 10·56‰ | Sea water | 35‰ |
| Mg | 1·30 | River water | 0·16 |
| Ca | 0·40 | Baltic Sea | 7·2 |
| K | 0·38 | Caspian Sea | 13 |
| Cl | 19·00 | Gulf of Kara Bogaz | 185 |
| $SO_4$ | 2·65 | Black Sea | 20 |
| $HCO_3$, $CO_3$ | 0·14 | Dead Sea | 192 |
| | | Great Salt Lake, Utah | 203 |
| Br | 0·06 | | |

(halite, NaCl) separates, accompanied by a little anhydrite, $CaSO_4$, and this continues almost to the end. The very soluble salts of potassium and magnesium do not begin to crystallize until the original volume is reduced to about 1/60th.

A geologically recent example of evaporation to dryness, with the resulting salt deposits still left uncovered at the surface, occurs across the boundary of the Ethiopian province of Eritrea, where the Red Sea begins to narrow. This is the *Piano del Sale*, a depressed region below sea-level which was originally part of the Red Sea and is now one of the hottest and driest places on the face of the earth. The isolation of this great evaporating dish was due to an accumulation of lavas from a series of volcanoes that are still not completely extinct. The depression is now lined with gypsum, representing an early phase of deposition. Within the inner rim of this 'saucer' of gypsum is a wide expanse of rock-salt, 32 km across, with intercalations of anhydrite. Each salt bed has also a saucer-like form, and because the level of the brine continued to fall with continued evaporation each 'saucer' is smaller than the underlying one. Potash salts appear only towards the heart of the depression, which drops to about 120 m below sea-level. Before their commercial exploitation, these were seen at the surface as a broad ring largely composed of carnallite, $KCl.MgCl.6H_2O$, thickening to about 60 m in the middle, where it was covered in turn by a metre or so of sylvite, KCl, over an area of 40 acres.

Although the original depth of the basin is not yet known, it could hardly have been more than 300 m. A column of sea water 300 m high would yield no more than about 5 m of salt deposits. In the *Piano del Sale* the thickness revealed by drilling are so great that they can only be accounted for by assuming that for long periods fresh supplies of sea water were continually added to the evaporating basin. Even today there are occasionally inundations from the sea. In the past, before the growing volcanic barrier reached its present size, it would necessarily go through a stage when it was less effective than now in keeping out the sea.

A process of continuous replenishment is manifestly responsible for the evaporites, mainly gypsum at present, now being precipitated on the floor of the Gulf of Kara Bogaz, a shallow embayment of the Caspian Sea (Figure 6.17). Because of the inflow from the Volga and Ural rivers, the level of the Caspian is always perceptibly higher than that of Kara Bogaz and consequently there is a continuous flow of water of medium salinity into Kara Bogaz through the narrow channel which is its only connection with the Caspian. Evaporation removes this water but leaves the salts behind. The concentration now reached is such that a little rock-salt is precipitated in warm seasons as well as gypsum. If the channel were closed and the Kara Bogaz evaporated to dryness the layer of salts (mainly rock-salt) would be little more than a metre thick on average, while if the channel remained open until Kara Bogaz filled up with salts the thickness would still be limited to 15 or 18 m, the maximum present depth of the water. To get greater thicknesses the floor of Kara Bogaz would have to subside while supplies of salt water continued to flow in from the Caspian. We can now realize how very extraordinary it is that in some parts of the world beds of rock-salt many hundreds of metres thick have accumulated in the past.

The salt mines of Wieliczka (near Krakow in south-west Poland) have been worked for over a thousand years and have long been celebrated as a tourist attraction because of the underground houses, churches and monuments, roads, railways and restaurants, all excavated far below the surface in a layer of salt over 3500 m thick. Neither the well-known Stassfurt deposits Germany, nor those of Alsace, are quite so thick, but they are commercially more important because valuable potash salts have been preserved under a protective cover of impervious clays and marls that prevented their removal and loss by subsequent solution. The thickest known salt deposits are those of Texas and adjoining states. About 3500 m have been proved in New Mexico. Even greater thicknesses are suspected in parts of Texas (see p. 158), but since the salt may have a cover of anything up to 9000 m

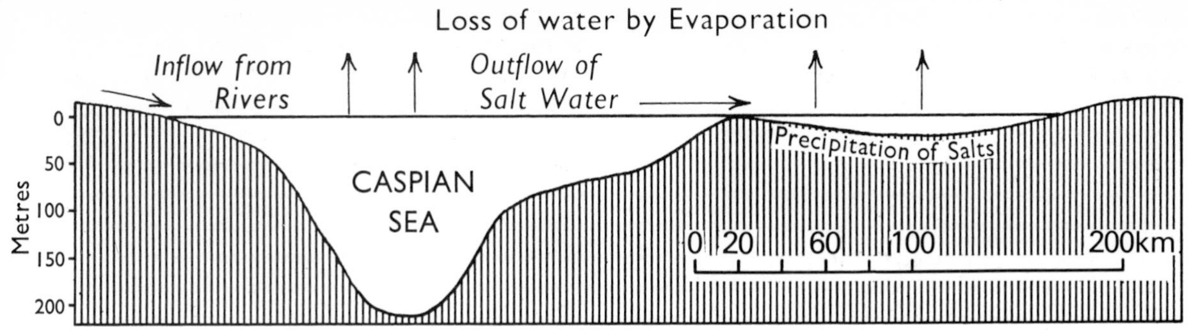


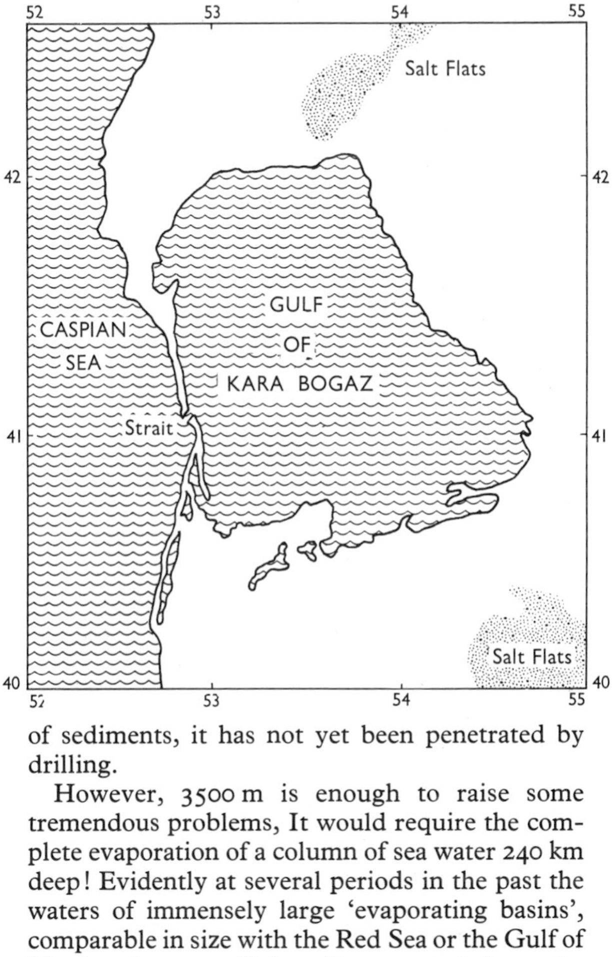


**Figure 6.17** Map and section to illustrate the accumulation of saline precipitates ('evaporates') in the Gulf of Kara Bogaz, Caspian Sea

of sediments, it has not yet been penetrated by drilling.

However, 3500 m is enough to raise some tremendous problems, It would require the complete evaporation of a column of sea water 240 km deep! Evidently at several periods in the past the waters of immensely large 'evaporating basins', comparable in size with the Red Sea or the Gulf of Mexico, became all but disconnected from the main body of the oceans, as the Mediterranean nearly is today. A coincidence of special conditions must have persisted for very long periods, the essential conditions being:

(*a*) a hot and arid climate to ensure the evaporation necessary to maintain a sufficiently high concentration for salts to crystallize out;

(*b*) an intermittent or continuous supply of salt: in part, perhaps, from rivers and inland seas as in the case of Kara Bogaz, but mainly from sea water, flowing into the basin through one-way channels;

(*c*) sinking of the floor of the basin to make room for the growing thickness of salt deposits;

(*d*) maintenance of the barrier between the sinking basin and the ocean, possibly by volcanic activity, upward growth of coral reefs, or earth movements;

(*e*) a final depression, carrying the barrier with it, to make room for the thick and widespread sediments that eventually covered and preserved the salt deposits.

More widespread than deposits of rock-salt and potash salts and precipitates of carbonates (limestone and dolomite) and anhydrite, which commonly occur as alternating layers. The anhydrite of this association characteristically occurs as nodules dispersed through carbonates, or as tightly packed nodules, Evidence as to how this may happen has been found in the desert region of the Trucial Coast, in the shoal water at the southern end of the Persian Gulf. Here evaporites of this variety are being precipitated today. Below low tide level calcareous sediments are being deposited. Mats of calcareous algae characterize the zone between high and low tide levels, and form traps for dolomite and gypsum. Above high tide level are extensive flats where anhydrite is precipitated as nodules from ground-water of marine origin that has become highly concentrated brine as a result of evaporation and lack of rain.

Trenches dug in these flats have exposed mineral sequences, from below upwards, that are respectively the equivalents of those below low tide level, between high and low tide levels, and above high tide level. The sections revealed in the trenches therefore indicate regression of the shore line. When traced laterally from land to sea, the evaporite of each facies will vary in age; each facies is diachronous.

The evidence from the Trucial Coast may have a bearing on the composition of some of the deposits of rock salt and potash salts, which differ chemically from evaporites obtained by experimental evaporation of sea water. Within some of these evaporites, for example the Middle Devonian potash deposits of Saskatchewan, not only is sulphate absent, but the proportion of magnesium is also lower than would theoretically be expected to be precipitated after rock-salt, and prior to and concomitantly with potash salt. Along the Trucial Coast, there is early fixation of magnesium through dolomitization of calcium carbonate. This releases calcium which fixes sulphate as gypsum and anhydrite. Hence the brines are initially depleted in magnesium and sulphate. It has been suggested that sea water gaining access to evaporating basins, within which rock-salt and potash salt are deposited, may be conditioned in this way whilst filtering through physical barriers such as barrier reefs (see R. C. Selley for discussion and references).

## Selected References


DUFF, P. McL. D., HALLAM, A., and WALTON, E. K., 1967, *Cyclic Sedimentation*, Elsevier, Amsterdam.

JOHNSON, J. H., 1951, 'An introduction to the study of organic limestones', *Quarterly of the Colorado School of Mines*, No. 46.

KRUMBEIN, W. C., and SLOSS, L. L., 1963, *Stratigraphy and Sedimentation*, Freeman, San Francisco.

OAKLEY, K. P., 1946, 'The nature and origin of flint', *Science Progress*, vol. 34, pp. 277–86.

PETTIJOHN, F. J., 1975, *Sedimentary Rocks*, Harper and Row, New York.

SELLEY, R. C., 1976, *An Introduction to Sedimentology*, Academic Press, London and New York.

# Chapter 7
# Pages of Earth History

These rocks, these bones, these fossil ferns and shells,
Shall yet be touched with beauty, and reveal
The secrets of the book of earth to man.

*Alfred Noyes*, 1925

## The Key to the Past

So far we have dealt with sedimentary rocks as materials which have been formed from pre-existing materials by the action of geological processes. Rocks are also the pages of the book of earth history, and the chief object of historical geology is to learn to decipher those pages, and to place them in their proper historical order. The fundamental principle involved in reading their meanings was first enunciated by Hutton in 1785, when he declared that 'the present is the key to the past', meaning that 'the past history of our globe must be explained by what can be seen to be happening now'. Rocks and characteristic associations of rocks, with easily recognizable peculiarities of composition and structure, are observed to result from processes acting at the present day in particular kinds of geographical and climatic environments. If similar rocks belonging to a former geological age are found to have the same peculiarities and associations, it is inferred that they were then formed by the operation of similar processes in similar environments.

We have already, as a matter of ordinary common sense, had occasion to apply this principle. The presence of fossil corals or of the shells of other marine organisms in a limestone indicates that it was deposited on the sea floor, and that what now is land once lay beneath the waves. The limestone may pass downwards or laterally into shale, sandstone and conglomerate. The last of these represents an old beach, and it indicates the shore line where land and sea came together. Elsewhere, old lava flows represent the eruptions of ancient volcanoes, and in places the vents that were active many millions of years ago still figure prominently in the landscape (Figure 7.1). Beds of rock-salt point to the former existence of inland seas that evaporated in the sunshine. Seams of coal, which are the compressed remains of accumulations of peat, suggest widespread swamps and luxuriant vegetation. Smoothed and striated rock surfaces associated with beds of boulder clay prove the former passage of glaciers or ice sheets. In every case the characters of older formations are matched with those of rocks now in the making.

Even the weather may be recorded in the structures of the rocks. A brief rain shower falling on a smooth surface of fine-grained sediment spatters it with tiny crater-like pittings known as *rain prints*. Sun cracks develop in the mud flats of tidal reaches or flood plains when the mud dries up and shrinks (Figure 5.5). Occasionally it happens that the polygonal cracks become filled with wind-blown sand before the next tide or flood sweeps over the area. Then, instead of being obliterated, they have a chance to become permanent. Thus it comes about that similar structures are preserved in older beds of corresponding origin (Figure 7.2). As far back as geological methods can be applied to the earth's history, such relics of 'fossil weather' prove that wind, rain and sunshine have always been much the same as they are today. Nevertheless, the distribution of climates over the earth's surface has varied in a most astonishing way.

In Great Britain the work of former ice sheets and glaciers is still written conspicuously in the form of the landscapes and in the boulder clays and other deposits left behind when the ice melted away. In striking contrast, the very much older clay through which London's underground rail-

**Figure 7.1** Arthur's Seat, Edinburgh, a Lower Carboniferous volcano that was active more than 300 million years ago. The bold escarpment of Salisbury Crags is the outcrop of a later sill. At least 14 lava flows can be recognized on Whinney Hill, with Lower Carboniferous sediments above and below. (*Airviews Limited*)

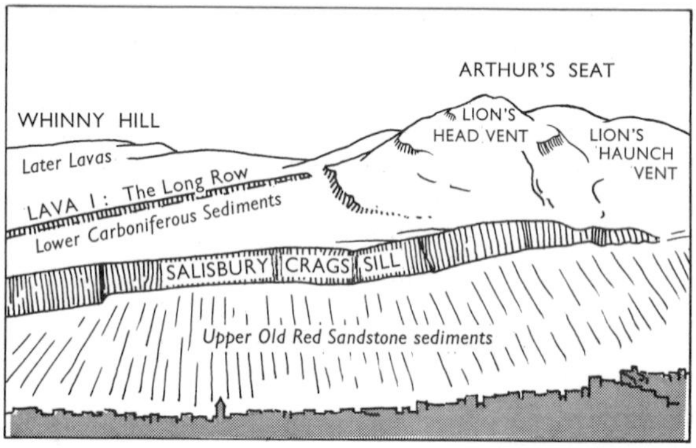


ways are bored contains remains of vegetation and shells and reptiles like those of the modern tropics. In the sandstones of still earlier periods there is evidence of desert conditions. Elsewhere the vicissitudes of climate are equally startling. In India and in central and southern Africa there is clear proof that while Britain was part of a region of swampy, tropical jungles (the time of coal formation) these lands were buried under great ice sheets like those of Greenland and Antarctica at the present day. In Greenland, however, there are sediments containing remains of vegetation that could have grown only in a warm climate. Similar discoveries have been made in various parts of Antarctica, beginning with Captain Scott's dramatic discovery of coal near the South Pole in 1912.

We may next turn to certain examples of Hutton's principle which can be applied to problems connected with earth movements. In the Alps and many other regions, including Britain, strata are found which have locally been turned upside down by overfolding. A small-scale example is illustrated in Figure 3.11, but it may happen that the overturn is on too large a scale to be seen as a whole. The reader may well wonder how such a structure, of which only a part may be visible or preserved, can then be recognized. Although no one has witnessed rocks being turned upside down by overfolding, Hutton's principle comes to our aid by suggesting various ways of determining which was the top and which was the bottom of a particular bed of rock at the time when it was deposited.

Current bedding (p. 81) provides one of the more easily applied tests, when it is present. Figure 7.3 shows that when the upper part of a sandbank, let us say, is removed by erosion, the

**Figure 7.2** Filled-in mudcracks in mudstone of Lower Old Red Sandstone age, shore east of Thurso, North coast of Scotland (*Institute of Geological Sciences*)

current bedding that remains is abruptly truncated at the top, whereas at the bottom it curves gently into the main stratification, or into an earlier surface of erosion. The truncated top and the original floor of the current-bedded layer of sandstone or quartzite can thus be recognized at a glance (Figure 7.4). This criterion was first used by the Irish geologists of a century ago to show that parts of the folded sandstones of the Dingle Peninsula in south-west Ireland are upside down.

Graded bedding, illustrated on p. 82, also provides a useful clue as to whether a particular bed is the right way up or upside down, especially if there are several graded beds all telling the same story. A single instance unsupported by other examples, might perhaps be otherwise explained, and so would remain ambiguous.

*Ripple marks*, like those seen on a beach after the tide has gone down, are often preserved in ancient sandstones (Figure 7.5). Desert sands are often beautifully rippled by the wind (Figure 3.4), but from the nature of the case wind ripples are very rarely preserved. Ripple marks formed by the to-and-fro movements of water have sharp crests and rounded troughs, and consequently the top and bottom of any bed of sandstone in which they occur can easily be recognized. Where such ripples remain in formations that have been disturbed by severe folding, they, too can be used to determine which was the original top of the bed (Figure 7.6).

There are, of course, certain processes and rocks to which Hutton's principle cannot be directly applied. It is impossible to observe granite in the making. Its origin can only be inferred from its internal structures and textures; its structural relationships to the rocks within which it is

**Figure 7.3** Sections to illustrate the value of current bedding in determining whether a bed is right way up or upside down: (a) Current bedding structure complete. After the upper part has been removed by erosion down to AB the original floor and the truncated top are easily distinguished; (b) is right way up; (c) is upside down.

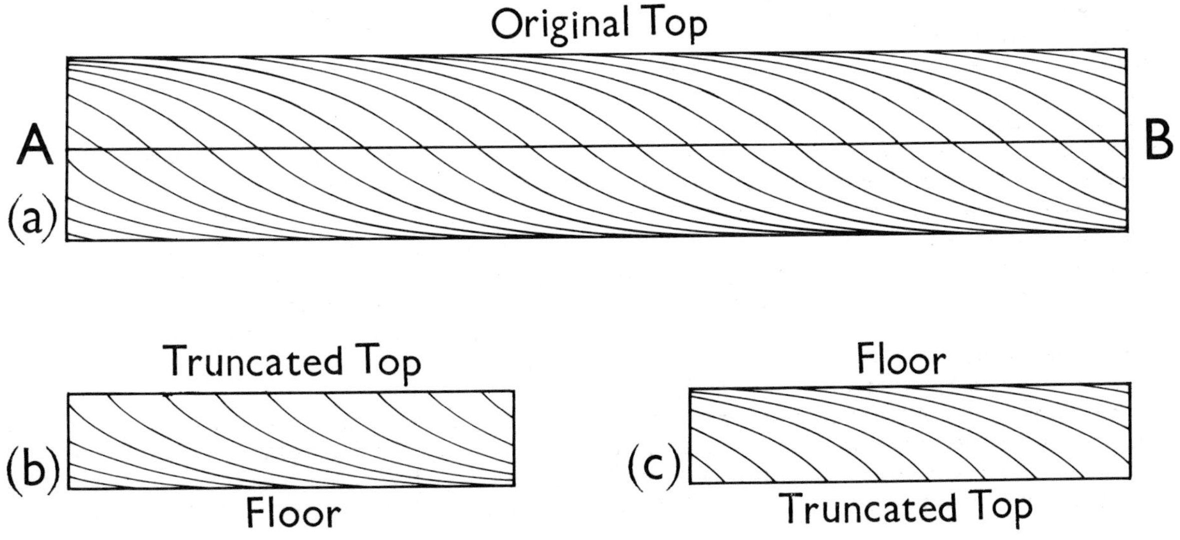

**Figure 7.4** Current bedding in metamorphosed biotite-quartzite (Glencoe Quartzite) at Rudha Cladaich, Loch Leven, South of Fort William, W. Scotland (*L. Weiss*)

emplaced; and from its metamorphic effects on the aureole rocks which, as observed long ago by Goldschmidt, a famous Norwegian geologist, are quite different from the metamorphic changes brought about by intrusions of other kinds of rocks. In such circumstances the best we can do is to suggest tentative explanations that are not inconsistent with either observation or experimental evidence. With every new discovery Hutton's key opens new doors into the past.

## The Succession of Strata

To place all the scattered pages of earth history in their proper chronological order is by no means an easy task. The stratified rocks have accumulated layer upon layer, and where a continuous succession of flat-lying beds can be seen, as on the slopes of Ingleborough (Figure 7.7), where there has been no inversion of beds by overfolding or repetition of beds by overthrusting, it is obvious that the lowest beds are the oldest, and those at the top of the series the youngest. The Grand Canyon of the Colorado (Figure 1.6) presents one of the finest successions of this kind to be seen anywhere in the world. However, where a sequence of beds has been tilted, as between the Welsh borders and London, the worn-back edges of layer after layer come in turn to the surface and it becomes possible to place a long succession of beds in their proper order (Figure 7.8).

**Figure 7.5** Ripple-marked surface of Triassic sandstone from Scrabo Hill, Co. Down, Northern Ireland (*Copyright Ulster Museum*)

Around Ingleborough the great limestone platform of the Pennines can be traced over a wide stretch of country, but where the streams have cut through to its base, the limestone is found to lie on the upturned edges of strongly folded grits and graywackes and slates, as shown in Figures 7.7 and 7.9. Here there is evidently a sudden break in the continuity of the record, a break that may imply a very long interval of geological time. The physical representation of the missing part of the record is

**Figure 7.6** Ripple-marked slab of Precambrian (Moine) gneiss, folded into a vertical position, with the lower side facing the observer. Glasnacardoch, south of Mallaig, Highland, W. coast of Scotland (*G. W. Tyrrell*)

an old erosion surface, and it is this that is described as an *unconformity*. The beds in continuous succession above the break in the sequence are said to be *conformable*. The lowest of the conformable beds rest *unconformably* on the underlying rocks. After the latter had been deposited on the sea floor as newly formed sediments, they were folded and uplifted, deep in the heart of an ancient mountain system—a series of ranges that extended throughout the length of Norway and across much of the British Isles. Because much of Scotland is carved out of its hard and contorted rocks, it is known to geologists the world over as the Caledonian mountain system or, more technically, as the Caledonian *orogenic* belt (Gr. *oros*, mountain; see p. 104). By denudation the folded grits and slates were gradually uncovered and ultimately reduced to an undulating lowland. Then the worn-down surface was submerged beneath the sea, to become the floor on which the horizontal sheets of the Pennine limestones were deposited. Successive stages of the events which occurred during the time gap represented by the unconformity are shown by the diagrams of Figure 7.10.

Another type of unconformity is exemplified by the Sphinx (Figure 2.1), which was sculptured out of a little hillock of limestone that rose abruptly above the desert sands. The sands of the Sahara rest unconformably on the surface of a landscape carved out of the underlying rocks—a landscape that is mostly buried out of sight, but of which the higher peaks still stand above the surrounding sands. In Charnwood Forest (Leicestershire) there is a similar unconformity of the buried-landscape type, made visible because the ancient landscape is now being exhumed by present-day erosion and quarrying. Here the desert sands are represented by Triassic sandstones and marls—characteristic rocks of the Midland plain of England—deposited at a time when desert conditions were widespread over much of Europe and North America. The desert deposits crept up against the flanks of an ancient mountain range and eventually buried it. The sharp and steep little crags of Charnwood Forest, rising abruptly above the plain, are only the summits of the highest ridges of the mountains now being exhumed.

In general terms, every unconformity is an erosion surface of one kind or another, representing a lapse of time during which denudation (including erosion by the sea) exceeded deposition at that place. If sediments were deposited there during the interval, they must subsequently have been removed. The time gap is likely to be represented by strata somewhere else (Figure 7.11), and out next problem is how to recognize such strata if we find them.

**Figure 7.7** Section across Ingleborough and its foundations, Yorkshire, showing the unconformity between the Carboniferous beds above and the intensely folded Lower Palaeozoic strata below. Length of section about 6 km (*After D. A. Wray*)

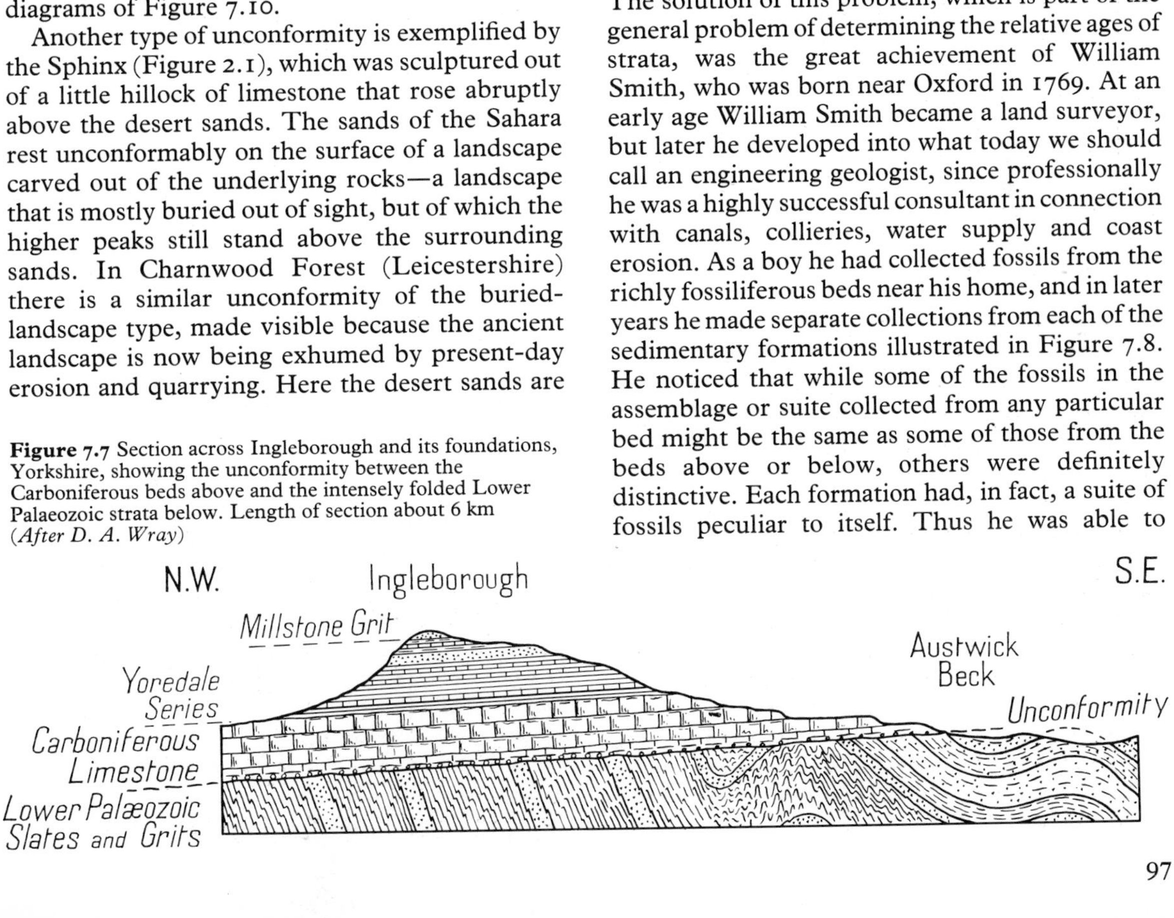


## The Significance of Fossils

The solution of this problem, which is part of the general problem of determining the relative ages of strata, was the great achievement of William Smith, who was born near Oxford in 1769. At an early age William Smith became a land surveyor, but later he developed into what today we should call an engineering geologist, since professionally he was a highly successful consultant in connection with canals, collieries, water supply and coast erosion. As a boy he had collected fossils from the richly fossiliferous beds near his home, and in later years he made separate collections from each of the sedimentary formations illustrated in Figure 7.8. He noticed that while some of the fossils in the assemblage or suite collected from any particular bed might be the same as some of those from the beds above or below, others were definitely distinctive. Each formation had, in fact, a suite of fossils peculiar to itself. Thus he was able to distinguish the different clay formations shown in Figure 7.8 (from the Lias to the Gault) by means of the fossils he found in them. By 1799, when his duties had taken him farther afield, he had examined all the formations from the Coal Measures to the Chalk, and everywhere he found the same types of fossils in the same formations and different suites of fossils in different formations. Smith had discovered that the special assemblage of fossils representing the organisms that lived during a certain interval of time never occurred earlier, and never appeared again. The relative age, or position in the time sequence, of a formation could thus be ascertained from its distinctive fossils.

In France, Cuvier (1769–1832) and Brongniart (1770–1847) made the same discovery by collecting fossils from the formations around Paris, which continue the sequence of strata upwards from the Chalk. By 1808 it became possible to correlate the older formations of England with those lying beneath the Chalk in France; and similarly to correlate the younger formations of France with those lying above the Chalk in England.

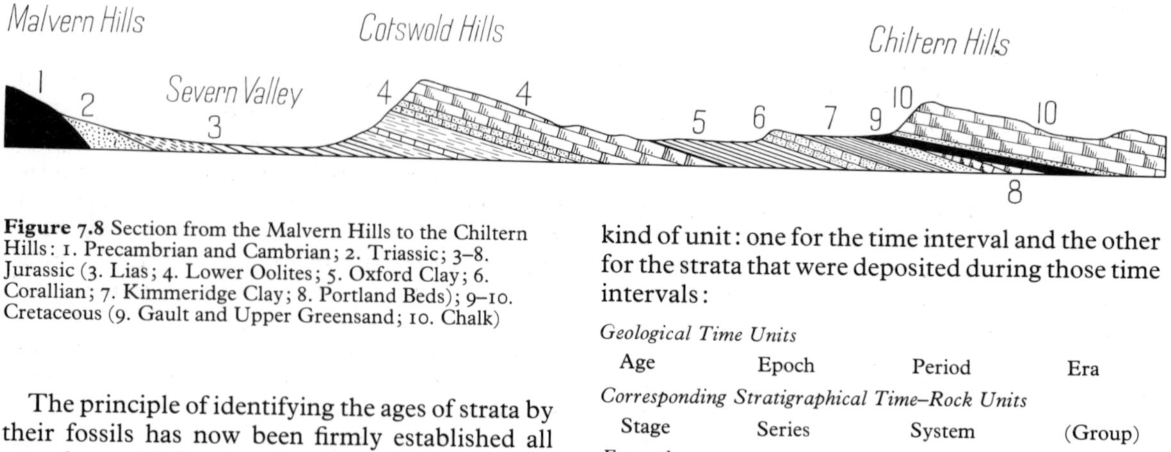


**Figure 7.8** Section from the Malvern Hills to the Chiltern Hills: 1. Precambrian and Cambrian; 2. Triassic; 3–8. Jurassic (3. Lias; 4. Lower Oolites; 5. Oxford Clay; 6. Corallian; 7. Kimmeridge Clay; 8. Portland Beds); 9–10. Cretaceous (9. Gault and Upper Greensand; 10. Chalk)

The principle of identifying the ages of strata by their fossils has now been firmly established all over the world. Strata in Europe and Australia, for example, are now known to be practically contemporaneous if they contain similar suites of fossils. The time required for the migration of a particular species from one region to another does not involve any practical difficulty, because the intervals represented by even the smallest divisions of geological time run into hundreds of thousands or even millions of years. In comparison the time required even for world-wide migration is relatively short. Everywhere the sequence of fossils reveals a gradual unfolding of different forms of life, and thus it becomes possible to divide the whole of the fossiliferous stratified rocks into appropriate divisions or units, each unit having its distinctive fossils and a definite chronological position.

## The Stratigraphical Time Scale

As the book of earth history is immensely long, it has been found convenient to divide and subdivide its contents in much the same way as a long book is divided into volumes, chapters, paragraphs, and sentences. If the book is read consecutively, then the order of the sentences etc. represents a time order. In building up stratigraphy on a firm basis it has been found desirable in the interests of clarity of thought to employ two sets of terms for each kind of unit: one for the time interval and the other for the strata that were deposited during those time intervals:

| | | | |
|---|---|---|---|
| *Geological Time Units* | | | |
| Age | Epoch | Period | Era |
| *Corresponding Stratigraphical Time–Rock Units* | | | |
| Stage | Series | System | (Group) |
| *Example* | | | |
| Portlandian | Upper Jurassic | Jurassic | Mesozoic |

Because of the abundance and variety of the fossils that characterize the Jurassic strata, the above scheme has been very successfully applied to the Jurassic Period and System. The System is divided into three Series, Lower, Middle and Upper, each of which in turn is subdivided into three or four Stages. In the example given the

**Figure 7.9** Carboniferous Limestone lying uncomformably on an erosion surface of steeply inclined Silurian slates, Arco Wood Quarry, 6 km north of Settle, North Yorkshire (*S. H. Reynolds*)

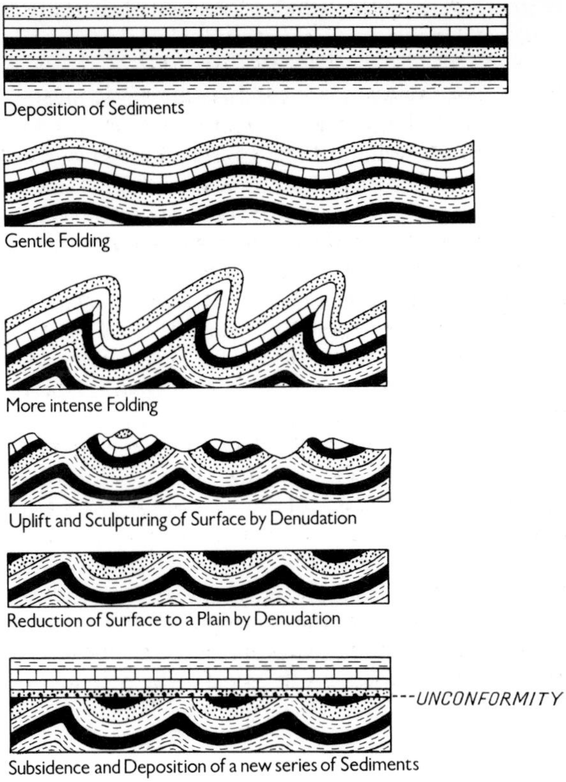


**Figure 7.10** Successive stages in the development of an unconformity

Portlandian Stage is named after the well-known Portland Limestone. It should be noticed that terms like *Jurassic* are commonly used as if they were nouns instead of adjectives. For example it is customary to speak of the Portland Limestone as belonging to the Jurassic (System); or of certain ammonites as having flourished during the Jurassic (Period).

The names for the Eras, Periods and Systems (as listed on p. 102), and for a few of the smaller units, have won general acceptance; but most of the names given to the smaller units have only limited application and vary from one country to another. The main difficulties in achieving the ideal that is aimed at are (*a*) that the field geologist does not map 'stages', but 'formations'; and (*b*) that the formations he maps may be lacking in the guide-fossils necessary to determine their exact ages.

A *formation* is a bed or assemblage of beds with well-marked upper and lower boundaries that can be traced and mapped over a considerable tract of country. It is, in fact, purely a *Rock Unit*; and only if it contains the appropriate fossils is it possible to say whether it represents a particular Age (i.e. a specific subdivision of geological time, defined by the relics of organisms that lived then), or part of an Age, or even more than one Age. In Britain it has long been common practice to use the term 'series' as an ordinary English word and so to speak of a succession of formations having some feature in common as a 'series'. The more technical use of the word, which then appears as 'Series', is for the strata deposited anywhere in the world during a particular interval of geological time known as an Epoch.

As examples of the non-technical use of the term 'series' we may consider the strata of the Carboniferous System. At the time when William Smith was first establishing his sequence of strata, the beds containing the chief English coal seams had long been known by the miners as the Coal Measures. A coal seam is a formation, associated with other formations of shale and sandstone, and it was natural to group the whole assemblage together as a series for which the obvious name was the Coal Measures. Underlying the latter comes another series consisting of massive sandstones and grits which had long been famous for the manufacture of millstones and grindstones. In consequence this series was given the quarrymen's name of Millstone Grit. Beneath this in turn are the massive limestone formations of the Pennines, originally called the Mountain Limestone, but now known collectively as the Carboniferous Limestone. As the limestone series was traced north through Northumberland into Scotland, it was found to pass into an increasing number of thinner beds of limestone, separated by intervening strata which included coal seams. The three series together were therefore styled the Carboniferous System, from the Latin words meaning 'coal-bearing'. It should be added that in North America the Carboniferous is divided into two Systems and Periods: the *Pennsylvanian*, including the Coal Measures, and the *Mississippian*, including the Carboniferous Limestone and its equivalents.

The tables on pp. 102–3 show the general scheme of classification by Eras and Periods that was gradually built up by the pioneer workers of the last century. It will be noticed that the Eras have names which broadly express the relations of the life forms then flourishing to those of the present. Underneath the oldest Palaeozoic beds there are enormously thick groups of sedimentary strata in every continent, passing down into or resting unconformably upon) widespread areas of crystalline rocks, metamorphic and igneous.

**Figure 7.11** Conglomerate of Old Red Sandstone age, deposited during part of the unrepresented interval of the Silurian/Carboniferous unconformity of Fig. 7.9. Gorge of the R. North Esk, Grampian/Tayside boundary (*Institute of Geological Sciences*)

Only rare and obscure forms of life have been found in the less altered strata of these ancient rocks, of no value for defining world-wide Systems. The only collective name for them all is *Precambrian*. Since classification into Systems and Periods is based on fossils, it cannot properly be extended into the Precambrian. However, with the development of methods of dating rocks in years (see p. 240) it is becoming possible to recognize a few Precambrian Systems as the rocks belonging to Periods of about the same duration as those of the fossiliferous rocks, say about 50 to 100 million years. So far, terms like 'formation', 'series' and 'system' have been used indiscriminately for Precambrian rock units of widely different ranks. At present the tendency is to use the term *Group* for all the formations belonging to a major Precambrian orogenic belt. The age of a Group is found to correspond to a time interval of a few hundred million years, that is, to an Era of several Periods.

It should be noticed that the duration of time represented by the Precambrian rocks that have already been dated—and probably there are older ones still to be discovered—was more than five times as long as the 600 million years that have elapsed since the fossil remains of a great variety of life forms first appeared in the sedimentary rocks of the Cambrian System.

## Crustal Movements and the Geological Time Scale

We can now return to the problem posed on p. 97 of how to recognize the strata corresponding to the time gap implied by the unconformity illustrated in Figures 7.7 and 7.9. From fossil evidence it is known that beneath the Pennines, and also around the English Lake District (Figure 19.9), Carboniferous limestones and associated sediments rest unconformably on the upturned edges of folded beds of Ordovician and Silurian age. The time gap is therefore the whole of the Devonian period, plus any part of the Lower Carboniferous that may be locally unrepresented, and possibly plus part of the Upper Silurian. At various times and places during this long interval the conglomerates (Figure 7.12) and sandstones known as the Old

Red Sandstone were being deposited, mainly in a series of deep depressions, sometimes occupied by lakes, that lay between the high mountain ranges of the Caledonian orogenic belt. In various parts of Scotland the 'Old Red' is itself found resting unconformably on an eroded surface of older rocks which had been intensely folded during the Caledonian orogeny (Figure 7.12). Obviously in the British area the Caledonian mountain building must have reached its climax at about the end of the Silurian or during the early part of the Devonian.

**Figure 7.12.** Upper Old Red Sandstone lying unconformably on vertical beds of Silurian slaty mudstones and graywackes, Siccar Point, Cockburnspath, Borders, Scotland. (*Institute of Geological Sciences*)

Figure 7.12 illustrates the celebrated unconformity, exposed at Siccar point, made classic by Hutton, who not only discovered it, but having already inferred its existence made a special expedition by boat with the confident expectation of discovering it. Beneath the unconformity he recognized what he called 'the ruins of an earlier world'; a world dating from a time so remote that he could only describe it as 'inconceivably long'; a vanished world that after passing through stages of mountainous landscapes had been worn down to a surface of upturned rocks such as can now be seen under the Old Red Sandstone at Siccar Point. Hutton had clearly realized that 'one land mass is worn down while the waste products provide the materials for a new one.' He recognized that the destruction of an old land by erosion, and the construction of a new land by upheaval of the resulting sediments (hardened and turned into a vertical position and in places invaded by granite) implied the existence within the earth of some agency sufficiently powerful to bring all this about. This agency he identified with 'subterranean heat'. To the effects of the earth's internal heat he ascribed (*a*) the general uplift (expansion); (*b*) the hardening and mineral changes suffered by the sediments (metamorphism); (*c*) the formation of granite and its forcible upward intrustion (additional localized expansion); and (*d*) the uptilting and dislocation of sediments on the flanks of the intrusive granites.

# The Stratigraphical Sequence

| Eras | Periods and Systems | Derivation of Names | |
|---|---|---|---|
| **CAINOZOIC** *Kainos or Cenos* = recent *Zoe* = life (Recent life) | QUATERNARY* | | 'cene' from *Kainos* = recent |
| | Recent *or* Holocene | *Holos* = complete, whole | |
| | Glacial *or* Pleistocene | *Pleiston* = most | |
| | TERTIARY* | | |
| | Pliocene | *Pleion* = more | |
| | Miocene | *Meion* = less (i.e. less than in Pliocene) | |
| | Oligocene | *Oligos* = few | |
| | Eocene | *Eos* = dawn | |
| | Paleocene | *Palaios* = old | |
| | *The above comparative terms refer to the proportions of modern marine shells occurring as fossils* | | |
| **MESOZOIC** *Mesos* = middle (Mediaeval life) | CRETACEOUS | *Creta* = chalk | |
| | JURASSIC | *Jura* Mountains | |
| | TRIASSIC | *Threefold* division in Germany | |
| | (New Red Sandstone = desert sandstones of the Triassic Period and part of the Permian) | | |
| **PALAEOZOIC** *Palaios* = ancient (Ancient life) | **UPPER PALAEOZOIC** | | |
| | PERMIAN | *Permia*, ancient kingdom between the Urals and the Volga | |
| | CARBONIFEROUS | *Coal* (carbon)-bearing | |
| | DEVONIAN | *Devon* (marine sediments) | |
| | (Old Red Sandstone = land sediments of the Devonian Period) | | |
| | **LOWER PALAEOZOIC** | | |
| | SILURIAN | *Silures*, Celtic tribe of Welsh Borders | |
| | ORDOVICIAN | *Ordovices*, Celtic tribe of North Wales | |
| | CAMBRIAN | *Cambria*, Roman name for Wales | |
| | **PRECAMBRIAN ERAS**: formerly described as | | |
| | PROTEROZOIC | *Proteros* = earlier | |
| | ARCHAEOZOIC *or* | *Archaeos* = primaeval | |
| | EOZOIC | *Eos* = dawn | |
| | *The term ARCHAEAN refers to the oldest known Precambrian crystalline rocks of a given region and has no other age significance* | | |

## Distinctive Life of the Geological Periods

| Era | Period | Distinctive life |
|---|---|---|
| CAINOZOIC | Recent | Modern Man |
| | Pleistocene | Stone-Age Man |
| | Pliocene | Great variety of Mammals<br>Elephants widespread |
| | Miocene | Flowering plants in full development<br>Ancestral dogs and bears |
| | Oligocene | Ancestral pigs and apes |
| | Eocene<br>Paleocene | Ancestral horses, cattle and elephants appear |
| MESOZOIC | Cretaceous | Extinction of Dinosaurs and Ammonites<br>Mammals and Flowering plants slowly appear |
| | Jurassic | Dinosaurs and Ammonites abundant<br>Birds and Mammals appear |
| | Triassic | Flying reptiles and Dinosaurs appear<br>First corals of modern types |
| PALAEOZOIC | Permian | Rise of Reptiles and Amphibians<br>Conifers and beetles appear |
| | Carboniferous | Coal forests<br>First Reptiles and winged insects |
| | Devonian | First Amphibians and Ammonites<br>Earliest trees and spiders<br>Rise of Fishes |
| | Silurian | First spore-bearing land plants<br>Earliest known coral reefs |
| | Ordovician | First fish-like Vertebrates<br>Trilobites and Graptolites abundant |
| | Cambrian | Trilobites, Graptolites, Brachiopods, Molluscs,<br>Crinoids, Radiolaria, Foraminifera<br>Abundant fossils first appear |
| | Late Precambrian | Scanty remains of primitive invertebrates<br>sponges, worms, algae, bacteria |
| | Earlier Precambrian | Rare algae and bacteria back to at least 3000<br>million years for oldest known traces of life |

The supreme genius of Hutton lay in his demonstration that the earth is a thermally and dynamically active planet, internally as well as externally; and that the earth's history can be regarded in terms of an overlapping succession of cycles. The later stages of one cycle necessarily imply the earlier stages of the next cycle. As Hutton expresses it: 'This earth, like the body of an animal, is wasted at the same time as it is repaired. . . . It is thus destroyed in one part but it is renewed in another.'

**Figure 7.13** Tectonic map of Europe. The Alpine orogenic belt is outlined by thick black lines with white arrowheads indicating the outward thrusts and overfolding towards the forelands.

The long belts of folded rocks which represent the various cycles referred to are described as *orogenic* belts to emphasize their relationship to past or present systems of mountain ranges. Mountain building by folding and uplift—as a purely structural or tectonic concept, without reference to the development of high peaks and deep valleys by erosion—is called *orogenesis*. The sum of the structural changes brought about during a time of mountain building is called an *orogeny* or an *orogenic revolution*.

Earth movements by which extensive regions are upheaved or depressed, with little folding, if any, apart from broad undulations, are distinguished as *epeirogenesis* (Gr. *epeiros*, land, con-

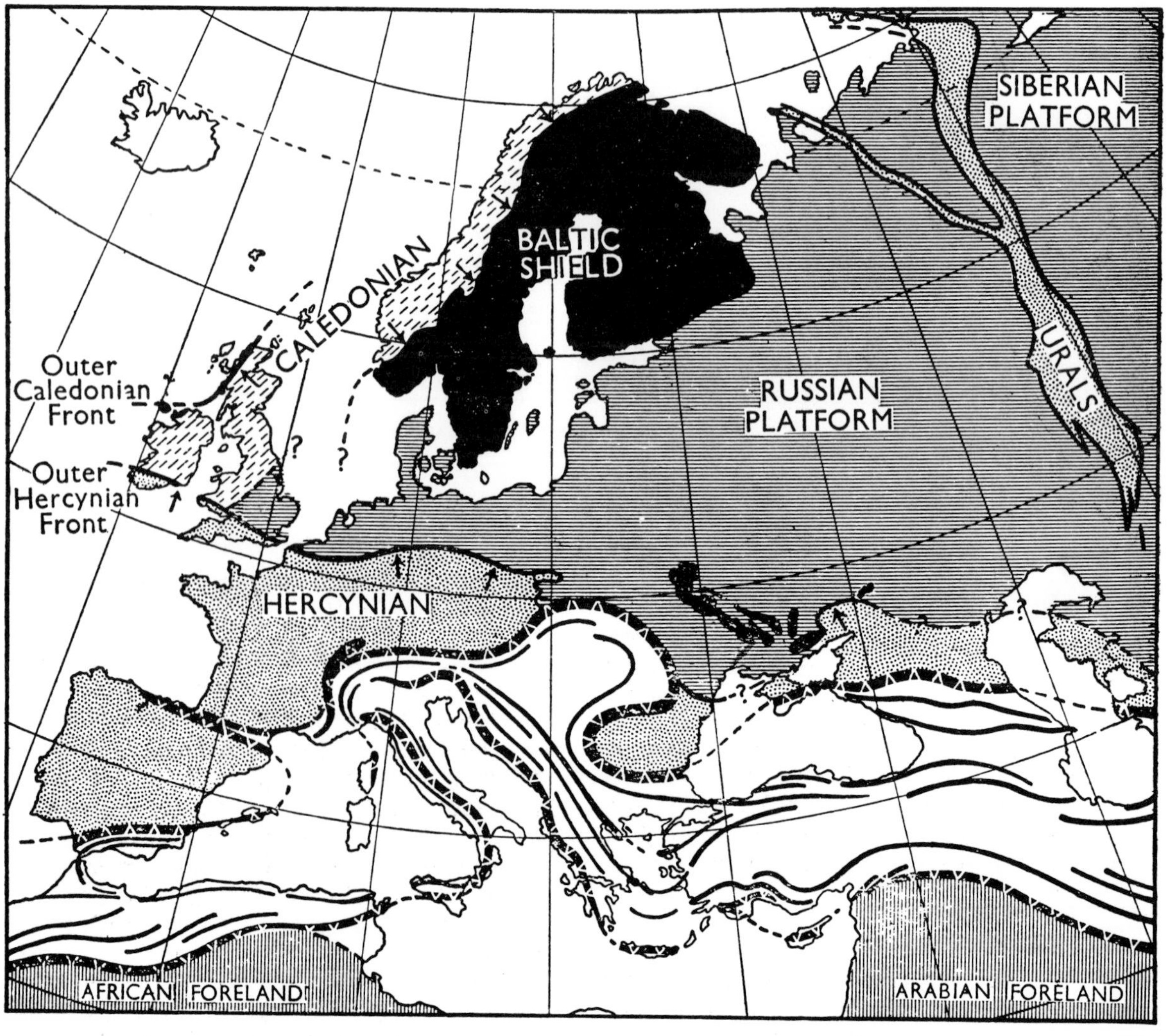

tinent). Plateaus result from epeirogenic movements of uplift. Some sunken regions, such as the Black Sea, involve more than epeirogenic depression—possibly thinning or even tearing apart of the crust.

The widespread marine sediments of former periods which now blanket much of the land clearly record fluctuating changes of level between land and sea. A geological Period is characterized by one or more invasions of the land by the sea, during which the marine beds of that particular System were deposited. Each invasion can be divided into (*a*) a phase of advance as the sea overflows the lands; culminating in (*b*) the phase of maximum flooding of the lands; which is followed by (*c*) a phase of retreat as the sea withdraws. Relative to sea level the lands may rise and fall many times during an era or major cycle.

Each revolution or minor orogeny is recorded by folding or tilting of the strata already deposited, and by the presence of an unconformity between these strata and the immediately overlying sediments. The geological age of the folding is obviously somewhere between the age of the oldest beds above the unconformity and that of the youngest beds below the unconformity. Only the orogenies of the last three major cycles in Europe can be relatively dated in this way by means of fossils. It might be thought that a geological Era could be defined in terms of physical events, i.e. as a cycle consisting of several periods of sedimentation together with the closing orogeny. But such a definition could apply only to particular regions. Wherever fossil evidence is available it is found that neither the major revolutions nor the minor orogenies are exactly contemporaneous in different parts of the world.

The Alps and Himalayas are examples of the last orogenic cycle, the *Alpine* cycle, in which we are, in fact, still living. The rocks of Cornwall and Devon and their continental continuations are amongst those formed in the immediately preceding European cycle—known to geologists as the *Hercynian* (after the Roman name for the forested mountains of Germany, typified by the Harz Mountains)—a cycle that reached its main mountain-building culmination towards the end of the Carboniferous. The rocks of most of Scotland, and of the English Lake District, Wales and Norway represent the still earlier *Caledonian* cycle. In Finland and Sweden, on one side of the Caledonian orogenic belt, and in the Hebrides and north-west Highlands of Scotland on the other side, the rocks of very much older cycles appear at the surface (See Figure 7.13).

So, going farther afield than Hutton, we can also travel backwards in time through a far longer series of these contorted volumes of earth history, And however far we penetrate into the past, we still, like Hutton, can find 'no vestige of a beginning'. Although Hutton was recording no more than the sober truth, his words were twisted by his critics to mean that, according to him, the earth never had a beginning and so was never created. Thus, far from being welcomed, Hutton's great discoveries were regarded by most of his contemporaries with righteous horror.

## Selected References


BENNISON, G. M. and WRIGHT, A. E., 1969, *Principles of Stratigraphy*, Edward Arnold, London.

DUNBAR, C. O., and RODGERS, J., 1957 *Principles of Stratigraphy*, Wiley, New York.

HEDBERG, H. D., 1961, 'The stratigraphic panaroma', *Bulletin of the Geological Society of America*, vol. 72, pp. 499–518.

OAKLEY, K. P., and MUIR-WOOD, H. M., 1960, *The Succession of Life through Geological Time*, British Museum (Natural History), London.

RAYNER, D. H., 1967, *The Stratigraphy of the British Isles*, Cambridge University Press, London.

TOMKEIEFF, S. I., 1962 'Unconformity—an historical study', *Proceedings of the Geologists' Association*, London, vol. 73, pp. 383–417.

INSTITUTE OF GEOLOGICAL SCIENCES, *British Regional Geology* (illustrated accounts of the geology of the natural regions of Britain), H. M. Stationery Office, London and Edinburgh (arranged from North to South):
*Scotland: The Northern Highlands; The Grampian Highlands; Scotland: The Tertiary Volcanic Districts; The Midland Valley of Scotland; The South of Scotland; Northern England; The Pennines and Adjacent Areas; The Central England District; East Anglia; The Welsh Borderlands; North Wales; South Wales; The Bristol and Gloucester District; South-West England; The Hampshire Basin; London and the Thames Valley; The Wealden District*
THE GEOLOGISTS' ASSOCIATION publish Geological Guides to many classic regions (e.g. *Arran; Snowdonia; Yorkshire Coast; Peak District; Shropshire; The Weald; Isle of Wight; Dorset Coast; Dartmoor*) and to areas around university and industrial centres (e.g. *London; Oxford; Durham; Hull; Sheffield; Manchester; Birmingham; Cardiff; Swansea; Belfast*), Benham and Co. Ltd., Colchester, Essex.
For the study of British Stratigraphy the following map will be found invaluable:
INSTITUTE OF GEOLOGICAL SCIENCES, 1977, *Geological Map of the British Isles*, 3rd edn, Scale ten miles to one inch (2 sheets), Ordnance Survey, Southampton.
1977, *Quarternary Map of the United Kingdom*, Scale ten miles to one inch (2 sheets), Ordnance Survey, Southampton.

# Chapter 8

# Metamorphic Rocks and Granitization

This tale has seven variations and all cannot be told if time be short.

*East African Saying*

On pp. 36–8 the chief processes responsible for the rock transformations to which Lyell gave the name *metamorphism* in 1833 were briefly summarized. Metamorphism includes all the processes of internal origin which bring about the recrystallization of rocks, with or without change of composition, while the rock remains essentially solid in the sense that it does not pass into the liquid state. A familiar example of recrystallization is the transformation of loose snow into compact ice, as it is buried beneath later falls. Snow and ice are aggregates of a mineral that happens to be near its melting-point. Recrystallization of rock-forming silicates and other minerals occurs just as easily at the appropriate temperatures and pressures. It cannot be over-emphasized that when rocks and minerals are buried to great depths their properties become very different from those with which we are familiar in everyday life. We shall meet with many illustrations of this as we pass in review some of the common rocks produced by the metamorphic processes—thermal, dynamic and geochemical, and their various combinations.

### Marble and Crystalline Limestones

Limestones are known commercially as 'marble' when they can be effectively polished and used for decorative purposes. Corals and the stems of sea-lilies give a variegated pattern commonly seen on polished slabs cut from the grey limestones of the Pennines. Many limestones have been fractured or brecciated by earth movements, and permeating ground waters have mottled the stone with patches of varied tints such as pink and orange, while the cracks commonly become white veins of calcite. Polished slabs and columns of such 'marbles' show in spectacular fashion how readily limestone responds to the effects of pressure, movement and percolating solutions.

To the geologist, however, the term *marble* is restricted to limestones that have been completely recrystallized by metamorphic processes during their burial in the earth's crust. Under the influence of heat, shells and fine particles of $CaCO_3$ are gradually reconstructed into crystals of calcite of roughly uniform size. All traces of fossils are eventually destroyed and the rock, when pure, becomes a white granular crystalline limestone like the well-known statuary marble from Carrara in Italy. On a smaller scale, limestones in the north of England locally pass into saccharoidal marble at their contacts with the Whin Sill (Figure 8.1). The similar metamorphism of chalk by the feeding dykes of the Antrim plateau basalts has already been mentioned (p. 36).

It may be asked how it is that $CO_2$ did not escape under such conditions, as it does when limestone is heated to make lime. The explanation is that when the heating takes place under pressure, as when the limestone is confined under a load of overlying rocks, say 45 m thick or more, the $CO_2$ is not liberated as a stream of gas; only a few dispersed molecules are temporarily freed. These mobile molecules act like tiny ball-bearings, lubricating the crystal boundaries and so facilitating the process of recrystallization. Conditions favourable to marble formation were successfully imitated in 1805 by Sir James Hall, an eminent friend of Hutton's, who tested experimentally many of the rival hypotheses around which controversy was

**Figure 8.1** Contact metamorphism of Carboniferous limestone at the base of the Great Whin Sill, Falcon Clints, Teesdale, Co. Durham. The limestone at the top of the Sill is similarly metamorphosed. (*E. J. Garwood*)

then raging. Hall enclosed pounded chalk in a porcelain tube, which in turn was fitted closely into a cylindrical hole bored in a solid block of iron, and securely sealed at the open end. On heating this 'bomb' in a glass furnace at Leith, the chalk was transformed into granular marble.

When dolomite is heated it loses part of its $CO_2$—the part associated with MgO—comparatively easily. Consequently dolomite-marble is formed only under the much higher pressure that is necessary to prevent this $CO_2$ from escaping. Under lower pressure $CO_2$ leaks away, leaving behind calcite and MgO. The latter soon combines with percolating water to form the stable mineral *brucite*, $Mg(OH)_2$. Considerable masses of the Cambrian dolomites of Skye were metamorphosed to *brucite-marble* when they were invaded by Tertiary granite and other intrusions.

If the original limestone or dolomite contains impurities of quartz grains or clay, various chemical reactions take place between the ingredients when they are heated up sufficiently. Here again, dolomite reacts more readily than calcite. New minerals, such as olivine or garnet, are then developed, and the $CO_2$ thus liberated is driven off:

$$\underset{\text{Dolomite}}{2CaMg(CO_3)_2} + \underset{\text{Quartz}}{SiO_2} = \underset{\text{Olivine}}{Mg_2SiO_4} + \underset{\text{Calcite}}{2CaCO_3} + \underset{\text{Carbon Dioxide}}{2CO_2}$$

When water is present during or after metamorphism of this kind it may react with the olivine and turn in into serpentine (p. 57). The beautiful green serpentine-marble of Connemara, familiar to all who know Ireland, probably originated in this way.

$$\underset{\text{Olivine}}{5Mg_2SO_4} + \underset{\text{Water}}{4H_2O} = \underset{\text{Serpentine}}{2H_4Mg_3Si_2O_9} + \underset{\text{(removed in solution)}}{4MgO+SiO_2}$$

The 'waters' associated with metamorphism are likely to be hot, highly compressed solutions of the kind generally described as *hydro-thermal fluids* or *emanations*. It is known that high-pressure steam readily dissolves silica and other rock-forming materials, especially if a little alkali is present. The solvent effect becomes greatly augmented if the steam is at a temperature above its *critical point*. The critical point of a gas is the temperature above which it cannot be liquefied, however great the pressure.

For water the critical temperature is 374°C, and the minimum confining pressure is 218 atmospheres, corresponding to the load of an overhead thickness of 835 m (just over half a mile) of rock of density 2·7. At the critical point water and steam have the same density, 0·324. Above that point water passes into the condition of a *supercritical* gas. This is nearly as dense as some liquids, but it is enormously more active and mobile, since it has all the molecular freedom of a gas. It can therefore search out the finest channelways, through fractured crystals or along the boundaries between adjoining mineral grains, so passing freely through every kind of rock and thereby promoting recrystallization and facilitating the growth of new minerals. Being an efficient solvent of a great variety of rock-forming constituents, especially when accompanied by other volatiles, it can transport new materials into a rock and carry away the unwanted by-products, like the MgO and $SiO_2$ in the above equation.

**Figure 8.2** Cleavage intersecting bedding at a high angle in grey banded slate. The estuary of the River Camel, West of Wadebridge, Cornwall (*Institute of Geological Sciences*)

## Slate

The world-famous roofing slates that are quarried from the rugged hills around Snowdon owe their value to a structure whereby they can be split along planes that are not necessarily parallel to the bedding, like the lamination of shales, but are usually inclined to it, and locally at a high angle (Figure 8.2). This structure is neither stratification nor lamination, but a fissility or cleavage, often of great perfection. It is distinguished as *slaty cleavage* (Figure 8.3). It must not be confused with the cleavage of crystals, for the latter is a property depending on the orderly arrangement of atoms, whereas slaty cleavage depends on the orderly arrangement of minute flaky minerals such as mica, clay and chlorite within the rock. A thin section of slate cut at right angles to the cleavage shows all the flaky minerals lying with their flat faces parallel to the cleavage planes of the slate. It is due to this orientation that the rock splits along these planes and along no others.

Traces of the original bedding planes can be made out in the quarry face wherever bands

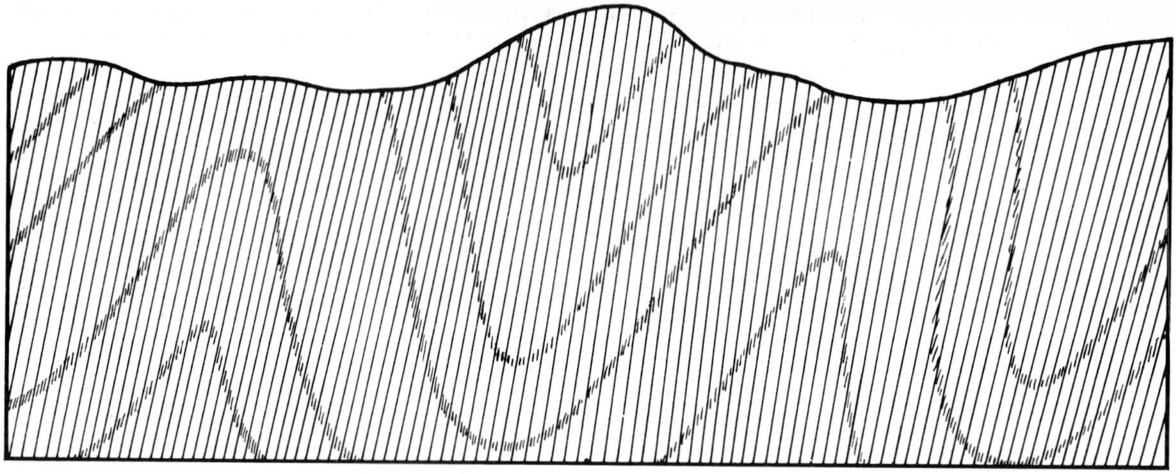

**Figure 8.3** To illustrate the relation of the slaty cleavage to bedding and folding, The cleavage planes are approximately parallel to the axial planes of the folds (see p. 135).

of contrasted colour or more gritty material are present; such bands are often badly crumpled and contorted. If by chance a fossil is found, it, too, is deformed and squeezed out of shape. In 1846 Daniel Sharpe studied the distortion of fossils in the old slate quarries along the Cornish cliffs near Tintagel Castle. He found that brachiopod shells had been shortened by at least 25 per cent in the direction at right angles to the cleavage, and correspondingly elongated parallel to the cleavage.

With the realization that beds of mudstone and shale had been compelled to flow at right angles to the direction from which the pressure came, it was thought that all the micas and other flaky minerals would shift around mechanically until their flat surfaces were also at right angles to the direction of maximum pressure. This mechanical explanation was experimentally tested by Sorby in 1853. He applied lateral pressure to a stiff mixture of clay and randomly distributed flakes of haematite. As expected, the resulting flow caused the flakes to orientate themselves with their flat surfaces at right angles to the direction of the pressure.

However this is not the whole story. We find that slates have also been made out of certain very fine-grained materials that contained no flaky minerals to begin with. The silvery-green slates of the English Lake District, for example, were formed from beds of fine volcanic tuffs. Their cleavage and silvery sheen is due, not to mechanical flattening of the original particles but to the development of new minerals from the volcanic materials. Although on no more than a microscopic scale, recrystallization has taken place during the flow movements induced by pressure. Minute shreds of mica and wisps of chlorite have grown in the rock, all with their film-like surfaces lying parallel to the cleavage planes. The slates formed from shales and mudstones also show that growth of new micaceous minerals has begun. Indeed, the new micas and chlorites, though still very thin, sometimes become so abundant that the cleavage surface has a characteristically lustrous sheen. The rock is then called a *phyllite*; it is intermediate in structure between slate and crystalline schist (p. 112).

### Kinds of Metamorphism

Slate is thus an example of a rock on which a new 'grain' has been impressed—partly by the mechanical effects of flow, partly by the growth of new minerals which have similarly accommodated themselves to the direction of flow. The original shale or volcanic ash has responded by becoming a slate, a new type of rock, characterized by *flow cleavage*. Since the main process is dynamic, slate is said to be a product of *dynamic metamorphism*.

The change from limestone to marble, on the other hand, is mainly brought about by the action of heat. It illustrates the effects of *thermal metamorphism*. The rocks in contact with igneous intrusions, like those of Figure 8.4, are commonly metamorphosed by heat and migrating emanations, and metamorphism of this kind is distinguished as *contact metamorphism*. The zone of altered rock surrounding the intrusion is described as the metamorphic *aureole*. Where shales and related rocks come into contact with intrusions like those of Figure 8.4 the mineral changes promoted by rise of temperature become visibly conspicuous. Traced from the edge of the aureole towards the granite, the rocks begin to be variegated by little spongy spots which as yet are hardly individualized as definite minerals. Nearer the granite these commonly develop into a glistening felt of tiny brown and white flakes, which further on become larger, and can be recognized as micas. Close to the contact other new minerals may appear. Metamorphic rocks of this kind are called *hornfels* (plural, *hornfelses*). They generally consist of a mosaic of grains, sometimes with a sprinkling of larger crystals; mica, when present, is usually arranged at random and there is no sign of cleavage, unless it is an inheritance from an earlier stage of metamorphism, as when a slate is incom-

**Figure 8.4** Aureoles of contact metamorphism around small plutons (batholiths and stocks), Galloway, south-west Scotland

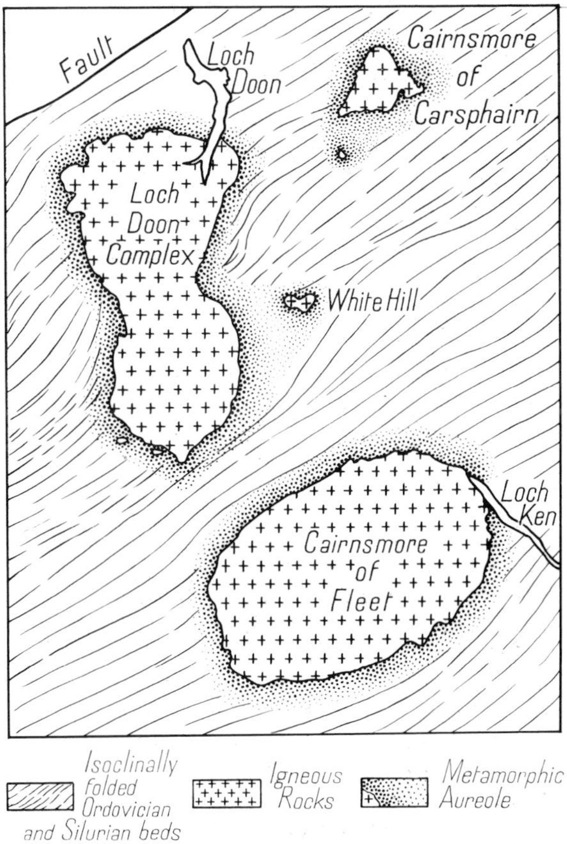


**Figure 8.5** An outlier of metamorphosed Lias on top of a Tertiary dolerite sill at Portrush, Co. Antrim, Northern Ireland. The soft clays and marly shales of the Lias are here hornfelsed to a hard splintery rock so much like porcelain that it is called porcellanite. The platform on which the outlier lies has a thin skin of ammonite-bearing porcellanite adhering to the top of the sill. This was one of the prize exhibits of the Neptunists, who, regarding the porcellanite skin as part of the sill, pointed to the ammonites as proof of the aqueous origin of basalt. (*R. Welch Collection, Copyright Ulster Museum*)

pletely metamorphosed into hornfels. There is rarely any evidence of much movement or of significant change of composition when a rock is hornfelsed, except for the driving off of unrequired water and carbon dioxide (Figure 8.5).

Except in and around certain volcanoes it is generally at greater depths (i.e. greater at the time when the metamorphism was taking place) that changes of composition are promoted. A notable example is the replacement of carbonate rocks and marls by silicate minerals such as iron-rich garnets, pyroxenes and amphiboles. Contact rocks of this kind are commonly known as *skarn*, from the Swedish word for rubbish, the implication being that the silicate minerals are merely so much waste in quarries that are being worked for limestone. Skarns result from what may be called *metasomatic metamorphism*, brought about by the passage of hydrothermal emanations. In some regions valuable deposits of iron ores accompany the silicate minerals of the skarns.

When all the agencies of metamorphism operate together, as they do in the heated depths of an orogenic belt while mountain-building and intrusive movements are in progress, the rocks throughout an extensive region become characteristically transformed, and the metamorphism is then described as *regional*.

## Crystalline Schists

When shale or slate is recrystallized under the conditions of regional metamorphism, heat and migrating hydrothermal fluids lead to the development of mica and other new minerals on a visible scale, as in contact metamorphism. At the same time the effects of shearing and flow movements impart to the rock a new structure, due to the streamlined arrangement of the platy and elongated minerals along planes of gliding. This structure is called *foliation*, a term based on analogy with that of tightly packed leaf-mould. Foliation may develop along an earlier cleavage, or it may follow the stratification of sediments; in other cases shearing or flow movements and the resulting foliation may follow a direction that is independent of pre-existing bedding or cleavage. The surfaces along which a foliated rock can be divided may be plane to undulating, wavy or contorted (Figure 8.6). When the foliation is closely spaced throughout the body of the rock, so that almost any part of it can be split into flaky sheets or wavy lenticles, the rock is called a *schist* (Gr. *schistos*, divided) or, when necessary to avoid ambiguity, a *crystalline schist*.

The ambiguity alluded to arises from the fact that the French word *schiste* means shale as well as schist, while the German word *Schiefer* can mean both of these and also slate. This is probably a hangover from the long-abandoned Wernerian notion that foliation was a result of sedimentation. It was Charles Darwin who first clearly distinguished foliation from stratification, as a result of his pioneer work on the metamorphic rocks of South America.

There is still some confusion as to the precise difference in meaning between the terms foliation and schistosity. Some writers regard them as synonyms. Others prefer to define *schistosity* as a variety of foliation in which the arrangement of flaky or platy or elongated prismatic minerals, concentrated in closely spaced planes, makes the rock fissile, so that it easily splits along those planes. This leaves *foliation* as a more general term to include related parallel and banded structures in which the planes of micaceous minerals are not closely spaced—as in gneiss—or in which micaceous minerals are rare or absent—as in granulites. For structures such as these the term 'schistosity' is obviously less appropriate.

Schists are named after the chief mineral responsible for the schistosity in any given case, e.g. mica, chlorite, hornblende, graphite, talc. Mica-schist is by far the commonest type, because it develops from argillaceous rocks (shales and slates etc.) which are themselves the most abundant sediments. It is also formed from tuffs. The commonest mica-schist is rich in muscovite, but biotite may also be present, other minerals being quartz, a little plagioclase and possibly garnet. Hornblende-schists are formed from basaltic rock; they contain hornblende, plagioclase, and possibly a little quartz.

Schists rich in green minerals, like hornblende and chlorite, are often called *green schists* or, if the foliation is only poorly developed, *greenstones*. These collective names are widely used as field terms. If the rocks are known to be metamorphosed basalts or dolerites the term *metabasite* is commonly used. But quite similar-looking rocks may represent impure dolomites or marls which have recrystallized and lost their original carbon dioxide in the process. More coarsely foliated

**Figure 8.6** Crumpled schist, north side of Fearna Bah, Kyles of Bute, Strathclyde Region (*Institute of Geological Sciences*)

rocks that are rich in hornblende and plagioclase are called *amphibolite*.

Sandstone and limestone do not characteristically form schists, the reason being that quartz and calcite are naturally granular minerals. Even though they may be recrystallized into elongated forms, the lack of cleavage in quartz and the three rhomboidal cleavages of calcite preclude the development of anything more than a very crude schistosity, unless micaceous or other flaky minerals are also present. The usual metamorphic equivalents of sandstone and limestone are quartzite and marble, respectively. Both are granular rocks in which foliation appears most commonly as banding.

## Granulites and Eclogites

When regional metamorphism becomes still more intense, rocks like amphibolites are transformed into 'basic' granulites. In these rocks hornblende has changed into pyroxenes, characteristically including hypersthene and diopside in place of the more complex augite that occurs in basalt. Granular interlocking minerals such as pyroxenes, feldspars and garnets predominate, and control the granulitic structure. Micaceous minerals are absent or extremely rare so that schistosity in the ordinary sense is lacking; but alternating bands or long-drawn-out lenticles varying in mineral composition or grain size invest the rocks with a foliation that may be very conspicuous.

As a result of similarly intense metamorphism, micaceous quartzites, mica-schists and igneous rocks of granitic composition become 'acid' granulites. Here the main mineralogical change is the transformation of mica and some of the quartz into garnet and microcline.

At still higher pressures the effect is to stimulate the development of new minerals with higher densities, so that the rock, literally strained beyond further endurance, accommodates itself by occupying less volume. The appearance of garnet in the granulites is the first symptom of this response to high pressure. Amongst the basic rocks the process is completed by their transformation into the beautiful red and green rock to which the appropriate name of *eclogite* has been given, from the Greek *ekloge*, meaning a choice selection—in this case, of colours (Figure 8.7). Here the partnership of albite and anorthite in pagioclase becomes unstable. Albite ($NaAlSi_3O_8$) is converted into jadeite ($NaAlSi_2O_6$), which usually unites with the original pyroxene to form a bright-green jadeite-augite. The liberated $SiO_2$, after uniting with FeO from ilmenite ($FeO.TiO_2$), also enters the new pyroxene; while the $TiO_2$ crystallizes as red prisms of rutile. Anorthite, on the other hand, combines with olivine or hypersthene to form a deep-red garnet. The transformation from a basaltic rock to eclogite may be represented in a simplified form as follows, the figures in brackets being the approximate densities:

**Basaltic Rock**
(2·9–3·1)
*Ilmenite* + *Pyroxene* + *Albite* + *Anorthite* + *Olivine*
(4·7) (3·25) (2·6) (2·75) (3·3)

**Eclogite**
(3·4–3·5)
*Rutile* + *Jadeite-augite* + *Garnet*
(4·2) (3·35) (3·25) (3·5)

The all-over reduction in volume is from 100 to about 87 or 85.

Eclogite has been described above as a product of high-pressure metamorphism of pre-existing basic rocks. At very great depths, however, e.g. in

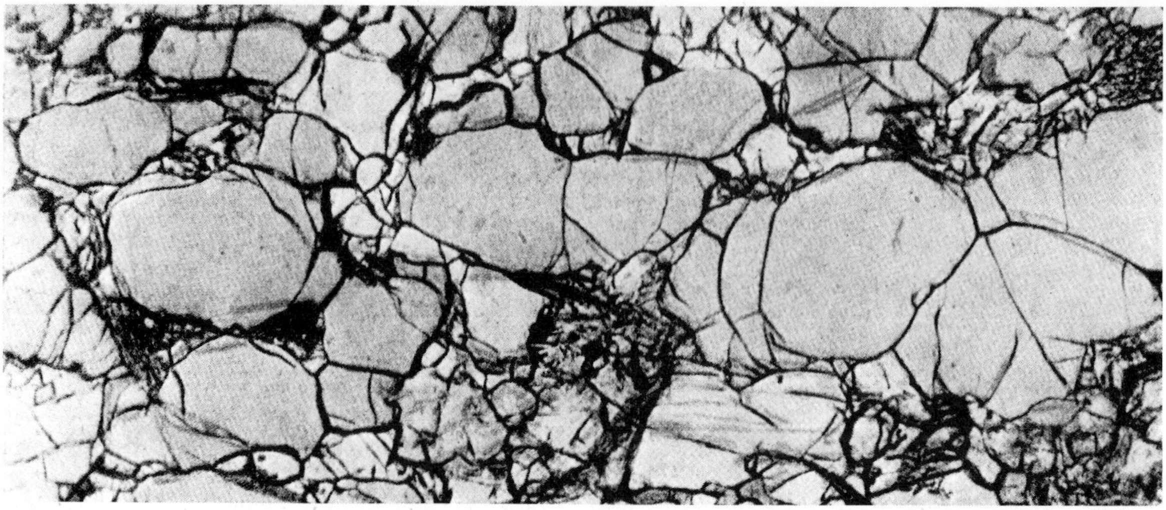

**Figure 8.7** Photomicrograph of eclogite, Dodoma diamond mine, Tanzania. Garnet, dark; pyroxene, light (*A. F. Williams*)

the mantle, such basic material might not be able to crystallize in any other form than eclogite, which could then be properly regarded as an igneous rock. What may be samples of this deep-seated variety are the eclogite nodules brought up to the surface by volcanic action of the kind responsible for the celebrated diamond pipes of South and Central Africa (see pp. 179–82). It is worthy of notice that neither these mysterious eclogite nodules nor the metamorphic eclogites formed in the crust are changed mineralogically merely by being brought to the surface. When they are so changed, it is because they have been metamorphosed afresh at some stage on the way up, in which case the usual product is amphibolite, a rock that requires the addition of water and fluorine to facilitate the growth of hornblende.

## Gneiss and Granitization

Returning to the older orogenic belts of the continents, where long-continued exposure to erosion has enabled us to see the metamorphic rocks that must originally have been formed at depths of three to five kilometres or more, we find the most characteristic types are schists and gneisses, intimately associated with granitic rocks. *Gneiss* is a venerable rock name, derived long ago from a Slavonic word meaning 'nest' and applied by the medieval miners of Central Europe to the foliated rocks in which the metalliferous veins occur: the rock, so to speak, being the 'nest' of the ores. Like schist, gneiss is a foliated rock, but its foliation is open and interrupted (Figure 8.8). Granular minerals, such as quartz and feldspar,

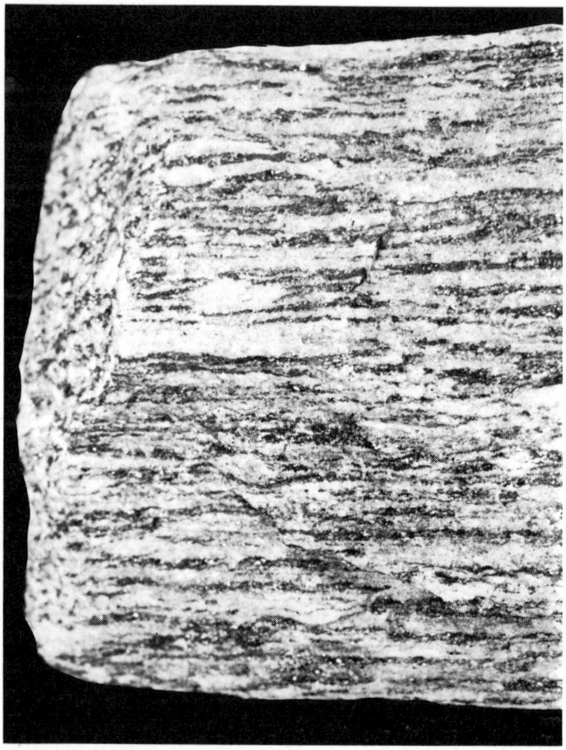

**Figure 8.8** Biotite-gneiss, Oxbridge, Mass., U.S.A. (*Ward's Natural Science Establishment, Inc.*)

and sometimes garnet, alternate with schistose or amphibolite layers. A common type is one with highly micaceous layers alternating with bands or lenticles that are granular in texture and often like granite in composition. These banded varieties sometimes pass into *augen gneiss*, the term *augen* referring to the presence of relatively large *eye*-shaped minerals (generally feldspars) or aggregates (generally feldspars and quartz) around which the foliation is wrapped in stream-lined fashion. Many gneisses, indeed are made of the same minerals as granitic rocks, though in different proportions. Common gneisses differ most characteristically from schists in containing feldspar as a dominant mineral.

Up to the middle of the last century, gneiss was commonly thought to be part of the original crust of the earth. This idea was based on the observation that schist and gneiss commonly underlie rocks of sedimentary origin. From this sequence in space, wrongly described as stratification, a time-sequence was inferred, and the schists and gneisses thus came to be considered as necessarily older than the overlying sedimentary rocks. Along these lines it was easy to take a further step and to suppose that the 'stratified crystalline rocks' were part of the original crust of the earth. Lyell (1833), knowing that rare fossils had been found in schists, recognized that they must have been formed from sedimentary rocks as a result of recrystallization during a later geological period, and he attributed the change to plutonic agencies, under pressure, in the depths of the earth, and described the process as metamorphism.

During the latter half of the last century, August Michel Lévy (1887) vigorously combated the still common belief that these rocks were part of the earth's original crust. Michel Lévy supplemented his field work by examining thin slices of rocks under the microscope, utilizing the techniques introduced by Sorby (p. 64). Up to this time, determinations of the optical properties of minerals had been made only from thick slices specially cut from large crystals in directions of known orientation. Michel Lévy realized that before progress could be made in the study of metamorphic rocks it was necessary to be able to identify, with greater accuracy, the minerals examined in random thin sections under the microscope. In collaboration with Fouqué he devised new optical techniques whereby it would be possible to distinguish the various types of feldspar under the microscope. Eventually, in 1879, descriptions of these new optical techniques, making use of both parallel and convergent polarized light, were published by the Geological Survey of France, with illustrations closely similar to those found in books of optical technique today.

Michel Lévy observed that the crystalline schists of the Massif Central are transformed to gneiss within contact aureoles around deep-seated intrusive granites. He also found, at still deeper structural levels, that the schists are similarly transformed to gneiss on a regional scale. This suggested to him that there is a genetic relationship between granite and gneiss. He observed that as granite is approached, newly formed feldspars grow throughout the fabric of the schist. At some distance from the granite contact these feldspars could only be seen and identified with the aid of the new microscope techniques. Nearer the contact they increased in size, becoming visible to the naked eye, and finally reaching lengths of 10 cm. So spectacular are these large feldspars that they were already well known to the country-folk of the Massif Central, who graphically referred to them as *dents de cheval*.

In the inner part of the granite aureoles Michel Lévy found banded gneiss characterized by the interlamination of layers of gneiss with narrower layers of fine-grained granite. This structure he described by the expression *lit-par-lit* (Figure 8.14), and he thought the granite *lits* to result from the intrusion of granite magma along structural planes in the newly formed gneiss. In using the expression 'granite magma' Michel Lévy clearly had an aqueo-igneous melt in mind, and he regarded the gneiss as having been formed through imbibation of a fluid hydrous melt.

Under the microscope Michel Lévy observed that the gneiss was composed of an alternation of micaceous laminae and more coarsely crystalline laminae consisting of intergrowths of quartz and feldspar. He observed that the micaceous laminae resembled the mica-schists both in mineral composition and in textural relationships, and microscopic investigation revealed that they were sometimes invaded and dislocated by quartz-feldspar laminae. Here a time sequence was indicated; the micaceous laminae had evidently existed before the intervening laminae of quartz and feldspar. Michel Lévy concurred with his friend A. de Lapparent, who had appropriately called gneiss a *petrographic hieroglyph* for, like the symbolic scripts of antiquity, gneiss had to be deciphered. Because the transformation of mica-schist to

**Figure 8.9** Wave-swept surface of schists, migmatites, and granite veins, Borgå (*E. Wegmann*)

gneiss is associated with granite and results from introduction of feldspar-forming materials. Michel Lévy described the transformation process by the term *granitization*. Keihau (1836) had previously used the term *granitification* for similar transformations in Norway.

Five or six years later, Barrois, while studying the granites and crystalline schists of Brittany, found spectacular proof of Michel Lévy's contention that gneiss was a product of granitization of mica-schist. In Brittany widespread regions of schist have been granitized and transformed to gneiss, but intercalated bands of quartzite have escaped the otherwise general feldspathization. Since these quartzite bands can sometimes be traced from the mica-schist into the gneiss, they provide striking evidence that the gneiss is a transformation product of the schist. In 1951 Read proposed the term *resisters* to describe bands like those of quartzite which have escaped granitization. 'Resisters' are rocks of extreme composition like basalt, quartzite and marble in which so many chemical interchanges are necessary for their transformation to a granitic composition that they remain as indicators of the initial structure.

The conclusion that gneisses are derived from schist was widely accepted in France. Pierre Termier (1904) attributed the transformation to the effect of percolating columns (*colonnes filtrantes*) of hot gas which carried alkalies, etc., and caused rise of temperature and chemical interchanges along their routes. To such chemical reactions Termier attributed both regional metamorphism and the formation of granite magma and its regional metamorphic aureoles. He thought granite magma to result from the melting of a eutectic mixture with a minimum fusion point.

## Opposed Views on Chemical Change

The observations of the French geologists were strongly opposed by the German petrologist, Rosenbusch. As a young man Rosenbusch (1877) investigated the contact aureole around the Barr-Andlau granite in the Vosges. This was an important piece of work, because it provided evidence of progressive metamorphic changes in argillaceous rocks of reasonably uniform type, as the granite was approached. Rosenbusch found that the aureole could be divided into a series of distinctive metamorphic zones. In the outermost zone the initial shales, the Steigerschiefer, were converted to spotted slate. Nearer the granite,

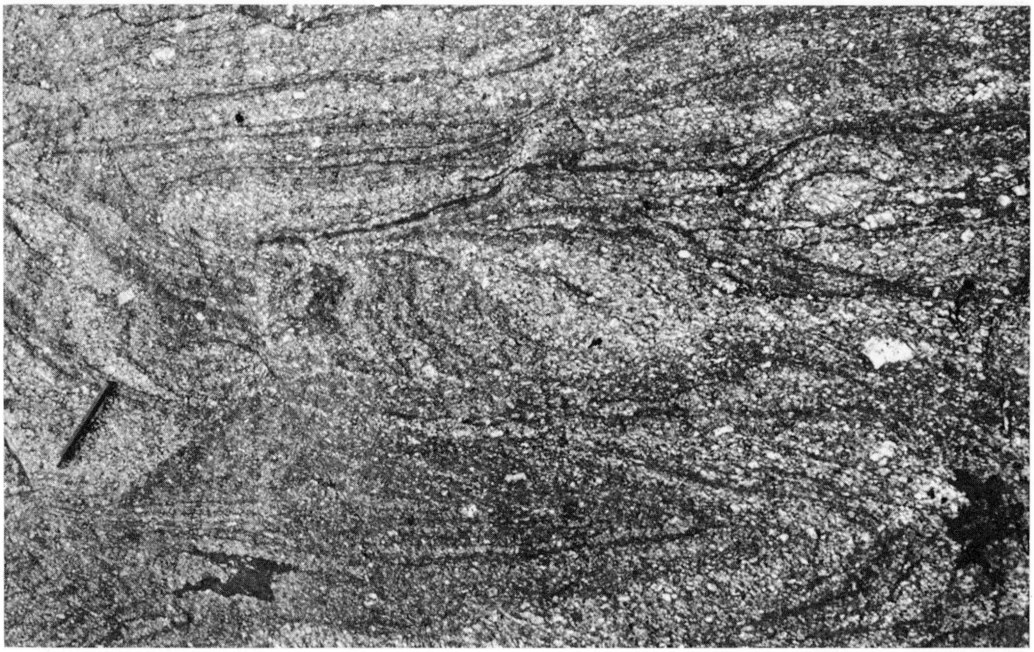

**Figure 8.10** Folded schists, in places almost completely granitized, Bodø, Pörtö (*E. Wegmann*)

biotite appeared, and within the innermost zone prismatic crystals of chiastolite ($Al_2SiO_5$) appeared in place of the spots, and changed to the clearer form andalusite adjacent to the granite. Chemical analyses showed that these progressive changes had been accomplished without any significant chemical change. It could therefore be inferred that the mineralogical changes were due solely to recrystallization of the initial rock-materials in response to rising temperature, the temperature being highest adjacent to the granite where andalusite crystallized. In later years, when Rosenbusch wrote what was destined to become the most influential textbook of the time on petrography, he made the sweeping generalization, from the one example of contact metamorphism that he himself had studied, that metamorphism adjacent to granite always took place without chemical change. Moreover, he insisted that all gneisses resulted from crushing and recrystallization—dynamo-metamorphism, as he called it. This mode of origin by itself involves no essential chemical change.

In 1921, V. M. Goldschmidt, who became a famous geochemist, working in the Stavanger area of southern Norway, supplied the chemical data that were necessary for the complete confirmation of the French geologists' observations that the deep-seated metamorphic rocks mantling intrusive granites become enriched in feldspar-forming constituents and progressively transformed to granite-gneiss. The normal country-rock in the Stavanger area is *phyllite*, i.e. a lustrous schistose rock that is midway in the coarseness of its crystallinity between slate and mica-schist. The intrusive granites, however, are each mantled by an aureole in which the phyllite is progressively transformed to granite-gneiss. At distances of from 1 to 4 km from the contacts the grain-size of the phyllite increases and it becomes enriched in albite. Because the eyes of albite are large compared with the grain-size of the schist matrix, just as porphyritic crystals are large in comparison with the groundmass crystals in porphyritic granite (Figure 4.12), Goldschmidt called the rock albite-porphyroblast-schist (Greek, *blastos* = a bud), a name used earlier by Beck to describe similar rocks occurring in the Austrian Alps.

Within the innermost part of the metamorphic aureoles, albite-porphyroblast-schist is progressively transformed to augen-gneiss, within which augen of potash-feldspar (microcline) are aligned parallel to the schistosity, and sometimes form the continuation of quartzo-feldspathic veins. Moreover, like the augen described by Michel Lévy, they enclose relics of earlier formed albite and quartz. Associated with the augen-gneiss is *lit-par-lit* gneiss consisting of granite layers alternating with relatively narrow layers of biotite-rich albite-schist.

With the aid of eleven chemical analyses, Goldschmidt confirmed his field and petrological inferences that phyllite—garnet-biotite-schist—albite-porphyroblast-schist—augen-gneiss form a gradational series. Moreover, from the chemical analyses it was apparent that $SiO_2$ and $Na_2O$ progressively increase through the series, so that there can be no doubt that these constituents were added to the phyllite ($Na_2O = 1{\cdot}26$) in increasing amounts as it was transformed to albite-porphyroblast-schist ($Na_2O = 3{\cdot}09$).

Goldschmidt considered five different ways in which the soda content of the phyllite could have been increased to that of the albite-porphyroblast-schist. He finally concluded that it could not have resulted from injection of magma corresponding to the adjacent granite, but had depended on the selective absorption of soda and silica from a permeating solution or vapour that resembled water-glass.

Ten years before his investigation of the metamorphic aureole in the Stavanger area, Goldschmidt, like Rosenbusch in the Barr-Andlau area, had found that the mineralogical changes within metamorphic aureoles of the Oslo area had taken place without chemical change. With great sagacity he now realized that this might mean that geologists would be tempted to enquire whether the results he had obtained in the Oslo area were correct and those in the Stavanger area wrong, or vice versa. But this, he added, would be asking the wrong question. From a scientific point of view it would be more fruitful to enquire under what conditions metamorphism *without* chemical changes takes place, and under what conditions metamorphism *with* chemical change takes places.

## Migmatites

Gneisses that *look* like mixtures of new granitic materials and older rocks were called *migmatites* (from Greek *migma* = mixture) by Sederholm during his fundamental studies of the Precambrian orogenic belts of Finland. They are now known within orogenic belts of all ages (Figure 8.13).

If we examine areas where migmatites are well exposed, as on the low wave-swept islands of southern Finland, where Sederholm examined them, we can see that the older rocks *seem* to have been impregnated with granite in every possible way. In some places thin sheet-like veins of granite lie between the folia of the schist, almost as if they had got there like water seeping between the pages of a book (Figure 8.9). Close by, the schists or other metamorphic rocks are strewn with crystals of potash feldspar (Figure 8.11) sometimes accompanied by quartz. Every veined gneiss and feldspathized schist can be seen to pass into other types of migmatite with more abundant and more uniformly distributed feldspars, so that much of the rock begins to look like granite, except for a shadowy background representing remnants or 'ghosts' of the original schists and their structures (Figure 8.10). And some of the large feldspars themselves contain microscopic inclusions that can be recognized as relics of the schists in which they grew. In these feldspathized and granitized rocks the pre-existing structures are commonly preserved, so that it is apparent that there has not been wholesale melting. Where broad areas of the rocks are clearly exposed, the forms of folds (Figure 8.10) and basaltic dykes (Figure 8.12) can easily be recognized. Finally, these ghosts of the older structures become fainter and vanish, leaving bands and patches of rocks that are indistinguishable from granite.

**Figure 8.11** Schists in various stages of replacement by granite, Hamnholmen, Pörtö (*E. Wegmann*)

**Figure 8.12** Sheared and deformed metabasite dyke cutting gneiss. The gneiss has been granitized, but the dyke remains as a resister. Gråskär, Sibbo Fjärd, east of Helsinki, Finland (*E. Wegmann*)

**Figure 8.13** Large boulder of highly complex migmatite associated with the Galway granite of Caledonian age; on the north shore of Galway Bay between Galway and Barna, Ireland (*Doris L. Reynolds*)

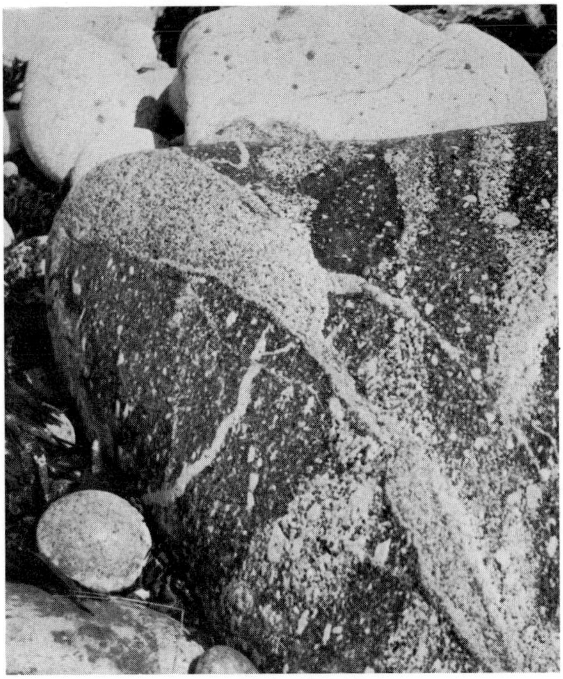

Sederholm ascribed migmatization to the circulation through the rocks of what he called 'granitic ichor', *ichor* being the etherial blood of the Greek gods. Of course this was an evocative word for something essentially unknown, that was endowed with a capacity for the most intimate penetration and circulation. Ichor can be equated with the mineralizers or *colonnes filtrantes* of the French geologists. Others have used the expression emanations, whilst others again have attributed migmatization to ionic diffusion. In the light of melting experiments (p. 69), many would now equate ichor with a water-rich melt or solution, resulting from partial melting of sedimentary or metamorphic rocks.

Many investigations of migmatites have now been made, including attempts to assess the kinds and amounts of materials that have been added to and subtracted from the initial country rock. Extreme views are (*a*) that all the granitic material was introduced as granite magma (Eskola), and (*b*) that migmatization is some form of metamorphic differentiation, with or without partial melting, but without addition or subtraction of material.

An excellent example of (*b*) above is illustrated by Figure 8.14, from the Black Forest, Germany, where a detailed investigation of migmatites has been made by Mehnert. The rock from which the banded gneiss has been evolved is biotite-plagioclase-gneiss (Figure 8.14(A)), and this parent rock has separated into two parts; one light in colour and of granitic composition is rich in plagioclase and quartz (B), whereas the other is dark in colour and richer in biotite than the initial gneiss (C). In Figure 8.15 the mineral compositions of the rocks concerned are plotted; it can be seen that the average composition of the granitic layers and their basic margins is closely similar to the average composition of the initial gneiss. Mehnert considers that separation of light from dark coloured minerals results from partial melting (anatexis) of the initial biotite-plagioclase-gneiss. On the basis of melting experiments in the presence of water, under pressure, he regards the light coloured layers as having crystallized from melt of granitic composition. The dark-coloured margins, on this interpretation, represent the unmelted residue of the initial rock.

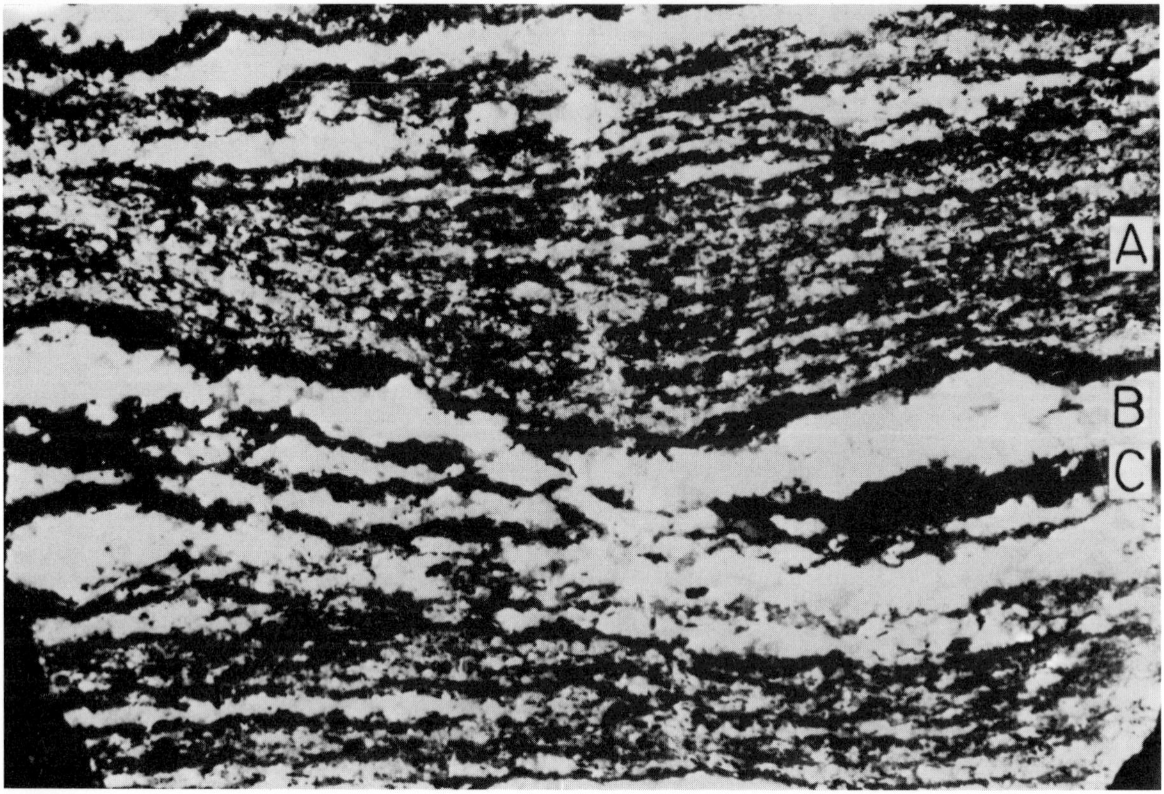


**Figure 8.14** Typical *lit-par-lit* gneiss from the Black Forest. A = initial biotite-plagioclase-gneiss: B = granite *lits*; C = biotite-enriched selvages of A, adjacent to B (*K. R. Mehnert, 1962*) (About 1·3 × natural size)

## Migmatization and Movement

In the evolution of banded gneisses, similar to that just described, other investigators have found that movement has played a dominant role. Misch (1949), for example, found similar banded gneisses, associated with augen-gneiss, outcropping over an area of more than 250 km$^2$, in the Nanga Parbat area, around the margin of a great body of granite-gneiss of batholithic dimensions that forms part of the crystalline core of the Himalayas. Misch found the banding to be related to closely spaced shear-zones; he concluded that migmatization and movement were synchronous. He rejected the idea that the granitic layers could have been emplaced in *lit-par-lit* style along shear zones, particularly where this direction dips gently, because any such *lit* would have to uplift and support the tremendous load of overlying rocks. The innermost core of granite-gneiss Misch regarded as having been formed in a similar way, by feldspathization of the initial phyllites and slates. Striking evidence for this conclusion is the presence within the granite-gneiss of layers of marble, up to 30 m thick, and commonly traceable for long distances. These layers are the metamorphic representatives of calcareous horizons within the initial sedimentary sequence, which remained as 'resisters' within the final migmatite.

B. C. King and A. M. J. de Swardt (1949), working in the Osi area, Nigeria, similarly found banded gneiss, consisting of granitic bands alternating with biotite-rich bands, to have formed concomitantly with shearing along closely spaced shear-planes. A related phenomenon, in the Malin Head Peninsula, Co. Donegal, Ireland, was investigated both chemically and microscopically by D. L. Reynolds (1961). Here the country rocks are quartzites and silvery mica-schists, with inter-layered sheets of metamorphosed basalt, now appearing as epidiorite. The rock-sequence is cut by shear-planes, and, as in the Osi area, this has resulted in banding at right angles to the strike of the initial metamorphic rocks. At one locality where a sheet of epidiorite overlies quartzite, the upper layer of the quartzite has been upthrust into the epidiorite (Figure 8.16), and transformed to a fine-grained granitic rock (trondhjemite aplite) rich in plagioclase and quartz, and poor in coloured minerals. The margins of the epidiorite adjacent to the quartzite layer are laminated by shearing and enriched in biotite. These basic selvedges of the epidiorite are about three or four times as wide as the granitic layer. Calculations based on six chemical analyses revealed that the transformation could be explained as a result of mechanical mixing, resulting from mylonitization (pp. 144–5), combined with metamorphic differentiation. The chemical and mineralogical changes were most likely implemented by percolating hydrothermal solutions; recrystallization being aided by the distortion of crystal lattices resulting from shearing. The crystal forms of some of the minerals indicate that recrystallization outlasted movement. It is important to notice that granitic intrusions are completely absent from the area concerned.

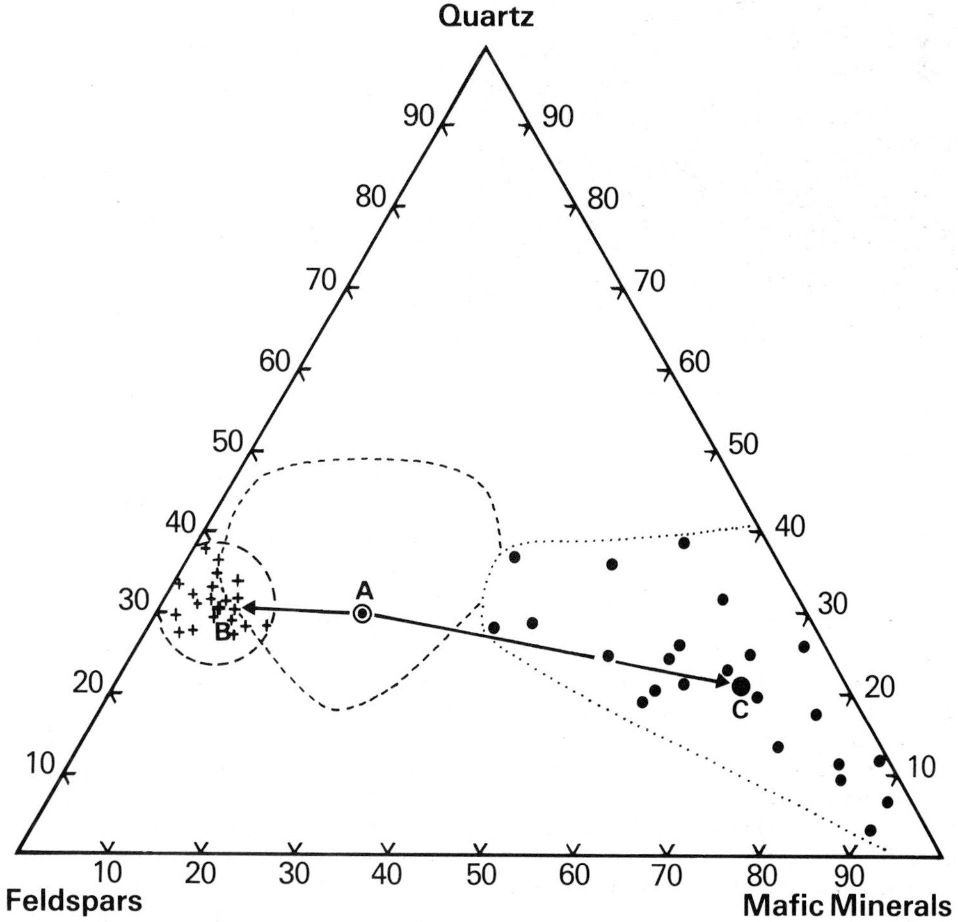


**Figure 8.15** Mineral composition of *lit-par-lit* gneiss from the Black Forest. The area A = the range of mineral composition of the initial biotite-plagioclase-gneiss of which the large circle represents the average. The area B = the range of mineral composition of the granite *lits* of which the large dot represents the average. The area C = the range of mineral composition of the biotite-enriched selvages, the large dot indicating their average. The fact that the average values lie on an approximately straight line indicates that A = B+C. This strongly suggests that B+C result from metamorphic differentiation of A, with or without differential melting. (*After Mehnert, 1962, 1968*).

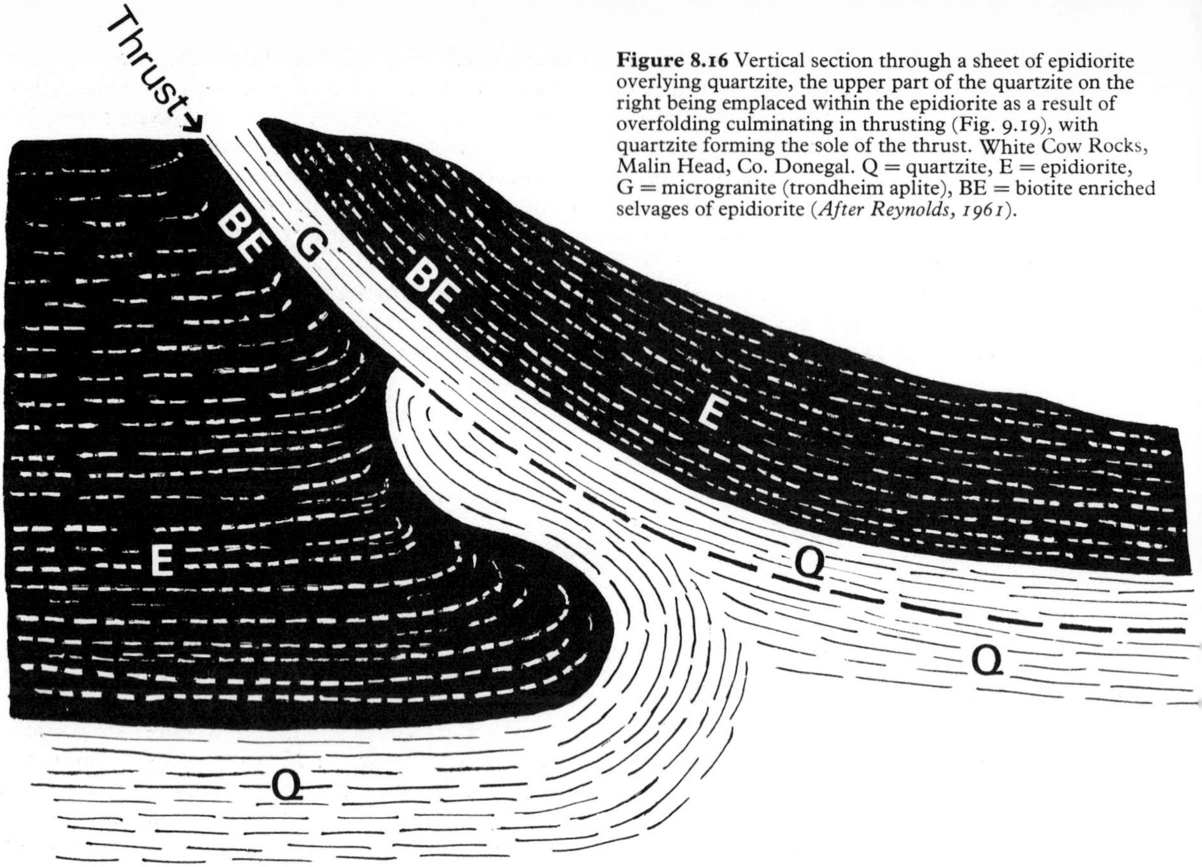


**Figure 8.16** Vertical section through a sheet of epidiorite overlying quartzite, the upper part of the quartzite on the right being emplaced within the epidiorite as a result of overfolding culminating in thrusting (Fig. 9.19), with quartzite forming the sole of the thrust. White Cow Rocks, Malin Head, Co. Donegal. Q = quartzite, E = epidiorite, G = microgranite (trondheim aplite), BE = biotite enriched selvages of epidiorite (*After Reynolds, 1961*).

In the light of knowledge derived from experimental melting of sedimentary and metamorphic rocks, probably nobody would deny that some degree of melting or solution may occur during migmatization, but that there has been wholesale melting is ruled out by the fact that 'resisters' commonly maintain the pre-migmatite structural pattern.

## The Source of Sodium

Since continental sedimentary rocks are deficient in sodium, in comparison with biotite-plagioclase-gneiss, migmatites, granites or even average igneous rock, it is apparent that there has been an overall addition of sodium either previous to or during migmatization. The problem is to find the source of this sodium. Is it of juvenile origin, having risen from the mantle and reached the continental crust for the first time, or is it recycled sodium of crustal origin?

When a balance sheet is drawn up for sodium in the crust and hydrosphere, it becomes apparent that there is a deficit of sodium in the oceans. Sodium liberated by weathering from continental rocks of all types is transported by rivers to the sea, but oceans do not contain nearly enough sodium to account for all the denudation that has gone on during geological time (see p. 253). Nieuwenkamp (1950, 1956) correlated sodium missing from the oceans with that trapped, in connate water, in the pore-spaces of sediments deposited on the sea and ocean floors. Shand (1943, p. 224) suggested, and Nieuwenkamp (1956) investigated the possibility that this connate water supplies the sodium necessary for migmatization and granitization, and found the amounts to be of the correct order.

According to the present-day concepts of plate tectonics, with oceanic lithosphere disappearing down into the mantle at ocean trenches, some of the adhering ocean-floor sediments, together with their connate water, will be likely to be carried down into the mantle. When the sinking lithosphere reaches a region of high temperature, the connate water will be driven off and rise upwards as hydrothermal solutions, rich in sodium. Here is a source for the sodium necessary to trigger off migmatization and granitzation, and also an explanation as to how sodium is abstracted from the oceans. Moreover, the sodium is supplied where migmatites form, in the cores of newly rising orogenic belts, where the leading edges of plates, surfaced with continental crust, meet plates with oceanic crust (Figure 31.3). According to Barth's (1962) calculations. sodium ions remain in the oceans for about 120 million years, a length of time that is of the correct order to match the rate of ocean-floor spreading.

## Classification of Regional Metamorphic Rocks

Many attempts have been made to classify regional metamorphic rocks. Rosenbusch (1898, 1901) dismissing the possibility of chemical change, divided gneisses into *ortho-gneiss*, derived from rocks of igneous origin, and *para-gneiss*, derived from rocks of sedimentary origin. This classification, still used by some geologists at the present time, obviously breaks down in face of the observations that rocks of sedimentary origin can acquire the composition of granite-gneiss.

Grubenmann (1904) classified regional metamorphic rocks into zones according to their depth of origin; recognizing with increasing depth an *epizone, mesozone* and *katazone*. This classification is still useful today. The index minerals: chlorite, biotite, garnet, staurolite, kyanite and sillimanite provide a method of mapping zones of increasing regional metamorphism. This method was first introduced by Barrow (1912) in the south-eastern part of the Grampian Highlands, Scotland, where he found the zones of higher grade curved round a central region of 'Older Granite', subsequently found by Read (1927) to be oligoclase-biotite-gneiss grading to typical migmatites. Eventually this method of zonal mapping was extended through the lower grade zones to the south-west, by Elles and Tilley (1930). Figure 8.17 illustrates the mineral zones, probably overlying and mantling a migmatite dome, and themselves intersected by a later zone of enrichment in albite.

More recently, two French geologists, Jung and Roques (1952), fully recognizing that with progressive regional metamorphism chemical changes occur, have divided schists and gneisses into two major groups: *ectinites* (*ekteneia* = tension) and *migmatites*. Schists and gneisses in which there is no obvious chemical change are ectinites. The name refers to the fact that they acquire their schistosity or foliation through crystallization under stress. Gneisses of dynamometamorphic origin belong to this group. If change of chemical composition is revealed by chemical analysis then the rocks are classified as metsomatic ectinites. Metasomatic ectinites include albite-schists, formed by the growth of albite porphyroblasts, commonly along structural breaks, within schists. They occur in south-west Scotland, north-eastern Ireland, Switzerland, Brittany, North Africa, Indian and many other places.

Jung and Roques divided migmatites into two groups: *embrechites* (*embrechin* = imbiber) and *anatectites*. Embrechites include augen-gneiss and banded gneiss (*lit-par-lit*), whilst anatectites are granite-gneisses that have flowed, and exhibit irregular fold-structures. Anatectites, like salt diapirs (pp. 153–7) characteristically intrude the overlying rocks as domes or nappes (French, sheet). Migmatite nappes, known from East Greenland since 1934 (Backlund and Wenk) are well exposed on the steep fjord walls (Figures 8.18 and 3.32), whilst migmatite domes are widely recorded (Figures 11.28 and 11.29). Wegmann (1935) classified migmatites as being either *autochthonous*, that is formed in situ, or *allochthonous*, that is formed elsewhere than *in situ*, having flowed and invaded higher structural levels as diapirs (Figures 8.18, 10.1 and 10.2). This distinction was adopted by Read (1957) as part of his granite series.

Selected References


BARTH, T. F. W., 1962, *Theoretical Petrology*, 2nd edition, Wiley, New York.

FOUQUÉ, F., and MICHEL LÉVY, A., 1879, *Minéralogie Micrographique. Roches Éruptives Françaises, Mem. Serv. Carte Gélogique France*, Text 3 Pts., Plates 2 Pts.

GOLDSCHMIDT, V. M., 1921, *Die Injectionsmetamorphose im Stavanger-Gebiete*, V. Geol.-Petrog. Studien im Hochgebirge des südlichen Norwegens, *Videnskaps. Skrifter I. Mat.-Nat. Kl.* 1920, No. 10.

JUNG, J. and ROQUES, M., 1952, 'Introduction à l'etude zonéographique des Formations Cristallophylliennes', *Bulletin du Service de la Carte Géologique de la France*, No. 235.

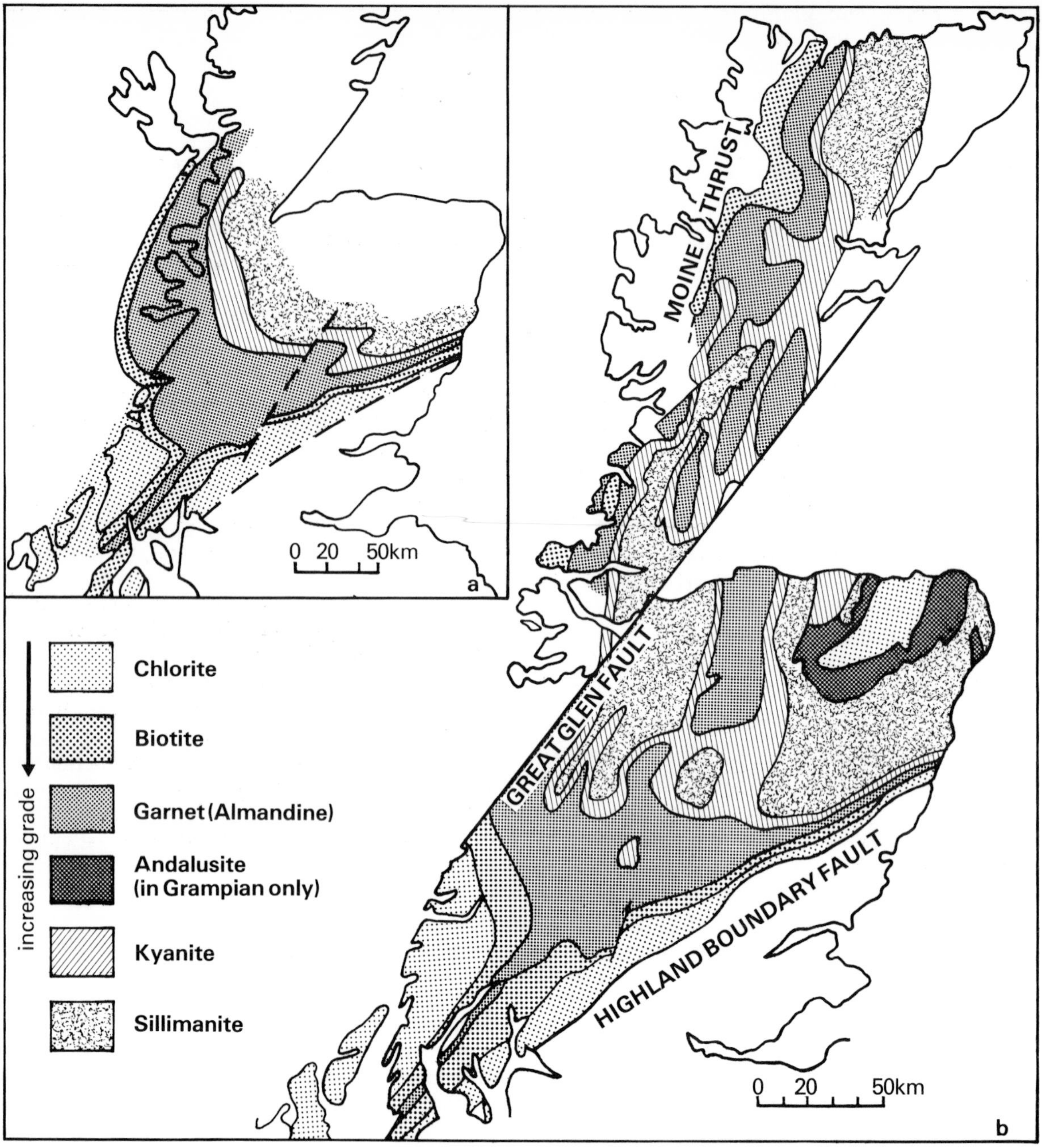


**Figure 8.17** The pattern of the regional metamorphic mineral zones of the Highlands of Scotland, with the subsequent displacement along the strike-slip fault of the Great Glen restored. (a) The mineral zones were mapped by Barrow (1893, 1912) in the south-eastern Highlands, and extended towards the south-west by Elles and Tilley (1930). After extending the mineral zones into the north-west Highlands, Kennedy (1948) recognized that when the region north of the Great Glen was moved north-eastward for 105 km along the Great Glen strike-slip fault (see Figs. 9.34, 9.41)—to its initial position the mineral zones depicted a thermal arch or anticline (see Fig. 9.12), plunging south-westward (see Fig. 9.17).
(b) A more detailed map of the metamorphic mineral zones by Winchester (1974), shows the thermal anticline to be more complex than was previously thought, as a result of post-metamorphic folding (see Chapter 30), and further metamorphic events. When the region north of the Great Glen fault is displaced about 210 km towards the north-east, along the strike-slip fault, the outcrops of the mineral zones become continuous across the fault, suggesting that the fault movement has been greater than previously thought.

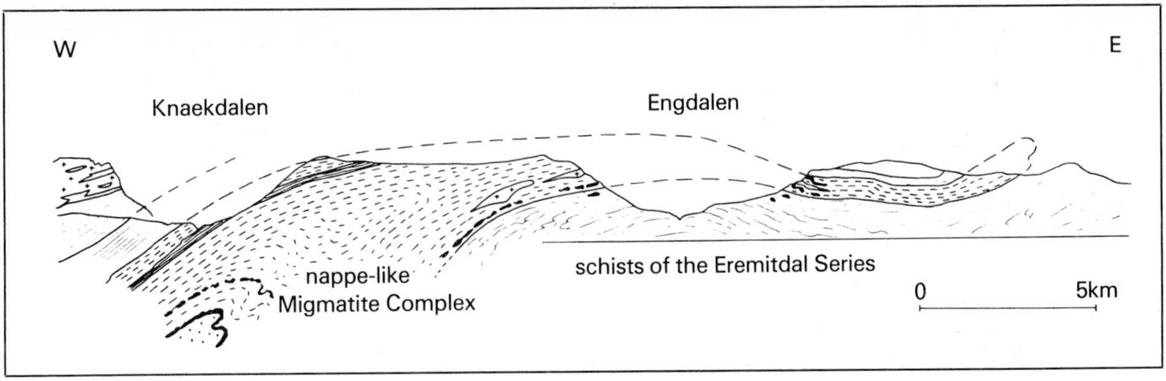


**Figure 8.18** An allochthonous migmatite complex that has flowed upwards as a diapir, and assumed a nappe-like form. Frankels Land, East Greenland (*Wenk and Haller, 1953*)

KING, B. C., and SWARDT, A. M. J. de, 1949, 'The geology of the Osi Area, Ilorin Province', *Geol. Surv. Nigeria Bull.* 20.
MENHERT, K. R., 1968, *Migmatites and the Origin of Granitic Rocks*, Elsevier, Amsterdam.
MISCH, P., 1949, 'Metamorphic granitization of batholithic dimensions', Part 1, *Amer, J. Sci.*, vol. 247, pp. 209–245.
NIEUWENKAMP, W., 1950, 'Geochemistry of sodium', *Report of the 18th International Geol. Congress, Great Britain, 1948*, pp. 96–100.
NIEUWENKAMP, W., 1956, 'Géochemie Classique et Transformiste', *Bull. Soc. Geol. France*, 6 série, vol. VI, pp. 407–429.
NORTH, F. J., 1952, *Slates of Wales*, National Museum of Wales, Cardiff.
READ, H. H., 1957, *The Granite Controversy*, Allen and Unwin, London.
REYNOLDS, D. L., 1946, 'The sequence of geochemical changes leading to granitisation', *Q. J. Geol. Soc.* London. vol. 102, pp. 389–446.
REYNOLDS, D. L., 1961, 'A "vein" of Trondhjemite aplite: granitization of quartzite forming the sole of a fold-thrust, Malin Head, Co. Donegal', *Geological Magazine*, vol. 98, pp. 195–209.
ROQUES, M., 1941, 'Les Shistes cristallins de la Partie sud-ouest du Massif Central Français', *Mém Serv. Carte Géologqiue France.*
SEDERHOLM, J. J., 1923, 'On migmatites and associated Pre-Cambrian rocks of Southwestern Finland', Part I, *Bulletin de la Commission Géologique de Finlande*, No. 58; 1926, Part II, No. 77; 1934, Part III, No. 107.
WEGMANN, C. E., 1935, 'Zur Deutung der Migmatite', *Geologische Rundschau*, vol. 26, pp. 305–350.

# Chapter 9

# Tectonic Features: Folds and Faults

Shall not every rock be removed out of his place?

*Book of Job* (*c.* 400 B. C.)

## Earth Stresses

The reality of earth stresses is forcibly demonstrated by the rock bursts that are one of the hazards of deep mining. If left inadequately supported, roof, walls and floor are liable to bulge inwards until the overstrained rocks suddenly give way, shattering and bursting into the excavation. By removing the confining pressure on at least one side, mining operations inevitably unbalance the stresses normally acting on confined rocks. One of the methods adopted to reduce the risk of disaster is to 'de-stress' the rocks for a metre or so ahead of a working face. Long holes are drilled, and after blasting at the ends of these mining can proceed more safely—or at least less dangerously—through ground that has been fractured in advance.

Any mass of rock in the earth's crust is subject to gravitational pressure corresponding to the load of overlying rocks, and also to a great variety of other stresses set up by such processes as tides, flow movements in the mantle, intrusions, expansion of some parts of the crust and contraction of others. Imagine an ideal plane situated anywhere within a stationary confined rock mass. The material on each side of the plane exerts a certain stress on the material on the other side. Stress is measured as force per unit area, but it must be remembered that the force is operating equally, but in opposite directions, on the two sides of the plane. The total 'stress field' in and around a given mass of rock can always be represented by three principal stresses, $P_1$, $P_2$ and $P_3$, acting mutually at right angles, $P_1$ being greater than $P_3$, while $P_2$ has an intermediate value (Figure 9.1). When one of the three is vertical, which is commonly the case because of gravity or rising intrusions, the other two are horizontal; but the possibility of departures from this arrangement ('inclined stress fields') should not be overlooked, especially on the flanks of rising batholiths.

If a stress is normal to a given plane it may be acting towards the plane, as implied in Figure 9.1, in which case it is a compressive stress or *compression* (→ ←); or it may be pulling away from the plane, in which case it is a tensile stress or *tension* (← →). For any other plane the stress can be resolved into a normal component acting at

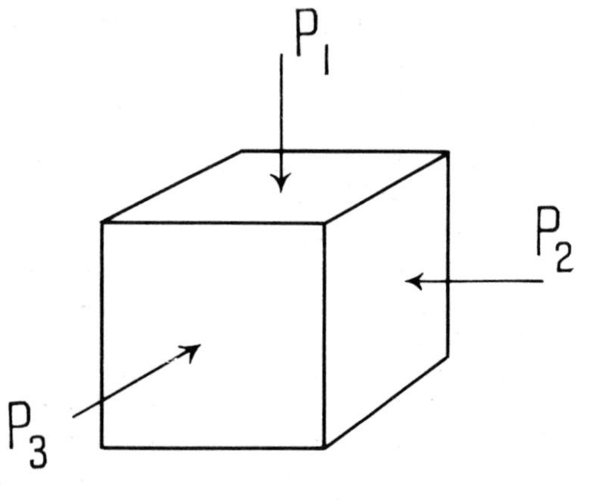


**Figure 9.1** Resolution of the stress conditions acting on a mass of rock into stresses acting along the three principal axes of stress, each of which is perpendicular to the other two. $P_1 > P_2 > P_3$. The stresses are here drawn as compressive and regarded as positive. Arrows in the opposite directions would represent tensile stresses, regarded as negative, It should be noted that it is usual engineering practice to regard tensile stresses as positive, compressions as negative.

right angles to the plane and a tangential component acting parallel to the plane (Figure 9.2). The tangential force exerted by the material on one side of the plane acts in the opposite direction to that exerted on the other side, as in the action of a pair of scissors. A tangential or lateral stress of this kind is called a shearing stress or *shear*.

If the maximum stress difference or deforming stress, $P_1$–$P_3$, acting on a mass of rock is gradually increased, the rock is elastically strained, that is, changed in shape or deformed, until a limit is reached at which the shearing or crushing strength is overcome. The rock then breaks, if it is brittle; or begins to 'creep' or flow, if it is plastic or viscous. Elastic strain is recoverable when the stress is removed, as when a spring balance rebounds when the weight is taken off. Change of shape due to plastic or viscous flow is not recoverable in this way when the stress is removed, although the elastic part of the total deformation may still be recoverable.

Solids that flow when the strength is overcome are commonly described as plastic, but except for certain kinds of clay this term has now too limited a technical meaning to be appropriate for the flowage induced in large masses of rock by long-continued geological processes. In plastic flow the deformation is proportional to the excess of stress over that corresponding to the elastic limit. It corresponds to the behaviour of dough and of many commercial substances while they are being moulded. A somewhat indistinct limit is reached as stress increases: indistinct because there are various little-understood transient effects between the end of elastic behaviour and the beginning of what can be recognized as viscous behaviour. Eventually, however, the material either cracks, if the extra stress is rapidly applied (as by a sudden blow), or it continues to flow, This time the flow is found to be continuous for a given stress, being proportional only to the time during which the stress is maintained. The flow is now like that of a highly viscous liquid. It is not yet precisely known at what stage this viscous flow really begins. It may be imperceptible in operations from the beginning, remaining unrecognized only because its first effects are negligible, compared with the elastic and plastic deformations, until it has broken through the successive 'strength' barriers. However, as we shall see on pp. 131–2 where the behaviour of rocks that flow is further considered, this unsolved problem ceases to matter when long periods of geological time are available.

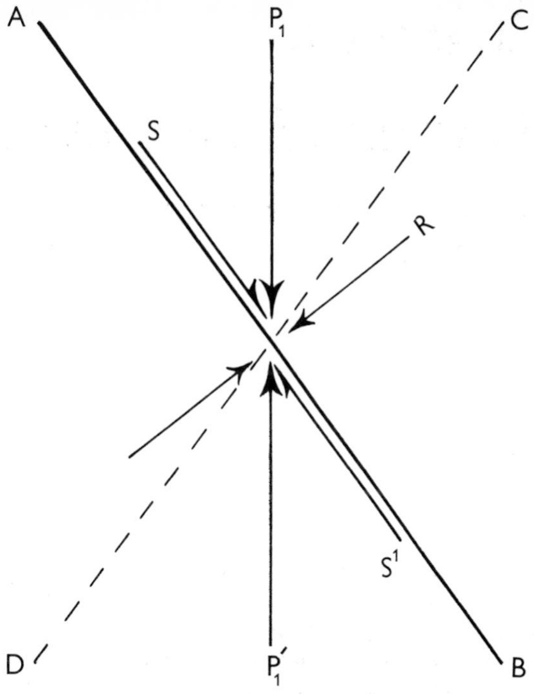


**Figure 9.2** Resolution of $P_1$ into S and R, and of $P_1'$ into S′ and R′; S and S′ being the shear along the shear-plane AB, and R and R′ being the component at right angles. The plane AB (and also the other possible shear-plane CD) is at right angles to the page. If the material yields along, say, AB, the layer on one side slips over the layer on the other side.

Brittle rocks would theoretically be expected to fracture along the planes of maximum shearing stress: that is, along planes parallel to $P_2$ and bisecting the angles between $P_1$ and $P_3$. However, rocks are far from being homogeneous; and the planes of maximum shearing stress rarely coincide with planes of weakness already present in the rock (e.g. bedding planes), which would offer the least resistance to shearing. In practice there is a compromise between highly complicated sets of conditions which vary from rock to rock. The actual shear fractures seen as joints and faults make an angle with $P_1$ that is generally less than 45° in hard rocks like quartzite, angles ranging around 25° to 30° being characteristic. There are, of course, two such planes, represented by AB and CD in Figure 9.2, and $P_1$ bisects the acute angle between them. Fracture may occur parallel to either of the two planes or to both, and what actually happens depends on the internal frictional properties of the rocks and the confining pressure. In softer rocks, like shale, an angle of more than 45° is not unusual.

When such rocks begin to flow, as they do when slaty cleavage is developing, the angle rapidly increases to 90°. This is because the behaviour is no longer elastic, but viscous or practically so: let us say *quasi-viscous*. Consequently flow occurs at right angles to the direction of compression, just as it does when tooth-paste is squeezed out of a tube.

Rocks, like other solid materials, also have a tensile strength, and this is generally considerably lower than the shearing strength. If $P_3$ is a tension adequate to overcome the tensile strength of a rock, the latter either breaks or contines to stretch. If it breaks it does so along fracture planes at right angles to $P_3$. Such tensile fractures are readily opened by invading fluids under high pressure, and some of them eventually come to be occupied by dykes or mineral veins.

## Fracture and Flow in Ice

In glaciers such tensile fractures tend to open into crevasses which gape at the surface and may penetrate the outer crust of ice to depths of as much as 45 m or more before dying out. They may, however, be hidden by snow-bridges and the treacherous surface is then difficult and dangerous to cross (Figure 9.3). But moving ice is not always or only torn into crevasses; it also responds to the appropriate stress changes by folding or shearing. As R. T. Chamberlain has said: 'the moving glacier affords far and away the most concrete illustration of rock deformation available.' At the conclusion of a highly stimulating paper on this subject (for reference see p. 442), he tells us that when 'perplexed by clashing theories' of rock deformation he 'has frequently found that the best remedy . . . is to turn to the glacier, a rock of simplest sort actually undergoing a deformation before our eyes.' We cannot do better than adopt this wise and apposite suggestion.

**Figure 9.3** The Rhône Glacier, showing crevasses and an ice fold (middle right), viewed from the ridges above the Belvedere Hotel, near Gletsch, Switzerland (*F. N. Ashcroft*)

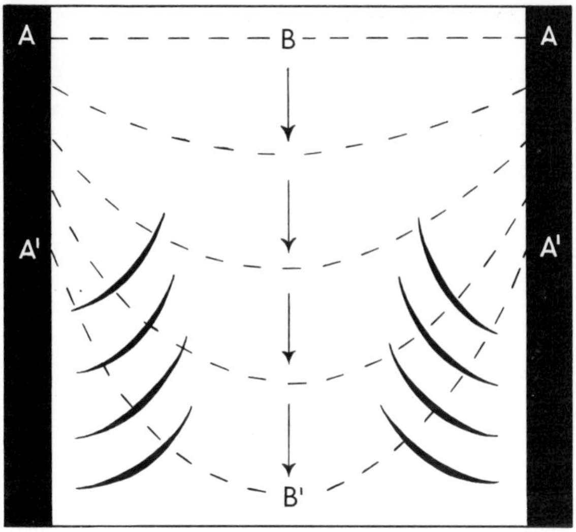


**Figure 9.4** Diagram to illustrate the development of marginal crevasses arranged *en échelon*, as a result of the differential flow of glacier ice

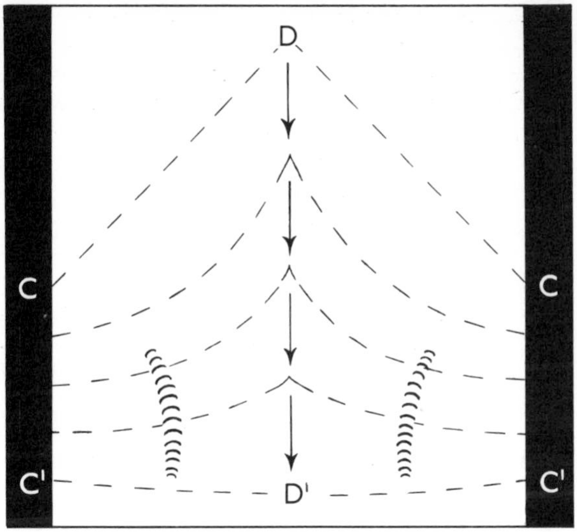


**Figure 9.5** Diagram to illustrate the development of folds as a result of the differential flow of glacier ice

The various types of crevasses caused by the stretching of moving ice are considered on pp. 413–14. Here we need only refer to one of them. Marginal crevasses, pointing or curving upstream from the sides of a glacier, develop as shown in Figure 9.4. It is well known that the middle part of a glacier travels more rapidly than the sides, where the ice is thinner and has to overcome the frictional resistance of the walls. Thus any line ABA is extended into a gradually lengthening catenary, such as A′B′A′. If the rate of extension is relatively slow, the moving ice can continue to adapt itself to the differential flow without cracking. But if the extension takes place more quickly, as is commonly the case, the ice will sooner or later crack. If A′B′ were a straight line the crack would be at right angles to it, but since A′B′ is curved the cracks and the resulting crevasses also tend to be curved. This curvature can be seen in Figure 9.3, which also shows a good example of a fold (snow-covered, on the right) cut across nearly at right angles by a series of crevasses. Evidently folds and crevasses can develop simultaneously.

Folds develop as shown in Figure 9.5. Assuming the same differential movement as in Figure 9.4, we can start with the lines CDC and follow the change of shape as the glacier advances. For a time the line CD continues to shorten and so to be thrown into compression. In Figure 9.5 the maximum shortening would be just before the position C′D′ is reached by the line considered. Folding occurs if the shortening in a given time is sufficient to provide the necessary compressive stress. Since ice cracks more easily than it folds, it follows that for one fold there may be many crevasses. This feature of the behaviour of ice is also clearly demonstrated by Figure 9.3.

Figure 9.6 illustrates one of a series of ice folds cut across by crevasses, the growth and decay of which were intensively studied as part of the International Geophysical Year investigations in Antarctica. The folds vary in height from 5 to 15 m above the general level of the ice, and in length from 600 to 1500 m. They occur in groups of several sub-parallel folds, each group occupying an area of about 2·5 km². The development of the folds implies that the ice is shortening in the direction at right angles to the folds. By repeated measurements of the length of a base line the rate of this shortening was found to be about 5 cm a day per 1·6 km. Another instructive observation is that the folds themselves eventually break up into irregular blocks by fracturing along shear planes. The sequence of phenomena shows that the ice can break under tension ($-P_3$) while it is flowing, and that it may break in this way before it begins to yield by folding under compression ($P_1$). However, it does not break by shearing until a later stage, when the maximum stress difference, $P_1-(-P_3)=P_1+P_3$, has increased to the limit necessary to overcome the shearing strength.

Evidence that ice can be folded by differential

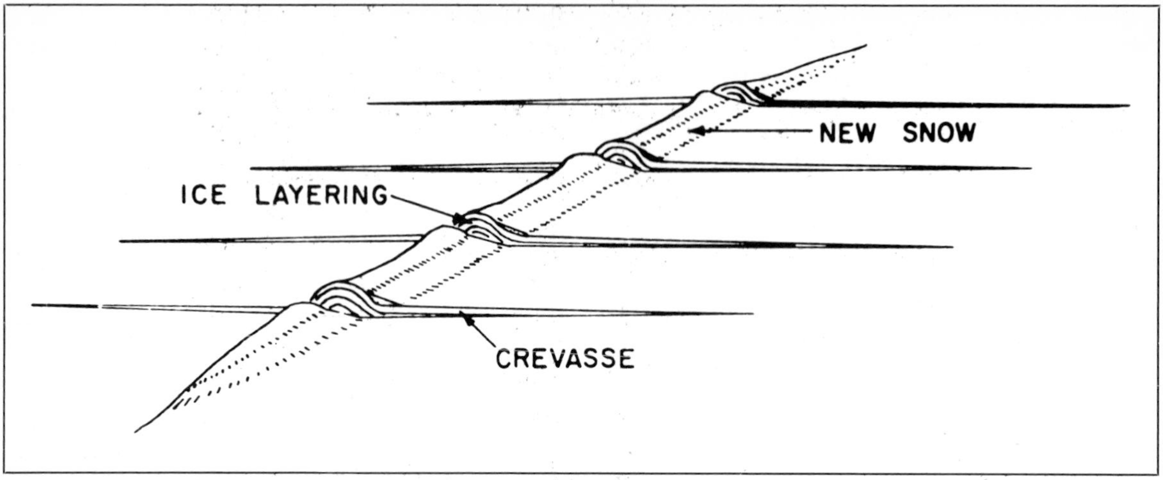


**Figure 9.6** Diagram of a typical anticlinal ice fold in the Ross Ice Shelf near Camp Michigan, one of the American International Geophysical Year bases (1957–8), Antarctica. The thickness of individual ice layers exposed by the transverse crevasses is only a few centimetres and is greatly exaggerated in the drawing. (*Transactions of the American Geophysical Union, vol. 39 (1958), p. 796*)

flow is illustrated by Figures 3.20 and 9.7, which show the contortions of bands of ice and morainic debris resulting from the struggle for space between confluent glaciers advancing at different speeds. Normally, however, fracture is the dominant response to excess of strain in the upper layers of most glaciers. This is visually demonstrated by the prevalence of crevasses. Below the depth at which crevasses die out the ice creeps down its valley by flowage, aided by a certain amount of sliding. Even if we define a superficial 'zone of fracture' in terms of the occurrence of crevasses, the top of the 'zone of flowage' is never far below the surface. Glacial ice differs from ordinary rocks in this respect, but only because it happens to be near its melting-point and is therefore under conditions favourable to rapid and continuous recrystallization.

Near the surface most ordinary rocks yield to earth stresses by fracture, but at greater depths deformation more commonly takes place by rock-flowage of one kind or another. For *each kind of rock* it may be said that fracture predominates above a certain depth. This 'zone of fracture' passes into a zone of transition in which both fracture and flowage are characteristic, according to circumstances. In turn, this zone merges into the region where rock-flowage is predominant: as, for example, where migmatites are generated. The migmatites we now see at the surface are, of course, in their zone of fracture; but while they were being formed, millions of years ago and several hundred metres below the surface of that time, they were in the zone that for them, and at that time, was the zone of flowage.

Whether or not flow movement occurs depends upon a wide variety of factors, such as (*a*) the kind of rock; (*b*) the prevailing conditions of confining pressure, deforming stress, temperature, and lubrication by interstitial liquids, vapours or gases; and (*c*) the duration of these conditions, and especially the duration of the deforming stress. It is easy to think of pressure and temperature as increasing with depth, and this has encouraged the misleading practice of relating the so-called zones of fracture and flowage to depth *alone*. But it must be realized that geologically the dimension of time is certainly no less important. The significance of duration has already been emphasized on p. 127. Now it can be approached from a different aspect.

## Rheidity—A Time Aspect of Rock Deformation

The three familiar states of matter—solid, liquid and gaseous—are clearly defined by reference to the melting- and boiling-points of the particular substance concerned. A liquid is at a temperature above its melting-point and flows in virtue of this fact. A solid is below its melting point, and yet it, too, is capable of flow given the proper conditions. This raises the difficult question of the nature of the conditions that determine whether a substance behaves as an elastic solid or a 'viscous' solid, or in some intermediate way. A whole new border-line science, known as *rheology* (Gr. *rheos*, flow) is now devoted to the investigation of these phenomena. Here it must suffice to say that the essential control depends on the relationship between two properties: *viscosity*, resistance to flow, and *rigidity*, resistance to elastic deformation.

S. Warren Carey has proposed the term *rheidity* for this relationship, which turns out to be expressible in terms of time. For given conditions of temperature, confining pressure and deforming stress, rheidity is arbitrarily defined by Carey as the time required for deformation by viscous flow to become more than 1000 times the elastic deformation. The factor 1000 may seem to be extravagant, but it ensures that the elastic and associated effects have become so trivial that they can be neglected, and the deformation regarded as if it were that of a viscous fluid. To refer to a substance which is behaving in this way (or to refer to it as it *was* during a former period of active deformation by flow) Carey proposes the term *rheid*. A rheid differs from a liquid in being below its melting-point and in having a certain amount of strength. It flows in virtue of the relatively long duration of the deforming stress.

Figure 9.8 is an attempt to indicate in a purely schematic way how rocks behave in relation to duration of stress as well as to temperature. Though in no way quantitative, the diagram may help to familiarise the admittedly difficult concept that solid rocks can have a double or triple personality, so to speak: being able to fracture or flow, or to fracture while they are flowing, according to circumstances.

Rheidity happens to have the dimension of time (and therefore to be of fundamental geological importance) because in the ratio of viscosity to rigidity the other dimensions, length and mass, cancel out. Taking ice as a familiar example for which data are available, and using *c.g.s.* units, the rheidity can be estimated as follows:

$$\text{Rheidity} = 1000 \times \frac{\text{viscosity}}{\text{rigidity}} \text{ seconds}$$

$$\text{Rheidity of ice} = 1000 \times \frac{10^{13}}{10^{10}} = 10^{6} \text{ seconds}$$

(about 12 days)

**Figure 9.7** Rheid folding of glacier ice and morainic bands as a result of the struggle for space between glaciers advancing into the same valley at competing rates of flow. The Susitna Glacier and 'tributaries', Alaska (*Bradford Washburn*)

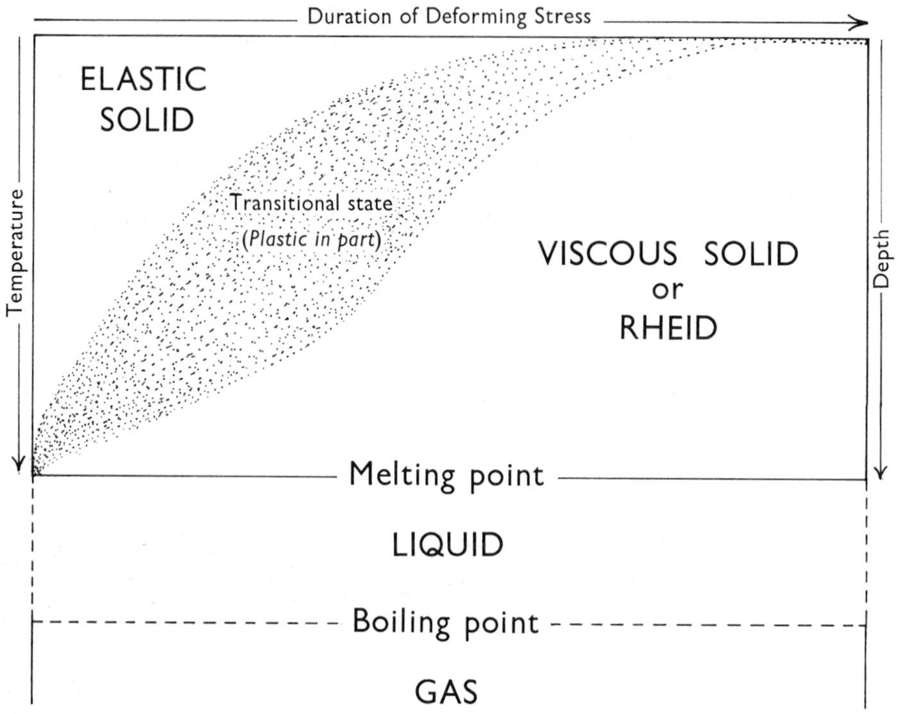


**Figure 9.8** Schematic diagram to illustrate the conditions under which a solid becomes a rheid and flows like a highly viscous fluid. Temperature increases downwards to suggest a rough correlation with depth. The diagram is in no way quantitative. 'Duration' may range from a fraction of a second on the left to anything between a few days and the age of the oldest-known rocks on the right. (See the paper by S. Warren Carey to which a reference is given on p. 41)

From this result we can easily realize that if a given deforming stress is applied for only a few minutes or less, ice behaves as an elastic and brittle solid. But if the stress continues for weeks or longer, as in the case of glaciers, the ice flows as if it were a highly viscous liquid. It should not be overlooked, of course, that the physical mechanisms of rheid flow are entirely different from that of liquid flow. Here, however, we are not so much concerned with the physics of the processes involved as with the geological results achieved over long periods of time.

For geological materials rheidity ranges from a week or two for ice, a year or two for gypsum, and a few years for rock-salt, to thousands of years for the sub-crustal materials of the mantle and a million times longer for ordinary rocks at ordinary temperatures. With rise of temperature rheidity rapidly decreases, approaching zero when fusion occurs, and the rheid becomes a liquid. Thus it is easy to understand that rocks undergoing metamorphisms, whether or not they are lubricated by hot gases, are able to flow in the depths of the crust as effectively as glacial ice can flow at the surface, the only significant difference being that they flow more sluggishly.

It should not be forgotten that while ice or any other solid body in the rheid state is slowly flowing, the deforming stress may temporarily increase to a point where the material is compelled to fracture, if only momentarily. Even water, the very symbol of fluidity, 'breaks' when a sea wave becomes a breaker and shatters itself against the cliffs, or when water falling freely from a tap, or as a waterfall, breaks into drops and spray. Given the appropriate time relations anything can break, and so far as rocks are concerned our experience leads us to expect them to break. If at first the idea that rocks can flow as well as break seems strange and perhaps unnatural, it is partly because we forget that glacier ice and rock-salt are also rocks, partly because ordinary rocks as we see them at the surface are destroyed by weathering or erosion long before they could possibly show signs of flow, and partly because our lives our so short. But geological time is long and ample, and one of the inevitable consequences is that the hardest and strongest rocks behave as rheids in the proper environment of temperature and time. At various periods and places in the history of the earth's crust, rocks have been mobilized and made to flow so that they finally display structures and intrusive forms that would seem incredible if they were not real (Figures 9.9 and 30.32).

The chief structural results of fracture and rheid flow to be briefly described in this and the following two chapters are listed below:

*Joints*: fractures along which practically no displacement of the rocks has occurred.

*Faults*: fractures along which the rocks on one side have been displaced relatively to those on the other side.

*Folds*: buckling, bending and contortion of rocks by a group of very complex processes involving fracture, sliding, shearing and flowage.

*Slaty cleavage* and *foliation* are rock-structures intimately related to folds that formed mainly by

**Figure 9.9** Late Caledonian overfold of migmatites formed from older (Precambrian) gneisses, east Greenland, about 100 km west of Shannon Island. The plateau, just seen at the top of the picture, is about 800 m above the floor of the Kildedalen, the valley in the foreground. (*Large Koch Expedition, 1947–55. See John Haller, 'Meddelelser om Grönland', vol. 154 (1956), Fig. 15*)

flowage while the rocks concerned were behaving as rheids.

*Diapirs* (Gr. *dia*, through; *peiro*, piercing): intrusion of relatively light rock material, while in the rheid state, through overlying rocks, so giving rise to such structures as *diapiric folds* (also known as *piercement folds*); the now familiar *salt domes* (or stocks or plugs); and *igneous diapirs*, including many plutons, stocks, bosses and batholiths, or parts of them.

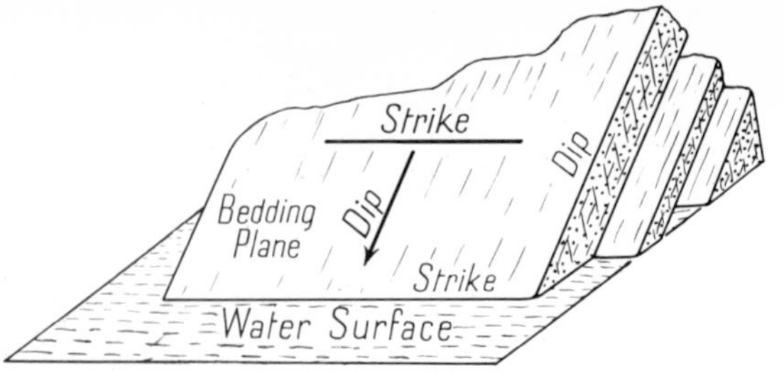


**Figure 9.10** Diagram to illustrate the meaning of the terms *dip* and *strike*

## Dip and Strike

Most sedimentary rocks were originally deposited on flat or very gently inclined surfaces. As a result of special circumstances certain beds may have started with an initial inclination, as in current bedding (pp. 80–1), but where we find great thicknesses of strata tilted into conspicuously inclined positions, perhaps for many kilometres, it is generally clear that the beds have been tilted by movement that occurred after the deposition. However, whatever the origin of the inclination, whether primary or due to slumping or to folding by earth movements, it is often important that the attitude of an inclined bed should be accurately determined. This is done by measuring what is called the *dip* of the bedding plane (Figures 9.10 and 9.11). The dip includes both the *direction* of the maximum slope down a bedding plane and the *angle* between the maximum slope and the horizontal. The dip direction is measured by its true bearing (e.g. as so many degrees east or west of north), the compass reading being suitably corrected for magnetic variation. The angle of dip is measured with a clinometer. The *strike* of an inclined bed is the direction of any horizontal line along a bedding plane; the direction, for example,

**Figure 9.11** Sea cliff of Torridonian Sandstone strata, showing dip and strike. Cailleach Head, Highland Region (Ross-shire), north-west Scotland (*Institute of Geological Sciences*)

**Figure 9.12** Anticline in Coal Measures, Sandersfoot, South West Dyfed (Pembrokeshire), South Wales (*Institute of Geological Sciences*)

**Figure 9.13** Syncline in Upper Carboniferous, north of Bude, Cornwall (*Institute of Geological Sciences*)

of the intersection of the bedding plane with still water or level ground. The dip direction is at right angles to the strike, but it must nevertheless be specified, even if the strike is known, since there are two such directions. If the strike is E–W, for example, the dip could be either to the N or S. The attitudes of other structural surfaces such as joint planes, fault planes, cleavage planes and foliation planes can be similarly described in terms of the angle and direction of dip. It should be noticed, however, that amongst mining men it is the traditional practice to record the *hade* of a fault, instead of a dip, hade being the angle between the maximum slope of the fault plane and the vertical. Hade and dip are complementary angles.

## Folds

When strata are upfolded into an arch-like form the structure is called an *anticline*, because the beds on either side then 'incline away' from the crest (Figure 9.12). When the beds are downfolded into a trough-like form the structure is called a *syncline*, because in this case the beds on either side 'incline together' towards the keel (Figure 9.13). The two sides of a fold are referred to as its limbs, the limb which is shared between an anticline and its companion syncline being called the middle limb. The plane that bisects the angle between the two limbs as nearly as possible is called the *axial plane*, and the line of intersection of this plane with a bedding plane (the *hinge-line*) gives the direction of the *axis* of the fold at that place (Figure 9.14). So long as this direction remains unchanged, which may be for a few centimetres or for many kilometres, the axis can be thought of as a line that would generate the form of the fold if moved parallel to itself. It should be noted that an axial plane has a definite position in space, whereas the axis is only a direction, although it is commonly thought of, and drawn on geological maps, as situated in the axial plane. If the axial plane is vertical and the axis horizontal, as in Figure 9.14, the fold is said to be upright and symmetrical. If the axial plane is inclined, as in Figure 9.15, the fold is also said to be inclined, and is described as an *overfold* or *recumbent fold* as the axial plane becomes increasingly inclined and approaches the horizontal (Figure 9.9).

The axes of folds are not infrequently found to be tilted instead of horizontal; the folds are then said to *pitch* or *plunge*. The angle of plunge is the angle between the axis and a horizontal plane. In Figure 9.16, which represents an anticlinal mountain, the plunge is towards the bottom left-hand corner in front; but towards the top right-hand

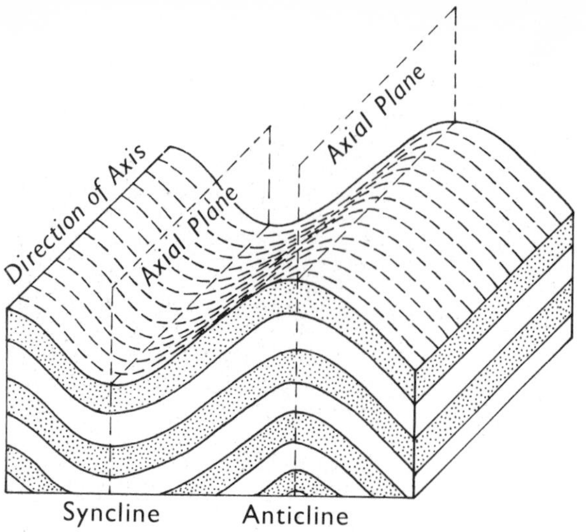


**Figure 9.14** Upright symmetrical folds, to illustrate the meaning of the terms *axial plane* and *axis*. The upper surface shown is not that of the ground but of one particular bed in a series of folded strata.

corner in the background, 18 km away, it is in the opposite direction. From this illustration it can be seen that at each 'end' of the mountain the U- or V-shaped outcrop of the outermost exposed bed 'points' or 'closes' in the direction of plunge. This would be true of the outcrop of any particular bed if the anticline had been worn down to a more or less level surface. An example of this kind is illustrated in Figure 9.17. Conversely in synclines the outcrop of a particular bed diverges or opens out like a horse-shoe in the direction of plunge. The 'end' of a plunging fold, as seen on the ground, is often described as a nose.

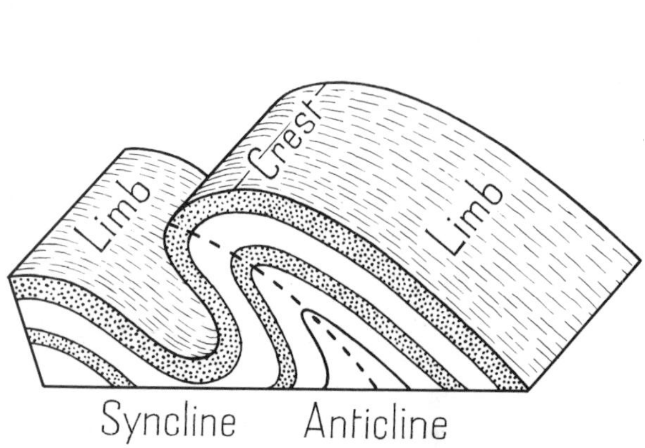


**Figure 9.15** Inclined folds, with the upper strata removed to show the surface of one particular bed. The trace of an axial plane is marked by the broken line.

The Asmari Mt anticline of Figure 9.16 could be described as an elongated dome. In an ideal structural dome the beds dip radially outwards from the centre. In a similarly perfect basin the beds dip radially inwards towards the centre. Actual examples are at best only roughly circular or crudely elliptical in outline, and other irregularities, commonly due to faulting, are usual. Domes are a natural result of the upheaval of overlying strata by ascending salt plugs or diapiric igneous intrusions (p. 153). Some basins are known to be collapse structures, but there are many others that are far from easy to account for.

From more or less elongated domes to anticlines that extend for very long distances there is every gradation, and there are similar transitions between basins and long synclines. Such structures must not be confused with hills and valleys, for they refer solely to the forms and attitudes of the bedrocks: not to the relief of the surface, Landscape forms may coincide here and there with the underlying structure for a time, as they are seen to do in Figure 9.16. But in the long run, when landscapes have been carved out of uplifted tracts of folded rocks by long-continued denudation, this relationship is sooner or later reversed. Snowdon in North Wales is a well-known mountain peak with a synclinical structure.

Folds range in intensity from broad and gentle undulations to tightly compressed plications in which the dips of the beds are almost parallel, except near the hinge-lines. Folding of this style is described as *isoclinal*, e.g. the isoclinally folded Silurian strata exposed along the Berwickshire coast (see page 101). In the cliffs shown in Figure 7.12, the limbs of the closely packed folds are found to be nearly parallel to the axial planes. Such folding exemplifies in a striking way one of the characteristic features of most folds—thinning along the limbs and complementary thickening towards the hinges, where the curvature is greatest (Figure 9.18).

When a pack of cards is folded into, say, an anticlinal form, each card slips a little over the underlying card and rises slightly above it towards the crest, so leaving a crescent-shaped opening between them, which represents the thickening referred to above. Similarly, when bedded rocks are folded, slipping along the bedding planes takes place. The necessary thinning and thickening is mainly achieved by weak, *incompetent* rocks, like shale, though stronger *competent* beds, like sandstone, are not immune.

**Figure 9.16** Air view from the NW of Asmari Mountain, Iran (26 km long by 5 km across in the middle). A relatively short anticline or elongated 'dome' of the Asmari Limestone (Oligocene and Lower Miocene) surrounded by younger sediments of the Fars Series (*British Petroleum Company Limited*)

**Figure 9.17** Air view of a plunging anticline in the Parry Island orogenic belt, looking from Peel Inlet to Erskine Inlet, Bathurst Island, Canadian Arctic. The core of the anticline is being dissected by a stream and its tributaries, and a small delta is being built in Peel Inlet. (*Royal Canadian Air Force*)

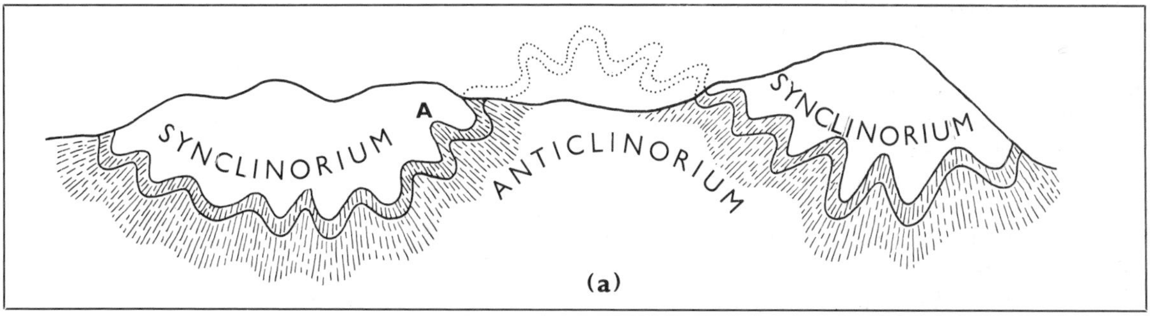


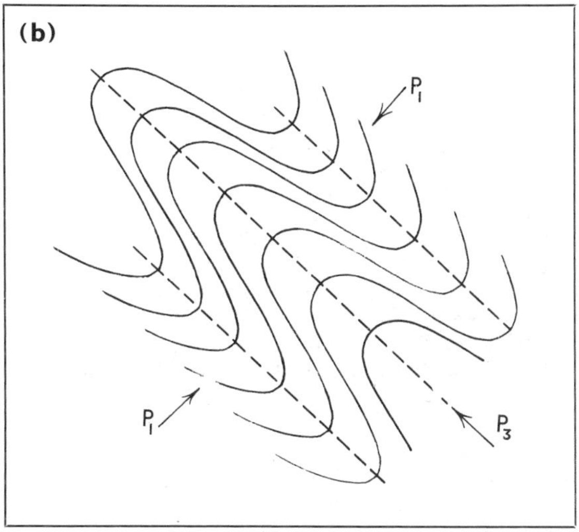


**Figure 9.18** *(a)* Schematic section through an anticlinorium bordered on each side by a synclinorium. The corresponding cleavage fans are indicated by the broken lines.
*(b)* Parallel folds at A in the above diagram showing thinning of limbs, cleavage planes (broken lines) and corresponding stress field

At considerable depths, generally in geosynclines, the folding of bedded rocks is accompanied by slipping along shear planes that are inclined to the bedding planes. As the folding becomes intense these shear planes become parallel to the axial planes of the growing folds. Shale, for example, adjusts itself by a process of recrystallization in which the new crystals all tend to be elongated or flattened parallel to the shear planes. In other words the shale becomes slate with the typical flow cleavage of such rocks. Paul

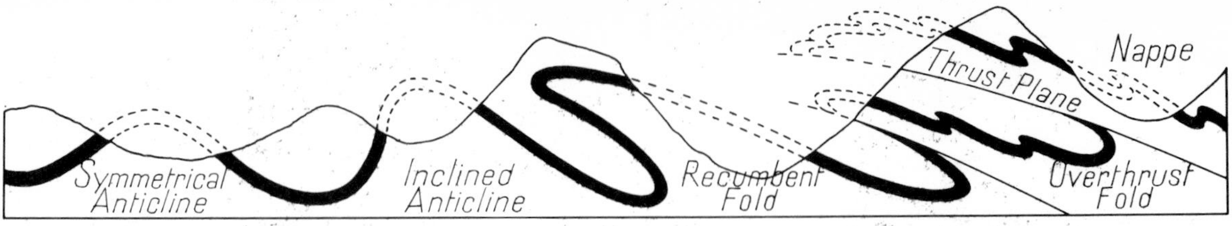


**Figure 9.19** Section (purely diagrammatic) to show various types of folding and thrusting

Fourmarier, a famous Belgian geologist who had spent much of his long life in studying cleavage and schistosity in many parts of the world, concluded that the weight of an overburden several kilometres thick is essential for the development of cleavage.

Associated beds of graywacke or other competent rocks adjust themselves partly by intergranular movements (akin to the flow of loose sand, but with extreme sluggishness) and partly by minute gliding movements along innumerable shear planes. The latter, being like closely spaced joints, give the rocks a fissility which is referred to as *fracture cleavage* to distinguish it from slaty (flow) cleavage. At higher levels, above the zone of slaty cleavage, folded and indurated shales commonly exhibit fracture cleavage.

**Figure 9.20** Chimney Rock, North Fork, Shoshone River Canyon, Wyoming, U.S.A. A pillar of sandstone bounded by vertical joints, separated by weathering and erosion from the vertical canyon wall (*U.S. Forest Service*)

It will have been gathered that slates are likely to be developed in regions of tight or isoclinal folding. The latter are usually parts of orogenic belts where small puckers, like the pleats of an accordion, are superimposed on anticlines and synclines of a larger order, which in turn may be integral parts of still vaster structures. Such a complex of folds of different orders is called an *anticlinorium,* if the whole appears to have been arched upwards; or a *synclinorium* if the generalized form is that of a corrugated depression. The actual relationships are generally much more complex than the simplified representation drawn in Figure 9.18(*a*). For example, faults have been omitted, but it should not be forgotten that in advances stages of deformation, when the limbs of the folds become drawn out and attenuated by flow, they easily become 'glide planes' along which mechanical displacements by thrust faulting are facilitated. However, Figure 9.18(*a*) adequately illustrates one very important feature: the fact that the cleavage planes of slates in an anticlinorium, far from being parallel throughout the structure, radiate inwards in a fan-like fashion. An excellent example occurs on the Clwyd moors, Wales, where the cleavage-dip increases from 40°S in the north to 80° or 90° in a central belt, beyond which it decreases again to 40°N in the south. In a synclinorium the 'cleavage fan' resembles a suspended fan. The importance of these fan-like arrangements is their implication that neither the cleavages nor the folds could have been produced by horizontal compression alone. The cleavage planes of slates are at right angles to the greatest principal stress of the stress field in which they were produced (Figure 9.18(*b*). Consequently the fans give us a clear picture of the directions of the principal stresses responsible for them.

**Figure 9.21** Flat-bedded Permian sandstones cut by two sets of well-developed vertical joints nearly at right angles; traces of a poorly developed third set, inclined to the others, can also be seen. The 'Tessellated Pavement' at Eaglehawk Neck, southern Tasmania (*Dorien Leigh Limited*)

**Figure 9.22** Characteristic jointing of a granite illustrated by the castellated cliffs of Land's End. Part of a large intrusion of granite (see Fig. 5.9) now exposed to weathering and marine erosion (*Aero Pictorial Limited*)

*Recumbent* folds have already been mentioned. In these the beds of the middle limb (the lower limb of the anticline and the upper limb of the underlying syncline) are then upside down. Parts of the Grampians are carved out of gigantic recumbent folds. In structures of this kind most of the middle limb may be sheared out altogether. Further development of the structure then results in forward movement of the rocks of the upper limb along the plane of shearing. The latter has become a *thrust plane* and the structure an *overthrust fold* (Figure 9.19). The sheet of rocks that has moved forward along the thrust plane is referred to as a *nappe*, this being the French word for cover or sheet.

## Joints

If we examine quarry, cliff or shore exposures of sandstone or limestone, we find that in addition to the bedding planes the rocks are transversed by fractures that are generally approximately at right angles to the stratification and therefore nearly vertical when the beds are flat (Figures 6.3 and 9.20). Joints frequently occur in sets consisting of two series of parallel joints (Figure 9.21). If the beds are inclined, one series approximates in trend to the dip direction and the other to the strike. Such joints are of great assistance to the quarryman in his task of extracting roughly rectangular blocks of stone, especially the 'master joints', which are often remarkably persistent and strongly developed. In many rocks, however, and especially in the older formations, subordinate joints cut across the main sets, thereby dividing the rocks into irregular angular blocks.

At the surface, joints of all kinds are very susceptible to the attack of weathering agents; they are readily opened up by the work of rain, frost, wind and plant roots. It is due to this opening of joints and bedding planes that cliffs and mountain scarps often resemble roughly hewn masonry (Figure 9.22). Along the shore, waves attack the rocks selectively along joints and the influence of the joint pattern is often clearly shown in the outlines of inlets, caves and skerries. The joint pattern may also control the course of rivers, the joint planes themselves commonly forming the walls of steep-sided gorges and canyons.

Most joints seem to be results of either shearing or tension. Some of the latter type (e.g. as in Figure 9.21) have been formed in response to the relief of elastic strain that accompanies uplift and the

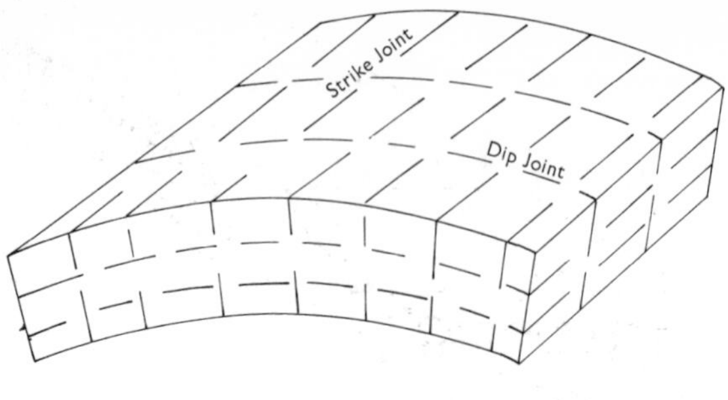


**Figure 9.23** Strike joints and dip joints in gently folded strata

possibility of extension over a greater area than before. Tension joints may have rough and irregular surfaces. Because of their tendency to open if tension continues, some of them provide easy passageways for natural waters from which minerals may be deposited, coating the sides or developing into mineral veins. The fissure-like spaces occupied by many dykes probably began as tension joints. Shear joints, on the other hand, tend to be clean-cut and in unweathered rocks they are tightly closed. As indicated by the two complementary shear planes represented by AB and CD in Figure 9.2, they occur in two sets, which together constitute what is called a *conjugate joint system.*

The polygonal cracks seen on dried-out mud flats (Figure 5.5) and the characteristic jointing of sheets of columnar basalt and ignimbrite (Figures 5.7 and 5.16) are familiar examples of tension joints that have already been described. In granite masses, however, there are generally three sets of joints. Two of these are upright or vertical and together they constitute a conjugate system of which one set, the *rift*, is more strongly developed than the other, which is known as the *grain*. The third set is responsible for the *sheeting* or *sheet* structure that is often seen in exposures of granitic and other plutonic intrusions. The parting planes between the sheets are more conspicuous and closer together near the actual surface, the undulations of which they tend to follow in gently curved or dome-like forms that are truncated by the ground only where the latter is steep. All three sets of joints are opened up by weathering, and the sharp edges of the resulting blocks are then rounded off. This effect is well seen in the tors of granite moorlands: piled-up tiers of gigantic blocks and rounded boulders, with a bold architecture suggesting a cyclopean fortress (Figures 9.22 and 19.50).

The sheet joints are amongst those referred to above as due to relief of elastic strain as the original overhead load is gradually stripped off by erosion. That this is essentially the correct explanation is sometimes demonstrated by the sudden upspringing of a sheet of rock from a freshly excavated quarry floor.

When the strata overlying a rising intrusion are domed, they are necessarily extended in all directions. The usual result is a conjugate system of two sets of tension joints, one being roughly radial, the other more or less concentric. If an anticline is thought of as an uplifted and greatly elongated dome, it is easy to see that folding in some cases involves extension of the strata both at

**Figure 9.24** A simple classification of faults and their stress fields. $P_2$ always lies in the fault plane. In (a) and (b) $P_2$ is horizontal; in (c) and (d) it is vertical. (Dip-slip without any strike-slip component)

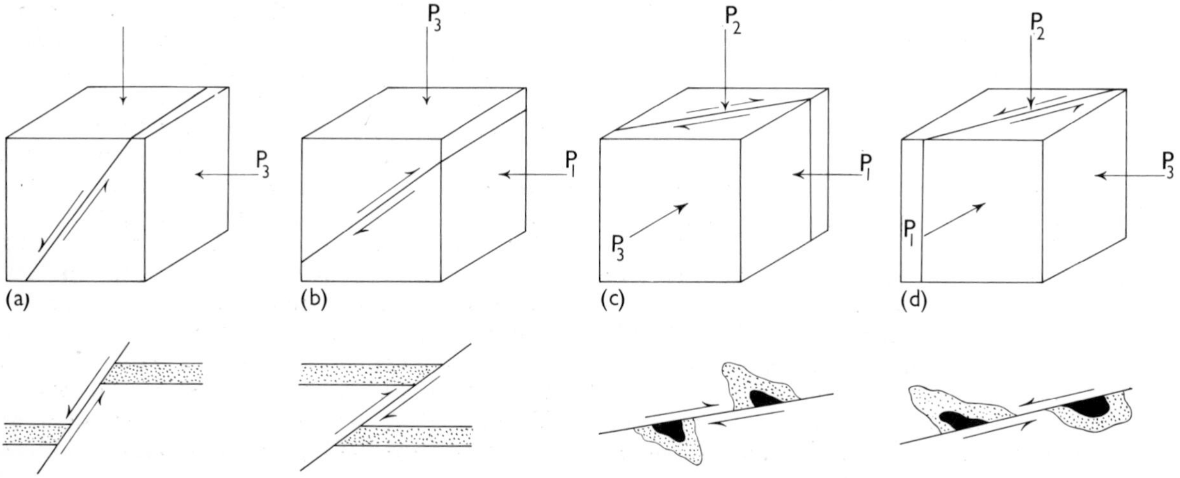

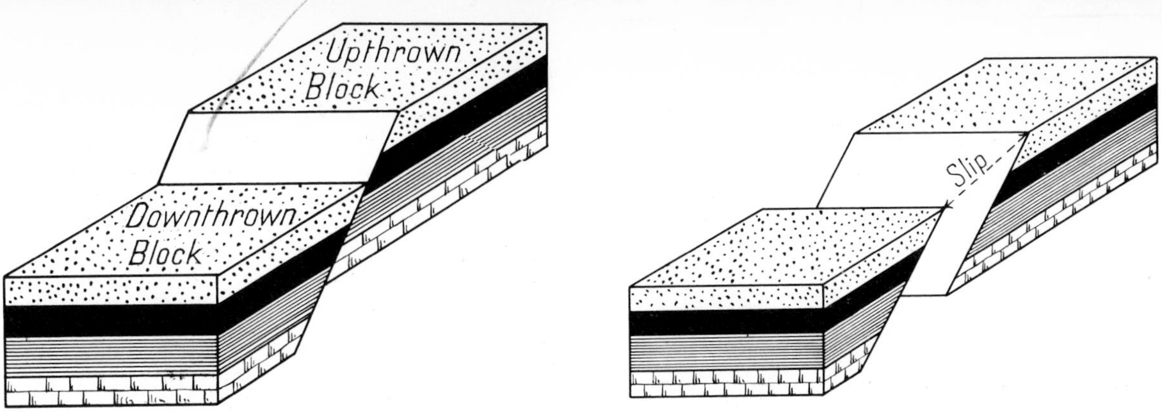


**Figure 9.25** *(left)* The relative movement involved in a normal fault. The visible part of the fault plane is left unshaded.
*(right)* Oblique-slip normal fault, illustrating the meaning of the term *slip*. Here the movement has involved both dip-slip and strike-slip, the former being the greater.

right angles to the axis and along the axis (Figure 9.23). Corresponding to these two directions of tension there may be joints parallel to the axis (strike joints, likely to be well developed in competent beds near the crests of anticlines, and similarly near the troughs of synclines), and joints at right angles to the axis (transverse joints, of which the crevasses shown in Figure 9.6 are a development).

## Faults

A fault is a fracture surface along which the rocks have been relatively displaced. Vertical displacements of several hundred metres and horizontal displacements up to 50 or more km are well known, but in none of these is there any reason to suppose that the total movement occurred during a single super-catastrophe. Earthquakes result from sudden movements along faults, but the displacements are rarely more than a metre or so at a time. even so, it is obviously of vital importance that engineering structures, such as great dams and lengthy bridges, should not be constructed across faults that are still active.

The attitude of a fault plane is described by the angle and direction of its dip (or *hade*, if the angle is measured from the vertical). The relative displacement of the rocks (or *slip*, as it is called) is more difficult to determine with accuracy, although the fact that the rocks on the two sides of a fault have been displaced can generally be easily recognized, given suitable exposures. However, corresponding to the standard distribution of the principal stresses, in which one is vertical and the others horizontal, it often happens that the slip approximates in direction either to the dip or to the strike of the fault plane. Remembering that $P_2$, the intermediate stress, is parallel to the fault plane, we can distinguish *dip-slip* movements and faults (in which $P_2$ is horizontal) and *strike-slip* movements and faults (in which $P_2$ is vertical). Each of these can be divided into two (Figure 9.24). In dip-slip faults the relative displacement of the upper block may have been either down the dip (giving a *normal* fault) or up the dip (giving a *reverse* or *thrust* fault). In strike-slip faults the relative displacement of the block on the far side of the fault, as viewed from the ground, may have been either to the right (giving a *right-lateral* or *dextral* fault) or to the left (giving a *left-lateral* or *sinistral* fault). Strike-slip faults are commonly described as *wrench* faults, *tear* faults or *transcurrent* faults.

Faults not coming readily under the four classes described above, and illustrated in Figure 9.25, are called *oblique-slip* faults; they can be more specifically described as combinations of, say, normal and wrench faults, or reverse and wrench faults. Except when precise details are called for, the name adopted is that of the dominant component.

Since a fault plane is a shear plane, it could theoretically lie on either side of $P_3$. When there is practically no movement, as when joints are formed, both directions are utilized and a conjugate system develops. But once movement has started along one of the two fractures, further movement is likely to continue along the same plane so long as the stress field is not radically changed. It should also be noted that the angle of dip of a fault surface may vary at different depths and in different kinds of rocks. Shearing surfaces, if curved, are generally concave towards places of least resistance, as is illustrated by landslides, When the Panama Canal was being excavated, the materials that had previously supported the

**Figure 9.26** A normal fault, Ballygally Head, Co. Antrim. Tertiary basalt, overlying Chalk in depth, has been downthrown against Chalk from which most of the overlying basalt has been removed by erosion.

**Figure 9.27** Normal fault between Upper Old Red Sandstone on the downthrow side (to the left) and Lower Old Red Sandstone on the upthrow side (to the right). The fault has served as a passage-way for ascending solutions from which a vein of barytes ($BaSO_4$) has been deposited. North side of Carlingheugh Bay, two miles NE of Arbroath, Tayside Region, Scotland (*Institute of Geological Sciences*)

hillsides were removed, and there was much downward slipping along concave surfaces that began steeply on the heights and curved down to the horizontal at the level of the canal floor.

The walls of a clean-cut fault may be polished by friction between the moving blocks, and traversed by grooves or striations. Such surfaces are called *slickensides*. The striations serve to indicate the direction of the latest movements, but it must be remembered that any similar evidence of earlier movements may have been entirely obliterated. Major faults, however, are generally accompanied by a great deal of grinding and crushing of the wall rocks, and there may be a more or less broad fracture belt, known as a *fault zone*, consisting of lenticles and sheets of shattered rock. The latter is called a *fault breccia* when it is largely made up of angular fragments.

## Normal Faults

The vertical stress is the greatest of the three principal stresses. The resulting fracture is generally inclined at an angle between 45° and the vertical. The rocks abutting against the fault on its upper face or 'hanging wall' are displaced downwards relative to those against the lower face or 'footwall'. The terms 'downthrow' and 'upthrow' for the two sides are, of course, purely relative. In faults recently active the footwall is exposed at the surface as a fault scarp, and survey measurements sometimes show that both sides were uplifted relative to sea-level; one, however, being heaved up more than the other. In other cases the block on one side is found to have subsided. Normal faults involve an extension of the faulted beds. A boring through the fault would fail to penetrate some particular bed altogether, passing between its ruptured ends. This lateral separation indicates that freedom of movement was possible, not only upwards or downwards, but also in the direction of $P_3$. Such freedom would be facilitated if $P_3$ were a tensile stress, involving stretching instead of compression, a condition that results, for example, when an area is extended by upheaval.

Regions that are divided by faults into relatively elevated or depressed blocks are said to be *block faulted*. The face of the earth of today is diversified by many boldly preserved topographic features which are primarily due to vertical movements involving normal faulting. Such movements have been unusually active during late geological time,

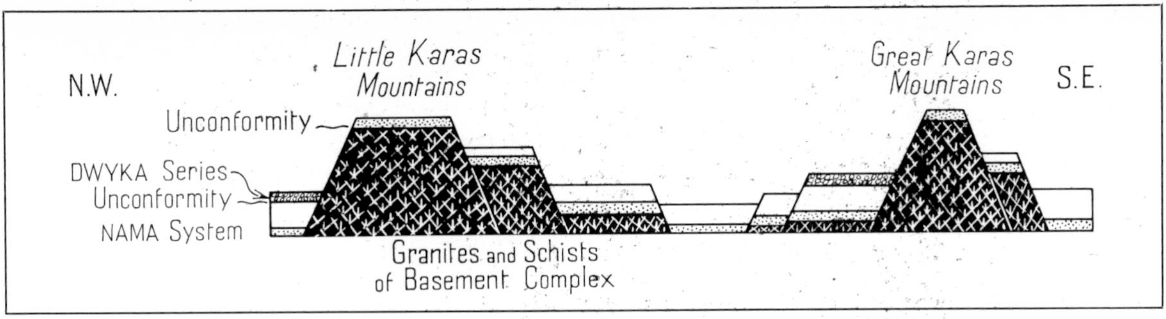


**Figure 9.28** Diagrammatic section to show the horsts and intervening rift valley or *graben* of the Kharas Mts, SW of the Kalahari Desert, South West Africa (Namibia). Minor modifications of the topography by denudation and deposition are omitted. Length of section nearly 100 km (*After C. M. Schwellnus*)

and they still are, right down to the present time. Fault scarps exposed at the surface (see Figure 25.1, p. 569) are being gradually worn back by the sculpturing hand of erosion, but long ages must elapse before such features as the block mountains illustrated in Figure 9.28 are levelled out. Upstanding fault blocks, which may be plateaus or long ridge-like block mountains, are called *horsts*. The Hercynian massifs of Europe, e.g. the Vosges and the Black Forest and the Harz Mountains, are horsts.

Fault blocks depressed below their surroundings form *fault troughs*, and here deposition of sediment from the adjoining uplands is favoured. A long fault trough—a tectonic valley bordered by approximately parallel fault scarps—is known as a *rift valley* or *graben*. Between the horsts of the Black Forest and the Vosges the Rhine flows through a rift valley (the *Rheingraben*, Figure 29.20). The river occupies the rift valley but did not excavate it. The most renowned systems of rift valleys are those which transverse the East African plateaus from the Zambesi to the Red Sea. These will be considered in chapter 29, but here it should be said that they may represent an early stage in the splitting apart of continental crust; a possible preliminary to the evolution of a new ocean.

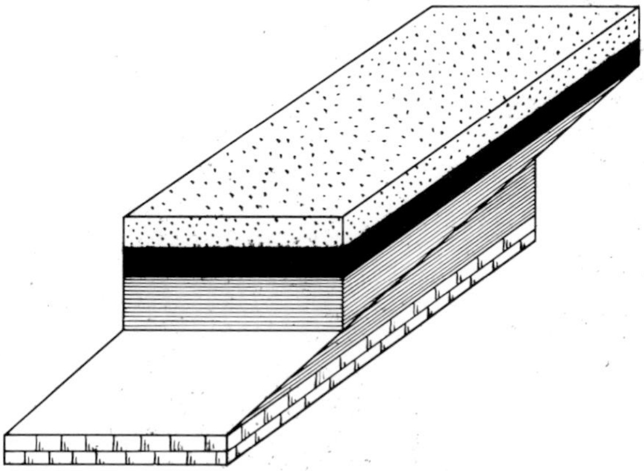

**Figure 9.29** Reverse or thrust fault

## Reverse or Thrust Faults

The vertical stress is the least ($P_3$) of the three principal stresses. When the resulting fracture is inclined at an angle between 45° and the horizontal, as it often is, the corresponding fault is described as an *overthrust*. High-angle thrusts or reverse faults are, however, far from rare. But whatever the angle of dip, the beds on the upper side are displaced *up* the fault plane relative to those below (Figure 9.29). Shortening of the faulted area is thus involved and the operation of compression seems to be implied. As already pointed out, overthrusts commonly develop along the middle limbs of recumbent folds. In the Alps and Himalayas and many other great mountain ranges overthrust folds can be seen on a spectacular scale (Figure 9.31), as well as nappes which have travelled forward for many kilometres from their original place of origin.

The celebrated Moine and associated thrusts of the North-West Highlands, where they mark the north-western boundary of the Caledonian orogenic belt in Scotland, are the classic examples of low-angle thrust faults (Figs 9.31 and 9.32).

**Figure 9.30** Reverse fault, Lebung Pass, central Himalayas. Carboniferous quartzites have been thrust over Permian shales. (*Geological Survey of India, Memoir xxiii*)

Between the major thrust planes high-angle thrusts are developed. At the base of the Glencoul thrust these are so closely packed as to give rise to what is called *imbricate structure*. Ridden over by a nappe of Lewisian Gneiss, the underlying Cambrian sediments have been sliced by reverse faulting into a large number of sloping thrust-sheets, which overlap like stacked roofing slates or tiles.

Along the sole of a major thrust severe crushing and grinding of the rocks is to be expected. The resulting product is a layer of hard, streaky or banded rock, consisting of pulverized materials which have been rolled out as if they had been passed through a rolling mill (Figure 9.33). Because of the highly characteristic structure and mode of origin, Charles Lapworth gave the name *mylonite* (Gr. *mylon*, a grinding mill) to rocks of this kind, which he was the first to discover.

**Figure 9.31** Section across the North-West Highlands of Scotland, showing the Moine and Glencoul overthrusts and the imbricate structure between the Glencoul trust plane and the underlying sole. Minor thrusts indicated by T

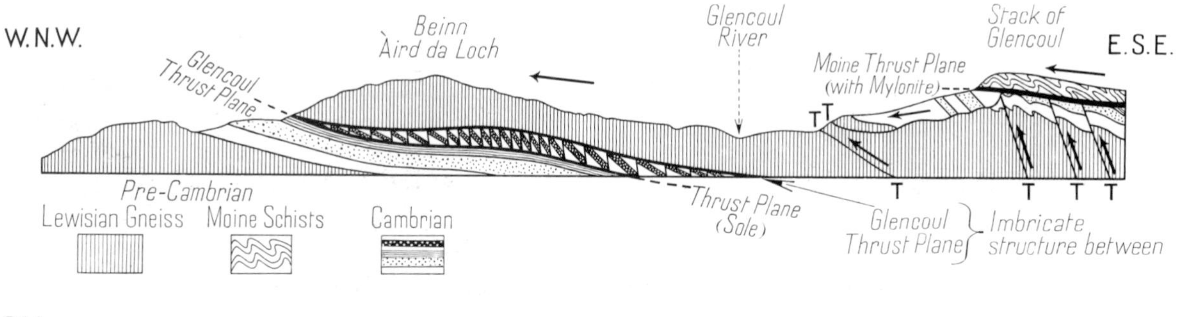

**Figure 9.32** Inclined sole of the Moine thrust exposed by stream erosion near Knockan, north-west Highland Region. Precambrian rocks above the stream on the right (Moine) have been thrust over Cambrian magnesian limestone. (*Institute of Geological Sciences*)

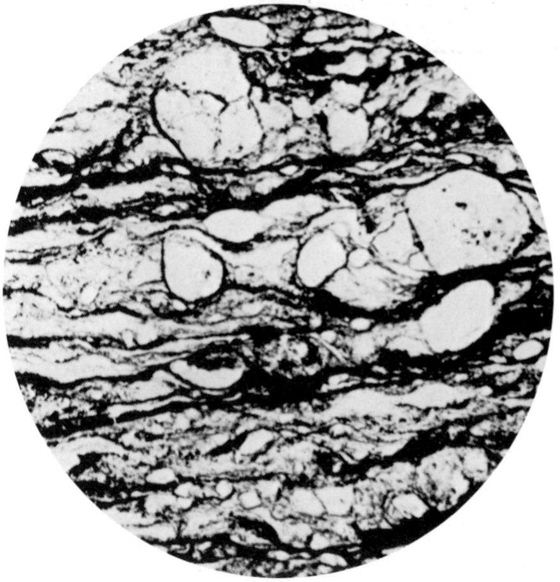

**Figure 9.33** Photomicrograph of mylonite formed from Lewisian Gneiss along a thrust plane near Laxford, North-West Highlands of Scotland (×32) (*J. J. H. Teall*)

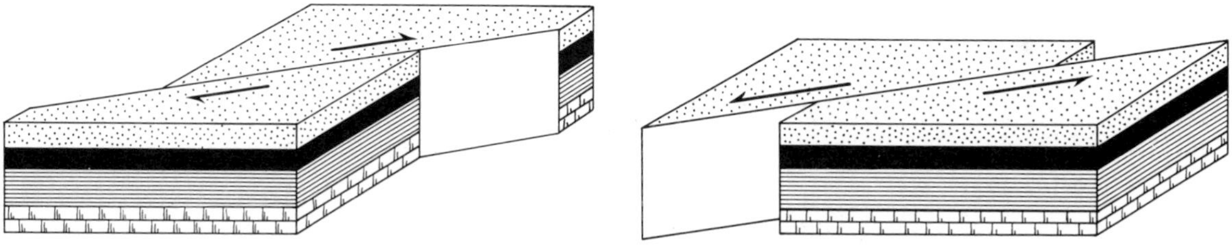

**Figure 9.34** Strike-slip or transcurrent faults: (*left*) right-lateral or dextral; (*right*) left-lateral or sinistral

## Strike-slip Faults

Faults known as strike-slip are characterized by movement essentially parallel to the strike of the fault-plane. The resulting movement is predominantly horizontal, and right-lateral or left-lateral, according to the rule indicated in Figure 9.34.

A variety of names for faults of this group have been fashionable in different countries, such as *transcurrent*, *wrench*, and *tear-fault*. In 1967 Tuzo Wilson recognized a completely new variety of strike-slip faults which he called *transform faults*. These faults are evolved either along plate boundaries or where new oceanic crust is formed, and were first recognized connecting apparently offset portions of mid-oceanic ridges. They are described in this connection on pp. 620–2. The San Andreas fault of California, The Alpine fault of New Zealand, and the Great Glen fault of Scotland are now thought to be major transform faults, the two former margining present-day plates.

The best known of all strike-slip faults is the great San Andreas fault of California, which has a known length of over 1200 km (Figure 9.35). The fault is really a fault zone, in places exceeding 1·5 km in width (Figure 9.36), consisting of nearly parallel or interlacing faults and intensely brecciated rocks.

A sudden right-lateral movement of only a metre or so along one of the constituent faults was responsible for the earthquake that wrecked San

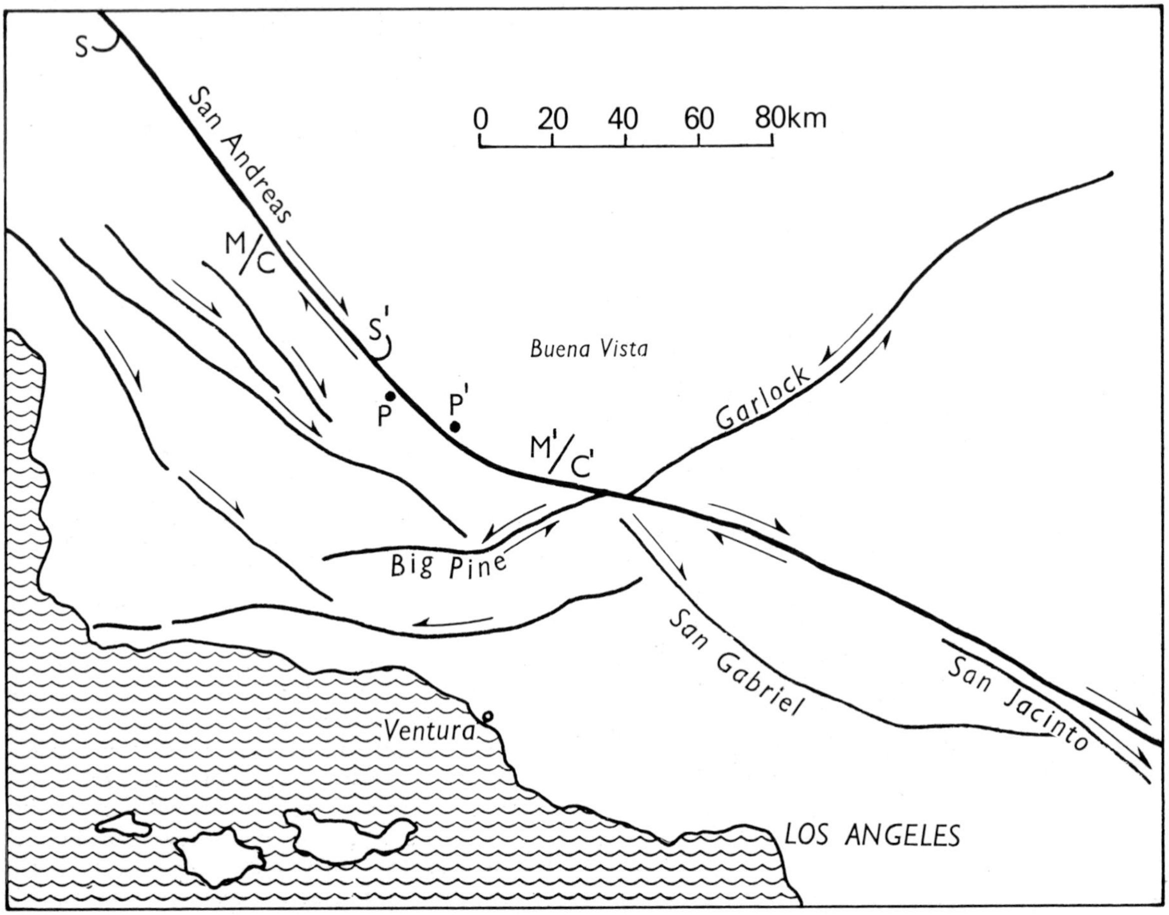


**Figure 9.35** The San Andreas and associated faults in southern California. Arrows indicate the directions of relative movement. Some cumulative displacements along the San Andreas fault zone are indicated by the following letters:
*Pleistocene*—P to P′ = 16 km: junction of shale-pebble gravels with granite-pebble gravels
*Upper Miocene*—M/C to M′/C′ = 105 km: M/C junction of blue-grey marine shales with red continental beds; M′/C′ same junction on NE side of fault zone
*Upper Miocene*—S to S′ = 105 km: outcrop of Margarita Shale

**Figure 9.37** *left* Looking NW along the San Andreas fault zone between the southern Temblor Range (visible in upper right) and the Caliente Range (out of sight below the lower left), SW corner of the Great Valley of California (*J. S. Shelton and R. C. Frampton, Claremont, California*)

**Figure 9.36** *above* Looking NW along the San Andreas fault zone in Cajon Pass. On the left (pacific side) the mountains consist of metamorphic rocks; those on the right are Miocene and older Tertiary sediments. (*J. S. Shelton and R. C. Frampton, Claremont, California*)

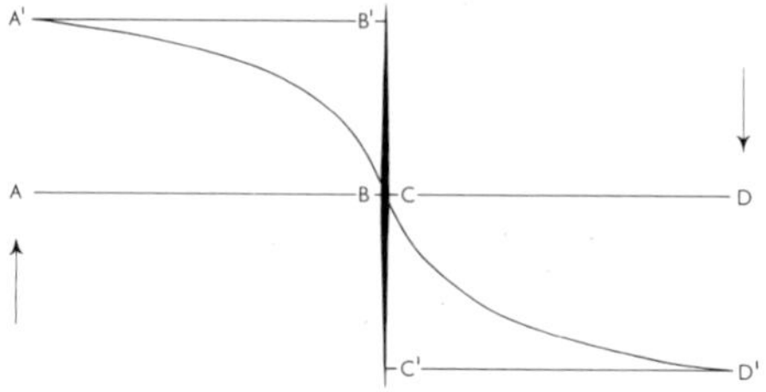


**Figure 9.38** Diagram to show how continuous lateral movement leads to intermittent earthquakes. Consider the deformation of any straight line ABCD across the fault. While A moves to A′, B moves to B′ until the strain is sufficient to overcome resistance to friction, fracture and movement within the fault zone. B then suddenly rebounds to B′, and C to C′. The resulting shock initiates an earthquake, which may be recorded all over the world.

Francisco in 1906. Roads, fences, and water and gas mains that crossed the fault were cut through and the ends displaced up to 6 m. On the Pacific side of the fault zone the direction of movement was north-westward, and on the continental side south-eastward. On its western side the Pacific plate moves north-westward, away from the east Pacific rise where new ocean floor forms at the southern end of the fault zone. On its eastern side

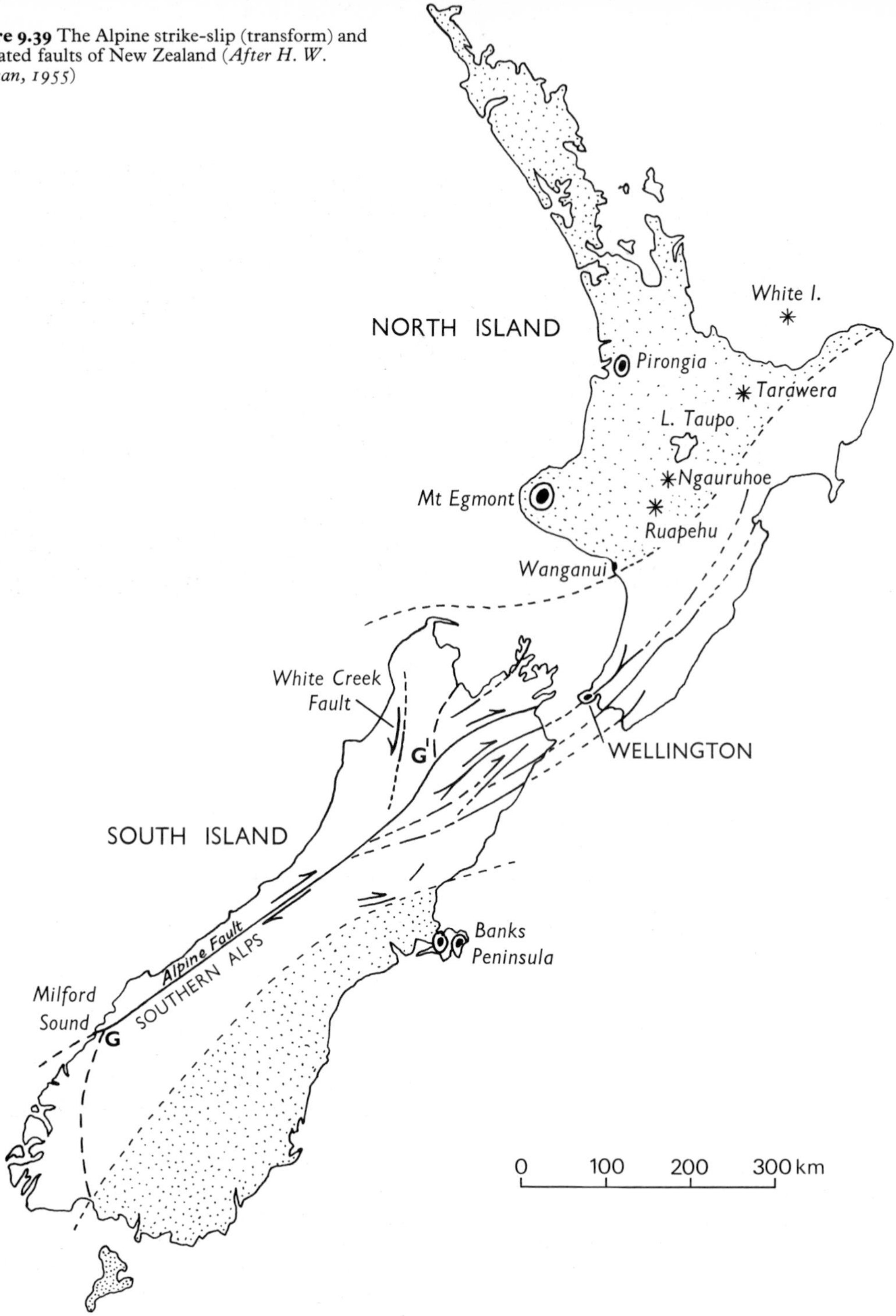


**Figure 9.39** The Alpine strike-slip (transform) and associated faults of New Zealand (*After H. W. Wellman, 1955*)

**Figure 9.40** The shatter belt of the Great Glen strike-slip fault, viewed from above Loch Oich, with Loch Ness in the distance (*Robert M. Adam*)

the North American plate moves south-eastward, away from the Gorda and Juan de Fuca ridges, where new ocean floor is evolved at the northern end of the fault zone (Figure 29.31).

Movements along the fault zone have been persistently right-lateral for a very long period—possibly many millions of years. This is established by measuring the cumulative displacements of easily recognizable geological features on both sides of the fault zone. Some of these are indicated on Figure 9.35. A distinctive sequence of Eocene strata also occurs near M′/C′, but its continuation on the other side is 360 km away. Displacements of Cretaceous and late Jurassic rocks amount to 515 and 560 km respectively.

The average rate of movement has been estimated in various ways. Seismic techniques indicate movement from 4 to 6 cm a year between the plates on either side of the fault-zone, but estimates made from the amount of displacement of stream courses, and easily recognizable rock formations, that can be related to earthquakes that have taken place during historical times, suggest movement of no more than 1·3 cm a year in northern and central California.

Systematic surveys have revealed that slow movement along the fault zone is continuously in progress. Sooner or later the growing strain of deformation now in progress will reach the elastic limit, and the sudden jerk of the resulting rebound (illustrated in Figure 9.38) will add yet another great earthquake to the hundreds of thousands that must already have occurred along this section of the boundary between the Pacific and North American plates.

Although most of the faults in the San Andreas fault zone strike approximately north-west, there are others, like the Garlock fault (Figure 9.35) that strike about east-west across the zone. A number of small upthrusts have been mapped in the region where the San Andreas 'rift' is deflected by the Garlock fault and its continuation on the western side (Figure 9.35). One of these minor thrust faults is known to be active at present. Oil wells in the

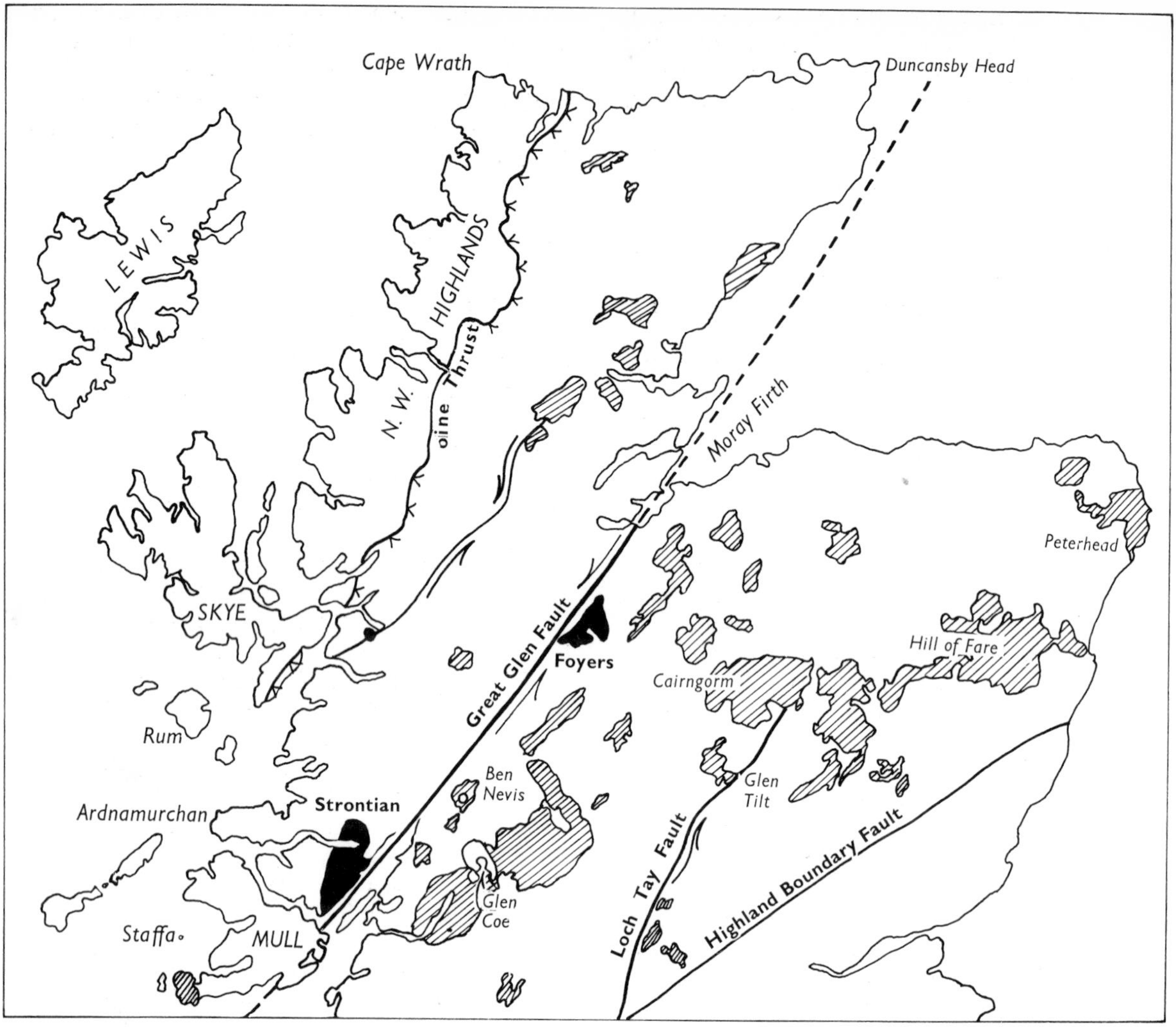


**Figure 9.41** Map of the Great Glen strike-slip fault (*After W. Q. Kennedy, 1946*)

Buena Vista Hills Oilfield had been inadvertently drilled through the fault, and by 1933 it was found that the well casings had been dislocated and cut. The ground above had crept up and the thrust plane and the forward edge of the advancing mass emerged at the surface as a low scarp. Between 1933 and 1947 the upper block advanced a further 30 cm, raising the height of the scarp by 18 cm. By 1957 the forward movement had reached 50 cm, equivalent to an average rate of over 3 m in a hundred years.

A few small thrusts are known on the western side of the San Andreas 'rift', but probably of more importance is the Ventura anticline (see Figure 9.35) for locality), which is remarkable for its phenomenal rate of growth. Since Middle Pleistocene time—that is within the last few hundred thousand years—Pleistocene sediments have been uparched by several hundred metres. Had it not been for exceptionally rapid erosion, which swept away the poorly consolidated sediments almost as rapidly as the crest of the anticline ascended, the region would by now have been part of a mountain range trending east and west.

Another great strike-slip fault, now thought to be a transform fault, is the Alpine fault of New Zealand, which forms part of the boundary between the Pacific and Indian plates (Figure

29.2). This extends along the north-west side of the Southern Alps (Figure 9.39). Although it has much in common with the San Andreas, it differs in having a greater vertical component of movement, and in sharing some of the features of a steep reverse fault (cf. Figure 3.10). In places old schists have been thrust over glacial moraines. But horizontal movement is dominant (right-lateral). All topographic features are cut through, and river terraces, deposited about 10,000 years ago, have been displaced 90 m horizontally and 15 m vertically.

The margin of the New Zealand geosyncline and a series of related structural features involving Jurassic and older rocks are cut off by the Alpine Fault in the south; they reappear on the other side of the fault 480 km away to the north-east (see G—G′, Figure 9.39). Thus the average rate of horizontal movement since the Jurassic (3 km per million years) has been of the San Andreas order.

A much more ancient strike-slip fault, still not entirely dead, cuts straight across Scotland from coast to coast. This is the fault zone of the Great Glen, a broad shatter belt of intensely crushed rocks, along which erosion has been relatively easy, so that it has become the site of a string of lochs (Figure 9.41). Horizontal displacement (left-lateral) of about 100 km is indicated by the various geological features that can be matched on the two sides (Figure 9.41). These include the Strontian and Foyers granite masses, which originated as a single pluton during the Caledonian orogeny. The vertical component of movement amounts to about 1·6 km near Inverness, where Middle Old Red Sandstone on one side comes against Moine schists on the other. Most of the displacement occurred during late Devonian and early Carboniferous times, but the fault zone is still a line of weakness. Its continued response to the straining earth is indicated by the fact that about sixty minor earthquakes have originated along the zone during the last two centuries.

The Walls boundary fault of Shetland is probably the continuation of the Great Glen fault to the NE. The continuation to the SW is represented by several parallel wrench faults cutting across Donegal. These follow the right direction, have left-lateral displacement and are of the right age. Across the Atlantic a narrow belt of similar faults, also of the same age, has also been traced from Newfoundland across the Cabot Strait and along the Bay of Fundy to Massachusetts. In 1962 Tuzo Wilson suggested that the whole belt, which he calls the Cabot fault, may be a further continuation of the Great Glen fault, and that it provides additional evidence that Newfoundland and the British Isles may once have been closely connected, that is, before the opening of the Atlantic ocean.

## Selected References


ANDERSON, D. L., 1971 'The San Andreas Fault', *Scientific American,* Freeman, San Francisco, vol. 225, No. 5, pp. 52–68.

ANDERSON, E. M., 1951, *The Dynamics of Faulting and Dyke Formation*, Oliver and Boyd, Edinburgh.

CAREY, S. WARREN, 1962, 'Folding', *Journal of the Alberta Society of Petroleum Geologists*, pp. 95–144.

CLOOS, E., 1955, 'Experimental analysis of fracture patterns', *Bulletin of the Geological Society of America*, vol. 66, pp. 241–256.

EDINBURGH GEOLOGICAL SOCIETY, 1959, *Geological Excursion Guide to the Assynt District of Sutherland* (with four coloured maps), Oliver and Boyd, Edinburgh.

HILL, M. L., and DIBBLEE, Jr., T. W., 'San Andreas, Garlock and Big Pine Faults, California', *Bulletin of the Geological Society of America*, vol. 64, pp. 443–58.

HILLS, E. S., 1963, *Elements of Structural Geology*, Methuen, London.

HUBBERT, M. K., and RUBEY, W. W., 1959, 'The role of fluid pressure in mechanics of overthrust faulting', *Bulletin of the Geological Society of America*, vol. 70, pp. 115–66.

KENNEDY, W. Q., 1946, 'The Great Glen Fault', *Quarterly Journal of the Geological Society of London*, vol. 102, pp. 41–76.

WELLMAN, H. W., 1955, 'New Zealand Quarternary tectonics', *Geologische Rundschau*, vol. 43, pp. 248–57.

— 1956, 'Structural Outline of New Zealand', *New Zealand Department of Scientific and Industrial Research, Bulletin 121.*

WILSON, J. TUZO, 1962, 'Cabot Fault, an Appalachian equivalent of the San Andreas and Great Glen faults and some implications for Continental displacement', *Nature*, vol. 195, pp. 135–8.

WILSON, J. TUZO, 1965, 'A new class of faults and their bearing on continental drift', *Nature*, vol. 207, pp. 343–7.

# Chapter 10

# Structural Features: Salt Domes and Plugs

And still the wonder grew *Oliver Goldsmith* (1728–1774)

## Piercement Folds and Diapirs

In the preceding chapter it was tacitly assumed that folding through great thicknesses of strata is made possible by thinning of the limbs and complementary thickening towards the hinge-lines. This is called *similar* or *parallel* folding, and it is depicted in an idealized form by Figure 9.18(*b*). In the actual conditions of the crust there are departures from this geometrical ideal towards two contrasted extremes according as the the thickening and thinning of a particular bed are (*a*) *less* or (*b*) *more* than is required for parallel folding.

(*a*) Strong competent beds of rocks like quartzite cannot readily change their thicknesses. If they are to retain their thicknesses unchanged, such beds must bend concentrically. However, any attempt to draw *concentric* folds soon shows that if one particular bed is folded in this way, the beds above and below cannot conform to the same pattern of folding while at the same time maintaining their original length (as they would do in parallel folding). Taking an anticline as an example, the bed below is too long to fit the concentric pattern; to shorten it involves thickening, which may be accomplished by thrust faulting, or flowage, or some combination such as small-scale folding within the bed. The bed above, however, is too short to fit the concentric pattern; to lengthen it involves extension by the opening up of fractures, or stretching and thinning by flowage. In a word, departures of type (*a*) in a particular bed require to be compensated by departures of type (*b*) in the beds above and below.

(*b*) Weak incompetent beds of rocks with low rheidity, like salt deposits and soft clays and marls, flow with ease and readily accommodate themselves to the requirements of any associated concentric folding. But this is not all. Such materials not only fill up the spaces beneath the more competent members of, say, a growing anticline, but where they come under sufficient stress they begin to burst out of their confined position by breaking-through the overlying beds, taking full advantage of cracks and fissures in the latter. The fold has now become a *piercement fold* or *diapiric fold*. If a structure of this kind were seen only below the level of the break-through, it would appear to be no more than a fold; but seen above this level, where the underlying core has pierced its roof, the structure is clearly an incipient intrusion.

The term *diapir* (through-piercing) was proposed in 1910 by the Rumanian geologist L. Mrazec (1867–1944) after his discovery in the Carpathians of elongated domes and anticlines with cores of salt that had locally broken through the crests or penetrated along faults. Such a fold becomes a *diapiric fold* after the mobile core has pierced the stronger mantling rock, but if the salt has continued to develop into a plug-like intrusion the structure is called a *diapiric intrusion* or, briefly, a *diapir*. *Diapirism*, a group name for the geological processes involved, is the development of an intrusion from a geological formation of low rheidity (p. 131) which comes under sufficient stress to deform, pierce and break through rocks of higher rheidity. There are many kinds of diapirs, ranging from pingos (made from ice, p. 277) and intrusive mudstones (Figure 10.1) to granite plutons in great variety (p. 125); but, thanks to the exploring activities of oil companies, by far the

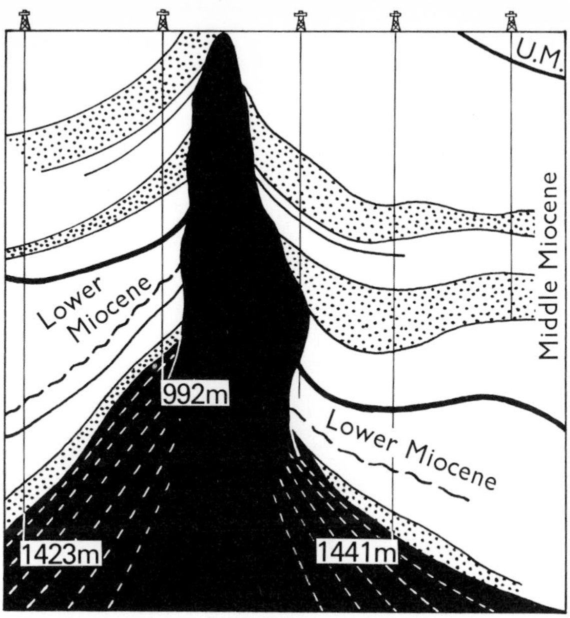


**Figure 10.1** Diapir breaking through to the surface from an anticlinal core of Upper Oligocene shale and mudstone. Depths (in metres) penetrated by some of the oil wells that have revealed the structure are given to indicate the scale (Horizontal = vertical). Barrackpore oilfield, Trinidad (*Simplified after G. E. Higgins, 1955*)

best known are those that ascend from thick deposits of salt and other evaporites. Salt diapirs are conventionally called *salt domes* in reference to the form of the overlying strata which they have uplifted, and which in consequence may reveal their presence underground. But since diapirs of salt may be columns 2 km or more across that have ascended from depths of several kilometres, they are referred to as *salt plugs* or *salt stocks* when their three-dimension structural forms are under discussion.

In the Carpathians, and elsewhere in orogenic belts where salt domes were first found, the motive force necessary for their development was ascribed to the compression thought to have been responsible for folding the associated rocks. However, after the significance of the celebrated oilfield around the Spindletop salt dome in SE Texas was fully realized, hundreds of salt domes were discovered in the thick sediments of the Gulf Coast States and the adjoining sea floor. Throughout this great salt-dome province the strata dip gently seawards, but nowhere do they show any sign of orogenic folding. The only folds are those occurring over the tops or around the flanks of the domes or plugs. Here, then, instead of a diapiric intrusion being a break-through resulting from excess of folding, it is the other way about: the folding has resulted from the active ascent columns of salt. Obviously, then, there are circumstances in which salt diapirs can develop without the aid of tectonic forces other than gravitation.

## Diapirism

As early as 1912 Arrhenius pointed out that if the average density of the surrounding strata is higher than that of the intrusive salt, an upward force of buoyancy would be acting on the latter, tending to lift it towards the surface. The density difference has been amply confirmed by the discovery of strongly negative gravity anomalies over actual salt domes. Such anomalies imply underlying deficiencies of mass (i.e. the presence of material of relatively low density) extending downwards to great depths, sometimes for many kilometres. Given a long column of salt the buoyancy effect might be adequate to promote further upward movement, but there remained the apparently insuperable difficulty of understanding how a salt dome ever got itself started, since the initial stress differences could only have been trivial in the absence of tectonic squeeze.

Let us consider Figure 10.2(*a*) illustrating a thick layer of salt overlain by an 'overburden' of sediments having an average density somewhat higher than that of salt. Freshly deposited sediment has a lower density than salt, but compaction during burial beneath younger sediments increases the density and brings it into equality with that of salt when the cover has reached a thickness of about 450 to 600 metres. As subsidence and sedimentation continue, the deeper sediments become steadily denser than salt.

Like much of the sea floor of today, the surface between salt and overburden is likely to have had considerable irregularities here and there, either fom the start or at some stage during the deposition of the sedimentary overburden (Figure 10.2(*a*)). Probably faulting would be a common cause of marked differences of level, but folding and slumping might also contribute. Suppose there is a projection of salt at B, relative to the surrounding surface AA. The projection may be a ridge or a mound. Because of the density differences the pressure $P$ at A is greater than the pressure $p$ at B. If the pressure difference $P-p$ is sufficient to overcome the resistance, then salt will begin to

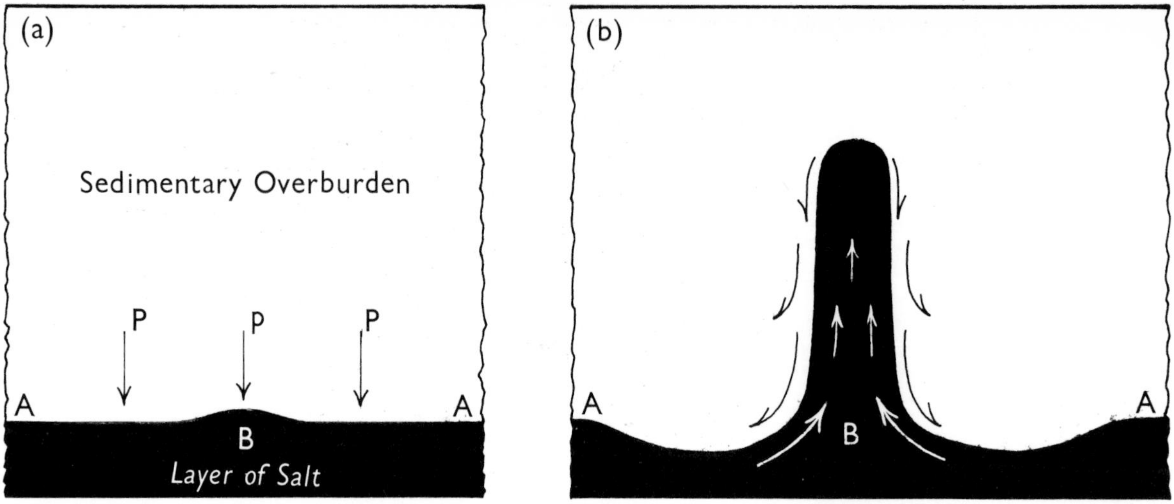


**Figure 10.2** Illustrating the development of a salt plug from a pressure difference, $P-p$, operating
(*a*) on a roughly circular protrusion at B
(*b*) with eventual production of a ring-shaped depression AA (the 'peripheral sink') around B

flow from A towards B, so that an initial mound tends to grown into a dome, and eventually into a plug. If B represents a ridge the latter develops into a salt anticline, part of which is likely to grow more rapidly than the rest, so becoming a plug on an anticlinal foundation (Figure 10.6).

The pressure difference is opposed by (*a*) internal friction in the salt itself; (*b*) friction between salt and overburden; and (*c*) resistance offered by the overburden to updoming, fracturing and gliding along bedding planes. The list seems to be a formidable one until it is remembered that both salt and the sediments through which it intrudes have relatively low rheidity. This means that they readily become mobile in response to long continued stress differences. The viscosity of rock-salt falls between $10^{16}$ and $10^{18}$, according to the temperature and water content, while the rigidity averages about $10^{11}$. The rheidity of salt is therefore about $10^8$ to $10^9$ seconds (as defined on p. 131), i.e. between 3 and 30 years. Consequently salt behaves as a viscous fluid in response to stress differences acting for periods of this order or longer. For gypsum and anhydrite the corresponding periods have a wider range, being shorter for wet gypsum and longer, up to 1000 years or so, for dry anhydrite. However, all these periods are brief compared with the many millions of years that are available. Accordingly the materials concerned eventually behave like liquids, following the principles of fluid mechanics, just as a layer of oil covered by a layer of water would do—only very much more sluggishly.

An inevitable result of salt flowage towards B is that the layer of salt becomes thinner in the surrounding region. If B grows into a plug, the latter is necessarily surrounded by a ring-shaped depression. This annular trough is technically called a *peripheral sink*; its volume at any stage is equal to that of the salt plug that has developed from the initial mound. Thus adequate space is available to accommodate the displaced sediments of the overburden. These glide into the peripheral sink and thus acquire a synclinal form which is described as a *rim syncline* (see Figures 10.1 and 10.2 (*b*)).

It is important to notice that the volume of material concerned in the diapiric mechanism remains unchanged. As suggested by Figure 10.1, there is merely an exchange of position between salt and sediments. Lighter material (salt) ascends as a diapir, while simultaneously the heavier displaced materials (compacted sediments) move towards and downwards into the peripheral sink. The behaviour is like that within a convection cell (see Figure 2.15, p. 23). Here, however, we have a 'cycle' operated by gravity that, in the nature of the case, cannot be completed; the overturn of material is necessarily limited to a half turn or very little more. An actively growing plug may reach or overshoot the half-turn stage in two ways.

If a salt plug rises from so great a depth that its buoyancy or head of pressure has become sufficient to carry it above the surface it will not

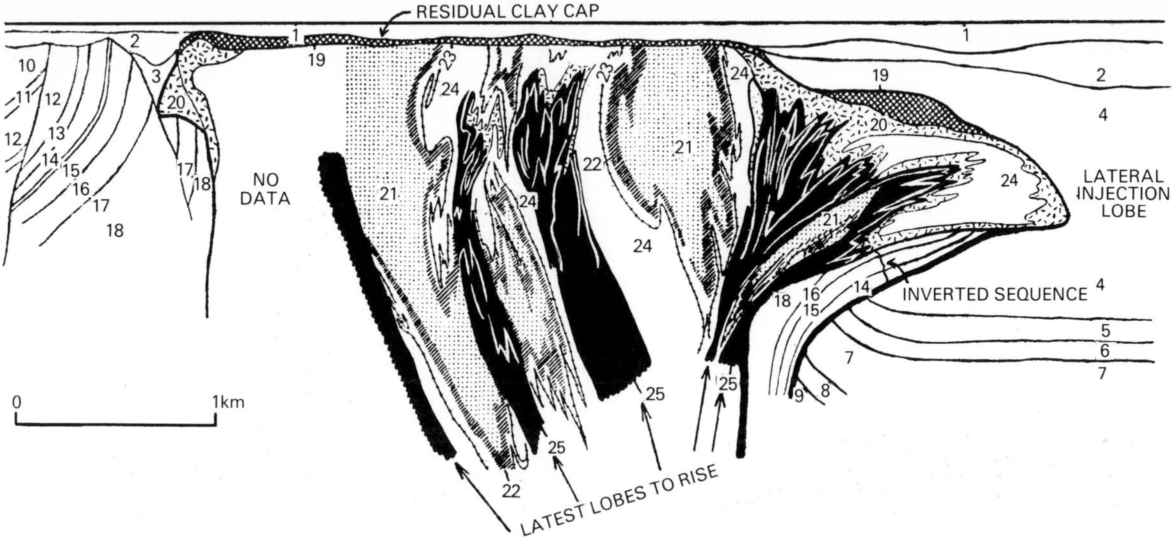


**Figure 10.3** Heide salt plug, near Hanover, W. Germany, showing lateral expansion at its head, with overturning and thrust faulting of the surrounding sediments, and intrusive lobes of older salt (black) into a younger saline series. 1. Recent and Pleistocene 2. Tertiary 3–9. Cretaceous 10–11. Jurassic 12–17. Triassic 18. Permian 19. Residual clay cap-rock 20. Anhydrite cap-rock 21–4. Younger saline series: Upper rock-salt (21); Red salt clay (22); Potash bed (23); Lower rock-salt (24) 25. Older rock-salt. The highly complex structures have been revealed by borings and mining operations. (*A. Benz, 1949*)

necessarily do so. By the time it reaches a level where the density of the surrounding sediments is lower than that of the salt, it may well be easier to spread along bedding planes and lift the overlying sediments. This will involve less work—expenditure of energy—than forcing its way upwards as a plug. For this reason sills of salt are sometimes encountered, but they do not generally proceed far from the feeding plug. More commonly the plug spreads out laterally to form an overhanging head, like a lop-sided mushroom. Sediments which might have ended steeply or vertically against a cylindrical plug may be turned back on themselves by the overhang (Figures 10.3 and 10.4).

Penetration of the surface is most easily achieved in regions where the lighter sediments have already been removed by denudation. Then, if the rainfall be not too great, a dome-shaped mountain of salt (with or without gypsum etc.) may be built up high above the surrounding country, until the salt is enabled to flow away under its own weight as a 'salt glacier'. Conditions of the kind described are characteristic of the salt-dome province of southern Iran, Oman and the Persian Gulf (Figure 10.5).

The salt plugs illustrated by Figures 10.3 and 10.4 are unconformably covered by Tertiary and later deposits, showing that upward growth came to an end long ago. This suggests that the supply of salt in depth became exhausted, as it would do if the peripheral sink reached the floor of the mother bed of salt. The remarkable fact that out of nearly 200 salt domes located around the Persian Gulf at least 80 still have outflowing 'salt glaciers' implies that the mother bed, which is here of Cambrian age, must be immensely thick. The plugs themselves have an average diameter of about 6 km. They break through Cretaceous and Tertiary limestones, which are sharply upturned around the margins. The salt carries up with it immense numbers of inclusions, large and small, of Cambrian and other strata and also of igneous rocks. The

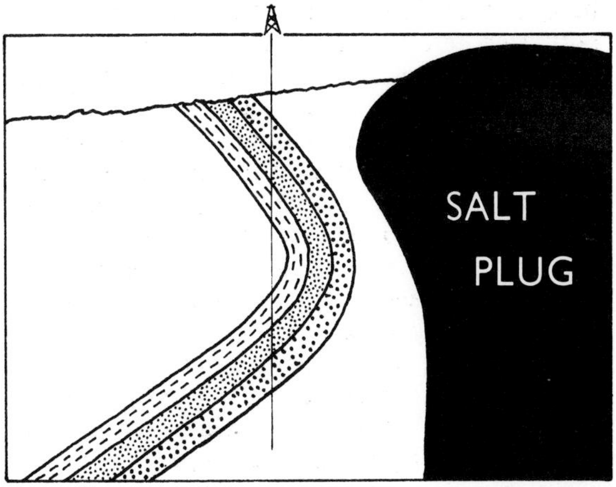


**Figure 10.4** Overturning of country rocks by a salt plug with an expanded head, San Filipe Dome, Texas (*After M. Y. Halbounty and G. C. Harding Jr, 1954*)

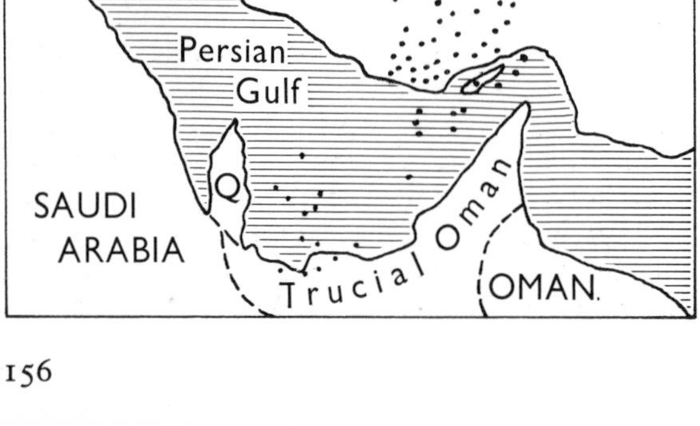


**Figure 10.5** (*above*) A mountain of salt 10 km across and 1310 metres high. Kuh-i-Namak salt dome, Iran, showing small outflowing 'salt glaciers'. The contact between salt and surrounding Cretaceous limestone is overturned in places. Steep outward dips of 85° are common, falling to 40° or 30° about 1·5 km from the contact. (*British Petroleum Company Limited*)

(*left*) Index map of the Persian Gulf province of salt domes which have ascended from a Cambrian mother-bed of evaporites. Miocene evaporites are the source of more complex forms of salt diapirism in the region around Asmari Mt (Fig 9.16) to the NW.

drilling cores from oil wells have revealed that some of the plugs formed islands in an Upper Cretaceous sea, to which they contributed sediments derived from the insoluble inclusion. Evidently then, as now, much more salt ascended than was necessary just to reach the surface. The present-day salt mountains reach heights of over 1200 m above the surrounding plains, at which level slow ascent compensates for the salt carried away in solution by the rainfall.

During Upper Permian times the North Sea formed part of an essentially land-locked sea which extended from Britain, across Germany to Poland. The German and Polish evaporites, deposited in this natural evaporating pan, have already been mentioned. The evaporites beneath the North Sea have played a dual role in the economic sense. They have formed an impermeable seal through which natural gas, accumulating within Lower Permian sandstone, has been unable to escape. Secondly, through their flowage into innumerable diapirs of various shapes and sizes, they have imparted movements to their covering rocks, and provided catchment zones for oil.

## True-to-Scale Model Experiments

All the leading features of salt domes and plugs have been remarkably well reproduced by experiments with approximately true-to-scale models designed to behave as nearly as possible as the natural beds of salt and sediment have done. Engineers have long been familiar with the advantages of making models that are dimensionally true-to-scale for testing and modifying their designs for large-scale constructions, e.g. of ships, aeroplanes and harbours. In recent years geologists have adopted similar methods in order to investigate the mode of origin and development of various tectonic and intrusive structures. In laboratory experiments designed to imitate in a short time the large-scale geological processes which extend over long periods it is essential that the properties of the materials employed to represent rocks should match the reductions in size and time. For example, if 50 km of country are to be represented in the model by 1 metre, the reduction in the dimension of length is 50,000 to 1. The strength of the materials used to represent the actual rocks must be scaled down in the same proportion: that is to say, they must be very weak.

In his fascinating paper on *Strength of the Earth* M. King Hubbert has demonstrated this point by considering that a man's ability to jump depends both on his weight (i.e. on his volume, or length cubed) and on his strength (i.e. on the cross-sectional area of his muscles, or length squared). In order that a model should be capable of performing the appropriate jump, the ratio of strength to weight must be kept constant. In a miniature working model of a man reduced from, say, 1000 to 1 in the dimension of length, the weight would be reduced by $1/1000^3$, but the strength would be reduced by only $1/1000^2$. Our minute Lilliputian would thus be 1000 times as strong as a man in relation to his weight and, like a flea, would be able to jump 1000 times as high. Clearly, in order to ensure the same relative performance as before, the materials used in the model would need to be only 1/1000th as strong as the materials of a man—the same reduction as in the length dimension.

The viscosity reduction in true-to-scale models has to be much greater than necessary for strength, because time is involved dimensionally as well as length. The viscosity of the materials used must be scaled down by the product of the length and time reductions (and also by the density reduction, but this is of minor importance). If a million years' natural growth of a salt plug is to be carried out in 1 hour, then the time reduction is from about $10^{10}$ to 1. With the same length reduction as before, 50,000 to 1, the viscosity reduction becomes $5 \times 10^{14}$ to 1. Allowing a factor of 2 for the density reduction, the viscosity of salt, $10^{17}$ to $10^{18}$, has to be reduced by $10^{15}$, i.e. to between 100 and 1000. Both glycerine and golden syrup are more viscous than this at ordinary temperatures.

It will be realized from the above discussion that it is not always easy to find suitable materials for use in experimental models. Viscous liquids were used in a long series of experiments carried out by L. L. Nettleton in 1936. It was not only shown that intrusion forms like salt domes and plugs could be faithfully reproduced, but that they developed from surprisingly small initial irregularities. Nettleton's results were later confirmed by other workers; notably T. J. Parker and A. N. McDowell, who used materials capable of fracture as well as flow, so that they were able to reproduce the characteristic fracture patterns that develop around and above growing salt plugs. They used weak muds for the sediment and asphalt for salt.

Experimental evidence suggests that the diameter of the plug that eventually develops is about the same as the original thickness of the mother

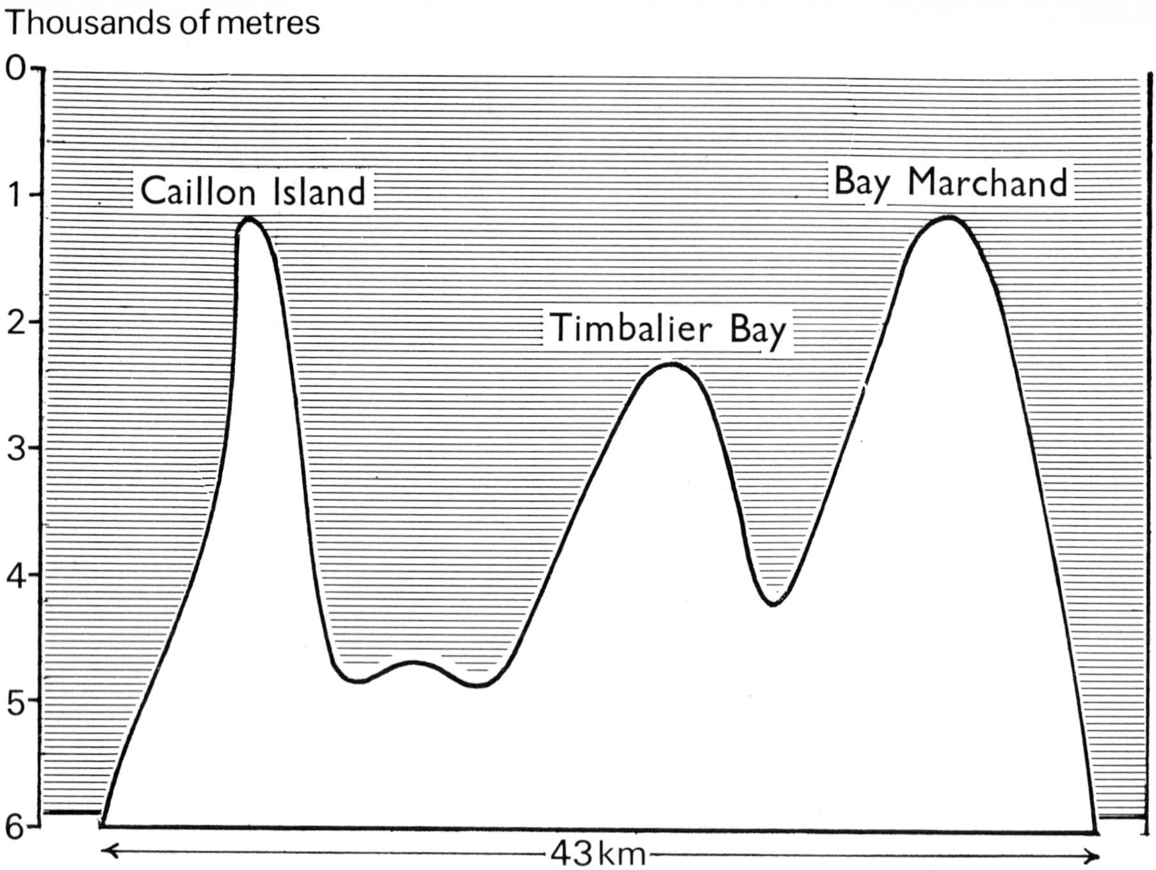


**Figure 10.6** Section through three salt plugs, coastal Louisiana, about 16 km west of the Mississippi Delta. Deep drilling has revealed that the plugs unite at a depth of about 4900 metres, below which the salt extends downwards as a large anticlinal structure. Down to 6000 metres the volume of diapiric salt is about 1105 cubic kilometres.

bed, and that the peripheral sink formed in the latter reaches a diameter about 6 to 8 times that of the plug. Moreover, as will be clear from Figure 10.2(*b*), the additional weight of heavy sediment on the salt underlying the peripheral sink not only increases the inward flow of salt into the growing plug, but also tends to start an outward flow of salt towards regions where the overlying pressure may be relatively low. New satellite or secondary domes then begin to grow, provided, as always, that the mother bed of salt is sufficiently thick to maintain the supply necessary for continued growth. All these and many other features elucidated by experiments have been found to conform to the actual findings of the oil companies in the course of their detailed exploration of salt domes and plugs by drilling and by geophysical methods. For example, Spindletop Dome is about 1·5 km in diameter; its peripheral sink is about 13 km across; and here and there satellite salt domes rise from the rim of the peripheral sink.

It should be added, however, that Spindletop is itself only a relatively small example of the 200 or more salt domes that have now been discovered in the Gulf Coast province. Some have spread laterally to as much as 11 km across, near the top; the average diameter of the more cylindrical forms is between 3 to 5 km. This suggests a thickness of 3 to 5 km for the deeply buried source of salt that underlies more than 259,000 $km^2$ of coastal plain and sea floor. Only on the outskirts of this province has the salt formation been reached and penetrated (p. 90). In Louisiana, where it has not yet been reached by drilling, it is suspected to have a thickness of about 5 km and to be covered by 15 or 16 km of sediments of Jurassic and later ages. Figure 10.6 illustrates one item of evidence that proves the existence of salt deposits on a scale that would have been considered incredible only a few decades ago.

## Cap Rock

As already mentioned, where salt domes have reached the surface the solvent effect of rainfall is to leave behind a mantle of insolubles. However, long before reaching the surface, a plug may encounter percolating ground-waters, which have a similar effect. Apart from any concentration of dragged-up wall-rocks, the evaporite minerals lose more rock-salt than anything else by solution, while anhydrite (or gypsum, to which it is commonly altered) and insoluble impurities such as clay form a mantle, which is called the *cap rock*, over the core of salt. Many of the cap rocks of Texas and Louisiana have been found to contain sulphur deposits, locally on a large enough scale to be of great commercial value. The sulphur is associated with a cavernous calcite rock ('lime-stone') and it has been generally supposed that petroleum, trapped against the flanks of a growing dome, reacted with $CaSO_4$ to produce $CaCO_3$ and sulphur. Experiments carried out at the Lamont Geological Observatory, New York, show that sulphate-reducing bacteria are the primary agents in the production of cap-rock sulphur. Wherever oil and a little water come into contact with sulphate minerals, such bacteria flourish by abstracting oxygen from $CaSO_4$, which is thereby reduced to $H_2S$ and CaO. Sulphur is produced by reaction between $H_2S$ and $CaSO_4$, with water and CaO as by-products. Omitting all the intermediate stages and their products, the sulphur-liberating process may be summarized as:

| $CaSO_4$ | $-\ 3O$ | $+\ CO_2$ | $=$ | $S$ | $+\ CaCO_3$ |
|---|---|---|---|---|---|
| *anhydrite or gypsum* | *abstracted by bacteria* | *supplied by bacteria* | | *sulphur* | *'lime-stone'* |

## Diapiric Roof Structures

Before leaving salt domes it is appropriate to refer here to the potential fissures, represented by dykes and faults, that may develop in the overburden above and around the crest of a rising dome.

The overlying sediments are necessarily subjected to a stretching force as the salt-dome rises. As a result radial and concentric or elliptical fractures appear at the surface above the crest of the dome (Figure 10.7). A few of the radial fractures become dykes of salt, while the rest fail to open because of the pressure exerted against their

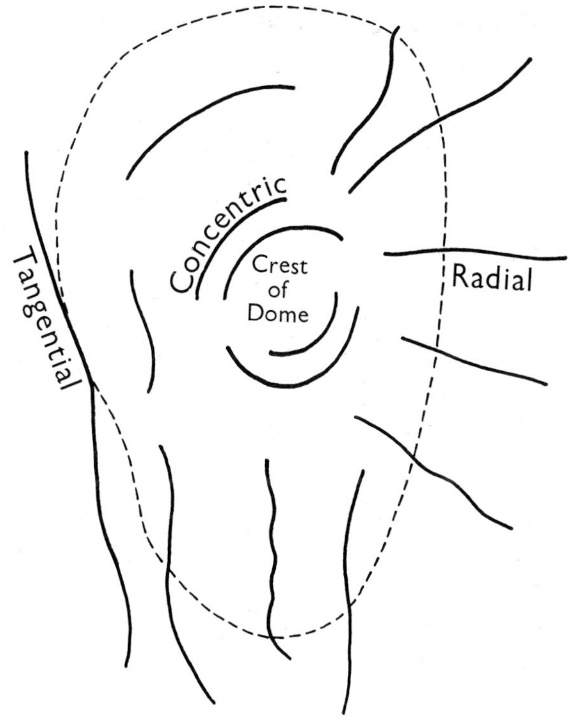


**Figure 10.7** Types of potential fissures developed in the overburden of a growing salt dome.

walls by neighbouring dykes, which continue to increase in width and length by a wedge-like process of growth. Dykes and sills of salt are especially notable in the salt province of SW France. Here, in front of the northern slopes of the Pyrenees, domes and long anticlinal arches of salt, some of them nearly 5 km across, have risen from evaporite formations over 2000 m thick through a sedimentary overburden of over 9000 metres. Similar conditions are found in Algeria, in front of the southern slopes of the Atlas Mountains.

Currie studied experimentally, simultaneous deposition and deformation during the development of a graben, above a rising salt dome. The concentric or elliptical fractures referred to represent the graben faults at the surface. Currie used a sectional model apparatus, consisting of two parallel glass plates two inches apart. Between the plates the crest of a rising salt dome was represented by a dome mounted on a screw, above which powdered material of different colours was deposited to represent stratified beds. As the domed top of the screw, driven by a motor, rose,

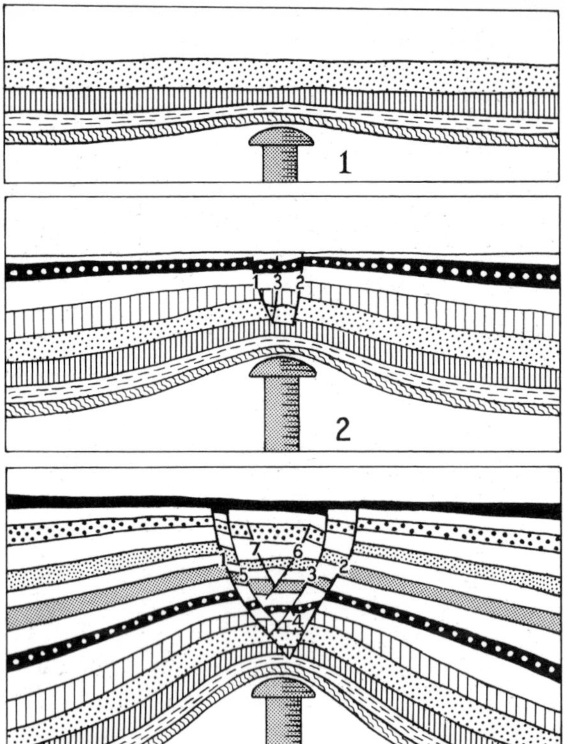


**Figure 10.8** Experimental true-to-scale sectional model illustrating how updoming of sedimentary strata, through the uprise of a deep-seated salt dome with concomitant sedimentation, results in graben formation. In the model the radius of curvature of the simulated dome was 3 inches (7·6 cm).

1 A layer of undeformed sediment deposited prior to updoming.

2 Further deposition with concomitant updoming has resulted in two faults (1 and 2) bounding a graben, within which a third fault appears.

3 With further deposition and updoming the graben widens and additional faults form within it.

4 Early formed faults 1, 2 and 5, bounding the graben, have become inactive. The general sequence of faulting progressing from the older faults margining the graben, to younger faults (10 and 11) in the middle.

*(Drawn from J. B. Currie's (1965) photographs of his experiment)*

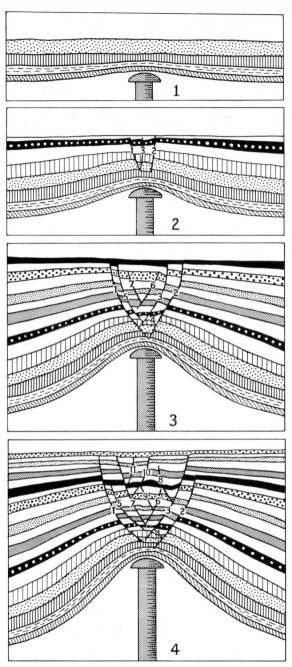

successive beds of powder were deposited. Figures 10.8 are drawn from photographs of one of Currie's experiments. Deposition of the stratified layers exceeded the rate of domical uplift by a factor of about 1·75. In Figure 10.8(1) an underformed bed, present before the beginning of the updoming, can be seen. Figure 10.8(2) represents an early stage of graben formation; two fractures, dipping inwards, define the margin of the graben. As the dome rose the graben increased in size (Figure 10.8(3) and (4)), and the number of faults increased. The older faults are located at the outer margin of the graben, and the younger faults towards its middle.

## Other Sedimentary Intrusions

Mudstone diapirs have already been mentioned and they will come up for further discussion later, as the feeders of mud volcanoes (p. 292). They also serve as the feeders of mudstone sills and dykes which originate in the same way as the sills and dykes of salt associated with salt plugs. All these minor intrusions occupy space that was made available by moving apart of the walls in the case of dykes, or of roof and floor in the case of sills. Here the space problem has been solved by *dilation*, i.e. the separation of boundary surfaces that were originally in contact.

There are many other examples of sedimentary dykes which occupy ready-made fissures and are therefore of dilation type. These can be of two main kinds. Fissures that open on the floor of a sea or lake, or wherever there is a cover of loose sediment, are likely to become filled by sediment that falls in from above. Around the Bristol Channel, for example, dykes of Triassic and Jurassic sandstones occur in the Carboniferous Limestone. The former distribution of Ordovician and Devonian sandstones over a wide area around Lake Chad is revealed by the occurrence of fossiliferous sandstone dykes in the Precambrian rocks south of the Sahara. Other sedimentary dykes (Figure 10.9) have been formed by the infilling of earthquake fissures from below, sand and silt having been carried up by outbursts of ground-water.

**Figure 10.9** Sandstone dyke, on the shore near Corriegills, Arran. (*T. M. Finlay*)

## Selected References


ATWATER, G. I., and FORMAN, McL. J., 1959, 'Nature and growth of Southern Louisiana salt domes, *Bulletin of the American Association of Petroleum Geologists*, vol. 43, pp. 2592–622.

BALK, R., 1949 'Structure of Grand Saline Salt Dome, Texas', ibid., vol. 33, pp. 1791–829.

— 1953, 'Salt structure of Jefferson Island Salt Dome, Louisiana', ibid., vol. 37, pp. 2455–74.

CLARK, J. A., and HALBOUTY, M. T., 1952, *Spindletop*, Random House, New York.

CURRIE, J. B., 1956, 'Concurrent deposition and deformation in development of salt-dome graben', *Bulletin of the American Association of Petroleum Geologists*, vol. 40, pp. 1–16.

DUPOUY-CAMET, J., 1953, 'Triassic diapiric salt structures, south-western Aquitaine basin, France', ibid., vol. 37, pp. 2348–88.

FEELY, H. W., and KULP, J. L., 1957, 'Origin of Gulf Coast salt-dome sulphur deposits', ibid., vol. 41, pp. 1802–53.

HARRISON, J. V., 1930, 'The geology of some salt plugs in Laristan', *Quarterly Journal of the Geological Society of London*, vol. 86, pp. 463–522.

JONES, R. W., 1959, 'Origin of salt anticlines of Paradox Basin', *Bulletin of the American Association of Petroleum Geologists*, vol. 43, pp. 1869–95.

KENT, P. E., 1958, 'Recent studies of South Persian salt plugs', ibid., vol. 42, pp. 2951–72.

NETTLETON, L. L., 1955, 'History of concepts of Gulf Coast salt-dome formation', ibid., vol. 39, pp. 2373–83.

O'BRIEN, C. A. E., 1957, 'Salt diapirism in South Persia', *Geologie en Mijnbouw*, The Hague, vol. 19, pp. 357–76.
(This number, September 1957, also contains papers on salt diapirs in Algeria.)

PARKER, T. J., and MCDOWELL, A. N., 1955, 'Model studies of salt-dome tectonics', *Bulletin of the American Association of Petroleum Geologists*, vol. 39, pp. 2384–470.

WALTON, M. S., Jr., and O'SULLIVAN, R. B., 1950, 'The intrusive mechanics of a clastic dike', *American Journal of Science*, vol. 248, pp. 1–21.

# Chapter 11

# Structural Features: Igneous Intrusions

Long is the way
And hard, that out of hell leads up to light.

*John Milton* (1608–74)

## Dykes and Sills

The forms and attitudes of igneous intrusions largely depend on their relation to the parting planes of the invaded formations. This is seen most clearly where the strata have remained horizontal or have been only gently tilted or folded. One of the commonest signs of former igneous activity is the familiar *dyke* or *dike* (Figures 3.15 and 11.1).

**Figure 11.1** North Star dyke, between Ballycastle and Fair Head, Co. Antrim. A Tertiary dyke, pointing due north. Rathlin Island in the distance. (*R. Welch Collection, Copyright Ulster Museum*)

Here the magma has ascended through an approximately vertical fissure, so that on cooling it became a vertical sheet of rock with roughly parallel sides cutting across the bedding planes. Such intrusions are said to be *transgressive* or *discordant*. If, however, the magma finds a channel of least resistance along a bedding plane, it will consolidate to a *concordant* intrusion with roof and floor parallel to the adjacent stratification. The *sill* (Figures 3.16, 11.7 and 11.8) is the type example, but it should be noted that few sills remain concordant throughout their extent. Like the Whin Sill (Figure 11.6) they commonly break across the bedding at intervals—locally becoming transgressive—to continue concordantly at a higher level. There may be some confusion in terminology where an intrusion happens to follow the bedding planes of beds that were vertical or steeply inclined before the time of intrusion. The essential distinction is that at the time of intrusion a dyke had walls, whereas a sill had a roof and floor.

When a dyke is more resistant to weathering and erosion than its wall rocks (as exposed at the surface), it projects as a more or less prominent ridge. On the other hand, if the dyke is less resistant it is worn away (e.g. by sea waves on the foreshore) into a narrow trench or cleft. In its ordinary usage the word 'dyke' may refer either to a ditch or to a wall (Figures 11.2 and 11.3).

Dykes vary greatly in thickness, from a few centimetres to many metres, but widths of two to seven metres are most common. There is also great variation in length, as seen at the surface, from a few metres to many kilometres. Dykes are very numerous in some regions of igneous activity. Along a 24-kilometre stretch of the coast of Arran,

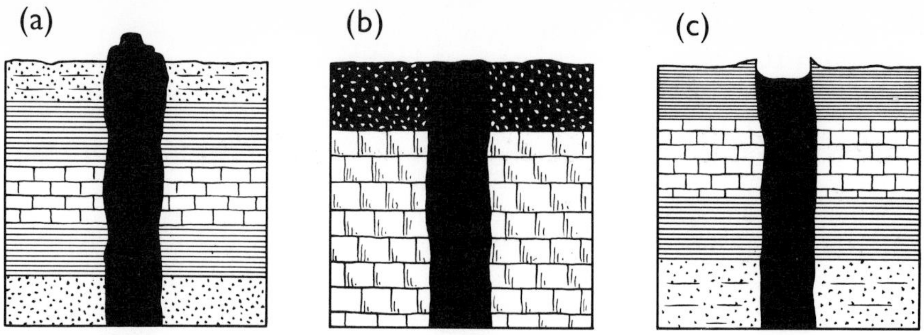


**Figure 11.2**
(a) Dyke more resistant to erosion than wall rocks (cf. Figs. 11.1 and 11.3)
(b) Dyke wall rocks equally resistant (cf. Fig. 3.15)
(c) Dyke less resistant than wall rocks.

**Figure 11.3** Dolerite dyke cutting Tertiary plateau basalts and projecting into the sea because of its greater resistance to marine erosion; height of cliffs, 24–26 m. Ardmore, Vaternish. (*D. Haldane*)

for example, a swarm of 525 dykes can be seen, the total thickness of the dykes being 1649 m. Here the local extension of the crust has been more than one metre in fifteen. Farther north, focused on Mull, there is another great swarm of dykes, some of which can be traced at intervals across southern Scotland into the north of England (Figure 11.4). Most of these dykes are dolerites of various kinds, except along the margins, where the rock is of finer grain, owing to chilling by the walls, or loss of gases, so that it merges into basalt or even tachylyte. The wall rocks themselves show the effects of thermal metamorphism.

Some of these dykes can be seen to die out upwards, indicating that the magma ascended by forcing aside the walls of a fracture until its internal pressure was insufficient to do so. These dykes failed to reach the surface. Others, however, represent the feeders of lava flows of plateau basalts and must, therefore, have been under sufficient hydrostatic pressure when molten not only to reach the surface, but to build up great thicknesses of lava flows. This contrast in behaviour suggests that the high-pressure magma came from a very great depth compared with that of the dykes that failed to reach the surface.

The contrast becomes even more striking when it is remembered that basaltic volcanoes in various parts of the world reach a height of well over 4500 m. The magma evidently ascends with the aid of a hydraulic pressure capable of supporting a column of this height with a density of about 2·6–2·7. Now, if such magma should enter a thick sequence of sediments, the density of which would probably average less than 2·4, it would find it much easier to penetrate laterally along bedding planes and so to lift the light sediments rather than itself. Accordingly it is found that sills are particularly abundant in basins of thick unfolded

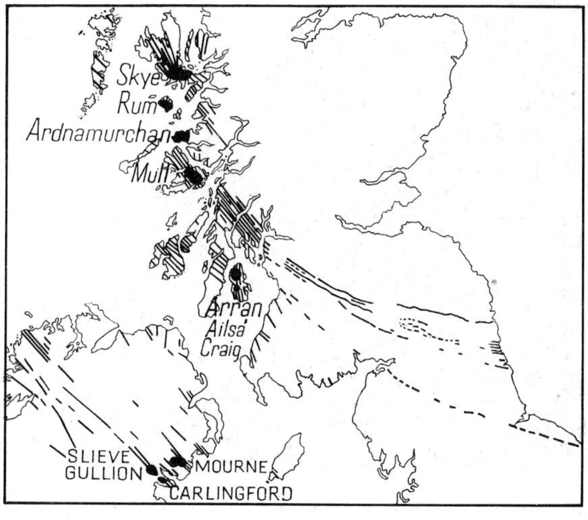


**Figure 11.4** Map showing the Tertiary dyke swarms of the British Isles and their relation to the intrusive centres of igneous activity.

sediments, such as those of South Africa, Brazil, Tasmania and Antarctica (Figure 3.16). In all these basins sediments had been accumulating from the Carboniferous until the early Jurassic, and it was then that the great invasion of basic magma took place.

On a smaller scale the Palisades and other Triassic sills of New York and New Jersey illustrate the same principle, and so also does the Great Whin Sill of the north of England (Figures 11.5, 11.6 and 11.7). *Whin* and *whinstone* are quarrymen's terms for any dark-coloured rock, such as basalt or dolerite, which can be used as road-metal. Beginning in the north of Northumberland, the surface outcrop of the edge of the Whin Sill swings round to the castled crags of Bamburgh and seawards to the Farne Islands. Appearing again on the coast below the ruins of Dunstanburgh Castle, it can be followed across Northumberland towards the River North Tyne. Then for many kilometres its tilted and weather-worn scarp boldly faces the north (Figure 11.7), and here, with an eye for the best defensive line, the Romans carried their famous Wall along its crest. To the south-west the Whin Sill outcrops in the valleys that notch the western margin of the Pennines. Inland from the exposed edge, the Upper Tees cuts through it in the waterfalls of Cauldron Snout and High Force. Near Durham it is encountered in depth, over 300 m below the surface. The thickness of the sill varies from 2 or 3 m, through a general average of nearly 30 m, to more than 60 m. In places it divides into two or more sills. These lobe-like offshoots are linked by oblique transgressions cutting across the strata from lower to higher levels. In this way, amongst others, it betrays its intrusive character. Moreover, as with other sills (Figures 8.5 and 11.8), the strata immediately above are thermally metamorphosed, as well as those below (Figure 8.1). Both sets of observations prove that the Whin Sill could not have been a lava flow that was afterwards buried beneath later Carboniferous sediments. It must have been intruded after the deposition of the Carboniferous rocks with which it comes into contact. An upper age limit is established by the presence of pebbles of the Whin Sill dolerite in a Permian conglomerate that occurs in the Vale of Eden. Evidently an outcrop of the Sill was exposed

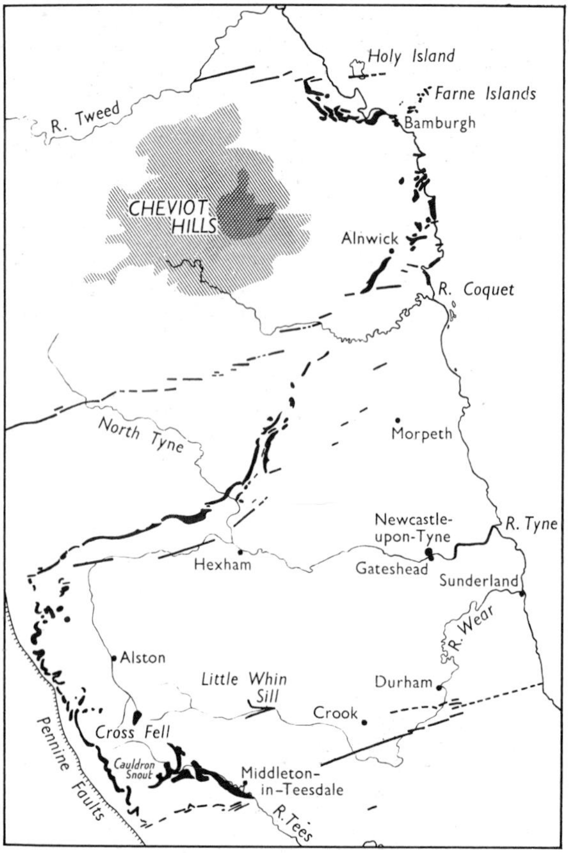


**Figure 11.5** Map of the Great Whin Sill of the north of England and its four echelon series of contemporaneous dykes (late Carboniferous). The Cheviot Hills are carved from the remains of an ancient volcano of Lower Devonian age; the original volcanic conduit and some of the adjoining lavas and ruffs have been replaced by an intrusion of granite (darker shading).

**Figure 11.6** Section across the northern Peninnes to show the thinning and up-stepping of the Great Whin Sill towards the west. T=Triassic; C=Carboniferous; S=Silurian and Ordovician.

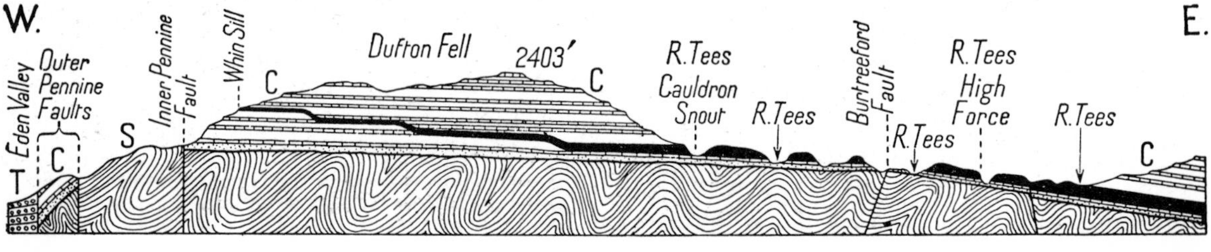

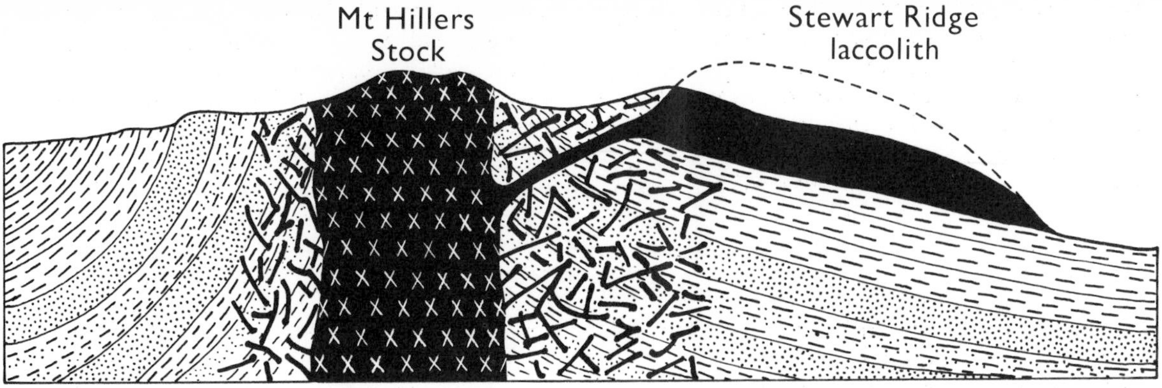


**Figure 11.10** Section through the Stewart Ridge laccolith, showing its relationship to the Mt Hillers Stock, Henry Mountains, Utah (*After C. B. Hunt*). For location see Fig, 19.32.

More detailed mapping has disclosed that each of the groups of laccoliths in the Henry Mountains (and elsewhere in the Colorado Plateau) occurs within a large structural dome. At the middle of each dome is a stock, around which the laccoliths are clustered like the petals of a lop-sided flower (Figure 11.9). The stocks themselves have all the characteristics of diapiric intrusions, and as some of them have concordant dome-shaped roofs, like many salt domes, they have in the past been themselves mistaken for laccoliths. The real laccoliths are tongue-like bulges which appear to have been fed sideways from the stocks (Figure 11.10). But as the latter are often surrounded by a zone of shattered strata which have been invaded by a plexus of sills and dykes, it is not always possible to trace a direct connection from stock to laccolith. There are no well-established laccoliths in Britain, many of those intrusions that were formerly so interpreted having turned out to be stocks or even volcanic necks.

## Lopoliths

Large intrusions, which on the whole are concordant and appear to have a saucer- or spoon-like form (Figure 11.12(*a*)), are distinguished as *lopoliths* (Gr. *lopas*, a shallow basin). The type example is the Duluth gabbro, which extends from the south-western corner of Lake Superior for 193 km to the north-east, and has an estimated volume of about 200,000 km$^3$. The sagging of such great sheets would seem to be an inevitable consequence of the transfer of enormous volumes of basic magma from the depths to the upper levels of the crust. But lopoliths are not merely very thick sills that have subsided into a spoon-like shape; their structure and history are incomparably more complicated.

The Sudbury Complex in Ontario (Figure 11.11), celebrated as the world's chief source of nickel, was originally thought to be a typical lopolith, with norite as its characteristic basic rock. Norite is a variety of gabbro with hypersthene as its chief pyroxene. The complex is now thought by Thompson and Williams (1956) to be a large caldera; a ring-complex or volcano-tectonic depression (pp. 170–1). In early Precambrian times a basin already existed, into which 'glowing avalanches' were erupted on a gigantic scale, from vents near the rim. Over 1500 m of pyroclasts of Peléan type (p. 203) were deposited and possibly caused further subsidence. Later these were covered by sedimentary rocks, followed by intrusion of the norite and later the micropegmatite (Figure 11.11). On the northern side of the basin the norite dips gently southward, but on the southern side it dips almost vertically or steeply southwards. Around the rim of the asymmetrical basin valuable sulphide deposits of iron, nickel and copper were introduced. Since the mines are located on the southern side of the southern limb of the norite, against what was initially thought to be its floor, it was for a long time supposed that the ores had been deposited from the basic magma during the course of its crystallization to norite. For this reason the norite has been misleadingly described as the 'nickel eruptive'. The association is not so much with the 'eruptive', however, as with later faults around the periphery of the basin, along which ore-bearing solutions found passageways towards the surface.

The norites and associated basic rocks of the

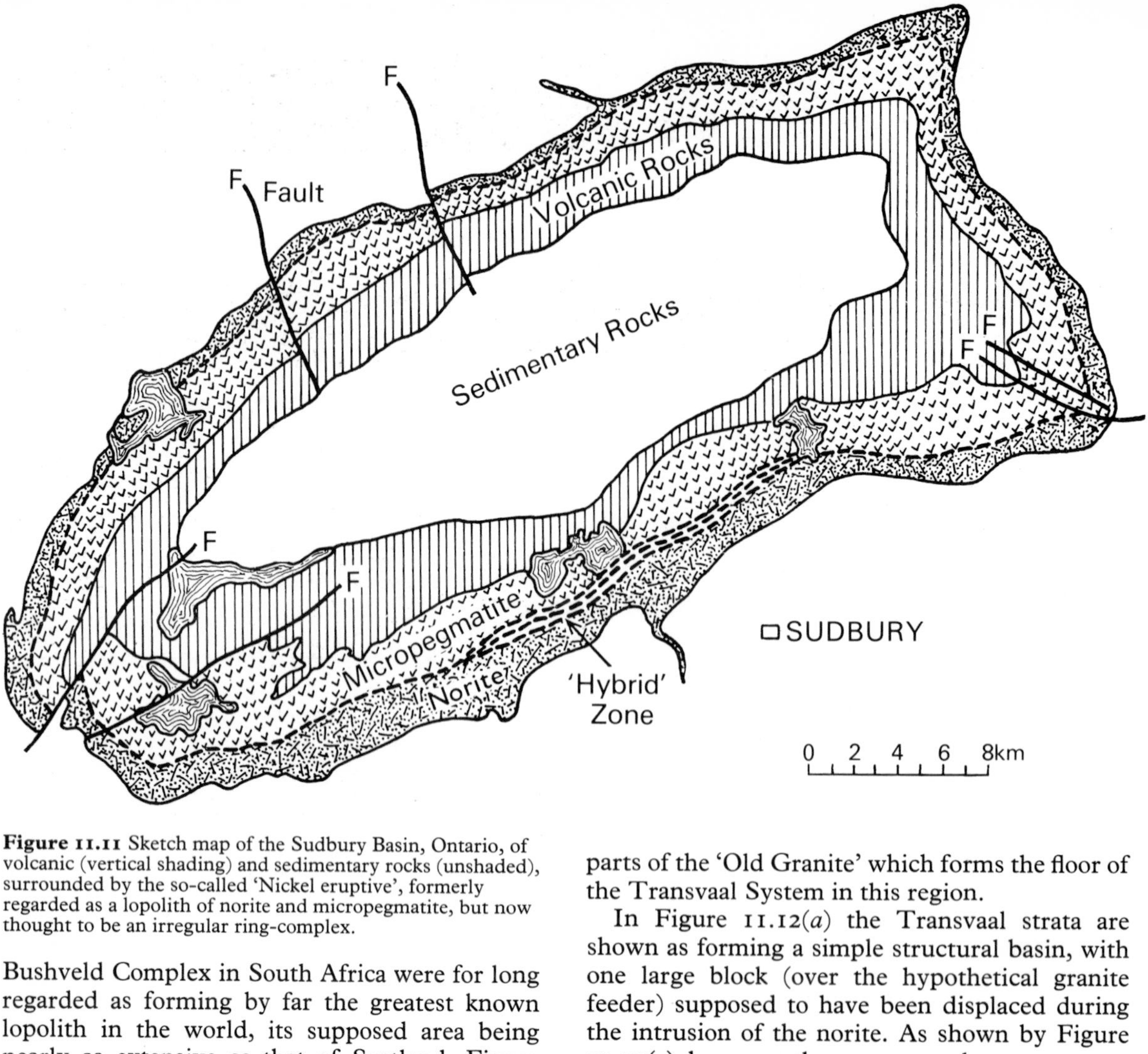


**Figure 11.11** Sketch map of the Sudbury Basin, Ontario, of volcanic (vertical shading) and sedimentary rocks (unshaded), surrounded by the so-called 'Nickel eruptive', formerly regarded as a lopolith of norite and micropegmatite, but now thought to be an irregular ring-complex.

Bushveld Complex in South Africa were for long regarded as forming by far the greatest known lopolith in the world, its supposed area being nearly as extensive as that of Scotland. Figure 11.12(*a*) illustrates the original interpretation. However, detailed mapping and gravity surveys make it quite clear that the Complex includes a cluster of several lopoliths, three of which are depicted diagrammatically in Figure 11.12(*b*). As at Sudbury, the history of the Complex begins with a vast outpouring of ignimbrites, represented by the Rooiberg felsites and associated volcanic rocks. These generally served as the roof of the great norite intrusions, the usual floor being the strata of the Transvaal System. Metamorphism of the Rooiberg volcanic rocks transformed some of them into granophyre and granite. There are also considerable masses of granite intrusive through the norite; these, presumably, represent mobilized parts of the 'Old Granite' which forms the floor of the Transvaal System in this region.

In Figure 11.12(*a*) the Transvaal strata are shown as forming a simple structural basin, with one large block (over the hypothetical granite feeder) supposed to have been displaced during the intrusion of the norite. As shown by Figure 11.12(*c*), however, there are many large exposures of Transvaal rocks in the middle of the Complex, and study of the dips and faults reveals that the interior of the Transvaal Basin has been updomed and upfaulted in several places, as schematically depicted in Figure 11.12(*b*). More recently a systematic gravity survey of the whole region has provided convincing evidence that the central part of the Bushveld (between B and C in Figure 11.12(*c*)) cannot be underlain by a thick, continuous sheet of heavy basic rocks, such as is shown in Figure 11.12(*a*). In contrast there are strong positive anomalies wherever the norite is exposed, and also south of C, where boreholes drilled through a cover of younger (Karroo) strata have penetrated basic rocks of Bushveld types. At

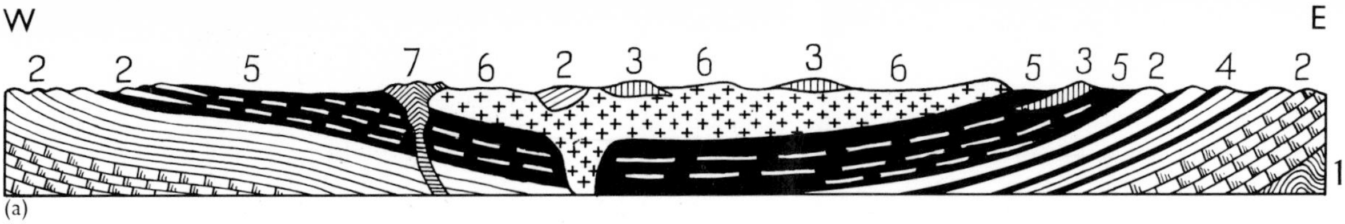


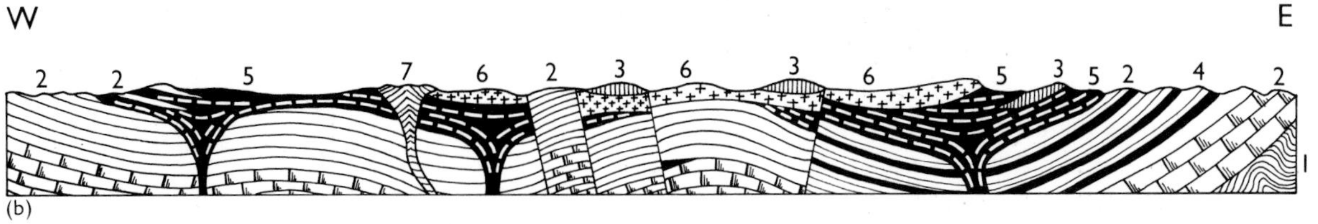


**Figure 11.12** The Bushveld Complex and its geological setting.
(a) Section to illustrate the original interpretation of the norite and associated basic rocks as a single lopolith.
(b) Section to illustrate a revised interpretation (1959) involving a cluster of lopoliths. Three of these (indicated by A, B and C on the map) are shown with their inferred feeders, as located by gravity 'highs'
(c) Sketch map of the Transvaal Basin and the Bushveld Complex.

A, B, C, D and E the gravity highs imply the occurrence of basic rocks locally extending downwards to very great depths. In other words they are evidence of thick dyke-like feeders such as are indicated in Figure 11.12(*b*).

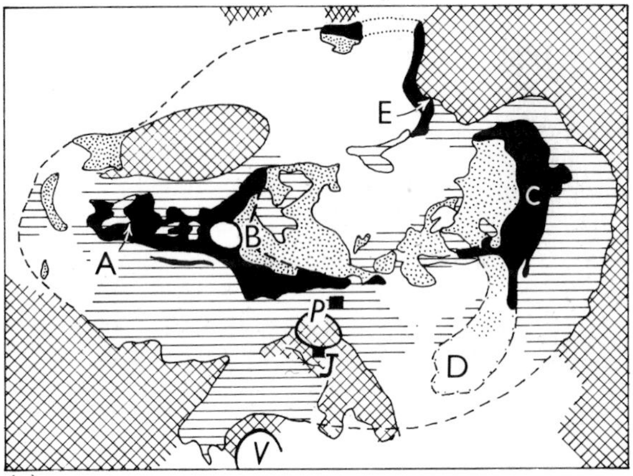


(c)

P Pretoria J Johannesburg V Vredefort Dome

| *Nos. on Sections* | *Sequence of Rocks* | *Shading on Map* | |
|---|---|---|---|
| | Younger formations | *unshaded* | YOUNGER ROCKS |
| 7 | Pilandsberg volcanic centre | | |
| 6 | Granophyre and Red Granite | *dotted* | BUSHVELD IGNEOUS COMPLEX |
| 5 | Norite and other basic rocks of the Bushveld lopoliths | *black* | |
| 4 | Basic sills in the floor rocks | | |
| 3 | Rooiberg volcanic rocks | *dotted* | |
| 2 | Transvaal and Ventersdorp Systems | *horizontal lines* | OLDER ROCKS |
| | Witwatersrand and Dominion Reef Systems | *criss-crossed lines* | |
| | 'Old Granite' and | | |
| 1 | Crystalline Basement Complex | | |

## Radial Dykes, Cone-Sheets and Ring-Dykes

Like the fractures developed over rising salt domes (Figure 10.7), many of those produced in the rocks near the top of a diapiric igneous intrusion (or above any centre of highly concentrated magmatic or gaseous pressure) are likely to be radial or concentric. If the roof rocks are gently upheaved and stretched into a broad domal form, the formation of radial fractures and their opening up and filling are easy to understand. Any circle around the apex increases in circumference as up-doming continues, and so is thrown into tension. Volcanoes and their foundations are often observed to bulge up before a great eruption and this explains the radial pattern of dykes found around certain old volcanic centres from which the superstructure of lavas and tuffs has been stripped by denudation (Figure 11.13). And even without the revealing aid of denudation a radial arrangement of fractures can sometimes be inferred from the distribution of the satellitic vents on the flanks of active volcanoes. The giant basaltic volcano Klutchevskoi, in Kamchatka, has twelve lines of vents, 6 to 18 km long, radiating down the slops from the summit crater, the number of vents along each line ranging from 3 to 11.

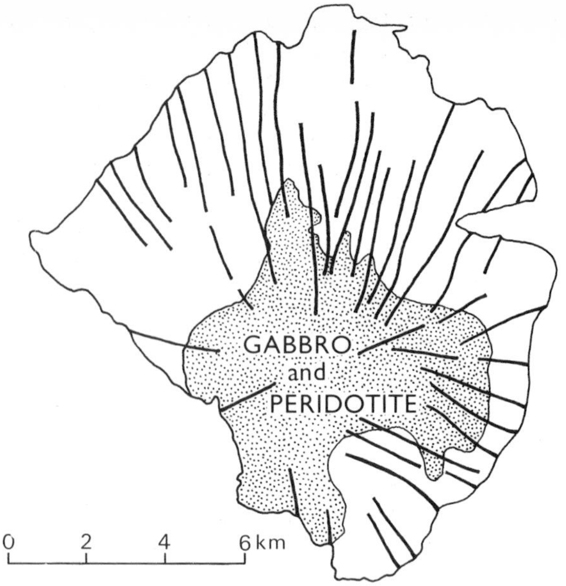


**Figure 11.13** Radial dykes around the Tertiary volcanic centre of the Island of Rum, between Ardnamurchan and Skye. See Fig 11.4 for locality

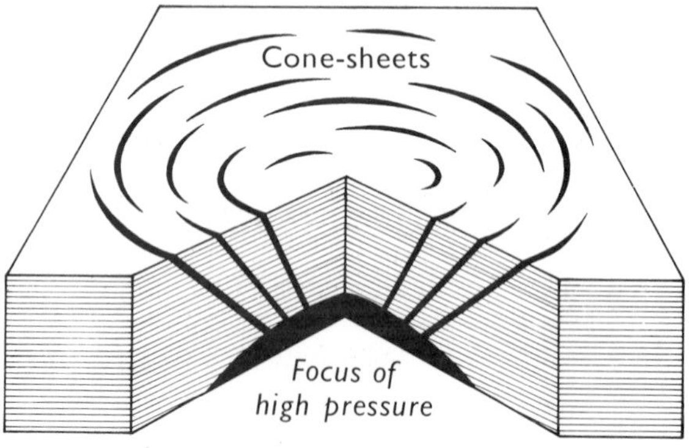


**Figure 11.14** Diagram to illustrate the form of a series of concentric cone-sheets and their probable relationship to an underlying focus of high magnetic pressure

Intrusions in the form of concentric arcs or rings are of two distinct types, forming ring complexes such as are exceptionally well developed around the Tertiary volcanic centres of NW Britain (Figure 11.4). One type consists of *cone-sheets* (Figure 11.14), which have the form of parts of inverted cones dipping inwards towards a common focus. The outermost sheets tend to flatten towards the surface, suggesting that their three-dimensional form may resemble the outward-opening end of a trumpet. Uplift of an inner cone relative to its outer neighbour makes space available for the injection of magma, which was presumably under sufficient hydraulic pressure to produce the necessary fractures in the roof rocks. Swarms of these intrusions (Figure 11.14) are to be seen in Mull and Ardnamurchan.

The other type of ring-intrusion, the *ring-dyke*, has a much broader outcrop, and is characteristically composed of microgranite, granophyre, felsite, or quartz-porphyry. At the time when ring-dykes were known only in Britain, from observations at Glen Coe and in the Tertiary volcanic centres in Mull, Ardnamurchan, and north-east Ireland, it was supposed that they dipped steeply outwards. The reason was that volcanic and foundation rocks, encircled by ring-dykes in Mull and at Glen Coe, were known to have subsided, and such *cauldron subsidence*, as it is called, would make room for the emplacement of a narrow ring-dyke if the fault bounding the cauldron dipped outwards. As Hawkes pointed out, however, in the case of broad ring-dykes the amount of subsidence would be embarrassingly great. In Ardnamurchan, for example, where there are ring outcrops up to more than 1·5 km in width, subsidence of many kilometres would be necessary for the emplacement of the whole ring-complex in this way. Nevertheless the idea that ring-dykes dip outwards has persisted.

In Britain, observations of the actual dips of ring-dykes are lamentably few and isolated, and for no British ring-dyke are there sufficient recorded measurements to determine the three-dimensional form. The most that can be said is that they dip steeply. Detailed investigations of ring-dykes in other countries have revealed that the dips at the surface are essentially vertical, and cauldron subsidence of the kind envisaged by the early British investigators would therefore provide no space for the emplacement of ring-dykes.

There is, however, a structural method by which the three-dimensional form of ring-dykes or ring-fractures, encircling cauldron subsidences, can be determined. At Glen Coe and Mull, in Britain; in the Glitrevann and Sande cauldrons in the Oslo area, Norway; in the Ossippe cauldron, New Hampshire, U.S.A.; and in the Messum cauldron, Namibia (South-West Africa) layers of volcanic rocks, variously consisting of basaltic, trachy-andesite and rhyolite lavas (including ignimbrites in the Oslo area), tuffs, and agglomerates have subsided within the respective cauldrons, and now appear as piles of saucers or bowls (Figure 11.15). The marginal inward dips indicate that in subsiding the volcanic strata have had to accommodate themselves to a narrowing space within downward converging ring-fractures. In the Glen Coe cauldron the saucer-like pile of volcanic rocks is locally marginally overturned. Intruded within the volcanic piles, which may initially have formed part of a volcanic cone or been erupted into a caldera (pp. 225–8), there may be thick sills of gabbro or dolerite, for example in the Messum and Slieve Gullion cauldrons (Figures 11.16 and 11.17).

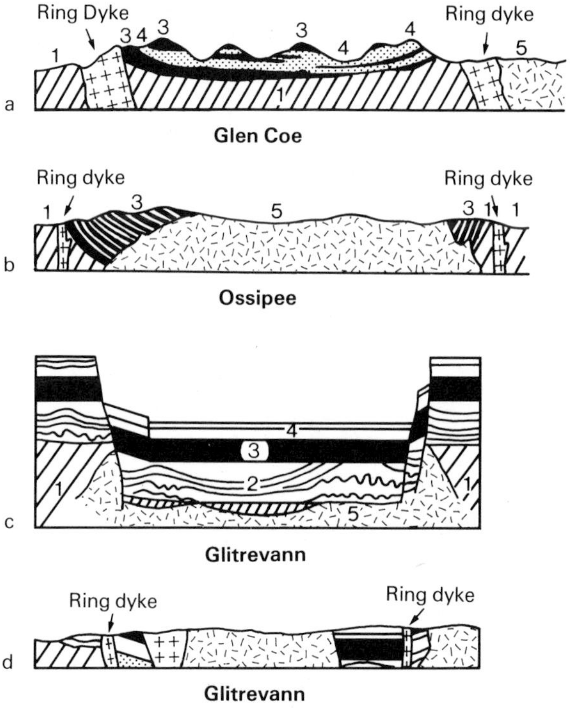


**Figure 11.15** Sections illustrating the saucer-like form of volcanic strata down-faulted within the cauldrons of ring complexes, and having ring-like outcrops. Glen Coe, Scotland, after Bailey and Maufe, 1916. Ossipee, New Hampshire, U.S.A., after Kingsley, 1931. Glitrevann, Oslo area, Norway, after Oftedahl, 1953. (c) represents the Glitrevann cauldron after the volcanic strata had subsided about 1500 metres within it, (d) the Glitrevann cauldron as it now appears after erosion.
1 = Foundation rocks
2 = Palaeozoic strata, the supercrustal rocks at the time when the Glitrevann cauldron was formed
3 = andesites, basalts, quartz-porphyry and pyroclasts within the Glen Coe and Ossipee cauldrons; interlayed basalts and rhomb-porphyry within the Giltrevann cauldron
4 = rhyolite in the Glen Co cauldron, and interlayered basalt and rhomb porphyry within the Glitrevann cauldron
5 = granite and granite aplite invading the saucer-form layers

Just as the graben overlying salt-domes (Figure 10.8) are associated with updoming, so the down-faulting within cauldrons is sometimes associated with updoming. Around one of the ring-complexes in Ardnamurchan, the country rocks dip peripherally outwards at about 30°. Richey (1930) interpreted this as evidence of early updoming, and estimated the vertical uplift to have been of the order of 4000 ft (1320 m). In central Arran, cauldron subsidence similarly followed updoming (King, 1955). At Slieve Gullion the fault pattern, like that over many salt domes, depicts updoming, yet on the inner side of the ring-dyke the presence of down-faulted Silurian rocks, that roofed the Caledonian granodiorite within which the ring-dyke is intruded, bears witness to cauldron subsidence.

## Volcanogenic Granitization Associated with Cauldrons

Within the Messum cauldron, South-west Africa (Korn and Martin, 1954; Mathias 1956), basaltic lavas, with interspersed acid lavas, and thick gabbro sills are marginally uptilted in the typical saucer-like manner (Figure 11.16). There are two ring-faults along which aplogranite has been intruded. One margins the cauldron, and the other

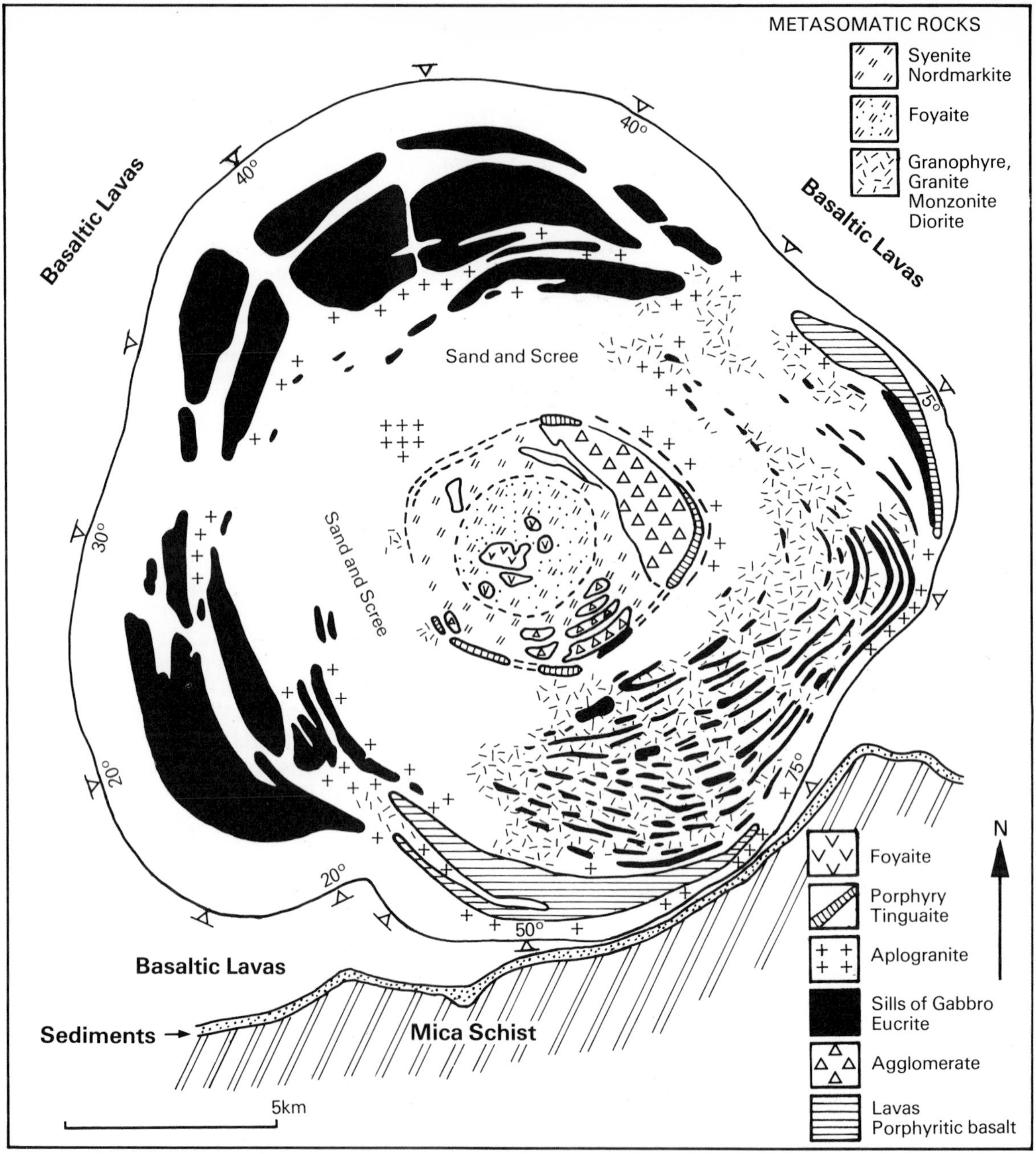


**Figure 11.16** The Messum Cauldron, Damaraland (*After Korn and Martin, 1954*). The directions and angles of dip reveal the conical form of the marginal ring-fracture, and the three-dimensional form of the concentric ring-outcrops is saucer-like.

**Figure 11.17** The Slieve Gullion cauldron, showing the ring-dyke and, within the area it encircles, down-faulted layers of a Tertiary volcano at Slieve Gullion, Foughill, and at Carrickcarnan where they are tilted. (*Reynolds, 1951*)
(a) Map of the Slieve Gullion area.
(b) Cross-section of the layered structure of Slieve Gullion itself, where layer 10 is a dolerite or gabbro sill with fine-grained edges.

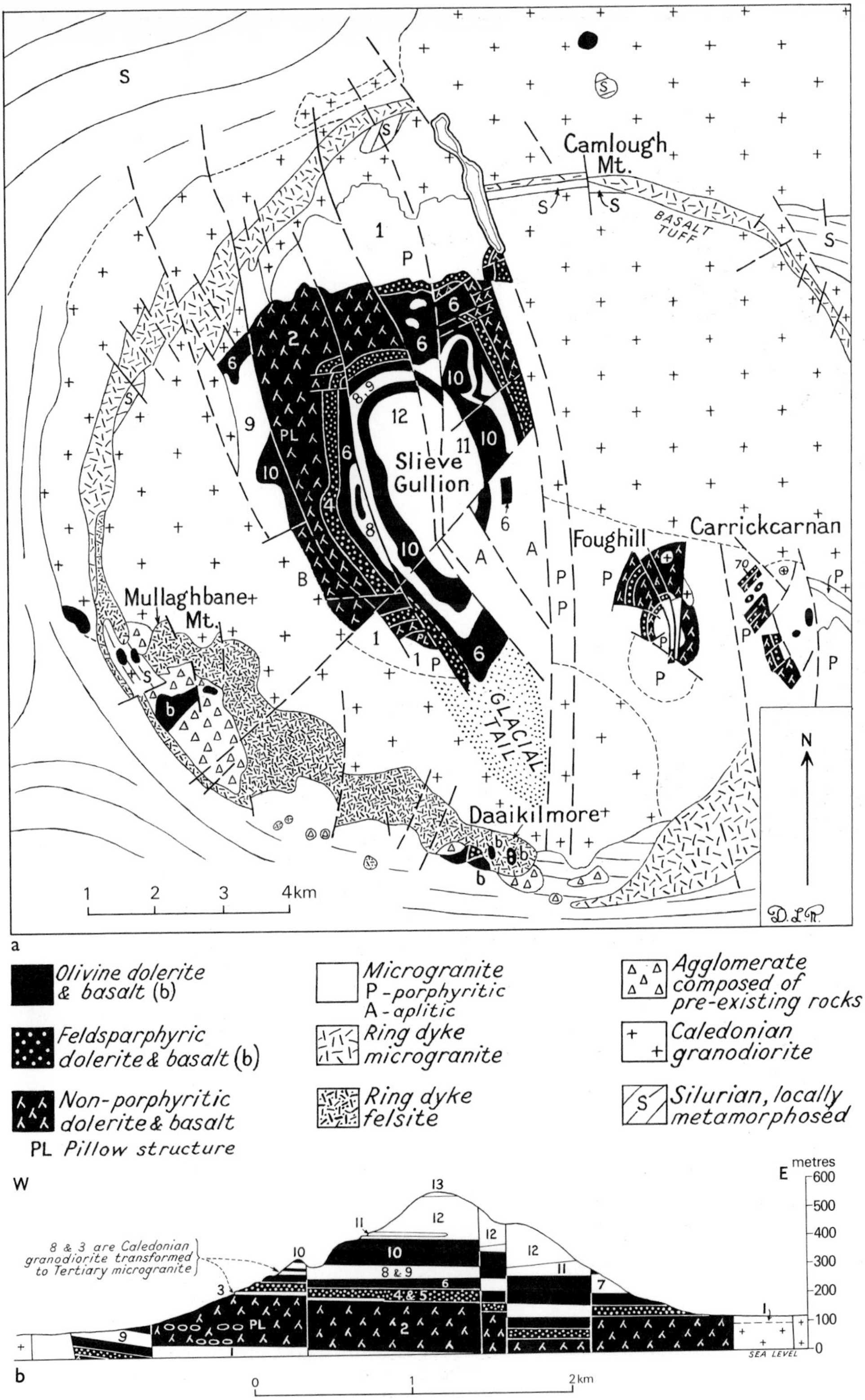
S
Camlough Mt.
BASALT TUFF
Slieve Gullion
Foughill
Carrickcarnan
Mullaghbane Mt.
GLACIAL TAIL
Daaikilmore
1 2 3 4km
N
D.L.R.
a
Olivine dolerite & basalt (b)
Feldsparphyric dolerite & basalt (b)
Non-porphyritic dolerite & basalt
PL Pillow structure
Microgranite
P - porphyritic
A - aplitic
Ring dyke microgranite
Ring dyke felsite
Agglomerate composed of pre-existing rocks
Caledonian granodiorite
Silurian, locally metamorphosed
W
E
metres
600
500
400
300
200
100
0
8 & 3 are Caledonian granodiorite transformed to Tertiary microgranite
SEA LEVEL
b
0 1 2km

margins an inner cauldron where there has been further subsidence in the middle of the complex. The saucer-like gabbro sills give rise to concentric scarp and dip slopes.

The acid lavas and pyroclasts have been metasomatically altered to granophyre, granite and monzonitic rocks, which are packed with layer-like inclusions of the basaltic lavas. The thick gabbro sills have chilled edges against the granitized rocks which they antedate and are themselves invaded by rheomorphic veins extending from the latter. Such seemingly contradictory evidence is characteristic of granitization (see p. 184). In the south-eastern part of the cauldron, where the gabbro sills are relatively thin, they also are intensely altered and eventually remain as ghost-like dark-coloured bands of diorite within the granitized rocks. Evidence that the granitization may have been implemented by pneumatolitic or hydrothermal action is provided by the presence of vugs filled with quartz and feldspar, with which tourmaline and calcite are sometimes associated.

The Slieve Gullion ring-complex, in County Armagh, west of the Mourne Mountains, also provides evidence of granitization (Reynolds 1950, 1951). Around Slieve Gullion a spectacular ring-dyke forms a ring of hills, the northern half of the ring-dyke being composed of granophyre and the southern part of felsite (Figure 11.17). It is possible that the felsite is partly extrusive, and microscopically it then resembles ignimbrite. Within the southern part of the felsite, or ignimbrite, fragments of basalt show various degrees of melting to brown glass.

Slieve Gullion itself, a steep-sided miniature mountain occurs in the middle of a region of Caledonian granodiorite, and is encircled by the ring-dyke (Figure 11.17). Both Slieve Gullion, and two smaller hills to the south-east, have a layered structure (Reynolds, 1951). On Slieve Gullion, where the layers are almost horizontal, this has given rise to marked terraced topography, each layer having a scarp face, and a flat upper surface leading back to the scarp of the next higher layer, The layers consist of highly metamorphosed lavas and pyroclasts. Only at the extreme north and south of the hill do basalts remain unaltered. The uptilted layers at the southern end are closely comparable with variolitic basalts in Mull, and one layer, like them, is of pillow lava.

The most spectacular metamorphic transformations are shown by the Caledonian granodiorite within an aureole about 275 m wide around the layered structure of Slieve Gullion; within an aureole around the minor vent of Carrickcarnan; and around small plugs of gabbro intruded into the Caledonian granodiorite. The layered structures of Slieve Gullion, Carrickcarnan and Foughill are downfaulted into the Caledonian granodiorite and thus conceal the sites of the major and two minor vents. Coarse agglomerates that form part of the layered structures exhibit similar transformations. From the transformed Caledonian granodiorite of these occurrences, rheomorphic veins penetrate

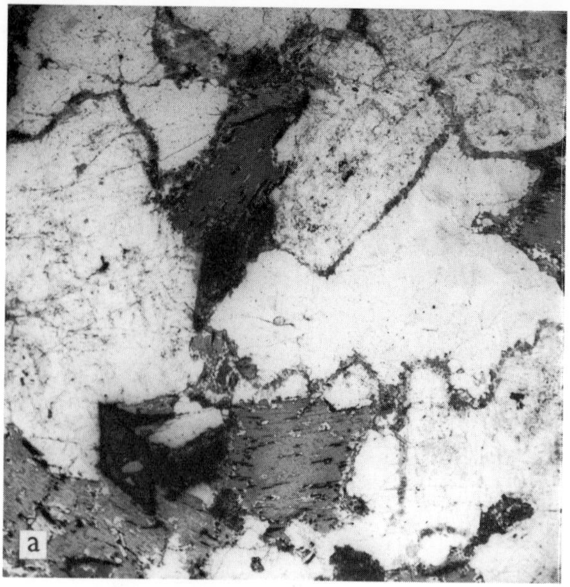
a

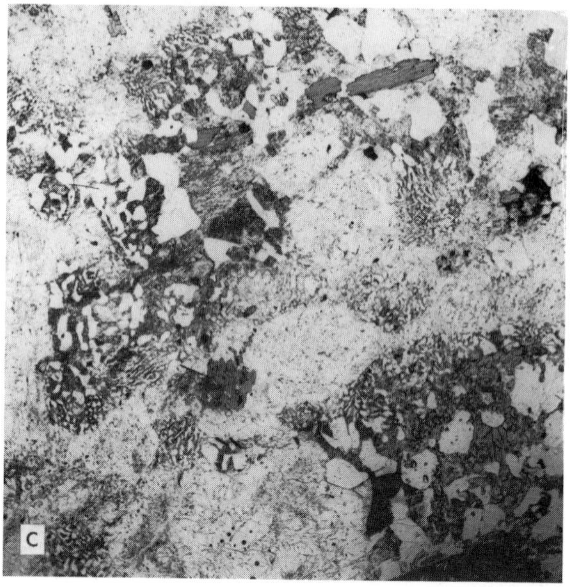
c

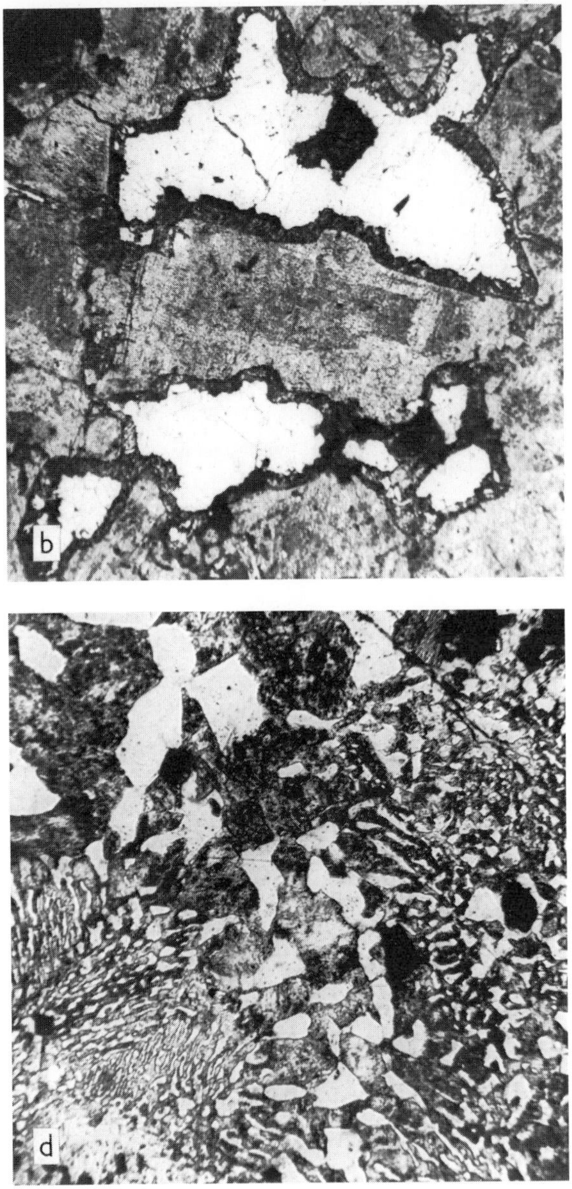


**Figure 11.18** Photomicrographs illustrating stages in the transformation of Caledonian granodiorite ('Newry granite') to Tertiary granophyre. Slieve Gullion, Co. Armagh, Northern Ireland (*Reynolds, 1937, 1950*)
(a) The rock has the typical appearance of Newry granodiorite, except that where quartz and feldspar were originally in contact they are now separated by a narrow rim of potash feldspar, formed at the expense of plagioclase. (×14)
(b) The rims widen and incorporate quartz, so forming the intergrowths known as micropegmatite. Being iron-stained, these are reddish brown and appear black in the photomicrograph. (×14)
(c) The transformation spreads along boundaries and cracks, and as more and more of the original rock is progressively changed the rims of micropegmatite widen into patches which are fine-grained where they replace plagioclase but coarse-grained where they replace quartz. (×14)
(d) Finally the patches of fine and coarse micropegmatite coalesce and the rock becomes a typical granophyre. (×50)

the Tertiary layered rocks (Reynolds, 1937).

The original granodiorite is a fresh glistening rock composed of crystals of feldspar, mainly plagioclase, groups of interlocking quartz crystals, and a sprinkling of biotite and hornblende. The first obvious change is a lack-lustre appearance. Under the microscope thermal alteration can be seen; hornblende being changed to aggregates of pyroxene, and biotite becoming bleached, and finally represented by ghost-like forms of iron ore granules. Around each group of quartz crystals a narrow rim of potash feldspar appears, formed at the expense of the adjoining plagioclase crystals (Figure 11.18(*a*)). At higher grades of transformation the rims gradually widen (Figure 11.18(*b*)), now partly at the expense of the quartz, and their composition changes to micropegmatite. Fingers of potash feldspar invade the quartz groups along the grain boundaries and cracks, and fingers and lobes of quartz invade the plagioclase crystals. Eventually the quartz groups are transformed to relatively coarse grained micropegmatite, and the pagioclase crystals to lace-like micropegmatite (Figure 11.18(*c*)). The replacement of Caledonian quartz and plagioclase by Tertiary micropegmatite is largely metasomatic, like the replacement of wood by silica in wood opal (Figure 6.15), since apart from the loss of lustre the rocks retain the outward appearance of Caledonian granodiorite almost to the end. Ten chemical analyses provide evidence of complete chemical gradation between Caledonian granodiorite and Tertiary granophyre (Figure 11.19). The chemical changes are progressive with increase of potash and silica and decrease of all other major constituents (Reynolds, 1950, 1951).

## Volcanic Necks and Plugs

When the less resistant rocks of an extinct volcanic cone, or even part of the foundation, have been removed by erosion, the more or less pipe-like intrusion that represents the conduit may continue to be exposed, often as a highly conspicuous feature of the landscape. Such intrusions have been given a variety of names. Those that consist

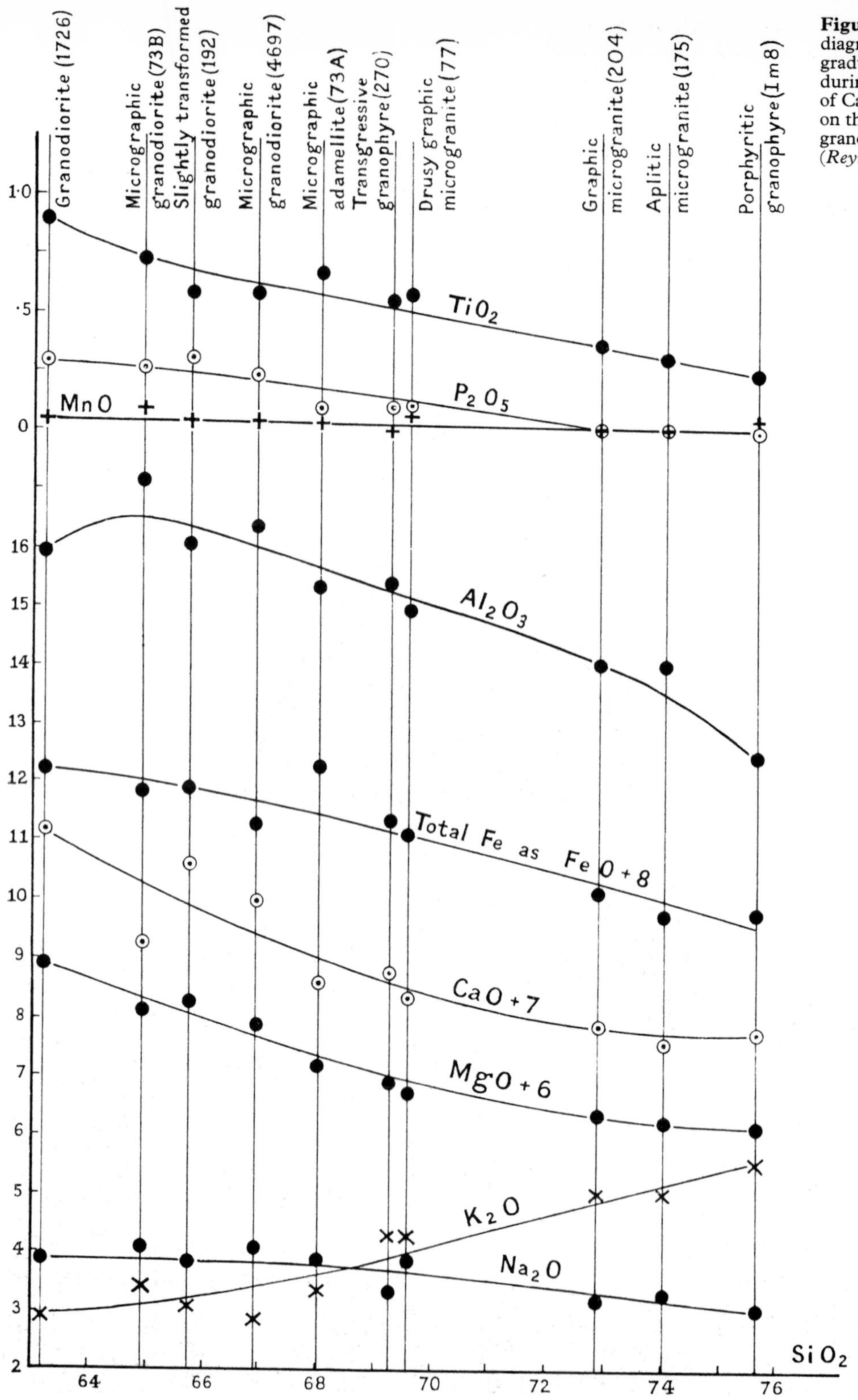


**Figure 11.19** Variation diagram showing the gradual chemical changes during the transformation of Caledonian granodiorite, on the left, to Tertiary granophyre, on the right (*Reynolds, 1951*)

of compact igneous rock, like the massive basalt of the Castle Rock of Edinburgh are commonly called *plugs*. The term *neck* is of more general application and includes volcanic pipes from which the broken-up fragments of the original rocks were not completely expelled (Figure 11.20), any follow-up of magma having been limited to intrusive tongues or the products of magmatic spray.

**Figure 11.20** A spectacular example of a volcanic neck consisting mainly of agglomerate. Le Puy, Velay, Haute Loire, France (*Burton Holmes from Ewing Galloway, N.Y.*)

At one time it was thought that all such volcanic necks had been formed by explosive outbursts. Whilst this explanation is satisfactory for vents filled with blocks of country rocks with haphazard arrangements, it is inadequate to account for others within which blocks of country rocks, embedded in 'tuff' retain their original orientation both with respect to the wall rocks and to one another, although they may have subsided or risen within the vent. The significance of vents of this kind were first recognized by Hans Cloos (1941) as a result of detailed structural mapping of the Tertiary tuff-pipes of Swabia, east of the Black Forest.

The Swabian examples contain blocks of Jurassic strata through which the pipes are drilled, embedded in 'tuff' consisting largely of Jurassic debris down to dust size, and to a lesser extent of tiny solidified droplets (*lapilli*) of lava. By mapping the dip and strike within the blocks, Cloos found the structure to be continuous from one block to another over considerable areas, even though the blocks have subsided some hundreds of metres, relative to their true stratigraphic horizon as seen in the wall rocks. Some of the blocks have been tilted at steep angles. The major blocks have subsided gently within the 'tuff' as if in quicksand (Figure 11.21).

Cloos found blocks of Jurassic rocks in all stages of detachment from the wall rocks and from one another, the separating medium being 'tuff' derived mainly from the adjacent rocks. Channels of 'tuff' similarly dissect and vein the blocks themselves. For 'tuff' of this kind, formed largely *in situ* from the adjacent country rocks, Cloos proposed the name *tuffisite*, to distinguish it from the tuffs normally deposited on the surface as volcanic ash. Cloos envisaged the process as beginning with streams of tuff-bearing gas rising upwards along pre-existing joints, fissures and cracks. The channels became widened with the active formation of tuffisite from the wall rock by a process of *tuffisitization* (German; *Tuffisierung*), until masses of detached rock, weighing up to several million tonnes, gently subsided within the tuffisite. As the masses sank they became dissected by tuffisite to which they were progressively converted. Cloos envisaged the tuffisite as having been in a mobile active state, able to penetrate cracks in adjacent rocks and convert them to tuffisite. He suggested that the process was a form of *pneumatoclasis* (fragmentation by gas action). He thought of through-flowing gas as causing 'a kind of ebullition' or perhaps of even causing boiling of water contained in the rocks through introduction of volcanic heat. Cloos, a great experimenter himself, expressed the hope that experimental investigations might eventually provide the explanation of this astonishing process. Volcanic explosions and their effects are familiar enough, But we are dealing here with phenomena that cannot be so easily explained. As can still be seen around some of the Swabian pipes, and even better around those near Ruwenzori (Figure 12.18) in Central Africa, the earliest tuffs consist of excessively fine materials. At the surface the vents are shallow ring-craters, rimmed by low mounds of ash or tuff.

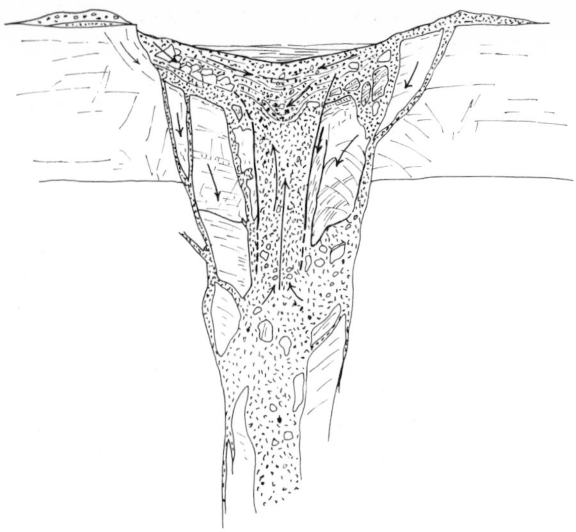

**Figure 11.21** Section through a Swabian tuffisite pipe. *White*: massive blocks of Jurassic strata, detached from the wall rocks and displaced by subsiding in fluidised tuff. *Grey*: tuffisite consisting mainly of micro-lapilli of melilite-basalt (olivine-melilitite) and small fragments and dust of Jurassic rocks (*H. Cloos, 1941*)

## Fluidization

Since the Second World War, an industrial process known as *fluidization* has become widely used by chemical engineers, and the results make it abundantly clear that the phenomena with which Cloos was dealing in the Swabian tuff pipes were purely dynamic, his idea of boiling being quite unnecessary. In the industrial process, gas is passed through a bed of fine-grained particles. At a particular rate of gas-flow, appropriate to the size and density of the particles, the bed expands upwards and the particles become separated from one another, buoyed up by the gas. When the rate of gas-flow is increased still further, gas not only flows uniformly through the bed; it also rises upwards as bubbles, which cause violent agitation and disturb the originally level surface of the bed, which now has the mechanical appearance of boiling, and for this reason is known as the boiling bed (Figure 11.22). The particles within it show turbulent movements and become thoroughly mixed. Moreover, because the particles serve as heat-carriers and make intimate contact with gas, an even temperature is maintained throughout the turbulent bed which provides ideal conditions for chemical reaction between gas and solid particles. Within the turbulent bed the particles are said to be fluidized. If there is further increase in the rate of gas-flow, the particles become entrained and transported by the gas, and the term fluidization is no longer applicable.

Fluidized systems erode the walls of their containing vessels, and because it is advantageous to reduce erosion of the vessels to a minimum, some of the factors that control erosion by fluidized systems have been evaluated experimentally. It has been found that the maximum erosion occurs when particles approach the wall at an angle of 20°. Furthermore it has been found that particles within a fluidized system become abraded and rounded, a spherical form being that which is most resistant.

Cloos knew nothing of the industrial process of fluidization, and it is amazing how well he deciphered the properties of such a system from examination of tuffisite and its activity within the Swabian tuff-pipes. The natural gas-tuff systems operative within the Swabian tuff pipes will have differed from industrial fluidized systems in that the rock particles concerned varied widely in dimensions. The particles used in industrial systems are all accurately matched for size in any one system. This difference will have meant that fluidization and entrainment went on side by side within the volcanic pipes. When the rate of gas-flow was just right for fluidization of particles of a particular size and density, smaller and/or less dense particles will have been entrained.

Evidence of fluidization is not restricted to the Swabian tuff pipes. British Tertiary volcanic rocks and pipes provide similar evidence, as do the Carboniferous vents of Fife, Scotland. Within an arcuate vent, associated with the felsite ring-dyke of Slieve Gullion, blocks of Caledonian grano-

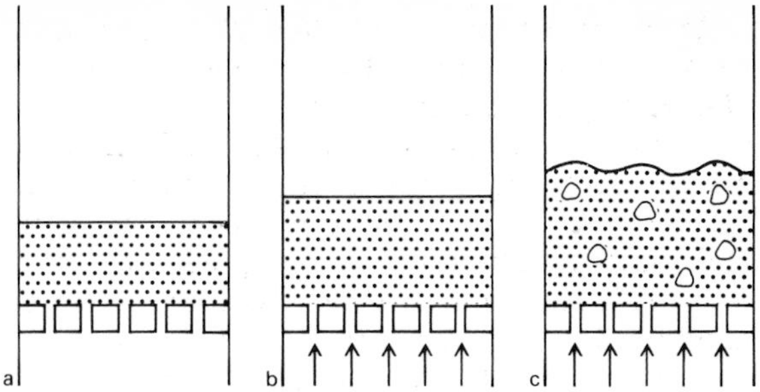


**Figure 11.22** Diagrammatic representation of the process of fluidization, which in industry provides an ideal method for chemical interchange between gas and particles.
(a) *The static bed* before gas is passed through the layer of fine-grained particles.
(b) *The expanded bed* Gas flows through the bed (indicated by arrows) and buoys up the particles, and the bed expands.
(c) *The boiling bed* The rate of gas-flow is increased until eventually it rises partly through the bed, and partly as bubbles. The result is turbulent flow within the bed which appears to be boiling. The particles are now said to be fluidized.

diorite set in a matrix of comminuted rock are so well rounded that in a different setting they could be thought to be boulders and pebbles rounded by marine or river action. Indeed geological evidence of fluidization is world-wide (Reynolds, 1954). It is appropriate here, however, to mention an ore-bearing pipe at Tsumeb, in Namibia (formerly South-West Africa), the contents of which appear to have been emplaced as a fluidized system. The pipe, which pinches and swells, contains a rock known locally as pseudoaplite. Although the latter is definitely intrusive, it is not of igneous origin, and there is no associated thermal metamorphism. The pseudoaplite consists of rounded and crushed grains of quartz and feldspar, associated with breccia in depth. Near the margin of the pipe are pods and veins of ores of lead, zinc and copper, with accessory germanium, silver and cadmium, but the metals are also disseminated through the so-called pseudoaplite. It would appear that mineral grains have been lifted from an underlying formation, and abraded and rounded in a fluidized system, developed by the through-passage of ore-bearing gas.

## Diamond Pipes

The most celebrated volcanic pipes, as well as the best known and most informative, are the late Cretaceous and now deeply eroded diamond-bearing pipes of southern Africa, some of which

**Figure 11.23** The Kimberley Diamond Mine, excavated in a kimberlite pipe to a depth of 1073 m before mining operations were discontinued in 1908 because of increasing costs. The mine is now being reopened and production of diamonds from greater depths is expected. (*The Aircraft Operating Company of Africa (Pty) Ltd, Johannesburg*)

have been worked for diamonds to great depths, e.g. 1073 m in the Kimberley Mine (Figure 11.23). The volcanic material of the pipes is a peculiar ultrabasic rock called *kimberlite*, which is probably a hydrothermally altered tuffisite. Near the surface it is thoroughly decomposed into a sort of yellowish brown mud, known to the miners as Yellow Ground. This passes downwards into Blue Ground, also altered but, as its colour indicates, with much less oxidation. The unweathered kimberlite occurring at deeper levels contains well-shaped crystals and rounded or angular fragments of olivine, occasional grains of pyroxenes and garnet, flakes of mica, and a variety of accessory minerals, the rarest and most valuable of which is diamond. These are embedded in a groundmass consisting mainly of serpentine and calcite, corresponding to which the unweathered rock contains about 10 per cent of water and carbon dioxide. This is a striking indication of the immense proportion of gas that must have been available when the pipes were being drilled to the surface.

In addition to the rounded or broken crystals already mentioned, kimberlite generally contains inclusions of an astonishing variety of rocks, varying in size from small particles to enormous masses. The inclusions can be conveniently divided into four groups:

(*a*) Rocks that can be matched with the formations exposed in the walls of the pipes by mining; ranging from Precambrian granite and metamorphic rocks to strata of the Karroo System.

(*b*) Rocks derived from *above* the present surface. These are samples from several hundred metres of strata, perforated by the pipes, which have been removed, together with the upper parts of the pipes, by the denudation of the last 70 million years or so (Figure 11.24). Fortunately the diamonds from the missing tops of the pipes have not all been lost. The sands and gravels of old watercourses, and of beaches and dunes along the coast near the mouth of the Orange River, have been the chief sources of diamonds for many years.

(*c*) Rocks older than those of group (*a*), brought up from *below* the formations disclosed in the deepest mines. Some of these can be matched elsewhere in southern Africa, but others are samples of deep-seated rocks of which we would otherwise have no direct knowledge. Most remarkable of all are well-rounded 'boulders' of eclogite (p. 113). Diamonds have been found in some of the eclogite inclusions. Although they differ from the kimberlite diamonds, and cannot be the source of the latter, they are equally indicative of crystallization under extremely high pressure.

(*d*) Coarse-grained ultrabasic rocks, also brought up from *below*. These provided kimberlite with the greater part of its load of rounded or broken crystals.

From the Cape to Zaïre, kimberlite pipes are of widespread occurrence, but only a small proportion of those discovered have proved to contain diamonds in sufficient abundance to be profitably worked. Inclusions of eclogite and ultrabasic rocks and their minerals seem to be invariably present, and this peculiarity is equally characteristic of kimberlite pipes in other parts of the world. Besides the relatively recent discovery of diamond pipes in the frozen plains of Siberia, examples have long been known in N. and S. America, Australia and India. Now it can hardly be imagined that kimberlite happened by accident to perforate the crust only in those places where it could be sure of passing through deep-seated masses of eclogite. We are clearly given a broad hint that layers or bands of eclogite, associated with the more abundant peridotites, are likely to be everywhere present at some depth or other beneath the continents. To account for the

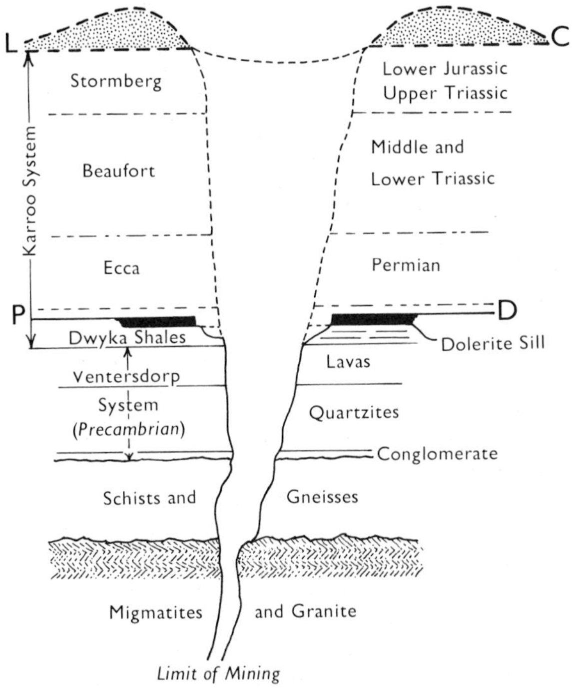


**Figure 11.24** Section through the Kimberley Diamond Pipe, showing the rocks exposed by mining and indicating the sequence of strata removed by erosion between the late Cretaceous surface L–C, when the pipe was formed, and the present-day surface P–D

diamonds found in eclogites the depth would have to exceed 110 km to ensure the requisite overhead pressure. It is therefore inferred that kimberlite pipes provide us with samples from the upper part of the mantle as well as from the crust. Since eclogite is of basaltic composition, its fusion at great depths would readily account for the eruption of great floods of plateau basalts and the intrusion of equally voluminous sills. Moreover, such a transformation would inevitably bring about vertical crustal movements on a considerable scale, as a result of the density changes to which reference has already been made on p. 113.

Here, however, after this introductory description, we are more concerned with kimberlite pipes as magnificent demonstrations of the effects of long-continued fluidization. The old idea was that the pipes were formed by a series of violent explosions, and that some of the shattered rocks blown into the sky fell back into the pipes, together with material that slumped in from the walls, all being churned up later by successive surges of kimberlite magma. This view admittedly failed to explain how inclusions of heavy, deep-seated rocks like eclogite came to be mixed with those of light sedimentary rocks from shallow levels. In one of the Transvaal pipes unmetamorphosed masses of coal, each weighing many tonnes, were found at depths of between 180 and 305 m below the coal seams from which they were derived.

These and many other difficulties vanish when the formation and infilling of the pipes is ascribed to fluidization. The picture is essentially one of a seething, turbulent 'expanded bed' which is continually being added to by masses of rock detached from the walls of the growing pipe, while at the surface a spray of gases and fine-grained materials is continually escaping and so making room for the gas-charged kimberlite and its load of inclusions coming in from below. The larger masses of wall rocks gradually subside through the 'expanded bed' in which they find themselves. In turn they are fragmented, veined and abraded. The fragments behave in various ways according to their sizes, shapes and densities:

(i) The larger ones continue to subside until by further disintegration they are reduced to smaller fragments; or until the gas streams wane and they become 'frozen in', as apparently happened to the masses of coal mentioned above, which must have been late-comers from the walls.

(ii) Smaller fragments are violently agitated by

the turbulent motions of the 'expanded bed', rising and falling in swirls and eddies, but rising on balance as they become smaller by abrasion; and in the case of the less friable materials acquiring the beautifully rounded and ellipsoidal forms that characterize so many of the inclusions.

(iii) The smallest particles are entrained by the swifter gas streams and transported towards or beyond the surface. These also include the materials that fill wedging veins in the wall rocks and larger inclusions—exactly as in the Swabian pipes.

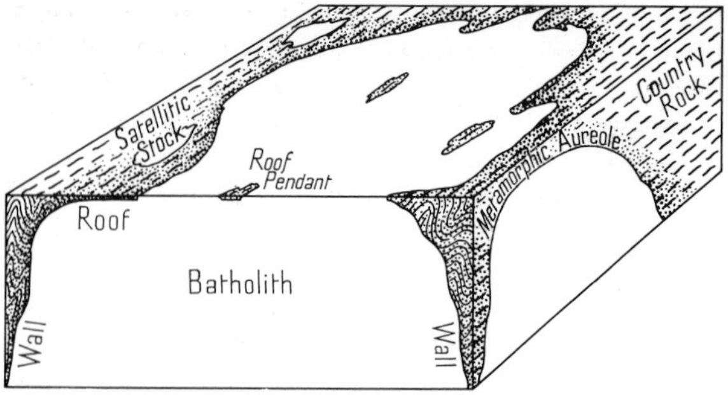


**Figure 11.25** Block-diagram illustrating the characteristic features of a batholith according to Daly (compare Figs. 30.16 and 30.23b where 'roof pendants' are formed as *pincées*—entrapments)

## Batholiths and their Emplacement

The term *batholith* was introduced by Suess (1888) to describe the major granitic intrusions emplaced within the cores of orogenic belts. At first he thought they represented cavities which were progressively filled with granite magma as they opened during orogenic movements. Subsequently, however, he concluded that they resulted from fusion of older formations, as if a red-hot poker had pierced the crust with the evolution of batholiths essentially *in situ*. He envisaged batholiths as being shield-shaped, with steep walls extending down to unknown depth.

Although for convenience the rocks of batholiths are referred to in a general way as 'granite', it must be remembered that this broad use of the term includes granodiorites of many varieties, and that on average these rocks are more abundant than true granites. Moreover, such related rocks as quartz-diorite and diorite, or monzonite and syenite, may occur as marginal varieties or as individual *stocks*, a term used for intrusions similar to but smaller than batholiths.

Daly (1914, 1933) defined batholiths as having a cross-section of 40 square miles or more (over 100 km$^2$), and as having no visible or inferable floor of older solid rock. In cross-section, as exposed at the surface, the long edges of batholiths are essentially parallel to the fold-axes of the orogenic belts within which they are emplaced. This implies some degree of parallelism between the boundaries of batholiths and the strike of the invaded country rocks, but cross-cutting boundaries are common (Figure 11.25).

As Suess recognized, batholiths pose a space problem. If country rocks were previously present within the space now occupied by granite, what has happened to them? Daly, who held the view that batholiths were emplaced as magma, solved the space problem by supposing the country rocks to have been removed through overhead stoping by the granite magma. Guided by the presence of angular inclusions of country rocks just within the margins of some granites, and by the presence of networks of granite veins within the adjacent country rocks, he postulated that batholiths were emplaced by engulfment of blocks broken out of their roofs and walls. To this process he gave the name *magmatic stoping*. The sunken blocks of country rocks he supposed to be assimilated in depth and consequently to have become part of the granite magma.

As to the mechanism by which blocks of country rocks might be detached from the walls and roofs of batholiths, Daly made three suggestions: (*a*) that granite magma exerted bursting pressure on the roof rocks and thereby shattered them, (*b*) that the roof rocks are shattered by differential heating, and (*c*) that updoming stretched and fractured the roof rocks, blocks of which, when completely surrounded by granitic veins, become available for stoping.

Daly himself recognized the inherent difficulty in these postulates; the difficulty of understanding how a magma, capable of exerting bursting pressure on its roof, and at the same time liquid enough for detached blocks to sink through it, was not completely poured out at the surface as lava flows. This was always a serious worry to him, and he thought the thick widespread rhyolite flows of Yellowstone Park might represent one such breakthrough—but, if one, why not hundreds of floods of rhyolite elsewhere? In fact rhyolite is not the most abundant of volcanic rocks, as it should be on this hypothesis, and this consideration suggests that either the bulk of actual melt was small, or alternatively it was too viscous to escape at the surface, and consequently too viscous for blocks of replaced country rocks to have sunk through it. It has now been recognized that the Yellowstone rhyolites are ignimbrites, so that if they were erupted from an underlying parental batholith they represent only such part as could be fluidized by escaping volatiles.

Daly, like Suess envisaged batholiths as being steep sided, and taller than they are wide, and as including downward directed wedges of the country rocks in the roof region, which he called 'roof pendants'. A batholith, according to Daly, would have a three-dimensional form much as depicted in Figure 11.25. When it is considered that the Coast Range batholiths of British Columbia are over 240 km wide, whilst the thickness of the crust there is of the order of 25 km, and certainly not more than 40 km, there is clearly something wrong with this idea of the dimensions of a batholith, the shape of which should rather be thought of as tabular.

One of the most formidable opponents of magmatic stoping was Hans Cloos. He devised a method of mapping planar and linear structures within granites, and so obtaining a three-dimensional picture of their form and flow movements (Figure 11.26). Planar structures result from the parallel orientation of layers of different composition, and of flattened schlieren, and tabular crystals. The word *schlieren* is an old German word denoting a flaw in glass, i.e. a patch of abnormal composition. Within granite, schlieren are commonly richer in ferromagnesian minerals than the normal rock. Linear structures result from the parallel orientation of prismatic crystals, such as hornblende, and of spindle-like schlieren. An account of Cloos's methods in English is recorded by his student and eventual collaborator Robert Balk (1937).

Linear structures within granite depict the direction of maximum extension, whilst planar structures depict the flow-planes. Cloos found that the flow planes within granite commonly define dome-structures, and he interpreted this form as depicting upward flowage. Where the walls of a granite mass dip steeply, the flow planes dip steeply. In the roof region, however, the flow-planes are commonly more or less horizontal, and at right angles to the direction of flow. Cloos interpreted these nearly horizontal planes, which are commonly characterized by linear structures, as depicting extension parallel to the roof. He inferred that batholiths, for which he preferred the

**Figure 11.26** Block diagram illustrating the structures within an intrusive granite and its contact rock (*After Grout and Balk, 1934, by courtesy of the Geological Society of America. Slightly modified from H. Cloos*)

The diagram illustrates flow vertically upwards, lineation on the domed surface depicting extension, joints and dykes at right angles to flow-lines depicting extension during cooling, diagonal joints and dykes, comparable with the marginal crevasses of glaciers (Fig. 9.4), resulting from differential flow.

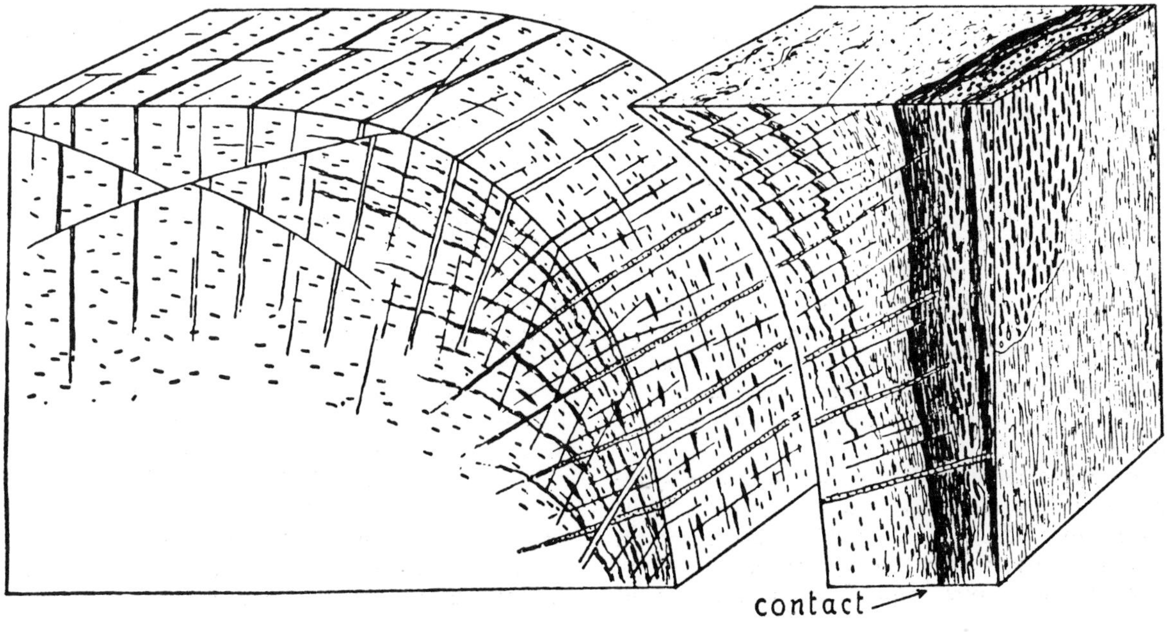

term *pluton*, are forcibly emplaced during orogenesis, updoming and stretching their roofs. Moreover, he discovered that granite plutons are sometimes thick updomed sheets, like that illustrated in Figure 8.18.

Cloos maintained that the sheet-form of some granite plutons, and the very existence of domed flow-planes, are incompatible with Daly's stoping hypothesis. The down-sinking of blocks of country rocks that were conjointly equivalent in volume to the volume of the emplaced granite mass, would inevitably have disturbed the flow planes. Moreover, Cloos added that (*a*) the number of observed inclusions (xenoliths) within granite plutons is too small for the requirements of the stoping hypothesis, and (*b*) the difference in specific gravity between xenoliths and granite is commonly too small for the stoped blocks to have sunk down to the abyssal depths before the granite became too viscous for them to sink through it.

To Cloos's criticisms of the stoping hypothesis, a geophysical objection can now be added. The stoping hypothesis implies that an enormous volume of light magma has exchanged places with an equal volume of heavier rock which was originally above the magma. The total mass of material in that particular stretch of crust has remained the same. Since the heavier rock has been displaced a few kilometres downwards, its gravitational attraction at the surface is slightly lowered and a small gravity 'low' would be expected. But the anomalies actually found are usually so strongly negative that they cannot be accounted for in this way. At best, stoping cannot be more than a small-scale process, mainly confined to the upper parts of batholiths and stocks.

Both Daly and Cloos thought of granite as being emplaced as magma, but it is now well known that the presence of flow-structures, as in salt diapirs (Figure 10.3) does not constitute evidence of liquid rather than of rheid flowage. Indeed Wegmann (1930) suggested that the term diapir should be extended to include granite plutons which, like salt diapirs, have flowed upwards and pierced through the dome-structures formed by their uprise (Figure 11.27).

Domes of granite gneiss have been mapped within the Karelian orogenic belt, at Pitkäranta, on the north-west side of Lake Ladoga, now in the U.S.S.R., and also in central Finland (Figure 11.28). Early in the century it was recognized that the domes near Lake Ladoga represent the foundation rocks upon which the overlying regional metamorphic rocks, of sedimentary origin, were deposited (Trustedt, 1907). In 1917, however, Laitakari found a marginal variety of the foundation granite gneiss to be intruded as dykes into the limestone overlying a dome, and he inferred from this that the granite-gneiss must be younger than and intruded into the overlying mantling rocks. Sederholm, on the other hand, concluded that the granite gneiss had been migmatized and, as it were, reborn during the Karelian orogenic movements, so that it had flowed again. Such rejuvenation of an old granite, with renewed ability to flow, he called *palingenesis*; Backlund's term *rheomorphic*, applicable also to mobilized rocks of sedimentary origin, is now more commonly used. Eskola at first agreed with Laitakari, particularly because the foliation of the gneiss is parallel to the margin of the domes, and to the dip of the mantling rocks. The gneiss domes near Joensuu, however, are directly overlain by conglomerates containing boulders and pebbles of the gneiss of the dome. There can therefore be no question that the gneiss of the dome is palingenetic gneiss, initially older than the overlying mantling rocks, and that it formed part of the land surface from which the materials of the overlying conglomerate were derived by denudation. Throughout a period of years Eskola (1948, 1951, 1952) became convinced that granitization, as Se-

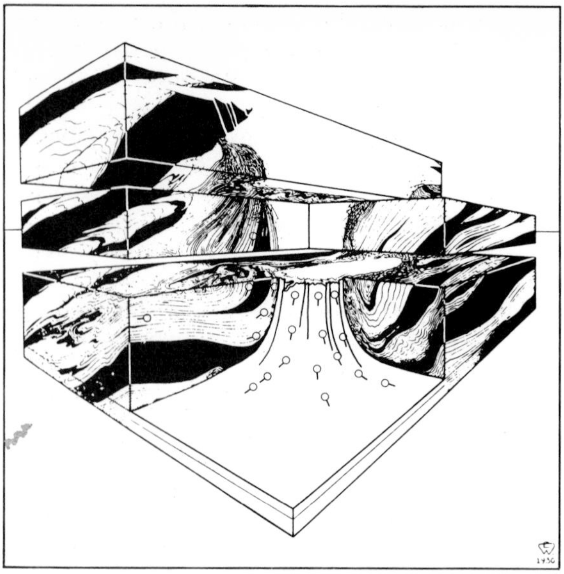

**Figure 11.27** Schematic illustration of the upward flow of a granite diapir (*Wegmann, 1930*). The symbols represent the uptilting of fold-axes, from near horizontal to vertical, both within the diapir and its mantling rocks.

derholm had thought, provided the clue to the enigma. From upward directed shearing movements at the margin of the Sumeria dome, and recrystallization of the resultant mylonite, Eskola eventually recognized that the foundation gneiss had moved upwards, and that concomitant metamorphic differentiation, combined with metasomatic granitization, had transformed the old gneiss marginally into a younger potash-rich granite, within which microcline replaces old minerals.

Subsequently, Wetherall, Kouvo, Tilton and Gast (1962) determined the radiometric ages of biotite and zircon in the Joensuu domes, considered to be free from granitization. Biotite was found (by the K–Ar and Rb–Sr methods, p. 235) to have an age of about 1800 million years. The zircons were found (by the lead method) to be much older, with ages of about 2600 million years. Biotite is the first mineral to be altered or destroyed during metamorphism, whether it be dynamic, thermal or metasomatic. During the former it recrystallizes along shear-planes and thus acquires the age of the movements. Zircon, on the other hand, takes no part in changes induced by movement or granitization, and its age therefore corresponds more closely to that of the initial rock.

Eskola described the rejuvenated gneiss domes as mantled gneiss domes (Figure 11.28), and he suggested that gneiss domes in other areas, previously described as being of magmatic origin, might similarly be granitized foundation rocks. In particular, he suggested that the gneiss domes of Maryland, investigated by Ernst Cloos and Broedel (1940), and those of New Hampshire, investigated by Billings and his co-workers, all within the Appalachian belt, might have originated in a similar way to those of Finland.

Another interesting gneiss dome is that of Agout, in the southern part of the Massif Central of France (Figure 11.29). The dome is historically of interest because under the influence of Werner, Marcel de Serre (1827) considered the gneiss to have been precipitated from a primeval ocean; the unmetamorphosed sediments, particularly the Coal Measures, supposedly being *Flötz* separated from the gneiss by Werner's *Transition* zone, the mantling mica-schists. Subsequently Dufrenoy, in the early days of his collaboration with Élie de Beaumont, in making the first geological map of France, held similar views. He observed the domal structure of the gneiss but, according to Werner's teaching, this could arise from precipitation of the gneiss on original protuberances from the earth's nucleus. Subsequently, in another part of France, he discovered that limestone occurs within gneiss, just as it is interstratified with sedimentary rocks.

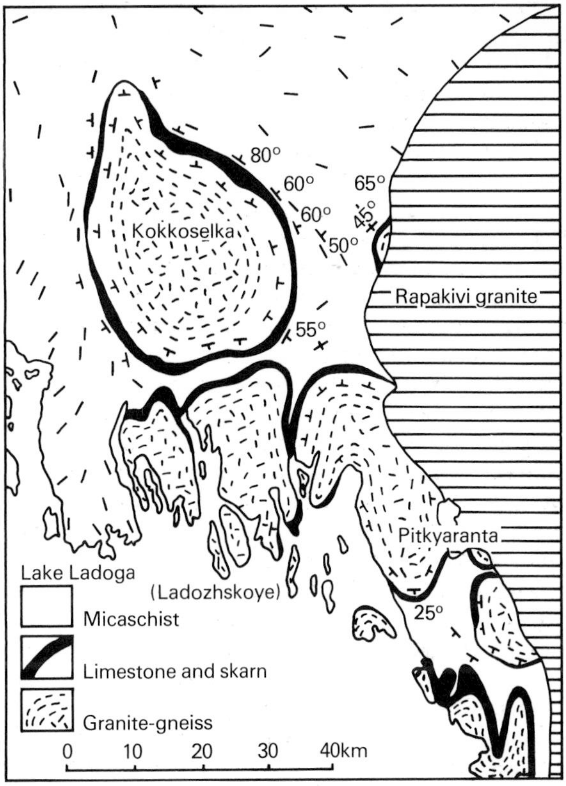


**Figure 11.28** Mantled gneiss domes of Pitkäranta, Finland (*After Eskola, 1951*)

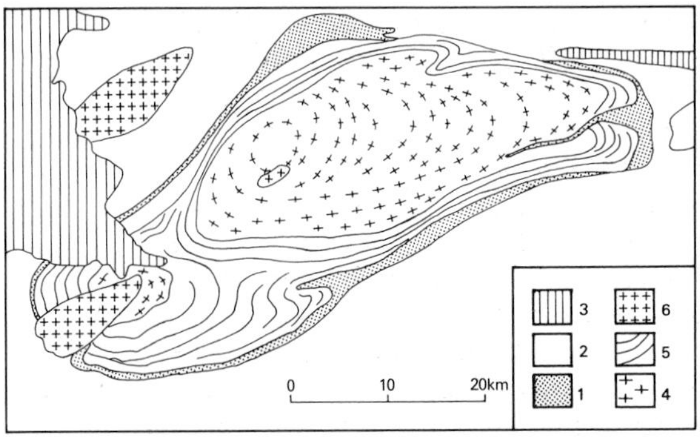


**Figure 11.29** Mantled gneiss dome of Agout, Montagne Noire, south of the Massif Central, France. Anatexite (4) is mantled by augen and banded gneiss (embrechite 5), in turn mantled by mica-schist characterized by regional metamorphic mineral zones (1). The country rock is mica-schist (ectinite 2), overlain by Tertiary and Carboniferous strata (3). 6=intrusions of granite. (*After M. Roques, 1941, flow structure mapped by R. Schuiling, 1960*)

This led him to recognize that schists and gneisses are transformation products of sedimentary rocks.

In 1941, Roques published a masterly memoir recording the results of an intensive geological, petrological and chemical investigation of the dome of Agout. The dome, mantled by mica-schists (ectinite), has a central core of anatexites with irregular, plastic style folds. This grades outwards to a zone of embrechites, consisting of augen- and banded-gneiss. The mica-schist and embrechites are updomed, like a carapace, over and around the anatexite core, the dome-structure being depicted by their respective schistosity and foliation. By microscopic examination Roques discovered that the orientated minerals of the mica-schist, mantling the dome, are commonly broken into pieces and their debris aligned parallel to the schistosity. From the fact that the broken pieces are commonly but little separated, so that they present the appearance of having been gently pulled apart, Roques recognized that the fragmentation resulted from gentle stretching over the uprising gneiss dome. Furthermore, he discovered that movement and crystallization were concomitant phenomena, the broken minerals having been healed by recrystallization only to be again pulled apart. The embrechites show similar microscopic evidence of extension over the gneiss dome at the time of its uprise. By mapping the linear structures and slickensides, Schuiling (1960) found that during emplacement the movement of the gneiss dome, relative to the mantling mica-schist, was upwards. Within the mica-schist, a succession of metamorphic zones appear, denoted by minerals comparable with those of the Grampian Highlands (Figure 8.17). The zone minerals range from biotite, through garnet and andalusite to staurolite; the sillimanite zone occurs within the outer part of the gneiss dome itself, both within the embrechites and anatexites. The central core of anatexite is characterized by cordierite.

Plutons which, from their surface outcrops appear to be simple domes may, if seen in three dimensions, be much more complex. As already mentioned on p. 123 allochthonous migmatites, in Eastern Greenland, are magnificently exposed on the steep and high fjord walls. After years of work, following up the pioneer explorations of Backlund and Wegmann, Haller has illustrated some of the forms of migmatite plutons which have flowed upwards, like salt diapirs, and intruded and infolded the overlying, and now regional metamorphic rocks (Figure 30.32). The migmatite plutons represent a thick sequence of Precambrian sedimentary rocks, which formed the foundation of the region. During the Caledonian orogeny the foundation-rocks became granitized and rose upwards by rheid flow, the flowage forms being made clearly visible by dark coloured bands of amphibolite. Figure 9.9. is a photographic illustration of one of the nappe-like tongues of migmatite.

## Selected References

ANDERSON, E. M., 1951, *The Dynamics of Faulting and Dyke Formation*, Oliver and Boyd, Edinburgh.

BALK, R., 1937, 'Structural Behavior of Igneous Rocks', *Geological Society of America*, Memoir 5, pp. 177.

CLOOS, H., 1941, Bau und Tätigkeit von Tuffschloten: Untersuchungen an dem Schwäbischen Vulkan, *Geologische Rundschau*, vol. 32, pp. 709–800.

DALY, R. A., 1933, *Igneous Rocks and the Depths of the Earth*, McGraw-Hill, New York, pp. 598.

ESKOLA, P., 1949, 'The problem of mantled gneiss domes', *Quarterly Journal of the Geological Society*, London, vol. 104, pp. 461–76.

— 1951, 'Around Pitkäranta', *Annals Academica Scientiarum Fennicae*, 3 Ser., A27 pp. 1–90.

GILBERT, G. K., 1877, 'Report on the geology of the Henry Mountains', *U.S. Geographic and Geologic Survey*, Rocky Mountain Region.

GROUT, F. F., and BALK, R., 1934, 'Internal structures in the Boulder batholith', *Bulletin of the Geological Society of America*, vol. 45, pp. 877–96.

HALL, A. L., 1932, 'The Bushveld igneous complex', *Geological Survey of S. Africa*, Memoir 28m pp. 554.

HUNT, C. B., 1953, 'Geology and Geography of the Henry Mountains Regions, Utah', *U.S. Geological Survey*, Professional Paper 228.

KING, B. C., 1955, 'The Ard Bheinn area of the central igneous complex of Arran', *Quarterly Journal of the Geological Society*, London, vol. 110, pp. 323–55.

KORN, H., and MARTIN, H., 1954, 'The Messum igneous complex in South-West Africa', *Transactions of the Geological Society of S. Africa*, vol. LIX, pp. 83–124.

MATHIAS, MORNA, 1956, 'The petrology of the Messum igneous complex South-West Africa', *Transactions of the Geological Society of S. Africa*, vol. LIX, pp. 1–35.
OFTEDAHL, C., 1953, 'The Igneous Complex of the Oslo Region, XIII—The Cauldrons', *Skrifter Det Norske Videnskaps-Akademi 1*, Oslo, 1 Mat.-Naturv. Klasse, No. 3.
PARK, C. F., and MACDIARMID, R., 1975, *Ore Deposits*, 3rd edn., W. H. Freeman, San Francisco.
REYNOLDS, D. L., 1937, 'Contact phenomena indicating a Tertiary age for the gabbros of the Slieve Gullion district', *Proceedings of the Geologists' Association*, London, vol. 48, pp. 247–75.
— 1950, 'The transformation of Caledonian Granodiorite to Tertiary Granophyre on Slieve Gullion, Co. Armagh, N. Ireland', *International Geological Congress*, 18th session, Gt. Britain, 1948, Part 3, pp. 20–30.
— 1951, 'The geology of Slieve Gullion, Foughill and Carrickcarnan: an actualistic interpretation of a Tertiary gabbro-granophyre complex', *Transactions of the Royal Society of Edinburgh*, vol. 62, pt. 1, pp. 85–143.
— 1954, 'Fluidization as a geological process and its bearing on the problem of intrusive granites', *American Journal of Science*, vol. 252, pp. 577–614.
— 1956, *Calderas and Ring Complexes, Gedenkboek H. A. Brouwer*, Mouton and Co., 'S-Gravenhage, Netherlands, pp. 355–79.
ROQUES, M., 1941, 'Les schistes cristallins de la partie sud-ouest du Massif Central français', *Mémoires Servir à l'explication de la Carte Géologique Détaillée de la France.*
SCHUILING, R., 1960, 'Le dome gneissique de l'Agout (Tarn et Herault)', *Mémoires de la Societe Géologique de France*, No. 91, pp. 1–59.
THOMSON, J. E., and WILLIAMS, H., 1956, 'Geology of the Sudbury Basin', *Annual Reports of the Ontario Department of Mines*, vol. 65, part 3, pp. 1–56.
WEGMANN, C. E., 1930, 'Über Diapirismus besonders im Grundgebirge', *Bulletin de la Commission Géologique de la Finlande*, No. 92, pp. 58–76.

# Chapter 12

# Volcanoes and their Products

A blast of burning sand pours out in whirling clouds.
Conspiring in their power, the rushing vapours
Carry up mountain blocks, black ash and dazzling fire.

*Lucilius Junior* (*c.* A.D. 50)

## General Aspects

As briefly indicated on p. 33, volcanic activity includes all the phenomena associated with the surface discharge of magmatic materials, solid, molten and gaseous, from pipes or fissures communicating with the heated depths. In addition to the eruption of hot gases and molten lavas from volcanoes, vast quantities of fragmental materials (pyroclasts) are produced by gas erosion and explosion. The great clouds of ash-charged gases and vapours which are the most conspicuous features of explosive eruptions may be luminous or dark, according as the fragmental material is incandescent or not. These 'fiery' and 'smoky' appearances, together with the glare reflected from glowing lavas beneath, were responsible for the formerly widespread notion that volcanoes are 'burning mountains'. Apparently supporting this description, the pyroclastic materials that drop from the volcanic clouds often resemble cinders and ashes, by which terms, indeed, they are still commonly described. There is, however, no implication that these volcanic products are combustion residues.

When eruptions take place through a vertical chimney the orifice is widened, by outward explosion and inward slumping, into a crater with

**Figure 12.1** Mayon, SW Luzon, Philippine Islands. A volcanic cone reaching a height of 2,421 m from a plain nearly at sea-level. Apart from radial irregularities due to rain gullies, avalanches and recent lava flows (e.g. in 1928, 1938, 1947 and 1953) the symmetrical form approaches mathematical perfection. Its rivals in this respect are Fujiyama (Japan) and Cotopaxi (Ecuador). The crater is about 500 metres across. (*J. P. Iddings*)

flaring sides. Volcanoes with the familiar cone-and crater structure (Figure 12.1) are said to be of the *central* type, because the activity is centralized about a pipe-like conduit. The lavas of a central volcano do not always issue from the crater; they may find it easier to break through at lower levels, forming 'satellitic' or 'parasitic' cones on the flanks of the main structure. Etna, the highest European volcano (over 3300 m), has more than two hundred of these satellitic vents and cones. This example makes Cotopaxi, the highest active volcano in the world (about 5974 m), all the more remarkable. It erupts only from its crater and has built up a perfect cone nearly as high as Etna, but founded on an uplifted platform of older rocks which itself stands 2743 m above sea-level (cf. Figure 12.3).

Other volcanic structures, more or less shield-like in form (e.g. Mauna Loa in Hawaii, Figures 12.26 and 12.27) are added to from fissures opening along the crest or on the flanks, or from chains of vents localized along such rift zones. This is particularly well illustrated by Heleakala on the island immediately NW of Hawaii (Figure 12.2). To some extent shield volcanoes are transitional to the Icelandic type of fissure volcano, exemplified by the Laki eruption of 1783 (p. 34).

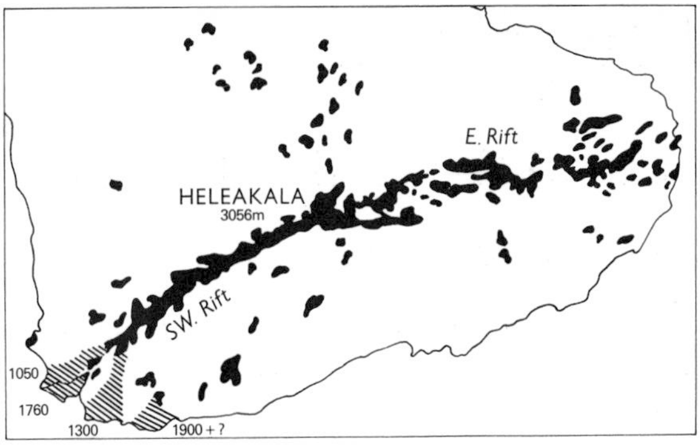


**Figure 12.2** Heleakala and its rift zones, eastern end of Maui, NW of Hawaii. Rift zones and cinder and spatter cones are in black. Lava flows of the last thousand years are shaded. (*After Stearns and Macdonald, 1942*)

The great fissure eruptions of former ages, which poured out prodigious floods of highly fluid basic lava and built up gigantic spreads of flat-topped plateau basalts, were probably similar to the Laki eruption except that the volume of lava was often greater and the proportion of pyroclasts generally

**Figure 12.3** Uplifted plateau in Chili, with volcanic peaks of the High Andes in the background (*American Geological Society: photo G. R. Johnson*)

very much smaller. It is thought that certain voluminous sheets of ignimbrite, like those of the volcanic district of New Zealand, may also have issued from fissures, but fortunately there have been no eruptions of this appalling kind during recorded historical times.

Some central volcanoes have gigantic depressions within the walls of their truncated summits, at first sight giving the impression that at some earlier stage of catastrophic activity the volcano concerned had not only blown off its own head but much of its body as well. Such a depression is called a *caldera* (Figure 12.24). The diameter of a caldera, which may be fifteen kilometres or more, is many times greater than that of an eruptive vent. It is essentially this contrast in size that distinguishes a caldera from a crater. Most of the calderas investigated, including many of the largest known, are found to have resulted from the collapse and engulfment of the original superstructure during great eruptions that covered the surrounding country with widespread sheets of 'ashes' and ignimbrites. 'Explosion' calderas that can properly be ascribed to the blowing-off of the upper part of a cone are relatively small. A crater so enlarged may be quickly restored by later eruptions to its normal proportions, as indicated by Figure 12.23.

A few volcanoes remain more or less continually in eruption for long periods, but intermittent activity is more usual, and some volcanoes, amongst them the more dangerous ones, have long intervals of repose, during which all external signs of activity either cease or are restricted to exhalations of steam and other gases and vapours from small vents called *fumaroles* or *solfataras*. As a volcano becomes extinct it passes through similar waning stages. Where the concentrated escape of hot exhalations takes place through an abundance of ground-waters, as in Yellowstone Park, favourable conditions are provided for geysers and hot springs (p. 271).

Long-dormant volcanoes have sometimes been considered extinct until a disastrous rejuvenation of activity proved otherwise. The classic example of this kind of tragic mistake was the unexpected outbreak of Vesuvius in A.D. 79. By that time the volcano had slumbered so long that no tradition of its prehistoric eruptions survived. When Pompeii and Herculaneum were founded, about the time when Nebuchadnezzar was king of Babylon, the fertile outer slopes of the mountain were clothed with woods and vineyards, and the rugged crags of the half-encircling collar, known as Monte Somma, were luxuriantly festooned with wild vines. For many centuries the volcano had been regarded only as a background of familiar scenery. In the year 79, however, after several years of local earthquakes, such as would now be clearly recognized as danger signals, the appalling eruption occurred that overwhelmed Pompeii in showers and blasts of fiery ashes, and obliterated Herculaneum in torrents of hot mud that set hard like cement.

Those who live far from volcanic districts sometimes express surprise that men and women should settle where their crops may be destroyed by vapours and ashes and their fields and vineyards blotted out by streams of lava, to say nothing of the danger to themselves. But volcanoes are not altogether unfriendly, nor are they uniquely so. Tornadoes, earthquakes and floods may be equally devastating. Calamities like the horrible doom that overtook Pompeii in 79 and St Pierre in 1902 (p. 220) are fortunately infrequent. Despite the fact that the energy liberated in the more catastrophic eruptions is equivalent to that of thousands of megaton hydrogen bombs, it is nevertheless true that, on average, all the volcanoes of the world fail to compete with motor transport as a menace to life. And compensating for the risk there is the irresistible attraction of the fertile soils for which volcanic districts are renowned. The decomposition products of the lavas, rich in plant foods, are carried down the rain-washed slopes to the plains beyond, and intermittent showers of ash, if not too heavy, rejuvenate the soil and add to its bulk. This double aspect of volcanic activity—destructive and beneficent—has been recognized from the earliest times of which legendary stories remain.

### Volcanic Gases

Steam is by far the commonest of the volcanic gases. It may be partly, and in some eruptions even wholly, derived from superficial sources, such as crater lakes, ground-waters and the sea. Steam blasts due to the heating of such extraneous water, without the eruption of incandescent materials, are often described as *phreatic* explosions, but it is generally far from easy to decide how much of the steam liberated in most eruptions is phreatic and how much is magmatic. Gases have been collected from cracks in small domes of spattered lava

**Figure 12.4** Collecting gases on the edge of the lava lake of Halemaumau, Kilauea, 28 May 1912. The pipe line penetrated about 30 cm into a crack through which gas was escaping from the spatter cone. See also Figs. 12.29 and 12.30 (*A. L. Day, Geophysical Laboratory, Carnegie Institution of Washington*)

formed over gas fountains on the edge of the lava-lake of Halemaumau (Figure 12.4). By this technique contamination by air is avoided as completely as is humanly possible. Besides steam (60 to 90 per cent) the gases are found to consist, in order of abundance, of carbon dioxide, nitrogen and sulphur dioxide, and smaller proportions of hydrogen, carbon monoxide, sulphur and chlorine. Similar assemblages of gases are liberated elsewhere from active lavas and fumaroles, together with various related compounds, such as sulphuretted hydrogen, hydrochloric and other acids, and volatile chlorides of iron, potassium and other metals. Incrustations of sulphur and chlorides, as well as of rare compounds, often in great variety, are deposited on cool surfaces.

Comparative study of the many gas analyses now available indicates that the samples considered to be least contaminated are richest in the elemental and less oxidized gases, such as hydrogen and carbon monoxide. Some of the steam results from oxidation of hydrogen, whence it follows that hydrogen and its highly reactive companions must be more abundant in the primary magmatic gases than in the samples that eventually reach the surface. Even the latter are still capable of reacting amongst themselves with generation of heat, and also with oxygen from the air or from the $Fe_2O_3$ of the lavas in which the gases are liberated or through which they are escaping. Striking evidence of such heat generation has been recorded and illustrated from Kilauea (p. 212 and Figure 12.30). From the spectra of the luminous gases emitted from Nyamuragira during its great eruption of 1934, Jean Verhoogen found that nitrogen was present in an 'active' form, i.e. in a more highly energized state than the relatively inert nitrogen of the atmosphere. This magmatic nitrogen must have acquired its additional energy in the depths and some of it, at least, still remained to be given up when the surface was reached. The source of magmatic gases is not known, but while they may have been acquired in part from pre-existing sedimentary and other crustal rocks, there are many reasons for inferring that additional gases have been contributed from deeper sources. The state of the nitrogen mentioned above, the isotopic constitution of carbon and oxygen in volcanic $CO_2$, and above all the fact of localized high temperatures, all suggest that the ascent of highly energized gases from the mantle may well be one of the fundamental causes of volcanic activity. A paroxysmal eruption is an overwhelming manifestation of the energy that can be liberated, and the propulsive power exerted, by volcanic gases.

The part played by gases in boring conduits through the crust and so providing passageways for the escape of lava has been discussed on p. 178. Recognition of the importance of fluidization, involving erosion and transport by gases, suggests that the proportion of gas to lava in volcanic activity may be much higher than is usually supposed. Observations at the basaltic volcanoes of Hawaii and Parícutin have suggested estimates ranging from 0·4 to 1·1 per cent of gas by weight. This seems trivial, although if we remember that 1 g of water (1 $cm^3$ at ordinary temperature and atmospheric pressure) yields 4500 $cm^3$ at 1000°C, the volume relationship becomes more impressive. In contrast, however, the gas output from the 1906 eruption of Vesuvius was estimated from the size of the conduit and the velocity of emission to be 'many times greater than the total mass of the ash and lavas'. Moreover, the high proportion of pyroclasts—mostly in the state of spray—produced during eruptions of the more acid lavas (p. 197), clearly implies an adequately high proportion of gas.

The density of magmas and molten lavas is reduced by the presence of dissolved gases, and is still further lowered when gas bubbles separate and turn the liquid into froth. Because of this effect, magma rich in gases is enabled to ascend to much higher levels than would otherwise be possible. The phenomenal height of Cotopaxi and other Andean volcanoes, perched as they are on high plateaus (Figure 12.3), is at least partly related to the exceptional abundance of the gases responsible for their devastating eruptions. Not only are the gases abundant, but even at that great height their propulsive powers are such that during the 1929 eruption of Cotopaxi a block weighing some 200 tonnes was hurled over 14 km from the crater, while clouds of self-explosive spray ascended several hundreds of metres above the summit.

## Lava flows

The temperature of freshly erupting lava is rarely much above the melting-point and, according to the composition and the gas content, it usually ranges between about 900° and 1200°C. Basic lavas, such as basalt, are generally the hottest, though not invariably so. The mobility of molten lavas depends on the same factors. While sufficient gas remains in solution a lava may continue to flow until the temperature falls to, say, 700°C. But when the same lava has crystallized and nearly all its gas has been expelled, the temperature necessary to soften it again is found to be several hundred degrees higher. It follows that loss of gases induces consolidation.

So long as they retain the properties of coherent liquids, basaltic lavas tend to flow freely for long distances, even down gentle slopes, before they come to rest. The speed depends on the mobility and the slope; when confined to a steeply descending valley a lava stream may locally advance at 45 to 65 km an hour. But such speeds are very rarely attained. Even 15 km an hour is unusual, and as the lava cools and congeals, or liberates expanding gases and becomes froth-like, its viscosity increases and movement becomes sluggish.

In contrast, silica-rich lavas (if coherent) are always so stiff from the start that they congeal as thick tongues before they have travelled far. But, as we have already seen (p. 73), the widespread formations that used to be accepted as lava flows of rhyolite or dacite are proving on more detailed investigation to be sheets of ignimbrite. These were erupted as clouds of incoherent particles of lava, swiftly transported by expanding gases with velocities far in excess of those attainable by coherent lava flows. It is important to remember that in the passage from liquid (with gas in solution) to froth or foam (with gas in bubbles) the viscosity increases very considerably, whereas if the mixture of gas and liquid bursts into spray, the viscosity suddenly drops to a very low value.

Newly consolidating flows of basic lavas assume forms of two contrasted types, dependent on whether or not they pass into a coarse froth. The two kinds are described in English as *block* lavas (Figure 12.5) and *ropy* lavas (Figure 12.6), but technically and internationally they are known by their Hawaiian names, *aa* (pronounced *ah-ah*; meaning rough or spiny) and *pahoehoe* (meaning satiny). *Aa* or block lava tends to form in partly crystallized flows from which the gases escape in sudden bursts. The congealing crust expands by gas-inflation and breaks into a wild assemblage of rough, jagged, scoriaceous blocks, often with dangerously knife-sharp edges. Through openings in the piled-up front the glowing pasty lava is revealed as it pushes between the clinkers. The latter come sliding down at intervals, to be rolled over and

**Figure 12.5** *Aa* or block lava: *above* Advancing front of a flow from Hekla; eruption of 1947 (*G. Kjartansson*) *below* Lava flow from Etna, near Nicolosi; eruption of 1886 (*Tempest Anderson*)

buried beneath the sluggishly advancing flow with its load of grinding blocks.

At Mauna Loa it has been observed that when gas is violently escaping from the summit vents the rapidily outflowing lava soon acquires the characters of the block variety. On the other hand lava issuing from fissures lower down the slopes is usually of the ropy type. This illustrates a feature of many volcanoes: the far more copious escape of gas from the summit crater than from satellitic vents on the flanks. Ropy lava cools more slowly and remains mobile for a longer period than the block variety. Gas escapes tranquilly and any bubbles that may eventually form are rarely of more than microscopic sizes. The flow congeals with a smooth skin that wrinkles into ropy and corded forms like those assumed by flowing pitch (Figures 12.6 and 12.29). At the front, before such a flow comes to rest, the still mobile lava advances by the protrusion of bulbous lobes and fingers through its own half-consolidated plastic skin. It sometimes happens that after the upper surface

**Figure 12.6** Cascades of *pahoehoe* (ropy) lava, Kilauea, Hawaii (*W. C. Mendennall, U.S. Geological Survey*)

**Figure 12.7** Pillow structure of a late Precambrian submarine lava, now exposed on a glaciated surface near Tayvallich (Loch Awe district), Strathclyde (*Institute of Geological Sciences*)

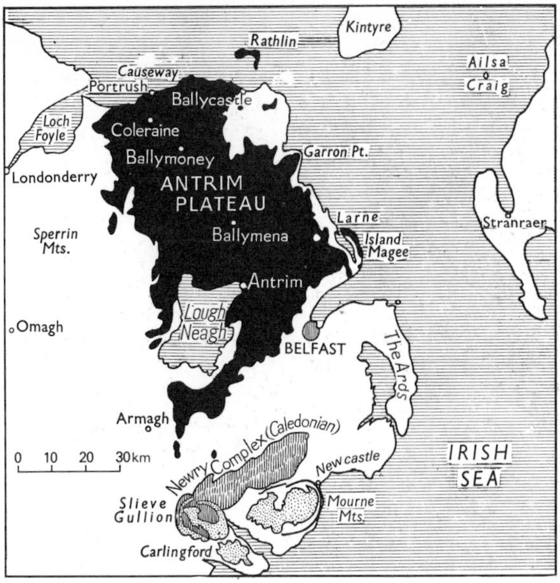


**Figure 12.8** Map of north-eastern Ireland showing the Tertiary plateau basalts (black), the tertiary volcanic and intrusive centres of Slieve Gullion (see Fig. 11.17), Carlingford and Mourne (dotted), and the Caledonian Newry Complex (broken vertical shading)

and edges of a flow of this kind have solidified, the last of the molten lava drains away, leaving an empty tunnel. Some of the lava caves of Iceland and Hawaii are famous for the shining black 'icicles' of glass which adorn their roofs.

When lava of the ropy type flows over the sea floor, or otherwise beneath a chilling cover of water, it consolidates with a structure like that of a jumbled heap of pillows and is then appropriately described as *pillow lava* (Figure 12.7). By the time each emerging tongue of lava has swollen to about the size of a pillow the rapidly congealed skin prevents further growth. New tongues which then exude from the base of the advancing front, or through cracks in the glassy crust, similarly swell into pillows, and so the process continues. Pillow structure has been seen actively developing where basaltic flows from Pacific island volcanoes entered the sea. Moreover, pillow lavas are extruded at mid-oceanic ridges, and are well known on the oceanic floors.

**Figure 12.9** Index map of Iceland showing the localities illustrated in Figs. 3.5, 3.12, 3.13, 3.14, 12.5, 12.15, 12.46, 14.3, 18.33, 20.56, 23.54

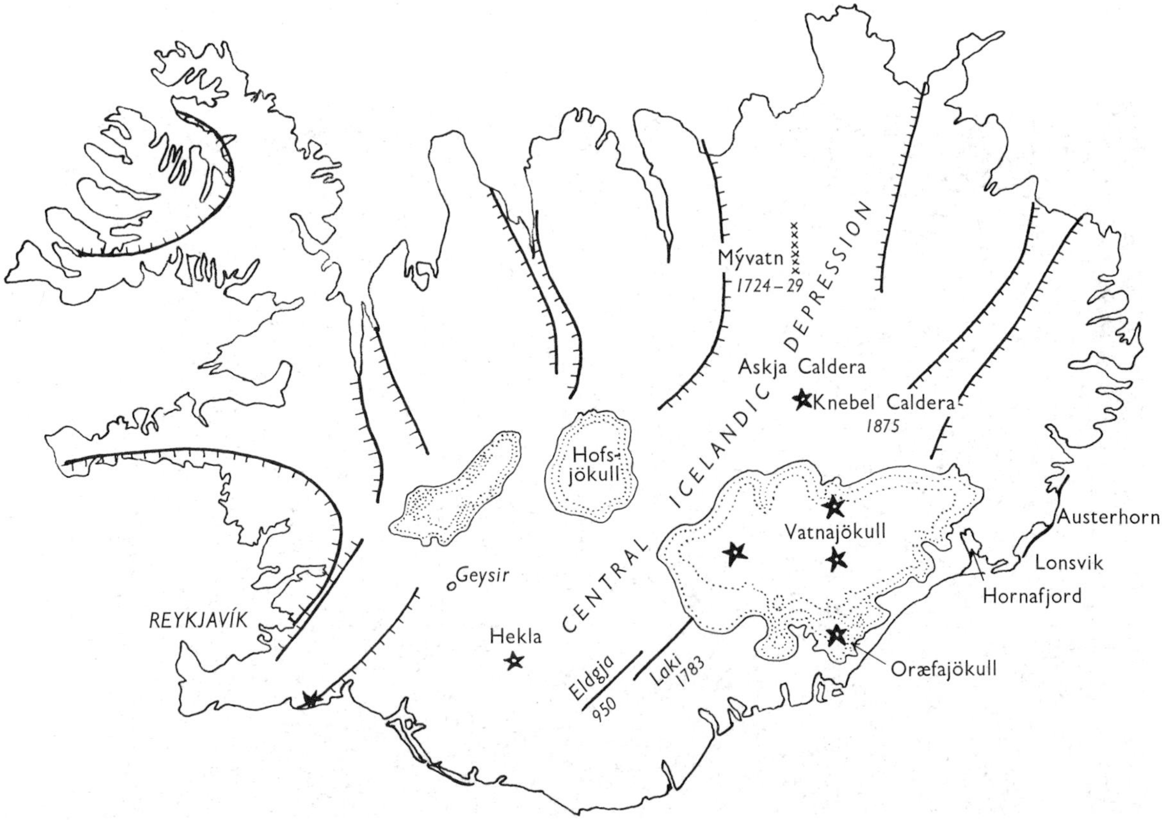

Columnar structure, already fully discussed on p. 65, develops across the lower parts of sheets of lava which have consolidated under stagnant conditions (Figures 1.1 and 1.2). It is especially characteristic of fine-grained plateau basalts of uniform texture, uninterrupted to any appreciable extent by vesicles. The largest area of plateau basalts in the British Isles is that of Antrim (Figure 12.8), but this is only a small part of the far greater Brito-Arctic province, extensive areas of which were flooded with basalts in early Tertiary times. Although considerable tracts have foundered and are now beneath the sea, over 150,000 km$^2$ still remain in Antrim, the Inner Hebrides, the Faröes, Iceland and East and West Greenland. Shield and other types of volcanoes arose later at certain localities, including Iceland, where activity was renewed during the Pleistocene and still continues as vigorously as anywhere in the world (Figure 12.9). In our own islands, from Slieve Gullion (Figure 11.17) to Skye (Figure 12.10) and St Kilda, we see only the roots of the ancient volcanoes exposed by millions of years of denudation.

Plateau basalts covering areas of over 500,000 km$^2$ occur in the Columbia and Snake River region of the north-western United States (Miocene to Recent), the Deccan of India (early Tertiary) and the Paraná region of South America (Jurassic). Other vast areas which were flooded with basalts in Jurassic or Tertiary times occur in Mongolia and Siberia, in Arabia and Syria, in many parts of Africa (e.g. Ethiopia, around the Victoria Falls and along the Drakensberg behind Natal) and in parts of Australia. Where the flows thinned out and denudation has since removed them, the underlying rocks are seen to be penetrated by swarms of dykes. In many other regions (as mentioned on p. 163), such feeding channels failed to reach the surface, but riddled the stratified rocks just below with innumerable sills. Altogether, more than 4 million km$^3$ of basalt have been transferred from the depths during the last 180 million years or so.

**Figure 12.10** Roots of the chief centre of Tertiary igneous activity in Skye. Red Hills (granophyre and granite) with Cuillin Hills (basaltic volcanic rocks and sills, now largely gabbro) in the left background (*G. P. Abraham, Keswick*)

## Volumes (in km³) of Lavas and Pyroclasts discharged during major eruptions

| *Lavas: mainly basaltic* | |
|---|---|
| Iceland (Prehistoric) | 43 |
| Laki, Iceland, 1783 | 12 |
| Kilauea, Hawaii, 1840 | 5·5 |
| Mauna Loa, Hawaii, 1955 | 5 |
| Nyamuragira, Zaïre, 1894 | 3·6 |
| Nyamuragira, 1938 | 2·5 |
| Hekla, Iceland, 1947 | 1·3 |
| Sakurajima, Japan, 1914 | 1·2 |
| Etna, Sicily, 1669 | 1 |
| Parícutin, Mexico, 1943–52 | 0·7 |
| Hekla, Iceland, 1845 | 0·5 |
| Mauna Loa, Hawaii, 1950 | 0·5 |
| Etna, Sicily, 1852 | 0·4 |

| *Pyroclasts: mainly rhyolitic-andesitic tephra* | |
|---|---|
| Tambora, Sumbawa, Indonesia, 1815 | 100 |
| Crater Lake, Oregon (Prehistoric) | 40–50 |
| Katmai and Novarupta, Alaska, 1912 | 20 |
| Krakatau, Indonesia, 1883 | 18 |
| Oraefajökull, Iceland, 1562 | 10 |
| Coseguina, Nicaragua, 1835 | 10 |
| Bezymianny, Kamchatka, 1956 | 2·8 |
| Parícutin, Mexico, 1943–52 | 1·3 |
| Hekla, Iceland, 1947 | 1 |
| Sakurajima, Japan, 1914 | 1 |

(In most eruptions the quantities are much smaller, e.g. for Vesuvius, 1906, the estimates are: lava 0·02, tephra 0·24 km³)

## Volumes (in km³) of Regional Assemblages

(Minimum estimates only)

| *Basaltic lavas: mainly plateau basalts* | |
|---|---|
| Deccan, India, Eocene | 700,000 |
| Columbia and Snake River, U.S.A. — Miocene and later | 300,000 |
| Siberia, Early Jurassic | 250,000 |
| Paraná, S. America, early Jurassic | 200,000 |
| L. Superior, Precambrian | 120,000 |
| Iceland (above sea-level) | 55,000 |
| Clyde Plateau, Scotland, Carboniferous | 2,500 |
| Antrim Plateau, Tertiary | 800 |

| *Pyroclasts: mainly tephra and ignimbrite flows* | |
|---|---|
| Great Plains, U.S.A. — Cretaceous | 15,000 |
| Great Plains, U.S.A. — Tertiary | 15,000 |
| Patagonia, Tertiary | 10,000 |
| Taupo-Rotorua, N.Z., Plio-Pleistocene | 8,000 |
| Andes of Peru, Pleistocene | 4,000 |
| Colorado, U.S.A., Miocene | 4,000 |
| Yellowstone Park, U.S.A. — Pliocene and later | 2,400 |
| Lake Toba, Sumatra, Prehistoric | 2,000 |

The estimates listed in the adjoining tables are of interest in showing the relative proportions of the different kinds of lavas and pyroclasts. Amongst lavas basalt easily takes first place, and it may be added that from most of the basaltic island volcanoes of the Pacific only about 1 per cent of the erupted materials are pyroclasts. On the other hand, the volumes of the tuff flows and other pyroclasts erupted from many of the andesite-rhyolite volcanoes of the Pacific borders and of Indonesia are immensely greater than those of the lavas, although at times there have been some notable exceptions to this 'rule'. The regional volumes of acid pyroclasts are very impressive, but even so they are no more than a small percentage of the gigantic volumes of granitic batholiths. The Coast Range batholiths of British Columbia, for example, occupy some 3,000,000 km³, and those of California (Upper and Lower) about half that volume.

## Pyroclasts

Pyroclasts have already been briefly described on p. 71, where fragmental volcanic *ejecta* or *ejectamenta* from three significantly different sources were distinguished; also on pp. 178–82, where some of the various fragmentation processes—explosion, shock waves and gas erosion—were considered. Pyroclastic materials from the three sources referred to are described (not very happily) as:

(*a*) *essential*, when they consolidated from 'live' lava during the eruption concerned;
(*b*) *accessory*, when they consist of fragments of the 'dead' lavas and pyroclasts of earlier eruptions; and
(*c*) *accidental*, when they consist of fragments of crustal rocks brought up from the foundations of the volcanic edifice.

In 1954 the term *tephra* (Greek for ashes) was introduced by Thorarinsson for loose pyroclasts and its use has rapidly spread from Iceland, particularly for volcanic products in which the essential materials predominate. To avoid the obvious ambiguity of the last few words, the term *tephra*, used without qualification, will be adopted for unconsolidated pyroclasts of the 'essential' class.

The fragmental materials blown into the air shower down at various distances from the place of eruption according to their sizes, the force and direction of propulsion, and the force and direction of the winds. Except in violent outbursts, the larger fragments, including bombs, lumps of scoria or pumice, or blocks of older rocks, commonly fall back near the crater rim and roll down the inner or outer slopes, forming chaotic deposits of *agglomerate* or *volcanic breccia*. Volcanic bombs represent clots of live lava most of which congealed, at least externally, before reaching the ground. Some of them have globular, spheroidal or spindle-shaped forms, acquired by rapid rotation during flight (Figure 12.11). Others are of less regular shape, either because they were stiff from the start or because, like *aa* lava, they lost their gases in explosive spurts. Some of these have a network of gaping cracks in a glassy crust and are appropriately called *bread-crust* bombs. It should be remembered that the larger masses of dead lava or foundation rocks are not bombs; they should always be referred to as ejected *blocks*. If such blocks are well rounded (as, for example, in the diamond pipes described on p. 180), their occurrence is an indication of long-continued attrition in the conduit, probably under conditions of highly turbulent fluidization. Explosion alone inevitably yields angular blocks.

**Figure 12.11** Volcanic bombs. (A) Basalt, Puy de la Vache, Auvergne ($\frac{1}{10}$ natural size); (B) Breadcrust bomb of dacite, Mt Pelée, 1901 ($\frac{1}{10}$); (C) Basalt, Patagonia ($\frac{1}{3}$) (*A. Lacroix*)

Smaller fragments, about the size of walnuts or peas or seeds, are called *lapilli* (Italian, 'little stones'), if they are not conspicuously vesicular. If they are, terms like 'pumice', 'scoria' and 'cinders' are used, without much reference to the size of the fragments, except that all the finer materials (whatever their structure) are referred to collectively as ash (Figure 12.12). These and the smaller lapilli and cinders etc., accompanied by occasional bombs and blocks, fall mainly on the slopes and beyond, and form deposits which, when more or less indurated, are known as *tuffs*. The parabolic path followed by large bombs sometimes carries them far from the crater and shows, moreover, that they acquired velocities far greater than could be attributed to the gas stream issuing from the volcanic vent. Possibly the escape of highly explosive gas from the bomb itself, if properly directed, would transform the bomb into a jet-propelled missile of natural origin. Showers of augite, feldspar and other minerals that had already crystallized from the magma (before its explosive expulsion) occasionally fall from the volcanic clouds and contribute layers of concentrated crystals to the ashes and tuffs, which then become attractive collecting grounds for mineralogists.

The finer particles, ranging down to dust size, and including shards and splinters of glass, often travel far beyond the cone before they descend. The most notable deposits of this kind are the great sheets of ignimbrite described on p. 73. The finest dust may be carried to such great heights—

**Figure 12.12** Country school buried by showers of volcanic ash erupted from Sakurajima, Japan, in 1914 (see Fig. 12.47) (*F. Perret, Carnegie Institution of Washington*)

20 to 50 km—that it takes part in the planetary circulation of the atmosphere. Microscopic volcanic dust from the catastrophic eruption of Krakatau in 1883 encircled the globe, and its dispersion through the atmosphere was responsible for the vividly coloured sunsets that were seen during the following months. In 1956 a violent eruption of the volcano Bezymianny, in Kamchatka (Figure 12.24), reached its climax with the expulsion of a stupendous jet of incandescent tephra. The finer particles were carried to a height of 45 km and formed a cloud well up in the stratosphere. Spiralling winds brought samples of the volcanic dust to Britain within three days. In this eruption the initial velocity of the ejected material was between 500 and 600 m/s, that is, nearly twice the velocity of sound.

## Types of Central Eruptions

It will be readily understood that eruptions vary widely in character and intensity according to the pressure and quantity of gas, and the viscosity of the lava from and/or through which it is released. Fissure eruptions, discharging floods of basaltic magma which flow nearly as freely as water and come to rest with almost horizontal surfaces, are presumably the type in which both gas content and viscosity are least. The nearest approach to this type in historical times was the Laki eruption of 1783, but since it finished up as a string of conelets (Figure 3.14), it is sometimes distinguished as the *Icelandic* type. The eruptions of central volcanoes depend still more on the nature of the gas activity. Several well-defined types (Figure 12.13) have been recognized, of which the following are the chief.

*Hawaiian Type* Effusion of mobile lava is dominant, either from 'lakes' of lava occupying pit-like craters or from fissures, gas being generally liberated more or less quietly. At times, however, fountains of incandescent spray are thrown up by the rapid emission of spurting gases (Figures 12.14, 12.30 and 12.32). Most of the spray returns to the molten lava lake from which it came and deposits of tephra are sparse. When caught by a strong wind the blebs of molten lava are drawn

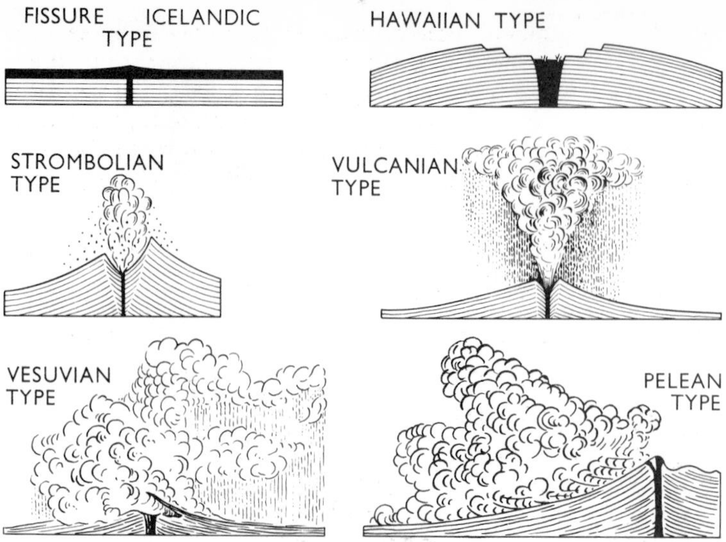


**Figure 12.13** Diagrams to illustrate the chief types of volcanic eruptions

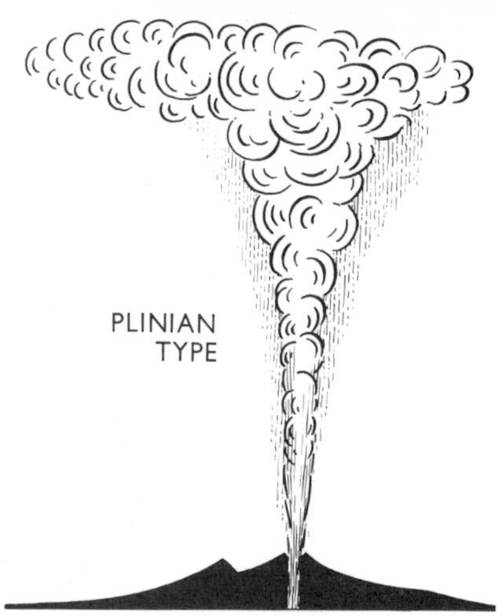


**Figure 12.14** Hawaiin type of eruption. Incandescent fountain of molten basaltic spray about 365 m high, Kilauea Iki eruption of 1959. Temperature between 1120° and 1190°C. Photographed at 5 a.m., 29 November. See also Fig. 12.32 (*Hawaiian Volcano Observatory, United States Geological Survey*)

out into long glassy threads known as *Pele's hair*, Pele being the Hawaiian goddess of fire.

*Strombolian Type* When less mobile lava is exposed to the air in a crater, the pent-up gases escape more spasmodically, with moderate explosions which may be rhythmic or nearly continuous. Clots of lava, characteristically incandescent, are blown out, to form bombs or lumps of scoria (Figure 12.15). Stromboli, one of the Lipari Islands north of Sicily, normally behaves in this way, its minor eruptions recurring at short intervals ranging from a few minutes to an hour or so. In phases of more intense activity there may be outflows of coherent lava, while violently expanding gases ascending from a deeper source tear their way through the lava of the conduit and give rise to luminous fountains playing above the crater.

**Figure 12.15** Strombolian type of eruption on 30 April 1947 from a crater near the south-western end of the 5 km long fissure along the summit ridge of Hekla, Iceland, (Fig. 2.9) from various parts of which lava and tephra were erupted between 29 March 1947 and 21 April 1948 (*P. Hannesson*)

*Vulcanian Type* (named after Vulcano, also in the Lipari group) The lava is more viscous and pasty, and quickly crusts over between eruptions. Gases accumulate and gather strength beneath the congealed cover and blow off at longer intervals, with correspondingly greater violence. The viscous lava and its crust are shattered to angular fragments of all sizes down to the finest dust. The resulting ash-laden volcanic clouds are dark and often black, incandescent bombs being rarely seen, even at night. The clouds assume a convoluted or 'cauliflower' shape as they ascend and expand. The major eruptions of many volcanoes begin with a Vulcanian phase whenever an obstructed vent has first to be cleared out. They may end in the same way, with outbursts continuing intermittently until the waning gas activity no longer has the power to eject the debris that avalanches into the vent from the unstable crater walls. If it is possible to make the distinction, eruptions of the latter kind are known as *pseudo-Vulcanian*. An important variety of this sub-type is the 'steam-boiler' or *phreatic* eruption, due to explosions beneath or within obstructing materials, where waters of surface origin have come into contact with hot lava and flashed into steam. An example is illustrated by Figure 12.31.

*Vesuvian Type* This is a paroxysmal extension of the Vulcanian and Strombolian types, the new and specific feature being the extremely violent expulsion of magma which has become highly charged with gas during a long interval of superficial quiescence or mild activity. Because of the concentration of streams of explosive gas through the conduit, while coherent lava flows are escaping from fissures and vents on the flanks (Figure 12.16), the conduit may be emptied of lava down to a considerable depth. As a result of this reduction of overhead pressure, the underlying magma bursts into an explosive spray and expels itself through the crater as vast clouds of 'cauliflower' form, which may be brightly luminous when seen

**Figure 12.16** Vesuvian type of eruption, Vesuvius, 26 April 1872 (*G. Sommer*)

at night. These ascend to great heights, and from them showers of ashes are widely distributed.

*Plinian Type* The most violent Vesuvian eruptions sometimes culminate in a stupendous blast of uprushing gas, which rises to a height of many kilometres, and there spreads out into an expanding cloud of globular masses of gas and vapour. The proportion of ash is low and is mainly confined to material eroded from the walls of the pipe and crater. The first recorded example of this type was observed by Pliny during the disastrous eruption of Vesuvius of A.D. 79.

*Peléan Type* Here a distinctly mysterious combination of high viscosity and more or less delayed explosiveness is reached. Upward escape is prevented by the growth of an obstructive dome above the conduit (Figure 12.40). Intermittent cloud-encircled masses of 'tightly compressed froth' force a passage or are squeezed through points of weakness near the base of the dome, and each of these sweeps down the slopes as an intensely hot avalanche of incoherent self-explosive lava, lubricated by constantly expanding gases and vapours. These downward-rolling explosive blasts, one of which wiped out St Pierre in 1902, are commonly referred to as *nuées ardentes* (Figure 12.17). Perret, who had observed them at close quarters, defined a *nuée ardente* as an 'avalanche of an exceedingly dense mass of hot, highly gas-charged and constantly gas-emitting fragmental lava, much of it finely divided, extraordinarily mobile, and practically frictionless, because each particle is separated from its neighbours by a cushion of compressed gas.'

*Nuées ardentes* are sometimes referred to as 'glowing avalanches', perhaps for lack of a better English term. They are not always 'glowing' as the French term suggests, and they commonly look menacingly black by day. But more important is the fact that the use of the word 'avalanche' is misleading without the qualification that Perret specifically introduced. The marked effect of

**Figure 12.17** Peléan type of eruption. *Nuée ardente* of 16 December 1902 approaching the sea down the valley of the Rivière Blanche, Martinique (*A. Lacroix*)

topography in guiding the course followed by *nuées* shows that their movements are dominantly controlled by gravity. In this respect they behave essentially like lava flows, but instead of being coherent they consist of self-expansive particles and are far more mobile than any lava flow could be. The movement of an avalanche is due to gravity alone. Unlike a *nuée* an avalanche has no internal resources of energy, no power of self-inflation save one that the *nuée* also gains during its development (and far more effectively because of the high temperature), i.e. the expansive force of drawn-in, entangled and highly compressed air.

It should be noticed that in eruptions of the Strombolian-Vesuvian-Plinian series the main escape of gas takes place vertically and centrally through the summit crater. Peléan activity presents a striking contrast. Here, what is erupted vertically and centrally is mainly lava that is relatively poor in gas and so viscous as to be nearly solid. This almost congealed lava is forced up the conduit to accumulate at the summit, or within the crater if there is one, as a highly obstructive dome. The gas-rich material which swiftly develops into a *nuée ardente* puffs or leaps out laterally, sometimes quietly, sometimes explosively, from a point of weakness that is generally low down on the flanks of the dome. The beginning of one of the 1902 *nuées* of Mont Pelée has been well described by Anderson and Flett. They 'saw a black cloud suddenly appear in the V-shaped break of the crater wall. It hung there for a moment as if suspended, then plunged downwards at extreme speed.' In the *nuée ardente* astounding quantities of gas come *out* of the erupted material *after* its arrival at the surface, whereas in the Vesuvian and related types most of the gases ascend from greater depths and burst *through* the molten lava in the conduit; much less gas escapes from the lava that issues from openings on the flanks.

The gas in the parental material of a *nuée ardente* seems to be present in at least two very different ways. In one of these states the gas is immediately expansive and on occasion violently explosive, as it might be if it were present as bubbles in hot froth; this is probably the gas mainly responsible for the ejection of the *nuée* material. Once liberated, however, the latter then behaves like a powerful rocket propellant that gives up its energy more or less evenly as successive surfaces are exposed, instead of exploding bodily in a single gigantic outburst. But if a flaw in a rocket propellant suddenly develops into an internal crack, the resulting increase of surface is liable to start a disastrous explosion. Some of the larger masses of *nuée* material have been seen to explode during their turbulent flight, and also when colliding with obstructions in their path. This behaviour and indeed the whole process of *nuée* development—the passing of a spray of foaming particles into ever finer and finer spray—strongly suggests the liberation of gas in proportion to the surface exposed, so long as an adequate supply remains. One wonders with Perret if this gas is in some unfamiliar latent state of high-internal-pressure solution of chemical combination. We still do not know.

It should be added that there are *nuée ardente* eruptions that differ in certain fundamental ways from those described as Peléan. In the latter the bulk of the material avalanches downwards under gravity, while the expanding clouds mount higher in successive bursts and leaps until the *nuée* reaches the sea at the foot of the mountain. In the *St Vincent* or *Soufrière* type explosions directed vertically from a domeless crater are accompanied or immediately followed by downward-rushing *nuées* which well out of the crater on all sides, though naturally most voluminously where there are breaks in the rim. Because of the initial vertical explosions the expanding clouds are highest above the summit. Apart from this effect, the behaviour of such *nuées* has been likened to the boiling over of milk. This analogy was originally suggested by C. N. Fenner to clarify his conception of the mode of emission of the 'incandescent tuff' deposits of the Valley of Ten Thousand Smokes in Alaska. These and the neighbouring tephra deposits were erupted from Katmai and Novarupta in 1912. The eruptions were probably like the St Vincent type in coming from what had originally been open craters, but otherwise there were notable differences, particularly in that the materials discharged had the composition of rhyolite instead of dacite. *Katmai* had a domeless crater before its eruption, and ended with a large caldera. The *Novarupta* eruptions are thought to have issued from neighbouring fissures as well as from the summit crater; the latter was found to be obstructed by a plug-like dome when the eruptions were over. A. G. MacGregor has admirably discussed the nature and mechanism of the various types of *nuée ardente* eruptions (see p. 229). To the types briefly described above he adds (*a*) avalanches made up largely of gas-generating blocks, blown out laterally from the exposed flank of a volcanic dome: a type—amongst others—

that has been observed in the eruptions of Merapi in Java (see Figure 17.18); and (*b*) a hypothetical type to include the 'incandescent tuff-flows' responsible for the more widespread flows of ignimbrite, which are referred to vertical discharges from innumerable fissures.

## Volcanic Cones and Related Forms

Eruptions that eventually leave a great caldera in place of the former volcanic edifice, like that of Krakatau in 1883, are likely to be extreme examples of the St Vincent or Katmai types. But this can only be conjecture, because an eruption on so catastrophic a scale hides its major operations behind an impenetrable curtain of suffocating darkness. Consideration of calderas is therefore deferred to a later section (p. 225). There are also certain volcanic structures, such as the ring-craters described below, which may have resulted from types of volcanic activity not yet witnessed or, at least, not yet recorded. The more familiar structural forms, however, result from types of eruption that are already well known, e.g. the building of typical cones by Strombolian-Vulcanian-Vesuvian eruptions, and of the more dome-shaped shield volcanoes by those of Hawaiian type.

fluidization; there are, of course, innumerable examples in which both processes have operated. *Explosion vents* are common enough, and present no particular problem. They are generally small craters with flaring sides, and the pyroclasts include a high proportion of coarse angular material.

*Fluidization craters* On the other hand there are wide shallow craters with nearly flat floors and low rims, sloping gently outwards to thin deposits of ash spread widely over the surrounding countryside. Most of the pyroclastic material is excessively fine, and invariably so at the base of the deposit, while the larger fragments are so well rounded that they have often been misleadingly described as 'bombs'. Fresh tephra is generally represented by minute lapilli of approximately spherical form, and these tend to be uniformly mixed with debris from the perforated rocks. The most recent ring-craters of this kind occur near Ruwenzori in the western rift valley of Africa (Figure 12.18). Although there are native traditions suggesting that some of them may have been formed as recently as a few hundred years ago, no actual eruption producing such ring-craters has ever been witnessed and described. Nevertheless, in the light of the treatment of volcanic pipes and fluidization on pp. 175–9, it is clear that copious gas streams and the attendant phenomena of

### THE CHIEF TYPES OF VOLCANIC FORMS

| *Mainly pyroclastic* | *Pyroclasts and lava flows* | *Mainly lava flows* |
|---|---|---|
| 1. Ring craters or ash rings—explosion vents and fluidization craters<br>2. Ash and cinder (scoria or pumice) cones | 3. Composite volcanoes and strato-volcanic cones | 4. Tholoids and cumulo-domes, of mainly internal growth<br>5. Domes of external growth—shield volcanoes |

1. *Ring-craters*, otherwise known as ash or tuff rings, are the surface expression of perforations of the crust through which gases and vapours escaped, carrying pyroclasts of various sizes according to the velocities of the gas streams. The floor of the crater and the low ring surrounding it consist mainly of pyroclasts in which older rocks (volcanic or foundation) are more abundantly represented than fresh volcanic material. Two contrasting types can be distinguished, owing their characteristic features to explosion or to fluidization must have characterized these unrecorded eruptions. The flat floors of the craters, which may be up to three kilometres across; their well-mixed materials; and the local occurrence of outcrops of the neighbouring country rocks veined with tuffisite; all suggest that during the eruptions the materials of the floor were in the states known in industry as the 'expanded' and 'boiling' bed. As with all volcanic activity, explosions (including those of phreatic origin) would be expected to occur from time to time. The occasional occur-

**Figure 12.18** Ring-crater, Lake Nyamununka, SE of Ruwenzori in the Western Rift Valley of Africa (*A. D. Combe*)

rence of well-rounded blocks, some of them far-flung, and of particular layers in which they are abundant, constitute evidence of explosive propulsion; but, apart from freshly shattered fragments, the propelled materials had already been rounded by a long subjection to gas-and-spray erosion of the sand-blast type.

Where the floor of a ring-crater is deep enough to fall below the ground-water level of the district it becomes the site of a shallow lake, which is then able to defy evaporation, as in Uganda (Figure 12.18). In the Eifel volcanic district a ring-crater occupied by a lake is called a *maar* (Figure 12.19).

2. An *ash or cinder cone* is built up when a sufficient supply of tephra is erupted. The profile is determined by the angle of rest of the loose material that showers down around the vent. Fine ash comes to rest at angles of 25° to 35°, while nearer the summit the coarser fragments may stand at 40° or more. In 1538 an eruption suddenly broke out in the country west of Naples and in a single Vulcanian outburst, lasting only a few days, Monte Nuovo, about 130 m high, was constructed. The first materials to be thrown out were angular fragments of the country rocks, but these were soon followed by ash and bits of pumice having the composition of trachyte.

One of the more recently born volcanoes, Barcena (also called El Bouquerón), on the island of San Benedicto off the Pacific coast of Mexico (south of Lower California), built a similar cone of trachytic ash and pumice which reached a height of 300 m in the first twelve days of August 1952. Vulcanian eruptions then diminished and a few weeks later the crater became plugged with highly viscous lava. The plug ceased to rise in December, when lava began to flow into the sea from the base of the cone (Figure 23.61). Since February 1953 there has been no further activity.

3. *Composite volcanoes.* Barcena introduces a type of composite volcano in which the growing cone remains pyroclastic, but is gradually surrounded by a gently sloping platform of lava flows that have issued from vents low down on the flanks or near the base. With this arrangement the later flows pile up on the earlier ones or on the deposits of tephra which may have covered them. This was the kind of structure built up by Jurollo in Mexico, long famous as one of the best known of new volcanoes, and by the now even more celebrated Parícutin, 105 km to the NW (Figure 12.21). In 1759 Jurollo burst out in the middle of a plantation with dramatic violence and soon constructed a shapely cone of basaltic scoria, surrounded by ash deposits that extended for 240 km in some directions. The first of a long series of lava flows appeared in 1764. Ten years later the volcano became quiescent, except for a few fumaroles, and by that time the scoria cone had reached a height of 400 m above its peripheral foundation of alternating ashes and lavas.

The birth of a volcano is rarely witnessed, but this happened on 20 February 1943, near the village of *Parícutin* (about 300 km west of Mexico City) after which the new volcano was named. For about a fortnight earthquake shocks had been felt with increasing frequency, totalling about 300 on 19 February. The next afternoon a fissure opened in a field that was being ploughed, and half an hour

**Figure 12.19** The Laacher See, a typical *maar* in the Eifel volcanic district west of the Rhine between Bonn and Coblenz. This ring-crater was formed at some time after an eruption of tephra that has been dated at about 11,000 years ago.

**Figure 12.20** Parícutin, 23 March 1943, as seen from the outskirts of San Juan Parangaricutiro looking south. The turbulent eruptive column reached a height of 6 km. In the face of advancing lava flows Parícutin village was abandoned in April, and the town of San Juan in June 1943. Both places were overwhelmed, as indicated by the locality map below, which shows the area eventually covered by lavas. See Fig. 12.22 (*Paul Popper Ltd*)

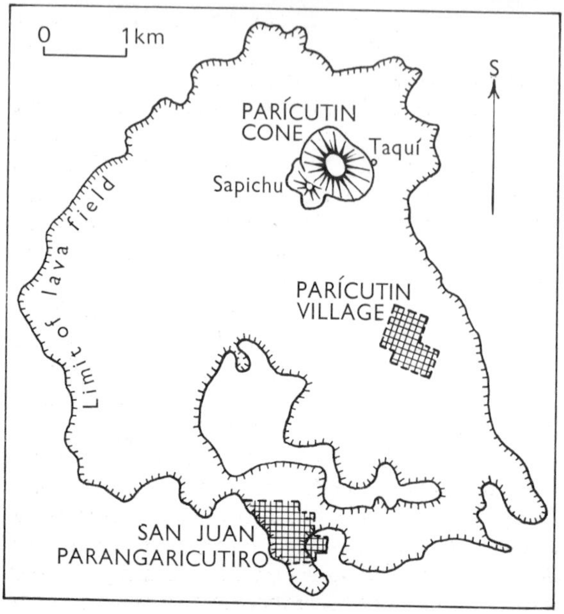


later a column of grey 'smoke' began to ascend from a vent only 30 cm or so across. As the vent gradually widened by slumping and eruption, its contents appeared to be 'vigorously boiling, while sparks and hot stones were thrown out'. Two of the latter were collected and found to be fragments of the basaltic rock perforated by the growing volcanic pipe. Towards midnight 'the activity of the new volcano became violent; incandescent rocks were violently ejected in great numbers, and a large eruptive column, accompanied by frequent lightning flashes and a tremendous roaring sound, arose from the newly formed vent. This change in activity suggests that the advancing gases of the initial phase were followed by the rising lava column and that it reached the surface at this time' (Foshag and Gonzalez: see p. 229 for reference). Next morning it was seen that a cone about 40 m high had been constructed. The eruption was now in full blast (Strombolian type) and basaltic bombs were being flung 1·5 to 3 km high, some of them exploding into incandescent fragments during flight. The first coherent lava flow appeared the next day (22 February), but it came from a fissure that opened at the base on the north side. It emitted white vapours more abundantly than any of the later flows, but this activity was negligible compared with that at the crater, where a vast and turbulent eruptive column rose continuously, with great explosions every few seconds throwing out masses of lava that became bombs or lumps of scoria. During the night the volcano presented a magnificent spectacle. In the words of F. M. Bullard: 'With each explosion literally thousands of red-hot fragments were thrown to a height of 2000–3000 feet above the crater rim, and, as they reached their zenith, they seemed to stop before beginning their fall', showering down like a giant firework display and covering the entire cone 'with interlacing fiery trails, which outlined it against the blackness of the night.'

The lava that reached the crater continued to be eviscerated during its expulsion (Figure 12.21) except on one occasion in June 1943, when a tongue of coherent lava succeeded in spilling over the rim. All the flows before and after came through fractures that developed as a result of preliminary up-swellings of the ash-covered ground near the base of the cone. One of the 1944 flows spread out in a lake-like mass over the site of Parícutin village, which had long before been buried in a shroud of ash and scoria. In June of the same year San Juan de Parangaricutiro was buried

**Figure 12.21** Parícutin at night, July 1943. Rhythmic explosions ejecting fountains of molten lava. The larger clots—volcanic bombs—are still brightly incandescent as they roll down the slopes. The glare in front of the onlookers is from a lava flow issuing near the base of the 500 metre cone. (*Three Lions Inc., New York*)

under a flow nine metres thick, leaving only the cathedral towers projecting above the jagged blocks of *aa* lava (Figure 12.22). After several more years of variable but on the whole decreasing activity, the volcano ceased to erupt in 1952, just after its ninth birthday. Although at this time the rim of the summit crater was about 400 m above the buried farmlands of 1943, it was barely 150 m above the platform of thick lavas and intervening ashes that had accumulated around the cone.

Izalco, west of San Salvador in Central America, is a volcano that was 'new' in 1770. Like Parícutin, it began as a scoria cone, but the basaltic lavas that soon followed included many that overflowed from the crater, so that almost from the start its cone was of the *strato-volcanic type*. The history of Izalco to date has included long periods of rhythmic Strombolian activity, interrupted by lava emission from the crater or the flanks, by short periods of repose, and occasionally by outbreaks of Vesuvian intensity. After nearly two centuries Izalco stands 670 m above the surrounding country and is still actively growing. In 1956 there were voluminous outflows of lava from the crater and several satellitic vents; the height of the volcano increased by 12 m during that year.

All the larger cones around the Pacific and, apart from the great shield volcanoes, most of those elsewhere are of strato-volcanic structure. With few exceptions they have histories extending over many thousands, in some cases millions, of years. Strato-volcanoes build accumulations of well-stratified tephra alternating irregularly with tongue-like lava flows (Figure 12.23).

4. *Cumulo-domes and tholoids* The shapes of volcanic structures built predominantly of lava flows depend on the fluidity of the lava concerned. The more silica-rich lavas, such as rhyolite, dacite and trachyte and the corresponding obsidians, are generally so highly viscous that they rarely flow far from the vent through which they are extruded. Steep-sided domes that form in the craters or calderas of certain strato-volcanoes are called *tholoids* (Gr. *tholos*, a domed building). Their growth commonly accompanies eruptions of *nuée*

**Figure 12.22** *Aa* laval flow from the base of Parícutin which overwhelmed San Juan Parangaricutiro in June 1944. (*Tad Nichols*)

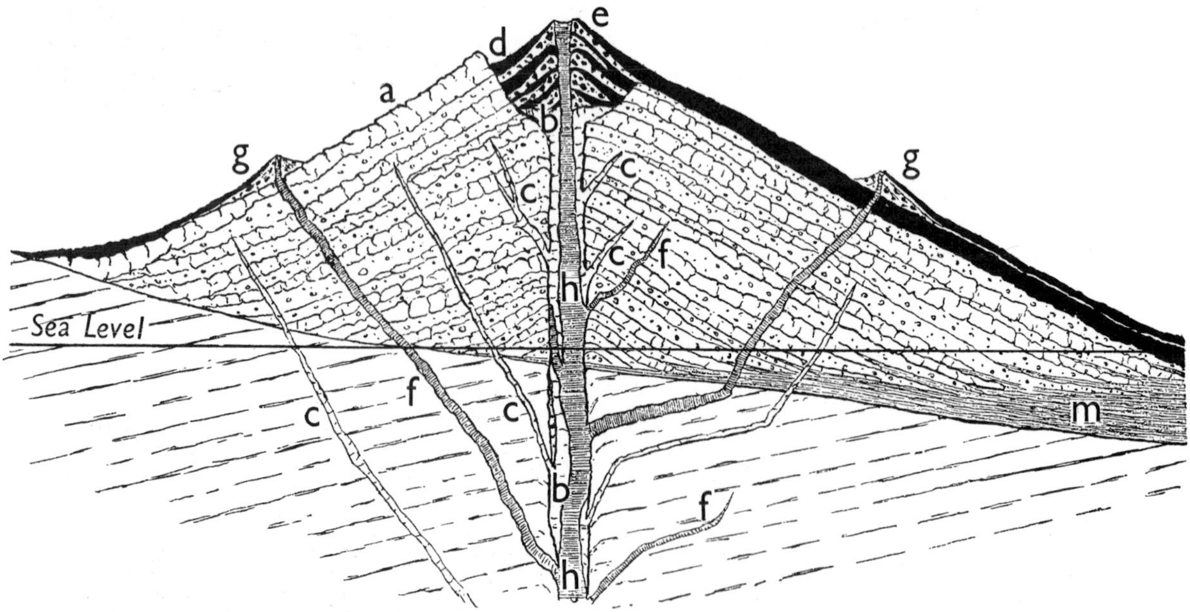


**Figure 12.23** Diagrammatic section through a strato-volcano. The main cone (*a*) is built of tephra deposits and tongue-like flows of lava fed from the conduit (*b*) and is braced by dykes (*c*). If (*d*) a large explosion crater (or a small explosion caldera) is formed at some stage it may be more or less filled up by continued growth of the cone (*e*) fed from the conduit (*h*). Some of the dykes (*f*) serve as feeders of lateral parasitic cones (*g*). Marine deposits interstratified with tuffs and lavas are indicated by (*m*). (*James Geikie and Messrs Oliver & Boyd*)

*ardente* types, and results from slow exudation of the extremely stiff lava then occupying the main conduit. Because of self-obstruction, continued growth takes place partly by internal additions which stretch and crack the outer congealed layers, and partly by the squeezing through the cracks of obelisks and spines of hot but nearly solid lava. These protrusions may reach considerable heights before they are demolished by gas expansion within them or by fumarolic activity around their margins. In this way the dome becomes mantled with debris, including many large blocks. The stupendous eruption of Bezymianny in March 1956 was followed during the next few months by the growth of a tholoid of this kind within the newly formed caldera (Figure 12.24). A similar dome began to develop in the crater of Mont Pelée before the disastrous eruption of 1902. Later, what appeared to be the plug of the conduit was forced bodily upwards through the dome a hundred metres or so, thus forming the celebrated 'spine' of Mont Pelée (see Figure 12.38 and p. 220).

Isolated domes with the form of an overturned bowl, such as the Puy Grand Sarcoui of Auvergne (Figure 12.25) and the 'mamelons' of Réunion (Indian Ocean, north of Madagascar), are distin-

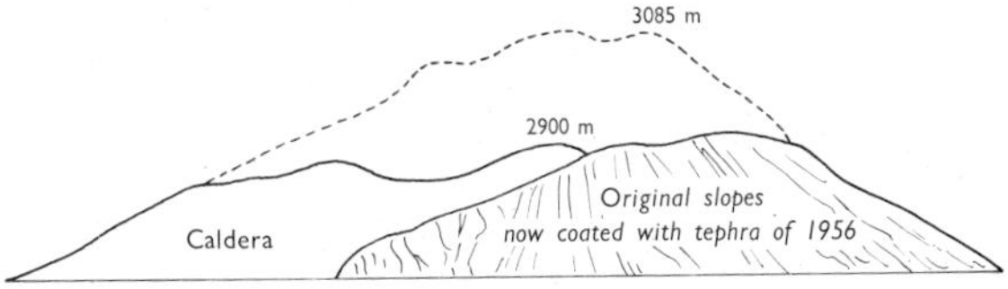


**Figure 12.24** Dome (tholoid) growing over the conduit within the caldera of Bezymianny, Kamchatka, August 1956. The caldera was formed concomitantly with the great eruption of 30 March 1956. The adjoining sketch shows the profile of Bezymianny before (dotted line) and after the March eruption (*G. S. Gorshkov*)

**Figure 12.25** Puy Grand Sarcoui, Auvergne: a craterless cumulo-dome of trachyte lava (*L. Hawkes*)

guished as *cumulo-domes*. They show no sign of a crater or orifice, and what little evidence bears on their mode of origin suggests that they, too, developed by internal as well as external accumulation of highly viscous lava over a hidden vent.

5. *Domes of external growth* Highly fluid basaltic lavas, from which gases escape so easily that explosive activity becomes subordinate, spread out as thin sheets for great distances. By the accumulation of successive flows in various directions a wide-spreading dome with gentle slopes, rarely exceeding 6° or 8°, is constructed. The classic examples of these *shield volcanoes* are those of the Hawaiian Islands; others occur in the Samoa group and in Iceland. Hawaii (Figure 12.26), the largest of the island chain to which it belongs, has been built up from the sea floor by the coalescence of several shield volcanoes. Mauna Loa is the highest, rising over 9000 m (about 4200 m above sea-level) from a broad base about 113 km in diameter (Figure 12.27). Kilauea lies on its flanks, about 30 km from the summit and 1200 m above the sea. The 'craters' of shield volcanoes are calderas of subsidence within which there are deep pit-craters, often occupied by swirling lakes of lava, open to the sky. These sometimes overflow, but at other times the lava drains away through deep fissures, sometimes to emerge lower down on the slopes, which may be below sea-level. As will appear from the following account of Kilauea, the plumbing and drainage systems of these volcanoes (to say nothing of others less well investigated) is complex beyond our present understanding.

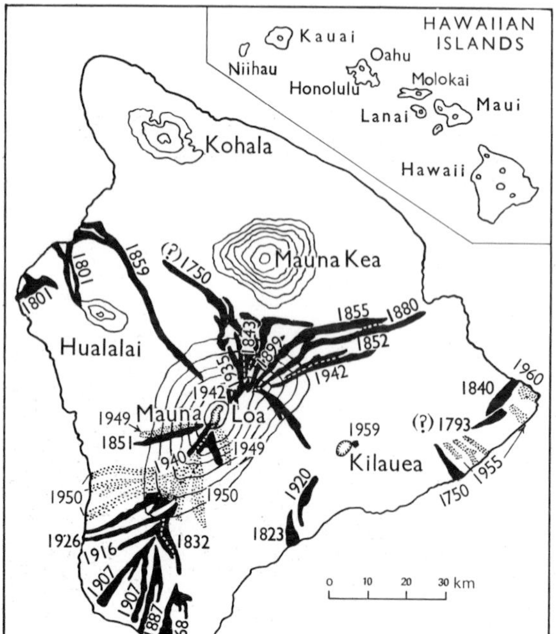


**Figure 12.26** Map of Hawaii showing its lava flows from 1801 to 1960

## Kilauea, Hawaii

On account of its easy accessibility and relative freedom from danger, Kilauea has become the most closely investigated of all volcanoes. After scientific observations began in 1823 lava-lake activity in various parts of the caldera floor was very vigorous for about a century. The floor was then over 300 m below the rim. By 1832 the accumulated overflows of lava from 'fire pits', fissures and occasional small cones had made a new floor 116 m nearer the rim. This then collapsed nearly to its 1823 level, after which it was similarly built up, only to subside again in 1840. Later changes of level have included bodily uplifts of the floor as well as minor collapses. The caldera is obviously bounded by ring-faults. At present the floor is about 150 m below the highest point on the rim and the only active pit-crater within the caldera is Halemaumau (Figures 12.28–30). During the last half-century the lava column has been alternately rising and falling, occasionally disappearing out of sight, sometimes crusting over, sometimes overflowing, sometimes forming a circulating lake with incandescent fountains spraying from the dazzling streaks that break through the usually duller surface.

In 1911, when the lava lake was being closely studied by Perret, it was covered by a thin glassy skin, channelled by cracks through which the brightly glowing lava could be seen. But the whole surface was moving towards the eastern end, where it broke up and sank into a long 'grotto' from which great bubbles of gas escaped and burst into splashes of golden spray. Clearly the congealed lava was dropping into the space vacated by

**Figure 12.27** Profile of Mauna Loa, the type example of a shield volcano. Length of section 34 km; natural scale

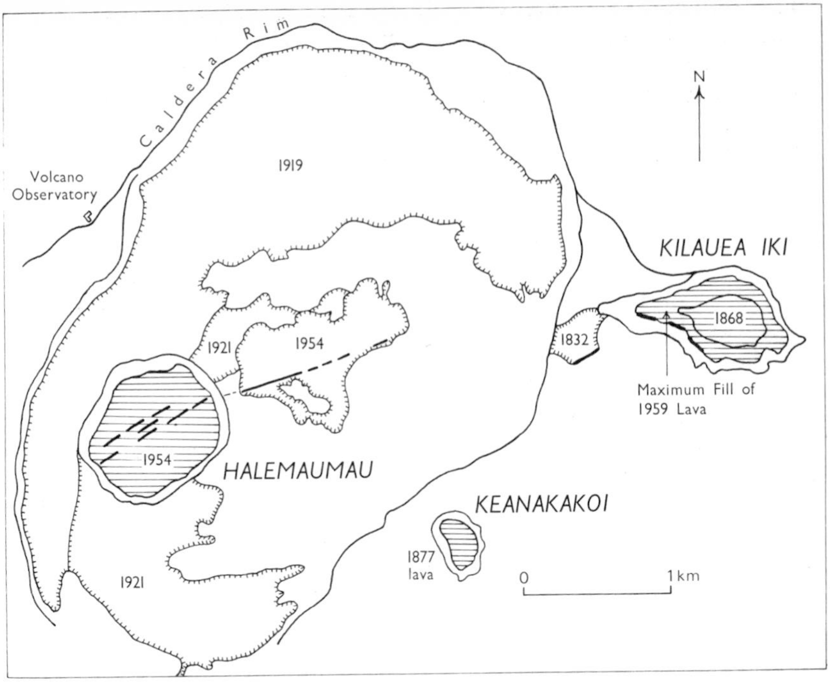


**Figure 12.28** Map of Kilauea caldera and associated pit-craters, showing the extent of lava flows. The 1954 eruptive fissures on the caldera floor and those of 1959 on the side of Kilauea Iki are indicated by heavy lines.

**Figure 12.29** The lava lake of Halemaumau on 23 May 1912. As the lava rises slowly above the ledge surrounding the lake it spills over and congeals, so building up a rampart within which the lake can rise still higher. Here the lava is seen breaking through the rampart in two places, right and left, with formation of typical pahoehoe in the foreground. The gas bubble bursting in the middle of the lake was about 9 metres in diameter. (*A. L. Day, Geophysical Laboratory, Carnegie Institution of Washington*)

ascending gases. The circulation was completed in depth by a return current towards the western end of the pit, where the rejuvenated lava rose again, so maintaining the lake at a constant level for several weeks. But for this gas-activated circulation the lava would very rapidly have solidified. At other times gases ascend near the middle (Figure 12.29), and exceptionally over the whole surface (Figure 12.30).

In 1912, during a period when the lake was about 60 m below the caldera floor, Day and Shepherd measured the temperature of the lava and found that in the course of 23 days it rose from 1070°C to 1185°C. Over the same period a steady increase in the rate of gas discharge was observed, culminating at the time of highest temperature in the maximum development of lava fountains (Figure 12.30). As the level of the lake had not varied, it was concluded that the increase of temperature was due to heat generated by chemical reactions, involving the uprising gases, the air, and the $Fe_2O_3$ of the lava itself.

As indicated on Figure 12.28, Halemaumau overflowed the caldera floor on a big scale during 1919 and 1921. For several years the lava lake had been rising and falling within a range of about 210 m. But in 1924 the lava withdrew to a far greater depth, clearly as a result of drainage through subterranean fissures. Earthquakes occurred 50 km to the east and beyond; and it is probable that the lava found an exit on the sea floor along a submarine continuation of the same eastern rift zone. The walls of Halemaumau, no longer sustained by lava, avalanched into the pit while it was emptying, thus enlarging it at the top and choking it at the bottom. Ground-water seeped into the debris and on passing into high-pressure steam expelled the successive obstructions by a series of phreatic explosions (Figures 12.31(*a*) and (*b*)) which continued until the supply

**Figure 12.30** View of the lava lake of Halemaumau during a period of maximum gas activity, 8 July 1912, photographed at night by the light of the incandescent fountains (*A. L. Day, Geophysical Laboratory, Carnegie Institution of Washington*)

**Figure 12.31** *above and below* Pseudovulcanian explosive eruptions from the pit of Halemaumau, May 1924, viewed from the rim of the caldera (*American Museum of Natural History, New York; below, H. T. Stearns, United States Geological Survey*)

of heat, residual or otherwise, ceased to be effective. It was then found that the avalanches and explosions of a few weeks had increased the surface dimensions of Halemaumau from 245 × 150 m to 1036 × 914 m. The bottom of the enlarged pit, now about 400 m below the rim, was really only a scree-encircled depression at the top of the immense accumulation of volcanic debris that had fallen in. Only a few hot spots and gas vents could be detected in the walls, and when lava began to return it first spurted through a scree on the lower slopes and fountained to a height of 53 m.

An earlier explosive eruption of the same kind—almost certainly of the 'steam boiler' or phreatic type—is known to have occurred in 1790, and is still recorded by a few outcrops of the pyroclastic debris of older lavas blown out at the time. Similar deposits preserved between lava flows and now exposed in the caldera walls of Kilauea show that there were at least 11 such eruptions during the time represented, which seems to be of the order 1000 to 2000 years.

Figure 12.26 clearly shows that Kilauea and its

adjoining rift zones to the south-west and east constitute an active belt which lies *across* the general axis of the Hawaiian chain of islands. Mauna Loa and its rift zones constitute a similar and nearly parallel belt. Between the two there can be no direct connection, or the Mauna Loa lava would drain out through Kilauea, which is 3000 m lower. Moreover, there is no sympathetic relationship of any kind between the eruptions of the two volcanoes and their rifts. Heleakala (Figure 12.2) also illustrates the significant fact that the main communications with the magmatic sources are transverse to the island chain and are, apparently, without direct interconnections. Even along each transverse belt connections seem to be temporary and variable.

These symptoms of a complex drainage system have been repeated during the more recent volcanic history of Kilauea. In 1927 the level of the Halemaumau lava slowly began to rise, but it remained low until the considerable rises of 1952 and 1954. During this time there were several voluminous eruptions from Mauna Loa and its rifts. It was not until 1954 that lava again spilled over the floor of Kilauea, but it was not an overflow from Halemaumau, where the level of the lava was still about 137 m below the rim. It was an outpouring from ENE fissures that opened on the caldera floor (Figure 12.26). Later, in 1955, activity migrated along the eastern rift, as indicated on the map.

**Figure 12.32** Incandescent fountain above the fissure from which lava is flowing into the lava lake occupying the pit crater in front, Kilauea Iki eruption of 1959. Photographed at 6 a.m., 19 November. See also Fig. 12.14 (*Hawaiian Volcano Observatory, United States Geological Survey*)

In 1957 it was found by measuring the tilting of the ground that the whole caldera region was bulging upwards. In the autumn of 1959 the rate of swelling accelerated and swarms of earth tremors, originating from a depth of about 56 km, were recorded. On the night of 14 November the threatened eruption suddenly began, but not where it had been expected. Dazzling lava fountains burst through fissures along the SW slopes of Kilauea Iki, a pit-crater that had remained inactive for nearly a century. One fountain played at intervals for a month or so (Figures 12.14 and 12.32), reaching a height of about 580 m in its most vigorous outburst. Kilauea Iki became a lake of molten lava which submerged the active vents and reached a depth of over 120 m when fountaining was particularly active. At times when the fountains stopped, lava drained away, like water from a bath, and it was noteworthy that the rate of withdrawal was two or three times as rapid as that of filling. This type of intermittent activity continued for five weeks.

Earthquakes then began along the eastern rift and in January 1960 heavily gas-charged lava spurted through a series of fissures near the most easterly point of Hawaii, forming an incandescent curtain from which a stream of lava flowed down to the sea. It was evident that lava was being withdrawn from beneath the caldera region, as the tilt measurements now showed rapid settling. On 7 February, while lava was still flowing into the sea, the flat floor of Halemaumau dramatically collapsed. First it became a saucer-shaped depression, about 46 m below its level of a few hours earlier. Then 'within a period of 10 minutes a small area, over 300 m in diameter in the centre of the large collapse feature, dropped an additional 60 m and a small volume of highly viscous lava oozed into the newly formed hole' (Richter and Easton). So ended the most remarkable volcanic episode that even Hawaii has staged.

Figure 12.33 Vesuvius half encircled by the breached caldera known as Monte Somma (*Fox Photos Ltd*)

## Vesuvius

During the sixteen centuries that followed the eruption of A.D. 79 Vesuvius broke into violent eruption only ten times, each outbreak being followed by a prolonged period of quiescence. With the eruption of 1631, after 130 years of repose, the volcano assumed its modern habit of cyclic activity, the cycles being marked by definite crescendos leading to outbursts of paroxysmal intensity at intervals which so far have varied from 11 to 40 years. The last three of these major eruptions occurred in 1872 (Figure 12.16), 1906 and 1944.

Throughout the 1906 eruption, Frank Perret, most courageous of vulcanologists, remained at the observatory on the western slopes, and his intimate record of the sequence of events has become one of the classics of geological literature. In 1905, when the eruption was already threatening, the cone fissured high up on the north-west slope (A, Figure 12.34), and thus led to an emission of lava which temporarily relieved the tense conditions in the crater. It was at this time that Perret had his first experience of witnessing the re-fusion of lava already congealed and cold.

**Figure 12.34** Map of Vesuvius and Monte Somma showing the points of lava emission in 1905 (A) and 1906 (B–G)

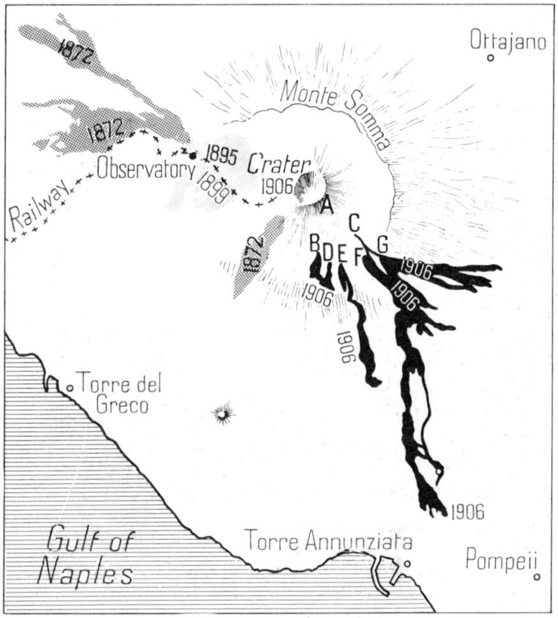

He records that he was standing with a group of students upon the hard lava surface of the cone when this gradually increased in temperature until the heat underfoot was unbearable. On moving away it was observed that a round area some 2 metres in diameter was becoming incandescent in full daylight, and this began to swell upward until, without a fracture, the material reached the point of complete fusion and a small, ephemeral lava stream started flowing down the slope.

Nearly a year later, on 4 April 1906, the cone was fractured low down on the slopes (B–G, Figure 12.34). Torrents of gas-saturated lava gushed out and thereby lowered the level of coherent lava in the conduit. The effect of the resulting relief of pressure beneath the crater was a swift emission of expanding gases, punctuated by sharply defined explosions which kept the whole mountain in a state of powerful vibration. White-hot lava, violently transported up the conduit, was torn to pieces of all sizes down to the finest spray and projected for miles into the overhanging ash cloud, like a fiery effervescing geyser. Intense electrical discharges added to the lurid glare. At one stage an inclined jet of molten projectiles shot over Monte Somma and brought disaster to the towns on the plains beyond.

Four days after the paroxysm—the Vesuvian phase—began, it culminated in a mighty uprush of gases—the Plinian phase—which continued for the greater part of a day, blasting out the throat of the chimney, tearing away the upper portions of the cone, and reaching a height of over 12 km before spreading out. It was the staggering intensity of this prodigious blast that impressed upon Perret the conviction that gas is the principal agency of volcanic eruption.

As the pressure and momentum of the gas eventually became less overwhelming, the walls of the widened crater began to fall in, creating temporary obstructions and thus leading to powerful explosions and the generation of pseudo-Vulcanian clouds heavily charged with dark ash. This concluding stage of the eruption persisted intermittently for several days, during which the intensity gradually declined. By 22 April the waning energy was exhausted.

During the next few years the deep funnel of the crater was gradually filled up by avalanches from the walls. The floor of debris first came into view in 1909. Apart from the emission of vapours, the volcano remained in repose until 1913, when incandescent lava perforated the floor, and began to build up a new eruptive conelet (Figures 12.36 and 12.37). It is probable that during these years of apparent inactivity the 'avalanche debris' that choked the conduit was already being remelted by ascending gases.

From 1913 Vesuvius was in a phase of almost continuous moderate activity of the Strombolian type, long periods of effervescence of the lava column within the growing conelet alternating

**Figure 12.35** The cone of Vesuvius before and after the 1906 eruption. The crater rim as it was on 20 March 1905 is indicated by *b*. Within this a terminal conelet was built up to a maximum height of 1335 m by 23 May 1905. The eruption of April 1906 lowered the crater rim to *c* by gas erosion and inward slumping, the maximum height then being 1220 m. (*F. Perret, Carnegie Institution of Washington*)

**Figure 12.36** Eruptive conelet in the crater of Vesuvius, 1918 (*Fox Photos Ltd*)

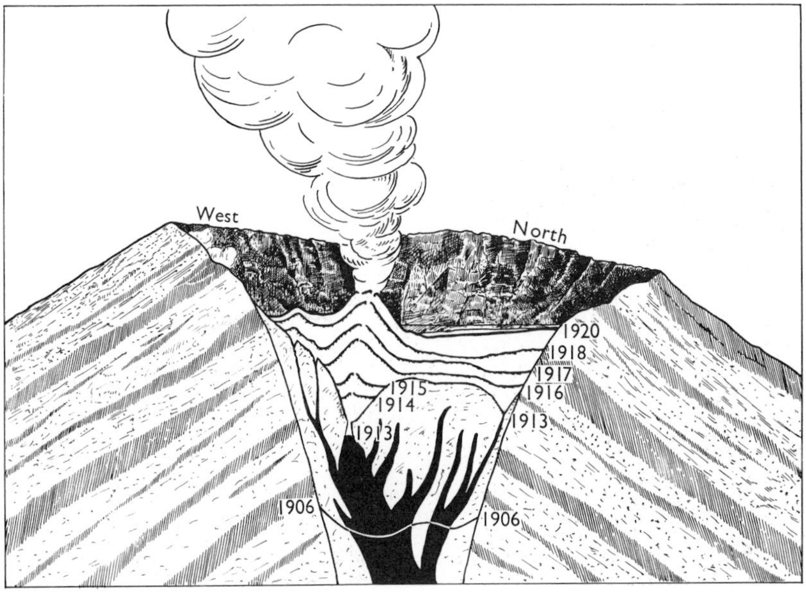


**Figure 12.37** Diagrammatic representation of seven years' growth of the active conelet shown in Fig. 12.36, following seven years of apparent repose after the great eruption of 1906

with occasional outflows which, on the whole, became more voluminous with the passing of the years. This long-drawn-out crescendo reached its climax with the eruption that began in March 1944, and lasted 17 days. On this occasion there were extensive lava flows from the crater. One of these advanced across the funicular railway and rolled through the town of San Sebastiano, little of which remained when the crunching front of *aa* lava finally came to rest. The phase of effusive lava was followed after nine days by typically Vesuvian paroxysms of increasing gas activity, with periodic production of luminous lava fountains reaching to about 1000 m above the summit. After a day and night of these displays, each of which lasted an hour or so, against a continually changing background of vast cauliflower clouds, the available lava was apparently exhausted. Gas activity then culminated with several hours of the Plinian uprushing gas phase. The 1944 eruption ended with a few days of pseudo-Vulcanian explosions, corresponding to the dark-ash phase of 1906. Avalanching of the inner walls then reduced the depth of the crater, sometimes raising impressive clouds of dust that were wrongly reported as eruptions.

## Mont Pelée, Martinique

Before the appalling catastrophe of 1902, Mont Pelée (Martinique, in the volcanic arc of the West Indies) had long been dormant, the only previous activity known to the inhabitants having been quite moderate Vulcanian eruptions in 1792 and 1851. In the early spring of 1902 Vulcanian activity again broke out, but this time the crater began to be filled and sealed over from within by the growth of a dome of extremely viscous lava. A great flood of hot mud poured down the Rivière Blanche, as the boiling lake-waters displaced from the crater escaped through a V-shaped gap or notch in the rim. This gap determined the points of least resistance, near the base of the growing dome, through which the parental material of nearly all the *nuées ardentes* subsequently issued: mostly gushing out in cloudy globular masses more or less quietly, but occasionally with explosive violence. Because the gap was situated at the head of the Rivière Blanche, the valley tended to confine the *nuées* as they rolled westwards with hurricane speed towards the sea. But on 8 May the

**Figure 12.38** Ruins of St Pierre after the 1902 disaster. Mont Pelée and its spine in the background (*A. Lacroix*)

**Figure 12.39** *Nuée ardente,* Mont Pelée, 7 January 1930
*above* 4 minutes after emission
*below* a few seconds later, showing swift expansion to a height of 4000 metres (*F. A. Perret, Carnegie Institution of Washington*)

first and by far the most powerful of all the *nuées*, heralded by an exceptional explosion, swept down the slopes of a much wider sector (extending from north-west to south) and so included St Pierre, the port and capital of Martinique, in the devastated area. Within a few minutes the whole city and its 30,000 inhabitants were utterly annihilated by the irresistible, asphyxiating blast (Figure 12.38). In the harbour ships burst into flame, turned turtle and sank in the boiling sea. After this unparalleled disaster *nuée* activity and growth of the dome continued for many months.

In October the plug of almost consolidated lava in the conduit began to be forced bodily upwards, and a gigantic steaming column rose through the crater-dome and projected high above it. Presently half of it broke away along a slanting crack from the summit, the part that remained thus assuming the shape of a spine. The latter reached a height of nearly 300 m above the dome in the course of seven months, but by July in the following year it had crumbled to pieces.

Only two days before St Pierre was wiped out a *nuée* eruption of somewhat different type (p. 204) occurred in the island of St Vincent to the south. After a series of violent earthquakes the crater lake of the Soufrière volcano boiled and overflowed, and a black *nuée*, heavily charged with incandescent masses of lava, descended from the summit in all directions and destroyed everything in its track. Here the death-roll amounted to about 1600.

From 1929 to 1932 Mont Pelée again became active, the preliminary symptoms being earth tremors and increased fumarole activity. During the first few months pseudo-Vulcanian explosions, gradually becoming more frequent and violent, established a crater on the site of the 1903 dome, most of which was blown away. Following the removal of this obstruction the first *nuée* appeared early in December 1929. Within a few days Perret improvised an observatory on the edge of the Rivière Blanche and kept the volcano under careful watch. Very soon, brave man, he felt able to reassure the terrified population—most of whom had fled after the threat of the first *nuée*—and the normal business of St Pierre was gradually resumed. In February 1930 a new dome began to appear.

Thereafter, the eruptions followed a remarkably rhythmic course, periods of *nuée* discharge from the same V-shaped notch as before (Figure 12.39) alternating with periods of dome formation, with very little overlapping of these two sharply contrasted types of extrusion. Innumerable spines of viscous lava were squeezed through the dome during its phases of growth (Figure 12.40), but within a short time, never more than a few days, the projections collapsed into large blocks and smaller fragments which slithered or avalanched down the ashy slopes. The *nuées* all poured down the valley track of the Rivière Blanche as swift torrents of self-inflating lava particles, transporting blocks of all sizes up to those of small cottages, mostly derived from the broken-down spines of the dome, and feeding overhead the ever-expanding, turbulent and billowing clouds of liberated gas and comminuted ash.

**Figure 12.40** View of the growing dome of Mont Pelée in 1930, showing small spines in various stages of protrusion and collapse (*Drawn from a telephoto by F. A. Perret*)

## Krakatau, Indonesia

For two centuries before its impressive awakening in 1883 (Figures 12.41 and 12.42), the old volcanic wreck of Krakatau (between Java and Sumatra) had been dormant. In May of that year the vent of Perboewatan became active, Vulcanian explosions being followed by eruptions of moderate Vesuvian type. During the next few weeks many new vents were opened around Danan, until by August at least a dozen Vesuvian eruptions were in progress, and steadily increasing in violence. The climax was reached during the last week of August. On the 26th formidable detonations were heard every ten minutes. Dense volcanic clouds reached a height of 27 km, and ashes, converted into stifling mud by the incessant rain, fell over Batavia (now Djakarta) which was plunged into thick darkness, relieved only by vivid flashes of lightning. On the morning of the 27th came four stupendous explosions, the greatest of which was heard about 5000 km away in Australia, and a vast glowing cloud of incandescent pumice and ashes rose 80 km into the air. Although Krakatau was uninhabited, the catastrophe did not forego its toll of life. Highly destructive sea waves (*tsunami*: see p. 576), one of them over 36 m high, swept over the low coasts of Java and Sumatra and 36,000 people were drowned.

When Krakatau again became visible, it was found that two-thirds of the island had disappeared. Subsequent survey showed that a deep submarine hollow had taken the place of 20 $km^2$ of land. It was originally thought that the greater part of the island—amounting to about 16 $km^3$—had been blown away by the colossal explosions. When the surrounding deposits of tuff came to be examined, however, it was found that they contain less than 5 per cent of material representing the vanished rocks. All the rest, consisting of glassy ash and pumice, is a product of the magma that was responsible for the eruption. Thus it was, for the most part, not the rocks of the volcanic edifice that were blown away, but the highly gas-charged underlying magma. The latter was evidently akin to the parental material of the Peléan *nuées ardentes*, but the quantity involved—about 18 $km^3$—was enormously greater. Evisceration of the molten volcanic roots on this gigantic scale left the superstructure without support, so that it collapsed and foundered, leaving a vast island-rimmed submarine caldera, 6 by 7 km across. The

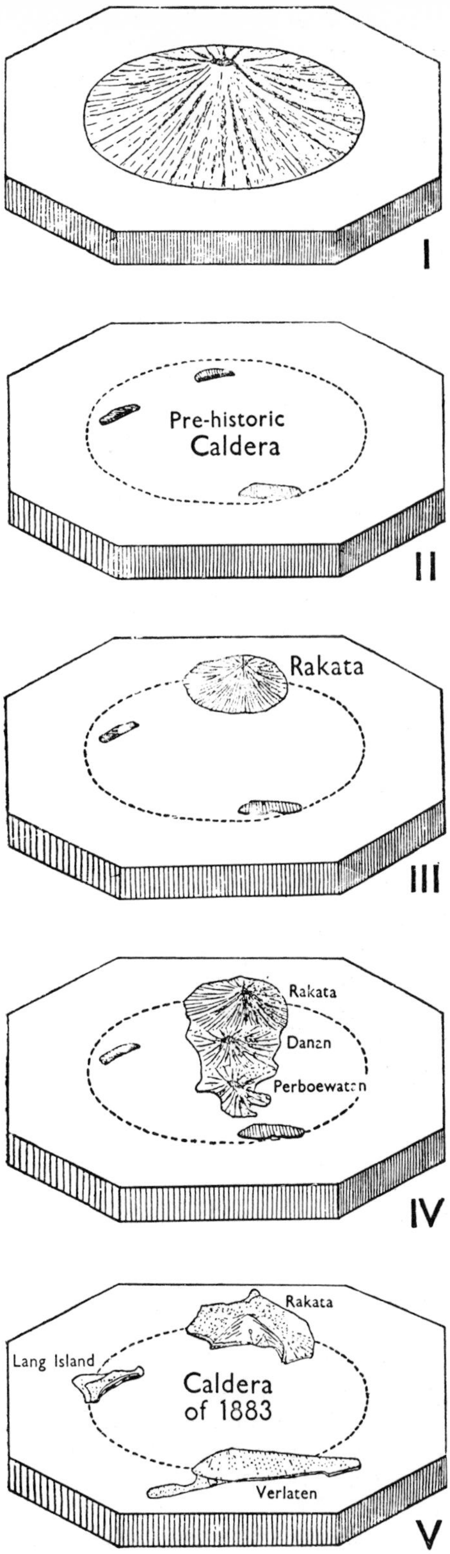


**Figure 12.41** Block diagram illustrating five stages in the history of Krakatau (*after B. G. Escher*)

I Original andesitic cone of Krakatau
II After explosive evisceration, probably accompanied by collapse of the superstructure, a great caldera was formed, rimmed by three small islands.
III Growth of Rakata, a basaltic cone
IV Krakatau before 1883, after two later andesitic cones had coalesced with Rakata
V Island remnants of caldera rim after the 1883 eruptions. Later eruptions have since built up the island of Anak Krakatau within the caldera.

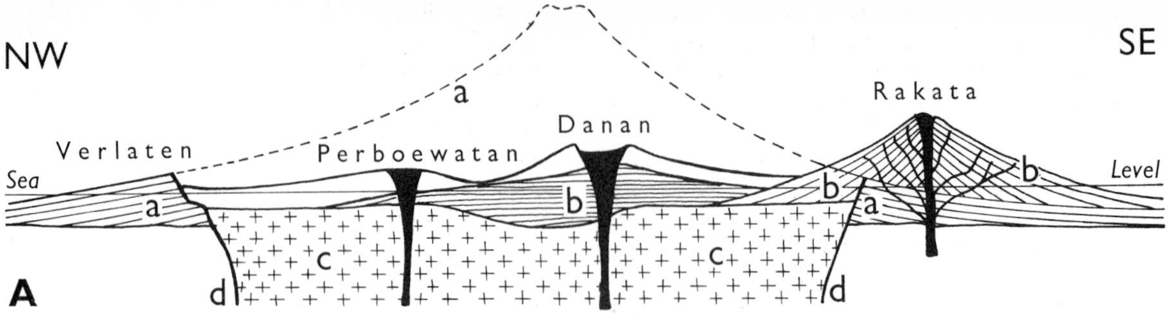


**Figure 12.42** (A) Profile of Krakatau before 1883, with dotted indication of the original cone, of which *a* remains on Verlaten and near the base of Rakata; *b* represents the lavas etc. of the later cones (stages III and IV of Fig. 12.41), *c* the collapsed rocks of the cone *a* and its foundations, which subsided between the marginal faults *d*; *c* is underlain by *e*, the highly gas-charged eruptible material (depth unknown) responsible for the 1883 outbreak.

(B) Profile of Krakatau after the 1883 eruptions, showing schematically some of the many conduits and vents through which *e* escaped, as a fluidized system, so providing conditions for the foundering of the superstructure

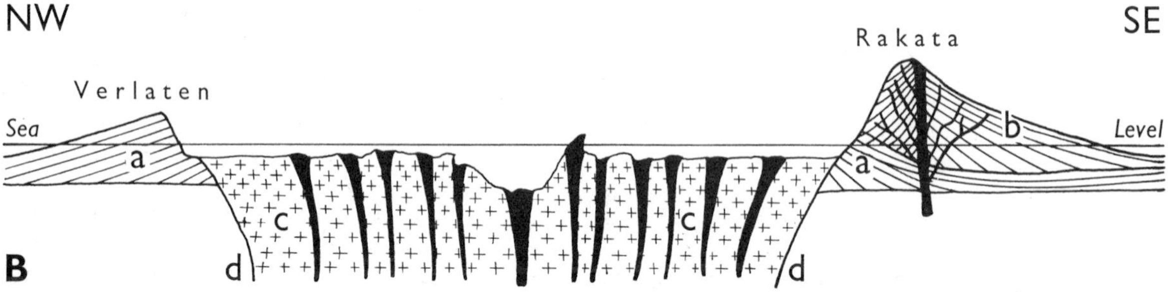

**Figure 12.43** Anak Krakatau (*anak*, child of), as seen in 1931, within the caldera of Krakatau (*Paul Popper Ltd*)

energy expended during this volcanic operation has been estimated at not less than could be liberated by 5000 megaton hydrogen bombs.

After 44 years of superficial quiescence, submarine eruptions broke through the caldera floor where the sea was about 300 m deep and built up a cone that first appeared above sea-level late in 1927 (Figure 12.43). Eruptions from the new vent occurred at intervals until 1933. It is of interest to notice that whereas the tephra discharged in 1883 has the composition of dacite, with about 65 per cent silicon, the bombs and lapilli from Anak Krakatau are of basaltic composition, with 52 per cent silica. Similar basaltic ejecta were again erupted in 1953 and 1959.

## Calderas

As already mentioned, if a caldera originated by an explosive eruption that blew the superstructure of a volcano to smithereens and showered the debris over the surrounding countryside, the resulting deposits would consist mainly of fragments of the missing part of the cone, i.e. of 'accessory' materials. Only a few minor calderas, generally less than 1·5 km in diameter, have originated in this way. In the tephra deposits surrounding the larger calderas, and many of the smaller ones, the proportion of accessory debris is so small as to leave no doubt that the vanished part of the cone must have foundered out of sight. Such wholesale engulfment implies a corresponding removal of support in depth and the provision of space into which the superstructure could subside. We have already had examples illustrating two possibilities. Kilauea demonstrates the removal of support by underground drainage of lava. Krakatau and Bezymianny demonstrate the paroxysmal discharge of gas-rich fluidized magma by eruptions which are probably akin to *nuées ardentes* of mega-Peléan intensity, but which may also include Vesuvian-Plinian phases of exceptional violence. There are, however, many calderas that greatly surpass these three in size. A selection is listed below, with the maximum dimensions in kilometres.

| | |
|---|---|
| Lake Toba, Sumatra | 50×20 |
| Buldir (submarine), Aleutian Is. | 43×21 |
| Aira (with Sakurajima), Japan | 25×24 |
| Lake Kutchaio, Japan | 26×20 |
| Tarso Yega, Tibesti, Sahara | 20×17 |
| Aso San, N Kyushu, Japan | 23×14 |
| Circus, Teneriffe, Canary Is. | 20×12 |
| Tarso Voon, Tibesti, Sahara | 18×14 |
| Santorin, Greece | 17×11 |
| Alban, N. of Rome, Italy | 11×10 |
| Aniakchak, Aleutian Is. | 11×9 |
| Crater Lake, Oregon, U.S.A. | 10×10 |
| Askja (including Knebel, formed since 1875), Iceland | 8×6 |
| Krakatau, Indonesia (1883) | 7×6 |
| Tambora, Indonesia (1815) | 6×6 |
| Kilauea, Hawaii | 5×3 |
| Bezymianny, Kamchatka (1956) | 2×1 |

Crater Lake, Oregon (Figure 12.44), occupies a caldera about 10 km in diameter, formed by the collapse of what was once a lofty composite cone. The cone must have been there not long ago, for it supported glaciers that have left abundant evidence of their former existence on the outer slopes of the caldera walls. Glacial deposits alternate with beds of tephra; and U-shaped valleys, not completely filled with tephra, can be followed up to the rim of the caldera, where they are abruptly truncated by the walls. Carbonized wood from trees killed during the eruption responsible for the collapse have been dated by the radio-carbon method at about 8000 B.P. (before the present). For the ancestral cone to have supported glaciers at that time, it must have been nearly 4000 m high, in which case more than 70 km³ of the original structure have disappeared. Howel Williams (1942) estimated that at least 1·5 cubic miles (6 km³) of the vanished cone were dispersed by explosive activity. The disappearance of the remaining 15·5 cubic miles (64 km³) can therefore only be explained as a result of engulfment by cauldron subsidence. Space for the subsidence Howel Williams attributed to withdrawal of underlying magma, and the creation of an actual or potential void, partly by eruption of 5 cubic miles (20 km³) of magma (liquid and crystals), and partly by lateral migration. A relationship to ring-complexes is indicated by a semicircular arc of vents, along the northern wall of the caldera, which are probably arranged along a ring-fracture.

The eruption of Tambora in 1815, long after the volcano was thought to be extinct, was probably the most violent of historic times. Although Krakatau and Pelée took a far greater toll of human lives, Tambora discharged at least 100 km³ of andesitic tephra and left a caldera 6 km across where there had formerly been a cone well over

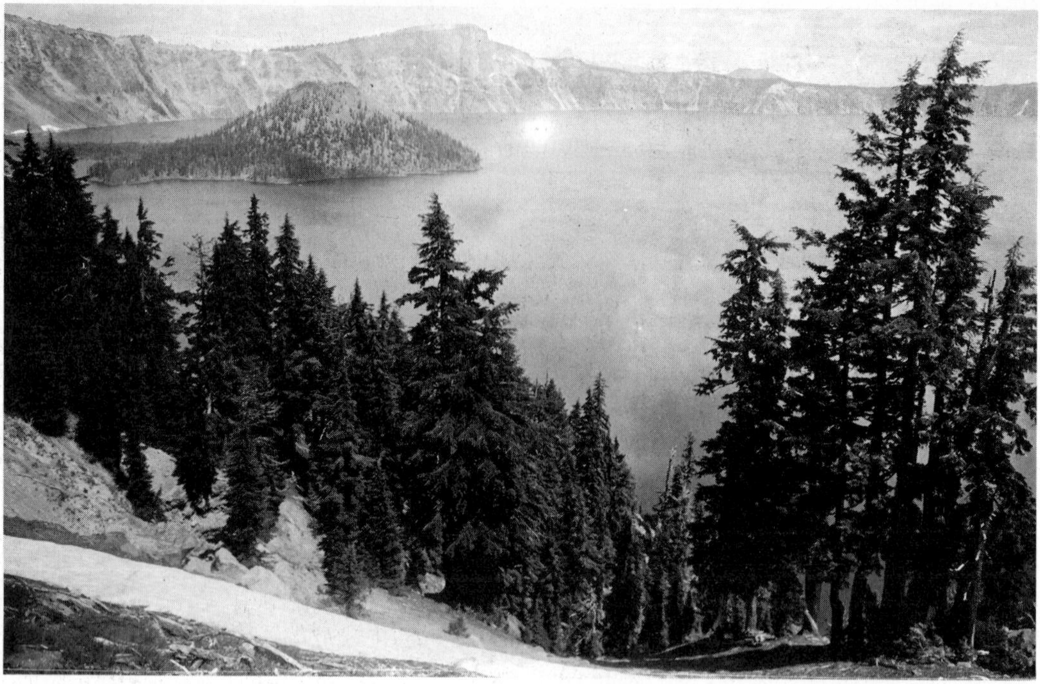

**Figure 12.44** Crater Lake, Oregon. A prehistoric caldera, with the cone of Wizard Island rising above the lake (*American Museum of Natural History, New York*)

3500 m high. At least 30 $km^3$ of the cone disappeared. From descriptions of the eruption it appears to have been an essentially ash-laden Plinian type. Figure 12.45 illustrates a possible subsidence mechanism that is consistent with cauldron subsidence within ring-complexes (Figure 11.15).

One of the greatest of known calderas is situated at the crest of the Barisan Highlands in NW Sumatra, and is partly occupied by Lake Toba, from which it takes its name. Its mode of origin has been studied by many geologists, notably by van Bemmelen. During the late Pleistocene there were paroxysmal eruptions of rhyolite tephra, formerly regarded as tuffs and lava flows of rhyolite, but

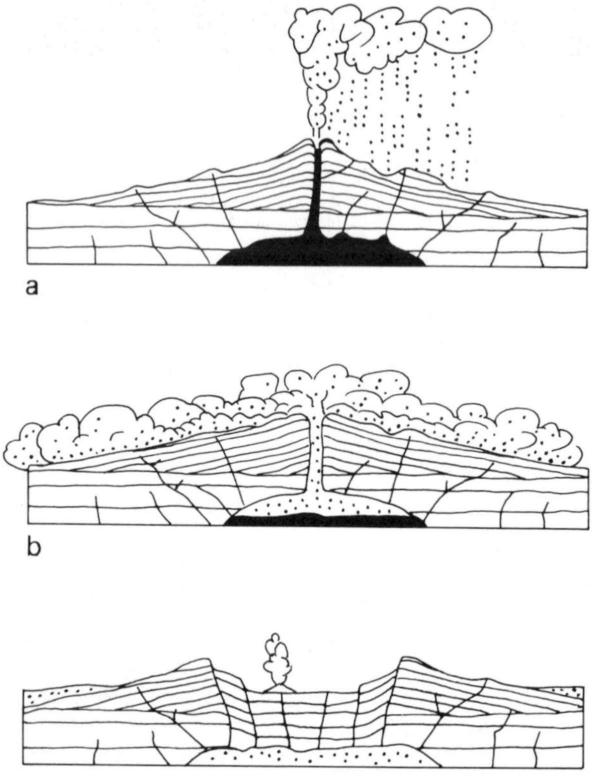


**Figure 12.45** Stages in the evolution of a caldera. Adapted from Howell Williams (1942) in the light of more recent knowledge of ring-complexes.
(a) Initiation of volcanic activity with eruptions of pumice, and magma rising high in the conduit
(b) Culmination of volcanic activity with outpouring of *nuées ardentes* and emptying of the magma chamber
(c) Collapse and subsidence of the cone as faulted saucer-like layers due to lack of support, with the formation of a caldera within which further eruptions occur

later recognized as being mainly ignimbrites. These tuffs and tuff flows are thickly distributed over 29,000 km$^2$ of Sumatra and they have been traced into Malaya, where their thickness is still 1·5 to 6 m in places that have escaped erosion. The estimated volume—allowing for sea-floor deposits—is of the same order as that of the caldera which came into existence as a result of the titanic outbursts.

It is thought that a diapiric granite pluton began to arch up the overlying crustal rocks during Pliocene times. The swelling culminated in what is now the Lake Toba region, presumably because a great head of highly energized gases accumulated and led to magma formation. The eventual fissuring of the roof by the expansion taking place beneath it allowed this dangerously explosive material to expel itself as a vast fluidized system. The actual eruptions—judging from their products—may have resembled giant *nuées ardentes*. As space became available, part of the up-arched roof rocks foundered in the rising fluidized system.

When a cauldron subsidence is preceded by marked up-arching or up-doming, the resulting

**Figure 12.46** The crater of Viti, the most active vent of the eruptions of 1875, which were responsible for the enlargement of the Askja caldera by the formation of the Knebel caldera. Viti itself originated by a sudden explosive eruption in 1724. The Knebel caldera is partly sunk in the Askja floor, as indicated in the accompanying sketch map. See Fig. 12.9 for locality (*Páll Jónsson, Reykjavik*)

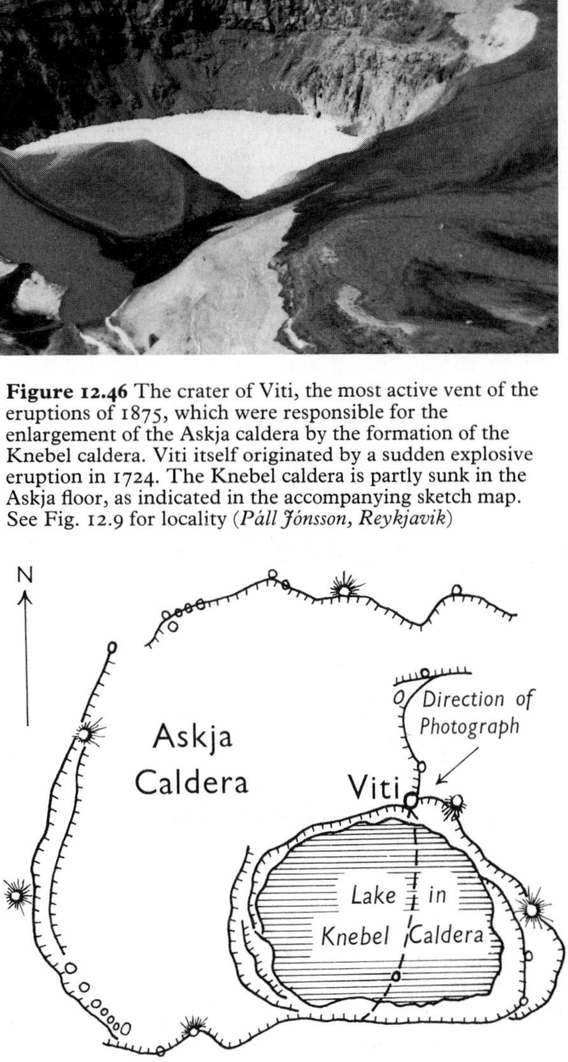

caldera is sometimes described as *volcano-tectonic*. The Askja caldera in Iceland, investigated by van Bemmelen and Rutten, is a down-faulted depression of the upper part of an up-domed region of plateau basalts. It is of particular interest (*a*) because of its numerous rim-craters, pointing to ring-dyke development in depth; and (*b*) because of its renewed development in and after 1875. Before that year the caldera was floored with basalt. In January of 1875 eruptions broke out and several small craters appeared along the SE edge of the Askja caldera. In March there were paroxysmal discharges of pumice and ignimbrite from Viti, the crater seen in Figure 12.46, and probably from smaller vents which were subsequently submerged by the lake. The resulting Knebel caldera took several years to reach its present size. The lake gradually increased in volume until 1907, since when it has not fluctuated much.

In certain volcanic regions, notably Hawaii (p. 212) and Japan, systematic surveys and precise levellings of volcanoes, calderas and their surroundings have revealed a significant association of upheaval before eruptions and subsidence immediately after. The case of Sakurajima on the southern rim of Aira caldera (Figure 12.47) is typical. Just after the 1914 eruptions (Figure 12.12) it was found that a nearly circular area enclosing the caldera had subsided. About the middle of the caldera floor the depression reached its maximum of about 2 m. Although it was so shallow, the volume of the depression was of the same order as that of the lava, pumice and ashes discharged during the eruptions (p. 197). After a few months it became evident that the same region was being slowly upheaved again. This continued until the swelling reached its maximum (again near the middle of the caldera) just before Sakurajima flared into a major eruption in 1946. As before, the evacuation of vast quantities of lava and gases was accompanied or followed by a widespread depression. The evidence strongly suggests that the upheavals result from the slow ascent of heavily gas-charged magma, and perhaps in part to the expansion involved in the formation of new magma with the aid of ascending gases. The eruption that follows involves a rapid withdrawal of support, with consequent down-sagging of the roof. The tremendous collapse implied by the caldera of Aira must, by analogy, have been preceded by eruptions that stagger the imagination.

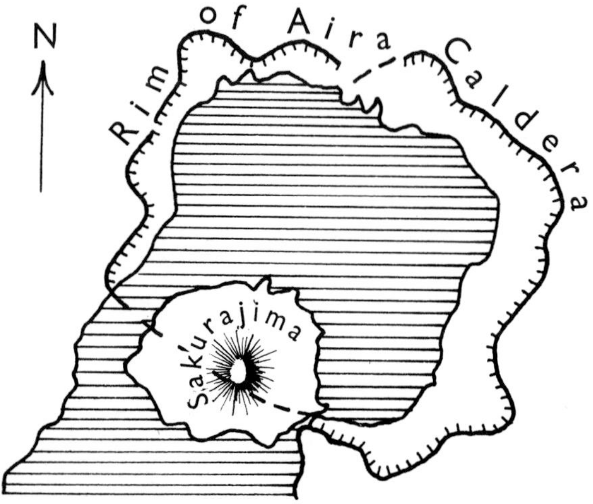


**Figure 12.47** Sakurajima volcano on the rim of the Aira caldera (24 × 25 km) at the heand of Kagojima Gulf in the extreme south of Kyushu, Japan (*After K. Mogi, 1958*)

## Selected References


ANDERSON, T. and FLEET, J. S., 1903, 'Report on the eruption of the Soufrière, in St Vincent, in 1902, and on a visit to Montagne Pelée, in Martinique', *Philosophical Transactions of the Royal Society of London*, A, vol. 200, pp. 353–553.

BARTH, T. F. W., 1950, *Volcanic Geology, Hot Springs, and Geysers of Iceland*, Carnegie Institution of Washington, Publication 587.

BAKER, P. E., GASS, I. G., *et al*, 1964, 'The volcanological report of the Royal Society Expedition to Tristan da Cunha, 1962', *Philosophical Transactions of the Royal Society of London*, A, vol. 256, pp. 439–578.

BEMMELEN, R. W. van, 1939, 'The volcano-tectonic origin of Lake Toba (North Sumatra)', *Geologie en Mijnbouw*, The Hague, Netherlands, vol. 6, pp. 126–40.

BROWN, M. C., 1962, 'Nuées ardentes and fluidization', *American Journal of Science*, vol. 260, pp. 467–70. Also McTaggart, K. C., ibid., pp. 470–6.

BULLARD, F. M., 1962, *Volcanoes: In History, in Theory, in Eruption*, University of Texas Press, Austin, Texas.

FOSHAG, W. F., and GONZALEZ, J. R., 1956, 'Birth and development of Parícutin Volcano, Mexico', *United States Geological Survey Bulletin*, 965D, pp. 355–485; see also Part B (Volcanoes of the Region).
LACROIX, A., 1904, *La Montagne Pelée et ses Éruptions,* Masson, Paris.
MACGREGOR, A. G., 1938, 'Royal Society Expedition to Montserrat, B. W. I. The volcanic history and petrology of Montserrat, with observations on Mt Pelée in Martinique', *Philosophical Transactions of the Royal Society of London*, B, vol. 229, pp. 1–90.
PERRET, F. A., 1924, *The Vesuvius Eruption of 1906*, Carnegie Institution of Washington, Publication 339.
— 1935, *The Eruption of Mt Pelée*, 1929–32, ibid., Publication 458.
— 1950, *Volcanological Observations*, ibid., Publication 549.
THORARINSSON, S., 1967, *Surtsey: The New Island in the North Atlantic*, Viking Press, New York.
WILLIAMS, H. M., 1941, *Calderas and their Origin*, University of California Press, Los Angeles.
— 1941, *Crater Lake: The Story of its Origin*, University of California Press, Los Angeles.

# Chapter 13

# Dating the Pages of Earth History

The terror with which men await the end of the world decides me to chronicle the years already passed, that thus one may know exactly how many have elapsed since the earth began.

*St Gregory the Great* (540–604)

### Hutton to Kelvin: the Great Controversy

Until the discovery of radioactivity geologists were in the same position as a historian who knew, for example, that the Roman invasion of Britain was followed by the Norman Conquest and that both these events occurred before the Battle of Waterloo, but who could not find any record of the dates of these or any of the other great events of history. The geologist had to be satisfied with putting the scattered pages of earth history into their proper order, with establishing a purely relative chronology, a chronology without years. He knew that our heritage of coal had accumulated long before Northern Ireland and the Western Isles were flooded with basaltic lavas, but he could do no more than vaguely guess at the ages in years of a coal seam or of a column of basalt from the Giant's Causeway.

Hutton himself, in the absence of any guiding data, quite properly made no attempt to estimate the rates of geological processes. He contented himself with having recognized that geological time was unimaginably long. Many of his successors, however, became unduly reckless in their extravagant claims. Kelvin (1824–1907), one of the great pioneers of geophysics, then entered the field with a dramatic counterblast against speculative estimates that assumed a background of almost unlimited time. Penetration of the earth's crust by bore-holes and mines shows that temperature increases with depth. This fact means that there is a flow of heat from the interior to the surface. True, the amount of escaping heat is only a minute fraction of the enormously greater amount of heat received from the sun (p. 298). But nonetheless the earth *is* losing heat, and Kelvin argued that it must therefore have been progressively hotter in the past. Beyond the dim horizon of the oldest known rocks, where Hutton could see no sign of a beginning, Kelvin, as a physicist, saw a beginning corresponding to a time when the earth was a molten planet, newly born, as was then generally supposed, from the sun. In 1862 he set himself the problem of calculating the time that had elapsed since the earth's consolidation from a molten state. Because of uncertainty in the data then available he allowed wide limits to cover the possible errors, and within those limits he found that the crust had solidified between 20 and 400 million years ago.

Kelvin's challenge initiated one of the great controversies that enlivened the scientific world in Victorian times. Despite energetic protests from the geologists, however, Kelvin felt justified by 1897 in narrowing his limits to 20 and 40 million years. And here it is of interest to notice that with the far superior data of the present day the solution to Kelvin's problem—as he posed it—is between 25 and 30 m.y. Archibald Geikie responded in 1899 by pointing out that the testimony of the rocks emphatically denied Kelvin's inference that geological activities must have been more vigorous in the past than they are today. In this Geikie was perfectly right. Evidently some unconsidered factor—or factors—governing the earth's behaviour remained to be recognized. And, strangely enough, although one such factor, *radioactivity*, had already been discovered, its bearing on the earth's thermal history continued to remain unrecognized for a few more years.

Looking back now, when it is easy to be wise after the event, the general nature of the flaw in

Kelvin's argument is quite obvious. Indeed, Kelvin himself was aware of it and pointed it out, but only as a theoretical possibility. The flaw lies in the assumption that the earth must be cooling because it is losing heat. We do not say an electric fire is cooling because it is losing heat. From the moment when it is first switched on it is losing heat—otherwise it would not begin to heat the room—but it is also getting hotter and it continues to do so until a balance is achieved between the heat electrically generated and the heat lost by radiation, conduction and convection. Only when the current is reduced or switched off does cooling begin. Kelvin's treatment of his problem was concerned only with the limiting case when the 'current' was switched off, or more accurately, when there was no 'current', i.e. no internal source of heat, at all.

The physical nature of at least one of the missing factor's in Kelvin's treatment was dramatically disclosed in 1906 when R. J. Strutt (later Lord Rayleigh) detected the presence of radium in a great variety of common rocks from many parts of the globe. This fundamental discovery, that the crustal rocks contain radioactive elements and are consequently endowed with an unfailing source of heat, showed that the earth is not living on a dwindling capital of internal heat inherited from the sun, as Kelvin assumed, but that it has an independent and regular source of heat-income of its own. Estimates of the age of the earth based on the outward heat-flow at once became valueless, and even now we cannot actually be sure whether the earth, as a whole, is cooling down or heating up. All that can be said is that *if* the earth is cooling, the process must be so slow that Kelvin's estimates are no more than a minute fraction of the 'real' age. But there is no certain evidence of cooling. The age-long recurrence of volcanic activity and its undiminished violence during Tertiary and Pleistocene time leaves no doubt that our planet still has heat resources that show no sign of waning even after thousands of millions of years.

## Varved Sediments

To measure geological time in years, the first essential is the recognition of a natural process that operates rhythmically or at a known rate from a defined starting-point and brings about a measurable result. The traditional geological method of estimating time from thicknesses of sediment is unreliable because of the fact that rates of sedimentation are extremely variable. However, here and there in the geological record there are rhythmically laminated sediments made up of pairs of laminae (e.g. of silt and clay) which can be recognized as representing a year, like the annual rings of a tree. Each pair is called a *varve* (Swedish, *varv*, a periodic repetition) and sediments characterized by this annual banding are said to be *varved*. By counting the successive couplets or varves in a given section, the time represented by the latter can be stated in years (Figure 13.1).

The most celebrated example of time measurement by counting varves is that provided by De Geer (1858–1943) and his co-workers, whose detailed work has made possible an exact system of geological chronology covering the period of retreat of the last Scandinavian ice-sheet up to the present day. As described on p. 457 the receding ice front terminated in a great marginal lake of which the present Baltic Sea is the descendant. Each spring and summer, as the ice thawed, the lake in front received a supply of sand, silt and clay from the streams that flowed into it. The coarser material settled down at once, but the finer particles remained longer in suspension and were not completely deposited until much later in the year. But during the late autumn and winter the glacial streams were frozen and the lake, itself frozen over, received no further sediment. The mud that still remained suspended slowly sank to the bottom, forming a thin layer of dark clay, easily distinguishable from the thicker layer of sandy silt beneath it. The following year the sediment liberated from the ice was again sorted out and deposited in two well-marked seasonal layers, sharply separated from the underlying pair. As this process continued year after year the area of deposit moved northwards with the receding ice, and the varves thus became superimposed after the fashion of wedge-shaped tiles on a roof.

In 1884 De Geer began to count the varves near Stockholm and in later years, when he had devised a method of correlation, the record was systematically carried backwards in time to the coast of Scania at the southern end of Sweden, and forwards to Lake Storsjön, 1000 km to the north. At any one locality the thickness of the varved deposit rarely exceeds 10 m, but this may contain several hundreds of varves, each representing a single year (Figure 13.1). Those near the top can then be matched against the lower varves at a neighbouring locality to the north, where the sequence

**Figure 13.1** Varved sediments in a clay pit, Uppsala, Sweden. The photograph shows 60 varves in a thickness of about a metre (*A. Reuterskiold, 1920*)

includes the varves of the immediately succeeding years. Thus, by tracing the overlapping varves through sections and borings at more than 1500 localities, and so carrying on the counting bit by bit, De Geer and his students suceeded, after thirty years of laborious work, in establishing an accurate chronology of the retreat stages of the last of the European ice-sheets. The Ice Age was conventionally regarded as having ended about 8700 years ago, when the ice-sheet had retreated to Ragunda and separated into two isolated ice-caps (Figure 21.15). Varves continued to be deposited in Lake Storsjön until 1796, when collapse of the morainic barrier holding back the water catastrophically drained the lake. The varves of the lake-floor sediments made it possible to complete the count up to a date known historically. About 13,500 years have elapsed since the ice-front stood along what is now the coast-line of southern Sweden.

The only geological period of which the duration has been estimated from varve investigations is the Eocene. W. H. Bradley discovered long sequences of varved sediments in the Green River formation, which consists of lake deposits about 650 m thick in Wyoming, Colorado and Utah. Here, the dominant type of varve has two laminae, one being markedly richer in organic matter than its associate, exactly as happens in the annual lamination found in many present-day lake deposits. By counting the varves wherever they occur, Bradley estimated the Green River epoch to have lasted some 5 to 8 m.y. (6·5 ± 1·5 m.y.). Assuming the rate of accumulation of the Eocene lake deposits above and below the Green River Series to have been the same as that of the latter, the duration of the whole Eocene period comes out at about 22 ± 5 m.y. This estimate compares well with the 16 m.y. allotted to the Eocene in Figure 13.6.

With few exceptions varved sediments record only short intervals of time. Apart from these, geological 'hour-glass' methods are necessarily based on little more than crude guesses.

In 1878 an Irish geologist, Samuel Haughton, introduced a principle that has often been used for lack of a better. In his own words 'the proper relative measure of geological periods is the maximum thickness of the strata formed during these periods'. The figures listed in Figure 13.6 add up to the astonishing total of 137,769 m for the fossiliferous systems alone. However, beyond emphasizing the immensity of geological time these figures are of little help in solving the age problem, because we do not know the rates at which the beds were deposited. To get a preliminary idea of the sort of time-scale that is implied we might take the sedimentation of the Nile—a fast worker—as an example. Since the reign of Rameses II, over 3000 years ago, 30 cm of sediment have been added to the neighbourhood of Memphis every 400 or 500 years. At this rate the fossiliferous strata of the geological column would represent about 200 m.y. But neither this nor any other estimate of its kind can be taken seriously, and it is not surprising that the geologists of last century should have failed to reach agreement in their efforts to assess the time elapsed since the beginning of the Cambrian period. Those who yielded to the cramping influence of Kelvin's authority favoured short estimates, like the 27·6 m.6. of Walcott (1893) and the 18·3 m.y. of Sollas (1900). A few steadfastly refused to be bullied into respectable conformity. Goodchild in particular deserves a salute for his 1897 estimate of 704 m.y.

It is easy to find fault with Haughton's principle, but it is only now, about 100 years later, that it can be dispensed with. It should be noticed, moreover, that the enormous maximum thicknesses that have been recorded for the various systems (Figure 13.7) imply that space must have been made available for their accumulation by long-continued down-sagging of the crust. In a very rough-and-ready way the maximum thicknesses are measures of the maximum crustal depressions during the periods concerned. The next task was to tie these maximum thicknesses to actual age determinations based on the phenomena of radioactivity. Sollas was obviously unhappy about his 1900 estimate, for he exclaimed: 'How immeasurable would be the advance of our science could we but bring the chief events which it records into some relation with a standard of time!' Today that hoped-for advance is in full acceleration.

## Radioactivity

The first of the radioactive methods to be developed was based on the generation of helium and lead fron uranium and thorium, and the accumulation of these stable end-products in radioactive minerals. The story begins in 1896 when the French physicist Becquerel most unexpectedly found evidence that uranium salts give out invisible rays. These rays first revealed their existence by their effects on photographic plates wrapped in black paper in a closed drawer. Madame Curie, then a young research student, followed up this epoch-making discovery by testing all the other elements then known. She was soon able to announce that thorium also possesses the astonishing property of spontaneously emitting radiations. For the new phenomenon she proposed the name *radioactivity*. But nothing impressed her more than the discovery that uranium minerals are far more powerfully radioactive than the uranium they contain. This clearly pointed to the presence in the minerals of a previously unknown and highly radioactive element, or possibly of more than one.

Working with pitchblende (a mineral consisting mainly of the oxides of uranium) Madame Curie and her husband, Pierre Curie, detected two new radio-elements before the end of 1898. The second of these was radium. Later, when radium was isolated, it was found to be continuously generating heat. This is true of every radio-element, but in the case of radium, which is about three million times as active as uranium, the heat was easily detectable with an ordinary thermometer.

Rutherford found that the radiations emitted are of three different kinds which he distinguished as $\alpha$-rays, $\beta$-rays and $\gamma$-rays.

$\alpha$-rays are beams of positively charged particles which were eventually shown to be the nuclei of helium atoms (2 protons + 2 neutrons), ejected from the nuclei of the disintegrating atoms with velocities of thousands of kilometres per second. The resulting atomic collisions are the main source of the heat generated by radio-elements that discharge $\alpha$-rays.

$\beta$-rays are beams of negatively charged particles, expelled even more swiftly. These were soon identified as electrons, but electrons from the nucleus, not from the surrounding orbital swarm (see p. 42). One of the neutrons of a radioactive nucleus spontaneously shoots out an electron (−) and turns into a proton (+).

$\gamma$-rays are like X-rays of very short wave length; they travel with the velocity of light.

Now clearly when the nucleus of a radium atom loses a helium nucleus (i.e. an $\alpha$-particle), the element that is left can no longer be radium. The new, residual, element turns out to be a gas; it is called *radon* and is also radioactive. Radon in turn loses helium and is transformed into another radio-element. There are several generations of these daughter elements, some losing helium nuclei ($\alpha$-particles) and others losing nuclear electrons ($\beta$-particles), until eventually the series comes to an end. Apart from the helium that has been generated on the way, the stable element that is finally left is an isotope of the familiar element lead.

Obviously, left to itself, radium could not continue to exist indefinitely. It loses helium at such a rate that after 1622 years half of it will have decayed into its daughter elements. This period is described as its *half-life* or *half-value period*. Obviously, unless radium were being continuously generated from something else, the world's supply—however great initially—would have dwindled away to practically nothing long ago. Actually, it was soon discovered that radium is constantly being renewed as a member of one of the families of radio-elements that descend from uranium. The parental element, uranium, as everyone knows nowadays, consists almost entirely of two isotopes, $^{238}U$ and $^{235}U$. Each is the progenitor of a large family of elements, all of which are radioactive except the end-products, helium and lead. A third family, also terminating in an isotope of lead, has thorium, $^{232}Th$, as its parent. From being a father of gods, Uranus has become, with Thor, a father of elements. From being a wielder of thunderbolts, Thor has become, with Uranus, a scintillating generator of high-speed atomic missiles (Figure 13.2).

## The Geologists' Timekeepers

Omitting the energy that is liberated and appears as heat, the ultimate results of these atomic transformations can be summarized as follows:

$$^{238}U \rightarrow {}^{206}Pb + 8\,{}^{4}He$$
$$^{235}U \rightarrow {}^{207}Pb + 7\,{}^{4}He$$
$$^{232}Th \rightarrow {}^{208}Pb + 6\,{}^{4}He$$

Some uranium minerals, of which pitchblende is the chief, contain little or no thorium, but most radioactive minerals contain both uranium and thorium, and consequently all three isotopes of lead are simultaneously accumulating in such minerals. The ordinary lead of commerce, obtained from common lead minerals like galena, is a mixture of the same three isotopes, together with a fourth, $^{204}Pb$, which, unlike the others, is not a product of radioactive decay. Throughout geological time the other isotopes have been steadily increasing in abundance, while $^{204}Pb$ has remained unchanged in amount. This is highly fortunate, because if $^{204}Pb$ occurs in the lead separated from a radioactive mineral, its amount provides a measure of the proportion of inherited lead which was present as an original constituent of the mineral at the time of its crystallization, Such lead, if present, must be subtracted from the total lead in order to arrive at the amount that has been generated within the mineral during its lifetime.

The term 'radioactive mineral' generally means one containing appreciable amounts of uranium or thorium or both. Several other natural elements are now known to be radioactive parents, but only two of these, potassium and rubidium, are of practical importance as geological timekeepers.

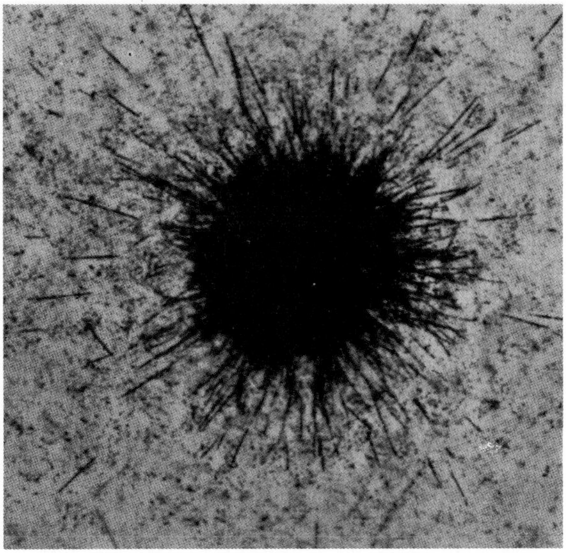

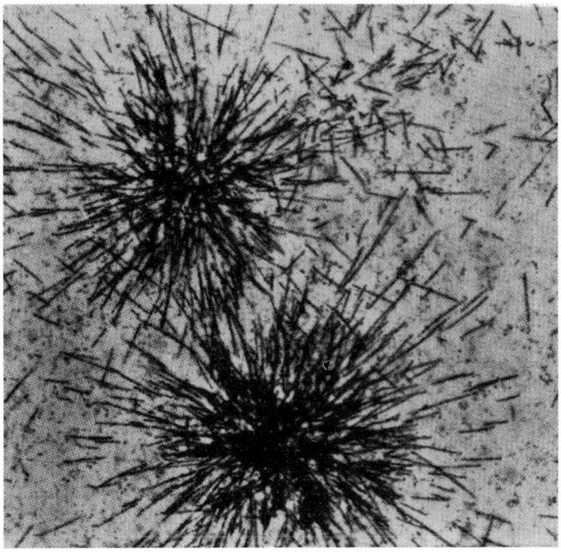

**Figure 13.2** Photographic demonstrations of radioactivity. *left* Auto-radiographs of monazite included in feldspar in a thin section of granite from Predazzo, Italian Alps. The section was placed on a photographic plate (unexposed to light) for 36 days. The radiation has blackened the area covered by the monazite, but beyond this the tracks of individual $\alpha$-particles are seen as darkened radial lines. *right* Zircons, containing minute granules of xenotime, included in feldspar in a thin section of Predazzo granite. Exposure 36 days, as in the photograph to the left. (*O. H. Merlin and E. Justin*)

They are, in fact, of quite special value, because they occur, often together, in common minerals such as potash feldspars and micas, and also in a most interesting bright green mineral called *glauconite* (a hydrous silicate of potassium and iron) which forms on the sea floor, where it is responsible for the striking colour of the green sands and muds that are now being deposited in many places on the continental shelves and slopes. Glauconite occurs in a great variety of fossiliferous strata of known geological age from the Cambrian to the present day. Being a potassium mineral it keeps a record of the time that has elapsed since it originated on the sea floor. In practice there are many difficulties to overcome, and these also apply to the dating of igneous rocks containing potash feldspars and/or micas. The geological age of igneous rocks can rarely be closely established except in the case of lava flows interbedded with fossiliferous strata. Dating by the rubidium method is generally more reliable than that based on potassium, and fortunately it can often be successfully applied to potassic rocks and their minerals.

The radioactive isotope of rubidium, $^{87}Rb$, decays in a very simple way. On disintegration an atom of this isotope loses a $\beta$-particle from its nucleus, which means that a neutron in the nucleus becomes a proton. The mass number remains the same, but the atomic number increases from 37, that of rubidium, to 38, which is that of strontium. An atom of rubidium has been transformed into one of strontium:

$$^{87}\mathrm{Rb} - \beta = {}^{87}\mathrm{Sr}$$

The radioactivity of potassium is more complex. The radioactive isotope is $^{40}K$, and of the atoms that disintegrate about 89 per cent behave like $^{87}Rb$, each losing a $\beta$-particle from its nucleus, and so changing to $^{40}Ca$, with an atomic number of 20 instead of the original 19:

$$^{40}\mathrm{K} - \beta = {}^{40}\mathrm{Ca}$$

Since ordinary calcium is a very common element and is itself mostly $^{40}Ca$, the minute additions contributed by potassium are generally lost beyond recognition, like raindrops in the sea. This change is therefore of very limited value for dating purposes.

However, the remaining 11 per cent of the $^{40}K$ atoms disintegrate by an opposite process, so to speak. Each nucleus captures an orbital electron, so that a proton becomes a neutron and the atomic number decreases from 19 to 18, which is that of argon. This transformation can be widely used for dating minerals:

$$^{40}\mathrm{K} + e^{-1} = {}^{40}\mathrm{Ar}$$

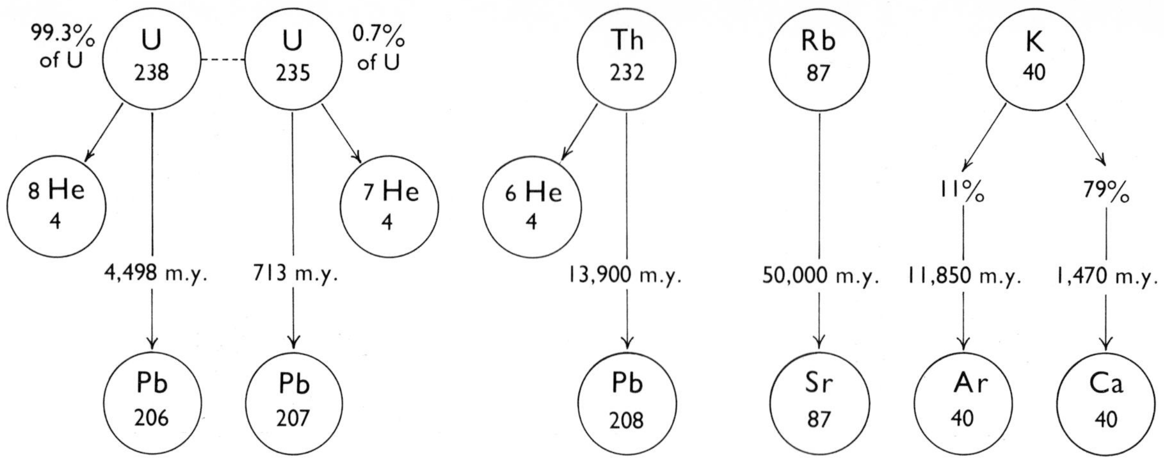


**Figure 13.3** Radioactive timekeepers: diagram showing the parent isotopes and their half-lives (in millions of years) and end products. The half-life of Rb is uncertain

With the recent development of extremely delicate methods of of analysis we have now six different kinds of radioactive 'clocks' (Figure 13.3):

(1, 2 and 3) three ticking out three different isotopes of lead;
(4) the same ticking out helium;
(5) another ticking out argon; and
(6) yet another ticking out strontium.

Of these, the accumulation of helium in ordinary radioactive minerals is of very limited value for dating purposes, since helium is a light gas, some of which readily escapes by outward diffusion from any mineral in which it has been generated in more than slight traces. Argon is also a gas, but a much heavier one. Though less liable to leak out than helium, it does tend to escape, especially from feldspars, and many 'argon ages' have turned out to be somewhat low when they have been checked by other methods.

## Reading the Radioactive Timekeepers

The rate of decay of all radioactive elements is found to correspond with a very simple law that was discovered and announced by Rutherford and Soddy as long ago as 1902. The number of atoms, $n$, which decay during a given unit of time (e.g. a second or a year, as may be convenient) is directly proportional to the number of atoms, $N$, of the radioactive element present in the sample concerned. It is important to notice that the rate of decay is quite independent of the length of time the atoms have already existed. Unlike human beings, amongst whom the chance of dying increases with old age, the atoms of a radioactive isotope are just as likely to live or die (i.e. to decay or disintegrate) whatever their age may be. The constancy of the rate of decay is purely statistical. Out of ten million atoms ($N$) of radium, for example 4273 ($n$) decay every year. The fraction $n/N$ is called the disintegration or decay constant, and is represented by the symbol $\lambda$. For radium $\lambda = 0{\cdot}0004273$ per year. It is generally easier to think of the half-life period, $T$, which is the time taken for any initial number of atoms to be reduced to one-half. $T$ always $= 0{\cdot}693/\lambda$. Thus for radium the half-life is 1622 years.

The fact that there is still a good deal of uranium left in the world shows that the half-lives of its isotopes must be enormously longer than that of radium. A near-coincidence worth remembering is that the half-life of $^{238}U$ is practically the same as the age of the earth's crust. That is to say there was twice as much $^{238}U$ in the world 4498 million years ago (to the value of $T$ for $^{238}U$) as there is today (Figure 13.4).

For those who are mathematically minded we may now briefly consider the general formula for calculating radiometric ages.

Let $N_p$ be the number of atoms of a parental radio-element now present in a given sample of a mineral that crystallized $t$ years ago, $t$ being the age of the mineral.

Let $N_d$ be the number of atoms of the end-product—the stable daughter-element—generated in the sample during the time $t$.

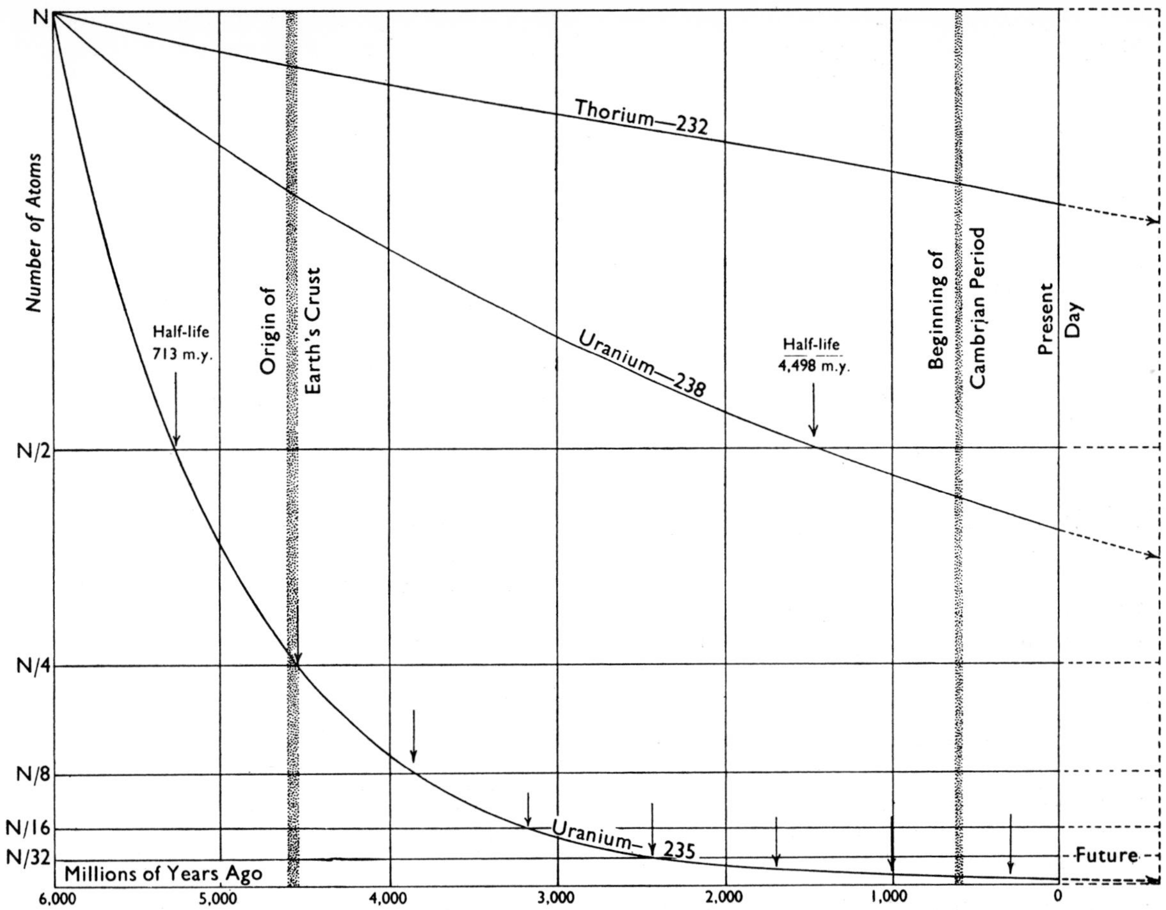


**Figure 13.4** Diagram to illustrate the radioactive decay of uranium and thorium, starting with N atoms of each isotope 6000 m.y. ago. It will be seen that while U-235 has passed through six of its half-lives (and is well into its seventh) during geological time, U-238 has passed through only one, and Th-232 through only a fraction of one.

$N_p$ and $N_d$ are known by analysis of the sample, and our problem is to calculate the value of $t$ from the analytical results. For this we need an equation relating $t$ to the rate of decay of the parent element, that is, to the disintegration constant, $\lambda$.

From Rutherford and Soddy's fundamental law of radioactive decay, the number of atoms of the parent element originally present in the sample when the mineral crystallized can be expressed as $N_p e^{\lambda t}$, where e is the base of Naperian logarithms. Thus we arrive at the equation

$$N_p e^{\lambda t} - N_p = N_d$$

which reduces to

$$e^{\lambda t} = 1 + N_d/N_p.$$

Taking Naperian logarithms of each side of this equation.

$$\lambda t = \log_e (1 + N_d/N_p),$$

and transposing to ordinary logarithms this becomes

$$t = (2{\cdot}303/\lambda) \times \log_{10} (1 + N_d/N_p)$$

or, in terms of the half-life, $T$,

$$t = 3{\cdot}323\ T \times \log_{10} (1 + N_d/N_p).$$

The age of a pegmatite containing, say, uraninite, and rubidium-bearing feldspars and micas can be estimated in five independent ways by substituting the appropriate values for $\lambda$ or $T$, and for the respective ratios corresponding to $N_d/N_p$. These are, as indicated in Figure 13.3, $^{206}Pb/^{238}U$; $^{207}Pb/^{235}U$; $^{208}Pb/^{232}Th$; $^{87}Sr/^{87}Rb$; and $^{40}Ar/(11\%$ of $^{40}K)$.

A value for $t$ is also given by the ratio $^{207}Pb/^{206}Pb$. This depends on the fact that $^{235}U$

decays more than six times as fast as $^{238}U$. Consequently, looking backwards in time, it is easy to see that the older the mineral the higher was the initial proportion of $^{235}U$, and the greater the proportion of $^{207}Pb$ afterwards generated in the mineral. The ratio $^{207}Pb/^{206}Pb$ increases with the age of the mineral as indicated by Figure 13.5.

It may be added here—in anticipation of a very common question—that atomic fission of the kind effected in atomic bombs and nuclear power-stations can be completely neglected in age determinations. Such fission does in fact occur naturally in rocks, as a result of accidental and extremely rare head-on collisions between high-energy neutrons (mainly derived from cosmic rays) and the nuclei of the heavy radio-elements. But the rate of this kind of atomic destruction is excessively slow compared with that of normal radioactive decay: between a million and a thousand times less. Not only is the effect on geochronology negligible, but the energy liberated by natural fission is far too minute to have the slightest geological significance; in particular, it has no possible bearing on the problem of volcanic activity.

**Figure 13.5** Graph giving the approximate age of a mineral for which the ratio of radiogenic Pb-207 to radiogenic Pb-206 is known. It will be noticed that for minerals younger than 400–500 m.y. the variation of the ratio with time is too slight to be of much value.

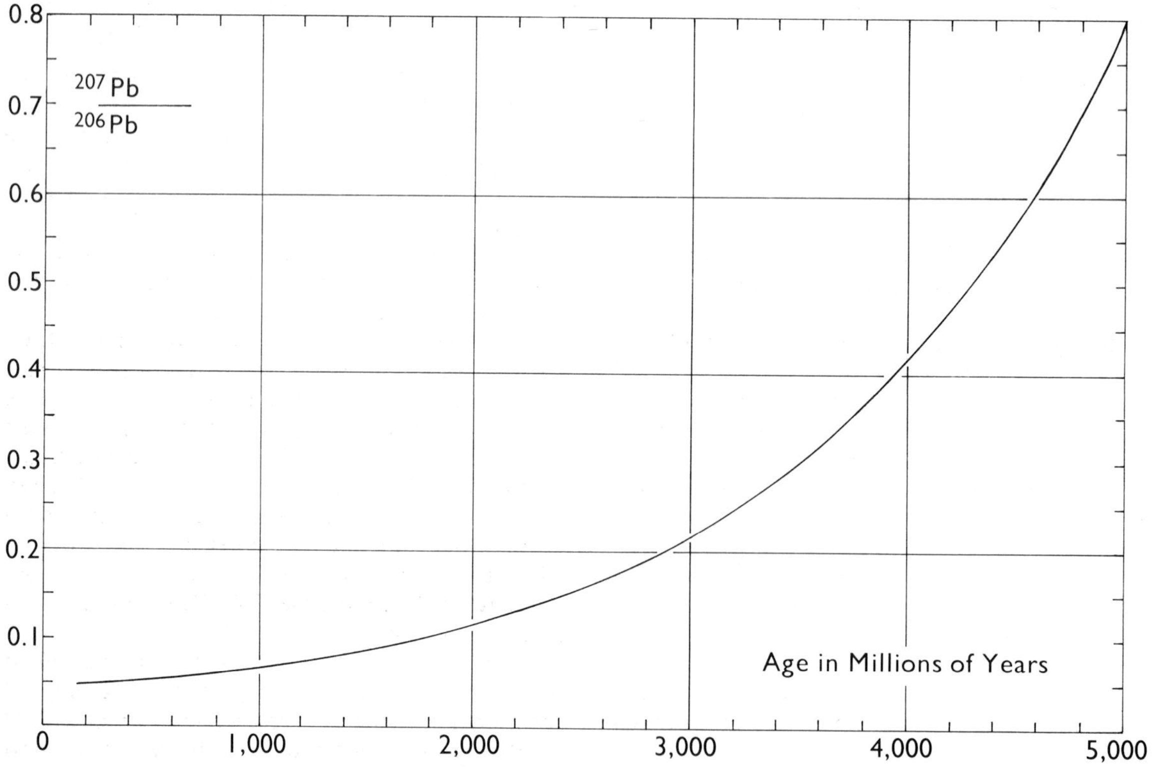


## Dating the Geological Periods

To construct an acceptable time-scale it is necessary to have reliable radiometric age measurements on minerals or rocks of known geological age, and also to have determinations which are evenly distributed, through the periods, in time. So far neither condition is fully realized. Many results have to be discarded either because the evidence as to stratigraphical age is not conclusive, or because the radiometric ages are not completely reliable. One difficulty at present is uncertainty as to the half-life of rubidium.

The rocks and minerals used for radiometric determinations are derived from one of the following varieties of rocks:

(*a*) Lava flows and pyroclasts interbedded with sedimentary rocks whose stratigraphical age is firmly established from fossil evidence. The radiometric ages of such lava flows or pyroclasts

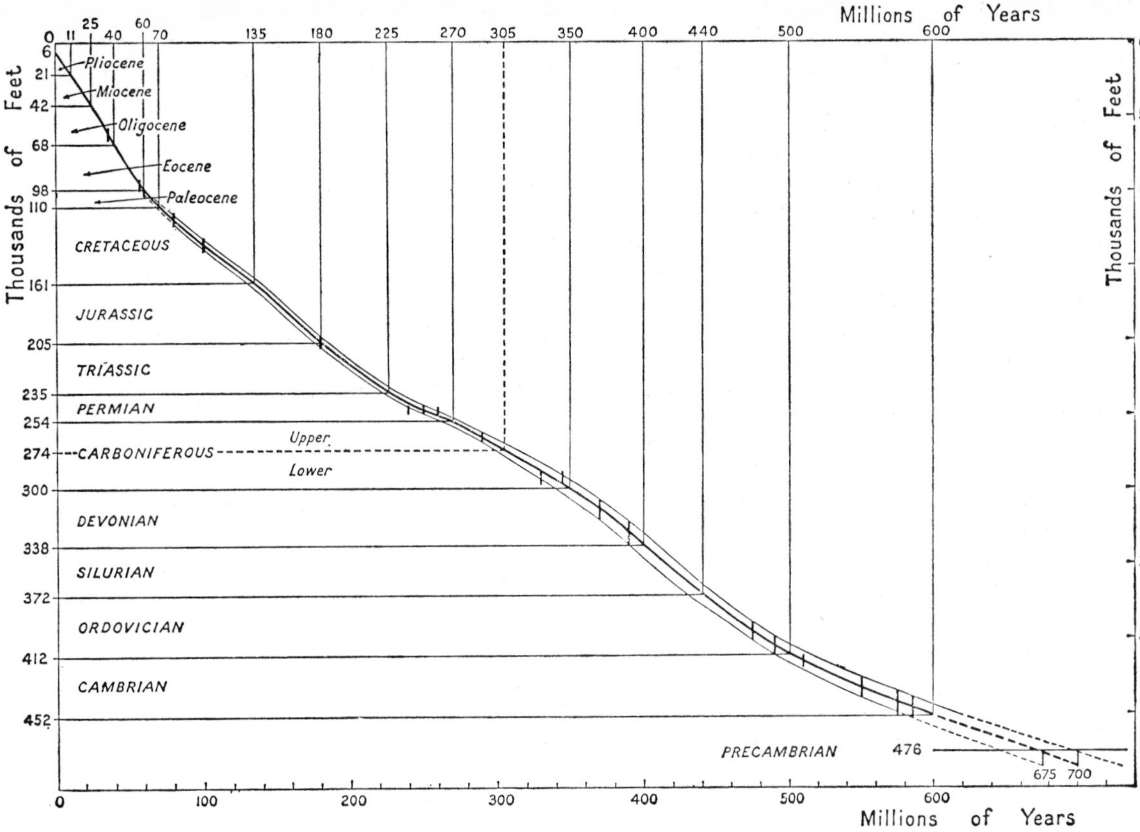


**Figure 13.6** Construction of a geological time-scale from the beginning of the Cambrian Period (*A. Holmes, 1959*)

will correspond very closely to the actual ages of the strata with which they are interbedded.

(*b*) Sedimentary rocks, accurately dated stratigraphically from fossil evidence, which contain authigenic (generated *in situ*) minerals, which have grown at the time of, or after, the deposition of the rock of which they form part. The radiometric ages are here the minimum age of the sedimentary rock. Glauconite has commonly been used for radiometric dating, particularly glauconite collected from the cores of deep bore-holes. Compared with specimens that have been exposed to near-surface conditions they are less likely to have lost some of their accumulated store of argon. From a survey of the available data, however, Lambert (1971) considers glauconite to be unreliable for dating rocks older than Oligocene.

(*c*) Intrusive igneous rocks, and minerals derived from them, that intersect sedimentary rocks of established stratigraphic position. The radiometric ages found in this way are minimum ages of the sedimentary rocks, which are necessarily older than the igneous rocks that cut them.

(*d*) Metamorphic rocks, in which event the radiometric age, particularly if determined from minerals, probably dates the time of metamorphism (see p. 241).

Many time-scales have been constructed for the Phanerozoic, that is from the base of the Cambrian upwards. Phanerozoic is a group name, derived from Greek and meaning 'plainly evident life'. When such time-scales were first constructed, radiometric data were extremely scarce. As more data accumulate it frequently becomes necessary for time-scales to be revised by geologists capable of assessing the value of both the stratigraphic and radiometric evidence concerned, so that only reliable data are used. Figure 13.6 is a radiometric time-scale for the Phanerozoic, constructed from data available up to 1968 (*The Phanerozoic Time-scale*, 1964, 1971).

In the early days of radiometric dating, when dates were lacking for many parts of the geological column, it was desirable to have some independent geological standard which, however rough and ready, would serve as a means of interpolating from one fairly well-dated part of the column to another. It is here that Haughton's principle (p. 233) continued to be of service. By plotting ages against cumulative thicknesses of strata, as in Figure 13.7, at least a provisional approximation towards an internally consistent time-scale could be achieved. From the resulting curve (actually the line lying in the middle of a sort of 'channel of uncertainty' which allows for errors either way) it is possible to read off the approximate duration of each of the periods and the corresponding dates of their beginning and ends. As reliable radiometric dates have increased, it has not only become possible but also preferable to construct time-scales from radiometric data alone, as exemplified by Figure 13.6.

## Dating the Precambrian and the Age of the Earth

In the early days of geology, Precambrian rocks were thought to represent parts of the original crust of the earth, formed during an 'Archaean' era of hot and turbulent conditions. What are now recognized as fold-structures, and more particularly folded folds (p. 693), were at that time thought to depict turbulent flow. Towards the end of the last century, whilst working in Finland, Sederholm found that the Precambrian could be divided into orogenic cycles, starting with sedimentation, proceeding to orogenesis, with associated regional metamorphism and granitization, and finally to intrusion of granites and emplacement of pegmatites. Subsequently Wegmann, whilst a young man from the school of Argand, devised special techniques for investigating and interpreting successive tectonic events depicted by the gneisses and migmatites of Finland. Sederholm's principle, that there is no obvious reason why orogenesis should be limited to the Cambrian and younger periods, has since been extended to Precambrian shields in other parts of the earth.

Sederholm made great use of swarms of metamorphosed basaltic dykes (metabasalts) as strati-

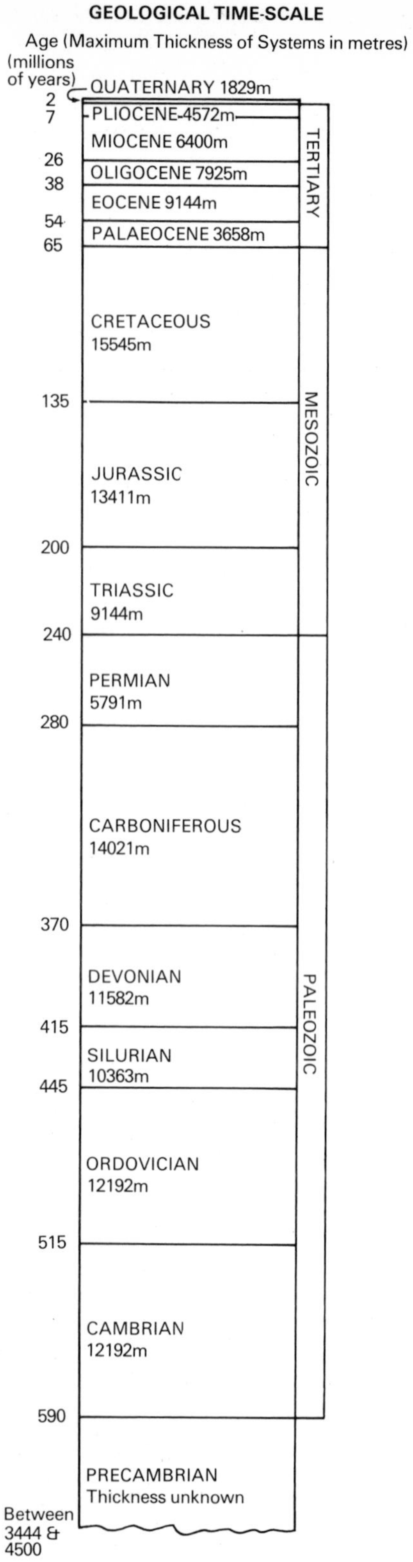


**Figure 13.7** The geological radiometric time-scale based on the Geological Society Phanerozoic time-scale (1964) with recommended amendments by R. St John Lambert, 1971 (*Geological Society Supplement to the Phanerozoic Time-scale*). The radiometric ages, in millions of years, relate to the base of the periods, of which the approximate maximum thicknesses in metres are given.

graphical time-markers, with which to separate successive events from one another. Long afterwards this same method was used by Sutton and Watson (1951) to divide the Lewisian complex of the North-west Highlands of Scotland into an older Scourian province or complex and a younger Laxfordian. The Scourian, having a radiometric age of from 2800 to 2200 m.y. is cut by a swarm of metabasalt dykes about 2000 m.y. old. Emplacement of the dyke-swarm was followed by the Laxfordian, 2200 m.y. to 1600 m.y. old, which is partly composed of metamorphosed and rejuvenated Scourian. This is revealed by two different radiometric ages for the Scourian, derived respectively from minerals and whole rocks. When biotite or hornblende from metamorphic rocks are dated by the K–Ar or the Rb–Sr methods, the date relates to the time of metamorphism, whether the metamorphism resulted from heat alone, or from heat and rock flowage that gave rise to schistosity or foliation. An approach to the age of the original rock is provided by Rb–Sr age determinations obtained from the whole rock. The reason for the two different results is that during metamorphism biotite and hornblende lose argon and radiogenic strontium. Radiogenic strontium does not diffuse far, commonly being accommodated in nearby plagioclase, within the crystal lattice of which it can substitute for calcium. Age determinations on the whole rock, by the Rb–Sr method are therefore likely to reveal the initial age. It is interesting to recall that early in the century, Lyell suggested that metamorphic rocks should have a hyphened name, consisting of the geological age of the original rock, and the time of the metamorphism. Scourian rocks, metamorphosed and rejuvenated in Laxfordian times, would thus be called Scourian-Laxfordian.

In Africa four Precambrian orogenic belts have been distinguished, as illustrated in Figure 13.8.

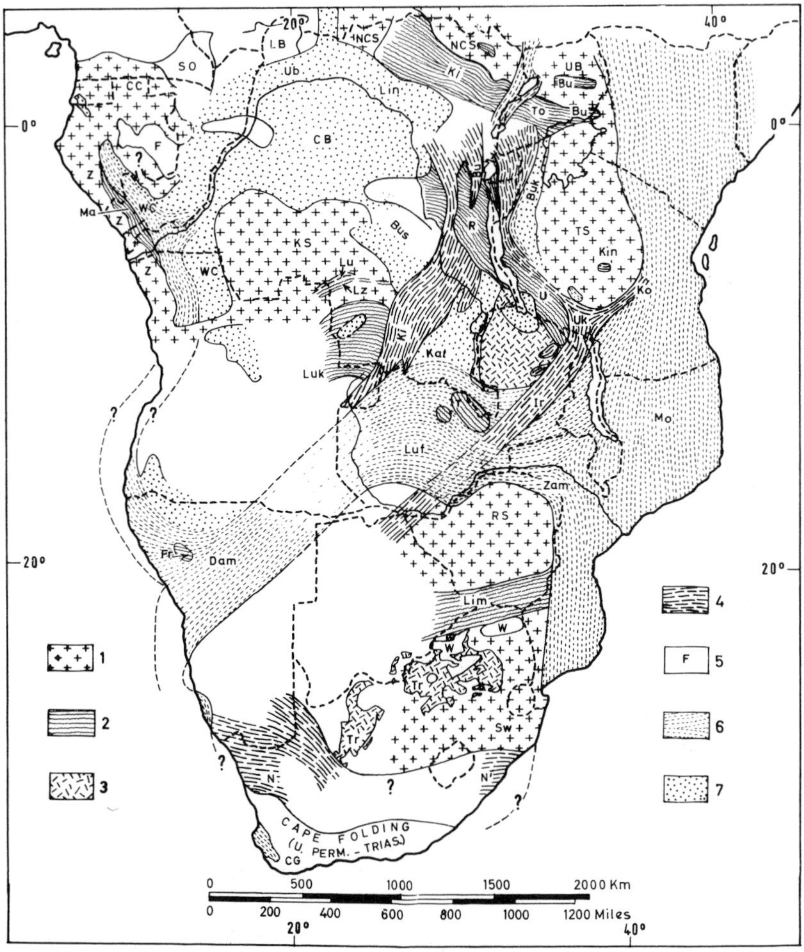


**Figure 13.8** Sketch map illustrating four radiometrically dated orogenic belts in equatorial and southern Africa. The belts were successively folded and metamorphosed about (1) 2500–2600, (2) 2100–1950, (4) 1300–1100, and (6) 730–600 million years ago (*L. Cahen and N. S. Snelling, 1966*)

Of these the Katangan cycle is of interest because it spans a very long period of geological time. Beginning with sedimentation, about 1200 m.y. ago, followed by an orogenic phase, 720 to 750 m.y. ago, the end of folding is evidenced by intrusive granites about 620 m.y. old and more. Finally post-tectonic pegmatites were emplaced from 525 to 510 m.y. ago, followed by mineral veins 510 to 455 m.y. old. The length of the Katangan orogenic cycle was therefore from about 7 to 8 million years.

In Africa there are regions of rock older than those recognized as forming parts of orogenic belts. The rocks of these shields, or cratons as they are sometimes called, are older than 2500 m.y. and locally as old as 3500 m.y. In 1906 Van Hise divided the Precambrian rocks of Minnesota into four groups, which were later divided into two major groups as follows:

| | |
|---|---|
| Keweenewan<br>Huronian | Proterozoic |
| Keewatin<br>Laurentian | Archaean |

This classification, of local significance only, was at one time widely applied to Precambrian rocks of other continents. In recent years the two major divisions have come back into use, the term Proterozoic being applied to rocks between 600 and 2500 m.y. old, i.e. older than the Cambrian and younger than the so-called Archaean. The term Proterozoic is thus used to indicate that part of the Precambrian that has been successfully divided into orogenic cycles, whilst the term Archaean is restricted to still older rocks, older than 2500 m.y., within which orogenic belts have not, at least as yet, been recognized, and which characteristically consist of granulites and syncline-like greenstone belts.

**Figure 13.9** Generalized distribution of the Precambrian provinces of the North Atlantic region as it was, according to Bullard *et al.* (1965), prior to the opening of the North Atlantic ocean. The numbers represent the approximate time, before the present, in millions of years, by which each province was firmly established. The Grenville province of North America correlates in age with that of south-west Scandinavia, but it is still doubtful whether this belt crosses the British Isles beneath the widespread younger rocks. (*After J. Watson, 1973*)

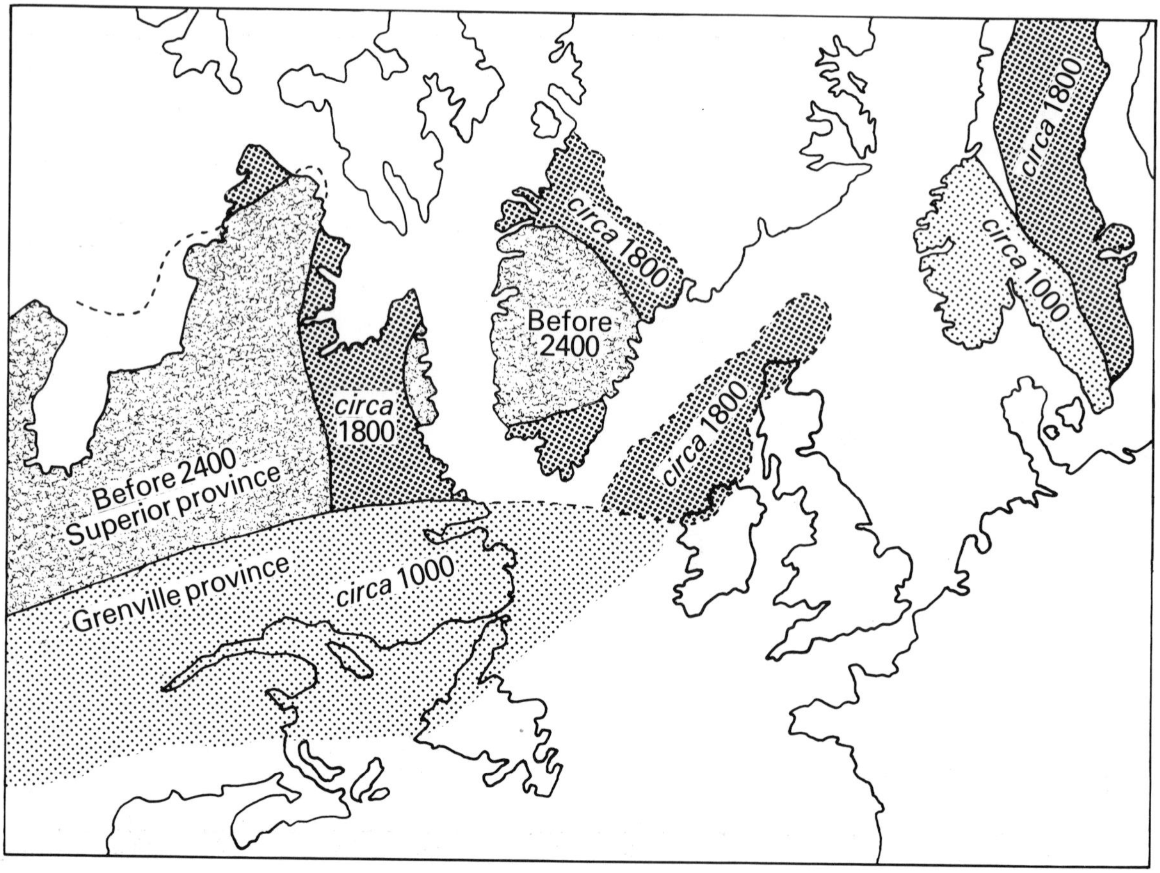

Many attempts have been made to correlate Precambrian rocks from continent to continent, but, in the absence of zonal fossils, this can only be done by reference to radiometric ages. Algae and bacteria have indeed been discovered within Precambrian rocks, even older than 3000 m.y., but these are useless for correlation purposes. Figure 13.9 depicts a recent attempt to correlate Precambrian belts on either side of the North Atlantic, the map representing the time before the North Atlantic Ocean opened.

The oldest rocks that have been found so far are gneisses, 3750 m.y. old, in the Godthaab area of south-west Greenland. Even older crustal rocks existed, however, because gigantic inclusions of metamorphic rocks occur within the gneiss, and appear to have originated as lavas and clastic sediments.

Obviously the earth must be older than 3750 m.y., but its actual age can only be estimated, and this has been done by reference to the three following factors:

(*a*) the age of meteorites, as determined by the U–Pb, and the Rb–Sr methods is about 4600 m.y.

(*b*) The oldest rocks brought back from the moon, with ages determined by the U–Pb and Rb–Sr methods, are again about 4600 years old.

(*c*) Through geological time radiogenic lead ($^{206}Pb$ and $^{207}Pb$) has increased relative to primeval lead ($^{204}Pb$). This increase, as determined from lead ores ranging in age from 500 to 4000 m.y., is shown in Figure 13.10, where the curve, extrapolated backwards through time, passes through the measured lead isotope ratios for a meteorite 4600 years old. This 'fit' indicates that the earth has a similar age.

It can thus be fairly concluded that the earth has an age of about 4600 m.y. It is important to notice that the time that elapsed before the

**Figure 13.10** Graph depicting the increase, through geological time, of radiogenic lead ($^{206}Pb$ and $^{207}Pb$) relative to primeval lead ($^{204}Pb$), as determined from major lead deposits, from localities indicated on the graph, the ages of which have been determined by other radiometric methods. When the curve is extended backwards through time it passes through the position of the measured isotope ratio for a meteorite, 4600 years old, from Canyon Diablo. (*After S. Moorbath, 1975*)

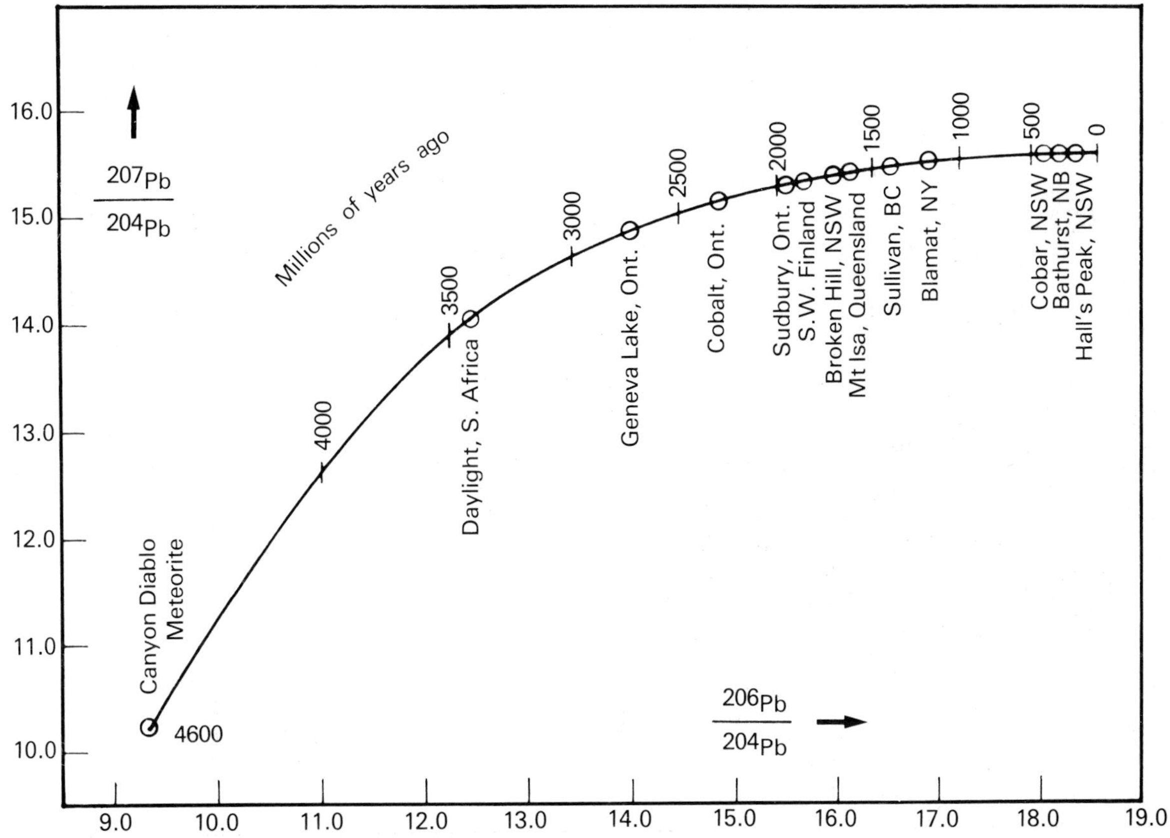

beginning of the Cambrian was about seven or eight times as long as the time that has elapsed since.

## Radiocarbon Dating

The parental radio-elements so far considered are those having half-lives comparable with the age of the earth, and therefore suitable for measuring long intervals of geological time. For measuring shorter intervals, such as those of historical and archaeological interest, no similar methods were available until 1951, when W. F. Libby discovered that minute amounts of a radioactive isotope of carbon, $^{14}C$, exist in air, natural waters and living organisms. Carbon consists mostly of $^{12}C$ (98·89 per cent) with a little $^{13}C$ (1·11 per cent), neither of which is radioactive. The proportion of radiocarbon, $^{14}C$, would not affect these percentages unless they could be expressed to eight places of decimals.

Radiocarbon is being continuously produced in the atmosphere by reactions caused by cosmic rays. When direct hits are registered on the nuclei of atoms of nitrogen or oxygen, the nuclei are shattered and high-speed neutrons appear amongst the by-products. In turn some of these collide with other atoms of nitrogen, transforming them into $^{14}C$. In due course, each $^{14}C$ nucleus loses a $\beta$-particle and is transformed back into $^{14}N$ (Figure 13.11). The rate of decay is such that the half-life is 5570 years. There is good evidence that the rate of production of $^{14}C$ in the atmosphere has not appreciably varied for many thousands of years, and consequently a *natural* state of equilibrium should have been achieved, in which the abundance of $^{14}C$ is such that the gain from new production just balances the loss by decay. The word *natural* is here emphasized, because human activities have upset the balance in recent years, as pointed out below.

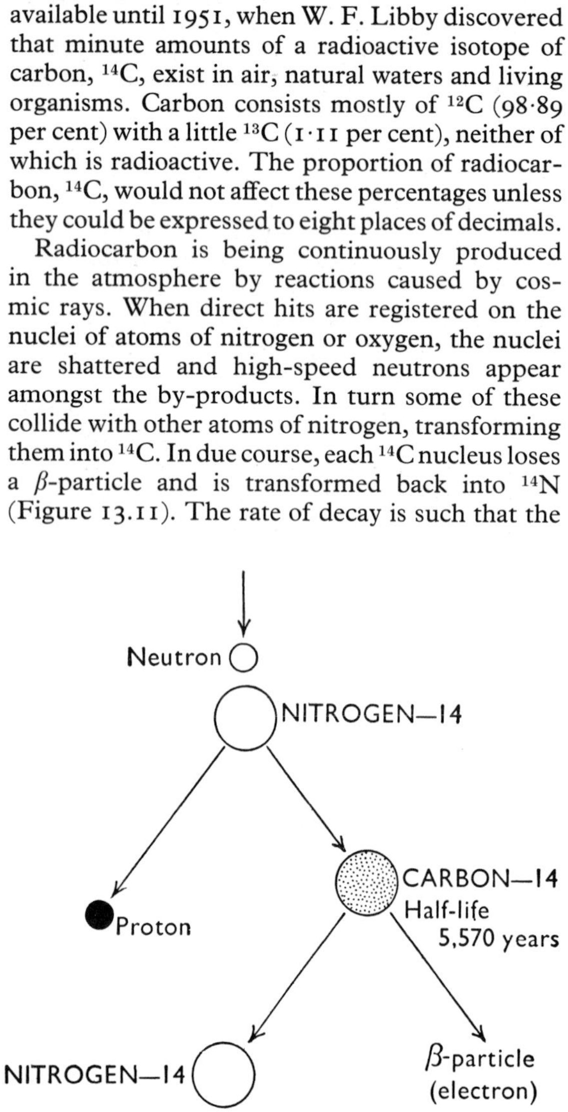


**Figure 13.11** Diagram to illustrate the mode of origin and the radioactive decay of radio-carbon (carbon 14)

The newly-born atoms of $^{14}C$ are speedily oxidized to $CO_2$, which soon becomes uniformly distributed by wind, rivers and ocean currents throughout the circulating carbon dioxide of the world, so that the ratio $^{14}C/^{12}C$ remains theoretically constant. Living creatures, both plants and animals, continually absorb carbon dioxide to replenish their living tissues, and consequently the proportion of $^{14}C$ in their carbon also remains constant. When the organisms die, however, there is no further replacement of carbon from fresh carbon dioxide and the amount of $^{14}C$ already present at once begins to diminish in accordance with the rate of decay. For example, if the carbon from prehistoric wood is found to contain only half as much $^{14}C$ as that from a living plant, the estimated age of the old wood would theoretically be 5570 years.

In practice there are sources of error against which special precautions must be taken. The human interference referred to above is a troublesome complication. The $^{14}C$ content of the atmosphere is being diluted year by year as a result of the combustion of coal and oil. The carbon dioxide so added to the air is made of carbon which long ago lost every detectable trace of $^{14}C$. On the other hand, one of the by-products of testing nuclear missiles is $^{14}C$, and in a few years this increased the natural abundance of $^{14}C$ by about one per cent. On balance the carbon from living plants contains less $^{14}C$ than the carbon from those that lived a century or two ago. It cannot therefore be used as a standard for comparison, and it is common practice to regard wood from the year 1850 as representing modern atmosphere. Another source of error arises from secular variation in the content of $^{14}C$ in the atmosphere. This variation was discovered from radiocarbon ages determined from dated annular rings in old wood. For the last 6000 years this error can be allowed for, and prior to this it was small. Errors also arise from the introduction of carbon from various sources carried by ground-water, and the resulting age determination may be too great or too low, dependent on the source of the introduced carbon.

In spite of the possible errors, the radiocarbon method has proved to be of great value for dating events during the later part of the Ice Age, including sediments brought up as cores from borings drilled into the deep-ocean floor.

## Selected references


CAHEN, L., and SNELLING, N. J., 1966, *The Geochronology of Equatorial Africa*, North-Holland Publishing Company, Amsterdam.

— 1974, 'Potassium–Argon Ages and Additions to the Stratigraphy of the Malagarasian (Bukoban System of Tanzania) of SE Burundi', *Journal of the Geological Society*, vol. 130, pp. 461–70.

DE GEER, G., 1940, *Geochronologia suecica Principles* (Text and Atlas), K. Svenska Vetensk. Handl. ser. 3, v. 18, no. 6, Almqvist and Wiksell, Stockholm.

FAURE, G., and POWELL, J. L., 1972, *Strontium Isotope Geology*, Springer-Verlag, Berlin.

HAMILTON, E. I., 1965, *Applied Geochronology*, Academic Press, New York.

JACOBS, J. A., RUSSELL, R. D., and WILSON, J. TUZO, 1974, *Physics and Geology*, McGraw-Hill, New York.

LAMBERT, R, ST. J., 1971, 'The pre-Pleistocene Phanerozoic time-scale—a review' *in* Part 1 of *The Phanerozoic Time-scale—a Supplement*, Special Publication of the Geological Society, No. 5, London, vol. 120 s.

MAGNUSSON, N. H., 1965, 'The Pre-Cambrian history of Sweden', Eighteenth William Smith Lecture, *The Quarterly Journal of the Geological Society of London*, vol. 121, pp. 1–30.

MOORBATH, S., 1975, 'Progress in isotope dating of British Precambrian', in *Precambrian: A Correlation of Precambrian Rocks in the British Isles*, Geological Society, Special Report, No. 6. Scottish Academic Press, Edinburgh.

— 1975, 'The geological significance of early Precambrian rocks', *Proceedings of the Geologists' Association*, vol. 86, pp. 259–79.

SHOTTEN, F. W., 1967, 'The problems and contributions of methods of absolute dating within the Pleistocene period (President's Anniversary Address, 1966), *Quarterly Journal of the Geological Society of London*, vol. 122 (for 1966), pp. 357–83.

WATSON, J. V., 1975, 'The Precambrian rocks of the British Isles—a preliminary review', in *Precambrian: A Correlation of Precambrian Rocks in the British Isles*, Geological Society Special Report No. 6, Scottish Academic Press, Edinburgh.

# Chapter 14
# Rock Weathering and Soils

The soil, considered as a rock, links common stones with the atmosphere, and the dead dust of the earth with the continuity of life.

*Grenville A. J. Cole*, 1913

## Weathering and Climate

Weathering is the total effect of all the various sub-aerial processes that co-operate in bringing about the decay and disintegration of rocks, provided that no large-scale transport of the loosened products is involved. The work of rain-wash and that of wind are both essentially erosional and are thus excluded. The products of weathering, however, are subject to gravity, and there is consequently a universal tendency on the part of the loosened materials to fall or slip downwards, especially when aided by the lubricating action of water (Chapter 17). It is, indeed, only through the removal of the products of weathering that fresh surfaces are exposed to the further action of the weathering processes. No clean-cut distinction between weathering and erosion can therefore be attempted.

The geological work accomplished by weathering is of two kinds: (*a*) physical or mechanical changes, in which materials are disintegrated by temperature changes, frost action and organisms; and (*b*) chemical changes, in which minerals are decomposed, dissolved and loosened by the water, oxygen and carbon dioxide of the atmosphere and soil waters, and by organisms and the products of their decay. The physical, chemical and biological agents actively co-operate with one another. Shattering requires stresses powerful enough to overcome the strength of the materials, but the strength is gradually reduced by the progressive action of decomposition. Shattering in turn provides increased opportunities for the further penetration of the chemical agents. Everywhere full advantage is taken of joints and of bedding and foliation planes which, together with the cracks newly formed, admit air, water and rootlets down to quite considerable depths (Figure 14.1). Thus, although the processes of weathering may be considered separately, it must not be forgotten that the actual work done is the resultant effect of several processes acting together in intimate co-operation.

The materials ultimately produced are broken fragments of minerals and rocks; residual decomposition products, such as clay; and soluble decomposition products which are removed in solution. The products of weathering differ widely in different places according to the climatic conditions and the relief and configuration of the surface. In general it may be said that disintegration is favoured by steep slopes and by the conditions characteristic of frost-ridden or desert regions, while decomposition and solution are favoured by low relief and by humid conditions, especially in tropical regions. In the temperate zones the weather is widely variable, and most of the leading processes are to be found in operation during one part of the year or another.

## Disintegration by Temperature Changes

Frost is an irresistible rock breaker. When water fills cracks and pores and crevices in rocks and then freezes, it expands by more than 9 per cent of its volume, and exerts a bursting pressure of about 150 kg to the square centimetre. The rocks are ruptured and fragments are wedged apart, to become loose when thaw sets in. Steep mountain slopes and cliffs are particularly prone to des-

**Figure 14.1** Stac Polly, North-west Highland, Scotland, a peak of Torridonian Sandstone lying uncomfortably on Lewisian Gneiss. The serrated skyline and the opening-up of joints and other parting planes illustrate the effects of weathering, dominated by frost-wedging (*L. S. Paterson*)

**Figure 14.2** Frost shattering of Ordovician igneous rocks, North peak of Tryfaen, North Wales (*G. P. Abraham Ltd, Keswick*)

**Figure 14.3** Screes from the Tertiary igneous rocks (granophyre and gabbro) of Austerhorn, Iceland. For locality see Fig. 12.9 (*L. Hawkes*)

truction in this way, especially where joints are well developed (Figures 14.1 and 14.2). The frost-shivered fragments fall to lower levels and accumulate there as talus slopes or screes (Figure 14.3). The treacherous nature of these loose aprons of angular debris is implied by the term *scree*, which is an old Norse word meaning rubble that slides away when trodden on.

In arid climates the rocks exposed to the blazing sun become intensely heated, and in consequence a thin outer shell expands and tends to pull away from the cooler layer a few centimetres within. Under perfectly dry conditions, however, the stresses so developed are insufficient to fracture fresh, massive rocks. Experiments leave no doubt about this. But experiments also show that in the presence of water repeated alternations of heating and cooling eventually lead to rupture. Under natural conditions massive rocks must first be weakened by chemical weathering before shells and flakes of rock can be set free and broken down into smaller fragments. Even in the desert chemical alteration is continually in progress; slowly, it is true, but time is long. The process is facilitated by the rapid chilling due to sudden rainstorms, for in the rare downpours of desert regions the rain may be near freezing-point, and even hailstones are not unknown. The resulting contraction opens up any cracks that may have developed at right angles to the surface, and water is thus given a temporary but effective entry into even the most stubborn of stones. Individual minerals swell and shrink and gradually crumble apart, especially in coarse-grained rocks like granite. The disintegration of pebbles on the desert surface is often conspicuous (Figure 14.4).

Where rocks are bedded or jointed parallel to the surface, actual separation of slabs of rock readily takes place. Here the sheet-joints due to pressure-release (p. 140) are of importance. However, such separation can occur on a spectacular scale, and in fact generally does, where the rocks affected are either massive or have structural planes of weakness (e.g. foliation) that are not far from being parallel to the existing surface. The scaling or peeling-off of flakes and curved shells of rocks, a phenomenon known as *exfoliation*, gives a highly

**Figure 14.4** Desert surface near the Pyramids outside Cairo, showing 'sun-cracked' pebbles. Shattering by temperature changes follows preliminary weakening by chemical weathering. (*J. J. Harris Teall*)

characteristic appearance to the surfaces and profiles of upstanding hills and peaks in the more torrid lands. The effects of exfoliation are especially well developed on the bare, steep-sided inselbergs (isolated 'island mounts') of Africa (Figure 14.5). Sharp edges, newly broken corners and projecting knobs soon become rounded off, because chemical and temperature changes can penetrate more deeply into the rock in such places than where the surfaces exposed are less irregular. The hills become dome-shaped with increasingly steep sides as time goes on. On the convex slopes (Figure 14.6) successive shells of rock can be seen, overlapping like the tiles on a roof, each ready to fall away as soon as it is liberated by cracks at right angles to the surface. Sometimes after sunset the loud report of a splitting rock and the noise of its fall can be heard.

Exfoliation seems to be the result of a complex group of processes which include and co-operate with the stresses set up by daily alternations of expansion and contraction, so that the latter eventually become far more effective agents of destruction than they appear to be when dealt with by themselves in the laboratory. This culmination

**Figure 14.5** A striking example of the inselberg type of landscape, north of Ribaue, Mozambique (*Sketch by E. J. Wayland, 1911*)

**Figure 14.6** Exfoliation in a ring-dyke of syenite, rising about 900 metres from the village seen in the foreground. Chambe Plateau Complex, Malawi. (*R. L. Kinsey, by courtesy of Federal Information Dept., Southern Rhodesia*)

results from the 'fatigue' set up by the many millions of repetitions of heating and cooling that a shell of rock must undergo before it finally springs loose. The principle involved is that stresses are strongly concentrated, and chemical activity greatly increased, at the inner feather edges of even the tiniest cracks. For this reason preliminary weakening by chemical weathering may be imperceptible. Recent researches on the growth of cracks in structural materials (e.g. for use in jet planes) have revealed that a microscopic scratch on the surface may be sufficient to initiate a crack that might lead to failure and disaster. At first the incipient crack opens and extends so slowly as to be hardly perceptible, but sooner or later, depending on the stresses involved, a certain critical point is reached and the crack then suddenly develops into a fracture at 10,000 or more times the previous rate.

## The Role of Animals and Plants

Earthworms and other burrowing animals such as rodents and termites play an important part in preparing material for removal by rain-wash and wind. Worms consume large quantities of soil for the purpose of extracting food, and the indigestible particles are passed out as worm-casts. In an average soil there may be 150,000 worms to the acre, and in the course of a year they raise 10 to 15 tonnes of finely comminuted materials to the surface.

The growing rootlets of shrubs and trees exert an almost incredible force as they work down into crevices. It is not usually realized that the cellulose of which all cell walls are made is actually stronger than many metals. Cracks are widened by expansion during growth (Figures 3.3 and 14.7) and wedges of rock are forcibly shouldered aside. Plants of all kinds, including fungi and lichens, also contribute to chemical weathering, since they abstract certain elements from rock materials and consequently liberate others in the process. Moreover, water containing bacteria attacks the minerals of rocks and soils much more vigorously than it could do in their absence. The dead remains of organisms decay in the soil largely as a result of the activities of bacteria and fungi. In this way carbon dioxide and organic acids, together with traces of ammonia and nitric acid, are liberated, all of which increase the solvent power of soil-water. The chief organic product is a 'complex' of brown jelly-like substances collectively known as *humus*. Humus is the characteristic organic constituent of soil, and water charged with humus acids can dissolve certain substances, such as limonite, which are ordinarily insoluble.

Another effect of vegetation, one which is of vital importance in the economy of nature, is its protective action. Rootlets bind the soil into a woven mat so that it remains porous and able to absorb water without being washed away. The destructive effects of rain and wind are thus effectively restrained. Grass roots are particularly effective in this way. Forests break the force of the rain and prevent the rapid melting of snow. Moreover, they regularize the actual rainfall and preclude the sudden floods that afflict more sterile lands. For these reasons the reckless removal of forests may imperil the prosperity of whole communities. Soil erosion is intensified, agricultural lands are impoverished and lost, and barren gullied wastes, like the 'badlands' of North America, take their place (p. 319). Except after heavy rainfall, the rivers run clean in forested lands, but after deforestation their waters become continuously muddy. Destruction of the natural vegetation by land clearing and ploughing, and the failure to replace forests cut down for timber of destroyed by fire, have had disastrous economic consequences in many parts of Africa and America. Soil devastation has been one of the major factors in the downfall of past civilizations. Man himself has been, and still is, amongst the most prodigal of the organic agents of destruction. But there is a growing awareness, now almost worldwide, of the urgent necessity for soil conservation and the increase of soil fertility.

**Figure 14.7** Tree roots exposed after wedging out the rock masses that originally formed the banks of a stream; cf. Fig. 3.3 (*S. H. Reynolds*)

## Chemical Weathering

The alteration and solution of rock material by chemical processes is largely accomplished by rain-water acting as a carrier of dissolved oxygen and carbon dioxide, together with various acids and organic products derived from the soil. The degree of activity depends on the composition and concentration of the solutions so formed, on the temperature, on the presence of bacteria, and on the substances taken into solution from the minerals decomposed. All natural waters are slightly dissociated into $H^+$ and $(OH)^-$ ions. The acidity of natural water is measured by the concentration of $H^+$ ions, pH as it is called. If the pH value is greater than 7 the water is alkaline; if less than 7 it is acid. Any increase in acidity increases the rate of weathering reactions. The pH of rain-water ranges from 4 to 7. Its acidity comes mainly from dissolved carbon dioxide, which ionizes water:

$$CO_2 + H_2O = H^+ + (HCO_3)^-$$

The lower values of pH may be due to lightning discharge, which produces nitric and other acids in minute amounts. Much lower values (down to 2·8) are also found in certain peat-bog waters, especially where *Sphagnum* and related mosses are actively flourishing.

The chief chemical changes that occur during weathering are solution, oxidation, hydration or hydrolysis, and the formation of carbonates. Only a few common minerals resist decomposition, quartz and muscovite (including sericite) being the chief examples. Others, like the carbonate minerals, can be entirely removed in solution. Most silicate minerals break down into relatively insoluble residues, such as the various clay minerals, with liberation of soluble substances which are removed in solution.

Limestone is scarcely affected by pure water, but when $CO_2$ is also present the $CaCO_3$ of the limestone is slowly dissolved and removed as calcium bicarbonate, $Ca(HCO_3)_2$. Limestone platforms like those around Ingleborough clearly show the effects of solution in their deeply grooved and furrowed surfaces (Figure 14.8). The joints are opened into 'grikes' with intervening ribs or 'clints' of harsh bare rock. The surface is free from soil except where a little wind-blown dust has collected here and there in crevices. On steep or verticle exposures of limestone, deep grooves, like hemi-cylindrical pipes, may be formed by the solvent action of rain-water that has been concentrated into narrow vertical channels, e.g. at joint intersections. These remarkable flutings are only a special variety of 'grikes', but they are known internationally as *lapiés* and they have become of special interest because they can also be formed on vertical surfaces of basic rocks other than limestones, provided that the water concerned has a sufficiently low pH value. Exactly similar vertical flutings which occur on the exposed joint planes of certain basaltic rocks are described below (see Figure 14.12).

When limestone contains impurities such as quartz and clay, these generally remain more or less undissolved, and so accumulate to form the mineral basis of a soil. The red earth, known as *terra rossa*, that covers the white limestones of the

**Figure 14.8** 'Clints' and 'grikes' produced by chemical weathering on a shelf of the Great Scar Limestone (Lower Carboniferous), above Malham Cove, North Yorkshire (cf. Fig. 6.8) (*Bertram Unné*)

Karst, a plateau behind the Adriatic coast of Yugoslavia, is a weathering residue rich in insoluble iron hydroxides derived by long accumulation from the minute traces of iron compounds in the original limestone.

Clay minerals are the chief residual products of the decomposition of feldspars. Under the hydrolyzing action of slightly carbonated waters the feldspars break down in the following way:

$$\underset{\textit{Water}}{6H_2O} + CO_2 + \underset{\textit{Orthoclase}}{2KAlSi_3O_8} = \begin{cases} Al_2Si_2O_5(OH)_4 \\ \textit{(a clay mineral)} \\ 4SiO(OH)_2 \\ \textit{(silicic 'acid')} \\ K_2CO_3 \\ \textit{(removed in solution)} \end{cases}$$

From plagioclase the products are similar, except that $Na_2CO_3$ is formed from albite and $Ca(HCO_3)_2$ from anorthite. Most of the clay is probably at first in colloidal solution, that is to say, it consists of minute particles dispersed through water, the particles being larger than ions but much smaller than any that can be seen with a microscope. The particles eventually crystallize into tiny scales or flakes. The alkalies easily pass into solution, but whereas soda tends to be carried away, to accumulate in the sea, potash is largely retained in the soil. It is withdrawn from solution by colloidal clay and humus, from which in turn it is extracted by plant roots. When the plants die the potash is returned to the soil. Analyses of river waters show that relatively little ultimately escapes from the lands.

As a result of certain obscure reactions, still not cleared up in detail, colloidal forms of clay break down still further:

$$\underset{\textit{Water}}{nH_2O} + \underset{\textit{Clay}}{Al_2Si_2O_5(OH)_4} \longrightarrow \begin{cases} Al_2O_3.nH_2O \textit{ (colloidal} \\ \textit{aluminium hydroxide)} \\ 2SiO(OH)_2 \\ \textit{(colloidal silicic 'acid')} \end{cases}$$

In the humid conditions of the temperate zone, aluminium hydroxide is not liberated to any important extent, but during the dry season of tropical and monsoon lands it is precipitated in a highly insoluble form, and so accumulates at or near the surface as bauxite (p. 257).

The decomposition of the ferromagnesian minerals may be illustrated by reference to a simple type of pyroxene:

$$\text{Water} + \text{carbon dioxide} + \underset{\textit{Diopside}}{Ca(Mg,Fe)(SiO_3)_2} \longrightarrow \begin{cases} 2SiO(OH)_2 \text{ (silicic 'acid')} + \\ \text{soluble bicarbonates of Ca,} \\ \text{Mg and Fe} \end{cases}$$

When $Al_2O_3$ and $Fe_2O_3$ are also present (as in biotite, augite and hornblende) clay and choritic minerals and limonite remain as residual products. In the presence of oxygen, limonite is also precipitated from solutions containing $Fe(HCO_3)_2$. For this reason weathered rock surfaces are commonly stained a rusty brown colour. Ordinary rust is, in fact, the corresponding product of the action of water and air on iron and steel. Rusting is, indeed, the reversion of iron to one of the common ores from which it might have been smelted; it is an economic menace that involves enormous wastage and expense.

The residual products of weathering are naturally more stable in the presence of air and water than the parent minerals from which they were derived. Corresponding to the fact that they are the result of oxidation and hydration, they contain a larger proportion of negative ions, $O^{2-}$ and $(OH)^-$, than the parent minerals. It should be noticed that when a mineral is 'hydrated' it is the ion $(OH)^-$ that is built into the new crystal lattice. The source of the hydroxyl ions is not water alone, but a reaction between water and oxygen ions:

$$H_2O + O^{2-} = 2(OH)^-$$

The resulting increase in the number of negative ions (anions) relative to positive ions (cations) makes for increased stability, i.e. resistance to weathering. If we represent the number of cations in the formula of a mineral by 100, then the corresponding proportion of anions is easily found. For quartz, $SiO_2$, it is 200. Now quartz is a very stable mineral during weathering, and so is muscovite, with 170 anions. Minerals with less than this number are weathered to more stable residual products with a higher number. The feldspars (160) yield successive products which fall into a series of increasing stability: hydromica (200) → clay (225) → bauxite (250–300). The ferromagnesian minerals, hornblende (160), augite (150), biotite (150) and olivine (133) yield chlorite (180) or serpentine (180) and limonite (200–300). Of these olivine, with the lowest proportion of anions, is the most rapidly weathered of all the common rock-forming minerals. Consistently with these figures, granophyre and most other granitic rocks withstand the attack of chemical weathering more effectively than basalts and dolerites, especially the olivine-bearing varieties.

Chemical weathering contributes to the disintegration of rocks (*a*) by the general weakening of

**Figure 14.9** Spheroidal weathering of dolerite sill, North Queensferry, Fife (*Institute of Geological Sciences*)

**Figure 14.10** Residual boulders of dolerite produced by spheroidal weathering, North Queensferry sill, Fife (*Institute of Geological Sciences*)

the coherence between minerals, so that the rock more readily succumbs to the attack of the physical agents; (*b*) by the formation of solutions which are washed out by the rain, so that the rock becomes porous and ready to crumble (e.g. the liberation of the grains of a sandstone by solution of the cement); and (*c*) by the formation of alteration products with a greater volume than the original fresh material, so that the outer shell swells and pulls away from the fresh rock within. In exfoliation the only alteration of this kind that has been detected is the appearance of chlorite in place of biotite.

The separation of shells of obviously decayed rock is distinguished as *spheroidal weathering* (Figure 14.9). It is best developed in well-jointed rocks which, like many basalts and dolerites, are readily decomposed. Water penetrates the intersecting joints and thus attacks each separate block from all sides at once. As the depth of decay is greater at corners and edges than along flat surfaces, it follows that the surfaces of rupture become rounded in such positions. As each shell breaks loose, a new surface is presented to the weathering solutions, and the process is repeated again and again, aided, according to the climate, by temperature changes with or without frost.

Each successive wrapping surrounding the core (if any) becomes more nearly spheroidal than its outer neighbour, until the angular block is transformed into an onion-like structure of concentric shells of rusty and rotted residual material. Cores of fresher and more coherent rock may eventually stand out like boulders when their soft outer wrappings have been washed away by rain erosion (Figure 14.10). Residual boulders of granite, sometimes of great size (Figure 14.11) probably result mainly from exfoliation but with the aid of spheroidal weathering.

By no means all basaltic rocks leave rusty shells of residual materials behind when exposed to weathering. Given the right conditions everything in the rock may be dissolved and washed away, even the iron and aluminium compounds, leaving outcrops with almost fresh-looking surfaces. In appropriate environments, vertical joint faces of compact basalts and dolerites are found to be fluted with the hemi-cylindrical vertical gutters

**Figure 14.11** Residual boulders of granite, produced by combined exfoliation and spheroidal weathering, perched on the summit of an exfoliated surface of granite, Matopo Hills, Rhodesia (*C. T. Trechmann*)

**Figure 14.12** Lapiés formed by chemical weathering and erosion along vertical joint planes in impervious olivine-basalt, Oahu, Hawaii (*H. S. Palmer 1927*)

known as lapiés—solution forms already referred to in the simpler case of limestones. They have been described from regions as far apart as Hawaii (Figure 14.12), New Zealand, and Slieve Gullion in Northern Ireland. Complete solution of basaltic minerals requires water of unusual acidity (with pH between 2·8 and 3·7), such as is available from the percolation through peat-moss or bogs, and from the drippings from dense vegetation. The waters responsible for spheroidal weathering have pH values ranging between 4 and 8.

### Weathering Residues

In dry climatic regions, on steep rock-slopes, and over massive crystalline rocks, the coating of chemically weathered material may not become more than a thin film. But where rain- and soil-water can soak deeply into the rocks, weathering may proceed to a considerable depth. In regions where the rainfall is heavy and evenly distributed, granite may be converted into soft friable earth to depths of 15–30 m. In tropical regions which have a heavy rainfall during the wet season, succeeded by a dry season, when the temperature is high and evaporation rapid, the weathering residues may be very different. Soil-water is removed by plants, and water from below is drawn up to make good the loss so long as the supply holds out. The weak solutions produced by leaching of the rocks during the wet season thus become concentrated by evaporation, and the dissolved materials are deposited, the least soluble being the first to be precipitated. The products include hydroxides of aluminium and iron, silica and various carbonates and sulphates. Most of these are re-dissolved by the rains of the next wet season, but the hydroxides of aluminium and iron are left in a highly insoluble state. They remain at or near the surface, and gradually accumulate as a reddish-brown deposit to which the name *laterite* has been given (L. *later*, a brick). In depth the material is variegated and paler in colour, and it is here that alumina tends to be specially concentrated. At greater depths the bedrock may be intensely decomposed, with abundant development of clay minerals.

Sheets of laterite rarely become thick—10 m is exceptional—because, being impervious, their very formation puts an end to the drainage which is essential to their continued growth. For this reason, too, the overlying soil becomes infertile, and then, no longer held together by roots, it is easily scoured away by torrential rains. Patches of the sterile surfaces of laterite then exposed may resist further erosion for very long periods. In certain regions, as in parts of India, laterite is found to be quite soft below its hard and slag-like crust. The soft laterite can be readily cut into bricks which set hard on exposure to the sun. This easily worked and valuable building material was called laterite not because it resembles brick, but because bricks are made from it.

Quartz, and clay minerals that have not suffered the ultimate loss of silica, remain cemented in the deposit in various proportions. There is every possible gradation between quartz (the chief mineral of sand), clay (mainly composed of clay minerals) and laterite, as indicated by the following scheme:

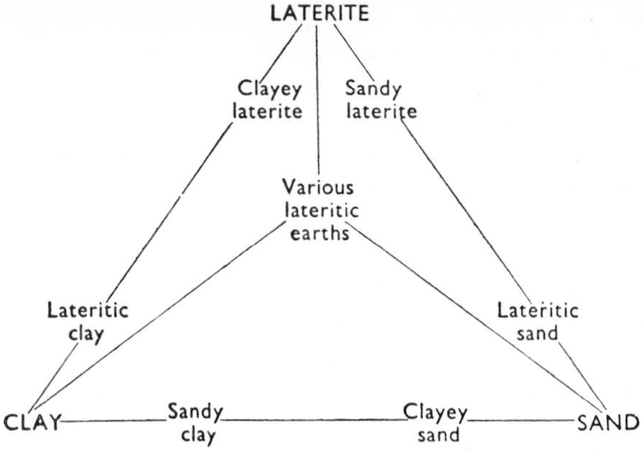


Of the laterites themselves, two important varieties are distinguished: those rich in iron and those rich in aluminium. The latter, when of high grade, constitute bauxite, the only ore of aluminium from which it is practicable to extract the metal on a commercial scale.

## The Mantle of Rock-Waste

The superficial deposits which lie on a foundation of older and more coherent bedrocks form a mantle of rock-waste of very varied characters. In this context it is now commonly necessary to speak of the *waste-mantle* to distinguish it from the *sub-crustal mantle* of the earth. In many places the waste-mantle lies directly on the bedrock from which it was formed by weathering. In this case quarry sections and cuttings of all kinds generally show a surface layer of *soil*, passing gradually downwards through a zone of shattered and partly decomposed rock, known as the *subsoil*, to the parental bedrock, still relatively fresh and unbroken by weathering agents. In the soil vegetable mould and humus occur to a varying extent, and under appropriate conditions they accumulate to form beds of peat which it is often convenient to include with the mantle. Soils develop, however, not only on bedrock, but also on a great variety of loose deposits which have been transported into their present positions by gravity, wind, running water or moving ice. These transported deposits will be considered in their appropriate chapters. Here they are summarized, according to their mode of origin, together with the untransported or sedentary deposits of the waste-mantle. It should perhaps be mentioned that in its engineering usage the word 'soil' refers to all unconsolidated sediments overlying bedrock. 'Soil mechanics' is the study of the behaviour of the ground—whether dry, wet or frozen—under the loads placed upon it by the erection of buildings and the stresses due to engineering structures generally.

## The Growth and Nature of Soils

The purely mineral matter of the residual or transported deposits is first colonized by bacteria, lichens or mosses. By the partial decay of the dead organisms, mould and humus begin to accumulate; lodgment is afforded for ferns and grasses; berries and winged seeds are brought by birds and the wind; and finally shrubs and trees may gain a

THE MANTLE OF ROCK-WASTE
(Continental deposits)

| *Mode of Origin* | | *Characteristic Deposits* |
|---|---|---|
| SEDENTARY | | |
| Essentially inorganic | *Residual* | Gravel, sand, silt and mud<br>*Terra rossa,* lateritic earths and laterite |
| Inorganic and organic | *Soils* | Including soils on bedrock (bedrock soils) and on mantle deposits (drift soils) |
| Essentially organic | *Cumulose* | Vegetation residues: swamp deposits and peat |
| TRANSPORTED BY | | |
| Gravity | *Colluvial* | Screes and landslip deposits |
| Wind | *Æolian* | Sand dunes, sand wastes, and loess |
| Ice | *Glacial* | Boulder clay, moraines, and drumlins |
| Melt-water from ice | *Glacifluvial* | Outwash fans, kames and eskers |
| Rivers (deposited in lakes) | *Lacustrine* | Alluvium and saline deposits |
| Rivers | *Fluviatile* | Alluvium, passing seawards by way of *Estuarine* or *Deltaic* deposits into *Marine* deposits |

footing. The rootlets work down, burrowing animals bring up inorganic particles, and the growing mass becomes porous and spongelike, so that it can retain water and permit the passage of air. Frost and rain play their parts, and ultimately a mature soil, a complex mixture of mineral and organic products, is formed. But though the soil is a result of decay, it is also the medium of growth. It teems with life, and as the chief source of food it is one of the three indispensable assets of mankind. Soil is the base of the pyramid of life on the lands. Its fertility depends on a proper balance of air and water, minerals and trace elements, humus, and a healthy population of bacteria and other lowly organisms.

From a geological point of view soil may be defined as the surface layer of the mantle of rock-waste in which the physical and chemical processes of weathering co-operate in intimate association with biological processes. All these processes depend on climate, and in accordance with this fact it is found that the resulting soils also vary with the climate in which they develop. Other factors are also involved: particularly the nature of the bedrock or other deposit on which the soil has developed, the relief of the land, the age of the soil (that is, the length of time during which soil development has been in progress), and the superimposed effects of cultivation.

The influence of the parental material is easily understood. Sand makes too light a soil for many plants, as it is too porous to hold up water. Clay, on the other hand, is by itself too impervious. A mixture of sand and clay makes a *loam*, which avoids these extremes and provides the basis of an excellent soil. A clay soil may also be lightened by adding limestone, and the natural mixture, known as *marl*, is also a favourable basis. Limestone alone, as we have seen, cannot make a soil unless it contains impurities. In most climates granite decomposes slowly and yields up its store of plant foods very gradually. Basaltic rocks, on the other hand, break down much more quickly. Volcanic ashes and lavas provide highly productive soils which, even on the flanks of active volcanoes, compensate the agriculturist and vine-grower for the recurrent risk of danger and possible destruction.

These differences are most marked in young soils and in temperate regions. As the soil becomes older, and especially when the climate is of a more extreme type, the influence of long continued weathering and organic growth and decay makes itself felt more and more. Certain ingredients are steadily leached out, while others are concentrated. Humus accumulation depends on the excess of growth over decay, and this in turn depends on climatic factors. The composition of the evolving soil thus gradually approaches a certain characteristic type which is different for each climatic region. The black soil of the Russian steppes, for example, is equally well developed from such different parent rocks as granite, basalt, loess and boulder clay. Conversely, a single rock type, like granite, gives grey soils in temperate regions (*podsol*), black soils in the steppes (*chernozem*), and reddish soils in tropical regions of seasonal rainfall (*lateritic earths*). The colours of soils are almost wholly due to the relative abundance (or paucity) of various iron compounds and humus.

Deeply cultivated soils may be more or less uniform throughout, but this is not the case in purely natural soils. A vertical cutting through an old natural soil reveals a layered arrangement which is called the *soil profile* (Figure 14.13). The successive layers of one type of profile are clearly developed in the grey soils of the more or less forested north-temperate belt of Canada, northern Europe and Asia. As the drainage is dominantly descending, iron hydroxides, and humus derived from the surface layer of vegetable mould, are carried down in colloidal solution. Thus a bleached zone is developed, and for this reason the soil type is called *podsol* (Russian, ashy grey soil). By the accumulation of the ferro-humus material at a depth of 5–30 cm, accompanied by particles of silt and clay washed down mechanically, a deep brown or nearly black layer of variable thickness is formed. This may develop into a hard, well cemented band, impervious to drainage, which is known as *hardpan*. One of the objects of ploughing is to prevent the growth of hardpan. Otherwise, waterlogged conditions may set in, and there will then be a marked tendency for peat to accumulate.

Farther south in the grasslands of the steppes and prairies, summer drought and winter frosts favour the accumulation of humus, largely provided by the grass roots which die each year. During the dry season ground-water is drawn towards the surface, and $CaCO_3$ is precipitated, often in irregular nodules, at a depth of 60–90 cm. Under the influence of the ascending calcareous solutions the humus becomes black and insoluble, Iron hydroxides are therefore not leached out as in the podsol. The upper layer of the soil profile is

## Climatic Soil Types and Associated Residual Deposits

| | *Climatic Regions* | *Characteristic Natural Vegetation* | *Mainly Organic* | *Mixed Organic and Inorganic* | *Mainly or Entirely Inorganic* |
|---|---|---|---|---|---|
| ARCTIC | Glacial | — | — | — | — |
| | Tundra (short mild summer) | Mosses and Lichens | Peat Soils Peat | Frost-shattered Stony Soils Polygonal Soils | |
| TEMPERATE HUMID | Boreal (cold winter) | Coniferous Forests | Peat Soils Peat | PODSOL (Grey) | |
| | Temperate (mild and humid) | Coniferous and Deciduous Forests | Peat Soils Peat | PODSOL (Grey) Grey Forest Earths | |
| | Mediterranean (short mild winter) | Evergreen Shrubs and Trees | | Brown Forest Earths | *Terra Rossa* |
| ARID and SEMI-ARID | Steppes (dry summer) | Grasses | | CHERNOZEM (Black) Chestnut-brown Soils | |
| | Deserts | Marginal Scrub — Marginal Scrub | | Grey Marginal Soils<br>Grey Marginal Soils | White Salt Encrustaceans Desert Sands White Salt Encrustaceans |
| TROPICAL RAINY | Savannahs (wet and dry seasons) | Grasses and Scattered Trees | | Brown Soils Black Soils | Laterite |
| | Monsoon Lands (wet and dry seasons) | Forests: Mixed, Open to Sub-tropical Evergreen | Swamp Soils Peat | Brown Lateritic Earths Red Lateritic Earths | LATERITE (Red—Reddish and Brown) |
| | Equatorial (continuously humid) | Evergreen Tropical Rain Forests | Swamp Soils Peat | Brown Earths Grey Bleached Earths | |

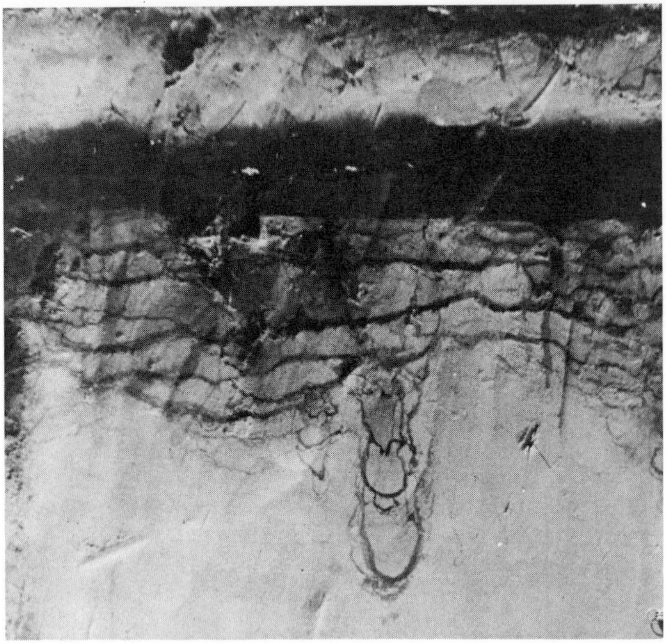


**Figure 14.13** Soil profile of podsol type formed from buff-coloured sandy alluvium C by descending soil-water enriched in humus by its passage through $A_1$ (vegetable mould and humus) here provided by a cover of heather; near Eindhaven, Netherlands. $A_2$ is the zone of bleached grey soil from which the podsol type takes its name; iron hydroxides have been leached out as ferro-humus materials and fixed, with washed down clay particles, in the dark and more compact zone $B_1$, which may eventually develop into hardpan. Continued downward migration leads to periodic precipitation of ferruginous materials in zone $B_2$, which is generally mottled, but sometimes banded, as here. Where there are favourable passageways, localized tongues of $B_2$ may invade the parental material C. (*U.S. Department of Agriculture, photograph by Roy W. Simonson*)

black, becoming brown in depth where there is less humus. For this reason the soil type is called *chernozem* (Russian, black earth). The black cotton soils of India and the 'black bottoms' of the Mississippi flood plains are of similar origin.

## Selected References

BRIDGES, E. M., 1970, *World Soils*, Cambridge University Press, London.
CARROLL, D., 1970, *Rock Weathering*, Plenum Press, New York.
CHAPMAN, R. W., and GREENFIELD, MILDRED A., 1949, 'Spheroidal weathering of igneous rocks', *American Journal of Science*, vol. 247, pp. 407–29.
FREDERICKSON, A. F., 1951, 'Mechanism of weathering', *Bulletin of the Geological Society of America*, vol. 62, pp. 221–32.
HAW, R. C., 1959, *The Conservation of Natural Resources*, Faber and Faber, London.
HUNT, C. B., 1972, *The Geology of Soils*, Freeman, San Francisco.
JACKS, J. V., 1954, *Soil*, Nelson, London.
PRESCOTT, J. A., and PENDLETON, R. L., 1952, 'Laterite and lateritic soils', *Technical Communication* No. 47. Commonwealth Bureau of Soil Science, Farnham Royal, Bucks., England.
REICHE, P., 1950, 'A survey of weathering processes and products', *University of New Mexico Publications in Geology*, No. 1.
REYNOLDS, DORIS L., 1961. 'Lapiés and solution pits in olivine-dolerite sills at Slieve Gullion, N. Ireland', *Journal of Geology*, Chicago, vol. 69, pp. 110–17.

# Chapter 15

# Underground Waters

Rivers do not rise with the first rainfall; the thirsty ground absorbs it all.

*Seneca* (A.D. 3–65)

## Sources of Ground-Water

There is abundant evidence for the existence of important underground supplies of water. At least from the time of the Babylonians there was a widespread and firmly held belief that not only springs and wells, but also rivers, were fed and maintained by water from vast subterranean reservoirs. The underground streams seen in limestone caverns supported this belief, and so did the spurting up of 'the fountains of the deep' through fissures riven in alluvial flats by earthquakes. Moreover, in arid countries like Mesopotamia and Egypt it was far from obvious that rivers could be maintained by rainfall. The author of Ecclesiastes remarked that although 'the rivers run into the sea, yet the sea is not full', and inferred that the balance was restored by a return circulation, underground from the sea floor, back to the sources of the rivers. It was only late in the seventeenth century that it first came to be realized, notably by Halley, that the circulation from sea to rivers was not underground, but through the atmosphere by way of evaporation and rainfall. Aristotle's erroneous conviction that the rainfall was quite inadequate to supply the flow of rivers was not dispelled until accurate measurements took the place of mere opinion. In 1674 Pierre Perrault completed the first quantitative investigation of the relation between rainfall and stream flow. He found that in the upper valley of the Seine the rainfall was actually several times greater than the stream flow, and so demonstrated for the first time a relationship that in humid climates seems now to be almost a matter of common sense.

The following scheme shows the various ways in which rain-water is distributed (see also Figure 3.1, p. 26):

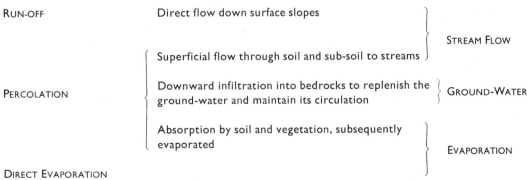

| | | |
|---|---|---|
| Run-off | Direct flow down surface slopes | Stream Flow |
| Percolation | Superficial flow through soil and sub-soil to streams | Stream Flow |
| Percolation | Downward infiltration into bedrocks to replenish the ground-water and maintain its circulation | Ground-Water |
| Percolation | Absorption by soil and vegetation, subsequently evaporated | Evaporation |
| Direct Evaporation | | Evaporation |

(During weathering a relatively trifling amount of water is fixed by hydration in clay minerals and other weathering products.)

Ground-water supplied by rain or snow or by infiltration from rivers and lakes is described as *meteoric*. Fresh or salt water entrapped in sediments during their deposition is distinguished as *connate*. During burial and compaction of the sediments, much of this 'fossil' water is expelled, and during metamorphism most of the remainder is driven out, carrying with it dissolved material which helps to cement the sediments at higher levels. Connate waters heated up during times of metamorphism and igneous activity contribute to the hydrothermal solutions responsible for many mineral veins, and to the hot springs that appear in many volcanic districts. It is commonly believed that part of the steam and hot, mineral-laden water liberated during igneous activity arises from great depths and reaches the surface for the first time. Such water is referred to as *juvenile*, for convenience in discussion, but much of it is likely to be of connate origin (see p. 666).

## The Water Table

Below a certain level, never far down in humid regions, all porous and fissured rocks are saturated with water. The upper surface of this ground-water is called the *water table*. The water table is arched up under hills, roughly following the relief of the ground, but with a more subdued surface. In general, three successive zones may conveniently be recognized (Figure 15.1).

(*a*) *The zone of non-saturation*, which is never completely filled, but through which the water percolates on its way to the underlying zones. A certain amount of water is retained by the soil, which yields it up to plant roots.

(*b*) *The zone of intermittent saturation*, which extends from the highest level reached by ground-water after a period of prolonged wet weather, down to the lowest level to which the water table recedes after drought.

(*c*) *The zone of permanent saturation*, which extends downwards to the limit beneath which ground-water is not encountered. The depths in mines and borings at which the rocks are found to be dry vary considerably according to the local structures, but a limit of the order 600 to 900 m is not uncommon. Juvenile and expelled connate water may, of course, ascend from much greater depths.

Wherever the zone of permanent saturation rises above ground level, seepages, swamps, lakes or rivers occur. When the zone of intermittent saturation temporarily reaches the surface, floods develop and intermittent springs appear. Conversely, many springs and swamps, and even the rivers of some regions, go more or less dry after long periods of dry weather when the water table falls below its usual level.

## Storage and Circulation of Ground-Water

Rocks through which water can pass freely are said to be *permeable*. They may be porous, like sand and sandstone; or they may be practically non-porous, like granite, but nevertheless pervious because of the presence of interconnected open joints and fissures through which water can readily flow. *Permeability* is the capacity of a rock to allow water or other fluids such as oil to pass through it; it is measured by the rate of flow defined in suitable units. The rate of movement of ground-water depends on the permeability of the rock through which it is flowing, and on the hydraulic gradient which provides the motive force. Both these factors vary widely, and even in rocks like sandstone rates of flow may be only a metre or so a year or as much as a metre or so a day; in limestones riddled with caves, or where pumping is resorted to, very much higher rates may be reached.

**Figure 15.1** Diagram to illustrate the relation of the water table to the surface and the variation of its level within the zone of intermittent saturation after prolonged periods of rain or drought. Water temporarily held or trickling down between the surface and the water table is described as *vadose*; that below as *phreatic*.

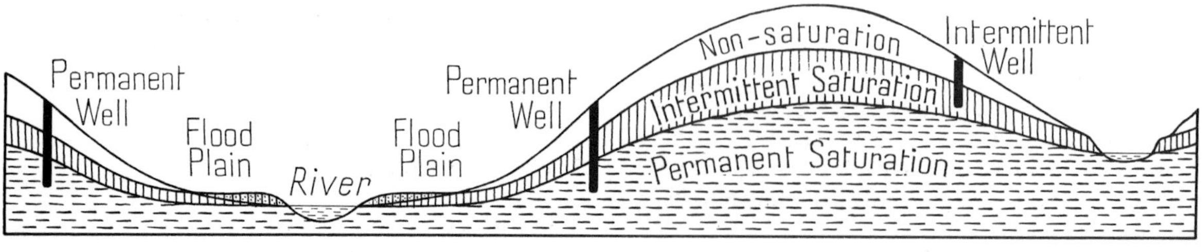

Impervious or impermeable rocks are those through which water cannot easily soak; they may be of two kinds: porous, like clay, or relatively non-porous, like massive unfissured granite. It should be noticed that porosity is not necessarily a sufficient condition to ensure permeability. The size and arrangement of the openings must also be such that continuous through-channels for the free passage of water are available. In clays there are no continuous passageways wide enough to permit the flow of water, except by slow capillary creep.

*Porosity* is expressed as the percentage of free space, or 'voids', in the total volume of the material concerned. Ordinary loose sand or gravel has a porosity of about 35 (i.e. the material in bulk is made up of 35 per cent of 'voids' and 65 per cent of 'solids'), but this drops to about 15 in common sandstones, according to the degree of compaction and the amount of cementing material.

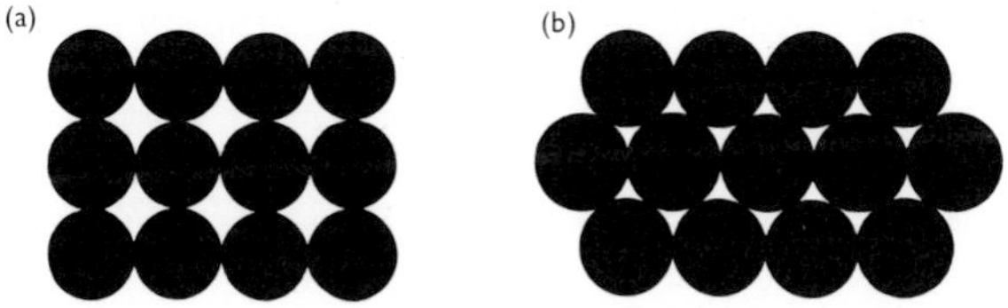


**Figure 15.2** Diagrams to illustrate how pore space varies according to the way the grains (here idealized as spheres) are packed together in a porous sediment:
(a) the most open packing: each sphere touches six others (in three dimensions) and the porosity is 47·6;
(b) the closest packing: each sphere touches twelve others and the porosity is 26.

Figure 15.2 illustrates two extreme ways of packing spherical grains. Loose granular materials, like sand, tend to shake down to an arrangement approximating to closest packing, even though they are not all of the same size and shape. The shape of a given mass of such grains cannot be changed without an increase in voids; and increase of voids can only mean increase of volume. This expansion of closely packed granular masses with change of shape is called *dilatancy*, and it has many geological applications. To take a familiar example of the phenomenon: if you walk along a beach of firm sand left wet by the outflowing tide, you will notice that as each foot is put down it is immediately surrounded by a halo from which the water has drained away. Your weight has changed the packing of the surrounding mass of sand and so has expanded it, thereby providing the empty space into which the surface water has been sucked.

A *quicksand* exemplifies the opposite state of affairs. Here there is a tract of sand through which water is flowing: perhaps imperceptibly, but sufficiently to maintain a loose packing. If you stand on this, the pressure due to your weight compresses the underlying sand into a state of closer packing. Consequently the surface under your feet is lowered and the expelled water wells up all around. A quicksand, in fact, as its name implies, behaves like a fluid. If the layer of quicksand is thick, and especially if it is activated by a rapidly inflowing or outflowing tide, it may be as treacherous and dangerous as many sensational works of fiction have suggested. Closer packing is also very effectively brought about by rapid vibration, and this is the reason for the sudden appearance of spouting sand-vents during earthquakes in alluvial tracts through which water is slowly percolating.

Clay, although it is impervious, may have a porosity of over 45 per cent. By compaction under pressure and the squeezing out of water the porosity drops gradually, falling to as little as 5 in some shales, and to 3 or less in slates. In limestones the porosity ranges from 30 in friable chalk to 5 or less in indurated and recrystallized varieties. Limestone, however, may carry a great deal of water in joints and other channels opened out by solution. The porosity of massive igneous and metamorphic rocks is generally less than 1, but here again water may circulate in appreciable quantities where passageways are afforded by interconnected joints and fissures.

Alternations of permeable and impervious strata, especially when folded, faulted and jointed, form underground reservoirs and natural waterworks of great variety. Where the catchment area is sufficiently high the water slowly migrates through the most permeable formations towards places at a lower level where the water can escape to the surface. It may emerge through natural openings (seepages and springs) or through artificial openings (wells), or it may feed directly into rivers or lakes or even discharge through the sea floor. The sustained flow of rivers which, like the Nile, successfully cross wide stretches of desert is to some extent due to supplies received from underground sources.

## Springs and Wells

When rain-water sinks into a permeable bed, such

as sandstone, it soaks down until an underlying impervious bed, such as clay or shale is reached by the ground-water. If the surface of the junction is inclined, the water flows down the watertight slope, to emerge where the junction is intercepted by a cliff or valley side (Figure 15.3(*a*)). A general oozing out of the water along the line of interception is called a seepage. More commonly a line of localized springs appears. The other diagrams of Figure 15.3 illustrate various examples of other structures favouring the development of springs. In (*b*) a fault brings permeable sandstone against shale which, being impervious, holds up the water. Springs are localized along the line of the fault, and the low ground on the left is marshy. In (*c*) water enters the joints in a massive rock, such as granite, and issues in appropriate places. In (*d*) water impounded by a dyke escapes along the outcrop of the junction. In (*e*) the upper spring is thrown out by a conformable bed of shale, as in (*a*); the lower spring appears at the outcrop of an unconformity, the underlying folded rocks being impervious. In (*f*) water enters jointed limestone, widens the joints by solution, forming caves and underground channels down to the impervious base of the formation. The latter holds up the water and allows it to drain out, sometimes as an actual stream, where a valley has been excavated through the limestone into the underlying rocks.

Wells are simply holes dug or bored or drilled into the ground to a depth at which water-bearing permeable formations or fissured rocks are encountered. Shallow wells, as shown in Figure 15.1, may dry up at certain seasons, unless they tap the zone of permanent saturation. Ground-water percolates into the bottom of the well, and rises to a level that depends on the head of pressure behind it. Pumping or lifting may be necessary to bring the water to the surface. In selecting sites for shallow wells special precautions must be taken to avoid pollution by germ-laden water which might drain into the source of supply from farmyards and cesspools. The ground-water from more deep-seated formations is preferable for human consumption, as it is more likely to be free from the dangers of surface contamination.

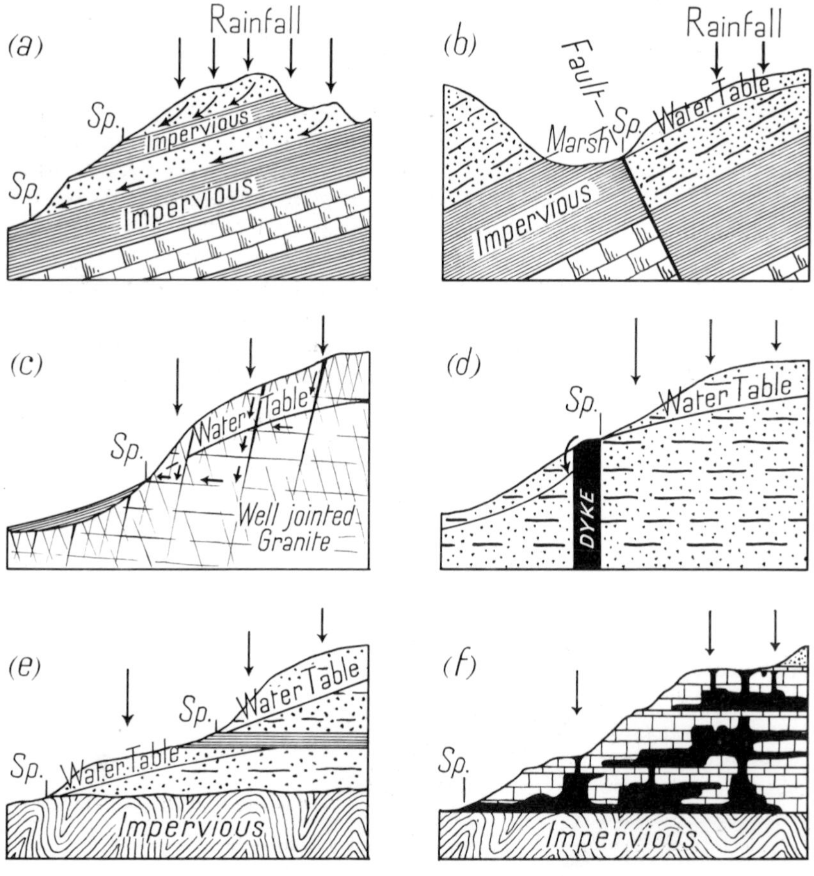


**Figure 15.3** Diagrams to illustrate various conditions giving rise to springs (see text)

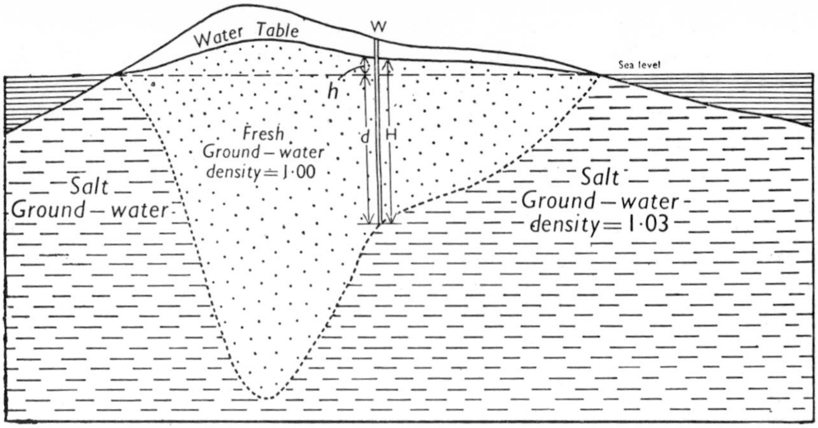


**Figure 15.4** Schematic section through an island or peninsula of permeable rock to illustrate the hydrostatic equilibrium between fresh ground-water received from the rainfall and the heavier salt ground-water that seeps in through the sea floor (not to scale). Ground-water continues fresh to a proportionate depth about three times as great as that shown (see text).

In wells sunk near the coast in permeable rocks which receive an adequate supply of fresh water from the rainfall, the water encountered at sea-level is not salt, as might at first be expected. The lighter fresh water floats on the heavier sea water in roughly hydrostatic equilibrium, so long as it is not actively flowing. Figure 15.4 illustrates the natural distribution beneath an island or peninsula under stable conditions. If $h$ is the height of the water table above sea-level at W (where a well is sunk) and $d$ is the depth below sea-level at which salt water is struck, then the total depth of fresh water, $H$, is $h+d$. For equilibrium,

$$H = h + d = 1{\cdot}03d;\ \text{whence},\ d = 33h.$$

If $h$ is 1 m, then $H$, the total depth of fresh water to be expected, is about 30 m. Such estimates, however, are subject to considerable modification when the water is actively flowing. If flow is towards the sea, the boundary between fresh and salt water is pushed outwards and $H$ increases. More commonly, however, the flow that matters is towards the well and is brought about by pumping. This has the effect of drawing salt water in, towards the bottom of the well, so that $H$ decreases. The boundary between fresh and salt water is naturally variable because of tides, and because of the diffusion, it is never sharp.

## Artesian Wells

*Artesian wells* are those in which the water encountered in depth is under a sufficient hydraulic pressure to force it up to the surface, as it does at Artois in France, from which the name comes. The necessary conditions are: (*a*) an inclined or broadly synclinal water-bearing formation, or *aquifer*, enclosed on both sides by watertight beds; (*b*) exposure of the rim of the aquifer over a catchment or intake area at a sufficient height to provide the ground-water with

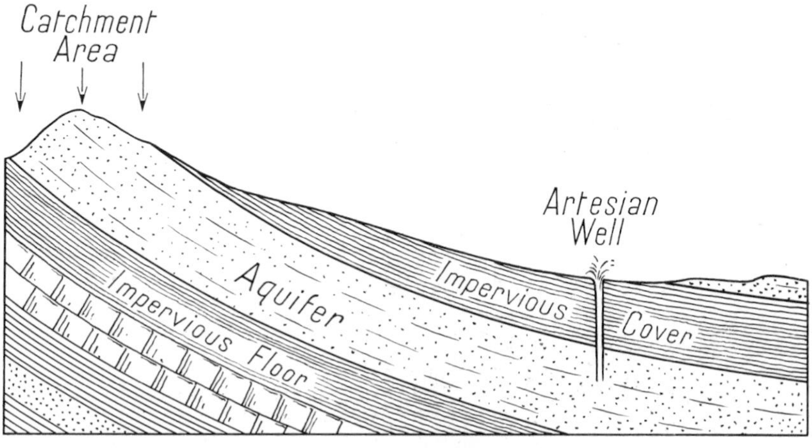


**Figure 15.5** Diagram illustrating the structural conditions favourable to artesian wells

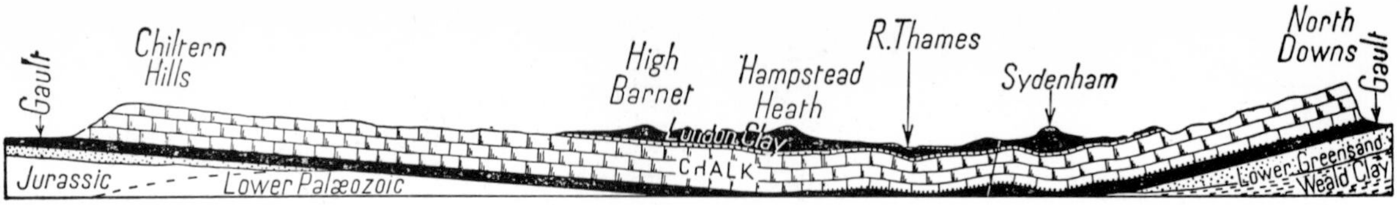


**Figure 15.6** Section 93 km long, through the London Basin

the necessary hydraulic head to raise it to the surface of the ground (or to overflow) where the wells are sunk; (*c*) a sufficient rainfall over the catchment area to furnish an adequate supply of water; and (*d*) absence of ready means of escape of the water except through the wells. The term *artesian* is sometimes extended to include deep wells in which the water approaches the surface but does not actually reach it.

The London Basin (Figure 15.6) formerly exemplified these conditions. The aquifer is the Chalk, with sandy beds above and, locally, below. The enclosing impervious formations are the London Clay above and the Gault Clay below. Water falling on the Chalk, where it is exposed along the Chilterns to the north, and the North Downs to the south, sinks into the basin and would accumulate there, were it not for the insatiable thirst of London. The water in the Chalk is tapped in the London area by more than a thousand wells sunk to depths of 180 to 210 m. Up to a century ago the Chalk was saturated, and when the fountains in Trafalgar Square were first constructed the water gushed out well above the surface. In more recent years the enormous supplies which have been drawn from the Chalk reservoir have far exceeded replenishment by rainfall on the rims of the basin. The level of the water in the Chalk under London has therefore fallen—as much as 120 m in places—and water can now be raised to the surface only by pumping. Near Deptford, where the Chalk is locally exposed, fresh water used to escape naturally into the Thames through springs that could be seen flowing at low water. But for many years the pressure has acted the other way. Salt water, brought in by the tidal rise of the Thames, flows back into the Chalk, and many wells have had to be abandoned because of this contamination.

In North and South Dakota an important aquifer dips off the edge of the Black Hills, and carries a copious supply of water beneath the plains to the east. Over an area of more than 38,000 km² the water can be tapped by artesian wells. The largest artesian basin in the world is that of Queensland and adjoining parts of New South Wales and South Australia. The catchment area is in the Eastern Highlands where porous sandstones of Jurassic and Cretaceous age come to the surface. These sandstones, with their accumulated stores of water, held in by a thick cover of clay, underlie an area of over 1·5 million km². Without the artesian wells, some of which are 1·5 km deep, much of this vast region would be a barren waste. The Desert Basin in the north of Western Australia and the Murray Basin in Victoria also provide invaluable supplies of artesian water.

## Oases

Many of the *oases* of the Sahara and other deserts owe their existence to the local emergence of artesian water at the surface. Fertilized by the escaping underground water, vegetation flourishes amazingly and makes 'a paradise in a setting of blazing sand and glaring rocks'. Between the Chad basin and the Libyan Desert the highlands from

**Figure 15.7** Schematic section across the Sahara to illustrate conditions favourable to the development of oases

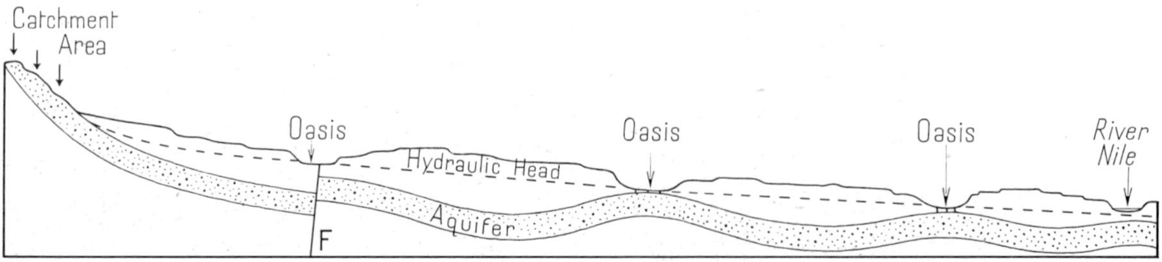

**Figure 15.8** Part of the great oasis of Colomb Bechar in the NW of the Algerian Sahara, as it was about 1930. Since then the abundant supplies of ground-water have been conserved by underground dams and reservoirs; coal has been discovered and, not far away, ores of iron and manganese. Colomb Bechar has now become the centre of a rapidly expanding industrial area. (*Paul Popper Ltd*)

**Figure 15.9** Valley oasis of Tafraouk, SW Morocco. The valley floor is habitable because it has been eroded down to the zone of intermittent saturation. The bordering mountains are elongated domes of Precambrian granite (*Polar Photos, London*)

Tibesti to Darfur constitute an important catchment area. There the occasional rains are readily absorbed by bare sandstones which continue underground far across Libya and Egypt. Many a traveller has died of thirst in the heart of the desert with water only about a hundred metres beneath him. Oases occur where this usually inaccessible water emerges through fissures or artesian wells, or is brought to the surface by anticlines, or where the desert floor itself has been excavated by the wind down to the level of the ground-water (Figure 15.8). South of Aswan artesian water contributes to the flow of the Nile where the river has cut its channel into an aquifer locally brought up by an anticline.

The general rate of flow of ground-water beneath the Libyan Desert has been found to be about 50 m a year, and since some of the Egyptian oases are 1500 km from the intake areas they must now be fed with water that has taken 30,000 years to reach them. Irrigation of this desert region and of the Sahara farther west must be carried out with due regard to the resources available.

## Swallow Holes and Limestone Caverns

The solution of limestone by rain-water charged with carbon dioxide has already been described (p. 252). In limestone districts water readily works its way down through joints and along bedding planes until it reaches an impervious layer, which may be within the limestone formation or beneath it. The water then follows the natural drainage directions until it finds an exit, perhaps many kilometres away from the intake. Once a through-drainage is established, but not until then, dissolved material is carried away and fresh supplies of water coming in from above continue the work of solution, principally along joints and bedding planes, until a labyrinth of interlacing channels and caves is dissolved out of the limestone (Figure 15.3(*f*)).

The surface openings become gradually enlarged in places where the contours of the ground favour a special concentration of the flow-off, and funnel-shaped holes, known as *swallow holes* or *sink holes* are developed, By continued solution beneath these, and the falling in of loosened joint blocks, the holes may be enlarged into roughly cylindrical shafts or chasms (Figure 15.10) communicating with great vaulted chambers, perhaps hundreds of metres below. One of the most impressive of these giant swallow holes in Britain

Swallow Holes in the Carboniferous limestone of Yorkshire
**Figure 15.10** A Yorkshire pot-hole, showing fallen roof-blocks; Great Dour, near Chapel-le-Dale (*Bertram Unné*)

**Figure 15.11** Gaping Ghyll, south-eastern slopes of Ingleborough (*Flatters and Garnett*)

is Gaping Ghyll (Figure 15.11), on the south-eastern slopes of Ingleborough. The shaft goes down for 111 m into a chamber 146 m long and 33 m high. The water escapes through an intricate system of passages into Ingleborough Cave, whence it emerges as the Clapham Beck.

The 'cavernous limestone' plateau of Kentucky has over 60,000 sink holes and hundreds of caves, including the great Mammoth Cave, which itself has over 48 km of continuous passages. Another famous American cave, the Carlsbad Cavern of New Mexico, has a 'Big Room' nearly 1220 m long, with walls over 180 m apart, and a ceiling rising to a height of 90 m (Figure 15.13). Cave explorers, known as speleologists, find a great fascination in the thrills and dangers of penetrating

**Figure 15.12** The Cheddar Caves (Carboniferous limestone, Mendip Hills) from a diorama illustrating stalactites, stalagmites, cave pillars and cave curtains (*Institute of Geological Sciences*)

**Figure 15.13** The King's Chamber, Carlsbad Cavern, New Mexico (*Paul Popper Ltd*)

the realms of subterranean waters. Some of their greatest achievements have been in the French Pyrenees and, more recently, in the French Alps, where underground waterfalls, rivers and lakes have been explored and mapped to depths of more than a kilometre from the surface.

In addition to the streams which flow through the underground network of passageways, there is generally a slow seepage of lime-charged water from innumerable joints and crevices in the roofs and walls of caves. $CaCO_3$ is deposited when a hanging drop of such water begins to evaporate or to lose part of its $CO_2$. When the drop falls on the dry floor of a deserted channel, the remaining $CaCO_3$ is deposited. Thus, long icicle-like pendants, called *stalactites*, grow downwards from the roof; and thicker columns, distinguished as *stalagmites*, grow upward from the floor (Figures 15.12 and 15.13). In time stalactites and stalagmites unite into pillars, and these are commonly clustered together in forms resembling organ pipes and other fantastic shapes that are often given fanciful names. Where the water trickles out more or less continuously along a roof joint, a fluted curtain or wavy screen may grow across the cave. When the water comes through a bedding plane it builds up encrustations from wall to floor which look like 'frozen cascades'. The internal decoration of caverns in which these varied structures have grown in profusion produces an underground scenery of unique and fascinating beauty.

Occasionally the roof of a cave collapses and leaves a corresponding depression at the surface. When the roof of a long underground channel falls in, a deep ravine, floored with limestone debris, further diversifies the irregular limestone topography (Figure 15.14). For a time one part of the roof may hold firm, thus forming a natural arch or bridge (Figure 15.15). Limestone regions such as those referred to above, having a roughly etched surface, pitted with depressions due to solution or roof collapse, and with underground drainage in place of surface streams, are said to have a *karst* topography, from the prevalence of these features in the Karst Plateau bordering the Adriatic coast of Yugoslavia.

**Figure 15.14** Trow Ghyll, on the slopes of Ingleborough, North Yorkshire. A dry valley due to roof collapse of a former limestone cavern (*A. Horner and Sons, Settle*)

**Figure 15.15** Natural Bridge, Virginia. Part of the roof of a former limestone cavern (*American Museum of Natural History, New York*)

## Hot Springs

Ground-water that has circulated to great depths in deeply folded rocks becomes heated, and if a sufficiently rapid ascent to the surface should be

**Figure 15.16** Terraces of travertine, Mammoth Hot Springs, Yellowstone National Park (*Dorien Leigh Ltd*)

locally possible, it will emerge as a warm spring. Such conditions are unusual, however, and the source of heat generally remains an unsolved problem except in regions of active or geologically recent vulcanism. It is possible that some hot springs, like those of eastern Tibet, may be a mixture of meteoric water with juvenile emanations and hot connate water ascending from underlying zones of metamorphism and granitization. This would be consistent with the exceptional height of the Tibetan plateau. Most hot springs result from the emergence of groundwaters, mainly of meteoric origin, which have passed through recently erupted volcanic rocks (such as ignimbrite) or with which hot volcanic gases or other hot emanations have mingled before they reached the surface.

There are three volcanic regions which have long been celebrated for their hot springs and geysers: Iceland, Yellowstone Park, and the North Island of New Zealand. Others on an imposing scale are now known in Morocco and Kamchatka. The waters are highly charged with mineral matter of considerable variety. The Mammoth Hot Springs of the Yellowstone Park are rich in calcium bicarbonate ($H_2Ca(CO_3)_2$) derived partly from neighbouring limestones, and partly from carbon dioxide which appears to be of juvenile origin, judging from the isotopic compositions of the carbon and oxygen. Calcium carbonate is deposited at the surface as mounds and terraces of *travertine* (Figure 15.16). In all the regions mentioned many of the springs are alkaline and carry silica in solution; this is similarly deposited as *siliceous sinter* or *geyserite* ($SiO_2$, with varying amounts of water, like opal) (Figure 15.17). As to the water itself, various investigations indicate that 80 or 90 per cent is of meteoric origin. In Iceland, for example, much of it comes from melting snow. However, the minor constituents carried in solution include many of the rarer elements, a peculiarity that points to a juvenile source and suggests their introduction by hot emanations. If so, one would expect the latter

**Figure 15.17** White terraces of Rotomahana, North Island, New Zealand; destroyed by explosions during the catastrophic Tarawera eruption of 1886 (*Paul Popper Ltd*)

to include steam of juvenile origin. Borings through the ignimbrite flows of Yellowstone Park encountered vast quantities of dry, high-pressure, superheated steam. In one case the temperature at a depth of 75 m was found to be 205°C. In New Zealand, Italy, California and Japan underground supplies of superheated steam are being increasingly tapped by bore-holes to provide heat and hot water for domestic and industrial purposes. At Wairakei the world's first geothermal power station is now feeding electricity into New Zealand's supply system.

## Geysers

*Geysers* are hot springs from which a column of hot water and steam is explosively discharged at intervals, spouting in some cases to heights of over 100 m. The term comes from *Geysir*, the Icelandic name (meaning 'spouter' or 'gusher') for the most spectacular member of a group situated in a broad valley north-west of Hekla. It will serve as a typical example. A mound of geyserite, built up by deposition from the overflowing water, surrounds a circular basin, about 21 m across and 1·2 m deep, filled to the brim with siliceous water at a temperature of 75° to 90°C. From the middle of the basin a pipe, also lined with geyserite, goes down about 30 m. At the bottom the temperature of the water is well above that at which the water would boil if it were not for the pressure due to the weight of the water column above it.

Following Bunsen it has been generally supposed that continued accession of superheated steam through cracks in the pipe gradually raised the temperature until eventually the boiling-point was reached far down in the pipe. But investigations in recent years have shown that, although the temperature of the water increases with depth, the maximum temperature reached at the depths where eruptions begin remains below the boiling-point corresponding to the pressure of the overlying column of water. Thorkelsson's explanation is that the observed 'boiling' results not so much from increase of temperature as from reduction of pressure. The waters of hot springs are charged with dissolved gases. Well below the surface these begin to be liberated as bubbles, and

**Figure 15.18** *Geysir*, the Great Geyser of Iceland, in eruption. See Fig. 12.9 for locality. (*Iceland Tourist Bureau*)

**Figure 15.19** Old Faithful in eruption, Yellowstone National Park. A fragment of wood preserved in the siliceous sinter (geyserite) of the mound surrounding the basin has been dated at about 730 B.P. by the radio-carbon method (*G. A. Grant, United States Geological Survey*)

being in contact with hot water they necessarily become saturated with water vapour. When the water is near its boiling-point the proportion of water vapour in the ascending bubbles becomes very large and some of the resulting hot foam overflows at the surface. This so relieves the pressure on the heated water in depth that it violently flashes into a vast volume of foam and spray, which surges up with irresistible force, hurling itself up to a height of as much as 60 m. A similar height is commonly reached by Old Faithful (Figure 15.19). The world record for height was made in 1901 by a geyser that broke out on one of the volcanic fissures of the Tarawera rift in New Zealand. During a few months of intense activity some of its intermittent fountains of hot spray shot up to a height of 450 m.

Barth has confirmed the relation between eruptions and gas bubbles. Describing Uxahver, a geyser in northern Iceland, he writes: 'In the lulls between eruptions the water is clear and one can see far down into the spring. Just before a new eruption one will see something like a whitish cloud from great depths (about 5 m) rapidly ascending. This is a swarm of tiny gas bubbles. As they rise they rapidly expand and, near the surface, become 2 to 5 cm in diameter. One must be careful not to become too absorbed in watching, though, for at this moment a new eruption starts, and the observer has to make a rapid retreat.'

In some geysers the amount of water discharged is many times greater than that contained in the pipe and basin. In these cases the pipe must therefore communicate with underground chambers into which continuous supplies of meteoric water and hot gases (including steam) have access. The cavities and tunnels which sometimes occur in basaltic lava flows would provide the sort of reservoir required (Figure 15.20). During each period of quiescence the whole system of cavities, communicating channels, pipe and basin rapidly fills up. Gas content and temperature gradually

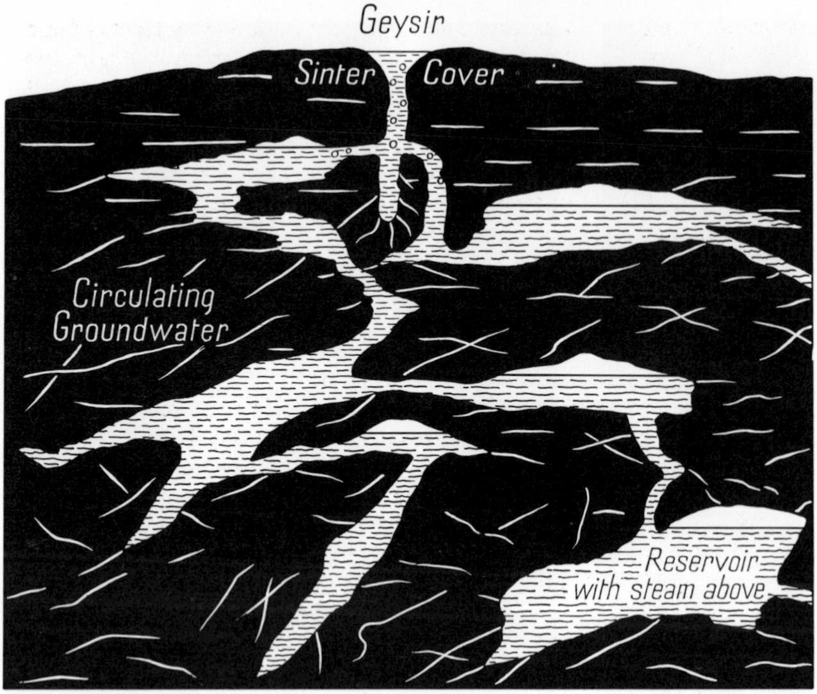


**Figure 15.20** Schematic section through *Geysir*, the Great Geyser of Iceland, to illustrate the conditions appropriate to intermittent eruption, showing subterranean reservoirs fed by groundwaters heated from below by the ascent of hot emanations, including superheated steam (*After T. F. W. Barth*)

rise until the quiet phase of the cycle is disturbed by the ascent and expansion of bubbles, and terminated by a roaring eruption of hot spray and steam.

### Deposition from Ground-Waters

In the preceding pages examples of deposition in limestone caverns and from the waters of hot springs and geysers have been given. In Chapter 6 the cementation of porous sediments, replacements of various kinds including petrified wood, and the development of nodules and concretions, such as those of flint and chert, were described.

Precipitation at the surface or in caverns is easy to understand, as the main factors concerned are evaporation, loss of carbon dioxide and cooling. Aragonite and calcite provide an instructive example of the effect of temperature in controlling the particular form of calcium carbonate that crystallizes. Calcite is the relatively 'low' temperature form and is being deposited today, except in caverns where the air is warm. Aragonite is the 'high' temperature form; at lower temperatures it becomes unstable and gradually turns into calcite. In the western U.S.A. there is a belt of caves in which calcite (still being deposited) has covered aragonite that formed in prehistoric times. In the caves to the south of this belt aragonite is still forming, while in those to the north the deposits, both old and new, consist only of calcite. During the time when aragonite was being precipitated in the central belt the temperature must have been some 8°C higher than today. From measurements of the present rate of deposition of calcite it is estimated that it has been accumulating for *c.* 5000 years. From other evidence it is known that a marked lowering of temperature brought the 'Climatic Optimum' to an end about 5000 years ago.

The return of dissolved material to the rocks through which ground-water is circulating involves the operation of a complex of delicately balanced processes which are still but little understood. Precipitation may be brought about by such factors as loss of gases and consequent decrease of solvent power; cooling while waters are ascending; changes of pressure during circulation; mingling of water from different sources; and reactions between solutions and the materials through which they pass.

Small *veins* of common minerals, such as calcite and quartz, may be formed from ground-waters in joints and fault fissures, or in gashes across the limbs of folds. In tightly folded rocks and in areas of regional metamorphism, irregular quartz veins are locally very abundant. Many of these have been deposited in tension clefts from siliceous water 'sweated out' of the original rocks during orogenesis. Most *mineral veins*, however, and especially those containing commercially valuable ores, have been deposited from hydrothermal solutions to which emanations of juvenile origin have contributed heat and chemically active gases, and in many cases a proportion of the mineral riches which gave them their name of 'mineralizing agents'.

## Freezing Ground-Water

In the tundra and adjacent parts of boreal regions the ground is permanently frozen, except for a surface layer which thaws out during the summer and varies from a few millimetres to 3 m in thickness, according to the climate. The frozen ground, known as *permafrost*, amounts to nearly one-fifth of the land area and is probably a relic from the last glaciation. In Arctic Canada and Alaska thicknesses of up to 400 m have been penetrated, and in Siberia a maximum thickness of 700 m is known.

*Mud and Stone Polygons* In flat-lying tracts where the top layer of thawed-out mud or morainic debris is badly drained, very remarkable polygonal patterns are developed (Figure 15.21). The origin of these patterns is clearly related to contraction and expansion, but the details are not well understood, probably for two main reasons. One is that contraction takes place not only when ice thaws into water, but also when it is cooled well below freezing point. At 0°C the density of ice is 0·92, but at −22°C it is 1·03, corresponding to a decrease in volume of 11 per cent. The other reason for uncertainty is that we cannot always be sure that the polygons were initiated under conditions like those of the present day.

The discovery of similar frost polygons on the flat morainic spreads of certain dry valleys in Victoria Land, from which the marginal Antarctic glaciers have retreated (Figure 15.22), throws new light on the problem. Here there is perennial frost, with little or no thaw, and the cracks outlining the polygons are due to the contraction that accompanies the severe drop in temperature from summer to winter. Snow and perhaps a little thaw water and wind-blown sand fill up the crack, which

**Figure 15.21** Mud polygons of two orders of size near Bruce City, Klaas Billen Bay, Spitsbergen. The mountain avens is growing in gutters that outline the larger hummocks. The smaller polygons, outlined by mud-cracks, are a seasonal result of summer desiccation and shrinkage. (*J. Walton*)

**Figure 15.22** Frost polygons developed in the morainic mantle covering the floor of a 'dry valley' in Victoria Land, Antarctica. Those illustrated average about 9 metres across; they are outlined by peripheral troughs about 60 cm deep. In areas of coarser moraine, with boulders up to 60 or 90 cm across, the polygons are smaller (4·5 to 6 metres across), but the peripheral troughs are deeper (up to 1·5 metres). (*P. N. Webb and B. C. McKelvey*)

consequently becomes an ice-wedge that grows wider and deeper each year. Examples up to 3 m deep are known, surrounding roughly polygonal mounds up to 12 m across. The annual expansion of the wedges in spring and summer is responsible for the dome-like hummocks within each polygon.

The diagram already used to illustrate the formation of columnar jointing will again serve to elucidate the development of the mud and stone polygons (Figure 15.23(*a*)) of the northern hemisphere, if we assume that they began in the same way as those of Victoria Land. At the present time, however, expansion is the dominant process in most of the northern regions: expansion of summer water to winter ice. Since the only direction of relief is upwards, the arrows now indicate actual movement of mud and ice, inwards and upwards towards each centre (Figure 15.23(*b*)). The bulging up of material reaches its maximum at each centre, and there the hummocks are highest. Eventually gutters outline the polygons and in these many Arctic flowering plants flourish in the late spring and summer (Figure 15.21). If the hummocks temporarily dry out, smaller polygons, outlined by mud-cracks due to shrinkage, develop in the ordinary way.

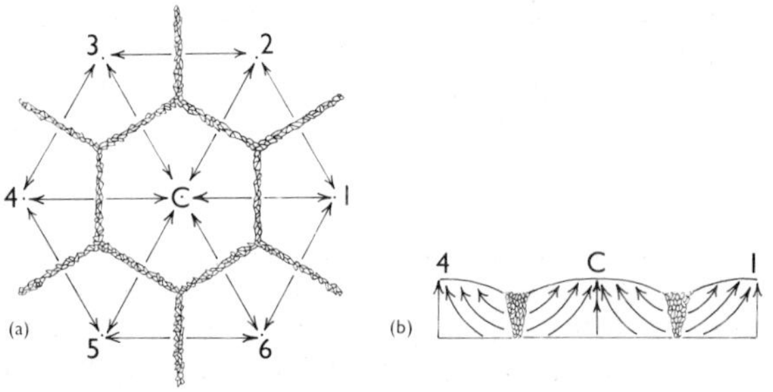


**Figure 15.23** Diagram to illustrate the formation of an idealized pattern of hexagonal hummocks

So far the active top layer has been regarded as consisting only of mud. More commonly, however, it is a mixture of fine and coarse debris. Conditions are then much more complex. The progressive tendency of stones to work their way up to the surface in the stony soils of cool and temperate climates is well known to gardeners. Stones behave in this curious way because of various differences between their thermal properties and those of the wet soil or mud in which they are embedded. When freezing is in progress and working downwards, the under surface of a stone is chilled more quickly than the surrounding wet mud. Hence the first ice to form at that level is immediately beneath the stone. Moreover, having once started, this ice continues to grow vigorously. The reason is that the freezing point of water is considerably lowered when it occupies capillaries like those of the pore-spaces of mud. The expansion due to ice formation beneath the stone tends to open up the capillaries of the underlying mud, so providing channelways through which water can creep towards the stone and be frozen there. All the conditions favour the growth of a layer of ice beneath the stone, and as the ice thickens it pushes the stone upwards. At the surface stones may sometimes be seen perched on top of a pedestal of ice. Uplift of stones—or of particles or fragments of any size—by the growth of ice is called 'frost-heaving'. This stage is followed during the next thaw by down draining of surface mud, some of which lodges under the stone and provides it with a new partial support before the whole of its underlying ice has melted. On balance, therefore, the stone continues to rise with every major frost and thaw. At the surface the stone topples off its pedestal, and on a slope tends to fall in a downhill direction.

In the tundra, frost-heaving is reinforced by the annual upward movement that results in the formation and maintenance of hummocks. As the stones reach the surface they are subjected to another frost effect. When the wet surface mud freezes it expands and pushes outwards, and any stones that are in the way are also thrust outwards. Assisted by gravity and a variety of minor processes (see p. 307) the stones slowly migrate towards the periphery of the hummock. There they accumulate as rings or polygons of stones.

On gently sloping ground the form taken by the stones is roughly elliptical, elongated in the downhill direction. On steeper hillsides alternating stripes of stone and finer debris are formed (Figure 15.24). The additional factors responsible for this modification are gravity and gullying by melt water. After thaw, waters spilling over from the lowest parts of the slightly inclined gutters near the top of the slope localize a series of parallel rills into which the drainage is concentrated. Subsequent ice-wedging in the little gullies so formed, and general expansion of the ground between, heaves up the latter into hummocky ridges. As stones reach the surface they tend to migrate down the steepest slope, and as this is no longer directly downhill, but partly towards the nearest gully, they eventually come to rest in the gullies, where they are now concentrated as long stone stripes. We see in these remarkable patterns the cumulative effects of many thousands of years of frost and thaw.

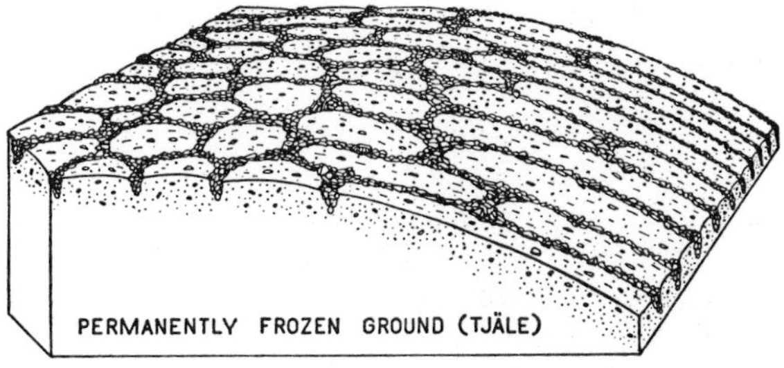


**Figure 15.24** Diagram illustrating the merging of stone-rings on a flat surface into stone-stripes on a slope (*C. F. Stewart Sharpe, 1938*; see p. 324 for reference)

For descriptions of patterned ground in general, and experimental investigations see Embleton and King, 1975.

*Pingos* In Arctic Canada, Greenland and northern Siberia there are sporadic mounds and cones, covered with deeply fissured layers of mantle deposits which have been upheaved by intrusions of ground-ice. They have dimensions twenty to a hundred times greater than those of the biggest mud polygons, their elevation above the surrounding country being anything up to 43 m. Many of them have a crater at the summit, sometimes containing a shallow lake during the summer months. These remarkable structures are called *pingos* (their Eskimo name), and they have also been referred to as 'hydro-laccoliths' and as 'cryo-laccoliths'. They appear to form where unfrozen ground-water (*a*) is trapped between an underlying zone of permafrost and an overlying frozen zone, or (*b*) has access to such a position from a high intake area which gives it a hydraulic head like that of artesian water.

East of the Mackenzie delta in NW Canada, pingos of the first kind can be seen in various stages of development. The example shown in Figure 15.25 was investigated by a Canadian team in the summer of 1954, when it had a crater 9 m across containing a shallow pool. A bore-hole drilled to a depth of 10 m adjacent to the pool passed entirely through ice after penetrating over a metre of raw soil. The following year the pingo of Figure 15.26 was similarly investigated by a member of a Danish team who had already made a detailed study of some East Greenland pingos. In this case a bore-hole was drilled through the floor of the crater. After passing through 14 m of frozen deltaic and other lake deposits, including peat, it traversed 55 m of clear, coarsely crystalline ice without bottoming the ice intrusion.

**Figure 15.25** Pingo, 30 metres high, east of the delta of the Mackenzie River (69°N; 134°25′W), North-West Territories, Canada

**Figure 15.26** 'Crater Summit' Pingo, about 40 metres high, and 290 metres across at its base, east of the delta of the Mackenzie River (69°50′N; 133°W) (*National Research Council, Division of Building Research, Ottawa*)

The zone of permafrost in this region is about 90 m thick. Each pingo stands in the middle of the site of a former lake, beneath which, it is thought, there was originally an extensive layer of ground-water, surrounded on all sides by permafrost. As the lake filled up with sediment and peat, the trapped ground-water was encroached upon by frost from all directions and thus came under steadily increasing pressure as a result of the 9 per cent expansion. Relief would be gained only by upheaval of the roof, and this would occur where the overhead pressure was least, which would be about the middle of the former lake. Water squeezed into this growing dome would be actively frozen from above, and by the time freezing was completed a considerable intrusion of ice would have formed. Drill cores from the margins of some of the ice bodies show that the roof and wall deposits have not only been upheaved and stretched, but also dragged up and penetrated by ice, indicating rheid flow of ice, like that of salt in a salt dome or, more familiarly, like that of ice itself in a glacier. The mechanism of upheaval and intrusion is evidently a variety of diapirism (p. 152) but with the motive force arising not from a difference in density between ice and mantle deposits, but from the expansion of trapped ground-water on freezing.

The East Greenland pingos grow from the freezing of ground-water that originates from the summer melting of snow on high surrounding mountains, so that it has sufficient hydraulic head to penetrate the permafrost zone at lower levels. The mechanism then proceeds as described above. In this region numerous dykes and sills of ice have been found on the margins of pingos.

## Selected References


ALLEN, E. T., and DAY, A. L., 1936, *Hot Springs of the Yellowstone National Park*, Carnegie Institution of Washington, Publication 466.

BARTH, T. F. W., 1950, *Volcanic Geology, Hot Springs and Geysers of Iceland*, Carnegie Institution of Washington, Publication 587.

BLACK, R. F., 1954, 'Permafrost: a review', *Bulletin of the Geological Society of America*, vol. 65, pp. 839–56.

CULLINGFORD, C. H. D. (Ed.), 1953, *British Caving: an Introduction to Speleology*, Routledge and Kegan Paul, London.

EMBLETON, C., and KING, C. A. M., 1975, *Periglacial Geomorphology*, Edward Arnold, London.

FENNER, C. N., 1936, 'Bore-hole investigations in Yellowstone Park', *Journal of Geology*, vol. 44, pp. 225–315.

HORBERG, L., 1949, 'Geomorphic history of the Carlsbad Cavern Area, New Mexico', *Journal of Geology*, vol. 57, p. 464.

MEAD, W. J., 1925, 'The geologic role of dilatancy', *Journal of Geology*, vol. 33, pp. 685–98.

MÜLLER, F., 1959, 'Beobachtungen über Pingos: with an English Summary', *Meddelelser om Grønland*, vol. 153, no. 3, pp. 1–127.

PIHLAINEN, J. A., BROWN, R. J. E., and LEGGET, R. F., 1956, 'Pingo in the Mackenzie Delta, Northwest Territories, Canada', *Bulletin of the Geological Society of America*, vol. 67, pp. 1119–22.

SWEETING, M. M., 1950, 'Erosion cycles and limestone caverns in the Ingleborough District', *Geographical Journal*, vol. 115, pp. 63–78.

WHITE, D. E., 1957, 'Thermal waters of volcanic origin', *Bulletin of the Geological Society of America*, vol. 68, pp. 1637–58; 'Magmatic, connate and metamorphic waters', ibid., pp. 1659–82.

# Chapter 16

# Life as a Fuel Maker: Coal and Oil

Organic matter equivalent in quantity to the weight of the earth has been created by living creatures since life originated on this planet.

*Philip H. Abelson*, 1957

## The Sources of Natural Fuels

Carbon dioxide, water and sunshine are the primary sources of the carbon compounds of all living organisms and of all those that have lived in past ages. Under the influence of the sun's rays green plants, including most of the bacteria, absorb radiant energy and utilize it to synthesize $CO_2$ and $H_2O$ into carbohydrates, such as cellulose $(C_6H_{10}O_5)_x$, starch $(C_6H_{10}O_5)_y$ and sugar $(C_6H_{12}O_6)$. Since these compounds are equivalent to carbon and water, their formation involves liberation of oxygen. Much research has been undertaken to discover how this *photosynthesis* is done, since many of the urgent problems of mankind—notably the need for ever-increasing supplies of energy—would be eased if it became possible to imitate the methods of nature economically. When the colouring matter of green plants (chlorophyll) is exposed to sunlight in water 'labelled' with traces of $H_2O^{18}$ (instead of the normal $H_2O^{16}$) it is found that most of the oxygen liberated comes from water and not from carbon dioxide as was previously supposed. Roughly speaking, what happens is that oxygen and hydrogen are separated from water-bearing compounds, and the hydrogen then combines with carbon dioxide to form the carbohydrates mentioned above:

$$2H_2O \longrightarrow 2H_2 + \underset{\textit{liberated}}{O_2}; \quad 2H_2 + CO_2 \longrightarrow \underset{\textit{carbohydrate}}{(CH_2O)} + H_2O$$

Some of the oxygen so set free combines with the carbon of organic matter to form carbon dioxide; another part is consumed in weathering processes; and the balance passes into the atmosphere or into the sea. The cycle of changes is schematically portrayed in Figure 16.1. If the chlorophyll reaction ceased to maintain the present supply of oxygen, the atmosphere would soon become unbreathable.

If all the decaying remains of dead organisms were completely oxidized there would be no free oxygen left over. Under waterlogged conditions, however, oxidation is not complete. The decomposition products of vegetation, for example, accumulate as humus in the soil and as deposits of peat in bogs and swamps. The buried peat deposits of former ages have been transformed into coal seams. In marine sediments a high proportion of the organic material of plant and animal life is either eaten or lost by oxidation, but some escapes complete destruction and is entrapped in muddy deposits to form the minute droplets of oil and bubbles of gas which are the source materials of the petroleum and natural gas concentrated in oil-fields. Life is thus responsible for all the natural fuels, including wood, peat, coal, oil and gas, and for the enormously greater amount of carbonaceous and bituminous matter that is dispersed through shales and other sedimentary rocks.

It is of interest to consider what has happened to the oxygen corresponding to the carbon fixed in organic matter. The figures at least give some idea of the prodigious amounts involved, and bear eloquent witness to the work of innumerable generations of plants and animals. Using all the data available, including recent work on the proportions of carbon isotopes in a great variety of carbon compounds, F. Wickmann estimates the fixed carbon in all sediments at about $3500 \times 10^{12}$ tonnes for the whole earth. This includes the

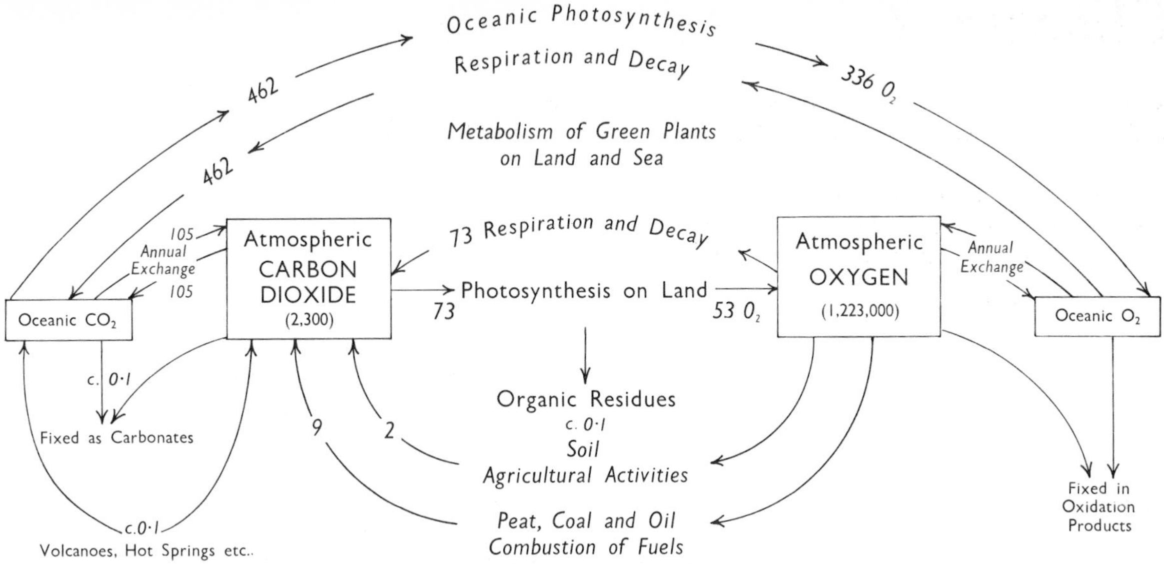


**Figure 16.1** Schematic summary of the Carbon dioxide-Oxygen cycle. All figures represent units of $10^9$ (a thousand million) tonnes. Figures in brackets are the total amounts now present in the atmosphere; those without brackets are annual amounts of $CO_2$, except in the specified cases of $O_2$ produced by photosynthesis.

carbon in coal and oil, which together with that in peat and soil amounts to only 1/400th of the total. Living matter is probably about 1/4000th of the total. 3500 m.m. tonnes of fixed carbon imply the liberation of 9300 m.m. tonnes of oxygen (in accordance with the equations on p. 280). Not all of this can be accounted for, as the following table indicates:

| OXYGEN | *Millions of millions (m.m.) of tonnes* |
|---|---|
| In the atmosphere | 1223 |
| Dissolved in the ocean and other surface waters | 12 |
| Consumed in oxidizing FeO to $Fe_2O_3$ | 1400 |
| Consumed in oxidizing S | 4100 |
| | 6735 |
| Unaccounted for | 2565 |
| Total amount liberated | 9300 |

The estimates for oxidation are maxima, and the small amounts of oxygen consumed in other weathering processes and in the formation of nitrates from nitrogen are therefore fully covered. It may be that much of the oxygen unaccounted for has been used up, directly or indirectly, in reactions with volcanic gases such as hydrogen (oxidized to $H_2O$) and carbon monoxide (oxidized to $CO_2$).

## Peat

In waterlogged environments, such as bogs and swamps, the decay of dead plant debris by bacteria and fungi is limited by the paucity or absence of oxygen and the generation of antiseptic organic acids which inhibit the bacterial activities. Under these conditions the softer and finely macerated plant debris changes into a dark-brown jelly-like humus. Part of this soaks into the cells of fragments of wood, bark, roots, twigs, etc., which are also being 'humified', and the cellular structures of these remains are in consequence often wonderfully well preserved. All the humified products, together with a variable proportion of the less destructible materials, such as resin and the waxy pollen cases and spores, accumulate to form deposits of peat.

The process of humification enriches the residue in carbon, as indicated approximately by the following equation:

$$\underset{\text{cellulose}}{2C_6H_{10}O_5} = \underset{\text{humified residue}}{C_8H_{10}O_5} + 2CO_2 + \underset{\text{methane}}{2CH_4} + H_2O$$

Methane, more familiarly known as marsh gas, is highly inflammable, and its pale flames are responsible for the 'will-o'-the-wisp' which is occasionally seen flickering over the surface of a bog. In coal-mines, where the gas sometimes escapes in disastrous quantities from the coal face, it is the chief constituent of firedamp, which all too easily catches fire and explodes.

The vegetation which contributes to peat formation ranges from mosses (bog peat—Figure 16.2) to trees (forest peat—Figure 16.3), and the environment may be a swampy lowland or a waterlogged upland with imperfect drainage. The climate must therefore be humid and the conditions such that growth exceeds wastage. In the bogs of cool humid regions the rate of decay lags behind because of the low temperature, whereas in the densely forested swamps of tropical regions the phenomenal rapidity of growth more than keeps pace with the high rate of decay.

A special variety of peat accumulates at the bottom of lakes and pools surrounded by marsh vegetation. Wind-blown pollen and leaves fall into the water, and all manner of organic particles drift into it. Year by year these settle down to form layer after layer of organic ooze. Fresh-water algae may add further contributions. Locally the spores and algal remains may predominate, giving rise to a deposit that is especially rich in the waxy and oily ingredients of vegetation. If streams are flowing into the water the ooze is likely to be contaminated by a certain amount of muddy sediment.

Many thousands of the shallow lakes that formerly occupied depressions in glaciated areas have been converted into peat bogs by the steady encroachment of marsh and swamp vegetation, and in others the process of infilling is still in progress. The rushes, reeds and pond weeds gradually advance over the dark gelatinous slime formed from the residues of earlier generations. Floating vegetation sometimes grows out in thick spongy rafts across the surface. Meanwhile the floor is being built up as organic ooze accumulates, and finally the site of the lake becomes a swamp. The treacherous surface may be covered with quaking tussocks of sphagnum moss, as in the bogs of Ireland. Where the drainage conditions are suitable, the plant sequence may culminate in a forest of trees with roots adapted to the precarious foundation through which they spread.

On a more extensive scale swamps are developed from the shallow lagoons and lakes of low-lying coastal plains, flood plains and deltas. The Dismal Swamp of the coastal plain of Virginia and North Carolina is an immense forested area, only a metre or so above sea-level, interspersed with stretches

**Figure 16.2** Bog peat, Dunmore Moss, ESE of Stirling, Scotland, showing cut peat lying out to dry and stacks of dried peat awaiting transport (*Institute of Geological Sciences*)

**Figure 16.3** Forest peat, showing remains of pine trees exposed by removal of peat, near Daless, Findhorn Valley, Moray Firth Region, Scotland (*Institute of Geological Sciences*)

**Figure 16.4** Aerial view of forest, swamps and creeks, Irrawaddy Delta, Burma (*From Burma Forest Bulletin No. 11*)

of open water (Figure 23.52). Here over 2400 km$^2$ have been covered with peat averaging 2 m thick. Along the north-east coast of Sumatra there are many scattered swamps supporting almost impenetrable tropical jungles. In one of these peat is known to have accumulated to a depth of 9 m.

The densely forested swamps of the Ganges and other tropical deltas (Figure 16.4) provide ideal conditions for peat growth and serve to illustrate the climatic and geographical conditions under which the coal seams of the Carboniferous period originated. Moreover, borings through the Ganges delta, and many others, reveal a succession of buried peat beds with intervening deposits of sand and clay. The sequence points to repeated alternations of sedimentation and plant growth, with the actual surface never far from sea-level.

As raw peat accumulates year after year, entangled water is squeezed out of the lower layers and the maturing peat shrinks and consolidates. It still contains a high proportion of water, however, and before being used as a fuel prolonged air-drying is necessary. In appearance it then ranges from a light brown fibrous or woody material to a dark brown or black amorphous substance.

## Coal and its Varieties

Peat becomes still further compacted when it is buried beneath a cover of clays and sands. As the overhead pressure increases, water and gases continue to be driven off, their composition being such that the residue is progressively enriched in carbon until it is transformed into a variety of coal (Figure 16.5). It has been estimated that 5 cm of peat are necessary to make 1 cm of ordinary coal. The essential conditions for the development of a coal seam are thus: long-continued growth of peat; subsidence of the area; and burial of the peat beneath a thick accumulation of sediments.

Considerable variation in the character of coal depends on the relative percentages of carbon, volatiles and moisture that it contains, with the exception of anthracite the percentage of fixed carbon determining the *rank* (Figure 16.5). The rank of coal depends primarily on the depth of burial, increasing overhead pressure and rise of temperature with depth, combining to drive off volatiles. This correlation has been well established in the Ruhr coalfield of West Germany where rank increases with depth. There are exceptions, however, in that elevated pressure by preventing the escape of volatiles may prevent the evolution of higher rank coals. It is thought, for example, that rise of rank in the Durham coalfield has depended primarily on rise of temperature (Jones and Cooper, 1972). The rank of coal may be influenced by additional factors, such as length of time of burial, i.e. the age of the coal; by tectonic deformation, and by thermal metamorphism induced by younger, neighbouring igneous intrusions. Such changes are purely local when induced by dykes or overlying sills. Sills underlying coal, particularly when they are thick, may cause regional changes in the grade of the coal. Rather unexpectedly, the nature of the initial plants, and the environment in which they accumulated have but small effect in determining the rank of coal.

In coals of *normal types* the remains of wood and bark predominate, indicating derivation from forest peats. Coal of the lowest rank, that is, the variety most like peat, is called *lignite*. It commonly retains visible vegetable structures, but there are also varieties, often known as *brown coal*, in which the woody tissues are obscure. Lignites and brown coals are common in the Cretaceous and Tertiary coalfields of Europe and North America, but are of no importance in Britain.

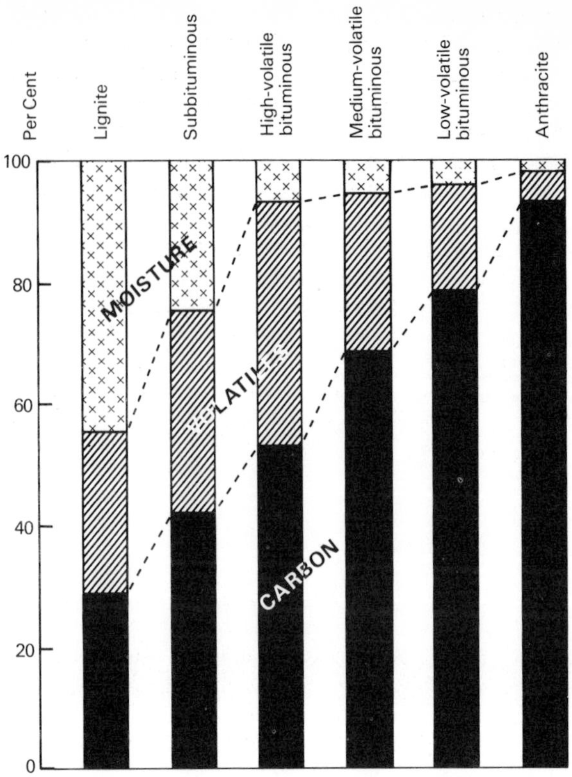


**Figure 16.5** Diagrammatic illustration of the variation in composition of coals of different rank (*After P. Averitt, 1974*)

The familiar shining black or dark-grey coals of the British and other Carboniferous coalfields belong to a group of fuels of higher rank known collectively as *bituminous coal*. This term does not imply the presence of the material properly called 'bitumen', but has reference to the fact that in the manufacture of coal gas and coke one of the distillation products, *coal tar*, is of a bituminous nature. Coal of the highest rank—called anthracite—is hard, brittle and stony-looking, does not soil the fingers, burns with a smokeless flame, and has a high heat-producing capacity.

Coals of *special types* are formed from organic oozes rich in spores or algal remains. These contain less carbon than the normal coals of corresponding rank, but are notably richer in hydrogen. When heated they give off abundant supplies of hydrocarbons and on distillation they yield oil. Varieties in which algal remains predominate are called *boghead coals*, or *bogheads*, after the name of the locality, west of Edinburgh, where they were first mined. They are close-grained and tough and often resemble dark brown or nearly black leather. They leave a considerable residue of ash, most of which is the muddy material that contaminated the original ooze. By increase of inorganic sediment the bogheads pass into *oil-shales*.

Between bituminous coal and the bogheads there are intermediate varieties known as *cannel coal*. In some of these, spores and blebs of resin are very abundant, while others contain algal remains in addition. Cannel coal is dull black and appears quite structureless to the unaided eye. It occurs as individual seams and also as lenticles and bands in seams of ordinary coal. The name refers to the fact that splinters of cannel can be burnt like a candle, a fact that in turn demonstrates the richness of the material in inflammable hydrocarbons.

## The Constitution of Coal

Almost any block of bituminous coal can be seen to have a well-marked banded or stratified structure (Figure 16.6). The commonest bands are composed of soft bright coal which readily breaks into approximately right-angled pieces with smooth brilliant surfaces. Many of the bands appear to be quite structureless, and since the material is not unlike black glass in appearance it has been called *vitrain*. Other bands are finely laminated and consist of shreds and films of vitrain in a very fine-grained matrix. This type of coal material is known as *clarain*. These bright bands are separated by layers of a dull grey-black type of coal which, being relatively hard and tough, is distinguished as *durain*. None of the three types already mentioned soils the fingers. The 'dirtiness' of coal is due to the presence, generally in quite small amounts, of a fourth type of material called *fusain*. It consists of thin flakes or lenticles of extremely friable 'mineral charcoal' which occur at intervals through the seam. Coal naturally splits very easily along these planes of weakness, and as fusain readily crumbles to powder the broken coal becomes dusty.

All types of coal are intensely opaque, and it is necessary to make sections ten times as thin as those of ordinary rocks before light can penetrate the material so that microscopic examination becomes possible. A translucent film of vitrain 0·0025 mm in thickness is no longer black, but has a rich golden brown or reddish colour. It is found to consist of bark or wood, each band representing a single fragment. Figure 16.7 shows a highly magnified sample of typical vitrain in which the cellular structure of bark is perfectly preserved. The material clearly corresponds to the humified bark found in peat.

**Figure 16.6** Polished surface of a block of coal (magnification about 1·7), Herrin No. 6 coal seam, Gallatin County, Illinois. C=clarain; D=durain; F=fusian; V=vitrain. (*Illinois State Geological Survey*)

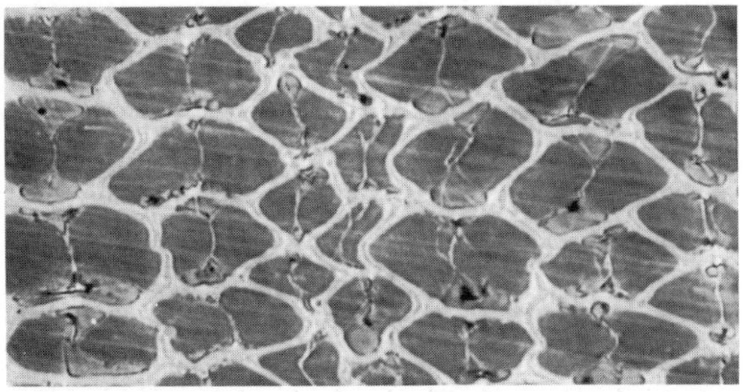

**Figure 16.7** Bark preserved as vitrain. ×317. (*C. E. Marshall*)

Fusain, similarly examined, also turns out to

be wood or bark, but only the carbonized cell walls remain, the cells being empty or occupied by ash. It represents woody and other fibres that escaped humification, though how they became altered into charcoal is a problem that remains unexplained. Some of the fusain may be the charred relics of forest fires, but most of it appears to have been formed after burial during the process of 'coalification'.

Unlike these two types, which are constituents of uniform composition, durain is found to be an assemblage of minute particles, like the organic ooze of peat. It contains the resistant coats of spores, more or less crushed and flattened, microscopic shreds of vitrain, lenticles and grains of fusain, and blebs of resin, in an obscure matrix of debris too finely macerated to be identified without an electron microscope. Figure 16.8 shows durain consisting of microspores (male) and macrospores (female) embedded in a dark matrix. Clarain contains the same ingredients as durain, but in very different proportions; abundant closely packed strips of vitrain are associated with extremely thin laminae of durain-like material. Cannel coal is essentially durain which is especially rich in spore cases and other waxy and resinous remains.

Lignites and anthracites are found to consist of the same structural types of material as bituminous coals. The variation in properties depends partly on the proportions in which the type of ingredients are present, and partly on the degree of alteration which they have suffered, i.e. on the rank of the coal. Fusain, for example, is highly inflammable, because its friability and high porosity make for easy oxidation. Coal-dust explosions—formerly a serious menace to mining before precautions were enforced—result from this dangerous property. Spore cases and resins are rich in hydrocarbons; consequently dull coals and cannels yield far more gas and tar than the bright coals of intermediate ranks. The latter, however, are excellent for household purposes. Steam coal, suitable for use in locomotives and ships, is of higher rank and is transitional towards anthracite, in which the rare spores that may still be detectable have been reduced to ghostly carbonized relics.

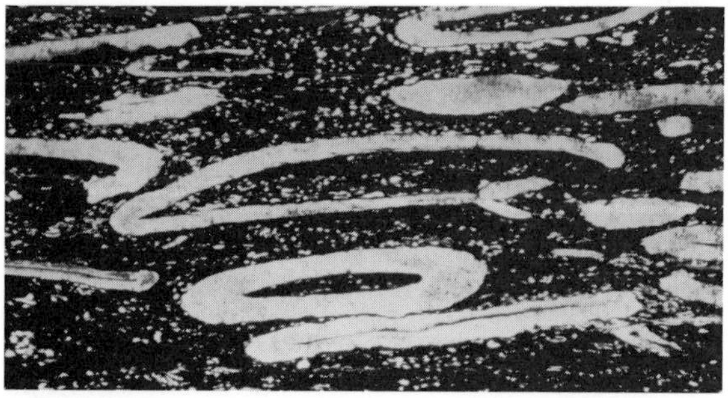

**Figure 16.8** Durain (dull coal) with spores. ×27. (*C. E. Marshall*)

## Coal Seams and Coalfields

The peak period of coal formation was the Upper Carboniferous. Hercynian earth-movements then provided the extensive, subsiding basins in which the sediments and coal seams of the Coal Measures were deposited; and provided them, moreover, along a belt running through North America (bordering the Appalachians), the British Isles and parts of Europe and Asia, where the climate was then hot and the vegetation luxuriant.

Land plants capable of preservation make their first feeble appearance only towards the end of the Silurian. By Carboniferous times, however, a rich and prolific flora had developed, and the fossilized remains of more than three thousand species are already known. The chief coal-makers were tall forest trees (*Lepidodendron*—Figure 16.9—and *Sigillaria*, with widely spreading roots known as *Stigmaria*) which grew to heights of as much as 30 m; and giant reeds called *Calamites* (the ancestors of the little horse-tails of today) which flourished in bamboo-like thickets to a height of 15 m or more; together with an undergrowth of smaller rushes and ferns, and slender plants of trailing or climbing habits (Figure 16.10). No flowers or birds enlivened these gloomy jungles, but insects—again of extravagant size—were abundant.

Practically all seams of bituminous coal and anthracite have certain characteristics from which it is inferred that each seam represents the actual site of the swamp in which the parental vegetation lived and died.

(*a*) The 'seat earth' which forms the floor of a seam may be a high-grade fireclay (useful for making refractory bricks). It is a fossil soil, riddled with innumerable rootlets of the plants which first colonized the swamp, and locally containing casts of stigmarian roots, often in the position of growth. The trunks either rotted away above water level or fell into the swamp to contribute to the lower part of the seam.

**Figure 16.9** Remains of a Carboniferous forest showing fossil stumps and forked roots of *Lepidodendron* rooted in shales, Fossil Grove, Victoria Park, Glasgow; discovered in 1882 and preserved as a natural museum (*Institute of Geological Sciences*)

(*b*) The roof of the seam sometimes contains casts of the trunks of great forest trees. These represent the generation that was drowned when the swamp conditions were brought to an end.

(*c*) Seams may be of very wide extent, sometimes covering areas of many thousands of square kilometres.

(*d*) Seams are locally interrupted by 'wash-outs', that is, by the sandstone-filled channels of streams that flowed through the forest swamps, like the creeks and distributaries of modern deltas (Figure 16.4).

(*e*) Coals of normal types contain no fish remains or other fossils of aquatic animals, and (except in certain bands of durain) are uncontaminated by muddy sediment. Such ash as remains when the coal is burnt is derived either from the vegetation itself or from carbonate minerals and pyrite that have been deposited in cracks by percolating ground-waters. Seams and bands of the durain type may, however, leave a little sedimentary ash. These dull coals seem to have accumulated in stretches of stagnant water into which a limited

**Figure 16.10** Reconstruction of a Carboniferous forest

amount of fine sediment was introduced while the delta streams were in flood.

In the special coals—the cannels and bogheads—muddy sediment is much more abundant than in durain. Moreover, these varieties contain the remains of fish and other aquatic organisms. The water in which the mud-contaminated ooze accumulated was therefore not stagnant, but was continually renewed and oxygenated. Evidently these coals were formed not *in situ*, but from plant debris that was drifted by wind and running water into lakes with a through drainage. Such conditions would also arise locally in the hollows of a peaty surface just beginning to subside. Matching expectation, thin lenticles of cannel of limited area are found to occur at the top of many seams of otherwise normal coal.

In the Upper Carboniferous coalfields each seam is a member of a characteristic succession of sediments which is commonly repeated dozens of times, and in some localities hundreds of times. A coal seam is usually roofed with shales considered to be of non-marine origin. Fossil leaves or non-marine bivalves frequently occur in them though very occasionally (and of considerable significance) a marine fauna is present. Followed upwards the shales become sandy and pass into sandstones, sometimes with shaly interruptions. Then follows the seat earth that underlies the next coal. The sequence is one that implies alternations of clastic deposition and plant growth at or about sea-level.

The subsiding regions eventually became deep tectonic basins lying between rising tracts of country which supplied the basins with sediments and kept them filled. The floor of part of the South Wales coalfield sank more than 3000 m in all. The South Wales and neighbouring coalfields were originally parts of an elongated basin that extended far across the south of England. To the north a persistent ridge of higher ground, often referred to as the Midland Barrier, separated this southern basin from a vast area of coalescing depressions which included the northern groups of coalfields (Figure 16.11).

At this time the British area and the adjoining parts of Europe were being warped into a pattern of alternating basins and swells as a result of vertical movements—complementary depressions and uplifts—associated in some way with the Hercynian mountain-building that was then in progress farther south. Later on the filled-up basins were themselves buckled, folded and faulted. The dominantly up-folded parts were at

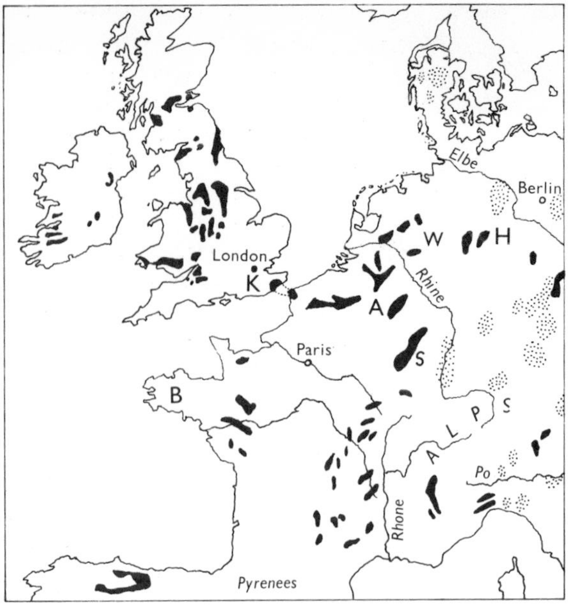


**Figure 16.11** Map of the coalfields (some concealed by later strata) of Western Europe. Carboniferous, black; Cretaceous, dotted. By middle Carboniferous time the northern front of a Hercynian mountain range extended from Brittany (B) through the Ardennes (A) to the Harz Mts (H) and beyond. Coal Measures were deposited in a series of outer basins to the north (e.g. from S Wales and Kent, to Westphalia, W, including the Ruhr) and in a series of intermontane basins to the south (e.g. the Saar, S).

once exposed to denudation and these areas gradually lost their Coal Measures wholly or in part. The down-folded and down-faulted parts, however, were buried to still greater depths beneath successive blankets of Permian, Triassic and later sediments. Thus the Coal Measures of the isolated coalfields of today were preserved. Subsequent uplifts, mainly during the Tertiary, have made possible the removal of much of the protective cover by erosion. In many places (Figure 16.11) the faulted and folded Coal Measures are exposed at the present-day surface. In others, as along the Durham coast and to the east of the Yorkshire and Nottingham field, the underground continuations of the exposed coalfields are concealed beneath Permian or Triassic strata. The little coalfield in the south-east corner of Kent is entirely concealed.

## Petroleum

Petroleum (Gr. *petra*, rock, L. *oleum*, oil) is the general term for all the natural hydrocarbons—

whether gaseous, liquid or solid—found in rocks. In common usage, however, it refers more particularly to the liquid oils. Gaseous varieties are distinguished as *natural gas*. Highly viscous to solid varieties are called *bitumen* or *asphalt*, but the latter term is also applied to the bituminous residues left when petroleum is refined, and to natural and artificial paving materials composed of sand, gravel, etc., with a bituminous cement.

Petroleum consists of an extremely complex mixture of hundreds of different hydrocarbons, generally accompanied by small quantities of related compounds containing nitrogen, sulphur or oxygen. The hydrocarbons fall into several natural series of which the paraffin series is the most familiar. Its members, all of which can be expressed by the formula $C_nH_{2n+2}$, range from light gases (e.g. methane, $CH_4$, the chief constituent of natural gas), through a long series of liquids (the chief ingredients of successive products of distillation such as petrol or gasoline, paraffin oil and lubricating oil), to paraffin wax (including $C_{20}H_{42}$ and higher members). Crude oils in which these hydrocarbons predominate are said to have a paraffin base; they are generally of pale colour with a yellowish or faintly greenish hue. The darker brown and greenish oils generally contain a high proportion of the naphthene series, each member having a composition of the type $C_nH_{2n}$. These furnish heavy fuel oils and, as they leave a dark asphaltic residue on being refined, they are said to have an asphaltic base. Intermediate varieties have a mixed base of wax and asphalt. In all crude oils there are also smaller proportions of several other series, including acetylene and its higher members, $C_nH_{2n-2}$, and a great variety of aromatic hydrocarbons, of which the benzene series, $C_nH_{2n-6}$, is an example.

To avoid confusion it should be clearly understood that neither oil shales nor the cannel and boghead coals contain petroleum as such. If they did, it could be dissolved out by carbon disulphide. They do, however, contain *pyro-bituminous* substances or kerogens which can be altered into oil and bitumen by heat. Such deposits can therefore be made to yield a group of petroleum products by destructive distillation. Petrol and related products can be obtained in commercial quantities from suitably powdered coals of ordinary types only by fluidization with hydrogen at high temperatures. Petrol can also be made from the heavier and less valuable oils by a similar but less elaborate process of hydrogenation. The table at the foot of this page summarizes the sources of oil and related products.

Being fluids, oil and gas behave very much like ground-waters. They occupy the interstices of permeable rocks, such as sand and sandstone and cavernous or fissured limestones, in places where these 'reservoir rocks' are suitably enclosed by impervious rocks, so that the oil and gas remain sealed up. Accumulations on a scale sufficient to repay the drilling of wells are referred to as oil or gas pools. The 'pool', however, is merely the part of a sedimentary formation that contains oil or gas instead of ground-water.

### The Origin of Petroleum

Unlike coal, petroleum retains within itself no visible evidence of the nature of the material from which it was formed. It has been suggested as a purely speculative possibility that oil may have been formed by volcanic or deep-seated chemical processes akin to the production of acetylene by the action of water on calcium carbide. But these hypotheses are quite incompatible with the geological distribution of oil and with certain peculiarities of its composition and properties. All the relevant evidence points convincingly to an organic origin.

(*a*) Some of the constituents of petroleum have the property of altering the direction of vibration of light rays. This 'optical activity' is a characteristic of many substances produced by plants and

| *Bituminous deposits of Petroleum* | *Pyro-bituminous deposits requiring destructive distillation* | *Carbonaceous deposits requiring fluidisation with hydrogen* |
|---|---|---|
| Natural gas | Special Coals: | Ordinary coals |
| Crude or mineral oil | Cannels | |
| Bitumen and mineral wax | Bogheads | |
| Tar sands and asphalt | Oil shales | — |

animals, but it is not shared by similar compounds which are generated by purely chemical reactions.

(*b*) The minor constituents of petroleum include *porphyrins* (formed from chlorophyll or from the corresponding colouring substances of animal origin) and other compounds that can be extracted from plants and animals by organic solvents. Most of these, and especially the porphyrins, are quickly destroyed in the presence of oxygen. Their persistence in oil indicates that the latter must have originated in an environment free from oxygen.

(*c*) The existence of oilfields in pre-Carboniferous sediments (as far back as the Ordovician) suggests that land plants were not essential to oil formation, and this inference is strengthened by the fact that no significant lateral connection between coal seams and oil pools has anywhere been traced. The two may occur in close association by some accident of faulting, and one may lie above the other in a sequence of varied strata, but in neither case has the association any bearing on the origin of oil. Drifting relics of land vegetation swept into the sea by great rivers may have contributed a little to oil formation, but such a source would be quite subsidiary to the contributions furnished by the organic remains of marine algae and diatoms and of similarly unconsumed animal matter. Cellulose and lignin, so important for the genesis of coal, are apparently of little importance as a source of petroleum. Most of the hydrocarbons extracted from marine sediments, whether recent or old, include constituents that are identical or similar to those made by marine plants and animals, but are not found in cellulose or lignin.

(*d*) Investigations on the organic debris of samples of sediment brought up from borings into the floors of the Gulf of Mexico and the Black Sea show that at least the early stages of oil formation are in progress at the present day. Long-continued bacterial activity is one of the essential conditions. Although no hydrocarbons can be detected in newly deposited muds, they gradually appear within little more than a metre below the sea floor. Age determinations of the organic matter (by the radio-carbon method) show that the earliest detectable stages of conversion into oil require from 3000 to 9000 years.

(*e*) Oil is not found in association with volcanoes or igneous rocks, except accidentally. West of Edinburgh, for example, oil shales have been invaded by intrusions and penetrated by volcanic pipes, with results comparable to those obtained when oil shales are distilled. Oil so liberated by metamorphism would naturally migrate into overlying sandstones, and there it has been occasionally found. No major oilfield, however, has originated in this way. About 70 per cent of the world's known oilfields have been located in marine sediments of Mesozoic and Tertiary ages, generally along the flanks and in the less folded portions of orogenic belts. Most of the remaining fields occur in Palaeozoic sediments (e.g. in North America and the U.S.S.R.), or in the uptilted beds around salt domes.

The various lines of evidence all lead to the conclusion that petroleum originates from the organic matter of muddy sediments deposited in depressed regions of the sea floor, where the water was stagnant and deficient in oxygen. Under such conditions anaerobic bacteria would be expected to abstract oxygen from the organic matter, so transforming it, molecule by molecule, into fatty and waxy substances. The lighter members of the paraffin and other hydrocarbon series appear to be later derivatives, possibly produced by a kind of natural refining brought about by increasing pressures and temperatures during deep burial, together with continued bacterial activity. Bacteria are known to exist in the deeply buried ground-waters of certain oilfields. Finally, time, which is geologically plentiful, is certainly a fundamental requirement.

## Migration and Concentration of Petroleum

Sediments in which petroleum had its origin are called *source rocks*, to distinguish them from the *reservoir rocks* in which oil and gas are now found on a commercial scale. The reservoir rocks carry far more oil than could possibly have originated within them and, moreover, they commonly contain fossils which indicate deposition in oxygenated waters where no appreciable quantities of organic matter could have survived. An oil pool is, in fact, a concentration of oil which has migrated from the source rocks into places where it could draw to a head and accumulate.

Source beds such as clay and shale are now compact and impervious. But while they were still unconsolidated they contained a high proportion of sea-water carrying dispersed globules of oil. During this stage, circulation of the mixed fluids would sooner or later become possible, in response to pressure differences set up by variations from place to place in the overhead load or, more effectively, by earth movements. As the source beds are compacted, the squeezed-out fluids pass into more coarsely porous formations, such as sands. Once the oil has been flushed into these permeable beds it may migrate through them for long distances, but it cannot again escape from them, unless the overlying impervious rocks are fissured. If the mixed fluids encounter a sediment with very fine pores, the water may filter slowly through, but, because of its very different surface tension, the oil is held back. Within the sand or other reservoir bed the oil trickles upwards through the water until it comes to an impervious barrier and collects there.

In general, then, oil migrates outwards and upwards from the source beds, passes into porous or fissured reservoir beds, rises to the highest possible level, and collects into an oil pool wherever the structure provides a trap which impedes further migration. Gas, if present in excess of the amount that the oil can hold in solution, bubbles to the top and forms a *gas cap* over the oil pool. Beneath the pool the pore spaces are occupied by ground-water (often salt) which is commonly under a very considerable head of pressure. If the pressure and gas content are sufficiently high, the oil gushes out like an effervescent fountain when the pool is tapped by drilling. But when the pressure is too low to drive the oil to the surface—or becomes so as the initial pressure falls off—pumping is necessary to bring it up.

In accordance with the principles of oil concentration a dome or anticline of alternating permeable and impervious sediments makes an efficient trap for oil migrating towards it (Figures 16.12 and 16.13(*a*)). Isolated open anticlines surrounded by extensive gathering grounds have a much better chance of being productive than more closely packed folds. Not only have the latter to

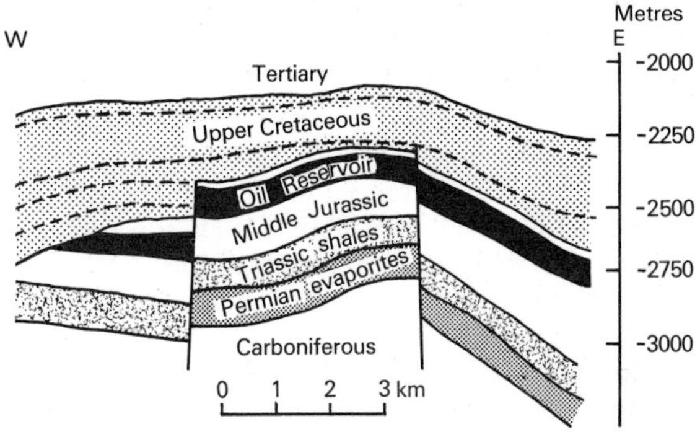


**Figure 16.12** Section across the Piper oil field, 200 km east of Wick, Scotland, illustrating the characteristic block faulted dome structure of North Sea oil fields. In the Piper oil field the block faults began to move during Lower Jurassic times or earlier, and ceased to move early in the Upper Cretaceous. The oil reservoir is here Upper Jurassic (Oxfordian) sandstone, and is sealed by Upper Jurassic (Kimmeridge) shales, and locally by the unconformably overlying and overlapping Cretaceous shales and marls. (*After P. E. Kent, 1975*)

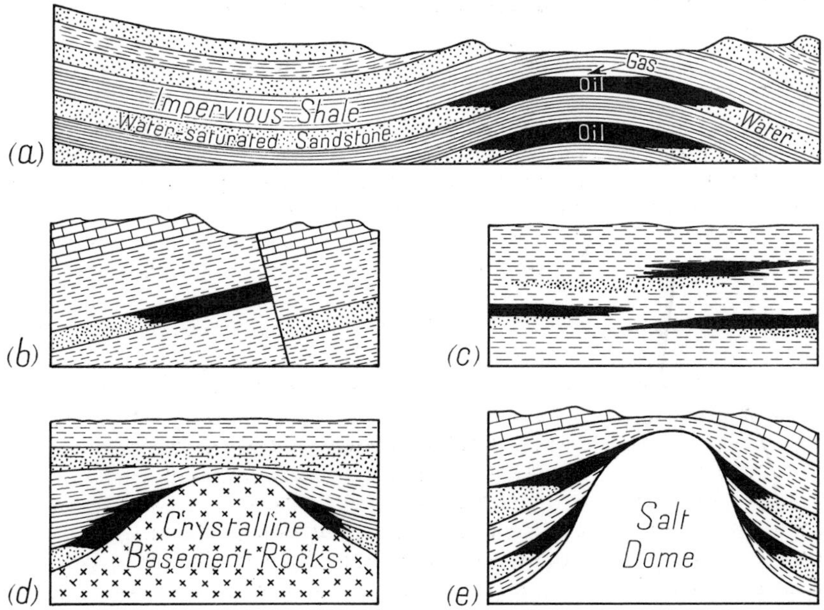


**Figure 16.13** Sections to illustrate various types of structural traps favourable to the accumulation of oil and gas (gas is omitted except in (*a*))

share a limited supply, but they are likely to be too much broken and fissured to retain any oil and gas that passed into them. The North Sea oil-traps include both open anticlines and fault block structures, and combinations of the two (Figure 16.12).

Although the 'anticline theory' of oil concentration dominated the search for oil for many years, it gradually came to be realized that anticlines are far from being the only structural traps. The early-discovered oilfields of Pennsylvania, for example, occur in a broad sedimentary basin in which the formations still remain practically horizontal over widespread areas. Here the oil pools occupy lenticular bands of porous sandstone which pass laterally, as well as vertically, into shales. The sandstone/shale boundaries served as a filter which held back oil while allowing water to pass through and escape. Thus the oil was concentrated and sealed within a series of impervious envelopes (Figure 16.13(*c*)). Oil pools occupying the upper ends of tilted reservoir beds are also of great importance. The tilted bed may pass laterally into shale, or it may be more abruptly cut off by an impervious barrier. The obstacle may be a fault throwing an impervious bed against the reservoir bed (Figure 16.13(*b*)); or a hill belonging to an ancient land surface which was unconformably overlapped and buried by a later series of petroliferous strata (Figure 16.13(*d*)); or a salt dome which has perforated and ascended through a thick series of sediments. Figure 16.13(*e*) shows oil accumulating against the walls of a salt dome. Oil may also be dammed back by faults produced in the surrounding rocks by the upward drag of the intrusive salt (cf. Figure 10.3), and finally, it may collect in the domed sediments over the roof.

Oil is not necessarily, or even generally, confined to a single reservoir bed in a given field. Any suitably placed formation may have been fed either from an outlying primary source, or via an underlying pool, the oil from which escaped upwards through fractures in the intervening impervious beds. Gas, in particular, readily migrates to higher levels, and in places vast quantities have accumulated as buried 'gas fields'. This reflects the natural tendency of some varieties of petroleum to differentiate into asphaltic and gaseous fractions. If the original reservoirs begin to leak after such fractionation has taken place, the mobile gas moves on and leaves the sticky asphalt behind. Most of the gas encountered by drilling was formerly discharged into the air and wasted. But now it is welcomed as an easily distributed source of power and illumination, and also because its heavier constituents can be condensed into petrol.

## The Discovery of Oilfields

Wherever fissures through the cover of a reservoir provide passageways to the surface, leakage of gas and oil becomes inevitable. Moreover, denudation may strip off the covering rocks until gas can force a passage through the roof and so open a way for the subsequent escape of oil. Thus the most obvious indications that a region is or has been oil-bearing are (*a*) seepages and springs of gas or oil; and (*b*) superficial deposits or veins of asphalt and other more or less 'solid' residues of petroleum left behind after loss of the volatile constituents by evaporation.

Inflammable gas may be found bubbling through the water of springs and wells or streams. In a few places gas escapes on a more spectacular scale. Blazing jets, like those of the Caspian coast near Baku, were long regarded with veneration by Fire Worshippers. The celebrated 'Eternal Fires' of Mesopotamia still burn today as they did in the time of Nebuchadnezzar (Figure 16.14). The 'fiery furnace' into which Shadrach, Meshach and Abednego were cast was probably a burning seepage of natural gas.

Where high-pressure gas breaks through water-bearing strata and clays it carries up wet mud and spatters it around the vent until a mound is built up with a crater at the summit. Groups and rows of these 'mud volcanoes', occurring above diapiric mud intrusions, or on the crests of anticlines, or along faults, are well known in Burma (Figure 16.15), Trinidad, California and various other localities. Eruption may be continuous or spasmodic, according to the build-up of gas pressure and the nature of the obstruction to be overcome. Apart from some similarity in gas mechanism, there is no connection with ordinary volcanic activity.

Flowing springs of oil emerge at the surface in some localities, but more commonly the exudations are sluggish, and the oil may be seen only as iridescent films on water. The largest surface 'shows' are sands cemented with residual 'tar' (bitumen), or localized deposits of more concentrated asphalt. These residues may eventually plug up the outlets and so prevent further losses

**Figure 16.14** Blazing seepage of natural gas, Iraq (*Iraq Petroleum Company Limited*)

**Figure 16.15** The largest mud volcano at Minbu, Upper Burma, showing two smaller ones to the left. Bubbles two or three feet across burst through the grey watery mud in the crater and spatter it over the slopes. The mud cones mark the top of a diapiric break-through of Miocene mud. (*E. H. Pascoe, Geological Survey of India*)

from depth; but, on the other hand, they may be all that remains of an otherwise dissipated oil pool.

Evidence of the use of bitumen in Mesopotamia dates back to very ancient times. Long before Noah caulked his Ark with asphalt, bitumen had been used for mortar and road making by the Sumerians (*c.* 5500 B.P.). Today it has become general knowledge that the Middle East is endowed with fabulously rich concentrations of oil, containing, as it does, almost two-thirds of the world's proved reserves.

Probably the best-known occurrence of asphalt is the celebrated 'Pitch Lake' of Trinidad (Figure 16.16). Although the island was discovered by Columbus in 1489, it is improbable that European travellers knew anything about the lake until it was seen by de Léry, who first described it in 1786. As a result of the removal of more than six million tonnes of asphalt during the present century, the surface has been lowered—but only by 6 m or so. Were it not for a slow process of replenishment from underlying sources, the lowering would have been about twice as much.

The first oilfields to be discovered were naturally found by digging wells in the neighbourhood of surface 'shows'. Many centuries ago the Burmese collected oil from surface exudations and later from shallow hand-dug wells. Yenangyaung is the classic example of the concentration of oil in the sands of an elongated dome. Oldham was the first geologist to survey the area, and as early as 1855 he pointed out the importance of anticlinal structures as oil traps. Nevertheless, little call was made on the services of geologists until after 1900, when the demand for oil began to accelerate, first for motor transport, then for aeroplanes, and eventually for war purposes and national security. It was then realized that the systematic discovery of oil involves the search for potential oil traps, that is to say, it demands the detailed geological survey of all the regions in which oil might conceivably lie hidden. The search for oil became a geological enterprise.

It is obvious, however, that in addition to the

**Figure 16.16** Convection cell of asphalt in the middle of the Pitch Lake of La Brea, SW Trinidad. The ashphalt is brownish black and contains a variable amount, up to a third, of sand and clay. Although it behaves like a solid and breaks when struck by a hammer, it nevertheless flows like a liquid under its own weight. The double behaviour is illustrated here by the tension cracks in the middle (light-coloured areas) due to upheaval and stretching over the ascending current; and the concentric folding towards the periphery, where the outward flow is slowed down as adjoining convection cells are approached (*H. H. Suter, 'Colonial Geology and Mineral Resources', vol. 2, 1951, p. 283*)

**Figure 16.17** Overcrowding of derricks, Beaumont Oilfield, Texas, illustrating the wasteful competition of the 1930s, when each owner of a plot tried to extract the maximum amount of oil in the least possible time. The field is still, however, a profitable producer. (*Ewing Galloway, N.Y.*)

favourable structures that can be located from outcrops at the surface, there must be some that cannot be detected in this way, and others that lie concealed by the sea and by tropical forests or swamps, by spreads of alluvium, boulder clay, desert sands or loess. Hidden anticlines and structures such as the buried hills and salt domes illustrated in Figure 16.13. can be detected by the gravity anomalies to which they give rise. Significant structures of all kinds can be explored by their effects on artificial earthquake waves (pp. 588–9). Seismic waves generated by exploding a charge of dynamite in the ground are reflected or refracted back to the surface by the rocks encountered in depth, and are there recorded by seismographs placed at suitable distances from the point of explosion. Many hundreds of salt domes, for example, have been successfully located by this method. Magnetic and electrical methods of exploring underground structures have also been devised. The search for oil is now a joint enterprise: geophysical as well as geological.

The discovery of a favourable structure does not, of course, guarantee that oil will be found. On the other hand, the absence of surface indications is not proof that oil will not be found; it may be that the oil, if there, is sealed in so efficiently that it cannot escape. Whether or not oil is present in commercial quantities can be determined only by drilling (Figure 16.17).

Important discoveries have occasionally been made by accident, as when bore-holes put down for water or coal happened to strike oil or gas. Indeed, the modern oil industry began a century ago as a result of a boring for water in Pennsylvania which brought up oil and salt water. This was in 1848 and the oil was sold for medicinal purposes. In 1859 the first well was drilled specifically for oil, and oil was duly found at a depth of only 21 m. The natural gas of western Canada, of which enormous reserves have since been proved, was first tapped in 1883 when it was hoped to find water for the Canadian Pacific Railway, then under construction. Moreover, there have always been optimistic operators willing to risk their capital and take a chance by sinking 'wildcat' wells on sites selected for no better reason than a personal hunch. Very few of these speculative ventures have ever proved successful, but some of those few yielded fabulous profits.

## Production and Reserves of Petroleum and Coal

The Middle East (1,083,240) is the largest oil-producing region, with North America (592,690) second, and closely followed by the Sino-Soviet area (537,580). The figures are estimates of the production of crude petroleum (including natural gas liquids) for 1974 in thousands of tonnes. Western Europe, which produced 20,350 thousand tonnes in 1974 lies at the bottom of the league. The world's total production for 1974 exceeded 2800 million tonnes. Recoverable reserves for the same year amounted to about 91,500 million tonnes, so that about 35 years supply is assured at the 1974 rate of consumption. As a glance at Figure 16.18 will show, the world's production and use of petroleum has persistently risen since 1940, but it has not risen nearly so steeply as the world's proven reserves, a glowing testimonial to the efficiency of geological mapping and geophysical probing. King Hubbert has estimated that oil reserves will continue to rise steeply until 1995, after which they will steeply decline, as the main sources gradually fail. On this basis it has been forecast that there will be an oil famine sometime during the early part of the next century. Albert Parker, in his summary of the information in the World Energy Conference Survey of Energy Resources, 1974, regards it as too pessimistic, however, to suppose that oil reserves will be exhausted within about 40 years. It will, of course, become increasingly profitable to search for and extract oil in difficult situations, an example being the location and recovery of oil from rocks beneath the floor of the North Sea. Other such enterprises are already being pursued. It will gradually become profitable to work oil shales, extracting the oil by distillation. The U.S.A. alone has a high reserve of oil in its widespread oil shales. Estimates of the total recoverable oil from world-wide oil shales and bituminous sands, indicate about 170,000 million tonnes.

Turning now to coal, the outlook for the future is not so cramping. Nevertheless, we are squandering our heritage of the solar energy stored up in coal several hundred thousand times as fast as it took to accumulate. Adopting the 1972 estimates (approximate only) of annual production in millions of tonnes, the leading coal-producing countries are: U.S.S.R. (580); U.S.A. (544); People's Republic of China (408); Western Europe (363). The total annual production for the world in 1972 was approximately 2866 million tonnes. Estimated reserves for the same year were over 8000 million tonnes. Although the use of coal as fuel has been decreasing relative to oil and gas, the total world consumption continues to increase because coal is itself an important source of gas and, like petroleum, of an infinite variety of by-products such

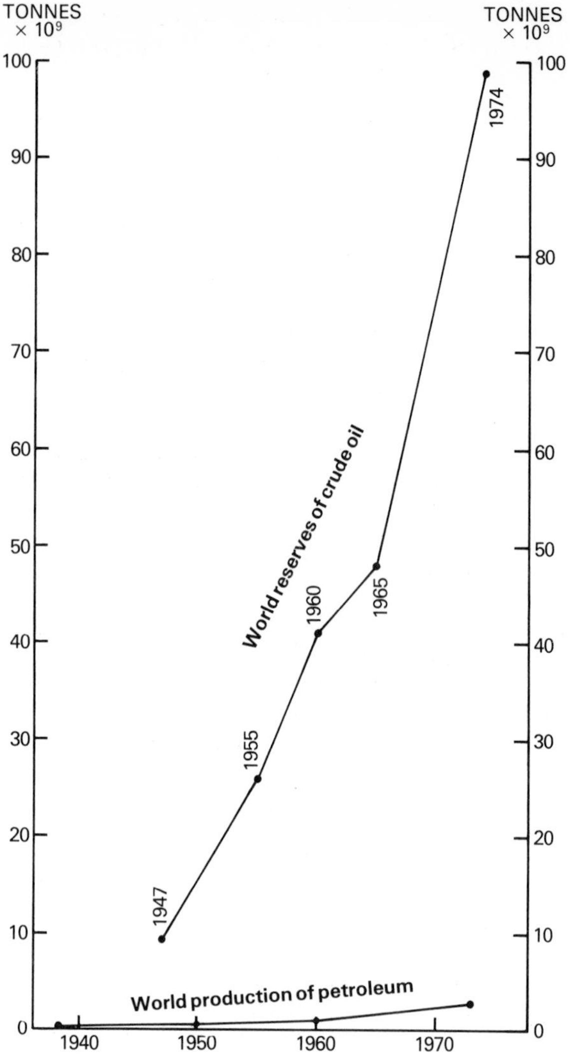


**Figure 16.18** Comparison between the world-wide use of petroleum and the known world reserves of crude oil up to 1974 (*Data from Institute of Petroleum, London*)

as plastics, dyes, and drugs. Moreover, in order to conserve oil it will have to be increasingly used for the production of electricity.

## Some Energy Comparisons

The sun is a gigantic nuclear power station which radiates energy in the form of sunshine. When we burn coal and oil we are using up part of the solar radiation which was received by the earth many millions of years ago, converted into chemical energy by photosynthesis, and in part stored up as natural fuels. We are, in fact, living at an alarmingly extravagant rate on an inheritance of capital that cannot be renewed. The present accelerating rate of consumption of energy per head of the world population, which itself is explosively increasing, stands high amongst the many causes of anxiety and embarrassment for the future well-being of our race.

Some terrestrial sources of energy which, like sunlight, can be regarded as income are enormous compared with human requirements, but unfortunately the extent to which they can be utilized to advantage is very limited. It is already possible to provide heat for a normal household from solar energy at great cost, but it is doubtful whether solar energy will ever provide adequate power for industrial purposes. A type of windmill has already been invented that could supply electricity if situated in a locality like the South Downs. To facilitate comparisons it is desirable to express the various manifestations of energy in the same units. One of the more readily grasped measures of energy is the kilowatt-hour (kWh), which corresponds to the electricity used by a small electric fire (1 kilowatt) in an hour.* It is appropriate, because electricity is generated from all natural sources of energy—fuels, sunshine, steam wells, tides, winds and nuclear fission.

The table on p. 298 brings out the supremacy of solar radiation. Amongst all other sources the most important in quantity, though the least important in practical value, is the heat received at the surface from the earth's interior. For a long time it was a popular belief that the 'Great Ice Age' was a symptom of the cooling of the earth. Yet in 1779, long before it was even suspected that there had been a 'Great Ice Age', Romé de l'Isle had shown that the outward flow of the earth's internal heat could have no climatic effects. So often has this been overlooked that it needs to be continually emphasized afresh. The radiation received from the sun is nearly 7000 times as much as the heat-flow from the earth's interior.

Volcanic activity is so localized that for the world as a whole its thermal contribution is only a minute fraction of the total outflow of internal heat—though this consideration does not make volcanic activity any easier to explain. In other respects, however, such as output of carbon dioxide and fine ash, volcanic activity may have significant climatic effects. Occasionally there is a great eruption, like that of Bezymianny in 1956, that liberates as much energy as the annual average for all the volcanoes of the world. Volcanic outbursts are altogether too ferocious to be tamed, but, as we have seen, geothermal power stations are now being established in favourable regions.

These geothermal installations are the more remarkable because, small though their total output of energy may be, it is enormously greater than has been obtained from the direct harnessing of sunshine. As yet man cannot compete with photosynthesis, which is by far the most efficient method of trapping the energy of sunlight. Merely as ensuring an important source of energy, forestry has a great future.

After combustion of wood and peat, hydro-electric stations come next in human importance, but despite their great number they contribute considerably less than 2 per cent to present requirements. It has been estimated that if the Mediterranean were turned to an inland sea by damming the Straits of Gibraltar and the Dardanelles, the level would eventually be lowered some 107 metres by evaporation. Power stations erected at Gibraltar, Gallipoli and the mouth of the Nile could then generate about $1000 \times 10^9$ kWh annually. This impracticable scheme merely serves to emphasize the limitations of water power, except for a few favoured countries like Norway and Switzerland.

It may be added in reference to nuclear energy that workable reserves of uranium ores are already

*Equivalent amounts of other energy units are as follows:
1 kWh $= 0{\cdot}86 \times 10^6$ calories $= 3{\cdot}6 \times 10^6$ joules $= 3{\cdot}6 \times 10^{13}$ ergs
1 erg is the energy required to make 1 gram move 1 cm per second faster than it moved during the preceding second
1 joule $= 10^7$ ergs
1 calorie ($= 4{\cdot}186$ joules) is the energy required to make 1 $cm^3$ of water 1°C hotter
1 joule per second $=$ 1 watt; 1 kilowatt (kW) $=$ 1000 watts $=$ 1·34 H.P.
The combustion of 1 tonne of average coal yields about 9000 kWh

ENERGY COMPARISONS

| *Sources of energy* | *Units of $10^9$kWh per annum* |
|---|---|
| Solar radiation received by the earth: | |
| at the outer limit of the atmosphere | 1,560,000,000 |
| at the surface, where it is mostly used in evaporating water | 930,000,000 |
| on land | 250,000,000 |
| Outflow of heat from interior of earth | 230,000 |
| Average volcanic activity | 650 |
| Geothermal power stations: Italy | 1 |
| New Zealand | 0·7 |
| Water power from waterfalls and elevated lakes (potential) | 20,000 |
| Tides (very little practically available) | 10,000 |
| Winds (very little utilized up to the present) | 84,000 |
| Lightning (an average flash dissipates 3000–5000 kWh and there are about 3000–6000 thunderstorms in progress at any given time) | 8,000 |
| Earthquakes | 3,000 |

known, e.g. in Canada, U.S.A., U.S.S.R., Africa and Australia, from which many times more energy could be released than from the whole of the world's reserves of conventional fuels. Given an adequate price there is sufficient recoverable uranium to yield something like 15,000 times the present annual consumption of energy. The immediate deterrents to the widespread use of nuclear power are the high cost of installation, and fear of serious accidents. Moreover, although the disposal of radioactive wastes has rightly become a matter for serious concern, it is not generally realized that the way in which we are disposing of our chief by-product, carbon dioxide, has already had climatic and economic consequences of world-wide importance. As we burn up our resources of coal and oil, millions of tonnes of carbon dioxide are poured into the atmosphere every day.

## Climatic Effects of Combustion

It was first pointed out by John Tyndall in 1861 that the carbon dioxide in the air must have a critical influence on temperature and climate because of its 'greenhouse effect'. The glass of a greenhouse is transparent to most of the sun's radiations. As the interior is warmed up by sunshine, the heat waves radiated back include infra-red wavelengths towards which glass is opaque. Accordingly a corresponding proportion of the heat is trapped and the temperature inside the greenhouse continues to rise. In the same way the atmosphere is practically transparent to the radiation that heats the surface of the earth. But the earth reflects back much of this energy on infra-red wavelengths towards which carbon dioxide is opaque. The more carbon dioxide there is in the atmosphere, the greater is the amount of radiation that is prevented from escaping into space. Consequently, if no other factors intervene, the temperature of the lower atmosphere rises as the content of carbon dioxide increases.

At present the atmosphere contains $2300 \times 10^9$ tonnes of carbon dioxide, about 0·03 per cent of its total mass. As indicated on Figure 16.1, the burning of fuels and various agricultural activities add $11 \times 10^9$ tonnes of carbon dioxide to this total every year. In the last hundred years about $400 \times 10^9$ tonnes have been added in this way. Only a small proportion can have been absorbed by the oceans in so short a time; allowing for this, human activities have raised the carbon dioxide content of the air by about 13 per cent. Theoretically this would be expected to cause a rise of temperature which would be most marked in higher latitudes. In northern Europe the average winter temperature has risen some 2° or 3°C, and higher increases are reported from Iceland and Greenland. Coal can be shipped from Spitzbergen for seven months in the year against three in 1900. Arctic pack ice now rarely reaches the coasts of Iceland as it did in the earlier years of the century. Farther south, countries on both sides of the Atlantic are now more exposed to cyclonic storms than formerly, and since the turn of the century the weather has deteriorated, and hurricanes and floods have become more frequent. And just as the belt of cyclonic storms has moved northwards into these latitudes, so the cod has migrated out of them far to the north of what were once profitable fishing grounds.

By the time all the reserves of coal and oil are exhausted, and allowing for oceanic absorption, it is estimated that there will be about ten times as much carbon dioxide in the atmosphere as there is today. On average, temperatures will then be about 12°C higher than they would otherwise have been. Later on, when equilibrium is established by continued oceanic absorption of carbon dioxide, this rise should gradually decrease to about 7°C. With the increase of warmth achieved during this century glaciers and ice sheets have been melting away faster than they have been replenished by fall of snow. A rise of temperature such as that estimated above would be amply adequate to lead to their complete disappearance.

Assuming that no counteracting processes come into effective operation during the thousands of years necessary to melt all the ice of Greenland and Antarctica, it is possible to estimate the effect on sea-level (see p. 15 for volumes of ice). If the melt-waters were simply restored to the oceans and nothing else happened the level would rise by about 76 metres. But in fact it would be less than this because (*a*) the weight of the restored water would depress the ocean floors isostatically; (*b*) the material so displaced in depth would flow under the continents (cf. Figure 3.18, p. 39) and uplift them; (*c*) Greenland and Antarctica would rise isostatically in response to the removal of the immense loads of ice; and (*d*) the inflow of material in depth implied by this isostatic recovery would mainly come from beneath the surrounding ocean floors, which would be correspondingly depressed. Each effect starts off others, like a decreasing chain reaction; but since they are all going on at once and at different rates, it is quite impossible to say what the sea-level would be at any given time in the future. If the ice were melted quickly, compared with the rate of isostatic response, the initial rise of sea-level might reach a maximum of 46–60 metres; at a much later time, as the isostatic adjustments restored equilibrium, this would gradually drop to about 40–37 metres. London and New York, together with all other ports and low land cities throughout the world, would be disastrously submerged. In partial compensation the higher parts of the rocky basements of Greenland and Antarctica would become lands with a genial climate.

However, the assumption of no counteracting processes is already turning out to be unjustified—at least temporarily. During the last two or three decades a few glaciers have ceased to melt away and have begun to advance again. We know that other factors besides the carbon dioxide content of the atmosphere are concerned with climatic changes, but what their relative effects may be still remains problematical. The carbon dioxide effect has been mentioned here only to illustrate the remarkable consequences of burning fuel hundreds of thousands of times as fast as it took to accumulate.

## Selected References


AVERITT, P., 1974, *Coal Resources of the United States*: United States Geological Survey Bulletin, No. 1416.

CARGO, D. N., and MALLORY, B. F., 1974, *Man and His Geological Environment*, Addison-Wesley, Reading, Mass.

JONES, J. M., and COOPER, B. S., 1972, 'Coal', Chapter 5 in Johnson, G., and Hickling, G., *Transactions of the Natural History Society of Northumberland, Durham, and North Tyne*, vol. 41, No. 1.

PLASS, G. N. 1959, 'Carbon dioxide and climate', *Scientific American*, vol. 201, pp. 41–7.

LEVORSEN, R. J., 1967, *Geology of Petroleum*, 2nd edn, Freeman, San Francisco.

STANTON, R. L., 1972, *Ore Petrology*, McGraw-Hill, New York.

*World Statistics 1975*, The Institute of Petroleum Information Service, p. 8.

# Chapter 17

# Surface Erosion and Landscape Slopes

The surface of the land is made by nature to decay. . . . Our fertile plains are formed from the ruins of the mountains.

*James Hutton*, 1785

## Rivers and their Valleys

The necessary conditions for the initiation of a river are adequate supply of water and a slope down which to flow. As we have already seen, Perrault was the first to discover that an adequate supply is provided by the rainfall. From the result of his pioneer work in the valley of the upper Seine he justly concluded that 'rain and snow waters are sufficient to make Fountains and Rivers run perpetually'. Rivers are partly fed from ground-waters, and some have their source in the melt-waters from glaciers, but in both cases the water is derived from the meteoric precipitation. In periods of drought rivers may be kept flowing, though on a diminished scale, entirely by supplies from springs and the zone of intermittent saturation. When these supplies also fail, through the lowering of the water table, as commonly happens in semi-arid regions, rivers dwindle away altogether. Even then water may still be found not far below the river floor, soaking slowly through the alluvium, where it is protected from evaporation.

The initial slopes down which rivers first begin to flow are provided by earth movements or, locally, by volcanic accumulations. Many of the great rivers of the world, e.g. the Amazon, Mississippi and Zaïre (Congo), flow through widespread downwarps of the crust which endowed them with ready-made drainage basins from the start. The majority of rivers, however, originated on the slopes of uplifted regions where, often in active competition with their neighbours, they gradually evolved their own drainage areas.

Most rivers drain directly into the sea. But in areas of internal drainage permanent or intermittent streams terminate in lakes or swamps which spread or contract so that evaporation from the exposed surface just balances the inflow, the conditions being such that the water is unable to accumulate to the level at which it could find an outlet. Notable examples occur in central Asia and Australia.

The development of a river valley depends on the original surface slope; on the climate, which determines the rainfall; and on the underlying geological structure, which determines the varied resistance to erosion offered by the rocks encountered. Where a newly emergent land provides an initial seaward slope, the rivers that flow down the slope, and the valleys they excavate, are said to be *consequent*. The valley sides constitute secondary slopes down which tributaries develop; these streams and their valleys are distinguished as *subsequent*. The resulting slopes and those developed later make possible the addition of further generations of tributaries. A main river and all its tributaries constitute a river system, and the whole area from which the system derives water and rock-waste is its drainage basin. Weathering continuously supplies rock-waste, which falls or 'creeps' or is washed by rain into the nearest stream. The latter carries away the debris contributed to it, and at times it acquires still more by eroding its own channel. Valleys develop by the removal of the material carried away by the streams which drain them. The load acquired by the main river is ultimately transported out of the basin altogether or deposited in its lower reaches. Deposits of gravel and alluvium are, of course, dropped on the way at innumerable places, but these are only temporary halts in the journey

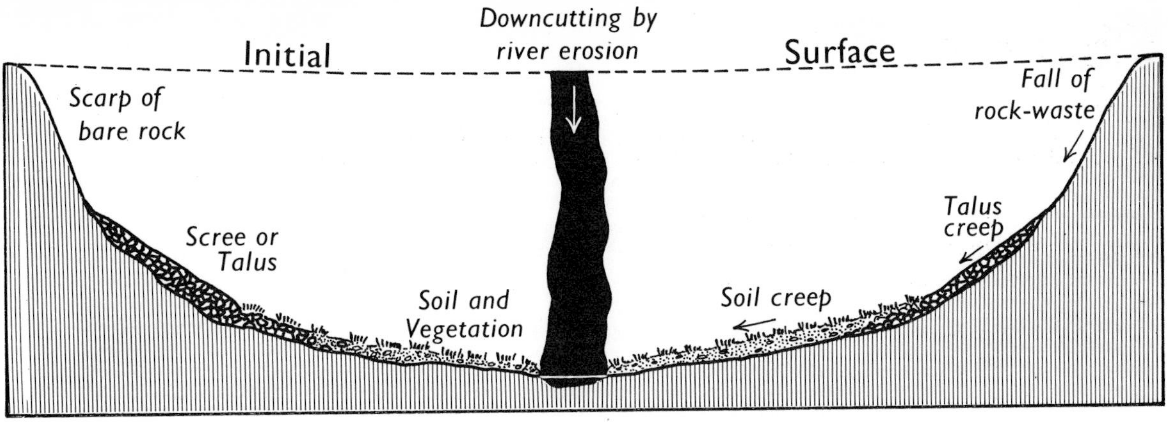

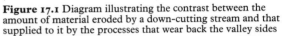

**Figure 17.1** Diagram illustrating the contrast between the amount of material eroded by a down-cutting stream and that supplied to it by the processes that wear back the valley sides

towards the sea. Rivers are by far the chief agents concerned in the excavation of valleys, not merely because of their own erosive work, but above all because of their enormous powers of transportation.

## Primary and Secondary Erosional Slopes

The widening of a valley such as that illustrated in cross-profile by Figure 17.1 clearly involves the delivery to the river of an amount of rock-waste that is many times greater than that liberated by the river itself during the down-cutting of its channel floor. If confined to down-cutting, the river would cut a valley in the form of a vertical-sided cleft. Deep gorges approximating to this form are found in mountainous areas and plateaus that have been raised high above sea-level, and where the rocks sawn through by the rivers are chemically resistant and mechanically strong (Figure 17.2). Down-cutting may then proceed a long way before the valley walls are worn back to any great extent by the successive operation of weathering, migration to the river of the resulting rock-waste under the action of gravity, and its eventual removal by the river. But sooner or later the rate of deepening slows down and widening begins to catch up, with production of a cross-profile that approximates to a V-shape (see Figure 18.19). In less resistant rocks widening on a more conspicuous scale accompanies deepening from the start.

While the valley is being widened by the wasting back of its sides, the river itself begins to widen its valley floor by under-cutting its banks, notably on the outer side of bends (Figure 17.3), where both down-cutting and lateral cutting are taking place

**Figure 17.2** Gorge of the Bhagirathi, one of the sacred headwaters of the Ganges, Garhwal Himalayas. The river issues from an ice tunnel at the end of the Gangotri glacier (Fig. 20.22) and traverses the range through a slit-like gorge about 185 metres deep. (*Prem S. Ray*)

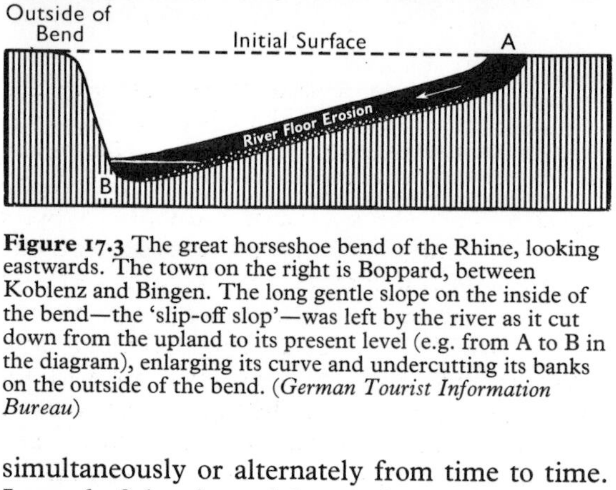


**Figure 17.3** The great horseshoe bend of the Rhine, looking eastwards. The town on the right is Boppard, between Koblenz and Bingen. The long gentle slope on the inside of the bend—the 'slip-off slop'—was left by the river as it cut down from the upland to its present level (e.g. from A to B in the diagram), enlarging its curve and undercutting its banks on the outside of the bend. (*German Tourist Information Bureau*)

simultaneously or alternately from time to time. Instead of the channel floor being sunk vertically into the rocks by river erosion, it digs in obliquely, as indicated by AB in the accompanying diagram. The relatively gentle slope on the inner side of the bend has been fashioned by the stream itself and carpeted with shingle or finer alluvium. This 'slip-off slope' is referred to again on p. 344. It is introduced here only in order to emphasize the fact that there are erosional slopes of two very different modes of origin:

(*a*) primary erosional slopes carved by the rivers themselves (e.g. the vertical walls of gorges and the slip-off slope in Figure 17.3 corresponding to the lower side of AB), or by the other chief agents of erosion: glaciers, winds, waves and currents; and

(*b*) secondary erosional slopes developed as a result of weathering and surface erosion of primary slopes, the latter including not only the erosional types (*a*), but also those of tectonic origin, such as fault scarps and tilted surfaces due to earth movements.

Here we shall consider the great variety of processes involved in downhill surface erosion and the chief results for which they are responsible. The specific activities of rivers, glaciers, sea waves and currents, and winds are dealt with in later chapters. These agents not only carry out the essential work of transporting the waste-products from the bases of the slopes that provide them; they also make characteristic types of slopes on their own account.

The valley widening that accompanies the excavation of river floors obviously implies the wearing back or *recession* of the slopes that lead down to the side or sides of the river where erosion and transport are taking place. Coastal cliffs similarly recede, so long as marine erosion undermines them and waves and currents continue to remove the fallen debris. The principles involved in all such cases are essentially the same. The land is consumed sideways, so to speak. For convenience, and to avoid undue repetition, we shall be dealing mainly with the slopes that constitute the sides of widening valleys or the scarps bordering extensive plains.

The slopes between two roughly parallel valleys

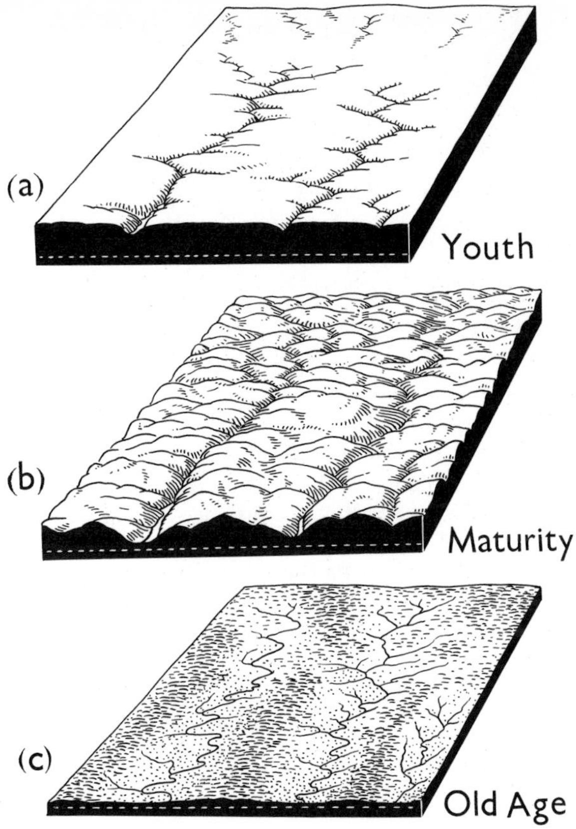


**Figure 17.4** Diagram illustrating the three main stages in the wearing down of an uplifted land surface according to the Davisian interpretation of the 'normal' cycle of erosion: *Youth*, while parts of the initial surface survive; *Maturity*, after most or all of the initial surface has vanished and the landscape is mainly slopes, apart from valley floors; *Old Age*, when the landscape becomes subdued and gently undulating, rising only to residual hills representing the divides between adjoining drainage basins. Eventually such hills are worn down and the region becomes a *peneplain* (*After V. C. Finch and G. T. Trewartha*)

necessarily approach each other as they recede from their respective valley floors. Eventually they meet and form a divide. The latter is then gradually lowered as the slopes continue to be worn back. Meanwhile, each main divide of this kind is itself being subdivided by the slopes leading down to tributaries of the trunk rivers. When there are several generations of tributaries, slope retreat comes into operation from several directions at once. Residual landforms are then developed in great variety: mountain peaks and hills; escarpments and ridges; inselbergs, tors and isolated pinnacles. Each uplifted area is dissected bit by bit into a slowly changing landscape. Some where or other every stage of landscape development is to be seen, from plateaus representing uplifted plains or sea floors, through mountainous or hilly regions of maximum relief where the country is practically all slopes, to those in which the land has again been reduced to a plain.

## Cycles of Erosion

These considerations suggest that from the time when the sculpturing of a newly emerged upland begins, and rock-waste as well as water streams away from every part of it in turn, the rivers, valleys and associated landforms should pass through a sequence of distinctive stages, provided that there has been no significant interference by earth movements, or by changes of sea-level or climate. This idealized concept of landscape evolution was introduced to geomorphology early in the century by W. M. Davis, who referred to the whole sequence of stages as a *cycle of erosion*. By analogy with the divisions of a lifetime he divided his evolutionary series into three main stages, metaphorically described as *youth, maturity* and *old age* (Figure 17.4).

Davis considered that as valleys are widened, the bordering slopes tend to become less steep during their retreat. During the stage of maturity, beginning when practically none of the original upland surface remains, the divides between adjoining valleys and drainage basins become rounded crests that gradually lose height. Finally, during a lengthy old age, the region is worn down and reduced to an undulating plain. Relief is faint, apart perhaps from an occasional isolated hill which owes its survival to the superior resistance of its rocks. Such residuals are sometimes called *monadnocks*, after Mt Monadnock in New Hampshire. The low-lying erosion surface which is the ultimate product of old age, Davis called a *peneplain* (L. *pene*, almost). This and the corresponding term *peneplanation* have thus come to be firmly associated with two leading ideas: the decline and flattening-out of hillside slopes during their retreat, and the accompanying down-wearing of divides and residual hills into forms presenting a broad and gently convexity towards the sky, but merging with equally gentle concavities into the surrounding plain.

This empirical treatment of a highly complex subject did not pass unchallenged, despite its persuasive virtues as a method of description and easy teaching. Erosion begins with the first appearance of a land above the sea and continues throughout the time of 'growing up', the time of intermittent uplifts culminating in the upland surface with which the Davisian cycle begins. Admittedly the whole cycle should include everything from plain to plateau or mountainous upland and back again from upland to plain. But one cannot deal with everything at once, as Davis, who was a superlative teacher, knew very well. No one doubts that his treatment provided an admirable starting-point. However, in 1924 Walter Penck tried to wrestle with the complications introduced by earth movements and, although in this he was not very successful, his new approach led him to a fertile idea which has since stimulated a great deal of lively discussion and active research. Penck argued that most hillside slopes should not flatten out as they are worn back, once they have attained an angle that is stable for the type of rock or waste-mantle concerned. Instead, he maintained, such slopes would be expected to recede without further change in declivity, i.e. each one should retreat parallel to itself (Figure 17.5). In a notable contribution in 1930 Davis agreed that this was so for the rocky and stony slopes of arid and semi-arid regions, but not for the forested and soil-mantled slopes of humid regions like New England and western Europe which he had previously been mainly considering.

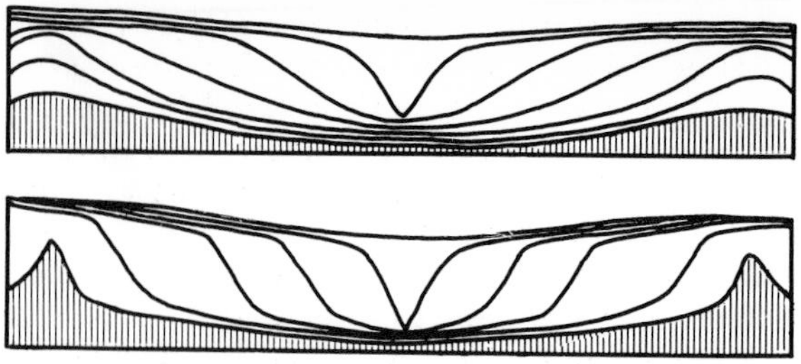

**Figure 17.5** Contrasted sequences of valley profiles from youth to old age—
*Above :* according to the wearing-down interpretation
*Below :* according to the wearing-back interpretation

Thus it came about that a 'cycle of arid erosion' was recognized as well as the original one, which was given a misleading prominence by being called the 'normal cycle of erosion'. The term *normal* in this context implied a standard of reference based on the landscapes now developing in the humid conditions of a temperate climate. It was agreed

**Figure 17.6** Ayers Rock, 320 km WSW of Alice Springs in the middle of Australia; an inselberg 2·6 km long of nearly vertical Precambrian strata rising in impressive isolation 335 metres above the surrounding pediplain. Mt Olga (left background) is 26 km to the west of Ayers Rock (*Australian National Travel Association*)

that in semi-arid regions a gently inclined surface—called a *pediment*—was left in front of the major slopes as a result of their parallel recession. Each pediment led down to the nearest stream or depression from bordering escarpments or from steep-sided isolated hills. In the latter case, well exemplified by the kopjes and inselbergs of Africa and the extraordinary landscape of the heart of Australia (Figure 17.6), it appeared that neighbouring pediments, encroaching from different directions, had coalesced to form a much more extensive *pediplain*.

Architecturally a pediment is a triangular feature crowning a portico of columns in front of a building in the Grecian style. In this sense the term is not an appropriate one for a gently sloping surface leading up to a scarp or butte or inselberg. It is worth remembering that the term in its geomorphological usage could equally well have been derived from the Greek *pedion*, which means flat open country or a piedmont.

## The Slope Problem

For some years the pediplain was regarded as no more than a climatic variety of the 'normal' peneplain. But the pediplain began to gain champions as geologists became more numerous and more widely travelled. In recent years attention has been mainly focused on what has come to be known as the 'slope problem'. In practice this means trying to discover the conditions under which major slopes (*a*) tend to remain essentially parallel to themselves as they recede in consequence of surface wastage and erosion, or (*b*) become progressively less steep. Which combinations of the many interwoven processes favour wearing-back and pedimentation, and which wearing-down and peneplanation?

On one point at least there is now general agreement. The Davisian scheme can no longer be properly described as 'normal'. The landscapes on which it was based are mostly hybrids, or 'fossil' varieties not yet brought up to date. Areas like New England and the British Isles passed through great fluctuations of climate during the Pleistocene. Most of their landscapes are products, not of the temperate conditions regarded as 'normal', but of repeated successions of glacial, permafrost and temperate-humid conditions. If we consider how the climates of particular latitudes have changed during the last 25,000 years or so (see Figure 17.7), we find that the regions now characterized by steppes and savannahs have been but little affected, if at all. In other words, the pediments and pediplains characteristic of these regions have been formed by the processes that are still operating there. These processes and their results can be and are being observed; and consequently the development of the pediment-inselberg type of landscape is becoming reasonably well understood. It is this type of landscape, which not so long ago was a profound mystery, that now provides us with an acceptable standard of reference.

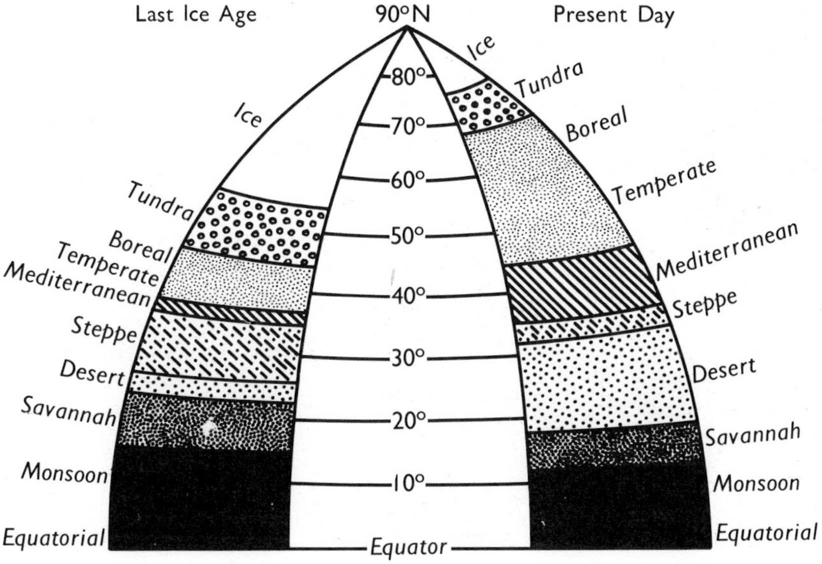


**Figure 17.7** The distribution of climatic zones in Europe and Africa during the last ice age (left) and at the present day (right)

From this standard there are two main deviations: (*a*) towards permafrost and glacial conditions; and (*b*) towards desert conditions, in which wind erosion plays a leading part. Our present landscapes of the temperate zones are composite products of all the varied processes involved in the first of these deviations. This means that the processes we now see modifying these landscapes are generally *not* those by which their major features were formed. Glaciated landscapes are easily recognized and are still well preserved in many parts of the temperate zone. And it is just because so many of us are familiar with landscapes of mixed origin that opinions diverge as to what purely temperate landscapes would look like.

Present-day landscapes have also been strongly affected by the many changes of level that have occurred since the beginning of the Pleistocene. On balance, as we shall see later (p. 449), these changes have been responsible for land emergence on a widespread scale in many parts of the world. In the south-east of England, for example, the land of today stands about 180 m higher, relative to sea-level, than it did a million years ago. Consequently a great many rivers are either still in the stage of youth or have been rejuvenated—endowed anew with youthful energy—so that their capacity for erosion and transport is still far from exhausted. This is a matter of primary importance in relation to slope retreat. Removal of the waste products from the base of a slope is essential, but by itself it is not sufficient to ensure parallel retreat of the slope. Another essential condition is that the base of the slope must be lowered or undercut, i.e. eroded downwards or laterally, to make room for the next 'slice' of waste to be removed from the surface of the slope. But we have still much to learn, not only from the researches now being carried out on surface erosion and slope profiles, but also from the practical experience of civil and hydraulic engineers and consulting geologists, who have to cope with the harsh realities of landslides, soil erosion and floods.

## Valley Sides and Hillside Slopes

A characteristic association of the hillside slopes that constitute the sides of many valleys is illustrated by Figure 17.8. This association forms

**Figure 17.8** Escarpment of Carboniferous Limestone, Eglwyseg Mountain, north of Llangollen, Clwyd, Wales. Below the free faces of the outcropping limestone beds the constant slope, determined by the angle of rest of the screes, is seen; this passes into the concave slope that leads down to the river. (*Institute of Geological Sciences*)

the basis of an illuminating analysis of slope development presented by Alan Wood in 1942. In addition to the initial upland and the flood plains of broad valley floors, the four slope elements recognized by Wood are shown in idealized form in Figure 17.9. From top to bottom these are listed below, the names given first being those adopted by Wood:

A. The *waxing slope*, the part of the upper surface which tends to become convex by being rounded off at, and towards, its edge with B; more commonly called the *convex slope*, sometimes the *upper wash slope*.

B. The *free face*, any outcrop of bare rock (e.g. a precipitous valley wall, scarp, bluff or cliff) that stands more steeply than the angle of repose of any scree or talus heap that may accumulate from its weathering products; it has also been called a *gravity* or *derivation slope*, because from it is derived the rock-waste that falls or rolls down under gravity.

C. The *constant slope*, which is that of the angle of rest of the scree-debris, whether it is a surface of scree or a bedrock surface on which a sprinkling of fragments may halt for a time. If scree is present its slope is often referred to as the *talus* or *debris slope*. Because scree-debris continues to be weathered, reduced in grain size and washed downhill by rain, the debris slope ceases to be constant towards its lower edge and merges into

D. The *waning slope* (*pediment, valley-floor basement* or *lower wash slope*), which stretches to the valley floor or other local base level with a diminishing angle, so that it is more or less concave upwards.

According to local circumstances certain of these slope elements may be repeated, or they may be undeveloped or worn out. In Figure 17.8, for example, and far more conspicuously in the pictures of the Grand Canyon of the Colorado (Figures 1.6 and 17.10), it can be seen that B and C are repeated several times, in accordance with the structural fact that the layers of strong limestone or sandstone responsible for the scarps, B, are separated by less resistant beds of shale. The latter have been worn back to a slope that is at least as steep as the constant slope, so ensuring that all the debris liberated from the sides of the canyon ultimately reaches the river at the bottom. On the other hand, in areas of low and gentle relief B is likely to be missing and C, if present at all, may be reduced to a short link between the summit convexity, A, and the concave waning slope, D. This case is familiarly illustrated by the smoothly undulating profiles that characterize the more subdued landscapes of parts of the British Isles and New England. The pediment-with-inselberg landscape is a strongly contrasted type in which B and D are the dominant elements (Figure 17.6). Where parallel down-cutting rivers have V-shaped valley sides that rise up to meet their neighbours in narrow ridges (see Figure 17.31), the constant slope, C, is highly developed. To cross country such as that between the Chindwin and Assam or between the Upper Irrawaddy and China is to toil laboriously up and down for hundreds of metres over and over again. It should perhaps be added that the valley walls adjoining an actively down-cutting river that has increased its rate of erosion will be steeper than the higher slopes, thus giving a profile that is convex upwards. However, this is a result of river erosion, to be dealt with in the next chapter. Here we are concerned with the surface erosion brought about over vast areas by the co-operation of gravity with weathering agents and rain.

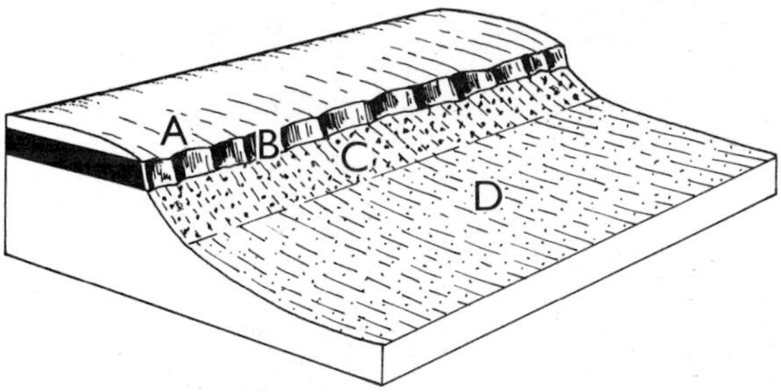

**Figure 17.9** Elements of hillside slopes (*After Alan Wood and Lester King*)

## Processes of Downslope Erosion

Downslope migration of the mantle of rock-waste and associated displacements of bedrock are commonly referred to as 'surface erosion' or as 'mass movements'. The movements range from falls of material and landslides resulting from slips, to soil creep and rain-wash resulting from various types of flow. In a masterly work on *Landslides and Related Phenomena* (see p. 324) C. F. S. Sharpe bases his classification on (*a*) the dominant kind of movement: slide or flow; (*b*) the relative rate of movement: fast or slow; (*c*) the kind of material: rock-waste or bedrock; and (*d*) the relative content of ice or water in the moving mass. Sliding

**Figure 17.10** Grand Canyon of the Colorado River, Arizona (*J. K. Hilliers, United States Geological Survey*) For location, see Fig. 19.32.

requires a slip plane between the moving mass and the underlying stable ground, i.e. an inclined surface on which the component of gravitational force exceeds the resistance due to friction and obstructions. Simple flowage requires no such slip plane, as the movement gradually dies out in depth. But flowage and sliding are often combined and give rise to complex transitional types of movement. Sharpe emphasizes the gradation between rivers as transporting agents and types of mass movement in which the debris load increases until water acts only as a lubricant or is absent, as in a dry landslide. There is a similar gradation between glaciers and movements of rock-waste in which ice, when present, occurs only in the spaces between the fragments like a sparse cement. Some of these relationships are indicated in the adjoining table. The work of wind and waves will be considered later. Here it need only be kept in mind that both promote rock-falls and landslides by their undercutting activities: waves by pounding at the base of cliffs, and wind by acting as a powerful sand-blast.

## Landslides

Falls of newly liberated fragments occur from vertical and overhanging outcrops or, by bounding and rolling, from slopes so steep that no slip plane is necessary (Figure 17.11). Such slopes include cliffs and fault scarps as well as the precipitous sides of gorges, glaciated valleys and fjords, and the heads of corries or cirques. A landslide may turn into a rock- or debris-fall if a precipice is encountered across its course.

The essential conditions for landslides are lack of support in front and a slip surface. Such conditions are liable to occur on the sides of undercut slopes and cliffs, or of road, railway and canal cuttings, particularly where heavy massive rocks (e.g. plateau basalts and basic sills) overlie weak and easily lubricated formations. Sliding

| | Ice | Bedrock, Rock-waste and Soil | | | Water |
|---|---|---|---|---|---|
| *Falling* | Icefalls | | Rock-fall<br>Debris-fall | | Waterfalls |
| *Sliding* | Glaciers<br>(in part) | | Landslides<br>Rock-slide<br>Debris-slide<br>Slump | | Rapids<br>(in part) |
| *Flow (relatively rapid)* | Avalanches | | Debris-avalanche<br>Earth-flow<br>Mud-flow<br>Lahars | Rain-wash<br>Sheet-wash<br>Rill-wash | Streams |
| *Flow (relatively slow)* | Glaciers | Rock-glaciers<br>Solifluction | Talus-creep<br>Soil-creep<br>Rock-creep | | Lake currents |

**Figure 17.11** Rock-fall on the Antrim Coast Road, Northern Ireland. Sliding of Tertiary basalt and Chalk on the underlying Jurassic clay (Lias), which becomes very slippery when wet, also causes temporary road blocks (*R. Welch Collection, Copyright Ulster Museum*)

(Figure 17.12) occurs when bedding or cleavage planes, master joints or fault fractures dip towards a valley or other depression at a dangerous angle. Slumping (Figure 17.13) takes place on curved shear planes. These are often spoon-shaped and leave an arcuate scar on the defaced cliff or hillside. Because of the rotational slip, backward tilting of the surface and of the dip of the beds often result (Figures 17.14 and 17.15).

Figure 17.16 shows the debris of a landslide that obstructed a tributary of the Ticino valley in 1927. During the preceding years a crack appeared near the top of a peak on the right-hand side and slowly widened until it became a gaping fissure 2 m

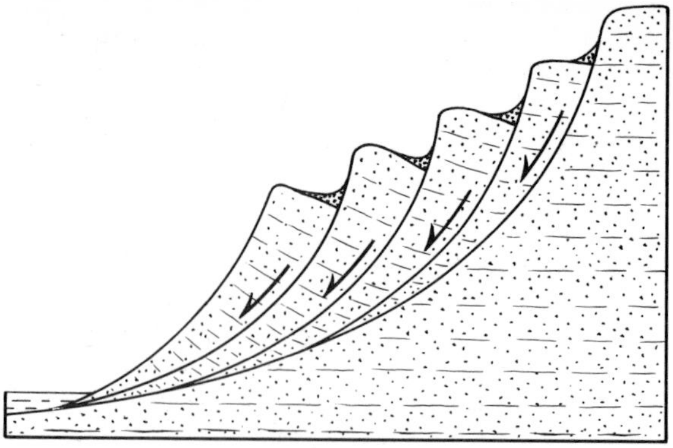

**Figure 17.13** Diagram illustrating slumping on curved surfaces in unconsolidated or other weak formations, with production of back-tilted beds that were originally flat. (cf. Fig 17.14)

**Figure 17.12** Diagram illustrating conditions leading to rock slides on bedding planes after lubrication

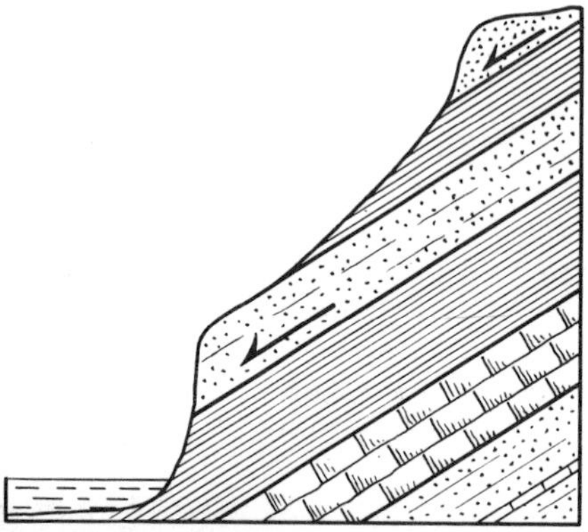

**Figure 17.14** A prehistoric landslide at Garron point, Northern Ireland, along the Antrim coast road, where Chalk overlain by Tertiary basaltic lavas has slumped over Jurassic (Lias) clay, resulting in back tilting of the strata as in Fig. 17.13. Differential movement during slumping (see Fig. 17.19) resulted in faulting of the sliding rocks essentially at right angles to the direction of slumping. This caused black basalt, seen on the right overlying white chalk. to be downthrown against chalk on the left. (*P. S. Doughty, Ulster Museum*)

**Figure 17.15** Landslide near Exmouth, looking west towards Sidmouth and the estuary of the Exe, SE coast of Devon. Chalk and Upper Greensand, seen *in situ* on the right, slipped towards the sea (to the left of the illustration) on a surface of Jurassic clay on Christmas Eve, 1839, after a prolonged period of heavy rain. A landslip of large continuous masses of rock was followed by slumping and back-tilting of narrower slices, as seen in the middle of the illustration. (*S. H. Reynolds*)

across. Subsequent movements were carefully measured by setting up a line of stakes and recording their positions every few hours. One day in 1927 an ominous slip of about 3 m alarmed the observers. Warnings were immediately telephoned and the danger zone was evacuated. Two days later the threatened landslide took place, fortunately without loss of life.

The north-western end of the Great Himalaya in Kashmir is a schist-migmatite-granite complex that is still rising and subject to occasional severe earthquakes. In December 1840 an earthquake shook loose part of the western spur of Nanga

**Figure 17.16** Landslide of 1927 which blocked the Valle d'Arbedo, near Bellinzona, Ticino Valley, above L. Maggiore, Italy (*F. N. Ashcroft*)

**Figure 17.17** Flat-surfaced deposits marking the site of a prehistoric lake formed by a landslide that dammed the Upper Rhine, Tobel Drun ravine, near Sedrun, Switzerland. The headwaters of the stream illustrate rapid widening by surface erosion (*F. N. Ashcroft*)

Parbat (8117 m), where the Indus has cut a gorge 4500 to 5000 m deep through the great mountain range. The gigantic landslide blocked the river and dammed back the water for 65 km. The resulting lake reached a depth of over 300 m before it overtopped the obstruction. The water then burst through with such violence that in less than two days the lake had emptied. A devastating flood tore down the valley, sweeping away a Sikh army encamped near Attock and carrying destruction for hundreds of kilometres.

The sediments of the flat terrace seen in the foreground of Figure 17.17 were deposited in a lake that formerly occupied a long stretch of the Upper Rhine and its tributaries. The lake was impounded by a prehistoric landslide that blocked the main valley near Flims, 50 km downstream from the place illustrated. About twelve and a half cubic kilometres of rock plunged down 1000 m or more. The forward momentum of this enormous mass carried it up the slopes on the other side of the valley to form a long tongue that extended 14 km from the deeply scarred mountain side and covered 50 km$^2$ with a chaos of smashed-up fragments. The sediments deposited in the lake and the great obstruction in front have now been cut through by the Rhine, and it can be seen that the landslide-dam that held up the lake for thousands of years was more than 820 m thick.

Van Bemmelen has described a most remarkable landslide which was an important early link in the chain of catastrophes that wiped out the ancient Hindu culture of central Java a thousand years ago. The south-western part of the cone of the Merapi volcano, including the crater and the upper part of the conduit, sheared off in a series of spoon-shaped slices bounded by crescent-like faults that curved down to a basement of soft, late Tertiary clays. On this weak and wholly inadequate foundation a great volcanic edifice had been built up until, after reaching a height of over 3000 m, parts of it began to slump (Figure 17.18), the movement being probably triggered off by an earthquake. As the colossal weight of displaced rock glided forward, it thickened and crumpled up in front. This obstruction created a dam behind which a flourishing countryside, famed for its temples and monuments, was submerged by a deep and extensive lake. But this was not all. The collapse of the summit of the volcano so reduced the pressure in depth that a cataclysmic eruption burst out through a new conduit and crater. The combination of landslides, earthquakes, floods and fiery ashes ruined the country for hundreds of years. The geological evidence still remaining adds dramatic detail to the brief stone inscriptions which have been found, recording a horrifying calamity that reduced Java to chaos in A.D. 1006.

The examples described above indicate that major landslides are often responsible for serious floods. Conversely, floods are generally accompanied and followed by a crop of minor slides and slumps. The banks are waterlogged when the streams are swollen; when the water subsides lateral support is withdrawn and the banks tend to fall in. The catastrophically heavy rainfall of August 1952 over Exmoor is chiefly remembered for the disaster it brought to the little town of Lynmouth (p. 331). It was also the cause of dozens of landslips, mostly slips of earth and debris into the Exe and neighbouring streams draining Exmoor. The wet summer of 1960 again saturated the

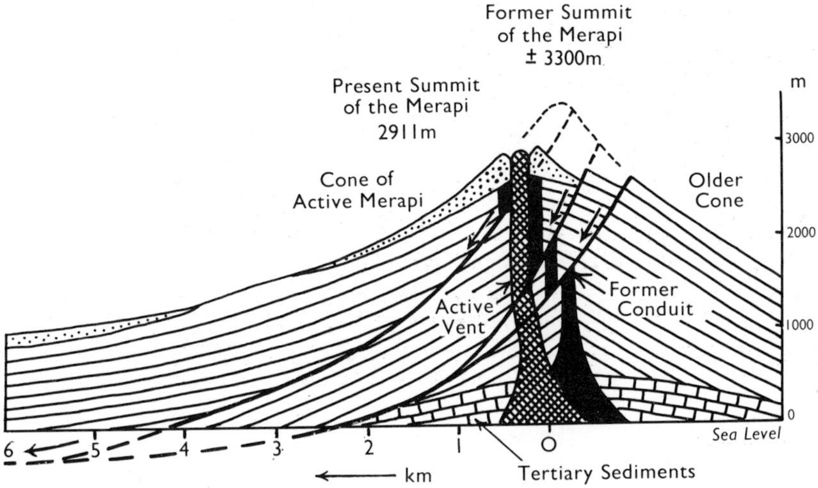


**Figure 17.18** Diagram to illustrate the slumping of part of the cone of the active volcano Merapi in central Java on its foundation of weak Tertiary sediments (*After R. W. van Bemmelen*)

ground so that in many areas the high rate of run-off accompanying the exceptional rainfalls of the autumn was repeatedly more than the rivers could discharge. Floods were inevitable and these in turn provoked landslips and debris-slides along the river banks of a wide region extending from Dartmoor and the New Forest to the Severn and the Trent.

It is a matter of common observation that steep grassy river banks are often scored at intervals with so-called 'sheep tracks' in the soil or waste-mantle. These features have nothing to do with sheep, except that sheep may use them. Technically described as *terracettes*, they are the surface expression of incipient landslips which play an important part in feeding streams with materials to be carried away, particularly where the slopes are kept steep by undercutting. The slip-rate is slow, except after very heavy rains. In the colder regions river-bank slumping is also promoted by frost and thaw, and by the rapid melting of thick snow. In hot humid regions the more frequently repeated landslips accompanying long-continued rains are largely responsible for the unusually steep slopes of many tropical mountains. Even forests may slide down bodily with their rooted foundations, and the jungle soon invades the scars left behind. Eventually the scars from adjoining valleys approach and meet, making sharp crests like the arêtes of glaciated mountains (p. 423). Many mass movements of these kinds are flow-slides, difficult to classify, since they merge into debris-avalanches and mud-flows.

### Earth-Flows, Mud-Flows and Lahars

Figure 17.19 illustrates a typical association between earth-flows and certain landslides of the slump variety. Here the earth-flow is a hummocky or crevassed 'toe', composed of materials that slumped down a curved slip-plane and thickened over the gentler slopes where the slip-plane flattened out or curved upwards. Other earth-flows start independently. When materials such as clay or shale are over-saturated with water and so dilated, any slight shearing movements (which might be due to sea waves, earthquakes or vibrations caused by traffic) tend to bring about a compaction of the 'solids' (see p. 263). The water potentially liberated cannot quickly escape by itself, because of the low permeability. Consequently the sodden material becomes 'quick', i.e. like a quicksand, and readily flows as a whole. Many of the so-called landslips along railway cuttings in the London Clay are of this type.

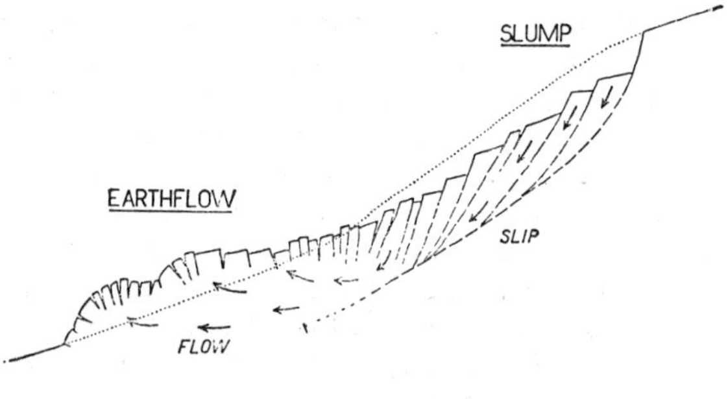


**Figure 17.19** Longitudinal section through earth-flow showing typical association of earth-flow with certain landslides of the slump variety (*C. F. S. Sharpe*)

*Mud-flows* tend to follow the stream of channels of arid or semi-arid regions, where dry accumulation of debris may be swiftly turned into a sort of 'porridge' by a sudden torrential downpour. If the run-off is high—i.e. if the bedrock near the surface is impermeable—the soaked debris flows down and out of the wadi or canyon (Figure 17.20) and spreads itself over the flatter country beyond. Debris of all sizes, including large blocks of fallen rock, may be transported by mud-flows, like the morainic load of glaciers and the great volcanic blocks carried along by *nuées ardentes*.

The bog-bursts of Ireland and similar regions of peat bogs are closely related to mud-flows. If the peat is the infilling of a former shallow lake, it may happen that water percolating along or near the underlying rock surface during long-continued rain finds that its previous underground exit has become blocked—probably by peat. It then accumulates at some level within the peat bog or between it and the bedrock. The bog then swells up until something gives way. If it bursts externally, as occasionally happens, a deluge of muddy peat flows over the surrounding countryside.

Deposits of ash and other pyroclasts on the slopes of a volcanic cone are obviously suitable material, all too favourably placed, for removal as mud-flows. When saturated by rain, melting snow or water from volcanic steam, mud-flows that may be highly destructive are inevitable. Volcanic mud-flows are widely known as *lahars*, this being their name in Java, where they are both frequent and dangerous. Hot and cold lahars are distinguished, the former being one of the special hazards of a volcanic eruption, as Herculaneum, buried in 79 A.D., still remains a grim witness. Every

**Figure 17.20** Mud-flow of 1930 at the mouth of Parrish Canyon, on the side of the Lake Bonneville flats, Utah (*United States Forest Service*)

gradation between a lahar and a flood heavily laden with debris may occur when a crater lake is ejected at the beginning of an eruption. In 1919 Java suffered a disaster of this kind when over a hundred villages and most of their inhabitants were overwhelmed. In Iceland the melting of glacier ice during eruptions and the resulting calamities have led to the abandonment of some of the farms and settlements of earlier centuries.

## Soil-Creep and Solifluction

Slow downward movement of soil on hillsides, known as *soil-creep*, is evidenced by tilted fences, piling up of soil against outward-bulging walls, and the curving of tree trunks near the ground. Creep is the sum of innumerable tiny displacements of grains and particles, with gravity in control at the steering-wheel to ensure that the cumulative effect is a downhill migration. Rain-splashing and rain-wash by surface run-off co-operate; but the effects of rain erosion—and also of wind erosion—are so powerful and distinctive that it is convenient to deal with them separately. Percolation of rain-water through the soil is one of the more important processes in soil-creep proper, and so is frost-heaving in regions subject to frost and thaw. Stones and particles of all sizes are heaved outwards by growing ice crystals (p. 276). This outward movement at right angles to the slope, followed by a verticle drop when thaw sets in, gives the fragment concerned a slight shift downhill. The return of fine particles to the surface by burrowing animals acts on balance in the same direction. Other less perceptible movements, all of which are dominantly downslope, result from expansion and contraction due to temperature changes; from swelling and shrinking by soaking and drying; from the growth and decay of roots; and from the filling of cracks and other openings, including those made by burrowing animals, and those made by the straining of roots as they respond to the swaying of trees in the wind.

Even the sub-soil and the upper part of the bedrock share in the slow migration downhill.

**Figure 17.21** Soil creep, showing the overturning of the cleavage of Cambrian slates, near St Davids, Dyfed, SW Wales (*W. Jerome Harrison*)

**Figure 17.22** Soil creep, showing overturning of Yeringian beds exposed in a road cutting in Melbourne, Victoria, Australia (*R. H. Clark*)

Rock-creep of this kind may lead to 'outcrop curvature', especially where the upper ends of steeply dipping or cleaved beds have been prised apart by frost or rootlets, so that they gradually curve over in the downhill direction (Figures 17.21 and 17.22). The apparent dips of superficial outcrops, exposed in hillside cuttings and gullies, may be very different from those of the undisturbed formations. Care must be taken in mapping not to be misled by such deviations.

The sub-soil activity may also be regarded as a by-product of the processes engaged in the formation of new soil. Where the soil survives, in spite of soil-creep and rain-wash, it means that soil removal is more or less balanced by soil formation in depth.

Many of the agencies responsible for soil-creep also operate in bringing about a similar downward migration of screes. However, the slow creep of screes, also known as *talus-creep*, is liable to sudden accelerations after thorough soaking and lubrication by excessive rain or melting snow. Increase of weight and decrease of friction may locally trigger off the slipping of a mass that was previously in uneasy equilibrium. Some debris-slides and rock-avalanches originate in this way. In permafrost regions screes may be permanently cemented by ice, except for a shallow spring thaw. With continuous replenishment by frost-shattered fragments from above, the increasing weight eventually forces the interstitial ice to flow. A sluggish *rock-glacier* then creeps down the slope and drapes the flatter ground in front with ridges and mounds of coarse rubble (Figure 17.23).

A rock-glacier has practically no interstitial mud. Where mud or a thin soil is present (with or without stones) on the gentler slopes of the permafrost lands, the process corresponding to soil-creep is almost entirely controlled by frost and thaw and is referred to as *solifluction* (soil-flow). The upper part of the waste-mantle creeps over the underlying frozen ground on slopes that may be nearly flat—no more than two or three degrees. Unless the migrating sludge is carried away at the foot of the slope, it accumulates and levels up any depressions that may lie in its path. Solifluction is clearly a slope-reducing process that generally leads to a landscape of low relief.

**Figure 17.23** A rock glacier, Copper River district, Alaska (*F. H. Moffit, United States Geological Survey*)

## Rain Erosion

Some of the sub-surface effects of percolating rain-water in promoting mass erosion have been indicated above. Interstitial rain-wash contributes to soil-creep; but where the soil has a strong cover of vegetation, and particularly if it is firmly bound together by a mat of interlacing grass roots, it is well protected against rapid surface erosion. On the other hand, where soil and poorly consolidated mantle deposits lose this reinforcement and are exposed to pelting rain and a copious run-off, the results may be disastrous. Raindrops splashing on wet soil are now known to have far more serious effects than was formerly suspected. From each miniature crater made by the impact hundreds of dislodged particles are scattered in a little fountain of spray, and on sloping ground most of these fall back a little farther downhill. Recent studies of splash erosion have revealed that during a heavy storm up to 100 tonnes of soil per acre may be shifted. This will seem less surprising if the energy involved is considered. In such a downpour thousands of millions of raindrops strike an acre of soil with velocities up to 30 km an hour or so.

The finest material of the top soil or earth is then readily transported to lower levels by rain-wash. The run-off may form a thin sheet of muddy water spread fairly uniformly over the slope (*slope-wash*), but generally it is concentrated into more or less intermittent trickles and rills (*rill-wash*). Observations on smooth sloping beaches show how quickly a thin sheet of water can develop into a network of shallow channels. The bearing on hillside soil erosion is clear. The good topsoil is the first to be carried away. If the process continues unchecked the deeper and inferior soil follows, long before it has had time to mature or to be replenished by the slow weathering of the sub-soil. Within a few years the soil may be first impoverished and then lost, either by being washed away down a surface furrowed by rills and gullies (Figure 17.24), or by being blown away (p. 474), or both.

Present rates of erosion are excessive because of human interference with the natural cover of the soil, e.g. by the widespread clearing of forests and the breaking-up of protective turf by ploughing. Where it was supposed that rain-wash was the major cause of soil erosion on hillsides, contour-terracing like that seen in the background of

**Figure 17.24** Badland type of erosion in the valley of the Ruindi, south of Lake Edward (L. Idi Amin) in the extreme east of Zaïre. Residual buttresses of Pleistocene sands and clays left between the gullies are themselves scored and fluted by rain erosion (*Félix, Inforcongo*)

Figure 17.26 was introduced to minimize the losses. Results often proved disappointing. The reason is that terracing fails to give permanent security against rain-splash erosion. Some additional safeguard, such as a covering of straw-mulch, is necessary to shield the vulnerable soil from the force of the rain.

But man is not always the culprit; he may be the victim. A localized cloudburst may concentrate the run-off into a violent torrent that bites deeply into the turf on sloping ground, sweeping the underlying soil to the foot of the slope, and leaving a long gash in the hillside. The gash is gradually deepened into a gully by recurrent rains, and as soon as the water table is tapped it begins to carry water and becomes a rivulet. This is one of the ways in which tributaries originate or extend their headwaters (see Figure 17.17).

In semi-arid regions, where the occasional rains

**Figure 17.25** Close up view of the turrets and spires of the Big Bad Lands, South Dakota, U.S.A.

are often exceptionally violent, rain-gashing reaches spectacular proportions on sloping ground underlain by clay or soft earthy deposits. Such land is sculptured into an intricate pattern of gullies and small ravines, separated by sharp spurs and buttresses. The gullies grow backwards into the adjoining upland, and the intervening ridges in turn are further cut up into smaller ribs and trenches (Figure 17.24). Tracts of the almost impassable country so developed are graphically described as *badlands* in North America, where they are widely scattered from Alberta to Arizona (Figure 17.25).

## Earth-Pillars

The remarkable residual forms known as *earth-pillars* are a standing proof of the efficacy of rain erosion. They are developed on the slopes of valleys from spurs and ribs of relatively impermeable material, which contains resistant boulders (or their equivalent) in a more easily eroded matrix: e.g. boulder clay (Figure 17.26), certain conglomerates (Figure 17.27), or tephra (not too porous) containing volcanic bombs or ejected blocks (Figure 17.28). Wherever boulders or other resistant masses, including concretions, large fossils and occasional hard layers, are encountered while the slope is being worn back by rain erosion, they act like umbrellas over the underlying less resistant materials. Where the slope is sheltered from strong winds, earth-pillars of surprising height may be etched out before their protective caps topple off.

The valley sides on which earth-pillars stand are eroded mainly by rain-wash. Their inclinations are generally between 25° and 35°, the angle being dependent on the properties of the materials of the slope and the nature of its cover of vegetation. The range of inclination is about the same as that of the constant slope (Figure 17.9), which is controlled by the angle of repose of screes. This equivalence implies that the slope of the valley side, eroded mainly by rain-wash, reaches an angle down which the boulders are continually being liberated can roll or slide, while the finer material is washed down by the run-off. If the slope were less steep the boulders would accumulate. But they do not accumulate, except temporarily in the river bed itself and on small adjoining flats, where the larger ones may rest for a time before being removed by an exceptional flood. This consideration indicates

**Figure 17.26** Buttress of boulder clay developing into earth pillars, Val d'Herens, tributary of the Upper Rhône (*G. P. Abraham Ltd, Keswick*)

**Figure 17.27** Earth pillar of soft conglomerate (Old Red Sandstone): one of many isolated stacks, some capped with boulder clay, on the valley side of the Allt Dearg, a tributary of the River Spey, above Fochabers, Grampian Region, Scotland. (*Institute of Geological Sciences*)

that a slope on which earth-pillars are developing must lead down to a river capable (*a*) of carrying away all the debris fed into it from the slope, and (*b*) of deepening its floor or widening its channel (or both) in order to make room for the successive slices of material eroded from the slope. Now we have already seen (p. 306) that parallel recession of a slope is not geometrically possible unless these conditions are fulfilled. Here, where they are fulfilled, preliminary observations confirm that the slopes on which earth-pillars stand do remain parallel, or nearly so, during their recession. And here we also know that the dominant process of surface erosion is rain-wash. The boulders are critical in providing the information that the slope of the valley side remains steep enough to ensure their easy migration to the foot, once they are liberated.

**Figure 17.28** Earth pillars in volcanic tuff erupted from Mt Argaeus (English Dagh), Gheureme valley, about 240 km SE of Ankara, Turkey. A lava flow of andesite forms the escarpment seen at the head of the slope. The valley is celebrated for the trogloditic dwellings and monasteries which have been excavated in the tephra. The lava caps have been removed from the earth pillars where practicable, partly for safety and partly to provide stone for the façades of the rock churches. (*Yan, by courtesy of Messrs Thames and Hudson*)

The slopes of the earth-pillars themselves range from nearly vertical to about 60°; they correspond to the free face of Figure 17.9. These also tend to recede parallel to themselves, since the sides of small pillars have about the same angles as those of the big ones where the materials and conditions are similar. The fact that many pillars develop a convex profile, as may be seen in the illustrations, results from the steepness of the sides, lack of vegetation and high run-off (factors that favour an increasing rate of rain-wash erosion from top to bottom), combined with removal of the debris at the base. However, where the run-off is being continually checked by small obstacles there can be little or no acceleration, and the slopes developed then tend to be straight.

## Measured Slopes and Erosion Rates

Figure 17.29 shows that two contrasted types of topography have developed under the same climatic conditions on the Brule and Chadron formations. The latter are of Oligocene age and consist of poorly consolidated clays and other sediments. The Brule, however, is relatively impermeable and dries out to a hard surface down which the run-off from subsequent rainfall is high. Erosion by rain-wash, accompanied by rill-wash, is dominant. As the volume of water increases downhill the flow tends to concentrate into tiny rills, most of which are temporary, as they fill up again with particles of soil or clay when the flow of water abates. Under these conditions small ∧ -shaped residuals left in front of the Brule escarpment continue to maintain steep, nearly straight

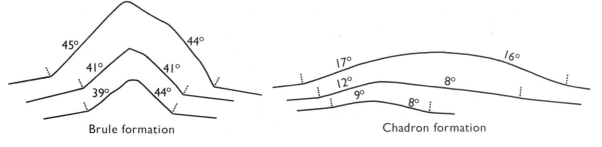

**Figure 17.29** Typical topographic forms developed in the Oligocene strata of Badlands National Monument, South Dakota. The pediment in front rises very gently to the rounded slopes of the Chadron formation (middle), above and behind which are the steeper, straight slopes of the Brule formation. Further along the escarpment the pediment locally truncates the Chadron and rises to make a sharp junction with the Brule. (*Stanley A. Schumm*)

**Figure 17.30** Series of slope profiles measured on erosion residuals, Badlands National Monument, South Dakota, to show the changes of angles as the slopes above the pediment become shorter. Dotted lines mark the junctions with the pediment. *Brule* (low permeability, high run-off): little variation in slope angles. *Chadron* (high permeability, low run-off): marked reduction in slope angles. (*After Stanley A. Schumm*)

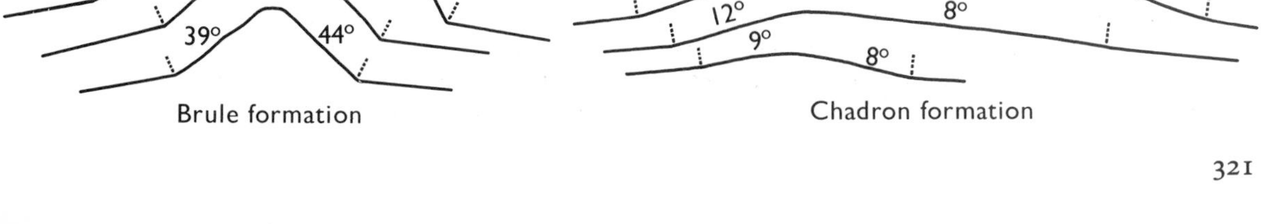

slopes. The latter meet in sharp divides except while a more resistant band is exposed at the summit, when the crest becomes rounded or even flat for a time, as seen in Figure 17.30. As material is eroded from the slopes their lengths from pediment to summit are shortened and the divides are lowered. Stanley A. Schumm has measured a variety of these residuals and finds no significant variation of slope angle from the highest and longest to the lowest and shortest, the average being between 44° and 45°. The slopes have remained parallel during their retreat.

The underlying Chadron formation, exposed by the more rapid retreat of the Brule escarpment, behaves very differently. It dries out to a surface of loose aggregates and, because of its permeability, much of the subsequent rain sinks in, so that the run-off is low. Here soil-creep, accompanied by small slumps and slides of wet mud, accounts for most of the erosion. Schumm's measurements on the Chadron residuals show a consistent decline in the slope angle from 33° for the longest slope, to 8° for the shortest (Figure 17.30). It is evident that convex summits and gently rolling topography are developed where surface creep and related processes predominate over rain-wash.

Measurement of the thickness of the material removed over a period of just over two years, during which 82 cm of rain fell on the area, revealed an exceptionally high rate of erosion, the thickness ranging from 2 to 3·8 cm at right angles to the surface. Figure 17.29 shows clearly that the Brule formation must have been worn back more rapidly than the Chadron, much of which still forms part of the landscape in front of the escarpment. Consistently, Schumm's observations indicate that the rate of present-day slope erosion on the Brule (mainly by rain-wash and rill-wash erosion) is nearly twice that on the Chadron (mainly by surface creep). In most areas, of course, both sets of processes are operative; but generally, as here, in widely different proportions. Parallel retreat of approximately straight slopes would appear to be favoured where rain erosion is dominant. Conditions appropriate to dominant surface creep, which favour slope retreat with declining angles, are of two kinds: (*a*) solifluction and related processes in the colder regions; and (*b*) a thick soil cover strongly protected by vegetation. The latter may be overcome in humid temperate and tropical regions by exceptional rainfall, especially if the underlying bedrock is such that infiltration is low. Figure 17.31 illustrates a case of parallel slope retreat in Sri Lanka, where protection by vegetation looks adequate, but where the rainfall is very high and the bedrock is a particularly massive quartzite. Schumm tentatively reached the conclusion previously suggested by C. D. Holmes (see p. 324) that 'the areas in which creep dominates over rain-wash, and vice versa, are end members of a continuous series ranging through all proportions of both processes' according to the prevailing conditions of vegetation, soils and climate.

### Sheet-Wash and Pediments

A curious feature that may have been noticed in Figures 17.29 and 17.30 is the sharp change of angle at the junction of hillside slope and pediment. This implies that all the material that reaches the base of the slope by rain- and rill-wash or creep is removed. Otherwise, as often happens, there would be some deposition at the base of the slope, resulting in a concave surface linking the steeper slope above to the more gently inclined pediment below. In Schumm's area the pediment in turn leads down to an actively eroding river which carries off all the debris supplied to it. This debris must therefore be transported across the pediment. Not only so, but Schumm found that during the period of his investigation the pediments themselves were lowered by nearly 2·5 cm in some places adjacent to a steeper slope. As each slope receded it left behind at its base a small extension of the pediment surface, the width of which ranged between 3·8 and 7·5 cm. Here, then, the pediment is a gently inclined surface that is gradually being extended at the expense of the steeper slopes as they recede. It is, moreover, a transportation surface.

The sharp break in angle between pediment and hill-slopes implies a sudden change in the processes responsible for the erosion and transportation, or in the properties of the materials being transported (or in some combination of both properties and processes). Schumm finds that during certain seasons of the year the hillside slopes have a relatively rough and permeable cover of aggregated soil, as compared with the smoother and less permeable surface of the pediment, where the soil aggregates are disintegrated by the greater energy of rain-splash, possibly assisted by changes in vegetation. On the steeper slopes these differences tend to reduce the run-off and its

**Figure 17.31** Heavily vegetated hill of quartzite with nearly straight steep slopes, heading in a slightly rounded crest of almost bare rock, near Sigiriya, Sri Lanka (Ceylon) (*Martin Hürlimnan*)

velocity, whereas on the pediments both run-off and velocity tend to be enhanced. If the relative smoothness of the pediment compensates for the gentle slope—given the right conditions of rainfall, run-off and level of the water table—a surface will be provided down which all incoming debris can be efficiently transported in the long run to the nearest stream. If the nature of the surface fails to compensate for the reduction of slope, a certain amount of deposition will occur at the foot of the hillside, thus giving a gradual change of slope down to the pediment (see p. 307). But if the conditions lead on balance to over-compensation, the pediment will be not only an efficient surface of transportation but also a surface subject to erosion from time to time by sheet-wash. Such conditions would favour the development and maintenance of an abrupt change of angle between hill-slope and pediment, and active investigations are now in progress to determine more precisely what these conditions may be.

The conditions under which pediments have developed on a large scale require a sudden change of process rather than of material. They have been studied chiefly by Lester King in parts of Africa where violent rainstorms quickly flood the ground over vast areas, so that the whole pediment surface acts like the floor of an extremely wide river channel. The resulting sheet-flood may be sufficiently vigorous and turbulent not only to pick up and transport any temporary deposits dropped during the waning stages of earlier floods, but also to accomplish a certain amount of sheet-wash erosion by removing rock-waste loosened from the bedrock by weathering. Between the steep sides of an inselberg and the gentle declivity of the surrounding pediment the change of angle is always impressive and often quite abrupt. At that point there is an obvious hydraulic 'break' in the behaviour of running water, such as occurs more familiarly in the plunge-pool at the bottom of a waterfall (p. 328). Heavy run-off, streaming rapidly down the smooth steep sides of an inselberg, suffers a sudden check at the bottom, where any accumulated debris is churned up and the fragments reduced in size. Even the transient rills that are largely responsible for the wearing-back of hillsides and escarpments are checked when they reach the bottom. The water then spreads out: perhaps to contribute to a sheet-flood if the rainfall and run-off are sufficient; perhaps to dwindle away, leaving a thin veneer of debris as they infiltrate into the soil and begin to co-operate with the processes of surface creep. The possibilities are both numerous and complex.

Selected References

DAVIS, W. M., 1930, 'Rock floors in arid and humid climates', *Journal of Geology* vol. 38, pp. 1–27 and 136–58.

— 1938, 'Sheet floods and streamfloods', *Bulletin of the Geological Society of America*, vol. 49, pp. 1337–416.

HOLMES, C. D., 1956, 'Geomorphic development in humid and arid regions', *American Journal of Science*, vol. 253, pp. 377–90.

KING, Lester C., 1953, 'Canons of landscape evolution', *Bulletin of the Geological Society of America*, vol. 64, pp. 721–52.

— 1957, 'The uniformitarian nature of hillslopes', *Transactions of the Edinburgh Geological Society*, vol. 17, pp. 81–102.

KRYNINE, D. P., and JUDD, W. R., 1957, *Principles of Engineering Geology and Geotechnics*, McGraw-Hill, New York.

SCHEIDEGGER, A. E., 1961, 'Mathematical models of slope development', *Bulletin of the Geological Society of America*, vol. 72, pp. 37–50.

SCHUMM, S. A., 1956, 'Evolution of drainage basins and slopes in Badlands at Perth Amboy, New Jersey', ibid., vol. 67, pp. 597–646; see also ibid 1962, vol. 73, pp. 719–24.

— 1956, 'The role of creep and rainwash on the retreat of Badland Slopes', *American Journal of Science*, Vol, 254, pp. 693–706.

SHARPE, C. F. S., 1938, *Landslides and Related Phenomena*, Columbia University Press, New York.

VARNES, D. J., 1958, 'Landslide types and processes', Chapter III of *Landslides and Engineering Practice*, Highway Research Board Special Report, Washington, D.C.

WARD, W. H., 1945, 'The stability of natural slopes', *Geographical Journal*, London, vol. 105, pp. 170–97.

WOOD, Alan, 1942, 'The development of hillside slopes', *Proceedings of the Geologists' Association*, London, vol. 53, pp. 128–39.

# Chapter 18
# The Work of Rivers

All these varied and wonderful processes, by which water mightily alters the appearance of the earth's surface, have been in operation since the most remote antiquity.

*Agricola*, 1546

## Processes of Erosion

Erosion is the cumulative effect of a great variety of processes. In general these can be conveniently divided into four groups: two involving the agent alone (chemical and mechanical activities); one combining the agent with its load of detritus, which arms it, so to speak, with 'tools' or 'teeth'; and one concerning the detritus alone. The four groups are listed on the following page for each of the chief agents of erosion, with the technical terms often used in describing them. Such a classification unduly emphasizes some of the distinctions, but it serves as a useful analysis provided that the general co-operation and overlapping that inevitably occurs in nature is not overlooked.

**Figure 18.1** The 'tools' of a river, left stranded in dry weather. Valley cut in boulder clay, Anglezark Moor, east of Chorley, Lancashire (*Institute of Geological Sciences*)

# Agents and Processes of Erosion

| *Type of erosion* | *Active agents* | *Agent alone* | | *Agent armed with detritus* | *Detritus alone* |
|---|---|---|---|---|---|
| | | Solvent and chemical action | Mechanical loosening and removal of materials passed over | Wearing of surfaces by transported materials | Mutual wear of materials transported |
| *Rain erosion* | Rain-water | Corrosion | Splashing Rain-wash Sheet-wash | Localized corrosion | Attrition slight, if any |
| *River erosion* | Rivers | Corrosion | Hydraulic lifting and scouring Cavitation | Corrasion | Attrition |
| *Glacial erosion* | Glaciers Ice-sheets | Corrosion limited to sub-glacial streams | Exaration (plucking and quarrying) | Abrasion (e.g. striated surfaces) | Attrition |
| *Wind (or Aeolian) erosion* | Wind | — | Deflation (blowing away) | Wind corrasion (sand-blasting) | Attrition |
| *Marine erosion* | Sea and ocean waves, tides and currents | Corrosion | Various hydraulic processes | Marine abrasion | Attrition |

*Note*—It should be noticed that the term *soil erosion* does not imply erosion *by* soil, but erosion *of* the soil, e.g. by wind, rain-wash, etc.

*Definitions*

*Abrasion* Wearing-away of surfaces by mechanical processes such as rubbing, cutting, scratching, grinding, polishing
*Attrition* Reduction in size of detrital fragments by friction and impact during transport
*Cavitation* Collapse of bubbles of water vapour in highly turbulent eddies of water; such collapse is like a negative explosion and sets up powerful shock waves which tend to disintegrate any adjacent rocks
*Corrasion* Cumulative effects of mechanical erosion by running water or wind when charged with detritus and so provided with 'tools' or abrasives
*Corrosion* Wearing-away of surfaces and of detrital particles and fragments by the solvent and chemical action of natural waters
*Deflation* Lifting and removal of dust and sand by wind (p. 473)
*Exaration* A term, now little used, for modes of erosion by glacial ice, akin to 'plucking' and 'quarrying' (p. 421)

## River Erosion

(*a*) *Corrosion*, as indicated in the table, here includes all the solvent and chemical activities of river water on the materials with which it comes into contact.

(*b*) Several *hydraulic* processes co-operate in bringing about the mechanical loosening, lifting and removal of materials by flowing water. Loose deposits are readily swept away, the initial lifting force being provided by turbulence, that is by eddies in which local velocities are rapidly changing and are often far higher than the rate of flow of the stream. Except when a river is actively downcutting its floor or undercutting its banks, it may not acquire much new material by erosion of its channel, but the coarser part of its load of debris is likely to be dropped again and again during transit (Figure 18.1), and each time the fragments have to be picked up afresh by the lifting forces before they can be transported farther downstream.

Changes of velocity during turbulent flow can produce some very remarkable effects. For example, a rapid increase of velocity is accompanied by a corresponding decrease of internal pressure. If the pressure falls below a certain critical amount, small bubbles of water vapour form and the water 'foams'. Sooner or later the velocity is decreased by friction against the floor or sides of the channel, the internal pressure increases again and the 'foam' becomes explosively—or rather implosively—unstable. The bubbles then suddenly and violently collapse, with production of shock waves which may shatter the adjoining surface with hammer-like blows, so liberating a crop of particles ready to be carried away. Wherever this process of *cavitation** occurs, the rate of erosion is greatly speeded up. Cavitation accounts for the hollows in stream beds that are later developed into pot-holes (Figures 18.2 and 18.3). It plays an important part in the erosion of plunge-pools below waterfalls (Figure 18.4) and it also helps to account for the extraordinary erosion achieved by turbidity currents on the ocean floor (p. 551).

A related application of the above principle is that water forcibly driven along a joint exerts a

**Figure 18.2** Pot-hole drilled in Tertiary plateau basalt, Glenariff, Co. Antrim, Northern Ireland. (*R. Welch Collection, Copyright Ulster Museum*)

**Figure 18.3** Pot-hole erosion of Carboniferous sandstone forming the bed of the River Taff, near Pontypridd, South Wales (*Institute of Geological Sciences*)

* To avoid possible misunderstanding it must be remembered that in ordinary foams the bubbles are occupied by a gas of different composition from the liquid, and cavitation does not then normally occur. Sea foam, for example, is mainly air in sea-water; its peculiar stability on certain beaches is due to the presence of protein in the films of water.

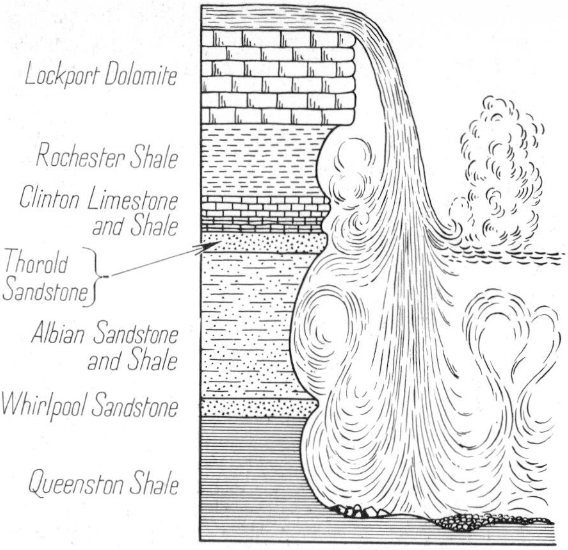


**Figure 18.4** Section across the Niagara Falls showing the sequence of formations, to illustrate the mechanism of recession by undercutting, and the erosion of the river bed in the 'plunge pool' beneath the falling column of turbulent water

very powerful pressure at the feather edge. The opening of the joint is extended and wedging action comes into play. Where the current is swift enough, as when a river is in full spate, the pressure of water driven along joints or bedding planes may be sufficient to liberate slabs of rock from the channel floor, or sides, and so prepare them for transport.

(*c*) *Corrasion* (an unhappily chosen term, too easily confused with 'corrosion') is the wearing away of the sides and floor with the aid of the boulders, pebbles, sand and silt which are being transported. With such tools even the hardest bedrocks can be excavated and smoothed. The drilling of *pot-holes* is one of the most potent methods of down-cutting. These develop in the depressions of rocky channels or from hollows started by cavitation and scouring. Boulders and pebbles, acting like drilling-tools, are rapidly swirled round by eddies (Figure 18.2). Vertical holes are cut deeply into the rock as the water plunges in and keeps the drilling-tools in action by its spiral motion. As the boulders wear away, and are swept out with the finer materials, new ones take their place and carry on the work. In front of a waterfall very large pot-holes may develop in the floor of the plunge-pool. This leads to deepening of the channel, and at the same time a combination of hydraulic and scouring activities undermines the ledge of the fall. The eddying spray behind the fall itself is particularly effective in scouring out the less resistant formations that underlie the ledge. Blocks of the overhanging ledge are then left unsupported and break off at intervals, thus causing a migration of the fall in an upstream direction, and leaving a gorge in front (Figures 18·5 and 18·6).

(*d*) *Attrition* is the wear and tear suffered by the transported materials themselves, whereby they are broken down, smoothed and rounded. The smaller fragments and the finer particles are then more easily carried away.

## Discharge and Transporting Capacity

The load carried by a river includes the rock-waste supplied to it by rain-wash, surface creep, slumping, etc., and by tributaries and external agents such as glaciers and the wind, together with that acquired by its own erosive work, as described above. The debris is transported in various ways. The smaller particles are carried with the stream in suspension, the tendency to settle being counterbalanced by eddies. Larger particles, which settle at intervals and are then swirled up again, skip along in a series of jumps—a process known as *saltation*. Pebbles and boulders roll or slide along the bottom, according to their shapes. Very large blocks may move along on a layer of cobbles which act like ball-bearings.

As the velocity of a river is checked, the bed load is the first to come to rest. With continued slackening of the flow the larger ingredients of the suspended load are dropped, followed successively by finer and finer particles. When the stream begins to flow more vigorously the finer materials are the first to be moved on again. In consequence a river begins to sort out its burden as soon as it receives it. The proportion of fine to coarse amongst the deposited materials tends on average to increase downstream, but there may be many local interruptions of this tendency because of additions of coarse debris from tributaries or from landslides and slumping of the banks.

It should be noticed that the term *load* does not specifically mean the maximum amount of debris that a stream *could* carry in a given set of conditions; that amount is referred to as the *transporting power* or *capacity*. When used by itself the term *load* is technically defined as the total weight of solid detritus transported in unit time

**Figure 18.5** General view of the Niagara Falls looking southwards. The America Falls on the left are separated from the Canadian or Horseshoe Falls by Goat Island. (*Photographic Survey Corporation Limited, Toronto*)

past the cross-section of the river at the place of observation. The corresponding quantity of water that passes is called the *discharge* and is usually represented by $Q$ and in English-speaking countries has been expressed in cubic feet per second. It includes all the material being transported in solution, such dissolved material being referred to as the *solution load* or *dissolved load*, to distinguish it from the load proper, which if necessary can be called the solid load. The latter consists of the suspended load plus the bed or traction load, which includes the coarser debris travelling by saltation, rolling and gliding. The traction load is difficult to measure and is generally taken to be not more than about a tenth of the suspended load. The following annual estimates for the Mississippi, averaged over a period of many years, are amongst those determined most accurately:

| | | *millions of tonnes per annum* |
|---|---|---|
| Dissolved or solution load | | 200 |
| Load or Solid load | Suspension load | 500 |
| | Traction or bed load | 50 |
| | *Total load* | 750 |

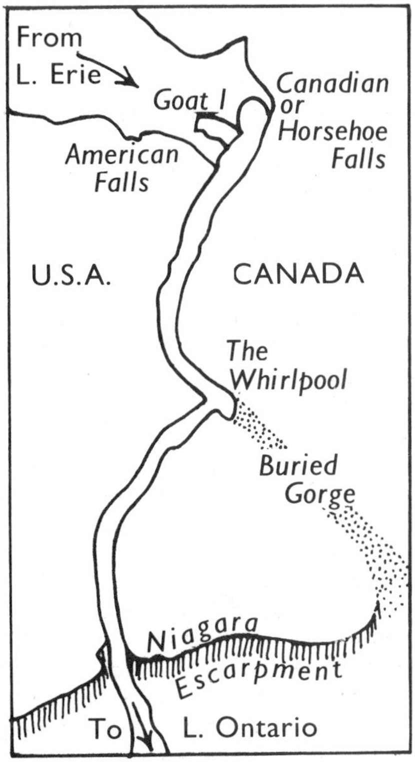


**Figure 18.6** Sketch map of the Niagara Falls and the 11 km gorge cut by headward recession since the Niagara River fell over the Niagara Escarpment about 12 000 years ago (see p. 338); looking southward as in Fig. 18.5. The Whirlpool is scoured out of glacial drift which now fills the Buried Gorge of an earlier 'Niagara River'.

The proportion of solid to dissolved load varies enormously from place to place and from time to time. Taking all the chief rivers of the world over representative periods of years, it appears that on an average about 8000 million tonnes of the products of rock-waste are transferred to the sea every year and that about 30 per cent of this total is carried in solution.

Both discharge and load depend on the climate and geology (lithology, structure and relief) of the river basin concerned, and both co-operate in carving out the channels down which the water and rock-waste are transmitted to the sea. The size and form of a river channel are determined by adjustments that are continually in progress between a bewildering variety of factors, the chief of which are the discharge, the mean velocity, the slope or gradient, the width, the mean depth, the load and the 'roughness' of the bed. 'Roughness' is a coefficient like that of friction, which expresses the resistance offered to flow; it depends on the varied sizes of the debris or other irregularities of the stream bed, being least when the bed is smooth. Indeed, there are so many factors—or variables—that it is only with the aid of experiments with models of rivers, and with the accumulated experience of the behaviour of real rivers, that river-control engineers can attempt to predict how a change in one factor is likely to affect the others.

It is found, for example, that on the brief occasions, generally seasonal but sometimes years apart, when the channel fills up to the brim (the 'bank-full' stage) and the stream overflows its banks in places, the changes brought about are far greater than those of the intervening months or years. The width in particular is likely to be increased by slumping and caving-in of the channel sides. Thus, in so far as the shape and gradient of a channel are determined by the discharge, it is the bank-full discharge that is the most effective agent, and the bank-full stage that provides the most satisfactory basis of comparison between one river and another, or for one river at different times. Now this is the condition associated with floods. The recurrent floods that afflicted many parts of Britain during the autumn of 1960 naturally provoked a widespread public outcry for systematic river control. One type of solution would be to try to forestall the possibility of bank-full discharge by increasing the size of the channels concerned. But such measures are extremely expensive. Moreover, if the width and depth are increased, the velocity is decreased; in consequence the new bed of an enlarged channel is liable to be silted up. Continual maintenance of the artificial interference is generally essential.

The Thames provides an instructive example. The mean discharge at Teddington is 2400 cubic feet (67 $m^3$) per second, but the bank-full discharge amounts to 10,700 (300 $m^3$). To protect London from floods the Thames Conservancy, amongst other measures, almost doubled the cross-sectional area of the channel for 12 miles (19 km) above Teddington, completing the work in 1939. This lowered the river by several feet (a metre or so): in fact within 2 or 3 inches (5 to 8 cm) of the predicted level. But to maintain this condition the Conservancy finds it necessary to excavate some 250,000 tonnes of sediment from the channel floor every year, and more than twice that amount from tributaries.

Another method of forestalling floods is to provide a new channel through which any dangerous excess of water can be diverted to a depression or an alternative exit. Mesopotamia, now inherited by Iraq, has a long record of floods dating back to the time of Noah. The melting of snowfields in the mountains to the north has been an almost annual menace to Baghdad. At the time of the great flood of 1954 a channel was already being dug from the Tigris to a large depression, 70 m deep, between the river and the Euphrates. In 1957 Baghdad would have suffered an even greater disaster had not the rising water been drawn aside before the danger level was reached.

The transporting capacity of a stream rises very rapidly as the discharge and the velocity increase. Experiments show that with debris of mixed shapes and sizes the maximum load that can be carried is proportional to something between the third and fourth power of the velocity. But for fragments of a given shape, the *largest size* that can be moved (not the total mass of mixed debris) is proportional to the sixth power of the velocity, provided, of course, that the depth of water is also adequate for the purpose. This explains how it is that enormous boulders, weighing many tonnes and for long periods looking like permanent features of a river bed, can be heaved up by exceptionally swollen torrents and carried a stage further downstream. When this happens (and at correspondingly lower velocities for smaller boulders) the whole bed load grinds and rolls along the channel floor. It is extremely dangerous to attempt to wade across a river with so treacherous a moving bed.

The almost incredible power of rivers in spate, especially when rushing down steep and narrow valleys, was tragically shown by the disaster that befell the north Devon seaside resort of Lynmouth in 1952. As a result of a downpour of about 22 cm of rain in 24 hours on Exmoor, when much of the ground was already waterlogged, the run-off became far greater than could be carried away through the existing channels. Part of the overflow from the West Lyn river concentrated itself into a new course and became a raging cataract that tore its way down the hillside and burst through the little town before rejoining the original channel (Figure 18.7 and see Figure 18.46). Roads, houses and bridges were demolished; service pipes and cables were destroyed; 40,000 tonnes of boulders, uprooted trees and soil, and great masses of collapsed masonry were churned up together in a bottle-neck entanglement (Figure 18.8). The sea wall was shattered to pieces and swept out to sea. After this grim account of a terrifying night of havoc, it would be unfair not to place on record that Lynmouth, rebuilt and adequately safeguarded, remained entirely unscathed during the phenomenal floods of 1960, referred to on p. 330.

**Figure 18.7** Showing on the right the new channel cut by the flood waters of the West Lyn as they swept down the hillside through Lynmouth on the night of 15/16 August 1952. The distribution of boulders and the vanished roads show the extent and violence of this devastating cataract when at its peak—only a few hours before the pictures Figs. 18.7 and 18.8 were taken. (*Syndication International Ltd*)

Figure 18.8 The maelstrom of destruction in the bottleneck of Lynmouth. (*Syndication International Ltd*)

## Rates of Denudation

The estimate of 8000 million tonnes for the materials annually transported to the sea by rivers is based on the data available for drainage basins having a total area of 104 million km². Averaged over this area the annual loss of rock-waste corresponds to 80 tonnes per square kilometre, which is equivalent to about 1 metre in 30,000 years. This estimate is practically the same as that arrived at for the whole of North America, and the data for the Mississippi drainage basin yield a result that is only a trifle lower. For individual drainage basins, however, the average rates of denudation vary enormously, as would be expected. The Irrawaddy-Chindwin rate is 1 metre in 1300 years, while for the low-lying areas draining into Hudson Bay the rate falls to 1 metre in over 150,000 years. In certain regions (see pp. 378 and 390) remnants of old uplifted erosion plains still survive; and except at their edges these may have remained almost untouched by denudation for millions of years. In the drainage basins themselves the maximum rate of erosion is on the steeper slopes, i.e. on the flanks of the bordering uplands, which may be conspicuous escarpments. Towards the coast, erosion again drops to a minimum and becomes negative at times when deposition of alluvium is the dominant process. It is, however, convenient for purposes of comparison to record rates of denudation as they would be if each were averaged over the whole of a given drainage basin.

Accurate measurements of the sediment accumulated in the great reservoir known as Lake Mead (Figure 18.37) show that during the years 1935–48 it amounted to 2000 million tonnes. To this has to be added another 150 million tonnes for the material removed in solution from the area drained by the Colorado river and its tributaries. In terms of unweathered rock the total corresponds to an average reduction of the whole basin (above the Hoover Dam) by about 1 metre in 7000 years. At the present rate of infilling the reservoir will last another three centuries.

For South Africa an average rate can be estimated for the period of about 75 million years which has elapsed since the diamond pipes were formed. As indicated on p. 181 the pipes originally penetrated some 900 m of higher strata before reaching the surface of that time. The average rate of denudation has therefore been about 1 metre in 80,000 years.

For the Mississippi basin an average rate for a still longer period can be arrived at from the volume of Mesozoic and later sediments deposited in the Gulf Coast Province, already referred to in connection with its salt domes (p. 158) and oilfields (p. 292). J. T. Wilson has estimated the volume accumulated since the early Jurassic, say during the last 175 m.y., as $5{\cdot}6 \times 10^6$ km$^3$. This corresponds to the removal of a layer 1·75 km thick from the whole drainage area of $3{\cdot}2 \times 10^6$ km$^2$; that is 1750 m in 175 m.y., giving an average denudation rate of 1 m in 100,000 years.

The above results, which are at least of the right order, leave no doubt that the present rates are far higher than those that can have prevailed during past ages. And this is what we should expect, for a variety of reasons.

(*a*) The continents are now more elevated and of greater area, relief and climatic contrasts than has been usual. The total energy of rivers available for erosion and transport is therefore abnormally high, and conditions also favour a vigorous circulation of ground-waters. During the long periods when vast areas of the continents were submerged and pre-existing mountains had been reduced to low-lying plains, rivers and ground-waters would be relatively sluggish and ineffective. There must have been times when the rates of both chemical and mechanical processes of weathering and erosion became almost negligible. Geological maps show clearly that, at least since the beginning of the Cambrian, immense areas that are now land must, on balance, have received deposits rather than supplied them.

(*b*) Extensive regions are now covered with glacial and glacifluvial deposits, most of which offer little resistance to erosion.

(*c*) Human activities have greatly increased rates of denudation by deforestation and agricultural activities; by excavations and other engineering projects; and by the addition to the atmosphere of carbon dioxide from factories and fires, together with various more corrosive gases.

In the light of these considerations it would seem likely that an average rate of 1 m in 100,000 years is more likely to be representative of the geological past than the average based on present rates. It must, of course, not be forgotten that any average of this kind may cover individual rates ranging between the extremes of no erosion at all and the swift cutting of a hillside gully during a sudden cloudburst. But even the adoption of the very low average here suggested leads to some astonishing conclusions. If one assumed, for example, that a rate of 1 m in 100,000 years had applied to the present area of land through the last 3000 m.y., it would mean the transformation into sediments of a slice of rock averaging 30 km thick, with the saline materials dissolved in sea-water as by-products. In other words the volume of material denuded during geological time would be about the same as the total volume of the continents, down to the Moho. Here we have a most intriguing situation that obviously calls for discussion in connection with the origin of continents and ocean basins. It may be said straight away, however, that the amount of sodium now dissolved in the oceans (plus the smaller amount of former oceanic sodium now imprisoned in evaporites and salt plugs) indicates that the thickness of the slice of rock so transformed—averaged over the present area of land—cannot have been more than 2·5 km. From this result (unless the amount of former oceanic sodium has been grossly underestimated) it is possible to draw either or both of two kinds of inferences:

(i) that the present area of land is several times as extensive as its average area since denudation first began; if this inference were taken by itself, 'several' would be at least 12, i.e. 30 km/2·5 km;

(ii) that the same material had been several times re-worked or re-cycled through the processes of denudation and deposition, with varying degrees of metamorphism and granitization in successive orogenic belts (see Figure 30.38, p. 695); here again, if this inference were taken by itself, 'several' would average at least 12.

Geochemical statistics suggest that re-cycling of the same materials has occurred on average about three times, in which case the 'several' of inference (i) would be about four. This combination of (i) and (ii) is much more probable than either of the extremes and is consistent with other evidence to which we shall come in later chapters. The conclusion that the present area of land is four times as great as the average area since denudation began may seem to be surprising, but it is well in line with the statement already made that 'since the beginning of the Cambrian, immense areas that are now land must, on balance, have received deposits rather than supplied them.'

Now, however, we can extend our conclusion to a time far beyond the Cambrian; perhaps to a time when no land had yet appeared; to a time when the continents were entirely submerged, as parts of them, like the North Sea and Hudson Bay, are submerged today. We catch a few fleeting glimpses of a fascinating moving picture reproducing the fluctuating 'separation of the dry land from the waters'. At first the lands emerge very slowly, but they become increasingly extensive as time goes on, despite repeated setbacks due to erosion and marine invasions, until today their area falls but little short of that of the continental regions as a whole. And today we see the rivers carrying away the land at such a rate that, had it continued throughout geological time—which of course it could not have done—the whole volume of the continents would have been poured into the oceans three or four times over. An impossible supposition, certainly; but nevertheless an impressive illustration of the immensity of geological time and a dramatic way of indicating the unusual vigour of present-day erosion and of emphasizing the work of rivers in removing the resulting burden of rock-waste.

## Base-Levels and 'Graded' Profiles

Since a river that flows into the sea must have a gradient towards the sea, the deepening of its valley is necessarily limited by sea-level. An imaginary extension of sea-level beneath the land surface is called the *base-level* of river erosion. The long profile of a river (i.e. the line obtained by plotting elevations—generally of the water surface—against their respective distances measured along the stream from, say, the mouth to the source) begins at sea-level, or just below if the profile is that of the river bed, and rises inland. The profile of a youthful river is likely to be more or less irregular, in conformity with the slopes and undulations of the initial surface, and the nature of the rocks being eroded. Characteristic features of youth are lakes and swamps, waterfalls and rapids. However, all but the greatest of these irregularities, such as very deep lakes, are destined to be smoothed out while maturity is being approached.

In humid regions the discharge of a river increases from the source to the mouth, where down-cutting, if any, is strictly limited. Starting on an initial surface with a general slope towards the sea, down-cutting is dominant along the middle reaches of the river. The effect is to steepen the gradient of the stretch leading down from the source, and to decrease the gradient from the middle reaches to the sea. Given ample time and no critical disturbances by earth movements or changes of climate or sea-level, it is not difficult to realize that the profile would be systematically modified until it approximated to a smooth curve, gently concave to the sky, practically flat at the mouth and steepening towards the source. When a river or, more commonly, a particular stretch of a river is found to have such a profile, it is conventionally said to be *graded* (Figure 18.9).

Admittedly this is only a rough-and-ready definition, but it gives the term *graded* an objective meaning that in practice has some useful applications. Any attempt at greater precision leads one into a maze of abstraction. The usual interpretation of the graded condition is that the gradient is continually being modified by erosion and deposition so that, allowing for seasonal and other fluctuations of discharge over a representative period of years, the stream is everywhere provided with just the right velocities for the eventual transportation of the rock-waste supplied to it. It must be clearly understood that 'the rock-waste supplied to it' is not restricted to the debris fed into the stream from its drainage basin, but includes all that may be liberated by the stream itself from the bedrock exposed to corrasion on the floor and sides of its channel when and where any temporary cover of alluvium has been swept away. Failure to recognize this part of a graded river's activities has sometimes led to the quite wrong idea that further lowering of a river bed must come to an end when a graded profile has been achieved.

It is now becoming generally realized that a smoothly curved profile cannot be interpreted in

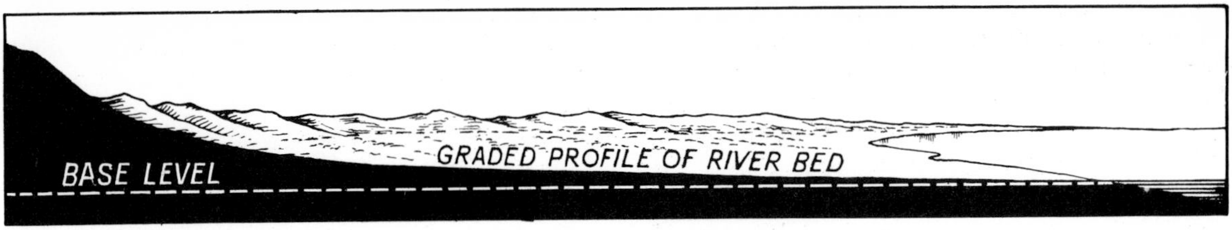


**Figure 18.9** Diagram to show the relation between base-level and an idealized graded profile (*After R. S. Tarr and O. D. von Engeln*)

terms of velocity and gradient alone. We have already seen (p. 330) that there are many other variable factors to be taken into account. For details of the most successful attempts to grapple with some of the complex and difficult problems involved in river hydraulics, reference should be made to the invaluable publications of Luna B. Leopold and his colleagues, listed on pp. 358–9. These workers have made it abundantly clear that 'the shape and pattern of the river channel are determined by the simultaneous adjustment of discharge, load, width, depth, velocity, slope, and roughness.' For a given discharge, the velocity depends not on the gradient alone, but also on all these other interrelated variables. The most one can say is that a river oscillates (with seasonal and other fluctuations) about a slowly changing condition of quasi-equilibrium. But this behaviour gives us no reason to suppose that it either requires or guarantees more than an approximation to the sort of idealized profile shown in Figure 18.9.

Some irregularities in the profile there are bound to be, if only because of abrupt changes of discharge and load introduced into a main river by its tributaries. Attempts to define grade in terms of a highly complex and theoretical state of equilibrium, which we could not recognize even if by chance it were temporarily achieved, cannot be correlated with the empirical definition of grade with which we started. The equilibrium concept of grade is a mathematician's dream that no river ever realizes. But if, on the other hand, we accept a profile approximating to a smooth curve as the criterion of grade, we are dealing with something real that can be readily recognized by field work or from the data recorded on accurately contoured maps, even though as yet we do not fully grasp its significance.

It is important to understand, as mentioned above, that erosional lowering of the channel floor does not cease when a river is graded, though it is negligible near the mouth and becomes slower and slower throughout the graded stretch as time goes on. We are still, for convenience, assuming that no major disturbances are introduced by earth movements, or by changes of climate or sea-level. Continued down-cutting is then ensured by two associated sets of circumstances: (*a*) the amount of debris and the sizes of the particles delivered to the river gradually decrease as the whole drainage area is being worn away; while (*b*) the stream flow or discharge remains about the same. With reduction of the available load, its finer state of comminution and the resulting reduction of the roughness factor, a considerably reduced velocity would be adequate to do the job of transportation, including that of the bed-load. The fact that the actual velocity is more than sufficient for this purpose means that the bed-load, which is the active agent of corrasion, can still abrade the bedrock wherever the latter is encountered on the river floor. Energy and tools are thus available for downward erosion and the gradient continues to be lowered. Consequently the graded profile can be slowly flattened out. Base-level, always being approached but never quite attained, is its only downward limit.

The level of the main river at the point where a tributary enters acts as a *local base-level* for that tributary. In the uninterrupted development of a river system, graded tributaries thus become so adjusted to the main stream that they join it tangentially or nearly so. When tributaries fail to behave in this way the absence of adjustment is a clear indication that the cycle of erosion has been interrupted by changes of slope or level due, as a rule, to earth movements or to glaciation. The waterfalls from 'hanging valleys' on the sides of deeply glaciated valleys (Figure 20.37, p. 426) are extreme examples of such lack of adjustment.

Various irregularities in a river channel may postpone the general establishment of grade, although above and below these features individual reaches of the river may be temporarily graded to the local base-levels controlling them. A lake, for example, acts as a local base-level for the streams discharging into it. Lakes that occupy

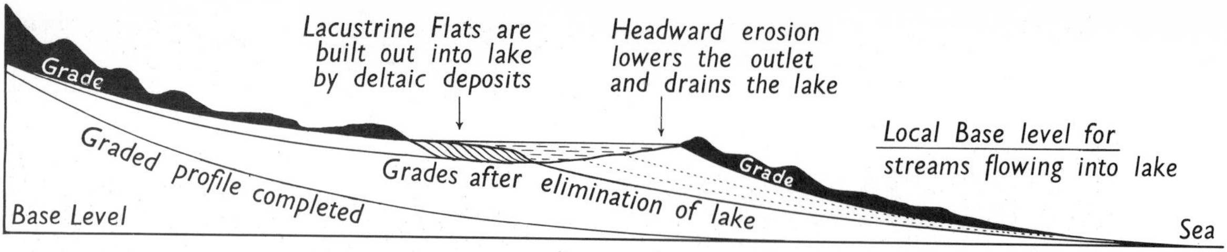


**Figure 18.10** Diagram to illustrate the elimination of a lake by sedimentation at the inlets and by headward erosion at the outlet. Successive positions of graded profiles before and after elimination are shown.

deep depressions have a very long life, but shallow ones are, geologically speaking, soon eliminated. A lake is a trap for sediments, destined to be silted up by deltaic outgrowths from inflowing streams. At the same time the outflowing water erodes the outlet and lowers its level, so that the lake is partly drained and its area reduced (Figure 18.10). Ultimately the lake is replaced by a broad lacustrine flat through which the river flows (Figures 18.11 and 18.12; see also Figure 17.17). The point where the lower graded profile intersects the upper one is called a *knick-point* (see Figure 19.20). Down-cutting through the lake sediments and underlying rock-floor then proceeds, and since this is most rapid where the slope is relatively steep in front of the knick-point, the latter migrates upstream until (in the absence of external interruptions) continuity of grade is established between the upper and lower reaches of the river.

A resistant formation encountered by a river also retards the establishment of grade and acts as a temporary base-level for the stream above until it is cut through by waterfalls and rapids. The latter persist so long as the outcrop of obstructive rock remains out of grade with the graded reaches in the softer rocks exposed above and below.

**Figure 18.11** Infilling of a lake by sedimentation. Head of Derwentwater viewed from the Wathendlath path, English Lake District (*G. P. Abraham, Ltd., Keswick*)

## Waterfalls

Figure 18.12 Alluvial flats occupying the site of a former lake, Borrowdale, looking north towards Derwentwater with Skiddaw in the background, English Lake District (*G. P. Abraham, Ltd., Keswick*)

Where an outcrop of resistant rock is underlain downstream by a weaker formation, the latter is relatively quickly worn down, and the resistant bed begins to be undercut. At the junction, and subsequently above it, the river bed is steepened and in this way *rapids* may be started. If the face of the resistant rock becomes vertical, the stream plunges over the crest as a *waterfall*. A waterfall is a knick-point of the most spectacular kind. The processes that bring about its recession upstream, leaving a gorge in front, have already been described (p. 328). Waterfalls eventually degenerate into rapids, which may persist for a long time before they are smoothed out and cease to make a break in the profile (Figure 18.13). A fall that descends in a series of leaps is sometimes

**Figure 18.13** Successive stages in the recession and elimination of a waterfall:
I  Profile of stream (drawn as graded) above an early position of the falls
II  Profile above present falls
III  Future profile after degeneration of the falls into rapids
IV  Future profile (graded throughout) after elimination of falls and rapids

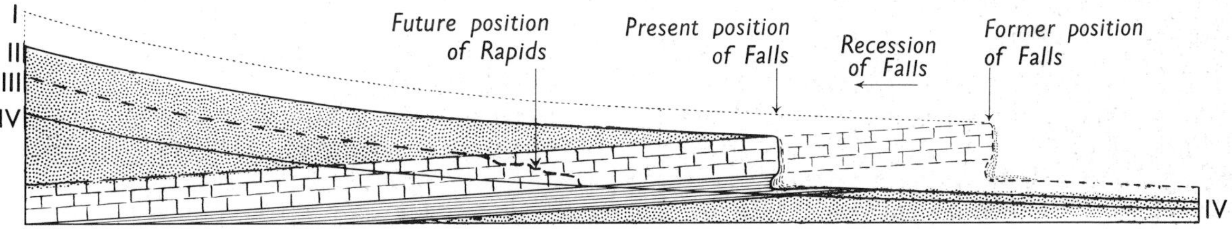

referred to as a cascade. An exceptional volume of water is implied by the term *cataract*, which may be applied either to waterfalls or, more commonly, to steep rapids. Rapids are characteristic throughout the wearing-through of an obstructive formation if its dip is downstream or steeply upstream.

Where a bed of strong rock, horizontal or gently inclined upstream, overlies weaker beds, the former is the 'fall-maker' and scouring of the softer beds underneath leads to undermining and recession. In the Yorkshire dales, for example, falls are well developed over ledges of Carboniferous limestone underlain by shales.

The *Niagara Falls* are the classic example of this type (Figures 18.4 and 18.5), the ledge being a strong dolomitic limestone of Silurian age. Before the diversion of much of the water to hydroelectric plants, the mean discharge over the falls (*c.* 5660 $m^3$ per second) was 85 times that of the Thames at Teddington. Most of the water passes over the Horseshoe Falls which, as their name implies, have been receding much more rapidly than the American Falls. After the last great ice-sheet withdrew from this region, and uncovered the pre-glacial Niagara escarpment about 12,000 years ago (p. 329), the Niagara river followed a course which descended from Lake Erie (174 m) to Lake Ontario (75 m) by way of a series of rapids and also one big vertical drop, where the river fell over the escarpment. Here the falls began and since then they have receded 11 km, leaving a gorge of which the rim is about 60 m above the river and on average about 110 m above the river floor (Figure 18.6). The rate of recession must have varied a good deal from time to time, but the measured rate for the Horseshoe Falls during the nineteenth century—about 1 m a year—is not far from the average of 0·9 m a year. Now, hydroelectric stations and the St Lawrence Seaway have greatly reduced the flow of water over the falls. Even so, the falls will recede until Lake Erie is reached and partly drained—an event too many thousands of years ahead to cause any present anxiety.

The *Kaieteur Falls* on the Potaro river in Guyana are also of this type, although in this case the ledge, over which the river makes a sheer leap of 244 m, consists of hard, flat-lying conglomerate and massive sandstones, underlain by softer shales (Figure 18.14).

The *Yellowstone Falls* (Upper, 33 m; and Lower, 94 m) are cutting through an immensely thick succession of ignimbrites, parts of which have been altered and weakened, and at the same time gorgeously coloured, by chemical changes due to thermal waters. Hot springs still emerge along the floor of the canyon, which in places is 780 m deep. The fresh layers of ignimbrite, being more resistant, are the fall-makers (Figure 18.15). Rugged spurs and pinnacles are left as the walls of the canyon are worn back, all vividly splashed with colours of every hue.

**Figure 18.14** Kaieteur Falls, Potaro River, Guyana (*John Smart*)

Uplifted areas of plateau basalts may provide the structural conditions for a long series of falls. The strong and compact internal parts of flows make ideal ledges, which are undermined by the more rapid removal of the far less resistant vesicular or amygdaloidal margins. The many falls of Glenariff, where the river is vigorously descending step by step through the Tertiary basalts of the Antrim plateau, would be hard to surpass for their

**Figure 18.15** Great Falls, Yellowstone River, Wyoming (*T. L. Bierwert, American Museum of Natural History, New York*)

**Figure 18.16** 'Tears of the Glen' waterfalls, Glenariff, Co. Antrim, Northern Ireland (*Northern Ireland Tourist Board*)

varied and exhilarating scenic attractions (Figure 18.16).

The *Victoria Falls* in Rhodesia, which have already cut back some 130 km through basalts of Karroo age, owe their unique features to structural controls of a quite different kind. Here the Zambesi drops 110 m, along a frontage of 2 km, from a nearly level basaltic plateau into a narrow gorge which is remarkable in being parallel to the falls (Figure 18.17). A break on the downstream side leads the river into a series of gorges through which it rushes as a powerful torrent with surging rapids at intervals (Figure 18.18). The reason for the astonishingly acute swerves is to be found in the zones of structural weakness that characterize the plateau basalts of this region. Along narrow zones in one direction the rocks are strongly jointed, while along shatter zones following another direction they are fractured and brecciated. Both directions lie athwart the general course of the river and the falls have receded alternately along one direction and the other, everywhere following the line that furnished the most easily liberated masses of basalt. Occasionally a third direction of special weakness and least resistance has been encountered by the river, like that between the present falls and the railway bridge (Figure 18.18). At the western end of the present falls the river is just beginning to cut back along another shatter zone. If this zone continues obliquely across the river, as it appears to do from the disposition of the islands, it will become the next stretch of the zigzag gorge.

Where rivers pass from uplifted highlands of metamorphic and massive igneous rocks to a coastal plain of weakly resistant sediments, waterfalls are initiated and often gain height as they recede. The rivers flowing from the hard rocks of

**Figure 18.17** Victoria Falls and zig-zag gorge of the Zambezi River, Rhodesia. Length of present falls 2 km; height 110–120 metres according to the discharge and depth of the river. While the successive zig-zags of the gorge have been cut by the recession of the Falls, practically no denudation of the surrounding erosion surface—and ancient pediplain—has taken place. (*Aircraft Operating Company, Johannesburg*)

**Figure 18.18** Another view of the Zambesi Gorge and Victoria Falls (*R. U. Light. American Geographical Society*)

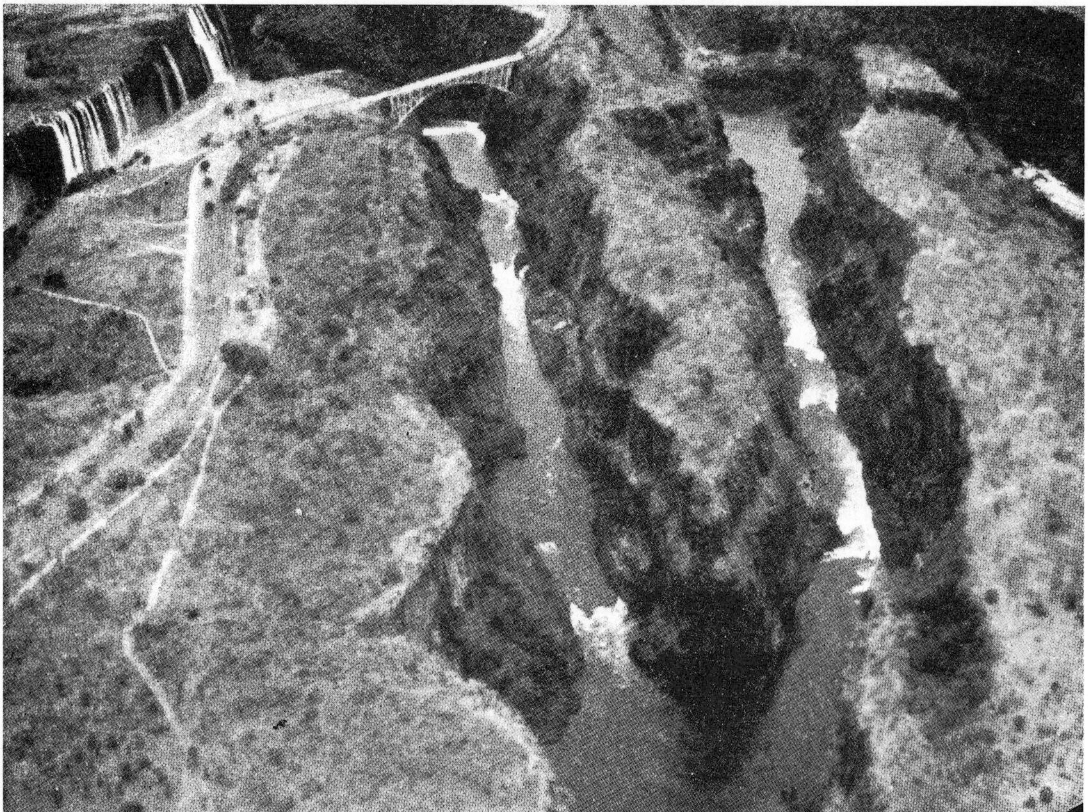

the Appalachian uplands have thus developed rapids and falls along the 'fall-zone' before they reach the softer sediments of the Atlantic coastal plain. More spectacular examples of this type are the *Aughrabies Falls* (140 m) of South Africa, where the Orange River passes into a grim and desolate gorge of naked granite and gneiss (Figure 19.44), and the *Gersoppa* or *Jog Falls*, on the frontier of Mysore and Bombay in the Western Ghats, which have a sheer drop of 250–260 m, according to the level of the water in the plunge-pool below. When in flood during the monsoon season the Jog Falls could be claimed as the greatest of the world's waterfalls, because of the combination of exceptional height with exceptional volume. In the dry season, however, the falls dwindle to a few ribbon-like trickles, some of which lose their continuity by passing into spray and mist before reaching the bottom of the rock face.

There are many falls of comparable or even greater height, but most of these result from relatively small tributaries or headwaters falling over precipieces already prepared for them by glacial erosion (Figure 20.36), by faulting, or by the long-continued recession of an escarpment at the expense of the uplands of a region that was rising while headward erosion was in progress (Figure 19.42).

## River Bends and the Widening of Valley Floors

In the last chapter the various processes that bring about the widening of valleys by slope recession were reviewed. Here we are concerned with the work of the rivers themselves in extending their channels sideways and so widening the valley floors at the expense of the pre-existing valley sides. Youthful streams rarely follow a straight course for any considerable distance unless they happen to be flowing rapidly down a steep gradient in a direction favoured by the structures of the rocks through which they are cutting their channels. A stream is likely to encounter belts of least resistance or rocks that are particularly obstructive, and these tend to predispose it to change its direction and so develop a winding course (Figure 18.19). But such obvious physical aids and restraints are not essential for changes of direction, as work with experimental streams in long tanks has clearly revealed.

**Figure 18.19** Steep-sided youthful valley with overlapping spurs, Crossdale Beck, Ennerdale, English Lake District (*Institute of Geological Sciences*)

An experimental stream which begins by flowing down a straight channel on a uniform slope of uniform sediment is always found to modify its channel by erosion and deposition so that a series of roughly symmetrical bends—like a sine curve—is developed (Figure 18.20). This curious behaviour is, in fact, a direct consequence of the general principle that when one medium moves over another the plane of contact tends to be shaped into a wave-like form. To be effective in this way the relative movement must be adequate to overcome such resistances as friction, viscosity and strength. There are many familiar examples. A light breeze blowing across a smooth sea ruffles the surface into ripples that quickly rise into waves as the wind increases. Wind blowing over loose sand builds up sand-dunes with rippled surfaces. The backwash of sea waves running down the slope of a beach leaves a characteristic surface of ripple

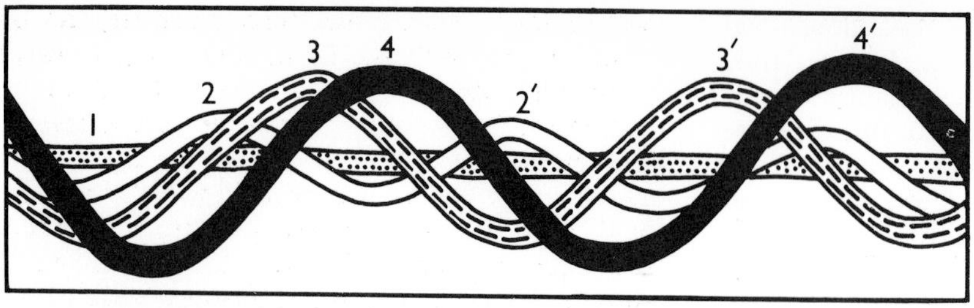


**Figure 18.20** Successive stages (1, 2, 3, 4) in the experimental development of incipient meanders in a model stream flowing in a bed of uniform sediment. The meanders migrate downstream and the wavelength (2 to 2′; 3 to 3′; 4 to 4′) increases until it is about ten times the width of the bank-full channel.

marks. Water falling freely through the air, as from a circular tap, at once begins to change its momentary cylindrical form, It oscillates in and out with rapidly increasing agitation until the pulsating column can no longer hold together, but breaks into drops. This example emphasizes the fact that the phenomena we are concerned with are three-dimensional in space; four-dimensional when time is included.

The oscillations of streams of running water on land are further complicated by the various factors listed on p. 335. Amongst these the roughness of the bed and the nature of the load are particularly important in controlling what happens. Suppose we start with an experimental stream flowing down a gently sloping straight channel floored with sand. Like the water from a tap the stream oscillates 'in and out', i.e. up and down and also sideways, but much more slowly and less obviously. However, clear manifestations of wave-like forms soon appear. Down the channel alternating shoals and hollows develop, corresponding to the up-and-down oscillation. Moreover, the shoals are alternately built up a little on one side of the middle line and a little on the other: an indication of the sideways oscillation. This change in the floor affects the flow of the stream, which concurrently begins to swing from side to side. The main current winds round the shoals, first on one side and then crossing over to the other side of the next shoal, and so on. Wherever the current impinges against the banks of the stream, lateral erosion occurs and bends are started. In this way the channel develops a serpentine form as seen in plan (Figure 18.20). The distribution of shoals and hollows is now controlled by the swinging habit of the current. As the curvature of a bend is increased by lateral erosion, the channel on the outside is deepened by vertical erosion, while on the inside it is shallowed by deposition (Figures 18.21 and 18.22).

It will be seen from Figure 18.20 that the successive wavelengths, 2-2′, 3-3′, 4-4′, gradually lengthen, and that the bends themselves migrate downstream. This migration steadily continues, but the wavelength adjusts itself to the width of the channel (bank-full stage) and does not permanently increase beyond a certain average limit, which is about 10 times the width of an average stream, the factor being less for narrower streams and more for wider ones (Figure 18.23). In their studies of natural rivers Luna B. Leopold and his colleagues find that straight reaches have scoured-out hollows in their floors alternating with shoals. Shoals (or hollows) are spaced at distances which bear the same relationships to widths as the

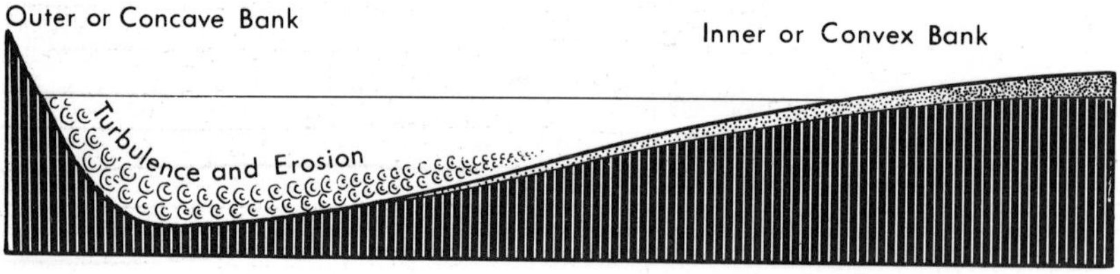


**Figure 18.21** Section across the bend of a river to show the asymmetrical profile due, on balance, to erosion on the outside and deposition on the inside. In floods turbulence and erosion extend to the inner bank.

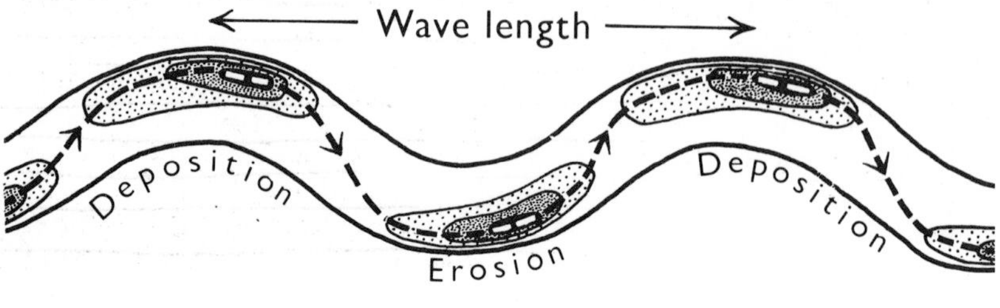


**Figure 18.22** Diagram to illustrate the effects brought about by water flowing down a meandering channel. Depth of water is indicated by depth of shading. The broken line indicates the axis of flow under normal conditions. With rising water the axis widens into a belt that may extend to the convex banks and remove some of the alluvium previously deposited there.

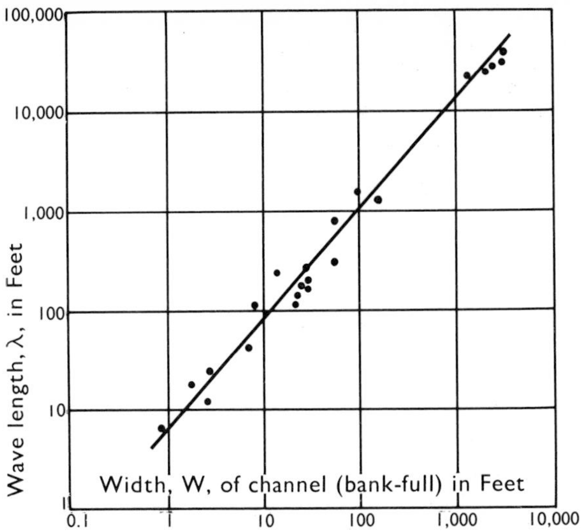


**Figure 18.23** Logarithmic graph of the wavelength of meanders, $\lambda$, plotted against the corresponding widths, W. The equation of the line is of the general form
$\log \lambda = \log a + b \log W$
($a$ and $b$ being constants) which is the logarithmic form of
$\lambda = aWb =$(in this case) $6{\cdot}5\ W^{1{\cdot}1}$
(*After L. B. Leopold and M. G. Wolman;* reference on p. 358)

wavelengths of bends described above. In such reaches the rivers are evidently on their way to becoming sinuous. Where they are beginning to develop meanders in neighbouring parts of their courses, the wavelengths correspond very closely with twice the distance between successive shoals (Figure 18.22).

Another discovery relating to the flow of water round a bend is that the position and extent of the belt of high velocity and turbulence depend on whether the discharge is (*a*) increasing towards bank-full conditions and a state of flood, or (*b*) decreasing from those conditions towards a state of low water. It is natural to suppose that the main current of the stream is deflected towards the outer bank, some way beyond the beginning of the bend, and that erosion will take place there while alluvium is being deposited on the inside of the curve. But this is only part of the story. With increasing rate of flow the maximum current impinges farther down the outside of the bend. The velocity also increases on the inside of bends until, during floods, it may be little less, if any, than on the outside. Shingle and sand previously deposited on the inside are then being scoured away. Some bedrock erosion may even occur locally. With decreasing flow the belt of high velocity contracts again towards the outside of the bend, and while the outside or concave bank continues to be eroded, the convex bank receives a new supply of sediment from upstream.

Under the varying conditions indicated above, the resulting changes in the river channel and its valley floor may be summarized in a general way as follows:

1. The channel is deepened towards the outer side of each bend (Figures 18.21 and 18.22), and particularly along the downstream part of the bend.

**Figure 18.24** Alluvial deposits built on the inside of a river bend, seen when the river is low; Glen Feshie, a tributary of Strath Spey, Highland Region, Scotland (*Institute of Geological Sciences*)

2. The outer bank is worn back and undercut by lateral erosion, with local development of river cliffs or 'bluffs', which are liable to cave in during or shortly after floods and thaws. At the same time banks of shingle or sand are built out on the inner sides (Figure 18.24).

3. As the bends are thus widened out by lateral erosion and deepened by downward erosion on the outside, and built out and shallowed on the inside, the river shifts its channel not only towards the outer bank but also downwards (Figure 18.25). Eventually this oblique migration of the channel leaves gently rounded slip-off slopes on the inside of the growing curves; classic examples are those of the Rhine (Figure 17.3) and of the Wye (Figure 18.26). In cross-profile the valley, as well as the channel floor, has become highly asymmetrical at every bend.

4. Since each bend is increasingly enlarged in the downstream direction, it gradually migrates downstream in serpentine fashion. It can now be realized that the changes in a river channel are three-dimensional: downwards (vertical erosion); sideways (lateral erosion); and along (downstream migration of bends). At an early stage, when there are overlapping spurs like those illustrated in Figure 18.19, each spur is eroded and undercut every time it is reached by the outer part of a migrating bend. Eventually, as bend after bend migrates downstream, the spurs are all trimmed off. Similarly the slip-off slopes are all cut back—apart from a few uplifted relics that have survived from earlier cycles of erosion (p. 380)—and a nearly flat trough-like valley floor is developed. The alluvial deposits extend into continuous stretches, with broad embayments bounded by bluffs or hills (Figures 18.27 and 18.28). The beginnings of a flood plain have been established.

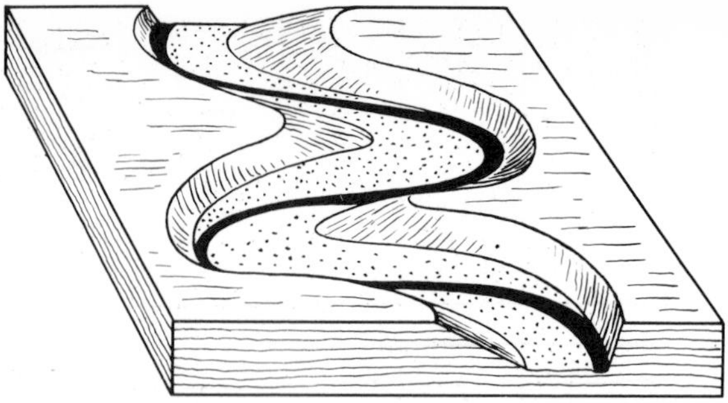

**Figure 18.25** Diagram to show the widening and deepening of a valley floor by oblique erosion (vertical and lateral) on the outside of river bends combined with the downstream migration of the bends. The swinging stream eventually trims off the overlapping spurs and slip-off slopes.

## Free Meanders and Meander Belts

As the river continues to swing from side to side of its valley floor, it undercuts the bordering bluffs wherever a bend impinges on them. So the widening of the valley floor proceeds, while the slopes above are slowly receding. The channel is

**Figure 18.26** The horseshoe bend of the River Wye at Wyndcliff, near Chepstow, Gwent, showing well-developed bluffs of Carboniferous Limestone and a gently inclined slip-off slope. Compare Fig. 17.3 (*Institute of Geological Sciences*)

**Figure 18.27** Meanders of the River Dee, near Braemar, Grampian Region (*The Scotsman, Edinburgh*)

**Figure 18.28** The meander belt of the River Glass; a flat-bottomed trough-like valley floor, veneered with alluvium, Strath Glass, Grampian Region (*Institute of Geological Sciences*)

now mainly in river deposits, bedrock (if any) being exposed only in and below the bluffs. Each part of the material deposited on the growing flood plain is worked over in turn during the downward sweep of the bends, fresh additions from above constantly making good the losses by erosion and transport. The bends, now free to develop in any direction, except where they encounter the valley side, are more quickly modified. Freely developing bends are called *meanders* from their prevalence in the classic river Meander, the Menderes of today, in the south-west of Turkey.

As we have already seen, bends develop into meanders in which 'the wave length is a function of the width' to a degree of approximation that is remarkable. As a direct result of this relationship, a freely meandering river flows in a *meander belt* (Figure 18.28; see also Figure 18.27), which is usually about 15 to 18 times the bank-full channel width. There are, as we have seen, many other factors besides width and discharge involved under natural conditions, and these are responsible for local and temporary irregularities. But as straight stretches soon begin to develop meanders (as indicated in Figure 18.20), there is always, barring natural 'accidents' or human interference, an approach to the relationships established experimentally. However, there is one type of 'crisis' or emergency that commonly arises in the development of freely meandering rivers. Erosion inevitably picks out the materials, and follows the structures, that offer least resistance. Consequently many meanders continue to swell out into broad loops with gradually narrowing necks, as illustrated in Figure 18.29. If a flood occurs when only a narrow neck of land is left between adjoining loops, the momentum of the increased flow is likely to carry the stream across the neck and thus short-circuit its course. On the side of the 'cut-off' a deserted channel is left, forming an *ox-bow* lake (Figure 18.29), which soon degenerates into a swamp as it is silted up by later floods. By making artificial cut-offs the Mississippi River Commission has straightened long stretches of the river and so shortened transport distances, both by road and river, by hundreds of kilometres.

The relics of many old ox-bows, indicating the positions of abandoned meanders, may no longer be obvious topographic features on the ground. But their outlines can be clearly seen from the air (Figure 18.30). This is because the soil and drainage of an infilled ox-bow lake differs from that of ordinary river alluvium, a difference to which vegetation visibly responds. Ox-bow lakes and their overgrown relics constitute the most convincing evidence of the reality and geological importance of the downstream migration of meanders.

As the bordering slopes of the valley are still cut back every time an individual meander impinges against them, the valley floor continues to be widened. After an immensely long period of time, with no important tectonic or climatic changes, it may become several times as wide as the meander belt. This indicates that as the meander belt becomes 'free', the belt itself proceeds to 'meander' to and fro across the widening flood plain, swinging down the latter towards the sea in curves of much longer and more variable wavelengths than those of the actual stream. It should be noticed that while lowering of the channel floor is limited by base-level, there is no such restriction on erosion of the banks. Widening can therefore continue long after significant lowering has ceased.

What appeared to be an interesting example of the meandering of a meander belt was provided by the discovery of the buried ruins of the ancient city of Ur of the Chaldees—the home of Abraham—in Mesopotamia. 5000 years ago the Euphrates flowed 8 km to the west of Ur; today it is 8 km east of the ruins. But there is another possibility to be considered. It is known that the full-bank width of a meandering river is roughly proportional to the square root of the discharge. If the rainfall were

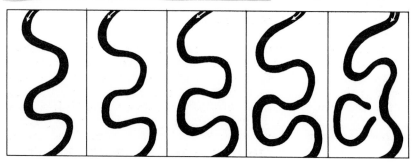

**Figure 18.29** Successive stages in the development of meanders, showing the formation of an ox-bow lake by the 'cut-off' of a loop

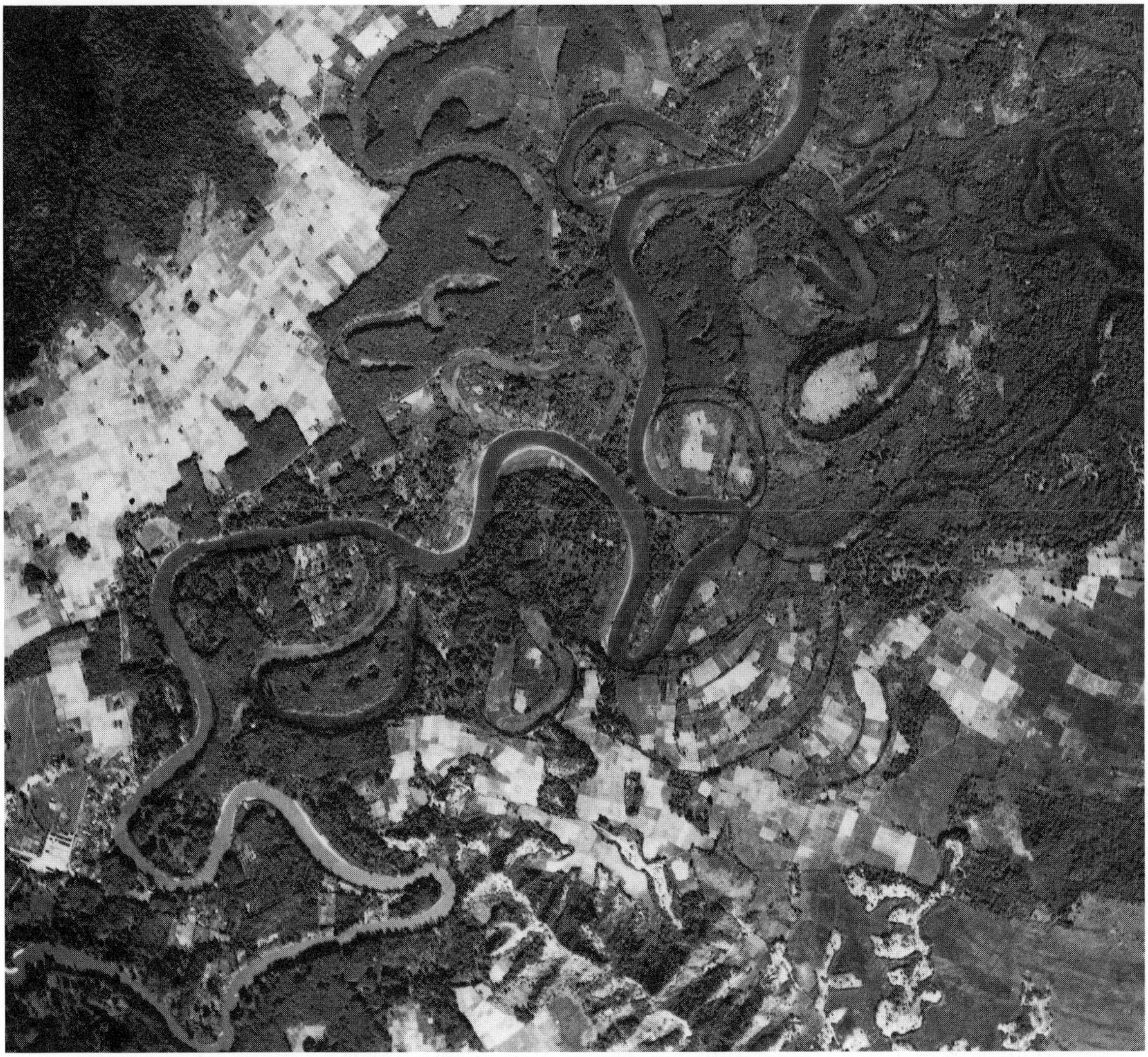

greater at some time in the past, the run-off would also be higher, and the discharge would therefore be very considerably increased. Doubling the rainfall would at least quadruple the discharge, so doubling the width of a river and consequently that of its meander belt. Some exceptionally wide valley floors are now ascribed to the meanderings of ancestral rivers which once had channels several times as wide as those of today. This interpretation, which has been supported by archaeological discoveries in deserts and semi-arid regions, is consistent with other evidence that at various times during the Pleistocene and even in the recent past, the rainfall in these regions was very much greater than it is today.

**Figure 18.30** Meandering stream with numerous ox-bow lakes and many that have been infilled with sediment and more or less obscured by vegetation; NE of Kota Kinabalu (Jesselton), Sabah, Malaysia (*R.A.F., Crown Copyright reserved*)

## Braided Rivers

Meandering very considerably lengthens the river concerned and so reduces its gradient or slope. This is only one of the ways in which a river can accommodate itself to prevailing conditions so as to achieve an approximation to equilibrium: that is, so as to become graded and remain graded. Another method is for the river to divide into an interlacing network of distributaries with shoals

**Figure 18.31** Braided headwaters of the Rangitati River, Canterbury, New Zealand, where they emerge from glacially eroded valleys in the Southern Alps, the floors of which, like that of the broader valley in front, have been built up by heavy deposition (*New Zealand Geological Survey*)

**Figure 18.32** Braided channel of the Rakaia River, incised in the alluvium of the Canterbury Plain (and locally superposed on the underlying bedrocks), between the Southern Alps of New Zealand and the sea (*V. C. Browne*)

and islands of shingle and sand between. The river is then said to be *braided* (Figures 18.31 and 18.32). The river channel as a whole is characteristically wide and shallow. In keeping with this the floor and outer banks are composed of incoherent sediments, generally consisting wholly or partly of the river's own deposits. Where a river, heavily laden with debris from a mountain range, emerges from a canyon or ravine on to the bordering plain, its velocity is suddenly checked by the abrupt change of gradient and a large part of its load of sediment, including all the coarser debris, is therefore dropped. Along the fronts of youthful scarps and ranges such debris is commonly spread out in the form of alluvial fans (p. 356). While thus alleviating the sudden change of slope by deposition, much of which is obstructive, the stream divides into branches which continually separate and reunite.

Another situation which favours braided drainage is in front of melting ice-caps (Figure 18.33) and of the snouts of certain glaciers (Figure 3.5). At the other end of a river, towards its mouth, braiding may be characteristic for long distances, like an inland anticipation of a delta. The Ganges and the Amazon are good examples.

In experiments designed to reproduce braiding the water fed into the channel at the head of a long tank is suitably charged with sediment from the start. This seems to have the effect of intensifying the up-and-down oscillation, while damping down the sideways oscillation. However this may be, elongated sandbanks are deposited at intervals down the middle of the channel and these divert

**Figure 18.33** Braided drainage through the outwash deposits south of Vatnajökull, Iceland. The two outlet glaciers are Skaftafells on the west (left) and Svinafells on the east; both are west of Oraefajökull (Fig. 3.5). See Fig. 12.9 for localities. (*Erlingur Dagsson*)

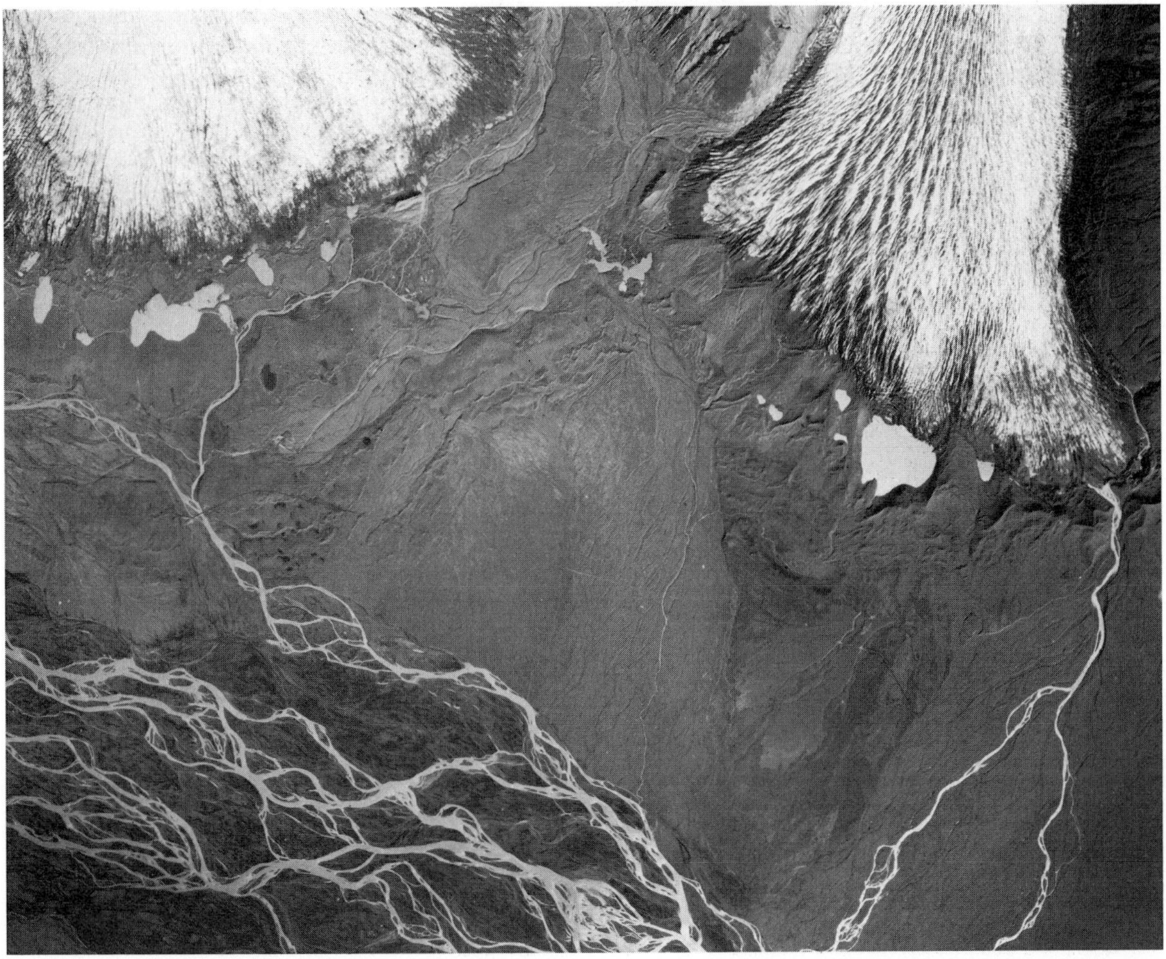

the flow of water to each side. Where the flanking channels become deepened by floor erosion the level of the water drops a little and the adjoining shoal may become an island. But lateral erosion of the outer banks also occurs, thus widening the channels and so providing materials for shallowing them and building up new sandbanks. Each sandbank migrates downstream as a result of erosion at its upper end and deposition at its lower end. The distribution of material changes, and some of it is continually being carried away, but the 'form'—the sandbank or island—may persist for a long time. Because of this migration the all-over widening process also migrates downstream, since each of the outermost banks continues to be eroded laterally by the stream directed towards it by the nearest sandbank.

Actual braided rivers follow this pattern of behaviour. During low water the channel is mainly sandbanks, as if the river had been choked with excess of sediment. But with high water, and especially during floods, new channels are cut and the islands, generally for the most part submerged, migrate downstream. A braided river becomes graded by depositing part of its load until the width, depth, slope and velocity are modified so that, on average, the given discharge and load can be transmitted with a close approach to equality of deposition and erosion. In saying this, however, it must be repeated that in the long run there always remains a slight balance in favour of erosion, unless this tendency is reversed by a change of climate or a rise of base-level. As Leopold remarks: 'Braiding does not necessarily indicate an excess of total load'.

Figure 18.34 embodies a remarkable range of evidence showing that meandering and braided channels can generally be distinguished in terms of only two of the many factors concerned. With a few exceptions, where the general rule is modified by one or more of these other factors, the diagram makes it clear (*a*) that for a given bank-full discharge braids usually require a higher gradient than meanders; and (*b*) that for a given gradient braids require a higher bank-full discharge than meanders. What is surprising is that there should not be more exceptions.

Amongst the factors not taken into account in Figure 18.34, the more critical ones are roughness of the floor and bed-load, and the ratio of all-over width to average depth, Bottom friction is greatly increased when a river is wide and shallow and has a rough floor. These handicaps must be compensated by a higher discharge and/or gradient to ensure the transmission of a given load. It is worthy of notice that a meandering river maintains a fairly steady width because erosion and widening of the channel on one side is balanced by deposition and narrowing on the other. In the development of braiding, however, a river increases its width because erosion and widening occur on both sides at once. The whole régime of sandbanks and inner channels eventually reaches an all-over width that meets the requirements of all but the very greatest floods. This width is likely to remain fairly stable until there is a really catastrophic flood. When this happens it is quite likely to upset the whole of the pre-existing régime by diverting the river into an entirely new course.

The bracketed B and M of the Kosi river in Figure 18.34 illustrate the effect of load and roughness in deciding the pattern of behaviour when gradient and discharge are essentially the same. The Kosi, also known as the Sapt-Kosi, because it is fed by the union of seven rivers, is the most powerful tributary of the Ganga (Ganges)

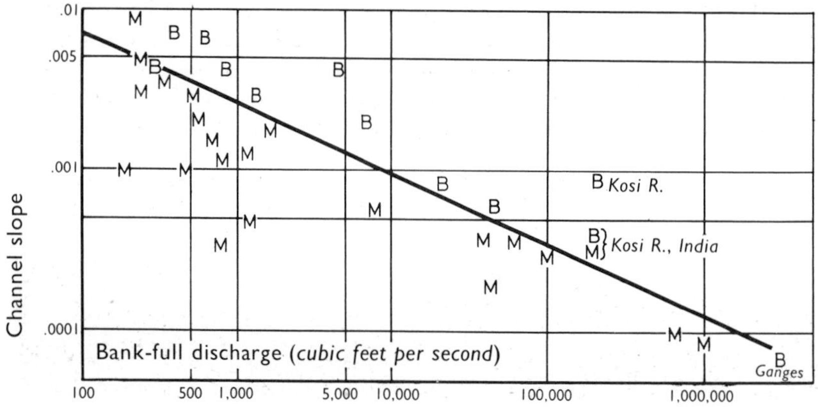


**Figure 18.34** Logarithmic graph of the gradient of a wide variety of rivers plotted against the corresponding discharges: M, meandering; B, braided. These types are separated by a straight line, except for one or two extreme examples at each end. For comparison the Thames at Teddington falls near the middle of the diagram in the field of meandering rivers. (*After L. B. Leopold and M. G. Wolman*; reference on p. 358)

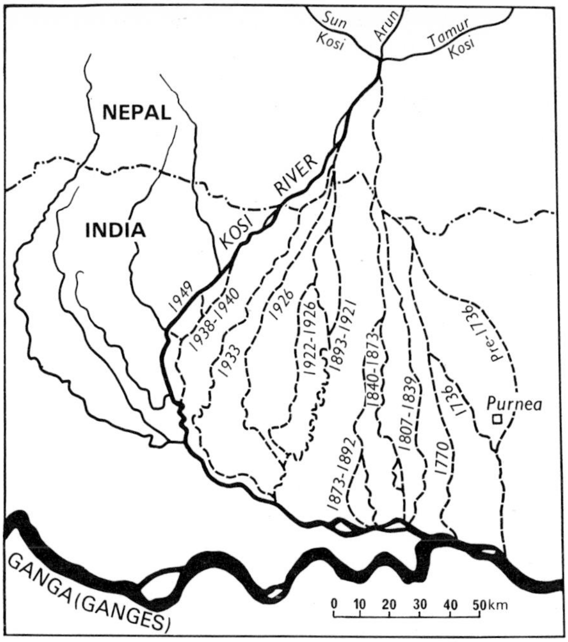


**Figure 18.35** Sketch map to illustrate the lateral migration of the Kosi River across its own alluvial deposits. The latter begin near the boundary between Bihar (India) and Nepal, which rougly corresponds to the southern edge of the Siwaliks, the foothills of the Himalayas.

(see Figure 19.36). Four major western Kosi rivers unite with the Sun Kosi, which in turn is joined by the Arun from the north and the Tamur Kosi from the east (Figure 18.35). The greatly enlarged river emerges on the plains near Chatra and flows for 120 km over a gently sloping alluvial fan of its own construction in a broad, conspicuously braided channel. By the time the Kosi has reached the foot of its alluvial fan most of its gravel and coarser sand has been deposited, and from there to its junction with the Ganga the river follows a meandering course. The Kosi is notorious for its frequent and often disastrous floods and some of these have altered the course of the river. Two centuries ago it flowed near Purnea (Figure 18.35), but now its main braided channel is 96 km to the west. By the construction of a barrage and 240 km of marginal embankments, it is hoped to tame the Kosi and utilize its immense water power.

## Flood Plains

The development of flood plains by the sweeping of meanders to and fro down the valley floor, and later, perhaps, by the swinging of the meander belt itself, has already been mentioned. As the valley floor is widened in this way the slip-off slopes become nearly flat and the flood plain may then have only a relatively thin veneer of mud and silt, generally with an underlying deposit of sand and gravel, covering a planed-off surface of bedrock. This rock surface results from the slight erosion that occurs each time the bedrock is uncovered at the deepest part of a meander loop which has reached the bordering bluffs (Figure 18.21). Braided streams provide thicker and coarser deposits. But the thickest spreads of alluvium are found where, for some reason, the level of the bedrock floor has been lowered. A common reason is crustal depression in front of mountain ranges, as exemplified by the Indo-Gangetic trough and its infilling of sediment. Another is the lowering of sea-level corresponding to the formation of great ice-sheets. This lowering of base-level stimulated the rivers to deepen their channels. Later, when sea-level rose again with the melting of the ice, these channels filled with sediment. In the Lower Mississippi valley the alluvium is 122 m thick where it fills a channel cut by the river when sea-level was 122 m lower than at present.

Deposits left on the flood plain by flood water outside the actual channel are described as *over-bank deposits*. These are generally very thin, except adjacent to the river banks; and if the flood is exceptional the alluvium swept away may exceed the amount that is afterwards left by the muddy waters as the flood subsides. Each time the river overflows its banks the current is checked at the margin of the channel, and the coarsest part of the load is dropped there. Thus, a low embankment or *levee* is built up on each side (Figure 18.36). Beyond the levees the ground slopes down, and in consequence is liable to be marshy. During floods, levees may grow across the junctions of small

**Figure 18.36** Schematic section across the flood plain of a stream bordered by natural levees. Vertical scale greatly exaggerated.

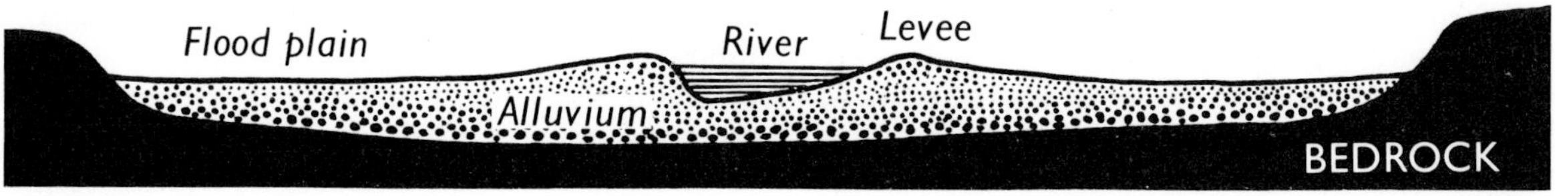

tributaries. The latter are then obliged to follow a meandering course of their own, often for many kilometres, before they find a new entrance into the main river. Depressions occupied on the way become ponds which degenerate into 'back swamps'. The characteristic features of the flood plain of a meandering river include ox-bow lakes and marshes, levees, back swamps and shallow lakes, and a complicated pattern of lateral streams. Levees are not constructed by typically braided rivers, as deposition from the latter tends to be concentrated within the channels.

Levees afford protection from ordinary floods, but it has been found by hard experience that confining the channel raises the water level (for a given discharge). Moreover, in some cases, e.g. the Mississippi, the river begins to silt up its floor with material that would otherwise have been spread across the plain as overbank deposits. This is possibly due to slight tilting of the region towards the sea, since it it known that many deltas are slowly subsiding. However this may be, the bank-full level is raised, and the danger from major floods becomes greater than before. For further protection artificial levees may be built, but these provide only temporary security, since they accentuate the tendency of the river to raise its level. In the flood plains of the Po in Italy and of the Hwang Ho and Yangtze Kiang in China the built-up levees are locally higher than the neighbouring housetops, and the rivers flow at a level well above that of the adjoining land. Such conditions are obviously extremely dangerous, as the swollen waters of a severe flood may breach the defences and bring disaster to agricultural lands over an enormous area.

Along the Mississippi and the Missouri and their tributaries the flood danger has become an increasingly serious menace. Little more than a century ago floods were easily controlled by levees about a metre high. The levels have since had to be raised several times. By 1927 they were three or four times as high, but nevertheless a great flood then broke through in more than 200 places, and devastated 50,000 square kilometres of cultivated land. Stronger and higher levees, up to 6 or 9 m, have now been constructed, but it is clear that by itself this method of flood control is far from satisfactory. It can, however, be supplemented by reafforesting the upland regions (to reduce the proportion of run-off) and by dredging, cutting through meander loops, and making vast storage reservoirs for flood waters.

The destructive flood of July 1951, in which over 200,000 people lost their homes, was due to an unprecedented rainfall of 47 mm in one night: like the Lynmouth disaster (p. 331), but over a much greater area. Most of the Mississippian floods, however, result from a combination of heavy rainfall with melting of the winter's snow. If, in addition, the snowfall has been abnormal and is followed by a widespread thaw, then the flood crests of all the tributaries may unite with that of the main river (instead of coming one after another in a drawn-out succession, as usually happens). An exceptionally heavy flood is then inevitable.

## Deltas

The essential condition for the growth of a delta is that the rate of deposition of sediment at or just beyond the mouth of a river should exceed the rate of removal by waves and currents. Where a river flows into a freshwater lake the current is usually checked and much of the load may be quickly deposited (Figure 18.11). Salt lakes provide still more favourable conditions. As river water mixes with saline water the finer particles of clay flocculate into aggregates that are too large to remain long in suspension. Rapid sedimentation is therefore promoted. A striking example of this effect is the Terek delta (between Baku and the Volga), which at one time was visibly advancing into the Caspian Sea at the rate of 1·5 km every 5 or 6 years. As the front has now reached much deeper water its outward growth is naturally slowing down, but even so the rate is still about 10 times that of the Rhône delta at the head of Lake Geneva.

These two examples illustrate another contrast that is of primary importance. Charged with suspended sediment, the Rhône has a milky appearance as it flows into the lake; and because of the higher density its waters visibly dive under the clear lighter water of the lake, forming what is called a *turbidity current*. This cloudy current flows through an 8 km trench towards the deepest part of the lake. Only a small proportion of the sediment—the coarser part—contributes to the outgrowth of the visible delta. Turbidity currents are frequently developed where the Colorado river enters Lake Mead (Figure 18.37). The murky inflowing water travels down the floor of the reservoir along the depression marking the submerged channel of the river, and builds a deposit of deltaic structure which extends as far as the

**Figure 18.37** Lake Mead 185 km long, held back by the Hoover (Boulder) Dam, the construction of which across the Colorado River was completed in 1935. The river, heavily laden with sediment, enters at the northern end as a turbidity current which can often be clearly seen from the air. By 1960 about 3500 million tonnes of sediment had been deposited. Left to itself the lake would be entirely infilled by the year 2250 (*Fairchild Aerial Surveys, Los Angeles*) For location, see Fig. 19.32.

Hoover Dam and so is extremely long compared with its width. In these examples and many others the river water, together with its suspended load, is 'heavier' during most parts of the year than the relatively still water into which it flows.

The Caspian water, however, is saline and has about the same density as that of the mud-laden Terek water. Here and in all similar cases the river water fans out on a broad and expanding front; it is rapidly slowed down by mixing with the relatively still body of water through which it is spreading. Sediment is therefore mainly deposited in a broad arc surrounding the mouth of the river. Having blocked its only means of exit, the river breaks through, so making new channels which then proceed to obstruct themselves in the same way. This is the essential mechanism by which deltas of the *arcuate* type are built up (Figure 18.38).

There is still a third possibility: when the river water is less dense than the salt lake or sea into which it is flowing. The fresh water, with a light suspended load of fine mud, then flows over the surface of the salt water, meeting little resistance in front, but rapidly losing velocity sideways. Deposition therefore occurs mainly along the sides of

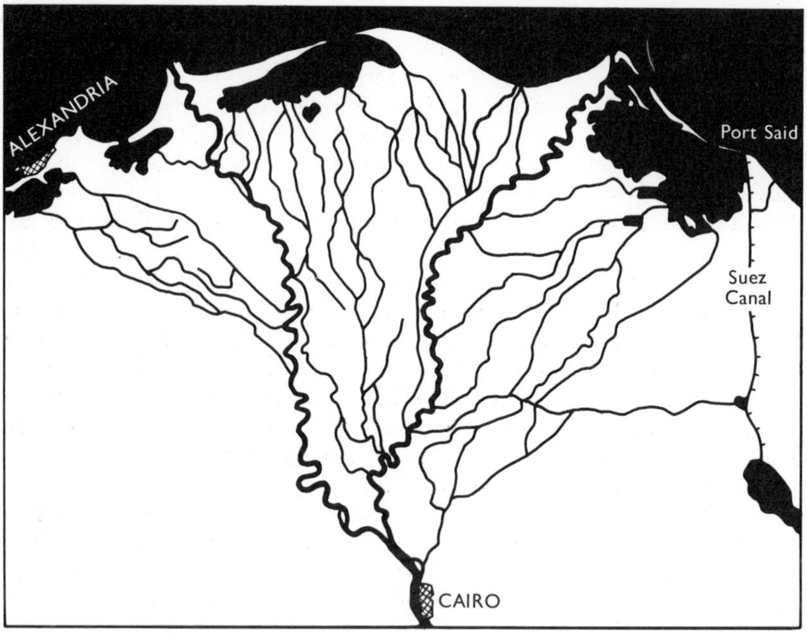


**Figure 18.38** The Nile Delta: a classic example of arcuate type.

the channel, until low levees are built up. Sooner or later these are breached and the resulting delta grows outwards by way of relatively deep channels resembling outstretched fingers. This is the *bird's-foot* type of delta (Figure 18.39).

The above summary is a very much simplified account of a hydrodynamical treatment by C. C. Bates (see p. 358) who regards the flow of river water into, let us say, the sea, as if it were akin to the discharge of a turbulent jet of fluid into a relatively still body of another fluid. There are then three possibilities:

(*a*) if the river water is the denser (because of its load of sediment) it flows along the bottom as a turbidity current and forms an elongated deposit of the deltaic type;

(*b*) if the river water has about the same density it spreads out fanwise and an arcuate delta is formed;

(*c*) if the river water is the less dense it makes a few confined channels for itself and a bird's foot delta is formed.

There are, of course, other considerations to be kept in mind. The coagulating effect of salt water has already been mentioned. Waves and currents may sweep away the incoming sediment before there is time for its accumulation above sea-level. The rise of sea-level that accompanied the melting of the last great ice-sheets of Europe and North America must have submerged a great many slow-growing deltas that have not yet had time to reappear. Moreover, the crust itself is sagging beneath many deltas, and not merely because of

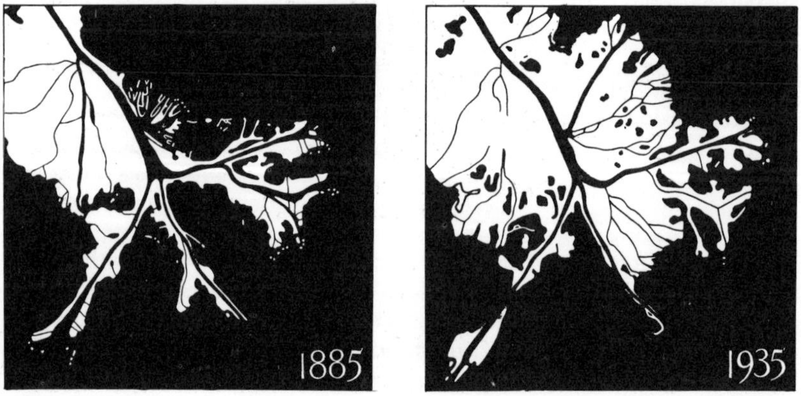


**Figure 18.39** Fifty years' growth of the Mississippi Delta: a classic example of the bird's foot type

the weight of sediment. Even when full allowance is made for the thick wedges of light sediment, it is found that the Mississippi, Nile, Indus, Irrawaddy and Ganga-Brahmaputra deltas are all areas of abnormally high gravity. Their tendency is therefore to subside and so to make room for sediments to accumulate and deltas to thicken—as indeed many deltas have already done to a most astonishing degree.

The delta of the Nile originally received this name because of the resemblance of its shape to the capital form of the Greek letter delta. It is the prototype of all deltas, but particularly of the arcuate type. It has an arc-like outer edge, modified by fringing sand-spits shaped by marine currents (Figure 18.38). After transversing 1600 km of desert, the Nile has comparatively little water left when it reaches the apex of the delta. Much of the remaining load is deposited before the sea is reached and frontal growth of the delta is correspondingly slow. Such continued accumulation of sediment is possible only because the region is subsiding. Subsidence must, indeed, have been in progress here for a very long time, as seismic exploration has revealed a thickness of at least 3000 m of poorly consolidated sediment. Frontal growth is therefore slow. The Po delta extends more rapidly. Adria, now 22 km inland, was a seaport 1850 years ago, the average rate of advance thus indicated being about 1 km in 75 years. Ostia, the seaport of ancient Rome, is now 6 km from the mouth of the Tevere (Tiber). The richly fertile delta of the Huang Ho in north China has grown across what was originally a broad bay of the Po Hai (Yellow Sea). A large island, now the Shantung peninsula, has been half surrounded. Since 1852 the main branch of the river has emptied to the north of Shantung, but before then it flowed across the southern section of the delta and reached the sea bout 480 km from its present mouth (Figure 18.40). In that year and again in 1887 there were calamitous floods in which the loss of life from drowning and famine amounted to many hundreds of thousands. Floods over the Ch'ang Chiang (Yangtze) delta, farther south, have also brought repeated disasters to the inhabitants. In contrast, the annual flood of the Nile was a disaster only when it failed. If, like those mentioned above, a river has a flood plain, the visible part of its arcuate delta is the seaward extension of the plain. Figure 18.41 illustrates the way in which a broad outward-sloping fan of sediments is gradually built up on the sea-floor.

The chief distributaries of the Mississippi delta (Figure 18.42), locally called *passes*, make a typical bird's-foot pattern. The sediment brought down by the river is rich in very fine mud and this confines the channels within impervious banks of

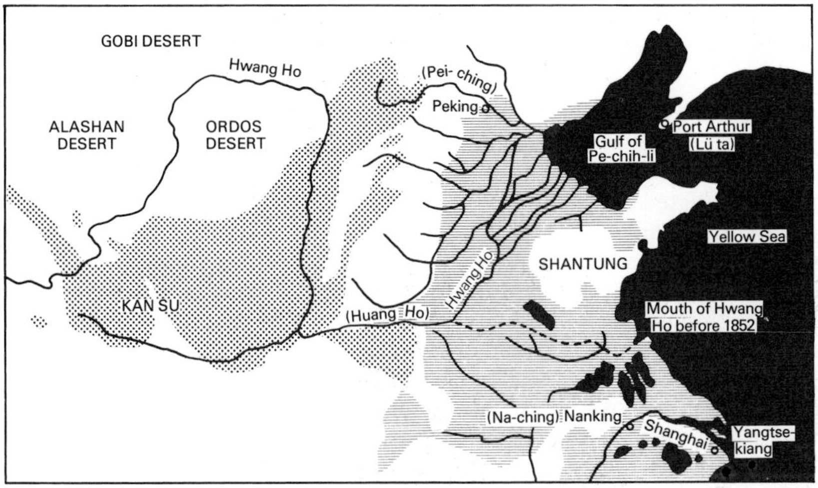


**Figure 18.40** Map of the Huang Ho (Hwang Ho) and its delta, showing the distribution of loess (dotted) and alluvium (horizontal shading) derived from loess (*After G. B. Cressey*)

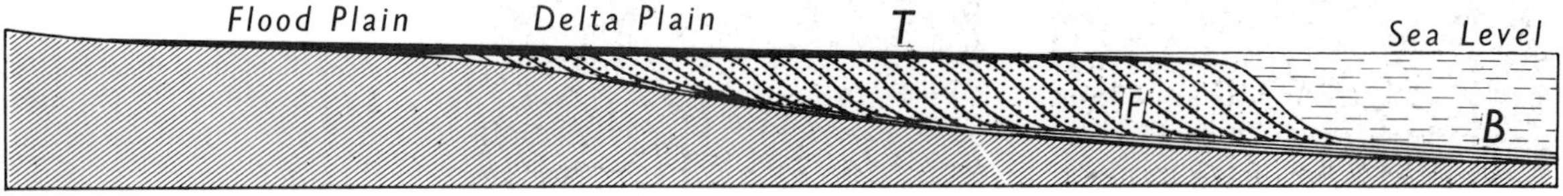


**Figure 18.41** Idealized section through the sediments of an arcuate delta: the *Foreset* beds, F, grow seawards (like an advancing railway embankment during its construction); finer sediment is deposited farther out on the sea floor as *Bottomset* beds, B; and a veneer of *Topset* beds, T, forms the seaward continuation of the flood plain.

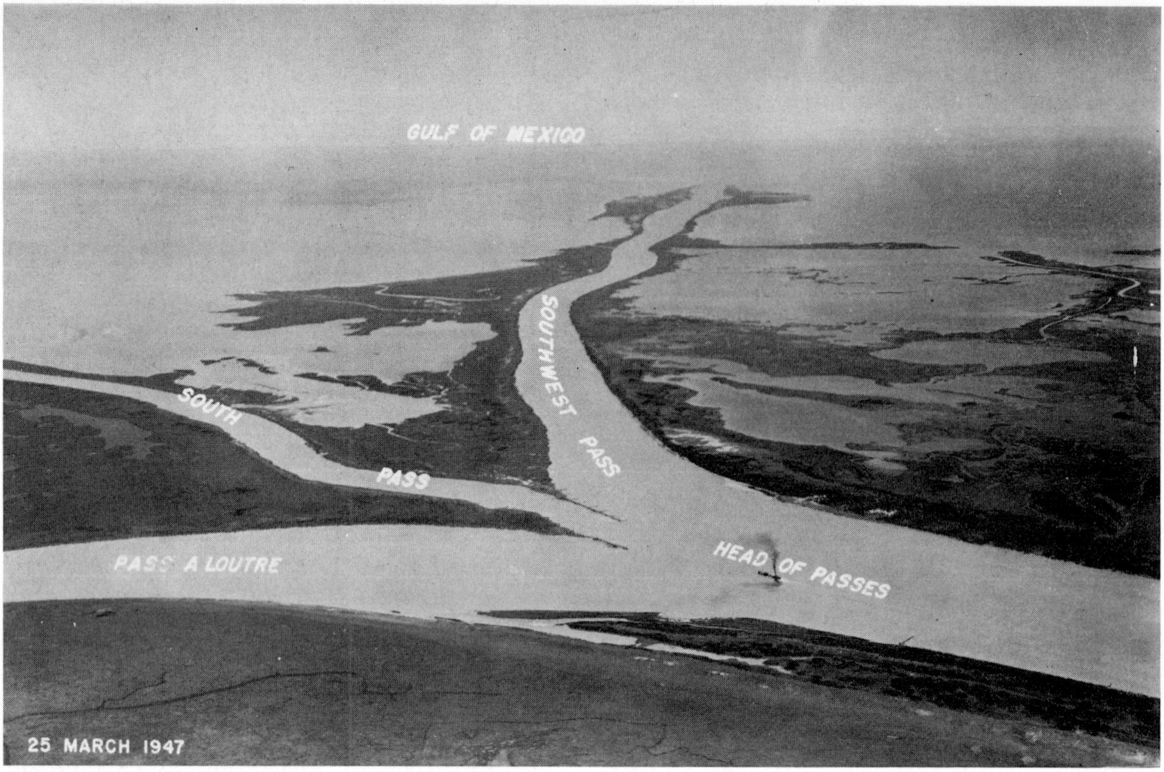


**Figure 18.42** Aerial view of part of the Mississippi Delta, showing the chief 'passes' and the levees by which they are confined (*Photograph by Corps of Engineers, U.S. Army, New Orleans District*)

clay. Because of the great depth of the Gulf of Mexico a good deal of sediment continues to be carried forward, so that slow deposition is in progress over a very wide area. The Mississippi delta is slowly subsiding and here, too, an immense thickness of deltaic sediments has accumulated. In references to the Gulf Coast salt-dome province, the region is often regarded as part of a modern geosyncline that has not yet passed beyond the stage of sediment accumulation. Though much of the region is now above sea-level, downwarping of the crust still continues on the seaward flank. Here the Mississippi is adding sediment to the floor of the Gulf of Mexico at the rate of 1 $km^3$ every 4 years or so. Were it not for the submarine downwarping, outward growth of the delta and of the continental shelf would have been on a far greater scale. There have been many interruptions due to changes of sea-level during the Pleistocene, but these can be dealt with more conveniently when the effects of glaciation are being considered.

## Alluvial Fans and Cones

Many youthful mountain ranges and plateaus descend steeply to the neighbouring lowlands; generally, but not always, because they are bounded by eroded fault scarps. Where a heavily laden stream reaches the plain after flowing swiftly through a ravine or canyon, its velocity is checked, it widens and much of its load of sediment is dropped. The stream may become braided and keep changing its course, like the notorious Kosi river (p. 351); or, as in delta formation, it may divide into branching distributaries; or it may behave in both ways. The deposited sediment is spread out as an *alluvial fan* (Figure 18.43). If the circumstances are such that most of the water sinks into the porous alluvium, while the rest evaporates—as commonly happens in arid or semi-arid regions—then the whole of the load is dropped. The structure gains height and becomes an *alluvial cone* (Figure 18.44). There are all gradations from wide fans 26–260 km across that are usually nearly flat (slopes less than 1°), through fans of moderate width and inclination (4–6°), to relatively small steep-sided cones (up to 15°) built

**Figure 18.43** Alluvial fan, eastern side of Death Valley, California (*J. S. Shelton and R. C. Frampton*)

**Figure 18.44** Alluvial cones at the mouths of canyons in southern Utah (*After C. E. Dutton, United States Geological Survey*)

of the coarser debris brought down by short torrential streams.

Where closely spaced streams discharge from a mountainous region across a *piedmont* (a mountain-foot lowland), their deposits may eventually coalesce to form a *piedmont alluvial plain.* Such is the vast crescent of boulder beds, gravel, sand, silt and clay known as the Indo-Gangetic Plain, which extends from the delta of the Indus to that of the Ganga-Brahmaputra. This has long been a 'foredeep' between the loftiest mountains in the world to the north and the Peninsular block to the south. Although the floor has sagged between 1800 and 3000 m during Tertiary times, the depression has been infilled by alluvium transported by the copious drainage from the Himalayas and supplemented on the south by minor contributions from the Peninsula.

Alluvial fans and cones are well displayed in western America: along the eastern base of the Rockies, where the Front Range faces the Great Plains; along the eastern edge of the Sierra Nevada, where the eroded fault scarp slopes down to the Great Basin; and in many other similar situations on the flanks of the block mountains of the Great Basin (Figure 29.36). They are also developed on a large scale at the foot of the various ranges of the Andes. Where rivers flow along or near the foot of the slope, many examples are to be found of river deflection due to the growth across their channels of alluvial fans deposited by tributaries from the mountains. Where there is a lake in a similar situation, an alluvial fan begun on the land passes into a delta which grows across the lake. Familiar examples which have divided what were originally single lakes into two are the fan-deltas of Buttermere in the English Lake District, separating Buttermere from Crummock Water (Figure 18.45); and of Interlaken in Switzerland, separating Lake Thun from Lake Brienz. A fan-delta that has spread into the sea at Lynmouth, where the East and West Lyn rivers rapidly descend from the plateau which they drain, is illustrated by Figure 18.46.

**Figure 18.45** Buttermere and Crummock Water illustrating the division of a lake into two by the growth of a fan-delta (*G. P. Abraham, Ltd., Keswick*)

## Selected References


BATES, C. C., 1953, 'Rational theory of delta formation', *Bulletin of the American Association of Petroleum Geologists*, vol. 37, pp. 2119–62.

BLENCH, T., 1957, *Regime Behaviour of Canals and Rivers*, Butterworth, London.

CRICKMAN, C. H., 1975, *The Work of the River*, Macmillan, London.

DURY, G. H., 1954, 'Contribution to a general theory of meandering valleys', *American Journal of Science*, vol. 252, pp. 193–224.

HACK, J. T., 1956, 'Studies of longitudinal stream profiles in Virginia and Maryland', *United States Geological Survey Professional Paper 294-B*.

KIMBALL, Day, 1948, 'Denudation chronology: the dynamics of river action', *University of London Institute of Archaeology, Occasional Paper No. 8*.

KIDSON, C., 1953, 'The Exmoor storm and the Lynmouth floods', *Geography*, vol. 38, pp. 1–9.

LEOPOLD, L. B., and WOLMAN, M. G., 1957, 'River channel patterns: braided, meandering and straight', *United States Geological Survey Professional Paper 282-B*, pp. 39–85.

**Figure 18.46** Lynton and Lynmouth, North Devon, showing the plateau drained and eroded by the East and West Lyn rivers, and the fan-delta built into the Bristol Channel at their common outlet. See also Fig. 18.7 (*Aerofilms, Limited*)


— 1960, 'River meanders', *Bulletin of the Geological Society of America*, vol. 71, pp. 769–94.

LEOPOLD, Luna B., WOLMAN, M. G., and MILLER, J. P., 1964, *Fluvial Processes in Geomorphology*, Freeman, San Francisco.

MACKIN, J. H., 1948, 'Concept of the graded river', ibid., vol. 59, pp. 463–512.

MENARD, H. W., 1961, 'Some rates of regional erosion', *Journal of Geology*, Chicago, vol, 69, pp. 154–61.

POWELL, J. W., 1895, *The Exploration of the Colorado Valley and its Canyons*, Dover, New York.

RASHLEIGH, E. C., 1935, *Among the Waterfalls of the World*, Jarrolds, London.

SCHEIDEGGER, A. E., 1970, *Theoretical Geomorphology*, Springer-Verlag, Göttingen.

SMITH, W. O., et al., 1960, 'Comprehensive Survey of Sedimentation in Lake Mead, 1948–49', *United States Geological Survey Professional Paper 295*, 254 pp.

WOLMAN, M. G., and LEOPOLD, L. B., 1957, 'River flood plains: some observations on their formation', ibid., *282-C*, pp. 87–109.

# Chapter 19

# Development of River Systems and Associated Landscapes

Straight my eye hath caught new pleasures
Whilst the landscape round it measures

*John Milton* (1608–74)

### Tributaries and Drainage Patterns

A consequent stream is one whose original course is determined by the initial slopes of a new land surface. From a volcanic cone or an uplifted dome the first streams flow off radially. A long upwarp or geanticline provides a linear crest—the primary watershed or divide—with slopes on each side. In many regions the uplifted area consists of a coastal plain backed by an older land already drained by rivers. These continue down the new surface as *extended consequents.* If there is no 'old land', the consequents begin some way below the crest, at each point where the drainage from above just suffices to initiate and maintain a stream. As the valley head widens and increased drainage is secured, each such stream is progressively lengthened by headward erosion. If uplift continues, the consequent streams are correspondingly lengthened by seaward extension.

Lengthening of a valley by headward erosion involves the same processes of surface erosion as those concerned in widening (p. 342). Nevertheless, the rate of recession of a relatively steep valley head may be much more rapid than that of the valley sides. Probably the main single reason for this difference is that the head has one great advantage over the sides. The valley head is concave (downstream) both in plan and in long profile, like the front half of a tilted spoon. Consequently rain-wash and rills from a wide area near the top converge towards the point where a definite stream channel begins. This concentration of the run-off towards the lower part of the slope tends to steepen it and so to promote slumping. If the rock is weak and impenetrable, like clay or shale, headward recession is favoured both by the high proportion of run-off and by the ease with which the infant stream can carry away the finely divided rock-waste. Another favourable condition is the emergence of springs or seepages; these tend to undercut the overlying permeable rocks, which fall away at intervals.

As the consequent streams dig themselves in, the valley sides furnish secondary slopes down which tributaries can flow. Subject to a general tendency to flow at right angles to the contours of the consequent valley, the pattern formed by tributaries and consequents depends largely on the nature and structure of the rocks which are being dissected. The latter may be homogeneous through a considerable depth, or they may consist of a series of alternating strong and weak beds.

Where the rocks have no conspicuous grain and offer nearly uniform resistance to erosion, the headward growth of a tributary is governed primarily by the regional slope, with modifications controlled by chance irregularities of surface and structure. Such streams are described as *insequent.* The regional slope, however, generally determines the prevalent direction followed by an insequent tributary; it commonly makes an acute angle with the upstream part of the consequent valley. As each insequent stream develops its own valley, it receives in turn a second generation of tributaries. The branching drainage pattern so established is tree-like in plan, and is described as *dendritic* (Figure 19.1). If the rocks are well jointed, however, a more nearly rhomboidal or rectangular pattern is likely to result.

Where, as commonly happens, the rocks consist of belts of alternately weak and strong beds,

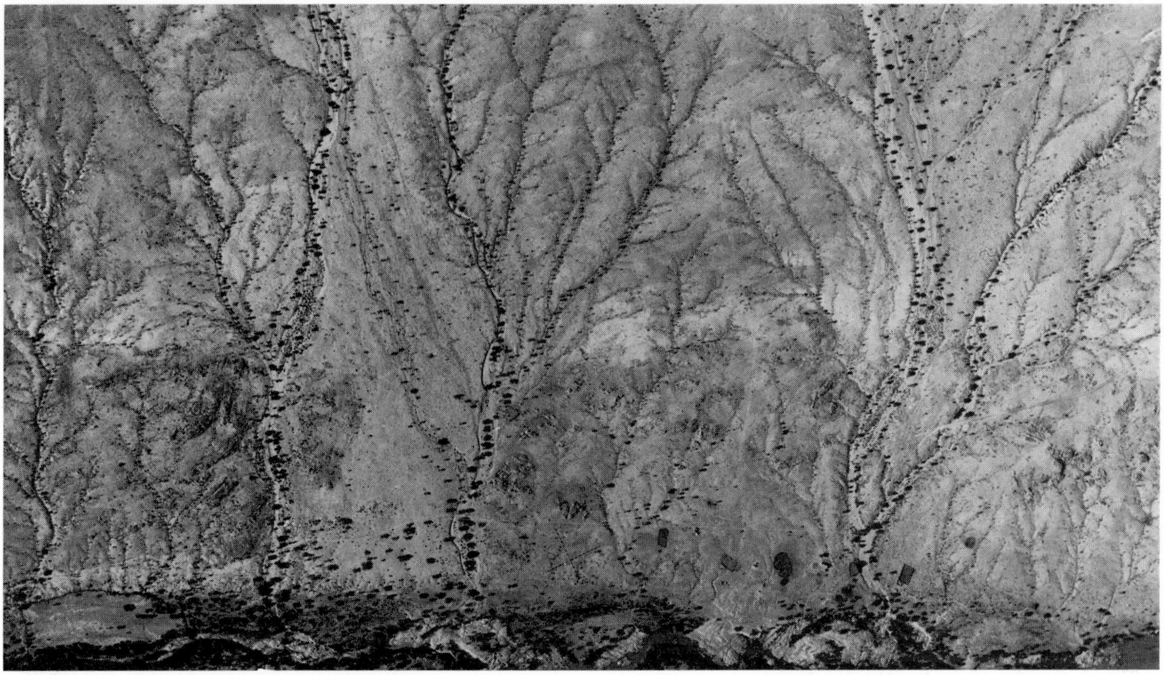

**Figure 19.1** Aerial view of the dendritic drainage developed by insequent streams on massive granite. Tributaries of the Orange River near the Aughrabies Falls (see Fig. 19.44) (*South African Air Force*)

dipping seawards, the consequent valley is narrow and steep-sided where it cuts through resistant beds (sandstones, limestones, lavas or sills), and broad and open where it crosses outcrops of clay or shale, or belts of rock structurally weakened by jointing or shattering. Tributaries naturally begin where the widening of the main valley has been facilitated in some such way. A tributary beginning on weak rocks has a great initial advantage in the struggle for space that depends on lengthening, since headward erosion unfailingly guides it back along the weak bed, parallel to the strike. Such a tributary is called a *subsequent* stream. The rectangular drainage pattern formed by consequent streams parallel to the dip and subsequent streams parallel to the strike is described as *trellised* (Figure 19.4). Later tributaries add further detail to the trellised pattern.

## Shifting of Divides and River Capture

The position of a divide remains permanent only if the rates of erosion are the same on each side, a state of affairs that is rarely achieved. When, as usually happens, the opposing slopes are unequally inclined, so that erosion is more effective on the steeper side, the divide gradually recedes towards the side with the gentler slope (Figure 19.2). Along primary divides this effect is most rapidly produced by headward erosion of the consequent valley heads. As some of the latter work back through the crest more vigorously than others, the divide becomes zigzag or sinuous, while the crest is notched and becomes increasingly varied in height. As deepening proceeds the dissection of the ridge is steadily elaborated until the more resistant rocks between the valley heads stand out as peaks. Where the headward migration of one valley head encroaches on a valley head at the other side, the notch in the crest is lowered

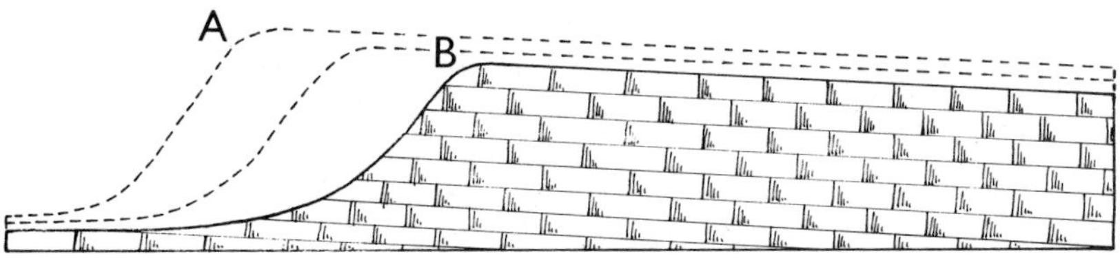


**Figure 19.2** Diagram to illustrate the recession of a watershed from A to B as a result of the effect of unequal slopes on rates of erosion

until it develops into a *col* which may become useful as a pass.

The new drainage area thus acquired by a headward-growing consequent stream is generally of little importance. Migration of the secondary divides between neighbouring consequent valleys leads to far more revolutionary changes. One of the original consequent rivers is likely to have a bigger drainage area than its neighbours: possibly because it occupies the largest hollow in what was an undulating initial surface, possibly because it is an extension of an older river in the 'old land' behind the uptilted coastal plain. In any case the major river will deepen and widen its valley more quickly than the neighbouring consequents. If the lateral divides are worn back until they reach these minor streams, the latter and their drainage areas are absorbed by the larger river—a process technically described as *abstraction*.

Capture of drainage on a still bigger scale becomes possible when the major river acquires vigorous subsequent tributaries, each working along a feebly resistant formation and each wearing back the secondary divide at its head (Figures 19.3 and 19.4). Endowed with a relatively low local base-level, a deeply entrenched subsequent, e.g. $S_1$ in Figure 19.5, cuts back towards a consequent $C_2$, which is still draining an area at a higher level. Eventually $C_2$ is intercepted, its headwaters are diverted into $S_1$ and its lower course is beheaded. This process is called *river capture*. The rectangular bend *e* at the point of diversion is known as the *elbow of capture*. The beheaded river, now deprived of much of its drainage, is described as a *misfit*, since its diminished size is no longer appropriate to the valley through which it flows. Its new source is some way below the elbow of capture, and the deserted notch $W_g$ at the head of its valley becomes a *wind gap*. A subsequent stream $S_2$, which originally entered the captured stream near or above the elbow, now has its local base-level lowered to that of $S_1$. It is then enabled to deepen its valley and to extend backwards until it in turn reaches and beheads the next consequent, $C_3$. In such ways a major consequent river, with the aid of its subsequents, may acquire a very large drainage area at the expense of its neighbours.

The rivers flowing into the Humber estuary illustrate the development of an actual river system by this process of capture (see Figure 19.7). The uplift of the Pennines provided the slopes down which a number of consequent streams flowed

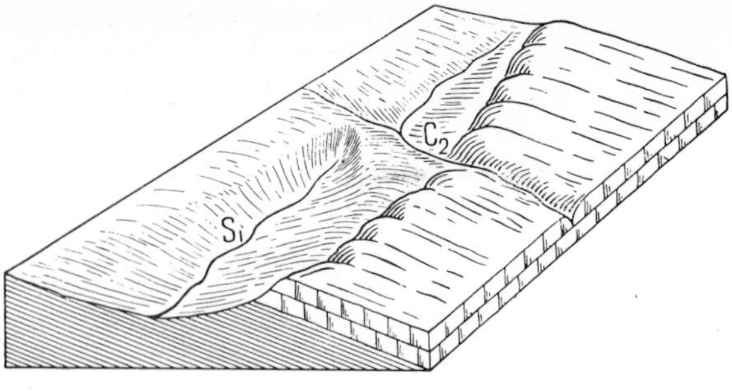


**Figure 19.3** The subsequent stream $S_1$ is cutting back by headward erosion from a relatively low level towards the consequent stream $C_2$. Capture of the headwaters of $C_2$ is imminent (cf. Fig. 19.6).

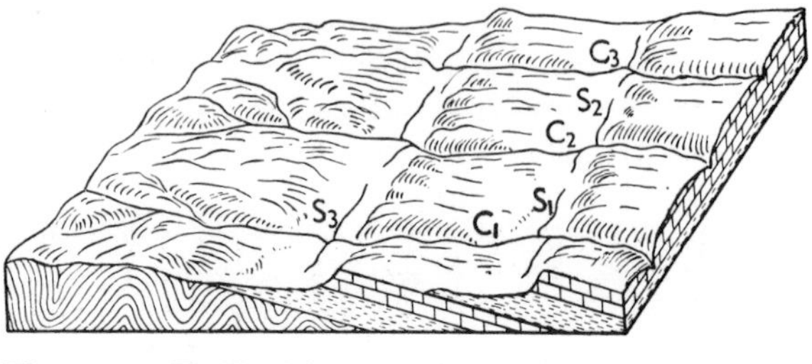


**Figure 19.4** Trellised drainage of consequent streams, C, and their subsequents, S, showing the dissection of a gently dipping series of hard and soft beds into escarpments and inner lowlands

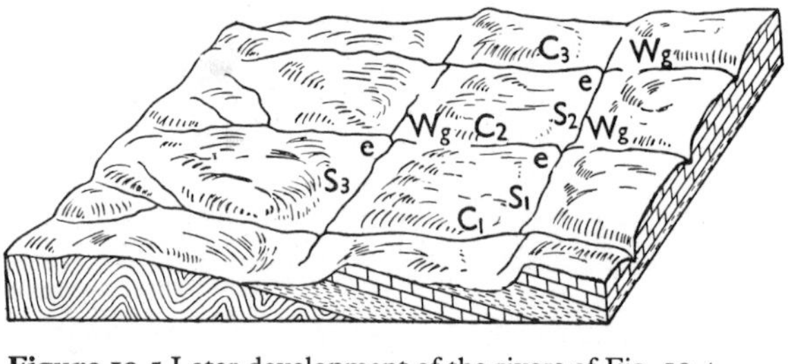


**Figure 19.5** Later development of the rivers of Fig. 19.4, illustrating river capture by the headward growth of the more vigorous subsequent streams; e, elbow of capture; $W_g$, wind gap

into the North Sea, which even then occupied a basin-like area with a long history of subsidence and sedimentation. The greater part of modern Britain probably emerged from the late Cretaceous sea with a widespread cover of Chalk, the exceptions being in the south-east, where subsidence with some fluctuations continued in the London and Hampshire basins; and locally in the north and west, where there may have been islands representing parts of the 'old land' that had remained unsubmerged. The first streams to which our present drainage can be related appeared on gentle, intermittently emerging slopes of Chalk. In mid-Tertiary times there were remarkable movements of uplift and folding in the south of England (see Figure 1.5) and of uplift subsidence and faulting in the volcanic north-west of our islands (see Figure 9.26). It is likely that the Pennines and the 'old lands' of the Highlands, the Southern Uplands, the Lake District and Snowdonia all shared in a second main uplift at about the same time.

However that may be, the ancestors of the Yorkshire rivers of today appear to have been eastwardly flowing consequents. But only the Aire-Humber still maintains an uninterrupted approximation to its original course. Even so, the headwaters of the Aire, including a lengthy subsequent tributary, were captured by the Ribble, which, flowing to the Irish Sea, had the advantage of a shorter run and a steeper gradient. The Wharfe, Calder and Don were probably tributaries of the Aire from an early stage in their history. The Nidd, Ure and Swale, however, have each been captured in turn by the Ouse, a powerful subsequent stream which lengthened itself northwards by headward erosion along the soft strata of the Trias (Figure 19.7). On the eastern side of the Ouse it is difficult to trace the former courses of the beheaded streams, because of uplift of the Cleveland Hills and obliteration of the older valleys by glacial deposits.

A more straightforward example is provided by the rivers of Northumberland. The three main streams, *a*, *b* and *c*, of the North Tyne system clearly correspond to the Wansbeck, *a'*, a tributary

**Figure 19.6** Brecon Beacon (886 metres), South Wales. Streams on the right drain to the Usk and the sea at Newport; their headward erosion threatens to capture the stream and lake on the left, draining to the Taff and the sea at Cardiff. (*Aerofilms Ltd*)

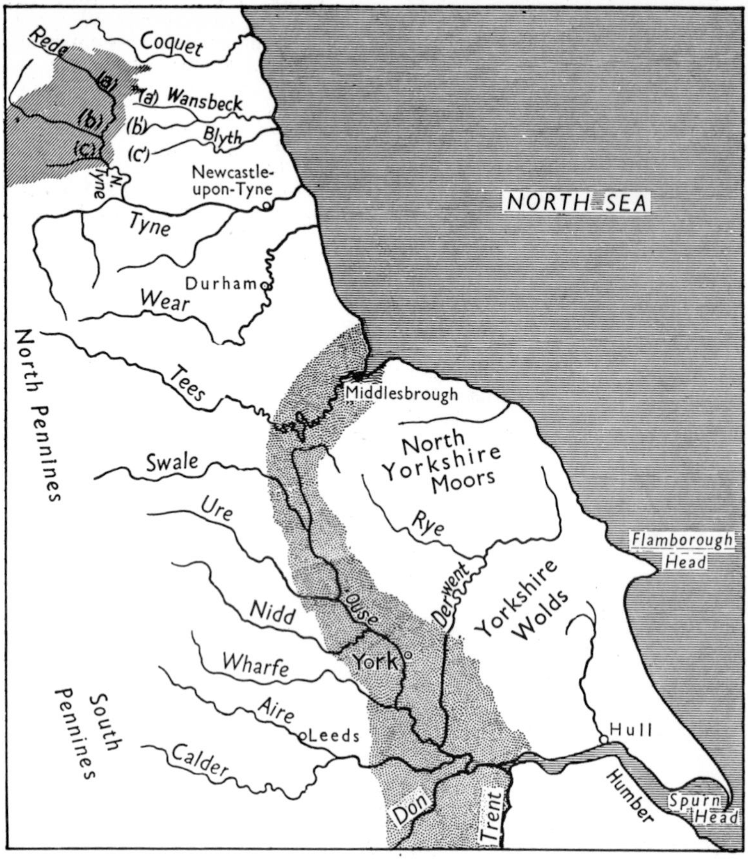


**Figure 19.7** river systems of NE England, to illustrate river capture by the Ouse—along the outcrop of soft Triassic beds (dotted)—and by the North Tyne—along the outcrop of the Scremerston Coal Group (inclined shading)

of the Wansbeck, *b′*, and the Blyth, *c′*. The headwaters of the forerunners of these were captured by the North Tyne, a subsequent of the Tyne, as it worked back along the soft beds of the Scremerston Coal series.

Tracing the ancestry of rivers and their competition for drainage areas has always had a special fascination for geomorphologists. Like every other branch of the earth sciences it springs from the main trunk of geology, detached from which it cannot flourish. And so we find that the history of rivers cannot be successfully investigated without detailed reference to earth movements and fluctuations of sea-level, and to radical changes of climate, including the highly complex interventions of the Pleistocene glaciations.

## Superimposed Drainage

In all the 'old lands' of Britain—representative of innumerable regions of similar history throughout the world—we see ancient folded rocks that were formerly hidden beneath an unconformable cover of later sediments. The Chalk, for example, that originally covered most of the British area has since been worn back to its present escarpment, which extends across England from Yorkshire to Dorset. The underlying Jurassic strata have been pared away almost as extensively. As the cover emerged as land from the late Cretaceous sea, rivers were initiated with a drainage pattern appropriate to the form of its surface. Sooner or later the deepening valleys were incised into the older, more resistant rocks, many of which have strong structural features trending across the river courses like barriers. As the covering strata were removed, the underlying rocks were gradually exposed over a steadily increasing area and, as uplift continued, were sculptured into a landscape of bold relief, with the rivers maintaining a close approximation to their original courses. The drainage pattern as we see it today has been *superimposed* on the older rocks as an inheritance from the vanished cover.

The clearest example of superimposed drainage

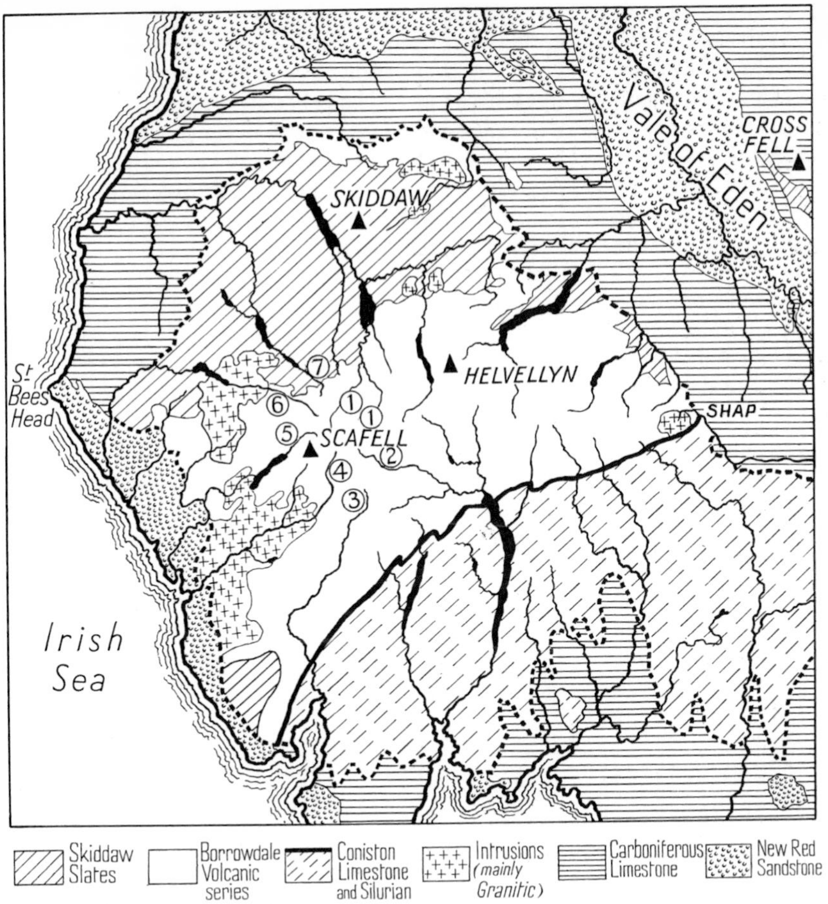


**Figure 19.8** Geological map of the English Lake District, showing the radial pattern of the superimposed drainage. Numbers refer to the headwaters of valleys and lakes: (1) Borrowdale and Derwentwater (2) Langdale and Windermere (3) The Duddon (4) Eskdale (5) Wasdale and Wastwater (6) Ennerdale (7) Buttermere and Crummock Water. To the east of (1) the streams leading into Thirlmere, Ullswater, and Haweswater flow off the northern side of the axis.

in Britain is afforded by the rivers and lakes of the Lake District. As illustrated in Figures 19.8 and 19.9, the Lake District consists of an oval-shaped area of Lower Palaeozoic rocks (folded during the Caledonian orogenesis and having a general trend from ENE to WSW) enclosed in a frame of Carboniferous Limestone and New Red Sandstone, the beds of which everywhere dip outwards. These strata originally covered the older rocks unconformably and were themselves covered in turn by later formations up to the Chalk. During Tertiary times the region was uplifted into a slightly elongated dome, with its main axis curving eastwards from the neighbourhood of Scafell towards the Shap Granite. The first streams were consequents that flowed radially down the slopes of the rising dome. After cutting through the cover and transporting all the younger strata out of the 'frame', they excavated their valleys deeply into

**Figure 19.9** Geological section across the Lake District (igneous intrusions omitted): (1) Skiddaw Slates (2) Borrowdale Volcanic Series (3) Coniston Limestone (4) Silurian strata (5) Carboniferous Limestone (6) New Red Sandstone

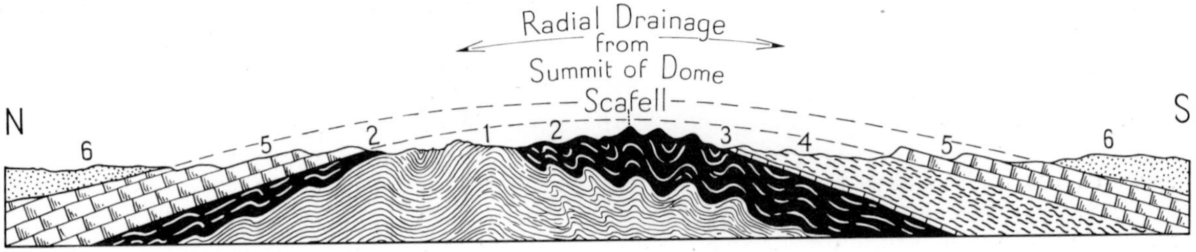

the underlying Lower Palaeozoic rocks. There they still persist with only minor adaptations to the very different structures through which they are now flowing. The radial pattern of the valleys and lakes, and of the mountainous ridges centred near Scafell, is particularly striking (Figure 19.8). The attractive contrast of lake and mountain for which the Lake District is renowned, is due to the finishing touches given to the landscape by the Pleistocene glaciations.

The Appalachian rivers of the United States are now interpreted as an example of superimposed drainage. Long after the folding of its Palaeozoic rocks, the Appalachian belt was reduced to a low-lying plain of erosion, much of which was invaded by the Cretaceous sea and buried beneath a cover of sediments, corresponding to the blanket of Chalk that then lay over Britain. Later the region was broadly uparched and the present drainage was then initiated by the consequent streams that flowed down the slopes of the cover: mainly into the Atlantic, but partly into the depressions that have developed into the Great Lakes and the Mississippi basin. The long course of this cycle of erosion has been interrupted by further arching and intermittent uplifts. The rejuvenated rivers have cut their valleys into the foundation rocks, but with a pattern largely determined by the cover, which has vanished except along the coastal plain. Powerful subsequent streams have dissected the Appalachians into long mountainous ridges, while the main consequents have continued to cross the latter through deep 'water gaps'. Examples of these transverse valleys, excavated from above through what would otherwise have been impenetrable structural barriers, are the Hudson gorge in the highlands of New York state; the Delaware gap (Figure 19.10) farther south; and the gap of the Potomac at Harper's Ferry, made famous a century ago as the strategic gateway into the broad interior valley of what is probably the most celebrated of all subsequent streams—the Shenandoah.

**Figure 19.10** Delaware water gap, cut by the superimposed Delaware River (Pennsylvania–New Jersey boundary) through a tilted formation of highly resistant conglomerate (*United States Geological Survey*)

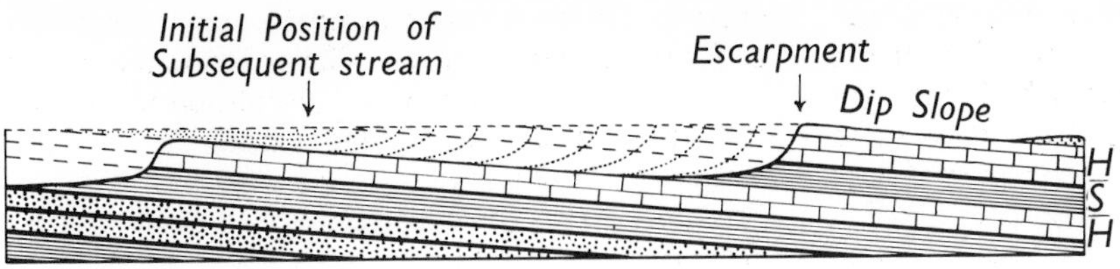


**Figure 19.11** Stages in the development and recession of an escarpment. H=relatively resistant formation S=easily eroded formations

## Escarpments and Interior Lowlands

The valley of a subsequent stream is widened and deepened between divides formed by the bands of more resistant rock on either side (Figure 19.11). As the intervening weak bed is gradually worn away, the upper surface of the underlying resistant bed is uncovered, and on this side the valley slope approximates to the dip. On the other side the overlying resistant bed is attacked at its base and soon begins to steepen into the free face of an escarpment. Until it is restrained by approaching base-level, the stream excavates its channel in the weak bed. As a result of combined lateral and downward erosion being specially favoured on one side, the stream shifts its channel obliquely in the direction of dip. The normal recession of the escarpment is thus widened, while a gentler slope—resembling a continuously guided slip-off slope—is left on the other side of the valley. This is generally referred to as the dip slope, unless it has been notably steepened or flattened after its exposure to erosion; the term *back slope* is then more suitable. Besides the temporary rills of surface wash, small tributaries called *secondary consequents* or *dip streams* flow down the dip slope. Others, known as *obsequent* or *anti-dip* streams, descend the escarpment. Both sets add detail to the trellis pattern of the drainage. As the main subsequent stream begins to meander, its valley becomes steadily wider and develops into an *interior lowland.*

Figure 7.8 illustrates the succession of escarpments and interior lowlands between Gloucester and the London Basin. From the Lias clays and marls of the Severn valley the escarpment of the oolitic limestones of the Jurassic rises to the crest of the Cotswolds. The Oxford Clay is responsible for the interior lowland occupied by the Thames above and below Oxford. A minor escarpment, that of the Corallian limestone, is then followed by the interior lowland of the Kimmeridge Clay. Beyond this the Chiltern Hills begin with the Chalk escarpment, the dip slopes of which lead down to the London Basin (Figure 15.6). On the south side of the basin the Chalk again emerges as the North Downs, with its well-known escarpment overlooking the Weald. The scarp continues through Kent until it reaches the coast near Folkestone. Followed in the other direction it swings round the western border of the Weald, where it makes a less conspicuous feature, and continues on the far side as the South Downs; at Beachy Head (Figure 6.14) it is again cut off by the sea.

The erosional history of the Weald was brought to light by a long series of detailed investigations (Figure 19.12). After the London Clay and other Tertiary formations had accumulated in the bordering depressions, the main broad upwarping of the region occurred. This was accompanied and followed by faulting and by localized buckling along a number of fold-axes. From the resulting elongated and slightly corrugated dome a framework of escarpments was developed by a first generation of subsequent streams. Eventually these early scarps were worn back until they disappeared, while the interior lowlands were widened at their expense into broad and more or less graded erosion plains sloping gently down to

**Figure 19.12** Section across the Weald, approximately north and south through Brighton, with an indication (dashed) of the Pliocene erosion surface from which the present topography has been developed. AB and CB represent the approximate extent of the Pliocene sea within the section illustrated. Length of section 58 km (*After S. W. Wooldridge and R. S. Morgan*)

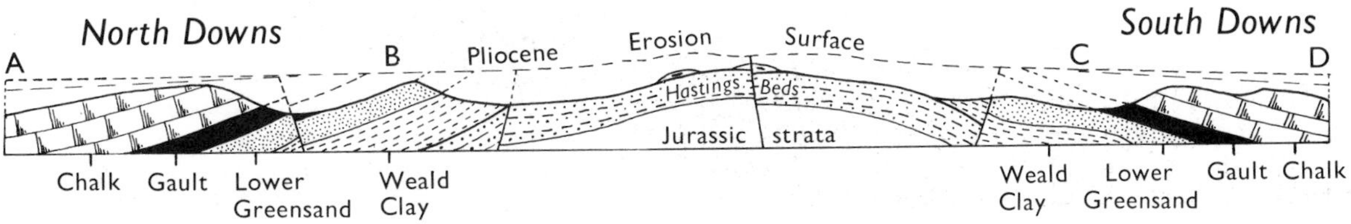

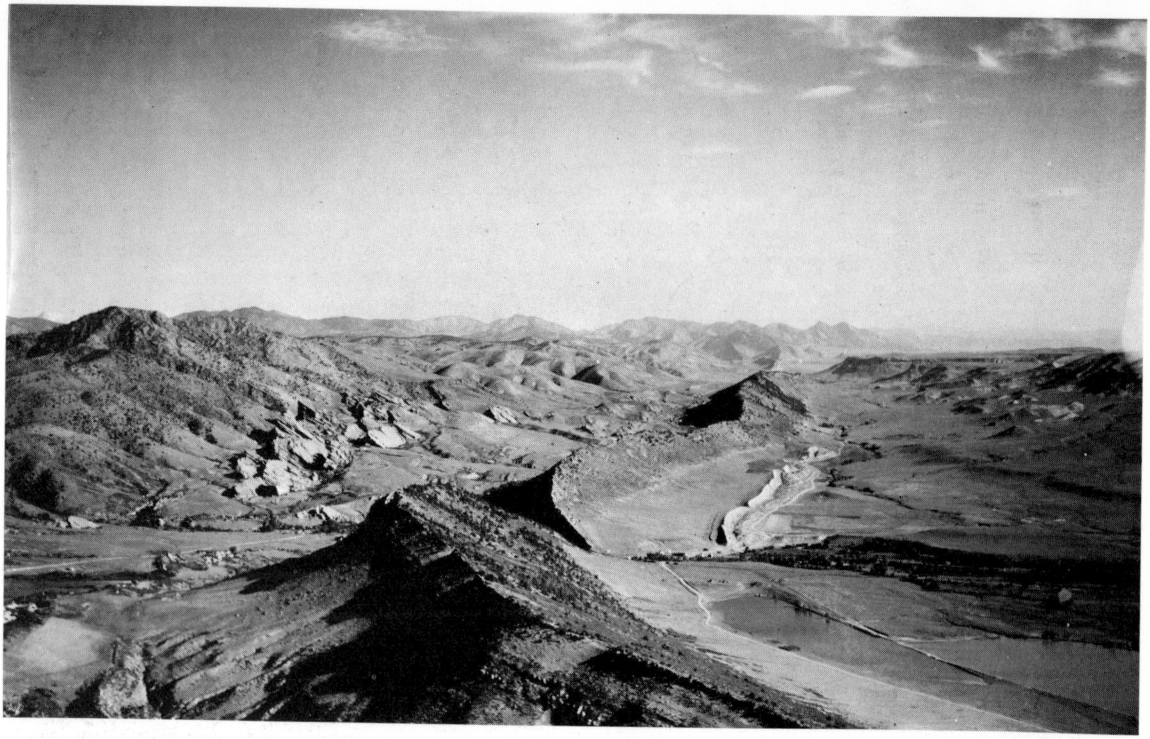

Figure 19.13 Hogback of Dakota Sandstone (Cretaceous) in the foothills east of the Front Range, Rocky Mts, Colorado (*T. S. Lovering, United States Geological Survey*)

the base-level of late Pliocene times. Parts of the area were then invaded by the sea. Remnants of the early Pleistocene sea floors can still be seen on what are now the upland surfaces of the Chalk along the North and South Downs, and also along the Chilterns. On some of them veneers of marine gravels have been preserved, although it may be 2 or 3 million years ago since they were deposited. At that time these bevelled hilltops were at sea-level. Today they stand about 180 m above sea-level. But on balance it is the level of the sea that has fallen by this amount, not the land that has risen.

The present drainage system and the associated valleys, scarps and interior lowlands of the Weald have all developed from a Pliocene erosion plain which had its margins trimmed by a Pleistocene sea. As the rivers were increasingly rejuvenated, scarps appeared on the emerging landscape and were worn back to the positions they occupy today, some 32 km apart in the case of the Downs. Recession is still going on, at widely variable rates that have been estimated to average about 5 to 8 cm per century. River capture has already occurred on a conspicuous scale. Examples can easily by recognized on geological maps by following the subsequent streams that extended themselves along the Gault Clay (at the foot of the Chalk escarpment) or the Weald Clay (at the foot of the Lower Greensand escarpment).

An escarpment together with its dip slope or back slope form a feature for which there is no English name. The Spanish term *cuesta* (pronounced *questa*) for this combination has been internationally adopted. If the beds dip at a high angle, so that the dip slope becomes about as steep as the escarpment, the feature corresponding to the cuesta is a ridge or *hogback* (Figure 19.13). At the other extreme, in horizontal beds, the cuesta becomes a *mesa* (Spanish for table), that is, a tableland capped by a resistant bed and having steep sides all round. By long-continued wearing-back of the sides, a mesa dwindles into an isolated flat-topped hill. In America such a hill is called a *butte*, from its resemblance to the butt or bole of a tree, and the term has been widely adopted. In western America buttes commonly occur where the beds dipping off the mountain flanks flatten out, as illustrated by Figures 19.14 and 19.15. Similar residual landforms in South Africa, many of which are capped by outliers of dolerite sills, are called *kopjes*.

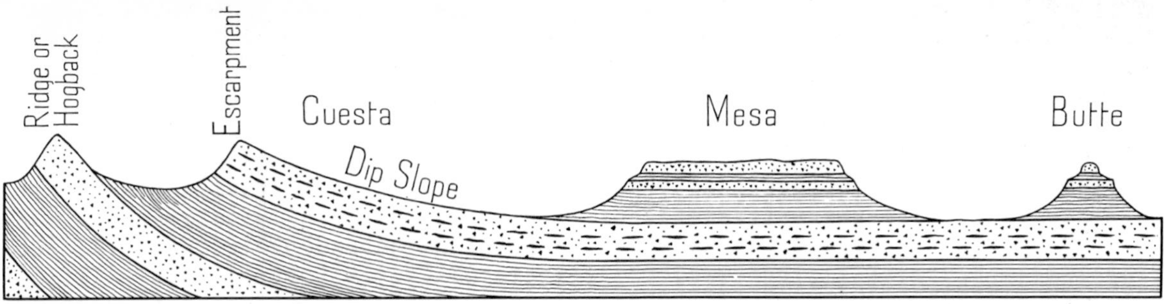


**Figure 19.14** Diagram to illustrate the relation of various erosional landforms to the structure and dip of the strata from which they have been carved

**Figure 19.15** Mesa topography with marginal 'badland' erosion due to gullying of the 'constant slope' Zion National Park, Utah (*Grant, U.S. Dept of Interior*)

## Youth, Maturity and Old Age

It will be clear from the last few pages that the original concept of an erosion cycle in terms of a major uplift, followed by prolonged denudation towards a stable base-level, fails to correspond with the real conditions that control the behaviour of rivers and the development of landscapes. The Davisian scheme was based on the opinion, then prevalent, that the uplift of an area was achieved in an interval of time that was short compared with the long millions of years required to reduce the area to a peneplain. Certain regions, like the Colorado plateau, have undoubtedly been upheaved by relatively rapid movements, but the total effect to date has been produced by a succession of uplifts with intervening periods of slowing down and standstill, and even of local reversals. These intervening periods of comparative stability have generally been too short for the completion of a cycle over more than a limited area. The distribution of river terraces, for example, and of scarps between upland and lowland plains, indicate that we have to do with a series of uncompleted cycles, overlapping in time. Such landscapes are described as *polycyclic*. Although the term 'cycle' and its derivatives are not strictly appropriate in this context, their adoption can be justified by the plea that no better terms exist.

The youth, maturity and old age of rivers and landscapes differ fundamentally from the successive stages of, say, a human life. A man may be old, mature or young, but he cannot be all three at once. Rivers and landscapes can. The headwaters of a river may still be youthful, while maturity may have been reached in the wider valley tract of the middle reaches; and nearer the sea a broad floodplain may have developed with all the characteristics of old age. Given time enough, the monotonous features of old age advance inland from the coastal plains, driving back the scarps and widening the valley floors.

The actual time required for a cycle to run its course over a given area naturally varies enormously with the rate and amount of uplift, the structure and past history of the emerging region, and the efficacy of the rivers. It is worth noting that it has taken about 60 million years for the Chalk escarpment to retreat some 725 km from the north-west to its present position north of London. Yet many parts of the area stripped of its cover of Chalk and other Mesozoic strata have mature landscapes now being incised by youthful rivers. Intermittent emergence has repeatedly interfered with the progression towards old age. Another point of interest in this uncovering process is that landscapes that were buried by Triassic deposits, and then shrouded by successive layers of Jurassic and Cretaceous sediments, are now beginning to reappear here and there. In Charnwood Forest, for example, a Triassic landscape that was young when it was buried is now being exhumed, so adding youthful features to the advanced age of its environment.

The stage of *youth* ideally begins with the dissection of a plateau or an undulating folded region. It is essentially the period during which the valley form is undergoing vigorous development, especially in depth and headward growth. The early rivers flow swiftly and have irregular gradients. Lakes, rapids and waterfalls, and gorges are highly characteristic features. In recently folded regions the main rivers occupy synclinal furrows. Tributary development proceeds rapidly during youth and river capture is common. The streams compete for space until the victorious ones acquire well-defined valleys draining large areas. Between the main valleys extensive tracts of the original surface, known as *interfluves*, may remain for some time. Except where these slope outwards or consist of soluble rocks (e.g. limestone) they are hardly affected by erosion. In regions of strong relief and widely spaced rivers the remnants may be very long-lived, even geologically speaking (see Figure 19.16). However, they necessarily diminish in area by attack from the sides, as the bordering escarpments or valley slopes continue to retreat at their expense.

**Figure 19.16** The head of the canyon of one of the Little Colorado tributaries, Arizona: an early stage of youth in the rejuvenation of an old erosion surface formed during an earlier cycle (*United States Army Air Force*)

**Figure 19.17** Kosciusko (2234 m) the highest peak of the Australian Alps, in the south of New South Wales. A maturely dissected plateau of the Gondwana cycle being invaded by the valley flats of the present cycle (cf. Fig. 19.45). (*E. O. Hoppé*) For location, see Fig. 19.32.

The landscape passes from youth into *maturity* when the relief reaches its maximum amplitude. Apart from the beginnings of flood-plains and any remnants of interfluves between the headwaters (Figure 19.17), the area is then 'all slopes'. Rivers or particular stretches of them may be said to have reached maturity when the irregularities of their long profiles have been approximately smoothed out, i.e. when they have become graded. It must not be assumed that landscapes and rivers attain maturity at the same time, or indeed that they are ever completely in phase. In the geological circumstances of the present time it is common for rivers that are obviously in a youthful stage to traverse a landscape that has advanced well into maturity or even reached old age. The contrast is due to emergence of an area that has already been dissected during an earlier but uncompleted cycle.

Transport of waste, lateral erosion and flooding are the major activities of mature rivers. The rate of lateral erosion tends to slow down as the flood plains being swept over by migrating meanders become wider and wider. Nevertheless the widening continues, until uplift or some other form of rejuvenation interrupts the process and makes it possible for a river to deepen its channel and start a new flood plain at a lower level. That this has frequently happened during recent geological time is made abundantly clear by the fact that the parts of mature landscapes that are nearly flat are by no means confined to the flood plains of the valley floors, but also include marginal terraces which are the relics of earlier flood plains. These valley flats, whether single or in flights of terraces, may abut against the steeper scarps and hillside slopes of the maturely dissected uplands; or they may be linked to the latter by another variety of erosion surface that becomes well developed during the stage of maturity. This is the pediment, the gently sloping surface (left by the retreat of steeper slopes) that is graded so that rain-wash and the occasional, but more effective, sheet-flood can convey rock-waste either to the nearest stream, which carries it away, or towards the nearest flood plain, on or near which it is deposited until it falls a prey to a meander loop. Because of recent rejuvenation, many flooded plains (e.g. those of the Mississippi) are flanked by bluffs. These tend to be weathered back and buried, both by their own debris and by waste from the pediment or other erosion surface, washed over from above. A wasted bluff may be steepened afresh by the lateral erosion of a passing meander, but this also widens the flood plain. The same processes are going on in the neighbouring flood plains and eventually two or more such plains will join up.

*Old age* may be considered to have set in, starting in the coastal regions, when coalescence of the flat and gently sloping erosion surfaces of neighbouring river systems (surfaces more or less mantled with rock-waste or alluvium) begins to extend up the valleys at the expense of the outlying bluffs and divides. The coalescing surfaces may be of different kinds. Neighbouring flood plains unite to form a type of old-age surface called a *panplain* by C. H. Crickmay, who ascribes its formation to long-continued lateral erosion by meandering rivers. One such river may invade and work over an area temporarily abandoned by its neighbour. In the area of overlap the divides are obliterated, but where there has been no overlap the residual hills—such as monadnocks—may be very persistent features of old age topography.

Similarly, as already indicated (p. 305) pediments may unite into pediplains; while inland from the coast and the major rivers the growing pediments encroach upon residual hills: e.g. the buttes of North America and the kopjes of southern Africa, or the inselbergs of areas where the granite rocks and migmatites and other resistant crystalline rocks have been long exposed. Finally, in many lands, each pediment on which such residuals stand can be followed up to a scarp which leads up to a higher pediplain, or other erosion surface, representing a cycle that had an earlier start but was interrupted before its completion.

Whatever it may be called, and however limited it may be, the end-product of a cycle of erosion over the area where completion was or has been achieved, is an erosion surface of faint relief. Landscapes that have been reduced to this stage of extreme old age may have approached it in different ways, corresponding to the various kinds of slope retreat already passed in review. Where the ground and bedrocks are permeable and the proportion of run-off low, where solution or soil-creep or solifluction becomes the dominant process of surface erosion, and where rivers fail to remove all the waste products, slopes gradually decline. The surviving interfluves then assume gently undulating forms with broad convexities towards the sky. This is the sort of surface that matches the conventional idea of a peneplain. As Lester King has pointed out, it suggests 'peneplanation by downwearing which has mistakenly been thought to have operated throughout the period of erosion'. As early as 1933 Crickmay raised several objections to the Davisian scheme of peneplanation, the most damaging being that 'most plains formerly quoted as peneplains occur in series, at successive levels, one above another. . . . With only the peneplanation theory as a basis for understanding such relationships, it is impossible to see how the lower 'peneplains' were formed without the higher ones being destroyed.'

Despite these and other criticisms, the term 'peneplain' is still widely used, particularly for older erosion plains that have been uplifted to become the initial surfaces of later, uncompleted cycles. When this term is encountered in the literature of geomorphology it should be kept in mind that the surface referred to may have been a pediplain or panplain or a plain or marine erosion. It is not always easy to decipher the mode of origin of a surface that is now represented only by a few remnants. It would save much confusion if every such surface was simply called an *erosion plain* until adequate evidence had been collected to justify a more specific descriptive name.

## The Isostatic Response to Denudation

The reduction of a region to an erosion plain involves the removal of an immense load of material, the mass of which is proportional to the height of the initial surface. While the crust was being thus unloaded by denudation, slow isostatic uplift must have been continuously in progress, thereby giving the rivers more work to do and delaying its completion. This effect has so far been tacitly ignored for the sake of simplicity of treatment, but it should not be overlooked.

Let us suppose that a thickness of 1000 m of rock having an average density of 2·6 has been removed from a region while isostatic equilibrium is maintained and no other change of level occurs. The mass lost is proportional to $2{\cdot}6 \times 1000$, and this must be made good by the inflow at depth of a thickness $h$ of material with a density of about 3·4. The condition for the maintenance of isostasy is that $3{\cdot}4h = 2600$ m; whence $h = 765$ m. This influx of sima raises the plan AB (Figure 19.18(*a*)) to A′B′, and the new surface is only 235 m below the original level of the denuded block of country. For the reduction to base-level of a plateau which

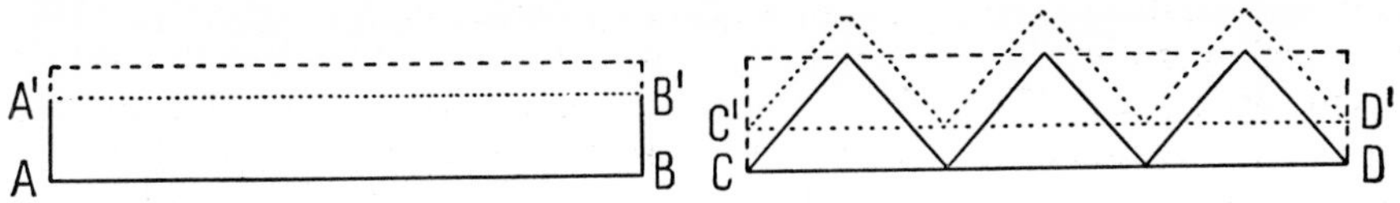


**Figure 19.18** Isostatic response to denudation: (a) Uplift of plateau (b) Uplift of mountain peaks

originally had an elevation of 1000 m, the thickness of rock to be removed is not 1000 m but about four times as much.

Such uplift must be considered a normal accompaniment of a cycle of erosion. It involves the curious effect that during late youth and early maturity the summits of peaks and divides become elevated above the initial surface. This is illustrated by Figure 19.18(*b*). When the cross-sectional areas of valleys and divides are equal, half the mass of the denuded block has been removed. The plane CD will by then have been raised to C′D′, i.e. by 765/2 m, assuming the summits to be 1000 m above the valley floors. The bearing of this remarkable result on the high altitudes of the Himalayan peaks is referred to on p. 388.

## Interruptions in the Course of Denudation

At any stage in an uncompleted cycle the normal sequence of changes may be slowed down by the effects of isostasy or by those due to a slow lowering of sea-level. More serious interruptions may be due to earth movements of uplift or subsidence, with changes of slope due to tilting, and commonly with production of fault scarps; by volcanic action; or by changes of climate leading to glaciation, increased rainfall or aridity. The distinctive landscape features developed under glacial and desert conditions are described in later chapters. Meanwhile, however, it must be remembered that glaciation involves (*a*) abstraction of water from the sea to form ice-sheets and its subsequent restoration when the ice melts away, with corresponding changes in sea-level and consequently in base-level; and (*b*) isostatic depression due to the loading of an area by an ice-sheet, followed by uplift when the ice-sheet retreats and disappears.

Volcanic activity may introduce local accidents, such as the obstruction of a valley by a lava flow. Youthful features are then temporarily restored while the river is regrading its course through the obstacle. On a larger scale whole landscapes may be buried beneath a thick cover of plateau basalts, in which case a new cycle then begins on the volcanic surface.

If a region is depressed by earth movements its surface is brought nearer to base-level, the work to be done by erosion is diminished, and the stages of the cycle then in progress are passed through more quickly. When a depression is localized across the course of a river a lake is formed. This also happens when a river is ponded back by the up-arching of a fold across its course (see p. 384).

When coastal regions subside—unless sedimentation keeps pace by growth of flood-plains and deltas—the sea occupies the lower reaches of valleys, and estuaries are formed. Tributaries which entered the valley before it was drowned now flow directly into the tidal waters of the estuary and become *dismembered streams*. Rivers like the Thames and Humber are sufficiently active—with some human assistance—to keep their channels open. More sluggish rivers, however, may be unable to prevent the growth of obstructive bars and spits across their estuaries, and the latter then become silted up and overgrown with peat-making vegetation.

The broads of East Anglia (Figure 19.19) were thought to be natural relics of the formerly widespread estuary of the rivers Bure, Yare and Waveney. However, it is now known that the Broads owe their origin not only to changes of level and climate, but also to human interference on a big scale. They are, in fact, the flooded sites of excavations (margining the rivers) dug in peat to depths of 3 or more metres. About 2000 B.C. the region was widely invaded by the sea, which then withdrew until about 1000 B.C., while thick peat deposits were accumulating on the emerging land. Another marine transgression then occurred, marked by a deposit of clay, which was again followed by emergence and further growth of peat until about A.D. 700. Since then submergence has been almost continuous, with only slight and temporary interruptions. By the year 900 a prosperous peat industry had begun to flourish and this continued to expand for nearly 400 years. During the thirteenth century severe storms and flooding, from the sea as well as from the rivers, not only checked the progress of this industry but began its enforced decline. The worst disaster was in 1287, when over 50,000 lives were lost by drowning in the Netherlands. At that time the

**Figure 19.19** Typical scenery of the Norfolk Broads, near Wroxham 6 km NE of Norwich. River Bure (looking east) with Salhouse Broad on the right (south) and part of Hoveton Great Broad on the left (north) (*Aerofilms Ltd*)

coastal region in the neighbourhood of Yarmouth stood 4 m higher above the sea than it does today. Even so, some of the pits left by the peat-cutters were flooded, and as they could not be drained they had to be abandoned. Later floods discouraged new enterprise and brought peat-digging practically to an end. By 1350 it had become more profitable to use the flooded pits—the Broads of today—as fisheries. The rivers now leave their flat meadows to flow along the irregular ramparts of peat and sediment which originally separated them from the adjoining peat-cuttings. Full details of this fascinating story, in which geology is linked to living history by archaeology, will be found in the memoir by J. M. Lambert *et al.*, listed on p. 401.

Geologically, the Broads may be said to owe the immediate possibility of their existence to a local subsidence, i.e. to a rise of base-level. If, on the contrary, base-level is lowered, the work to be done by erosion is increased and the river is endowed with renewed energy to begin the task of regrading its profile to the new base-level. The river is *rejuvenated*. The change begins with the restoration of youthful features where the gradient is steepened, and gradually works upstream. During the process of regrading there is a more or less marked change of slope at the place of intersection (the *knick-point*) of the newly graded profile with the older one (Figure 19.20). The knick-point is particularly conspicuous when the uplifted region is bounded by a growing fault scarp, or in any other circumstances favourable to the development of rapids or a waterfall. The newly deepened part of the valley may be excavated as a gorge or narrow V in the wide V- or trough-shaped floor of the pre-existing valley. Consequently the cross-profile also shows a steepening of slope below the point where the earlier valley form is intersected by the new. The valley sides tend to become convex as a result of an increased rate of down-cutting.

The remaining sections of this chapter are devoted to features of special significance, such as river terraces, canyons and entrenched meanders, and polycyclic (or multicyclic) landscapes in general, developed in response to changes of base-level which have been relatively rapid compared with the slow isostatic uplifts induced by erosion.

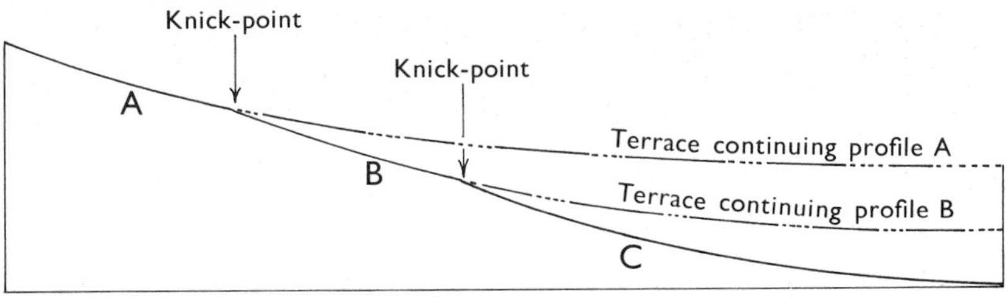


**Figure 19.20** A long profile of graded reaches, A, B, and C, with knick-points at their intersections. Profiles A and B may be indicated by terraces preserved on the valley sides.

## River Terraces

When a river that has already established a flood plain is rejuvenated, it cuts through its own deposits into the underlying rocks. The margins of the original valley floor are then left as terraces above the new level of the river. In the course of time the new valley is widened and a second flood plain forms within the first one, of which only local remnants may survive. By subsequent rejuvenation a second pair of terraces may then be left on the valley side. The sides of many of the valleys of Britain (Figure 19.21) and western Europe—and indeed of many other parts of the world (Figure 19.22)—are bordered by a series of such river terraces, each corresponding to a phase of rejuvenation and valley deepening followed by one of rising base-level and valley widening. A typical terrace is a platform of bedrock mantled with a sheet of gravel and sand passing upwards into finer alluvium. As indicated in Figure 19.20, such terraces slope gently seawards and so provide a means of reconstructing earlier profiles and of estimating the sea-levels towards which they were graded. In some places they can even be correlated with 'raised' beaches along the neighbouring coasts. But correlation of terrace remnants is not quite so easy as Figure 19.20 might suggest, either

**Figure 19.21** Alluvial terraces of the River Findhorn, flowing into the Moray Firth, Scotland, cut through glacial deposits and marking successive stages in the erosion of the valley (*Institute of Geological Sciences*)

**Figure 19.22** Alluvial terraces of the Fraser River, British Columbia (*A. M. Cockburn*)

along a single valley or from one river system to another. However, the immense amount of detailed work that has now been carried out does establish two points of extraordinary interest and importance:

(*a*) The sequence of terraces and 'buried channels' corresponds to the fluctuations of climate during at least the later part of the Pleistocene. During a cold phase, when the growth or advance of an ice-sheet farther north caused the sea-level to fall, rivers near the coast cut deep channels in their flood plains. When the ice melted and the sea-level rose again, these channels were rapidly choked with glacial debris and with waste supplied by solifluction while the frozen ground was thawing out. But during the warm phase that followed, the sea failed to recover its former level, and the rivers then regraded themselves to the new base-level, with flood plains at a lower level than before.

(*b*) The oscillations of sea-level that have been recognized in this way—and also from the corresponding evidence of 'raised' beaches—show that, superimposed on the falls and rises accompanying the growth and shrinkage of ice-sheets, there has been a progressive lowering of sea-level. As we have seen already (p. 299) the melting of all the ice-sheets and glaciers of today would raise the sea-level only a fraction of the 180 m that has been lost since the beginning of the Pleistocene. We have here a preliminary indication of a world-wide phenomenon of profound significance, since it can only mean that the ocean basins have increased in depth or in area, or in both these ways.

Since the early Pleistocene there have been at least a dozen well-marked changes of level in the London Basin, recorded by terraces and also by buried channels which can be discovered only by borings and excavations. Until glaciation has been dealt with, only the last few of these oscillations will be considered. In early Pleistocene times, after leaving the Henley-to-Marlow stretch, the Thames flowed ENE through the Vale of St Albans. Later it was diverted by advancing ice, and the present valley through London dates from the time when all the earlier exits to the North Sea had been completely blocked by ice and boulder clay (Figure 21.8). As illustrated in Figure 19.23, the Thames valley is shown with three terraces that can most easily be recognized in and near London:

(1) The *Boyn Hill Terrace*, named after a locality near Maidenhead, reaches a height of about 60 m, but at Swanscombe (near Gravesend, 80 km nearer the sea) it is much lower and corresponds to a sea-level of about 30 m. At Swanscombe it is

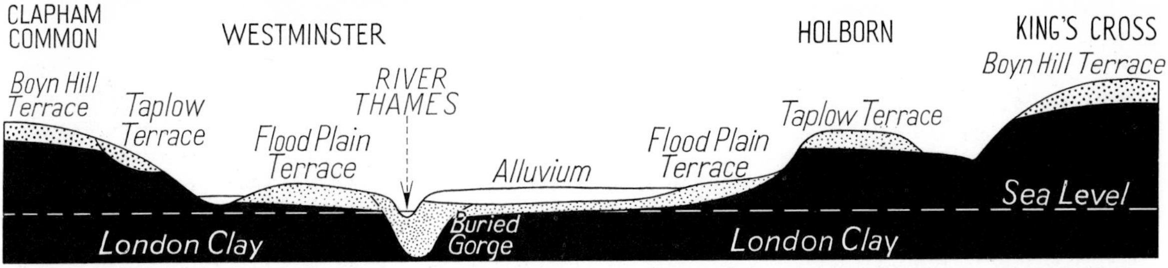


**Figure 19.23** Section across London to show the paired terraces and one of the buried 'gorges' of the Thames valley (*Institute of Geological Sciences*)

particularly well exposed on a platform of Chalk. The sequence of gravel and the loamy clay locally called 'brickearth', together with evidence from borings, corresponds to a long history of a warm and genial climate followed by cold and the cutting of a deep channel, after which more temperate conditions were again restored. The gravels, mostly composed of flint, contain the fossil remains of extinct species of elephant, hippopotamus and rhinoceros. Palaeolithic flint implements are also found.

(2) The step down to the *Taplow Terrace,* which would originally have been a fairly steep bluff, is generally hidden by downwash from the terrace above. This is illustrated by Clapham Common (Figure 19.23), and the slopes in its vicinity. The type locality, Taplow, is farther upstream, near Maidenhead. The most familiar example of this terrace is Hyde Park and Kensington Gardens. Bones and skulls of lions and bears have been found in the lower gravels, but nearer the surface these are absent and remains of the hairy mammoth appear, indicating the oncoming of much colder conditions.

(3) The step down to the *Flood-Plain Terrace* has also become a gentle slope in most built-up localities; e.g. South Kensington Station and the museums are on the Flood Plain, while most of the Imperial College buildings and the Albert Hall are on the slope leading up to the Taplow Terrace of Kensington Gardens. In the other direction the gradual descent towards the Thames takes us through Chelsea, mostly built on the gravel and alluvium of modern times. Between the formation of the Upper Flood Plain and the alluvium of today there were three important oscillations of sea-level. During the times of 'low water' the river cut deep channels with floors far below the present level of the sea. The first was infilled and followed by the formation of the Lower Flood Plain. The second was probably the deepest and is graphically described as the *Buried Gorge.* At Charing Cross the gravels filling the 'Gorge' have been found by borings to continue to a depth of at least 30 m below the river; and elsewhere there is evidence that the sea-level of the time was lower than −120 m. In turn the Lower Flood Plain was constructed over the parts of the Upper that had been washed away. The third deep channel was similarly filled as the sea rose to a level not very different from that of today, and so made possible the present alluvial plain, now largely 'made ground' instead of the marshes and meadows of earlier days.

The earlier deposits of each part of this composite Flood-Plain Terrace contain occasional fossils, all indicative of a frigid climate. In the oldest of these, remains of the mammoth have been found, accompanied by reindeer and elk. In the Lea valley there is a celebrated 'Arctic-Plant Bed'. Excavation of the London docks has revealed the presence of several beds of peat, those in the later deposits containing relics of oak and beech. The latest of these fossil forests is followed by the first signs of the Bronze Age, which began about 4000 years ago. Here it may be added that London owes its remarkable growth not only to its position on the great estuary that is nearest to Europe, but also to the geological conditions that provided an adequate water supply: in the early days from the terrace gravels, and later from the artesian wells that tapped the Chalk and afterwards the Lower Greensand. It is noteworthy that Paris and Brussels have also shared in the geological advantages provided by Tertiary basins and terraced rivers.

## Uplifted River Terraces and Slip-off Slopes

World-wide changes of sea-level, whether due to the growth and decay of ice-sheets, to displacement of sea water by deposition of sediments, or to changes in the volume of the ocean basins, are said to be *eustatic*. In general, as we have seen, these are invariably involved with slow crustal movements due to *isostatic* readjustments. Moreover, in many places, independent earth movements may be simultaneously affecting the level of the crust. The term *diastrophism* is used for all movements of the solid crust which result in a relative change of position (including level and attitude) of the rock formations concerned, i.e. for the movements commonly referred to as *orogenic* and *epeirogenic* (p. 104). Strictly speaking, the term also includes isostatic movements, but it is convenient to consider these separately, so far as it is possible to do so, which in fact is never very far.

For the above reasons the recent changes of level indicated by land emergence—e.g. river terraces and 'raised' beaches—and land submergence—e.g. 'buried' gorges and submerged forests or peat beds—can be referred to eustatic changes only over limited areas and even then only *mainly* so. The terraces of the Thames fall into the latter group, but, as we have already seen in describing the Broads (p. 374), there has been a notable subsidence of East Anglia during recent centuries, while farther north isostatic uplift seems still to be taking the lead. On the other hand, the 180-metre sea-level of the early Pleistocene can be traced from the bevelled Chalk hills of SE England into Devon (see Figure 18.46) and Cornwall, and still farther afield into Wales, where marine platforms at a similar height are found at intervals behind the coastal regions. This strongly suggests—though it does not prove—that the levels of a considerable part of south Britain have been controlled during the last 2 or 3 million years essentially by eustatic changes. In general, however, it must be remembered that the actual displacement at any place during the time under consideration is the algebraic sum of the changes of level due to (*a*) eustatic changes of sea-level; (*b*) isostatic readjustment; and (*c*) independent diastrophic movements. Even the 180-metre sea-level is not free from these complications, since in practice the level is found to vary from place to place between 165 and 195 m.

When the relative movement between land and sea is no more than a hundred metres or so during a relatively short period of geological time, such as the Pleistocene, it is often difficult or even impossible to disentangle the separate effects of these three factors. In dealing with the major effects of diastrophic movements, however, where changes of level may amount to several hundred metres, this difficulty is not seriously felt. Moreover, wherever faulting or tilting has occurred it is clear that earth movements have operated, since eustatic changes are necessarily of world-wide uniformity. From the Thames, with its essentially eustatic terraces, we may therefore pass to examples where the land has been genuinely uplifted and not merely left behind by a falling sea.

A most remarkable case has been described by Crickmay (see 1959 reference on p. 401) from the thickly forested Cariboo district of British Columbia, between the upper Fraser river, itself celebrated for its river terraces (Figure 19.22), and the Canadian Rockies. This is a mountainous country of steeply dipping metamorphosed sedimentary rocks. It is now being vigorously dissected by streams that are cutting downwards from headwaters at 1800 m or higher. Around Nugget Mountain (Figure 19.24), the summit of which is a nearly flat terrace, there are a great many other 'flats' and gently sloping slip-off surfaces at a variety of levels, connected by steeper valley sides which have an average declivity of about 26°, corresponding to the 'constant slope' determined by talus and slipping fragments. Apart from roughening by frost, the terrace and slip-off remnants show no significant modifications by denudation. Many are still covered with river sands and water-worn pebbles. Others, less well protected, have a cover of frost-disintegrated fragments like some of the upland flats of Scotland (Figure 19.25). But the most revealing feature is the preservation, almost unaltered, of small narrow canyons across some of the uplifted flats. Overthrow Canyon traversing Nugget Mountain is typical (Figure 19.24). Open at both ends, it is the only surviving vestige of a tributary to one of the streams on the northern side: a stream that long ago was widening a flood plain that now has vanished or can no longer be identified in the maze of surrounding remnants. The little 'canyon' is 12 m deep; like the others it is floored with alluvium that was abandoned when it was cut off from its drainage system. Only the northern outlet and the walls have wasted a little and the debris from the latter encumbers the original floor. The slopes of the walls have declined in consequence, but otherwise there has been no retreat. Yet the time required to carve the existing relief of the district, allowing for the wide flood plains that must have been repeatedly formed and all but destroyed, must have been very long. A million years, or several millions? It is hard to estimate,

**Figure 19.24** Nugget Mountain and its environs, Cariboo District, British Columbia. From a photograph of a small, natural-scale model of the area investigated, looking northwards. Vegetation is omitted in order to show the summit flats and other erosion terraces. Heights in feet. (*C. H. Crickmay*)

**Figure 19.25** A typical 'block field' on the plateau-like summit of Aonach Beg (about 1200 m), Ben Nevis Range, Scotland; illustrating the splintering action of frost on schists. The white trails seen in the distance are the shattered tops of quartz veins (*Institute of Geological Sciences*)

but the Colorado river provides a little guidance (p. 382).

Nugget Mountain (1682 m) is not the highest of the flats and slip-off surfaces in this district. Within a few kilometres there are similar platforms and summits that have survived unharmed at 1737, 1828 and 1890 m. Below the 1682-metre level there are well-defined terraces at 1554 and 1463 m, notably between Nugget Gulch and Victorian Creek (Figure 19.24). There are also several slip-offs sloping from 1463 to 1432 m, and many benches at lower levels. These surfaces can have been produced only by lateral and oblique river erosion. The side-to-side migration of the streams responsible must have varied in rate and distribution beyond hope of reconstruction. Figure 19.26 is intended to do no more than suggest the nature of the process in a greatly simplified way; the amplitude of the migrations was probably over a far greater width of country than is indicated in the diagram. The intervening steeper slopes correspond to times of dominant down-cutting (as at present), made possible by a long series of successive rejuvenations.

In a 'stepped' landscape, like that of the Nugget Mountain district, the elevated flat or gently sloping surfaces are capable of long survival except along their edges, where they are consumed by the retreat of the steeper slopes that lead down to an actively eroding river. Such a landscape of platforms and steeper slopes—like the steps and risers of a flight of stairs—indicates a long succession of uplifts 'separated by periods of standstill during which denudational progress was made'. Crickmay describes as *stagnant* the uplifted steps which have so long escaped the lowering effects of denudation, in contrast to the *active* valley floors and sides. It is this contrast between geological repose and normal activity (to say nothing of the violence of landslides, cloudbursts and flash-floods) that has led Crickmay to draw attention in the title of his monograph to the nominal inconsistency of the 'geological principle of uniformity' (see pp. 29–30). The difficulty does not arise if one thinks of *actualism* instead of uniformitarianism. Much more important are Crickmay's conclusions from his study of this and other stepped landscapes. He writes, 'In such a history there can be no stages comparable to those of the Davisian scheme, for each of a hundred-and-one rejuvenations brings back some slight semblance of Davisian youth, and the result is—at least around Nugget Mountain—a perpetual blending of youth and age, a blending of increasing complexity as time passes.' As he also points out, both the number of these uplifted erosion plains and their well-established mode of origin rule out any possibility that they could be surviving relics of peneplains. Nevertheless, such is the momentum of familiar ideas that relics far less well preserved than these are still often regarded as parts of former peneplains, especially if they can be traced over a considerable area. But how large must that area be? Wales? Africa? We shall return to this question on p. 392.

## Entrenched Meanders

Incised meanders with well-developed slip-off slopes have been described on pp. 302, 344. These dig themselves in by oblique erosion while their loops are being enlarged, and are said to be *ingrown*. If, however, erosion is mainly vertical, the existing loops have less opportunity to enlarge

**Figure 19.26** Approximately N–S cross-section through Nugget Mountain to illustrate the to-and-fro migration of the main streams (Fig. 19.24) regarded as the minimum necessary to account for the origin of the summit flats and terraces. (*Modified after C. H. Crickmay*)

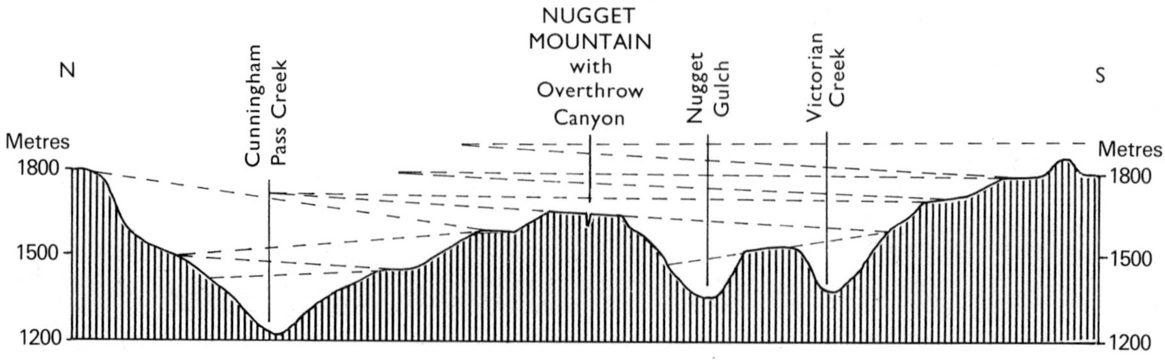

themselves and the resulting incised meanders are distinguished as *entrenched*. There is, of necessity, every gradation between these two, entrenchment being favoured by relatively rapid rates of uplift and down-cutting. If, for example, at the time of rejuvenation, a stream was freely meandering on a floor of easily eroded deposits underlain by more resistant formations, the deepening channel would soon be etched into the underlying rocks with a winding form inherited from the original meanders. The 'hair-pin gorge' of the river Wear at Durham is a familiar British example (Figure 19.27). A well-protected site within the loop was selected for the Cathedral, which is thus enclosed

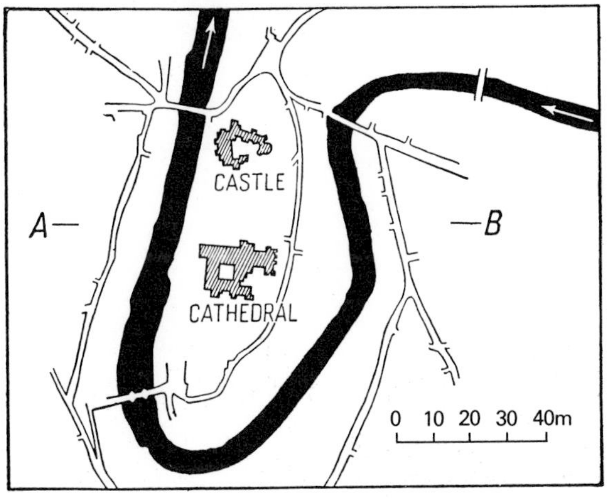


**Figure 19.27** The entrenched meander of the River Wear at Durham (*Aerofilms Ltd*) with explanatory map and section

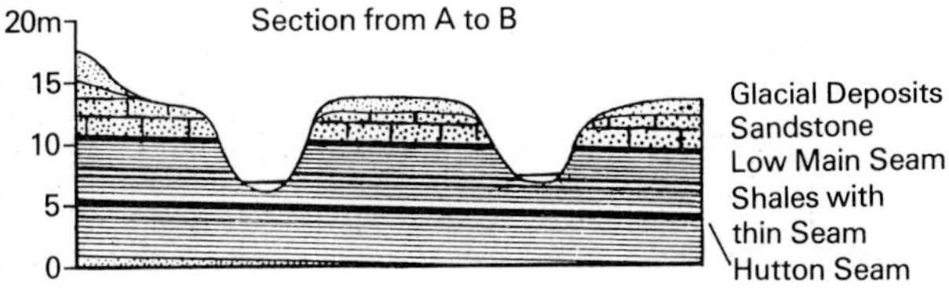

by the gorge on three sides. The fourth and easily vulnerable side was safeguarded by building a castle which is now the home of the senior college of Durham University.

The change of form of entrenched meanders and the wearing back of the confining walls are relatively slow processes controlled by lateral undercutting of the river banks. Localized undercutting on both sides of the narrow neck of a constricted loop sometimes leads to the formation of a natural bridge. On each side a cave is excavated, especially if the rocks at river level are weaker than those above. Eventually the two caves meet, and the stream then flows through the perforation. The stronger rocks of the roof remain for a time as an arch spanning the stream, and the loop-shaped gorge at the side is abandoned. In Utah, where recent uplift has made possible the development of many deeply entrenched meanders (Figure 19.28), there are several examples of such arches. The most impressive of these is Rainbow Bridge (Figures 19.29 and 19.30), a graceful arch of sandstone which rises 94 m over Bridge Creek in a span of 85 m.

## The Canyons of the Colorado River

Bridge Creek is a tributary of the Colorado river, and is thus related to one of the world's most awe-inspiring scenic wonders—the Grand Canyon of the Colorado river (Figures 1.6, 19.32 and 19.33). At the end of Cretaceous times the region which is now the Colorado Plateau was near sea-level, eastern Utah and Arizona being a coastal plain (possibly a panplain or a pediplain) which extended towards an inlet of the sea in Colorado and New Mexico (Figure 19.31). To the west were the early mountains of the Basin and Range province, while to the east the sea was already bordered by the Rockies. After the sea withdrew, the Uinta Mountains arose in the north of Utah, while much of the southern part of the region became the Green River Lake, occupying a downwarp that developed during the Eocene and became the depository of about 3000 m of sediment. Uplift, upwarping, faulting and igneous activity then began more vigorously and have continued at

**Figure 19.28** Entrenched meanders of the San Juan River, Monument Valley, Utah (*Dorien Leigh, Ltd*)

Figure 19.29 Rainbow Bridge, Bridge Creek, Utah. Water from Lake Powell, impounded by Glen Canyon Dam (1964), now fills and obscures the miniature canyon beneath the arch. See Figs. 19.30 and 19.32 (*Ewing Galloway*)

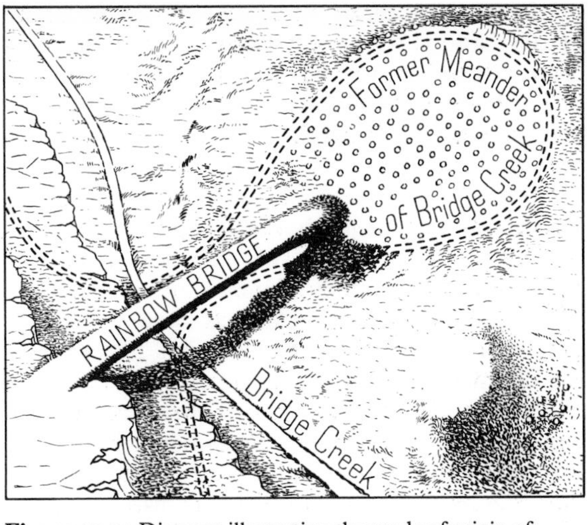


**Figure 19.30** Diagram illustrating the mode of origin of Rainbow Bridge

intervals and at varying rates ever since.

Most of the Mesozoic and later formations have been removed from the Grand Canyon region during the Miocene, when the cutting of the present canyons may be said to have begun. J. W. Powell, the intrepid explorer who was the first (in 1869) to pass through the gorges of the Colorado river, suggested that the river had been able to maintain its course by deepening its valley while

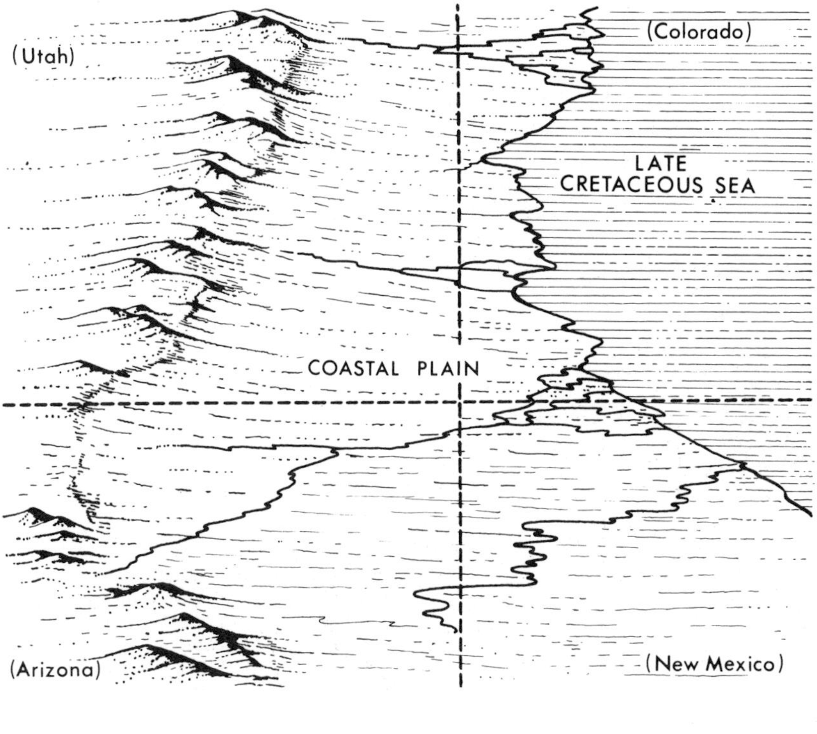


**Figure 19.31** Diagrammatic representation of the Colorado Plateau area as it was in late Cretaceous times (*After C. B. Hunt*)

uplift was in progress, so becoming permanently entrenched in the rising landscape. A rejuvenated river which thus succeeds in maintaining a downward slope from a source behind the uplands to the plains in front, is called an *antecedent* river, to express the inference that the river must have been in existence *before* the upheaval of the land through which it had to cut its channel.

Although mainly correct, this idea does not cover the whole history of the Colorado river. Of the various upwarps that lie across the course of the river, one was formed early in the Plicoene near the site of the present Lake Mead. Here the river failed to keep pace with the rate of uparching of the ground beneath it, and in consequence its waters were ponded back, as they are today by the Hoover Dam (Figure 18.37). The Pliocene lake deepened until it could spill over the lowest point on the rim, i.e. the old canyon floor, which thereupon resumed its former function of draining the basin upstream. In due course the lake was eliminated in the usual way by sedimentation from the inlet and headward erosion from the outlet. This combination of superposition (p. 364) and antecedence has been called *anteposition* by C. B. Hunt, who has made a comprehensive study of the Plateau (see p. 186). The Colorado river is in part an *anteposed* river.

Hunt estimates that the average thickness of strata removed from the Plateau amounts to about 3000 m. At the present rate of denudation (see p. 332) this would have taken about 21 million years, which agrees very well with the geological evidence. Canyon cutting was renewed on a big scale during the Pleistocene, in response to further uplift which raised the surface to heights of about 2000–3000 m. Corresponding to the infilling of the 'buried gorges' of the Thames and other rivers, there is some evidence that there were intervals of heavy deposition in the canyons during the Pleistocene, but on balance deepening has predominated. During recent time erosion has been more vigorous than weathering. Much of the region is bare rock, and what little soil there is remains poor and thin. Although the river has to cross hundreds of kilometres of desert country, it receives sufficient water from the Rockies to carry it successfully through.

The Grand Canyon has now reached a maximum depth of about 1900 m. The river drops 457 m in the 322 km stretch above L. Mead, which is 380 m above sea-level. A narrow inner gorge has been cut through 300 metres of crystalline rocks (Figure 19.33). The walls above, carved through a nearly flat Palaeozoic cover of strong

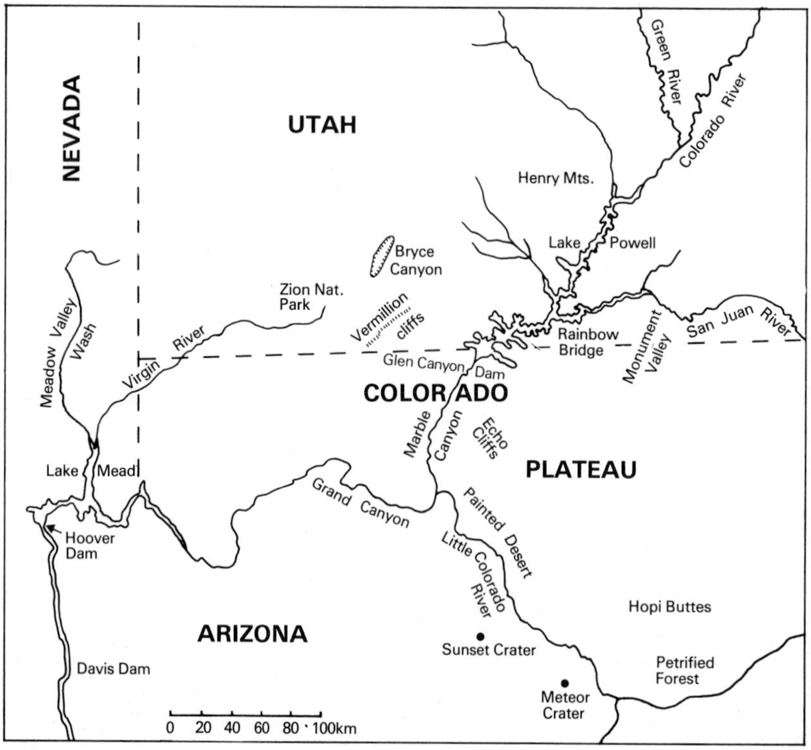


**Figure 19.32** Sketch map of the Colorado Plateau to show the localities illustrated in Figs. 1.6, 6.15, 6.16, 11.9, 11.10, 17.10, 18.37, 19.16, 19.28, 19.29, 19.31, 19.33 and 19.34

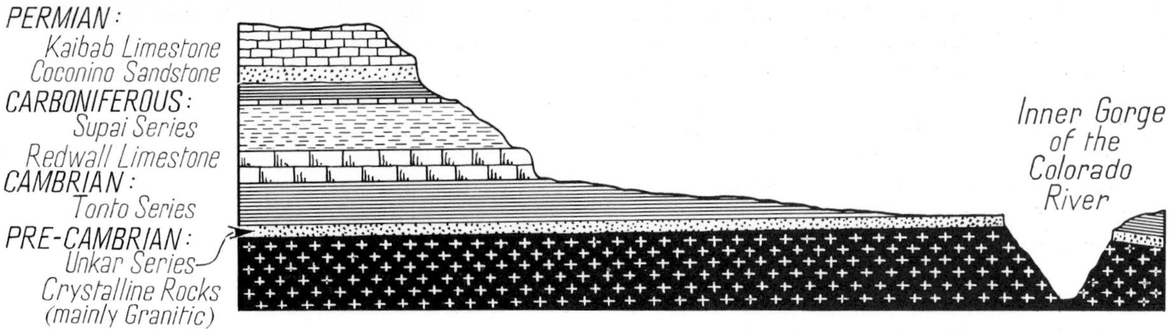


**Figure 19.33** Section across the Grand Canyon of the Colorado River at Grand Canyon Station, Arizona. Scale approximately 1 in 63,000 (horizontal and vertical) (*After Darton*)

sandstones and limestones alternating with weak shales, rise by a succession of steps and slopes of varied colours which add to the architectural grandeur of the scene. As a result of differential erosion the width of the Grand Canyon from rim to rim ranges between 8 and 24 km. During the recession of the walls bold spurs between the bends have been carved into pyramids and isolated pillars. The plateau is trenched by several tributary canyons (Figure 19.16), but otherwise the general surface is but little dissected. Northwards from the Grand Canyon the surface rises by successive cliffs and terraces (Figure 19.34) to over 3350 m in the high plateau of Utah, the edge of which has been sculptured by erosion into such

**Figure 19.34** Aerial view of part of the Grand Canyon region: looking over the north-western end of the Painted Desert and the Marble Canyon to the Vermillion Cliffs and the Plateau beyond (*G. A. Grant, United States National Park Service*)

fantastic landscapes as those of Zion and Bryce Canyons (Figure 17.25). North-east of Bryce Canyon are the famous stocks and laccoliths of the Henry Mountains (Figure 11.9). To the south-east of Bruce Canyon, on the other side of the Colorado river, is situated Rainbow Bridge (Figure 19.29); and farther east, south of the San Juan river, is Monument Valley, so called because of its obelisks and towers and other castellated erosion remnants, carved under arid conditions from red Triassic rocks. The celebrated Painted Desert stretches from the Marble Canyon (Figure 19.34) to the Petrified Forest (Figure 6.16) and beyond. All round the edge of the Colorado Plateau there are extensive spreads of lavas and tephra. South of the Grand Canyon the great cone of the San Francisco volcano rises near the edge of the Plateau and is surrounded by a vast volcanic field dotted with hundreds of small vents, many of them so perfectly preserved that they are obviously very youthful features. Some of the eruptions are known to have been quite recent. One that occurred at Sunset Crater in the year 1160 brought disaster to a Pueblo settlement. Not far to the south-east is Meteor Crater, a great depression over 1200 m across and 150 m deep, caused by the explosion which followed the impact of a giant meteorite. This brief outline gives but a faint idea of the variety and interest of the magnificent scenic and geological features for which the region is so justly renowned.

## Himalayan and Pre-Himalayan Rivers

By far the most remarkable examples of rivers that are generally accepted as antecedent are the Brahmaputra, and the Indus and its tributary the Sutlej. It should be noticed that the term *antecedent* is not applied to the rivers of drainage basins in areas that have been bodily uplifted. If it were, a great many rivers which have inherited their courses from an earlier cycle would have to be described as 'antecedent' and the term would lose its distinctive value. The significant criteria in the case of the Himalayan rivers are (*a*) that the river has sawn its way through ranges with peaks that now rise high above the level of the source; and (*b*) that the deep valleys or gorges traversing the ranges have not resulted from headward erosion (e.g. by waterfall recession), which would have made possible the capture of streams flowing on the far side (Tibet) of what would have been the original watershed. As usual in such cases it may not be easy in practice to dispose of (*b*) with any confidence.

All the above-mentioned rivers have characteristics favouring a pre-Himalayan origin. They rise in Tibet well to the north of the highest peaks. The Brahmaputra flows eastwards (and there it is called the Tsangpo), while the others flow to the north-west: all three for long distances before they abruptly change direction, turning *towards* the mountain barriers and passing through them by way of stupendous gorges, some of which are deep slit-like gashes cut in the bottom of V-shaped valleys. This latter feature points to an increased rate of rejuvenation during Pleistocene and recent time. As the Indus, for example, leaves Kashmir in the neighbourhood of Nanga Parbat (8117 m), the river itself is about 900 m above its delta, but the precipitous walls by which it is confined rise almost vertically at first, and then by a series of steps, to 6000 m. Like a gigantic saw the river has cut through over 5000 m of rock, keeping pace with an uplift of the same order.

The Himalayan tributaries of the Ganges (or Ganga, as the river is now called in India) nearly all descend from glaciers on the southern side of the Great Himalayan crest line (Figure 19.35), but two or three have their sources in Tibet, i.e. on the northern side, and so may be antecedent. Of these, the Arun, one of the seven rivers that unite to form the Kosi (p. 351), is the best known, and yet there is still controversy as to whether it should be regarded as antecedent or not (Figure 19.36). The Arun rises at a height of about 6706 m, and after dropping to about 4267 it becomes a braided river flowing eastwards through a valley flanked with terraces of gravel. At about 3962 m it abruptly turns and plunges into the heart of the mountains, first through the relatively short Yo Ri gorge and then through a much longer one, between the great mountain groups of Everest and Kanchenjunga, which carries it down to about 1219 m.

Some Indian geologists think that the gorges have been cut by headward erosion, thus enabling the Lower Arun to capture the Tibetan streams in the rear, which are now its headwaters. A good deal of river capture must inevitably have been involved, but there are serious objections to the headward-erosion interpretation of the gorges. If the Lower Arun had been able to cut back through the crest at its highest point, one might reasonably expect several other Himalayan rivers to have been equally successful where the crest line was not so high. In 1937 L. R. Wager discussed the problem

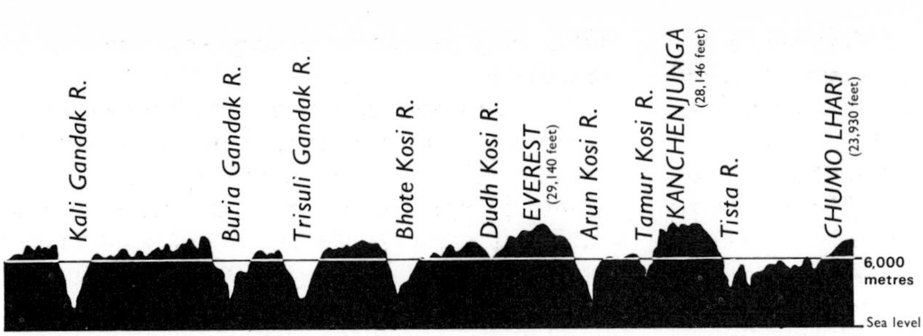


**Figure 19.35** Section along the Great Himalaya of Nepal, showing the dissection of the Range by rivers, some of which have their sources on the Tibetan side, nearly 100 km to the north. Length of section about 685 km (*After E. H. Pascoe*)

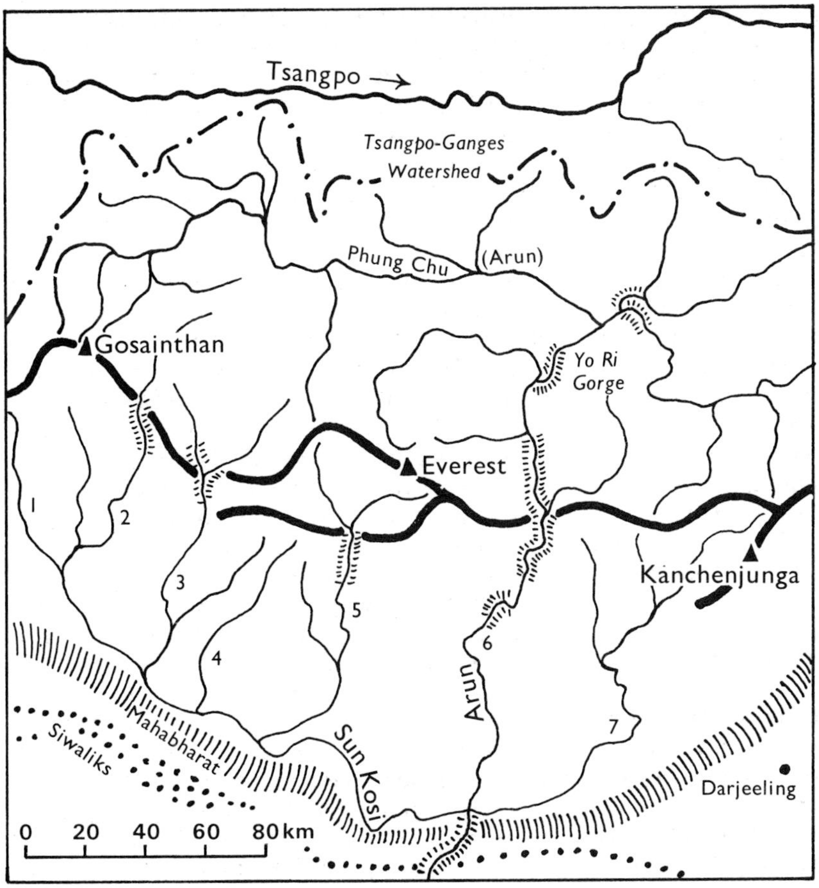


**Figure 19.36** Sketch map to show the gorges of the Arun River, which is one of the seven feeders of the Sapt Kosi (see Fig. 18.35): (1) Indrawati (2) Sun Kosi (3) Bhote Kosi (4) Tamba Kosi (5) Dubh Kosi (6) Arun (7) Tamur

after exploring the critical region during the Everest expedition of 1933. He found that if the Arun had extended itself through the Yo Ri region by headward erosion it could most easily have done so by working along a belt of weak schists. In fact the Yo Ri gorge has walls of highly resistant gneiss. The river has deeply entrenched itself in the hardest rocks of the district and it could have done this only by accidentally coming upon these rocks from above, at a time when its downstream course had already been firmly established. For this and other reasons Wager concluded that the course of the Arun was established long ago on a 'mountain slope which formed the descent from the Tibetan Plateau to the Ganges plain, in those days perhaps an arm of the sea. No such surface exists nowadays because since that time the range of the Himalaya has risen across the middle part of the river's course.'

More recent French explorations suggest that the mountains are probably still rising against the down-cutting Arun, or at least have been doing so quite recently. On both sides of the valley to the north of the Great Himalaya the gravel terraces slope up towards the mountains. It has not yet been recorded whether the higher terraces slope more steeply than the lower ones. However, such *is* the case on the southern side of the mountains, where the upper terraces flanking the Ganges alluvium have been tilted up towards the foothills (the Siwaliks) slightly more than the lower ones. That movements are still in progress is indicated by the occurrence of earthquakes.

Like the Colorado river the Arun has not continuously maintained its course against the rising ground. Locally and at times it has been what Hunt would call an anteposed river. About half-way between the great gorge and the Ganges plain the Arun has cut a channel over 150 m deep through a broad plateau of sediments. These appear to be the infilling of a lake formed at a time when the uparching of one of the Middle Himalaya branches was sufficiently active to pond back the river until it could spill over and resume its downward erosion.

The 'Valley' of Katmandu (the capital of Nepal) has had a similar history. The valley is floored with deltaic lake sediments which everywhere slope inwards towards the city, thus forming a kind of amphitheatre surrounded by spurs of the Middle Himalaya (see Figure 19.37 for heights) and, on the south, by the Mahabharat range. There is only one exit for the Bagmati river, which drains the amphitheatre, and that is by way of a deep gorge through the Mahabharat, cut after the latter had risen to form a natural dam. Apart from basins of purely internal drainage, the Upper Bagmati and its tributaries make what must be the world's most remarkable example of centripetal drainage. Along the line of exit the Bagmati flows southwards towards the Ganges, while less than 2 km to the east the Nakhu Khola flows due north, to join the Bagmati just before reaching Katmandu.

Wager has suggested that the great peaks of the Himalaya owe part of their additional elevation over the average height of the Tibetan Plateau to the effect of isostatic readjustment of the kind discussed on p. 372, i.e. to the uplift following the relief of load where deep valleys have been eroded. Instead of a 300 m block carved into hills and valleys, we have here a 5000 m block. Given a proportionate relief of load, the resulting isostatic

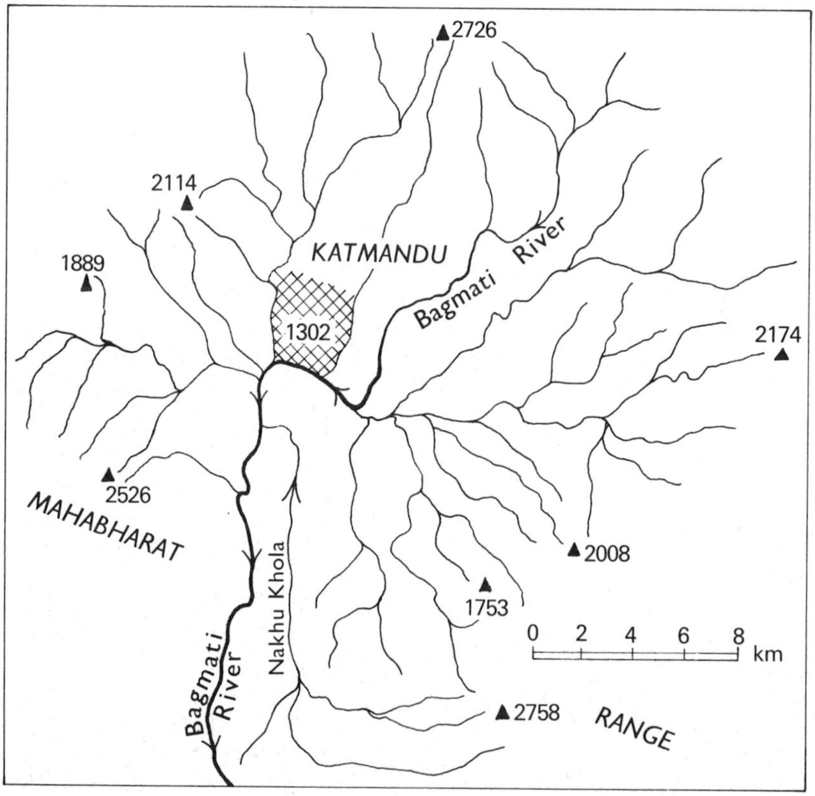


**Figure 19.37** Centripetal drainage pattern of the Valley of Katmandu, Nepal. Heights in metres (*After H. L. Chhibber*)

reaction would theoretically upwarp the serrated margin of the Plateau, so elevating the ridges and peaks of the Himalaya to heights some 1900 m higher than they would otherwise have been. However, while this little exercise emphasizes the importance of isostasy in raising a mountain range along the deeply eroded edge of a plateau, it also indicates that some other process or processes must be at work in the depths to account for the exceptional heights of the Everest and Kanchenjunga groups of peaks, and of the Karakoram giants.

## Uplifted Erosion Plains

Erosion plains representing the partially completed cycles of former periods, but since uplifted to form the initial surfaces of later sub-cycles, can be detected in many landscapes. In the Grampian Highlands an old erosion plain, now dissected into a landscape of late youth or early maturity (though lately rejuvenated by glaciation and uplift), is easily recognized by the even skyline seen from one of the higher peaks (Figure 19.38). This widespread uniformity is due to the abundance of 'accordant' summit levels at about 600 m. The plane through these summits—the 'summit plane'—is commonly referred to as an ancient peneplain; while the occasional peaks that rise to higher altitudes have been interpreted as the relics of the monadnocks that diversified its surface. As already indicated on p. 372, this nomenclature would be better replaced by terms free from hypothetical implications. Whatever kind of erosional origin the Grampian summit plane may have had, we can at least be certain that it was produced later than the Tertiary dykes, since some of these have been truncated by it. This observation is in harmony with our earlier conclusion that, apart from the effects of glaciation, the British landscapes are largely the products of erosion stimulated by a series of Tertiary earth movements, accompanied by eustatic changes of sea-level. The factors concerned are so numerous that their

**Figure 19.38** The sky-line of the Grampian Highlands of Scotland, looking south from Ben Nevis. An example of summit levels representing an uplifted erosion plain which has been deeply dissected by rivers and former glaciers (*Institute of Geological Sciences*)

individual influence is not easily recognized in the 'old lands', such as Wales, the Lake District and the Highlands, which have lost whatever cover of Chalk etc. they may once have had. It is notable, however, that in all these areas, and also in the Southern Uplands of Scotland and in various parts of Ireland, a summit plane roughly corresponding to the 600 m level can be detected, as well as others leading down like a staircase to the 180-metre marine platform that appears to herald the beginning of the Pleistocene in so many places.

For a variety of reasons the summits and hilltop flats belonging to a particular erosion plain cannot be expected to be more than roughly accordant. Graded valley floors and pediments of the present day rise progressively as they are followed inland, and so do sea floors and marine platforms as the shore is approached. Genuine accordance would therefore demand a similar regional slope. But then there are monadnocks and inselbergs of all sizes; there are variations of level due to structural inequalities in the rocks exposed to erosion; and there are earth movements that may have warped the surface or sliced it through with fault scarps that can be particularly misleading. The most one can reasonably expect to establish in tracing a surface is that each remnant should be consistently placed with regard to its neighbours. What is wanted is 'an eye for the country' as the older geologists used to say.

One of these, A. C. Ramsay, an early Director of the Geological Survey, wrote: 'To the eye of one who appreciates the physical features of Wales there is . . . nothing more striking than the average flatness of the tops of many of the hills; . . . a flatness, be it remembered, not connected with anything like a horizontal position in the beds, for everywhere they are contorted and often stand on end.' He had recognized the '2000-foot (*c.* 600 m) summit plane' as early as 1846, and realizing that it cut across a great variety of rocks and structures he correctly interpreted it as an old erosion surface (Figure 19.39). However, he thought it represented the remnants of a wave-cut platform and this idea prevailed for some sixty years. Then Davis came and claimed that Ramsay's platform was a peneplain. It is now generally agreed that this and a number of other hilltop surfaces in Wales (at levels down to 200 m or so) owe their form to sub-aerial erosion. Despite some growing doubts they are still called 'peneplains' by most writers, including E. H. Brown, who has given us an inspiring and coherent synthesis of the landforms of Wales and its borders (see p. 401). Brown concludes 'that the lower or coastal plateaux are in all probability the work of the sea, i.e. wave-cut platforms, whilst the higher dissected plateaux are the result of long-continued erosion by rain and rivers', all more or less modified by ice erosion since they were formed. 'Together', he writes, 'they form a gigantic staircase from the highest upland plain, which impressed Ramsay so much, down to the lowest raised beach around the coast. Each tread in the staircase is a plateau separated from its neighbours below and above by breaks in slope, i.e. more steeply sloping portions of the landscape.'

The treads of the staircase from top to bottom may be summarized as follows:

1. A high-level *Summit Plain* along the crests of the 'monadnock' group of peaks, including those of Snowdonia, rising to well over 900 m; Snowdon being 1073 m. This surface was domed at the beginning of the Tertiary, at which time it had a cover, probably mainly consisting of Chalk. By the removal of this cover it was exhumed and subjected to further erosion.

2. The *High Plateau*, a series of flat-topped summits (*c.* 500–600 m) left by a cycle of erosion that was initiated by mid-Tertiary earth movements and ran its course during the Miocene.

3. The *Middle Erosion Plain* (*c.* 360–500 m), well represented on the Longmynd and the Denbighshire Moors (Clwyd).

4. The *Low Erosion Plain* (200–335 m), also recognized around Dartmoor and up its valley floors, and traced eastwards into Hampshire. Then followed the 180-metre wave-cut platform (*c.* 150–200 m), which was succeeded in turn by others at levels of approximately 120, 90 and 60 m, and so on through the Pleistocene.

**Figure 19.39** Projected section through Wales to show the summit levels representing the uplifted erosion plain known as the '2,000-foot High Plateau'. Length about 280 km (*After E. H. Brown*)

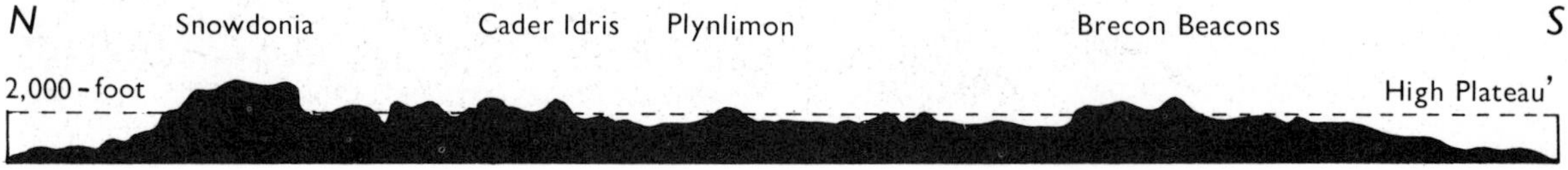

## Erosion Surfaces of Southern Africa

By far the most spectacular examples of the co-existence of landscapes of successive cycles of erosion are to be seen in Africa. They have been made familiar by the writings of Frank Dixey and Lester King, both great geological explorers of the continent in which most of their work has been done. King, like Crickmay, has insisted that such polycyclic landscapes are incompatible with pene-planation, but are a natural consequence of scarp retreat. Almost flat remnants of old erosion surfaces dating from the Jurassic have been preserved almost intact for something like 140 million years, not merely because their bedrocks are highly resistant types or are strongly protected with an armour of laterite, but essentially because they are vulnerable to denudation only from their flanks, downward erosion having been negligible. The retreating scarps by which they are bounded remain youthful throughout their active existence (Figure 19.40), while the summit flats, which are being slowly consumed by the scarps, continue to represent extreme old age and may, indeed, be practically dead. However, the lower 'treads' of the erosion 'staircase' continue to advance inland, provided that the run-off and the drainage system are adequate to maintain at least some retreat of their boundary scarps. Any outlying parts of these lower surfaces that become isolated (e.g. the summit flats of mesas or kopjes) may also remain stagnant for long periods. As we have seen, this condition of stagnancy is not peculiar to Africa or its climates, since remnants of high-level erosion surfaces are also well preserved in Britain and in British Columbia. But, while not forgetting these accidents of isolation, King envisages the African landscape in its most dynamic aspect as 'a procession of scarps sweeping inland' (Figure 19.41).

**Figure 19.40** The Great Escarpment of the Natal Drakensberg, showing gullies in the Jurassic basaltic lavas (over 1200 metres thick) on the precipitous walls leading up to the 30,000 metre plateau of Lesotho (Basutoland). (*South African Air Force*)

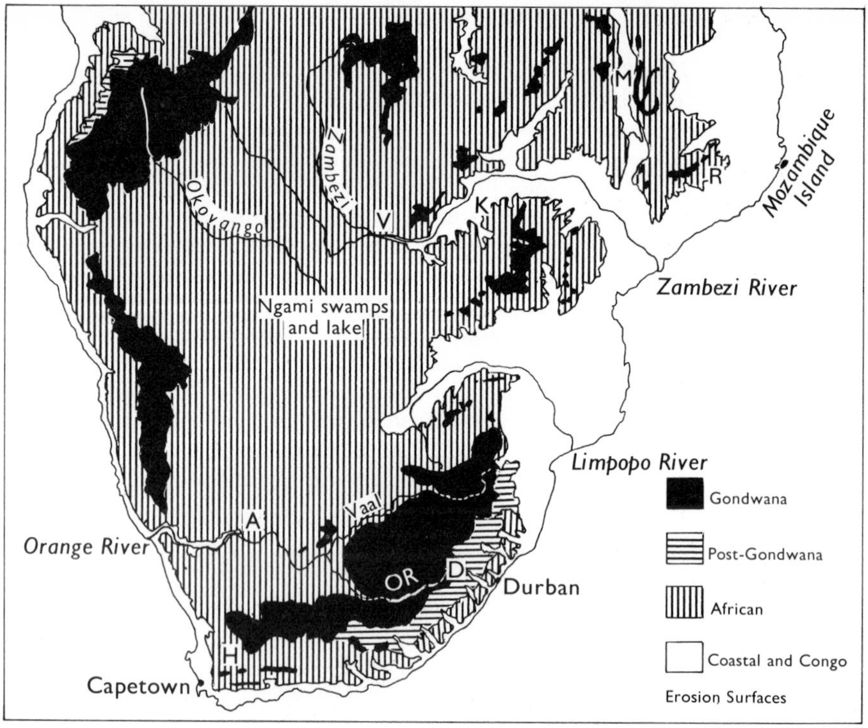


**Figure 19.41** Erosion surfaces in southern Africa. A, Aughrabies Falls (Fig. 19.44); D. Drakensberg (Fig. 19.40); K, Kariba Dam; M, Lake Malawi; OR, Orange River headwaters; R, Ribaue Peaks (Fig. 19.47); H, Hex River (Fig. 19.45); V, Victoria Falls (Figs. 18.17 and 18.18)

At the close of the Triassic the long period of accumulation of sediment in the Karroo basins and troughs was brought to an end by the intrusion of innumerable sills and dykes, and the outpouring of floods of plateau basalts, which continued into the early Jurassic. The basalts form the boldest crags of the Drakensberg (Figure 19.40) and cover large areas around the Victoria Falls (Figure 18.17). Ever since this 'blaze of volcanic fury' the geological history of much of Africa has been characterized by a series of intermittent uplifts on a continental scale. There have been great depressions, too: making room for over 3000 m of Cretaceous sediments in the Mozambique Channel and over its shores, and for over 12,000 m—also marine Cretaceous—in southern Nigeria. But here we are concerned only with the main upheavals and the cycles of erosion that each one initiated.

Sequences of five well-defined erosion surfaces have been recognized in Africa from Madagascar to Angola and Zaïre, and from Uganda to the Cape. Detailed work in limited areas has already shown that some of the 'standard' surfaces are composite and that all of them have local features corresponding to minor pulses and warpings, to say nothing of the major interruptions by rift valleys and such giant uplifts as Ruwenzori. Nevertheless the five main surfaces serve well as a provisional basis for discussion. The main difficulty in correlating erosion surfaces over vast distances lies in determining the geological ages or events to which they should be referred. A surface begins at the coast in Cretaceous times, let us say, and extends itself half-way across Africa, to end at the foot of a scarp or of a waterfall if it has been progressing up a river. There, where it is still working inland, its age is the present. But if it can be traced back towards the coast, it may possibly be found bevelling middle Cretaceous sediments or lavas; and nearer the sea it may be seen emerging from an unconformable cover of, say, Eocene beds. This evidence would enable us to date that spread of the surface as approximately late Cretaceous. In all such cases it is clearly desirable to date a surface from the earliest time when some part of it first became really distinctive. So far as evidence is available, dating on these lines is generally found to be consistent with the geological age of the main uplift responsible for initiating the erosion cycle concerned.

The five surfaces referred to are known by the following names, the uppermost being the oldest:

| **Main erosion surfaces** | **Sequence of main geological events** |
| --- | --- |
| | *Outpouring of Lower Jurassic plateau basalts* |
| **Gondwana*** **surface** | Erosion during and since Jurassic |
| | *Break-up of Gondwanaland*: pronounced faulting; downwarping of coastal regions* |
| **Post-Gondwana surface** | Erosion from early or middle Cretaceous (according to locality) |
| | *Middle to late Cretaceous uplift* |
| **African surface** | Erosion from late Cretaceous or early Tertiary (according to locality) |
| | *Middle Tertiary uplift* |
| **Coastal Plain surface also called Victoria Falls surface** | Erosion during and since Miocene |
| | *End-Pliocene to recent uplifts (including fluctuations due to oscillating sea-levels)* |
| **Zaïre surfaces** | Erosion from Plio-Pleistocene to present |

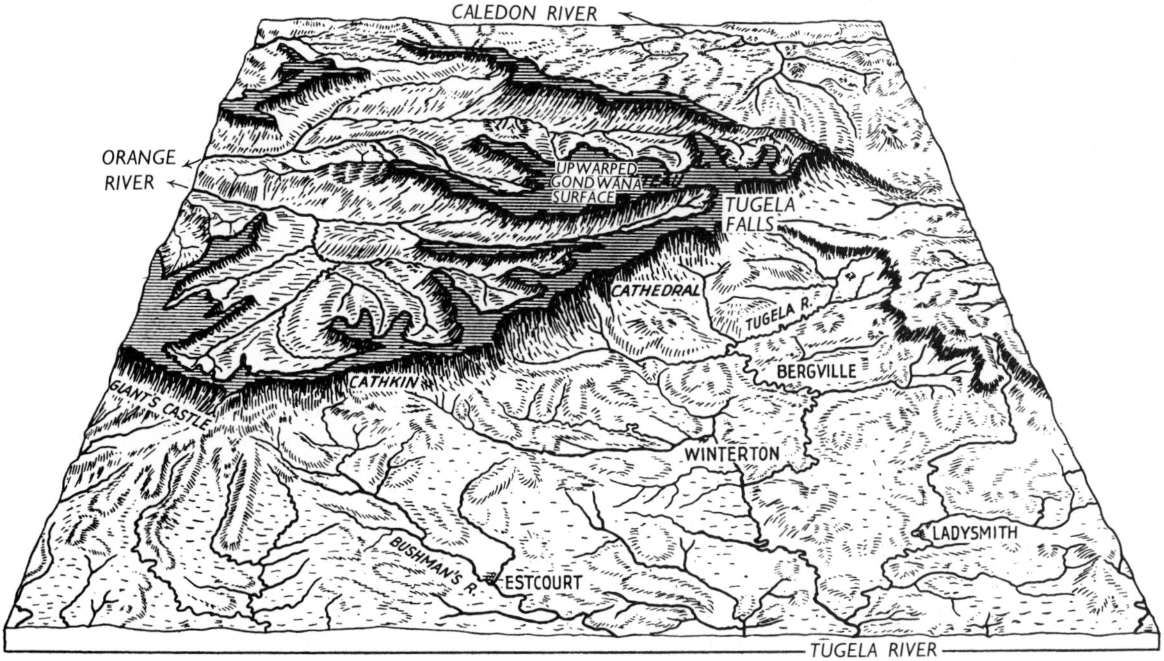


**Figure 19.42** Block diagram of western Natal and Lesotho (Basutoland), looking a little S of W showing remnants of the upwarped Gondwana erosion surface above the Great Escarpment (*After Lester C. King*)

* *Gondwana* or *Gondwanaland*: a collective name proposed by F. E. Suess for a great Southern Continent comprising South America, Africa, India, Australia and Antarctica. Soon after the close of Palaeozoic times Gondwanaland began to break up into gigantic crustal fragments that gradually drifted apart into the positions they now occupy (Chapter 28).

The *Gondwana* surface now much dissected by the headwaters of the Orange river (Figure 19.42), rises to its greatest height (3292 m) in the plateau of Lesotho (Basutoland), which is separated from Natal by the great castellated wall of the Drakensberg (Figure 19.40). This basalt-capped scarp is the most spectacular part of the Great Escarpment, which faces the stepped landscape leading

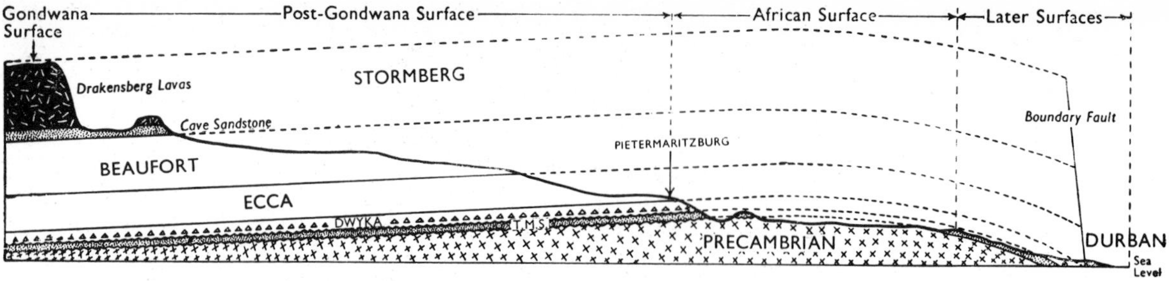


**Figure 19.43** Generalized section from the Drakensberg on the west to the coast of Durban on the east, to illustrate the recession of the Great Escarpment since Jurassic time. T.M.S. = Table Mountain Sandstone (late Silurian or early Devonian), resting uncomfortably on Precambrian rocks and covered unconformably by late Carboniferous tillites and other strata known as the Dwyka Series. Late Karroo sills and dykes, and minor faults and flexures omitted. Length of section 200 km (*After Lester C. King*)

down to the coast with but little interruption here and there, from the Limpopo southwards and westwards towards the Cape and then northwards across the Orange river into Damaraland and Angola (Figure 19.41). The Drakensberg, which now forms the high western boundary of Natal, affords a magnificent example of scarp recession (Figure 19.43). It began in a small way near the present coast about 140 million years ago and during that time it has been intermittently uplifted and warped, probably to a considerable extent by the isostatic response to unloading by denudation. At the coast a thickness of over 3900 m of rock has been removed, corresponding very nearly to an average rate of 1 cm in 350 years; but over the whole of Natal—from the coast to the escarpment—this figure is reduced to 1 cm in about 700 years. The average rate of scarp recession has been 225 km in 140 million years, or 1 cm in 6 years. However, a glance at Figure 19.43 shows that this estimate covers an enormous range from place to place, and the rate must also have greatly varied from time to time.

On the plateau side of the escarpment the headwaters of the Vaal and Orange rivers begin their long journey to the Atlantic, while the shorter rivers of Natal begin as mere trickles which descend the escarpment. Not far from its source the Tugela river plunges over the brink of a precipitous amphitheatre in the escarpment. Five clear leaps with intervening cascades make up a total drop of 856 m. Though of insignificant volume, the Tugela Falls are probably the highest in the world.

The *Post-Gondwana* surface is in many places not yet clearly distinguished from those above and below it. However, it is well established in Natal and the Cape Province (Figure 19.41), and also in Angola, where there is an equally well-developed series of steps rising inland from Benguela. It is probable that the Zaïre and Kalahari Basins began as downwarped depressions of this surface.

The *African* surface, as its name suggests, forms the most extensive plateaus of southern and eastern Africa. Figures 18.17 and 19.44 both show the surface admirably: one at the Victoria Falls, where the Zambezi makes its abrupt descent into the gorge which it has cut deeply back through the African surface; the other at the Aughrabies Falls, where the Orange river has a much more difficult task in sawing through the tough and compact granites and migmatites of a grim and desolate semi-arid land. Along the borders of Uganda and Kenya the African surface supplied the foundation for Elgon and other great volcanic cones. Here an additional surface has been recognized, a good 300 m below the other, and it is a gentle downwarp of this lower surface that is occupied by the shallow Lake Victoria. Rift-valley faulting, however, has brought about changes in level amounting to several hundred metres. One of the 'African' surfaces has been uplifted far above the snow line to form the glaciated summits of Ruwenzori. And one has been depressed far below sea-level, to form the bedrock floor of Lake Tanganyika. Some of the rift-valley fault scarps have been notched by gorges with waterfalls at their heads. Near the SE end of the Tanganyika rift, for example, the Kalambo Falls have cut back 5 km from the lake, and there a stream flowing placidly across the African surface suddenly takes an unbroken leap of 215 m over the brink. On the western side of the rift the lake finds an outlet into the Congo by way of the Lukuga river, which carries the Coastal surface into the very heart of the continent, where it is represented by the lake surface and its narrow shores. Lake Malawi (Nyasa) is similarly rimmed by the Coastal surface, which slopes down the valley floor of the Shire river towards the Zambezi.

**Figure 19.44** Aerial view of the Aughrabies Falls and Gorge of the Orange River, South Africa. Here the nearly flat African erosion surface is deeply incised by the Coastal erosion surface, which extends inland as the Falls slowly recede. (*South African Air Force*)

The African surface is itself a broad valley floor in some places, where it is bordered by uplands whose summits represent a higher surface, generally the Gondwana (Figure 19.45).

The *Coastal* surface, already mentioned above, surrounds the basin of the Zaïre. It makes deep inroads into the continent up the Zambezi as far as the Victoria Falls; and up the Limpopo where, in striking contrast to the Zambezi gorges, the surface is wide and monotonous and has all the characters of extreme old age. The Limpopo only begins to be vigorous near the Messina copper mines, where it descends by way of a long stretch of cataracts to the Zaïre surfaces which correspond to the real coastal plains. Around south and south-west Africa the coastal plains include the Zaïre and Coastal surfaces, but the latter are too narrow to be separately distinguished on a small-scale map.

**Figure 19.45** Aerial view across the Hex River Valley and Mountains, NE of Cape Town (see Fig. 19.41). The summit level in the background represents a remnant of the Gondwana erosion surface, now largely isolated by the advance into the mountains of the African erosion surface (the valley flats). (*Aircraft Operating Company of Africa*)

The *Zaïre* surface, as its name indicates, covers a vast extent of a country in the basin of the Zaïre. It also makes a broad belt inland from the shores of Mozambique, particularly south of the Zambezi, up which river it has progressed as far as the recently damned Kariba Gorge. The Zaïre 'surface' includes a group of minor surfaces, many of which are represented by river terraces of the kinds already well illustrated from northern lands.

### Inselberg Landscapes

The inselberg type of landscape was first described (1900) under this name by W. Bornhardt, one of the early explorers of Tanzania. W. M. Davis later suggested that residual peaks having the characteristics of inselbergs should be called 'bornhardts'. This name is sometimes used, but it is more generally realized that if the peaks have the characteristics of inselbergs they might as well be called inselbergs. In 1904, after seeing the equally remarkable inselbergs of parts of the Kalahari and Namibia (South-West Africa), S. Passarge convinced himself and many others that they were the typical residual hills of desert regions where resistant crystalline rocks had long been exposed. However, since it is now known that inselbergs and pediments also occur in humid regions, and that the Kalahari and some other deserts have been humid regions in the past, the Passarge hypothesis has become untenable. In 1911 the inselberg landscape was described by J. D. Falconer from Nigeria; and in the same year A. Holmes and his fellow explorers E. J. Wayland and D. A. Wray independently discovered it—with growing astonishment—in Mozambique. Here is how two of them afterwards described it:

> 'Proceeding westwards from the coast, the country begins to be diversified with isolated peaks or clusters of hills which rise abruptly from the surface of the

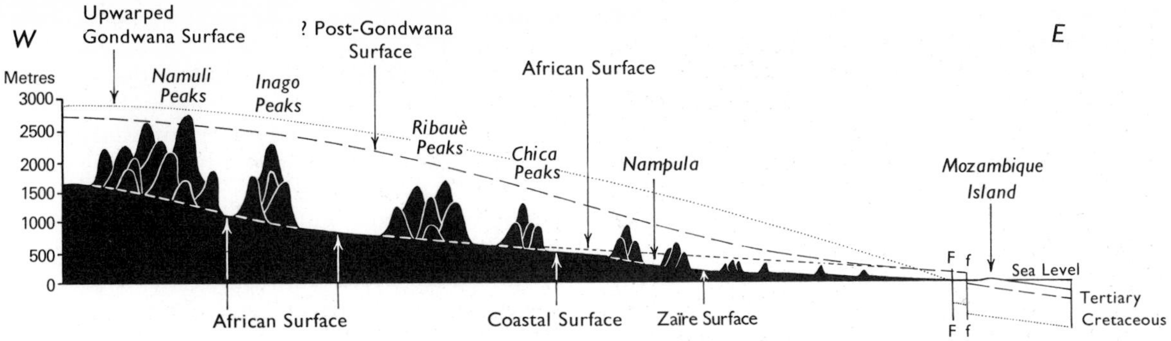


Figure 19.46 Diagrammatic section across Mozambique to illustrate the relation of the inselberg peaks to the erosion surfaces, which here are well developed pediments. FF and ff represent faults near the coast. Length of section 450 km.

plateau and exhibit the most remarkable outlines. Within 30 miles of the coast these island mountains are not numerous, and only rise a few hundred feet above the surrounding country. Further inland, however, they gradually increase in numbers and bulk, until, assembled together in picturesque chains or hurled up, peak upon peak, in towering blocks of rugged gneiss, they cooperate to form fantastic and impressive mountain systems, which arrest the attention by the colossal scale of their daring architecture. The summits of the peaks are of every conceivable shape, varying from gracefully rounded domes of smooth and naked gneiss to irregular knobs and pinnacles. In general, the outlines approximate to smooth curves above, dropping precipitously to the plateau surface below. . . . The most striking feature of these detached groups of hills lies in the abruptness of their discontinuity with the plateau.' (A. Holmes and D. A. Wray, 1913, *The Geographical Journal*, London, vol. 42, pp. 143–52)

Wayland's sketch (Figure 14.5), and Figure 19.48 give a better idea of these fantastic peaks than any description in words. Occasionally summits are seen that approximate to angular or rectangular knobs, with only slightly rounded edges. These result from the fall of rock masses, liberated by fracture along foliation planes or the opening-up of joints, too recently for exfoliation to have restored the normal convexities. At that time (1911) all had to admit that the origin of this type of scenery was not properly understood. But there was general agreement that inselbergs were best developed from tough and massive crystalline rocks, offering more than average resistance to chemical weathering, and so capable of standing up to considerable heights with precipitous sides; many gneisses and granitic rocks fulfil these conditions. In Mozambique it was noticed that on the flanks of some of the inselbergs the gneissic foliation dips radially outwards with increasing angles from the crest, in places being roughly parallel to the surface. This domical structure may continue for some distance over the surrounding pediment, where the dips are steep. In some examples, e.g. those of Figure 14.5, the inselbergs consist of granitic rocks around the crest or farther out, passing into migmatite and gneiss on the slopes or on the adjoining pediment. But others are completely independent of this dome-like or 'mantled dome' structure, the banding of the gneiss being gently inclined and sharply truncated by the steep sides, as seen in Figure 19.47.

**Figure 19.47** Embayment pediment rising from the African pediplain through the Ribaue group of inselberg peaks, Mozambique. A. Holmes is sitting on a pile of boulders representing the wastage of a former midget inselberg. In the background, is East Peak, over 300 metres above the pediment. (*R. L. Reid, 1911*)

An inselberg peak is not merely a core of rock which has been preserved in virtue of its structural soundness while the weaker surrounding rock has fallen a prey to erosion. The surrounding rock, now part of a pediment, is generally just as tough and massive as the inselberg that rises above it. Probably the inselberg itself is only a small remnant of an original large block of resistant rock which became separated from a retreating escarpment after the manner already described. With the history of the erosion surfaces of Africa made familiar by Dixey and Lester King, the mode of development of the inselberg landscape has become clear. Figure 19.46 is a schematic attempt to show the relationship of the Mozambique inselberg peaks and groups of peaks to the various erosion surfaces, all of which are well-defined pediments. The highlands on each side of Lake Malawi (Nyasa) and the Shire river have been carved from remnants of the upwarped Gondwana surface. Corresponding to the Namuli peaks in Mozambique, rising more than 1500 km above the African surface, is the Mlanje plateau of southern Malawi, which rises 2100 m above the African surface. The Gondwana and African surfaces disappear beneath marine sediments near the coast: sediments which are only the feather edge of immense thicknesses that accumulated in the Mozambique Channel. The latter subsided to great depths while the mainland on the west and Madagascar on the east were rising and providing rock-waste. The erosional history of Madagascar, as Dixey has shown, resembles that of the mainland almost as closely as if it were a mirror image.

With each major uplift a new cycle of erosion started at the coast, encroaching on the pediment of an earlier cycle (now raised to a plateau), advancing up the rivers and their tributaries, wearing back the escarpments, but leaving massive bastions and projections behind to be slowly worn into groups of inselbergs and eventually into isolated peaks (Figures 14.5 and 19.48). Figure 19.47 shows a pediment embayment penetrating the Ribaue group of inselbergs. Here the slope is steeper than that of the well-developed African pediplain of which it is an extension. This particular intermontane pediment serves as a pass, since it rises to meet a similar pediment which has advanced into the heart of the group from the other side. The pile of irregular blocks and rubble seen in front of the hut on Figure 19.47 represents a very late stage in the wastage of a tiny inselberg that once interrupted the downward sweep of the pediment. The hut was part of the base camp of the expedition and before it was erected the pediment was cleared of its thin cover of thorn and scrub. The vegetation, however, was insufficient to

**Figure 19.48** Miniature finger-like inselbergs, Katsina, Northern Nigeria. (*Dorien Leigh Ltd*)

prevent the descent of occasional sheet-floods that scoured the pediment during periods of exceptional rainfall. At such times water streams down the steep sides of an inselberg in closely packed rills, or even like a thin waterfall. The inevitable hydraulic break (p. 327) at the base is responsible for the sharp angle often seen between inselberg and pediment. But, as the illustrations show, a smooth gradation from steeply convex to gently concave slopes is not uncommon. A sudden change of slope seems to be favoured by torrential seasonal rainfalls and by the liberation of only minute amounts of fine debris which can be readily swept away by sheet-wash over the pediment. A gradual change of slope is favoured by the liberation of coarser debris (e.g. by exfoliation) which accumulates for a time at the base, so providing interspaces for the retention of smaller particles and opportunities for the growth of vegetation and soil. But, as indicated on p. 322, investigation of the headward development of pediments has only just started. Much more detailed work in a variety of climates remains to be done.

In terms of slope nomenclature, the *free face* controlled by gravity and the *pediment* controlled by sheet-wash are the dominant features of the inselberg landscape. The essential requirements are strong rocks and repeated uplifts over sufficiently long intervals of time to enable erosion to provide the necessary relief between the summits of the inselbergs and the lower levels of the intervening pediments. The latter may be no more than broad valley floors, as in the Matopo Hills of Rhodesia, which have been described as 'an archipelago of inselbergs'; or they may be surprisingly extensive plains, like those of central Australia (Figures 17.6 and 21.25). The inselberg landscape of Brazil is of interest in two respects: it has developed in a tropical humid climate; and along the coast it has been partially drowned by recent downflexing of the land (Figure 19.49).

**Figure 19.49** The Sugar Loaf, Rio de Janeiro, Brazil, an inselberg peak that has been partly submerged (*Dorien Leigh Ltd*)

If the rocks exposed are not sufficiently strong and cohesive through a sufficient thickness to allow inselbergs to be sculptured from them, the residual hills left between successive pediments generally take the form of kopjes and buttes (p. 368). Here the rocks are usually stratified, perhaps armoured on top by laterite, or capped by indurated sediments or by resistant volcanic rocks or sills. Where the rocks are weak and impermeable, and rainfall is adequate, the conditions are favourable for the development of badlands between uplands and adjoining plains (Figures 17.24 and 22.24): commonly, that is, between pediments at different levels. In terms of drainage density the badland and inselberg types of landscape stand at opposite extremes. *Drainage density* is the sum of the stream-channel lengths in a given drainage area divided by that area. The increase of run-off that follows removal of forests leads to accelerated erosion and gullying, and so to the high drainage densities—500 to 1000—characteristic of badlands. On the other hand, where rocks are strong and/or effectively protected by vegetation, the drainage density may fall to a single figure. In many typical inselberg landscapes it is no more than 3 to 5 km of stream-channel per 3 km$^2$.

## The Origin of Tors

The boldly upstanding piles of joint-blocks of highly resistant rocks known as *tors*, such as the granite tors of Dartmoor (Figure 19.50), have commonly been regarded as residual features left behind by the discriminating hand of superficial denudation. This view is unsatisfactory because, as D. L. Linton and others who have studied the problem have found, tors are now being more rapidly weathered and eroded than the rock platforms on which they stand. Some authors have

**Figure 19.50** Bowerman's Nose, Manaton, Devon—Dartmoor tor; a residual of massive granite resistant to weathering and with widely spaced joints (*Fox Photos Ltd*)

claimed that tors are near relatives of inselbergs and have a similar origin. This view is also unsatisfactory, because few tors are as high as 15 or 18 m, whereas inselbergs may be ten times as high. It is interesting, however, to notice that a long-discarded hypothesis proposed by Falconer to account for the inselberg landscape of Nigeria now finds a successful application to the problem of tors. Faloner pictured a region of granite and gneiss rotted by chemical weathering to considerable but unequal depths, according to the composition and structure of the various rocks. 'When elevation and erosion ensued, the weathered crust would be removed, and an irregular surface would be produced, from which the more resistant rocks would project.' It is well known that certain granites suffer deep sub-soil decomposition and lose their coherence where closely spaced joints facilitate the downward migration of groundwater, acidified by its passage through the soil. But locally, where the joints are far apart, massive blocks and rounded cores of relatively fresh rock may remain, as in the early stages of spheroidal weathering (p. 255). These are the tors of a future time, ready to be exhumed by a process of erosion that removes the rotted material (Figure 19.51).

This is the basis of the hypothesis developed by Linton in 1955. He defines a *tor* as 'a residual mass of bedrock produced below the surface level by a phase of profound rock rotting effected by groundwater and guided by joint systems, followed by a phase of mechanical stripping of the incoherent products of chemical action. . . . Tors may rise sharply from a *basal platform* of bed-rock which may be flat or inclined and is interpreted as representing the position of the water-table during the period of rock-decomposition, so that tor height cannot exceed the depth of the vadose zone of that period.' This hypothesis is supported by the facts:

(*a*) that quarrying operations have revealed small 'tors' still embedded in incoherent rock, but not yet naturally exhumed because of their low-lying position in the landscape;

(*b*) that the more conspicuous tors rise from summits and spurs, where the operative depth to the water table would originally be greatest;

(*c*) that there were long pre-glacial and interglacial periods in the past (Pliocene and Pleistocene) when the climate was warmer and more humid than today, thus favouring deep weathering in the appropriate places; and

(*d*) that solifluction was very active in British and similar tracts in front of ice-covered regions, and would be a highly effective process for the removal of loose material from uplands and slopes. In tropical lands where tors occur, rejuvenation by uplift to base-level would provide favourable conditions for their exhumation.

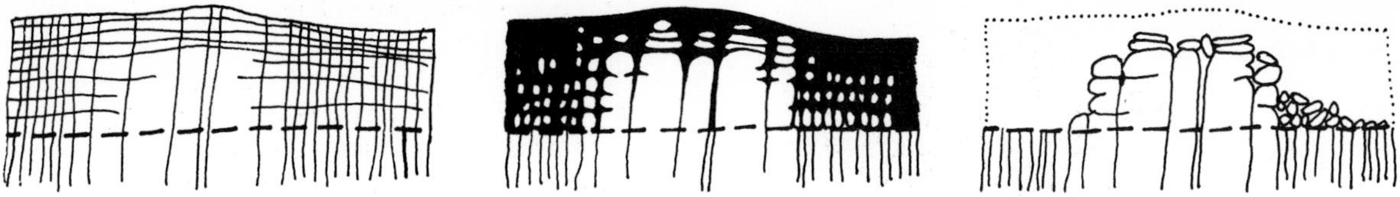


**Figure 19.51** Stages in the evolution of a granite tor:
(a) vertical section of fresh granite with varied spacing of joints
(b) after a period of rotting by percolatry ground-water down to the level of permanent saturation; decomposed rock, black
(c) decomposed rock removed leaving a tor rising abruptly from the surrounding surface

## Selected References


BROWN, E. H., 1960, *The Relief and Drainage of Wales*, University of Wales Press, Cardiff.

CRICKMAY, C. H., 1933, 'The later stages of the cycle of erosion', *Geological Magazine*, Vol. 70, pp. 337–47.

— 1959, *A Preliminary Inquiry into the Formulation of the Geological Principle of Uniformity*, Evelyn de Mille Books, Calgary, Alberta.

DIXEY, F., 1956, 'Erosion surfaces in Africa', *Transactions of the Geological Society of South Africa*, vol. 59, pp. 1–16.

GEORGE, T. Neville, 1955, 'British Tertiary landscape evolution', *Science Progress*, vol. 43, pp. 291–307.

HORTON, R. E., 1945, 'Erosional development of streams and their drainage basins', *Bulletin of the Geological Society of America*, vol. 56, pp. 295–370.

KING, Lester C., 1951, *South African Scenery* (2nd edn), Oliver and Boyd, Edinburgh.

— 1967, *Morphology of the Earth*, 2nd edn, Hafner, New York.

LAMBERT, J. M., JENNINGS, J. N., SMITH, C. T., GREEN, C. and HUTCHINGSON, J. N., 1960, 'The making of the Broads', *Royal Geographical Society Research Memoir*, No. 3, 153 pp., Royal Geographical Society, London; John Murray, London.

LEOPOLD, Luna B., and MILLER, J. P., 1956, 'Ephemeral streams—hydraulic factors and their relation to the drainage net', *United States Geographical Survey Professional Paper 282-A*, pp. 1–37.

LINTON, D. L., 1955, 'The problem of tors', *Geographical Journal*, vol. 121, pp. 470–87.

MITCHELL, G. H., 1956, 'The geological history of the Lake District', *Proceedings of the Yorkshire Geological Society*, vol. 30, pp. 407–63.

PALMER, J., and NEILSON, R. A., 1962, 'The origin of granite tors on Dartmoor, Devonshire', *Proceedings of the Yorkshire Geological Society*, vol. 33, pp. 315–40.

SMITH, K. G., 1958, 'Evolutional processes and landforms in Badlands National Monument, South Dakota', *Bulletin of the Geological Society of America*, vol. 69, pp. 975–1008.

WAGER, L. R., 1937, 'The Arun River drainage pattern and the rise of the Himalaya', *Geographical Journal*, vol. 89, pp. 239–50.

WOOLDRIGE, S. W., and LINTON, D. L., 1955, *Structure, Surface and Drainage in South-East England*, George Philip, London.

# Chapter 20

# Glaciers and Glaciation

Mais où sont les neiges d'antan?
(But where are the snows of yesteryear?)

*François Villon* (1431–85)

## Snowfields and the Growth and Decay of Glaciers

Glaciers are masses of ice which, under the influence of gravity, flow out from the snowfields where they originate. Permanent snowfields occur in every continent except Australia. The level up to which the snow melts in summer, i.e. the lower edge of a permanent snowfield (if present), is called the *snow line*. Its height varies with latitude from sea-level in the polar regions to 610 m in S Greenland and S Chile, 1525 m in S Norway and S Alaska, 2750 m in the Alps, 4420 m (Assam) —5800 m (Kashmir) in the Himalayas, and 5200–5500 m on the high equatorial peaks of Africa (Figure 20.1) and the Andes (Figure 12.3). It is of interest to notice that the higher summits of the Scottish Highlands, e.g. Ben Nevis, just fail to reach the level of the snow line as it would be in Scotland.

**Figure 20.1** Kilimanjaro, Tanzania viewed from the plateau (African erosion surface) over 900 m showing the ice cap on the summit of Kibo (5895 m) *(Dorien Leigh Ltd)*

**Figure 20.2** Avalanche descending the slopes of the Jungfrau, above the Giessen Glacier, Switzerland (*Paul Popper Ltd*)

Low temperature alone is not sufficient to ensure the growth of a snowfield. Although the permafrost region of Siberia is one of the coldest parts of the globe, it is kept free from perpetual snow because the scanty winter falls are quickly dissipated in the spring. It is where the winter snowfall is so heavy that summer melting and evaporation fail to remove it all, that snowfields are formed and maintained. Snow may also be swept away by the wind, or lost from steep slopes by avalanching (Figure 20.2). Wherever a balance of the snowfall is left over to accumulate from year to year, a snowfield grows in depth and surface area until pressure on the ice that has formed in depth is sufficient to start the outward flow of a glacier.

The transformation of snowflakes into a glacier ice is a kind of low-temperature metamorphism. As the loose feathery snow is buried by later falls it gradually passes into *névé*, a closely packed form with a density of about 0·8, air-free ice being 0·917. Névé still retains a white colour because of the presence of entangled air. With further compaction more air is squeezed out, melt-water from above seeps in and freezes, until the deeper layers are transformed into ice. Minute bubbles of air still remain and as the bubbles become smaller, fewer and more dispersed, the colour of the ice changes from the usual opaque white to the clear blue that is commonly restricted to particular bands. Glacier ice in bulk is a granular aggregate of interlocking grains, each grain being an individual crystal of ice. Between the grains there remains an extremely thin *intergranular film* consisting of an aqueous solution containing chlorides and other salts, mainly of sodium. The presence of these ions lowers the freezing-point and so maintains the film in a liquid state. The same phenomenon is put to practical use when salt is sprinkled over an icy pavement to 'melt' the ice. In a glacier the intergranular film plays an important role in regulating the flow of the ice.

Glaciers originating in and around the heads of valleys creep slowly downwards as tongue-like streams of ice, the material and pressure responsible for the flow being maintained by the yearly replenishment of the névé fields above. By the operation of various wasting processes, mainly melting, evaporation and calving of icebergs (Figure 20.3), glaciers lose volume, the annual loss of ice being described as the *ablation*. Glaciers advance until their fronts or 'snouts' reach a position where, over a period of years, accumulation is more or less balanced by ablation. This position may be several hundred metres below the snow line. The areas of accumulation and ablation naturally overlap and they also vary with time, but supply dominates in the upper regions and wastage at the lower levels. In response to seasons of heavier snowfall, accumulation exceeds ablation, so that the glacier thickens and its rate of flow increases. When the effects of such a resurgence reach the snout, which may take some years, the latter becomes notably steeper and the glacier advances farther down its valley. Conversely, in response to a falling-off in the rate of supply relative to that of wastage, the snout eventually tapers and recedes. The glacier may be said to retreat or recede, but this is only a manner of speaking. It is not the ice that moves backwards, but the position of the terminus or snout (Figure 20.4).

**Figure 20.3** Front of the Miles Glacier, near Cordova, Alaska. Photograph taken a few seconds after the calving of an iceberg from the 46-metre high cliff of ice (*Bradford Washburn*)

**Figure 20.4** Tidewater terminus of the John Hopkins Glacier in 1940, John Hopkins Inlet, Glacier Bay, Alaska. From 1894 to 1929 the terminus retreated about 12 km after which there was a slow advance amounting to just over 800 m by 1940. The advance has since continued. In 1907, when the area was mapped, the glacier was about 610 m thicker in the locality of the 1940 terminus. The high peak on the left is Mt Grillon, 3879 m. (*Bradford Washburn*)

**Figure 20.5** The Rhône Glacier, showing the area of accumulation in the high snowfields and the zone of wastage (*ablation*) terminating in the snout. For crevasses and folds in the middle reach of the glacier, see Fig. 9.3. (*Aerofilms Ltd*)

The capacity of powerful valley glaciers to extend far below the snow line before they melt away is due not only to the immense supplies of ice which are drained from the upland gathering grounds, but also to the fact that the area exposed to wastage is small relative to the great volume of the ice. Because a glacier moves extremely slowly, it occupies its valley to a very great depth. On an average slope, with gravity as the only motive force, water can flow about 100,000 times as fast as ice. To drain a given area the cross-section of a glacier has therefore to be enormously greater than that of the corresponding river, and in accordance with this comparison the streams that suffice to carry off the summer melt-water from the snout of a glacier always appear small and insignificant (Figures 20.5 and 20.6).

When glaciers overflow the land and terminate in sea water sufficiently deep to allow the ice to float, huge masses break away from the front and become *icebergs*. Loss of ice by this 'calving' process may be a high proportion of the total ablation. Theoretically about 0·9 of an iceberg would be submerged if it were made of pure ice.

**Figure 20.6** A close view of the snout of the Rhône Glacier, near Gletsch, Switzerland, showing the source of the Rhône, which issues from an ice cave as a narrow milky-looking stream (*F. N. Ashcroft*)

Figure 20.7 Tabular iceberg from the Ross Barrier; Ross Sea Antarctica, February 1959 (*P. N. Webb and B. C. McKelvey*)

The actual proportion, however, is subject to variation according to the proportion of entangled air and the load of rock debris present in the ice. Some of the vast tabular icebergs liberated from the Antarctic ice float with as much as 0·2 of their total height above the sea (Figure 20.7).

## Types of Glacier

Glaciers fall naturally into three main classes:

(*a*) *Ice-sheets* and *ice-caps* that overspread continental or plateau regions of supply, where the snow line is low, creeping with a slow massive movement towards the margins.

(*b*) *Mountain* or *valley glaciers*, occupying the pre-existing valleys of mountain ranges that rise above the snow line.

(*c*) *Piedmont glaciers*, consisting of sheets of ice formed by the coalescence of several valley glaciers which have spread out below the snow line—like lakes of ice—over a lowland area of wastage.

There are many gradational and subsidiary types and some of these are referred to below.

Greenland and Antarctica provide the only examples of *continental ice-sheets* that still exist. Covering nearly one-tenth of the land area of the globe, they lock up an immense amount of water that would otherwise be in the oceans. Melting of the ice-sheets would raise sea-level by about 60 m.

The Greenland ice-sheet, about 1,300,000 km$^2$ in extent, is largely enclosed within a mountainous rim. Near the middle of the sheet the ice has been shown by seismic methods to be over 3000 m thick. The immense weight has isostatically depressed the rock-floor, which now has a saucer-like surface, parts of which are below sea-level (Figure 20.8). Towards the edge the higher peaks and ridges of the mountains project through the ice as *nunataks* (Figure 20.9). The ice itself overflows through passes in the mountain zone and terminates in the sea or in the valleys leading down to the green coastal belt. Recent expeditions have made it possible to give a rough estimate of the present rate of wastage:

| | |
|---|---|
| Annual loss by melting, evaporation, etc. | 315 km$^3$ water |
| Annual loss by calving of icebergs | 215 |
| Total ablation | 530 |
| Annual accumulation from snowfall etc. | 446 |
| Net annual loss | 84 |

The volume of the Greenland ice-sheet and the outlet glaciers is estimated at $3{\cdot}7 \times 10^6$ km$^3$,

Figure 20.8 Profile across central Greenland from Disco Bay on the west to Franz Josef Land on the east. Rock basement, black (*After P. E. Victor, Lauge Koch 1949–54 Expedition*)

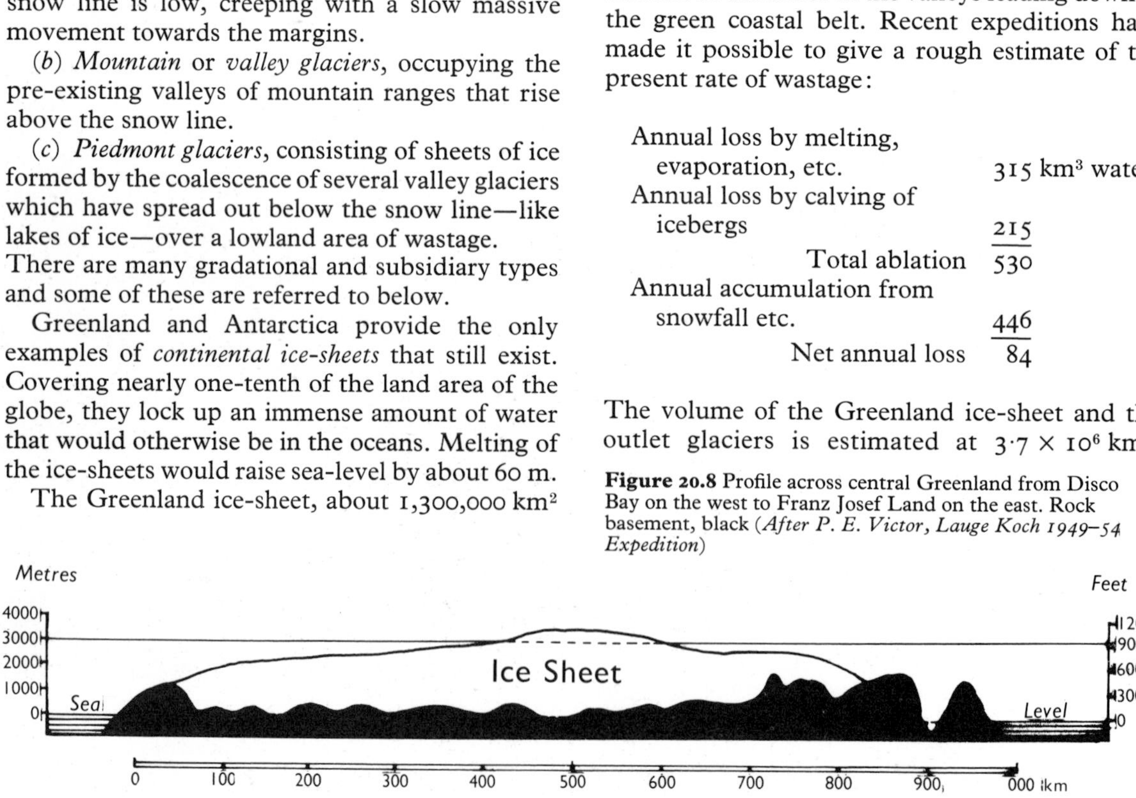

corresponding to $3{\cdot}4 \times 10^6$ km$^3$ of water. Thus it would take about 40,000 years for all the ice to disappear, if the present rate were to continue unchanged—which, of course, is most unlikely in view of the astonishing fluctuations now known to have occurred during the last 40,000 years (see Chapter 21).

The continental ice-sheet of Antarctica, more than seven times as extensive as that of Greenland and having an estimated volume of $28{\cdot}5 \times 10^6$ km$^3$, forms a great plateau rising to over 4300 m (Figure 20.10). Except in a few localities of fringing mountains and coastal strips, interrupted by powerful outlet glaciers, the ice-sheet overruns the coast and spreads over the sea as vast masses of more or less floating shelf ice. The best known of these is the Ross Ice Shelf, which reaches a thickness of over 390 m. The shelf terminates in sheer cliffs of floating ice rising 30 to 50 m above the Ross Sea and known as the Great Ross Barrier. The shelf ice is worn away by submarine thawing and marine erosion, in addition to the normal processes of ablation, of which the most important is the breaking away of gigantic icebergs (Figure 20.11). For many years after James Clark Ross discovered the barrier in 1841 wastage exceeded supply and the high ice cliffs receded several kilometres to the south. However, photographs show that here and elsewhere along the route to the South Pole followed by Scott there has been little change since 1912.

The Filchner Ice Shelf of the Weddell Sea is another example, known for the treacherous behaviour of its outlying pack ice. Marie Byrd Land (facing the Pacific) is mainly a variable thickness of ice creeping out over a deeply depressed sea floor which appears to be margined with mountain peaks and islands that are detectable only by geophysical means. At one point on

**Figure 20.9** Nunataks, consisting of Precambrian quartzite with basic sills and dykes (black), rising 200–300 metres above the ice sheet, Carlsbergfondet Land, East Greenland (about 76°N, 26° 15′W); looking NW. The effects on the ice of the relative movement between ice and the large nunatak are well seen—as if the blunt end of the nunatak were ploughing through the ice towards the left. (*John Haller, Lauge Koch 1949–54 Expedition*)

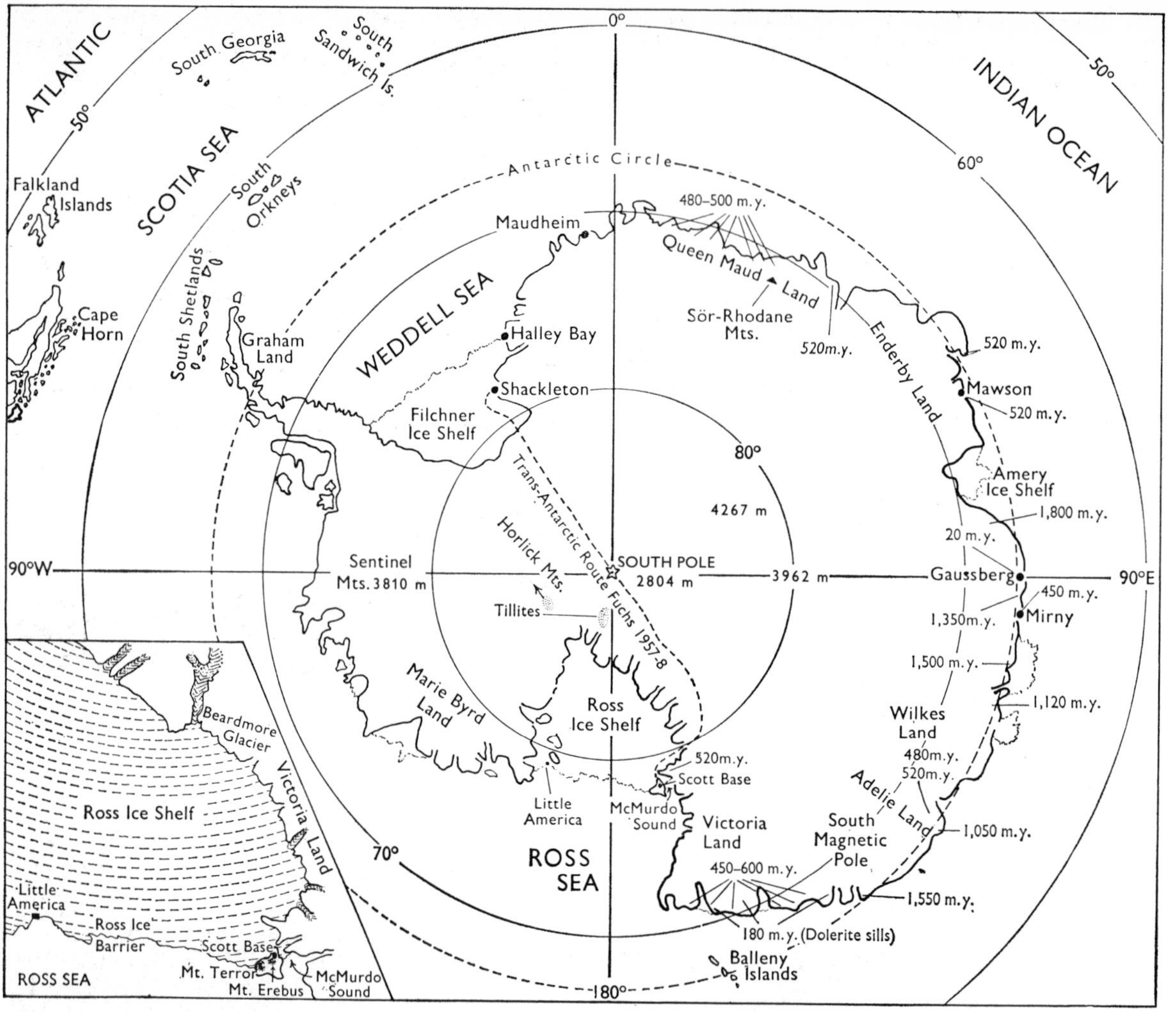


**Figure 20.10** Map of Antarctica; radiometric ages of coastal rocks in millions of years; Permo-Carboniferous tillites near the South Pole, dotted

the overlying plateau of ice, at a height of 1510 m, seismic reflections revealed an ice thickness of about 3000 m. If the region were to be freed of ice, the isostatic response would probably restore the original coastal ranges of Marie Byrd Land and disclose a series of off-lying islands with deep channels between.

Glaciers are thus very sensitive registers of climatic changes. For most of the last hundred years ablation has exceeded accumulation and most glaciers—there have been a few local exceptions—have shrunk and become conspicuously shorter (Figure 20.4). The average thickness of the Arctic ice cover has diminished about 150 cm since it was measured by Nansen during his *Fram* expedition, 1893–6. The Antarctic Astralobe Glacier in Adélie Land has receded over 3 km since it was first seen in 1840. In the tropics the glaciers of Ruwenzori, Kilimanjaro (Figure 20.1) and Mt Kenya (Figure 21.3) have greatly diminished since they came under observation at the beginning of the century. Corresponding to the world-wide melting of the ice, water has been restored to the oceans, particularly from Antarctica and Greenland, and the mean sea-level has risen 7·7 cm since 1880. As pointed out on p. 299, these and related changes can be accounted for by the rise of temperature that has already resulted from the daily addition of millions of tons of carbon dioxide to the atmosphere. However, there are counteracting influences at work, since the last few years have made it increasingly clear that many glaciers have reversed their previous

**Figure 20.11** Tabular iceberg from the Ross Barrier, off Beaufort Island at the entrance to the Ross Sea, Antarctica; the visible part above sea level is about 37 m high. (*New Zealand Geological Survey*)

habits by readvancing over the debris-laden ice of their old and stagnant snouts.

Smaller ice-sheets, distinguished as *plateau glaciers* or *ice-caps*, cover large areas in Iceland, Spitzbergen and the Arctic islands of northern Canada (Figure 20.12); areas from which they emerge through marginal depressions as blunt lobes or large valley glaciers. The tips of the underlying mountains project from certain less continuous caps of highland ice. Where the supply of ice is rather less, these interrupted caps pass into a network of connected glacier systems, the ice of each valley system overflowing the cols into neighbouring valleys and smothering all the lower

**Figure 20.12** Axel Heiberg Island, Canadian Arctic, looking eastwards. The hanging glaciers on the far side of the valley drain the Schei ice-field. The protruding peaks and ridges are over 2000 m above sea-level. The ice tongues coming from the right drain the similar Krueger ice-field. (*Royal Canadian Air Force*)

**Figure 20.13** Flow-contorted moraines of the Susitna Glacier, Alaska, looking NW towards Mt St Elias (5489 m); spurs are well illustrated. (*Bradford Washburn*)

divides. Such transitional types grade to the more familiar valley glaciers.

Trunk glaciers and their tributaries occupy the upper parts or the whole of the valley system of a single drainage area (see Figures 3.5 and 20.20). Smaller valley glaciers tend to be confined to single valleys. They characteristically originate in deep armchair-shaped hollows, called *corries* or *cirques*, situated at the valley heads (Figure 20.5), and so also do the feeders of larger glaciers, as many of the illustrations clearly show. Small isolated glaciers occupying cirques or hanging valleys that are now perched high on the side of a deeper valley are referred to as *cirque* or *corrie glaciers* (Figure 20.13) or as *hanging glaciers* (Figures 20.12 and 20.13).

Where a glacier passes from a restricted channel to a more open lowland, it fans out into an *expanded foot*; and where several neighbouring glaciers so emerge, a *piedmont glacier* results. The outstanding example of the latter type is the great Malaspina Glacier of the coastal plain of SE Alaska (Figure 20.14). Maintained by the confluence of the glaciers from the St Elias Range along the Canadian frontier, it has an area of about 2200 km$^2$ and locally reaches the sea. As a result of surface melting, much of the marginal ice is covered with morainic debris and soil which here and there support dense forests of pine (see Figure 20.15). Seismic reflections from the floor of the Malaspina show that midway between mountains and sea the ice attains a thickness of over 600 m. In this region the floor is from 180 to at least 250 m below sea-level. Since the glacier terminates a little above sea-level, it follows that the floor is like a shallow saucer, sloping upwards, not only towards the mountains to the north, but also towards the ice margin to the south.

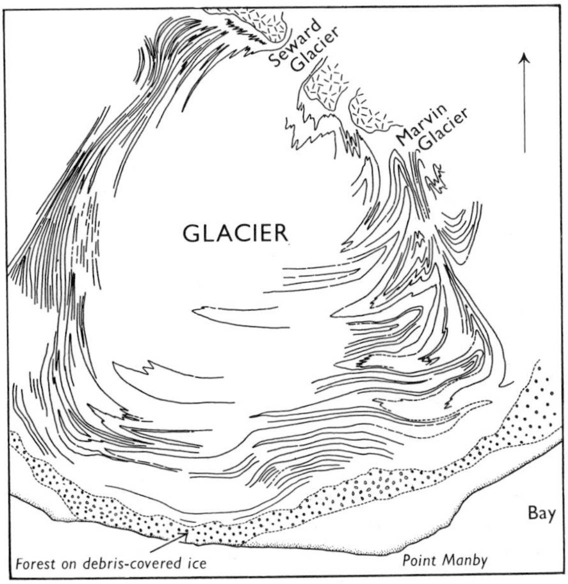


**Figure 20.15** Map of the Malaspina ice-and-debris folds: constructed by R. P. Sharp (reference p. 443) from a mosaic of air photos taken in 1948 by the U.S. Navy

**Figure 20.14** Photograph to illustrate the flow of ice and its morainic load of rock waste; eastern part of the Malaspina Glacier, Alaska, looking NW towards Mt St Elias (5489 m); the high peak on the right is Mt Augusta (4289 m). The accordion-like folds are primarily due to differences in the rate of flow of debris-laden ice fed into a shallow basin by the Seward and other mountain glaciers. (*Bradford Washburn*)

## The Movement of Glaciers

Were it not for the fact that ice in bulk behaves as a rheid and flows under its own weight, the world would now present a very different appearance. Practically all the water of the oceans would be locked up in gigantic circumpolar ice-fields of enormous thickness. The lands of the tropical belts would be deserts of sand and rock, and the ocean floors vast plains of salt. Life would survive only around the margins of the ice-fields and in rare oases fed by juvenile water.

The most rapidly moving glaciers are those of Greenland, some of which advance as much as 18 m a day in the summer. In general, however, a metre or so a day is a more characteristic rate. The

Beardmore Glacier of Antarctica, the greatest in the world, moves at less than a metre a day.

By observing the changes in position of lines of stakes driven into the ice it is found that the middle of a glacier moves more rapidly than the sides. The rate of flow increases with the steepness of the slope, with the thickness and temperature of the ice, and with constriction of the valley sides. Movement is retarded by the presence in the ice of a heavy load of debris and by friction against the rocky channel. These facts suggest that the flowage of glaciers is controlled mainly by stress differences and temperature.

It is well known that increase of pressure lowers the melting-point of ice, and thus stimulates its transformation into water. More important, however, is the fact that stress differences are very much more effective in liberating molecules of water from rigid grains of ice. The mechanism of skating provides a clue to the problem of glacier flow. A skater really glides in a narrow groove of water formed momentarily under the intense stress applied to the ice by the thin blades of his skates. As he passes, the water immediately freezes again.

In a glacier or ice-sheet the shear stresses due to weight increase with depth and tend to squeeze out the ice in the direction of least resistance. Lowering of the melting-point at the base, where pressure and stress are greatest, and the rise of temperature that accompanies friction and erosion may partially offset the braking effect of frictional resistance. By means of a number of tests carried out in deep boreholes it has been shown that part of the motion of cirque and valley glaciers is accomplished by slipping and sliding, especially where there is a well-smoothed floor. But movement is mainly achieved by distortion of the ice, brought about by a complex group of processes akin to those responsible for the recrystallization and flowage of ordinary rocks.

The intergranular film of unfrozen saline water, already mentioned on p. 403, facilitates minute movements of the grains of ice amongst themselves. Wherever the interlocking crystals are subjected to stresses that vary from point to point, mobile molecules of water are temporarily liberated at points where the strain is most severe. The temporary excess of water diffuses through the intergranular film into places where the stress is lower, and there an equivalent trace of water crystallizes on to the adjacent grains. In this way—roughly speaking by gaining in front what it loses behind—an average granule of ice may advance about 0·0001 of its own diameter in the course of a day. The migration is cumulative from points of high pressure and stress to points where these are lower. In the case of valley glaciers this commonly means that the direction of flow approximates to that of the downward slope of the valley. But flow does not necessarily come to an end if the floor happens to slope upwards. We have already seen that the Malaspina Glacier completes its journey to the sea over a rising floor. For this to be possible a necessary condition is that the upper surface of the ice must have a downward slope, as indeed it has across the apparently stagnant expanse of the Malaspina.

There is no doubt that ice can move uphill, provided that the ice is sufficiently thick to give an adequate downward slope at the surface. Scandinavia and North America both provide striking proofs. Blocks of recognizable rocks from the Swedish lowlands were transported by Pleistocene ice over mountains 1800 m high and deposited again on the coastal plain of Norway. Here the surface slope was adequate, since over the Gulf of Bothnia the thickness of the great ice-sheet was at least 3000 m and may have approached 4000 m at times. A similar thickness of ice was responsible for the transport of boulders of Precambrian rocks from near Hudson Bay up to heights of about 1400 m in the foothills of the Canadian Rockies of SW Alberta.

Movement in the same general direction—from high stress to low—also takes place by mechanical slipping along planes of fracture or dislocation in the individual grains of ice within a glacier. Studies with the electronic microscope show that in stressed crystals of ice the dislocations (imperfections in the lattice structure) commonly migrate to the edges, in much the same way as the ruck in a carpet can be swept out. Thus every part of a grain of ice is provided in turn with an internal plane on which slipping can occur, and this process goes on repeatedly so long as there is sufficient stress to generate dislocations, and an intergranular film to accommodate the change of position implied by each minute slip. In all these various ways each grain changes its shape and position under the influence of stress differences, waxing on one side and waning on the other, so that the whole behaves as a rheid and flows continuously forward.

The parallelism with rock flowage and the simpler forms of rock metamorphism is complete. The chief factors concerned are stress differences,

**Figure 20.16** Mer de Glace, from near Montanvert, France
*(G. P. Abraham Ltd., Keswick)*

relatively high temperatures and migrating fluids. In the interior of a glacier stress differences are provided by the weight of overlying ice and the pressure of upstream ice; the temperature is already 'high' in the sense that the ice is never far from its melting point; and the migrating fluid is the intergranular film with its continually varying content of saline water.

## Crevasses

A glacier has generally an outer crust, sometimes as much as 60 m thick but usually much less, which readily cracks and so behaves like an elastic solid rather than as a rheid. This more or less passive and brittle crust is carried along by the flowage of the deeper parts; and the strains so induced within it are applied too rapidly for the processes of flow described above to come into effective operation. If the ice becomes overstrained and gives way by stretching, as it commonly does, it breaks into the gaping fissures known as *crevasses*. It must not be forgotten, however, that differential flow may be responsible for superficial folds as well as for crevasses (Figure 9.3). Similar crumplings and *pressure ridges* of ice tend to form when a glacier is subjected to lateral confinement while passing through a constriction in its course.

Stretching of ice, witnessed by the formation of crevasses, takes place where a glacier passes round a bend (Figure 20.16) or over a declivity in its floor, or spreads into an open valley or plain. *Marginal crevasses*, resulting from differential drag against the valley sides, have already been described on p. 129. *Transverse crevasses* develop very readily across a glacier wherever there is a marked convexity in the shape of its floor (Figure 20.18(*a*)). The limiting case is the *icefall* (Figure 20.17). Here the whole thickness of the glacier is torn into a chaos of blocks which nevertheless soon reunite and flow on again, as before. *Longitudinal crevasses*, roughly parallel to the direction of flow, form where ice spreads out laterally.

**Figure 20.17** Lower part of the icefall of the Khumba Glacier, which splits into a maze of crevasses and seracs after leaving the cirque encircled by the great peaks of Everest, Lhotse and Nuptse (Fig. 2.5). (*Royal Geological Society and Alpine Club of Great Britain*)

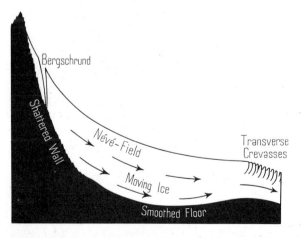


**Figure 20.18** (a) Schematic section through a cirque occupied by the head of a glacier, showing the bergschrund near the top and transverse crevasses above the threshold

**Figure 20.18** (b) Bergschrund crevasses in an ice-filled cirque at the head of a glacier (*Drawn from an aerial photograph of the Gelmerhorner, Switzerland*)

Since crevasses are a by-product of flow they are not permanent features; some are closing up while others are opening. This adds to their danger; but the treacherous surface becomes most perilous to cross after a snowstorm, when the crevasses are likely to be hidden by snow bridges. Difficult and sometimes impassable surfaces may result from the intersection of two closely packed sets of crevasses (Figure 20.3). The upper part of the glacier is broken into a jumbled mass of jagged ice pinnacles known as *seracs*.

A special type of crevasse, which may be very wide and deep, is the *bergschrund*. This opens in summer near the top of the névé field of a cirque, where the head of a glacier is pulled away from the precipitous walls or from the ice adhering to them (Figure 20.18(*a*)). It frequently happens that several such fissures are formed instead of an especially large one (Figures 20.18(*b*) and 20.19).

In sunny weather small pools and rills may diversify the surface of a glacier, gathering into streams which mostly fall into crevasses. By a combination of melting and pot-hole action (aided by sand and boulders) deep cauldrons called *glacier mills* or *moulins* are worn through the fissured ice, and the water may escape to the snout through a tunnel (Figures 20.6 and 20.22).

## Moraines

Rock fragments liberated from the steep slopes above a glacier, mainly by frost shattering, tumble down on the ice and are carried away. Thus the sides of a glacier become edged with long ribbons of debris described as *marginal* or *lateral moraines*. When two glaciers from adjacent valleys coalesce, the inner moraines of each unite and form a *medial moraine* on the surface of the united glacier (Figure 20.20). A trunk glacier fed by many tributaries may thus come to be ridged with a series of medial moraines composed of materials from different

**Figure 20.19** Bergschrund crevasses in the cirque glaciers of the Breithorn (3782 m) in the Bernese Oberland due south of Interlaken. The high peak in the middle distance is the Bietschhorn (3934 m). Beyond the Pennine Alps are seen along the skyline, with the Weisshorn and the Dent Blanche rising high above the general summit level. (*Swissair-Photo A. G., Zurich*)

**Figure 20.20** Barnard Glacier, Alaska, showing lateral moraines and their coalescence into a multiple series of medial moraines. The high peak on the skyline is Mt Natazhat (4109 m) on the Alaska–Yukon boundary, 40 km away. (*Bradford Washburn*)

**Figure 20.21** Gangotri Glacier in the Central Himalaya, looking ESE from a point about 24 km from the snout with its ice tunnel (see below) from which issues the Bhagirathi River, one of the two sacred sources of the Ganges. The other, the Alakananda River, issues from a similar glacier on the far side of the heavily glaciated peaks here seen. The ice is smothered beneath a thick cover of morainic debris, which itself is bordered with conspicuous lateral moraines (*Prem S. Ray*)

**Figure 20.22** One of the two ice caves at the snout of the Gangotri Glacier, Central Himalaya (4206 m), from which the Baghirathi River issues. The ice near and between the two tunnels is vertically banded as a result of the flow structure imposed on the glacier where it has reunited after its bifurcation immediately upstream by a rock barrier. (*Prem S. Ray*)

parts of the area of supply, thus providing samples of rocks that might otherwise be unobtainable. Figure 20.21 illustrates a celebrated glacier that is entirely covered with morainic debris.

Sooner or later part of the debris is engulfed by or washed into crevasses. Material that is enclosed within the ice is referred to as *englacial moraine*. A certain proportion reaches the sole of the glacier, and there, together with the material plucked or scraped from the rocky floor, it constitutes *subglacial moraine*.

If the lower part of the ice becomes so heavily charged with debris that it cannot transport it all, the excess is deposited as *ground moraine*, which is then overridden by the more active ice above. All the varied debris that finally arrives at the terminus of the glacier, ranging from angular blocks and boulders to the most finely ground rock flour, is dumped down haphazardly when the ice melts. Where the ice front remains stationary for several years an arcuate ridge is built up, called a *terminal* or *recessional moraine* (Figures 20.23 and 21.10). If, however, the snout is continually receding summer after summer, no piling up of a ridge is possible. The load liberated from the retreating front then forms an irregular sheet which rests on the ground moraine already deposited.

Thin isolated slabs of rock or patches of debris on the surface may be sufficiently heated by the sun to melt the underlying ice. Larger blocks, however, act as a protection from the sun's rays, and as the surrounding ice melts away they are left as *glacier tables* perched on a column of ice. Even morainic ridges may stand out for a time on thick walls of ice.

Examples of folds, outlined by moraines, due to the rheid flow of ice have already been discussed (pp. 129–30). The most spectacular flow structures

**Figure 20.23** A succession of arc-shaped recessional moraines left by the North Iliamna Glacier, Alaska, during the recent retreat of its snout. Braided streams of melt water are breaching the moraines and depositing glacifluvial 'drift' on a characteristic outwash plain (cf. Fig. 3.5). (*Bradford Washburn*)

of this kind are to be seen on the surface of the Malaspina Glacier at times when the snow cover has been removed by ablation. As this great 'expanded foot' of ice spreads out from the mountain valleys towards the sea, it becomes deformed into a kind of natural model of Alpine folding. The folds look like plunging anticlines and synclines, truncated by erosion, but as indicated by Figure 20.15 they are more like tightly packed recumbent folds and overfolds passing in places into thrusts. R. P. Sharp (see p. 411) has traced some of the moraines back to their sources through fifteen pairs of anticlines and synclines. It is a point of special significance that these structures have developed as a result of flow, spreading and extension. Convincing proof of extension is provided by the occurrence of immense numbers of crevasses at right angles to the axial planes of the folds and to the foliation-strike of the ice, which has been continually recrystallized during flowage.

## Glacial Erosion

As we have seen, a valley glacier soon acquires a load of morainic material. Moreover, loose debris on the floor and sides is quickly dislodged and engulfed by actively advancing ice. Blocks from protruberances of jointed bedrocks are firmly grasped by the ice and withdrawn from the downstream and unsupported side by a quarrying process referred to as *plucking*. The ice works its way into joints, bedding planes and other fractures and closes around each block in turn with sufficient cohesion and frictional drag to carry it away from the parent mass. The rugged surface left behind is readily susceptible to further plucking, and the process continues until the obstruction is removed or the glacier wanes.

Thus, even pure ice, which by itself would be a very ineffective tool for eroding massive rocks, is sooner or later transformed into a gigantic flexible file with embedded fragments of rock for teeth. *Abrasion* is the scraping and scratching of rock surfaces by debris frozen into the sole of a glacier or ice-sheet. The larger fragments cut into and groove the floor and sides (Figures 20.24 and 20.25), and may themselves be worn flat and

**Figure 20.24** Snout of the Woodworth Glacier, Tasnuna Valley, Alaska, which by its retreat has revealed a heavily striated rock surface, crossed by a conspicuous esker (see Fig. 20.53). (*Bradford Washburn*)

striated. The finer materials act like sandpaper, smoothing and polishing the rock surfaces and producing a further supply of powdered rock, or *rock flour*, in the process.

The rate of glacial erosion is extremely variable. Theoretically, the maximum rate of abrasion for a given kind of rock is approximately proportional to the cube of the velocity of the ice against its channel. Thus a powerful glacier in northern Greenland may be 30,000 times more effective than the sluggish glaciers of the Alps. A continental ice-sheet moves so slowly that it cannot be expected to do much more than remove the soil and smooth off the minor irregularities of the buried landscape. In such a case the broader features of the pre-glacial relief are, on the whole, protected from denudation, though the surface is modified in detail into an irregular pattern of hummocks and hollows which reflect the varied resistances offered by the rocks to abrasion. But

**Figure 20.25** Glacially striated surface of a slate formation, Kilchiarin, Islay, Scotland (*Institute of Geological Sciences*)

when the outflowing ice, or a valley glacier in a mountain district, is concentrated in a steeply descending valley, the erosive power reaches its maximum, and the pre-glacial relief is strongly accentuated. Beyond the region of steep gradients and rapid movement the rate of erosion gradually falls off and gives places to deposition as the ice becomes overloaded and reaches the zone of wastage. The three realms of accumulation, movement and wastage are clearly seen in Figure 20.5.

## Recognition of Former Glaciations

The growth and decay of glaciers and ice-sheets, together with their geological work of erosion and deposition and the consequent modifications of landforms, are collectively known as *glaciation.* The sculpturing of the surface beneath existing glaciers can be studied directly only by such dangerous methods as the exploration of ice tunnels and the descent of crevasses. Fortunately a great deal can be learned by taking full advantage of the fact that most of the familiar glaciers of today—quite apart from the revealing recessions that have occurred during the last hundred years—are but the shrunken descendants of immense ice-caps that covered the Alps and similar mountainous regions, or of even greater continental ice-sheets that spread over most of north-western Europe and half of North America during Pleistocene times. And this happened not once, but repeatedly.

One of the earliest observers to suspect that glaciation was once far more widespread than in his own day was de Saussure, the first scientific explorer of the Alps. In 1760 he noticed that kilometres below the snouts of the Alpine glaciers the rock surfaces were scratched and smoothed—in striking contrast with the frost-splintered peaks above—and strewn with morainic material exactly like that still being carried and deposited by the ice. He concluded that the glaciers had formerly extended far beyond their then limits. But finding gigantic blocks and boulders of Alpine granite lying about on the slopes of the Jura Mountains, 80 km away from their source, he ascribed their transport on more traditional lines to a catastrophic flood. Hutton, however, reading of this remarkable discovery, at once grasped its significance. Picturing the Alps surmounted by ice, this is what he wrote in his *Theory of the Earth* (1795, vol. 2, p. 218): 'There would then have been immense valleys of ice sliding down in all directions towards the lower country, and carrying large blocks of granite to a great distance, where they would be variously deposited.' This far-seeing inference, for which Hutton has only lately begun to receive due credit, was either ignored or regarded as 'contrary to revealed truth'. In his *Illustrations of the Huttonian Theory* (1802) Playfair emphasized this remarkable example of Huttons's genius, but it was conveniently overlooked by the geological 'establishment' of the time and, eventually, all but forgotten.

A few observers in the Alpine valleys independently reached similar conclusions during the next few decades, but the 'glacial theory', when heard of at all, was ridiculed as a kind of lunacy. At first Noah's Deluge was its most powerful competitor. When this was found to be inadequate as a scientific hypothesis and had reluctantly to be abandoned, its place was taken by the idea that the far-travelled boulders had indeed been carried by ice, but by ice in the form of floating icebergs. Enormous areas of land were thought to have been submerged for a short time by the latest of a long series of marine invasions. So it came about that the glacial theory met with little or no success until after 1840, when the great naturalist Louis Agassiz awakened more general interest in the subject by the publication of his classic studies on the glaciation of the Alps. In later years Agassiz recognized that the similarly striated rock surfaces and morainic deposits of Scotland were also due to the former passage of ice. Even then there were many who opposed the idea for another twenty years. However, with a band of enthusiastic supporters he convinced the scientific world that such features could be accounted for in no other way. It is now familiar knowledge that the landscapes of vast areas of Europe, North America and many other regions bear the unmistakable hallmarks of glaciation. Thus it happens that in many countries the characteristic effects of ice erosion and deposition modified but little by subsequent weathering and river action, can be seen and studied close at hand.

## Evidences of Glacial Erosion

Among the evidences of erosion by continental and valley glaciers, striated surfaces (Figure 20.25) and ice-moulded hummocks of the more resistant

**Figure 20.26** Ice-sculptured surface of *roche moutonnée* type, east of Sanna Bay, Ardnamurchan, Scotland (*Institute of Geological Sciences*)

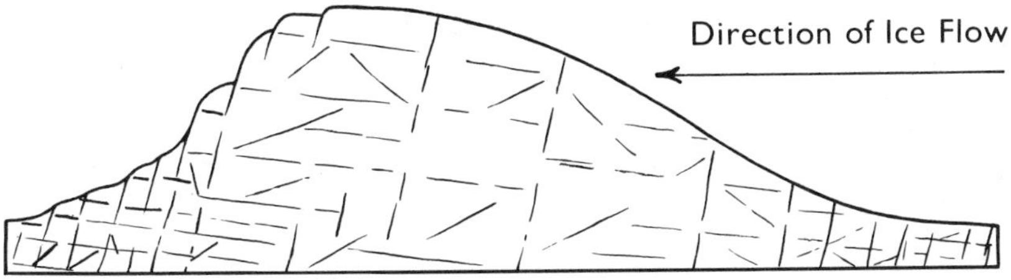


**Figure 20.27** Section through a typical *roche moutonée*, showing the effect of ice abrasion where the rock is sparsely jointed, and of plucking where the jointing is closely spaced

bedrocks (Figure 20.26) are of widespread occurrence. The residual hummocks vary widely in size, and have a characteristically streamlined form which is developed in relation to the direction of ice movement so that the resistance presented by each obstacle is gradually reduced. The side up which the ice advanced rises as a smoothly abraded slope, while the lee side falls more steeply, sometimes as an abraded slope, but often by a step-like series of crags and ledges obviously due to the plucking out of joint blocks (Figure 20.27). Seen from a distance the more isolated examples resemble sheep lying down, or wigs placed 'face' downwards. They are, therefore, described as *roches moutonnées*, a term first used by de Saussure in 1804 in reference to the sheepskin wigs, styled *moutonnées*, which were then in vogue.

Highly resistant obstructions on a larger scale, such as old volcanic plugs, that lay in the path of the ice are responsible for features known as *crag-and-tail* (Figure 20.28). The *crag* boldly faces the direction from which the ice came, while the *tail* (bedrock with or without a covering of boulder clay) is a gentle slope on the sheltered side, where the softer sediments were protected by the obstruction from the full rigour of ice erosion. The classic example is provided by the Castle Rock of Edinburgh, from the eastern side of which the High Street follows the sloping crest of the tail. The massive basalt plug diverted the ice-flow, and a deep channel was excavated in the sediments on each side of the crag-and-tail feature, one being now occupied by Princes Street Gardens and the other by the Grassmarket. Figure 20.29, showing

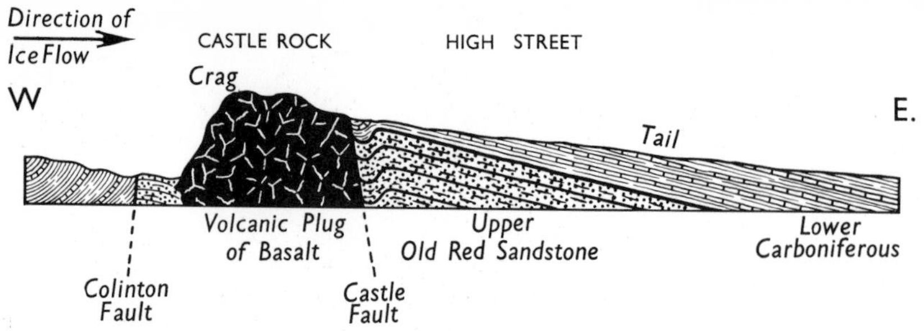


**Figure 20.28** Crag and Tail, Edinburgh. The Castle Rock probably represents an early outbreak of the Arthur's Seat volcano (see Fig. 7.1).

**Figure 20.29** Slieve Gullion, the core of a Tertiary volcano, with its 'Tail' on the right, extending southwards like a sloping railway embankment; viewed from Forkhill, Co. Armagh, Northern Ireland (*Doris L. Reynolds*)

the celebrated tail of Slieve Gullion, illustrates an example where glacial debris has been concentrated along the sheltered belt.

In mountainous and upland coastal regions with well-developed valley systems the topographic modifications superimposed on the landscape by glacial erosion include U-shaped valleys with truncated spurs and hanging tributary valleys; corries or cirques surmounted by sharp-edged ridges and pyramidal peaks; and rock basins and fjords. Waterfalls descending the precipitous valley sides, and lakes occupying the over-deepened hollows of the valley floors, add variety to an assemblage of features that can be easily distinguished from the landscapes of unglaciated regions (Figures 20.30 to 20.42 provide a variety of illustrations of these features).

## Cirques (Corries) and Associated Features

It has been observed that chance hollows occupied by persistent snowbanks are steadily cut back and deepened by (*a*) disintegration of the marginal and underlying rocks by frost and thaw, (*b*) removal of the shattered debris by falling, avalanching, solifluction and transport by melt-water. By this process of snow-patch erosion or *nivation* the surrounding slopes are kept steep as they recede (Figure 20.32). The larger hollows grow more rapidly than the smaller ones, especially near and above the snow line, until the mountain slopes and valley sides are festooned with deep snowfields, the largest of all being at the valley heads. Eventually these nourish small glaciers which carry away the debris and begin more active excavation of the floor. Headward erosion of the walls continues, not only by frost-sapping at the exposed edges of the snowfield, but also by a process of sub-glacial disintegration which comes into play whenever the bergschrund allows surface water to reach the rocks behind or beneath the ice. Draining into cracks and joints, the water freezes and breaks up the rocks until they are gripped by the ice, and carried away as ground moraine. The amphitheatre-like hollows that are eventually

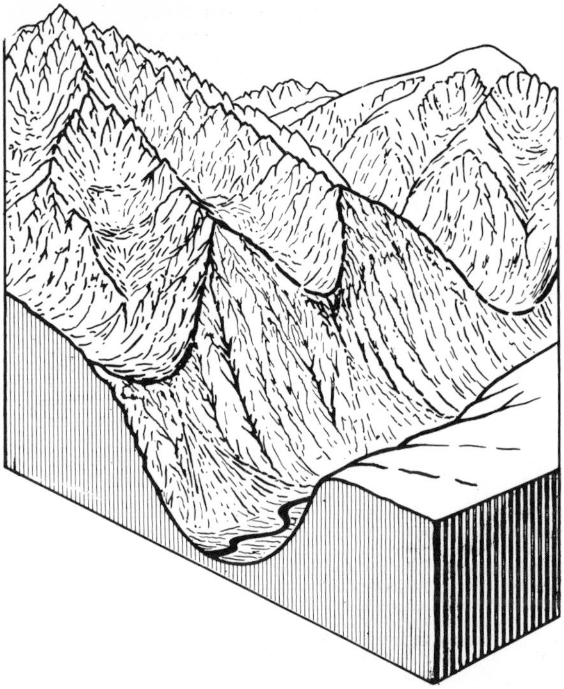

**Figure 20.30** Block diagram illustrating some of the characteristic landscape features of glaciated valleys and mountains: U-shaped valleys; truncated spurs; hanging tributary valleys; cirques, arêtes and horns. The bench across the lower right-hand corner is depicted as unglacicated. (*Modified after W. M. Davis*)

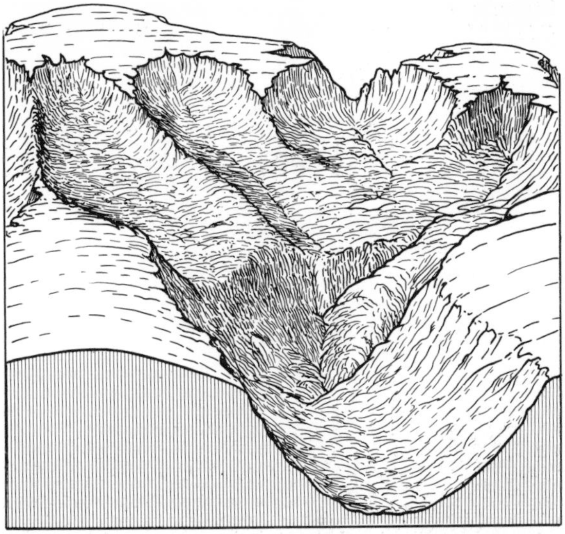

**Figure 20.31** Block diagram to illustrate the 'trough-end' at the head of a valley glaciated by tributary glaciers from several confluent cirques. The hill summits and lower interfluves are depicted as unglaciated. (*Modified after W. M. Davis*)

hollowed out are known as *cirques* in the Alps, *corries* in Scotland, and *cwms* in Wales. For international purposes the term *cirque* has come into general use.

A small glacier in a large cirque is characteristic of the waning stages of glaciation. During the maximum of glaciation most of the cirques we now see must have been deeply buried, or at least full to the brim with overflowing ice. This can sometimes be proved by the preservation of striated surfaces above the head walls. At this stage the floor and sides of a growing cirque would be subjected to vigorous abrasion, in the course of which a shallow rock basin would commonly be excavated by the downflowing and outflowing ice, to become the site of a mountain tarn or lake after the disappearance of the ice (Figure 20.33). Such lakes may also be held back by arcuate ridges of morainic material left stranded by the waning ice during its recession.

Two adjoining cirques may grow at each other's expense until only a sharp-edged precipitous ridge, known as an *arête*, remains between them. When the ice has gone the steep rocky slopes fall

**Figures 20.32** Cirques developed by 'snow rotting' on the cliffs of Spitzbergen (*Sketch by F. Nansen*)

**Figure 20.33** Snowdon (1085 m) and Glaslyn, a tarn occupying an ice-eroded rock basin, viewed from the long arête of Crib Goch, North Wales, (*G. P. Abraham Ltd, Keswick*)

**Figure 20.34** The Matterhorn (4482 m), highest peak of the Pennine Alps, along the Swiss–Italian frontier (*G. P. Abraham Ltd, Keswick*)

a ready prey to frost action, and soon become aproned with screes. Many an upland region has been eaten into by cirque-erosion from several sides at once, and so reduced to a series of arêtes radiating like a starfish from a central summit. Snowdon (Figure 20.33) and Helvellyn are good examples. At a later stage the arêtes themselves are worn back, and the central mass, where the heads of three or more corries come together, remains isolated as a conspicuous pyramidal peak. In this way the *horns* of the Alps have been formed, the world-famous Matterhorn being the type example of its class (Figure 20.34).

### Modifications of Valleys by Glacial Erosion

By the passage of a vigorous glacier through a pre-existing river valley the mantle of rock-waste is removed, the overlapping spurs are trimmed off into facets, and the floor is worn down. The valley is thus widened and deepened, and is eventually remodelled into a U-shaped trough with a broad floor and steep sides and a notable freedom from bends of small radius (Figure 20.35). Flat floors are not uncommon, however, where the bottom of the trough has been levelled up by subsequent deposition of alluvium. In the less severely glaciated regions, where valleys have not been entirely filled by ice, the upper slopes may remain as high-level benches which meet the ice-steepened walls in a prominent shoulder.

Tributary valleys have their lower ends cut clean away as the spurs between them are ground back and truncated. The floor of a trunk glacier, moreover, is deepened more effectively than those of feeders from the side (Figure 20.30) or at the head (Figure 20.31). Thus, after a period of

**Figure 20.35** Val Giuf, a tributary of the Upper Rhine, viewed from the slopes near Milez, Switzerland. The lower part of the U-shaped valley is heavily strewn with debris washed down from the steep hanging valleys seen on the left, where they are carved out of the granitic rocks of the Aar Massif. The rocks on the skyline, with cirques partly occupied by snow, are schists. (*F. N. Ashcroft*)

Truncated Spur of El Capitan    Half Dome    Hanging Valley and Bridal Veil Falls

**Figure 20.36** Yosemite Valley, Sierra Nevada, California (*R. H. Anderson, U.S. National Park Survey*)

prolonged glaciation such valleys are left hanging high above the main trough. The streams from such *hanging valleys* plunge over their discordant lips in cascades or waterfalls, some of which are amongst the highest in the world. The Yosemite Valley in the Sierra Nevada of California is renowned for its impressive examples of these and other spectacular features due to glacial erosion (Figures 20.36 and 20.37). A remarkably similar glacial trough—the finest of its kind in Europe—is the Lauterbrunnen valley, with its celebrated falls, between Interlaken and the Jungfrau.

Glacially excavated floors are deepened very unevenly, the effect at each point depending on the thickness and rate of flow of the ice, and on the nature and structure of the bedrocks. Poorly consolidated strata are scoured out more rapidly than resistant rocks, and tracts of well-jointed rocks are selectively quarried away by plucking. Thus, where the ice encounters a sequence of rocks of varied resistance the floor is excavated into a series of successive steps, often with abrupt descents from one tread to the next, so that the long profile of a heavily glaciated valley may

**Figure 20.37** Yosemite Falls, Yosemite Valley, Sierra Nevada, California (*R. H. Anderson, U.S. National Park Service*)

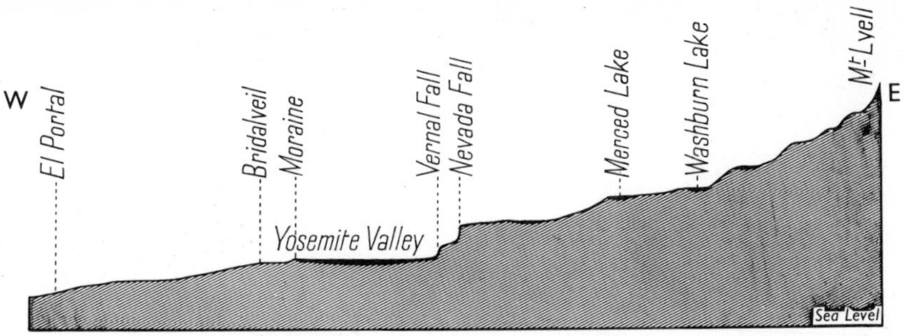


**Figure 20.38** Longitudinal profile along the Yosemite Valley, 240 km E of San Francisco. A typical 'glacial stairway' developed by selective ice erosion. Length of section=58 km.

resemble a giant stairway (Figure 20.38). A tread may even be hollowed into a basin with a barrier of resistant rock in front. Such rock basins are now occupied by lakes, or by tracts of alluvium representing the sites of shallow lakes that have since been silted up. Though referred to as 'basins', these depressions are generally conspicuously elongated, and some of them have been excavated to very impressive depths. It was at one time objected that the ice at the bottom of a basin could not flow out of it, that is, up the slope at the lower end. However, as we have seen (p. 412), the objection is not valid. All that is necessary is that the ice had a means of exit and that the downward gradient of its *surface* towards the exit was adequate to maintain its flow beyond the barrier.

The following are examples of lakes occupying 'over-deepened' basins and troughs. Morainic deposits on the bedrock barrier add slightly to the depth of water in some of these lakes. See Figure 20.46 for an illustration of such a double barrier.

In the examples cited and in the fjords described on p. 429 the ice flowed approximately parallel to the valleys concerned. When sufficiently thick, however, a sheet or cap of ice may have crossed a pre-glacial watershed. Flow would then be concentrated in the cols or passes of the time, and some of these would be deepened into rock basins, now occupied by lakes, which breach the former watershed and modify the previous drainage system. Many of the long lakes lying across the boundary between Norway and Sweden (called

| Lakes | *Maximum depth (in metres)* | *Height of surface above sea-level (in metres)* | *Maximum depth of floor below sea-level (in metres)* |
|---|---|---|---|
| ENGLISH LAKE DISTRICT | | | |
| Windermere | 67 | 39 | 28 |
| Wastwater | 79 | 61 | 18 |
| SCOTLAND | | | |
| Loch Coruisk (Fig 20.39) | 38 | 8 | 30 |
| Loch Lomond | 199 | 6 | 193 |
| Loch Ness (Fig. 9.40) | 235 | 16 | 219 |
| Loch Morar | 310 | 9 | 301 |
| SWISS-ITALIAN ALPS | | | |
| Lake Maggiore | 372 | 194 | 178 |
| Lake Como | 410 | 198 | 212 |
| NORWAY | | | |
| Hornindals-vatn (fjord lake of the Nord Fjord) | 514 | 53 | 461 |
| NORTH AMERICA | | | |
| Lake Michigan | 300 | 177 | 123 |
| Lake Superior | 397 | 183 | 213 |
| Great Slave Lake | 614 | 150 | 464 |

**Figure 20.39** Loch Coruisk, Skye. A double rock basin excavated by ice with the Cuillin Hills in the background (*Institute of Geological Sciences*)

*glint* lakes from the Norwegian for 'boundary') occupy elongated basins gouged out by the Scandinavian ice as it ground its way to the North Atlantic over the mountains and through the pre-glacial passes. Some of the Scottish lochs and Irish loughs represent deeply breached watersheds. Figure 20.40 illustrates a Donegal example. Powerful accumulations of ice over the Derryveagh Mountains flowed to the north and north-west across the well-known quartzite belt of Errigal, Aghla Mor, Aghla Beg and Muckish, and scoured out the cols between these high peaks. One of the cols was excavated into the trough now occupied by Altan Lough. As a result the Owenbeg has lost its main headwater, the Owenwee, which now drains into the lough and so, northwards, to the Atlantic.

## Fjords

*Fjords* are greatly over-deepened glacial troughs that reach the coast below the present-day sea-level, so that instead of forming elongated lake basins they have become long arms of the sea stretching inland between steep rocky walls. A terminal rock barrier (with or without a cover or moraine) wholly or partly submerged, characteristically occurs near the seaward entrance. This is the *threshold* of the fjord. Along the west coast of Scotland there are all gradations from the submerged thresholds of the sea lochs (fjords) to the more or less exposed barriers that separate fresh-water lochs from the sea. The terminal rim of Loch Morar is close to the sea and rises only 9 m or so above it. The threshold of Loch Etive is lower and is uncovered only at low tide; that of Loch Leven is not uncovered but becomes a 'race' at low tide. Twenty-three of the remaining sea lochs have thresholds near their entrances which, at the present sea-level, are permanently submerged.

Fjords have been developed during the intense glaciation of dissected coastal plateaus and mountains of appropriate structure in countries such as Scotland, Norway, Greenland, Labrador, British Columbia, Alaska, Patagonia and New Zealand. In plan (Figure 20.41) they everywhere have a

**Figure 20.40** Altan Lough and Aghla Mór, viewed from the lower slopes of Errigal, Co. Donegal, Ireland; looking north to the Atlantic, seen in the distance 13 km away. The index map shows the breached quartzite belt, dotted; the Derryveagh Mts are mainly granitic (*Bord Fàilte Photo : Irish Tourist Board*)

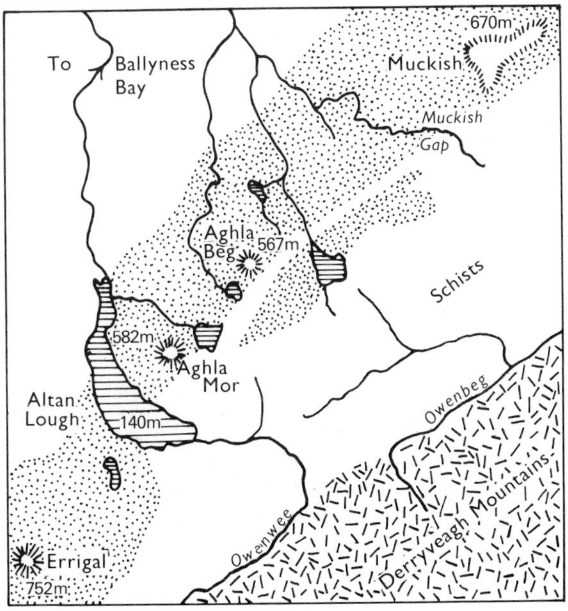


characteristic rectilinear pattern which is clearly determined by the distribution of belts of structural weakness. The latter may be synclines of relatively weak sediments or schists enclosed by massive crystalline rocks (as in the Sogne and Hardanger Fjords of Norway), but more commonly they are belts of closely jointed rocks or of shattered rocks along zones of faulting. The preglacial rivers carved their valleys along these lines of least resistance. The valleys in turn confined the ice and guided its flow, and because the structure facilitated plucking, the valley floors were steadily deepened, often to an extraordinary degree. In some of the fjords of Norway and Patagonia the sea is over 1200 m deep. Neighbouring fjords, however, vary enormously in depth, in accordance with the varying resistance of the excavated rocks. The distribution of fjords is thus conditioned by (*a*) appropriate tectonic structures in upland regions near the sea, (*b*) pre-existing valleys which followed these structures, and (*c*) heavy glaciation by seaward-moving ice of sufficient thickness and surface slope to ensure that the main valleys were over-deepened up to or beyond the present coast.

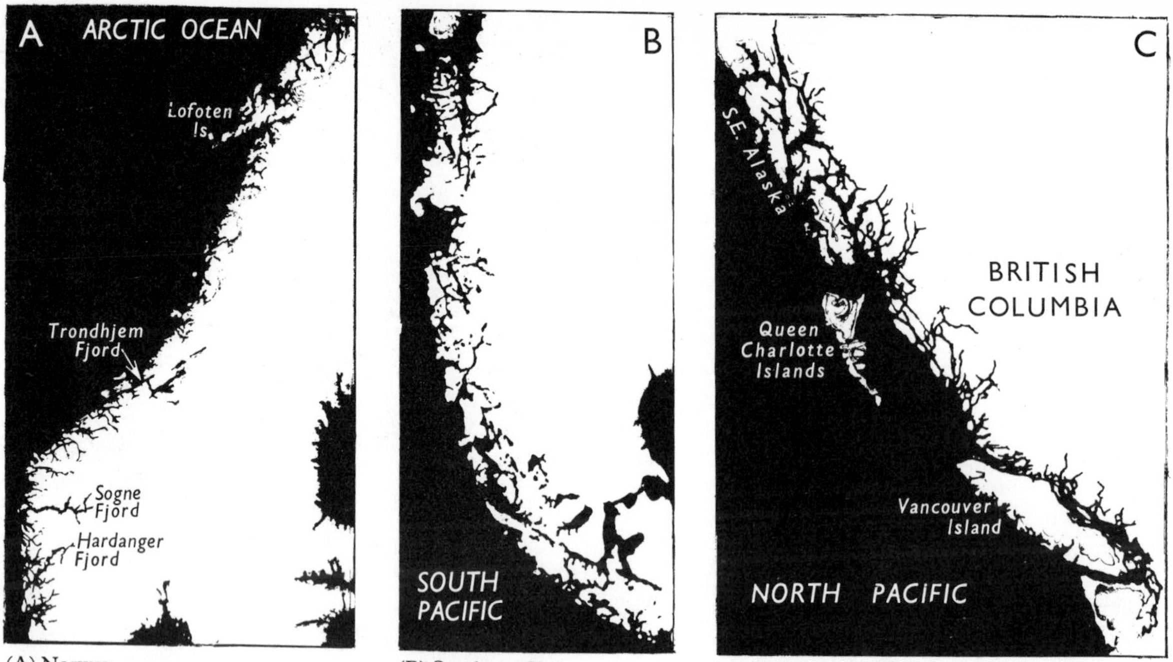


(A) Norway

(B) Southern Chile

(C) SE Alaska and British Columbia

**Figure 20.41** Three examples of fjord coasts

**Figure 20.42** Naeröyfjord, on the south side of Sognefjord, Norway (*Mittet Foto, Norway*)

## Glacial Deposits

As glaciers and ice-sheets reach the zone of wastage beyond the region of active erosion they become overloaded and begin to drop their burden of debris. During the subsequent recession of the ice the zone of deposition retreats with the ice front. The *glacial* deposits thus left stranded on the landscape, and also the *glacifluvial* sands and gravels transported and deposited by the associated melt-waters, have long been grouped together under the general term *drift*. At one time the vast spreads of drift occurring in Europe and North America were thought to be vestiges of the Noachian Deluge and, consequently, to be a proof of the world-wide extent of the Flood. It was this belief that delayed acceptance of the 'glacial theory' for so long. Eventually, however, it was recognized that the commonest type of drift, the haphazard assemblage of materials known as *boulder clay* or *till*, could not possibly have been deposited by water.

A distinction is sometimes drawn between the terms 'boulder clay' and 'till', the latter being restricted to varieties in which a stiff clayey matrix is predominant. The term *tillite*, however, is in general use for the indurated boulder clays and tills of earlier periods of widespread glaciation: e.g. late Carboniferous and Precambrian.

Boulder clay has obviously been dumped down anyhow, in a completely unsorted and unstratified condition (Figure 20.43). Its constituents range from the finest ground-down rock flour to stones and boulders of all sizes up to masses of rock that are occasionally of immense bulk. It usually consists of a varied assortment of stones embedded in a tenacious matrix of sand, clay and rock flour. Most of the stones, like those of screes, are irregular fragments showing little or no sign of wear or tear, but a few can generally be found which have been rubbed down and scratched and grooved, clearly by scraping along the rocky floor over which they were dragged by the ice.

A characteristic feature of glaciated regions is the occurrence of scattered boulders of rocks that are foreign to the place where they have been dropped. These ice-transported blocks, carried far from their parent outcrops, are called *erratics* (Figure 20.44). The largest ones commonly rest on abraded surfaces where the normal drift is thinly scattered or confined to hollows. Some have been stranded in exposed and precarious positions. Such *perched blocks* are striking monuments to the

**Figure 20.43** Boulder clay on glaciated rock surface (bottom right-hand corner), Borrowdale, English Lake District (*S. H. Reynolds*)

**Figure 20.44** Glacial erratic block of Silurian grit perched on a terrace of Carboniferous Limestone at Norber Brow, near Austwick, south of Ingleborough, Yorkshire. Since the ice retreated from this region probably about 100,000 years ago (see Fig. 21.15), a thickness averaging 45 cms of limestone has been removed by chemical weathering, except where blocks such as this protected the underlying limestone from solution and so came to be perched on a plinth of natural origin. (*Institute of Geological Sciences*)

former passage of ice and, as we have seen, they were amongst the first evidences of glaciation to be recognized.

Long trails of erratics of easily recognized rocks afford an unfailing guide to the direction of movement of the ice that carried them. Boulders of Shap granite, for example, can be traced from their original home in Cumbria (Westmorland) across the Pennines by way of the Stainmore Pass into the Vale of York. Ailsa Craig in the Firth of Clyde is an upstanding mass of finely speckled granitic rock which can be identified with certainty. Erratics of this rock, as isolated blocks and as stones embedded in boulder clay, are found in Antrim, Galloway and the Isle of Man, and on both sides of the Irish Sea as far as Wicklow and South Wales, and show that Ailsa Craig lay in the track of a great southward-flowing glacier. Familiar Norwegian rocks from the Oslo district, such as rhomb-porphyry and larvikite (a shimmering blue syenite much used for shop fronts), occur as erratics along the Durham and Yorkshire coasts, and prove that the Scandinavian ice-sheet crossed the North Sea and at times overran the British shores. At other times these rocks reached northern Germany, showing that the directions of ice dispersal were not always the same.

At or near the maximum extension of a glacier or ice-sheet the front may have remained stationary, or nearly so, sufficiently long for a ridge-like *terminal moraine* to be heaped up. Similar, but later, *recessional moraines* mark the sites of halting stages during the shrinkage of the ice (Figures 20.45 and 20.46). In certain lowland regions the ice appears to have fanned out and become stagnant. Such 'dead' ice simply melts away from the top and sides, and also from the edges of crevasses, and liberates its debris without forming terminal moraines.

The terminal and chief recessional moraines formed at successive stands of the last European ice-sheet are shown in Figure 21.15. Similar features, traversing the country in broad loops to the south of the Great Lakes, mark the various pauses in the recession of the North American ice. The terminal moraines left by mountain glaciers cross their valleys as crescent-shaped embankments (concave upstream) which in some cases continue along the valley sides as less conspicuous lateral moraines (Figure 20.23).

In the tracts between the morainic embankments the spreads of boulder clay naturally vary widely in character and thickness from place to place. In certain regions where boulder clay is thickly plastered over a floor of low but probably irregular relief, it has been moulded by the former ice-sheets into swarms of whale-shaped mounds called *drumlins* (Gaelic, *druim*, a mound). Being distributed more or less *en échelon* the mounds give rise to what has been aptly described as 'basket of eggs' topography (Figure 20.47). In the intervening depressions drainage is poor and confused, and is responsible for such features as ponds, marshes and waterlogged meadows. One of the most densely packed drumlin belts stretches across northern Ireland from Co. Down to Donegal Bay,

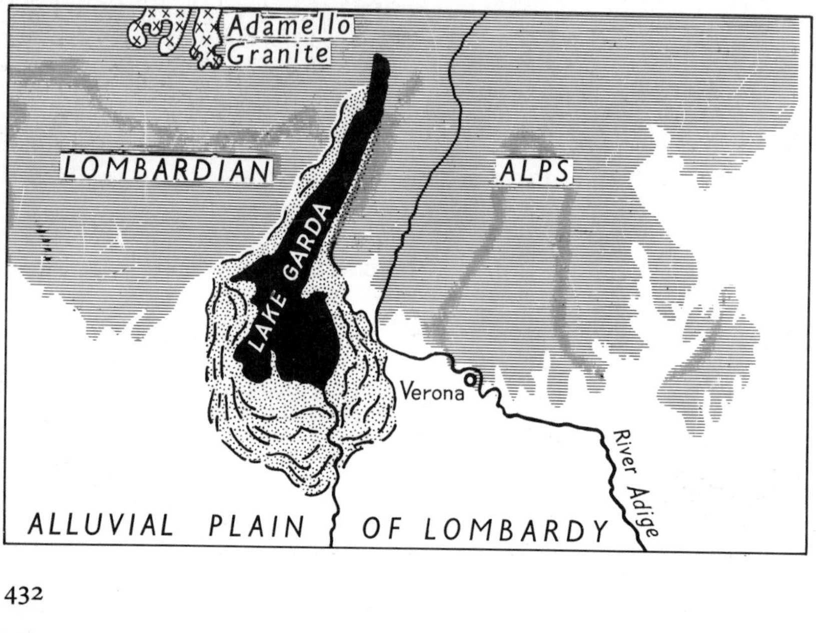


**Figure 20.45** Lake Garda at the foot of the Italian Alps, with its bordering lateral and terminal moraines. The lake owes 149 m of its depth to the thicknesses of the morainic barrier. Below that level (down to a maximum depth of 350 m) the lake occupies an ice-excavated rock basin.

**Figure 20.46** Lake on the Dontouz-Oran Pass, Caucasus Mts, occupying an ice-eroded rock basin with a terminal moraine on the threshold (*Planet News, Ltd*)

**Figure 20.47** 'Basket of eggs' topography. A typical landscape of drumlins which have been moulded into streamlined forms by the passage of ice moving from right to left

Figure 20.48 Map of part of a drumlin tract in Co. Down, Northern Ireland (*After J. K. Charlesworth*)

and contains tens of thousands of these streamlined mounds (Figure 20.48).

Drumlins are commonly 400 to 800 m long, but there is every gradation from low swells to enormous examples two or three kilometres in length and 30 to 60 m high. Most of them are elongated in the direction of ice movement, and the end facing upstream is relatively broad and blunt compared with the tapering downstream end or 'tail' (Figure 20.49). This is a typically streamlined form; implying that the surface between the moving ice and the subglacial drift was moulded by both erosion and deposition towards the form offering the least resistance to the advancing ice. A blunt nosed fish or a whale swims with the blunt end in front; that is to say, remembering the relativity of motion, it faces the direction of flow of the water streaming past it, so avoiding the setting-up of energy-consuming eddies. In the same way drumlins and crag-and-tail features present their blunt ends *towards* the advancing ice-flow. Roches moutonées are an exception to this rule because they are only partially streamlined, the end past which the ice was flowing being steep and irregular because it resulted from the plucking-out of blocks of strong but well-jointed rock.

From the regional distribution of drumlins (Figure 20.50) it is clear that they developed under deep ice at a distance of many kilometres from the front towards which the ice was advancing.

Figure 20.49 Typical drumlin country around Wigtown in the SW of Scotland (*Institute of Geological Sciences*)

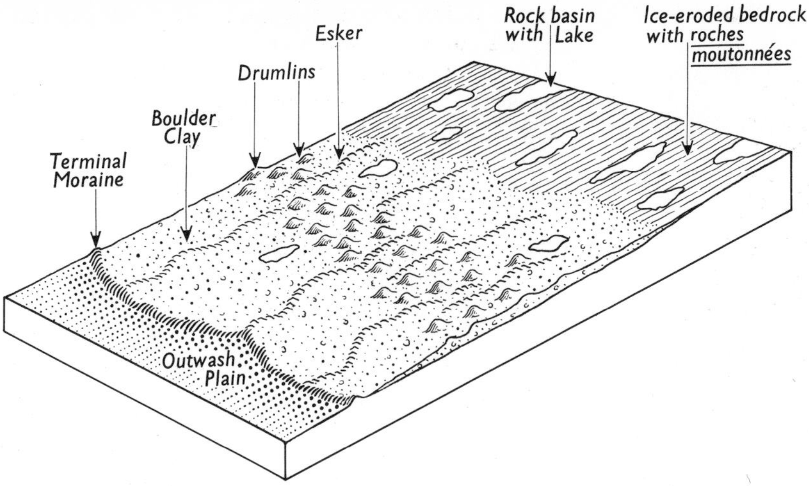


**Figure 20.50** The characteristic assemblage of features seen on a recently glaciated area of low relief

Consequently there is no possibility of seeing drumlins in course of formation, and one can only be guided as to their origin by analogy with related phenomena produced when one medium is flowing over another. Wind blowing over sand produces sand-dunes. Water flowing over beach sand or river sand produces sandbanks. In both cases material is continually eroded from one side of the structure and deposited on the other, so that the structure as a whole migrates slowly in the direction of flow. It seems probable that drumlins are formed in an essentially similar way, the high viscosity and slow rate of flow of ice being compensated by an abundance of time.

## Glacifluvial Deposits

The drainage from the long front of an ice-sheet escapes by way of an immense number of more or less temporary and constantly shifting streams (Figures 3.5 and 20.51). These carry off a great deal of sediment and, as the velocity is checked, low alluvial fans or deltas are deposited, according as the ice terminates on land or in standing water. On land the fans spread out and coalesce into gently sloping *outwash plains* (Figure 20.50) of irregularly stratified drift, ranging from coarse gravels near the source to sand farther out, and finally to clay. Valley floors are choked with spreads of similar deposits; mainly coarse, however, because the finer materials are rapidly washed downstream.

Beyond the terminal moraines of ice-sheets, where the supply of debris is abundant, outwash plains may extend for many miles. Vast areas of the

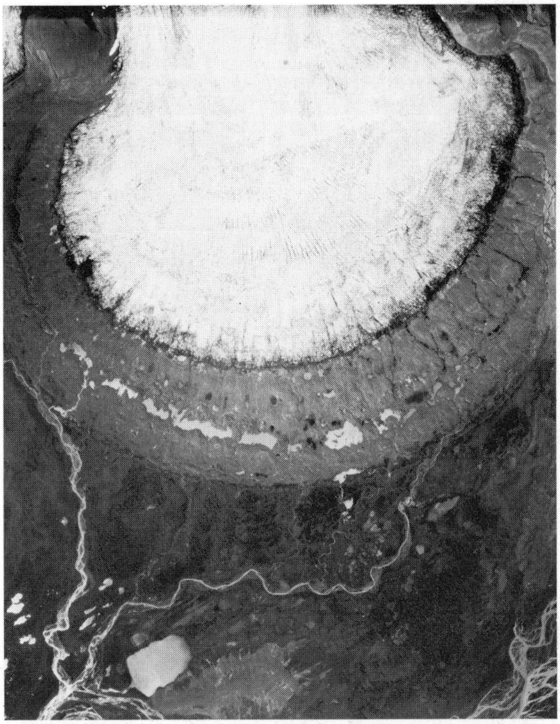

**Figure 20.51** Outwash plain of glacifluvial drift deposited by braided streams from the melt-water of an ice lobe on the SW of Vatnajökull, Iceland (*Erlingur Dagsson*)

North American prairies have been smoothly veneered with sediment in this way. Between successive moraines the outwash drifts rest on previously deposited boulder clay. Masses of stagnant ice, left stranded between deep crevasses as the main front melted back, are often surrounded and even buried by drift, and as they melt

**Figure 20.52** Kettle hole near Finstown, WNW of Kirkwall, Orkney (*Institute of Geological Sciences*)

**Figure 20.53** Esker at Holylee, Tweed valley, 19 km W of Melrose, Scotland (*R. J. A. Eckford, Institute of Geological Sciences*)

away they leave the surface pitted with depressions known as glacial *kettles* or *kettle lakes* (Figure 20.52).

Such drift-covered regions are further diversified by mounds (*kames*), long winding ridges (*eskers*, Figures 20.53 and 20.54), and relatively short and straight ridges (*crevasse infillings*). All of these are built of crudely bedded gravel and sand, showing that they are features for which glacial streams were responsible.

*Kames* are isolated or clustered mounds, each of which represents a steep-faced localized alluvial cone or delta built up by a stream emerging at a high level from a temporarily stagnant ice front. As the front receded, the unsupported back or sides of the accumulation slumped down, leaving a mound with slopes corresponding to the angle of repose of gravel or sand.

If such a stream, instead of being short and temporary, were long and persistent, then the deposit continuously formed at its mouth would grow backwards as the ice retreated, thus extending into a winding ridge that would reproduce the course of the stream. Some eskers may have originated in this way. But such a stream would also deposit sediment while flowing through its tunnel in the ice, thus gradually raising its floor. Most eskers are therefore regarded as the infillings of the tunnels of unusually long subglacial streams. In some cases, where later outwash-drift has lapped against them, it is obvious that they originated within the ice before it receded. Eskers characteristically disregard the underlying topography, which they cross like long railway embankments, this form resulting from the outward slumping of the original sides. Their courses, though sinuous, are generally aligned roughly at right angles to the ice front. See Figure 20.24 for examples exposed in recent times. In glaciated lands riddled with lakes and marshes, like Finland, Sweden and Canada, eskers provide natural causeways across many districts where road and railway construction would otherwise be difficult. Ridges that differ from eskers in being short and straight are interpreted as the infillings of crevasses that remained in a frontal or isolated sheet of ice when it became stagnant during the last stages of its melting away.

Where the glaciers and active volcanoes occur together, devastating floods of melt-water or of mud-flows are produced by eruptions. Cotopaxi

**Figure 20.54** Esker, Punkaharju, Finland. The lake occupies the depressions in an irregular surface of glacifluvial sands and gravels. (*Finnish Tourist Association*)

has an evil reputation for such encounters between fire and ice, and so have some of the Iceland volcanoes. A minor eruption may not liberate sufficient energy to melt and break through a thick covering of ice. In such a case the melt-water escapes along the floor of the ice-cap and the latter sags into a temporary depression. But, writes Barth, in reference to Iceland: 'If the entire ice cap above and around the volcano is melted, a lake will form in the glacier. The huge amounts of water thus concentrated will finally break through the dam of ice, and from the rim of the glacier an unimaginable mass of water mixed with icebergs, volcanic products, and rocks, will sweep down the mountain side, burying everything in its way, and flood the lowlands.' During an eruption of this kind in 1918 at Katla, the most terrifying of all the Icelandic subglacial volcanoes, the discharge of water reached twenty times that of the Amazon at the peak of its flood season.

**Figure 20.55** The Marjelen See, an ice-dammed lake held up by the Aletsch Glacier, between the Jungfrau and the Upper Rhône, Switzerland (*Hans Steiner*)

### Ice-dammed Marginal Lakes and Spillways

A glacier occupying a main valley may obstruct the mouth of a tributary valley and so impound the drainage and make a lake. The Marjelen See is a well-known Alpine example (Figure 20.55) and there are many others, large and small, in regions that still maintain glaciers. Where the ice barrier is sufficiently high and massive the impounded lake rises to the col or pass at the head of its valley and escapes through an *overflow channel* into the valley on the other side of the divide. During the degeneration of an ice-cap into valley glaciers (Figure 20.56), the higher ridges of a divide between two neighbouring valleys may be uncovered, while the ice still extends across the divide at a lower level. Melt-water then accumulates along the margin of the ice against the flanks of the hills, and if it overflows from one side of the ridge to the other, a channel is cut in the ridge itself, which thus becomes notched. Notching by marginal overflow channels may be repeated again and again at successively lower levels

**Figure 20.56** Ice-dammed lake held up by the Hoffell Glacier, Hornafjord, SE Iceland (*L. Hawkes*)

**Figure 20.57** The Parallel Roads of Glen Roy (*Institute of Geological Sciences*)

while the ice is retreating.

During the recession of the Pleistocene ice sheets ice-dammed lakes came into temporary existence on an extensive scale. Some of these marginal lakes overflowed at successively lower levels before they eventually disappeared, each stand of the lake being determined by the height of the lowest outlet available at any given stage during the wasting away of the ice barrier. The features and deposits left behind by such lakes (thereby providing evidence of their former existence) include: (*a*) overflow channels or spillways at the outlets, often eroded into conspicuous valleys and gorges (now dry) situated at the heads of the valleys from which the lake waters escaped, or across the ridges and spurs between neighbouring valleys; (*b*) shore-line deposits and terraces, formed by the action of waves and currents; (*c*) deltas deposited by streams flowing into the lake; and (*d*) lake-floor sediments, including the *varved* sediments described on p. 231.

The most celebrated lake-shore terraces of this kind are the Parallel Roads of Glen Roy (Figure 20.57). These are beaches about 12 to 15 m wide which follow the contours at the levels shown on

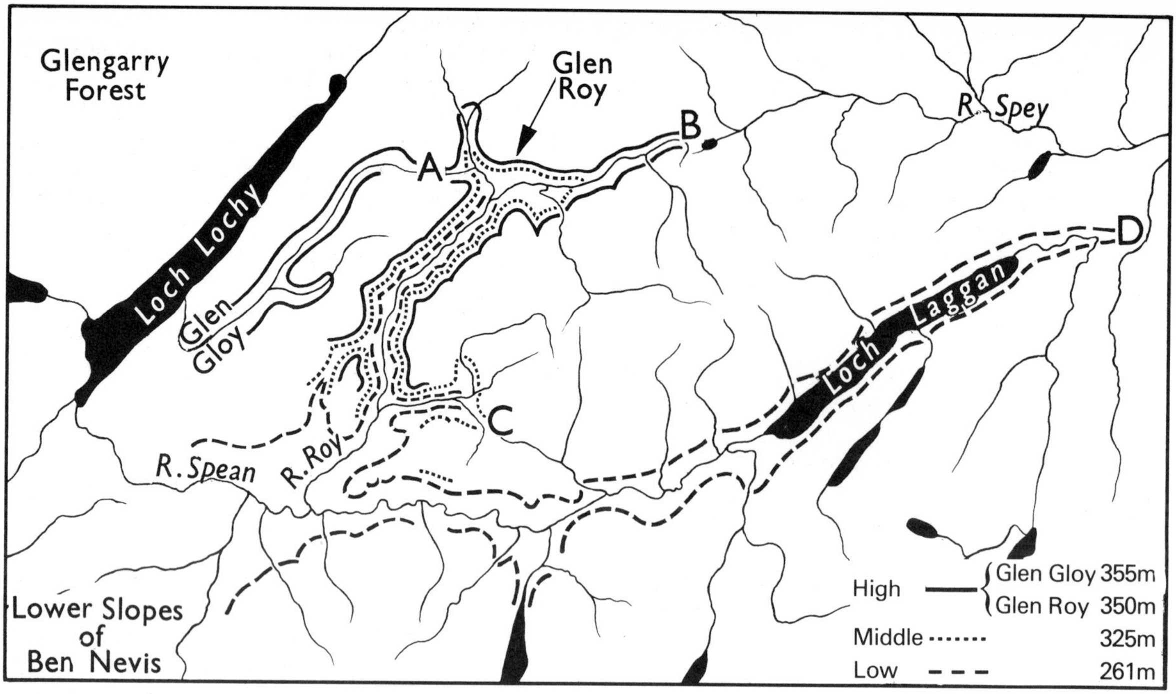


**Figure 20.58** Map of the Parallel Roads of Glen Roy and their overflow channels, South-west Highland Region, Scotland

Figure 20.58. Each beach can be traced to the head of a valley, where a spillway is found corresponding to the level at which the lake overflowed while the beach was being formed. Ice extending from Ben Nevis across the valleys to the north blocked the entrances to Glen Roy and Glen Gloy. The highest lake (355 m), that of Glen Gloy, discharged across the watershed at A into the Glen Roy lake (350 m) which overflowed into the river Spey at B. Later, an outlet into a tributary of Glen Spean was uncovered at C (325 m), and the lake rapidly drained to the level so determined. This stage lasted until further withdrawal of the ice allowed the lake to extend along Glen Spean, whence it overflowed through D (261 m), the outlet at its head. In due course the lobe of ice that blocked the lower part of the valley dwindled away sufficiently to allow the lake to drain towards the sea (Loch Linnhe) and so, finally, to disappear.

### Lakes and Lake Basins: a General Summary

It will already have been gathered that lakes are amongst the most characteristic features of the landscape of glaciated regions. Finland is renowned for its innumerable lakes, 55,000 of which have been mapped. Very appropriately, the Finns call their country *Suomi*—the Land of Lakes. Many parts of Ontario and the neighbouring Provinces and States are riddled with a comparable network of lakes and waterways (Figure 20.59). The extraordinary abundance at the present time of lakes of glacial origin—they are far more numerous than all other types put together—is a result of two circumstances: (*a*) immense numbers occupy hollows excavated in the less resistant rocks by ice scouring, or irregular concavities in the drift surface left behind by the retreating ice; and (*b*) they originated so recently that only some of the shallower ones have since been silted up and replaced by lacustrine flats (Figures 18.11 and 18.12).

Given a supply of water in excess of the amounts lost by evaporation or by seepage through the floor and sides, a lake continues to exist so long as the floor of its basin remains below the lowest part of the rim. Lakes are therefore conveniently classified according to the modes of origin of their basins.

Lakes of glacial origin may occupy:

(*a*) ice-eroded hollows in areas of varied structure (Figure 20.59); rock basins in valleys or valley heads (cirques, Figure 20.33); and rock basins in the deepened cols or passes of former watersheds (Figure 20.40); all with or without morainic fringes;

(*b*) valleys obstructed by morainic barriers (Figure 20.46);

(*c*) depressions due to surface irregularities of glacial drift;

(*d*) kettle holes left by the melting of masses of stagnant ice after burial or partial burial by glacifluvial drift (Figure 20.52);

(*e*) valleys obstructed by ice barriers (Figures 20.55 and 20.56).

Among lakes with a more varied history, involving both glacial erosion and deposition, with modifications due to isostatic rebound after removal of the load of ice (and perhaps to independent earth movements), the Great Lakes of North America (p. 451) are the most remarkable. In Europe, Lakes Ladoga and Onega had a similar origin and history.

Lake basins owing their origin to geological processes other than glacial are described in the appropriate chapters, but for convenience the following summary is added here. For an admirably complete classification, with examples from all over the world, the reader cannot do better than refer to the first chapter (pp. 1–163 of G. E. Hutchinson's great work on *Limnology* (see p. 443).

EARTH MOVEMENTS

Tectonic depressions (relative to surroundings) are responsible for the largest of the world's lakes (Caspian Sea), the deepest (L. Baikal), the lowest (Dead Sea) and, amongst those of notable size, the highest (L. Titicaca, Peru/Bolivia), as well as for many shallow lakes, both large (L. Victoria) and small (local subsidences accompanying earthquakes, e.g. sag ponds). In terms of origin the chief types are due to:

*Crustal Warping* (Caspian Sea, p. 90; L. Victoria, p. 652; Lough Neagh, Figure 12.8) and the back-tilting of valley systems (L. Kyoga, Figure 29.23).

*Folding* across pre-existing valleys (pp. 384 and 388).

*Differential Faulting*, especially in the African rift valleys and in the Great Basin of the western United States.

*Strike-slip Faults* or *Thrusts* across a pre-existing valley, whereby it may be obstructed by a hill range (Lac de Joux, Jura Mts; Fählensee, Säntis Alps). Sag ponds occupy some of the smaller topographic depressions formed along

**Figure 20.59** The intricate waterways of Sturgeon Lake and its neighbours, west of Lake Nipigon, north of Lake Superior, Ontario; characteristic summer scenery of much of the ice-eroded surface of the Canadian Shield (*Royal Canadian Air Force*)

active faults, where higher ground has been brought against lower, or where there has been local subsidence.

VOLCANIC ACTIVITY

*Maars* (Figures 12.18 and 12.19), *Craters* and *Calderas* (Figures 12.44 and 12.46) of extinct or dormant volcanoes.

*Collapse* of crust of hollow lava flows (Myvatn, Iceland).

*Lava Flows forming Barriers* across pre-existing valleys (Sea of Galilee; L. Kivu).

DEPOSITION OF SEDIMENTS

Obstruction of valleys and of river channels may be brought about by:

*Landslides* etc. (p. 313) and occasionally by avalanches and screes.

*River Deposits*: Alluvial fans from vigorous side streams. The sealing-off at both ends of abandoned meander loops (ox-bow lakes, Figure 18.30). Levee building in general (flood-plain lakes and swamps, and delta lagoons (p. 346 and Figure 18.42).

*Glacial and Glacifluvial Deposits* (p. 441).

*Wind Deposits*: Coastal sand dunes enclosing lagoons and marshes (as in the Landes of south-west France).

*Marine Deposits*: Closed bars and barrier beaches enclosing coastal lagoons.

DENUDATION

*Solvent Action of Ground-water*: Swallow holes, of which the outlets have been clogged by residual clays. Surface subsidences due to underground solution of limestone (p. 270) or of rock salt (the meres of Cheshire).

*Solvent Action of Rivers*: Expansion and deepening of river beds by surface solution of limestone (some of the Alpine lakes; Lough Derg, an expansion of the river Shannon).

*Glacial Erosion* (see p. 440).

*Wind Deflation*: Hollows excavated in arid regions to a depth where an adequate supply of ground-water is tapped.

ORGANIC AND HUMAN ACTIVITIES

*Growth of Coral Reefs*: Lagoons cut off from the sea by a continuous ring of coral rock, or by the emergence of atolls or barrier reefs above sea-level (p. 556).

*Beaver Dams* (Beaver Lake, Yellowstone Park).

*Dams built by Man* (L. Mead and L. Powell, Colorado river, Figures 18.37, 19.32; L. Kariba, Zambezi).

*Excavations made by Man*: Some reservoirs; many abandoned peat diggings (Norfolk Broads, Figure 19.19); abandoned diamond mines in kimberlite pipes, South Africa (p. 180).

IMPACT OF LARGE METEORITES

Many of the resulting craters are dry (e.g. Meteor or Barringer Crater, Arizona Desert) but some have become lakes, e.g. the meteorite crater of New Quebec (Figure 20.60); the largest known is the Ashanti Crater (Lake Bosumtwi) in Ghana, with a diameter of over 9·5 km. These scars in the earth's crust caused by the impact of giant meteorites have been called *astroblemes* (Gr. star-wounds) by R. S. Dietz (*Scientific American*, August, 1961, pp. 51–81).

UNKNOWN

*Carolina Bays*: A group of about 150,000 shallow elliptical depressions scattered along the coastal plain of the SE United States, mainly in the Carolinas and Georgia, and orientated in a NW to SE direction. They cannot be accounted for by any known process; it may be that they represent the effects of a glancing encounter with a comet, but this remains a purely hypothetical suggestion since the effects of a collision with a comet have not been observed.

## Selected References

CHAMBERLAIN, R. T., 1936, 'Glacial movement as typical rock deformation', *Journal of Geology*, Chicago, vol. 44, pp. 93–104.

CHARLESWORTH, J. K., 1957, *The Quaternary Era with Special Reference to its Glaciation*, vol. 1, pp. 1–592; vol. 2, pp. 593–1700, Edward Arnold, London.

DURY, G. H., 1957, 'A glacially breached watershed in Donegal', *Irish Geography*, vol. 3, p. 171.

EMBLETON, C., and KING, C. A. M., 1975, *Glacial Geomorphology*, Edward Arnold, London.

— 1975, *Periglacial Geomorphology*, Edward Arnold, London.

FLINT, R. F., 1971, *Glacial and Quartenary Geology*, Wiley, New York.

**Figure 20.60** A meteorite crater and lake (frozen over), in the extreme north of Quebec between Hudson Bay and Hudson Strait. The crater is almost circular, with a diameter of about 3 km, in striking contrast with the irregular outlines of the ice-eroded depressions occupied by the other lakes. (See *Scientific American*, May 1951 for its discovery and description.) (*Royal Canadian Air Force*)


HUTCHINSON, G. E., 1957, *A Treatise on Limnology*, vol 1; Geography, Physics and Chemistry, Wiley and Sons, New York; Chapman and Hall, London.

LEWES, W. V., (Ed.), 1957, *Norwegian Cirque Glaciers*, Royal Geographical Society Research Series, No. 4, pp. 104.

JAMIESON, T. F., 1862, 'On the ice-worn rocks of Scotland', *Quarterly Journal of the Geological Society*, London, vol. 18, pp. 164–84.

NYE, J. F., 1952, 'The mechanics of glacier flow', *Journal of Glaciology*, vol. 2, pp. 81–93; see also pp. 182–3.

— 1959, 'The motion of ice sheets and glaciers', *Journal of Glaciology*, vol. 3, pp. 493–507.

RAACH, G. O. (Ed.), 1961, *Geology of the Arctic*, vol. I, pp. 1–732, vol. II, pp. 733–1196, *University of Toronto Press*, Toronto.

SHARP, R. P., 1954, 'Glacier flow: a review', *Bulletin of the Geological Society of America*, vol. 65, pp. 821–38.

— 1958, 'Malaspina Glacier, Alaska', *ibid.*, vol. 69, pp. 617–46.

SIMPSON, F. A., (Ed.), 1961, *The Arctic Today*, A. H. and A. W. Reed, Wellington, New Zealand.

SPARKS, B. W., and WEST, R. G., 1972, *The Ice Age in Britain*, Methuen, London.

# Chapter 21

# Ice Ages and their Problems

For Hot, Cold, Moist, and Dry, four champions fierce,
Strive here for mastery. . . .

*John Milton* (1608–74)

## The Quaternary Ice Age

The evidences of former glaciations on a widespread scale are overwhelmingly conclusive, as indicated in the preceding chapter. The chief of these are abraded, often striated, bedrock surfaces of wide extent, with *roches moutonnées* and related ice-sculptured forms; spreads of boulder clay (till) or tillite and morainic ridges; and various types of outwash drift, locally including varves. At the present time just over one-tenth of the land surface is covered with ice (p. 15), but during the more widespread glaciations of the Pleistocene this proportion rose to nearly one-third. The maximum extent of the great ice-sheets in the northern hemisphere is shown in Figure 21.1. The total area buried in ice approached 30 million km$^2$. Half of this was in North America, where the ice radiated from three main regions: (*a*) Labrador; (*b*) Hudson Bay; and (*c*) the Cordilleran ranges of the west. As the ice thickened, (*a*) and (*b*) coalesced into the *Laurentide* ice-sheet. From the directions of ice movement, determined by striations and erratics etc., it appears that the ice 'divide' curved from Labrador around the northern part of Hudson Bay and then followed a more southerly direction through Keewatin province. Along this belt the ice surface reached its highest altitude. The subsequent isostatic rebound that began as the ice melted away has already uplifted former shore lines as much as 300 m above the present level of Hudson Bay, and uplift is still in progress. Both to the north (e.g. the northern coast of Ellesmere Island) and to the south (e.g. the area of the Great Lakes) the elevation and its present rate, and also the negative gravity anomalies that still remain, are all less than the corresponding values along the divide. It is a reasonable inference that in these regions the ice was thinner, and that it was thickest along the divide. The position of the divide—so much farther north than might have been expected—is an important indication that the Arctic Ocean was a major source of moisture and precipitation for the development and replenishment of the great ice-sheet. To fulfil that function the water surface of the Arctic must have been exposed to sunshine and wind during at least the summer months. Today the Arctic is permanently ice-covered and leaves northern Canada, like northern Siberia, intensely cold but with insufficient snowfall to nourish an ice-sheet.

Greenland and Iceland were independent centres of ice accumulation and dispersal. The case of Greenland is of particular interest, because it is largely bounded by the open Atlantic, which supplies sufficient moisture for the snowfall required to sustain its present state of glaciation. Canada, at the same latitudes, has scanty precipitation and few glaciers. As Maurice Ewing and W. L. Donn have pointed out: 'An open Arctic Ocean during the Pleistocene seems to be the only geographic condition which could have produced glacial conditions in northern Canada equivalent to those in Greenland today'.

The European ice-sheet and its continuation beyond the Urals covered about 10 million km$^2$. Scandinavia was the main region of accumulation, but at times of maximum extension the greatest height and thickness of the ice plateau lay not over the mountains but over the Gulf of Bothnia. At such times the Scandinavian ice approached and sometimes crossed parts of the eastern coast of

**Figure 21.1** Map showing the maximum extent of the Pleistocene ice sheets and ice caps in the Northern Hemisphere (*After E. Antevs and R. F. Flint*)

Britain, forcing back the British ice, which radiated from a number of highland areas. The Alps formed an important but independent centre of ice dispersal. From its successive ice-caps long valley glaciers extended down to low altitudes, far from their mountain-bound relics of today. The snout of the Rhône glacier, for example (Figure 20.5), has receded all the way from Lyon, more than 350 km beyond its present termination.

There were considerable ice-sheets over NE Siberia, recent work having shown that they were much more extensive than was formerly suspected. Here again, an open Arctic has to be invoked to help in providing the heavy precipitation of snow that was required to sustain them. The Karakoram and Himalayan ranges, much of eastern Tibet and the high mountains of central and eastern Siberia were also very heavily glaciated.

In the southern hemisphere the Kosciusko plateau of New South Wales (Figure 19.17) and a considerable part of Tasmania were glaciated. New Zealand, where glaciers still persist, was largely shrouded in ice, and so were extensive tracts of Patagonia and southern Chile. The Antarctic sheet, like that of Greenland, was thicker and more extensive than it is today. In central Africa moraines occur more than 1500 m below the ice that still remains on Ruwenzori and on the volcanic peaks of Kilimanjaro (Figure 21.2) and Mt Kenya (Figure 21.3). The climatic changes evidently involved a general lowering of the snow line and, as we shall see, they were world-wide in their effects and roughly contemporaneous.

## Stages of the Quaternary Ice Age

Even before the fact of widespread glaciation had become generally accepted it was realized by a few observers, in Britain as well as in the Alps, that the drift deposits could not be accounted for by a single great advance and retreat of the ice. In many of the valleys on the northern side of the Alps four or more successive sets of boulder clays, moraines, outwash gravels, etc. have been recognized, showing that there was at least a fourfold repetition of the glacial cycle. The intervening warmer stages are represented by soils and peat beds, by lake and river deposits and by local screes. Such stratigraphical evidence is preserved only towards the margins of glaciated regions, where the ice was at or near its outer limit and therefore unable to sweep away the superficial deposits in which earlier events were recorded. An admirable example of this kind is illustrated in Figure 21.4. Beyond the ice-fronts, i.e. in the *periglacial* regions, the corresponding climatic oscillations can be recognized by alternations of deposits of loess (wind-blown dust or rock flour) representing

**Figure 21.2** Aerial view of the dwindling ice cap on the Kibo summit (5895 m) of Kilimanjaro, Tanzania. See also Fig. 20.1 (*Aircraft Operating Co. of Africa*)

**Figure 21.3** Summit of Mt Kenya (5199 m), central Kenya, just south of the Equator, showing ice-filled cirques and deep scouring down to 4389 m by former large glaciers (*R. V. Light, American Geographical Society*)

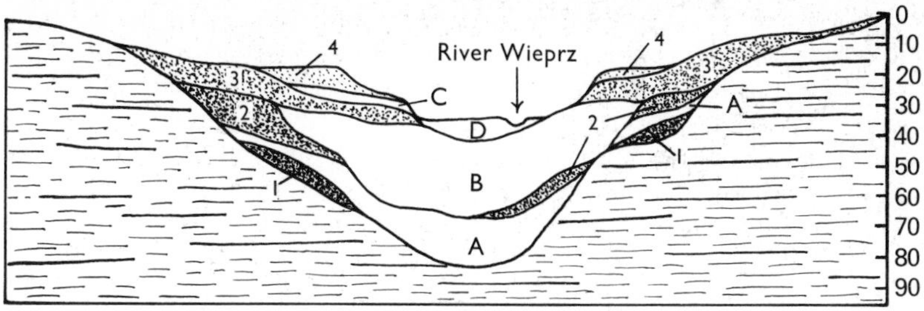


**Figure 21.4** A schematic section across a tributary of the Vistula, Lublin district of eastern Poland, showing the sequence of Quaternary deposits in a Pliocene valley. Boulder clays and glacifluvial deposits, marked 1, 2, 3, 4, represent glacial stages or sub-stages; these alternate with river deposits and peat beds, marked A, B, C, D, indicating interglacial stages or sub-stages. (*After H. Jahn*)

cold intervals, with bands of soil indicative of weathering under warmer conditions.

Each *glacial stage* was originally regarded as the time of active life of a major continental ice-sheet, the intervening stages, recognized by the remains of warmth-loving plants and animals, being distinguished as *interglacial stages*. The four major Alpine glaciations and the inter-glacials between each pair are known by the names listed in the following table. The names, selected to be in alphabetical order (Gunz, Mindel, Riss, Würm), are those of Bavarian tributaries of the Danube, where the glacial deposits so distinguished are well preserved. A fifth glacial stage, older than the Gunz, has since been discovered and is called the Donau or Danube. For easy reference the names of what are tentatively considered to be the corresponding stages in North America, Britain and northern Europe are also listed. Since the last glacial stage is mainly known by names beginning with 'W', it is convenient to refer to it by that initial letter. The other Alpine initial letters are often similarly adopted for general use, but it must be remembered that the correlations so implied (as in the accompanying scheme) remain hypothetical, unless they have been established by the finding of fossil plants and animals, or of human implements, or by actual dating in years.

The whole assemblage of alternating glacial and interglacial stages constitutes an *Ice Age*. Since we are living in a still unfinished Ice Age, it is desirable to refer to it as the Quaternary Ice Age in order to include the last 11,000 years as well as the succession of Pleistocene glaciations that characterized the preceding two million years or so. This is only a matter of definition and usage. What is far more important is that the rhythm of climatic change within an Ice Age is much more complex than this introductory idea of a simple alternation would suggest. Every glacial and interglacial stage

## **Pleistocene GLACIAL and *Interglacial* Stages**

From the beginning of the Donau Glacial, about 1,800,000 years ago, until about 11,000 years ago, when Holocene ('Recent' or 'Post-Glacial') time began.

| *Central North America* | | *British Isles* | *Alpine Region* | *Northern Europe* |
|---|---|---|---|---|
| WISCONSIN | | NEWER DRIFT | WÜRM (W) | WEICHSEL |
| *Sangamon* | | *Ipswichian* | *R/W* | *Eemian* |
| ILLINOIAN | OLDER DRIFT | GIPPING TILL | RISS (R) | SAALE |
| *Yarmouth* | | *Hoxnian* | *M/R* | *Needian* |
| KANSAN | | LOWESTOFT TILL | MINDEL (M) | ELSTER |
| *Aftonian* | | *Cromerian* | *G/M* | (Confused sequence |
| NEBRASKAN | | Weybourne Crag* | GUNZ (G) | of 'cold' and |
| | | *Norwich Crag* | *D/G* | 'warm' pre-Elster |
| (? Pre-Nebraskan | | Red Crag* | DONAU (D) | deposits) |

* Marine beds, some of which contain fossils indicating cold conditions

is itself made up of two or more sub-stages or phases. Each glacial stage has been interrupted by interglacial phases or *interstadials*, during which the ice-front receded. Similarly each interglacial stage has been interrupted by glacial phases or sub-stages, during which ice-sheets formed afresh or made a notable advance from their shrunken condition. By putting together all the bits of evidence from different glaciated regions it is possible to recognize at least a score of such phases, and an even greater number of minor oscillations.

## Oscillations of Pleistocene Sea-Levels, River Terraces

Eustatic changes of sea-level which, being uniform all over the earth, are not due to localized earth movements, may have a variety of causes besides glaciation. Displacement of water by deposition of sediments on the sea floor could raise the level by no more than 250 m, even if all the land were reduced to the resulting sea-level; so far as we know this has never happened everywhere at once. Cooling and heating of ocean water by, say, 10°C throughout, would result in a fall or rise of only 7 or 8 m. Gain of water from juvenile volcanic sources is negligible in this respect, like the loss of water bound in weathering products. There remain crustal and sub-crustal processes involving surface changes of level; for example, the isostatic rise and fall of ocean floors and continents in response to overlying load, i.e. ice on land during glacial periods and water from melt-ice added to the oceans during interglacial periods. In spite of the complexitities it is obvious that by far the greatest downward and upward swings of sea-level during the Pleistocene have been due to the abstraction of water piled up as ice on the land, and to its restoration by melting. The greatest fall during the last maximum accumulation of ice was 135 m. Today's level was reached about 6000 years ago. But there still remains a good deal of ice to be returned to the oceans. If melting were rapid, the further submergence might be 50 or even 70 m, but after isostatic adjustments (p. 299) the sea-

**Figure 21.5** The eustatic rise of late Pleistocene sea-level (with superimposed oscillations) to its present level is indicated by dated samples which were deposited close to the sea-level of their time. (*Modified after R. W. Fairbridge, 1960;* see also *Nature*, 1958, vol. 181, p. 1518)

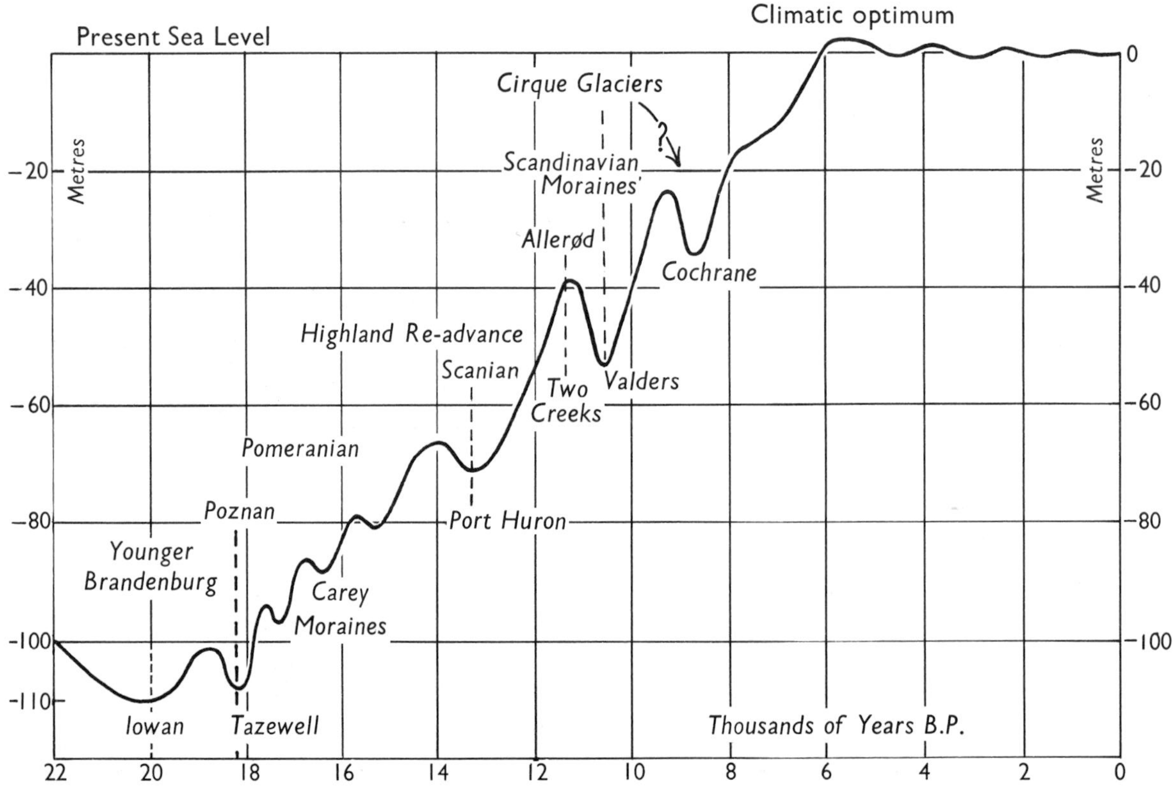

level would ultimately settle down at about 35 m above today's. The eustatic rise of sea-level since the water locked up in the last great ice-sheets was restored to the oceans is shown by the data plotted in Figure 21.5.

Zeuner (1959) and Fairbridge (1961, 1962) both found evidence that the high sea-levels of interglacial periods, prior to 100,000 years ago, regularly fell for a period of several hundred thousand years. Zeuner suggested that there was a fall in high sea-level of 0·12 m in a thousand years. Evans (1971) after re-examining all the data suggests that the fall was more probably between 0·15 and 0·2 metres in a thousand years. In other words, sea-level was about 200 m higher, a million years ago, that it is at present. Figure 21.6 illustrates this regular fall in sea-level upon which the oscillations of Pleistocene sea-level were superimposed.

The heights and ages of river terraces are closely tied to the heights and ages of high sea-levels during the Pleistocene. As illustrated by Figure 21.7, in harmony with the progressive lowering of sea-levels, related river terraces became successively lower. As King and Oakley (1936) expressed this, in discussing the terraces of the Lower Thames:

> 'In general the movements of base-level seem to have been like the waves of an ebbing tide. After each period of aggradation the river cut down rapidly; and generally did not return to quite such a high altitude ever again; at the next down-cutting it went lower than before, never to make up all its lost height, and so on time and again'.

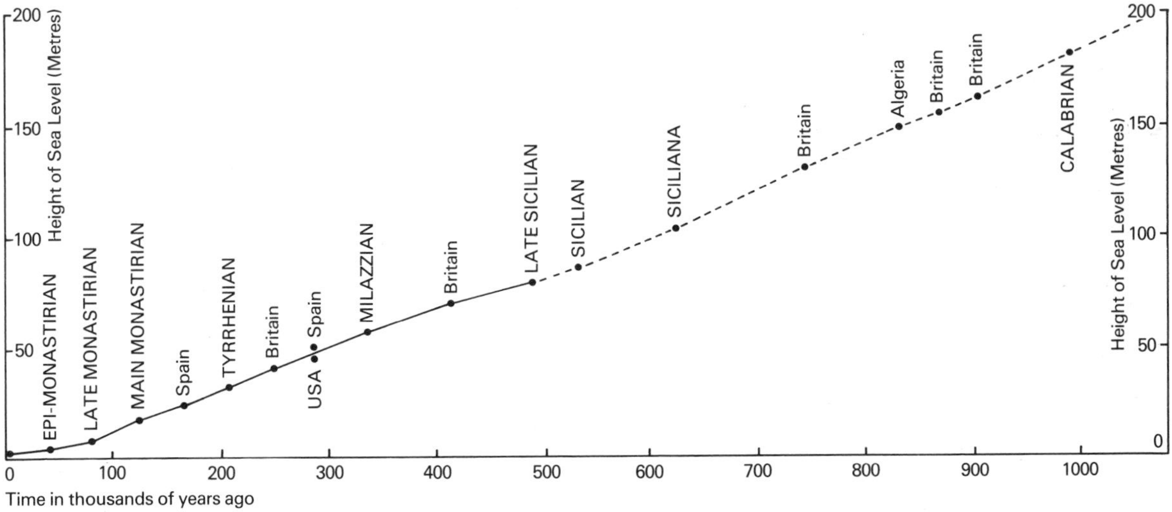


**Figure 21.6** The regular fall of high sea-levels upon which the oscillations of Pleistocene sea-levels were superimposed (*After P. Evans, 1971*)

**Figure 21.7** Diagrammatic section to illustrate the chief erosion platforms and gravel terraces occurring in the Thames Valley from the North Downs on the southern side, or from the Chiltern Hills on the northern side. Pleistocene sea-levels reached during successive interglacial stages are given in brackets. For terraces at lower levels than the Taplow Terrace, see Fig. 19.23.

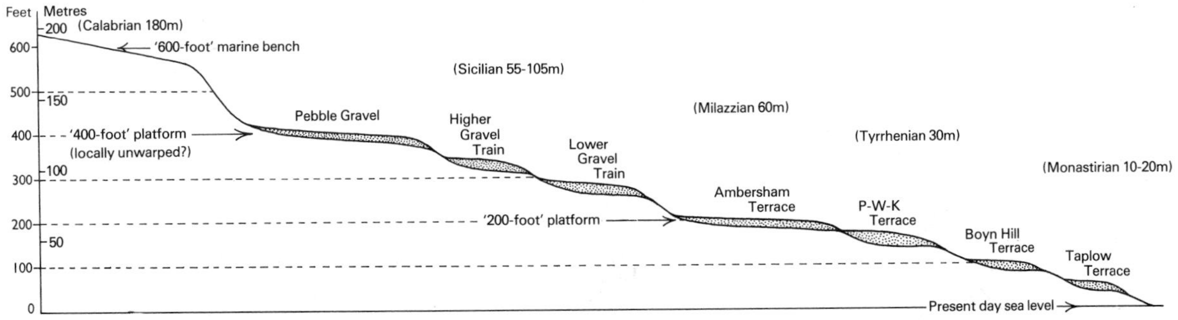

The base levels referred to in the above quotation are, of course, the height of sea-level during successive glacial and interglacial periods.

The chief erosion platforms and gravel terraces of the Thames valley are illustrated in Figure 21.7. The Boyne Hill terrace is of particular interest for several reasons. Its fossil remains and human implements indicate that it belonged to the Great Interglacial between Mindel and Riss. Its correlation, step by step down the valley sides, with the Swanscombe terrace, which is 80 km nearer the sea, shows that the Thames was then graded to a sea-level of about 30 m, which corresponds both in height and age to the *Tyrrhenian* sea-level recorded by emergent beaches and strand lines in many parts of the Mediterranean and elsewhere (Figure 21.18). The age of the terrace appears to be about 210,000 years. Moreover, the very existence of the Boyne Hill–Swanscombe terrace along the sides of the valley through which the Thames now flows indicates that the terrace was formed long after the river had been diverted from its original course along the northern edge of the London Basin (Figure 21.8).

## Pluvial Periods

In regions that are now arid or semi-arid, shrunken or dried up lakes evidence increased rainfall during the glacial periods. These climatic phases are referred to as *pluvial* (L. *pluvia*, rain). Pluvial conditions may be marked by terraces left stranded on the slopes of lake basins, or by widespread lacustrine deposits where the relief of the surrounding country is low. The ancient and now vanished Lake Bonneville in Utah, an enormous ancestor of the present-day Great Salt Lake, Provo and Sevier Lakes, had a maximum area of about 52,000 km². It was here that the discovery was made, by reference to the association of lacustrine deposits with glacial drift, that the maximum extension of the lake was synchronous with a glacial period. Lake Bonneville extended during glacial periods and contracted during interglacial periods several times, as evidenced by wave-cut platforms and deltas that were formed along its shore-lines. One of these platforms, more than 300 m above Great Salt Lake, is cut along the side of the Watsatch Mountains.

Lake Chad has several times spread from its present vague boundaries far over Nigeria and the Sahara, only to retreat again before the inter-

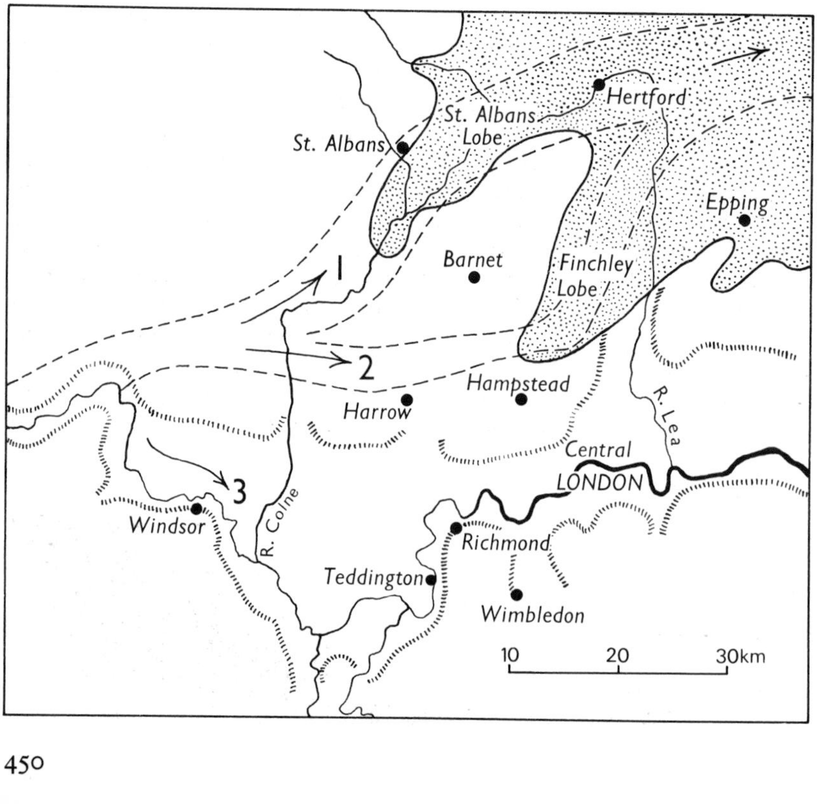


**Figure 21.8** Successive courses of the River Thames from early Pleistocene times: 1, through the Vale of St Albans; 2, through the Finchley Loop; 3, through the present valley tract. The diversion from 2 to 3 resulted from the blocking of the Finchley gap by advancing ice during the M glaciation. The area covered by the resulting Older Drift is dotted. (*After S. W. Woolbridge and D. L. Linton*)

pluvial advance of the desert. Wherever correlation by stone implements or other means is practicable, it is found that high lake levels and the rainy periods responsible for them correspond with glacial stages. One particularly impressive example has been recorded by Russian geologists. Each glaciation of the Caucasus Mountains was accompanied by a high level of the inland Caspian Sea, while the marine Black Sea (after its origin in mid-Pleistocene times) was at a low level, corresponding to the general glacial lowering of sea-level. It should be added that the Black Sea originated in a tectonic depression that sank below sea-level at the same time as the Sea of Marmara and parts of the Aegean Sea. All three were soon inundated from the Mediterranean, which then itself happened to be standing at a fairly high level, corresponding to the R I/II interglacial phase. Though less deep than it has since become, the original Black Sea covered a wider area, as shown by its abandoned strand lines.

## Lakes Margining Continental Ice-Sheets

During the recession of the continental ice-sheets of Europe and North America conditions were highly favourable to the development of widespread marginal lakes (Figure 21.9). At the time of maximum extension of the ice the less mountainous parts of the underlying floor were depressed into a shallow bowl by the isostatic effect of the load of ice. The thickness of the ice reached 2400 m or more, tapering off towards the margins. The corresponding subsidence of the crust, where the load was greatest, would therefore be over 600 m; sufficient, that is, to depress vast areas of the rock surface well below sea-level. Such is the condition of Antarctica today. During the retreat of the ice the crust was gradually unloaded and isostatic recovery worked in from the margins, though with a considerable lag. Consequently, for thousands of years there were large tracts, abandoned by the ice, that sloped towards and beneath the receding ice front. Many of these became giant lakes, while others were invaded by the sea. The isostatic recovery already achieved since the disappearance of the ice is clearly demonstrated by the emergence of beaches now locally preserved at various heights above sea-level, and by the tilted attitude of many lake terraces. Moreover, the fact that the shores of Hudson Bay and the Gulf of Bothnia are steadily continuing to rise shows that the process of restoring isostatic equilibrium is still going on. The deeper depressions within the areas that were formerly inundated are still occupied by lakes or by the sea; the Great Lakes of North America (Figure 21.11) and the Baltic Sea (Figure 21.12) are outstanding examples.

*The Great Lakes of North America*

The immense basins occupied by the Great Lakes have been gouged out by the grinding passage of ice, many times repeated, mainly along tracts of Palaeozoic shales and other relatively weak rocks. Because they were easily eroded, these tracts were already broad lowlands in pre-glacial times. The capacity of the lake basins has been further increased by the morainic barriers looped around their southern margins (Figure 21.10). The lakes appear to have left geological records of their history—such as shore terraces and beaches—dating back only some 17,000 years or so. During the many earlier occasions when ice was advancing or retreating over the same areas there were doubtless similar lakes, but no evidence of these that may have escaped destruction by the overriding ice has yet been detected. Some of the successive stages represented by the scraps of history that remain are illustrated in Figure 21.11. These reveal changes in the outlines and outlets of the lakes that cannot be accounted for merely by

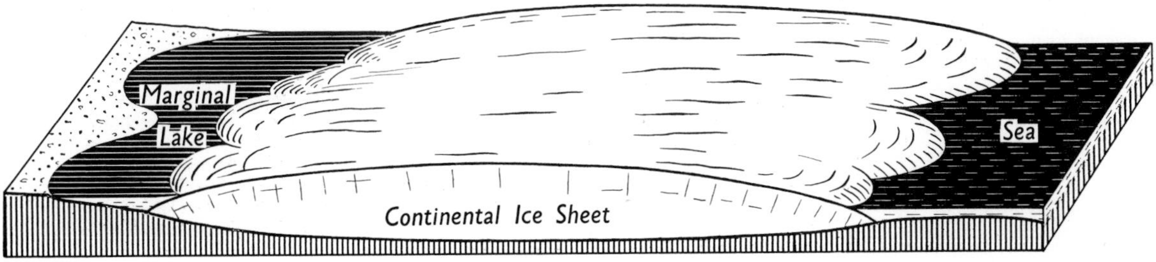


**Figure 21.9** Diagram to illustrate the isostatic depression of a land surface loaded by a continental ice sheet, and the consequent development of marginal lakes during the recession of the ice. Vertical scale and slopes are greatly exaggerated.

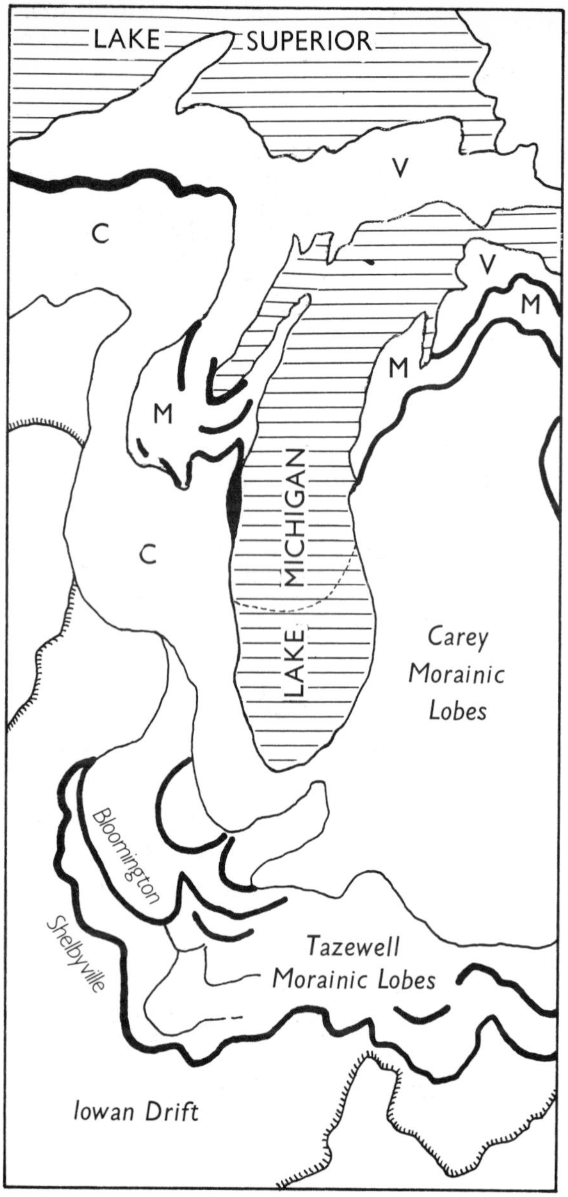


**Figure 21.10** Map of the drift and recessional moraines representing the main glacial sub-stages of the later part of the Wisconsin (W II) glacial stage in and around Lake Michigan. C. Carey; M, Mankato or Port Huron; V, Valders. (*After L. Horberg and R. C. Anderson, Journal of Geology, (vol. 64, 1956); and M. M. Leighton, ibid., vol. 68, 1960*)

fluctuations in the uncovering of depressions between the ice-front and the bordering moraines. Varying degrees of isostatic readjustment from time to time and from place to place introduced additional changes of level and tilting. Some independent crustal warping is also suspected to be a further source of complication.

The dates attached to the substages depicted in Figure 21.11 have all been determined by the radiocarbon method (p. 244). One particularly well-dated temperate interval is represented by the Two Creeks 'forest bed', which is an extensive layer of peat containing abundant remains of spruce, pine and birch trees. The forest flourished for hundreds of years but was then obliterated by a short-lived advance of the ice known as the Valders sub-stage. After this the recession continued with only minor fluctuations.

Figure 21.11 shows (*a*) the approximate position of the ice-front when the Cary advance was at its maximum. This was the latest occasion on which all the lake basins were entirely buried by the Laurentide ice-sheet. The outlines of the modern lakes are indicated by faint lines and dots.

(*b*) The marginal lakes that appeared during the Cary/Port Huron interval of recession. At first the main outflow was to the Mississippi by way of an outlet near the site of Chicago. Later, when L. Superior began to appear its waters drained into another tributary of the Mississippi by an outlet near Duluth. By this time the retreating ice had uncovered an outlet into the Hudson, and the other lakes drained into this, as its level was below that of the Chicago outlet, which temporarily fell into disuse. It is possible that the buried St David Gorge (Figure 18.6) began to be cut when the ice-front retreated to the north of the Niagara escarpment during this interval.

(*c*) The Port Huron advance at its maximum, when the basin of L. Superior was again overrun by the ice. The other marginal lakes emptied through the Chicago outlet. Only the drainage of the Finger lakes escaped to the Hudson. The ice now overrode the Niagara escarpment for the last time, and St David Gorge was choked with drift.

(*d*) The maximum retreat of the ice during the Two Creeks interval. At this time drainage was partly by overland flooding, and entirely eastwards into an enlarged St Lawrence. During the retreat the Niagara river again spilled over the escarpment—well to the east of St David Gorge—and started cutting its present gorge. Samples of

**Figure 21.11** Some successive stages in the development of the Great Lakes of North America, with approximate radio-carbon dates. The lake basins or the parts uncovered are indicated by the initial letters: S, Superior; H, Huron; M, Michagan; E, Erie; and O, Ontario. The Finger Lakes of northern New York State are marked F; C and D are the Chicago and Duluth outlets to the Mississippi. (*After J. L. Hough, 1958, with later modifications of radio-carbon dates*)

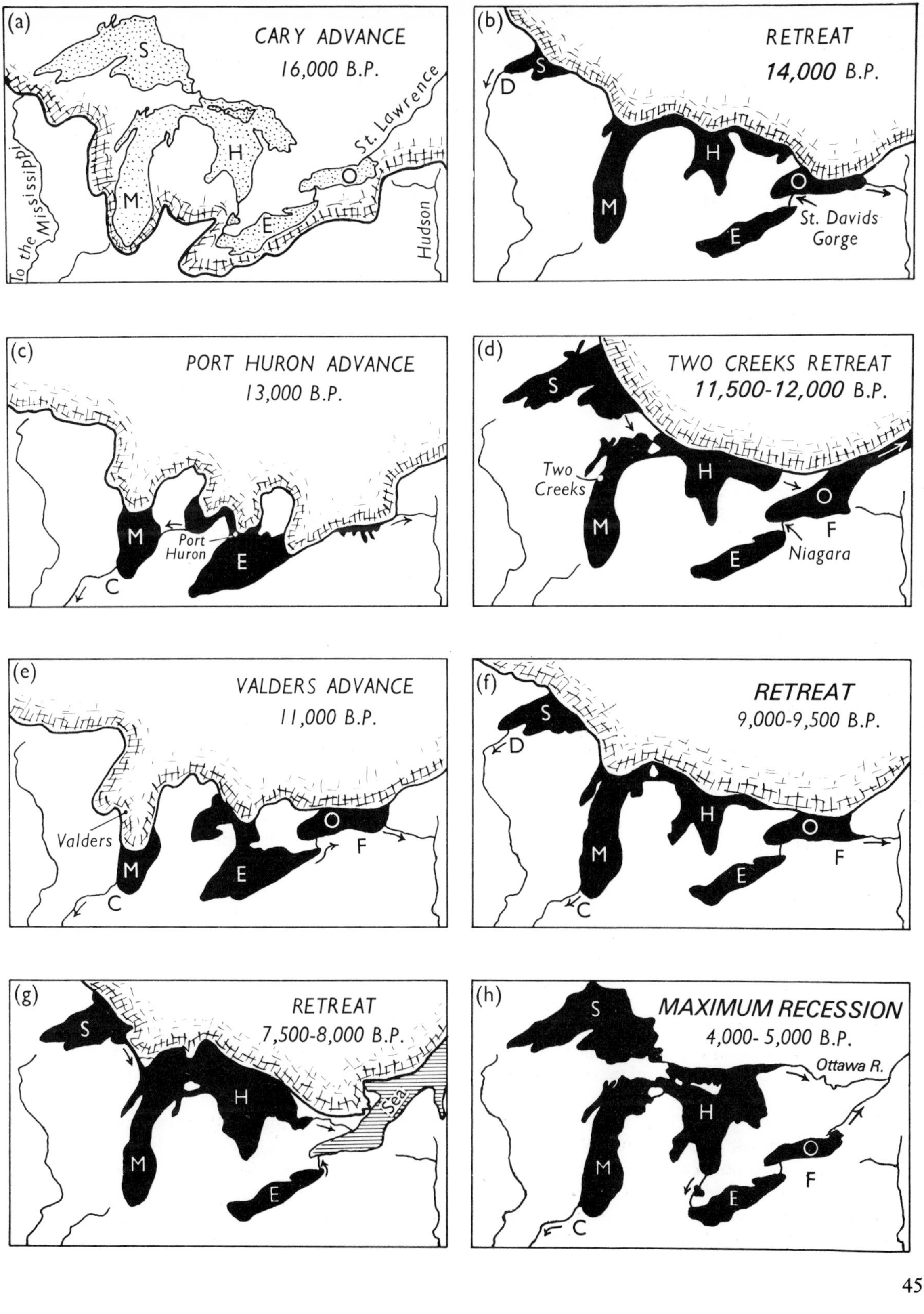
(a)
CARY ADVANCE
16,000 B.P.
S
M
H
E
O
St. Lawrence
To the Mississippi
Hudson
(b)
RETREAT
14,000 B.P.
S
D
H
M
O
E
St. Davids Gorge
(c)
PORT HURON ADVANCE
13,000 B.P.
M
E
Port Huron
C
(d)
TWO CREEKS RETREAT
11,500-12,000 B.P.
S
Two Creeks
M
H
O
F
E
Niagara
(e)
VALDERS ADVANCE
11,000 B.P.
Valders
M
C
E
O
F
(f)
RETREAT
9,000-9,500 B.P.
S
D
H
M
O
F
E
C
(g)
RETREAT
7,500-8,000 B.P.
S
H
M
E
Sea
(h)
MAXIMUM RECESSION
4,000- 5,000 B.P.
S
Ottawa R.
H
M
O
F
E
C

wood from lake deposits north of the escarpment give radiocarbon ages averaging 12.080 ± 300 years. As these deposits have not been covered by drift or otherwise disturbed by ice, it is inferred that the Valders advance that followed could not have reached as far as the escarpment. The age of the Niagara Falls and Gorge can thus be estimated to round figures at about 12,000 B.P.

(*e*) To the west of Niagara the Valders advance again overwhelmed the basin of L. Superior. The Two Creeks forests were cut through; and the greatly diminished L. Michigan drained through the Chicago outlet.

(*f*) Conditions as they were at one stage of the subsequent recession. About 9500–9000 B.P. there were three sets of marginal lakes, each with its appropriate outlet, as indicated.

(*g*) A later stage in the recession, when a long marine embayment, known as the St Lawrence Sea, extended over the site of L. Ontario. All the other lakes drained into this seaway, which did not withdraw until about 5500 B.P.

(*h*) The lakes as they were about 5000–4000 B.P., when rainfall was heavy and the lakes were swollen. The main outlets were into the St Lawrence, now a river again, not only by the present-day route, but also by a spillway (since abandoned) that led through L. Nipissing into the valley of the Ottawa river. At times when the level of the water rose sufficiently the old Chicago outlet carried off the excess overflow.

It will be realized from Figure 21.11 that the three upper lakes are particularly delicately balanced. A gentle tilt of their basins towards the south would suffice to send their waters to the Gulf of Mexico instead of to the east, as at present. Isostatic uptilting towards the north is still in progress. If this were to continue at its present rate, without human interference or some natural counterbalancing process, L. Michigan would again be draining to the Mississippi in about 1500 years.

To the north-west another series of lakes originated in much the same way as the Great Lakes. Their known history begins about 12,000 B.P. when a number of growing marginal lakes coalesced into a gigantic spread of water that lasted for several thousand years. This ancestral lake is known geologically as L. Agassiz. The flat-lying sediments deposited on its floor provide much of the soil of the rich wheat lands of North Dakota and Manitoba. At first the lake drained into the Mississippi by way of the valley of the Minnesota river. Lower outlets were uncovered later, draining at one time into L. Superior, but later into Hudson Bay. The remnants of this vast lake include L. Winnipeg and the Lake of the Woods. Farther north-west, still along the edge of the Canadian Shield, L. Athabasca, Great Slave Lake and Great Bear Lake are the descendants of other great marginal lakes.

*The Baltic Sea*

Just as a string of great lakes is situated along the border of the Canadian Shield, so in Europe L. Ladoga and L. Onega border the Baltic Shield, together with the Gulf of Finland, the Baltic and the White Sea. All these depressions have had a similar history, but that of the Baltic is the best-known. During the last fluctuating recession of the Scandinavia ice-sheet the Baltic began as a series of marginal lakes which coalesced and drained through whatever overflow channels were available from time to time, including one that led from the Gulf of Finland through L. Ladoga to the White Sea. At that time the sea-level was still low (see Figure 21.5), but the depressed land in front of the ice-sheet was beginning to rise. Eventually the receding ice uncovered an outlet at the point marked 'O' on Figure 21.12(*a*), and the Baltic Ice Lake then began to drain through an ice-marginal river (between the sites of L. Vättern and L. Vänern) into the North Sea. This was 10,275 years ago (varve date), a date that corresponds to the time of ice withdrawal from the second of the great Scandinavian moraines (S II) marked on Figure 21.15.

The sea was now rising more rapidly than the land. The outlet soon widened and deepened as the L. Vänern depression was submerged. The heavier sea water was then able to creep in along the floor of the broadening gap, and the Baltic Lake became increasingly saline, especially in the west and south. By the time the ice-front had withdrawn to the site of Uppsala (varve 9800 B.P.) the Baltic was definitely marine: an arm of the sea known as the *Yoldia* Sea (Figure 21.12(*b*)) after the name of a marine mollusc that left its shells in the beaches or other strand lines that survive. Today, *Yoldia* lives only along the shores of Arctic seas. After another few hundred years, during which the land rose more rapidly than the sea, the broad connection through L. Vänern to the North Sea was narrowed and reduced to a short river. The Baltic now became the *Ancylus Lake* (Figure 21.12(*c*)), named after a small freshwater snail.

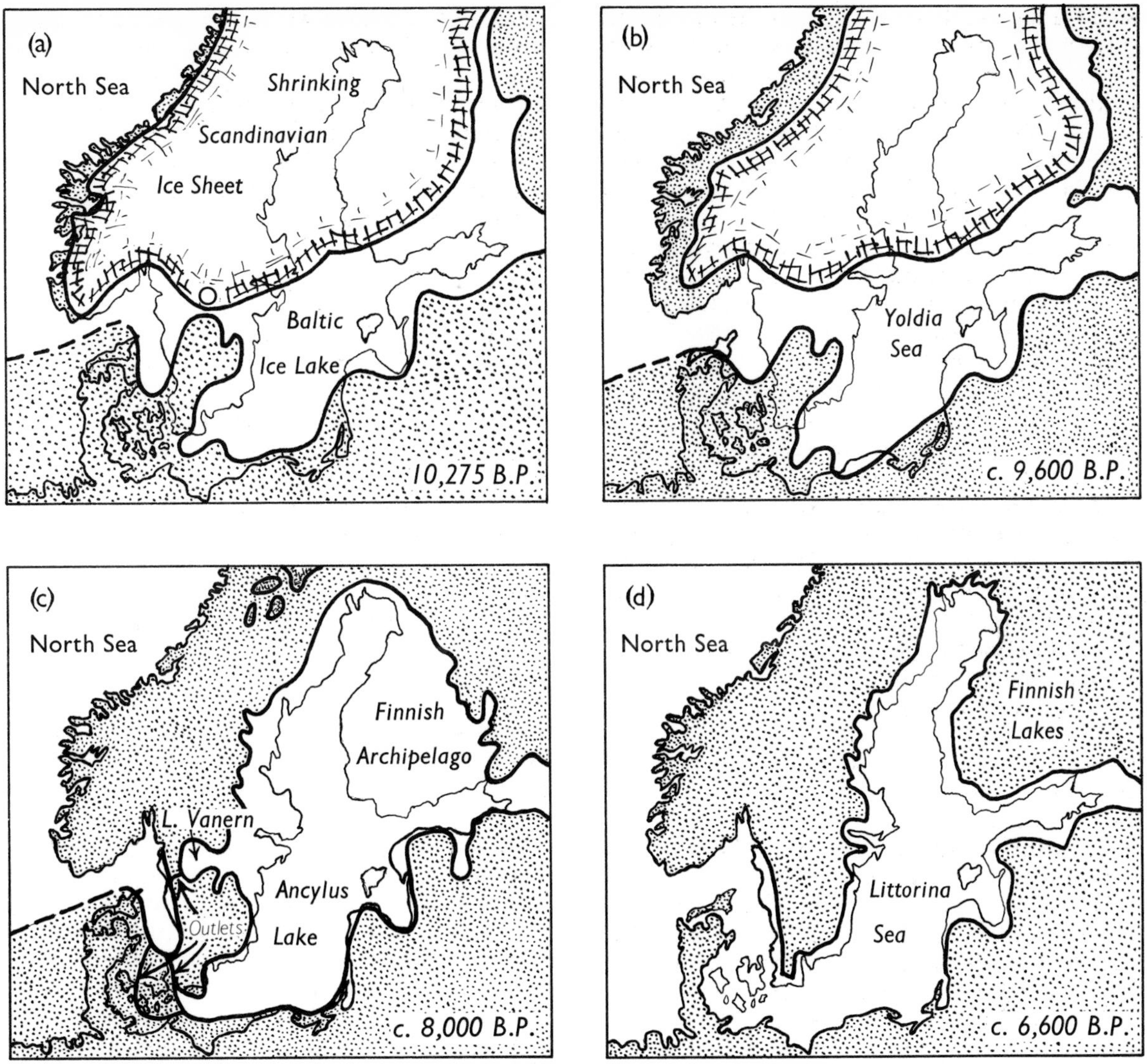


**Figure 21.12** Stages in the late ancestry of the Baltic Sea. Yoldia Sea is named after a marine mollusc. Ancylus Lake after a small fresh-water snail, and Littorina Sea after the common periwinkle. (*After E. Fromm, 'Atlas över Sverige', 1955*)

This lake lasted some 2000 years, but as land-, lake- and sea-levels were all rising at different rates, a time came when the outlet across southern Sweden was abandoned in favour of lower outlets through the Sound (between Copenhagen and Sweden) and the Great Belt (the main channel between the Danish islands to the west). Eventually, however, the isostatic recovery of Denmark began to lag behind, so that the eustatic rise of sea-level caught up again. About this time Britain became separated from the Continent. By about 7500 B.P. the Sound and the Great Belt were submerged and widened, and the Ancylus Lake developed into an arm of the sea. This ancestral Baltic Sea had originally a considerably bigger area than its present-day descendant—which is still shrinking as the land rises in the North (Figure 21.13)—and it is therefore convenient to refer to its earlier stages as the *Littorina Sea* (Figure 21.12(*d*)), after the common periwinkle that still flourishes along the shores of the Baltic and the North Sea.

**Figure 21.13** Ice-moulded rock surface still emerging isostatically at a rate of 40 cm per century. Söderskär Archipelago, near Helsinki, south coast of Finland (*E. Wegmann*)

## Quaternary Glaciations in the Southern Hemisphere

The growing demonstration that the pluvials of the Sahara and E Africa are synchronous with the glacial stages farther north raises the question whether the latter are also synchronous with the glacial stages of the southern hemisphere. Present-day conditions suggest that they are. The Andes are equally glaciated on both sides of the equator. Greenland and Antarctica are both glaciated, with the two ice-sheets in comparable states of recession from their last major advances. Going back over a longer period, the immense swings of sea-level that we have passed in review would have been impossible had not a fair proportion of the Antarctic ice melted away at about the same time as the waning of the northern ice-sheets. A few indications of this have already been mentioned. Here it may be added that gravity measurements made during the Commonwealth Trans-Antarctic Expedition led by Vivian Fuchs in 1958 revealed very strong negative anomalies, suggesting that Antarctica was formerly covered by a greater thickness of ice, and that isostatic recovery, as in N Scandinavia and Canada, is still far from being completed.

In the McMurdo Sound region (Figure 20.10) well-exposed boulder clays and moraines represent at least four major glaciations and a few minor advances. The earliest deposits are over 600 m above the present valley floor. Algal remains, dated at 6000 B.P., have been found on the latest terminal moraines, indicating that the Climatic Optimum of the north had its counterpart in the south. The formerly greater extent of the Antarctic ice is proved by the presence on outlying islands of erratics of granite and gneiss that can have come only from the mainland. The formerly greater thickness of the ice-sheet is spectacularly illustrated by Figure 21.14 from Queen Maud Land, where evidences of a general recession of the glaciers are to be seen wherever there is ice-free land. Farther round the coast, gneiss erratics occur at the summit of the extinct Gaussberg volcano, i.e. well over 300 m. Nearer the pole, lateral moraines of the Beardmore Glacier have been left at heights of 300–1200 m above its present surface.

All the available evidence is consistent with the conclusion that throughout the Quarternary the glaciations of the two hemispheres have been at least roughly contemporaneous. As to when and

**Figure 21.14** North-east face of Birger Bergensenfjellet, Monts Sør-Rondane, Queen Maud Land, Antarctica. The rocks are migmatites heavily veined with pegmatites. The needle-like peak on the right is an intrusion of fine grained granite. The vertical walls, characteristic of wind erosion, rise 800 metres above the snow-covered ice-field. The immense scale of this landscape is indicated by the sledge and its team of dogs—to which the arrow is pointing. (*E. Picciotto, Expédition Antarctique Belge 1957–1968*)

where the first glaciation began, there is as yet no clear indication. There was, however, a long and world-wide decline of temperature throughout the Tertiary, and this evidently heralded the initiation of the latest of the Great Ice Ages.

## Dating Events in the Pleistocene

An early method of dating events in the Pleistocene was devised by De Geer (1912) who, with his collaborators, counted the varves within the Scandinavian varved clay, as described on pp. 231–3. In this way he constructed a time-scale for the last recession of the ice across his native land (Figure 21.15). De Geer fixed his 'zero'—the end of the Ice Age and the beginning of Recent or Holocene time—by a thick varve near Ragunda, which he took to represent the separation of the Scandinavian ice-sheet into two isolated ice-caps depicted in Figure 21.15. There is, however, another thick varve, some 84 years older, which is equally likely to represent the bipartition of the ice, and has the practical advantage of being more firmly connected with historical time. In round figures, 8000 years before the present is widely adopted for the date of bipartition, in place of De Geer's 8700 quoted on p. 232. What is more important is that dating by the counting of varves has been extended into Denmark, Finland and north Germany. Moreover, many of the numbered varves have also been dated by the radiocarbon method.

Since about 1950 the radiocarbon method has been applied to dating events in the Pleistocene not older than 70,000 years. The reliability of the method, however, has been found to decrease for ages greater than 25,000 years. For Pleistocene events older than 70,000 years B.P. the potassium-argon method has been widely used. A difficulty in applying this method to the Pleistocene, however, is that determinations have to be made on igneous rocks or their minerals, and preferably on lavas, and these are only rarely associated with the Pleistocene deposits.

Other radiometric methods have been applied to dating the Pleistocene. One of these is based on the fact that protoactinium, $^{231}Pa$, and ionium, $^{230}Th$, both daughters of uranium, decay at different rates, their half-lives being 32,000 and 75,000 years respectively. Consequently the ratio $^{231}Pa/^{230}Th$ varies with time and can therefore be used as a measure of time. This method is applicable to clays brought up in deep-sea cores. Another method, known as the ionium-excess method, depends on an assumption that modern

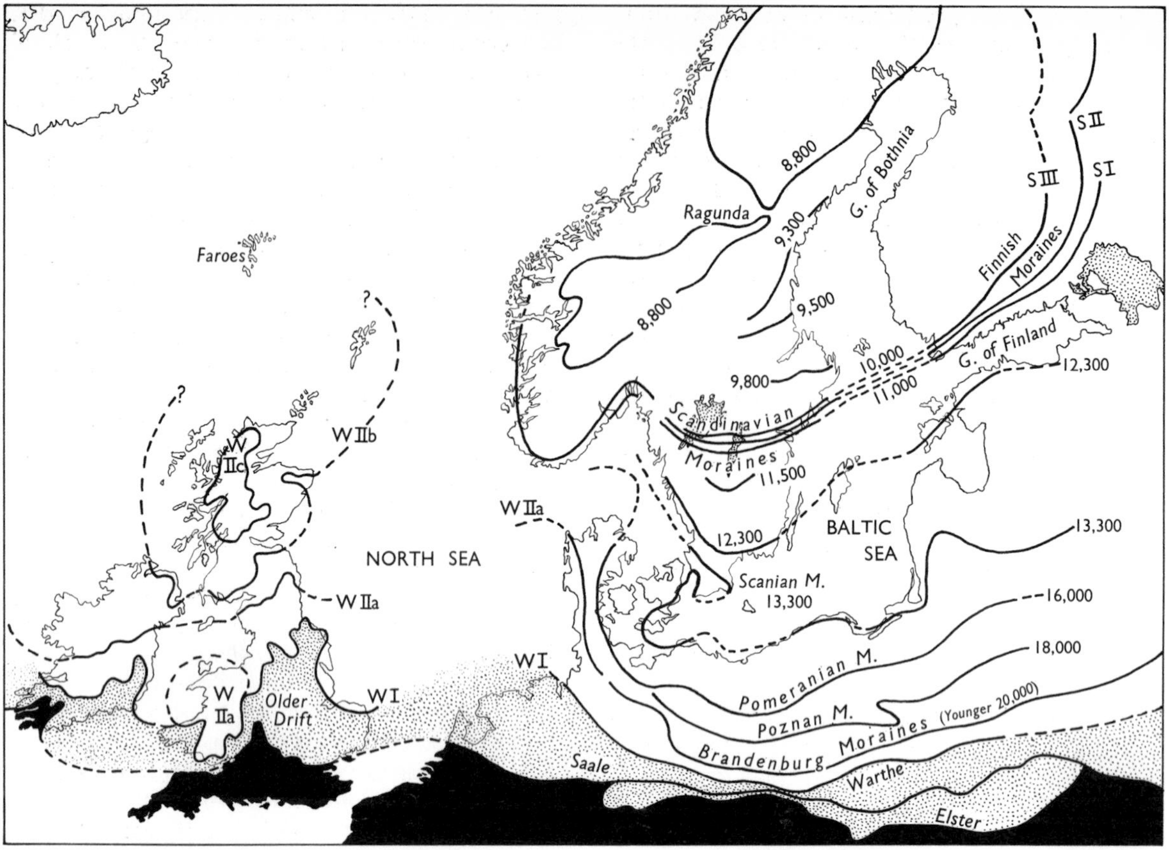


**Figure 21.15** Map illustrating the extent of Pleistocene glaciations over NW Europe and the British Isles, and the chief moraines and drift borders marking stages in the recession of the last great ice sheet (W). The Older Drift corresponds to the Saale (R) and the Elster (M). The Elster front was overrun by the Saale in the Netherlands, but in England it reached the site of north London. *Elster* and *Saale* are named after tributaries of the Elbe. *Warthe* is named after a town in Poland on a river of the same name that passes through Poznan on its way to join the Oder. The Warthe front probably corresponds to the outer border of the Newer Drift, both being here correlated with the maximum spread of WI. In the British area this is the Irish Sea–York–Hunstanton glaciation. W IIa marks the Scottish and Welsh re-advance, and possibly corresponds with the older of the Brandenburg moraines. W IIb is the Highland re-advance, corresponding to the younger Brandenburg and the Poznan, Pomeranian and Scandinavian moraines. W IIc is a British phase of minor cirque and valley glaciers of the same general age as the Scandinavian moraines.

deep-sea sediments contain a constant amount of $^{230}Th$, derived from sea water. A sediment at some depth in a deep-sea core that contains only half the amount of $^{230}Th$ in the surface sediment is 75,000 years old. For other amounts of $^{230}Th$ present, the age of the sediment can be calculated as illustrated on p. 237. For an assessment of the various methods of radiometric dating, as applied to the Pleistocene, see Shotten (1967).

An important advance in dating early Pleistocene events was made by the application of dating methods to deep-sea cores. Ages within cores can be estimated, as was first done by a team from Lamont Geological Observatory, from *(a)* the known depth of the layer to be dated, and *(b)* the known rate of deposition of accurately dated layers of identical types occuring higher up in the sequence. This extends the usefulness of radiometric dating beyond the levels to which they actually apply. Furthermore, sediments within deep-sea cores can be dated by reference to the palaeomagnetic time-scale (pp. 614–16).

So slowly have ocean sediments been deposited that some of the cores represent the accumulation of several million years. These, and some of the shorter cores, contain a clear record of all the glacial and interglacial stages and of the more important sub-stages. The contrast of 'cold' and 'warm' bands is particularly well marked in cores from the North Atlantic. The glacial layers are represented by a predominance of foraminifera belonging to species that now live only in the colder waters of the polar regions; whereas the species of the interglacial layers imply that the surface waters in which they lived were as warm as, and sometimes even warmer than, they are today. From the proportions of the two kinds of foraminifera in a given layer of the core it is possible to make a fairly close estimate of the average temperature of the surface waters from which the $CaCO_3$ of the shells was being abstracted by the organisms during their lifetimes. The temperature at the time of growth can also be estimated to within 1° or 2°C by an elegant method devised in 1947 by W. D. Urry. This is based on the discovery that the proportion of $^{18}O$ to $^{16}O$ in the oxygen of the $CaCO_3$ secreted from sea water by shell-builders depends on the temperature of the water. By measuring the isotopes with a sensitive mass spectrometer the temperature can be determined from the ratio $^{18}O/^{16}O$. Finally, the same shells—if not older than about 70,000 years—provide material for age determinations by the radiocarbon method.

Pollen has also been used together with diatoms for recognizing climatic conditions. The method relies on recognition of the relative proportions of pollen from warm dry and cold moist climates, and the percentage of arctic and subarctic diatoms.

From the above methods of dating Pleistocene events, and of distinguishing glacial and interglacial periods within deep-sea cores, Pleistocene time-scales have been linked to variations in the heat received from the sun, at any given latitude, as a result of cyclic changes in the orbit and axis of rotation of the earth. The axis is not at right angles to the plane of the orbit (i.e. to the *ecliptic*), but is inclined to it: a familiar fact which accounts for the seasonal alternations of summer and winter as shown in Figure 21.16. At present the inclination is nearly $23\frac{1}{2}°$, but it varies periodically between limits of just over $24\frac{1}{2}°$ and just over $21\frac{1}{2}°$. The angle is the same as that between the earth's equatorial plane and its orbital plane and is known as the *obliquity of the ecliptic*. The orbit itself has the shape of an ellipse, the *eccentricity* of which passes through a cyclic variation between extremes of almost zero, when the orbit is nearly a circle, to a maximum value when the ellipse reaches its greatest elongation. Because of the attraction of sun and moon on the earth's equatorial bulge, the direction of the axis in space (relative to the fixed 'stars') slowly changes, periodically describing, as it were, the surface of an imaginary cone, like the axis of an inclined spinning top. This terrestrial 'wobble', or *precession* in astonomical language, would by itself require 26,000 years for the completion of a cycle. But completion is hastened, i.e. reduced to 21,000 years, because the orbit itself also gradually swings round. The net result is described as the *precession of the equinoxes*.

These three cyclic disturbances of the earth's planetary movements have the following periodicities:

| | |
|---|---|
| Obliquity of the ecliptic | 40,000 years |
| Eccentricity of the orbit | 92,000 years |
| Precession of the equinoxes | 21,000 years |

At intervals during the last hundred years attempts have been made to calculate the variations in the heat received by the earth as a result of the interaction of the three cycles during the last few hundred thousand years. In such calculations it is necessary to assume that the sun's radiation has remained constant, apart from short-term sunspot effects. The most notable results have been those of M. Milankovitch, a Yugoslav geophysicist, who produced detailed curves showing the variations at various latitudes for the last 600,000 years. He found a conspicuous alternation of (*a*) long periods

**Figure 21.16** Diagram to illustrate seasonal changes in the Northern and Southern Hemisphere

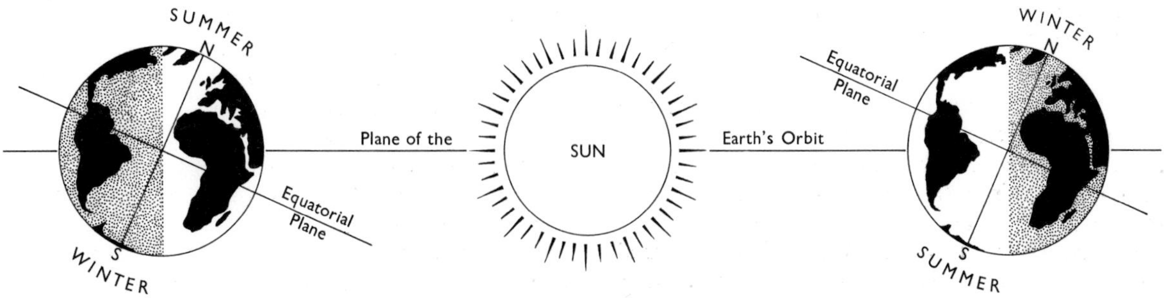

of cool summers which, he considered, would fail to melt all the winter snowfall in regions of appropriate latitude or elevation, and so would initiate permanent snowfields and glaciers; and (*b*) periods of warm summers which would hinder the accumulation of snow. More recently the variations for 65°N latitude have been recalculated, and extended to cover the last million years. The resulting curve, copied from the more detailed original published by C. Emiliani in 1955 (see p. 468), is here reproduced as Figure 21.17.

Many attempts have been made to correlate the calculated *Milankovitch* or *astronomical* sequence of cool summers with the actual *geological* sequence of glacial phases. Zeuner, as indicated in Figure 21.17, fitted all the glacial phases, from Gunz I onward, with the lows of the astronomical sequence starting at about 600,000 years ago. Others chose different 'fits', as illustrated by the interpretations of Fairbridge, and by Emiliani indicated on Figure 21.17, as *a* and *b* respectively. As a glance at Figure 21.17 will show, the interpretation of Milankovitch's insolation curves in terms of a Pleistocene time-scale has resulted in a wide range of results for individual events. Indeed, a date for the beginning of the Pleistocene was not in doubt by only a few thousand years, but by a million or more.

Percy Evans (1971), in a Special Publication of the Geological Society of London (see p. 468) has reviewed the situation and analysed the chronological evidence for Pleistocene events older than the limit of radiocarbon dating. From detailed analysis and correlation of (*a*) all the geochronological results arising from the various methods of dating sediments within deep-sea cores, including palaeomagnetic dating, and (*b*) the various methods of estimating climatic variation, he constructed climate-time curves. These curves reveal cyclic climatic changes for the last million years of the Pleistocene, with a period averaging 40,000 years (Figure 21.18) comparable with Milankovitch's curve. Evans, in comparing the astronomical and Pleistocene climate curves, used a refinement of Milankovitch's curve based on Woerkkom's revision of the insolation curve for 65°N and 65°S, and by taking into account the percentage of solar radiation that is reflected back (known as the *albedo*), and the latent heat of ice and snow. The albedo is increased when large areas are covered with ice and snow. Moreover, although dating methods applied to deep-ocean cores go back no further than 400,000 years, by reference to climatic and other evidence older than this, Evans extrapolated the climatic cycles back to a million years, in comparison with the insolation cycles. He numbered the climatic cycles from 1 to 25 for ease of reference, and the climatic cycles therefore form a dated framework within which to fix Pleistocene events.

Figure 21.18 is abstracted from Evans' diagram illustrating a tentative chronology for the Pleistocene based on comparison of the insolation and climatic cycles during the Pleistocene. Figure 21.18 shows tentative dates for high sea-levels in the Mediterranean; for the Thames Valley ter-

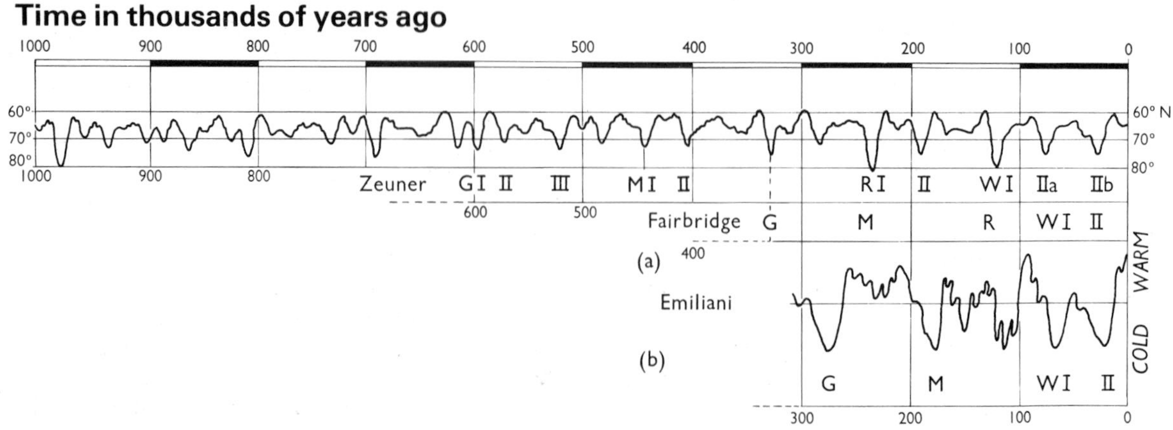


**Figure 21.17** The 'Milankovitch' curve representing the amount of solar radiation received at the earth's surface at latitude 65°N, expressed in terms of the amounts now received at higher (colder) and lower (warmer) latitudes; from data recalculated by A. J. J. van Woerkom for the last million years. Correlations favoured by F. E. Zeuner and (a) R. W. Fairbridge are indicated. (*Curve after C. Emiliani, 1955, Fig. 14*)
(b) C. Emiliani's correlation between the 'Milankovitch' curve and a generalized climatic variation curve based on deep-sea cores (*After C. Emiliani, 1955, Fig. 15*)

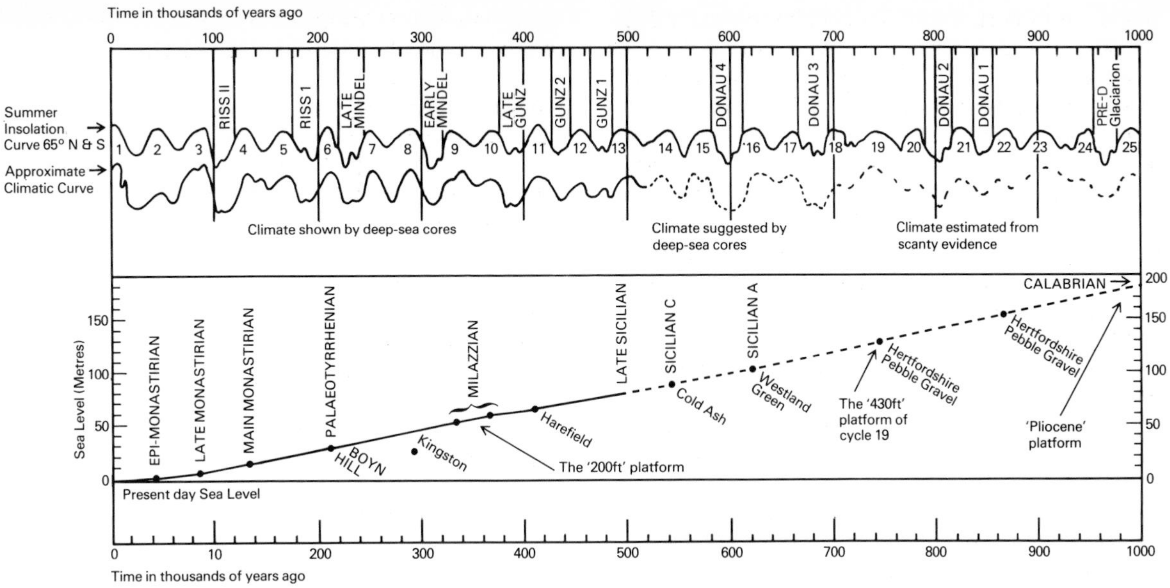


**Figure 21.18** *Above* The summer isolation curve depicting cyclic climatic changes throughout the last million years, is compared with a climatic curve (beneath it) constructed from climatic data recorded in sediments.
*Below* A curve based on the same time-scale depicting the dates assigned to high sea-levels during the last million years (*After P. Evans, 1971*)

races; and for the Alpine glacial epochs. There is no evidence for a sudden major deterioration in climate, that can be used to date the beginning of the Pleistocene, in deep-ocean cores. On the basis of palaeontological evidence it is dated at a little over two million years ago.

It should be pointed out that there is agreement that the cool summers represented on the insolation curve are insufficient by themselves to explain the onset of glaciation. 'Some unknown factor', writes Zeuner, 'created conditions favourable for glaciation during the Pleistocene'. There have been many hypotheses as to what this factor may be, but there is still wide diversity of opinion as to how glacial climates were initiated.

## The Permo-Carboniferous Climatic Zones: A Geological Dilemma

During the last few decades the discoveries made by geologists and geophysicists have included many unexpected surprises. But probably none has created so embarrassing a dilemma as Blandford's much earlier discovery in tropical India (at Talchir, in Orissa) of a late Carboniferous boulder bed which he recognized as being of glacial origin. This was in 1856. Before Blandford's Talchir tillite had been officially confirmed by a detailed survey of the area, completed in 1874, tillites of about the same age had been recorded in 1859 from South Australia and in 1870 from South Africa (Figure 21.19). South America was added to the list in 1888, when Carboniferous tillites were found in Brazil. As these astonishing discoveries were followed up it became unmistakably clear that Gondwanaland (see p. 466) had been repeatedly glaciated on a gigantic scale at the very times when most of Laurasia, i.e. North America, Europe and Asia (excluding Arabia and peninsular India), had tropical or hot-desert types of climate.

The history of science presents many examples of dilemmas of a similar kind. A current one that remains unresolved is inherent in our concepts of radiation: waves or particles? Whenever the challenge of a scientific dilemma is overcome, it leads to a broad advance on a wide front, to a 'break through', in the current jargon. A common attitude towards a dilemma, however, is either (*a*) to deny the evidence, or (*b*) to ignore it, or (*c*) to compromise. An amusing example cited by Anatol Rapoport is Zeno's paradox, which has been dramatized as a race between Achilles and a tortoise which is given a considerable start. Against the party who accept the obvious evidence that Achilles does, in fact, win the race, there is a theoretically minded party who maintain with equal confidence that Achilles could not possibly win the race, since to do so he must overtake the tortoise. This feat they declare to be impossible because Achilles must first reach the point from which the tortoise started, by which time the tortoise has again reached a point ahead . . . and so on *ad infinitum*. Rapoport suggests that if you had a compromise solution—the usual sort of thing in cricket or political life—the race would probably be pronounced a draw. Actually there was no hope of resolving the paradox until Newton and Leibnitz invented the infinitesimal calculus for dealing with the logical contradictions that arise when any portion of space or time is regarded as the sum of an infinite series of diminishing parts.

The Permo-Carboniferous climatic dilemma could only be resolved by realizing that the deep-rooted 'common sense' belief in the fixity of the continents relative to each other and to the earth's axis of rotation was now in direct conflict with the evidence of the chief witness—Earth herself. In other words, polar wandering and/or continental drift had to be taken seriously. But mathematical physicists declared both to be impossible and most geologists accepted their verdict, forgetting that their first loyalty was to the earth and not to books written about the earth. Consequently the prevailing attitudes until about 15 years ago were:

(*a*) to deny the evidence, e.g. by pretending that because a few local mud-flows or lahars had been wrongly interpreted as tillites, therefore all formations claimed as tillites, however widespread, had also been wrongly identified; or

(*b*) to ignore the evidence, as if it were irrelevant to the subject; or perhaps more commonly to ignore its geological significance, in the vain hope that meteorologists would presently discover some magic combination of phenomena that would explain simultaneous equatorial glaciation and warm polar regions.

Now, however, the mathematicians of a younger generation have found flaws in the classic treatment of the earth as a gyroscope. Given time enough and the rheid properties that depend on time, large-scale polar wandering is no longer declared to be theoretically impossible. Now, too, the physicists of a younger generation are finding more and more evidence every year that, relative to the earth's magnetic field, the continents have changed their positions by thousands of kilometres since Precambrian times (Chapters 27 and 28). Continental drift has at last become scientifically 'respectable'. But let us not forget that the geological evidence has been clear and conclusive for nearly a century, though few there were who recognized its meaning.

**Figure 21.19** Dwyka Tillite (Upper Carboniferous), exhumed by long-continued erosion, together with its glaciated pavements of Ventersdorp basalt. Nooitgedacht, near Riverton on the Vaal River, Kimberley District, South Africa (*A. M. Duggan-Cronin, Alexander McGregor Memorial Museum, Kimberley*)

## Permo-Carboniferous Ice Ages

*Africa* The widespread *Dwyka* tillite of South Africa has been partly obliterated by erosion, and is partly hidden by later formations. But innumerable exposures still occur at intervals from the Transvaal towards the Cape and from Namibia (South-West Africa) to Natal. In many places it can be seen resting on a glaciated floor (Figure 21.19), characteristically scored with striations (Figure 21.20). *Roches moutonnées*, excavated rock basins, drumlins and varved clays have been uncovered. In the walls of the gorge below the falls of the Cunene river in Namibia a beautifully U-shaped valley filled with Dwyka tillite is to be seen. Elsewhere, similar valleys, though less well revealed, are being exhumed by stream erosion. The tillite itself contains grooved and ice-faceted boulders and erratic blocks (Figure 21.21), some of which have been transported for hundreds of kilometres from the north. In some localities several tillites are known, with intervening interglacial deposits, including varved shales. The successive glaciations were not all from the same centre, but migrated from west to east (Figure 21.22). The associated deposits show that the glaciated region was one of moderate relief, and for

**Figure 21.20** Striated glaciated pavement of Ventersdorp basalt, after removal of formerly overlying Dwyka Tillite, Nooitgedacht, Kimberley District South Africa (*A. M. Duggan-Cronin, Alexander McGregor Memorial Museum, Kimberley*)

**Figure 21.21** Erratic block (transported from the Transvaal) in Dwyka Tillite, Kimberley District, South Africa; part of dispersal **C** in inset of Fig 21.22 (*C. T. Trechmann*)

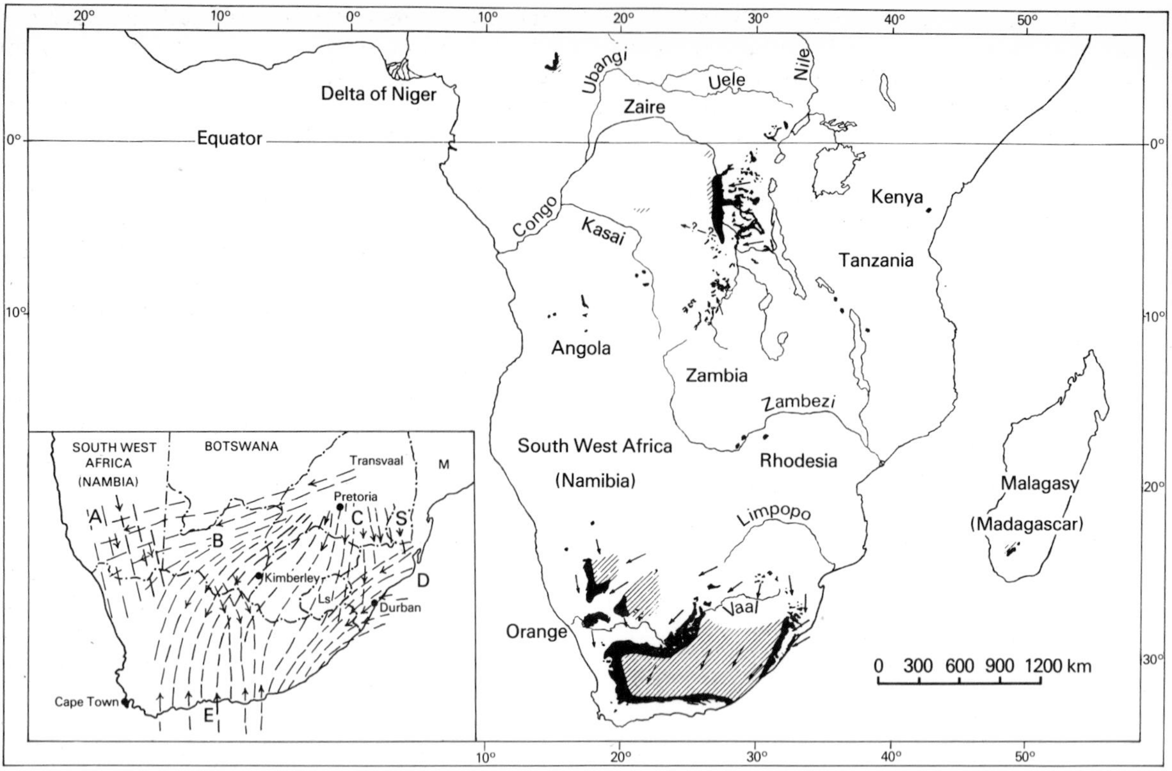


**Figure 21.22** Distribution of the Dwyka Series in Africa south of the Sahara. *Black*, outcrops of tillite and associated glacifluvial deposits; *diagonal shading*, Dwyka hidden by later formations; *arrows*, directions of ice flow (*After J. Lepersonne, 1961*)

*Inset map*; directions of successive ice radiations in South Africa (M Mozambique; S Swaziland; Ls Lesotho

**A** from Namaqualand highlands of SW Africa to south of Orange River (and suspected to South America)

**B** from highlands NW of Kimberley (a minor centre of dispersal) to S and SSE

**C** radially from the Transvaal to Natal, the Cape and SW Africa

**D** from a region that lay NE of the present coast of Natal

**E** from south of the Cape Province (position in the sequence A–D not known, but probably contemporaneous with C or D)

(*After A. L. du Toit and S. H. Haughton*)

the most part low-lying. At the margins the ice terminated in shallow water, marine, brackish or fresh, which followed up the ice-front as it retreated. No high mountain range or plateau lay to the north, from which great valley glaciers (or mud-flows) might have descended. Nor could the glacial phenomena be explained away as the work of lahars, for there was no volcanic activity in south and central Africa at this time. The glaciations were the work of thick continental ice-sheets that spread outwards like the Antarctic ice of today.

The ice came from centres lying far to the north of South Africa, and in the latest of the glaciations from beyond Natal, outside the present continent. Since the ice must have crept outwards in all directions and not only towards the south, it follows that the Dwyka tillites should be only part of a once continuous ring of such deposits, surrounding the region of ice dispersal. Confirming this deduction, tillites of the same age have been found in the north of Angola, in Katanga and in Uganda, and in each of these territories it has been established that the ice moved towards the north. Along the western side of Madagascar several tillites alternate with varved shales and other glacifluvial sediments, making up a thickness of over 300 m. More limited relics have recently been found in Tanzania and Kenya. As indicated in Figure 21.22, the ice-sheets advanced beyond the equator.

*India* In India, far to the north of the equator, the original *Talchir* tillite of Orissa, with its grooved and faceted boulders, has since been found to rest on a striated pavement. Similar evidence of heavy glaciation, exhumed by the removal of overlying deposits, shows that late

Carboniferous ice-sheets covered Rajasthan and Madhya Pradesh and spread northwards to what are now the Himalayan foothills, beneath which the tillites are buried by the frontal *nappes* of the mountains. In the Salt range (Pakistan, between the Indus and the Jhelum) erratics occur that have been carried 80 km or more from the south. Nearer the Himalaya at least two well-marked stages of glaciation are recorded by tillites separated by 460–760 m of varved shales. Near Simla, boulders of the earlier tillite occur in the later one. 1126 km to the east at least one bed of tillite has been uncovered by the Kosi river. Still farther east, where Gondwanaland is last seen before vanishing underground on the far side of the Brahmaputra, the Talchir tillite has been recognized in Assam. In India the centre of ice dispersal lay far to the south. Evidence in that direction has either been completely lost by erosion or is hidden beneath the plateau basalts of the Deccan. The evidence that does remain was provided by the margins of the great ice-sheets, where the ice radiated in what would now be a northerly direction, away from the present equator.

*Australia* In Australia, we again find evidence of glaciation from what is now the south. In New South Wales 6 glacial stages have been preserved, in Victoria 11 (with at least 51 beds of tillite) and in Tasmania 5, all with intervening interglacial deposits, including many long runs of varves. In South Australia striated pavements and *roches moutonnées* are notable features. Very large erratics occur in the associated tillites, some of which have been transported 480 km to the north. In Western Australia and S Queensland the corresponding deposits are largely marginal and glacifluvial. In SE Australia the tillites fall into two sets of markedly different ages: one belonging to the late Carboniferous; the other to the Middle Permian. The Lower and Upper Permian strata contain workable coal seams. The nature of the plant remains points to a temperate climate, very different from that of the tropical rain forests that provided the Carboniferous coals of Laurasia. The interval between the two Ice Ages was evidently of the order of 20 million years—too long to be regarded as a long interglacial stage.

*South America* In South America there are also two well-separated groups of tillites and associated varves, but both are older than those of Australia. In the north-west of Argentina the earlier group appears in the Lower Carboniferous. On the eastern side of the continent the later tillites range from about the Middle Carboniferous into the Upper, most of which is represented by the coal measures that followed. Several of these younger tillites occur near Buenos Aires and in Uruguay; while still farther north they cover enormous areas in Brazil. Glacial and interglacial deposits make up a thickness of more than 900 m around São Paulo. The ice advanced from land that is now the site of part of the South Atlantic. At the same time the Falkland Islands were overwhelmed by ice from land that lay to the north—now also part of the South Atlantic. Was this missing land part of Africa?

There were at least three Ice Ages, not a single Permo-Carboniferous Ice Age. The earliest was in NW Argentina during Lower Carboniferous (Viséan) times, say about 335 m.y. ago. The next was much more widespread. It extended from South America across Africa and India to Australia during the late Carboniferous; but it clearly began and ended first in the west and last in the east. Then, after a long, more temperate epoch represented by coal measures, the third Ice Age seems to have affected E Australia alone. This was during the Permian, say about 260 m.y. ago. As Lester King had pointed out, the wavelike succession of frigid climates from west to east (followed everywhere by coal measures) is itself a confirmation of 'the majestic drift of the Gondwana continent . . . through the climatic girdles' of the time.

*Antarctica* As indicated on Figure 20.10 tillites, containing striated pebbles, have been discovered in the Horlick Mountains, lying between Devonian sediments and a series of permian shales with numerous thin coal seams containing fossil plants like those of corresponding age in other parts of Gondwanaland. What was an ambiguous gap in the evidence relating to Gondwanaland has now been at least partially filled.

## Gondwanaland in the Late Carboniferous

A glance at Figure 21.23 shows that the glaciated lands of the late Carboniferous now occupy considerable areas on both sides of the equator. With the continents in their present positions such a distribution of ice-sheets at the same, or nearly the same, time is hopelessly inexplicable. The suggestion that Gondwanaland rose from sea-level to a plateau so enormously high that it was above the snow line is negatived by ample evidence that in

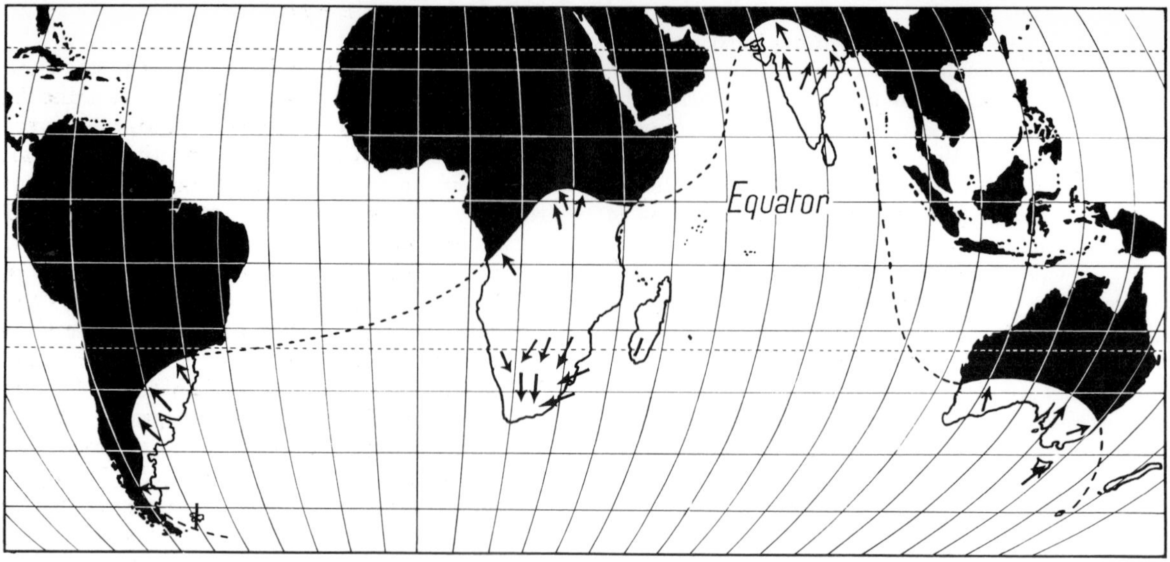


**Figure 21.23** Map showing the distribution of the late Carboniferous glaciations of Gondwanaland with the continents in their present positions; arrows indicate directions of ice flow. Neither the Lower Carboniferous glaciation of NW Argentina nor the Permian glaciation of E Australia is included.

most places it was low-lying. But whether it was or not, the tropics could not have been glaciated down to sea-level without the development of still greater ice-sheets over the northern lands. The only evidence of Carboniferous glaciation in the north is that of mountain glaciers, found in Alaska, which has probably never been far from the north pole since the Precambrian. The notorious Squantum 'tillite', near Boston, Mass., is no longer thought to be of Carboniferous age. Another local and doubtful 'tillite' occurring in the Tarim basin (north of the Tibetan plateau) is now known to be of Cambrian—not Carboniferous—age.

On the other hand great coal forests were flourishing in tropical swamps extending from North America across Europe to China, while much of Gondwanaland lay under ice. Moreover, deposits of laterite and bauxite that could only have formed in a tropical climate are found in the Upper Carboniferous of the United States (Kentucky and Ohio), Scotland (Strathclyde Region, in the former Ayrshire), Germany, Russia (south of the Moscow basin) and China (Shantung). The inference that the equatorial zone of the time is roughly indicated by this belt of coalfields and laterites is irresistible.

By itself no amount of polar wandering, even if so much could be admitted in so short a time, would give a distribution of climatic girdles corresponding to the picture outlined above. Wherever the south pole is imagined to have been in order to account for any one of the glaciated regions, it would still have been too distant from the others to account for more than one of them. The problem, indeed, remains an insoluble enigma, unless continental drift is added to any polar wandering there may have been. The straightforward inference can then be accepted that all the continents except Antarctica lay well to the south of their present positions, and that the southern continents were grouped together around the south pole.

Alfred Wegener, who was the pioneer in attempts of this kind, placed Antarctica between Australia and South America, approximately as illustrated in Figure 21.24. Other reconstructions are considered in Chapter 28, when much more evidence has been passed in review. In all of them, however, the late Carboniferous ice-sheets fall within an area comparable with that glaciated in the northern hemisphere during the Pleistocene. Moreover, the lateritic belt referred to above then comes into line with what would have been the equator of the time, and other known climatic details also fall into appropriate places. The corresponding site of the north pole would have been in what is now the NW Pacific, which is consistent with the fact that the lands of the northern hemisphere were not glaciated by Carboniferous ice-sheets.

Now for the north pole to have 'wandered' into

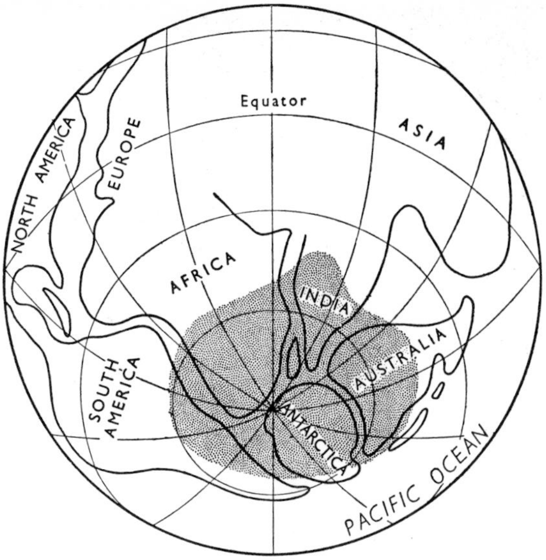


**Figure 21.24** Map showing the distribution of the late Carboniferous glaciations of Gondwanaland with the continents reassembled, though not quite so closely, as interpreted by Alfred Wegener in 1915. The laterite belt and the coalfields of the time coincide with what would then have been the equator.

the Arctic Ocean it seems most probable that NE Asia drifted *over* the pole during the course of the Mesozoic era. Yet no evidence of pre-Quaternary glaciation has been recorded from that quarter, although Mesozoic coals are common and are said to indicate a boreal climate. Thus is appears that a continental region does not automatically receive a shroud of ice merely because it happens to lie over or near one of the poles.

## Precambrian Ice Ages

Late Precambrian tillites have been reported from north-east and east Greenland, Scandinavia, Spitzbergen, northern France at Granville, in the Yangtse valley, North Korea, and the Transvaal and Witwatersrand in South Africa.

There were two important Ice Ages in Australia in late Precambrian times. From the Musgrave mountain belt, almost in the middle of Australia, late Precambrian strata of great thickness extend to the Flinders Ranges and beyond. Besides the Marinoan tillites, these include the massive Sturtian tillites, which are well exposed in Mt Olga (Figure 21.25), where they are gently dipping, and in Ayers Rock (Figure 17.6), where they are almost vertical. Here we have a remarkable display of dramatic contrasts, not only in tectonics, but of past and present climates (glacial deposits in a land of drought and heat), and of the strange topography developed during successive cycles of erosion (isolated inselbergs rising abruptly from a nearly flat pediplain).

From the occurrence of supposed tillites in Angola, Zaïre, and the Republic of Congo, late Precambrian glaciation has been thought to be widespread in equatorial Africa. The origin of some of the rocks originally identified as tillite has now been questioned, however, and they have come to be known as *mixtites*, since their fine-grained matrix

**Figure 21.25** Mt Olga, an impressive inselberg group rising above a semi-arid pediplain and carved out of gently dipping Sturtian (late Precambrian) tillites. See also Fig 17.6 for a neighbouring example in central Australia. (*Australian National Travel Association*)

is of non-glacial origin, although some of the pebbles they contain are faceted and striated as if of glacial origin. A reappraisal of the characteristics and origin of the mixtites of Lower Zaïre, and comparison with those of Angola and the Republic of Congo, has recently been made by Cahen and Lepersonne (1976). The lower mixtites are interstratified with lavas of basaltic and andesitic composition, sometimes characterized by pillow structure, and locally spilitic. They are therefore subaqueous lavas. The upper mixtites are characterized by larger pebbles, sometimes striated, and amongst them are crystalline rocks, indicating that they are of northern or eastern origin and have been transported for a considerable distance. Cahen and Lepersonne conclude that the mixtites are not of glacial origin, and that they have been deposited subaqueously within shallow basins. The striated pebbles of glacial origin appear to have been derived from neighbouring highlands characterized by glaciers.

A well-known Dalradian 'boulder bed' outcropping at intervals across Donegal and Scotland has been variously interpreted. Although now regarded as a tillite, it was formerly thought that the boulders had been transported from the north by floating ice.

## Selected References


ANDERSEN, S. T., 1960, 'Climatic change and radiocarbon dating in the Weichselian glaciation of Denmark and the Netherlands', *Geologie en Mijnbouw*, vol. 39, pp. 38–42.

BROECKER, W. S., EWING, M., and HEEZEN, B. C., 1960, 'Evidence for an abrupt change in climate close to 11,000 years ago', *American Journal of Science*, vol. 258, pp. 429–48.

CAHEN, L., and LEPERSONNE, J., 1976, *Les mixtites du Bas-Zaïre : mise au point intérimaire*, Rapport annual pour l'année 1975, Départment de géologie et de minéralogie du Musée Royal de l'Afrique Centrale, Tervuren, Belgium

COLEMAN, A. P., 1941, *The Last Million Years*, University of Toronto Press, Toronto.

EARDLEY, A. J., *et al.*, 1973, 'Lakes cycles in the Bonneville Basin, Utah', *Bulletin of the Geological Society of America*, vol. 84, pp. 211–16.

EMILIANI, C., 1955, 'Pleistocene temperatures', *Journal of Geology*, vol. 63, pp. 538–78.

EMILIANI, C., and GEISS, J., 1959, 'On glaciations and their causes', *Geologische Rundschau*, vol. 46, pp. 576–601 (in English).

EVANS, P., 1971, 'Towards a Pleistocene time-scale', Part 2 of *The Phanerozoic Time-scale—a Supplement*, Special Publication of the Geological Society, no. 5, London, pp. 123–356 (with a full Bibliography).

EWING, M., and DON, W. L., 1956, 'A theory of ice ages', *Science*, vol. 123, pp. 1061–6, and vol. 127, pp, 1159–62.

FAIRBRIDGE, R. W., 1960, 'The changing level of the sea', *Scientific American*, vol. 204, pp. 70–9.

FARMER, B. H., (Ed.), 1966, *Vertical Displacements of Shorelines in Highland Britain*, Institute of British Geographers, Transactions no. 39, George Philip, London.

FLINT, R. F., 1971, *Glacial and Quaternary Geology*, Wiley, New York.

GODWIN, H., 1961, 'Radiocarbon dating and Quaternary history in Britain', *Proceedings of the Royal Society*, B, vol. 153, pp. 287–320.

HARE, F. K., 1947, 'The geomorphology of part of the middle Thames', *Proceedings of the Geologists' Association*, vol. 58, pp. 294–339.

KUNEN, PH. H., 1955, *Sea Level and Crustal Warping*, Geological Society of America Special Paper 62, pp. 193–204.

SHOTTEN, F. W., 1966, 'The problems and contributions of methods of absolute dating within the Pleistocene period', *Quarterly Journal of the Geological Society of London*, vol. 122 (for 1966), 1967.

URRY, W. D., *et al.*, 1951, 'Measurements of palaeotemperatures and temperatures of the Upper Cretaceous of England, Denmark, and the Southeastern United States', *Bulletin of the Geological Society of America*, vol. 62, pp. 399–416.

WEGENER, A., 1967, *The Origin of Continents and Oceans*, English translation of the 4th edn (1929), Methuen, London.

WRIGHT, H. E., Jr., 1961, 'Late Pleistocene Climates of Europe: a review', *Bulletin of the Geological Society of America*, vol. 72, pp. 933–84.

ZEUNER, F. E., 1959, *The Pleistocene Period*, Hutchinson, London.

# Chapter 22

# Wind Action and Desert Landscapes

Where for many centuries only the camel has been able to penetrate, the helicopter . . . is now dropping geologists.

*Georg Gerster*, 1959

## Circulation of the Atmosphere

The circulation responsible for the winds is primarily a response to the familiar fact that air is cold over the polar regions and hot over the equatorial belt. If the earth did not rotate, heated air would ascend at the equator and blow towards the poles where, having become chilled and heavy, it would descend and return towards the equator, becoming steadily warmer as it followed the meridians into lower latitudes. This would be a simple convection system like that illustrated by Figure 2.15. But because the earth is rotating, a powerful deflecting force is at work, called the *Coriolis force*, after the physicist who first recognized it in 1835. If anything (e.g. air, water, projectile) moves relatively freely from south to north in the northern hemisphere, it starts from a place where it shares in the earth's rotational velocity to the east (over 1600 km/hour at the equator) and passes through places where that velocity is much lower, according to the latitude (e.g. about 1300 km/hour for New York; 1000 km for London; 0 for the north pole). Consequently the moving mass or object tends to travel eastward faster than the earth immediately beneath it does, and the farther north it goes the more it turns towards the east. Similarly if the movement is towards the equator, where the rotational velocity is higher, the tendency is to be left behind by the earth, i.e. to turn increasingly westward. Put more generally, the deflection is always to the right in the northern hemisphere, to the left in the southern one.

The Coriolis force is one of the minor factors concerned in river erosion, the channel being more effectively undercut on the side towards which the force is directed. The persistent westward migration of the southward-flowing Kosi river (Figure 18.35) can be partly ascribed to this force. The less confined movements of projectiles and aeroplanes, and of the winds themselves, are far more powerfully affected. The high-altitude winds that blow from the equator to the poles are deflected to the east. The return winds which complete the convective circulation near the ground would therefore be expected to be deflected to the west, i.e. to be 'easterlies'. And so they are across a broad belt on each side of the equator (the NE and SE trade winds) and, less regularly, in the polar regions. But we find that in each hemisphere there is a belt of disorderly 'westerlies' separating the polar easterlies from the tropical easterlies (the trade winds).

To account for this complication we must consider another global effect. Heated air ascending from the equatorial 'doldrums' turns polewards at a height of about 10–13 km, and passes into latitudes that are shorter than the equator. Latitude 30°, for example, is 13 per cent shorter than the equator. At about this position, but ranging between 25° and 35° (N or S according to the season), the crowding of the air has become sufficient to raise the pressure so that the air is obliged to descend towards the surface. Here, then, are the high-pressure subtropical calm belts that came to be known as the 'Horse Latitudes' in the old days of sailing-ships. At the surface the descending air divides into (*a*) the trade winds that blow towards the equator (so completing the tropical convection cells A and A′ of Figure 22.1), and (*b*) the disorderly westerlies that spiral *towards*

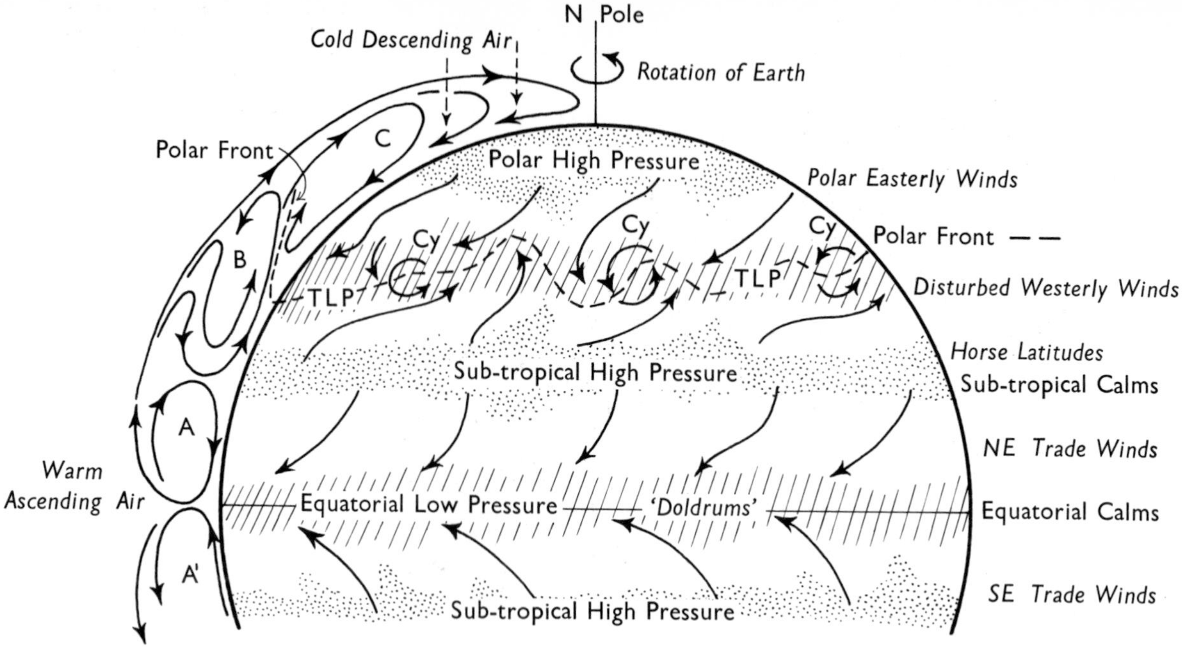


**Figure 22.1** The general planetary circulation of the lower atmosphere in three main cells, A, B and C in the Northern Hemisphere (and similarly A¹, B¹ and C¹ in the Southern Hemisphere). TLP, Temperate low pressure belt; Cy, Cyclones

the poles in cell B. But surface winds are already blowing *from* the poles, in cell C, and where the two opposing wind systems B and C meet, the weather becomes very disturbed and variable. The cold polar air tends to wedge itself southwards, while the warm moist air from lower latitudes flows up the surface of this wedge and so becomes cloudy and a source of rain or snow, often accompanied by strong winds. The surface of the cold wedge is called the *polar front*.

In the northern hemisphere the polar front advances far to the south in winter and retreats to the north in the summer, its range over land being much wider than over the oceans. The latter have a stabilizing effect because of the relative slowness with which water gains or loses heat. A corresponding winter advance to the north and summer retreat to the south takes place in the southern hemisphere. We now have the basic scheme of a threefold tandem arrangement of convection cells in each hemisphere (A, B, and C in Figure 22.1). But there are many further complications. The high-altitude westerlies are found to be concentrated into jet-like streams between the tropics and the poles. Instead of maintaining a nearly uniform direction as they blow towards the poles, the air-streams follow sinuous courses, alternately surging far to the north and south of their mean directions. These surges are probably mainly responsible for the atmospheric eddies familiar as cyclones (with lowest pressure at the centre) and anticyclones (with highest pressure at the centre). Each cyclone is like an enormous vortex with winds spiralling round in an anticlockwise direction in the northern hemisphere, and clockwise in the southern. For anticyclones these directions are reversed, the general rule being a direct consequence of the Coriolis force. It is because of the continual recurrence of these broad eddies that the westerlies are described as 'disturbed' or 'disorderly'. In the British Isles the most settled type of weather is, paradoxically, 'unsettled'.

Over tropical oceanic areas that have been abnormally heated, rapidly ascending streams of air generate the devastating vortices of winds known as *hurricanes* in the Atlantic and its great Mexican, Caribbean and Mediterranean embayments. The very term *hurricane* comes from a Caribbean word meaning 'the spirit of evil' (Figure 22.2). In the W Pacific (e.g. between Australia and Japan) similarly violent whirlwinds are called *typhoons*, after the malevolent monster Typhon of Greek mythology. No special term refers to those originating in the Indian Ocean,

**Figure 22.2** Mass of coral-rock thrown up by a cliff about 12 m high, carried 30 m from the cliff and turned upside down by a hurricane; east coast of Barbados, Windward Islands (*C. T. Trechmann*)

where they are commonly referred to as *tropical cyclones*. Whereas the comparatively mild cyclones of temperate regions are usually 1600 km or more across, the hurricane and typhoon systems of rotary winds may be only 300 or 400 km in diameter. But this means that from outside, where the pressure is highest, to the relatively calm but menacing 'eye' in the middle, where the pressure is lowest, the pressure gradient is correspondingly steeper. The resulting winds of hot moist air drawn into the spiralling updraught reach speeds of 120 to 200 or even 300 km/hour, and as the rising air expands and cools, its water vapour condenses and falls in an overwhelming deluge of rain. After passing their climax, hurricanes tend to broaden out into ordinary cyclones as they travel away from the tropics. Some cross the Atlantic and bring stormy weather to W Europe.

The *tornado* (Spanish *tornar*, to twist or turn), which often begins on land, often as a satellite to a severe hurricane, is a very much narrower column or funnel of swiftly spinning air rarely a kilometre and a half across and generally much less. The fiercely twisting winds create such havoc that any instruments that might have measured their speeds are inevitably destroyed. Estimates based on the fantastic effects of the winds sound incredible. The general destruction is due not only to the extreme violence of the winds, but also to the phenomenal reduction of pressure that takes place in the heart of a tornado. This results from the intensity of the surrounding rotation, just as happens in the vortex that develops in the water above the plug-hole when a bath is being quickly emptied. When a house is suddenly struck by a tornado it may be literally burst open by the excess of its internal air pressure, although the latter is likely to be something less than normal. A characteristic feature of a tornado is the long black funnel-shaped or snaky cloud that stretches towards the ground from the great thunder cloud riding above the storm. From this sinuous column the rainfall is catastrophic and wherever the column itself touches the ground the ruination of trees and structures of all kinds is most severe. Over the sea the pendant column swings about like an elephant's trunk and finally joins up with leaping peaks of water and spray to form a *waterspout* (Figure 22.3). The sea makes its contribution where the pressure is so low that a kind of temporary isostasy comes into play. In response to the relief of pressure the water spurts upwards to complete its swirling union with the cloud from which the bulk of the waterspout is derived.

In addition to the various planetary complications introduced into the basic scheme illustrated by Figure 22.1, there are further modifications dependent on the distribution of land and sea. Of these, only the Asiatic monsoon circulation

**Figure 22.3** Waterspout off the Island of Rhodes, Greece, October 1930 (*Syndication International Ltd*)

need be mentioned here. This is essentially due to the intense winter cold of Siberia and the high plateaus and ranges of central Asia, alternating with the summer heating up of the continent, which leads to extremes of high temperature over vast areas. In winter the outflowing monsoon strengthens the NE trades and brings cool dry air over India. In summer the directions of the monsoon winds are reversed. Moisture-bearing winds then blow in from the oceans (e.g. from the SW over India) and so control the rainy season.

Before proceeding to consider the direct action of winds on the land surfaces over which they blow, there are some general aspects that should not be overlooked. Because the wind distributes moisture over the face of the earth it is one of the primary factors responsible for weather and the weathering of rocks, and for the maintenance of rivers and glaciers. The more northerly and southerly of the atmospheric surges mentioned above introduce streams of moist oceanic air far into Greenland and Antarctica respectively, so providing sources of precipitation for the nourishment of the ice-sheets. Moreover, in blowing over the oceans and other bodies of water, the wind transfers part of its energy to the surface waters and so becomes responsible for waves and their erosive work. Hurricanes and typhoons in particular generate unusually high waves and locally increase the height of high tides by driving water towards the land, so adding much coastal flooding to the catalogue of their capacity for destruction.

## The Geological Work of Wind

As an agent of transport, and therefore of erosion and deposition, the work of the wind is familiar wherever loose surface materials are unprotected by a covering of vegetation. The raising of clouds of dust from ploughed fields after a spell of dry weather and the drift of windswept sand along a dry beach are known to everyone. In humid regions, except along the seashore, wind erosion is limited by the prevalent cover of grass and trees and by the binding action of moisture in the soil. But the trials of exploration, warfare and prospecting in the desert have made it hardly necessary to stress the fact that in arid regions the effects of the wind are unrestrained. The 'scorching sand-laden breath of the desert' wages its own war of nerves. Duststorms darken the sky, transform the air into a suffocating blast and carry enormous quantities of material over great distances (Figure 22.4). Vessels passing through the Red Sea often receive a baptism of fine sand from the desert winds of Arabia; and dunes have accumulated in the Canary Islands from sand blown across the sea from the Sahara.

By itself the wind can remove only dry incoherent deposits. This process of lowering the land surface is called *deflation* (L. *deflatus*, blowing away). Armed with the sand grains thus acquired, the wind near the ground becomes a powerful scouring or abrading agent. The resulting erosion is described as *wind abrasion*. By innumerable impacts the grains themselves are gradually worn down and rounded. The winnowing action of the wind effectively sorts out the transported particles according to their sizes. This is well illustrated in the desert wherever gravel, sand and mud are worked over by the wind. Supplies of such materials are slowly liberated by weathering and by the occasionaly short-lived torrents that flush out the wadis. Particles of silt and dust are then picked up by the wind, whirled high into the air and transported far from their source, to accumulate beyond the desert as deposits of loess (p. 485). Sand grains are swept along near the surface, travelling by leaps and bounds, until the wind drops or some obstacle is encountered. The dunes and other accumulations of wind-blown sand thus come to be composed of clean and uniform grains, the finer particles having been sifted out and the larger fragments left behind. It follows that pebbles and gravel are steadily concentrated on the wind-swept surfaces of the original mantle of rock-waste.

As a result of wind erosion, transport and deposition three distinctive types of desert surface are produced:

(1) the rocky desert (the *hammada* or *hamada* of the Sahara), a desolate surface of bedrock with local patches of rubble and sand (Figure 22.5);
(2) the stony desert, with a surface of rubble, gravel or pebbles (the *reg* of the Algerian Sahara; the *serir* of Libya and Egypt); and
(3) the sandy desert (the *erg* of the Sahara).

Complementary to these is the loess of the bordering steppes, deposited from the dust-laden winds that blow outwards from the desert.

## Wind Erosion

The most serious effects of wind deflation—from the human point of view—are experienced in semi-arid regions like the Great Plains of the United States, where during this century vast quantities of soil have been blown and washed away from thousands of formerly productive

**Figure 22.4** Dust storm approaching Port Sudan, west coast of Red Sea (*Paul Popper Ltd*)

**Figure 22.5** The rocky wastes of Ahmar-Kreddou, viewed from the Col de Sfa, Algerian Sahara (*Paul Popper Ltd*)

wheat-growing farms. Originally an unbroken cover of grass stabilized the ground, but long-continued ploughing and over-exploitation finally destroyed the binding power of the soil and exposed it as a loose powder to the driving force of the wind. This national menace became critical during a period of severe droughts, culminating in 1934–5, when great dust-storms originating in the 'Dust-bowl' of Kansas and adjoining States swept eastwards towards the Atlantic. Rain-wash and the creeping disease of badland erosion extended the devastation. Widespread measures of reclamation and protection have been undertaken to minimize the growing wastage and to conserve and improve the soil that remains. Wherever rain is deficient, deforestation, overgrazing or other misuse of land inevitably leads to soil erosion. In the Mediterranean area, with its long dry summers, the green lands have been shrinking for centuries, especially in Spain and Algeria, while the population has been increasing. Only recently has a start been made to break this vicious circle by afforestation and irrigation, and the application, enforced if necessary, of the 'balanced' use of land. The over-hasty development of agriculture in parts of the U.S.S.R. quickly ran into similar difficulties, dramatically indicated by the great dust-storms of a few years ago, blowing out of the semi-arid lands east of the Caspian and Aral Seas. These troubles are now being successfully corrected, and the lesson is being rapidly taken to heart in the new countries of Africa and elsewhere.

A characteristic result of deflation, especially over regions where unconsolidated clays and friable shales are exposed, is the production of wide plains and basin-like depressions. The excavation of hollows is limited only by the fact that even in deserts underground water may be present. Once the desert floor has been lowered to the level of the ground-water, the wind cannot easily loosen and pick up the moistened particles, though it may drive pebbles across the surface when blowing hard and continuously. The base-level for wind action is that of the water table, which may be far below sea-level. The 'pans' of S Africa and the Kalahari, and the depressions of the N African and Mongolian deserts, have all been excavated by ablation.

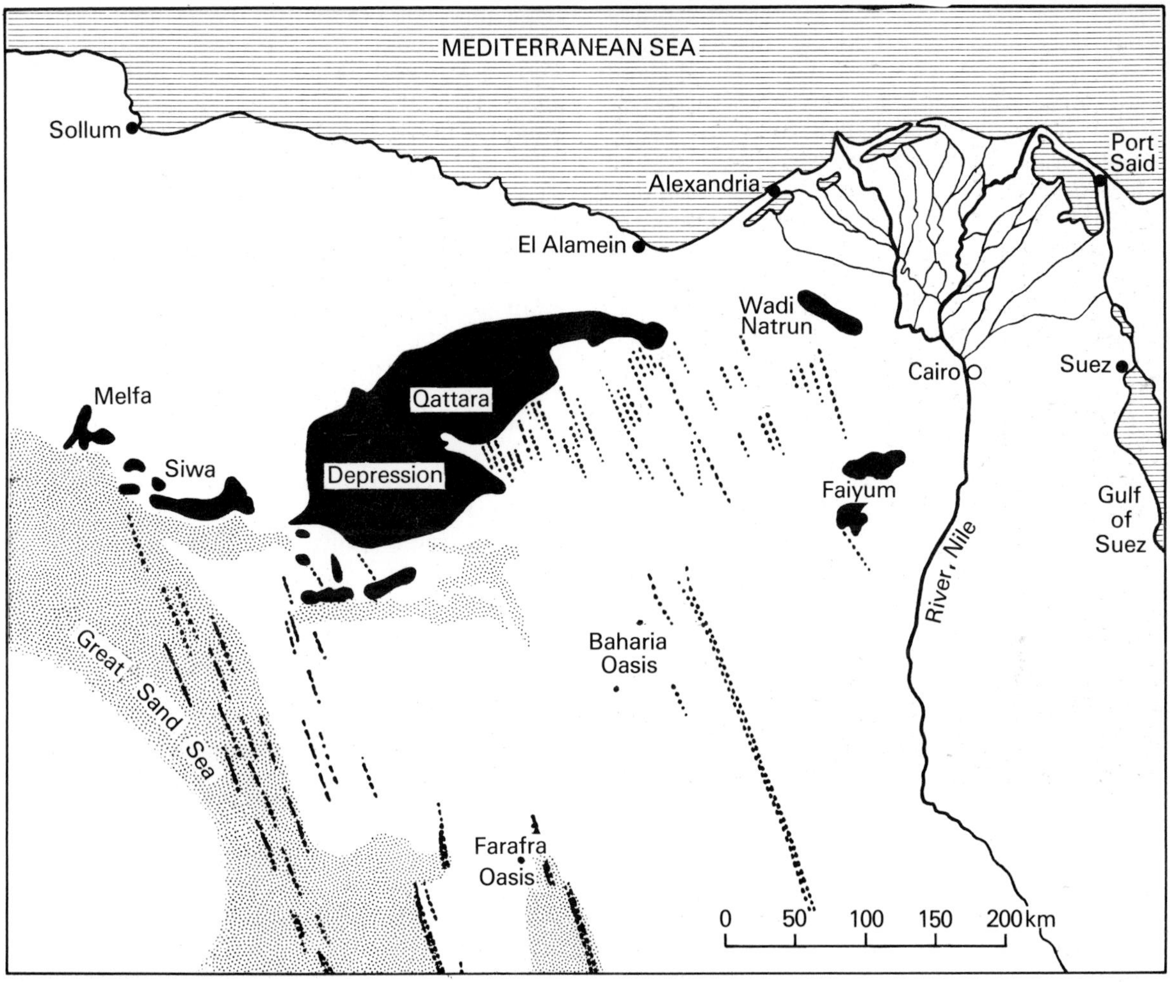


**Figure 22.6** Map showing depressions, sand seas and lines (*seifs*) of dunes in the Egyptian Desert

Westward from Cairo there is a remarkable series of basins with their floors well below sea-level (Figure 22.6), reaching −125 m in the salt marshes of the immense Qattara depression. Some of the smaller basins tap a copious supply of ground-water at depths of −15 to −30 m, and have become fertile oases. To the north, the surface rises by abrupt escarpments to terraced tablelands formed of hard sandstones and limestones which formerly extended across the softer rocks of the depressions. To the south, following the direction of the prevailing wind, long stretches of sand dunes represent part of the removed materials. The other well-known oases of Egypt—Baharia, Farafra, Dakhla and Kharga—are above sea-level, but they have originated in the same way. All of them are margined by steep escarpments of resistant strata, underlain by shales in which the floors have been excavated. These depressions are not crustal downwarps, like the shotts of Tunisia and Algeria, or Death Valley in California. Nor have they been hollowed out by water, for the sheet-floods due to rare cloudbursts tend to fill them up with debris. The wind has been the sole excavator.

The effects of wind abrasion are unmistakably expressed in the forms and surfaces of the desert bedrocks. Just as an artificial sand-blast is used to clean and polish building-stones and to etch glass, so the natural sand-blast of the wind attacks destructively everything that lies across its path. Cars driven against wind-blown sand may have their windscreens frosted and their paint scoured off. The action on exposed rocks is highly selective. Like a delicate etching-tool, the sand-blast picks out every detail of the structure. Hard pebbles, nodules and fossils are left protruding from their softer matrix until they fall out. Variably cemented rocks are fretted and honeycombed like fantastic carvings. Where there are thin alternations of hard and soft strata the soft bands are scoured away more rapidly than the hard, which thus come to stand out in strong relief, like fluted shelves and cornices with deep grooves between. Where the wind blows steadily in one direction over strata of this kind, especially if the beds are tilted rather than flat, the softer materials are excavated into long passageways between deeply undercut overhanging ridges. Such fantastically carved 'cockscomb' ridges are common in parts of the Asiatic deserts, where they are called *yardangs*.

Undercutting is everywhere a marked feature of wind abrasion, owing to the fact that the process is most effective within 30 to 60 cm of the surface, where the sand is most abundant (Figure 22.7).

**Figure 22.7** An illustration of undercutting by wind erosion: Balanced rock, Garden of the Gods, Colorado (*United States Geological Survey*)

Telegraph poles in sandy stretches of desert have to be protected by piles of stones against the cutting action of the sand grains hurled against them. Along the base of an escarpment alcoves and small caverns may be hollowed out. As always, the effect of undercutting on slowly weathered formations is to maintain steep slopes. Joints are readily attacked and opened up and these commonly determine the outlines of rock towers and pinnacles, left isolated like detached bastions in front of the receding wall of an escarpment (Figure 22.8). A spectacular example of an Antarctic landscape that has been remodelled by intense wind erosion in a 'cold desert' has already been illustrated (Figure 21.14). The volcanic necks of the Hoggar (Ahaggar) and Tibesti in the Sahara exhibit similarly wind-steepened walls and towers, but they rise from a sun-scorched pavement of basalt instead of protruding through a plateau of ice.

Where the bedrock of the desert floor is exposed to blown sand it may be smoothed or pitted or furrowed, according to its structure. Compact

**Figure 22.8** 'Cathedral Spires', pillars, detached by erosion along steeply dipping joint planes, remain in front of a receding escarpment on the left. Garden of the Gods, Pike National Forest, Colorado (*F. E. Colburn, Courtesy of United States Forest Service*)

limestones become polished, massive granites are smoothed or pitted, and gneisses and schists are ribbed and fluted, especially where their foliation approximates in direction to that of the dominant winds. Where pebbles have been concentrated by removal of finer material they become closely packed, and in time their upper surfaces are worn flat. In this way mosaic-like tracts of *desert-pavement* are devloped. Isolated pebbles or rock fragments strewn on the desert surface are bevelled on the windward side until a smooth face is cut. If the direction of the wind changes seasonally, or if the pebble is undermined and turned over, two or more facets may be cut, each pair meeting in a sharp edge. Such wind-faceted pebbles, which often resemble Brazil nuts, except that their surfaces are polished, are known as *dreikanter* or *ventifacts* (Figure 22.9).

As a result of continual attrition due to the friction of rolling and impact the sand grains themselves are gradually worn down and rounded. The prolonged action of wind is far more effective in rounding sand grains than that of running water, because of (*a*) the greater velocity of the wind; (*b*) the greater distances traversed by the grains as they bound and roll and collide with each other backwards and forwards across wide stretches of desert; and (*c*) the absence of a protective sheath of water. Some of the *millet-seed sands* of the desert are almost perfect spheres with a matt or frosted surface like that of ground glass. It is also noteworthy that visible flakes of mica, such as are commonly seen in water-deposited sands and sandstones, are very rare in desert sands and dunes. The easy cleavage of mica facilitates constant fraying during the wear and tear of wind action. Mica is thus reduced to an impalpable powder that is winnowed away from the heavier sand grains. These contrasts between water-laid and aeolian sands are of great value in deciding whether ancient sandstones were accumulated in deserts or under water. The Penrith Sandstone of

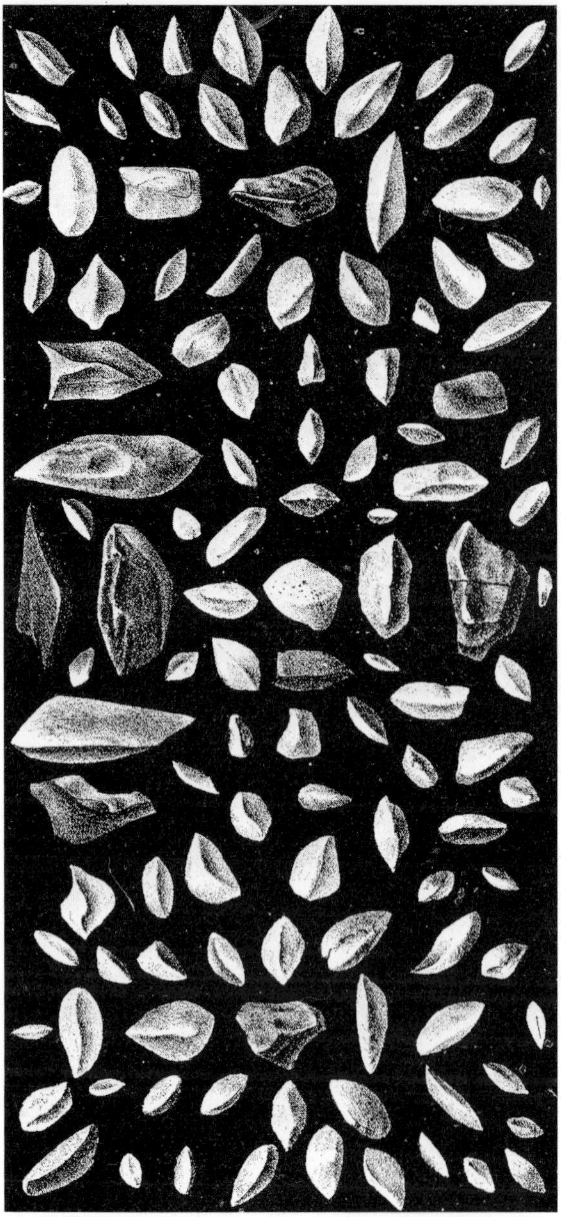

**Figure 22.9** Pebbles from the Egyptian Desert faceted by sand-blast (dreikanter or ventifacts) (*M. V. Binosi*)

the Eden valley, between the Pennines and the Lake District, is a well-known example of a Permian desert sand. Its rounded grains, the absence of mica and the cross-bedding of the formation all testify to the desert conditions of the time. Figure 22.10 illustrates an American example of millet-seed sand grains, rounded by Ordovician winds when Iowa, Illinois and the

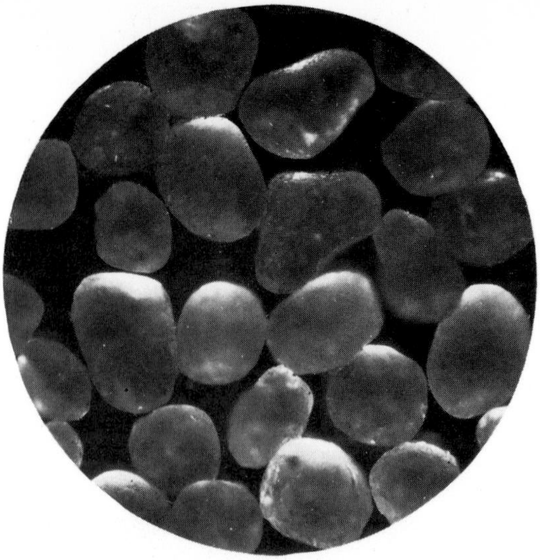

**Figure 22.10** Frosted 'millet seed' sand grains, St Peter sandstone (Ordovician), upper Mississippi States of U.S.A. Photomicrograph ×28 (*P. G. H. Boswell*)

southern part of the Canadian Shield were desert lands.

## Coastal Dunes and Sandhills

Along low-lying stretches of sandy coasts and lake shores, where the prevailing winds are onshore, drifting sand is blown landwards and piled up into dunes which form a natural bulwark of sandhills. Any mound or ridge of sand with a crest or definite summit is called a *dune*. Deposition begins wherever the wind is deflected by surface irregularities, including grasses and trees. In humid regions the conditions governing growth and removal are very complex. The wind varies in strength and direction. Vegetation and moisture tend to trap and fix the sand, but fixation is often incomplete. During severe gales old dunes may be breached and scooped out into deep 'blow-outs'. The resulting confused assemblage of hummocks and hollows gives coastal sandhills a characteristically chaotic relief. Where the water table is reached the ground becomes marshy.

An ideal dune has a long windward slope rising to a crest and a much steeper leeward slope (Figure 22.11). The latter is determined by the fact that sand blown over the crest falls into a wind shadow, and comes to rest at its natural angle of

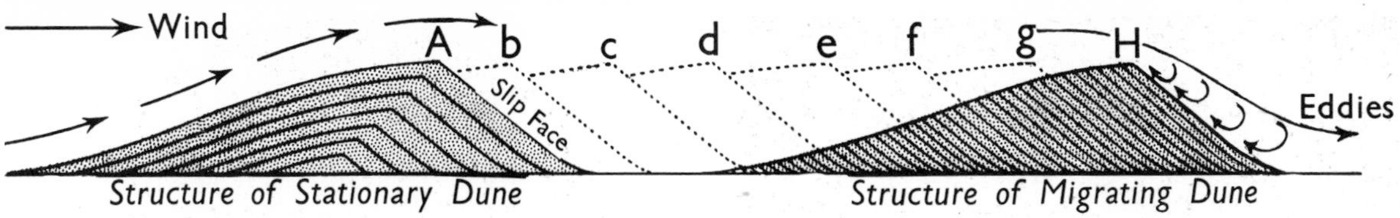


**Figure 22.11** Sections to illustrate the growth, migration and stratification of sand dunes. A stationary dune, A, grows in height with a forward and upward advance of the crest. When the sand supply and wind velocity involve migration of the dune, the crest advances to successive positions such as *b, c, . . . . . g,* and H.

repose—about 30° to 35° for dry sand. In situations where dunes are not effectively arrested by vegetation, or kept within bounds by winds from opposing quarters, they slowly migrate in the direction of the prevailing wind. When the wind is not fully loaded with newly acquired sand it sweeps up more from the windward slope, and drops it over the crest, where it streams down the 'slip-face'. By subtraction of sand from one side and addition to the other the dune travels forward.

As one belt of dunes migrates inland away from the beach, another arises in its place 'so that a series of huge sandy billows, as it were, is continually on the move from the sea-margin towards the interior' (Figure 22.12). Wide expanses of sandhills have spread inland in this way from low-lying coasts well supplied with sand, such as those of Holland and N Germany, the Landes of Gascony adjoining the Bay of Biscay, and the coastal desert of Namibia. Landward migration from the sea floor is convincingly demonstrated by the existence of beaches and dunes that are largely composed of ground-up marine shells. Excellent examples occur at St Ives and Perranporth along the north-west coast of Cornwall (Figure 22.13). The abundance of land snails in coastal dunes is a clear indication that the sand is at least partly composed of calcaereous fragments.

**Figure 22.12** A stage in the landward drift (from upper right toward lower left) of the Culbin Sands, Moray Firth area. Most of this area has now been stabilized by plantations. (*Institute of Geological Sciences*)

**Figure 22.13** Penhale sand hills, north of Perranporth, Cornwall, largely composed of minute fragments of marine shells; illustrating migration from the sea floor and beach, partial fixation by marram grass, and 'blow-outs' by severe gales (*Institute of Geological Sciences*)

**Figure 22.14** Advancing sand overwhelming a plantation of Scotch firs; Maviston Sands (SW continuation of the Cublin Sands, Fig. 22.12 (*Institute of Geological Sciences*)

The Culbin and Maviston sandhills, near the mouth of the Findhorn on the southern shore of the Moray Firth, furnish a classic example of the destruction of cultivated lands and habitations by advancing sand. Prior to 1694 the sandhills had already reached the fringe of the Culbin estate. In that year a great storm started a phase of accelerated encroachment which finally led to the complete obliteration of houses, farms and orchards, and even to the burial of fir plantations (Figures 22.14 and 22.15).

In most threatened regions of agricultural value, active measures are now taken to restrain the advance of the dunes. Tough binding-grasses such as bent or marram are excellent for this purpose, and since they thrive best where they are most needed, they commonly acclimatize themselves naturally and can easily be added to, if necessary. The harsh tufts check the wind, trap oncoming supplies of sand and continue to grow upwards as the entangled sand accumulates, leaving the underlying sand fortified with an intricate network of long roots. Such protected dunes become locally turfed over, and their subsequent growth tends to be seawards rather than landwards. If severe blow-outs undo or delay the work of stabilization when a great storm occurs, dunes can be anchored more securely by starting plantations of suitable trees on the landward side and gradually extending them across the sand already partially fixed by grasses. During the last few decades, large areas of the Culbin and Maviston sandhills have been successfully afforested with conifers in this way.

## Desert Dunes and Sand Sheets

About a fifth of the land surface is desert, and on an average about a fifth of the desert areas is mantled with sand. A high proportion of the desert floor is an erosion surface of bedrock, locally strewn with coarse rock-waste (Figure 22.5). Regions of shale and limestone provide little or no sand, but where sandstone is being disintegrated or mixed alluvium is being deflated, the wind picks up the loose grains and concentrates them into vast sand wastes (Figure 22.16) and long chains of dunes.

**Figure 22.15** Remains of an exhumed plantation formerly buried by sand which has now migrated farther on; Maviston Sands (*Institute of Geological Sciences*)

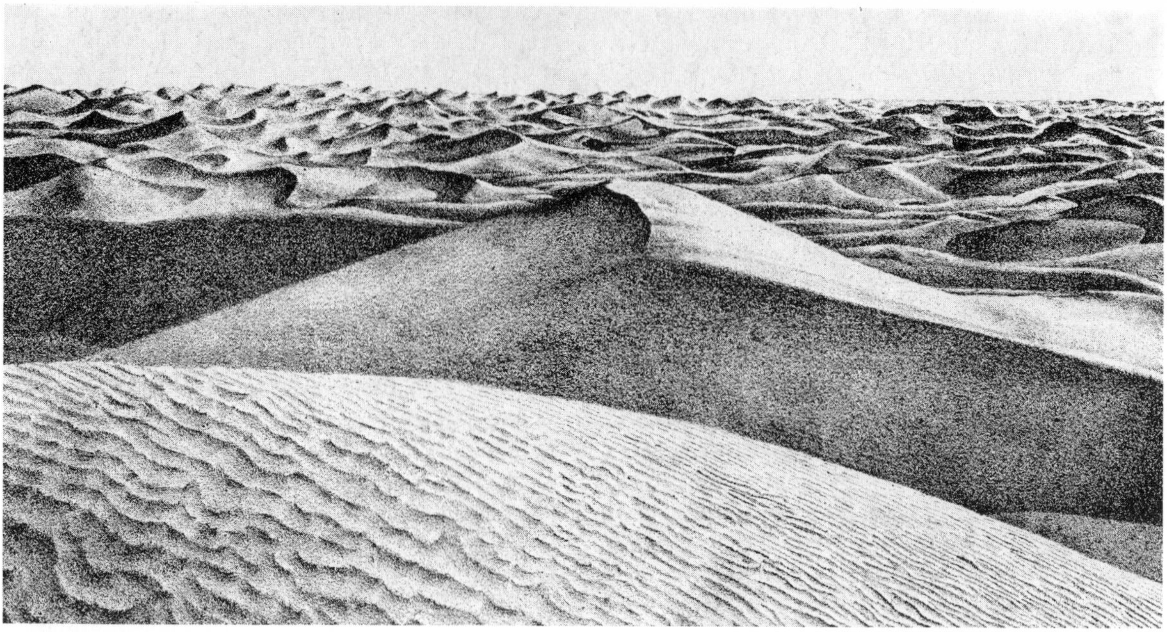

**Figure 22.16** Characteristic surface features of a sand sea in the Egyptian Desert (Fig. 22.6) (*Drawing based on photographs by M. V. Binosi*)

Complications due to vegetation and moisture arise only around oases or in the transition zones where the desert merges into steppe or savannah country. In the heart of the desert the wind has free play. But nevertheless, as a masterly study by R. A. Bagnold has made abundantly clear, the factors controlling the form of the sand accumulations are far from simple. They include the nature, extent and rate of erosion of the source of supply; the sizes of the sand grains and associated fragments; the varying strength and direction of the wind; and the roughness or smoothness of the surface (e.g. the presence or absence of pebbles) across which the sand is drifted and deposited. Of the resulting sand forms four main types can be distinguished:

(*a*) *Sand drifts*, which form temporarily in the wind-shadows of protruding rocks or cliffs.

(*b*) *Crescentic dunes* or *barchans* (a Turkestan name which has been generally adopted), which occur as isolated units (Figure 22.17), either sporadically or in long chain-like swarms, or as colonies, more or less linked together laterally, which advance across the desert like gigantic but irregular ripples (Figure 22.19).

(*c*) *Linear ridges* or *longitudinal dunes* (known as *seifs* in the Sahara), which commonly occur in long parallel ranges, each diversified by peak after peak 'in regular succession like the teeth of a monstrous saw'.

(*d*) *Sand sheets* of wide extent, which may be flat or undulating or diversified with more or less crude examples of (*b*) or (*c*) (Figures 22.16 and 22.18).

Dunes arise wherever a sand-laden wind deposits sand on the windward slopes of a random patch. The mound grows in height until a 'slip-face' is established by avalanching on the sheltered

**Figure 22.17** A typical barchan (drawn from a photograph)

**Figure 22.18** Sand dunes of Death Valley, California. Cottonwood Mts in the background (*United States Department of the Interior*)

leeward side. As the dune migrates, the extremities, offering less resistance to the wind than the summit region, advance more rapidly, until they extend into winds of such a length that their total obstructive power becomes equal to that of the middle of the dune. The resulting crescentic form then persists with only minor modifications of shape and size, so long as the wind blows from the same quarter. The width of a barchan is commonly about a dozen times the height, which ranges up to a maximum of 30 m or so. Winds blowing continually from nearly the same direction are essential for the growth and stability of barchans. The NE trade winds (e.g. in N African deserts) are those that best satisfy this condition. Given such winds and an adequate supply of sand, colonies and elongated swarms of barchans march slowly forward, like stream of vehicles in a one-way street (Figures 22.19 and 22.20). The rate of progress varies up to 6 m a year for high dunes, and up to 15 m a year for small ones.

Where the prevailing wind is occasionally interrupted by strong cross-winds which drive in sand from the sides, the conditions are like those of a one-way street which becomes densely crowded and choked by the inflow of traffic from every cross-road. Instead of a chain of barchans, a long seif dune is developed, a continuous but serrated ridge, parallel to the direction of the prevailing wind, with crests that may reach a height of over 100 metres. South of the Qattara depression (Figure 22.6) there is a long tract of parallel seifs with corridors of bare desert floor between. South of the area shown in Figure 22.6 the seifs come into the belt of NE trades and, accordingly, they wheel round towards the south-west, and pass progressively into barchans.

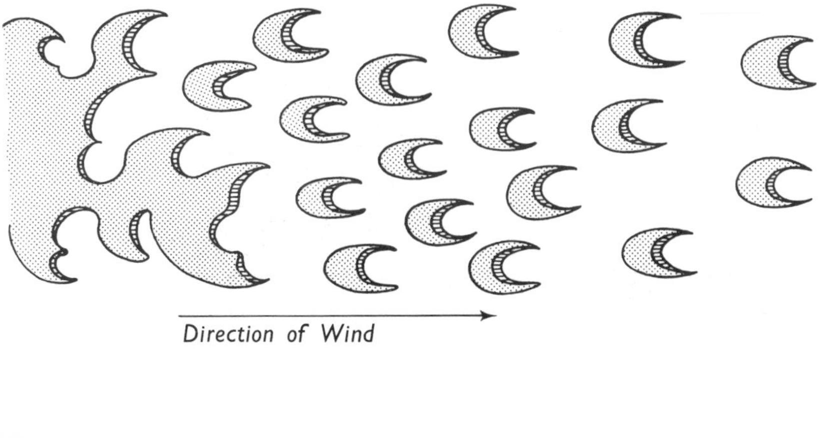


**Figure 22.19** Plan of a typical procession of barchans in the Libyan Desert

**Figure 22.20** A colony of barchans in Mauritania, West Africa, advancing towards the south-west (*I. F. N. A.*)

One of the most remarkable features of desert dunes is their apparent power of collecting all the sand in their neighbourhood. The explanation appears to be that the wind exerts a shepherding effect. If the surface between the dunes is fairly smooth, the drag on the wind there is less than it is along the edge of the dune. Eddies are thus set up which blow towards the dune, and so keep the intervening surface swept clean (Figure 22.21).

In the great 'sand seas', like that of Figure 22.16, other factors come into operation, dependent on the quantity and nature of the materials on which the wind has to work. The surface may be variegated with poorly developed seif dunes or, more commonly, with groups of irregular barchans. Where the dunes are not too crowded, they rise from a platform of coarser sand. The finer sand has always a better chance to be driven up the slopes. If a sprinkling of pebbles should be present, the wind disperses chance mounds of sand instead of building them into dunes, and wide, almost featureless, sheets of sand then accumulate. A patch of pebbles increases the drag on the wind to such an extent that the velocity near the ground is less than it is over a patch of clean sand. The resulting eddies therefore blow from the patch towards the pebbly surface until the sand is again evenly distributed between the pebbles. Widespread sand-sheets are also characteristically developed on the borders of deserts, where a scanty vegetation diversifies the surface (Figure 22.22). Grass and scrub break the force of the wind and inblown sand is more or less evenly distributed.

Among the curiosities of the desert, including 'dust devils' and mirages, perhaps the most mysterious is the sudden booming noise that

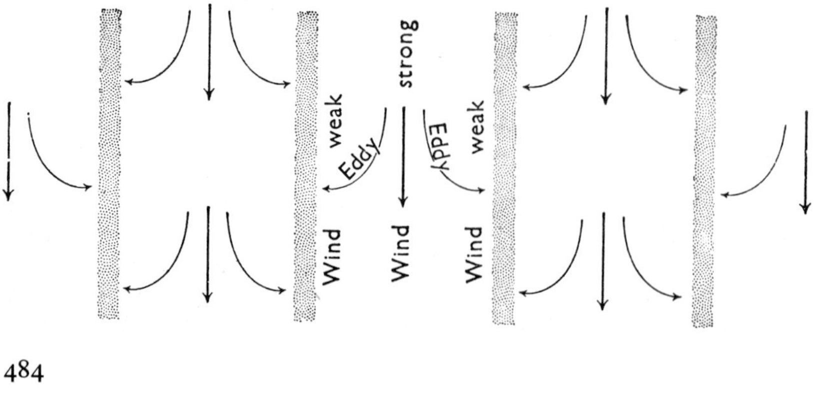


**Figure 22.21** Diagram to illustrate the shepherding effect of wind on sand ridges. The wind is strongest between the ridges and is retarded by friction against them; eddies are therefore set up, as shown.

**Figure 22.22** A pediplan of semi-arid 'desert scrub', with basalt-capped mesa behind; NE of The Solitario, Texas (*G. C. Gilbert, United States Geological Survey*)

occasionally interrupts the silence, apparently when the surface layer of sun-baked sand on the steep side of a dune becomes unstable and begins to slide down. Referring to the statement in the first edition of this book (1944) that no satisfactory explanation of these sounds was yet forthcoming despite much investigation and experiment, Dr D. W. Ramsden suggested that the boom is the sound of electrostatic discharges occurring when very dry sand slips down the slope of a dune. In a personal communication (1959) he wrote: 'A colleague who was recently in the Trucial States of Oman [extreme south of the Persian Gulf] created this noise every time he drove a truck down the slip face of the usual crescent-shaped dunes. The truck would sink deeply into the sand and slide down, producing a noise like rolling thunder.' Recently, Criswell, Lindsay, and Reasoner investigated the Sand Mountain dune near Fallon, Nevada. They placed a geophone to monitor vibrations transmitted through the sand, and a microphone for receiving sounds transmitted through the air. They found that the most effective way of causing the sand to boom was to dig a trench in it with a flat spade. The sound was like a low note on a cello that lasted for a few seconds and could be heard 30 m away. When the sand was pulled down by hand a strong vibration was caused, said to be reminiscent of an electric shock, which seems to confirm Ramsden's suggestion.

Examination of sands by the scanning electron microscope revealed that the sand grains of booming sand are more highly polished than those of silent sand. Criswell *et al.* found that out of 31 known booming dunes in U.S.A., 29 are composed essentially of quartz sand, the two exceptions, in Hawaii, being composed principally of calcite sand.

## Loess

We must next consider what happens to the vast quantities of dust which have been winnowed from the deserts and exported by the wind. From the deserts of Asia the wind carries the dust to the south and south-east where it is deposited over vast areas of the grassy regions of China, as a thick blanket of *loess* (Figure 22.23). Farther west, loess accumulates against the foothills of the Pamir plateau, where it provides a narrow belt of well-irrigated and fertile soil which has recently become one of the most densely populated agricultural regions of the world. From the deserts of N Africa much of the dust reaches the countries to the south and west, such as the S Sudan, northern Zaïre, Nigeria and Mali, where it adds to and is retained by the soil. Some of the dust blows over the Atlantic, some is trapped by the Mediteranean,

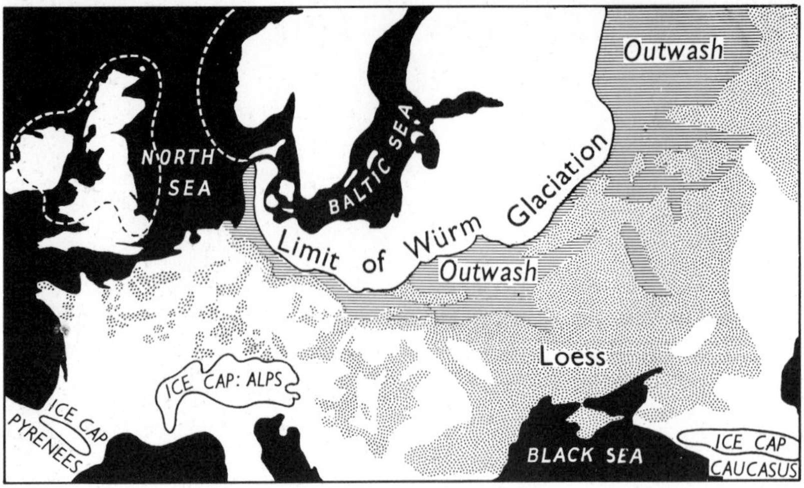


**Figure 22.23** Map showing the European distribution of loess and loamy soil formed from it, and the marginal relation to the last major glaciation (*After S. von Bubnoff*)

and the Red Sea receives passing contributions from Arabia.

Other very important supplies of fine silt and dust were formerly provided by the rock flour of glacial and glacifluvial deposits. During and after the retreat of each of the successive ice-sheets, the finer material was sifted out by the wind and deposited over the surrounding country (Figure 22.23). Thus it happens that a long belt of loess, mainly derived from glacial material in the west and from desert waste in the east, stretches from France to China. The term *loess* comes from a town of that name in Alsace. Beginning as local patches in France and Germany, the deposit becomes thicker and more extensive as it is traced across Russia and Turkestan, until in Shansi and the adjoining provinces of China it reaches its maximum development.

Loess is an accumulation of wind-borne dust and silt, washed down from the air by rain, and retained by the protective grip of the grasses of the steppe. Each spring the grass grows a little higher on any material collected during the previous year, leaving behind a ramifying system of withered roots. Over immense areas many metres have accumulated, whole landscapes having been buried, except where the higher peaks project above the blanket of loess. The material itself is yellowish or light buff, very fine-grained and devoid of stratification. Although it is friable and porous, the successive generations of grass roots, now represented by narrow tubes partly occupied by calcium carbonate, make it sufficiently coherent to stand up in vertical walls which do not crumble unless they are disturbed. The passage of traffic along country roads loosens the material, clouds of dust are removed by the wind, and the roads are worn down into steep-sided gullies and miniature canyons.

In the loess provinces of China rain and small streams carve the surface into a maze of ravines and badland topography (Figure 22.24). The larger rivers flow in broad and fertile alluvial plains bordered by vertical bluffs. Here and in the lowland and deltaic plains to the east most of the alluvium is simply loess redistributed by water. In the loess uplands cultivation of the slopes is made possible by terracing. The steep-sided cliffs and walls, whether natural or artificial, are often riddled with entrances to cave dwellings, many of which have their chimneys opening into the fields above. This mode of habitation has occasionally led to great distasters. In 1556, for example, widespread landslides and floods were started by a catastrophic earthquake and nearly a million peasants lost their lives.

In the Midwest of the United States there are deposits of loess, locally called *adobe*, that correspond in all essentials to those of Europe and Asia. In Kansas and Nebraska much of the wind-blown silt has come from the semi-arid lands of the west; but elsewhere it consists largely of rock flour, winnowed by the wind from the outwash and temporary lake deposits left by the shrinking ice-sheets of the Pleistocene, to find lodgment on grass-protected surfaces like the prairies. By subsequent erosion much of the original loess of the upper Mississippi basin has contributed to the alluvium of the downstream flood plains. When these dry up in times of drought they form an important source for a second generation of loess.

**Figure 22.24** 'Badland' erosion of loess in Kansu province, North China, brought about largely by destruction of timber and over cultivation in the past. Terrace conservation has checked this wastage in the valley, which is drained by a tributary of the upper Huang Ho (Hwang Ho) (see Fig. 18.40).

### Weathering and Stream Work in the Desert

The belt of deserts between the Atlantic and the Persian Gulf lies mainly to the south of latitude 30°N. Here the high-pressure descending air is dry to begin with (Figure 22.1) and the resulting trade winds become increasingly desiccating so long as they are blowing over land. This belt continues north of latitude 30° from Iran to the Gobi of central Asia, though with several interruptions, rainfall being low because of distance from the oceans or the intervention of mountainous rain barriers. High mountains near the western coasts are mainly responsible for the deserts of North and South America. In all these arid regions the rainfall is rare and sporadic, both temperature and wind intensity are subject to violent fluctuations—daily and seasonal—and vegetation is extremely scanty or entirely lacking.

Under these conditions mechanical weathering is dominant, involving the splitting, exfoliation and crumbling of rocks by alternations of scorching heat and icy cold (Figure 14.4). Nevertheless chemical weathering, though extremely slow, plays a far from negligible part. By decomposition and solution, rocks that would otherwise successfully resist the stresses set up by temperature changes are gradually weakened until they can be shattered. By evaporation minute quantities of dissolved matter are brought to the surface. The loose salts are blown away, but oxides of iron, accompanied by traces of manganese and other similar oxides, form a red, brown or black film which is firmly retained. The surfaces of long-exposed rocks and pebbles thus acquire a characteristic coat of 'desert varnish'.

Although most desert localities remain entirely unvisited by rain for years on end, no part can be regarded as permanently free from rain. Taken over a long period of years the rainfall averages only a few centimetres a year, 25 cm being reached only on the fringes of the desert. In the adjoining semi-arid regions the annual rainfall averages between 25 and 50 cm, but even here long periods of drought are usual.

The capacity for evaporation far exceeds the rainfall in the desert. Growth of lakes is effectively prevented. Nor can permanent streams originate

**Figure 22.25** Gorge of the Wadi Barud, Egyptian Desert (*O. H. Little*)

**Figure 22.26** Gorge of the Wadi Gasab in the plateau of Ma'aza between the Nile and the southern end of the Gulf of Suez (*W. F. Hume*)

under such conditions, although well-nourished rivers, like the Nile, with adequate sources in humid regions, may cross the desert without entirely dwindling away. Outflowing streams are otherwise short and intermittent, and confined to coastal districts where, moreover, the rainfall is less scanty. The desert drainage is almost wholly internal, and directed towards the lowest parts of the many depressions which, owing to earth movements and wind erosion, characterize the desert surface. The poetical generalization that 'the weariest river winds somewhere safe to sea' does not apply to desert regions.

The gorges and steep-sided wadis that dissect the desert uplands (Figures 22.25 and 22.26), some with extinct 'waterfalls'; the alluvial spreads that floor the depressions; the salt deposits and terraces of vanished lakes (Figure 22.27); the buried soils that only require turning over and irrigating to blossom afresh; the rock carvings and paintings of prehistoric artists, and other relics of ancient man from stone implements to actual habitations: all these point to pluvial climates when running water was far more actively concerned than it is today in developing the landforms of the desert which we actually see.

The Sahara has been described as 'the corpse of a once well-watered landscape'. But whereas man originally retreated before the spreading aridity, now he is returning for oil and gas and iron ores, and above all for the essential artesian water that alone can restore the long-vanished fertility and make the desert locally habitable. Between the Hoggar and Mauritania there is a vast expanse of desert-varnished *reg* or gravel desert known as the Tanezrouft and hardly heard of until the French exploded their first atom bombs. Here, during the preliminary quest for water, French geologists discovered a thick fossil soil just below the surface. Pollen from the soil is of vegetation of the kinds that now flourish in a Mediterranean climate, and radiocarbon dating shows the soil to be about 7000 years old. Drilling has revealed deep-lying water-bearing sandstones fed by the rain that falls on the Atlas mountains to the north, where the intake area extends for hundreds of kilometres along the upturned edges of the strata. On the northern side of the Sahara this great aquifer already supplies artesian water to all the newly developing oilfields, turning the growing settlements into market gardens of amazing fertility (Figure 22.28). Throughout this region official policy—not everywhere strictly adhered to—is that the rate of withdrawal of water for irrigation, industrial and social purposes must not exceed the rate of replenishment along the intake.

Despite the multiplication of artificial oases, the

**Figure 22.27** Shore terrace of former pluvial lake, Colorado Desert, California (*W. C. Mendenhall, United States Geological Survey*)

**Figure 22.28** El Oued, 'the city of domes' in South-east Algeria, where deep artesian wells now support a population of over 100,000, and basins dug through the surrounding dunes enable the roots of palms to tap a shallower source of flowing ground-water (*Ritchie Calder*)

extreme desiccation of the desert is still its most striking natural feature; and it remains something of a puzzle that rare rainstorms and the resulting spasmodic stream action should occur at all. When they do occur they are probably a result of cyclonic, funnel-shaped breakthroughs from the moist air of high-level westerlies (Figure 22.1). The chief characteristics of the rare desert rainstorms are their erratic distribution and brief duration, and—apart from occasional light showers—their intense violence. Houses of dried salt mud are turned into a miry pulp and washed away when a sudden 'cloudburst' descends on an oasis. Travellers have been drowned in the floods that race down the dry wadis with little or no warning. Such torrents, swiftly generated in a distant upland rainstorm, carry a heavy load of mixed debris, prepared for them by years of weathering and wind erosion. At the foot of the mountains or escarpments, where carrying-power is reduced by seepage or loss of gradient, the load is dropped to form alluvial fans (Figure 22.29) and 'deltaic' deposits: first the coarser and later the finer debris. Between the pediment or sediment-filled depression that commonly surrounds a desert mountain group or flanks an escarpment there may be a more or less continuous slope of waste formed by the coalescence of alluvial fans; this feature is called a *bahada* or *bajada* (Figure 22.30).

Choked by their own deposits the short-lived streams subdivide into innumerable channels and spread out laterally across the plain, which thus becomes covered with a veneer of finer sediment. If the rain falls on the gentle slopes of a depression a shallow sheet-flood may carry a mud flow towards the middle. If the water reaches the lowest part before it is lost by further seepage and evaporation a temporary lake is formed. There the dissolved material is concentrated by evaporation and finally deposited to form salt muds or glistening white sheets of rock-salt, gypsum and other

**Figure 22.29** Alluvial fan, east wall of Death Valley, south of Badwater, California, passing into a *salina* of salt and mud in the foreground. Notice the miniature delta near the middle in front, formed by a temporary stream from the other side of the depression. (*J. S. Shelton and R. C. Frampton*)

**Figure 22.30** The outer part of a bahada that originally continued across the foreground to the gullied escarpment on the Pacific side of the San Andreas fault zone, above which the photograph was taken, and against which the coalesced alluvial fans originally reposed. The valley in the foreground was formed at the expense of the bahada by the headward erosion of a stream draining into the Pacific. Now only a thread of water, the stream must have been much more voluminous during the pluvial phases of the Pleistocene. The view shows one of the beheaded channels leading down towards the playas and salinas of the Mohave Desert. (*J. S. Shelton and R. C. Frampton*)

**Figure 22.31** Salt marsh (salina) in Death Valley, California (*Dorien Leigh Ltd*)

salts. In W America the alluvial plains of arid or semi-arid enclosed basins are called *playas* (Figure 22.30), the more saline tracts being distinguished as *salinas* (Figure 22.31). So long as more material is brought in by rare torrents and sheet-floods than is removed by wind, the depressions continue to be filled.

### The Cycle of Erosion in Arid Regions

A peculiarity, still unexplained, of the present wide extension of deserts is that the spread of arid conditions occurred not only during the retreat of the last great ice-sheets, but continued long afterwards, and may still be doing so. One of the most striking historical proofs of the climatic deterioration within the last 2300 years is the account of the prodigious campaign of Alexander the Great (356–323 B.C.) to the Punjab and back through countries that were then sufficiently well watered and forested to provide the needs (including shipbuilding) of an army of 110,000 men, although today they include the Thar desert of Rajasthan and are 'so desolate that they cannot support passing caravans of even 100 men and animals' (D. N. Wadia, 1960).

Just as ice-sheets extended themselves for a time by a self-accelerating process, so, it appears, do desert areas. Wind erosion strips away the fertile soil formed during a preceding pluvial phase and turns the adjoining grasslands into sand-covered wastes. The destructive activities of mankind have powerfully contributed to the loss of fertile land in many places; but elsewhere unaided natural changes in the air and water circulations of the globe have promoted increasing desiccation. But how this is brought about we do not understand.

In these circumstances it is natural that there should be considerable doubt about the relative importance of the effects of wind and water in deserts that have remained arid sufficiently long to have passed through an arid cycle of erosion. Most present-day deserts have had so short or so interrupted a life that some of their landscape features are of hybrid origin. However, the following outline, though brief, will serve to indicate how the processes known to operate in the desert could ultimately reduce a given region of uplands and depressions to a desert plain.

The stage of *youth* is on the whole characterized by decrease of the original relief. Given sufficient time, the rare and short-lived flash-floods cut gullies in the uplands. These develop into wadis and canyons, from which rock-waste is spread over the depressions, by way of alluvial fans where the uplands are bordered by escarpments, The latter gradually recede by wind erosion, and because the 'sand-blast' operates most effectively at the base, desert scarps are characteristically steep. So long as more material is carried into the depressions by intermittent streams than is removed by wind, the depressions continue to be filled, so passing into playas or salinas. The only features that offset the prevalent reduction of relief are (*a*) hollows excavated by the wind in beds of soft and dry materials, and (*b*) sand-dunes.

Between the uplands and the playas the growing alluvial fans coalesce into bajadas which steadily increase in thickness as they encroach on the wind-swept escarpments. The stage of *maturity* may be said to begin when the mountains become like islands half submerged in their own waste products. Now the lower depressions begin to tap the higher ones by incipient wadis that slowly extend themselves headwards. While they are being eroded away, the soft deposits of a higher basin are likely for a time to develop typical badland topography, with the 'constant slope' much in evidence. The more resistant rocks of the escarpments, however, flanked by outlying buttes and inselbergs, rise abruptly with the 'free face' of bare rock as their most prominent slope element.

During *old age* the higher 'basins' contribute materials to the lower ones until they all unite, and the upland residuals dwindle away, provided the cycle is not interrupted by earth movements or a change of climate, or brought to an end wherever the surface is reduced to the base-level of wind erosion—which is the water table. And so the region is reduced to an erosion surface, closely related to a pediplain but of more complex structure, being composed not only of bare rock, thinly strewn with debris, but of smooth tracts of mosaic- or desert-pavement and vast wastes of sand.

## The Permian Desert Winds of Britain

We have already seen that crescentic dunes (barchans) are developed where the stronger desert winds blow from the same quarter throughout the year. This condition is fulfilled in deserts across which the trade winds blow. Since the

**Figure 22.32** Dune bedding of barchan type in desert sandstones of Permian age, Mauchline Quarries, Strathclyde, Scotland (*Institute of Geological Sciences*)

barchans with their slip-faces and horns advance in the wind direction, their cross-bedding preserves the pattern given them by the wind and therefore embodies a record of the direction of the wind (see Figure 22.11). Thus, by making a statistical study of the dips of the cross-bedding of a group of barchans of ancient origin, it is possible to ascertain the mean direction of advance and therefore of the wind direction at the time concerned. This ingenious method of determining the directions of 'fossil winds' was initiated by F. W. Shotton and successfully applied in 1937 to the aeolian sands of the New Red Sandstone (mainly Permian) of Britain (Figure 22.32).

Shotton has found that the various remnants of these British desert sands (e.g. in Shropshire and the adjoining Midlands, N England, SW Scotland, Arran and near the Moray Firth) are accumulations of barchans, indicating that they most probably formed under the influence of the NE trade winds of the time. As shown in Figure 22.33, the mean directions of dune advance determined by Shotton confirm this highly significant inference. The evidence indicates that in Permian times the British area lay within, say, 30° of the equator. If so, this area has not only changed its latitude by moving 25° or more to the north, relative to the present position of the north pole, but it has rotated about 40°.

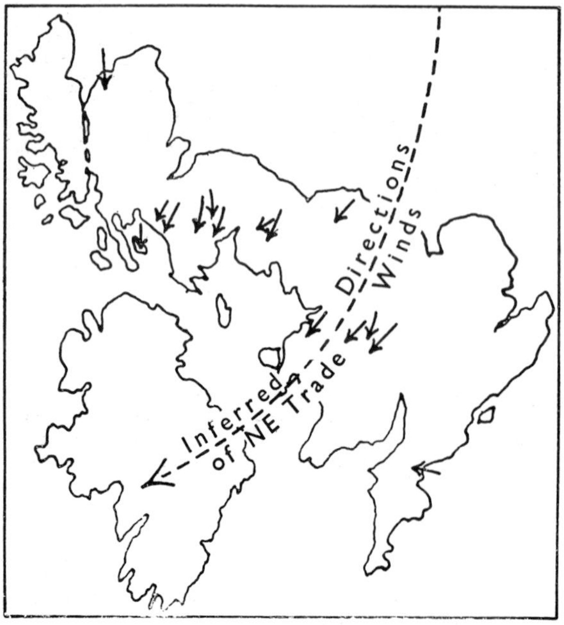


**Figure 22.33** Map showing the directions of dune advance in the Permian desert of Britain. The map is orientated so that the corresponding wind directions would be those of the North-East Trades of the time. If this interpretation is right, the British area was then situated about 15°–30°N of the Equator; since when, during its northward drift it has also rotated through an angle of about 40°. (*After F. W. Shotton and S. K. Runcorn*)

N. D. Opdyke and S. K. Runcorn have followed up Shotton's work, both in Britain and the western United States. From various localities in the States they find the inferred wind directions to have been nearly constant over a distance of about 1600 km during Upper Carboniferous and Permian times, when desert sands accumulated over an immense area. The results indicate that, provided the winds were trade winds like those of today, the Arizona-Utah-Wyoming area was situated about 30° nearer the equator than it is today.

## Salt Deposits and Ancient Climates

It is appropriate here to recall that on p. 91 it was pointed out that a hot and arid climate is one of the conditions necessary for the accumulation of *thick* deposits of evaporites. By far the most important salt deposits of Europe and the United States are of Permian age, and their existence is consistent with other evidence that the climate of these regions was then hot and arid. Vast spreads of evaporite occur across the southern States, and through Britain and Germany to the south-eastern part of the Russian Platform and the western foothills of the Urals. The high temperatures and excess of evaporation required for the formation of these deposits indicate that the great 'evaporating basins' in which they were precipitated were situated within or near what were latitudes of 15° to 40° during the Permian period. The immense salt deposits of the Kama river basin (west of the Urals), culminating in the thickest-known layers of potash salts, can hardly be supposed to have been formed in latitudes around 60°N, where they are now situated. Like Britain, Russia appears to have drifted in a direction which has carried it 25° or more to the north in, say, 250 m.y., corresponding to an average rate of not less than 11 km per million years.

Even more striking is the evidence from the Queen Elizabeth Islands of Arctic Canada. There, salt domes have been discovered, one of which (Figure 22.34) is 6 km across. These imply the presence in depth of immensely thick evaporites ranging in age from late Cambrian to late Silurian.

**Figure 22.34** Isachsen salt dome (78°N Latitude), Ellef Ringnes Island, one of the Queen Elizabeth group of Arctic Canada. The core of the dome has a diameter of 6 km and consists of late Silurian evaporites which penetrate Cretaceous strata. Away from the Isachsen and other domes of the Island these beds lie nearly flat, but within 3 to 5 km of a dome they begin to be upturned, outcropping as concentric rings which become nearly vertical against the core (*Royal Canadian Air Force*)

The very existence of these deposits across latitudes of 75° to 78°N suggests that at the time of their accumulation the Canadian Shield could not have been far from the equator. Here the northerly component of drift is of the order 40° in 450 m.y., giving nearly the same average rate of about 10 km per million years.

In the light of the above examples, it is of interest to notice that the present geographical distribution of salt deposits of various geological ages in America, and in Europe and Africa (Figure 22.35) points on balance to a northerly drift of these continental masses. The data suggest that the direction may have been temporarily reversed. But since many deposits of evaporite have been lost by denudation and others may remain to be discovered, our information is not yet sufficiently complete to give more than a general picture.

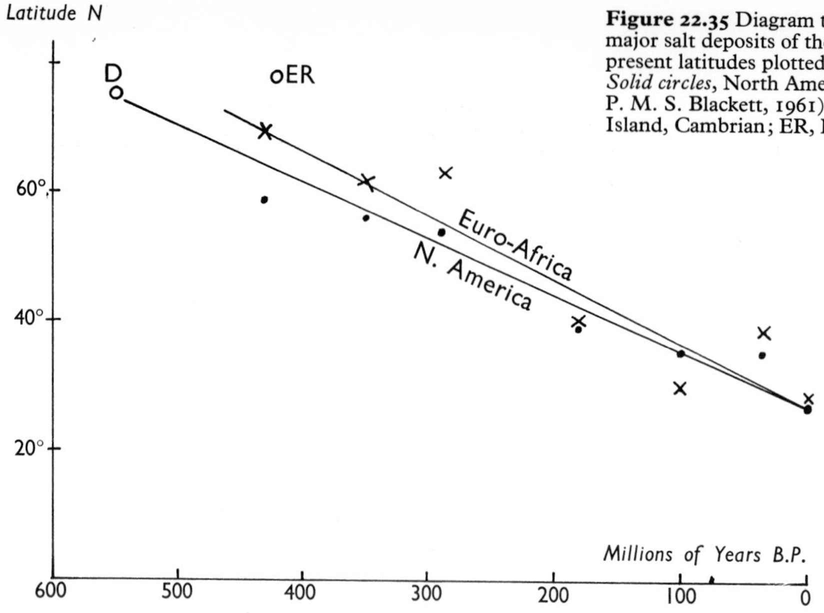


**Figure 22.35** Diagram to illustrate the distribution of the major salt deposits of the Northern Hemisphere; mean present latitudes plotted against ages. *Crosses*, Euro-Africa; *Solid circles*, North America (data from F. Lotze, 1957, and P. M. S. Blackett, 1961); *open circles* added: D, Devon Island, Cambrian; ER, Ellef Ringnes Island, late Silurian

## Selected References


BAGNOLD, R. A., 1973, *The Physics of Blown Sand and Deserts Dunes*, Chapman and Hall, London.

BLACKETT, P. M. S., 1961, 'Comparison of ancient climates with the ancient latitudes deduced from rock magnetic measurements', *Proceedings of the Royal Society*, A, vol. 263, pp. 1–30.

CALDER, Ritchie, 1958, *Man Against the Desert*, Allen and Unwin, London.

COOPER, W. S., 1957–8, 'Coastal sand dunes of Oregon and Washington', *Geological Society of America Memoir*, No. 72.

ELIAS, M. K., *et al.*, 1945, 'Symposium on loess', *American Journal of Science*, vol. 243, pp. 225–303.

FORESTRY COMMISSION, 1949, and later editions, *Britain's Forests: Culbin*, H.M. Stationery Office.

GERSTER, Georg, 1960, *Sahara* (translated from the German edition of 1959 by S. Thomson), Barrie and Rockcliff, London.

JAEGER, E. C., 1957, *The North American Deserts*, Stanford University Press, California; Oxford University Press, London.

LAMING, D. J. C., 1958, 'Fossil winds', *Journal of the Alberta Society of Petroleum Geologists*, vol. 6, pp. 179–83.

OPDYKE, N. D., and RUNCORN, S. K., 1960, 'Wind direction in the western United States in the Late Palaeozoic', *Bulletin of the Geological Society of America*, vol. 71, pp. 959–72.

RUNCORN, S. K., 1961, 'Climatic change through geological time in the light of the palaeomagnetic evidence for polar wandering and continental drift', *Quarterly Journal of the Meteorological Society*, vol. 87, pp. 282–313.

SHOTTON, F. W., 1956, 'Some aspects of the new Red Desert in Britain', *Liverpool and Manchester Geological Journal*, vol. 1, pp. 450–65.

WADIA, D. N., 1960, 'The post-glacial desiccation of Central Asia', *Monograph of the National Institute of Sciences of India*, New Delhi.

WHITE, G. F., (Ed.), 1956, *The Future of Arid Lands*, Washington.

# Chapter 23
# Coastal Scenery and the Work of the Sea

I with my hammer pounding evermore
The rocky coast, smite Andes into dust,
Strewing my bed and, in another age,
Rebuild a continent for better men.

*Ralph Waldo Emerson* (1803–82)

## Shore-Lines

Nearly all coast-lines have been initiated by relative movements between land and sea. Rise of sea-level or subsidence of the land leads to the submergence of a landscape already moulded by sub-aerial agents. The drowning of a region of hills and valleys gives an intended coast-line of bays, estuaries, gulfs, rias, fjords and straits, separated by headlands, peninsulas and off-lying islands. The Rio coast of Brazil (Figure 23.1) is one of the most striking and familiar examples of this type. In contrast very broad bays, like the Great Australian Bight, result from the submergence of plains. Coasts that have originated in these ways are called *coasts of submergence*. Because of the rise of sea-level that followed the last glacial period most coasts belong to this class. Indeed without that rise harbours would have been few and far between. The occurrence of 'raised beaches' along some of these coasts looks like a contradiction. But it only means that what we now see is the algebraic sum of a long history of ups and downs. The last major eustatic rise of sea-level has ensured that a majority of shore-lines still retain the characteristics of submergence.

**Figure 23.1** A celebrated coast of submergence: Rio de Janeiro, Brazil (*Brazilian Government Public Relations*)

There are, however, formerly glaciated regions that have been rebounding more rapidly than the rise of sea-level, and have continued to emerge during the last few thousand years of nearly stable sea-level. The results, such as can be well seen at Stockholm and Helsinki and along the neighbouring coasts, can be described as submerged glacial topography that is now emerging (Figure 23.2). More typical *coasts of emergence* occur where tectonic uplift outstripped the rising sea (or has since done so), e.g. along the Pacific coast of South America.

Other varieties of coastline include those determined by volcanic activity (Figure 23.61), faulting (Figure 24.22), outwash plains (Figure 23.54) and the growth of coral reefs and atolls (p. 555 et seq.). From a structural point of view Suess was the first to recognize two contrasted types which he distinguished as *Atlantic* and *Pacific.*

**Figure 23.2** An emerging coast due to post-glacial isostatic uplift. Örö Island and neighbouring parts of the Finnish Archipelago, Hitis, south-western Finland (*Copyright: Ilmavoimat—Finnish Air Force. Reproduced by permission*)

Coasts of Atlantic or *transverse* type are determined by fractures and subsidences that characteristically cut across the strike or 'grain' of the folded rock formations (Figures 9.17 and 23.3); they characteristically border relatively young oceans that are widening as a result of sea-floor spreading. Coasts of Pacific or *longitudinal* type border or lie within mountain chains, including island festoons like those of Asia, and follow the general 'grain' of the land. When partially drowned such coasts are said to be of *Dalmatian* type (Figure 23.4).

The general outlines of a newly formed coast are soon modified by marine erosion and deposition, with development of a wide variety of shore features and coastal scenery. By the incessant pounding of waves, which break up the rocks and wear back the cliffs, the sea cuts its way into the land like a horizontal saw. The liberated rock fragments are rounded by innumerable impacts and continual grinding as the line of breakers is carried backwards and forwards over the foreshore by the ebb and flow of the tide. The worn-down

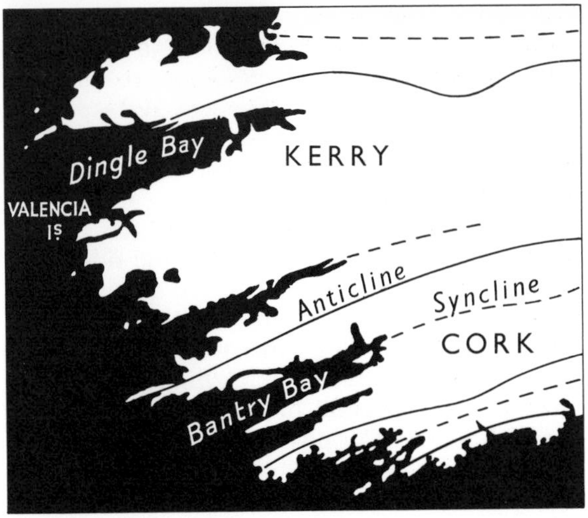


**Figure 23.3** An example of submerged 'Atlantic' or transverse coast. Old Red Sandstone crops out along anticlines which have remained as uplands or broad ridges that jut out as promontories. Less resistant Carboniferous strata outcrop as synclines in the valleys, which pass seawards into long bays or *rias*.

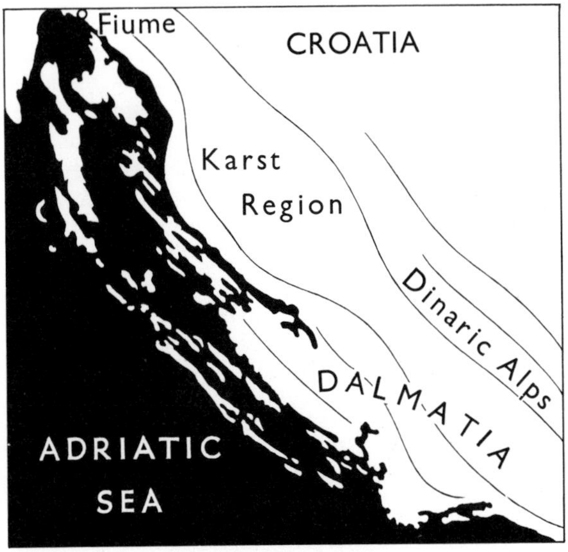


**Figure 23.4** An example of a submerged 'Pacific' or longitudinal coast (Dalmatian type), Yugoslavia

material is partly dealt with by the waves themselves and partly by currents. Most of the finer sediment—including contributions from rivers and glaciers and the wind—is carried into deeper water coming to rest on the sea floor. The coarser sediment is swept towards or drifted along the shore to form shoals and beaches or to build out spits and bars (long embankments of sand and shingle), where the coastline abruptly changes its direction. Inlets and lagoons sheltered from the sea in this way develop into marshes, which in time are silted up by contributions from the landward side, or blanketed with sand dunes that advance from the seaward side. In these and other ways new land is added to the fringe of the old in compensation for the losses suffered elsewhere.

The waters of the seas and oceans readily respond by movement to the brushing of the wind over the surface; to variations of temperature and salinity; to the gravitational attraction of the moon and sun; and to the Coriolis force. The work of erosion, transport and deposition carried out by the sea depends on the varied and often highly complex interplay of the waves, currents and tides that result from these movements.

It may be noted in passing that lakes, especially the larger ones, behave in much the same way as enclosed seas. In consequence the shore features of lakes and seas have much in common. A lake formed by obstruction (e.g. the lava-dammed Lake Kivu) drowns the surrounding land and so acquires a shore-line of the submergent type. A lake which has shrunk from its former extent in response to climatic or other changes (e.g. the Great Salt Lake of Utah) is margined by flats and terraces of sediment (Figure 22.27), and so acquires a shore-line of the emergent type. Tides are negligible in lakes, but seasonal variations of rainfall may cause the water to advance and recede over a tract of shore which is alternately covered and exposed, though much less frequently than the tide-swept foreshore of a sea coast. Waves and currents operate exactly as in land-locked seas of similar extent and depth, and are responsible for erosion and deposition on a corresponding scale. There are of course important biological contrasts. The swamps into which shallow lakes degenerate, with their luxuriant growth of aquatic vegetation and accumulations of peat, are very different from the mangrove swamps and tidal marshes that locally border the sea. On the other hand lake shores have nothing to be compared with the coral reefs of tropical seas.

## Tides and Currents

The tide is the periodic rise and fall of the sea which, on an average, occurs every 12 hours 26 minutes. Tides are essentially due to the passage around the earth, as it rotates, of two antipodal bulges of water produced by the differential attraction of the moon and sun. The bulges are the crests of a gigantic wave, low in height but of enormous wave length. It is easy to understand that the water facing the moon should bulge up a little, but it is less obvious why there should be a similar up-bulge in the opposite direction on the other side of the earth. The basis of the explanation is that the water centred at A (Figure 23.5) is attracted towards the moon more than the earth, centred at E, while the earth in turn is attracted more than the water centred at B. The water at the far side is thus left behind to almost the same extent as the water on the near side is pulled forward. From places such as C and D the water is drawn away and low tide results. As the earth rotates, each meridian comes in turn beneath the positions of high and low tide nearly twice a day; not exactly twice, because allowance must be made for the forward movement of the moon. Nor are these positions exactly in line with the moon (as for simplicity they have been drawn in the diagram), because the tides are affected (*a*) by the earth's rotation; (*b*) by the great continental obstructions met with during their circuit of the globe; and (*c*) by friction against the sea floor, especially in shallow seas.

The effect of the sun is similar to that of the moon but considerably less powerful. When the earth, moon and sun fall along the same straight line, the tide-raising forces of sun and moon help each other, and tides of maximum range, known as *spring tides*, result. The moon is then either new or full. When the sun and moon are at right angles relative to the earth, the moon produces high tides where the sun produces low. The tides are then less high and low than usual and are called *neap tides*.

In the open ocean the difference in level between high and low tide is only a metre or so. Enclosed basins have still weaker tides—only 30 cm or so in the Mediterranean and no more than 10 cm in the Black Sea. In shallow seas however, and especially where the tide is concentrated between converging shores, ranges of 6 to 9 m are common and tidal currents are generated. The tides around the British Isles are of special interest. After passing up the western coasts the crest of the tidal wave swings round into the North Sea and proceeds southwards. The Coriolis force (p. 469) drives the water towards the right, that is, the British side, which in consequence has far higher tides than Norway and Denmark. In passing northwards up the Irish Sea the Coriolis force results in the tides being at least twice as high on the Welsh and English coasts as on the Irish side. Similarly the tides running up the English Channel are also forced to the right, giving the French coasts higher tides. It will be readily appreciated that conditions of exceptional complexity arise in the Straits of Dover and the southern part of the North Sea.

A current of about 2 knots accompanies the flood tide as it advances up the English Channel. In the Bristol Channel however, which is like a wide funnel leading into the Severn, the inflowing tide is forced into a rapidly narrowing passage. It therefore rises in height. Spring tides may reach as high as 12 or 13 m and the inward current may attain a speed of 10 knots. In these circumstances, and especially if the wind co-operates, the overcrowded tidal waters eventually travel bodily up the river with a wall-like front of roaring surf. This is the Severn *bore*, a vigorously advancing flood of powerful waves and breakers (Figure 23.6), which may ride up the river almost as far as Gloucester before subsiding completely.

Near the shore and between islands, tidal currents are often sufficiently powerful to trans-

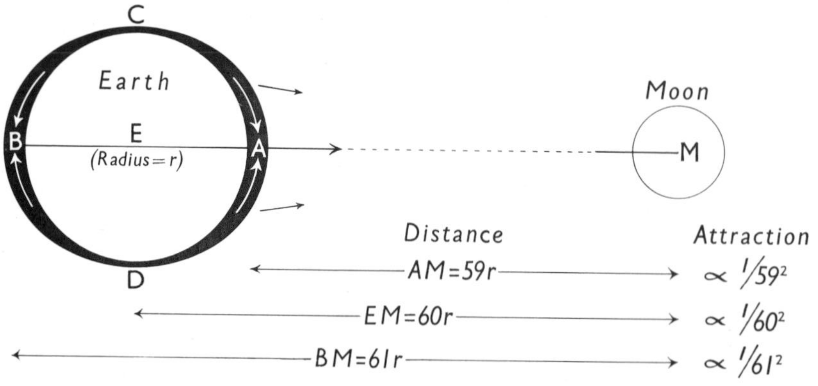


**Figure 23.5** Diagram to illustrate the generation of tides

**Figure 23.6** The 'bore' of the Severn advancing up-river at the peak of a spring tide (*Keystone Press Agency Ltd*)

port sand and even shingle, and so to scour and erode the sea floor. In estuaries, where the outward flow of river water is added to the ebb current, transport is dominantly seaward. But since the fresh river water, carrying a load of silt and mud, tends to slide out over the heavier salt water that has crept in along the bottom, it is the upper suspended load that is mainly swept out to sea, while the coarser debris is stranded and tends to accumulate as sand bars.

Powerful currents are generated by differences of salinity (see p. 89), such as depend on (*a*) inflow of rivers, rainfall and the melting of ice, all of which freshen the water; and (*b*) evaporation, which increases the salinity.

Evaporation over the Mediterranean lowers its surface and increases the salinity and density. Surface currents therefore flow into the Mediterranean through the Dardanelles from the Sea of Marmara and the Black Sea (where the evaporation is more than balanced by the inflow of rivers), and through the Strait of Gibraltar from the Atlantic (Figure 23.7). In each case undercurrents of higher salinity flow outwards from the Mediterranean. Shore deposits are affected by the surface currents, while farther out the floor is scoured by the deeper current. A similar interchange of water takes place between the highly saline Red Sea and the Indian Ocean, and also between the comparatively fresh Baltic and the North Sea.

The main current systems of the oceans are primarily of a convectional nature, brought about by density differences due to heating in the tropics and cooling in the polar regions, and also to variations in salinity. They are greatly modified by the configuration of the continents and the dominant winds. Apart from the superficial movements charted in every atlas, cold water from the Arctic spreads south at depths between 1000 and 1500 fathoms, is warmed by the undercurrent from the Strait of Gibraltar, and continues south until it rises over the cold currents moving northwards from Antarctica. The latter can be traced into all the oceans, including the North Atlantic, while deep Arctic and Atlantic water also passes into the Pacific and Indian Oceans.

Oceanic currents have important climatic and biological as well as geological effects, and will be referred to where appropriate. More localized currents, due essentially to interactions between wind, tides and waves (e.g. surges, longshore drift and rip currents), or to slumping and churning up

**Figure 23.7** Salinity currents of the Mediterranean and Black Seas. Water of high salinity (about 37–38; thick arrows) flows out of the Mediterranean, while water of lower salinity (thin arrows) flows in.

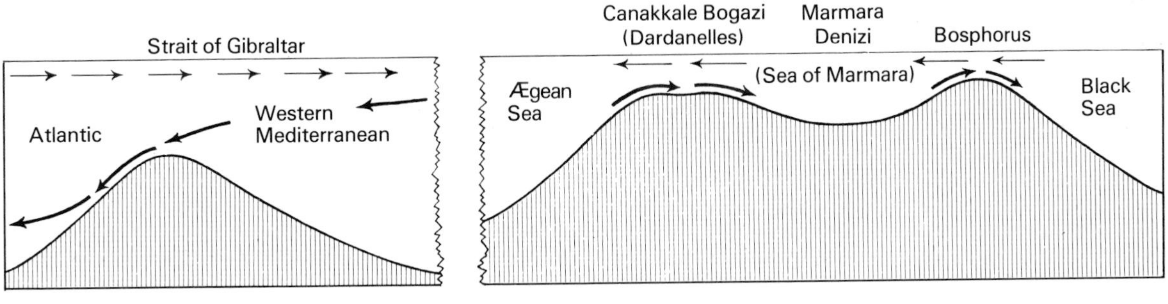

of bottom sediment on submarine slopes down which muddy water can flow (turbidity currents), are also dealt with on subsequent pages.

## Waves

The ordinary waves seen on the sea are almost entirely due to the sweeping of winds over the surface of the water. The exceptions are the far-travelled descendants of giant waves generated by catastrophic 'accidents' which suddenly displace immense volumes of ocean water. Such occurrences may result from earthquakes, volcanic eruptions, submarine landslides and avalanches of rock from high cliffs or steep mountain sides into deep water. Here we are concerned with waves that derive their motion and energy from the wind. To begin with, the surface is ruffled into undulations that move forwards and gradually increase in height and speed. The *height* of a wave is the vertical distance from trough to crest (Figure 23.8). The horizontal distance from crest to crest—or from trough to trough—is the *wave length*. This is measured at right angles to the wave front or crest line. Although the latter may be very long it should not be confused with the *wave length*, which is parallel to the direction of wave advance. The height ultimately attained by a wind-driven wave, where it is not restricted by shallowing water, depends on the strength, duration and *fetch* of the wind, the fetch being the length of the open stretch of water across which the wind is blowing. When the loss of energy involved in the propagation of the waves through the water is just balanced by the amount of energy supplied from the wind, the height reaches its maximum. The height cannot however be calculated from the speed and wave length of the waves. According to a very rough numerical rule adopted as a guide by mariners, the limiting height in feet is about half the speed of the wind in miles per hour.

**Figure 23.8** Profile of an ideal wave of oscillation from crest to crest, showing the directions of movement of water particles at various points

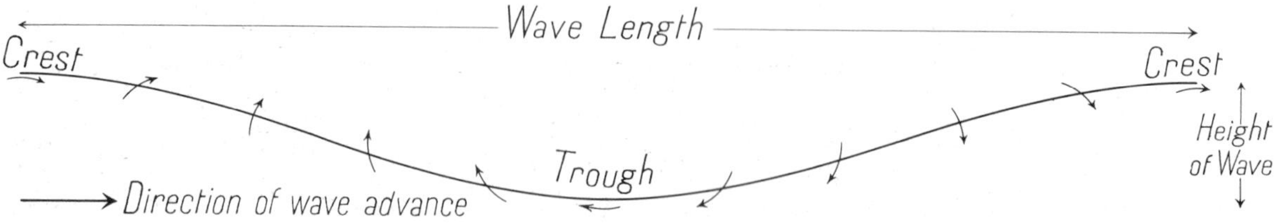


Where waves are being generated by the gusty winds of a storm new waves are continually forming. The resulting *sea*, as the assemblage of waves is called (e.g. a stormy or choppy sea), is likely to be a jumble of many different wave lengths. Here and there crests and troughs coincide and cancel out. Elsewhere crests may coincide and rise to form a high peak, and where two troughs coincide the sudden drop may be perilous to a passing ship. From the generating area the longer and higher waves grow at the expense of the shorter ones. Thus it happens that a series of more uniform waves is eventually sorted out of an initial chaotic 'sea'. Such waves may travel into regions of calm weather far beyond the fetch of the wind where they originated and started their progress towards the bordering shores. The height gradually declines when waves are no longer maintained by the wind; they are then described as a *swell* or *groundswell*.

A set of uniform waves (called a 'train') progressing through deep water can be described in terms of wave length $\lambda$, height $h$ and period $T$. The period is the time taken for two successive crests to pass a given point. The speed $v$ is then given by the simple formula $v = \lambda/T$. An approximate relationship that is often useful is $\lambda$ (in metres) $= 1{\cdot}56T^2$ (in seconds). This means that a deep-water wave with a period of 10 seconds has a wave length of about 156 metres, the corresponding speed ($v$) being 56 km per hour. In the open ocean heights of 1·5 to 5 metres are common, increasing to 12 or 15 m in high seas. The corresponding wave lengths range from 60 to 245 m, but as they flatten out into swell they may reach 300 metres or more, about 760 being the maximum recorded, with a speed of nearly 130 km per hour.

The sea waves we watch from the shore are generally a mixture of swells from distant storms and 'seas' from local winds. In other words they can be described as a *spectrum* of 'trains' of waves of different lengths, just as sunlight is a blend of light waves of different lengths which, when separated (as in the rainbow), appear as a spectrum of different colours. This is the origin of the traditional belief that 'every seventh wave is the biggest'. Actually the heights can vary irregularly within wide limits, according to the happenings in various parts of the globe (see pp. 576–9).

It is important to realize that in the open sea—apart from wind drift—it is only the wave form that moves forward: the shape, not the water itself. The wave is a device for the transport of energy. Each particle of water moves round a circular orbit during the passage of each complete wave, the diameter being equal to the height of the wave (see Figure 23.9). This is demonstrated by the behaviour of a floating cork under which a train of waves is passing. Each time the cork rises and falls it also sways to and fro, without advancing appreciably from its mean position. Such waves are called *waves of oscillation*. If the wind is strong, however, each water particle advances a little farther than it recedes and the waves may become strongly asymmetrical (Figure 23.10). Similarly in shallow water, where friction against the bottom begins to be felt, each particle recedes a little less than it advances. In both cases the orbit, instead of being a closed circle, resembles an ellipse which is not quite closed, and a certain proportion of the water then slowly drifts forward in the direction of wave advance.

When gusts of wind brush over a field of corn the stalks repeatedly bend forward and recover, and waves visibly spread across the surface. Here it is obvious that the wave motion is not confined to the surface, since it is shared by the stalks right down to the ground. In the same way the energy contributed by the wind to a body of water is transmitted downwards as well as along the surface. Owing to friction the diameters of the orbits rapidly diminish in depth until at a depth of the same order as the wave length they become negligible. Indeed, at a depth of half the wave length the orbital diameter is only 4 per cent of that at the surface. The greatest depth at which sediment on the sea floor can just be stirred by the oscillating water is called the *wave base*. This depth was formerly thought to be about 180 m, but it is now known to be considerably less, at least for surface waves. There is much conflicting evidence. Ripple marks have been photographed at depths of

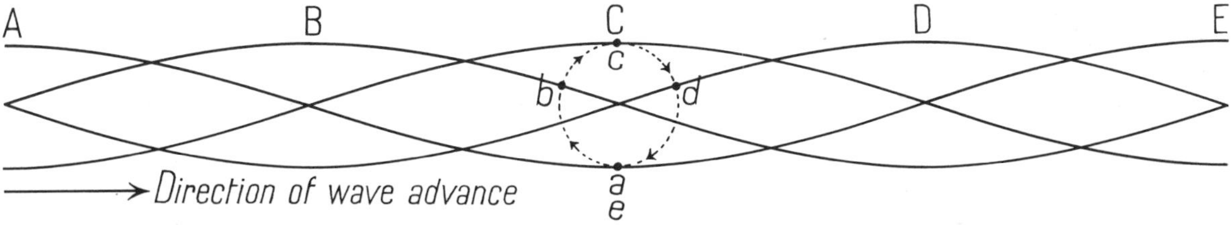


**Figure 23.9** Diagram showing the orbit of a water particle during the passage of a wave of oscillation. A, B, C, D, E mark successive positions of the crest as the wave moves forward; *a, b, c, d, e* are the corresponding positions of the particle. AE=wave length; C*a*=height of wave.

**Figure 23.10** An asymmetrical wave driven by a high wind. Here the orbits are nearly elliptical, but with a net advance for each rotation, so that a little of the water moves forward in the direction of the wave-motion. (*T. M. Finlay*)

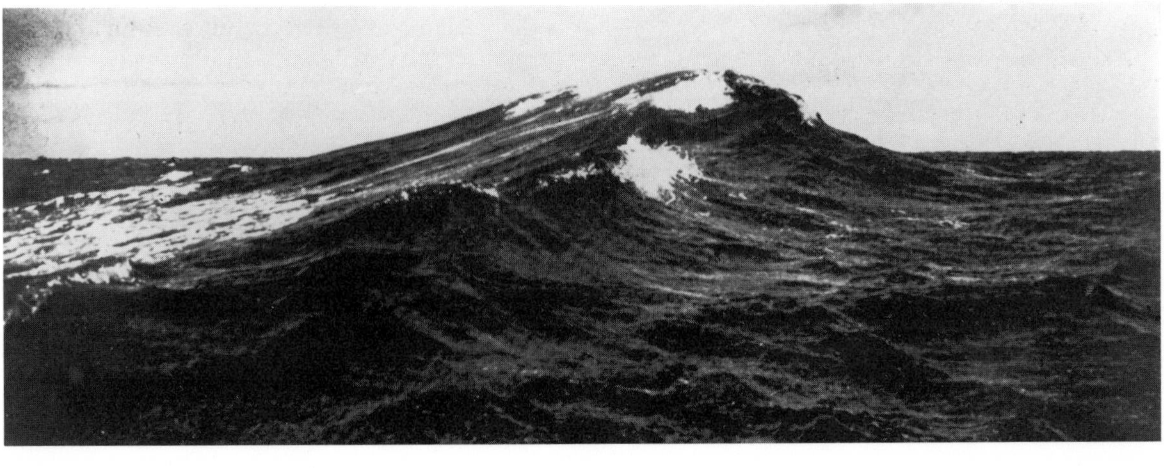

hundreds of metres, but such effects are far beyond the influence of surface waves. There are, however, deeper waves generated at the boundaries of currents travelling in opposing directions. It has also been suggested that deep-sea ripple marks may be the work of gentle currents flowing over loose sediment while the particles are kept vibrating to and fro by the passage of earthquakes.

### Waves in Shallow Water

As waves approach the shore and pass into shallow water, several changes of great importance occur. 'Shallow' in this context means a depth of water less than half the wave length of the approaching waves. After the waves begin to 'feel bottom' they gradually slow down as the water shallows, each progressing more slowly than the one behind. The only characteristic of a train of waves that remains constant is the period $T$. Remembering the formula $v = \lambda / T$, it follows that as the speed falls off the wave lengths become correspondingly shorter. Indeed the tendency of the wave fronts to crowd together as the shore is approached can often be clearly seen. Consequently when waves approach a shelving shore obliquely, as in Figure 23.11, the crest lines swing towards parallelism with the shore, This change of direction with change of speed is called *refraction*; it is essentially the same phenomenon as the sudden bending of a ray of light when it passes from air into, say glass, in which its speed is reduced by 30 per cent or more. Refraction of oblique waves around a headland and islet into a bay is well illustrated by Figure 23.12.

**Figure 23.11** Wave refraction: diagram to illustrate the swing of oblique waves in shallowing water towards parallelism with the shore. While the crest at *a* advances to $a^1$, the crest at *b* advances a shorter distance to $b^1$ . . . and so on. The crest lines *abcd* thus become curved, as shown.

The effects of wave retardation off an indented coast are illustrated by Figure 23.13. The waves advance more rapidly through the deeper water opposite a bay than through the shallower water opposite a headland. Thus the practically straight crest of a wave at *a* moves to *a′*, while the crest at *b* moves only to *b′*. Each wave crest in turn begins to approximate to the curves of the shore-line. Consequently when the shore is reached by a wave such as *abcde*, all the energy from the long stretch *ac* converges on the headland AC (and that from *de* on DE). In contrast the very much smaller amount of energy from the short stretch *cd* is dispersed around the shores of the bay from C to D. Thus while headlands are being vigorously attacked by powerful waves, deep embayments are not unduly disturbed and their waters provide safe anchorage for vessels sheltering from a storm. In the same way waves entering a harbour between or over piers spread out and rarely do more than ruffle the water inside.

In a train of waves advancing at right angles to a straight shore-line, each metre of the crest line of a wave represents about the same amount of energy. The wave energy from a deep column of water is concentrated into a shallow one. Since energy is proportional to $mV^2$, it follows that as $V$ is reduced, as it is on a shelving sea floor, $m$ must be increased (where $m$ is the mass of water per metre of crest line). $mV^2$ turns into $Mv^2$. Each wave therefore increases in volume and becomes higher as it passes into shallower water. At the same time the wave front becomes correspondingly steeper. Finally a critical point is reached where the volume of water in front is insufficient to fill up the wave form as required by the orbital movement. The crest of the uncompleted wave is then left unsupported, and because the orbit is broken the wave itself breaks. The depth in which waves break (i.e. the mean depth as it would be in still

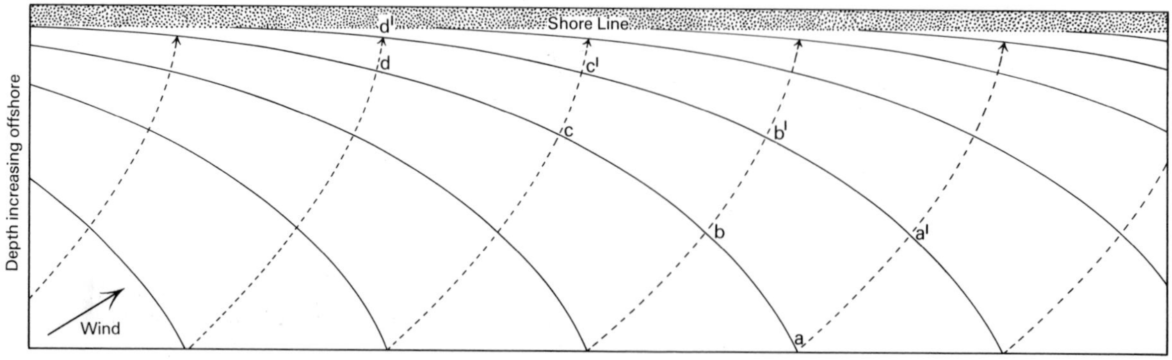

water) ranges between 1·25 and 0·75 times the height of the unbroken wave following on behind.

Dependent on a variety of factors, chief of which is the steepness of the wave front, two contrasted types of breakers are recognized. Between the two extremes there are naturally all gradations, according to the nature of the shore and the waves, and the strength and direction of the wind. Advancing

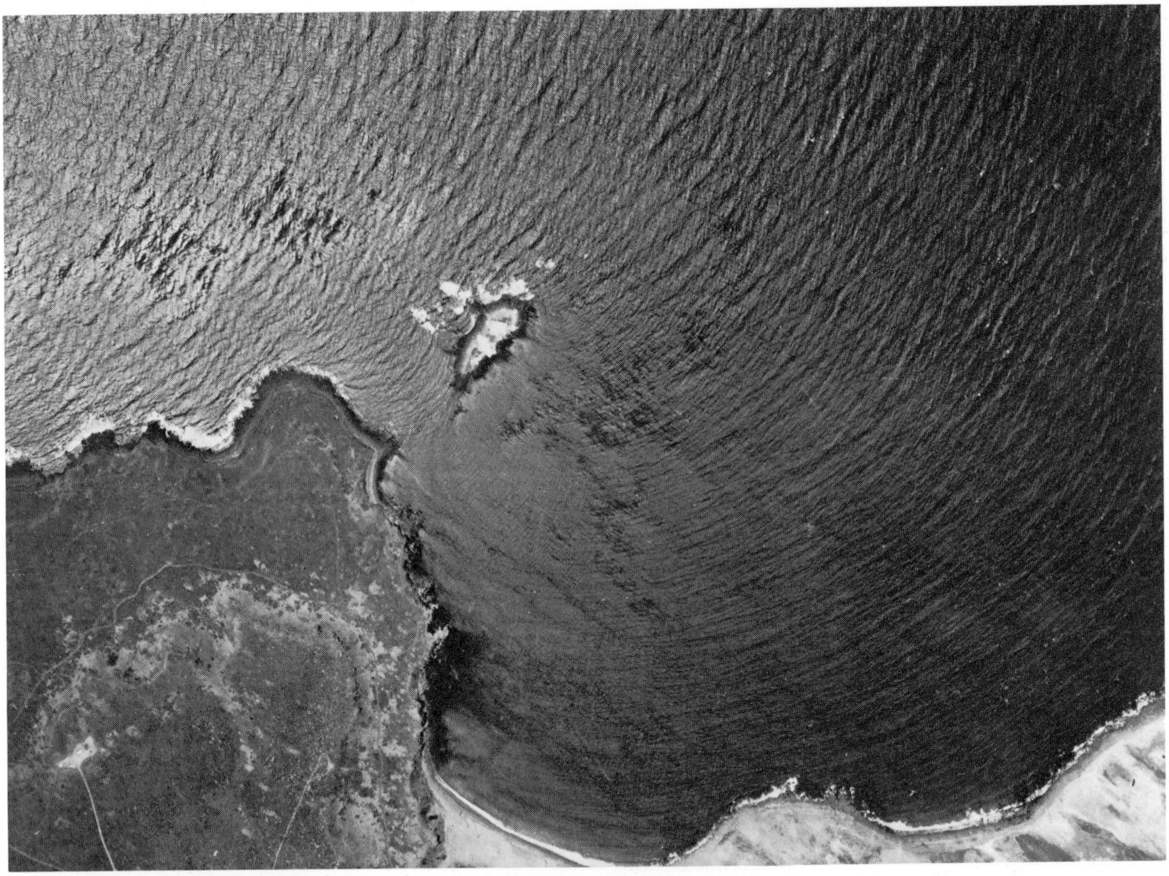

**Figure 23.12** Aerial photograph showing the refraction of waves around a headland and islet into a bay at San Clemente Island, California. The pattern of short-crested waves approaching from the upper left results from the interference of swells from two different directions. (*Official U.S. Navy photograph, by permission*)

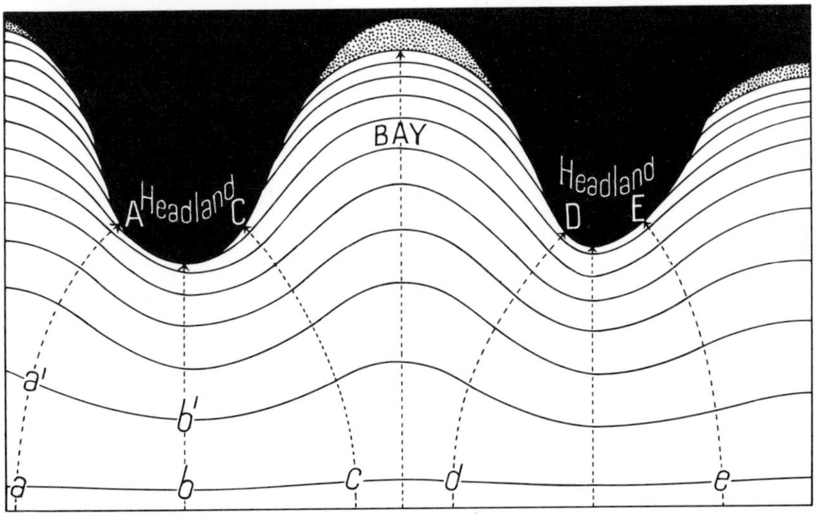


**Figure 23.13** Diagram to show the effect of wave retardation off an indented coast

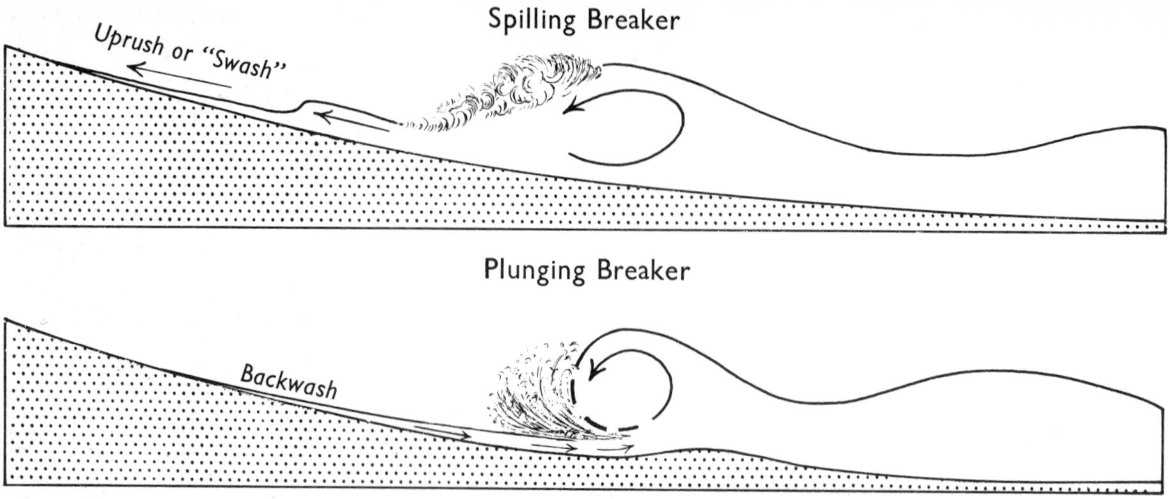


**Figure 23.14** Diagrammatic indication of the difference between a spilling breaker with an effective uprush (constructive type) and a plunging breaker with an effective backwash (destructive type) (*Modified after W. V. Lewis*)

over a gently shelving floor a wave front steepens gradually, until the crest merely spills over into the trough in front. There is a *spilling breaker*, a foaming mass of water that surges forward as a turbulent sheet of surf (Figures 23.14 and 23.15). But where the floor shallows rapidly, and especially if the waves are large, the rise of the crest is greatly augmented. The wave front steepens until an unstable-looking hollow appears in front. Over this the crest begins to curl, sometimes hovering a little while the wave continues its advance. But finally it plunges down, often with great violence. This is the *plunging breaker* (Figures 23.14 and 23.16).

What were *waves of oscillation* before they began to 'feel bottom' have now passed largely or wholly into *waves of translation*, in which the water advances bodily up the shore. This sheet of surf may re-form into smaller waves that break again higher up, so that there is a zone of breakers rather than a single break-point. The final translation of water up the beach is called the *uprush* or *swash*. The return of water down the slope is called the *backwash* (Figure 23.14).

Considerable volumes of air may be rapidly trapped by plunging breakers and compressed so strongly that the air reacts with explosive violence. It escapes by 'atomizing' the water into masses of foaming spray which reach astonishing heights in severe storms. Geyser-like displays may sometimes be seen when high waves are breaking on

**Figure 23.15** Small breakers of the spilling type, coast of New Jersey, U.S.A. (*Paul Popper Ltd*)

steep beaches, but uprushing jets of spray are more familiarly associated with the crashing pressure of breakers against rocky obstructions (Figure 23.17) and breakwaters (Figure 23.18).

**Figure 23.16** A plunging breaker viewed from above, approaching the coast of New Jersey. The marked break in the crest is an indication of a rip current (Fig. 23.38) flowing seawards (*Paul Popper Ltd*)

**Figure 23.17** Storm wave from the Atlantic breaking against the cliffs of St Ives, Cornwall (*Fox Photos Ltd*)

## Marine Erosion

The sea operates as an agent of erosion in four different ways:

(*a*) by the *hydraulic action* of the water itself, involving the picking up of loose material by currents and waves, and the shattering of rocks as the waves crash, like giant water-hammers, against the cliffs (Figure 23.17).

(*b*) by *corrasion*, when waves, armed with rock fragments, hurl them against the cliffs and, co-operating with currents, drag them to and fro across the rocks of the foreshore;

(*c*) by *attrition*, as the fragments or 'tools' are themselves worn down by impact and friction; and

(*d*) by *corrosion*, i.e. solvent and chemical action, which in the case of sea water is of limited importance, except on limestones and rocks with a calcareous cement.

The destructive impact of breakers against obstructions is often far greater than is generally realized. The pressure exerted by Atlantic waves averages over 9700 kg per $m^2$ during the winter, while in great storms it may exceed even 30,000. Thus not only cliffs but also sea walls, breakwaters and exposed lighthouses are subjected to shocks of enormous intensity. Cracks and crevices are

**Figure 23.18** A breaker crashing against the promenade at Hastings and bursting into jets of spray from the explosive expansion of entrapped air (*Judges Ltd*)

quickly opened up and extended. Water, often in the form of high-pressure spray, is forcibly driven into every opening, tightly compressing the air already confined within the rocks. As each wave recedes the compressed air suddenly expands with explosive force, and large blocks as well as small become loosened and are sooner or later blown out bodily by pressure from the back. The combined activity of bombardment and blasting is most effective as a quarrying process on rocks that are already divided into blocks by jointing and bedding, or otherwise fractured, e.g. along faults and crush zones.

The undercutting action of waves is illustrated by Figures 23.20 and 23.21. Cliffs originate and are maintained by similar undermining of the seaward edge of the land. By falls from an over-steepened rock-face or by collapse of rocks overhanging the notch which may have been excavated at the base of the cliffs, the latter gradually recede and present a steep face towards the advancing sea. But where the cliffs are protected for a time by fallen debris, and especially if they are composed of poorly consolidated rocks, the upper slopes may be worn back by weathering, rain-wash and slumping. At any given place the actual form of the cliff depends on the nature and structure of the rocks there exposed, and on the relative rates of marine erosion and sub-aerial denudation. (Figure 23.19).

The most striking evidence of undermining is provided by caves. There are few stretches of coast along which the rocks are equally resistant to wave attack. Caves are excavated long belts of weakness of all kinds, and especially where the rocks are strongly jointed (e.g. Fingal's Cave, Figure 1.2). By subsequent falling-in of the roof and removal of the debris, long narrow inlets are developed. In

**Figure 23.19** The Needles: stacks of Chalk in line with the cliffs at the western extremity of the Isle of Wight. The boldness of the cliffs reflects the unusual resistance of the Chalk, here due to folding and induration, and contrasts strongly with the low shoreline of Cretaceous sands and clays seen in the background (*Aerofilms Ltd*)

**Figure 23.20** Pedestal or 'mushroom' rock, of tough metamorphosed basalt, deeply undercut by marine erosion; near Maloy on the north side of Nordfjord, about 145 km N of Bergen, Norway (*Mittet Foto, Oslo*)

**Figure 23.21** Pedestal of calcareous sandstone (Jurassic) supported by an undercut pillar of shale on the foreshore at Sheepstones, Yorkshire. The sandstone is pitted with corrosion hollows caused by spray from breaking waves at high tide and related to the lapiés of Fig. 11.12 (*Institute of Geological Sciences*)

Scotland and the Faroes a tidal inlet of this kind (Figure 23.22) is called a *geo* ('g' hard— Norse *gya*, a creek). The roof of a cave at the landward end of a geo—or indeed of any sea cave—may communicate with the surface by way of a vertical shaft which may be some distance from the edge of the cliff. A natural chimney of this kind (Figure 23.23) is known as a *blow-hole* or *gloup* (a throat). The opening is formed by the falling-in of joint blocks loosened by the hydraulic action of wave-compressed air already described. The name *blow-hole* refers to the fact that during storms spray is forcibly blown into the air each time a breaker surges through the cave beneath.

When two caves on opposite sides of a headland unite, a natural arch results, and may persist for a time (Figure 23.24). Later the arch falls in, and the seaward portion of the headland then remains as an isolated stack. Well-known examples of stacks are the Chalk pinnacles at the western extremity of the Isle of Wight, known as the Needles (Figure 23.19), and the impressive towers of Old Red

**Figure 23.22** Sea cave and development of inlets by roof collapse in cliffs of Old Red Sandstone. The Wife Geo, near Duncansby Head, extreme north-east coast of Highland, Scotland, looking seawards. (*Institute of Geological Sciences*)

**Figure 23.23** Blow-hole or 'gloup', due to inland collapse of roof of sea cave. The part of the roof that still remains is known as The Devil's Bridge. Holborn Head, north-east Highland, looking seawards (*Institute of Geological Sciences*)

**Figure 23.24** Arch cut through a headland of Dalradian quartzites; 'Great Arch', Doaghbeg, north of Portsalon, Co. Donegal (*Irish Tourist Board*)

**Figure 23.25** Stacks and cliffs of Old Red Sandstone, looking north. The Stacks of Duncansby, near John o'Groats, Highland (*Institute of Geological Sciences*)

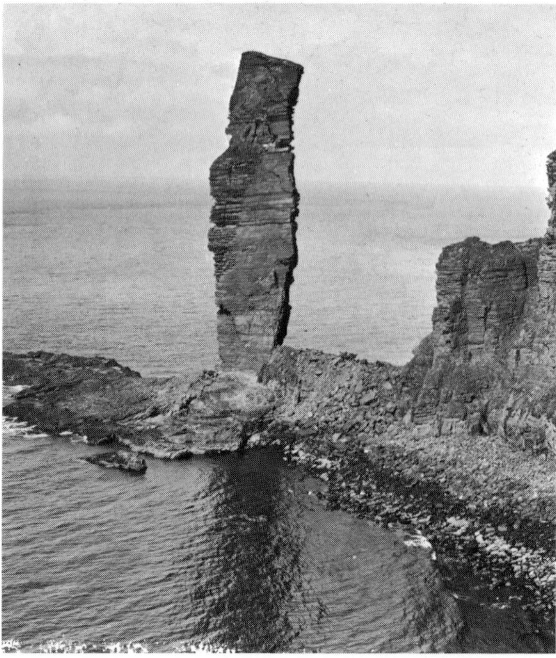

**Figure 23.26** The Old Man of Hoy, Orkney Is., north of Caithness (Highland). A stack of Old Red Sandstone, 137 m high, rising from a platform of Devonian lava (*Institute of Geological Sciences*)

Sandstone near John o'Groats (Figure 23.25) and in the Orkneys (Figure 23.26).

As the cliffs are worn back a *wave-cut platform* is left in front (Figure 23.27), the upper part of which is visible as the rocky foreshore exposed at low tide (Figure 23.28). There may be patches of sand and pebbles in depressions, and beach-like fringes strewn with fallen debris along the foot of the cliffs, but all such material is continually being broken up by the waves and used by them for further erosive work, until finally it is ground down to sizes that can be carried away by currents. The platform itself is abraded by the sweeping of sand and shingle to and fro across its surface. Since the outer parts have been subjected to scouring longer than the inner, a gentle seaward slope is developed. In massive and resistant rocks this is an extremely slow process. Consequently, as the cliffs recede and the platform becomes very wide, the waves have to cross a broad expanse of shallow water, so that when they reach the cliffs most of their energy has already been used in transporting the abrading materials. Thus the rate of coast erosion is

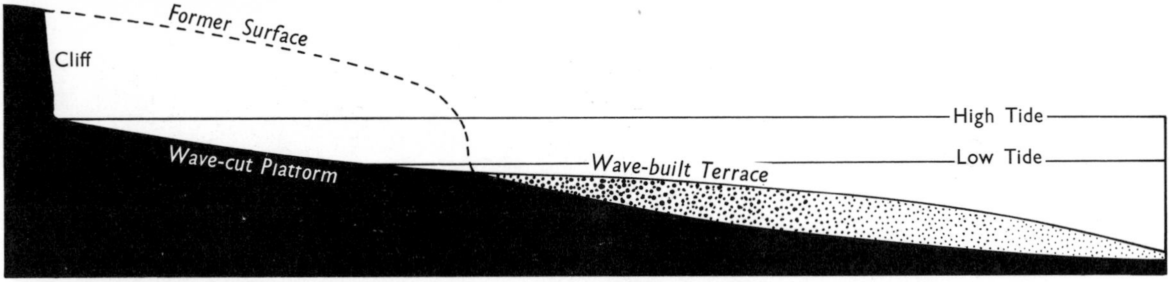


**Figure 23.27** Idealized section to illustrate a temporary stage in the development of a sea cliff, wave-cut platform, and wave-built terrace

**Figure 23.28** Wave-cut platform cut across steeply dipping Silurian strata, near St Abb's Head, Borders, south-east Scotland (*Institute of Geological Sciences*)

**Figure 23.29** Section across the *strandflat* north of Bergen, Norway. Length of section 52 km (*After Fridtjof Nansen*)

automatically reduced. In high latitudes, however, the cliffs may still continue to be worn back by frost and thaw, provided that the waves are able to remove what would otherwise become a protective apron of scree (cf. Figure 14.3). The wave-cut platform off the rocky coast of west and north-west Norway—there known as the *strandflat*—has reached an exceptional width by this co-operation of processes, locally up to as much as 60 km (Figure 23.29). The most extensive level now stands 15 to 18 m above sea-level because of recent isostatic uplift. Above this platform innumerable stacks and skerries, mostly flat topped, rise to heights of about 30 m, suggesting an older strandflat that has since been dissected. Similar strandflats have been described from parts of Spitzbergen

Figure 23.30 Coast erosion by the North Sea, south of Lowestoft, Suffolk; showing the disastrous effects of rapid recession of cliffs of glacial deposits due to destructive storm waves in 1936 (*Institute of Geological Sciences*)

and Greenland and other fjord coasts.

Where the sea is encroaching on a coast of poorly consolidated rocks, the platform in front is quickly abraded and normal coast erosion proceeds vigorously (Figure 23.30). In some localities the inroads of the sea reach alarming proportions. In Britain serious loss of land is suffered in parts of East Anglia and along the Humberside coast south of Flamborough Head, where the waves have the easy task of demolishing glacial deposits of sand, gravel and boulder clay. Since Roman times the Holderness coast has been worn back 4 or 5 km, and many villages and ancient landmarks have been swept away (Figure 23.31). During the last hundred years the average rate of cliff recession has been 1·5 or 2 m per year. The rate is not uniform, however, for severe storms and localized cliff falls are more destructive in a short time than the normal erosion of several average years.

As an extreme example of rapid coast erosion the great North Sea surge of 1953 may be cited. A *surge* is an abnormal rise of sea-level that occurs when spring tides, wind and storm waves all act together in the same direction in a more or less confined and shallow sea. Low-lying and easily erodable coasts are particularly vulnerable to these disastrous coincidences. On the last day of January 1953 very low atmospheric pressure over part of the North Sea caused a rise of sea-level which, though only 30 to 60 cm, meant the inflowing of immense volumes of additional water from the Atlantic, where the atmospheric pressure was higher. At the same time a violent gale from the north continuously drifted still greater quantities of surface water southwards. Whereas the usual flood tides hug the British coasts, this huge swell of water began to spread across the North Sea, particularly south of the latitude of the Firth of Forth. Only a limited amount of the excess could escape through the Straits of Dover. The peak of the growing surge had no alternative but to swing round towards the Netherlands. There the piling up of

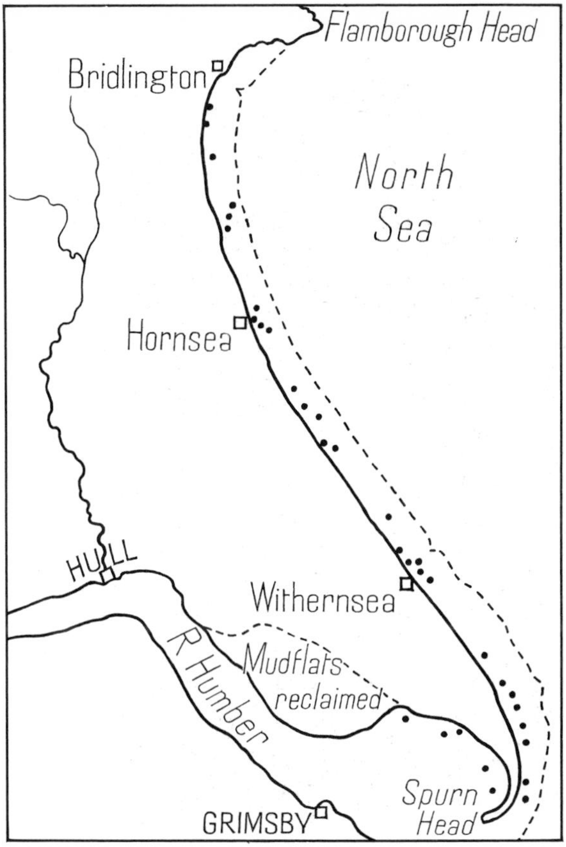


**Figure 23.31** Sketch map showing the loss of land and villages by marine erosion along the Holderness coast, Humberside. The broken line indicates the approximate position of the coast in Roman times; former settlements are shown by black dots. (*After T. Sheppard*)

the storm waters reached its maximum. The sea poured in through countless breaches in the coastal defences and the most ruinous floods of salt water for hundreds of years devastated the low-lying areas previously reclaimed from the sea.

Along the lowland coasts of East Anglia and the Thames estuary the disaster, though less widespread, was locally just as destructive. Around the time of high tide the theoretical height (which is normally quite reliable) was exceeded by 2·4 m in the Thames estuary (Figure 23.32) and by 2·7–3·4 m in the Netherlands. Apart from the breaching of protective barriers by enormous waves, erosion was most serious where low cliffs of glacial drift bordered the shore; 11 km south of Lowestoft, for example, a 7·6 km cliff was cut back nearly 11 m in two hours. Not far away, where the height of the cliff was less than 2 m, the frontal attack by undercutting was powerfully reinforced at high tide by the shattering blows of 4-metre breakers crashing down from above, and 27 m of land was lost to the sea in a fraction of a single night.

The great storm of February 1962 would have generated an even more serious surge had it not occurred during the moon's third quarter. As it was, the main atmospheric depression and the resulting winds directed the peak of the surge towards Hamburg and the adjoining coasts. The sea rose 6 m above the predicted height and the flooding of Hamburg was on an unprecedented scale.

## Shore Profiles

As indicated in Figure 23.27, sediment in transit across a wave-cut platform commonly tends to accumulate in the deeper water beyond, to form a wave-built terrace in continuity with the platform,

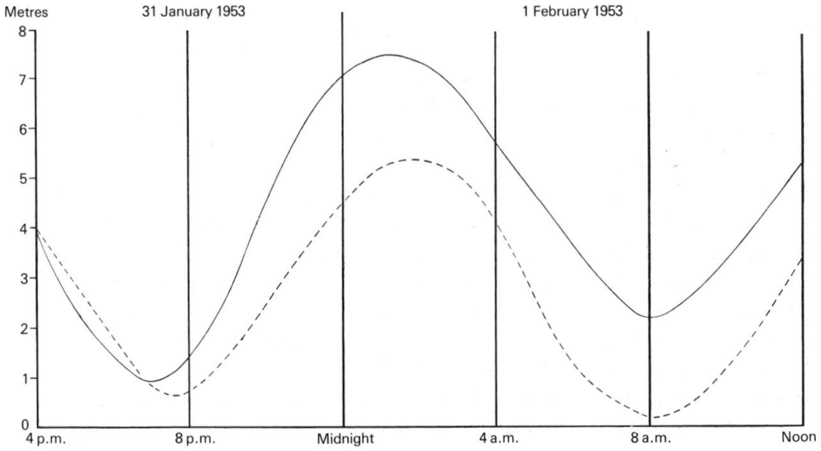


**Figure 23.32** The North Sea surge of 1953. Tidal curves for Southend, near the mouth of the Thames: predicted, broken line; and actual, continuous line (*After W. W. Williams*)

though generally with a less uniform surface than is suggested by the figure. The combined foreshore and offshore surface is a product of the joint action of erosion and deposition, each of which varies considerably from time to time and from place to place. The supply of sediment, for example, is irregular both in rate and distribution, since contributions are received from rivers and currents as well as from cliff wastage and platform abrasion—all widely varying sources of income. The processes concerned in the removal of sediment (also widely variable) are themselves largely controlled by the slope of the shore and its seaward continuation—that is to say, by the profile of the surface taken at right angles to the shore. A relatively steep slope favours destructive waves and removal of sediment from the landward side, so that the slope becomes less steep. Conversely a relatively gentle slope favours constructive waves and beach deposition on the landward side, so that the slope becomes steeper. The surface is therefore being continually modified, and in such a way that at each point it tends to acquire just the right slope to ensure that incoming supplies of sediment can be carried away just as fast as they are received. A profile so adjusted that this fluctuating state of balance is approximately achieved is called a *profile of equilibrium*. It is akin to the graded profile of a river, but along the shore the variable factors are still more numerous and difficult to evaluate. Theoretically speaking there must be a profile of equilibrium for any given set of conditions, but because of the alternating changes from tide to tide and from season to season, and the particularly violent changes from calm to storm, only a shortlived approximation to equilibrium can ever be achieved. The actual profile is continually being modified, especially along shores fringed with sand or shingle—loose materials which are easily moved.

The concept of the ideal profile of equilibrium has its uses however. At any given time the seaward slope at a given place may be either steeper or gentler than this ideal. Suppose AB in Figure 23.33 represents a relatively steep initial slope on a shore of submergence. In transforming this as nearly as possible into a profile of equilibrium CD, the waves cut a cliff-backed platform, while the resulting sediment is deposited as an offshore terrace, as already illustrated in Figure 23.27. Next, suppose *ab* in Figure 23.34 represents a relatively gentle initial slope, as when a broad valley floor becomes a bay by submergence. In transforming this into the profile of equilibrium *cd*, waves and currents build up a beach around the shores of the bay. Initial surfaces that may be almost flat are provided by the drowning of extensive alluvial plains, and on a still more widespread scale by emergence of the sea floor. In these cases, as described on p. 529, barrier beaches of sediment driven landwards are built up by the waves.

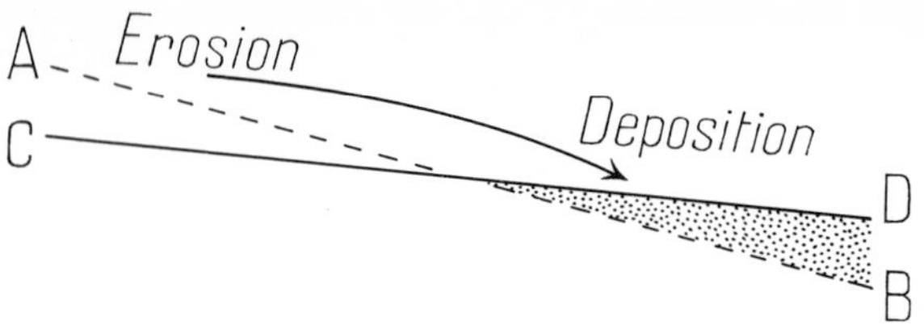


**Figure 23.33** The development of a profile of equilibrium CD from a more steeply sloping initial surface, AB

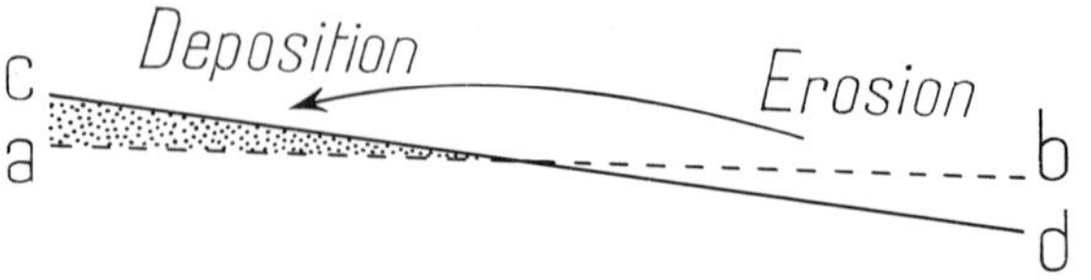


**Figure 23.34** The development of a profile of equilibrium cd from a more gently sloping initial surface, ab

Seasonal and other short-term periodic changes lead to alterations of the effects portrayed in Figures 23.33 and 23.34. Shingle accumulations under the cliffs or upbuilding of beaches by the constructive waves of, say, the summer, result in a slope like *cd*. In winter however this is likely to be reduced to the gentler slope CD. Destructive storm waves bring this about by transporting material seawards, so that beaches become thin or even disappear for a time. These considerations cover various intermediate cases: e.g. parts of a wave-cut platform may sometimes be covered with a temporary veneer of beach material, such as can be seen in Figures 23.21 and 23.28.

It must be clearly understood that Figures 23.33 and 23.34 are merely diagrams to illustrate the basic principles. The real profiles are commonly far from simple. Figure 23.35 attempts to depict a more realistic winter profile for a sandy shore.

## Beaches: Landward and Seaward Transport

Leaving the longshore drift of sediment to be

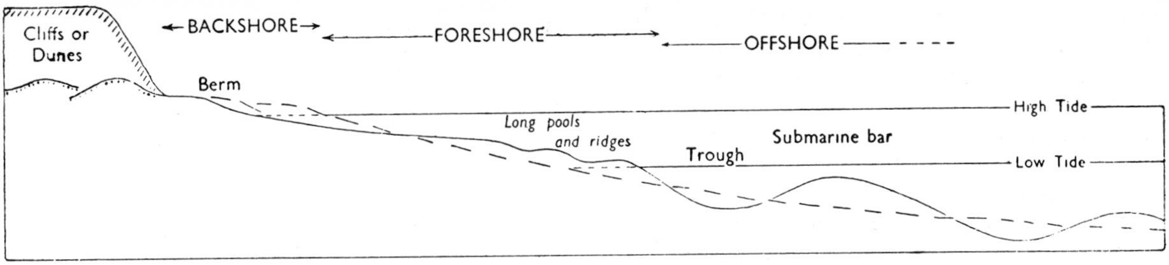


**Figure 23.35** Diagrammatic indication of the subdivisions of a shore, with characteristic beach and offshore profiles in summer (broken line, showing berms built up by constructive waves) and winter (full line, showing berms cut back by destructive storm waves)

considered later, let us now review the migration of beach and terrace material towards the coast and away from it. From the outset it has to be realized that the actual processes are so complicated that satisfactory explanations in terms of physical principles are still lacking for most of the phenomena observed. A good deal can be learned from scaled-down experimental models, but the application of these is limited by the fact that sediment sizes and hydrodynamical processes cannot simultaneously be scaled down in the same proportion as, say, the dimension of a beach to be investigated or a harbour to be designed. The preparations for D-Day, during the Second World War, emphasized the critical importance of knowing what really happens on the Normandy beaches under varying conditions. Since then many other types of shores have been studied over long periods of changing winds and currents. Offshore, direct observations are made by frogmen, and radioactive tracers added to pebbles and sand grains show how sediment migrates under natural conditions.

In a classic paper of 1931 W. V. Lewis distinguished between constructive and destructive waves (Figure 23.14). *Constructive waves* are those in which the backwash is relatively weak, so that it does not seriously obstruct the uprush from the next breaker. Such waves include those that give spilling breakers with a fairly long period: say about 8 to 10 seconds, corresponding to a frequency of about 6 to 8 per minute. The elliptical shape of the orbit ensures a relatively strong horizontal component of motion, i.e. a strong uprush. The long period gives the backwash time to return to the trough that helps to feed the next oncoming wave; moreover some of the backwash may sink into the shingle or sand. When a mixed assemblage of sediment is swept up the breach the coarser material is left stranded at the top. Taking the ebb and flow of the tides into account, the result of a long spell of constructive waves is the building-up of a beach headed with a broad *berm* of coarse sand or shingle. The berm commonly has a more or less steep front leading down to the more gently graded profile of the foreshore. This rapid change in slope tends to correspond with a break in the grain sizes of the sediment; e.g. if the berm is shingle the foreshore is likely to be sand, the latter tending to become progressively finer towards low-tide mark. The jump in grain size probably results from the accumulation of the coarser debris well above the reach of ordinary waves and tides, while the rest is dragged to and fro with the tides and subjected to long periods of attrition.

There are several definitions of the grain sizes of sedimentary fragments and particles, including beach materials. The British Standard recommendations are summarized in the table overleaf:

*Destructive waves* are those in which the backwash is strong. Such waves include plungers with a somewhat short period: say about 4 to 5 seconds, corresponding to a frequency of about 12 to 15 per minute. These waves break on the backwash from the previous breaker, and because the orbit is nearly circular the main component of motion is downwards. The plunging breaker churns up the beach material, scouring out pools and troughs like those shown in Figure 23.35. Part of the sediment may be swirled back a little way, building up temporary structures like the ridges and submarine bar also shown in Figure 23.35. Another part is carried forward in suspension by the surf of the uprush. The latter is relatively weak from the start and is further obstructed by having to advance over the returning backwash. Thus it is the backwash that is mainly in contact with the beach materials, and the dominant transport of sediment for the time being is seawards. Plunging breakers at high tide

Predominant Grain Sizes in Millimetres

| | *Coarse* | *Medium* | *Fine* |
|---|---|---|---|
| SHINGLE | | | |
| Boulders | 200 | | |
| Pebbles | 200–60 | | |
| Coarse gravel | 60–20 | | |
| GRAVEL | 60–20 | 20–6 | 6–2 |
| SAND | 2–0·6 | 0·6–0·2 | 0·2–0·06 |
| SILT | 0·06–0·02 | 0·02–0·006 | 0·006–0·002 |
| CLAY or MUD | | | < 0·002 |

attack the berm, carrying part of it away and leaving the rest with an unstable and weakened front, which is readily susceptible to further 'combing down' during the next spell of high tides.

However there is one constructive activity which, although rarely coming into operation, is of major importance over a long period. Violent splashes of spray from the more powerful plungers—generally during spring tides and winter storms—may fling pebbles and boulders to the highest part of the berm or even beyond, so building up a strong coastal defence against all but highly exceptional attacks from the sea. Chesil Beach (Figure 23.36 and p. 527) is one of the finest examples of this kind of protective barrier.

Figures 23.15 and 23.36 both show that the shoreward edge of the uprush is conspicuously indented at roughly equal intervals, instead of being smoothly curved, as might have been expected. This pattern reveals the presence of shallow troughs and low ridges on the upper part of the beach. These are rudimentary *beach cusps*, which in appropriate circumstances may become deeply scalloped features with a relief of 3 or 6 m, especially along the seaward slopes of a berm built largely of shingle overlying less permeable materials. Tapering ridges or 'horns' of shingle point seawards and are separated by embayments floored by beach-sediments of finer grain. It has been generally considered that the height and width of the cusps increase with the height of the waves, but this is only true within certain limits. M. S. Longuet-Higgins and D. W. Parkin have made a careful study of the conditions favouring cusp building along parts of the south coast of England, including Chesil Beach. They find that the dimensions of the cusps are more closely related to the length of the swash or uprush than to the wave height. Moreover, particularly high and powerful waves may wash over the horns and disperse the shingle into the depressions, so that the cusps are soon destroyed. Oblique waves have a similar effect. Consequently cusps are formed and maintained only by waves advancing with their crest-lines parallel to the shore. Under these conditions each advancing wave divides at the seaward extremities of the cusps, dropping the coarser shingle there, while water carrying the rest of its load floods each embayment from both sides, so that the backwash there is strong and results in deepening the embayment. Within the limiting conditions the horns are built up and the depressions are eroded. The writers mentioned above demonstrated this by conspicuously dyeing a number of pebbles occurring in cusp-depressions; it was found that the dyed pebbles were washed out and that those which returned were deposited on the horns. After the beach cusps have been smoothed out, the problem arises how they develop again. A possible explanation is that, starting with a smooth slope, the backwash acts like a sheet-flood and produces a series of rills or gullies. Of these the bigger ones grow at the expense of the smaller, until as they develop into cusp depressions they all come to have a similar size. The size changes from time to time in accordance with the length of the swash and the height and direction of the waves.

The contrasted conditions of constructive and destructive waves referred to in previous paragraphs tend to be correlated with quiet and stormy periods respectively, and therefore in a general way with summer and winter. The operation of 'combing down'—thinning or removal of beaches

**Figure 23.36** Chesil Beach, showing the highest part or 'berm' at Chesilton, which is 13 m above high-water level. (*Aero Pictorial Ltd*)

by destructive waves—is sooner or later followed by restoration of the material by constructive waves. In some places the alternation amounts to little more than a seasonal exchange of material between the berm at the top of the beach and the offshore submarine bar or bars. But beach removal by a catastrophe like the North Sea surge of 1953 naturally takes much longer to restore, if indeed the damage should not be beyond hope of recovery. On that occasion, for example, the beaches of Lincoln County were completely scoured away in one night, and it was not until 1959 that their former condition was restored.

Both thinning and thickening of beaches may be effectively assisted by winds in the appropriate directions. Gales blowing strongly offshore cause a surface drift of water away from the coast. To balance this an undercurrent is directed towards the shore and, though feeble, it co-operates with the waves of construction and so helps to reinforce the rather slow processes of beach growth and restoration. On the other hand, onshore gales not only whip up destructive waves, they also raise the hydraulic head of water along the shore and so strengthen the backwash and the outward currents that necessarily result from the piling-up of water against the coast. It should be noticed that the direction of sedimentary transport on the sea floor that results from onshore or offshore winds is opposite to that of the wind. The growth and maintenance of beaches are therefore favoured by offshore winds.

But if onshore winds blow some of the sand landwards to form sand dunes, the profile is altered so that the backwash is weakened. Constructive waves are then assisted in restoring the appropriate profile by washing in materials from the sea floor. Proof that this happens has already been presented in Figure 22.13. Offshore winds then further assist in beach maintenance by returning some of the dune sand. Where this has become part of the fluctuating balance of nature, artificial removal of the dune sand (e.g. for building etc.) can only promote increased erosion along that part of the coast.

Similarly, artificial interference with natural profiles by offshore dredging introduces a factor that will have dangerous results if it causes one-way changes outside the normal limits of thinning and thickening. Look back at Figure 23.33. Suppose an offshore terrace has been *deposited* at DB from sediment transported seawards to give the appropriate profile. *Erosion* on a highly abnormal scale is then suddenly introduced by dredging. This increases the slope from the shore and therefore promotes beach and cliff erosion at AC, so providing material to make good by deposition the loss from DB. The actual changes in the processes at work are mainly (*a*) refractive concentration of wave energy from the dredged area towards the coast; (*b*) strengthening of breakers; and (*c*) strengthening of backwash and seaward currents. In 1897, for example, dredging was begun off the coast north of Start Point (Figure 23.37) to furnish shingle for use in the harbour works then in progress at Plymouth. Thinning of the beach at Hallsands then also began to be a one-way process. By 1902 the beach had been lowered nearly 4 m and cliff erosion had become a steadily increasing menace. The licence to dredge offshore shingle was cancelled. But it was too late. Storm waves were already attacking the little fishing village of Hallsands, built on a post-glacial raised beach, and by 1917 only the walls of a few ruined cottages still remained.

The seaward currents which have been mentioned require some discussion. When breakers and onshore winds pile up water against the coast, the raising of sea-level must be counterbalanced by currents flowing away from the shore. These were formerly grouped together as the *undertow*, against the dangers of which bathers were warned, particularly in the deeper water over offshore depressions, where the seaward pull of the 'undertow' was recognized to be a source of peril, even to strong swimmers. It is now known that most of the water that would otherwise accumulate inside the zone of breakers finds an exit through occasional depressions or 'lows' in the offshore bars. These localized outflowing currents are much more than concentrations of undertow, since near the shore not only the bottom water flows outwards, but most of the water from sea floor to surface. Farther out these *rip currents* (Figure 23.38) weaken and

**Figure 23.37** Start Point, Devon, a promontory of Precambrian rocks. The remaining walls of the ruined cottages of the village of Hallsands can just be seen on the cliff edge in the background, on the left of the photograph, in front of the new white cottages. (*Fox Photos Ltd*)

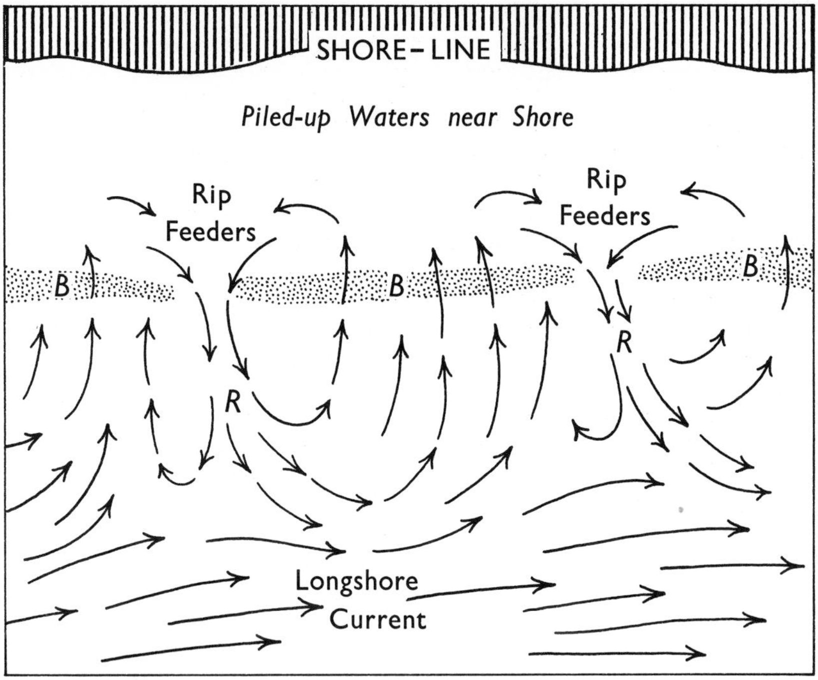


**Figure 23.38** Schematic indication of the relation of rip currents, R, to depressions in the submarine offshore bars, B. The currents here shown are purely diagrammatic, since at best they represent only a momentary stage in continuously changing systems. (*Simplified from diagrams by F. P. Shepard*)

die out in depth, but at the surface they may either continue for long distances or merge with other currents—longshore or shoreward—which they may encounter. Seen from the cliffs rip currents sometimes appear as long lanes of foamy or turbid water stretching far out to sea. From the shore they are most easily located by noticing places where the wave crests are lower than usual and the breakers interrupted and less active. A good example of the latter effect on a plunging breaker can be seen in Figure 23.16. Swimmers caught in a rip should not attempt to free themselves by swimming towards the shore against the current. By turning parallel to the shore and swimming at right angles to the rip they should soon escape the danger zone and pass into water where the currents are directed towards the shore.

It will have been noticed that rip currents are the only ones that are easily capable of transporting sediment seawards. They are far more effective than the average ebb-tide currents and the largely discredited undertow. At best the latter may extend the backwash movement a little way in special circumstances. In the other direction there is a much stronger tendency to return sediment landwards. Once ordinary waves 'feel bottom' the sea-floor sediment is stirred to and fro, and as the 'to' is greater than the 'fro' there is a slight shoreward movement on balance, which increases as the zone of breakers is approached, and increases again as the breakers turn into waves of translation. The landward movement is reinforced by inflowing tides, locally on a powerful scale.

Thus, on the whole, any migration of material seawards at the expense of the wasting land is extremely slow and is confined to the finer grades of sediment. So far as the British Isles are concerned, gain of land—mainly mud-flats, shingle banks, sand dunes and salt marshes—slightly exceeds the loss. But the latter gets more publicity, being destructive of land and property, and sometimes of life.

However, all along the shore freshly broken fragments are being steadily abraded and reduced in size to boulders, pebbles and grains. Some of the fine products of attrition may be entrapped for a time in the interstices of the coarser materials, but much is carried seawards in suspension by the rip currents. This is eventually deposited on the sea floor beyond the range of normal wave and current action. Thereafter, only a lowering of sea-level or a rise of the sea floor, or a sudden incorporation into a turbidity current (p. 551) is likely to disturb its rest.

Given a stable sea-level, one would expect the sediments on the sea floor to become finer and finer as the depth increases, the full outward sequence being shingle, sand, silt and mud. Corresponding to this sequence, in strata of the same geological age traced laderally, we often find conglomerates against old shore-lines, passing outwards through sandstones into shale or mudstones. But in Recent geological time the sea-level has *not* been stable; sea-level 20,000 years ago was over 100 m lower than today's (Figure 21.5). Nearly every part of the continental shelf has in turn been the coast-line of the past and berms and ridges of shingle, and the coarser sediments of glacier-fed rivers, have been left at various depths, to puzzle the oceanographer of former years who often found them far from the coast, where they were least expected. Obviously the present-day distribution of sediment on the continental shelves is highly abnormal, and must have been so throughout much of Pleistocene time. These 'abnormalities' however are consistent with what we have learnt already of the vicissitudes of Pleistocene history.

### Beaches: Transport along the Shore

Longshore drift of sediment is brought about in two ways: by beach drifting, due mainly to oblique waves on the foreshore, and farther out by transport due to longshore currents. When waves are driven obliquely against the coast by strong winds, debris is carried up the beach in a forward sweeping curve. The backwash may have a slight forward movement at the start, owing to the swing of the water as it turns, but otherwise it tends to drag the material down the steepest slope, until it is caught by the next wave, which repeats the process (Figure 23.39). By the continual repetition of this zigzag progress, sand and shingle are drifted along the shore.

The direction of drifting may vary from time to time, but along many shores there is a cumulative movement in one direction, controlled by the dominant or most effective winds. A subsidiary factor which aids or hinders beach drifting is the direction of the advancing flood tide. Farther out from the shore oblique winds and waves generate intermittent and fluctuating currents, since both have a component parallel to the general direction of the coast. These are the longshore currents, and again they are strengthened or weakened by the set of the tides. Longshore drift is by no means confined to the immediate shore-line.

In the English Channel the dominant winds and the inflowing tides both come from the south-west, and the cumulative direction of drift is up-Channel and though the Straits of Dover almost to the estuary of the Thames. Down the east coast of Britain the drift is mainly southwards, the dominant winds being from the north-east and the flood tide advancing from the north. There are a few exceptions: for example, the north coast of Norfolk, west of Cromer, stands athwart the main drift which, striking the coast obliquely, is deflected westward towards the Wash.

Wherever it is thought desirable to check the drift of sand and shingle, barriers known as *groynes* are erected across the beach. On the side from which the beach drift comes, sediment accumulates, and sometimes reaches a height a metre or so above the beach level on the other side, where sediment is being washed away, to be retained in turn by the next groyne (Figures 23.40 and 23.41). Groynes, like other structures designed to interrupt the natural flow of sediment, such as moles and piers constructed to protect harbours and river mouths, interfere with the replenishment of beaches farther along the coast by robbing them of their former sources of supply. Drifting does not stop along these starved beaches until they disappear. The coast behind and beyond them, which they had been protecting from erosion, is thereupon exposed to increasing attack by the waves. To the east of Brighton and Newhaven, for example, the wastage of the Chalk cliffs (Figure 23.58) has greatly increased since the erection of groynes.

In the Netherlands these problems are of vital importance. A supply of sand at the appropriate rate is found to give better coastal protection than groynes and sea-walls. To take a simple case: if a river mouth lies across the drift direction it was formerly protected by groynes on the up-drift

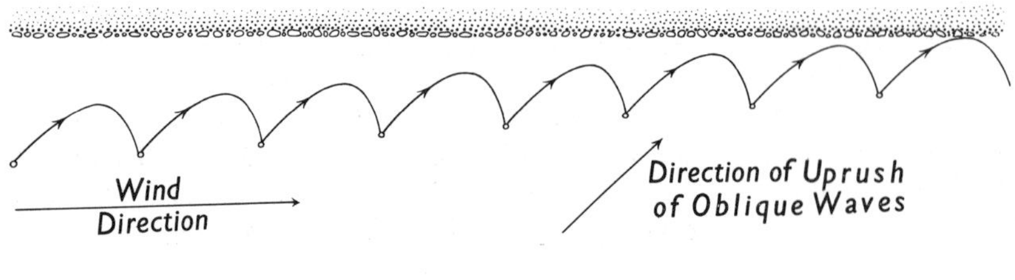


**Figure 23.39** Diagram to illustrate beach drifting, showing the path followed along a sloping beach by a pebble or sand grain under the influence of the uprush and backwash of successive oblique waves during an advancing tide

**Figure 23.40** Beach drift impeded by groynes at Eastbourne, NE of Beachy Head, Sussex. The direction of drifting is to the NE i.e. up-Channel. (*Aerofilms Ltd*)

**Figure 23.41** Beach drift impeded by groynes, St Margaret's Bay, NE of Dover, showing that the direction of drifting continues up-Channel through and beyond the Straits of Dover (*Aerofilms Ltd*)

side. This however weakened the coastal defences on the other side of the river mouth. Now, using suction pumps, sand on the up-drift side is delivered into pipe-lines through which it is conveyed beneath the river to the down-drift side. There, having left the river mouth clear, it provides the coast beyond with its natural income of sediment to balance the losses by beach drifting.

## Sand and Shingle Spits and Bars

Where shore drift is in progress along an indented coast, spits and bars are constructed as well as beaches. Where the coast turns in at the entrance to a bay or estuary the sediment transported by beach drift and longshore currents is carried more or less straight on, and some of the coarser material is dropped into the deeper water beyond. The shoal thus started is gradually raised into an embankment. This grows in height by additions from its landward attachment until a ridge or mound of sand or shingle is built above sea-level in continuity with the shore from which the additions are contributed. The ridge increases in length by successive additions to its end, like a railway embankment, until waves or currents from some other quarter limit its forward growth.

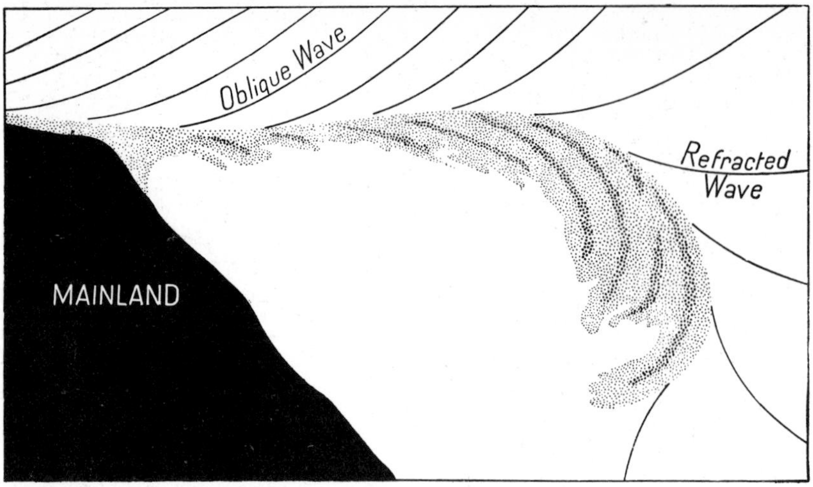


**Figure 23.42** Diagram to illustrate the development of a hooked spit by the refraction of oblique waves

If the ridge terminates in open water it is called a *spit*. Storm waves roll and throw material over to the sheltered side, especially when they approach squarely. Spits thus tend to migrate landwards, often becoming curved in the process. Curvature is also brought about by the tendency of oblique waves to swing round the end (i.e. to be refracted) in places where the sea floor beyond slopes rapidly into deeper water. A spit may thus be developed into a *hook*, as indicated in Figure 23.42. Cross currents may assist or modify hook formation, and it is usual for spits to lengthen by the addition of a number of successive hooks. Hooked spits of simpler structure occur in large lakes where tides are absent and currents are negligible (Figure 23.43). This suggests that the dominant winds and waves are the agents essentially responsible for the curving of spits.

A good example of a curved spit is Spurn Head (Figure 23.44), which extends into the Humber in streamlined continuity with the Holderness coast. The latter is almost everywhere fringed with sand and shingle that drifts steadily from north to south, fresh supplies being continually furnished by the rapid erosion of the coast. Nearly all (97 per cent) the transported material is carried beyond Spurn Head, cumbering the estuary with shoals on its way towards the Lincolnshire coast, where most of it is added to the seaward-growing coastal flats.

Southward drift is also very active along the east coast of Norfolk and Suffolk. Ten centuries ago the Yarmouth sands had already spread across the estuary of the Yare, forming an obstruction which deflected the river towards the south (Figure 23.45). The spit then continued to grow southwards, hugging the coast as closely as possible, with the river confined between itself and the mainland. By 1347 the end of the spit and the outlet of the river had reached Lowestoft. Since 1560, however, an artificial outlet has been maintained at Gorleston, where the spit now terminates. The truncated part has long ago drifted south. At Aldeburgh, half-way between Lowestoft and Harwich, the longest spit on the east coast has similarly diverted the outlet of the Alde (Figures 23.46 and 23.47).

**Figure 23.43** Hooked spit, Duck Point, Grand Traverse Bay, Lake Michigan (*I. C. Russell, U.S. Geological Survey*)

Figure 23.44 Spurn Head, built by beach drifting into the estuary of the Humber, in continuation of the Holderness coast, Humberside; looking NNE (cf. Fig. 23.31) (*J. K. St Joseph*)

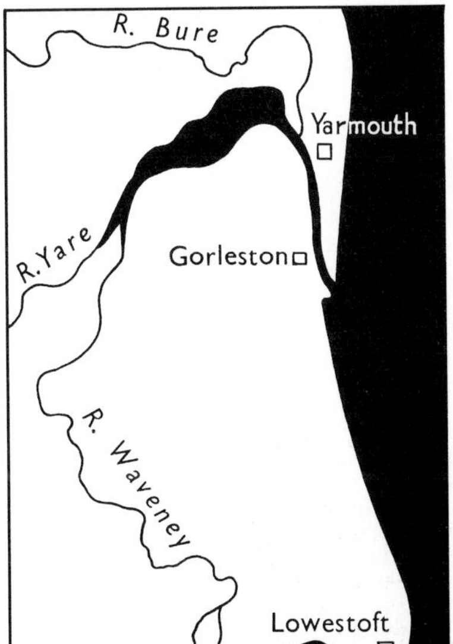


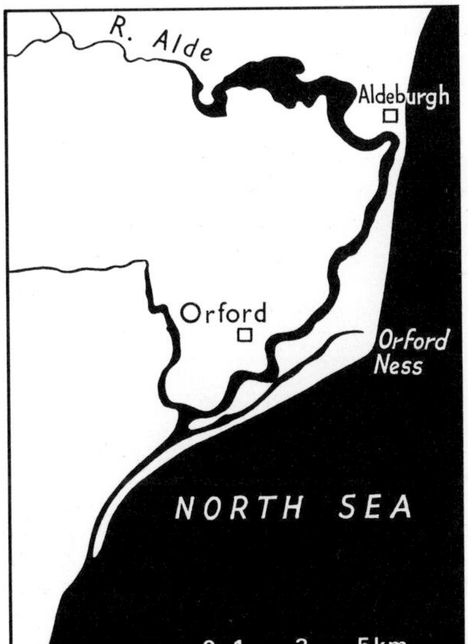


**Figure 23.45** River Yare, Norfolk
**Figure 23.46** River Alde, Suffolk
Examples of river deflection in East Anglia by the southerly extension of sand and shingle spits. Both drawn to same scale

Figure 23.47 Deflection of the River Alde south of Aldeburgh by the southerly growth of Orford spit, Suffolk (*Aerofilms Limited*)

A *bar* or *barrier beach* extends from one headland to another, or nearly so. When the bay inside is completely enclosed it becomes a marsh, or, if it receives streams from the mainland, a shore-line lake. More usually, however, outflowing drainage escapes through a deep and narrow channel kept open by vigorous tidal scour. Between Gdansk and Memel (Klaipéda), on the south-east Baltic coast, there are two very long bay-mouth bars, surmounted by sand dunes, with extensive tidal lagoons on the landward side (Figure 23.48).

A bar connecting an island to the mainland or to another island is called a *tombolo*. The name comes from an Italian example of a double tombolo, 130 km north-west of Rome, where a high rocky island is tied to the adjoining coast by a pair of shingle banks with a broad lagoon between.

By far the most impressive barrier beach of shingle in Britain is Chesil Beach (Figure 23.49).

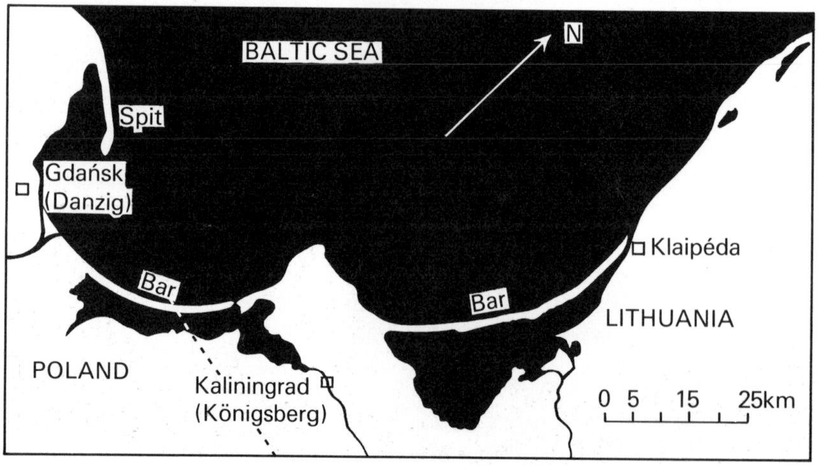


**Figure 23.48** A spit N of Gdansk (Danzig) and two smoothly curved bay-mouth bars enclosing broad lagoons along the SE Baltic coast of Poland and Lithuania (Königsberg and Memel are now called Kaliningrad and Klaipéda by the U.S.S.R.)

For 10 km south-east of Bridport the beach fringes the shore. Near Abbotsbury the shore recedes and for the next 13 km the beach continues in front of the tidal lagoon of the Fleet as a bar well over 6 m in height (Figure 23.50). Finally, becoming a tombolo, it crosses 3 km of sea to the Isle of Portland, which is thus tied to the mainland (Figure 23.36). Chesil Beach is a composite structure, its shingle having accumulated from local sources as well as by drift from each end. At the north-west end there are pebbles or rocks from Cornwall and Devon; at the south-east end the shingle includes larger pebbles, supplied from the Portland promontory. Between the extremities the pebbles are mainly relics from an eroded land, now submerged, that formerly stretched in front of the Fleet when the sea-level was much lower. Although towards the Portland end the beach rises to the quite exceptional height of 13 m, the sea sometimes bursts over it during storms and pours through breaches into the low-lying area beyond. Since two villages were demolished in 1824, the worst disaster of this kind occurred late in 1942, when the railway between Portland and Weymouth was partly washed away and the lower parts of Portland itself were seriously flooded. The beach is in fact very slowly migrating towards the Fleet, as a cumulative result of its alignment almost exactly at right angles both to the direction of the average dominant winds and to that of the maximum fetch (Figure 23.49). This alignment is one of great stability, since the beach ridges squarely face the more powerful storm waves and longshore drift is reduced to a minimum.

Because of the tendency of spits tied to the mainland to continue in the direction of the latter, or to curve inwards towards the coast if the water is shallowing in that direction, as it usually is, it is exceptional for wave-built structures to swing sharply away from the general trend of the coastline. When they do so the change of direction may be due to shallowing of water seawards, e.g. towards an island or shoal. But in some examples the change appears to be at least in part a reaction to the average direction from which the ridge-building storm waves approach. This in turn

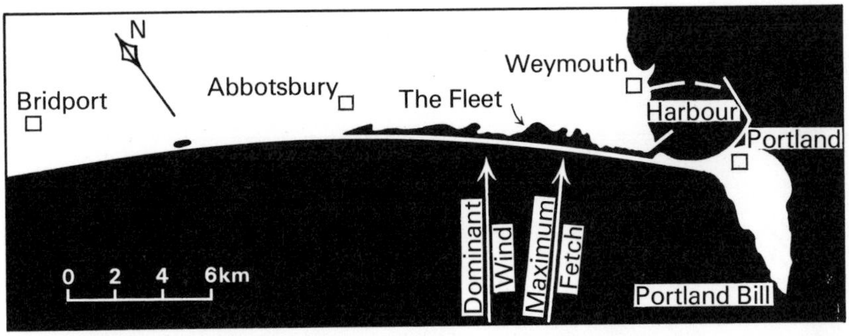


**Figure 23.49** Map of Chesil Beach, Dorset

**Figure 23.50** Chesil Beach, Dorset, viewed from the West Cliff, Portland (see also Fig. 23.36) (*Institute of Geological Sciences*)

depends on the directions of the dominant winds and of the maximum fetch. But the particular mechanism involved often remains a puzzle, as in the case of Dungeness.

Dungeness is a triangular outgrowth of shingle ridges forming the seaward margin of Romney Marsh, which, now silted up and reclaimed, was once a broad bay between the end of the cliffs near Fairlight (east of Hastings) and Hythe (Figure 23.51). During the Danish invasion of the year 893 a fleet of 250 vessels sailed past Lydd to Appledore. An old map of about 750 shows that Lydd was then at least 1·6 km to the west of the sea, suggesting that the invaders took advantage of a gap in the shingle barrier, possibly the outlet of the river Rother, which at that time probably reached the sea near the New Romney of today. In 750 Dungeness was already clearly outlined, as indicated on Figure 23.51, and can be recognized as having had, then as now, the outward form of the peculiar type of construction known as a *cupsate foreland.*

Most cupsate forelands (e.g. Cape Hatteras, Figure 23.52) are the points of accumulation where two curved spits happen to meet. In accordance with this view Dungeness was originally supposed to represent the meeting-place of up-Channel beach drift with that coming down the east coast and through the Straits of Dover. But this idea is easily shown to be untenable. The two seaward sides of the triangle behave quite differently: while the southern side has been retreating before the attack of the waves, the eastern side has been conspicuously built out. The foreland itself has advanced nearly 2·5 km during the last twelve centuries. Moreover, the piling of shingle against the groynes seen in Figure 23.41 shows that up-Channel drifting remains dominant beyond the Straits of Dover and continues along the east coast of Kent to accumulate between Deal and the Isle of Thanet—an 'isle' which in consequence has ceased to justify its ancient name.

The oldest storm ridges now to be seen are those on the west. W. V. Lewis supposed the first stage to be represented by a spit or bar that extended from the end of the cliffs as they were at that time, e.g. from A somewhere south of Fairlight to A′

**Figure 23.51** Map of Dungeness showing the pattern and sequence of the shingle ridges up to 1957. The outline of the shingle revealed by the 1967–8 Ordnance Survey suggests that Dungeness is still extending eastward. The scale is about 6 times that of the regional map on the left. The latter shows Romney Marsh and (horizontal shading) the higher ground bordering it on the north and west. It also illustrates the Lewis (1932) hypothesis of the development of Dungeness as described in the text. (*After W. V. Lewis and W. G. V. Balchin*)

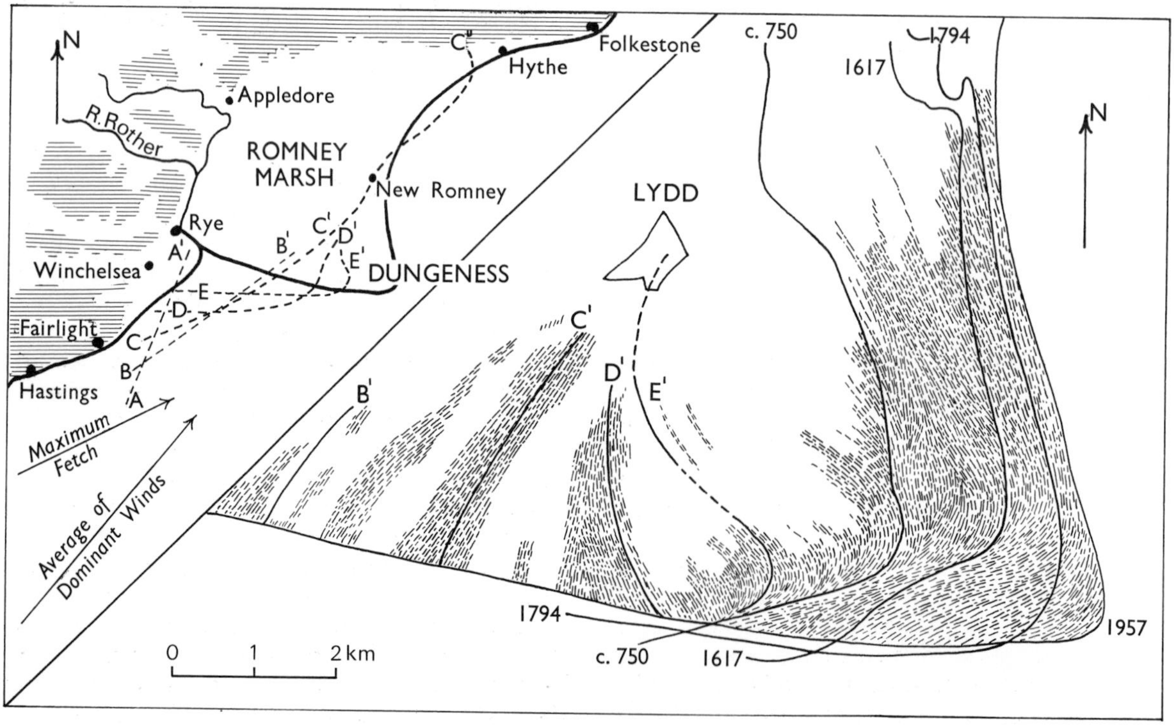

near Rye (see small-scale mp of Figure 23.51). At later stages this direction AA′ began to swing round, through BB′, CC′ and so on, towards the present-day direction, which is much more nearly at right angles to the dominant direction of advancing storm waves, and so is approaching stability. The oldest spit probably dates from Neolithic times, when the Flandrian rise of sea-level reached its maximum. Later there was slight emergence, and while the cliffs were being eroded back the lengthening spit became a bar that may have extended from CC′ through the shingle ridges near New Romney to those near Hythe marked C″, with a tidal outlet somewhere for the outflow of the Rother and other inland streams. But since no ridges are now exposed between Lydd and New Romney, it is equally possible that D′ or E′ were connected with C″. One of these stages, most probably near that marked E′, may have been reached about the time when Julius Caesar landed near Hythe. From then a series of successive stages can be dated from old maps, leading up to the latest survey of 1957.

Unfortunately the pre-Roman history outlined above is purely hypothetical. The numerous explanations given for both earlier and later developments fail to carry conviction, particularly as there are innumerable bars that have *not* developed into structures like Dungeness. Guilcher indeed can only conclude that Dungeness is one of the 'true wonders of nature'. And so it will remain until the missing factors concerned in its evolution have been discovered.

## Offshore Bars and Barrier Islands

The Atlantic and Gulf coasts of the United States are bordered by long stretches of barrier beaches which are separated from the mainland by lagoons or expanses of sea, except where they are locally tied to headlands (Figure 23.52). These are known as *offshore bars* or, if they should be discontinuous at both ends, as *barrier islands*. Their mode of origin was not easy to understand while they were thought to be diagnostic features of emergent shore-lines. A nearly flat coastal plain passing into a wide offshore zone of shallow water was first envisaged. In accordance with the principle illustrated in Figure 23.34, loose sediment from the sea floor would then be driven landwards in order to restore an appropriate profile of equilibrium. To bring this about the waves would begin

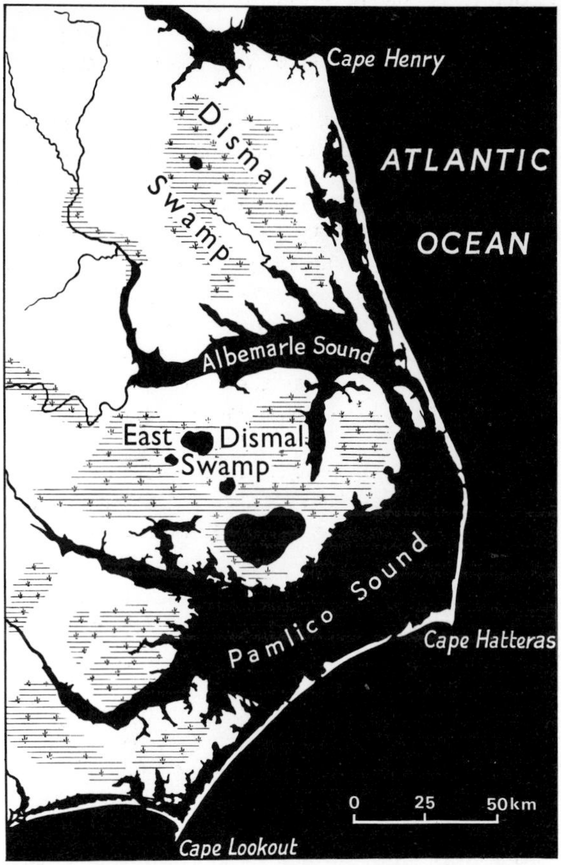


**Figure 23.52** Offshore bars and barrier islands, with lagoons, sounds and swamps behind, along the coast of North Carolina

to drag the bottom several kilometres from the low-lying mainland, and as they would lose much energy in crossing the shallows, the coarser parts of the stirred-up sediment would be dropped before the shore was reached. The foundations of a mound or shoal could thus be laid offshore, but then the difficulty remains to explain how it could be built up into a bar with its crest above sea-level. As soon as any such submarine feature reaches a certain height, it is swept by breakers and material is transferred from the seaward to the landward side. The submarine bar therefore advances towards the shore. But it cannot become other than a submarine bar until it finally joins the shore or a headland, by which time it is a beach or a bay barrier, and is no longer an 'offshore' feature.

Figure 23.52 shows that the coast depicted is clearly one of submergence—not emergence. A borehole drilled at Cape Hatteras passed through about 3000 m of marine sediments (from Recent to

**Figure 23.53** Miami, Florida, U.S.A., with its offshore bar and narrow lagoon (*Fairchild Aerial Surveys Inc.*)

Lower Cretaceous) before penetrating a pre-Cretaceous land surface of Precambrian rocks. This offshore belt has been subsiding almost continuously for well over 100 million years. But keeping in mind the Pleistocene changes of sea-level the difficulties can be overcome. 20,000 years ago the shore-line lay far out on the continental shelf, where it was probably margined by beaches with storm ridges already well above sea-level. As the latter rose these ancestral beaches continued to advance, maintaining contact with the coast where the hinterland was high enough, but losing contact wherever the surface on the landward side happened to be below the sea-level of the time—as in many places it is today (Figure 23.52).

The famous Florida beaches of Daytona, Palm Beach and Miami are offshore bars which are now nowhere far from the land (Figure 23.53) and in many places make contact with it. One of a long series of offshore bars and coastal barriers along the south-east coast of Iceland is shown in Figure 23.54. This coast is one of some complexity because it is being rapidly built out by glacifluvial outwash from the glaciers of Vatnajökull (see Figures 3.5 and 12.9). The offshore bars represent the return from the sea floor of some of the waste delivered from the land when sea-level was lower.

## Classification of Coasts

Valentin (1951) classified coasts, as *retreating* and *advancing* coasts; this has the practical advantage that it is primarily based on the criterion of whether land is being gained or lost. Loss can result from either submergence or erosion, and coastal retreat tends to reach its maximum when these two act in combination. Similarly gain of land can result from either emergence or outbuilding by constructive processes, such as deposition, and coastal advance tends to reach its maximum when these two co-operate. But there are many coasts in which the operative processes are in opposition; it is then the dominant one that determines whether on balance, i.e. over a suitably long period of years, the coast advances or retreats.

Valentin expresses his classification graphically by a diagram of which Figure 23.55 is a modification. The vertical axis through O represents emergence in the positive (upward) direction; submergence in the negative (downward) direction. The rate of coastal advance or retreat is proportional to distance from O. Similarly the horizontal axis represents outbuilding in the

**Figure 23.54** Offshore barrier beach, with lagoons and outwash plains from Vatnajökull, behind Lón Bay, SE Iceland (*L. Hawkes*)

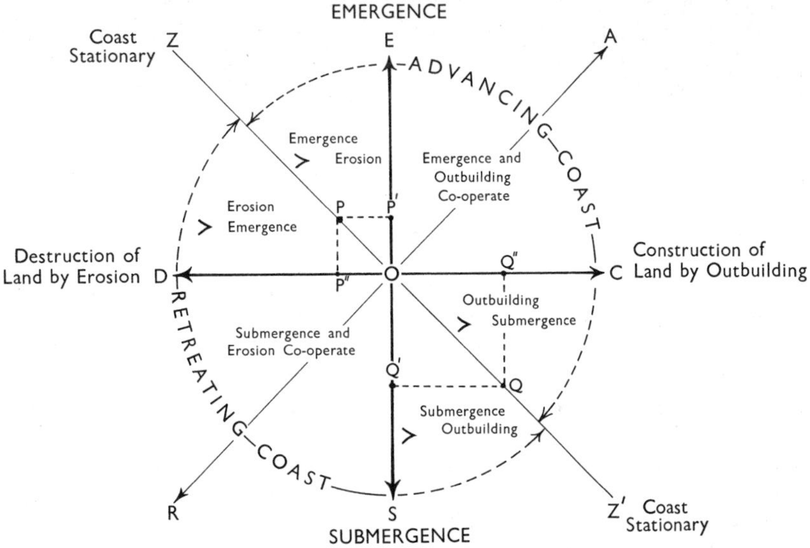


**Figure 23.55** Graphical representation of H. Valentin's classification of coasts

positive direction (to the right); erosion in the negative direction (to the left). Advance and retreat reach their maxima at points such as A and R on the diagonal AOR. The diagonal at right angles, ZOZ′, represents a 'zero' line, along which points like P have equal components of gain by emergence, P′, and loss by erosion, P″; or along which points like Q have equal components of loss by submergence, Q′, and gain by outbuilding, Q″. ZOZ′ represents balanced conditions in which the coast remains stationary on average over the period concerned.

The importance of taking time into consideration is illustrated in a remarkable way by the intermittent outward growth of the coastal plain near Cayenne in French Guiana. In the tropics mangroves spread vigorously over the tidal zone of low muddy shores—particularly of estuaries—forming an impenetrable thicket of interwoven roots which provide shelter for fearsome hordes of organisms. Practically all the sediment washed in by the rising tide is trapped and the mangrove swamp therefore continues to extend itself offshore across what had previously been shallow sea. In French Guiana this outward growth has added 10 to 16 km to parts of the coastal plain since the first surveys were made in 1751. But progress has not been uninterrupted. Over the course of the years a mysterious and still unexplained periodicity has been recognized. The north equatorial current flows from the south-east almost parallel to the coast, bringing from the mouth of the Amazon the silt and mud that is trapped by the mangroves. The curious feature here is that outward growth proceeds for eleven years, most rapidly during the middle of the period; but then the outermost mangroves begin to wither away, leaving the unprotected mud to be removed by the backwash from the waves and carried farther along the coast. This phase also waxes and wanes over a period of eleven years. The periodicity has rarely varied by as much as a year. There was a phase of silting up and coastal advance from 1947 to 1958, but by 1962 the sea was gaining at the expense of the fringes of the mangrove swamps. Possibly the alternation may have something to do with the cyclic changes in the waters of the equatorial current. These might be variations of salinity, or of the abundance of trace elements acting like fertilizers; or periodic 'bloomings' of minor organisms injurious to the mangroves. But why the periodicity? So much remains to be discovered that it would be futile to guess.

### Retreating Coasts

The widespread occurrence of coasts of submergence at the present time has already been emphasized with numerous illustrations (e.g. Figures 23.1, 23.4, 23.56, 23.60). On a regional scale one of the most remarkable examples of a submerged land surface, with drowned valleys that can still be recognized on the sea floor, is that which formerly connected Malaya with the islands of Sumatra, Java and Borneo, and innumerable smaller ones. The betrunked and dismembered modern valleys can be traced down to depths of about 100 m in the South China and Java Seas, and as shown in Figure 23.56 they join up into a small number of river systems. Such land connections must have existed several times during the Pleistocene.

Another important land bridge connected Asia and North America during phases of lowered sea-level. Siberia and Alaska were joined across the Bering Straits when the sea was 46 metres or more below its present level: that is, up to about 10,000 years ago on the latest occasion. It has generally been supposed that man, as well as a great variety of plants and animals, first migrated into North America by this route. Now that it is known that the deepest parts of the straits and the Chukchi Sea to the north are no more than 55 m beneath the present surface, there is no doubt that the land bridge—of tundra type, but not glaciated—was well over 1600 km wide for long intervals of Pleistocene time.

The first effect of marine erosion on a newly formed coast of submergence is usually to intensify the initial irregularities of outline (see Figure 23.3). Where the rocks vary in resistance the

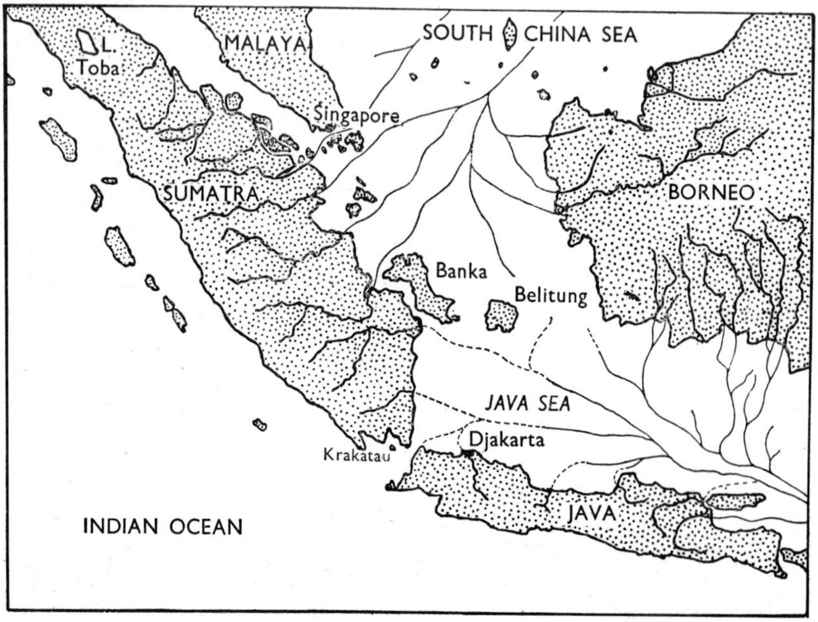


**Figure 23.56** Map showing the drowned river system of SE Asia, still identifiable across the shallower parts of the Java and South China Seas (*After J. H. F. Umbgrove*)

waves pick out all the differences. Soft and fissured rocks are worn back into coves and bays, while the harder and more massive rocks stand out conspicuously. The Dorset coast north-east of Portland shows this process in active operation. Here there is a long coastal strip of soft Lower Cretaceous beds, backed on the landward side by an upland of Chalk, and formerly protected from the sea by a continuous rampart of hard upfolded Jurassic limestones. The sea has breached the latter in places and excavated the softer rocks behind, until slowed down by the more resistant Chalk. The Stair Hole (Figure 1.5) illustrates the breaching stage. Lulworth Cove (Figure 23.57) is a beautiful example of a bay scooped out by waves and rip currents, and backed by Chalk.

Eventually, however, erosion and deposition co-operate to smooth out the intricate outlines of a youthful shore-line. Bay-head beaches and deltas, fed by lateral currents from the headlands and by additions from streams, locally extend the shore in the seaward direction. Meanwhile the headlands recede before the concentrated attack of the waves, and stretches of cliff become longer and straighter (Figure 23.58). Spits and bars bridge across the inlets where oblique waves favour beach difting, and these structures generally advance landwards to keep in line with the retreating cliffs. Embayments protected in this way are rapidly shoaled up by contributions from the land, aided by wind-blown sand and the growth of salt-marsh vegetation. Finally the bars and sand-dunes encroach on, and coalesce with, the lagoon and marsh deposits, and coast of smoothly flowing outline is evolved. Theoretically it might be supposed that if this cycle of marine erosion were to continue without interruption, the shore-line would slowly retreat as a whole. Lengthening lines of cliffs would join up at the expense of the deposits representing the embayments that originally

**Figure 23.57** Lulworth Cove, a beautifully curved bay scooped out by the sea after the breaching of the resistant barrier of Portland beds still forming the cliffs on both sides of the entrance to the Cove. (*Aerofilms Ltd*)

**Figure 23.58** The Seven Sisters, SSE of Newhaven Sussex; looking towards Beachy Head (Fig. 6.14) from the wave-cut platform and shingle beach of Cuckmere Haven. Coastal erosion has produced a continuous line of cliffs with the valleys of former tributary streams left hanging at different heights according to their original gradients. (*Institute of Geological Sciences*)

separated the headlands. But as we have seen, the latest rise of sea-level has been so recent that in most places only the earlier stages of the cycle have been completed. The later stage of slow retreat outlined above has been reached only locally and it is doubtful if it could continue far, given a stable sea-level. Retreat of cliffs implies a complementary broadening of the wave-cut platform in front. This in turn reduces the energy of the waves that reach the cliffs until it becomes negligible. Coastal retreat by marine erosion is probably a limited process in the absence of continued submergence.

## Advancing Coasts

Typical coasts of emergence are not common at present. Finland for example is steadily rising isostatically, and the south and south-west coasts are fringed with tens of thousands of islands as a result of the emergence of the higher parts of a hummocky ice-moulded surface (Figure 23.2). But although this archipelago owes its existence to emergence it is merely part of a drowned land surface that has become progressively less drowned than it was a few thousand years ago. Nevertheless the coasts are advancing because the emerging crystalline rocks are massive, and have been streamlined in form by previous glaciation, so that they resist erosion and are not easily cliffed. Similarly fjords, though fundamentally submergence features due to deep glacial erosion and invasion of the sea, are mostly along coasts that are still actively rising. The structurally weaker belts of rock having been already gouged out, marine erosion is extremely slow, and on the whole land is being gained.

A really typical shore-line of emergence is one in which the sea floor with its veneer of sediments has been uplifted to form a nearly flat coastal plain with a smoothly flowing shore-line margined by widespread stretches of shallow water. New Zealand furnishes ideal examples of uplifted sea floors (Figure 23.59), which are now fertile coastal plains that have lost little by cliffing, as well as others which illustrate submergence (Figure 23.60). Similar emergent plains occur along the south coast of Honshu (Japan), although to the

**Figure 23.59** The Motunau coast of North Canterbury. The coastal plain has an almost undissected cover of recent shelly deposits, indicating emergence by relatively rapid upheaval. Indented cliffs of older rocks are seen at the bottom right (*V. C. Browne*)

**Figure 23.60** A typical coast of recent submergence, Queen Charlotte Sound, Marlborough (*R. H. Clark*)

west submergence has reduced a former landscape of hills to an offshore archipelago. At the present time, however, this and much of the Honshu coast is rising at rates ranging between 13 and 20 cm per century. Corresponding figures for Taiwan (Formosa) are 18 cm, and for Hong Kong 15 cm. These rates are slow compared with the post-glacial rise of sea-level, which averaged about 80 to 85 cm per century up to 6000 years ago. It follows that where the sea floor has become a coastal plain by uplift the rise must have been both considerable and relatively rapid.

Figure 14.3 shows an uplifted sea floor on the south-east coast of Iceland, backed with screes which for the time being are protecting the cliffs against renewed attack by the sea. But for the uplift the scree materials would have been abraded and removed soon after their fall, while at the same time being used by the sea as ammunition in undermining the cliffs.

Examples of coastal advance by sedimentary outbuilding include the accumulation of beach-drifted materials (e.g. along the shore of Lincolnshire); the outward growth of storm ridges into forelands (e.g. Dungeness); offshore deposition of glacifluvial outwash (e.g. Iceland); and retention of washed-in mud by the intergrown roots of mangrove-swamps (e.g. French Guiana and Surinam coast of South America). Deltas also come here and many of them exemplify the case where new land continues to be built out despite the opposing submergence due to the isostatic depression that results from the growing weight of sediment.

Outbuilding of coasts by lavas and other volcanic products that flow or fall into the sea is not uncommon when eruptions occur on the seaward side of volcanoes situated near the shore (Figure 23.61). Advancing volcanic coasts are naturally

**Figure 23.61** Coastal advance by formation of a 'delta' of lava; Barcena (also known as El Boquerón) at the S end of San Benedicto Island, Mexico. This new volcano originated on 1 August 1952, as a cinder cone which reached a height of 300 m in 12 days. In November lava flowed into the crater. In December a fissure opened at the base, which was already cliffed, and lava flowed out into the sea until February 1953. Since then only fumarole activity has been reported. (*Paul Popper Ltd*)

most characteristic of growing island volcanoes that have succeeded in establishing their craters above sea-level. The eruption on Tristan da Cunha in 1961, which led to the evacuation of the settlement, culminated in the seaward flow of lava from the central cone. Before the end of the year the lava field had extended the coast-line by about 365 m, with a nearly vertical front in 15 m of water. Older volcanoes, like those of Hawaii, present many examples of lobe-shaped extensions of the coast where voluminous outpourings of lava have advanced over the sea floor.

On the other hand many more volcanic coasts are of the retreating type: (*a*) when the sea occupies a newly formed caldera (Figures 12.41 and 12.47); and (*b*) when an island volcano becomes extinct and no longer makes good the land it loses by marine erosion and isostatic subsidence. This item is referred to here, although it belongs to the class of retreating coasts, because it serves to introduce atolls and coral-reef coasts which, in tropical seas, are important members of the advancing class of coasts. The coral reefs of today have grown upwards and outwards during conditions of submergence. Since coral communities can grow upwards at a rate of about 300 cm per century, they can easily keep pace with all normal rates of either subsidence or rise of sea-level. Many atolls are built on foundations provided by ancient volcanoes that have subsided to considerable depths while the corals and their associates were maintaining a living front at sea-level. Coral reefs are in fact involved in so great a variety of the problems of oceanic geology that—notwithstanding the fascination of their coastal scenery—they can be more appropriately dealt with in the chapter that follows.

## Selected References


BASCOM, W., 1959, 'Ocean waves', *Scientific American*, vol. 201, pp. 75–84.

— 1960, 'Beaches', ibid., vol. 203, pp. 81–94.

COTTON, C. A., 1954, 'Tests of a German non-cyclic theory and classification of coasts', *Geographical Journal*, vol. 120, pp. 353–61.

— 'Deductive morphology and genetic classification of coasts', *The Scientific Monthly*, vol. 78, pp. 163–81.

GRIEVE, HILDA, 1959, *The Great Tide: The Story of the 1953 Flood Disaster in Essex*, County Council of Essex, Chelmsford.

GUILCHER, A., and KING, C. A. M., 'Spits, tombolos and tidal marshes in Connemara and West Ireland', *Proceedings of the Royal Irish Academy*, vol. 61B, pp. 283–338.

HAAG, W. G., 1962, 'The Bering Strait land bridge', *Scientific American*, vol. 206, pp. 112–23.

KING, C. A. M., 1972, *Beaches and Coasts*, Edward Arnold, London.

LEWIS, W. V., 1931, 'Effect of wave incidence on configuration of a shingle beach', *Geographical Journal*, vol. 78, p. 131.

LEWIS, W. V., 1932, 'The formation of Dungeness Foreland', ibid., vol. 80, pp. 309–24.

— 1938, 'Evolution of shoreline curves', *Proceedings of the Geologists' Association*, London, vol. 49, pp. 107–27.

LEWIS, W. V., and BALCHIN, W. G. V., 'Past sea-levels at Dungeness', *Geographical Journal*, vol. 96, pp. 258–85.

LONGUET-HIGGINS, M. S., and PARKIN, D. W., 1962, 'Sea waves and beach cusps', *Geographical Journal*, vol. 128, pp. 194–201.

MCKENZIE, P., 1951, 'Rip current systems', *Journal of Geology*, Chicago, vol. 66, pp. 103–13.

ROSSITER, J. R., 1954, 'The North Sea Storm surge of 31 Jan. and 1 Feb. 1953', *Philosophical Transactions of the Royal Society*, A, vol. 246, pp. 371–99.

RUSSELL, R. C. H., and MACMILLAN, D. H., 1953, *Waves and Tides*, Philosophical Library, New York.

SHEPHARD, F. P., 1972, *Submarine Geology*, 3rd Ed., Harper and Row, New York.

STEERS, J. A., 1969, *The Coastline of England and Wales*, Cambridge University Press, London.

STEERS, J. A., (Ed.), 1971, *Applied Coastal Geomorphology*, Macmillan, London.

— 1971, *Introduction to Coastline Development*, Macmillan, London.

TUREKIAN, K. K., 1968, *Oceans*, Prentice-Hall, Englewood Cliffs, N.J.

VALENTIN, H., 1952, 'Die Küste der Erde', *Petermanns Geographische Mitteilungen Ergänzungsheft* 246.

(For excellent summary and critical review see COTTON, above.)

# Chapter 24
# Marine Sediments and the Ocean Floor

Learn the secret of the sea? Only those who
brave its dangers comprehend its mystery.

*H. W. Longfellow* 1807–88

## Life as a Rock Builder

In earlier chapters many aspects of the geological work accomplished by living organisms have been reviewed. These include the growth and protection of soils; the fixation of sand dunes; and the differentiation of water and carbon dioxide into the fuels we burn and the oxygen we breathe. And besides the vast accumulations of peat and coal there are immensely greater deposits which are largely composed of the shells or other protective and supporting structures of once-living organisms. Most of these hard parts consist essentially of calcium carbonate secreted from the sea (*a*) by animals such as molluscs, sea urchins, corals, and the tiny single-celled foraminifera (Figure 24.7); and (*b*) by plants, of which algae, familiarly represented by seaweeds, are the chief. After death, the hard parts, if not dispersed through a superabundance of silt or mud, accumulate as shell deposits (Figure 3.6), coral reefs, algal reefs (Figures 24.1 and 24.2), deep-sea oozes and the like: all raw materials of limestones in the making. Other organisms, such as freshwater mussels, and snails and green algae, contribute calcareous material to deposits forming in freshwater lakes and lagoons. The biochemical precipitation of calcium carbonate as aragonite by the ammoniacal products of bacterial activities has already been mentioned.

Silica is extracted from sea water by minute single-celled plants (diatoms, Figure 24.8) and animals (radiolarians, Figure 24.10), both of which encase themselves within microscopic shells of opal. Siliceous deposits in which one or the other of these predominante constitute two important groups of deep-sea oozes. Siliceous deposits formed by freshwater diatoms, which abound in the lakes of recently glaciated districts, are distinguished as *diatomaceous earths*. In many of these

**Figure 24.1** Crusts of coralline algae exposed at unusually low tide along open 'surge channels' on the windward side of Bikini Atoll, near Bikini Island, NW Marshall Islands (*H. S. Ladd and J. L. Tracey, U.S. Navy*)

**Figure 24.2** Algal reef (*Cryptozoon*) of late Cambrian age in the 'Petrified Sea Gardens', nearly 5 km West of Saratoga Springs, New York (*New York State Museum and Science Service, Albany, N.Y.*)

lakes precipitation of limonite is brought about by iron-fixing micro-organisms: varieties of bacteria and algae that coat their cells with ferruginous filaments. Since the iron-fixing and the silica-fixing organisms tend to flourish at different seasons, and to have phenomenal outbursts of fertility in certain years, it comes about that some of the deposits are made up of alternations of finely banded iron ore and silica.

Similar alternations have been found on the sea floor. In an inspiring study of iron ores, H. G. Backlund has suggested that the observed biochemical deposition of limonite and silica may be the clue towards understanding some of the most enigmatic of the Precambrian rocks (Figure 24.3). In every continent these ancient rocks, usually metamorphosed and crystalline, include what are variously described as haematite-quartzites, banded iron formations, banded jaspers, or calico-rock. All of these have bands of iron ore alternating with bands of quartzite or jasper, the latter being a compact quartz-rock flecked with ferruginous inclusions which gave it a variety of bright colours, frequently brick red or yellow, and occasionally green. In some occurrences haematite has been changed to magnetite, and bands of black (magnetite) and white (quartz), or of black and red (jasper), or of all three colours—or more—may be seen.

Strongly supporting this idea is the discovery by Harvard biologists of structures that can be identified as the fossilized spores and filaments of primitive fungi and algae, still preserved in the old banded iron ores of Ontario and Minnesota. These rocks are known to be older than granites that have been dated at about 1800 million years. Even more remote ancestors of reef-building algae have been detected in Rhodesia. It is now well established that algae have flourished on the earth for at least 3000 m.y. Life must have existed long before then for algae to have been evolved, and as the age of our planet is about 4600 m.y. there was obviously no shortage of time. Probably life began its development on our planet at a very early stage of the earth's long history. Though as yet we are far from knowing how life did begin, we can all claim an ancestry that is older than any of the rocks on which we live.

Phosphates are of great value because of their vital importance as fertilizers, and because work-

**Figure 24.3** Polished section showing folded bands of quartz (black) and haematite (grey) in Precambrian banded iron ore from Norberg, Bergslagen, Central Sweden. ×1·4 (*N. Hjort*)

able deposits are far less common than we could wish. Most of them have resulted directly or indirectly from organic activities. Calcium phosphate is particularly concentrated in the bones, teeth, and excrement of vertebrates, especially fishes. Vast numbers of fish are sometimes killed almost simultaneously by the phenomenal outbursts of flagellates responsible for 'red water', which leads to catastrophic mortality of nearly all forms of coastal life, or by the shock of earthquake vibrations of unusual intensity. In such circumstances the remains accumulate on the sea floor as *bone beds*. Otherwise they are so rapidly disposed of by other organisms that fish bones are normally very rare as marine fossils.

Another interesting store of phosphates has been provided on rocky coasts and islands by the droppings of countless generations of fish-eating birds (Figure 24.4). The inorganic matter of the fresh deposit is a mixture of phosphates, nitrates and other compounds of calcium and ammonium. Warm climates with intermittent rain favour the removal of the soluble constituents and reaction products, until eventually a residual crust of *guano* is formed. In some localities crusts up to 30 m thick originally blanketed the rocks. But most of the thicker and therefore more valuable deposits have already been fully exploited and little now remains. Geologically it is of interest to note that where the base of the guano has been uncovered parts of the underlying rocks—generally volcanic—are found to have been metasomatically replaced by phosphate.

## The Floors of the Seas and Oceans

It is important not to be misled by the smooth curves of Figure 24.5, which is merely a diagram to illustrate the nomenclature used in describing marine environments. The results of echo-sounding, magnetic surveys, coring, bathyscaphe descents and a variety of geophysical techniques have upset practically every hypothesis based on the scanty knowledge of little more than a generation ago. A few submarine valleys carved into the slopes of continental margins have long been known, but echo-sounding has revealed the presence of great numbers of deep submarine canyons with their tributary systems. Oceanic islands, mainly volcanic peaks and coral atolls, rising from great depths, were once thought to be surrounded by almost featureless plains of enormous area. These regions of the ocean floors are now known to be traversed by great submarine mountain ranges,

**Figure 24.4** Asia Island, off the coast of Peru, thickly coated with guano, but now almost deserted by birds (*G. R. Johnson, American Geographical Society*)

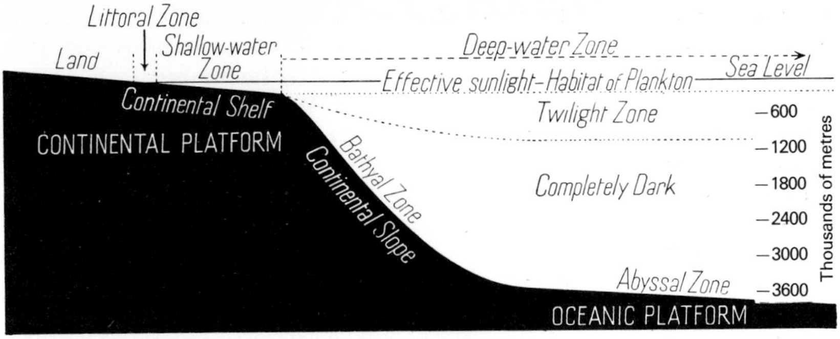


**Figure 24.5** Schematic section to show the chief zones of marine sedimentation

built of basalt, which encircle the earth and are known as mid-oceanic ridges (Figure 2.6). Great numbers of isolated peaks, volcanic cones and their relics, commonly occurring in rows or clusters, characterize the ocean floors (Figure 26.10). Some of the highest peaks rise above sea-level to form islands, but the vast majority fall short of the surface. These are known as *seamounts* or, if they have nearly flat tops, like truncated cones, as *tablemounts* or *guyots* (Figure 24.30). Deep ocean trenches, like those margining the Pacific garland of Island Arcs (Figure 2.7) have been familiar features since the first *Challenger* Expedition of 1872–6, but their significance remained a mystery until the 1960s (pp. 598–601). As a result of numerous geophysical investigations much has been learnt about their topography. They are sharply V-shaped in cross-section, sometimes free from sedimentary infilling, commonly more than 8000 m deep and sometimes over 3000 km long. Some trenches contain sediments of continental origin, and their floors are then flat and up to 8 km wide. Echo-sounding has revealed that their infilling sediments may be bedded and undisturbed (Figure 24.6). Innumerable faults, characterized by earthquakes, have been discovered intersecting and seemingly offsetting the mid-oceanic ridges (Figure 27.16). The ocean floor is no longer thought to be a static feature of the earth; newly formed at the mid-oceanic ridges, it travels away on either side, and disappears into the mantle at the deep ocean trenches (pp. 24 and 601, Figure 2.16).

The transition from abyssal plain to continental land—except where interrupted by the trenches mentioned above—is known as the *continental margin*. This is usually divided into three parts:

(*a*) the *continental rise*, which links the ocean basins to

(*b*) the *continental slope*, which is locally quite

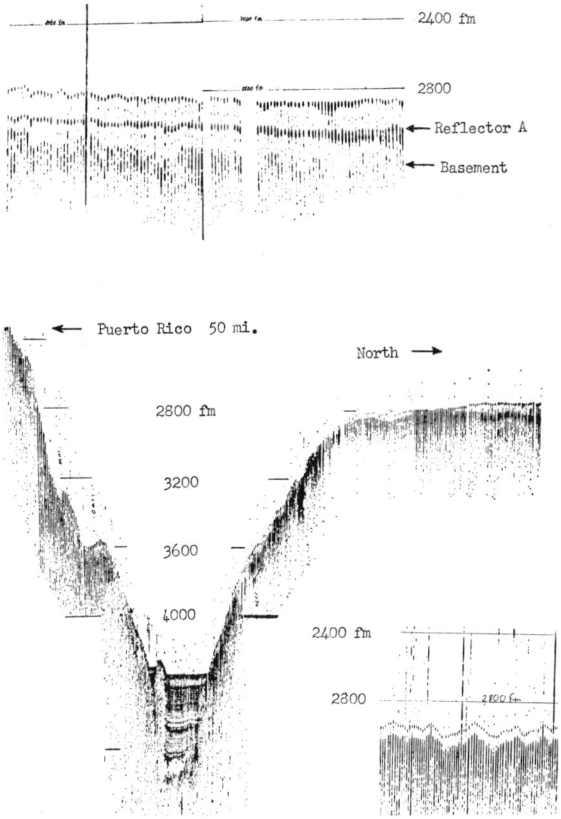


**Figure 24.6** Seismic reflection profiles; echo sounding records (*M. Ewing and J. Ewing, 1964*)

*The upper section* from south-west of Bermuda, at the edge of the Hatteras abyssal plain, illustrates the horizontal bedding characteristic of the ocean basins. The strata marked 'reflector A' was found to be widespread both within the north and south Atlantic.

*The middle left section* is a cross-section of the western end of the Puerto Rico trench, West Indies, where it contains 2 km of horizontal well-bedded sediment. The average thickness of the sediments elsewhere in the trench is less than 1 km.

*Section at bottom right* illustrates a uniform homogenous layer of sediments, probably pelitic, draped over a rough basement. This relationship between sediments and underlying basalt is characteristic of the mid-Atlantic ridge where the relief is not high (For high relief see Figure 26.5.)

steep and shares some of the features of an escarpment; it continues to the edge of

(*c*) the *continental shelf*, which extends shorewards at a very much gentler slope. The depth of the outer edge of the shelf is traditionally remembered as about 100 fathoms. Actually it varies considerably from place to place, like the width of the shelf. On average the steepening that marks the change from shelf to slope occurs at about 70 fathoms.

## Marine Deposits

According to their location on the sea bed or ocean floor, the marine deposits forming today are classified as follows:

*Littoral deposits*, which are formed between the extreme levels of low and high tides; they include beaches and other bay deposits.

*Shallow water deposits*, which collect on the continental shelf and at similar depths on the flanks of oceanic islands. These are now called *neritic deposits* by many marine geologists, although the term 'neritic' was originally applied only to shelf deposits of organic origin, such as shell beds or coral reefs. Nereus, a son of Oceanus, was one of the minor sea gods of the ancient Greeks; his domain was the shallow seas. In this sense 'neritic' could apply to shallow water deposits of any kind. But *nerita* became the Latin word for a sea-snail, and this secondary derivation suggested a restriction of the term to organic deposits. Confusion is easily avoided by referring to the latter as 'organic neritic' wherever there might otherwise be ambiguity.

Beyond the edge of the continental shelf are the muds and oozes referred to as *deep-sea deposits*. The muds, etc. of much of the continental slope, and of similar environments around oceanic islands, belong to the *bathyal zone*. Depending on currents and temperatures, the depth to which the bathyal zone extends varies considerably from place to place, the average in round figures being about 4000 m or 2000 fathoms. At greater depths lies the *abyssal zone*. Its characteristic deposits are the red clay or lutite and most of the deep-sea oozes. By the exploration of the deep oceanic trenches by photography and bathyscaphe descents, previously unsuspected forms of living organisms have been discovered on their floors, and it is becoming necessary to distinguish a still deeper marine environment. For this the name *hadal zone* (from Hades) has been proposed. It is provisionally regarded as beginning at depths of about 6500 m or 3500 fathoms.

According to the source of the dominant materials, three main groups of marine deposits are recognized:

(*a*) *Terrigenous*: derived from the land by rivers, glaciers, wind, and coast erosion. (Note that deposits *on* land are referred to as 'terrestrial'.)

(*b*) *Chemical* and *biochemical*: derived wholly or in part from ocean water, with or without the cooperation of organisms.

(*c*) *Organic*: comprising accumulations of calcareous and siliceous shells and other remains of marine organisms.

Terrigenous deposits are naturally found in the greatest bulk bordering the lands and especially off the mouths of great rivers. Sediment that is gently swept over the edge of the continental shelf, or carried more vigorously by turbidity currents, comes to rest on the slopes and plains beyond, or in the submarine canyons that diversify these regions. It must not be forgotten that during the Pleistocene stages of low sea-level, sand and gravel were deposited on the outer and deeper parts of the continental shelves, where the shore lines were then situated. The discovery of coarse sediments far out on the shelf seemed to be anomalous until it was realized that at the times when they were deposited there the depth of water was appropriately shallow. Recognition that coarser sediments and heavier minerals might be found on the deeper parts of the shelves has led to discoveries of great economic value. Gravels containing concentrates of cassiterite (tinstone, $SnO_2$) have long been dredged from the floor of the Java Sea between Borneo and Java (cf. Figure 23.56), and rich finds of diamonds (1961) have been made off the desert coast of Namibia (South-West Africa). For 80 km north of the mouth of the Orange River the gravels beneath the sand dunes constitute what is at present the most prolific source of diamonds of gem quality (one modest gem for every 10–20 tonnes of gravel). The newly found fields on the sea floor are the beaches of several thousand years ago, when sea-level was about 18 m lower than today. Sucking up diamond-bearing gravels with equipment something like a giant vacuum cleaner has already begun on a commercial scale, while greater depths, both here and off the coasts of other diamond-producing countries (e.g. Sierra Leone),

MODERN MARINE DEPOSITS

| *Dominant Kinds of Material* / *Zones of Deposition* | Terrigenous Deposits | Chemical (authigenic) and Biochemical Precipitates | Organic Deposits | |
|---|---|---|---|---|
| | | | Neritic (Mainly *Benthos*) | Pelagic (Mainly *Plankton*) |
| Littoral Zone<br>and<br>Shallow Water or Neritic Zone | Shingle<br>Gravel<br>Sand<br>Mud | Oolite sands<br>Calcareous muds<br>Evaporites<br>Cementing materials | Shell gravels and Shell sands<br>Coral reefs and Coral sands | (rare: usually overwhelmed by other materials) |
| | (8%) | | | |
| Bathyal Zone | Deep-Sea Muds and Sands<br>(With varied amounts of plankton remains)<br>(15%)<br>Green, black and blue muds<br>Volcanic muds<br>Deposits from Turbidity Currents | Glauconite<br>Pyrite, etc.<br>Cementing materials | Coral muds | Deep-Sea Oozes<br>(47%) |
| Abyssal Zone | | | | Pteropod ooze<br>Globigerina ooze<br>Diatom ooze<br>Radiolarian ooze |
| | Red Clay (30%) | | | |
| | Mainly terrigenous | Manganese nodules | (rare) | (variable) |

The figures in brackets represent the approximate areas covered by the various groups of deposits, expressed as percentages of the area of the ocean floor.

are being prospected.

The only really *continuous* supplies of land detritus that reach the abyssal ocean floor come from airborne dust and, in high latitudes, from melting icebergs. The normal supplies are so scanty that the rate of accumulation is extremely slow. But *discontinuous* supplies of typical terrigenous sediment are brought in by slumping and turbidity currents. It has been found in many of the core samples from the ocean floors that layers of shallow-water deposits alternate with pelagic oozes and red clay. In the early days of this unexpected discovery it was interpreted by some to mean that parts of the ocean floor must have been subjected to a succession of sudden but temporary uplifts, amounting in places to 3000 m and more. But most geologists hesitated to accept the possibility of such violent changes. The hypothesis became untenable when radiocarbon dating revealed that there had been many such alterations during the last few thousand years. The rival explanation that turbidity currents were responsible has gradually won general acceptance. We shall return to this important topic on p. 551.

## Marine Organisms

The marine organisms that contribute most conspicuously to the sediments of the littoral and shallow-water zones belong to a group known collectively as the *Benthos* (bottom dwellers). This includes seaweeds, molluscs, sea urchins and corals, and other forms that live on the sea floor. Many of them are firmly attached to the bottom. Deposits of shells or of their wave-concentrated fragments accumulate in favourable situations, while elsewhere similar remains are dispersed as fossils through the terrigenous deposits. The North Sea is mainly floored with terrigenous material, but between Kent and the Netherlands there are patches of several square kilometres consisting almost entirely of large shells. The shelly sands of some of the Cornish beaches have already been mentioned (p. 479). More extensive accumulations occur off limestone coasts and in other situations where organic remains are not smothered by sand and mud. The reefs and atolls constructed by corals and their associates illustrate limestone-building on a particularly spectacular scale.

The organic oozes and red clay of the abyssal zone are distinguished as *pelagic* deposits (Greek, *pelagos*, the open sea). The oozes are largely composed of the remains of marine organisms belonging to a group called the *Plankton* (the wanderers). This includes unicellular marine plants (diatoms) and animals (foraminifera and radiolarians); certain floating molluscs known as 'sea butterflies' or pteropods; most of the eggs and larvae of the benthos and other marine organisms; and all other forms which, unlike fishes, have no means of self-locomotion. The pteropods are blown along the surface by the wind, but the others, nearly all microscopically small, are passively suspended in the water.

**Figure 24.7** A picked assemblage of shallow water benthic foraminifera. ×35 *(By courtesy of the trustees of the British Museum—Natural History)*

Diatoms, being plants, cannot live below the depth of effective sunlight penetration, which in the open ocean reaches a maximum of about 100 m. Though individually quite invisible to the unaided eye, the diatoms are present in such prodigious numbers that they turn the sea in which they live into a kind of thin vegetable soup. This forms the main food supply of the rest of the plankton, whose habitat is therefore similarly confined to the sunlit or *photic* zone. Much of the silica devoured in this way goes into circulation again by being redissolved. Diatom ooze (Figure 24.8) is therefore restricted to cold regions where other forms of plankton do not flourish. There are, however, some exceptions, of which the Gulf of California—a veritable trap for silica—is a well-known example. Large amounts of silica are carried in solution into the Gulf by the Colorado River and also by bottom currents of ocean water flowing in from the Pacific. Only a small proportion of this silica is lost, because the warm surface outflow into the Pacific contains much less silica than average sea water, while the enormous amount of water removed from the Gulf by evaporation contains none. Most of the deposition of silica is concentrated in the northern part of the Gulf, mainly as a result of the 'explosive' reproduction of diatoms that occurs at certain times and seasons.

From the prolific surface waters penetrated by sunlight (the *photic zone*) the sea floor receives a slow but steady 'snowfall' of plankton shells which have escaped destruction. In the shallow-water zone the tiny shells are generally hidden in an overwhelming abundance of terrigenous materials and remains of benthos. In the bathyal zone, where the supply of terrigenous sediment is less overpowering and benthos forms begin to be scanty, they make a bigger show and can be readily found in the blue and green muds, both of which are characteristically calcareous varieties. In the abyssal zone, however, the plankton shells accumulate over vast areas with but little contamination from other sources, to form the deep-sea oozes which, together with the red clay, constitute the pelagic deposits.

Fishes, whales, and other marine animals which go actively after their food supply are grouped as the *Nekton* (swimmers). These contribute to all the marine deposits on a limited scale, but concentrated remains, such as the bone beds already mentioned as sources of phosphates, are quite rare.

The table on p. 543 summarizes the leading types of deposits now forming on the sea and ocean floors.

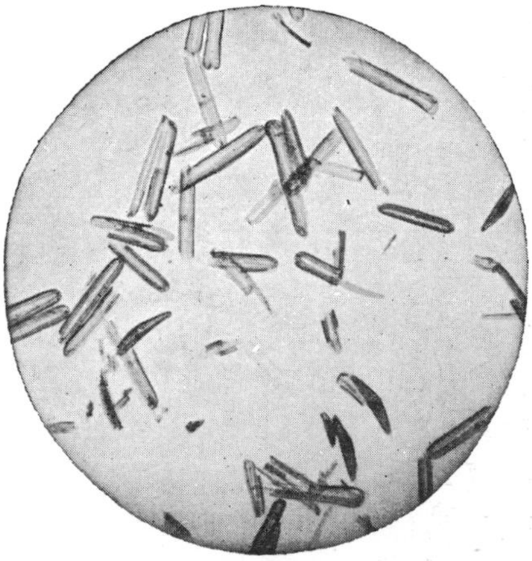

**Figure 24.8** Diatom cases of opaline silica. ×125 (*Jean Tarrant*)

### Pelagic Deposits

Until a century ago nothing was known of the deep-sea deposits. Globigerina ooze, dredged up by one of the cable-laying steamers in 1852, was the first to be discovered. A systematic exploration of the ocean floor was carried out by the famous *Challenger* Expedition of the years 1872–6. The thousands of samples then dredged up are described in one of the fifty bulky volumes in which the scientific results of that great enterprise are recorded. Since then relatively little further progress was made until echo-sounding and core-sampling began the accelerating advances of the last forty years.

The composition and distribution of the deep-sea oozes depend on the temperature of the photic zone and on the depth and circulation of the oceanic waters. It is convenient to start with the main food supply—the diatoms. Although they flourish wherever there is sunlight and water, diatoms make their greatest contribution to the oozes where conditions are least favourable to the organisms that feed upon them, i.e. in the cold regions around Antarctica and in the north of the Pacific. Minute box-like shells, and others of more intricate design, accumulate there as *diatom ooze*. The deposit usually contains the remains of certain species of foraminifera that are adapted for life in cold water. Contamination by mineral matter is usual, but whereas most of this consists of red clay ingredients in the north Pacific area between Japan and Alaska, it is mainly derived from floating ice around Antarctica and is relatively more abundant.

Radiolarians flourish most actively in warm tropical waters. *Radiolarian ooze* is essentially a variety of red clay that is notably rich in the remains of radiolarians, together with a sprinkling of diatoms and sponge spicules. It occurs principally along an east-west strip between Central America and Christmas Island in the Pacific, and also in two small south-equatorial patches in the Pacific and Indian Oceans (Figure 24.9). These are all places where the westward-flowing equatorial currents have become sufficiently warm, and where only a scanty supply of calcareous remains

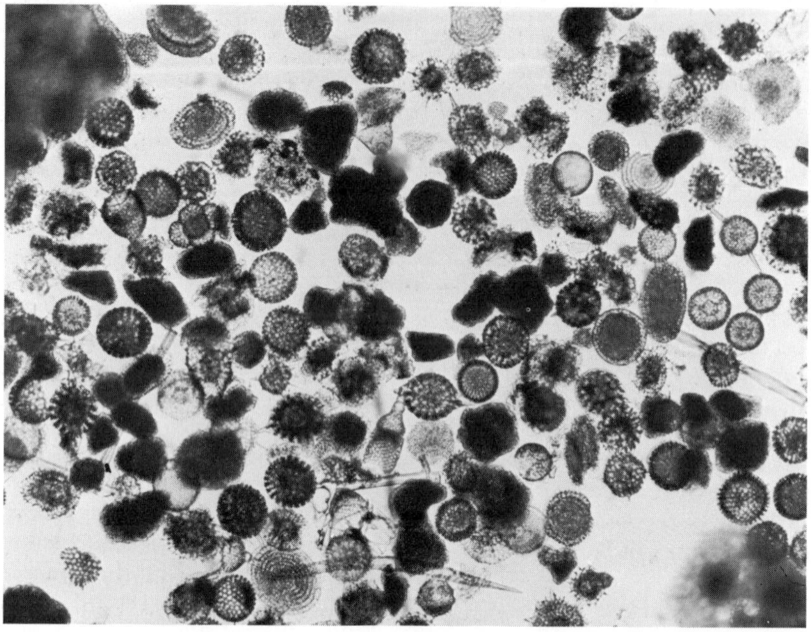

**Figure 24.9** Radiolaria from radiolarian ooze, North Pacific. ×91 (*By courtesy of the Trustees of the British Museum—Natural History*)

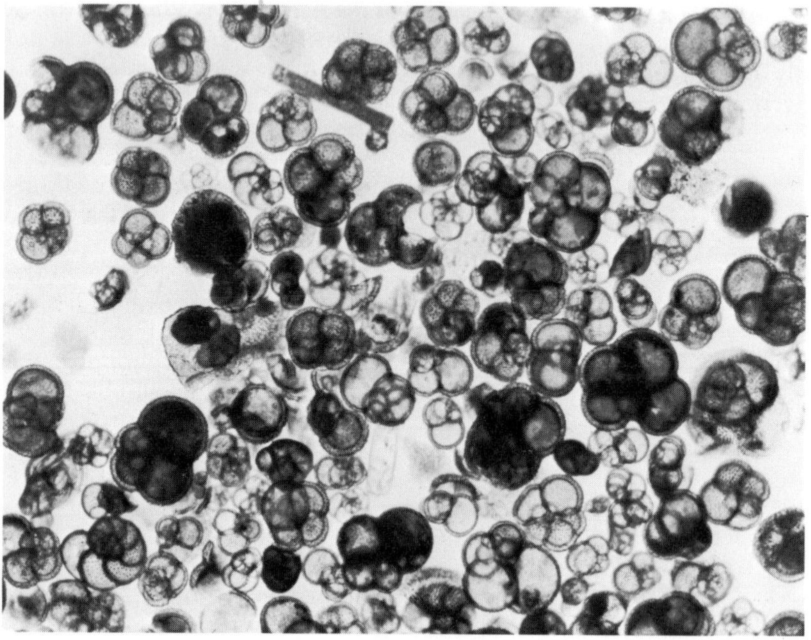

**Figure 24.10** Planktonic foraminifera from globigerina ooze, North Atlantic. ×78 (*By courtesy of the Trustees of the British Museum—Natural History*)

succeeds in reaching the bottom. The latter are mainly foraminifera, of which the shells of *Glogigerina*, each consisting of several globular chambers, are the commonest (Figure 24.10).

Of all the varieties of ooze, *globigerina ooze* is by far the most widespread, because foraminifera abound in the seas of both tropical and temperate regions, while a few distinctive species live only in very cold waters. Globigerina ooze is especially characteristic of the Atlantic, from the neighbourhood of Bear Island in the north to that of South Georgia in the south. It also makes extensive spreads in the western Indian Ocean and in parts of the SW and SE Pacific. In the shallower depths, over subtropical and tropical submarine swells and ridges, the shells of pteropods locally become abundant—mainly along the mid-Atlantic ridge. Where they predominate the deposit is distinguished as *pteropod ooze*.

In the following table the general composition, area and average depth of each of the chief pelagic deposits are summarized:

The tiny calcareous and opaline shells, being mostly extremely thin and of delicately embroidered forms, readily lend themselves to attack by solution as they sink towards the bottom. The solvent power of sea water tends to increase with depth, partly because of the increasing pressure and partly because the proportion of gases in solution, mainly $CO_2$, also increases. These conditions particularly favour the solution of calcareous remains, and consequently very few relics of even the thicker-shelled foraminifera reach depths of, say, 5000 m. The siliceous remains persist to greater depths, on average, as indicated in the above table. But depth and the time taken to reach the bottom are not the whole story. Diatom ooze in the southern hemisphere becomes enriched in silica because cold water from the Antarctic sinks to the bottom and spreads northwards, dissolving calcareous material as it goes, both from sinking shells and from the original deposit. As the water of these cold currents approaches saturation with $CaCO_3$, more of the abundant foraminiferal shells from the warmer surface waters succeed in reaching the floor, provided the latter is not too deep. Still farther north, and indeed wherever sufficient organic matter decays with the aid of oxygen dissolved in the water, $CO_2$ increases so that practically all the slow calcareous 'snowfall' is dissolved, and only the constituents of the red clay reach the bottom. Thus there is a lateral gradation of globigerina ooze into red clay. For similar reasons the latter merges into radiolarian ooze where the surface temperatures and depth of water are appropriate.

## The Red Clay or Lutite

Roughly 100 million square kilometres of the ocean floor are known to lie beyond the reach of more than small amounts of the plankton remains. These are the areas carpeted with the *red clay*.

| *Main constituents and areas* / *Type of ooze and average depth in metres* | *Calcareous* | | *Siliceous* | | |
|---|---|---|---|---|---|
| | Pteropod Ooze 2000 m | Globigerina Ooze 3600 m | Diatom Ooze 3900 m | Radiolarian Ooze 5300 m | Red Clay 5400 m |
| Calcareous remains | 45–98 | 30–97 | 2–36 | tr.–20 | 0–29 |
| *Average %* | 74 | 65 | 23 | 4 | 8 |
| Siliceous remains | tr.–20 | tr.–5 | 20–60 | 30–80 | 0–5 |
| *Average %* | 2 | 2 | 41 | 54 | 1 |
| Mineral matter (Av. %) | 24 | 33 | 36 | 42 | 91 |
| Area in millions of square kilometres | 2–3 | 127 | 32 | 6 | 100 |

tr. = trace (0·5% or less)

They include most of the North Pacific (except for the diatom belt in the extreme north and the radiolarian belt along the equator), the middle part of the South Pacific, the eastern part of the Indian Ocean and the deeper basins of the Atlantic. The chief ingredients of this remarkable deposit are:

(*a*) clay minerals derived from the finest wind-borne dust and from the alteration of the next three items;
(*b*) wind-borne volcanic ash, and pumice fragments that may have floated far from their sources before sinking;
(*c*) volcanic materials from submarine eruptions;
(*d*) rock-waste dropped from far-travelled icebergs or floating trees;
(*e*) insoluble organic relics such as sharks' teeth and the ear bones of whales;
(*f*) dust from meteors and meteorites, and very rarely larger fragments that have fallen from the sky;
(*g*) 'manganese' nodules, composed mainly of manganese and iron oxides, with more or less contamination by clay.

The red clay is chocolate coloured rather than red. Genuine red varieties occur locally, but are far less common than the original name would suggest. The colour is due to staining or filming of the ingredients by ferruginous matter, sometimes manganese-bearing, mostly derived from volcanic sources, including submarine fumaroles. 'Cosmic dust' is produced from the fusion of minute meteorites and of the surface skin of larger ones as they plunge through the outer atmosphere, and is constantly settling down over the earth's surface. But it is only high up in the atmosphere or where the red clay is accumulating very slowly (about 1 mm per 1000 years) that it is not smothered beyond recognition by more abundant varieties of 'dust'. The larger 'cosmic dust' particles are tiny spherules, rarely more than 0·1 of a millimetre across. These little pellets are magnetic and a score or so were separated from a sample of red clay, and correctly identified, by John Murray in 1876, shortly after his return from the *Challenger* Expedition, of which he was the leader. Renewed interest in these spherules has been aroused by the discovery of their occurrence in deep-sea cores of red clay from depths representing ages of 10 to 15 m.y. A meteoritic origin is confidently ascribed to these, as to those now accumulating, because they contain proportions of iron, nickel, cobalt, copper, etc. that match those found in iron meteorites. The corresponding proportions in volcanic rocks are quite different.

The manganese nodules also contain notable amounts of the same curious assemblage of elements, although it must be remembered that most of their manganese and iron is of volcanic origin. The nodules—commonly only 2 or 3 cm across, but sometimes 30 cm or more—have been familiar ocean floor curiosities ever since the first samples were dredged up during the *Challenger* Expedition. Now submarine photography has shown them to be so abundant in parts of the deep-sea floor that they may eventually become valuable as metallic ores. A possible clue to their mode of origin is the recently discovered fact that some of the rare foraminifera found in red clay are coated with a shiny veneer of dark mahogany-coloured material having the same composition as the manganese nodules. In turn, a few of these 'varnished' foraminifera support minute organisms which live on traces of protein and other organic matter that still remained attached to the shells. It is a provisional inference that the protein provides these microscopic 'bottom dwellers' with the vital energy that somehow enables them to extract from ocean water the queer assemblage of elements dissolved from the spherules and to concentrate them in film-like deposits. Long continued deposition of film on film would account for the growth and onion-like structure of the nodules.

## Cores from the Ocean Floors: Rates of Abyssal Sedimentation

At one time it was planned to drill through the oceanic crust and across the Moho right down into the mantle; this great enterprise was known as the *Project Mohole*. In 1961, using the drilling barge *CUSS I* (Figure 24.11), a preliminary trial was made to the east of the Island of Guadalupe, west of the peninsula of Baja California. The first boring ran into trouble after passing through 71 m of fine clay. A second boring then penetrated through 170 m of sediment, mostly Upper Miocene, and reached the basaltic crust, but after penetrating 13 m of basalt the drilling bit had to be withdrawn. It thus soon became apparent that the cost of boring through the whole oceanic crust would be too expensive, and the project had to be abandoned. It had been demonstrated, however,

**Figure 24.11** Aerial view of the drilling barge *CUSS I*, March 1961, a 3000-ton converted U.S. Navy barge with an oil-drilling rig installed amidships. The name represents the initials of the Continental, Union, Shell and Standard Oil Companies, who jointly owned the vessel. (*Global Marine Inc.*)

that the sediments on the ocean floor could be bored through without replacing the drilling bit.

In 1964 several American institutions combined in a Joint Oceanographic Institutions' Deep Earth Sampling Programme, commonly known as JOIDES, the object being to drill through the oceanic sediments in the Pacific and the Atlantic, and examine the cores. In 1968 the programme was extended as the Deep Sea Drilling Project (DSDP), and the ship *Glomar Challenger*, with a drilling rig installed midships, as in the *CUSS I* (Figure 24.11), began to drill through oceanic sediments. One method of coring is by driving a metal tube into the ocean-floor. Another is by actually cutting into the ocean floor. Initially cores up to 20 or 30 m long were collected, but with improved techniques cores up to 300 or even 500 m long have been obtained.

Information of fundamental geological significance is obtained by collecting and investigating cores. The record of glacial and interglacial periods found within cores has already been described on p. 458. When some of the upper layers of a core have been dated by the radiocarbon method, it is possible to estimate the rates of deposition during successive intervals within the core. However, in some of the cores the pelagic

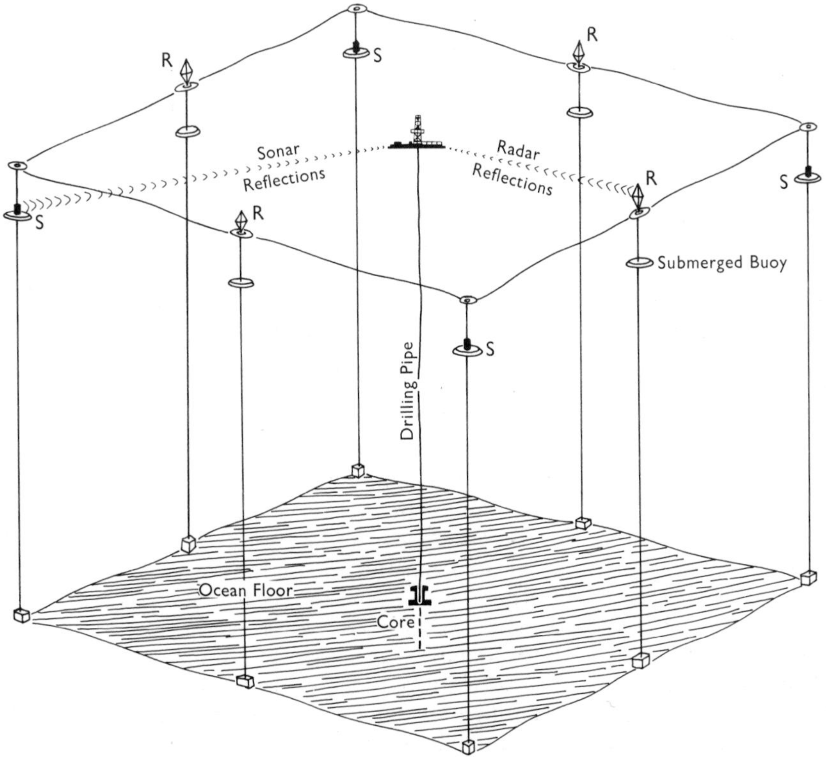


**Figure 24.12** Diagram showing how the drilling barge is held in position while floating. At each of the four corners there are diesel engines and propellor screws which can be rapidly speeded up or slowed down to drive the barge in any of the constantly changing directions required to counteract the effects of winds and waves. Manoeuvring is effected by a single handle which the steersman manipulates to keep the radar signals as closely as possible over the standard positions marked on the radar screen. The radar reflectors, R, are on floats attached to deeply submerged buoys, which are tautly moored to the ocean floor. Another set of buoys, S, provides an independent acoustic system of signals. Even though winds reached 40 km an hour and passing waves a height of 4 m, the drill-pipe did not deviate more than 1° from the vertical and was in no danger of being broken by undue bending.

clays and oozes are interrupted by layers of coarse sediment of obvious terrestial origin, often consisting of muddy sand with graded bedding. These interruptions are ascribed to turbidity currents (p. 551). For measuring sedimentation rates, it is essential to select samples free from all but climatic interruptions. Cores of this type were collected by Heezen, about 20 years ago, from both sides of the mid-Atlantic ridge, and detailed investigations of the deposits were later carried out by his colleagues at the Lamont Geological Observatory (see p. 566 under Broecker *et al.*). The results summarized in the following table show a definite relationship to the degree of continental glaciation. Rates of deposition of both carbonate and clay fractions decreased abruptly about 11,000 years ago.

*Rates of Pelagic Deposition in the Atlantic*

| | GLACIAL *Between* 45,000 *and* 11,000 *B.P.* | POST-GLACIAL *Since* 11,000 *B.P.* |
|---|---|---|
| Clay fraction | 0·82 | 0·22 g/cm²/1000 years |
| Carbonate fraction (mainly coccoliths and foraminifera) | 2·80 | 1·34 g/cm²/1000 years |
| *Total deposit* | 3·62 | 1·56 g/cm²/1000 years |

The above rates are expressed in terms of the mass of salt-free, dried material. In terms of thickness, e.g. in cm per 1000 years, the corresponding results depend on the density adopted for the material. Numerically the thickness rates are not very different from the above figures unless the sediment has been well consolidated. Hundreds of measured rates of recent Pacific and Atlantic pelagic sediments (from tropical regions well out of range of glacial debris) vary between 0·07 to 0·2 cm/1000 years for red clay, and up to 1·6 cm/1000 years for globigerina ooze.

## Submarine Canyons

Nearly a century ago it was discovered by soundings that the Hudson and Zaïre (Congo) valleys continue over the sea floor as submarine trenches which cut through the supposedly featureless floor of the continental shelf and become comparable in their dimensions with deep canyons where they traverse the continental slope beyond. 190 km SE of New York the Hudson submarine canyon is 10 km across from rim to rim, and over 1100 m deep (measured from the rim). The Zaïre (Congo) example is on an even vaster scale. these submarine canyons were naturally a source of great perplexity, and as others were continually being discovered the mystery of their significance and mode of origin presented geologists with still another challenge of increasing urgency.

About 1920 sounding ceased to be a slow and tedious process with the invention of echo-sounding. Since then the technique has been greatly improved by making use of supersonic waves, thus following what is now known to be the method of 'hearing' where things are, as practised by bats. These short, inaudible waves can be more sharply focused on the sea floor beneath the ship and the reflections returned to the receiving equipment are correspondingly clearer. Continuous soundings are now automatically recorded on a chart which shows a profile of the bottom—called an *echogram* (see Figure 24.30)—passed over by the ship while it is travelling along an accurately known course. Although vast areas still remain to be systematically surveyed in this way, many hundreds of submarine canyons are now known in considerable detail. The margins of every continent provide examples. A very few begin as gorges within the estuaries of great rivers, like that of the Zaïre. A few are in line with rivers, but discontinuously, like that of the Indus, which begins about 24 km from the land. Possibly the gap in such cases has been filled with sediment. But a large majority of the submarine canyons have their heads on the continental shelf or slope without any obvious relation to the drainage of the land. All of them become features of vigorous relief on the steeper slopes of the continental margins, which thus turn out to be far more rugged than anyone could have expected. A dendritic system of tributary valleys, as illustrated in Figure 24.13, is characteristic and the resulting submarine topography closely resembles that of a land surface youthfully dissected by river erosion. This similarity is strikingly exemplified by Figure 24.14. The trunk canyons range between broad valleys with V-shaped profiles and gorges with steep or even vertical sides. Some have been traced across the gentler slopes of the continental rise until they die out in the sediments of the abyssal plains beyond at depths of anything up to 3000 m. Cores brought up from the canyon mouths and beyond include layers of shallow-water types of sediment and fossils which could only have got there by

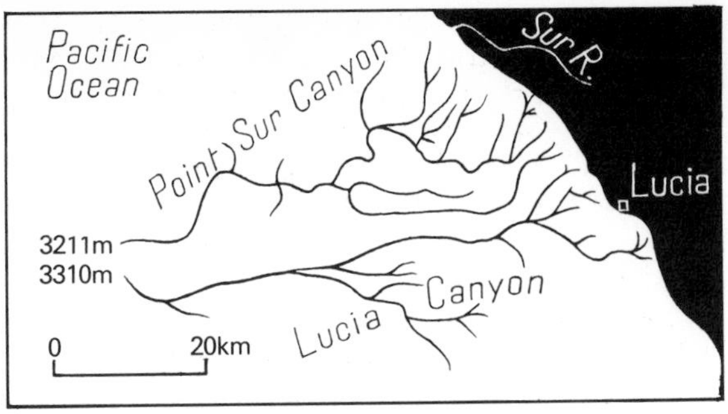


**Figure 24.13** Chart of the Point Sur and Lucia submarine canyons, off the Californian coast between San Francisco and Los Angeles, showing their tributaries and the depths to which they extend

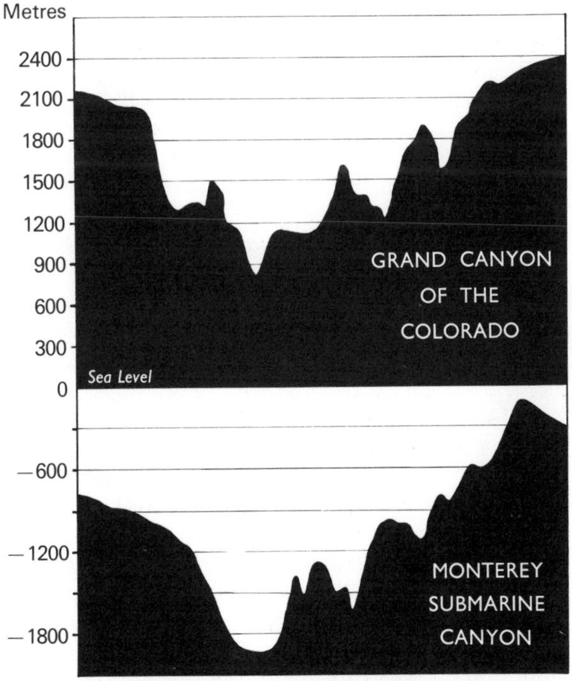


**Figure 24.14** Profiles 20 km across and drawn to the same scales to show the remarkable similarity of form and dimensions between the Grand Canyon of the Colorado and the submarine canyon of Monterey, off the Californian coast 56 km NW of Point Sur submarine canyon (*After F. P. Shepherd*)

being transported through the canyons themselves. This implies that the canyons were cut by the erosive effects of some kind of current.

From samples dredged from the walls it has been argued that some of these canyons must be geologically quite young. For example, the edges of Pliocene marine beds are exposed on the sides of the Monterey and many other canyons, suggesting that the latter must have been cut *after* the Pliocene, i.e. during the Pleistocene. But such evidence may only mean that the canyon has continued to develop since the Pliocene, e.g. by landslipping of the steep walls into the depths of a canyon that was already there. As we shall see presently, conditions were often especially favourable for the formation of submarine canyons during the Pleistocene Ice Age, but there is no reason to suppose that the Pleistocene had a monopoly of this submarine erosion. It is only the latest stage of development of a canyon that is demonstrated to be young, not necessarily the whole canyon.

The canyons must have been excavated either above or below sea-level. Before the very great depths reached by some of the longer canyons had been discovered it was not unreasonable to suppose that they must have been formed by river erosion, like their counterparts on the continents, i.e. that the submarine canyons were drowned river valleys. However, with the progress of submarine exploration it became evident that the relative changes of level required by the sub-aerial hypothesis (involving emergence and submergence of land and/or lowering and raising of sea-level) amounted to 3000 m or more, *all over the continental margins*. This was beyond belief. Moreover, the general application of the hypothesis became preposterous when it was later discovered by radiocarbon dating of the sediments in deep-sea cores that such extravagant changes of level would be required to occur repeatedly, with only short intervals of a few thousand years or less between the ups and downs. The only remaining possibility, that most submarine canyons were formed beneath the sea, presumably by the erosive action of turbidity currents, had perforce to be seriously considered, even by its former opponents.

## Turbidity Currents

Turbidity currents (Figure 24.15), also known as density or suspension currents, were introduced on p. 352 in connection with the treatment of deltas. There is no longer any doubt about their efficacy as agents of transport and deposition, and also of erosion when they take the form of relatively heavy underflows moving rapidly along the floor of a lake or reservoir, or down the submarine slopes of the continental margins. Sea water containing sediment in turbulent suspension

**Figure 24.15** A small turbidity current flowing into the Pacific from the Santa Clara River, near Ventura, California, during a springtime flood (*U.S. Department of Agriculture: Soil Conservation Service*)

has a higher density than the clear surrounding water and is therefore capable of such flow. It was Daly in 1936 who first suggested that submarine canyons could have been excavated by suspension currents, and particularly, he thought, during the glacial stages of the Pleistocene. The continental shelf was then exposed down to nearly 200 m below the present level of the sea. Waves and ordinary marine currents, then especially strong because of the prevalence of stormy weather, churned up the loose sediment of the outer parts of the shelf. The mud- and silt-laden bottom water would begin to flow downslope, like a heavy undertow or a turbulent rip current, with an accelerating speed determined by the density of the suspension and the angle of slope. By guiding and concentrating the flow of the loaded bottom water, chance depressions would become selectively eroded into furrows which thereafter would canalize and accentuate the currents. On reaching the continental slope (with an average gradient of about 1 in 15, a hundred times steeper than that of the shelf) the currents would gather speed and thus gain additional erosive power. Once started, such submarine flows would be self-accelerating until the slope began to flatten out. The inference is inevitable that on the continental slopes erosion would become much more vigorous than on the shelf above. Farther out the currents would slow down, probably spreading out over a broad front, and depositing sediment which would become increasingly fine as the abyssal plains were reached.

By carefully designed tank experiments, carried out in the Netherlands by Ph. H. Kuenen and, independently, by H. S. Bell in the United States, it has been convincingly demonstrated that turbidity currents do, in fact, behave in the ways that Daly had predicted (Figures 24.16 and 24.17). Subsidiary processes, locally capable of effective co-operation, include loosening and undercutting of the canyon sides and floors by springs, and faulting and fissuring of the sea floor by earth movements. But by themselves none of these could

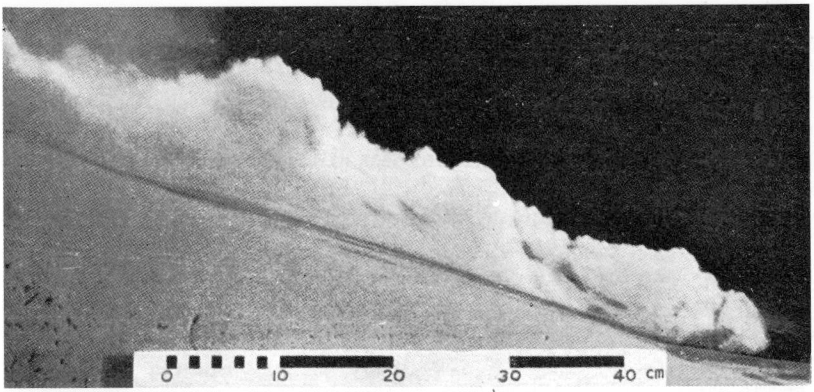


**Figure 24.16** An experimental turbidity current of fine sand and water flowing down a model valley excavated in fine sand in a laboratory tank. (*Ph. H. Kuenen*)

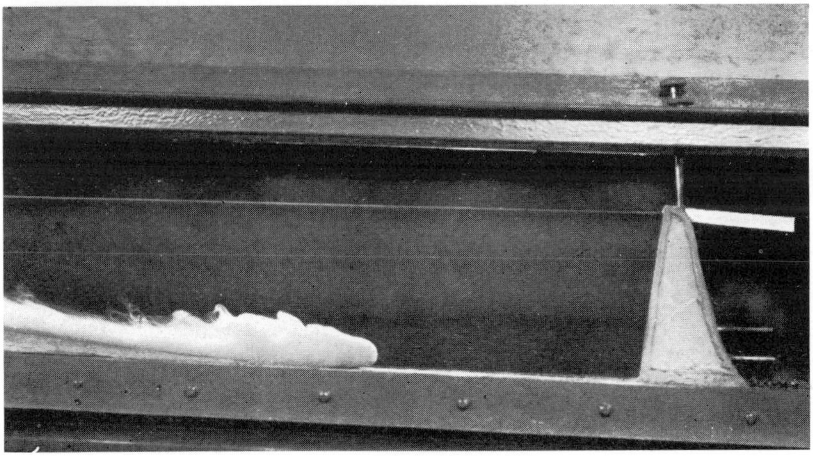

**Figure 24.17** An experimental turbidity current approaching the dam in a laboratory tank. The shape of the front seen in these two illustrations is characteristic of all turbulent suspension flows of higher density than the surrounding medium, e.g. dust storms (Fig, 22.4) and snow avalanches (Fig. 20.2) (*H. S. Bell, U.S. Department of Agriculture*)

bring about a submarine topography that bears all the hall-marks of erosion by a flowing medium somewhere between swiftly running water and avalanching mud flows. The only remaining doubt is to what extent turbidity currents could erode resistant crystalline rocks, such as have been dredged from the deeper walls of certain canyons. We do not know with any certainty, but it may well be that the currents have picked out fissured zones and shatter-belts, as rivers and glaciers are known to do on land.

As to the causes of turbidity currents: given a sub-aqueous slope, the essential requirements are a source of sediment and a trigger mechanism to start off the flow. A river in flood with a heavy bed-load at its mouth provides both sediment and flow. This combination would account for submarine canyons that are direct extensions of rivers. Such rivers, like the Zaire (Congo), may be said to have their deltas at the ends of their submarine canyons, hundreds of kilometres from where they pour into the sea. Another example is the Magdalena River, which flows northwards into the Caribbean Sea from the great highland valley of Colombia between the E and W Cordillera of the Andes. The real trigger required by these rivers is whatever causes a sufficiently heavy rainfall to yield the high discharge needed to ensure a heavy bed-load; thus a powerful hurricane might start a turbidity current. A more usual trigger is an earthquake that sets off submarine landslides or shakes loose unstable masses of sediment which then slump down-slope. Moving sediment and water mix together and rapidly develop into a turbidity current. Slumping may also be started by exceptional tides and waves over slopes of freshly deposited sediment that are just on the point of becoming too steep for stability. Submarine eruptions of *nuée ardente* type might develop into turbidity currents of volcanic ash.

What was probably the most gigantic turbidity current of modern times—and one that provided convincing evidence of its reality—was triggered off by the Grand Banks earthquake of 1929. This powerful earthquake not only devastated the southern coast of Newfoundland, it also broke twelve submarine telegraph cables in at least twenty-three places. Nearly all the breakages occurred in pairs, near the edges of a trough-like submarine canyon, i.e. where the cables were already in tension under their own sagging weight. The trough is in line with Cabot Strait, which is itself a deep submarine valley cut into the continental shelf. The trough reaches a depth of about 6400 m along the site of the cable marked C on Figure 24.18. This is more than 400 m below the abyssal plain on the western side and about 1200 m on the eastern side. Shortly after the earthquake and the disaster to the telegraph services, J. W. Gregory (*Nature*, 21 December 1929) interpreted the canyon as a submarine rift valley and considered the earthquake to be a probable indication of a fresh subsidence of its floor. At that time, the wholesale breaking of cables was attributed to the earthquake itself but, as shown below, this 'explanation' failed to account for the extraordinary timing of the breaks. About twenty years later Heezen and Ewing had a hunch that turbidity currents might be strong enough to break the cables. If so, the breaks should have occurred in a definite time sequence, depending on the speed of the current concerned. They therefore looked up the records of the Grand Banks earthquake and to their great delight found the evidence to be just what they had anticipated. In the words of Heezen:

> 'A study of the timetable of the breaks discloses a remarkable fact. While the cables lying within 60 miles of the epicenter of the quake broke instantly, farther away the breaks came in a delayed sequence. For more than 13 hours after the earthquake, cables farther and farther to the south of the epicenter went on breaking one by one in regular succession. . . . It seems quite clear that this series of events must indicate a submarine flow: the quake set in motion a gigantic avalanche of sediment on the steep continental slope, which broke the cables one after another as it rushed down-slope and flowed onto the abyssal plain.'

From the time intervals between the successive breaks and the intervening distances it is easy to estimate what the speed of the turbidity current must have been at various points along its course. the distance between A and B is 210 km and the interval between the breaks was just over three hours (Figure 24.19), indicating that this particular stretch was covered at an average speed of 70 km an hour. The speed was greater at A, slowing down towards B and beyond, until the current deposited its load in the form of fine or graded sediment and eventually faded out. The next step was to verify that the predicted deposit was actually there. Distinctly favourable evidence had been reported by the repair crew of the cable marked E, who described the material adhering to one of the broken ends as 'sharp sand and small pebbles'. Unfortunately, attempts to take cores from the sea floor south of the cable area had been frustrated by hurricanes in 1950 and again in 1951. Meanwhile, when Heezen and Ewing announced their results in 1952 (see p. 567), Kuenen accom-

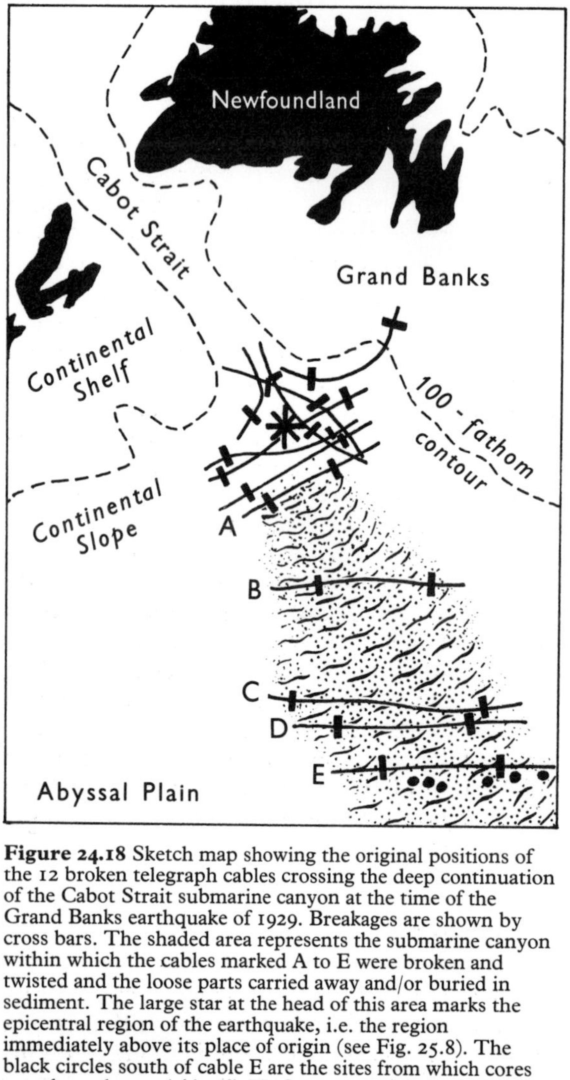


**Figure 24.18** Sketch map showing the original positions of the 12 broken telegraph cables crossing the deep continuation of the Cabot Strait submarine canyon at the time of the Grand Banks earthquake of 1929. Breakages are shown by cross bars. The shaded area represents the submarine canyon within which the cables marked A to E were broken and twisted and the loose parts carried away and/or buried in sediment. The large star at the head of this area marks the epicentral region of the earthquake, i.e. the region immediately above its place of origin (see Fig. 25.8). The black circles south of cable E are the sites from which cores were brought up. (*After J. W. Gregory, Nature, 1929; and B. C. Heezen and M. Ewing, 1952*)

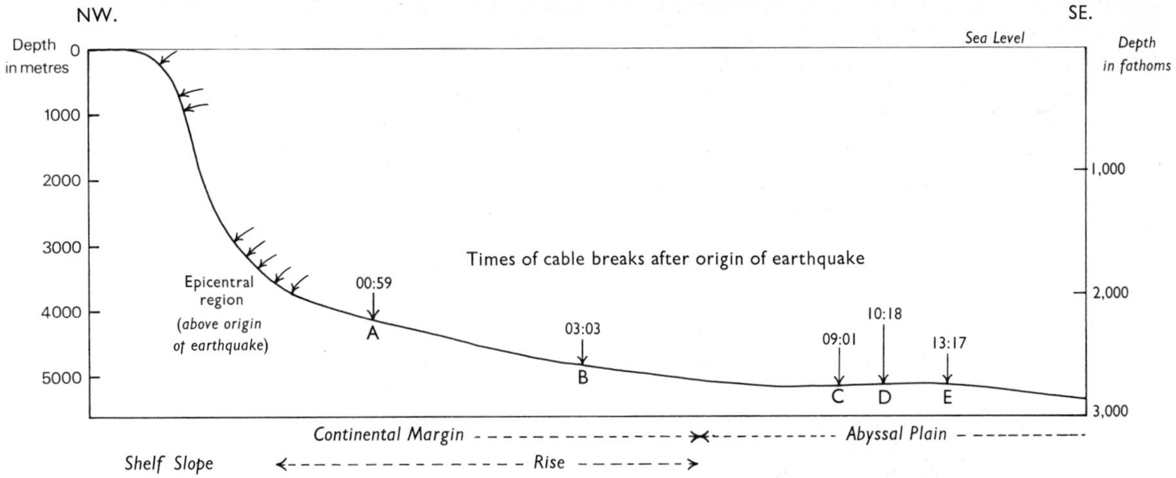


**Figure 24.19** Profile of floor of Cabot Strait submarine canyon, with arrows showing places where the cables marked A to E were broken and the corresponding time intervals after the earthquake, which originated at 20:32 G.M.T. on 18 November 1929. Length of section approximately 980 km (*After B. C. Heezen and M. Ewing, 1952*)

panied them with an estimate that the layer would probably have an average thickness of 40–100 cm over perhaps 260,000 km² of the abyssal plain. The dramatic conclusion to this story of discovery followed swiftly. Before the 1952 papers were published, Heezen had succeeded in bringing up six cores from the south of the southernmost cable. The first two were examined early in 1953 and found to consist of foraminiferal clay of abyssal type, with a top layer of graded silt, 130 cm thick in one core (Figure 24.18) and 70 cm in the other.

In 1954 an earthquake originating near El Asnam (Orleansville) about half-way between Oran and Algiers, was followed by the breaking, twisting and local burial of five cables which were spaced out on the continental slope of the Mediterranean floor. In this case the breaks occurred where the slope was steep, and far outside the area within which the ground was fissured (Figure 25.6) and underground pipes ruptured. No cables were broken directly by the earthquake shock, but the passage of earthquake waves was sufficient to release slumps of unstable sediment. These quickly developed into turbidity currents that rushed like express trains down the steep slopes. The times when telegraph messages were suddenly interrupted showed that the cables were broken one after another in order of their distances from the coast.

In recent years loose sediment from the lower reaches of the Magdalena River has several times contributed to turbidity currents which come to rest only far out on the abyssal plain of the Caribbean Sea. Jetties constructed at the river mouth have been shattered by the slumping of their foundations and considerable lengths of them have disappeared without trace. On the sea floor to the north a cable has been repeatedly broken at distances of 24 to 56 km from the land. The sediments of the Caribbean abyssal plain vary in thickness between limits of about 1000 and 2000 m—several times thicker than is usual—and a high proportion is obviously turbidity-current sediment, easily recognized by its richness in the decayed remains of marsh grasses that grow abundantly in the swampy lowlands between the Cordillera and the mouth of the Magdalena.

Before leaving this topic it should be mentioned that the surprisingly thin layer of sediment on the Pacific floor would be much thicker but for the fact that it is almost everywhere barred off from turbidity-current supplies by the deeps around the margins of the great ocean. The deep trenches associated with island arcs (Figure 2.7) are particularly obvious traps for sediment carried into them by turbidity currents, which may flow into them from the continental side either sideways or lengthwise.

## Coral Reefs and Atolls

In favourable situations in tropical seas, corals, together with all the organisms to which they give shelter and attachment, grow in such profusion that they build up reefs and islands of very considerable size. Clothed in vivid green, crowned

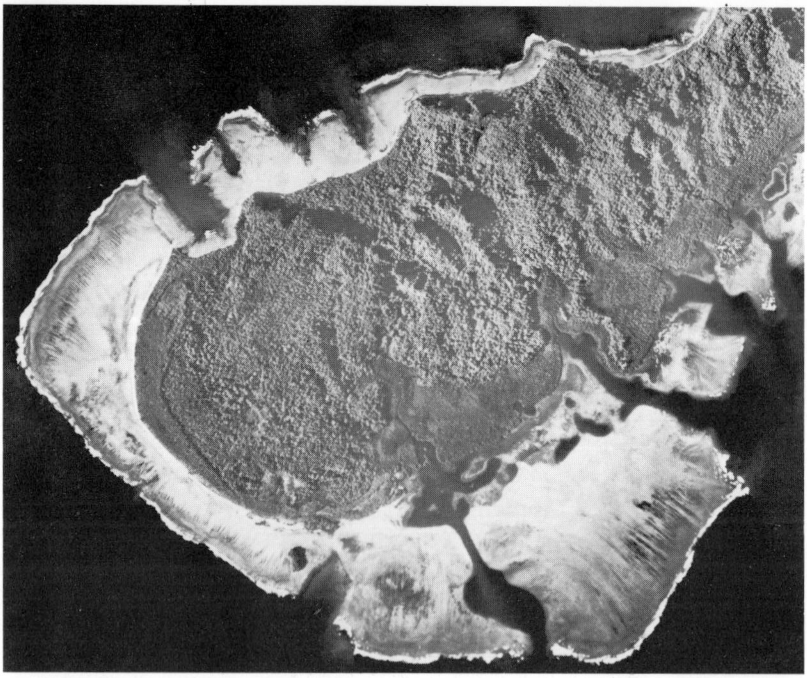

**Figure 24.20** Fringing reef, British Solomon Islands (*Overseas Geological Surveys*)

by the coconut palm, and fringed with the white foam of the ceaseless surf, the 'low islands' of the Elizabethan mariners have a reputation for dazzling but treacherous beauty. Dangerous to navigation and difficult to explore, they have been equally tantalizing to geologists who sought to account for their existence. Darwin was the first to face the problems in a scientific spirit and by him coral reefs were divided into three main classes:

(*a*) *Fringing reefs* consist of a veneer or platform of coral which at low tide is seen to be in continuity with the shore, or nearly so. The width may be 800 m or more, and where the reef is facing the open sea (Figure 24.20) the windward side may have a protective algal ridge (as in Figure 24.1) before sloping steeply down to the surrounding sea floor, which is commonly deep. Many fringing reefs, however, have grown on the inner side of relatively shallow lagoons enclosed by barrier reefs (Figure 24.21).

(*b*) *Barrier reefs* are situated up to 300 km offshore, with an intervening lagoon. The 2000 km complex of reefs known as the Great Barrier Reef, which forms a gigantic natural breakwater off the north-east coast of Australia, is by far the greatest coral structure in the world (Figure 24.22). Most barrier reefs, however, of which there are countless examples, are island-encircling structures

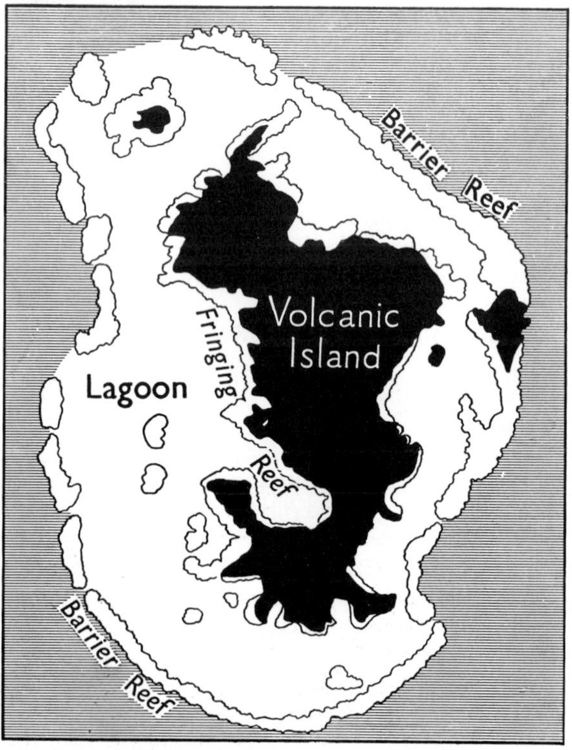


**Figure 24.21** Fringing and barrier coral reefs of mayotte, Comores archipelago, at the northern end of the Mozambique Channel. The embayed outlines of the islands indicate cumulative subsidence.

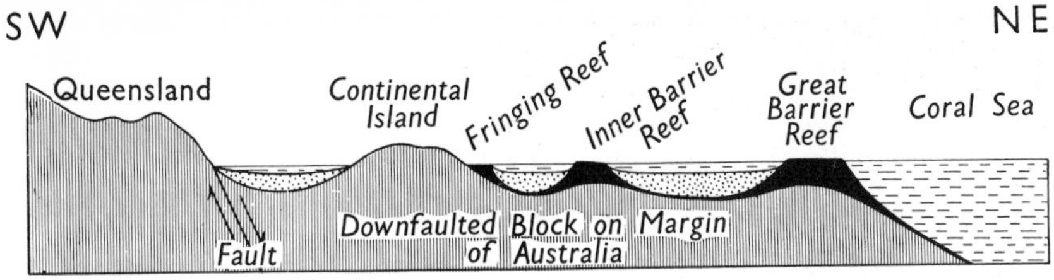


**Figure 24.22** Schematic section to show the tectonic relationship of the Great Barrier reef and its associated islands to the mainland of Queensland. reef rock, black; lagoon and channel sediments, dotted (*After J. A. Steers*)

forming irregular rings of variable width, more or less interrupted by open passages on the leeward side (Figure 24.21).

(*c*) *Atolls* resemble barrier reefs, but are without the central island (Figures 24.23, 24.24 and 24.26). They are essentially low-lying ring-shaped islands enclosing a lagoon which again is generally connected with the open sea by passages on the leeward side.

Reef-building corals live in colonies of thousands of tiny individuals (polyps), each occupying a cup-shaped depression in a calcareous framework which is common to the whole colony. As the successive generations of corals grow upwards and

**Figure 24.23** Hao Atoll, Tuamotu Archipelago (half-way between Fiji and Peru) (*J. P. Caplin, American Museum of Natural History, New York*)

**Figure 24.24** Ifaluk Atoll looking west, western Caroline Islands, western Pacific (*U.S. Navy*)

outwards through the restless waters in their competition for food, the stony frameworks also branch upwards and outwards and simulate the forms of certain plants, some being like shrubs and others like cushioned rock-plants (Figure 24.25). The interspaces between the dead structures are cemented and bound together by coralline algae called nullipores. These precipitate calcium carbonate within themselves, and still more as incrustations which coat their surfaces and cover the coral growths to which they may be attached. Where waves are strong the outer ridge-like margin of a reef is likely to be largely constructed by purplish red coralline algae, because these organisms can withstand exposure better than corals, so long as they are kept moistened by spray. Other calcareous contributions are made by shelled molluscs, foraminifera, worms and bacteria. The whole assemblage forms a white porous limestone which gradually becomes more coherent as it is buried and subjected to prolonged saturation by sea water.

The development and maintenance of coral reefs depend upon the conditions that favour a vigorous growth of the living colonies. A thriving reef has to contend not only with the waves, but also with boring organisms and voracious crustaceans that feed on the bodies of the individual corals. The reef represents the margin of success in a never-ceasing struggle against death and extinction. Not only have the corals and nullipores to supply material to maintain a flourishing living face; they have also to provide the broken masses of coral rock and other debris that accumulate to form the low land surface of the reef and the voluminous seaward foundations required for continued outward growth. The seaward face of the living reef passes downwards into a talus slope that may descend steeply to very great depths. On the lagoon or landward side of the growing face is the *reef flat*. This may be a platform with isolated clumps or patches of coral that remain submerged except at low tide, or it may be an irregular built-up surface consisting of material thrown up by breakers to a height of about 3 to 5 m. A certain amount of debris is also washed into the lagoon when the reef is swept by heavy seas.

Reef-building corals flourish best where the mean temperature of the surface water is about 23° to 25°C, with 18°C as the lower limit of tolerance—and that not for long. Reefs and atolls are therefore restricted to a zone lying between

**Figure 24.25** Coral growths exposed at very low tide on a fringing reef off the coast of Queensland, near Port Denison (*American Museum of Natural History, New York*)

latitudes 30°N and S, except locally where warm currents carry higher temperatures beyond these limits. The reefs of the Bermudas, for example, are dependent upon the warmth of the Gulf Stream. Along the torrid belts of the oceans the equatorial currents drift towards the west, becoming warmer on the way, and consequently reefs flourish far more successfully in the western parts of the oceans than on their colder eastern shores.

The water must be clear and salt. Opposite the mouths of rivers, where the diluted sea water carries suspended silt and mud, corals cannot live and no reefs appear. Conversely, reefs grow best on the seaward side of the reef, where splashing waves, rising tides, and warm currents bring them constantly renewed supplies of oxygen and food, and where, on the windward side, they are protected from violent breakers by the algal ridge. Corals cannot long survive exposure above water, nor coralline algae above the level continually reached by spray. Consequently living reefs can never grow much above low-tide level. Dead reefs are found above sea-level, but they have been uplifted by earth movements, to which they are therefore a most reliable index.

On the other hand, reef-building corals require abundant sunlight and do not grow freely at depths greater than about 30 m. Coralline algae are similarly restricted to shallow water, because they, too, are dependent on an adequate supply of radiant energy. Light intensity diminishes rapidly with depth and also with distance from the tropics, becoming negligible during the long nights of the polar regions. A few single corals—some still living—have been dredged up from the cold sea floor between South Patagonia and the Falkland Islands, so raising the problem of how these peculiar species manage to exist at all. Certainly they come from an environment quite incapable of providing the energy and food required by the vast communities of reef builders. Another clear indication that the latter can flourish only in warm, shallow, sunlit seas lies in the fact that reefs and atolls are easily 'drowned' if their upward growth cannot keep pace with any submergence (eustatic or tectonic) that may be in progress. Many reefs and atolls killed off in this way have been found on the sea floor. The average rate of coral growth varies with the species, the range being from 6 to 45 mm per year. Allowing for the necessity to provide material for the reef-flats and for the talus slopes on the seaward flanks, the average rate of reef growth is about 14 mm a year.

## The Origin of Reefs and Atolls

It follows from the above considerations that an essential condition for the initiation of a coral reef is the pre-existence of a submarine platform not far below sea-level. The origin of fringing reefs is easy to understand. The minute larvae of corals and the spores of coralline algae are drifted along by ocean currents, and those that reach suitable shores find attachment and start new reefs that gradually grow upwards and outwards towards the surface and the sunlight.

Barrier reefs and atolls, however, are remarkable in that they generally rise from depths where no reef-building corals or coralline algae could live. There are two possibilities: either (*a*) the reefs could have grown upwards from submerged banks within 30 or perhaps 60 m of the surface; or (*b*) they could have grown upwards and outwards from fringing reefs during the submergence of the land or island to which they were originally attached. Another feature that calls for explanation is that the lagoons have nearly flat floors (Figure 24.26), and depths that vary with the widths, the range being from about 45 fathoms for the larger

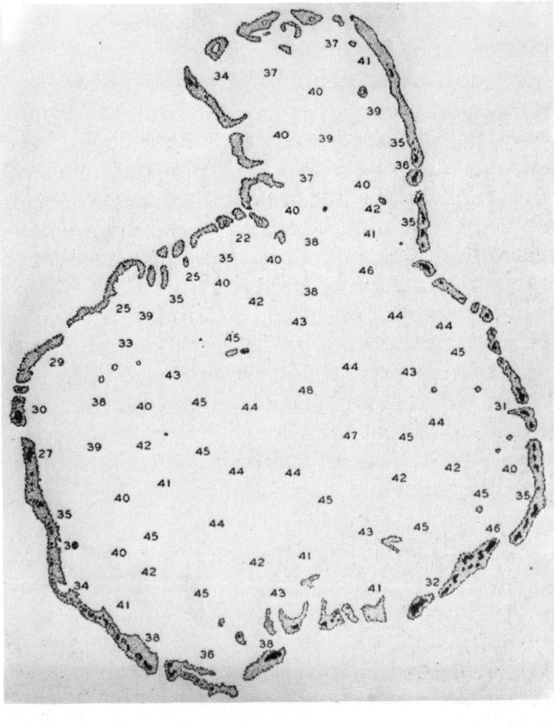

**Figure 24.26** Map of the Suva Diva Atoll (about 80 by 60 km), Indian Ocean, showing the depth of the lagoon floor in fathoms (*R. A. Daly*)

examples (80–160 km across) to 25 fathoms for the smaller ones.

The first general explanation was offered by Darwin in 1837 as a result of the observations he made during his celebrated voyage in the *Beagle*. He visualized all reefs and atolls as different stages in a single process (Figure 24.27). Growth begins with the building of a fringing reef around, let us say, a volcanic island. Subsidence of the island combined with continuous growth converts the reef into a barrier reef. Since reefs can grow upwards at a rate of about 30 cm in twenty years it will rarely happen that they are unable to keep pace with the movement. The submerged area between the island and the rim of coral rock forms the lagoon. By further subsidence the summit of the central island sinks out of sight, and the barrier reef becomes an atoll.

Darwin's simple theory has not passed unchallenged, but it satisfactorily accounts for most of the features associated with reefs, including several that he had not considered. The reality of subsidence—or at least submergence—is proved by the embayed shore lines and drowned valleys of land areas within the lagoons of barrier reefs. Moreover, the shore lines are not cliffed, as they would have been but for the protection from breakers afforded by the barrier reefs. Yet cliffs there must have been when the drowned valleys were being eroded, because then the sea would have been slightly turbid and there could have been no reefs to break the force of the breakers. Any original cliffs, like the valleys, must have been drowned. The theory does not, however, make it clear how the lagoons of the present day have come to be so remarkably uniform in depth. Figure 24.27 shows the enormous quantity of lagoon sediment necessary to fill in the 'moat' around a subsiding volcanic island. Alternatively, the flat lagoon floors of atolls and island-encircling barrier reefs suggest that the corals grew upwards from the seaward faces of platforms worn down by marine erosion and since submerged.

In 1910 Daly showed that these features are an inevitable result of the Quaternary oscillations of climate and sea level. He had already noticed the narrowness—and therefore the youthfulness—of the reefs fringing the Hawaiian Islands. Connecting this observation with the discovery that a former glacier had left its traces on the flanks of Mauna Kea, he inferred that corals could not have flourished along those shores during the glacial stages and that the existing reefs must have grown there while the sea gradually rose as a result of the melting of ice stored up during the last glaciation. Daly thought that during the glacial stages most of the pre-existing reef-builders would be killed off, leaving only a few living reefs in sheltered spots from which the reefs of the interglacial stages and those of the present day could be colonized as the seas grew warmer. However, it is now realized that only the marginal belts of the coral seas were colder, a fact that can be directly inferred from Figure 17.7. On the whole, the Pleistocene reefs suffered much more from the oscillations of sea-level than from lowered temperatures.

The most important contribution made by Daly in his 'Glacial Control' theory was the recognition of the effects of these oscillations. Pre-Pleistocene volcanic islands and successive generations of coral reefs must have been vigorously attacked by the waves through the wide range of levels illustrated on p. 448 by Figure 21.5. Whenever sea-level was falling, the tops of the then existing reefs would be exposed and so become a ready prey to marine erosion. As the sea rose the reefs would build themselves up again. Daly was mainly concerned with the last of these upbuildings. This began some 20,000 years ago, at which time there must have been innumerable platforms of marine erosion, planed down to levels more than 100 metres lower than the ocean level of today. From the outer slopes of these foundations, which now began to be submerged, the surviving corals would rapidly grow upwards, keeping close to the surface as the sea-level rose. There were setbacks now and

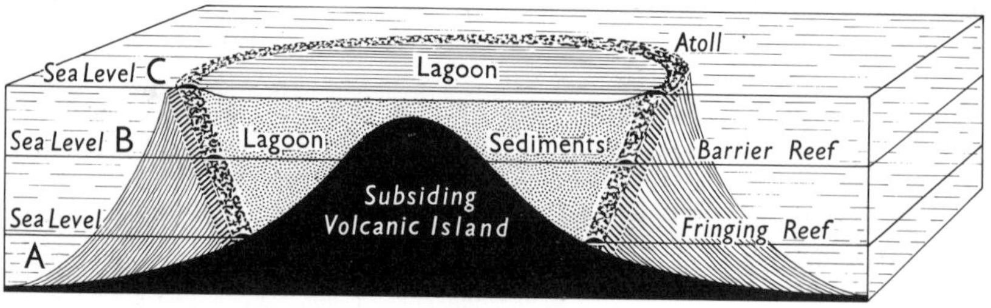


**Figure 24.27** Diagram to illustrate Darwin's theory of the successive development of fringing reef, barrier reef and atoll around a subsiding island. No account is taken of the interruptions caused by the Pleistocene fluctuations of the sea level.

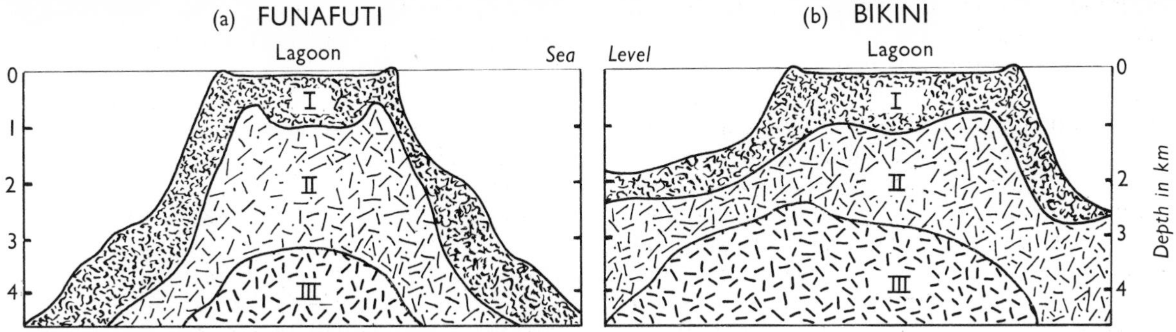


**Figure 24.28** Cross sections of Funafuti and Bikini Atolls showing the submarine structures inferred from the velocities of seismic waves (see p. 588) through the foundation rocks at various depths.
I represents reef and lagoon calcareous deposits.
II might be in part firm carbonate rock, but is more probably the equivalent of Layer 2 and mainly composed of volcanic rocks
III corresponds to the main refracting layer of the oceanic crust, composed of massive basaltic rock; its upper surface descends laterally to the normal depth of about 7 km. To the SW of Bikini the base of the crust (the Moho) was found at a depth of 13 km. Under both atolls III is abnormally thick, as would be expected in a volcanic foundation.
*Note:* the horizontal scales of (a) and (b) are not the same. Funafuti lagoon is 13 km across. Bikini lagoon is nearly 45 km across (*Simplified from R. W. Raitt et al.*)

then as the sea fell back for a time (cf. Figure 21.5), but the rate of growth of corals and their associates is such that these losses would soon be repaired. The time required for the upward growth of the reefs presents no difficulty: 14,000 years would provide roughly 200 m and so would amply allow for the setbacks. That growth was indeed unusually rapid is indicated by the spongy and porous nature of the first hundred metres or so of reef-rock penetrated by borings.

Incidentally, Daly's theory accounts for the rarity of lagoons with floors of abnormal depth, these being found only where marked subsidence due to earth movements has occurred. As already noted, lagoons must have been shallowed by deposition, the smaller ones more so than the larger because of the proportionately greater length of reef across which debris could be washed by the waves. This consideration is consistently matched by the fact that the depths of submerged platforms outside the coral seas are correspondingly greater, because where there were no corals or algae to build up and maintain an encircling rim the platforms would receive little sediment, and in many examples that little has been swept away by currents.

## The Foundations of Atolls

Darwin's subsidence theory refers essentially to submergence by earth movements, but it left the origin of platforms unexplained. Daly's glacial control theory makes good this deficiency and refers to submergence by a rising sea-level. Together, as a complementary pair, they account for all the major characteristics of coral reefs, but the results of recent borings and geophysical explorations have indicated that glacial control is only of the nature of an amendment to the more fundamental fact of subsidence, which has now been abundantly confirmed. In 1881 Darwin wrote to his friend Agassiz: 'I wish that some doubly rich millionaire would take it into his head to have borings made in some of the Pacific and Indian atolls.' Acting on this suggestion a few years later, the Royal Society of London sponsored the first of a series of drilling expeditions to Funafuti, one of the atolls of the Ellice Islands, north of Fiji. Coral rock and reef talus were penetrated to about 400 m. Some of the coral rock was thought to be in the position of growth and so to support Darwin's case, but much of the core material came from the talus slope, thus raising doubts that left the problem unsettled. More recently seismic exploration has revealed the presence of a foundation with the seismic properties of basalt underlying the whole of the lagoon at a minimum depth of about 900 m (Figure 24.28(*a*)).

In 1936 a hole was drilled through the dry lagoon floor (Figure 24.29) of an uplifted atoll known as Kita Daito Jima or North Borodino. This is a small island south of Japan and west of the Ryukyu arc. A depth of nearly 560 m below sea-level was reached in reef rock and reef sediments, the lowest strata being of Upper Oligocene or

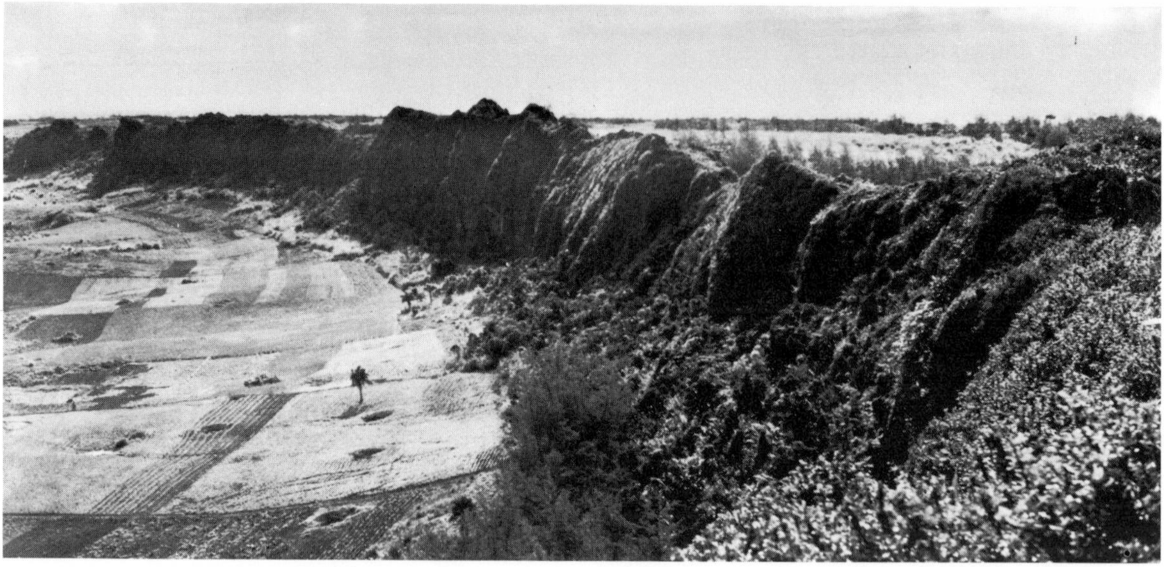

**Figure 24.29** The raised atoll of Kita Daito Jima (or North Borodino) showing part of the old lagoon floor and the surrounding rim of what was once the living reef (*R. A. Saplis, U.S. Geological Survey*)

Lower Miocene age. Similar results were obtained by drilling through a lagoon NE of Borneo (429 m), again without reaching a foundation for the reef.

Bikini, an atoll at the northern end of the Marshall group, became notorious when it was decided to test nuclear weapons there. Since 1946 it has been studied in great detail. Several holes were drilled, showing that corals and other organisms of shallow-water types occur in all the cores and chips brought up, the maximum depth reached being 779 m, in strata low down in the Miocene. This convincing evidence of subsidence was followed by seismic exploration (Figure 24.28(*b*)). The results indicated that the sediments were supported by a basement of hard rock, probably basaltic, at depths of between 2400 and 3000 m beneath the lagoon. It will be noticed that the peak of the foundation is not under the middle of the atoll. This lack of symmetry is probably due to coral growth having been more vigorous on the windward side, i.e. towards the south-east. The suspected nature of the foundation was confirmed in 1950, when basaltic rocks, some of them pyroclasts, were dredged from the slopes of Bikini at various depths between 1800 and 3600 m.

In 1952 two borings were put down in Eniwetok, an atoll to the west of Bikini. One on the northern part of the rim passed in turn through reef-rock and coralline sediments (rich in foraminifera and algae, as well as in corals) of Recent, Miocene and Eocene ages. Hard basement rock was encountered at 1405 m, but no samples of this were recovered. The absence of Pliocene sediments probably means that they were planed off by marine erosion during the exceptionally low sea-levels of the Pleistocene. Sediments of Oligocene age have not been recognized with certainty. The second boring, on the SE part of the rim, passed through a similar sequence, but reached a basaltic foundation at 1283 m. This time a total of 4 m of olivine-basalt core was recovered. Thus it was definitely established that the basement supporting this atoll is an old volcanic edifice, the summit of which is still more than 3300 m above the surrounding ocean floor. To this level it has subsided during the last 50 m.y. or so, while a rim of coral reef was successfully maintained at about sea-level by upward growth. The reef also supplied sediments for the lagoon floor and the outer talus slopes. After a long period of doubt and controversy Darwin's explanation of coral reefs can now be hailed as one of the very few Victorian 'theories' that has not become a casualty as a result of modern discoveries.

## Seamounts and Guyots

The evidence that the summits or flanks of worn-down volcanic cones served as the foundations of early Tertiary coral reefs, and afterwards subsided

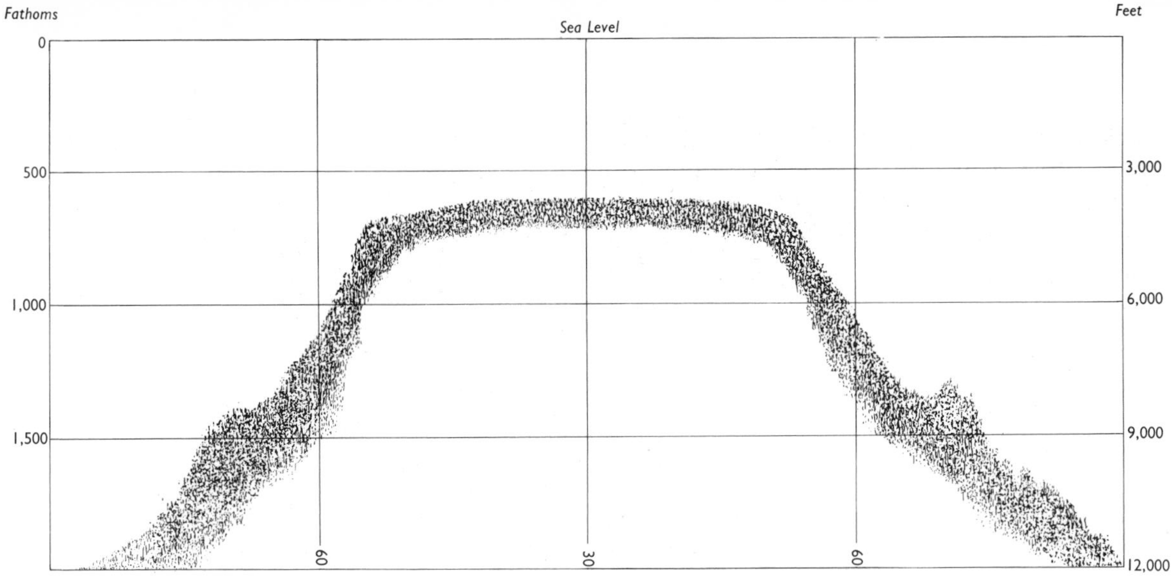


Figure 24.30 Echo-sounding record across a typical guyot south of Eniwetok Atoll. The truncated platform at a depth of 620 fathoms is nearly 15 km across, with a gently sloping rim down to 700 fathoms. The slopes then become abruptly steeper until the guyot is 35 km across at 2000 fathoms, after which they flatten out to the surrounding ocean floor at 2000 fathoms. (*After H. H. Hess, 1946*)

a thousand or more metres while the reefs grew upwards, at once suggests that there should be similarly drowned islands outside the coral seas but uncrowned by coral reefs. And even within the coral seas examples might be expected, provided that the islands subsided so rapidly through the critical zone that the coral community could not keep pace. These inferences would have been triumphant predictions had they not already been fulfilled. While on wartime service in the Pacific, H. H. Hess saw from the echo-sounder records that at least twenty flat-topped seamounts had been crossed during his voyages, the depths of their summits ranging from 900 to 1800 m. below sea-level. In a celebrated paper published in 1946 Hess interpreted these remarkable features as volcanic islands which had been truncated by wave erosion and deeply submerged (Figure 24.30). For the flat-topped seamounts he proposed the name *guyot* (after Arnold Guyot, a Swiss geographer of the eighteenth century) to distinguish them from other seamounts, such as volcanic cones that had failed to reach the surface. This work was followed up by several expeditions organized by various governments and private institutions, culminating in the International Geophysical Year. Among many surprises, the unexpectedly high speeds of currents at various depths was not the least. The reality of these currents is impressively demonstrated by the ripple marks seen on many of the photographs that have been taken by cameras lowered to the appropriate depths. Figure 24.31, illustrating conditions at a depth of well over

Figure 24.31 Deep-sea photograph of calcareous sands with pronounced ripple marks (averaging 20 cm from crest to crest), taken at a depth of 1710 fathoms (3127 metres) over the platform of a guyot rising from the Atlantic floor west of Portugal; lat. 41° 12′N, long. 15° 14′W. Area photographed about 1·5 by 2·5 metres (*The National Institute of Oceanography*)

3000 m, is an excellent example from the Atlantic, the exploring vessel being a later H.M.S. *Challenger*. Other pictures taken over the same guyot show part of the bedrock margined by a thin cover of ripple-marked sediment. The half scoured-out boulders seen in Figure 24.31 are probably wave-worn fragments of the same volcanic bedrock.

In addition to some 400 islands charted in the Pacific, well over 1000 deeply submerged seamounts have been recorded and at least 100 of these in sufficient detail to be recognized as guyots. Half of the guyots have been dredged, as well as some of the sharp-peaked seamounts, and in every case the materials brought up have been mainly volcanic.

In 1950 the Scripps Institution and the U.S. Navy joined in a Mid-Pacific expedition to determine the structure of Bikini Atoll and to investigate some of the guyots west of Hawaii. The guyots were found to be truncated peaks on a gigantic submarine range extending for 2400 km from Necker Island in the Hawaiian Chain to near Wake Island and now known as the Mid-Pacific Mountains (Figure 24.32). A typical example is Hess Guyot with a flat top, 19 × 13 km, standing about 1700 m below sea-level, but 3000 m above the surrounding ocean floor. Dredging from just below its top, and also from the edge of Cape Johnston Guyot, brought up reef-corals dating from the Middle Cretaceous, about 100 m.y. ago. Hollows and fissures in the flat tops of four of the guyots yielded globigerina ooze of Palaeocene and Eocene ages. But in all cases by far the commonest material of the hauls was olivine-basalt debris, ranging from rounded boulders and pebbles to erosional products of smaller sizes. The evidence is clear that the guyots formed a chain of basaltic islands that were eroded down to sea-level in Cretaceous times. Corals found lodgment here and there and many small reefs seem to have started, but the rate of submergence was such that they never developed into atolls.

E. L. Hamilton, who has graphically described the mid-Pacific range (Figures 24.32 and 24.33) compares it with the younger Hawaiian Chain, each being a complex linear series of volcanic ridges and peaks rising from a broad swell on the ocean floor. Another great range of seamounts begins just beyond the WNW end of the Hawaiian Chain, but is aligned northwards towards the junction of the Aleutian and Kamchatka-Kurile trenches. This is known as the Emperor Range because its guyots and seamounts were first discovered by the Japanese, who named them after their former emperors. Like the higher mid-Pacific peaks, those of the Emperor Range pro-

**Figure 24.32** Simplified chart to show the sunken seamounts of the Mid-Pacific Mountains. The greater part of the Hawaiian Chain appears in the NE and the northern group of the Marshall Islands in the SW. (*After E. L. Hamilton, 1956*)

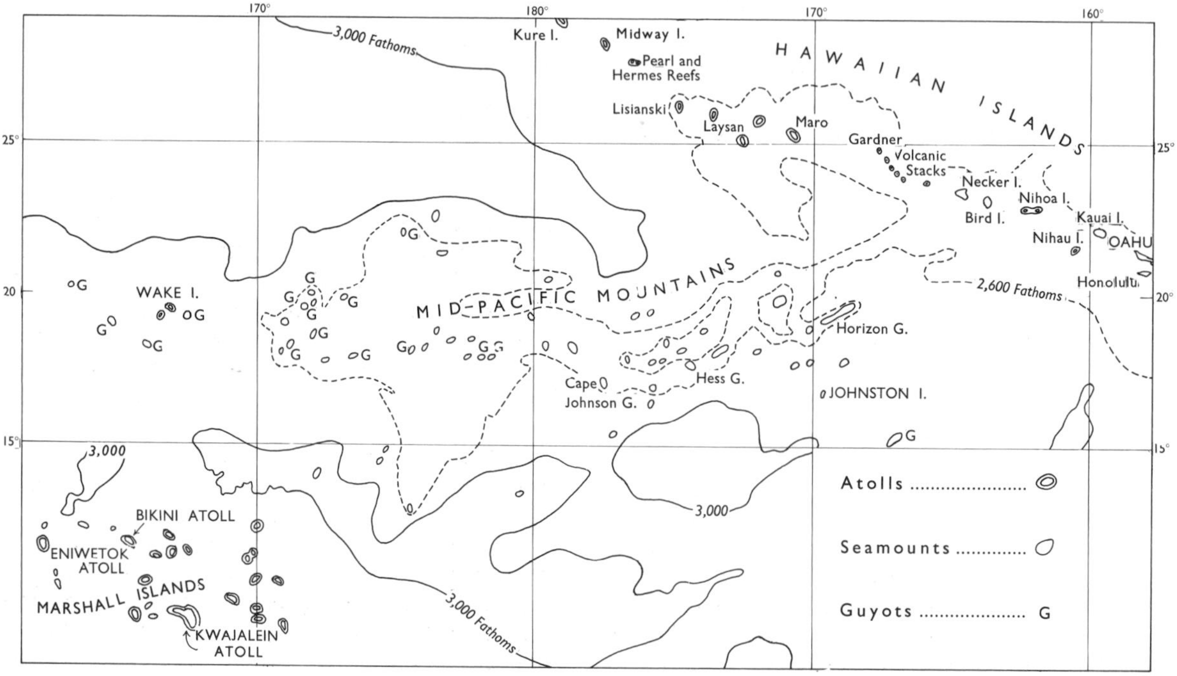

**Figure 24.33** An artist's impression of the Mid-Pacific Mountains. 'If the Pacific Ocean were drained away, the mile-deep sunken islands would emerge as truncated volcanic cones. The original oil painting is by the distinguished scientific illustrator Chesley Bonestell and is based on part of the bathymetric chart of the Mid-Pacific range.' (*Hamilton, 1956. Reproduced by permission of E. L. Hamilton*)

bably passed through their sub-aerial history long before the Hawaiian volcanoes began their activities.

## Fossil Reef-Corals and Ancient Climates

As indicators of past climates the distribution of fossil reef-building corals has a clear bearing on continental drift. Several geologists have pointed out that in Greenland and the Arctic Islands of Canada such corals flourished well within the Arctic Circle during Lower Palaeozoic times, and that since then the broad belt outlining their distribution has moved southwards towards the tropical zone of coral reefs of the present time. If reef-building corals were restricted in the past to the conditions which favour them today this means that the drift of Europe and North America had a strong northerly component. Alternative suggestions have been made: e.g. that Palaeozoic reef-building corals may have preferred cooler water, like some of the solitary species of today; or that the Arctic seas were then sufficiently warm, but that the tropical seas were too hot for coral growth. However, such *ad hoc* hypotheses are easily ruled out. Whether corals could survive in cold waters or not, they could only flourish with a continual and plentiful supply of sunshine. This condition could not be fulfilled during the long winter nights of the polar regions—unless, indeed, its equivalent was supplied by marine organisms radiating so powerfully that they could belong only to the realms of science fiction. When corals were living in the

| GEOLOGICAL PERIODS | RANGE OF AGE *in m.y.* | MEAN AGE *in m.y.* | MEAN LATITUDE of REEF-CORAL DISTRIBUTION *America* | *Europe and Africa* |
|---|---|---|---|---|
| Cretaceous, Jurassic, Triassic | 70–225 | 150 | 16°N | 35°N |
| Permian, Carboniferous | 225–350 | 290 | 38°N | 61°N |
| Devonian, Silurian | 350–440 | 395 | 50°N | 47°N |
| Ordovician | 440–500 | 470 | 53°N | 61°N |

Lower Palaeozoic seas that covered the regions now known as the Queen Elizabeth Islands and the north of Greenland, those regions were surely situated far to the south of the Arctic Circle. Independent evidence of this from the occurrence of salt domes has already been presented (p. 494). It seems fair to infer, as P. M. S. Blackett has done, that the mean latitude of the coral reefs of a given geological era or period may be taken as giving a first approximation to the average position of the equatorial belt during that interval. Blackett discusses the data very thoroughly and summarizes it in the table above (see next column for reference).

Russian and Chinese geologists have arrived at similar conclusions by measuring the convexities of the cushion-like forms of banded algal reefs of known ages. Here the principle involved is the well-known response of plants to sunlight, the turning towards the light technically called *heliotropism.* In the case of living algal reefs the axes of the convexities point towards the direction of maximum receipt of radiation from the sun. Algal limestones carry the climatic record far back into Precambrian time. A most interesting example comes from a region ENE of Peking, where algal reefs occur at intervals in unfolded Sinian strata. The time period represented is from 1200 to 600 m.y. ago. The convex upper surfaces of the bands are of the same kind as those formed by algal communities now living. But detailed measurements of the directions of the axes of the convexities reveal a progressive change in the apparent angle of the sun's maximum radiation. This can be interpreted either in terms of continental drift to mean that during the last 600 m.y. of Precambrian time the region was displaced southwards through 40° of latitude; or in terms of polar wandering to mean that the N pole migrated northwards by the same amount. The real explanation is likely to be a complex mixture of both.

## Selected References


BACKLUND, H. G., 1952, 'Some aspects of ore formation, Pre-Cambrian and later', *Transactions of the Edinburgh Geological Society*, vol. 14, pp. 302–35.

BLACKETT, P. M. S., 1961, 'Comparison of ancient climate with the ancient latitudes deduced from rock magnetic measurement', *Proceedings of the Royal Society*, A, vol. 263, pp. 1–30.

BROECKER, W. S., TUREKIAN, K. K., and HEEZEN, B. C., 1958, 'The relation of deep sea sedimentation rates to variations of climate', *American Journal of Science*, vol. 256, pp. 503–17.

DARWIN, CHARLES, 1842, *The Structure and Distribution of Coral Reefs*, Smith Elder, London.

EMERY, K. O., TRACEY, J. L., and LADD, H. S., 1954, 'The geology of Bikini and nearby atolls', *U.S. Geological Survey, Professional Paper*, 260 A, Washington, pp. 1–291.

HAMILTON, E. L., 1956, 'Sunken islands of the Mid-Pacific Mountains', *Geological Society of American, Memoir* 64.

— 1959, 'Thickness and consolidation of deep-sea sediments', *Bulletin of the Geological Society of America*, vol. 70, pp. 1399–1424.
HEEZEN, B. C., 1959, 'Dynamic processes of abyssal sedimentation: erosion, transportation, and redeposition on the deep-sea floor', *Geophysical Journal of the Royal Astronomical Society*, vol. 2, pp. 142–63.
HEEZEN, B. C., ERICSON, D. B., and EWING, M., 1954, 'Further evidence for a turbidity current following the 1929 Grand Banks earthquake', *Deep-Sea Research*, vol. 1, pp. 193–202.
HEEZEN, B. C., and EWING, M., 1952, 'Turbidity currents and submarine slumps, and the 1929 Grand Banks earthquake', *American Journal of Science*, vol. 250, pp. 849–73.
HESS, H. H., 1946, 'Drowned ancient islands of the Pacific Basin', *American Journal of Science*, vol. 244, pp. 772–91.
KUENEN, Ph. H., 1952, 'Estimated size of the Grand Banks turbidity current', *American Journal of Science*, vol. 250, pp. 874–84.
— 1958, 'No geology without marine geology', *Geologische Rundschau*, vol. 47, pp. 1–10 (in English).
LADD, H. S., INGERSON, E., *et al.*, 1953, 'Drilling on Eniwetok Atoll, Marshall Islands', *Bulletin of the American Association of Petroleum Geologists*, vol. 37, pp. 2257–80.
MACNEIL, F. S., 1954, 'Organic reefs and banks and associated detrital sediments', *American Journal of Science*, vol. 252, pp. 385–401.
MOORE, J. R., (Ed.), 1971, *Oceanography: Readings from* Scientific American, Freeman, San Francisco.
PETTERSSON, HANS, 1960, 'Cosmic Spherules and Meteoric Dust', *Scientific American*, vol. 88 (2), pp. 123–32.
RAITT, R. W., 1954, 'Seismic Refraction Studies of Bikini and Kwajalein Atolls', *U.S. Geological Survey, Professional Paper*, 260-K, pp. 507–27.
TRACEY, J. I. Jr., *et al.*, 1961, 'The natural history of Ifaluk Atoll', *Bernice P. Bishop Museum, Bulletin* 222, p. 75, Honolulu.

# Chapter 25

# Earthquakes and the Earth's Interior

Man is learning to harness for his enquiring use the very wrath of the earth; the tremblings of our vibrant globe are used to 'X-ray' the deep interior.

*Reginald A. Daly*, 1928

## The Nature of Earthquakes

When a stone is thrown into a pool, a series of waves spreads through the water in all directions. Similarly, when rocks are suddenly disturbed, vibrations spread out in all directions from the source of the disturbance. An earthquake is the passage of these vibrations. In the neighbourhood of the disturbance itself the shaking of the ground can be felt and the effects may be catastrophic, but farther away the tremors die down until they can be detected only by delicate instruments called seismographs (Gr. *seio*, to shake; *seismos*, an earthquake).

Vibrations are set up in solid bodies by a sudden blow or rupture, or by the scraping together of two rough surfaces. Corresponding causes of earthquakes in the earth's crust are volcanic explosions, the initiation of faults and the movements of the rocks along fault planes. Perceptible tremors are set up by the passage of trains and tanks, by avalanches and landslides, by rock falls in mines and caverns and by explosions of all kinds. When a munition factory explodes, the intensity of the resulting earthquake may be comparable with that of volcanic earthquakes, while that of the shock waves through rocks set up by the underground testing of H-bombs may be far more severe. The majority of natural earthquakes, however, including all the most disastrous examples, are due to sudden earth movements, generally along faults (Figure 25.1); these are distinguished as *tectonic* earthquakes. The term *tectonic* (Gr. *tekton*, a builder) refers to any structural change in rocks brought about by their deformation or displacement.

The cause of tectonic earthquakes is thus the building up of stresses in rocks until they are strained to breaking-point, when they suddenly rupture and move. The fault movements themselves, as already described on p. 141, may be vertical, horizontal or oblique. After the great Alaskan earthquake of 1899, it was possible from the presence of barnacles clinging to the uplifted rocks of Disenchantment Bay to measure the uplift, which in this case reached an exceptional maximum of 14 m. Crustal blocks often move obliquely, both vertical and sideways movements being observed. Surveys carried out after the Sagami Bay earthquake of 1923, when Tokyo and Yokohama were wrecked, showed that the floor of the Bay and the surrounding mainland had been twisted round in a clockwise direction, the measured shift of the volcanic island Oshima being over 3·5 m (Figure 25·2).

When the rocks are nearly at their breaking-point, an earthquake may be triggered off by some extraneous agent such as a high tide, a heavy rainfall or flood, a rapid change of barometric pressure, the tremors from an independent earthquake originating far away, and now, it is suspected, the shock waves from an exploding hydrogen bomb. The principal shock, which generally lasts only a few seconds, or at most, and rarely, a few minutes, may be preceded by fore-shocks and is invariably followed by a series of after-shocks. The fore-shocks represent the preliminary shattering of small obstructions along a fault plane or zone. When these have been overcome the main movement occurs. But complete stability is not restored immediately. The sudden jolt, or swift succession of jolts, often

**Figure 25.1** One of two parallel scarps, 2–3 km apart, representing the surface expression of the movement and the resulting major earthquake that originated at a depth of 35 km beneath Quiches in the Andes of Peru on 10 November 1946. The scarp illustrated has a vertical throw of 4 m and an unbroken length of 5 km through country at an elevation of 3600–3900 m. The zone between the two scarps subsided about 3 m on average. (*Arnold Heim*)

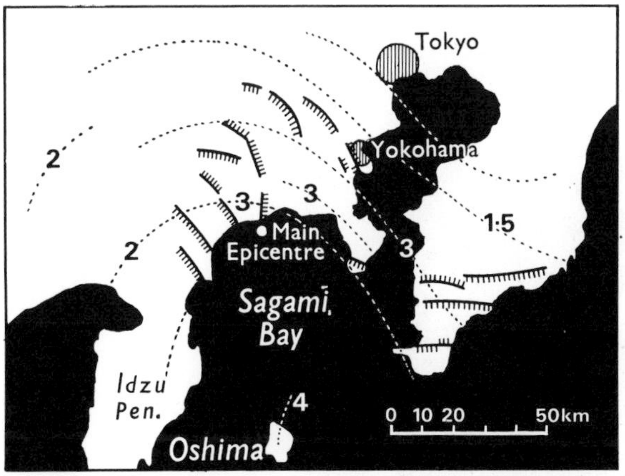


**Figure 25.2** Sketch map of the Sagami Bay area, Japan, to indicate the surface movements associated with the disastrous earthquake of 1 September 1923. Vertical displacements ranging from several centimetres to a metre or two are shown by shading on the downthrow side of the fault lines. Horizontal rotation (clockwise) is indicated by dotted lines, with numbers representing the approximate displacements at various localities in metres.

disturbs adjoining fault-blocks. The after-shocks represent a long series of minor movements that accompany the gradual settling down of the region. Considering the whole earth, earthquakes of one kind or another are known to take place every few seconds, but many of these are so slight that without nearby instrumental records their occurrence would not have been known. Really severe earthquakes, which would be disastrous or catastrophic (see p. 575) in populated areas, take place every two or three weeks on average. Most of these originate beneath the continental slopes and cause little damage from the human point of view.

## Effects of Earthquakes

One of the most alarming features of a great earthquake experienced on land near the place of origin is the passage of ground or land waves which throw the surface into ever changing undulations. These may be only 30 cm or so high and 6 to 9 m from crest to crest, but the effect is so universally terrifying that accounts of the writhing of the ground are usually wildly exaggerated. But the reality is quite bad enough. During the main shock of the earthquake swarm that ravaged Chile in May

1960, the motion of the ground was described by scientific observers as 'slow and rolling like that of the sea during a heavy swell. Parked cars in Concepción rolled to and fro through a distance of half a metre and bobbed up and down'. The period of the waves was 10 seconds or more and they continued to pass for 3 or 4 minutes before ceasing to be visible.

Apart from the dislocation of structures built across a fault when it suddenly jumps into action, the most spectacular and perilous effects are due to the passage of land waves. Fissures gape open at the crests, only to close again as the crests turn into troughs. Sometimes a broken undulating surface is preserved as it was when the land waves suddenly ceased to have effect (Figure 25.3). Water pipes and gas pipes are cracked open as well as roads (Figure 25.4). Railways are buckled and twisted (Figure 25.5); bridges collapse and buildings crash to the ground. Land waves are a little-understood local manifestation of the earthquake vibrations usually experienced, which are of a much smaller order. Frequent shaking through a few millimetres suffices to wreck most buildings not specially

**Figure 25.3** Deformation of the quayside, Yokohama, left after the passage of land-waves; Sagami Bay earthquake of 1923

**Figure 25.4** Gaping fissure left in a country road near Yokohama; Sagami Bay earthquake of 1923

**Figure 25.5** Buckling of railway lines near Tokyo; Sagami Bay earthquake of 1923

constructed to withstand earthquake shocks. A to-and-fro tremor of as little as 0·016 mm can be distinctly felt if one is sitting or standing still. In a strong earthquake the effect of the up-and-down vibrations on the feet has been described as like 'the powerful upward blows of a monstrous hammer'.

In the region of destruction, landslides are set moving on valley sides, and avalanches are started in snowy mountains. Glaciers are shattered and where they terminate and break off in the sea icebergs become unusually abundant. Vast masses of loose sediments may be so disturbed by submarine shocks that they slump for many kilometres down the continental slope. In Sagami Bay in 1923 landward parts of the floor were thus uncovered by 300 to 450 m, seaward parts being correspondingly built up where this enormous bulk of sediment came to rest. The passage of earthquake vibrations through water-filled sands, especially in alluvial districts, brings about the compaction of the deposit. The resulting decrease of volume obliges the water to escape, which it often does with sufficient violence to ascend through fissures as powerful sandy jets. These often issue at the surface like isolated fountains around which sand-craters develop (p. 160).

Ground-water and its circulation may be greatly disturbed in other ways by earthquakes. Old lakes may be drained off through open cracks, and new ones formed in depressions. In 1935 Lake Solar was engulfed by a great fissure that opened across its floor in the Kenya rift valley. Not all fissures close up again, as Figure 25.6 clearly demonstrates. Through a dramatic coincidence it has been discovered that broad fissures also open

**Figure 25.6** One of many permanent fissures opened during the Orleansville earthquake, Algeria, of 1954 (*Paris Match*)

up on the ocean floor. In July 1958, during a bathyscaphe descent by O'Byrne and Kumagori off south-east of Japan, one of the stabilizers struck a rocky cliff at a depth of about 3000 m. It was supposed that the bathyscaphe had been driven against a scarp on the sea floor by an unexpected current. However, on cautiously resuming the descent more jolts were felt, and presently the bathyscaphe was found to be scraping rock on both sides. It had wedged into a deep and wide fissure. Both stabilizers were badly cut, but fortunately the ascent was safely accomplished.

The appalling losses of human life that accompany great earthquakes in highly populated areas are mainly due to secondary causes such as the collapse of buildings, fires, landslides and the giant waves called *tsunami*[1] by the Japanese (*tsu*, harbour; *nami*, waves). Gas mains are torn open and, once started, fires rapidly spread beyond control, since the water mains are also wrenched apart. In San Francisco in 1906 far more damage was done by fire than by the earthquake itself. The Sagami Bay earthquake of 1923 occurred just as the housewives of Tokyo and Yokohama were cooking the midday meal. Fires broke out in all directions and completed the toll of death and destruction. At least 250,000 lives were lost and twice as many houses destroyed. In the loess country of Kansu in China, 200,000 people were killed in 1920 and another 100,000 in 1927 by catastrophic landslips of loess which overwhelmed cave dwellings, buried villages and towns and blocked river courses, so causing calamitous floods.

### The Lisbon Earthquake of 1755

The catastrophic earthquake that ruined Lisbon during the morning of All Saints' Day in 1755, one of the greatest on record, probably originated by sudden fault movements and fissuring of the sea floor to the west and south. The movements gave rise to two major shocks, about 40 minutes apart, of which the second was the greater, although the first had already brought down most of the buildings in avalanches of masonry. A third major shock about an hour later originated in the neighbourhood of Fez in Morocco, where the devastation was also on a catastrophic scale. The destruction of Lisbon by a triple earthquake and the accompanying fires and tsunami shocked Western civilization more deeply than any other natural calamity before or since (figure 25.7). In Portugal itself it seemed incomprehensible that such cruel blows should have occurred on a holy day, at the very time when the city churches were crowded with worshippers, many thousands of whom were crushed to death by the collapsing masonry. Abroad, the sense of horrified awe was intensified by a panic-born rumour, destined to become a widespread belief that still persists, to the effect that the broad marble quay alongside the Tagus, together with all the people who had fled to it in the vain hope of finding safety, had vanished for ever into a fissure that opened and closed again during the passage of the second and greatest of the main shocks. Engineers afterwards showed this story to be false. Although the marble blocks were loosened and dislodged they were still there. But the loss of life was no less real, for the quay and indeed all the lower parts of the city were swept by gigantic tsunami. Waves up to 9 m high rushed in and completed the ruin and destruction. The people on the quay were washed away and their disappearance added plausibility to the more dramatic story of their engulfment. Moreover, there was a similar report from a town in Morocco where many thousands of the population were said to have been 'swallowed up by the earth'. However exaggerated such accounts may be, they illustrate the undoubted fact that the possibility of engulfment—though dependent on accidental coincidence—has always been the most terrifying of the many perils associated with the great earthquakes. In Christendom the horror was magnified by fear of a sudden descent into the fiery torments of hell.

The Lisbon survivors only described what they had been taught to expect to happen, as witnessed by the appalling account of an earthquake recorded in the sixteenth chapter of the Book of Numbers. There we read that Korah and his associates and 250 followers had the temerity to accuse Moses and Aaron of taking too much upon themselves. Moses 'was very wroth'. Not content with rebuking them, he threatened them with the dire consequences of provoking the Lord:

'And it came to pass, as he had made an end of speaking all these words, that the ground clave asunder that was under them:

And the earth opened her mouth and swallowed them up, and their houses, and all the men that

[1] It should be noted that this word is plural, but is also used in the singular, like our English word *sheep*.

**Figure 25.7** Old wood engraving telescoping the story of collapsing buildings, fires, tsunami, and 'engulfment' of the marble quay that accompanied the Lisbon earthquake of 1 November 1755 (*Scribner, Walford and Co.*)

appertained unto Korah, and all their goods.

They, and all that appertained to them, went down alive into the pit, and the earth closed upon them: and they perished from among the congregation.

And all Israel that were round about them fled at the cry of them: for they said, Lest the earth swallow us up also.'

This last verse well expresses the general consternation felt throughout Europe after the Lisbon catastrophe. An unprecedented outburst of public discussion followed. The earthquake was variously attributed to the Devil and his legions of evil spirits; God's anger against the scandalous behaviour of his worshippers in the Lisbon churches; the need for frightening sinners into repentance; the need to punish Portugal for the undue severity of its Inquisition; and the need to remind humanity of the flames of hell-fire within the earth. John Wesley was particularly eloquent on the last of these themes. He boldly asserted that earthquakes were punishments for sin, and that any attempt to explain them away as natural consequences of the earth's internal heat was itself a sin. In a word, trying to account for earthquakes was doomed to produce more earthquakes.

The philosopher Kant, who was only thirty in 1755, took a very different line. He firmly condemned all attempts to interpret God's purpose as acts of outrageous impertinence. He regarded the Lisbon earthquake as a purely natural occurrence and made the sensible suggestion that we should find out where earthquakes were likely to happen and then take care not to build cities in such places. Rousseau also regarded earthquakes as a necessary part of nature and pointed out that we cannot expect to prevent an earthquake at a particular place merely by building a large number of churches there.

In the then narrow world of science the most notable contribution stimulated by the Lisbon earthquake came from John Mitchell (1724–93). Mitchell had already invented a torsion balance for determining the mean density of the globe and had done pioneer work in terrestrial magnetism and structural geology. Having collected what information he could about the effects of the great earthquake and their distribution, he published his tentative conjectures in 1760, but not without a salutary warning that he considered them to be still insufficiently supported by facts. What impressed him most was the enormous area over which the earthquake was felt and the evidence that it was something that travelled outwards from the neighbourhood of Lisbon, gradually dying out in all directions. He knew, for example, that distant lakes were set swaying, like water in a tilted bath, not only in Switzerland but as far away as Loch Ness, where the surface continued to rise and fall through a range of 60 to 90 cm for about an hour. This long-distance tilting he contrasted with the limited effects of volcanic earthquakes, which shake the surrounding country 'for only 10 or 20 miles'. Ascribing volcanic eruptions to the escape of vapour generated when large quantities of water come into contact with 'Subterranean Fire', he asks himself, *What is to be expected when the vapours are confined?* It was Mitchell who originated the idea that volcanoes act as safety valves in so far as they allow the upward escape of vapours. But, he thought, if the vapours originate too deeply to find a passageway to the surface, they must give rise to an explosive shock followed by a wave-like migration through the rocks in all directions. Mitchell was thus the first to recognize that from the region of the initial shock an earthquake spreads out in waves to distances so great that they can no longer be perceived. This was a remarkable achievement for two centuries ago. Mitchell may justly be hailed as the originator of the wave theory of earthquake transmission, and indeed as the founder of the vast subject that has since developed into geophysics.

## Earthquake Intensities and Isoseismal Lines

Mitchell's recognition that the destructiveness of an earthquake—or what we should now call its *intensity*—decreases outwards from the source of the disturbance stimulated attempts to define the degree of intensity in terms of information that could be supplied by people living in the area where the earthquake could be felt. From the human point of view damage depends on population density, building standards and the nature of the ground, weak alluvium being more susceptible than strong bedrock. Later it became possible to add the information recorded on seismographs to the effects described by observers. Intensity is now expressed by reference to an arbitrary scale of 12 degrees, which is summarized in the adjoining table. The present scale is a modification of a similar one (of 10 degrees) devised by the Italian seismologist, Mercalli.

In physical terms intensity is determined partly by the duration and number of jerks and tremors, but mainly by the maximum rate of change of these movements of the ground, i.e. by its maximum acceleration. This can be estimated from seismograph records. Approximate values of the accelerations have been added to the table for comparison, to indicate the range within which each empirical degree of intensity falls. In the same units the average acceleration of gravity ($g$) is 9800 mm/s$^2$. When and where this value is exceeded (intensity XII), the hammer-blows coming up from the ground throw everything that is loose into the air. Trees may be shaken loose and uprooted, and even well-casings are sometimes ejected.

It will be noticed that the divisions of the empirical scale are far from being equal. Each numeral indicating a degree of intensity is roughly proportional, not to the numerical value of the corresponding acceleration, but to its logarithm. In 1935 C. F. Richter devised a different type of logarithmic scale for comparing the magnitudes of Californian earthquakes. Since then his method has been widely extended and fruitfully developed. The *magnitude* of a tectonic earthquake is now defined so that it is closely related to the total amount of elastic energy released when the overstrained rocks suddenly rebound and so cause a shock (Figure 9.38). The relationship between the magnitude $M$, and the energy released, $E$, is given by the equation

$$\log E = 11{\cdot}8 + 1{\cdot}5M$$

where the energy, $E$, is expressed in ergs. For the definition of the *erg* as a unit of energy see p. 297. For a magnitude of 8·6, only three times reached and once exceeded during the present century, $E$ amounts to $10^{24{\cdot}7}$ erg. The average annual release of energy from all earthquakes ranges from about $10^{25}$ to $10^{27}$ erg and most of this, 80 per cent or more, is generally accounted for by a few really major shocks. For convenience of easy reference some numerical values of magnitudes have been added to the table. It must, however, be clearly realized that the magnitude assigned to a given earthquake corresponds only to the highest intensity of that earthquake. A *disastrous* earthquake, for example, spreads outwards from intensity X through *all* the lower intensities; but it has only *one* magnitude, which refers to the total energy set free by the shock. Since 1904, when seismograms

# Scale of Earthquake Intensities with Approximately Corresponding Magnitudes

| *Intensity* | *Description of characteristic effects* | *Maximum acceleration of the ground* | *Magnitude corresponding to highest intensity reached* |
|---|---|---|---|
| I | *Instrumental:* detected only by seismographs | | |
| | | 10 | |
| II | *Feeble:* noticed only by sensitive people | | 3·5 |
| | | 25 | to |
| III | *Slight:* like the vibrations due to a passing lorry; felt by people at rest, especially on upper floors | | 4·2 |
| | | 50 | |
| IV | *Moderate:* felt by people while walking; rocking of loose objects, including standing vehicles | | 4·3 |
| | | 100 | to |
| V | *Rather Strong:* felt generally; most sleepers are awakened and bells ring | | 4·8 |
| | | 250 | |
| VI | *Strong:* trees sway and all suspended objects swing; damage by overturning and falling of loose objects | | 4·9–5·4 |
| | | 500 | |
| VII | *Very Strong:* general alarm; walls crack; plaster falls | | 5·5–6·1 |
| | | 1000 | |
| VIII | *Destructive:* car drivers seriously disturbed; masonry fissured; chimneys fall; poorly constructed buildings damaged | | 6·2 |
| | | 2500 | to |
| IX | *Ruinous:* some houses collapse where ground begins to crack, and pipes break open | | 6·9 |
| | | 5000 | |
| X | *Disastrous:* ground cracks badly; many buildings destroyed and railway lines bent; landslides on steep slopes | | 7–7·3 |
| | | 7500 | |
| XI | *Very Disastrous:* few buildings remain standing; bridges destroyed; all services (railways, pipes and cables) out of action; great landslides and floods | | 7·4–8·1 |
| | | 9800 | |
| XII | *Catastrophic:* total destruction; objects thrown into air; ground rises and falls in waves | | >8·1 (maximum known, 8·9) |

first provided information from which magnitudes could be calculated, only a few shocks, including those of 1977 in China, have exceeded magnitude 8·4:

| | | |
|---|---|---|
| 1906 | Andes of Colombia/Ecuador | 8·6 |
| 1906 | Valparaiso, Chile | 8·4 |
| 1911 | Tien Shan, Sinkiang, China | 8·4 |
| 1920 | Kansu, China | 8·5 |
| 1933 | Japanese Trench | 8·5 |
| 1950 | North Assam, India | 8·6 |
| 1960 | Chile (three major shocks) | 8·3–8·9 |
| 1964 | Alaska | 8·6 |

The San Francisco earthquake of 1906 had a magnitude of 8·25; that of the Sagami Bay earthquake of 1923 (pp. 569–71) was 8·2.

Returning now to the earlier work, based on intensity, we can learn a good deal from this simpler concept, as did the pioneer seismologists. A line drawn through all places with the same intensity is an *isoseismal line* (Figure 25.8). Each one generally encloses a roughly circular or elliptical area, according as the place of origin of the earthquake is point-like or elongated. The place of origin is called the *origin* or *focus*, and the point or line on the surface vertically above is the *epicentre* or *epicentral line*. From the intensities at the epicentre E and at a point G on an isoseismal line at a known distance from E, R. D. Oldham showed, at least in principle, how the depth of focus could be estimated (Figure 25.9).

From the focus the intensity (expressed in terms of acceleration) theoretically decreases outwards inversely as the square of the distance. Thus to a first approximation we have $n/m = h^2/r^2 = (\sin\theta)^2$. The angle $\theta$ being thus determined, $d \tan\theta = h =$ the depth of the focus.

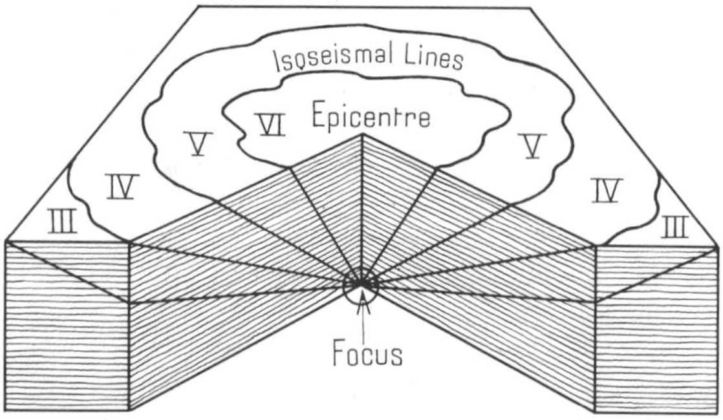


**Figure 25.8** Block diagram showing isoseismal lines and their relation to the epicentre and to the wave paths radiating from the focus of an earthquake. The Roman numerals indicate intensities.

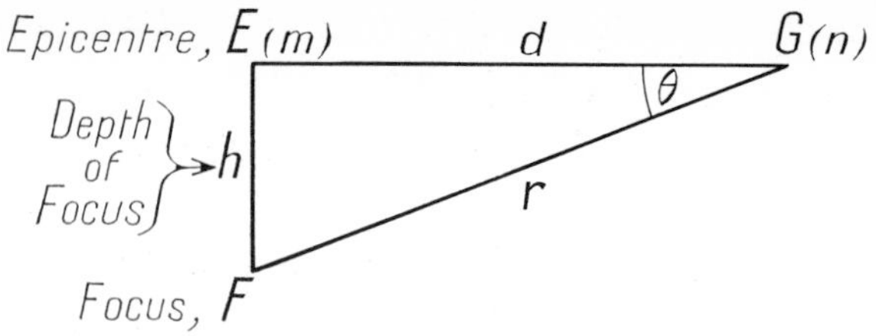


**Figure 25.9** To illustrate Oldham's method for estimating to a first approximation the focal depth ($h$) of an earthquake
Intensity at E=$m$ (known)
Intensity at G=$n$ (known)
Distance between E and G=$d$ (known)
*Problem:* to find $h$, the depth of the focus

From the records of 5605 shocks in Italy, Oldham found that 90 per cent of the earthquakes originated at depths of less than 8 km; nearly 8 per cent at depths between 8 and 30 km; and the rest at greater depths. Tectonic earthquakes are now classified as:

*Shallow:* when the depth of origin is less than 60 km (or 70 km, the depth adopted by some seismologists).

*Intermediate:* when the depth of origin is between 60 (or 70) and 300 km.

*Deep:* when the depth of origin is more than 300 km, the maximum depth so far recorded being about 720 km. Most deep earthquakes originate between 500 and 700 km.

Volcanic earthquakes are caused by phenomena such as gas explosions; or the updoming and fissuring of volcanic structures (figure 25.10) and their foundations during the period preceding an eruption; or the formation of calderas. They commonly occur in 'swarms' and are generally of shallow origin. For this reason the area of sensible disturbance is correspondingly small (a few hundred square kilometres being rarely exceeded), although the intensity may be high near the volcano itself. In contrast, tectonic earthquakes may be felt over areas a thousand times as great.

## Tsunami

The giant sea waves often associated with earthquakes of high magnitude are caused by displacements of water due to sudden large-scale changes of level of the sea floor, e.g. by the fault movements responsible for the earthquake or by submarine slumping set off by the earthquake. Related to the latter is the landsliding into deep

**Figure 25.10** Fissure opened during a swarm of earthquakes preceding a volcanic eruption, south-west rift zone of Kilauea, Hawaii (*Alex. Cockburn*)

water of immense volumes of rock from the oversteepened, glaciated sides of certain fjords. One of the most impressive giant waves generated in this way occurred in an inlet of the Gulf of Alaska, SE of the Malaspina Glacier, immediately after an earthquake in 1958. Heavy rainfall following frequent repetitions of frost and thaw had already weakened the highly fractured belt of rocks at the head of the inlet, and the earthquake shook loose a rock-avalanche so vast in volume that the displaced water started a wave with a steep front that rose to a height of 30 m or more and reached a velocity of 200 km an hour. The forest was destroyed for several kilometres along the shores and in places the momentum of the surging water carried it up to at least 525 m, as proved later by the height to which trees had been stripped of their bark and the bedrock of its covering of soil.

Tsunami are usually the result of topographic changes on the sea floor brought about by powerful earth movements occurring along deep-water coasts, such as those on the sides of the great Pacific trenches. Much less commonly the displacement of the sea floor results from the formation of a submarine caldera (e.g. the Krakatau tsunami of 1883), but in all cases the same general principles apply. If the displacement causes a large depression of the ocean surface, water is drawn in from all sides and throughout the whole depth, which may be very great. One manifestation of this inward flow is the menacing withdrawal of the sea from neighbouring coasts that commonly portends the onset of a dangerous tsunami. If the sea-floor displacement is upwards and causes a widespread upheaval of the ocean surface, water flows outwards in all directions and at all depths. However, whether the displacement is up or down, the momentum of the vast volume of moving water carries it far beyond its position of rest or equilibrium. The resulting ebb and flow takes the form of a series of gradually subsiding, oscillating waves. In deep water these waves have wave lengths of hundreds of kilometres and travel at speeds of hundreds of kilometres an hour, but they are no more than a metre high, so that they pass unnoticed by ships in the open ocean. Nevertheless the energy being transmitted is immense, since the whole depth of water is involved. Consequently, when these waves reach shallowing water and narrowing bays and inlets they rise to heights that swiftly become terrifyingly destructive near their source, and may again become dangerous after travelling and dispersing through thousands of kilometres (Figures 25.11 and 25.12).

**Figure 25.11** Sandy Beach, near Hilo on the NE coast of Hawaii, a few minutes before a tsunami, resulting from a powerful earthquake that originated beneath the Aleutian Trench, tore across the beach and the highway, with the results seen in Fig. 25.12 (*Honolulu Advertiser : Photo Y. Ishii*)

**Figure 25.12** Vehicles passing along the coastal highway were swept up the slopes by the sudden onset of the tsunami. A man fleeing for safety can be seen on the left. The position of the highway can be recognized by the road sign, right centre. (*Honolulu Advertiser : Photo Y. Ishii*)

Theoretically the velocity $v$ in deep water is given by $v^2 = gd$, where $g$ is the acceleration of gravity (about 960, or say 1000, cm/s$^2$) and $d$ is the depth of water. If $d$ is 4 km=400,000 cm, then $v^2$ is 400,000,000, and $v$=20,000 cm/s=720 km/h. This was first verified in practice by the study of the Krakatau tsunami, which were picked up by tide gauges in all the chief harbours of the world. Speeds across the Indian Ocean varied between 565 and 720 km an hour, according to the differing depths of water along the paths traversed between Krakatau and the tidal stations, e.g. Aden and Cape Town.

Tsunami originating with a major earthquake on the western slopes of the Japanese Trench in 1933 (magnitude 8·5) raised mountainous waves, with surges up to 27 m high, along the shores of some of the Pacific bays and inlets of Japan, where thousands of people were drowned. The waves were recorded at San Francisco about ten hours later, having traversed the Pacific at 756 km an hour. The waves took nearly twenty hours to reach Iquique in the north of Chile, the average speed being a trifle less because the San Francisco route is somewhat deeper. Iquique and other parts of the Pacific coast of S America have their own long history of destructive earthquakes and tsunami, some of the latter having been strong enough to be seen in Hawaii and recorded in Japan. Outside the Caribbean the Atlantic shores have not suffered seriously from tsunami since the outstanding Lisbon example of 1755, which was sufficiently powerful to raise visible waves in the North Sea after passing up the Channel.

Following powerful earthquakes that originated beneath the deep trenches of the western Pacific, about 150 disastrous tsunami have afflicted the Japanese coasts since records of earthquakes first began to be kept there. In recent years it has been noticed that submarine earthquakes with magnitudes over 8 have invariably been followed by tsunami. Despite the enormous number of earthquakes of lesser magnitudes very few of these have been associated with tsunami. It is therefore suspected that those few, and probably some of the others, were started by large-scale slumping and turbidity currents unloosed by the sudden movements. It should also be noticed that movements along strike-slip faults would be unlikely to cause tsunami by themselves, since the vertical crustal displacements would be negligible. But avalanching of thick layers of sediment down the steeper sides of the offshore trenches might cause—and in some cases are known to have caused—sea-floor changes of level amounting to a hundred metres or so.

Surprise is often expressed that the inhabitants of the 'low islands' of the Pacific and other mid-oceanic atolls should escape being swept away by the passage of tsunami. Their safeguard is twofold. These island homes rise steeply from the sea floor, like pinnacles with diameters that are very short compared with the wave lengths of the tsunami in deep water. It is the energy that is transmitted, not the water. Only where the ocean floor gradually shallows over a broad area, as for example across the great swell from which the Hawaiian Islands rise, is the low wave-front sufficiently retarded to steepen appreciably. Then, where the shore topography is appropriate, it may quickly become a wave of translation, like a bore (cf. p. 500). Figures 25.11 and 25.12 illustrate the swift growth of such a wave near Hilo, from a tsunami that originated over 3600 km away in the Aleutian Trench.

## Distribution of Epicentres: Earthquake Zones

Earthquakes may be recorded anywhere, but the places where they originate are confined to regions where earth movements are in operation, or where volcanoes are active. Figure 25.13 shows in a generalized way the distribution of known epicentres during the present century. Most earthquakes originate within three well-defined zones: a *Circum-Pacific zone*, a *Mediterranean* and *Trans-Asiatic* zone, and a zone follows the *mid-oceanic ridges*, with an extension along the African rift-valleys.

The Circum-Pacific zone closely follows the deep ocean trenches and associated island arcs, the quakes being located on the continental sides of the troughs. From Alaska, down the western side of the Pacific, it lies parallel to the Kurile, Japan, Marianas and Philippine trenches. Then one branch crosses to the Keramac-Tonga trenches to the north-west of New Zealand, whilst another swings round to the west parallel to the Java or Indonesian trench.

On the eastern side of the Pacific, the earthquake zone follows the west coast of North America, being particularly characteristic of California, although there is no ocean trench associated with it there. It continues southwards parallel to the

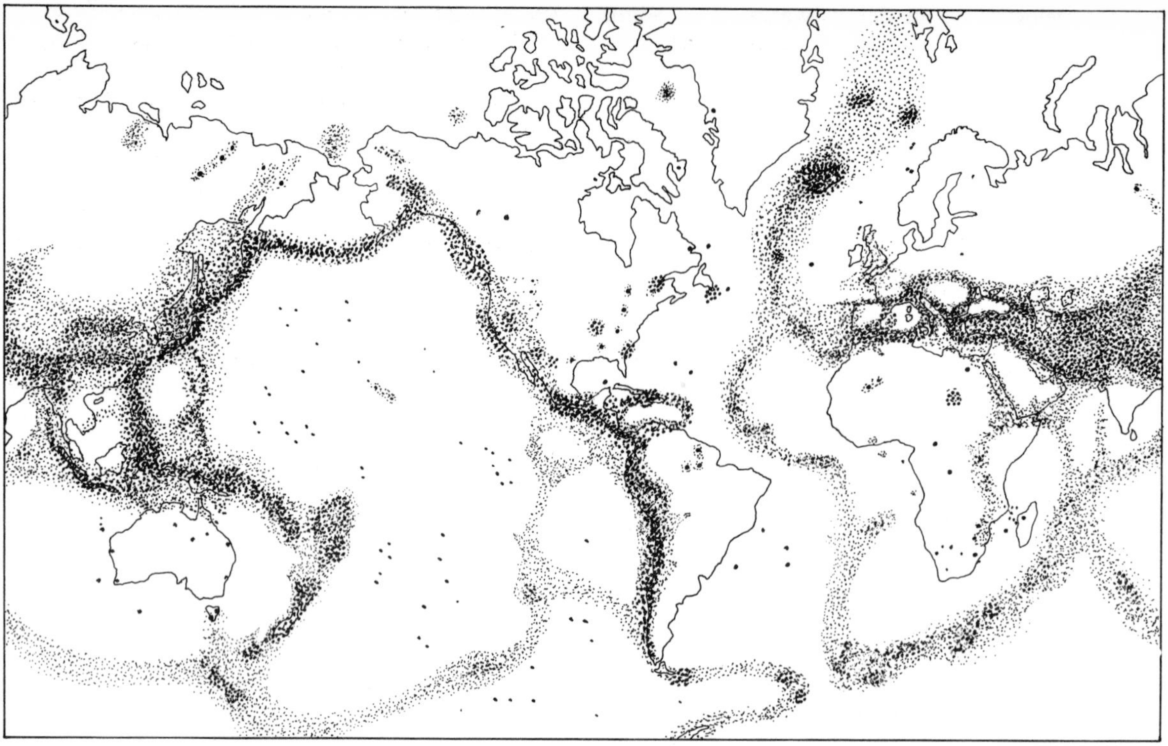

**Figure 25.13** Map showing the distribution of the epicentres of earthquakes during the present century. Heavy shading indicates the occurrence of frequent earthquakes, including most of those of great magnitude. The density of shading is relatively exaggerated along the mid-oceanic and rift-valley belts in order to bring out more clearly the significant pattern of distribution in these regions.

Middle American trench, and on down the Andes where it is associated with the Chilean trench.

The earthquakes along the Circum-Pacific zone include both shallow, intermediate and deep earthquakes. Indeed deep earthquakes are practically restricted to the Circum-Pacific zone, the foci deepening from shallow, through intermediate to deep in a landward direction from the margins of the trenches. This correlation between the deep ocean trenches and the earthquake belt obviously must have some high significance. As we shall see in the next chapter, the evidence suggests that the earthquake zone depicts the routes along which slabs of oceanic lithosphere move downward into the mantle along the trenches.

The second earthquake belt, that of the Mediterranean and Trans-Asiatic zone, extends along the Alpine mountain arcs of Europe, and North Africa, through Asia Minor and the Caucasus, Iran and Pakistan, to the Pamirs, Himalayas, Tibet and China. This zone is characterized by most of the larger quakes of shallow origin, and others of intermediate origin. For a long time this earthquake zone was thought to be free from deep earthquakes, but in 1954 a powerful earthquake originated at a depth of 630 km under the southern slopes of the Sierra Nevada in Spain. The Mediterranean and Trans-Asiatic earthquake zone loosely follows belts of Tertiary and Recent mountain-building. There are no associated deep-ocean trenches that are visible; indeed this earthquake zone lies inland, but attempts are being made to discover whether now-vanished trenches once existed.

Most of the remaining earthquakes are of the shallow variety, and are located along the mid-oceanic ridges and transform faults that intersect them. Along the transform faults they are confined to those parts that lie between and connect seemingly off-set portions of the ridges. On Figure 25.13, this earthquake zone, which continues along the rift-valleys of Africa, is greatly exaggerated. This exaggeration, however, emphasizes the fact that the earthquake zones jointly outline areas of the earth's surface that are relatively inert, and form a mosaic pattern. These inert areas are the plates of lithosphere of Chapter 29, the relative movements of which are the fundamental cause of the earthquakes.

## Seismographs

From the focus of an earthquake, waves are propagated through the earth in all directions, and when they arrive at a seismological station they are recorded on seismographs, provided they are not too vigorous to put the instrument out of action. Instruments for detecting earthquake waves have been devised in great variety (Figure 25.14). Between the two world wars most of those in common use were delicately poised pendulums designed to record either the horizontal or the vertical components of the vibrations of the ground. A good seismological station would have two horizontal seismographs mounted at right angles—usually one to respond to N–S movements and the other to E–W movements—and one vertical instrument; the three giving a complete record of the movements in three dimensions.

Here it must suffice to indicate the principles from the kind of apparatus that provided most of the data of geological interest up to thirty years ago. The essential requirement for recording the passage of earthquake waves is to have a point that remains stationary, or nearly so, while each earth tremor shakes the whole instrument, including its recording system. The horizontal pendulum illustrated in Figure 25.15 consists of a weighted boom pivoted against a massive support which is firmly attached to the ground so that it shares in the vibrations of the latter. Because of its inertia the weight tends to remain stationary. It is the relative movement between the end of the boom and the rest of the instrumentation that is recorded, after being suitably magnified.

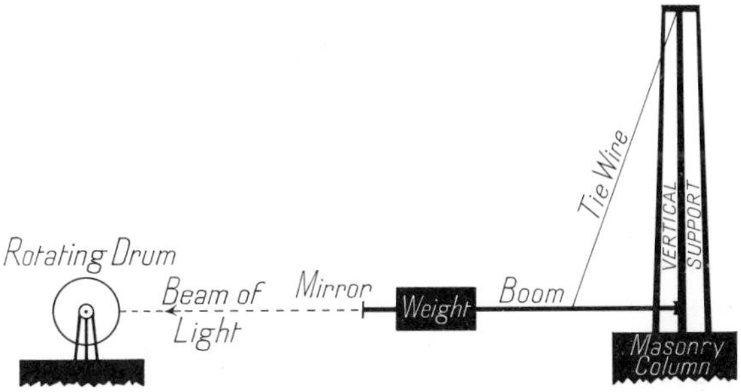


**Figure 25.15** To illustrate the essential parts of a horizontal seismograph of the Milne-Shaw type

A small mirror attached to the end of the boom reflects a beam of light on to a sheet of photographic paper wrapped around a cylindrical drum which rotates on a long screw; while rotating, the drum carries the paper along at right angles to the beam of light. The length of the beam determines

disturbed by an earthquake, and operated a mechanism which knocked a ball from the nearest dragon's head into the mouth of the frog beneath. (*The Science Museum, London*)

and the acute necessity for distinguishing natural earthquakes from nuclear explosions have stimulated the design of instruments making use of electromagnetic properties that are highly sensitive to relative movements. Examples are the change of electrical resistance in certain crystals in response to minute changes of pressure caused by the passage of vibrations; and the feeble current set up in a weighted coil suspended in a magnetic field when there is relative movement between the two. Small portable detectors of the latter kind—called *geophones*—are part of the standard equipment employed in seismic prospecting. One highly successful type of geophone has a powerful permanent magnet that is rigidly attached to the container (itself perhaps no bigger than a 5 cm cube), which in turn, when in use, is firmly buried in the ground. When a vibration arrives the current generated in the suspended coil is fed to a recording truck in which the rest of the instrumentation is housed. There the current passes into a valve amplifier connected with a galvanometer, the deflections of which are photographically recorded.

On a larger scale a very efficient system of detection and recording was installed in 1962 at Eskdalemuir, near Dumfries in the south of Scotland. This consists of a battery of instruments in steel cylinders, buried in the ground at intervals along two 8-kilometre stretches at right angles, and capable of picking up vibrations from any part of the globe. Each detector is connected to the observatory, where the vibrations are automatically recorded, and the records speedily interpreted with the aid of electronic computers.

### Seismic Waves

Records of distant earthquakes are generally complex and difficult to interpret without long and varied experience. In the simpler seismograms, such as that illustrated by Figure 25.16, three conspicuous pulses can be picked out (P, S and L), particularly when the earthquake originates at a distance of about 10° to 50° from the recording station. The first arrival is the *p*rimary or P wave, which is followed by a train of irregular oscillations. Then comes a *s*econdary pulse or S wave also followed by a train of oscillations. P and S are body waves that have travelled through the earth by routes like that shown in Figure 25.17. The concave form of the path towards the surface indicates that their velocities increase with depth. Although the existence of P and S waves in solids was predicted from the theory of elasticity by Poisson in 1829, it was not until 1899 that Oldham first recognized them on earthquake records.

P or '*p*ush-and-*p*ull' waves are *compression-dilatation* (or *compression-rarefaction*) waves like those of sound, in which each particle vibrates to and fro in the direction of propagation (Figure 25.18). We are all familiar with the transmission of sound through air and water as well as solids, and this exemplifies the more general fact that P waves pass through gases, liquids and solids alike. S or '*s*hake' waves are *transverse* or *shear* waves, in which the motion of each particle is at right angles to the direction of propagation, like the loops in a taut rope shaken at one end. Shear waves pass only through solids: not through liquids. Through rocks of the same kind P and S waves travel at different speeds because they depend on different properties. The velocity of P depends on density and compressibility (resistance to compression), while that of S depends on density and rigidity (resistance to distortion or shearing). As a rule P waves travel at about 1·7 times the speed of S waves.

**Figure 25.16** Seismogram recorded at Pulkovo Observatory, Russia, of an earthquake originating in Asia Minor on 9 February 1909. The time interval S–P is 3 minutes 43 seconds, corresponding to a distance of 2253 km from the epicentre. (*After B. Galitzin*)

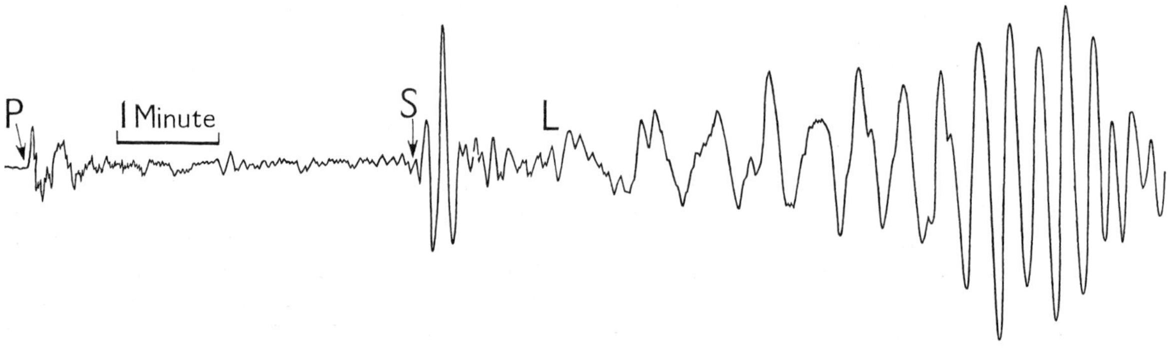

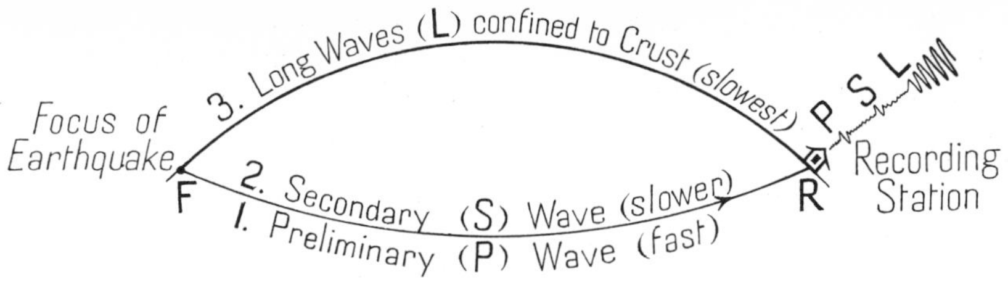


**Figure 25.17** Section through a segment of the earth, showing the paths followed by the P, S and L waves of an earthquake originating at the focus F and recorded at a station R

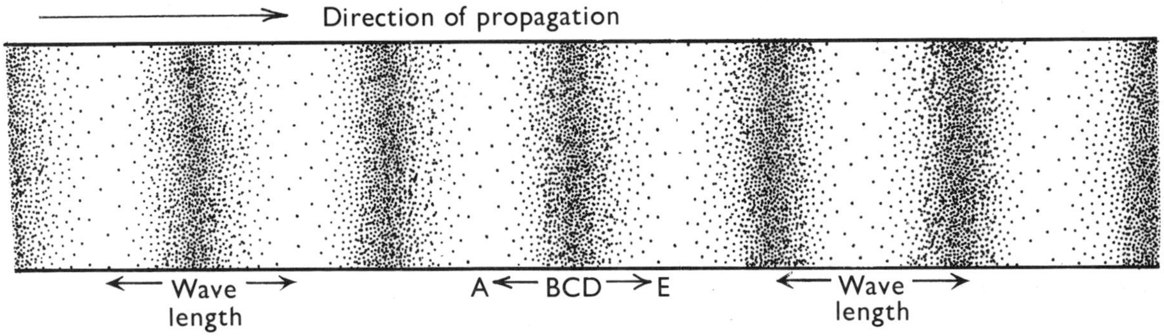


**Figure 25.18** Diagrammatic representation of compression-rarefaction waves (P waves). The initial shock sets up a zone of compression or rarefaction which rapidly moves outwards (speed in granite about 5·6 km/s), although the to-and-fro movements of the individual particles are very slight.
A zone of rarefaction behind C
C zone of maximum compression
E zone of rarefaction in front of C
B compressed particles moving back towards A
D compressed particles moving forward towards E

Besides transmitting P and S waves the earth can transmit two other kinds of waves which are guided around the neighbourhood of the surface by continuous reflections at the top and bottom of the surface layers. These 'surface waves', as they are called, are responsible for the third train of pulses, the onset of which is marked L in Figure 25.16. They are collectively known as L waves because they have *l*ong periods. Their prominence on seismograms recorded beyond a certain distance from the source is due to the fact that they spread out and disperse their energy nearly according to $1/d$ ($d$ being distance from the source), whereas the body waves, P and S, disperse their energies according to $1/d^2$, which rapidly becomes much smaller than $1/d$ for increasing values of $d$.

L waves arrive later than P and S because of the greater complexity of their paths through the crustal layers. One of the two kinds of L waves was predicted by the third Lord Rayleigh in 1887, twenty years before it was recognized on seismograms. In *Rayleigh waves* the motion of particles is in elliptical orbits in the plane of propagation—analogous to the nearly circular orbits of water particles during the passage of waves over lakes or the sea. In the second kind of L waves the motion is horizontal and at right angles to the direction of propagation. These waves were recognized on seismograms before they had been accounted for. It was an Oxford mathematician, A. E. H. Love, who explained them by an extension of Rayleigh's theory, and since then they have been known as *Love waves*. Both Rayleigh and Love waves fade out at various depths according to their periods and consequently they provide invaluable information for distinguishing between continental and oceanic types of crust (cf. Figure 26.4).

As a background to the P, S and L waves and their many associates there are small irregular earth tremors and quiverings going on all the time, like the noise of traffic that may be heard in a concert hall whether a concert is in progress or not. These *microseisms*, as they are called, set a limit to the degree of magnification that can usefully b employed, since they only confuse the earthqua record if they are made too big. Some of the m conspicuous microseisms are caused by d traffic, others by the pounding of breakers or coasts, while others have been traced to ch atmospheric pressure and especially to h and typhoons. But after all such region of microseismic activity have been ac there still remains a world-wide backg chaotic seismic 'noise'. Don and Florence have suggested that these microseisms are caused

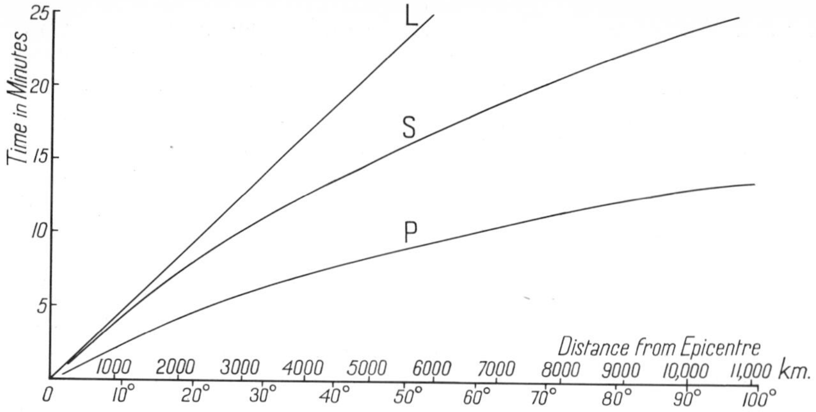


**Figure 25.19** Time-distance curves for P, S and L waves (the curve for L waves refers only to those that have traversed the continental type of crust).

by the strained condition of the crust, which 'hums' or 'sings' like a highly strained piece of steel.

When the *times* taken by P and S waves to travel to seismological stations in various parts of the world are plotted against the respective *distances* from the epicentres, they fall on smooth curves which are nearly the same for all distant earthquakes except those originating at great depths. Figure 25.19 shows at a glance that the time intervals between the arrivals of P and S steadily increase with distance. Thus, when an earthquake is recorded at a distant station, measurement of the interval between the arrival times, S–P, serves to determine to a first approximation both the distance from the epicentre and the time of origin of the shock. At least three such distance determinations at well-spaced stations are necessary for a first indication of the position of the epicentre (Figure 25.20). Greater accuracy is attainable when several determinations are available, and especially when the depth of the focus has also been calculated from the records, since S–P really refers to the distance of the recording station from the focus. When the focus is deep the difference between focal distance and epicentral distance may be considerable (cf. Figure 26.12). In practice there is close international co-operation for effective exchange of information. Some of the larger observatories act as regional clearing-houses, e.g. in Japan, New Zealand, France and Britain, while the U.S. Coast and Geodetic Survey serves this purpose on as nearly a world-wide scale as the political situation permits. For every well-recorded shock they receive many hundreds of cabled sets of readings, and within two days or less the position of the epicentre and the time of origin can be cabled back to the co-operating stations.

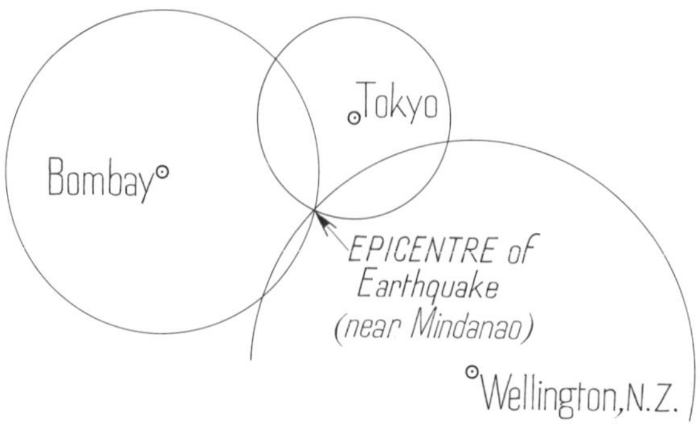


**Figure 25.20** To illustrate how the epicentre of an earthquake can be fixed when its distances from three suitably placed stations are known. A circle is drawn on a globe around each of the three stations (*e.g.* Bombay, Tokyo, and Wellington), with a radius corresponding in each case to the respective distance of the epicentre.

The epicentre lies at the point of intersection of the three circles.

## The Internal Zones of the Earth

The idea of 'X-raying' the globe with its own earthquake waves was introduced on p. 12. Seismograph records of distant earthquakes provide information of some of the properties of the interior at different depths, just as rays of light bring to our eyes the colours of, say, the internal parts of an intricately designed glass paperweight. We see the internal structure of the latter, and similarly, by a comparative study of seismograms of the same earthquake recorded at stations well distributed over the earth, we can 'see' at least the dominant features of the internal structure of the globe. Whether an earthquake is regarded as 'near' or 'distant' depends on the distance of its epicentre from the station where it is recorded. If the station is within a few hundred kilometres of the epicentre and the earthquake originated within the crust, then the P and S waves that are recorded there include those that have travelled only through crustal rocks. The results are of great importance in working out the thickness of the crust and its variation from place to place. Moreover, near earthquakes can be generated by means of artificial explosions and picked up by a whole battery of small instruments at suitable distances. Some of the details of crustal structure, continental and oceanic, are reviewed later. Here we shall first consider the P and S waves of distant earthquakes, which descend through the Moho into the mantle before coming up to the surface again (Figure 25.17).

The travel times of the P waves indicate that their velocities increase with depth from about 7·8–8·1 km/s near the top of the mantle (where there are some significant variations) to about 13·6 at the bottom, i.e. at a depth of nearly 2900 km. For S waves the corresponding increase is from 4·35 to 7·25 km/s. The waves that just attain this depth emerge at the surface at places about 11,500 km (105°) from the epicentre of the earthquake concerned. At stations beyond this distance there is a striking and critical change in the records. No P or S waves appear—although the L waves come along at the appropriate times—and this type of record is all that is received by stations within the next 4500 km. At distances greater than 16,000 km (143°) from the epicentre the P wave reappears, and continues to do so right up to the antipodes of the epicentre (the anticentre, as it is called). Corresponding to each earthquake of sufficient magnitude to be recorded so far from its source there is a ring-like shadow, free from P and S waves, as illustrated in Figures 25.21 and 25.22. This shadow zone was first noticed by Oldham in 1906. He at once inferred that the earth has a core which acts like a spherical lens by refracting the deeper waves inwards, so concentrating them in the antipodal region at the expense of a surrounding zone of shadow.

In accordance with this strong refraction the velocity of the P waves is greatly reduced when they pass from the mantle into the core (Figure 25.23). But the S waves are not transmitted at all, so indicating that the discontinuity between mantle and core is also a boundary between states of matter that behave towards seismic waves as if they were solid and fluid respectively. It was at first expected that the core would be fluid throughout.

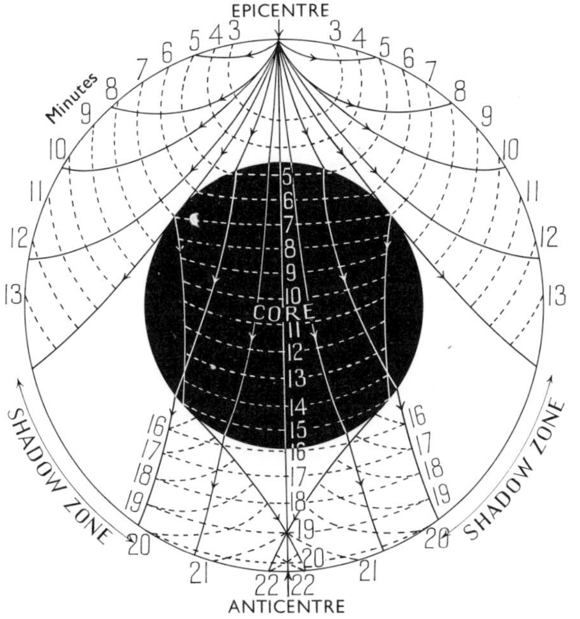


**Figure 25.21** Section through the centre of the earth showing the wave paths (firm lines with arrows), wave fronts (dotted lines), and arrival times (in minutes reckoned from the zero time of the shock). Since there is a shadow zone free from P and S waves for each such earthquake, it is inferred that the earth has a core which refracts the deeper waves as shown in the diagram.

**Figure 25.22** The shadow zone cast by the earth's core in the case of an earthquake originating in Japan

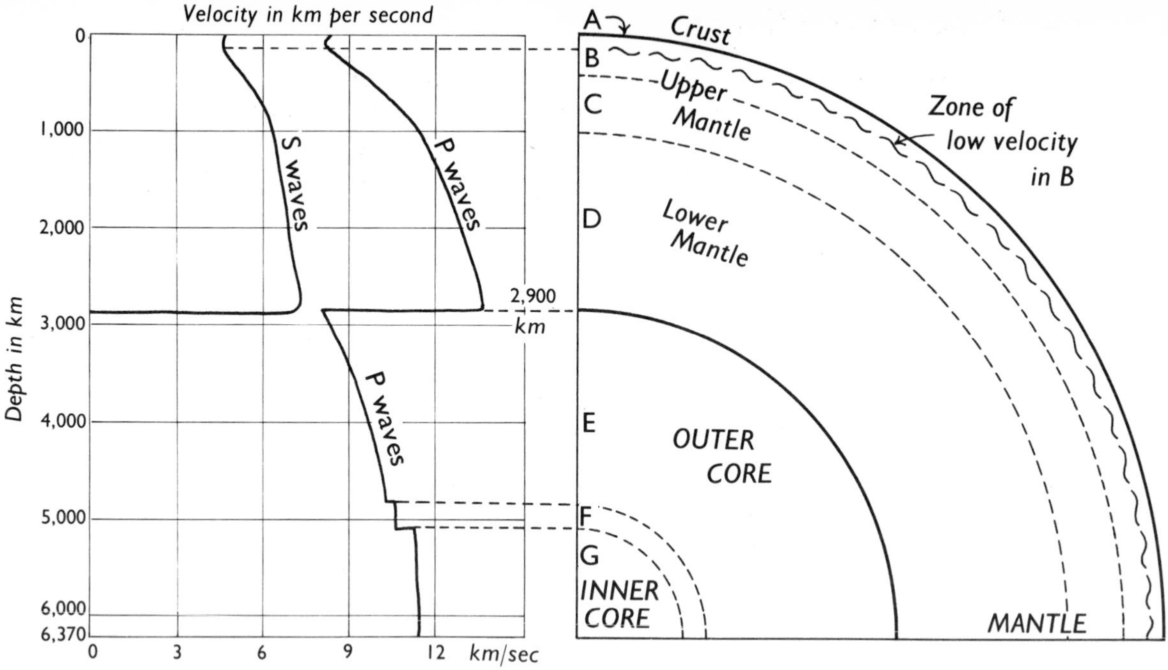


**Figure 25.23** Diagrams showing the variations in velocity of P and S waves, from the surface to the centre of the earth; and the inferred structure of a quadrant of the earth, lettered in accordance with K. E. Bullen's classification of the successive zones

However, in 1936, Lehmann, a Danish seismologist, detected a still deeper discontinuity enclosing an inner core which she suspected to be solid. Seismic records of underground H-bomb explosions (in which the time and place of origin are exactly known) have confirmed her tentative conclusions. Until 1962 the inward passage from fluid to solid was thought to be a gradual change spread over a transitional zone (F in Figure 25.23) about 160 to 320 km across. B. A. Bolt of Sydney then showed that agreement between observation and theory is greatly improved by regarding F as an independent zone with fairly sharp boundaries against the outer (E) and inner (G) parts of the core. The change from mantle to core is known to be sharply defined: i.e. no transition zone can be detected, as it should be if a zone were more than 2 or 3 km thick. It is of interest to notice in passing that the planet Mars has nearly the same radius (3332 km) as the earth's core (3473 km). Mars, however, has probably no core, its mean density being little more than 3·6.

When the velocities of P and S waves at various depths within the earth are known, it is possible by making certain reasonable assumptions to estimate the distribution of density with depth. The results are acceptable only if they provide the earth with the correct moment of inertia as well as with the correct mean density, both of which are accurately known. The calculated density distributions that survive these severe limitations are all very similar and they all involve a heavy core. There is still some uncertainty about the nature of density changes within the mantle. K. E. Bullen has found evidence of slight departures from a regular downward increase at depths of 413 and 984 km, suggesting a division of the mantle into three parts. These are labelled B, C and D in his lettered classification of the earth's internal zones, A being the crust. The table on p. 587 summarizes the data corresponding to our present state of knowledge. Densities and pressures are thought to be correct within 2 or 3 per cent.

## Selected References

Birch, F., 1960, 'The velocity of compressional waves in rocks to 10 kilobars: Part I', *Journal of Geophysical Research*, vol. 65, pp. 1083–102.

Bullard, E. C., 1957, 'The density within the earth', in *Gedenkboek F. A. Vening Meinesz*, pp. 23–42, Mouton, The Netherlands.

# Internal Zones of the Earth

(From data computed by Harold Jeffreys, K. E. Bullen, E. C. Bullard and B. A. Bolt)

| *ZONES* | | *Depth (and radius) of boundaries in km* | *Velocity of P waves km/s* | *Density gm/cm³* | *Pressure in $10^6$ bars: approx. in atmospheres** |
|---|---|---|---|---|---|
| | | Sea-level (6371) | | 2·7 | |
| CRUST (continental)† | A | | | 2·8 | |
| | | | | 2·9 | |
| *Mohorovičić Discontinuity* | | 33 (6338) | | | 9000 |
| | | | 7·9–8·1 | 3·32 | |
| | | 50 } Low velo- | 7·8 | | |
| | | 250 } city zone | 8·1 | | |
| | B | | | | |
| UPPER MANTLE | | 413 (5958) | 8·97 | 3·64 | 140,000 |
| | C | (720: deepest earthquakes) | 10·70 | 4·29 | 270,000 |
| | | 984 (5387) | 11·42 | 4·64 | 382,000 |
| LOWER MANTLE | D | | | | |
| | | | 13·64 | 5·66 | |
| *Oldham or Gutenberg Discontinuity* | | 2898 (3473) | | | 1,368,000 |
| | | | 8·10 | 9·71 | |
| OUTER CORE | E | | | | |
| | | 4703 (1667) | | 11·76 | 3,180,000 |
| | | | 10·31 | | |
| 'TRANSITION' | F | | | | |
| | | 5154 (1216) | | c.14 | c. 3,300,000 |
| | | | 11·23 | | |
| INNER CORE | G | | | | |
| | | 6371 (Centre) | | c. 16 | c. 3,600,000 |

*Standard pressure of 1 atmosphere = 1,031,250 dyne/cm²
1 bar = 1,000,000 dyne/cm²

†In oceanic areas Zone A extends from the ocean floor at a mean depth of 4·8 km to the Moho at a depth of about 10–11 km below sea-level.


Bullen, K. E., 1954, *Seismology*, Methuen, London.

— 1963, *An Introduction to the Theory of Seismology*, Cambridge University Press, London.

Davison, G., 1936, *Great Earthquakes*, Allen and Unwin (Murby), London.

Iida, K., 1956, 'Earthquakes accompanied by Tsunami', *Journal of Earth Sciences, Nagoya University*, vol. 4, pp. 1–43.

Kendrick, T. D., 1956, *The Lisbon Earthquake*, Methuen, London.

Matuzawa, Takao, 1964, *Study of Earthquakes*, Maruzen, Tokyo.

Oakeshott, G. B. (Ed.), 1955, 'Earthquakes in Kern County, California, during 1952', *Bulletin 171, Division of Mines*, San Francisco, California.

Richter, C. F., 1958, *Elementary Seismology*, W. H. Freeman, San Francisco.

Tsuboi, C., *et al.* (Earthquake Prediction Research Group), 1962, *Prediction of Earthquakes*, Earthquake Research Institute, Tokyo.

Wartnaby, J., 1957, *Seismology: Historical Survey and Catalogue of Exhibits, Science Museum, London*, H.M. Stationery Office.

# Chapter 26

# The Crust, Mantle, Moving Lithosphere, and Core

The Earth has a spirit of growth

*Leonardo da Vinci* (1452–1519)

## Seismic Exploration of the Crust

As already indicated, records of P and S waves from near earthquakes have been extensively used for estimating the nature and thicknesses of the dominant layers that make up the earth's crust. The waves travel with velocities that depend on the densities and elastic constants of the rocks through which they pass. For P waves these range from very low speeds, such as 2 km/s for loose or poorly consolidated sediments; through a wide range of higher values for indurated sediments; to 5·2–6·2 km/s for sialic crystalline rocks, including granite and granodiorite; and to 6·2–7·2 km/s for basic rocks such as gabbro and its common metamorphic equivalents. Eclogite, however, goes up to 7·9–8·1 km/s. The velocity for a particular kind of rock is found experimentally to vary not only with its composition and structure, but generally to increase with rising pressure (i.e. with depth), while decreasing only slightly with rising temperature.

The principles involved in seismic exploration of the crust can be most easily illustrated by considering the simplest type of case that arises in seismic prospecting. Here the vibrations are provided by artificial explosions, which have the great advantage over natural shocks that they are under complete control and can be used wherever and whenever happens to be convenient. A charge of dynamite is lowered into a shot hole previously drilled to a depth of a metre or so. The charge is detonated electrically at an exactly recorded time. The place of origin is exactly known and the depth of focus is negligible. The problem is to determine the thickness $h$ of an upper layer of rock (through which the P waves travel at velocity $v_1$) resting on a lower layer in which the velocity $v_2$ is considerably greater than $v_1$ (Figure 26.1).

The waves from the explosion spread out in all directions from the shot point. Some (*a*) reach the surface directly through the upper layer. Of those that reach the boundary between the two layers, some (*b*) are reflected back to the surface; some (*c*) are refracted along the boundary, where they travel with the higher velocity $v_2$ and are bit by bit returned to the surface from each point along the boundary; and some (*d*) are refracted into the lower layer until they in turn are reflected or refracted at the next boundary. At the surface the time of arrival of the P wave can be recorded wherever a suitable instrument is placed for its reception. In practice geophones $G_1$, $G_2$ . . . up to, say, $G_{20}$ are fixed in the ground in line with the shot point at known distances, and all are linked with the recording van. The geophones may be spaced to pick up the reflected waves (*b*) to best advantage, as in echo-sounding at sea. This method, known as 'reflection shooting', is in common use in the exploration for oil-bearing structures. A different spacing of the geophone is adopted for 'refraction shooting', designed to pick up waves (*c*), and also, waves (*d*) that penetrate more deeply before returning to the surface.

Figure 26.1 refers to the refraction method in its simplest form. The waves (*a*) are the first to reach the geophones ($G_1$–$G_3$) situated near the shot point. Beyond a certain distance, $l$, the refracted waves from the top of the upper layer arrive first because of their greater speed while travelling along the lower side of the boundary. $l$ is the distance at which both direct and refracted waves

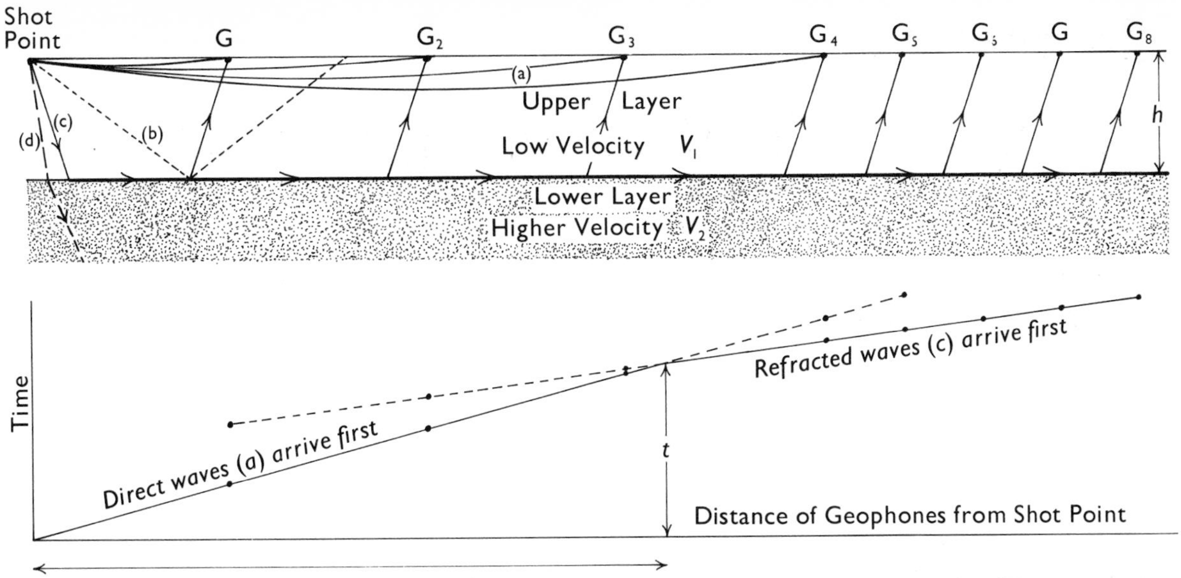


**Figure 26.1** Wave paths from a shot point to a series of geophones, G, with *(below)* the corresponding time-distance graphs for the first arrivals or direct and refracted waves. For details see text.

arrive at the same time $t$. As indicated in the lower part of Figure 26.1, $l$ is determined by the intersection of the graphs obtained by plotting the times of the first arrivals against the respective distances. In each case the slope of the graph gives the corresponding velocity, e.g. $v_1 = l/t$. Knowing the two velocities, the thickness $h$ of the upper layer is given by the relationship

$$h = \frac{l}{2}\sqrt{\left(\frac{v_2 - v_1}{v_2 + v_1}\right)}.$$

## Continental Crust

The large-scale structure of the crust, briefly outlined on pp. 11–13, was originally investigated by studying the records of near earthquakes; records made, not by a series of geophones, but by seismographs permanently established in observatories within a few hundred km of the epicentre. The first major boundary surface to be recognized was the discontinuity between mantle and crust, now familiarly known as the Moho. In the course of a study of an earthquake that occurred in NW Yugoslavia in 1909, A. Mohorovičić found well-defined pulses of P and S waves that had travelled directly through the crust in addition to those of P and S waves that had been refracted into the mantle before coming up to the surface again. The next advance was made by V. Conrad when studying the records of an earthquake that occurred in the Austrian Alps in 1923. Besides recognizing the four pulses just mentioned, Conrad found a fifth—a P wave intermediate in velocity between the other two. This he referred to a crustal layer lying between the upper part of the crust and the mantle. Jeffreys identified the corresponding S pulses in the records of shocks that originated in Jersey and near Hereford, both in 1926. The boundary between the upper and lower crustal layers (then thought to be respectively of sialic and basic composition) was called the Conrad discontinuity. But this has never won the celebrity attached to the Moho, because there are many areas in which it cannot be found.

Much of the work of crustal and upper-mantle exploration is now carried out by utilizing the effects of artificial explosions just below the surface—as in ordinary geophysical prospecting, but on a larger scale. The focal depth is then zero and the time and place of origin are exactly known. This leads to considerable simplification, but it also leads to the unexpected result that the wave velocities are often found to be higher than those based on records of natural earthquakes. One explanation of this discrepancy is lack of exact knowledge as to the time and focus of an earthquake, another is that it is not everywhere true that the wave velocities increase with depth. One such exception is the occurrence of metamor-

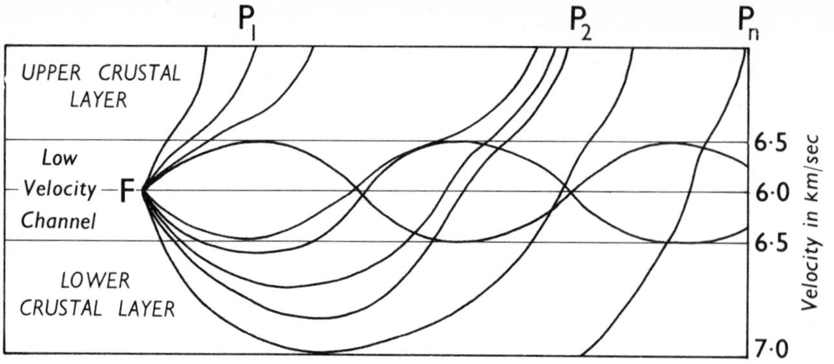


**Figure 26.2** Diagrammatic illustration to show how a 'low-velocity' channel tends to guide and confine some of the waves from an earthquake focus at F. Other waves take longer to reach the surface than where there is no low-velocity channel.

phic rocks, enriched in basic minerals such as biotite and hornblende, overlying granitic rocks through which seismic waves travel more slowly. As may be gathered from Figure 26.2, waves originating from a focus within such a 'low velocity channel' take longer to reach a given station, so that their apparent velocity is less than it would have been if the focus had been at the surface, or if the velocity had continued to increase downwards without interruption.

Another cause for velocity variations, whether the waves are generated naturally or artificially, lies in the structure of the rocks through which the waves are propagated. Transmission through schistose metamorphic rocks, for example, is faster across the schistosity than parallel to it. This is well illustrated by an earthquake of 1946 that had its focus in the Haut-Valais, between the Mt Blanc and Aar massifs (Figure 30.11). P waves recorded at Coire, at the NE end of the Aar massif, travelled parallel to the structural directions, the average velocity being 5·97 km/s. Those recorded at Neuchâtel travelled at right angles, at an average rate of 6·3 km/s.

Whereas in many regions there is seismic evidence for a division into upper and lower crust, there are others, where such evidence is lacking. Moreover, where upper and lower crustal layers have been determined they are not always sharply divided, and their relative thickness varies and appears to be dependent on the geological structure. There is, however, a general increase in the velocity of seismic waves with depth through the continental crust. Generally speaking, the values for the upper part suggest rocks of granitic composition, whereas those of the lower crust are more basic.

Probably the longest continental path over which surface waves have been recorded was between the Aleutian trough, where a great earthquake (epicentre marked A on Figure 26.3) occurred on 20 March 1958, and the seismic station of Lwiro in Zaire, where both Rayleigh and Love waves were not only recorded but later were recognized as such and carefully analysed. The direct path was mainly continental, 13,240 km out of a total great circle distance of 14,240 km, with an average elevation of about 600 m. The path taken in the opposite direction, through the antipodes, was mainly oceanic, 21,790 km out of a total of 25,790 km, with a mean depth of water of 4·5 km. From the two sets of results it is possible to correct each for its 'contamination' by the respective oceanic and continental interruptions.

Figure 26.4, together with a great deal of other data, indicates an average crustal thickness of nearly 40 km along the path across Asia and Africa, as shown in Figure 26.3; and an average oceanic crustal thickness of 5·3 km (probably basaltic) plus 0·8 km of sediment. This gives 10·6 km for the average depth of the Moho in the eastern Pacific and south Atlantic. This is very near the world oceanic average of 11 km for the depth of the Moho. The continental crustal estimate of nearly 40 km may be on the high side. The crustal thickness of the Baltic Shield averages 37·5 km by the method based on Rayleigh waves, whereas P waves from artificial explosions give 34 km, which is probably more accurate. For Africa as a whole, from Algeria to Cape Town, the Rayleigh method gives 35 km. It is, indeed, remarkable that all the old shield areas have nearly the same crustal thickness: 36 km for the Canadian shield; 35 for the Australian; and 35 for that of East Antarctica (i.e without its carapace of ice).

Although, generally speaking, the continental crust has an average thickness of about 35 km, it varies between 20 km and 50 km. The maximum of about 50 km occurs beneath mountain ranges; in other words, the mountains have roots.

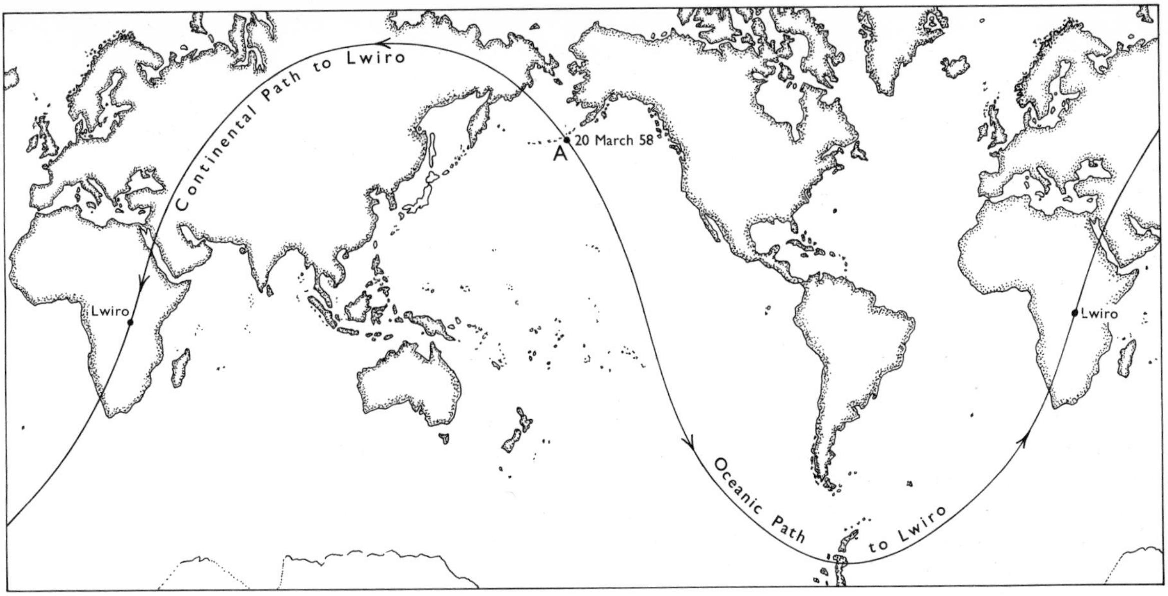


**Figure 26.3** Diagram showing the continental and oceanic paths followed by Rayleigh waves recorded at Lwiro in Zaïre from an earthquake with its epicentre at A in the Aleutian arc

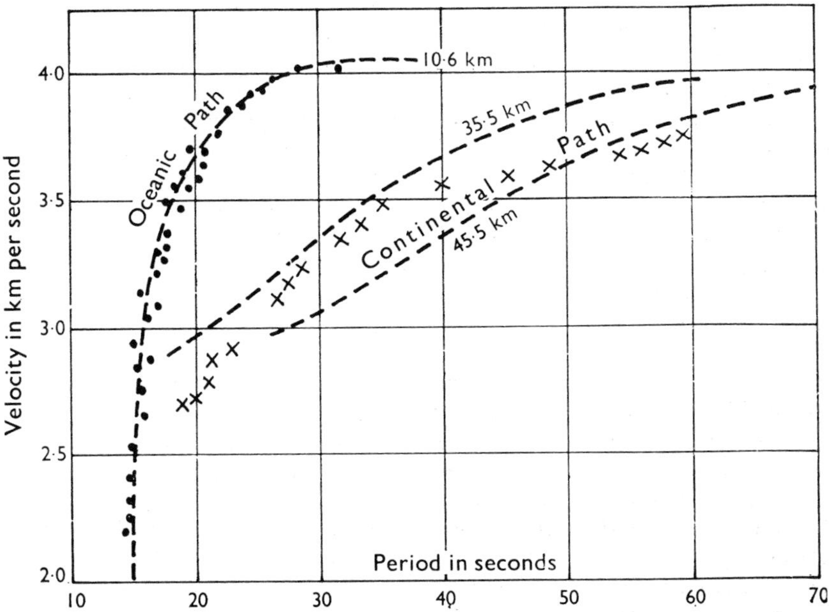


**Figure 26.4** Velocities of Rayleigh waves plotted against their periods along the continental and oceanic paths of Fig. 26.3. The broken lines indicate the results theoretically to be expected for the depths of the Moho below sea level shown by the attached figures. The actual results are plotted as dots (oceanic) and crosses (continental). The thickness of the continental crust is obviously variable, but averages about 40 km. *(Both after R. L. Kovach, 1959, Journal of Geophysical research, vol. 64, pp. 805–813)*

## Oceanic Crust

In principle the method of seismic profiling at sea is the same as that illustrated in Figure 26.1, except for the presence of sea water, the depth of which must be determined. Instead of explosives the sound sources are commonly air guns, or sparkers, and hydrophones are sometimes arranged in lengths of tubing maintained at a depth of about 50 m behind and at some distance from the ship. The echo times are continuously recorded and appear as a profile of the ocean floor complete with the interfaces between the various layers of rock (Figure 24.6).

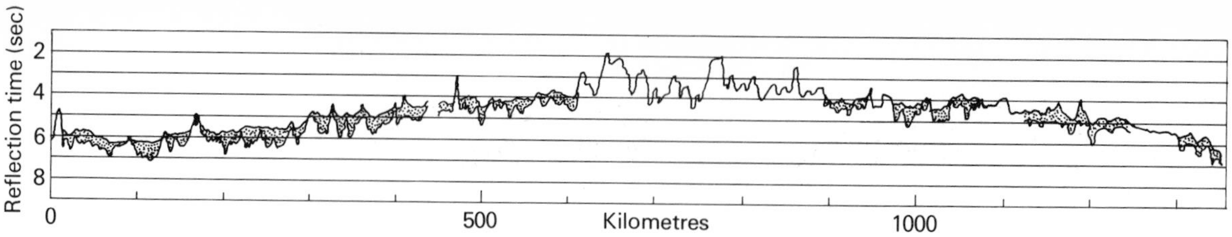


**Figure 26.5** Drawing of a seismic profile at about 40°N across the mid-Atlantic ridge, illustrating the absence of sediment from the median part of the ridge, and its collection within pockets, where the topography is strongly diversified, on the flanks of the ridge (*After J. and M. Ewing, 1967*)

Where the ocean is 4 to 5 km deep, that is away from swells and ridges on the one hand, and deeps and trenches on the others, the Moho is found at a depth of 10 to 11 km. It is overlain by the three following crustal layers:

*Layer 1*, the uppermost layer, is sedimentary with a thickness ranging from zero up to several km. The thickness varies according to distance from continental margins and the degree of transport by bottom currents, and in the case of pelagic sediments according to food supply and ocean depth.

Over mid-oceanic ridges sediment is sparse or absent or occurs in thin isolated pockets where the relief is high (Figure 26.5), a fact confirmed by dredging and more recently by observations from highly mobile submersible craft that can dive to 3000 m and more. The sediment gradually thickens on the sides of the ridges (Figure 26.5). At distances of from 100 to 400 km away from the ridge crests, varying according to geographical position, the sedimentary layer abruptly thickens, after which it maintains the same thickness away from the mid-ocean ridges, or thickens only slightly (Figure 26.6). This abrupt thickening of Layer 1 was found by J. and M. Ewing (1967) to characterize the North and South Atlantic, the South and equatorial Pacific, and the Indian Oceans. Assuming continuous and uniform spreading rates for the parts of ocean floors concerned (p. 23 and p. 619), they explained the abrupt increase in the thickness of Layer 1 as indicating that about 10 million years ago the rate of sedimentation was, for a short time, many times greater than the rates before or since (Figure 26.7). Alternatively they suggested that there might have been a change in the rate of ocean floor spreading with a pause between. By reference to more recent palaeomagnetic evidence the spreading rate is now thought to be essentially constant.

The thickest sediments occur close to the foot of continental slopes, and where large rivers have built sedimentary fans and cones that extend hundreds of kilometres out to sea. A relatively thick sedimentary layer also characterizes the equatorial belt which is a rich breeding ground.

*Layer 2* is about 1 to 1·5 km thick with P velocities that would fit many kinds of sedimentary as well as basaltic rocks. However, wherever samples of this layer have been brought up by dredging or drilling it has been found to be basaltic. Seismic profiles show Layer 2 to have a rough upper surface, and from submersibles working within the north Atlantic rift it has been seen to consist of pillow lava, and part of a pillow weighing over 180 kg has been collected.

*Layer 3* the lowest and main oceanic crustal layer has an average thickness of about 5 km, with P velocities indicative of a basic composition. It is variously thought to be serpentinized peridotite derived by hydration of the upper part of the mantle (Hess, 1965) or amphibolitized basalt representing metamorphosed basaltic crust (Cann, 1968). There is some evidence that Layer 3 may consist of two layers.

## The Mantle

It has long been thought probable that meteorites, stony and iron, might be direct clues to the nature of the earth's mantle and core. Stony meteorites are like our terrestrial peridotites in many ways and, moreover, a few varieties have a composition not unlike some of our basalts. For these and other reasons a similar range of composition for the mangle, with ultrabasic materials predominating, has seemed to be a plausible guess. The sudden change of seismic velocities at the Moho indicates either a change of composition (e.g. from basic rocks above to ultrabasic below) or a change of state (e.g. from, say, gabbro or amphibolite to a high-pressure modification which, for convenience, we may refer to as eclogite) without much change of composition.

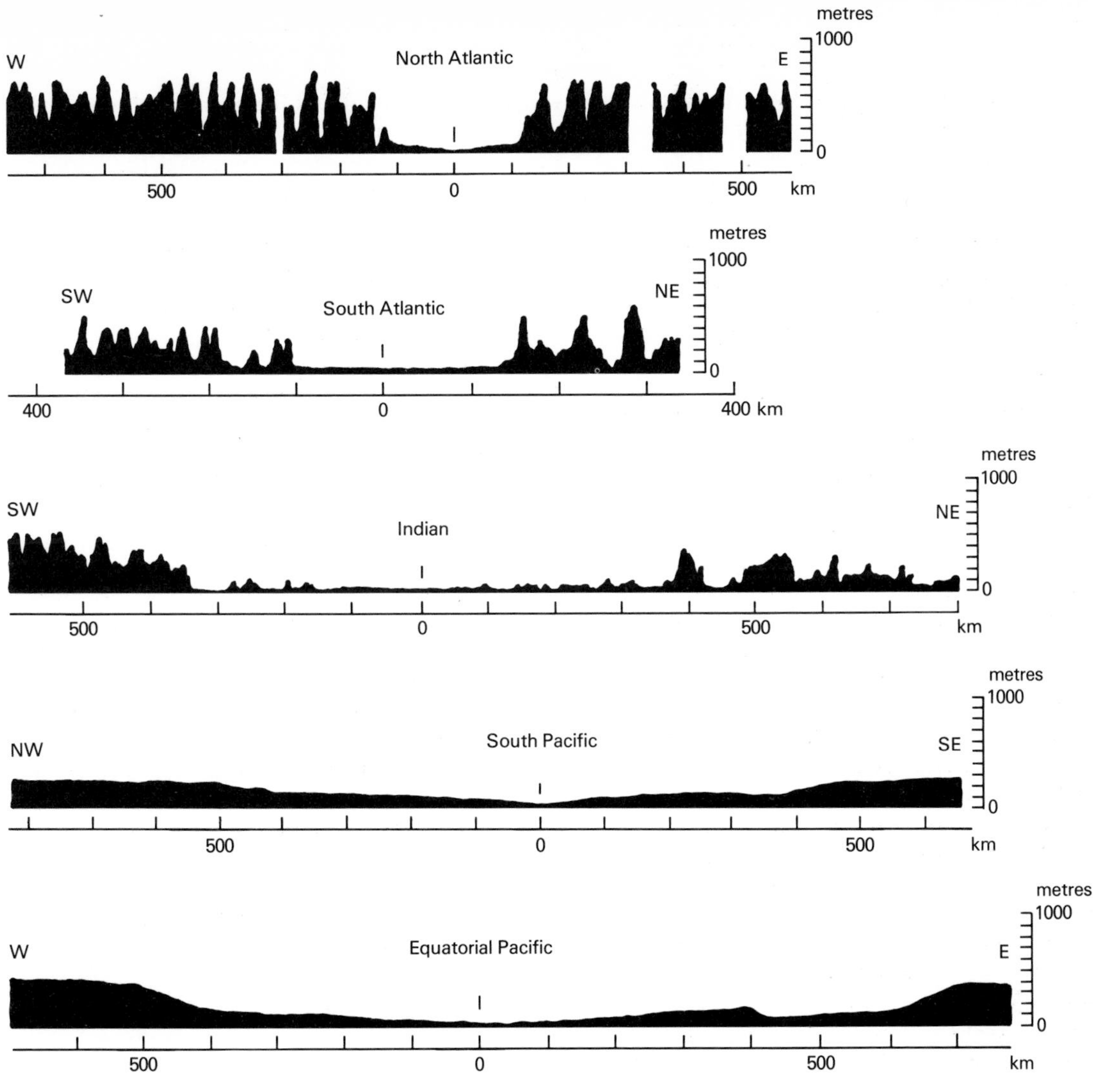


**Figure 26.6** Graphs depicting the measured thicknesses of sediments plotted against distance from mid-oceanic ridges in various oceans. In each case there is an abrupt thickening of the sediments at distances from the ridges varying from 100 to 400 km. The bottom graph represents three traverses across the equatorial Pacific rise and the thick equatorial sediments on either side. (*After J. and M. Ewing, 1967*)

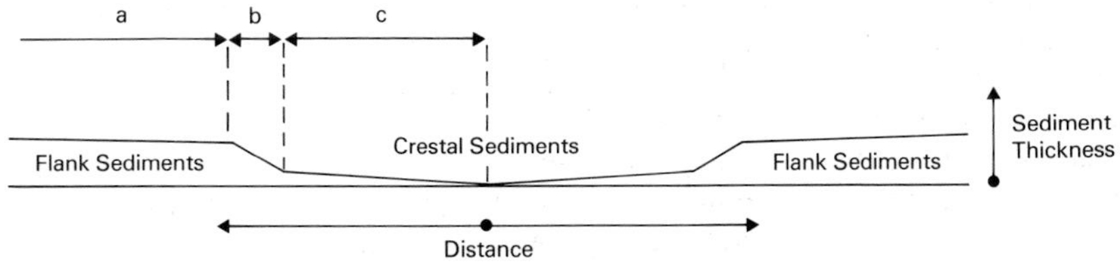


**Figure 26.7** Calculation of the changes in rate of sedimentation based on the lowest graph of Fig. 26.6, assuming the rate of sea-floor spreading has remained constant. a=an early cycle during which the rate of sedimentation was 3 mm in 10,000 years, followed by b=more rapid sedimentation for a relatively short space of time during which 100 mm of sediment were deposited in 10,000 years. c=the most recent cycle of sedimentation, commencing about 10 million years ago, during which the rate of sedimentation has been 13 mm per 10,000 years (*After J. and M. Ewing, 1967*)

Olivine-rich peridotites and garnet-bearing varieties, at the appropriate temperatures and pressures, give seismic velocities that are about right for most parts of the upper mantle. Eclogite would do equally well, and various associations of these rocks would match the seismic and density requirements. That eclogite does occur in the mantle is proved by the fact that it occurs in diamond pipes as inclusions, some of which themselves contain diamond. There cannot be much eclogite in comparison with ultrabasic materials, however, because eclogite contains too high a proportion of the heat-generating radio-elements to be an abundant constituent of the mantle. Further evidence that the mantle is composed of various types of peridotite is provided by olivine and peridotite nodules that occur as inclusions within basaltic lava, including that of Layer 2 of the oceanic crust. The basalt has evidently carried these xenoliths up from the mantle of which it is probably itself a differentiate, perhaps a first formed melt.

## The Low-Velocity Zone of the Upper Mantle

As long ago as 1926, the celebrated Californian geologist, Berno Gutenberg, suspected that there was a low-velocity zone in the upper mantle. As may be gathered from Figure 26.2, waves originating from a focus within a low-velocity zone take longer to reach a given station, so that their apparent velocity is less than it would have been at the surface, or if the velocity had continued to increase downwards without interruption. Thirty years after Gutenberg's suggestion, the reality of a low-velocity zone within the upper mantle was established by world-wide records of the blasts from underground nuclear explosions. Remembering that there is considerable variation from place to place, the velocity decrease begins on average at a depth of about 100 km and, after passing the minimum and resuming the normal downward increase, the velocity, as it was just below the Moho, is only regained at an average depth of about 150 km or more. It should be noted that the actual depths are lower for oceanic regions and higher for continental.

The existence of a low velocity zone within the upper part of the mantle is thought to depend on the temperature of the material in relation to the melting point or range, the velocity being least where the actual temperature makes its nearest approach to the melting temperature. Both temperature and pressure increase with depth, but they have opposite effects on properties such as rigidity and rheidity. In the upper part of the low-velocity channel the temperature effect is dominant; that is, as the depth increases the temperature rises more rapidly than the melting point of the material that happens to be there. Evidence of at least partial melting has been recorded by Gorshkov, an eminent Russian vulcanologist. In 1957 he found that S waves from earthquakes originating in the Japanese or Aleutian arcs failed to arrive at seismic stations so situated that the P waves, which did arrive, had passed through the mantle at depths of 55 to 60 km beneath the volcanic belt of Kamchatka and the Kurile Islands (see Figure 26.12). Since S waves cannot pass through liquids, whereas P waves can, Gorshkov concluded that the volcanic belt was underlain at these levels by pockets or layers of molten rock material. Judging from the lavas now being erupted the molten material would be of basaltic composition.

The low-velocity zone in the upper mantle is thus a kind of lubricated zone, making relative movement between the overlying layers and the interior possible. As a result, it has become usual to refer to the crust and the upper part of the mantle above the low-velocity zone as the *lithosphere*, and the low-velocity zone as the *asthenosphere*.

## Oceanic Ridges and Rises

An interconnecting network of submarine mountain ranges extends through the oceans, generally as mid-oceanic ridges. This extensive mountain system differs from continental mountains, not only in its greater length of about 64,000 km, but also because it is composed of basaltic lavas, and is, in fact, an outcrop of Layer 2 of the oceanic crust (see Figure 26.5).

The Atlantic mid-oceanic ridge (Figure 26.8) has been known since 1855, being recognized by the *Challenger* Expedition. In 1953, *Discovery II*, working along the ridge north of the Azores, observed that it had a double crest with a deep and steep-sided trough between. There is indeed, as shown graphically in Figure 26.9, a close topographic resemblance between this high-level submarine rift valley and those of the high-level plateaux of Africa. Soon afterwards, the Lamont geologists, in their research vessel *Vema*, disco-

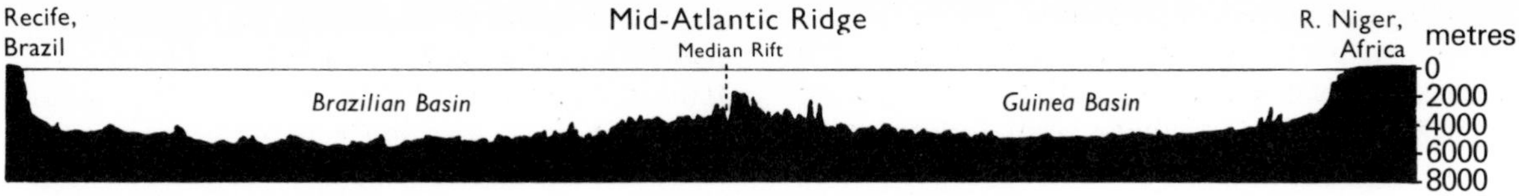


**Figure 26.8** Topographic profile across the Atlantic floor to show the Mid-Atlantic Ridge and its median rift. Length of section about 4800 km (*After B. C. Heezen, 1959*)

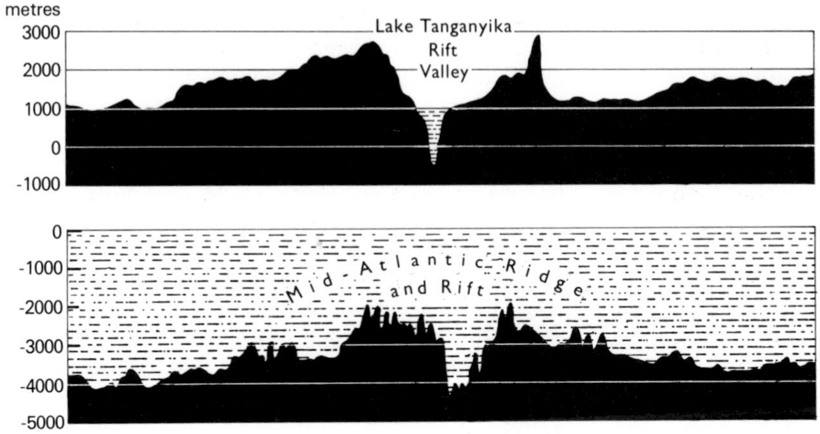


**Figure 26.9** Diagram to illustrate the similarity of form and scale between the Tanganyika rift of Africa and the submarine rift of the Mid-Atlantic Ridge. Each section is 725 km long. (*After B. C. Heezen, 1959*)

vered that the mid-Atlantic rift continues along the crest south of the Azores. The rift has been detected at many other places along the crest, but there are long stretches where, instead of a conspicuous central rift, a number of parallel ridges and troughs occur.

In Iceland, which is a culmination of the mid-Atlantic ridge, the rift is well represented by the central graben (Figure 12.9), which bisects the island. Topographically, the graben is shallower than the submarine rift, because it is largely filled with the volcanic products of Pleistocene and recent eruptions. Indeed the volcanic activity displayed along the ridge by Iceland and San Mayen in the North Atlantic, and by Tristan da Cunha in the South Atlantic are vigorous reminders of the active growth of the ridge. In 1970 Project FAMOUS (French-American Mid-Ocean Undersea Study) was founded. The intention was to use submersibles, specially constructed for manoeuvrability and ability to dive to depths of the order of 3000 m or more, and so descend into mid-ocean rifts and make detailed observations with the aid of specialized instruments and techniques. Starting in 1973, the mid-Atlantic rift was examined about 600 km south-west of the Azores. It was found in detail to be asymmetrical, both topographically, thermally and magnetically. The western side 300 m high is bounded by a steep cliff that rises by narrow steps backed by nearly vertical fault scarps, whereas the eastern wall rises gently in broad steps backed by short steep slopes. Fresh pillow lavas, thought to be no more than 100 years old, were viewed at close quarters, and collected from fissures located in the rift floor. Subsequent laboratory examination showed the mineralogy and chemical composition of the lavas to vary outwards from the middle of the rift and to do so more gradually eastward than westward.

In the east Pacific a broad submarine bulge, 2000 to 4000 km wide, known as the east Pacific rise, lies close to North America. Compared with the mid-Atlantic ridge, its topography is relatively smooth, with only occasional ridges and troughs parallel to its crest. As a submarine feature the crest of the east Pacific Rise passes into the Gulf of California and at the landward end is intersected by the San Andreas fault, only becoming submarine again to the south-west of the Island of Vancouver.

The mid-oceanic ridges are characterized by shallow earthquakes. Moreover, they are intersected at high angles, by strike-slip faults of a variety known as transform faults (p. 620). The ridges are off-set against the faults, and those parts of the faults connecting the off-sets, are characterized by shallow earthquakes.

## Migrating Oceanic Volcanoes

Before there was formal evidence (p. 616) that the oceanic lithosphere is moving away from the mid-oceanic ridges, where new basaltic crust is formed, and disappearing into the mantle at the trenches, Dietz (1961) and Hess (1962) had suggested that this was so. The idea was based on the thinness of oceanic sediments and the absence of Palaeozoic sediments from the ocean floors. The oceanic lithosphere was envisaged as moving as if carried along on a conveyor belt, now equated with the rheid low-velocity zone of the mantle. To test the hypothesis of sea-floor spreading, as it was then called, Tuzo Wilson (1963) ascertained, for the volcanic islands of the Atlantic, (*a*) the oldest age reported for each island (by interbedded fossiliferous beds or by radiometric dating) and (*b*) the distance between each island and the ridge. The results, summarized below with slight modifications, show that distance definitely tends to increase with age, apart from those volcanic islands that have remained on the ridge. The correlation is not perfect, as Wilson remarked, but in view of the roughness of the data the figures are sufficiently striking to support the interpretation that the volcanic islands originated on or near the ridge and have since been carried away from it.

Turning to the Pacific Ocean, it was long ago recognized by Dana (1890), in relation to the origin of coral reefs and atolls, that the Hawaiian chain shows a sequence of forms, starting from the active volcanoes of the ESE end, through volcanic domes showing various stages of erosion and increasing indications of submergence (like Pearl Harbour), followed by volcanic islets and stacks (some with fringing reefs), to atolls and seamounts at the WNW end. Since this first example was recognized, other chains and belts of islands and seamounts have been described from the SW Pacific by L. J. Chubb (1927–57) and Tuzo Wilson (1963), all of which illustrate a similar sequence of volcanic activity followed by extinction and increasing signs of submergence (e.g. drowned valleys, barrier reefs and atolls, and guyots and other seamounts). The chains all trend approximately ESE to WNW, are nearly straight and, with the exception of the Samoan chain, they all have their active or more recently extinct volcanoes at the ESE end and their oldest and most deeply submerged members at the WNW end, as illustrated in Figure 26.10. In the case of the Hawaiian chain, Miocene fossils have been discovered on Oahu, lava from the same island has been dated at not less than 20 m.y., and Cretaceous fossils have been brought up from one of the guyots where the range of the Emperor seamounts begins (p. 564). These discoveries indicate that the evolution of the chain has taken not less than 100 m.y. to reach its present state. Relative movement

## Volcanic Islands of the Atlantic

| *Islands (*active)* | *Oldest ages Geological* | *m.y.* | *Distance from Ridge (km)* | *Max. rate of motion (cm/yr)* |
|---|---|---|---|---|
| Jan Mayen* | Pliocene | 10 | — | — |
| Iceland* | L. Tertiary | 50 | — | — |
| Bouvet* | Pleistocene | 1 | — | — |
| Tristan da Cunha* | —(radiometric) | 18 | 0–350? | 0–1·9? |
| Ascension (? dormant) | Pleistocene | 3 | 150 | 5·0 |
| Gough | ?Mid-Tertiary | ?20 | 400 | 2·0 |
| Azores* | Miocene | 20 | 400 | 2·0 |
| Faroes | L. Tertiary | 50 | 600 | 1·2 |
| St Helena | Miocene | 20 | 800 | 4·0 |
| Madeira | pre-Miocene | 30–60 | 1500 | 2·5–5·0 |
| Canary* | Oligocene | 35 | 1800 | 5·1 |
| Bermuda | Eocene/Oligocene | 40 | 2100 | 5·2 |
| Cape Verde* | L. Cretaceous | 120 | 2200 | 1·8 |
| Fernando de Noronha | L. Cretaceous | 120 | 2400 | 2·0 |
| Bahamas | L. Cretaceous | 120 | 3000 | 2·5 |

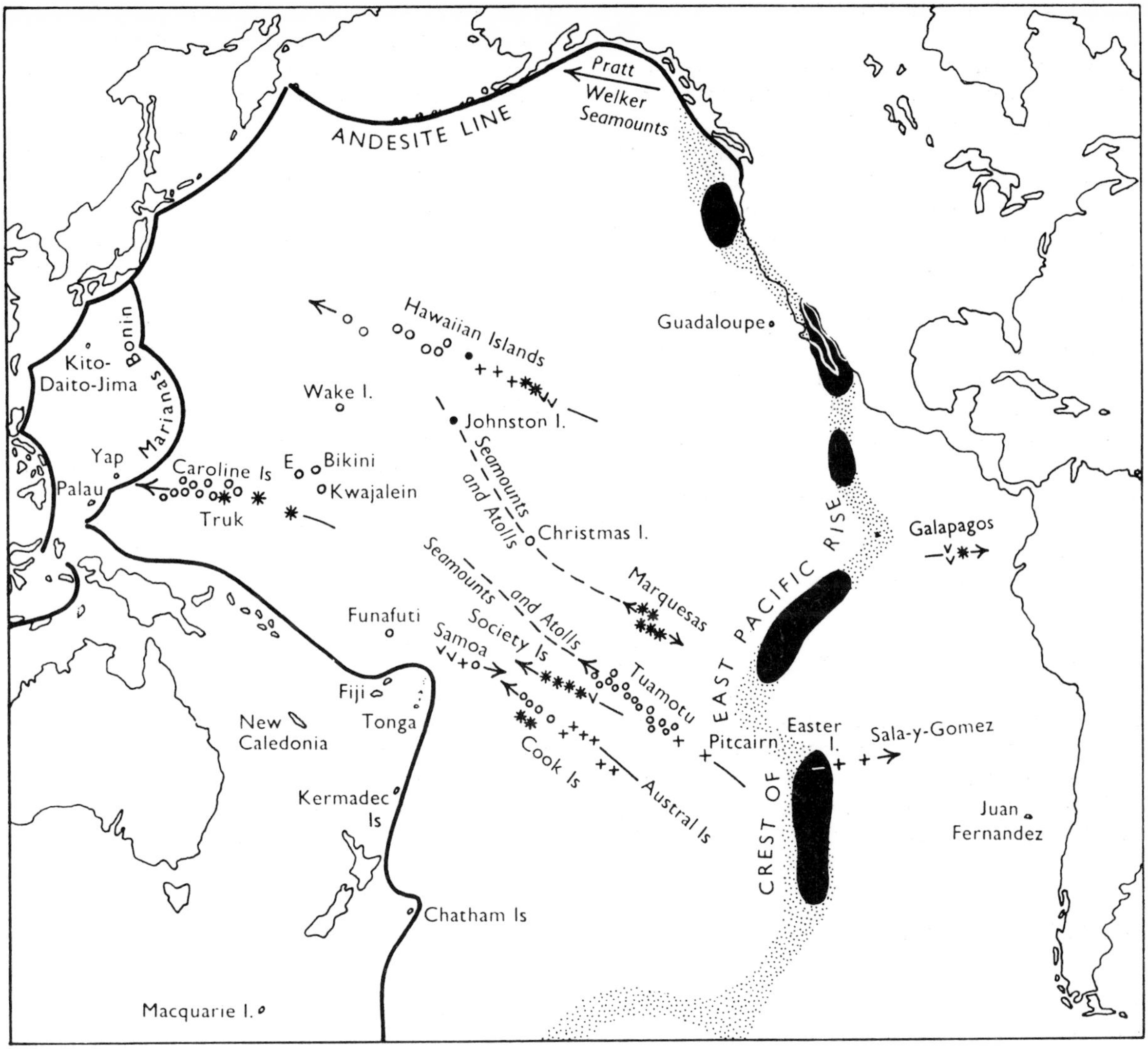


**Figure 26.10** Map of the Pacific Ocean to show the Andesite Line; the crest of the East Pacific Rise (black areas have heat-flow values $> 3$, ranging from 3·06 to 8·09); and linear chains of islands and seamounts, illustrating progressive increase of age and submergence in the directions indicated by arrowheads (v=active volcano; star=extinct but little eroded volcano; cross=deeply dissected and embayed volcanic island; solid circle=basaltic relics; open circle=atoll or seamount). The localities of atolls that have been drilled are named (Funafuti, Fiji, Bikini, E=Eniwetok, and Kito-Daito-Jima).

between the volcanoes and the magmatic source in depth has produced a chain about 3000 km long in 100 m.y. That is, there has been an average rate of movement of about 3 cm/year. There are two possible explanations:

(*a*) the sub-crustal magma source slowly migrated linearly from the end of the chain with atolls, where it began millions of years ago, to the end with volcanoes, where it is still operating;

(*b*) the magma source remained in its present position while the volcanic islets and underlying oceanic crust have migrated towards the WNW. In this connection Figure 26.11 is of special interest. The floor of the Gulf of Alaska slopes gently towards the Aleutian trench and is variegated by seamounts and guyots that rise abruptly to heights of 1060 m to nearly 4000 m. The most significant feature of this chain of submarine mountains is the exceptional position and depth of guyot GA–1. All the other guyots have their platforms at depths of about 760 m. GA–1 is unique because it lies within the Aleutian trench and has its platform at a depth

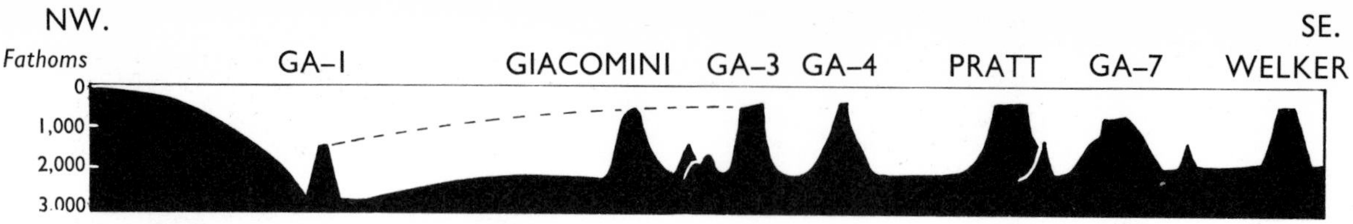


**Figure 26.11** Profile of the Gulf of Alaska floor from the continental shelf SE of Kodiak Island across the Aleutian trench (with guyot GA–1) and along the Pratt-Welker chain of seamounts and guyots. Length of section 628 km

of over 2500 m, over 700 m deeper than the others. Clearly this guyot was carried down to its abnormal depth during the formation of the trench (Menard and Dietz, 1951). Figure 26.11 gives the impression of the oceanic crust approaching the gigantic obstruction of the Aleutian Arc and subsiding in front of it, just as the surface of a great ice sheet subsides into a trough-shaped form in front of an obstructing nunatak (cf. Figure 20.9).

## Ocean Trenches: Subduction Zones

Deep-focus earthquakes occur beneath the continental sides of the deep ocean trenches; commonly on the continental side of island arcs which, like those of the Pacific, have their convex sides directed towards the ocean. Like the island arcs, the trenches that margin them are also arcuate. A trench about 4800 km in length, margins the west coast of South America, where there are no arc-like festoons of islands, and related deep-focus earthquakes occur beneath the Andes. The deep-focus earthquakes in both these settings are associated with shocks from intermediate and shallow foci. These all originate within a zone that typically extends from the ocean trench and dips towards the neighbouring continent down to a depth of 700 km or so.

The Kuriles are an excellent example of a *single* arc, with a submarine trench or trough on its outer convex side (Figure 26.12). Beneath the trench, and mainly to the landward side, there is a broad zone of shallow earthquakes with foci down to 60 or 70 km. This also roughly coincides with a belt of negative gravity anomalies, commonly referred to as a 'negative belt'. The arc itself is a line of active or recently extinct volcanoes, usually rising from an erosion surface of Tertiary or older sedimentary rocks which have been folded, metamorphosed and granitized. This orogenic belt is well exposed around the volcanoes of Kamchatka to the north, and those of Japan to the south-west. Beneath the line of volcanoes the earthquake foci (apart from shallow volcanic shocks) fall mainly within the range of the intermediate class, i.e. at depths of 70 to 300 km. Since on average the depth of foci increases with distance from the open ocean, the epicentres of deep earthquakes occur much farther inland, or within marginal seas such as the Sea of Okhotsk between the Kurile arc and the Asiatic mainland.

Hugo Benioff (1954), an American seismologist, assembled the following data for a number of earthquake zones which dip beneath the continental margins:

| *Marginal Region* | *Intermediate Foci* | | *Deep Foci* | |
|---|---|---|---|---|
| | *Depths in km* | *Dip* | *Depths in km* | *Dip* |
| Aleutian Arc | 70–175 | 28° | | |
| Kurile-Kamchatka Arc | 70–300 | 34° | 300–700 | 58° |
| Japanese Arc (Bonin-Honshu) | 70–400 | 38° | 400–550 | 75° |
| Indonesian (Sunda) Arc | 70–300 | 35° | 300–700 | 61° |
| New Hebrides | 70–300 | 42° | | |
| Chile | 70–290 | 23° | 550–650 | 58° |
| Peru-Ecuador | 70–250 | 22° | 600–650 | 47° |
| Central America (Panama-Acapulco) | 70–220 | 39° | | |
| | *Average dips* | 33° | | 60° |

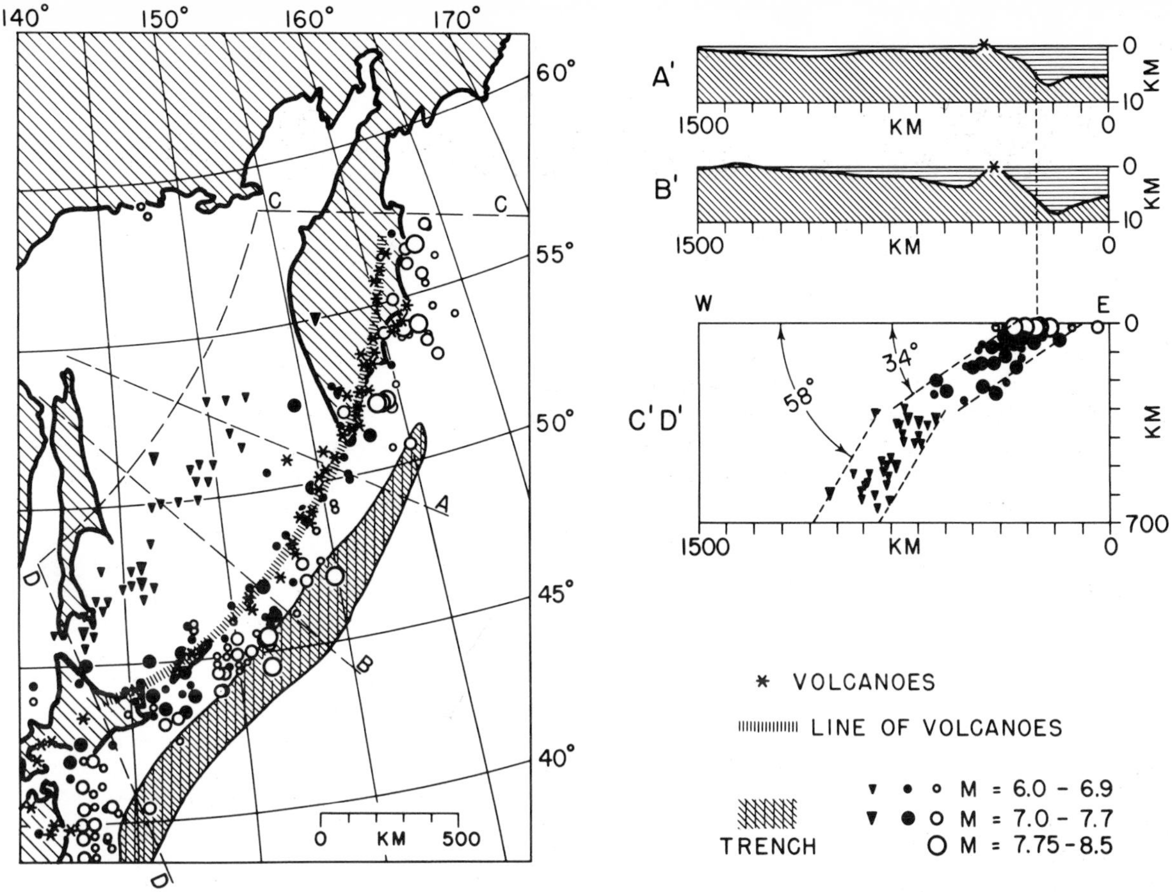


**Figure 26.12** Map of the Kamchatka–Kurile volcanic arc with topographic profiles and a composite seismic profile. Epicentres (map) and foci (seismic profile) of shallow, intermediate and deep earthquakes are represented by circles, circular dots and triangular dots respectively. These symbols are of different sizes according to the magnitudes (M), as indicated below the profiles. A′ and B′ are vertically exaggerated profiles along the lines A and B on the map. C′D′ is a composite seismic profile of the whole arc from CC across Kamchatka to DD across northern Japan. (*Hugo Benioff, 1954*)

Such earthquake zones, associated with ocean trenches have come to be known as *Benioff zones*.

Certain island festoons, of which the Indonesian arc is an excellent example, differ from single arcs like those of the Kurile and Aleutian Islands in having an additional feature consisting of a sedimentary outer arc which has developed from within the submarine trench. The single Aleutian arc shows a gradual transition towards the *double* type where it links up with the Alaskan peninsula. The volcanic line of the islands is continued by the volcanoes of Alaska, but a second chain of islands appears on the landward side of the trench, which is thus divided by a ridge into two troughs, the outer one remaining the deeper. In the Indonesian arc this outer ridge is a submarine feature for long distances, e.g. south of Java (Figure 26.13), but in places it has risen above sea-level forming rows of islands. The Mentawai Islands, off the coast of Sumatra, are a good example (Figures 26.13 and 26.14). There the strata can be seen to have been strongly folded early in the Tertiary and again during the Miocene. The islands have recently been rising, as the terraces of uplifted coral reefs clearly prove; and that they are still doing so is indicated by the effects of shallow-focus earthquakes. Farther north the ridge can be traced by way of the Nicobar and Andaman Islands into Burma, where it becomes a continuous land feature and passes into the high mountain range of the Arakan Yoma, west of the Irrawaddy. Far to the east of Java the ridge emerges conspicuously above sea-level as the Island of Timor, which is celebrated for the great heights to which its successive coral reefs have been upheaved, some of

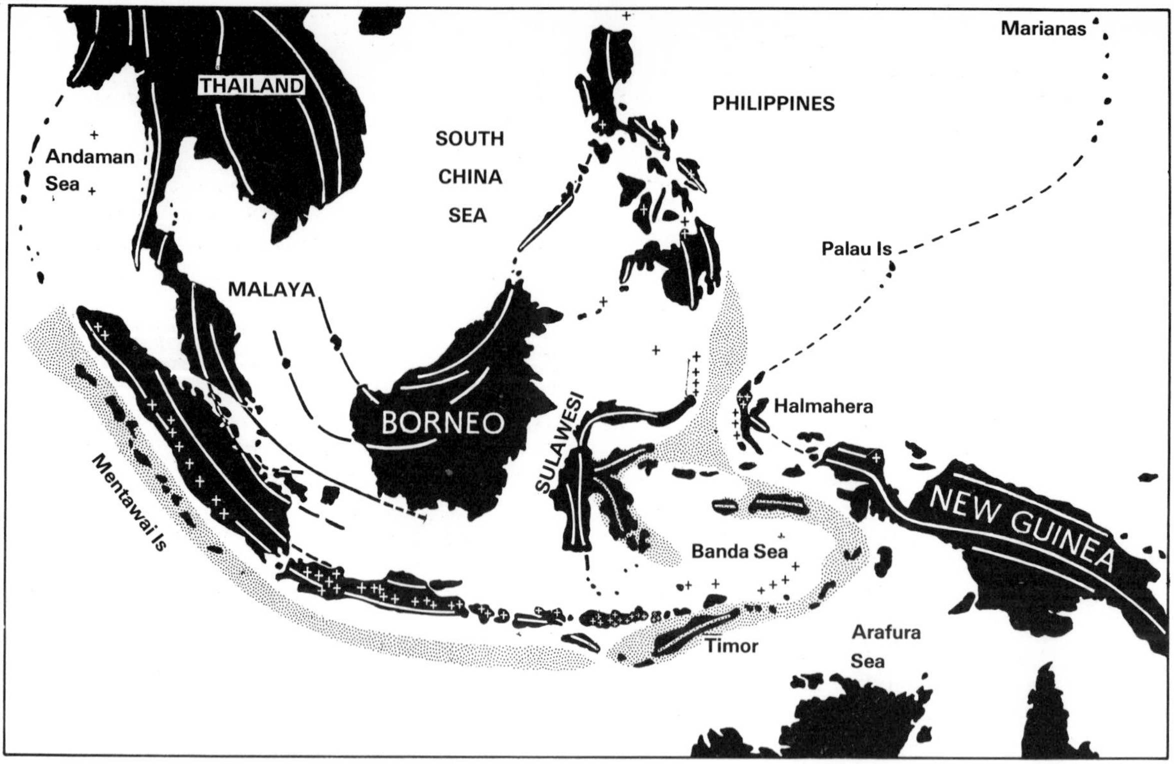


**Figure 26.13** Tectonic map of the Indonesian arc and adjoining territories. The belt of negative anomalies of gravity discovered by Vening Meinesz is indicated by a shading of fine dots; this can now be continued through the Andaman Islands into Burma. The line of active volcanoes is shown by black or white crosses. The deepest part of the trench, south of Java, is known as the Java Trench.

**Figure 26.14** Map showing the generalized distribution of intermediate and deep earthquakes associated with the trench bounding the southern side of the Indonesian archipelago. The seismic isobaths depict the Benioff zone dipping towards the continental area of south-east Asia. The line of active volcanoes lies above the foci of earthquakes of intermediate depth. (*Adapted from T. Hatherton and W. R. Dickinson, 1969*)

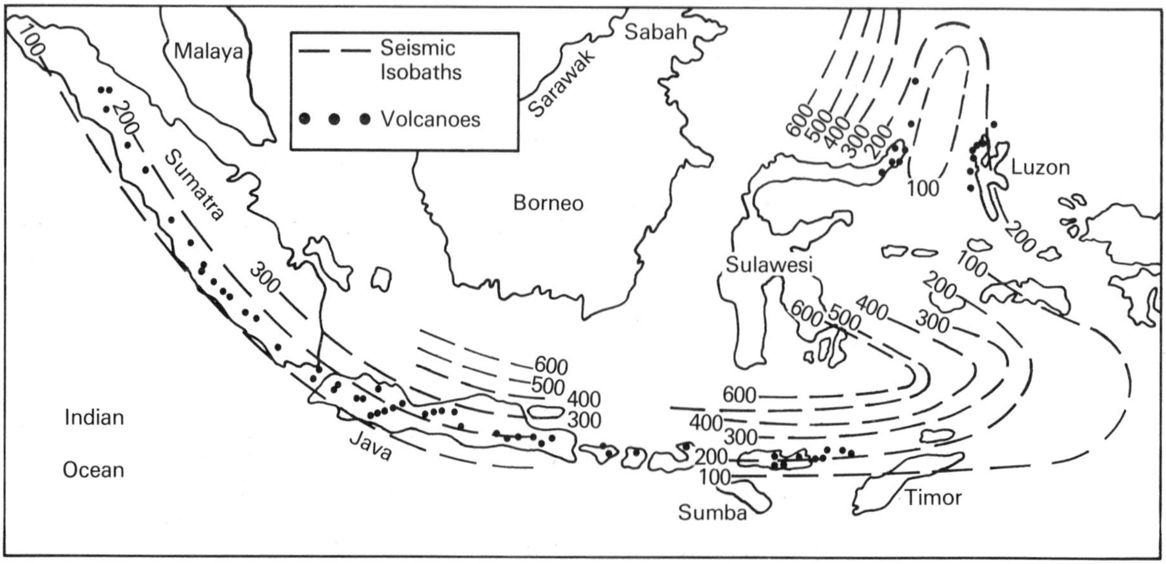

them now standing over 1200 m above sea-level.

The unity of this long arc of mountains, rows of islands and submarine ridges is indicated not only by its topographic continuity and association with a Benioff zone (Figures 26.13 and 26.14), but also by its coincidence with a belt of negative anomalies of gravity. These are particularly strong where the arc is still under water. The discovery of the major part of this negative belt (Figure 26.13), the first of its kind to be made, was the great achievement of F. A. Vening Meinesz. In the course of a series of confined and tedious expeditions (1923, 1926, 1929–30) he carried out his pioneer work on the measurement of gravity at sea, using submarines lent by the Royal Netherlands Navy. These could be submerged to depths where the effects of waves on swinging pendulums became negligible. The 'Diving Dutchman', as Vening Meinesz came to be called, was eventually able to chart a well-defined negative belt from the ridge NW of the Mentawai Islands to the Mindanao Trench off the Philippines, with only a short break near Sumba (west of Timor), 4000 km in all. In 1951 a gravity map for the whole Archipelago and its extension to Burma, made possible by including the later results of oil-company geophysicists, was published by J. W. de Bruyn.

From the belt of negative gravity anomalies that characterize the trench of the Indonesian arc, Vening Meinesz inferred a down-buckling and consequent thickening of the oceanic crust to form a root. Since then, however, the addition of seismic to gravity surveys both here and in other regions, notably the Puerto Rico and Peru–Chile trenches, have shown that although all of them are characterized by negative gravity anomalies, and are out of isostatic equilibrium, the oceanic crust beneath them is of the same order of thickness as elsewhere in the transitional region between continental and oceanic margins. Moreover, the sedimentary deposits within trenches are characteristically thin except for that part of the Peru–Chile trench south of Valparaiso where they are at least 2 km thick and form a flat floor to the trench (Figure 24.6). The negative gravity anomalies characteristic of trenches are now thought to be due to displacement of mantle resulting from subduction of oceanic lithosphere at the trenches; the only substitution for the volume of high density mantle being low density sea water filling the trench.

In studying data from several hundred earthquakes recorded from the Fiji region, and associated with the Tonga–Kermadec Island arc and trench to the north-east of New Zealand, Oliver and Isacks (1967) found evidence of oceanic lithosphere turning down into the mantle. They discovered an anomalous seismic zone in the mantle, the upper surface of which is roughly coincident with the upper surface of the Benioff zone that dips beneath the island arc down to a depth of about 700 km. Seismic body waves propagated in the anomalous zone are less attenuated than are waves of the same type propagated from similar depths in the mantle elsewhere. Oliver and Isacks suggested that if low attenuation can be equated with strength then the anomalous zone can be equated with the lithosphere, and thus depicts oceanic lithosphere travelling down into the mantle (Figure 26.15). In other words, the anomalous zone related to the Tonga trench is a subduction zone revealing the disappearance of the moving oceanic lithosphere down into the mantle. The zone is at most 100 km thick and probably only 50 km thick.

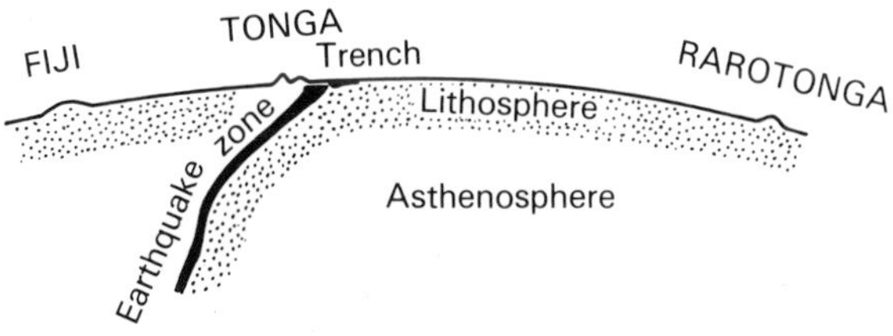


**Figure 26.15** Diagrammatic section across the Tonga trench illustrating the subduction of oceanic lithosphere based on seismic correlation of the oceanic lithosphere with an anomalous seismic belt, within the mantle, characterized by low attenuation and high velocity of seismic waves relative to those of the adjacent asthenosphere (*After J. Oliver and B. Isacks, 1967*)

## The Core

From the analogy with meteorites the earth's core has traditionally been supposed to consist essentially of an alloy of iron and nickel, but this hypothesis has now been challenged from several directions. Iron meteorites are more abundant in museum collections than the stony types because, being highly unusual natural objects, they are easily recognized and so have been found in far greater numbers than the stones, which weather like ordinary rocks and generally fail to attract attention unless they are being specially looked for. For this reason most of the irons are prehistoric, whereas nearly all the recent falls are of stones. Comparing the total masses of all the meteorites collected from observed falls, the irons make up less than a tenth of the whole. But the core makes up about a third of the earth's mass. Against this discrepancy it can be shown that irons are more readily captured than stones, which means that the source of meteorites (the belt of asteroids?) has been progressively depleted in iron during geological time. But this argument can be countered by a more significant consideration. The planet Mars has too low a density to have any core worth mentioning, in the sense that Earth has a core. It is very unlikely that Earth, in an intermediate orbit, could have captured a high proportion of iron from the primordial materials while Mars, so much nearer the belt of asteroids, got little or none (Figure 26.16).

| *Planet* | *Mean distance from Sun (in $10^6$km)* | *Rotation period (in days)* | *Volume (Earth=1)* | *Mean density* |
|---|---|---|---|---|
| Mercury | 58 | 88 | 0·66 | *c*5 |
| Venus | 108 | *c*200 | 0·876 | 5·15 |
| Earth | 150 | 1 | 1·000 | 5·52 |
| Mars | 228 | 1·03 | 0·154 | 3·6 |

A more direct objection to the 'iron core' hypothesis is based on ultra-high pressure experiments. Compressibility measurements on iron subjected to the shock waves of a high explosive, detonated in contact with the sample, show that its density reaches 11·8 when the pressure rises momentarily to 1·4 million atmospheres. There is every reason to conjecture that the density would become very much higher if such a pressure could be maintained for a long period, as it would be in the earth's interior. Yet the density at the top of the outer core is less than 10, and this is the region where the above pressure is reached (see table on p. 587). The discrepancy suggests that the outer core contains a considerable proportion of elements with lower atomic numbers than iron.

Another serious difficulty arises out of the current hypothesis that the liquid outer core acts like a self-exciting dynamo in virtue of its internal circulations and so is responsible for the greater part of the earth's magnetic field. If this view is correct in principle—and it has no rival—a source of energy is necessary to maintain the circulations that in turn maintain the earth's magnetism. H. C. Urey has speculated that metallic iron, distributed through the mantle, may have been migrating downwards and adding to the core throughout the earth's history, and that the process may still be going on. The core would be growing at the expense of the mantle on this view, and the earth would be contracting, so releasing gravitational energy for a variety of mechanical and thermal purposes. But if the earth is not systematically contracting, a more promising approach would be to consider how energy might be liberated if the mantle were growing at the expense of the core.

The first step in this direction was taken by W. H. Ramsey in 1948, when he suggested that the core consists of material having the same ultrabasic composition as the mantle, but in a high-pressure phase with a correspondingly high density. As first presented, this attractive hypothesis had to face a storm of adverse criticism. But with a little modification it has survived. And combined with P. M. Dirac's inference that the force of gravity continually weakens as the universe grows older, it has been developed by Egyed into a dynamic model of the earth which, though still speculative,

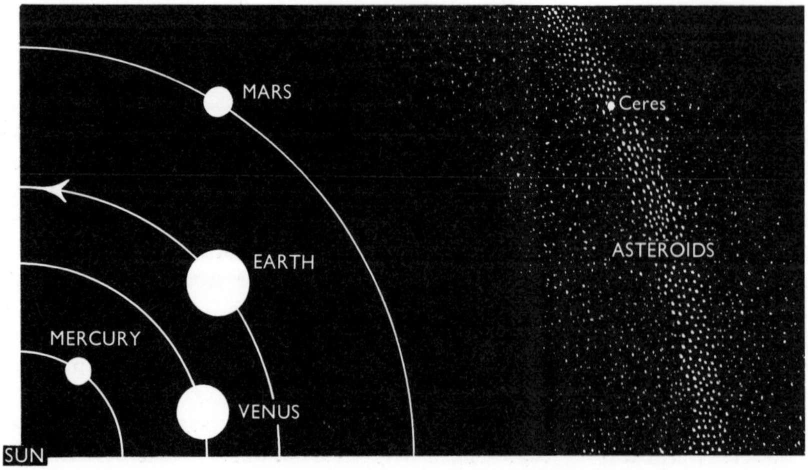


**Figure 26.16** The inner planets, showing the relative distances from the Sun and the main part of the belt of Asteroids; some of the orbits of the latter are much more eccentric than is depicted here.

leads to an expanding earth and yields a surplus of energy over and above that involved in expansion. This surplus, with contributions from other sources (e.g. radioactivity), could amply provide the energy required for earthquakes, volcanoes, mountain building, and other tectonic and plutonic activities.

## High-Pressure Transformations

Examples such as the metamorphism of basic rocks into eclogite and the artificial production of hard translucent diamond from soft, black graphite (p. 113) have already illustrated the effect of high pressure in reorganizing the lattice structures of minerals so that the atoms are more closely packed than before. Phase transitions of the graphite/diamond type are now known to be common throughout the mineral world. Silica, for example, passes through a number of phases ranging from the familiar quartz (density 2.65) to the much denser and harder coesite (3.01), which forms in the temperature range 500–800°C at pressures of 35,000 bars or more, corresponding to conditions near the top of the mantle in undisturbed continental regions. Coesite is evidently not a normal crustal mineral, though it might be expected to occur in the roots of high mountains, if there were any free silica there to be so transformed. The only known natural occurrences of coesite are in the walls and exploded fragments of certain large craters where pressures and temperatures have been high. Two kinds of such craters have been recognized: (*a*) those due to the searing and explosive impact of giant meteorites (Figure 20.60); and (*b*) those due to the explosive eruption of hot volcanic gases which had to overcome not only the weight of the overlying rocks, but their strength (e.g. the late Miocene Ries Basin in western Germany, ENE of the Black Forest) before developing the vents by fluidization. At pressures corresponding to the weight of overlying rocks at a depth of 300 km coesite is transformed into a still denser phase which was first made artificially and given the name *stishovite*. This man-made mineral is nearly twice as dense as quartz. It has the hardness of diamond, but is immensely more costly to make. In 1962 it, too, was detected in the glassy material formed from the shattered and fused granitic rocks from the floor of the Ries Basin.

At what may be called the ultra-high pressures produced by shock waves, still more remarkable changes occur. At a pressure corresponding to a depth of 1300 km, diamond ceases to be stable and turns into a metallic form of carbon not unlike iron. As the pressure of the shock waves dies down, the metallic carbon quickly reverts to graphite. The metallic state is one in which the ions are much more tightly packed than in the crystals of rock-forming minerals; so tightly, in fact, that some of the electrons are forced out of their orbits and become detached, so that they are free to move through the crystal lattice. It is for this reason that metals are good conductors of electricity. An electric current along a wire is the swift rush of detached electrons through the interstices between the ions of the metal. When non-metallic elements are subjected to a pressure sufficiently high to strip off some of the outer electrons, they too are turned into metals. With further increase of pressure the process of 'atom squashing' or *pressure-ionization*, to give it its technical name, continues until more and more electrons are forced out of their orbits: first from the outer shells and then from the inner ones in turn (see p. 44). Finally, when the intensity of pressure-ionization reaches its limit—which it does in the stars known as white dwarfs—all the atoms are crushed like hollow eggshells and matter becomes completely *degenerate*. It then consists of tightly crowded atomic nuclei immersed in a homogeneous medium of electrons.

The above brief introduction to a difficult branch of physical chemistry will suffice to show that the properties of matter under the high pressures and temperatures of the earth's deep interior differ fundamentally from everything we are familiar with in everyday life. The degree of pressure-ionization reached in the core is not precisely known. But from various magnetic phenomena it is inferred that the core is an exceptionally good conductor of electricity. This property implies the presence of free electrons. The earth has indeed a metal core, but the metal need not be mainly iron. Any of the elements of ordinary minerals, like those of peridotites, and even hydrogen if it happens to be there, will be metallic because of pressure-ionization.

These considerations are in accordance with Egyed's model of the earth, in which the inner core, outer core and mantle have the same all-over composition, but with the ingredients in very different states and phases. The seismic discontinuities are the boundaries between successive high-pressure phases. In the mantle continuous rise of density with depth depends on the effect of pressure in compressing the lattice structures of crystalline minerals. At certain critical depths a discontinuous jump to a higher density may occur either because the lattice has been modified into a different and tighter pattern, or because the minerals have absorbed sufficient energy to break down into their constituents (e.g. MgO, FeO, $SiO_2$, etc.), which recrystallize in denser forms, as illustrated by the behaviour of $SiO_2$. The comparatively sudden density jump and the change to a liquid-like state at the boundary of mantle and core are indications that this discontinuity marks the most critical transition stage of all. It is interpreted as an indication that the crystalline state has collapsed into a crowd of ions in the early stages of degeneracy. All chemical bonds have broken down and the mutilated ions are free to wander amongst the much more mobile electrons which were originally attached to them. Such freedom of movement would readily account for the fluidity of the outer core. On this hypothesis the fluidity is not regarded as a result of melting due to high temperatures. The liberation of electrons accompanying the breakdown of chemical bonds and lattice structures also makes the whole assemblage much more compressible than before, so accounting for the very high density of the outer core.

The nature of the inner core is more difficult to understand on any hypothesis. Since it behaves towards seismic waves as a solid, it may be presumed that the effect of still higher pressures is to decrease the mobility of the degenerate ions, until what was like a liquid in the outer core (Zone E, Figure 25.23) becomes more like glass in the inner core (Zone G). This would be consistent with the existence of a broad transition between them (Zone F), since such a change would be expected to take place gradually. But we do not know.

## Selected References


BENIOFF, H., 1955, 'Seismic evidence for crustal structure', in *Crust of the Earth*, Special Paper 62, Geological Society of America, pp. 61–73.

BOTT, M. H. P., 1971, *The Interior of the Earth*, Edward Arnold, London.

BULLARD, E. C., 1954, 'The interior of the Earth', in *The Earth as a Planet*. vol. 2, pp. 57–137, University of Chicago Press, Chicago.

CANN, J. R., 1968, 'Geological processes at mid-ocean ridge crests', *Geophysical Journal*, vol. 15, pp. 331–41.

CHUBB, L. J., 1957, 'The pattern of some Pacific island chains', *Geological Magazine*, vol. 98, pp. 170–11, and 1962, vol. 99, pp. 278–83.

DIETZ, R. S., 1961, 'Continent and ocean basin evolution by spreading of the sea floor', *Nature*, vol. 190, pp. 854–7.

HEIRTZLER, J. R., and BRYAN, W. B., 1975, 'The floor of the Mid-Atlantic rift', *Scientific American*, vol. 233, No. 2.

HESS, H. H., 1962, 'History of the ocean basins', in *Petrologic Studies: A Volume in Honour of A. F. Buddington*, pp. 599–620, Geological Society of America.

— 1965, 'Mid-oceanic ridges and tectonics of the sea-floor', in *Submarine Geology and Geophysics*, pp. 317–33, Colston Papers, vol. 17, edited by Whittard, W. F., and Bradshaw, R., Butterworths, London.

MENARD, H. W., and DIETZ, R. S., 1951, 'Submarine geology of the Gulf of Alaska', *Bulletin of the Geological Society of America*, vol. 62, pp. 1263–86.

OLIVER, J., and ISACKS, B., 1967, 'Deep earthquake zones, anomalous structures in the upper mantle, and the lithosphere', *Journal of Geophysical Research*, vol. 72, pp. 4259–75.

SYKES, L. R., OLIVER, J., and ISACKS, B., 1968, 'Earthquakes and tectonics', in *The Sea*, vol. 4, *New Concepts of Sea Floor Evolution*, Pt. 1, 1970. General editor: Arthur E. Maxwell, Wiley-Interscience, New York.

TOKSÖZ, M. N., 1975, 'The subduction of the lithosphere', *Scientific American*, vol. 233, No. 5, pp. 88–98.

WILSON, J. TUZO, 1963, 'Evidence from islands on the spreading of ocean floors', *Nature*, vol. 197, pp. 536–8.

# Chapter 27

# Magnetism, Palaeomagnetism, and Drifting Continents

There is always more chance of hitting on something valuable when you aren't too sure what you want to hit upon.

*Alfred North Whitehead*, 1945

## The Earth's Magnetic Field

Although Thales, earliest of the Greek philosophers, was familiar with the lodestone (magnetite) and its apparently magic properties, and the compass was used as a direction-finder by Chinese navigators well over 2000 years ago, it was not until the year 1600 that William Gilbert, Physician to Queen Elizabeth I, published his celebrated book *De Magnete*, in which for the first time it was shown that 'the terrestrial globe behaves like a giant magnet'. Gilbert experimented with spheres of magnetite and found that iron filings placed on the surface aligned themselves in the same directions as freely suspended magnetized needles situated in corresponding positions on the earth's surface. He concluded that the earth's magnetism, being like that of a uniformly magnetized sphere, is therefore of internal origin. This was a remarkably accurate first approximation. In 1839 Gauss rigorously proved that by far the greater part of the magnetic field originates below the earth's surface, but that there is a small and variable part that originates outside. The latter—the *transient* disturbances—include the magnetic storms which are often accompanied by widespread auroral displays. They originate high up in the atmosphere, where electric currents are generated by cosmic rays and by radiation and charged particles emitted from sunspots, solar flares and the like.

While the main part of the magnetic field is like that of Gilbert's magnetized sphere, it is equally like that of a powerful bar magnet (a 'dipole') placed near the middle of the earth along the axis joining the north and south geomagnetic poles (Figure 27.1). Gauss determined the dipole field that most closely matched the earth's actual main field, and found this left over certain irregular deviations, even when local anomalies due to the magnetic effects of iron ores in the crust have been allowed for. These minor irregular parts of the main field can be represented by weak dipole magnets situated some hundreds of kilometres from the centre of the earth. The whole field can thus be subdivided as follows:

- EARTH'S MAGNETIC FIELD
  - *Transient field*, originating in the upper atmosphere
  - *Magnetic anomalies*, due to concentrations or deficiencies of magnetic minerals and rocks in the crust
  - MAIN FIELD *of internal origin*
    - *Regular* major field equivalent to that of a strong centrally placed axial dipole (Figure 27.1)
    - *Irregular* residual fields equivalent to those of weak non-axial dipoles

The main field is continually changing at easily measurable rates. The *secular magnetic variations*, as they are called, are recorded at various observatories where the details necessary to determine the total magnetic field and its parts are continuously measured, like the elements that make up records of the weather. Three magnetic elements are essential and those in common use are the following:

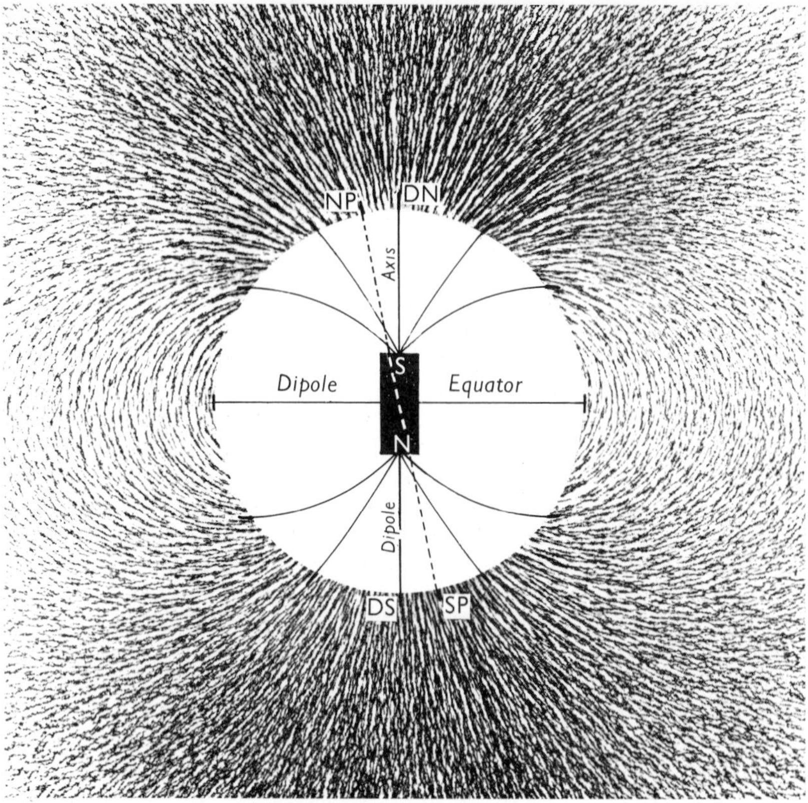


**Figure 27.1** Iron filings sprinkled on a sheet of glass over a strong bar magnet represent the dipole part of the earth's magnetic field. DN and DS are the north and south geomagnetic poles; NP and SP are the geographical north and south poles.

The *intensity*, which is the total magnetic force, F; the horizontal and vertical components are referred to as H and Z respectively.

The *magnetic dip* or *inclination*, which is the angle I, measured between the horizontal and the direction taken by a freely suspended magnetized needle; i.e. the angle between the directions OH and OF in Figure 27.2. At the geomagnetic N and S poles the dip is vertical.

The *magnetic declination* is the angle D between the direction of H and the geographical north.

The rapidity of the changes is well illustrated by the declination. In 1546 a compass in London pointed 8°E of true north. Since then the direction has varied year by year, becoming true north in 1665 and reaching 24°W in 1823. The direction of change then reversed, as shown on Figure 27.3, which also records how the inclination varied over the same period. The variation in the intensity, F, is both more rapid and more irregular. During the last hundred years the strength of the field in the London district diminished by 5 per cent, but recently it has been increasing again. About a century ago the intensity at Cape Town rapidly increased to about a third more than its present value. These remarkable changes are phenomenally fast compared with the sluggish processes occurring in the mantle. This highly significant fact leaves no alternative to the inference that the main field has its origin in the outer core, where there is the possibility of fluid motions such as might result from thermal convection, and where the material is metallic and capable of carrying electric currents.

Taken by itself, it is found that the pattern of the irregular part of the main field is continually changing and that it bears no relationship of any kind to the distribution of land and sea or to geological structures or rocks. This is another fact pointing to the core as its place of origin. Moreover, it drifts slowly westward at about 0·18° per annum, which implies that the magnetic field rotates more slowly than the earth as a whole. In orther words, the source of the field, i.e. the core, rotates more slowly than the mantle. The resulting shearing along the discontinuity would thus favour the generation of eddies. The latter, together with vortices in other parts of the outer core, might well be responsible for the irregular part of the field.

## The Origin of the Earth's Magnetic Field

The magnetic field cannot be a permanent property of the material of the core. Above a certain temperature known as the Curie point all magnetic materials lose their magnetism. For iron at atmospheric pressure the Curie point is 770°C; for nickel, 330°C; and for magnetite, 580°C. With increasing pressure the Curie point drops. Thus the high temperature and the liquid-like state of the outer core both render permanent magnetization impossible. The magnetic field must therefore be continuously produced and maintained, and this again suggests the generation of electric currents in material of high electrical conductivity and capable of internal movements. Only the outer core, being what we can best describe as a metallic liquid, fulfils these conditions. The suggestion that the outer core acts like a dynamo has been investigated in great detail by W. Elsasser in the United States and Edward Bullard in England.

The principle on which the dynamo hypothesis is based is that of the dynamo itself: namely, that when an electric conductor is moved through a magnetic field an electric current is generated in the conductor. Current is generated in an ordinary dynamo by the rapid rotation of wire coils in a strong magnetic field. In the outer core of the earth the rotating coils are represented by thermal convection cells. Given a magnetic field to start the system working (a very weak field would do) convection provides the motion necessary to generate electric currents in the convecting material. The electric currents increase the intensity of the magnetic field, which in turn increases the strength of the electric currents . . . and so on until convective movement, electric current and magnetic field reach an approximate state of equilibrium. For this reason the system is described as a 'self-exciting dynamo'.

Two problems remain to be solved before the hypothesis can claim the status of a reasonably complete theory. One is the source of the original magnetic field. The other is the source of the energy that keeps the convection and the associated vortices going without a breakdown. The need for an initial magnetic field sounds like assuming what has to be explained. But a weak field could have started in a variety of ways at the time when the earth was being formed as a planet. Since we do not known how the earth originated, the matter is of no more than theoretical interest at present.

Three sources of energy for maintaining the field have been considered. The one usually appealed to is radioactivity. But it is difficult to see how there could be sufficient radioactivity in the core to supply the heat necessary to maintain convection currents; if there were, the mantle would also be endowed with so much radioactivity that the crust could not have cooled down to its present temperatures. Another hypothetical

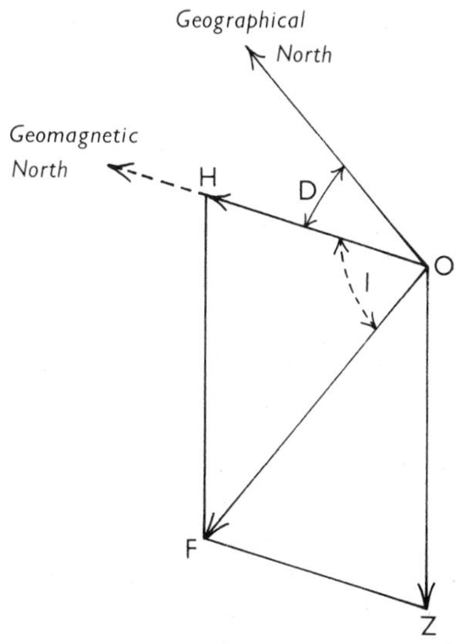


**Figure 27.2** For description see text.

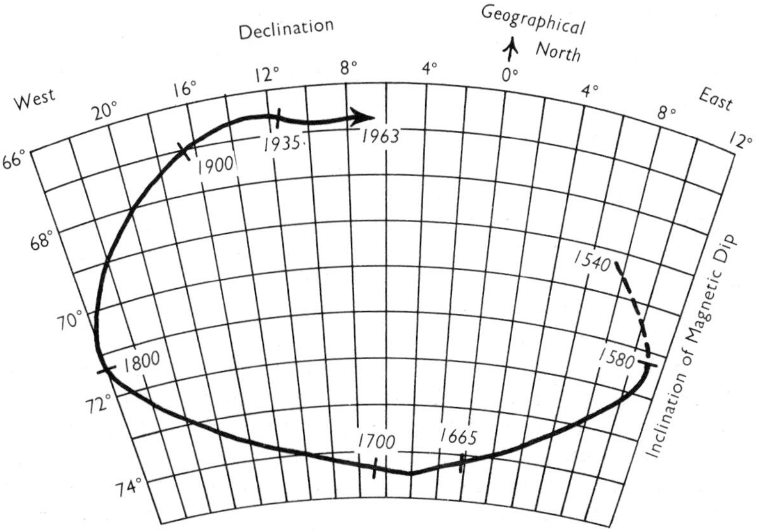


**Figure 27.3** The change in the direction of the mean magnetic force, F, in the London district since 1540, shown by records of magnetic declination and magnetic dip

process, already referred to, is downward migration of iron from the mantle to the core, a process which would release gravitational energy and generate heat. This hypothesis bristles with difficulties. If the process happens at all, it should have come to an end long ago, or have now become too slow to supply any significant amount of heat. But it leads to a contracting earth, and that is hardly compatible with the lengthening of the day to which Palaeozoic corals bear silent witness, or with sea-floor spreading (pp. 705–7).

The third source of energy to which appeal has been made is that liberated (like latent heat) when phase changes occur as a result of decrease of pressure. This is a hypothesis based on the work of Dirac and Ramsey, and developed by Egyed. Here convection in the outer core is particularly easy to understand (Figure 27.4), because material in the phase appropriate to the inner core is gradually changing into that of the outer core. The latter is necessarily being heated from below on this hypothesis, just as the mantle is being heated from below by the passage of material from the outer-core phase into the mantle phase. This explanation for convection in the outer core received unexpected support from the discovery that Venus has no detectable magnetic field. The American space-probe, *Mariner II*, passed 34,745 km above the sunlit side of Venus on 14 December 1962. Delicate instruments of three different kinds transmitted no evidence of a detectable magnetic field around the planet. Had there been one with only a thousandth of the strength of the earth's it would easily have been detected. Venus is sufficiently like Earth in size (Earth volume 1, Venus 0·876) and density (Earth density 5·52, Venus 5·15) to have a core. But assuming its materials to be the same as ours, a little arithmetic shows that Venus cannot have anything corresponding to our inner core. This suggests that the core of Venus cannot now act as a self-exciting dynamo, although it might have done so long ago, when the force of gravity and internal pressures may have been sufficiently high for Venus to have had an inner core. But if so, that inner core has ceased to exist, and Venus no longer has the internal energy to maintain a magnetic field.

## Palaeomagnetism: Rocks as Fossil Compasses

In describing the earth's magnetic field on p. 605 it was stated that the main part of the field is nearly equivalent to that of a powerful bar magnet (a dipole) placed near the earth's centre and orientated along the axis joining the N and S magnetic poles (Figure 27.1). This *geocentric axial dipole field* accounts on average for about 95 per cent of the whole. Most of the balance is made up of feeble and irregular fields equivalent to a number of weak non-axial dipoles not situated centrally. At the present time the magnetic axis departs considerably from the axis of rotation. However, the departure is continually changing, and even during the few hundred years in which records have been kept (see Figure 27.3) the average divergence between the magnetic and geographical north poles has been very much smaller than the value at any given time. But to make it a reasonably sound assumption that the mean magnetic field is that of a geocentric dipole orientated along the axis of rotation, the period over which the average is taken should be not less than several thousand years. As noted below, this consideration has an important bearing on the

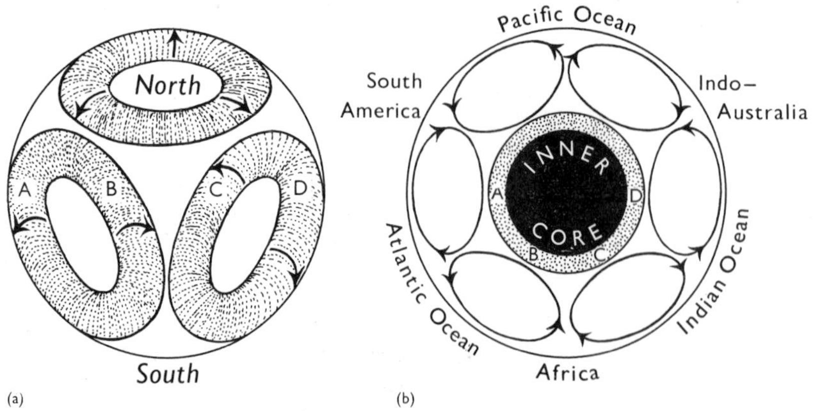


**Figure 27.4**
(a) A purely diagrammatic representation, within a spherical shell, of a tetrahedral pattern of convection cells in the earth's outer core, dotted in (b). The fourth cell, being entirely at the back, is not seen.
(b) An equatorial section through the earth, passing through ABCD in the outer core. Hypothetical circulation directions in the mantle are shown, together with the corresponding idealized positions of the continents and oceans.

collection of rock specimens from formations likely to give information about the geomagnetic fields of earlier times.

In rocks such as basaltic lava flows and basic sills and dykes every crystal of ferromagnetic material—mainly oxides of iron and titanium such as magnetite and ilmenite—acquires a stable magnetism which is 'frozen in' as it cools through its Curie temperature. The latter is about 575°C for magnetite, but may range between 700°C and 200°C, according to the mineral concerned. The newly acquired or *remanent* magnetism has the same direction (the declination, D) and the same dip (the inclination, I) as the local geomagnetic field at the time of consolidation. This can be demonstrated experimentally. It has also been proved for natural flows erupted at known dates from volcanoes situated where magnetic records for the same dates are available (notably Etna, Hekla, and some of the volcanoes of Japan).

Measuring the remanent magnetism is carried out with special types of highly sensitive galvanometers. The samples on which the measurements are made are usually slices cut from cylindrical cores drilled from the hand specimens (or *in situ* from the rock itself); non-magnetic tools being used. The specimens must be unweathered and must be accurately orientated at the place of collection: i.e. marked with the horizontal plane and dip (if any) as well as with the local direction of the magnetic north or south. To ensure that a statistically reliable mean value of the remanent magnetism can be calculated it is rarely enough to sample a single flow. Specimens from a series of flows, estimated to cover at least a few thousand years, are required to provide the data for calculating the approximately equivalent axial dipole field. A series of specimens taken across a sill or dyke from one margin to the other will provide an adequate time span if the intrusion is thick enough. Samples from the margins, which cooled quickly, will have passed through the Curie points thousands of years before those from the middle, where cooling was slow. Other samples of interest are those from the country rocks near their contacts with sills or dykes. Shortly after the time of intrusion such a rock will have been heated up well above its Curie points. It will then have lost its original magnetism and gained a new one on cooling, the latter correspondingly closely to that of the margin of the intrusion.

Similar considerations apply to the sampling and orientation of sedimentary rocks collected for geomagnetic data. Detrital grains of magnetite, for example, inherit a remanent magnetism from their parent rock. Thus, when they finally come to rest in a semi-consolidated sediment, they tend to align themselves like compass needles along the direction of the geomagnetic field in which they then find themselves. Probably more important is the fact that ferruginous alteration products, e.g. haematite, which form coatings to the detrital minerals and act as cementing materials to newly deposited sediments, also become magnetized in line with the local geomagnetic field. Such 'chemical magnetization' as it is called, makes red sandstones and shales, like those of many well-known Permian and Triassic formations, particularly rewarding sources of palaeomagnetic data. Sampling of a given formation should be through a thickness of 100 m or more, and over as wide an area as possible, in order to give an ample time range for effective averaging. Measurements of dip and strike must be carefully made, so that correction for the effects of folding or tilting can be properly applied. The importance of freedom from weathering is obvious, since ferruginous weathering products become 'chemically magnetized' at the time of their formation, which for all practical purposes is the present time.

Apart from some earlier work in France, which was not systematically followed up—probably because of the World Wars—the study of palaeomagnetism was begun about 1950 by P. M. S. Blackett, who soon attracted an enthusiastic team of co-workers, some of whom, notably S. K. Runcorn, soon started teams of their own. Because of the obvious bearing of the results on the geological problems of continental drift and the physical problems of the origin of the earth's magnetic field, the subject rapidly spread throughout the scientific world as one of the most fertile and actively developing branches of geophysics.

### Palaeomagnetic Results: Drifting Continents

The fundamental hypothesis on which palaeomagnetic investigations are based is that the geomagnetic field, averaged over an appropriate time, has always been very nearly that of a geocentric axial dipole. In other words, the geomagnetic poles have always, on average, coincided approximately with the geographical poles. If this coincidence were exact, then for rocks of a particular age from a given locality the averaged value of $I$ (the palaeomagnetic inclination or dip) would give the latitude $\lambda$ of the locality at that time from the relation

$$\tan I = 2 \tan \lambda$$

shown graphically on Figure 27.5. The latitude fixes the *distance* of the pole. The *direction* of the pole is given by the averaged value of $D$ (the palaeomagnetic declination). Knowing both the distance and direction of the geographical pole from a given locality fixes the position of the pole on the globe at the time when the measured rocks received their magnetization.

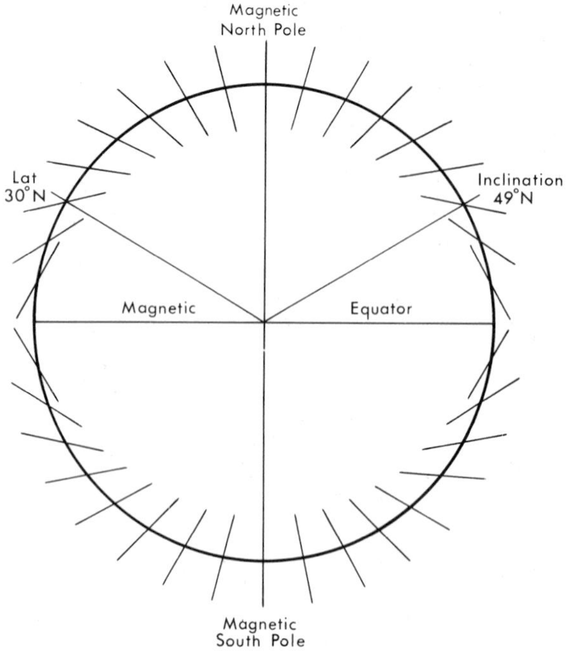


**Figure 27.5** Idealized diagram to illustrate the relationship between magnetic inclination and the corresponding latitude for a geocentric axial dipole field. An inclination of 49°, for example, corresponds to a latitude of 30°; tan 49°=1·15=2 tan 30°.

In practice such pole determinations can never be accepted as exact: because (*a*) the minor components of the earth's field (p. 605) have been neglected; (*b*) the rocks tested may have suffered magnetic changes that have not been recognized and allowed for; and (*c*) slight orientational and experimental errors inevitably creep in. Individual pole determinations may be correct within about 15°, and possibly less if based on a dozen or more reliable samples. It should also be noticed that considerably greater variations may be found in the pole positions relative to a given continent and recorded for a given geological Period (e.g. the Permian, as illustrated in Figure 27.7). These differences are not necessarily errors. Many of them may represent genuine changes in the position of the continent during the very long interval of time involved, or in the positions of certain regions that have been displaced by great strike-slip faults, or have gone through the severe disturbances of orogenic movements. The effects of some of these difficulties have been reduced as increasingly detailed work has been accomplished year by year, and especially as it has become more generally possible to combine radiometric dating with magnetic measurements.

The dipole hypothesis is already well supported for the last 20 m.y. or so. Statistically, the mean position of the Quaternary magnetic pole, as determined for a large number of specimens from various parts of the world, does not depart significantly from the geographical pole of today (Figure 27.6). The same is true for the Pliocene (Figure 27.7). But going backwards in time from then, the positions increasingly diverge until by the Permian, for example, the difference for European rocks amounts to 40° or 50° of latitude,

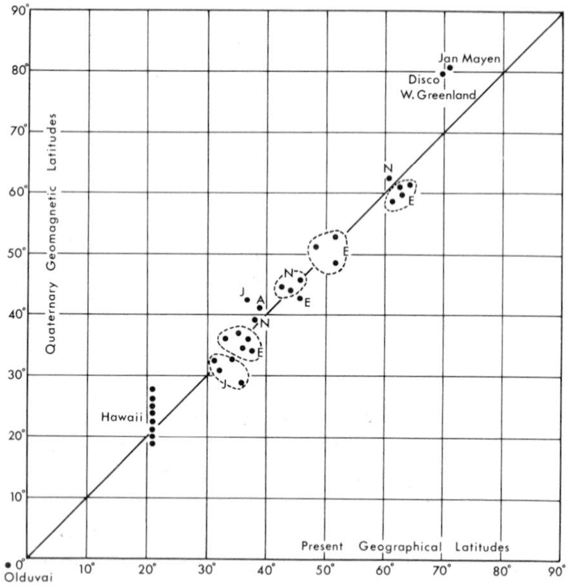


**Figure 27.6** Geomagnetic latitudes, determined from Pleistocene and Recent lavas, plotted against the present latitudes of the latter. A, Australia; E, Europe; J, Japan; N, North America. The points nowhere depart far from the diagonal, which implies that within the limits of error the mean positions of the Quaternary magnetic poles were not significantly different from the geographical poles of today.

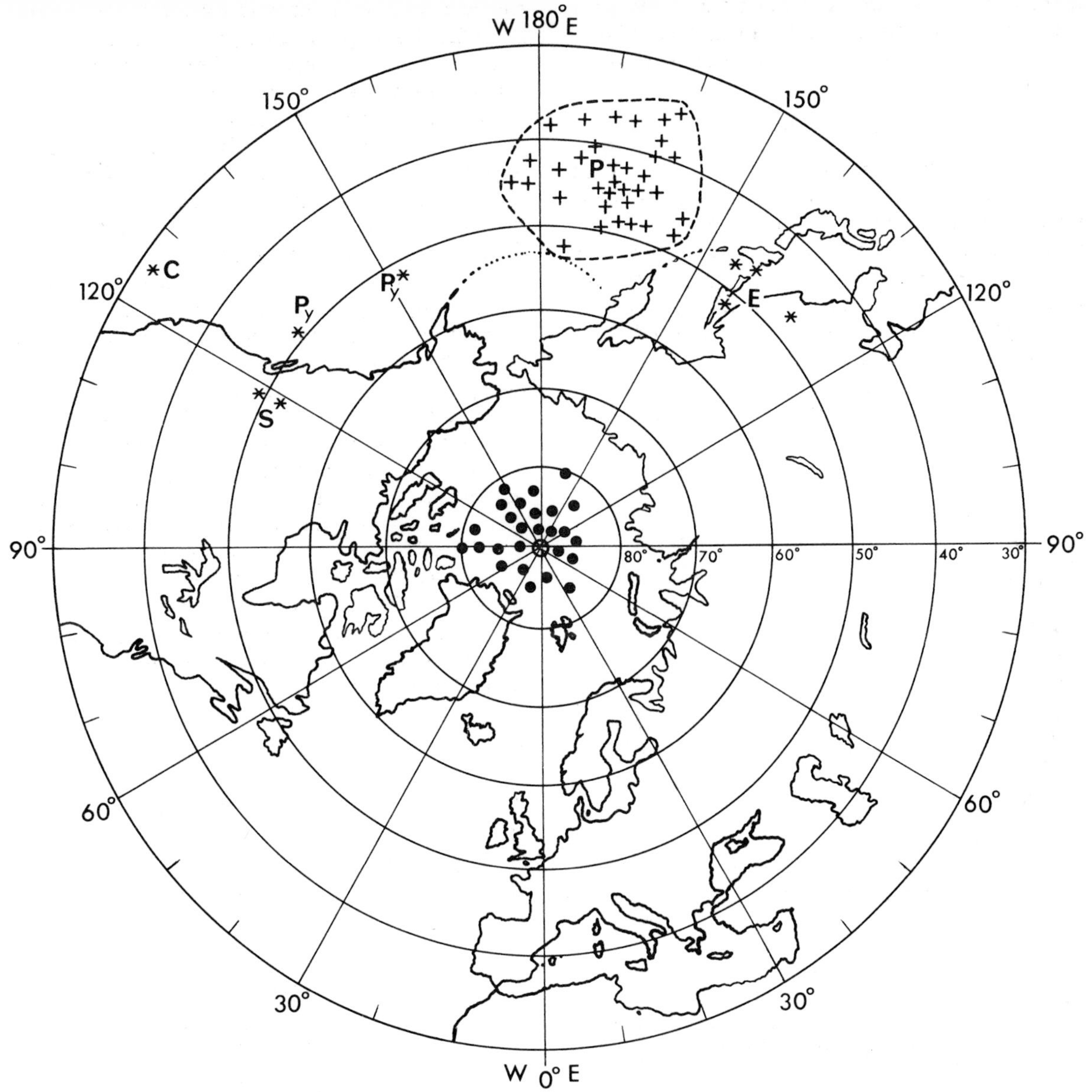


**Figure 27.7** Stereographic projection of the northern hemisphere (from the N Pole to Latitude 30°). *Black dots* clustered around the N Pole indicate the positions of Pleistocene and Pliocene geomagnetic poles. *Crosses* clustered around P indicate the positions of Permian geomagnetic poles determined from European sites undistributed by Alpine orogenies (British areas include Ayrshire, Durham and Devon). The mean position of these poles, P, is Lat. 43°N and Long. 170°E. *Stars*, dispersed on both sides of the cluster around P, indicate the positions of Permian geomagnetic poles from European sites within the Alpine orogenic belts; C, Corsica; E, Esterel (SE France); Py, Pyrenees; S, Southern Alps (*Data mainly from R. R. Doell and A. Cox, 1961; and D. van Hilten, 1964*)

which is well beyond observational and experimental errors (Figure 27.7). For the Lower Palaeozoic the divergence is still greater, and it reaches 90° or more in the late Precambrian. Similar results are obtained from the rocks of other continents, except that the 'polar-wandering curves' are found to be conspicuously different for the different continents (Figure 27.8).

To fix the position of a continental area (regarded for the moment as a rigid slab) on the globe requires three items of information, of which the palaeomagnetic data yield approximate estimates

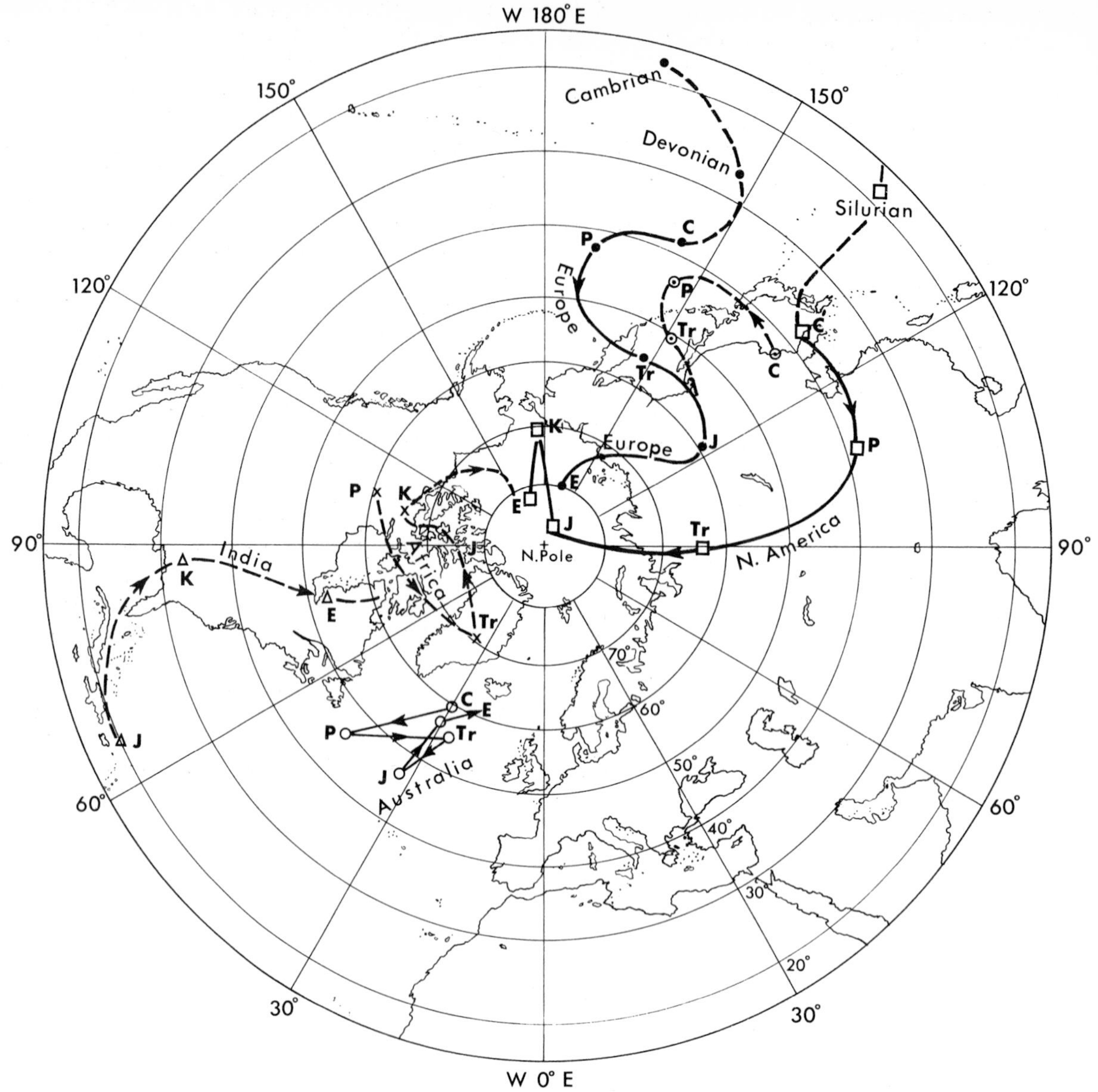


of only two. The three requirements are:

(*a*) the orientation relative to the contemporary poles (determined by the declination of the remanent magnetism);

(*b*) the latitude of each locality tested (determined by the inclination or dip of the remanent magnetism); and

(*c*) the longitude of each locality tested (*not* provided by palaeomagnetic measurements).

The third item, the longitudinal position, can be judged within certain limits since, at any given period of the past the continents of the present day

**Figure 27.8** Stereographic projection of the northern hemisphere from the N Pole to Latitude 16°, showing palaeo-magnetic polar-wandering curves for Europe (solid black circles) with an offshoot for Siberia (open circles with a central dot); North America (open squares); Africa (crosses); India (open triangles); and Australia (open circles). Geological ages of samples from which the data were obtained are indicated by letters: E, Eocene; K, Cretaceous; J, Jurassic; Tr, Triassic; P, Permian; and C, Carboniferous. It should be noticed that to avoid congestion these curves are drawn through the mean positions of clusters covering considerable areas (*cf* the P cluster in Fig. 27.7). To represent the data more accurately the curves should be broad bands instead of lines.

must not overlap, and they must occupy positions consistent with geological evidence of the kind appealed to before the data provided by palaeomagnetism were available. The Tibetan plateau is regarded as the result of an overlap between Asia and a nothern extension of India, but that northern extension is a hypothetical inference, not a 'continent of the present day'.

For convenience the longitudinal position is often judged relatively to some particular continent. Africa is commonly adopted for this purpose, following a convention started by all the pioneers, including Snider as well as Wegener. The meridian through Greenwich is similarly adopted as the conventional standard relative to which present-day longitudes are reckoned. There is no natural standard of longitude to be discovered; but the positions of the equator and poles and the intervening circles of latitudes can be discovered, and this applies equally to ancient times, with the aid of palaeomagnetism. The polar-wandering curve for Europe drawn in Figure 27.8 implies that the British Isles lay within 20° or so of the equator in Carboniferous and Permian times. But a more direct and clearer picture of the palaeomagnetic results is given by plotting the equator for a given pole and constructing *magnetic isoclines* (lines of equal magnetic inclination) between the two. The isoclines are small circles parallel to the equator of the time, which is the isocline for which $I = 0°$. As shown on Figure 27.5, the inclination increases towards the poles, where it is 90°. This very convenient method of representing the palaeomagnetic results has been extensively used by D. van Hilten. Figure 27.9 is an example of one of his isoclinal maps. It shows at a glance that, during the Permian, Europe and the British Isles lay thousands of miles south of their present positions and also that the subsequent drift towards the north was accompanied by a clockwise rotation.

On the basis of palaeomagnetic determinations it can thus be inferred that in past geological ages the continents have occupied positions relative to the poles and to each other, very different from those of today's familiar geography. Until about 1956 most geophysicists seem to have favoured *polar wandering* (without continental drift) as a sufficient explanation for the changes of latitude disclosed by the palaeomagnetic data then available. The polar-wandering hypothesis is generally understood to mean that an outer shell of the earth, involving the crust and probably part of the mantle down to the low-velocity zone (p. 586), has shifted *as a whole* relative to the axis of rotation, which remains almost fixed, relative to the stars. If this were an adequate explanation by itself, then the geographical positions of the poles at any given time would be the same for all the continents. Assuming the continents to have remained fixed

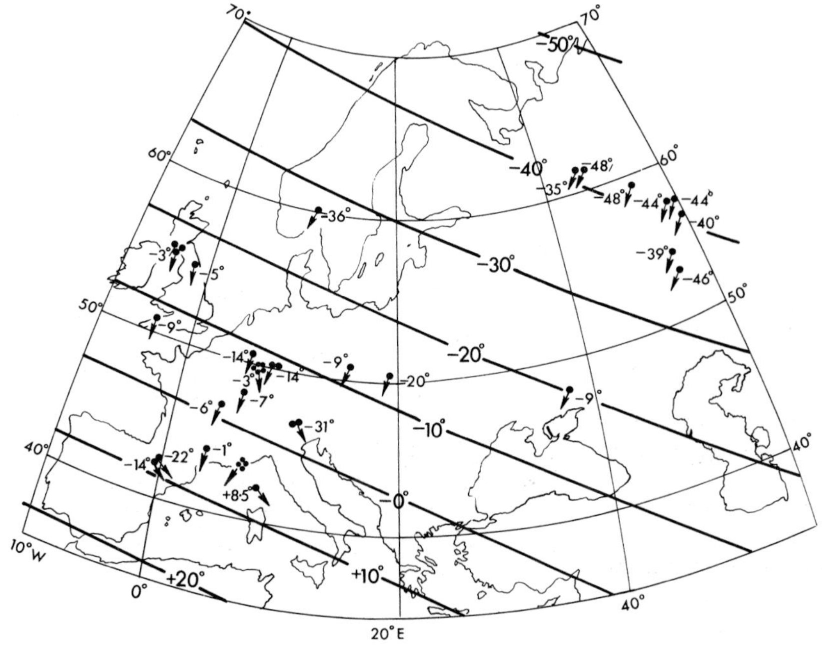


**Figure 27.9** Isoclines in heavy black lines drawn about the mean Permian pole position (43°N; 170°E) for Europe north of the Alpine front. The source-localities of measured samples are indicated by dots, with inclinations in degrees and orientations shown by arrows. Negative signs north of the Permian equator (and positive to the south) imply reversal of the normal magnetic field (*D. van Hilten, 1964*)

relative to one another, there would be a single 'polar-wandering' curve'. But every continent has its own 'polar-wandering curve' and these are so widely different that it is obvious that the continents have changed their relative positions through time.

## Reversed Magnetism: the Palaeomagnetic Time-scale

In a long sequence of rocks that have been adequately tested for remanent magnetization, two well-defined groups commonly turn up, within one of which the stable remanent magnetization is in the opposite direction to that of the other set. About half the rocks are N-seeking the remainder being S-seeking. Nearly all Permian samples, for example, are S-seeking. Initially it was thought that this phenomenon of *reversed magnetization* could be due either to alternating reversals of the earth's polarity (*field reversal*) or to certain processes of *self-reversal* within the rock itself, such as might be brought about by physical or chemical

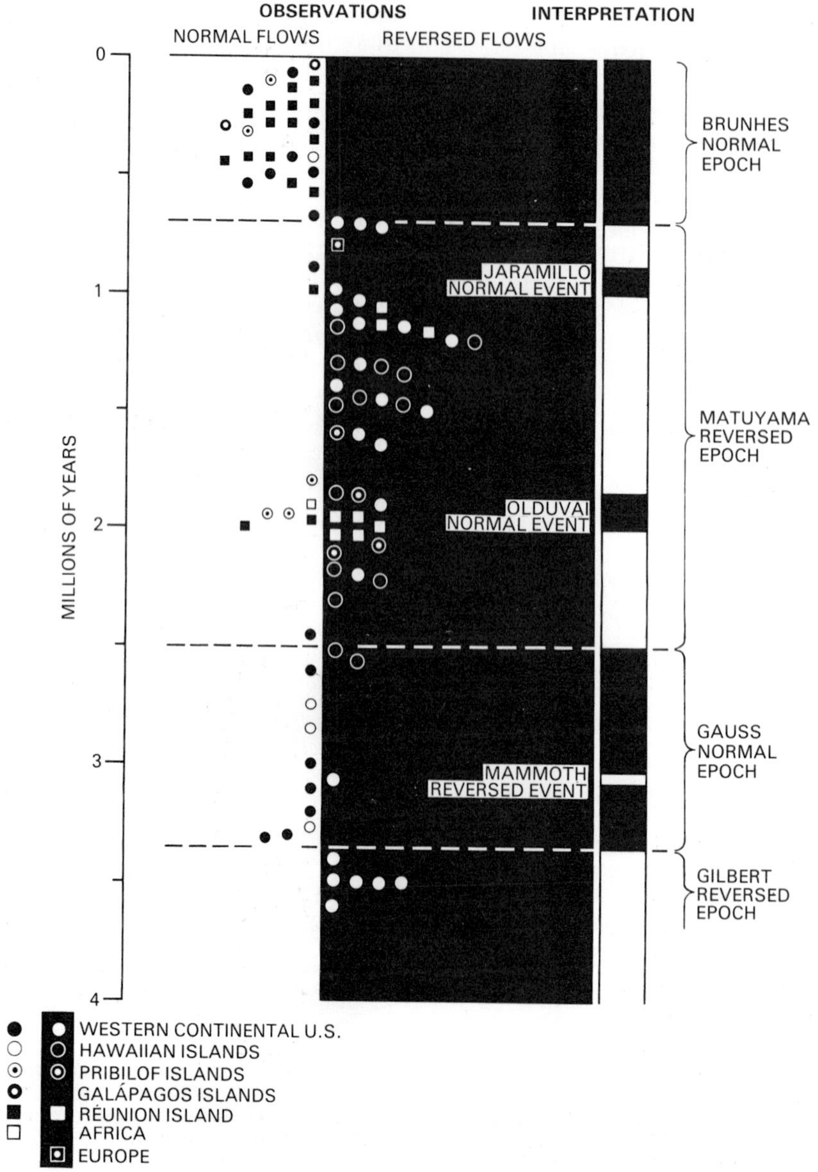


**Figure 27.10** A time-scale for reversals of the earth's magnetic field, established from palaeomagnetic data with matching radiometric ages derived from nearly 100 volcanic formations located in both hemispheres. The data group themselves into 4 major *polarity epochs*, in which shorter *polarity events* are superimposed. (*A. Cox, G. B. Dalrymple, and R. R. Doell, 1967. Copyright by Scientific American, Inc. All rights reserved*)

changes in the magnetic minerals during or after the rock's formation.

Self-reversal is evidenced by the fact that it can be brought about by laboratory experiments—but only on certain specimens constituting a small percentage of the number tested. Moreover, a few cases are known (e.g. in the Tertiary lavas of Mull) where both normal and reversed magnetization are found within different specimens from the same flow, so indicating that self-reversal can be a natural phenomenon. In the Mull example the chief magnetized mineral in a *normal* specimen is titaniferous magnetite in well-formed crystals; whereas in a reversed specimen there is a considerable variety of ferro-magnesian minerals, representing the products of subsequent alteration of a kind involving oxidation.

As long ago as 1929 Matuyama found that rocks with reversed magnetization in Japan were all early Pleistocene in age, whilst younger rocks showed normal magnetization. Complete evidence that magnetic reversals are not fortuitous, since they occur at the same time all over the earth, was not established, however, until 1964. By then there were sufficient determinations of remanent magnetization correlatable with radiometric age determinations, by the potassium-argon method, to establish a time-scale of normal and reversed magnetism extending backwards in time for 3·6 million years. The relevant data was obtained from close on 100 magnetized volcanic formations, occurring within both hemispheres. The investigations were mainly made by Cox, Dalrymple, and Doell, in the U.S. Geological Survey Laboratory in California, and by McDougall, Tarling and Chamabaum at the Australian National University. By the late 1960s the time-scale had been extended back to about 4·5 million years. The potassium-argon method, in this context, has insufficient precision to carry the time-scale farther back in time.

As can be seen from Figure 27.10, during the last 3·6 million years there have been two periods with normal polarity, as at the present day, and two with reversed polarity. These main periods are known as *geomagnetic polarity epochs*, and they are named after investigators, of various nationalities, who have made important contributions to knowledge of the earth's magnetism. Within these epochs are shorter *polarity events*, to which place names have been given, commemorating the localities where they were first recognized. When the geomagnetic time-scale was first established, three polarity events were known. By 1969, many more determinations having been made, eight polarity events were known (Figure 27.10).

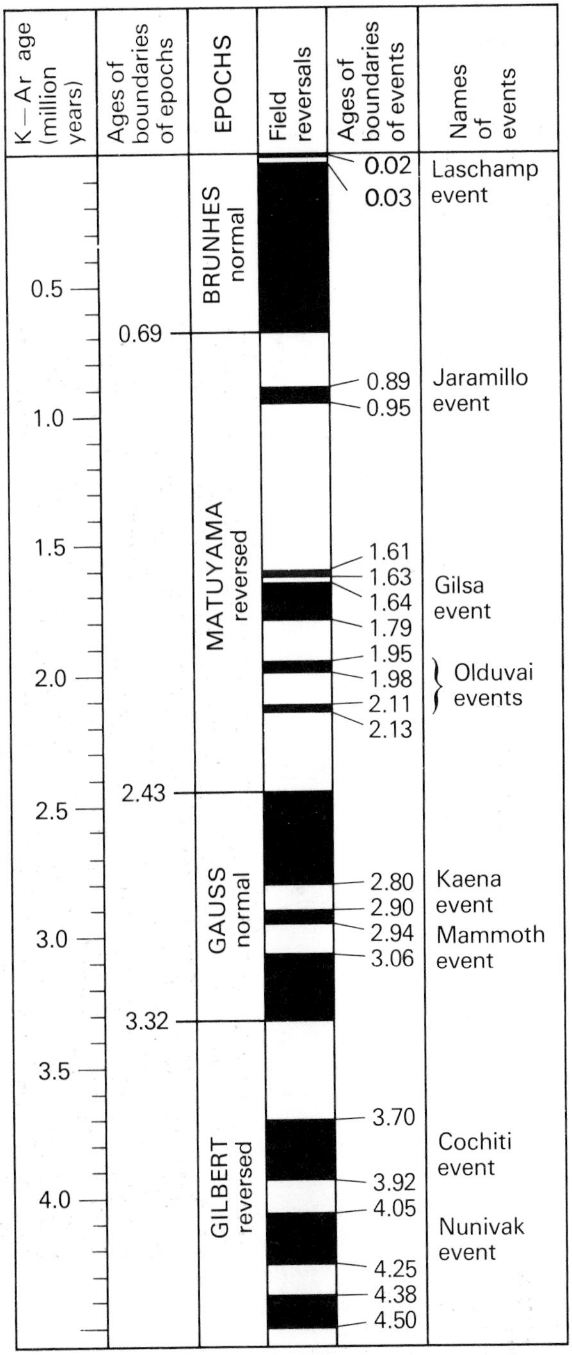


**Figure 27.11** Time-scale for geometric reversals in which the time-span of events is based partly on data from sediments (*Adapted from Cox, 1969*)

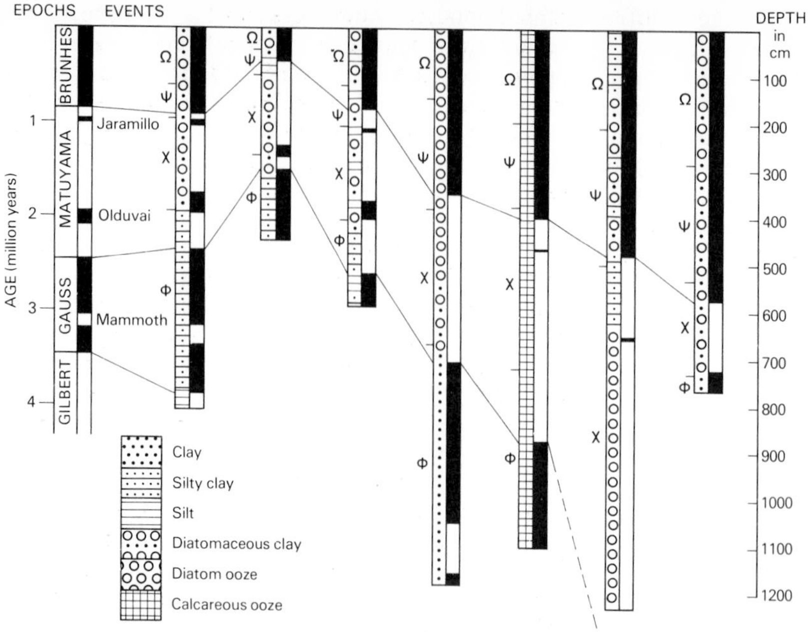


**Figure 27.12** 'Magnetic stratigraphy' established within deep-sea cores from the Antarctic, and correlated with radiolarian faunal zones (Greek letters) (*After N. D. Opdyke, B. Glass, J. D. Hayes and J. Foster, 1966. Copyright the American Association for the Advancement of Science*)

The magnetic reversal time-scale has an important application in determining the ages of deep-sea sediments. In 1966, Opdyke, Glass, Hayes and Foster established what might be called the magnetic stratigraphy within seven cores from the Antarctic, and they correlated this stratigraphy of normal and reversed epochs and events with faunal zones as defined by radiolaria. Here then was a method of correlating deep-sea sediments from widely separated regions, and also of determining the rate of deposition (Figure 27.12).

## Magnetic 'Stripes': Sea-floor Spreading

In 1952, Mason, of the Scripps Institute of Oceanography, towed a magnetometer, tied to the stern of a research ship, halfway across the Pacific Ocean. The results proved to be of so much interest that he and his colleagues subsequently surveyed a broad strip of the Pacific floor offshore from Mexico to British Columbia. When the magnetic anomalies were plotted on a chart, like contour lines, it was found that they revealed a roughly north–south pattern of stripes of which Figure 27.18(*a*) is a sample. Magnetic anomalies are simply the difference ($b-a$) between the calculated strength or intensity of the normal magnetic field ($a$), and the actual intensity ($b$) as measured by the magnetometer.

At the time when they were discovered, the cause of the magnetic anomalies on the Pacific floor could be no more than a matter of inference by anology. As Raff records 'a single glance at the map was enough to show that we had something quite new in geophysics'. Since the anomalies are sharply and steeply bounded, showing that whatever is responsible for them comes up to the ocean floor, or nearly so, it appeared at the time that they might be caused by structural changes as, for example, by the presence of lava flows, or dyke-like basic intrusions, or by variations in the topography of the main crustal layer.

In 1963 Vine and Matthews, after a detailed magnetic and bathymetric survey of the middle part of the Carlsberg ridge in the Indian Ocean, as part of the International Indian Ocean Expedition of 1963, computed the magnetic profile across the Carlsberg ridge, from the bathymetric profile, assuming normal magnetization. They found it differed completely from the observed anomaly profile. On the other hand, when they assumed that the basaltic crustal layer was divided into blocks 20 km wide, with alternately normal and reversed magnetization, the computed profile closely resembled the observed profile (Figure 27.13).

Assuming spreading of the oceanic floors, according to the hypothesis of Dietz and Hess, and periodic reversals of the earth's magnetic field (Figure 27.14), Vine and Matthews suggested that if the basaltic crustal layer was formed by rising

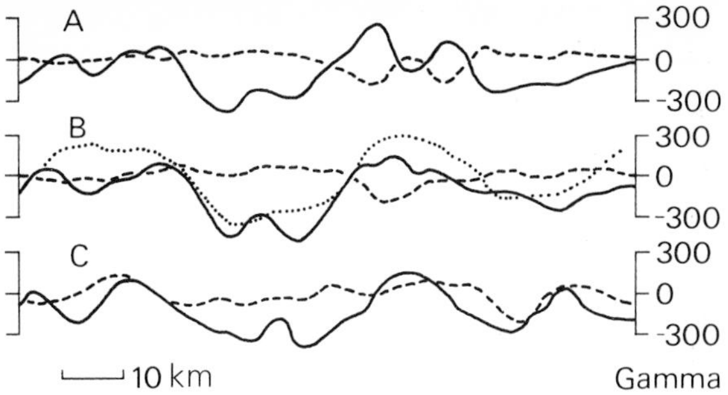


**Figure 27.13** Profiles perpendicular to the trend of the Carlsberg Ridge illustrating (a) the lack of resemblance between the observed (solid lines) and computed (dashed lines) profiles, assuming normal magnetism, and (b) the resemblance between the observed profile (solid line) and the dotted profile in B, which was computed for a model in which the basaltic layer is divided into parallel blocks with alternating normal and reversed magnetization (*After F. J. Vine and D. H. Matthews, 1963*)

convection currents along the middle of the oceanic ridges, and the resultant extrusions and intrusions of basalt each assumed the direction of magnetization of the earth's magnetic field at the time when it consolidated, then the striped pattern of magnetic anomalies, like those of the North Pacific Ocean, could be explained. The idea is that each new intrusion and extrusion of basalt, along the middle of the oceanic ridges, splits the previous intrusion into two, and the two halves move away on either side. The striped pattern should therefore be bilaterally symmetrical on either side of the median rift. In other words, according to the Vine–Matthews hypothesis, the magnetic anomalies might provide a magnetic record of ocean-floor spreading that could be read from each ocean floor outwards from the mid-oceanic rift.

At the time when Vine and Matthews proposed their hypothesis there was much resistance to its acceptance, because it was not then generally agreed that magnetic field reversals were a reality. Many earth scientists still regarded it as possible that reversals might be self reversals. The following year, however, Cox, Doell and Dalrymple (1964) established the reality of reversals of the earth's magnetic field (Figure 27.10). Since the reversals took place at the same time in various parts of the earth it would be too much of a coincidence to consider them to be other than field reversals, and the Vine–Matthews hypothesis gained favour. It gained complete acceptance after Opdyke *et al.* (1966), as recorded in figure 27.12, found a similar magnetic timescale for the last 4 million years, fossilized within the stratigraphic columns of sedimentary cores collected from the Antarctic. The reliability of palaeomagnetic epochs and events, whether determined from basaltic or sedimentary rocks, was thus accepted as a proven means of dating events during the last 4·5 million years.

In 1966 the results of an aeromagnetic survey by Heirtzler *et al.*, over an area of 350 km² over the Reykjanes ridge, part of the mid-oceanic ridge, were published. This investigation not only showed a sequence, from the ridge outwards on either side, of normally and reversed magnetized stripes conforming to the newly established sequence, but it also showed that the anomalies are symmetrical about the crest of the ridge, as the hypothesis of ocean-floor spreading requires (Figure 27.15).

Since teams from Lamont–Doherty Observatory, manning research ships, had collected magnetic data as a matter of routine, on their many cruises since the early 1960s, a wealth of magnetic

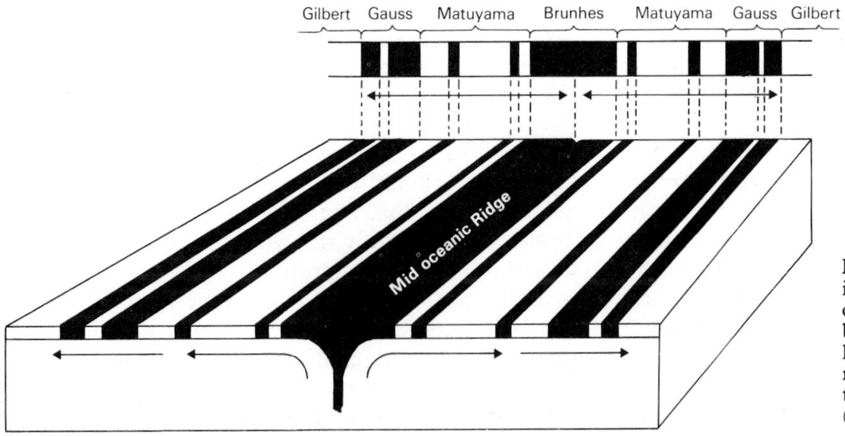


**Figure 27.14** Block diagram illustrating how 'magnetic stripes' on the ocean floors can be explained by spreading of the oceanic lithosphere away from oceanic ridges on either side, according to the Vine–Matthews hypothesis (*Adapted from A. Cox, et al., 1967*)

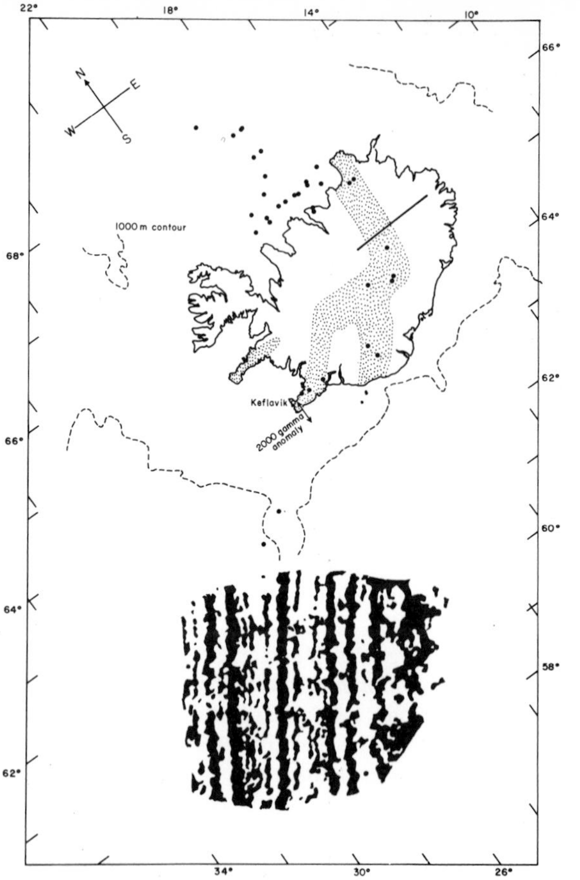


**Figure 27.15** Magnetic anomalies over the Reykjanes Ridge, south-west of Iceland, showing 'magnetic stripes' (positive anomalies black) and their bilateral symmetry (*After J. R. Heirtzler et al., 1965. Reproduced from Heirtzler, 1969, with permission of John Wiley and Sons*)

data already existed, but it had to be deciphered. Records across mid-oceanic ridges were not directly available, because the research ships had changed course and speed, according to the prime object of their investigations. Cox (1973) records that Heirtzler undertook the task of organizing and correlating the existing data and, with the aid of a computer, constructed magnetic profiles across the Atlantic. By 1968, profiles across the mid-oceanic ridges had been published for the South Atlantic, the South Indian Ocean, and the North Pacific. It was a proven fact that the striped patterns of magnetic anomalies characterized extensive areas of these three oceans. For the middle parts of the oceanic ridges the ages of the magnetic anomalies had by now been estimated by comparison with the known sequence of polarity events for the past 4·5 million years (Figure 27.10). By assuming constant spreading rates for individual profiles, and using the relative widths of magnetic 'stripes' as a guide to the rate of spreading, it was found that the calculated ages of the marine magnetic 'stripes' matched those that were ascribed by analogy with the ages of polarity epochs and events determined from lava-flows on land. Conversely, dating the marine 'stripes' by analogy, provided a means of estimating the rates of spreading in the various oceans. The widths of the magnetic 'stripes' vary according to the spreading rate, being wider where the spreading rate is rapid, as in the equatorial Pacific.

There is no direct way of dating marine stripes older than 4·5 million years. Heirtzler *et al.* (1968), however, by assuming constant spreading rates for the individual oceans, constructed a time-scale for the magnetic 'stripes' back for the past 75 million years, that is back into the upper Cretaceous. It was then possible to construct an isochron map of the ocean floors (Figure 27.16), by mapping known magnetic anomaly 'stripes' at time intervals of 10 million years. Isochron lines marked 10 lay on the mid-oceanic rift 10 million years ago; isochron lines marked 20 lay on the mid-oceanic rift 20 million years ago and so on. Spreading rates are calculated from the ages of the isochron lines and their distances apart. The slowest rates of 1 or 1·5 cm a year occur in the Atlantic and Indian Oceans, and the most rapid, 6 cm a year, on the east Pacific rise. If the rate of separation of oceanic crust at the mid-oceanic ridges is under consideration, the spreading rates have to be doubled. For example new oceanic crust with a width of 2 or 3 cm is emplaced in the Atlantic Ocean each year.

In 1970, the result of coring across the South Atlantic Ocean confirmed the calculated magnetic time-scale of Heirtzler *et al.* The cores penetrated down to the basaltic layer (Layer 2), Upper Cretaceous being the oldest stratigraphic level penetrated. The age of the oldest sediment immediately overlying the basaltic layer was determined from microfossils within each core, and this age was found to increase with increased distance from the mid-oceanic rift on either side. (Figure 27.17). The data reveal a constant spreading rate of 2 cm a year extending backwards in time for about 80 million years.

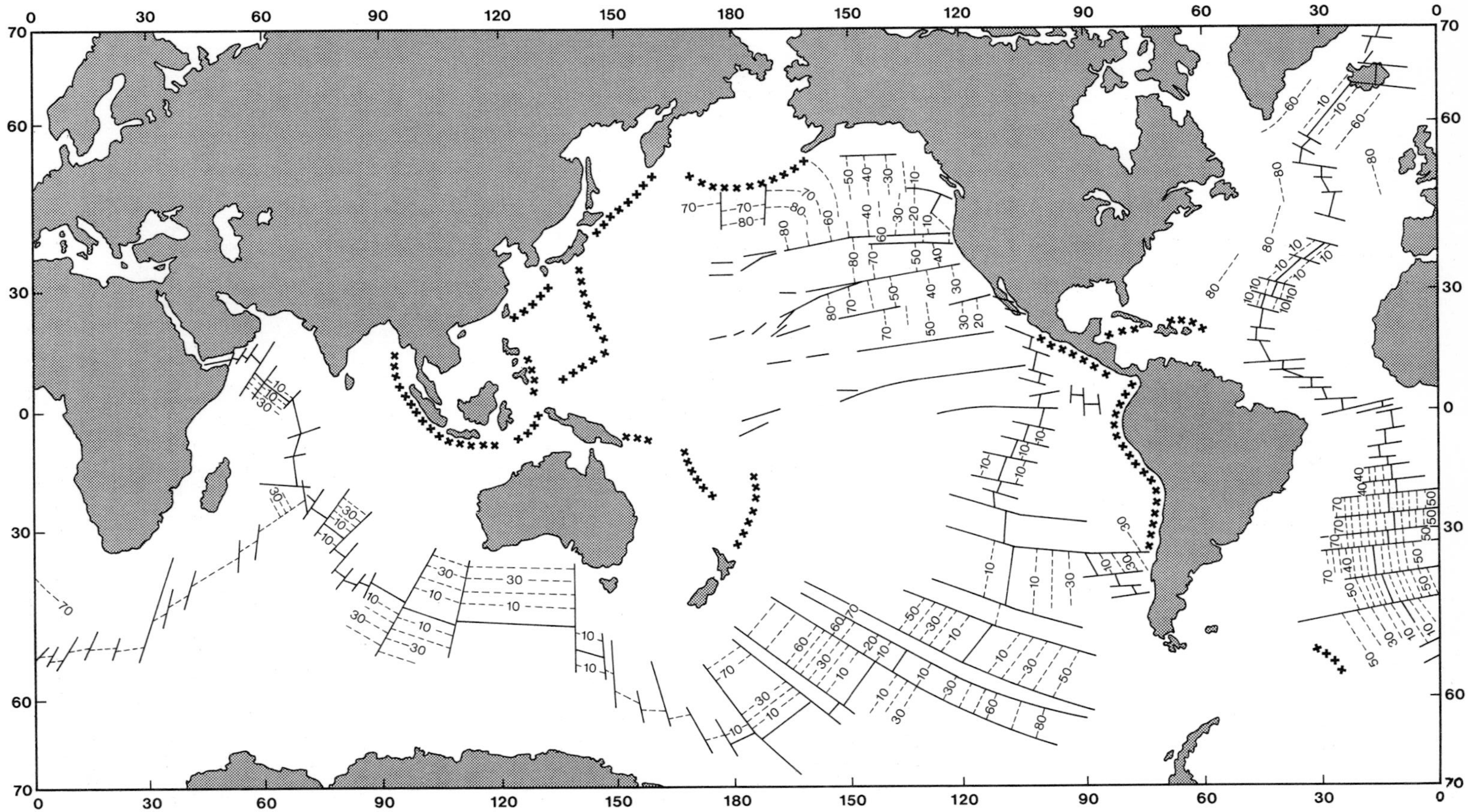


**Figure 27.16** Isochron map constructed from the 'magnetic stripes', by assuming that the spreading rates for individual ocean floors remain constant. The numbers, against the isochrons (dashed), represent time intervals of 10 million years. The solid lines represent mid-oceanic ridges when parallel to the isochrons, and transform faults when they intersect them. Crosses indicate the positions of deep ocean trenches (*After J. R. Heirtzler et al., 1968*)

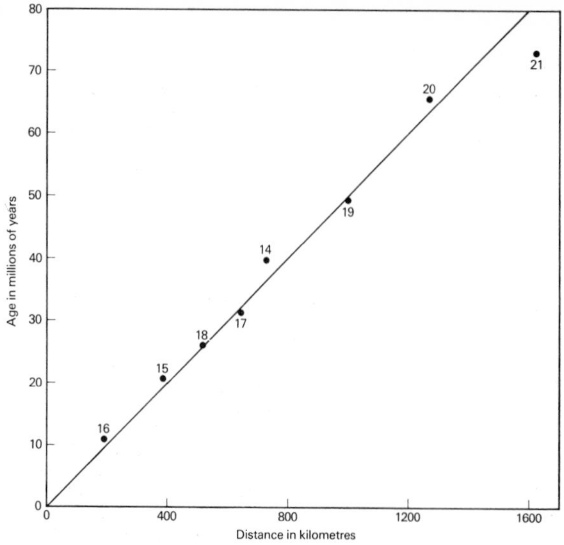


**Figure 27.17** Graphic construction revealing a constant spreading rate for the oceanic lithosphere of the South Atlantic. The ages, in millions of years, of sediments (from numbered sites) immediately overlying the basaltic layer are plotted against their distance from the South Atlantic ridge. The close agreement of the plotted points with a straight line respresenting a spreading rate of 2 cm a year is strong evidence of a constant spreading rate. (*After A. E. Maxwell et al., 1970*)

## Transform Faults

Associated with and commonly striking approximately at right angles to the mid-oceanic ridges are fracture zones that appear to off-set the ridges (Figure 27.18). Amongst the earliest to be studied in detail were those off the western coast of North America, which have sliced the rise into long crustal slabs, strikingly roughly E–W. These fractures were first recognized as faults from the vertical displacements which have produced great submarine scarps rising hundreds of metres above the ocean floor on the deeper side. Because they are pratically straight they were at once suspected to be strike-slip faults. The Mendocino, Murray, Clarion and Clipperton faults were first described by Menard in 1955, and since then many more have been discovered as indicated on Figure 27.16.

The 1600 km long Mendocino fault, for example, has a southward facing undersea cliff rising in height from 1500 to as much as 3000 m. The block to the north has been only slightly tilted as a whole, but in detail it is considerably fractured. The relative horizontal displacement along the Mendocino fault, and similar faults to the south, were determined by reference to the magnetic anomaly 'stripes', which are cut through by the faults, and in each case are quite different on the two sides of the faults (Figure 27.18(*b*)). The amount of lateral displacement along the Murray fault was the first to be measured. It was found that the east-west profile of the magnetic anomalies on the northern side (plotted as in Figure 27.18(*c*)) could be matched with the corresponding profile on the southern side when the latter was shifted 154 km to the east, so that an anomaly E (Figure 27.18(*c*)) was found to have its continuation at D.

At that stage no matching could be recognized along the two sides of the other great faults. The survey had not, in fact, been carried far enough west. In 1958 Vacquier and Raff continued the magnetic survey westward until they eventually found a convincing match along the Pioneer fault,

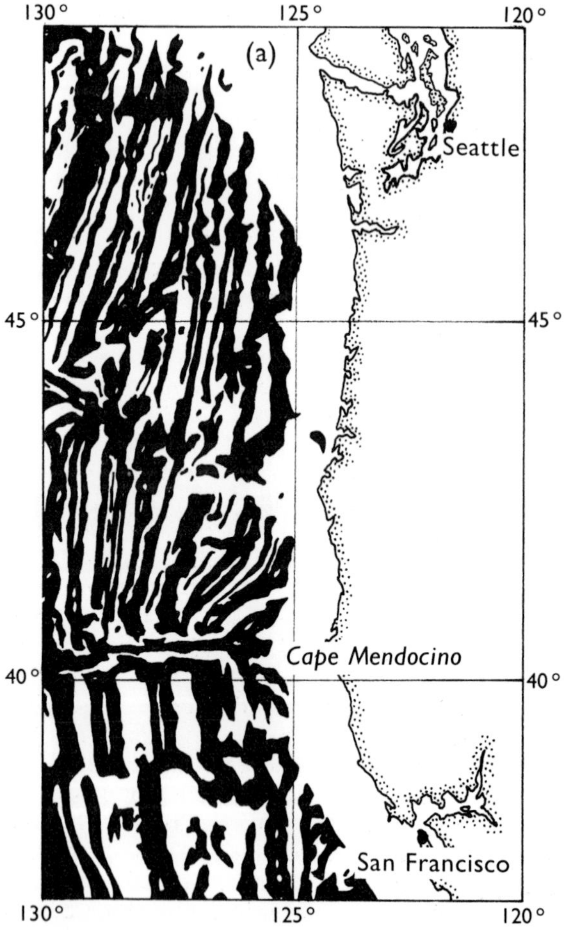


**Figure 27.18**
(a) Pattern of positive magnetic anomalies discovered in the Pacific floor off the coast between Seattle and San Francisco

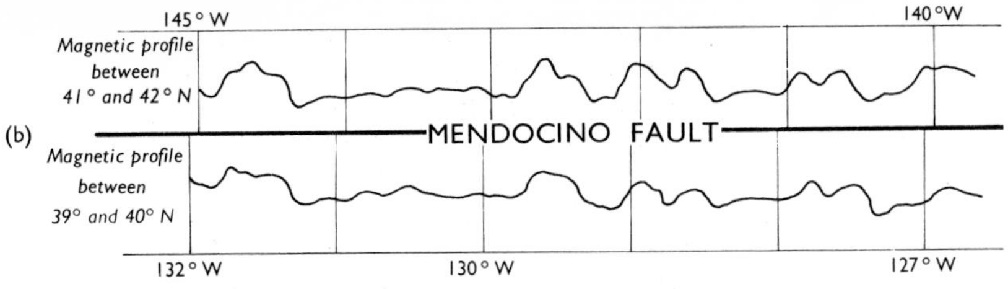

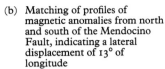

(b) Matching of profiles of magnetic anomalies from north and south of the Mendocino Fault, indicating a lateral displacement of 13° of longitude

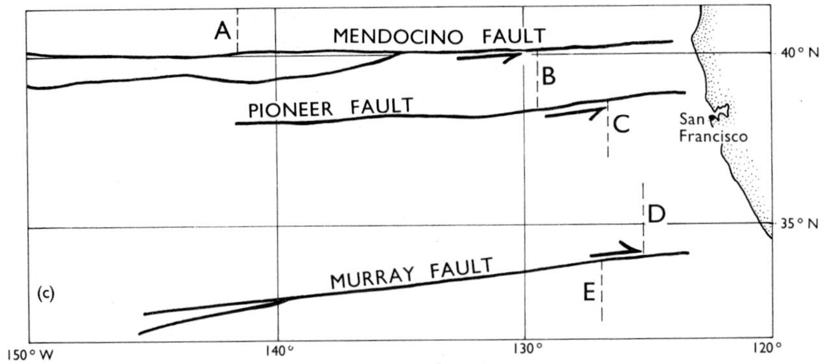


(c) The magnetic anomalies on the north side of the Mendocino Fault can be matched with those on the south side only when two anomalies such as A and B, 1160 km apart, are brought together. Similarly, the displacement along the Pioneer Fault is the distance between B and C, 265 km; and that along the Murray Fault the distance between D and E, 154 km. (*After A. D. Raff, 1961*)

which indicated a displacement of 265 km. But along the Mendocino fault no match was detected until the Scripps men had patiently extended their survey farther west throughout two seasons of work. In 1961 they succeeded in locating a pattern on the north side that matched the pattern on the south side—but 1160 km to the west (Figure 27.18 (*b*) and (*c*)).

Similar faults off-setting the mid-oceanic ridges have been found to be a world-wide phenomenon. Prior to 1965, faults characterized by horizontal displacements like those just described, and the San Andreas fault in North America and the Great Glen fault in Scotland, were thought to be transcurrent faults, and inferred to be younger than the phenomena, such as the mid-oceanic ridges, which they off-set. The great difficulty about this interpretation was that it seemed impossible for great strike-slip faults like the San Andreas fault to terminate. Yet there was no evidence that they continued as small circles right round the earth. For many years geologists had recognized that earth movements, expressed by earthquakes, ranges of young fold mountains and major faults, like the San Andreas with great horizontal displacements, are restricted to narrow belts, sometimes described as mobile belts. In 1965 Tuzo Wilson recognized that these mobile belts form a continuous network over the earth, dividing its surface into several large rigid plates, like a mosaic. Faults with major horizontal displacements can be traced into movement belts of other types, such as fold mountains and island arcs, or mid-oceanic ridges. Wilson suggested that the junctions between movements of different kinds should be called *transforms*, and that the faults concerned be called *transform faults*.

Transform faults differ from transcurrent faults in that they are intimately related either to the creation of new crust, as at mid-oceanic ridges, or to the annihilation of crust where it disappears into the depths at ocean trenches, or becomes narrowed in fold mountain belts. The faults that strike across mid-ocean ridges connect portions of the ridge crests that are off-set in relation to one another. With great intuition, Wilson recognized that the apparent off-sets of the ridges might be original features formed when continental crust split apart and new oceans were born (Figure 27.19). The mid-Atlantic ridge, including its off-sets, for example, is essentially parallel to the shores of the bordering continents (Figure 27.16). The direction of horizontal movements along transform faults, formed in this way, would be in the opposite direction to that required if the faults were transcurrent faults responsible for off-setting the ridges after they were formed (Figure 27.20). The directions of movements along transform faults intersecting mid-oceanic ridges have now been amply established by seismic studies, so that there is no doubt that the directions of movement along them conform with Wilson's concept of transform

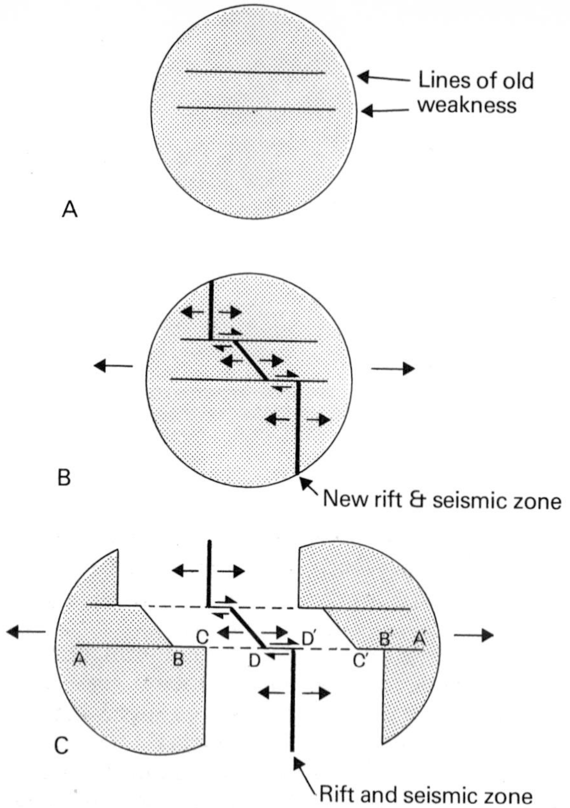


**Figure 27.19** Diagrammatic illustration of three stages in the rifting of a continent into two parts, and their separation resulting in the evolution of new ocean floor. Exemplified by the separation of South America and Africa with the opening of the South Atlantic ocean (*After T. Wilson, 1965*)

faults. Moreover, the associated earthquakes, which are all shallow, are confined to the ridges themselves and to the portions of transform faults between the off-sets. Where the faults extend away from the ridges, beyond the off-sets, no earthquakes occur; the faults are here dead.

## Selected references

BLACKETT, P. M. S., 1961, 'Comparison of ancient climates with the ancient latitudes deduced from rock magnetic measurements', *Proceedings of the Royal Society*, A, vol. 263, pp. 1–30.
BULLARD, E. C., 1970, 'Geomagnetic dynamos', In *The Nature of the Solid Earth*, 1972, International Series in the Earth and Planetary Sciences, McGraw-Hill, New York.
COX, A., DOELL, R. R., and DALRYMPLE, G. B., 1964, 'Reversals of the earth's magnetic field', *Science*, vol. 144, pp. 1537–43.
COX, A., DALRYMPLE, G. B., and DOELL, R. R., 1967, 'Reversals of the earth's magnetic field', *Scientific American*, vol. 216, pp. 44–54.
COX, A., 1973, *Plate Tectonics and Geomagnetic Reversals*, Readings, selected, edited, and with introductions by Alan Cox, W. H. Freeman, San Francisco
HEIRTZLER, J. R., 1969, 'Magnetic anomalies measured at sea', *The Sea*, vol. 4, Pt. 1, Wiley-Interscience, 1970, New York.
HILTEN, D. VAN, 1964, 'Evaluation of some geotectonic hypotheses by palaeomagnetism', *Tectonophysics*, vol. 1, pp. 3–71.
MCDOUGALL, I., and TARLING, D. H., 1963, 'Dating of polarity zones in the Hawaiian islands', *Nature*, vol. 200, pp. 54–6.
OPDYKE, N. D., GLASS, B., HAYS, J. D., and FOSTER, J., 1966, 'Palaeomagnetic study of Antarctic deep-sea cores', *Science*, vol. 154, pp. 349–57.
WILSON, T. J., 1965, 'A new class of faults and their bearing on continental drift,' *Nature*, vol. 207, pp. 343–7.

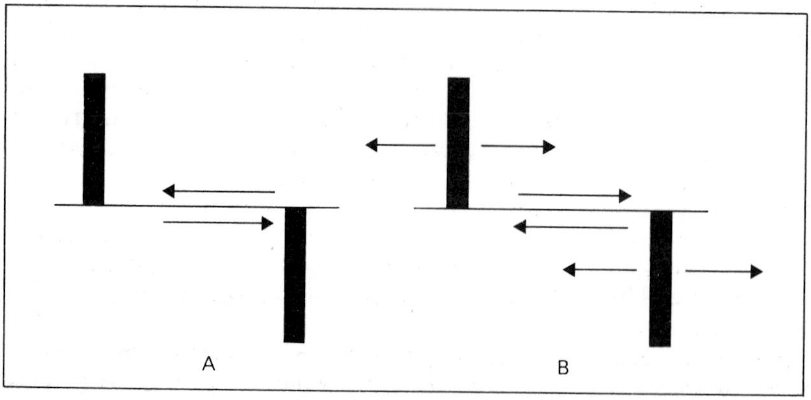


**Figure 27.20** Diagrammatic illustration of:
A The directions of movement along a transcurrent fault responsible for the off-setting of the black band
B The directions of movement along a transform fault connecting off-set portions of a mid-oceanic ridge (black)

# Chapter 28

# Reassembling the Continents

The latitude's rather uncertain,
And the longitude also is vague.
WILLIAM JEFFREY PROWSE (1836–70)

## Changing Views of Continental and Oceanic Relationships

The continents are essentially slabs of sial, distributed to form a northern group, known as *Laurasia*, and a more scattered southern group, known as *Gondwanaland*. The outer peripheries of the members of each group are defined by the Alpine orogenic belts (Figures 28.1 and 28.2), and the coastlines, generally backed by mountains are of Pacific type (Figure 23.4). The inner margins of the members of each group, against the Arctic, Atlantic, and Indian Oceans, are fractured, the coastlines being of Atlantic type (Figure 23.4).

To what extent these primary units and their arrangements have been stable or otherwise during geological time has been one of the fundamental problems of geology. For nearly a century there was a vigorous debate as to whether continents and oceans could interchange as a result of vertical or radial movements. During the last century it was tacitly assumed that if interchanges between continent and ocean had to be postulated then the movements involved could hardly be other than vertical, except for the slight lateral movements in the crust of a supposedly contracting earth.

The widespread occurrence of marine sediments over the lands suggested that the continents could sink to oceanic depths and the ocean floors rise to become dry land. It was gradually recognized, however, that with only a few minor exceptions these deposits proved no more than temporary flooding of the lands by shallow seas. For this reason, amongst others, Dana expressed the view in 1846 that continents and oceans have never changed places, and that the general framework of the earth has remained essentially stable. Nevertheless, Edward Forbes, tackling the subject from the biological side in the same year, found it impossible to explain how certain animals and plants had migrated from one continent to another unless some parts of the oceans had formerly been land. Thus began the long controversy regarding the permanence of the continents and ocean basins.

Support for permanence was found in the fact that deep-sea deposits, like those now forming on the ocean floors, had been discovered only in a few marginal islands, where they could be accounted for by the vertical movements involved in mountain building. Moreover, with the growing recognition of isostasy and its implications it became increasingly difficult to conjure up processes that could turn ocean floor into land or land into ocean floor. This difficulty became an impossibility when seismic exploration revealed the absence of sial from the ocean floors. Vulcanism, of course, was available to construct land-bridges like Panama, which joins North and South America, or island stepping-stones like those of the West Indies, which also link the Americas together for many biological purposes. Nevertheless, the main problem remained, and, only a few decades ago hypothetical land-bridges were being invented on a most irresponsible scale to provide biologists with the migration routes they required for the dispersal of land animals that could not withstand a marine environment, or for shallow water bottom-dwellers that could not cross the floor of an ocean.

Another problem was provided by the evidence that certain regions which supplied detrital sedi-

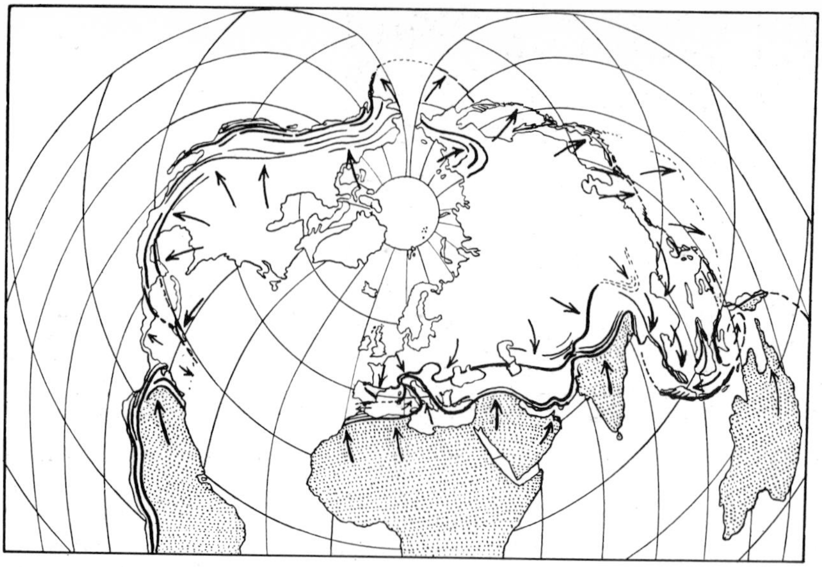

**Figure 28.1** Map showing the interrupted orogenic ring peripheral to the continental masses (unshaded) of Laurasia. The supposed movements of the continents directed outwards towards the Pacific and the Tethys are roughly indicated by arrows.

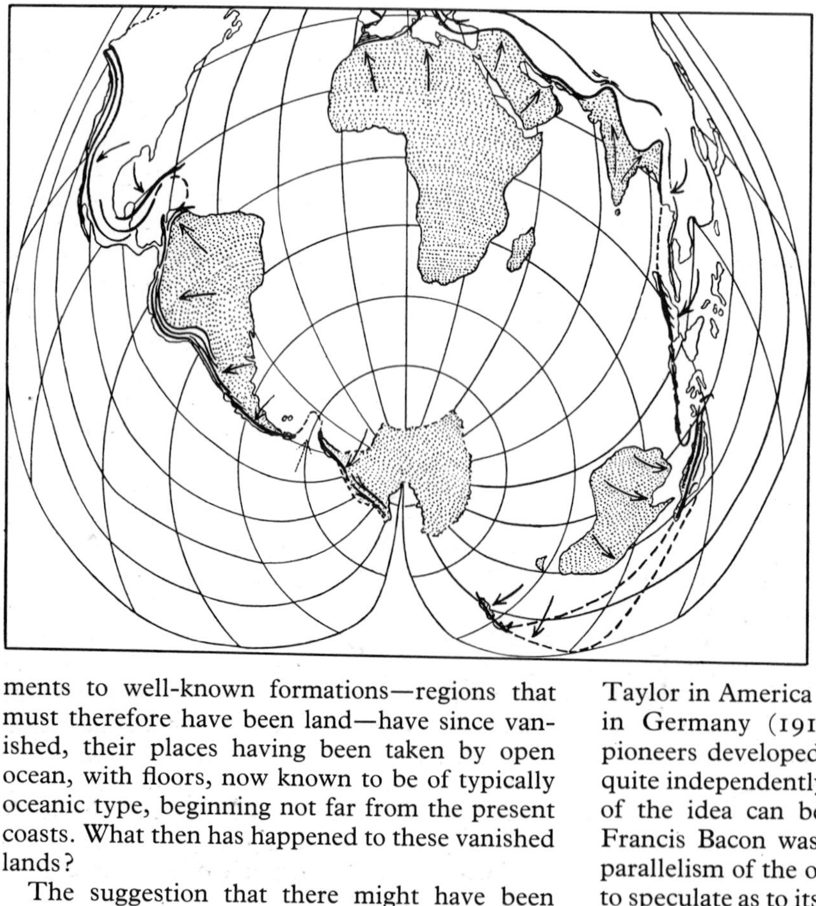

**Figure 28.2** Map showing the interrupted orogenic ring peripheral to the continental masses (dotted) of Gondwanaland. The supposed movements of the continents towards the Pacific and the Tethys are roughly indicated by arrows.

ments to well-known formations—regions that must therefore have been land—have since vanished, their places having been taken by open ocean, with floors, now known to be of typically oceanic type, beginning not far from the present coasts. What then has happened to these vanished lands?

The suggestion that there might have been lateral displacements of the continental masses on a gigantic scale is generally ascribed to F. B. Taylor in America (1908) and to Alfred Wegener in Germany (1910). For several years these pioneers developed their unorthodox hypotheses quite independently. Actually, however, the germ of the idea can be traced back to 1620, when Francis Bacon was sifficiently impressed by the parallelism of the opposing shores of the Atlantic to speculate as to its meaning. But he did not go so far as a Frenchman, one P. Placet, who published a fantastic memoir in 1668, entitled *La corruption du*

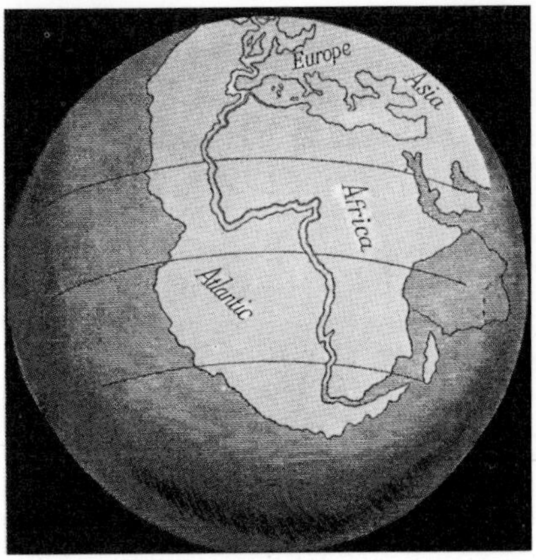


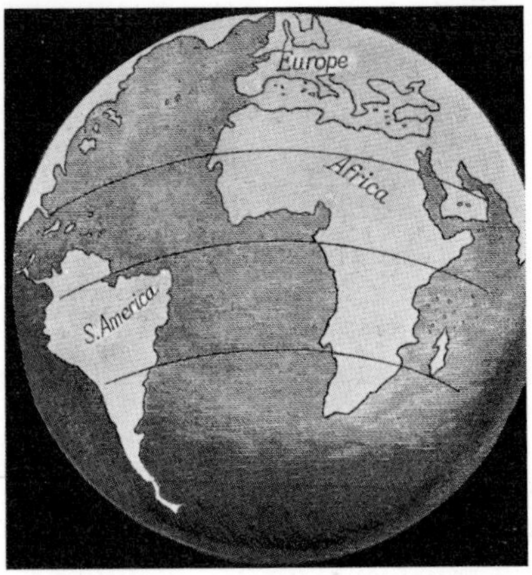


**Figure 28.3** Maps published by A. Snider in 1858 to illustrate his conception of continental drift. The left-hand map represents his reassembly of the continents for late Carboniferous times.

*grand et du petit monde, où il est montré que devant le déluge, l'Amérique n'était point séparée des autres parties du monde.* Nearly two centuries later Antonio Snider united the continents in much the same way as Wegener did. In a book with the optimistic title *La Création et ses Mystères dévoilés* (Paris, 1858) he published the two maps here reproduced as Figure 28.3. Snider's reconstruction of Carboniferous geography was intended to explain that fact that most of the fossil plants preserved in the Coal Measures of Europe are identical with those of the North American Coal Measures. Although the two diagrams reappeared in J. H. Pepper's popular and highly entertaining *Playbook of Metals* (London, 1861), the idea they embodied was evidently regarded as too outrageous to be worthy of serious scientific attention. For all practical purposes it was soon forgotten.

## Taylor's Concept of Continental Drift

It was not until Wegener published his famous book on the subject in 1915 that the possibility of continental drift began to be widely discussed. But Taylor must be given credit for making an independent and slightly earlier start in this precarious field. In a pamphlet privately printed in 1908 he described the continents as 'huge landslides from the polar regions towards the equator', but he did not publish his ideas until 1910. His immediate object was to account for the distribution of mountain ranges. He pictured the original Laurasia as being a continuous sheet of sial and supposed it to have spread outwards towards the equator, more or less radially from the polar regions, much as a continental ice sheet would do. Wherever the resistance was least, the crust flowed out in lobes, raising up mountainous loops and arcs in front (*cf.* Figure 28.1). Such movements, of course, would be impossible without complementary stretching and splitting in the rear. And, indeed, there is ample evidence of down-faulting and disruption in the coastal lands and islands of the Arctic and North Atlantic, and especially in the highly fractured region between Greenland and Canada, the map of which looks like a jig-saw puzzle with the separate bits dragged apart. In the southern hemisphere the originally continuous Gondwanaland similarly spread out, breaking up into immense rafts which also migrated towards the equator and raised up mountains in front (Figure 28.2). The basins of the South Atlantic and Indian Oceans were interpreted as the stretched and broken regions left behind or between these drifting continents.

For two reasons Taylor's hypothesis received scant attention. A considerable amount of lateral continental movement was thought to be implied by the structures of orogenic belts, but it seemed to be unnecessarily extravagant to invoke thousands of kilometres of horizontal displacement when from thirty to seventy according to the contraction hypothesis would have sufficed. Secondly, Taylor's attempt to explain the alleged movements was quite unacceptable. He postulated that the Moon first became the earth's satellite during the Cretaceous, and that at the time of its capture it was very much nearer to the earth than it is today. The resulting tidal forces were supposed to be sufficiently powerful not only to alter the rate of the earth's rotation, but also to drag the continents away from the poles.

Apart from the improbability that the earth was without a moon before the Cretaceous, there are two fatal objections to this conjecture:

(*a*) If the late Cretaceous and Tertiary mountain building is to be correlated with the supposed close approach and capture of the moon, then we are obviously left with no explanation for all the earlier orogenic cycles.

(*b*) If the tidal force applied to the earth by the newly captured moon had been sufficient to displace continents and raise mountains on the scale required, then, as Jeffreys convincingly pointed out, the friction involved would have acted like a gigantic brake, bringing the earth's rotation practically to a standstill within a year.

Taylor's 'explanation' is untenable, but from the criticisms one very important conclusion may be drawn. The fact that the earth continues to rotate shows that neither tidal friction nor any other force applied from outside the earth can be responsible for mountain building or for continental drift. Whatever the primary cause for the latter may be, it must be looked for *within* the earth.

## Wegener's Concept of Continental Drift

It is tempting to suppose that the seed destined to grow into the great concept of continental drift may have been implanted in Wegener's mind during his first expedition to Greenland in 1906–8. Off the NE coast he could not have failed to observe both the splitting up of the polar ice pack into gigantic floes, as undulating ocean swells surged beneath, and the formation of lanes of open sea as wind or currents drew some of the adjacent floes apart. It appears, however, that the analogous idea of the drifting apart of Africa and South America was not consciously inspired until the Christmas of 1910, when he had an opportunity of examining at leisure the beautiful maps of the André *Handatlas*. The near-parallelism of the opposing coasts of the South Atlantic at once fired his imagination. despite warnings of the wrath to come, he shortly afterwards announced his hypothesis of 'moving continents' to a gathering of German geologists. Their hostile reaction foreshadowed the fierce international controversy that followed the publication of Wegener's first papers on the subject in 1912.

Wegener's highly ingenious concept of the evolution of the continents and their distribution is graphically illustrated by his own maps, strange and fantastic on first aquaintance, but now widely familiar (Figure 28.4). His picture of the world in Carboniferous times is somewhat similar to Snider's, except that India and Antarctica are tucked in between Africa and Australia, with the horn of South America wrapping around West Antarctica. For this combination of Laurasia and Gondwanaland, making up the whole land area of the globe, Wegener proposed the name *Pangaea* (Gr., all earth). Snider had urged that the forests of the Carboniferous Coal Measures were tropical, and that in consequence Europe and North America were then near the equator, thus implying that South Africa lay near the south pole. Similarly, but because of the distribution of the Permo-Carboniferous glaciations, Wegener inferred that the Carboniferous south pole occupied a position just off the present coast of South Africa (Figure 28.5).

The present distribution of the continents was regarded as a result of fragmentation by rifting, followed by a drifting apart of the individual masses. The southern continents began to unfold during the Mesozoic era by being dragged away from wherever the south pole happened to be at any given time during the progress of the outward movements. Somewhat later North America began to break loose and to drift away to the west, Greenland being the last to go. The Atlantic is the immense gap left astern, filled up to the appropriate level by the inflow of sima from the mantle. By the time the continents had reached their present positions Antarctica found itself stranded over the south pole; Africa lay athwart the equator; India had been tightly wedged into Asia; and Australia and New Guinea had advanced far into the Pacific, by-passing the Banda arc and eventually separating it from the Pacific arcs.

The drift of the continents away from the poles was dramatically described by Wegener as the *Polflucht*—the flight from the poles. He ascribed the force involved to the gravitational attraction exerted by the earth's equatorial bulge. The force is a real one, but it is many millions of times too feeble to drag the continents from their mooorings.

Wegener also postulated a general drift towards the west. As the Americas moved westwards against the resistance of the Pacific floor, their prows, he supposed, would be crumpled up into great mountain ranges. Between the two immense rafts a trail of fragments lagged behind and formed the islands of the West Indies. The stretched-out isthmus connecting South America and Antarctica similarly lagged behind, forming the horns of

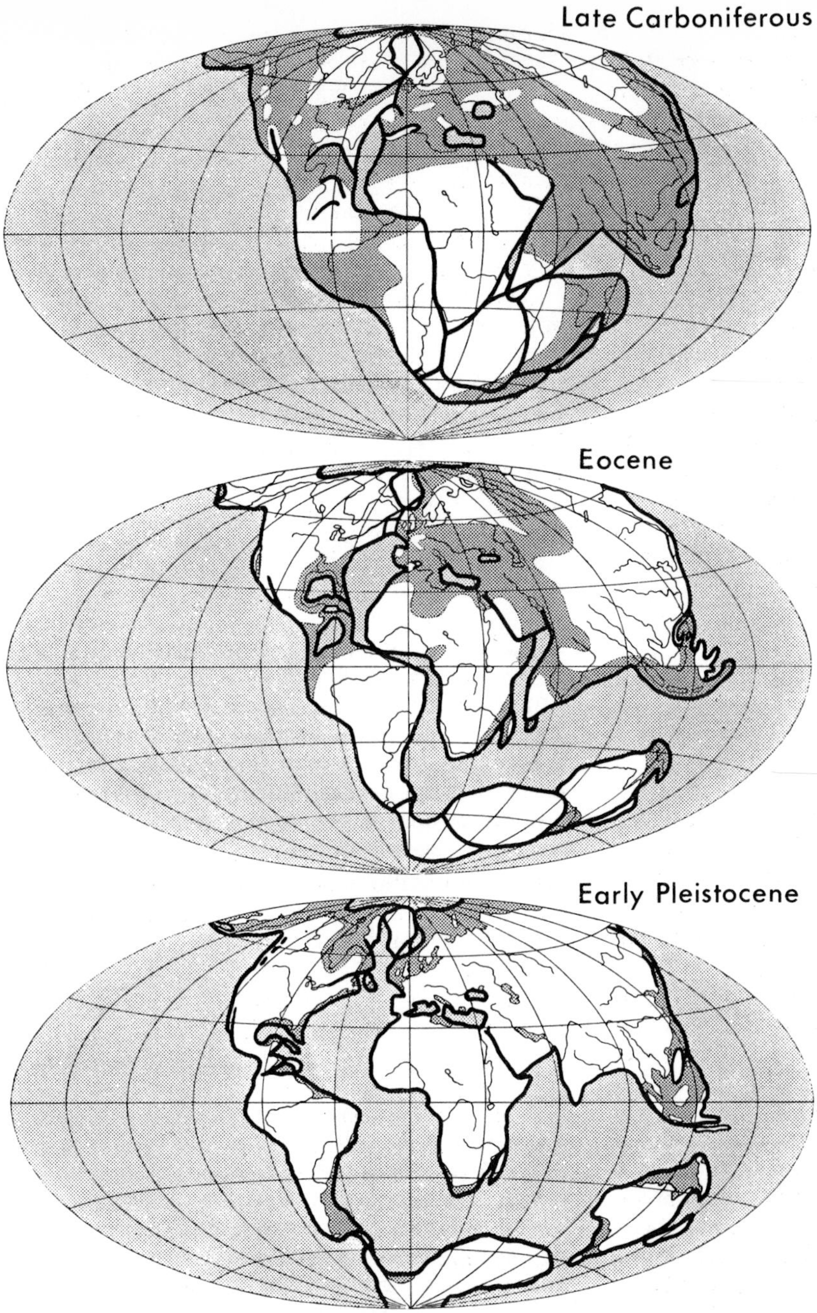


**Figure 28.4** Wegener's reconstruction of the distribution of the continents during the periods indicated. Africa is placed in its present-day position to serve as a standard of reference. The more heavily shaded areas (mainly on the continents) represent shallow seas (*From A. Wegener, Die Entstehung der Kontinente und Ozeane, 1915*)

the two continents and shedding bits of sial that now remain as the island loop of the Southern Antilles. The supposed effects of the westerly drift of Asia are less happily conceived. The great oceanic deeps were regarded as gaping fissures torn in the Pacific floor and not yet fully healed, while the island festoons were interpreted as strips of sial that remained attached to the mainland only at their ends.

The westward movements were ascribed to the differential attractions of the moon and the sun on the continents. Tidal friction acts like a brake on the rotating earth, and as the effect on protuberances is greater than that on lower levels of the crust the continents tend to lag behind. If they *did* lag behind, they would appear to drift to the

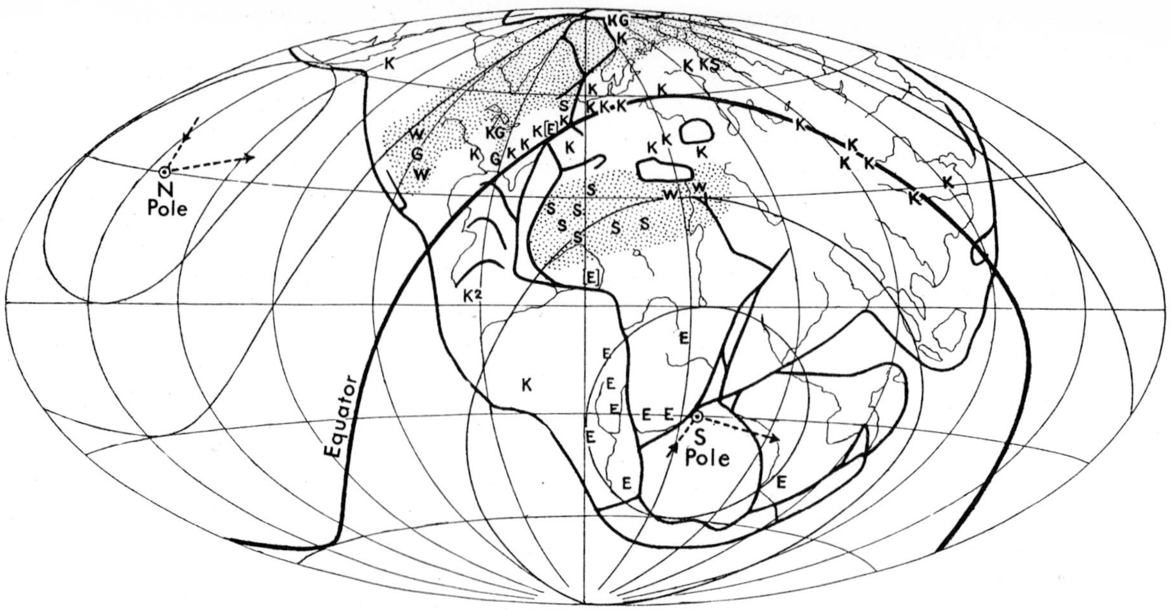


**Figure 28.5** Reassembly of the continents for the late Carboniferous, based on the palaeo-climatic evidence available in 1924. Glaciated regions are indicated by E (Eis, ice). K shows the distribution of coal, mainly near the equator of the time (heavy line). The dotted areas on both sides of the Carboniferous equator were arid regions: with W, desert sandstone; S, salt, and G, gypsum. The S Pole is placed near Durban and the N Pole in the Pacific, NE of Hawaii. It should be noticed that [E] on the equator near Boston, Mass. (the Squantum tillite), of doubtful age in 1924, is now ascribed to the Lower Palaeozoic. (*After W. Köppen and A. Wegener, Die Klimate der geologischen Vorzeit, 1924*)

west. But here again the force invoked is hopelessly inadequate to overcome the enormous resistance that opposes actual movement. The tidal force barely affects the earth's rotation, and is about ten thousand million times too small to move continents and raise up mountains.

In support of the case for continental drift Wegener marshalled an imposing collection of facts and opinions. Some of his evidence was undeniably cogent, notably the distribution of the Permo-Carboniferous glaciations, already discussed on p. 465. But so much of his advocacy was based on inadequate data and speculation that a storm of adverse criticism was provoked. It was all too easy to demolish some of Wegener's particular views. Moreover, following the lead of many influential geophysicists, most geologists were reluctant to admit the possibility of continental drift, because no recognized natural process seemed to have the remotest chance of bringing it about. The 'flight from the poles' and the westerly tidal drift have both been permanently discarded as operative factors. Polar wandering originally seemed to be impossible on a scale of geological importance, but it has since become dynamically respectable and henceforth it can be taken into account as fully as may prove necessary. Continental drift, now embracing crustal separation and ocean-floor spreading and renewal, is known to be in operation at the present time. And geophysicists have not only been won over by the testimony of palaeomagnetism, but they are leading the way with the concept of *plate tectonics* (p. 640).

## Geological Criteria for Continental Drift

The chief geological criteria for continental drift are based on the following considerations:

(*a*) If the continents formerly occupied widely different positions on the earth's surface, then the distribution of climatic zones, as inferred from geological evidence, should have correspondingly changed. Ample evidences of such changes have already been summarized in earlier chapters: e.g. dealing with the Permo-Carboniferous glaciations (p. 463); desert sands and wind directions (p. 493); the distribution of salt deposits (p. 494); and of ancient coral reefs (p. 565).

(*b*) If two continents, now far apart, were formerly united, it should be possible to detect the fact by the recognition of certain geological features that were originally shared in common;

e.g. orogenic belts of which the truncated ends can be naturally joined up; specific and unusual details and sequences of geological history as recorded in the sedimentary and other rocks of the separated lands; and the identity of the fossil remains of animals and plants (especially those of land and freshwater species) which could migrate freely across united continents but not across an intervening ocean.

(*c*) Indications from strike-slip (transform) faults, like the San Andreas fault, with cumulative lateral displacements sometimes amounting to hundreds of kilometres, that movements implying continental drift have long been in progress.

Finally, however, it is geophysical evidence that has provided the proof of continental drift. From rocks which can be regarded as palaeomagnetic compasses, since they have retained a record of the positions of the poles as they were at the time when they were formed, it is proved beyond doubt that in past geological periods the continents occupied positions, relative to the present poles and to one another, very different from those of today (pp. 608–14).

## The Opposing Lands of the Atlantic

Ever since Wegener thought the Atlantic Ocean to be an enormously widening rift, with sides still matching 'as closely as the lines of a torn drawing would correspond if the pieces were placed in juxtaposition', the parallelism of its opposing shores has been discussed. Figure 28.6 shows the result of a direct and accurate comparison made in 1958 by Carey between South America and Africa, not along the actual shore-lines which have been modified by marine erosion and deposition, but offshore at a depth of 200 m, i.e. towards the top of the continental slope. Carey made this comparison by sliding transparent caps over the surface of the globe until he attained the best fit. A similar comparison at a depth of 2000 m, i.e. about half way down the continental slope, is, if anything, slightly better. Subsequently, Bullard (1964), feeding all the relevant data into a computer, which was asked to find the level at which the 'fit' is most nearly perfect, received an answer between these two depths. The 'fit' between the opposing lands of the Atlantic is illustrated by Figure 28.7,

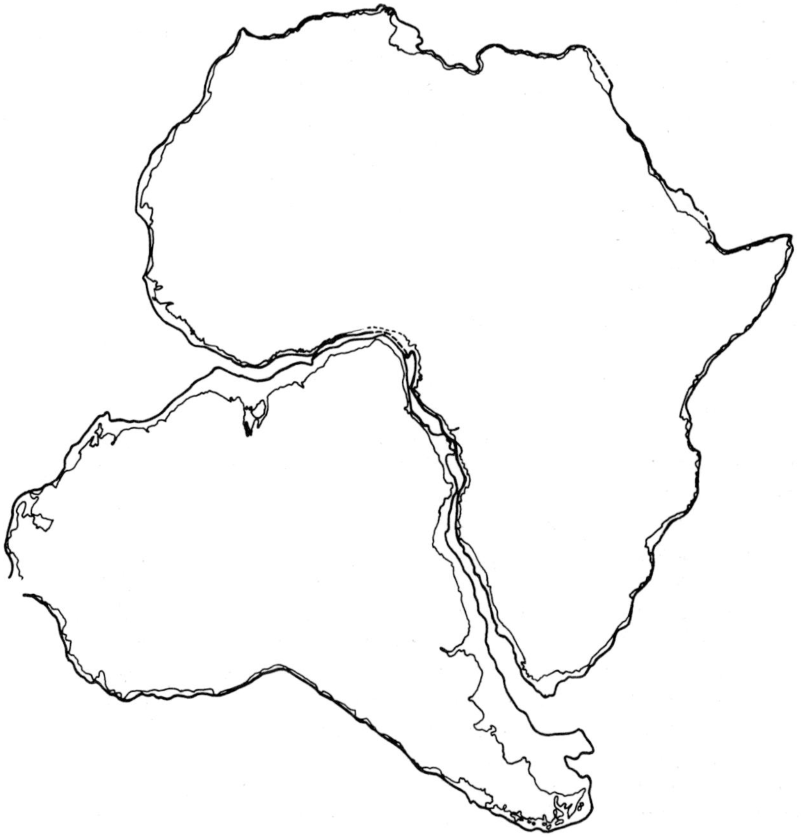

**Figure 28.6** Comparison of Africa and South America by placing the 200-metre isobaths (depths below sea-level) in juxtaposition (*S. Warren Carey*)

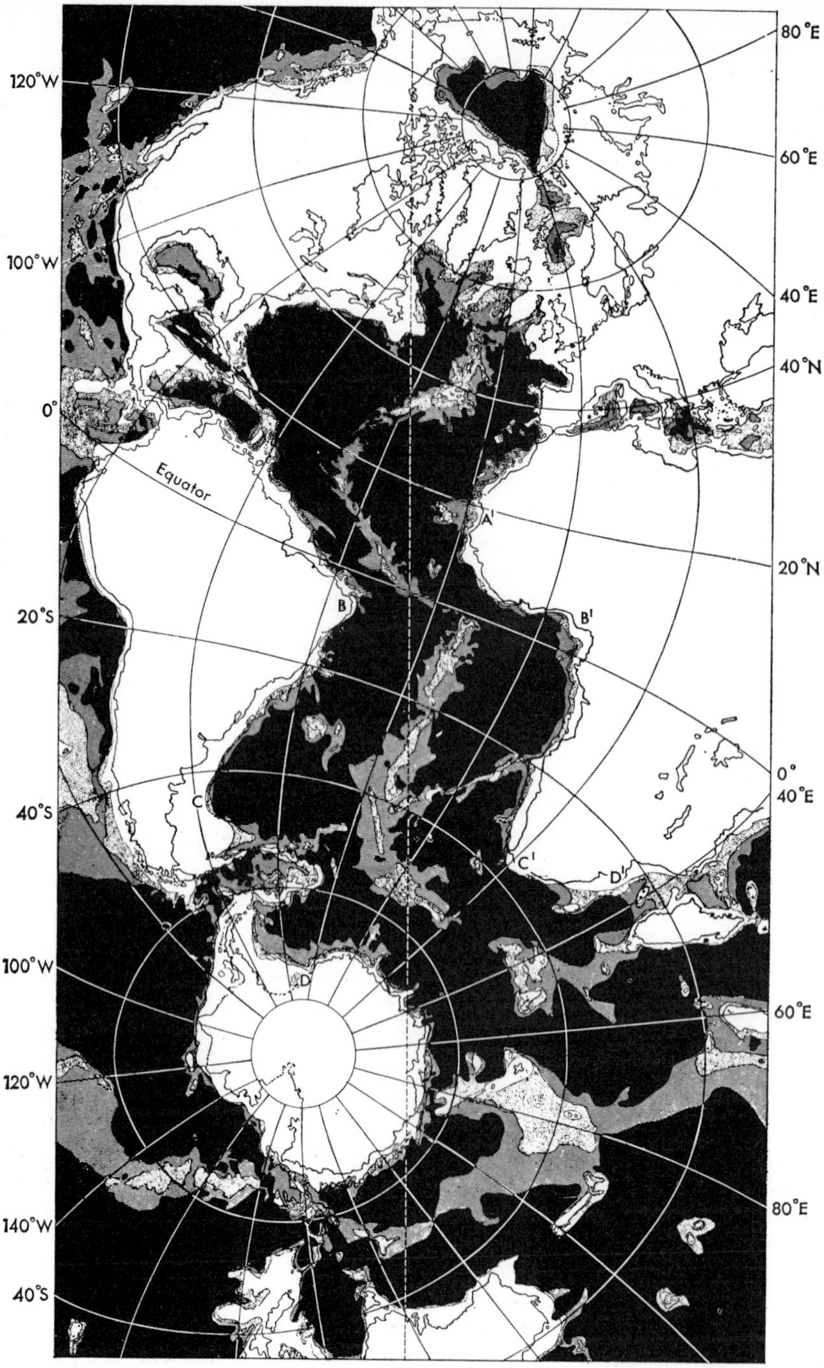


**Figure 28.7** Map illustrating the morphological fit of the opposing lands of the Atlantic (*S. W. Carey, 1958*). Carey envisaged N and S America, Antarctica and even Australia as having been swung as a whole, towards Eurasia and Africa, about the hinge (bend) in the Alaskan orogenic belt, suggesting that such motion through 20° or 30° would close the Arctic, Atlantic and Southern oceans. He proposed the term *orocline* to describe such an orogenic belt that appears to have been bent or flexed.

prepared by Carey (1958) for his Symposium on Continental Drift, that did so much to reawaken interest in the subject. Of this 'fit' Carey wrote: 'Thus the east coast of North America is the mould of the great western bulge of Africa (AA′), the Gulf of Guinea is the negative of the bulge of Brazil (BB′), and the embayment of the South American shelf to the Falkland Islands is the mould of the Cape of Good Hope and most of South Africa (CC′).'

In 1965, Bullard, Everett and Smith published a best 'fit' map of the continents margining the

Atlantic Ocean (Figure 28.8), constructed by using purely objective arithmetic methods. They moved the continents over the surface of the earth with the aid of Euler's theorem, according to which a spherical surface displaced over a sphere can be moved to any other position by a single rotation about an axis passing through the centre of the sphere. The point where the axis intersects the surface of the sphere is known as the centre or pole of rotation. In the calculations the earth was treated as being spherical. Movement of an area of an ellipsoid or geoid, like the earth, over itself involved distortion, but since this distortion was calculated to be of the order of only 0·2 km, it was considered to be negligible in comparison with the misfits occurring on the best 'fit' map, Figure 28.8.

With the aid of a computer, Bullard, Everett and Smith found the best 'fit' to be at 500 fathoms, rather less than 1000 m below sea-level. The closeness of the 'fits' between South America and Africa, on the one hand, and between North America, Greenland and Europe on the other, not only exceeded their hopes but fully confirmed the results of Carey's earlier 'fits'.

In fitting together the continents margining the Atlantic, Bullard, Everett and Smith found, as Carey had done before them, that when Africa was fitted to North America it overlapped Spain. In order to achieve a 'fit' without overlap, Spain had to be rotated so as to close the Bay of Biscay and bring its northern shore against the 500 fathom contour to the west of France.

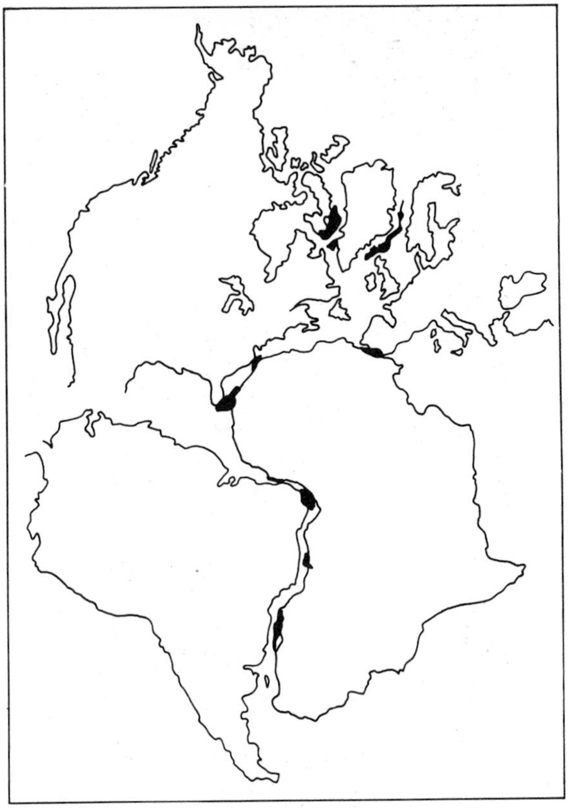

**Figure 28.8** The best 'fit' of continents surrounding the Atlantic at the 500 fm contour. The black areas indicate overlaps. (*After E. Bullard, J. E. Everett and A. G. Smith, 1965*)

## Geological Similarities between the Two Sides of the Atlantic

For many years South Africa's eminent geologist, Alexander du Toit, was indefatigable in assembling the evidence bearing on continental drift. In his well-known book *Our Wandering Continents*, published in 1937, he showed that a striking series of correspondences can be recognized in the sediments, fossils, climates, earth movements, and igneous intrusions of the two sides of the Atlantic. Both regions had essentially the same geological history during the Palaeozoic and early Mesozoic times, and the combined evidence points very persuasively to the high probability that they were then very much closer together than now. Du Toit considered it possible that the original distance between the present opposing shores may have been as little as 400 km. But this was a minimum estimate. In 1961 H. Martin reviewed the subject afresh, with special reference to Namibia (South-West Africa) and Brazil. Discussing whether the more detailed knowledge now available has tended to increase or decrease the similarities between the two sides of the South Atlantic, he writes: 'There is not the slightest doubt that from the Silurian to the Cretaceous every correction of the statigraphy and lithologic columns has increased the similarities. The stratigraphic and lithologic columns for this period of some 200 m.y. have become almost identical. I do not think that, for a comparable length of time, a similar likeness between parts of any two continents can be found.'

Just as striking evidence is the fact, recognized by radiometric dating, that a long and complex orogenic cycle reached its closing stages during the late Precambrian and Cambrian. The trend lines are sub-parallel to the coasts and appear equally on both sides of the South Atlantic (Figure 31.6).

While the Permo-Carboniferous glaciations of

Gondwanaland still remain the most convincing line of geological evidence for continental drift (see pp. 463–5), there can now be added the discovery of the Iapo tillite WSW of São Paulo in Brazil, in the same stratigraphical position as the Table Mountain tillite, near Cape Town. Both are late Silurian or early Devonian. Martin recorded that the direction of ice movement near Cape Town, and the directions of transport of Devonian sediments in the Paraná Basin both indicate the existence of elevated areas in regions occupied today by the South Atlantic. He also drew attention to the striking contrast between Table Mountain and Brazilian glacial deposits and the strata of early Devonian age in Australia, which contain a warm-water fauna with corals. This climatic contrast had disappeared in the late Carboniferous, when Australia was glaciated like the other southern continents.

It should also be added that the widespread indications of late Precambrian glaciations (averaging about 650 $\pm$50 m.y.) in southern Africa (Katanga, Lower Zaïre, Namibia) have been matched by the discovery of similar deposits in Lavras (Brazil), the age of which lies between limits of about 600 and 800 m.y. It is probable, though not rigidly proved, that all of those correspond approximately with the Sturtian tillites of South and Central Australia (see p. 467) and Figure 21.25), and others that are correlated with them in Tasmania and Northern Australia.

Amongst a host of similarities too numerous to mention, attention should again be drawn to the great floods of basalt that inundated the Triassic deserts of both continents about the beginning of Jurassic time. In the Paraná Basin basalts cover an area of a million $km^2$ or more and reach a thickness of 1·5 km. In southern Africa a smaller area of basalts now remains, but originally, judging by the criss-crossing swarms of basic dykes and sills in the Karroo Basin and by the enormous thickness of flows still preserved on the eastern side from the Drakensberg to the Lebombo, the volume of lava erupted may have been still greater than in South America. Such spectacular signs of unusual activity in the depths would seem to be a significant preparation for the major break-up of Gondwanaland and the appearance of seaways between the separating lands. Palaeomagnetic evidence based on these basaltic rocks indicates that separation of Africa from South America began while this igneous activity was in progress. But no actual proofs of the existence of the South Atlantic before the Lower Cretaceous have so far been detected. Marine beds of that age occur on terrestrial Jurassic in the oilfields of Luanda, Angola, and also in those of Bahia, Brazil.

Palaeontological evidence of the former union of South America and Africa is provided by the bones of a small fresh-to-brackish water reptile, called *Mesosaurus*, in a band of lacustrine deltaic clay near the top of the Carboniferous. Since remains of this little animal, about 45 cm long, have not been found anywhere else in the world, they suggest that South America and Africa were united towards the end of the Palaeozoic.

One of the early indications that there was no North Atlantic in Palaeozoic times was the recognition that similar fossils, including the Cambrian trilobite *Olenellus*, occur both in Scotland and Newfoundland. Moreover, the mid-Cambrian trilobites of Sweden are the same as those of Utah. Such evidence implies the former existence of shallow water between the two continents, such as the Caledonian geosyncline would provide. When North America and Europe are placed together so that their Permian palaeomagnetic poles coincide and the corresponding isoclines, including the equator, fall continuously into line, as shown in Figure 28.9, the Caledonian fronts of Greenland and the North-West Highlands, and the corresponding front of the Appalachians are automatically joined up in practically the same way as was thought most probable before palaeomagnetic data were available as a guide. Moreover, the western convergence across Europe of the outer Caledonian and Hercynian fronts, until they almost meet in the south-west of Ireland, is continued in North America, where the fronts eventually cross. Radiometric dating has shown that much of the northern part of the Appalachians was affected by orogenesis that would be called Hercynian or Variscan in Europe. Tuzo Wilson has suggested that the Great Glen Fault has its natural continuation in the Cabot Fault and this correlation is again found to be consistent with the palaeomagnetic reassembly of Figure 28.9. Both in Scotland and Newfoundland there are similar volcanic centres associated with dyke-swarms and granodiorite complexes of Lower Devonian age, like Ben Nevis; and the Old Red Sandstone type of Devonian, so characteristic of Scotland and the rest of Caledonian Britain and Ireland, is also found in Norway, Greenland, and Spitzbergen, as well as in parts of eastern North America.

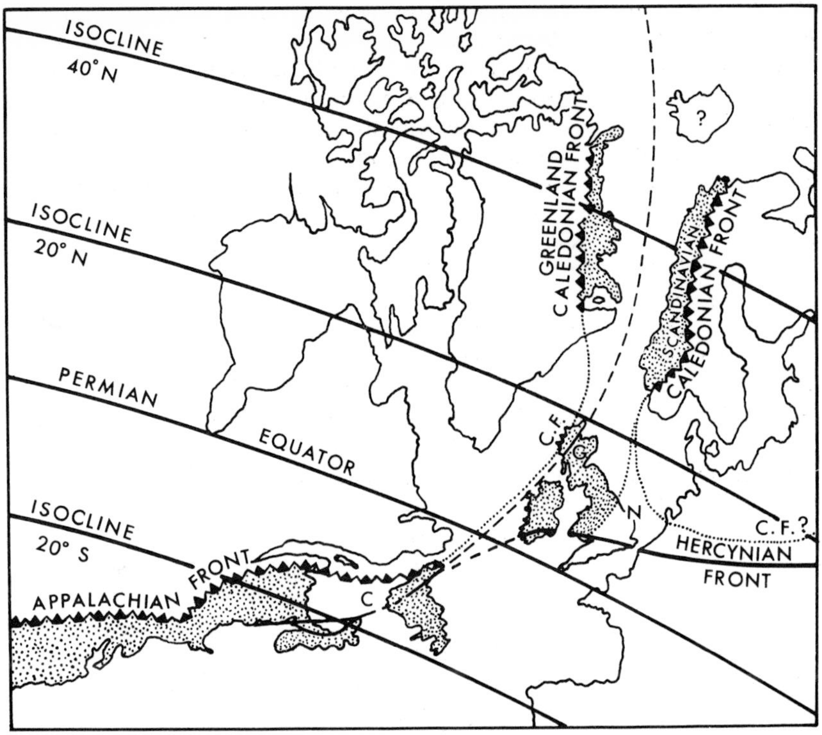


**Figure 28.9** Tentative reassembly of the opposing lands of the North Atlantic, based primarily on moving Europe and North America so that the mean positions of their Permian poles coincide and the isoclines come into line as shown. The Caledonian and Appalachian orogenic belts are shaded, C.F., the Caledonian Front of the North West Highlands of Scotland; G, the Great Glen Fault of Scotland; C, the Cabot Fault of Newfoundland and Nova Scotia and its probable continuation along the New England coast; N, a possible median area (including part of the North Sea, East Anglia and the Netherlands) if the Scandinavian Caledonian Front should continue through Brabant and Poland.

## Attempts to Reassemble Gondwanaland

Du Toit, like Taylor, thought it more logical to suppose that a double land-mass existed during the Palaeozoic, rather than the single land-mass, *Pangaea*, envisaged by Wegener. However this may be, prior to the Tertiary Alpine–Himalayan orogeny, Laurasia and Gondwanaland were separated from one another by a long seaway that was called the Tethys by Suess. The Tethys, interspersed between Eurasia and Indo-Africa was bordered by, and perhaps locally consisted of, the various geosynclines from which the Alpine-Himalayan mountain system originated.

Reconstruction of Laurasia is mainly a matter of closing the Atlantic Ocean, as discussed in the previous pages. Reconstruction of Gondwanaland is in many ways more difficult. Many attempts have been made, using all the geological evidence, and most of the results have a remarkable family resemblance. The greatest disagreement relates to the positions of India and Madagascar. The Talchir boulder bed indicates that in Permo-Carboniferous times India must have been situated far from the equator. Du Toit (1937) tucked it in between Africa and Antarctica, and he has been followed by many others. In 1961, however, Ahmad established a good Permian 'fit' between India and Western Australia, which was adopted by Krishna (1969). Amongst a host of similarities one notable detail is the occurrence at Umari (just above the Talchir beds) of four thin beds of limestone made up amost entirely of *Productid* shells. Fossils of the same group are thickly packed in the corresponding limestones on the west coast of Australia. Crowded colonies of these organisms were probably destroyed by the effects of violent earthquakes. The same reconstruction also brings the Precambrian banded iron ores of Yampi Sound, Australia, opposite the similar ores of Singhbhum and Bastar in India, both of which have radiometric ages in the range 2000–2200 m.y. The Dharwar orogenic belt of India, with its celebrated Kolar gold mines, is also brought nearly into parallelism with the gold-bearing Kalgoorlie belt of Western Australia. In both regions the gold was introduced about 2400 m.y. ago.

Petrascheck (1973) has drawn attention to the Precambrian ore province of Africa, Southern India and Western Australia, which unite to form one province when Australia, India and Africa are fitted together as illustrated in Figure 28.10. Petrascheck suggests that the ore province of Madagascar fits this assemblage better when it is

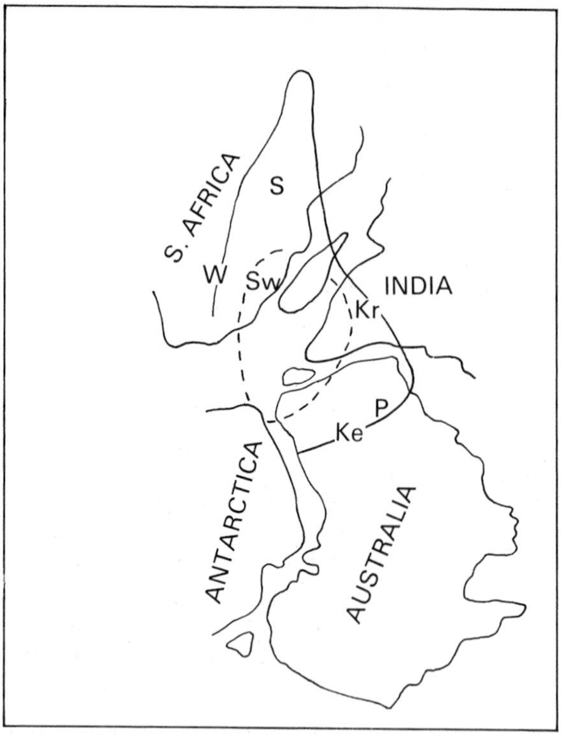


**Figure 28.10** The 'fit' of the ore provinces of South Africa, Madagascar, India and Australia. The dashed line outlines a pegmatite province (600 to 1000 million years old) characterized by niobium, tantalum, beryllium, zirconium, some tin, and uranium. The firm line outlines the older gold province of Kolar (Kr) in India, Kalgoorlie (Ke) and Pilbara (P) in Australia, and of Witwatersrand (W), Swaziland (Sw) and Salisbury (S) in South-East Africa. The ore provinces of Madagascar fit within this province when Madagascar is placed opposite South-East Africa as in Fig. 28.12(b) and (c). (*After W. E. Petrascheck, 1973*)

placed opposite south-east Africa, rather than opposite Kenya as it is commonly depicted. Another Gondwanaland ore province, extends between South America and western Africa (Petrascheck, 1973).

Figures 28.11 and 28.12 present some examples of Gondwanaland based on geological evidence. In these reconstructions the difficulty has commonly arisen that some item of seemingly favourable evidence has had to be sacrificed in order to achieve a reasonable 'fit' of all the parts.

Figure 28.11(*a*) was a valiant attempt by du Toit (1937) based on his idea that the Cape folds of South Africa continued into South America at one end, and along the eastern side of Australia at the other. He called this belt the Samfrau Geosyncline. The Indo-Australian link was not known to du Toit, the real sacrifice he was obliged to make was the unnatural severance of Graham Land from the main body of Antarctica. It is of interest to note that the more recent investigations of Antarctica have justified du Toit's faith that one day this continent would provide evidence of Permo-Carboniferous glaciation.

Du Toit made two inferences that are of interest today:

(*a*) He saw in the rift-valleys of Africa, the Red Sea and the Rhine graben, early stages in the separation of land masses. He also compared the escarpment-like edges of the tilted blocks, that margin these rift-valleys, with the fault-like scarps that form the shore-lines of continents like Africa and South America which have been separated by rifting. Furthermore, these escarpment-like coast-lines, tilted gently inland, provided du Toit with an explanation of rivers which rise close to such high shore-lines and flow inland, sometimes right across a continent before reaching the sea. Amongst others he cited as examples in South Africa, the Orange and Limpopo rivers, and farther north the Congo, now the Zaïre. Similarly the Paraná, Uraguay and São Francisco rivers in South America rise close to the shore-lines and flow for long distances parallel to the coast before reaching the sea.

(*b*) Du Toit left spaces between the present-day continents in his reconstruction of Gondwanaland. In doing so he was guided by the sedimentary facies. For example, in comparing the Cape ranges of South Africa with the Sierra de la Ventana, Argentina, he concluded that the lateral variation of the Palaeozoic sediments and their structures required that they should be separated from one another by a distance of from 500 to 600 km. He verified this distance by the intersections of other lines of evidence, and he thought the gap to have been occupied by land at least during the Jurassic. Eventually, however, by the entry of the sea, du Toit envisaged the gap between South America and Africa to have become the proto-South Atlantic.

Figure 28.11(*b*) is a reconstruction of Gondwanaland by Warren Carey (1958). Here Ahmad's Indo-Australian link is adopted, but Antarctica gets badly in the way, and again Graham Land is unnaturally displaced.

Figure 28.11(*c*) is a reconstruction by Tuzo Wilson (1963). He adopts the Indo-Australian link and uses the mid-oceanic ridges as a guide. But this leaves no room for Antarctica; in consequence the rule forbidding overlapping is broken.

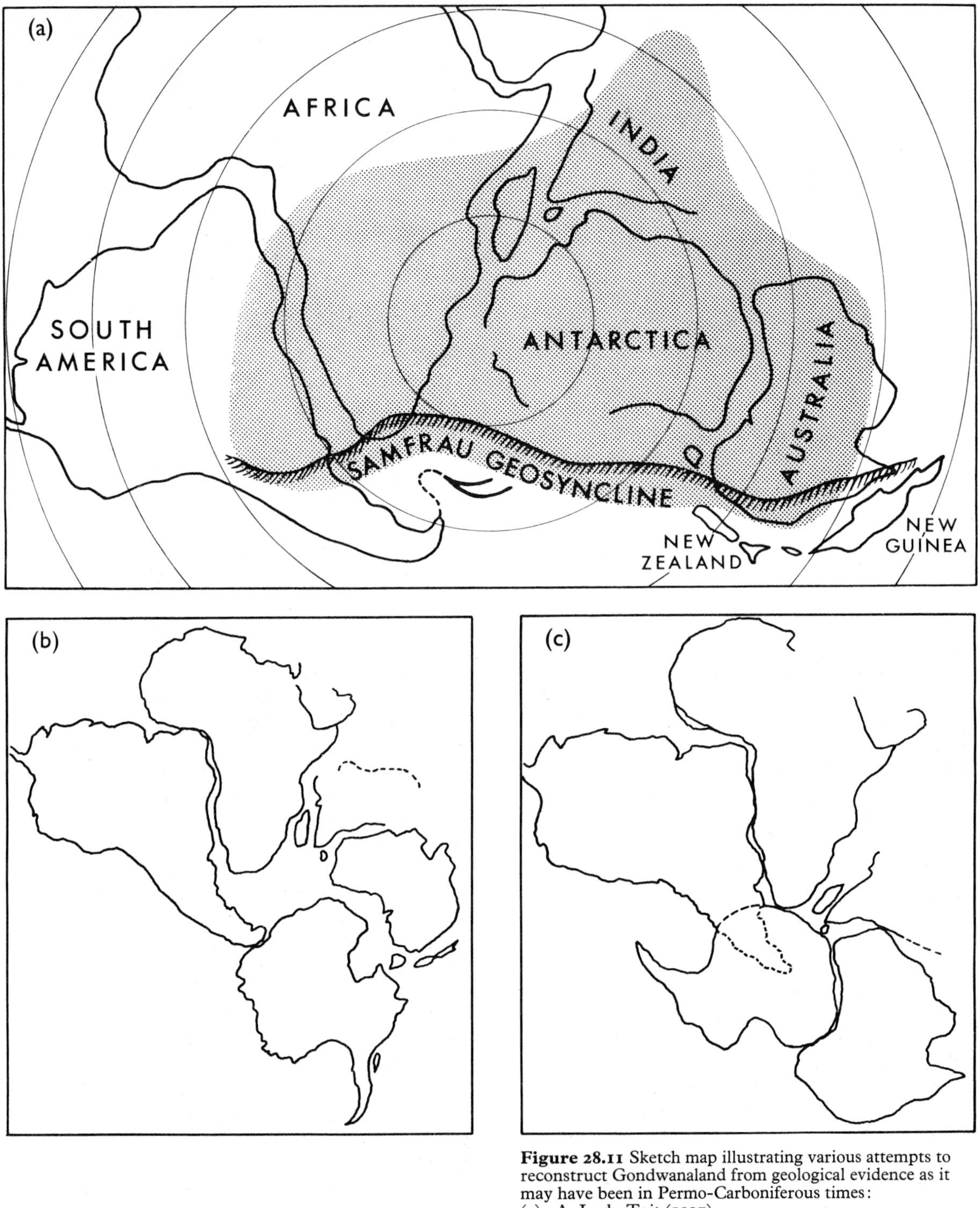


**Figure 28.11** Sketch map illustrating various attempts to reconstruct Gondwanaland from geological evidence as it may have been in Permo-Carboniferous times:
(a) A. L. du Toit (1937)
(b) S. W. Carey (1958)
(c) J. T. Wilson (1963)

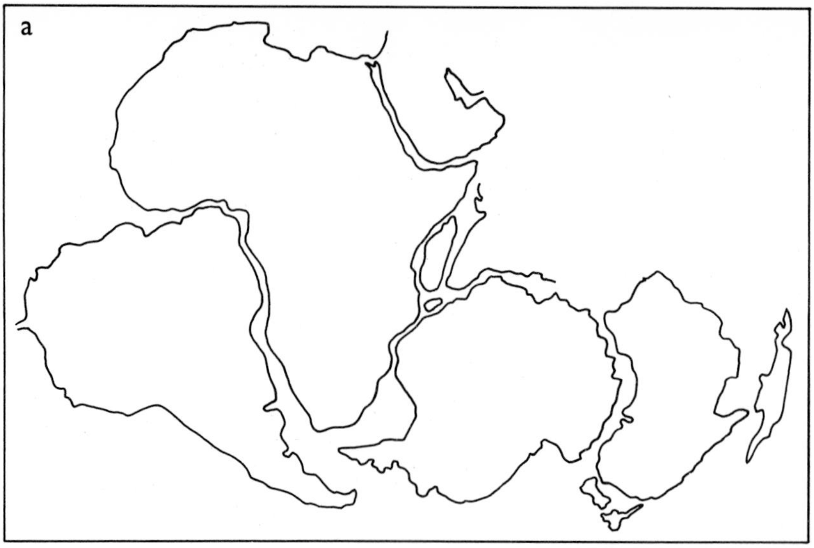


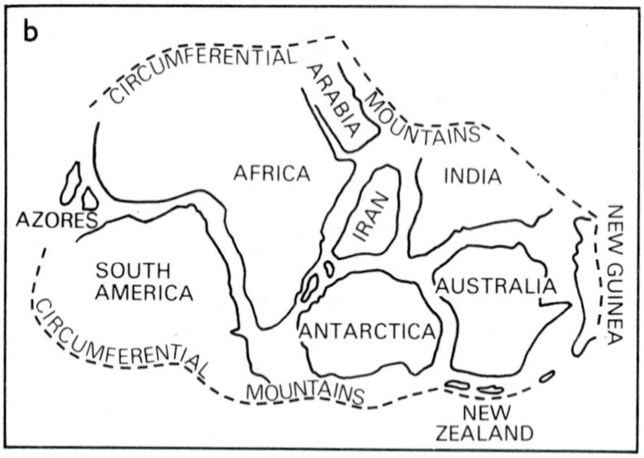


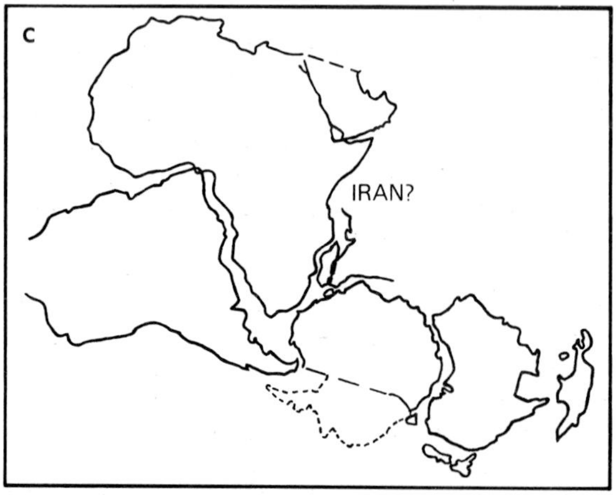

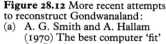

**Figure 28.12** More recent attempts to reconstruct Gondwanaland:
(a) A. G. Smith and A. Hallam (1970) The best computer 'fit' for Gondwanaland at 500 fathoms, supported by geological evidence
(b) Lester King (1973) A reconstruction based on geological evidence
(c) D. H. Tarling (1973) A geometric reconstruction tested against palaeomagnetic evidence.

Figure 28.12(*a*) is a reconstruction by Smith and Hallam (1970). With the aid of a computer they found the best 'fit' for Gondwanaland at 500 fathoms, using only those 'fits' that they judged to be geologically sound. They found Ahmad's 'fit' for India against north-west Australia to be geometrically sound, but they discarded it because they found that India–Australia–Antarctica could not be fitted against Africa without leaving large gaps between them. From the stratigraphic evidence they considered that such gaps did not exist.

Figure 28.12(*b*) is a reconstruction by Lester King (1973) showing Gondwanaland as 'a simple ovoid body surrounded by a girdle of tectonic mountain ranges'. King adopts Ahmad's Indo-Australian link, and finds compelling evidence for doing so in the fact that this reconstruction shows continuity between the Tertiary orogenic belts of the Himalayas and New Guinea. For geological reasons King includes the Iran–Afghanistan block in Gondwanaland, thus filling the space referred to by Smith and Hallam when the Indo–Australian link is adopted. King places Madagascar opposite south-east Africa, in contrast with the reconstructions by du Toit and Smith and Hallam. From geological evidence, Kent (1972, 1973) and Tarling (1973) agree that Madagascar was situated in the Mozambique region of south-east Africa at the time when Gondwanaland existed.

## Testing the Reassembly of Pangaea by Reference to the Palaeomagnetic Pole Positions

Creer (1965) calculated the positions of the palaeomagnetic poles for various geological per-

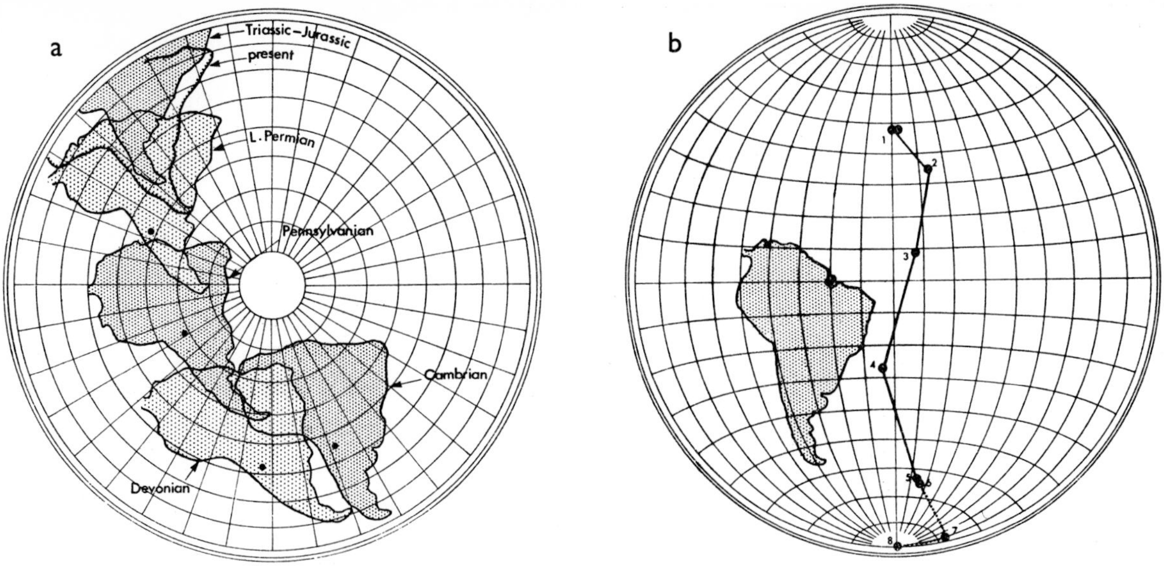


**Figure 28.13** (a) The palaeolatitudes and orientations of South America for periods ranging from early Palaeozoic to early Mesozoic, constructed from palaeomagnetic data
(b) Palaeomagnetic data expressed as a polar wandering curve for the South Pole, with South America fixed, for the Palaeozoic era (*K. M. Greer, 1965*)

iods from reliable palaeomagnetic data of the time, and showed that it is possible to discover whether or not there has been movement between two continents by comparing their 'polar wandering' curves. If the curves have the same shape, so that they can be superimposed one upon the other, then there has been no relative movement.

From the relevant palaeomagnetic data for South America, and by choosing a sequence of related longitudes, Creer found evidence for the movement of this continent across the south pole during the Palaeozoic and early Mesozoic eras (Figure 28.13(*a*)). This result can also be expressed as a 'polar wandering' curve (Figure 28.13(*b*)). Creer similarly constructed a 'polar wandering' curve for Africa, and found that the Palaeozoic parts of the 'polar wandering' curves for South America and Africa can be superimposed, the two continents at the same time being fitted together, indicating that there was no relative movement between them during this era. (Figure 28.14). Furthermore, for the Palaeozoic era, the 'polar wandering' curve for Australia can be superimposed on those for South America and Africa, indicating that there was no relative movement between these three continents during this era, and that their relative positions were approximately as indicated in Figure 28.15.

More recently Creer (1973) has plotted the positions of the Palaeomagnetic poles on a map of *Pangaea* for the periods Cambrian to Cretaceous inclusive. The particular form of *Pangaea* tested is composed of Laurasia reconstructed from the closing of the Atlantic according to Bullard *et al.*, and Gondwanaland according to Smith and Hallam. He found the poles for the Ordovician period to be so widespread as to be inconsistent with the existence of *Pangaea* at that time. Clustering of the poles for the Lower Carboniferous to the Jurassic periods, indicates the existence of *Pangaea* in a form similar to that tested. For the Cretaceous period the poles are more dispersed, indicating the breaking up of *Pangaea* during this period. A point of interest is that Creer found the position of the pole for Madagascar during the Cretaceous suggested that Madagascar is wrongly placed in the Smith–Hallam reconstruction.

In 1972, Tarling made a new geometric construction of Gondwanaland, which differs from that of Smith and Hallam in that he placed Madagascar against Mozambique, and India against Madagascar (Figure 28.12(*c*)). Australia he fitted against Antarctica as in the reconstruction by Smith and Hallam, but placed them farther south relative to Africa. Tarling then tested this new 'fit' against the most reliable palaeomagnetic data available. He found, for the Mesozoic, the scatter of palaeomagnetic poles to be less than they are for the reconstruction by Smith and Hallam.

From the above review of the evidence it is apparent that there is as yet no unique construction of *Pangaea*, or even of Gondwanaland, acceptable to all earth scientists. Palaeomagnetic evidence, however, does prove that *Pangaea* did exist in some form at the end of the Palaeozoic era.

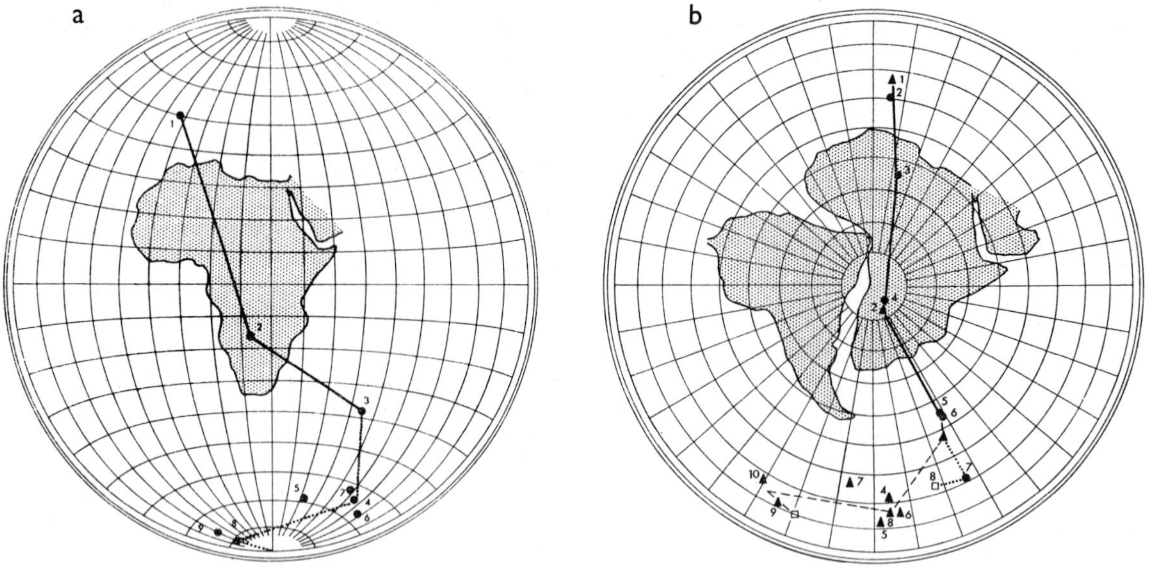


**Figure 28.14** (a) The polar wandering curve for the South Pole, relative to a fixed Africa, constructed from palaeomagnetic data derived from African rocks

(b) The fact that the Palaeozoic parts of the polar wandering curves for South America and Africa have the same shape and can be superimposed indicates that there was no relative movement between the two continents during that time. (*K. M. Creer, 1965*)

**Figure 28.15** The relative positions of South America, Africa and Australia during the Palaeozoic era, determined by superimposing their polar wandering curves for that time (*K. M. Creer, 1965*)

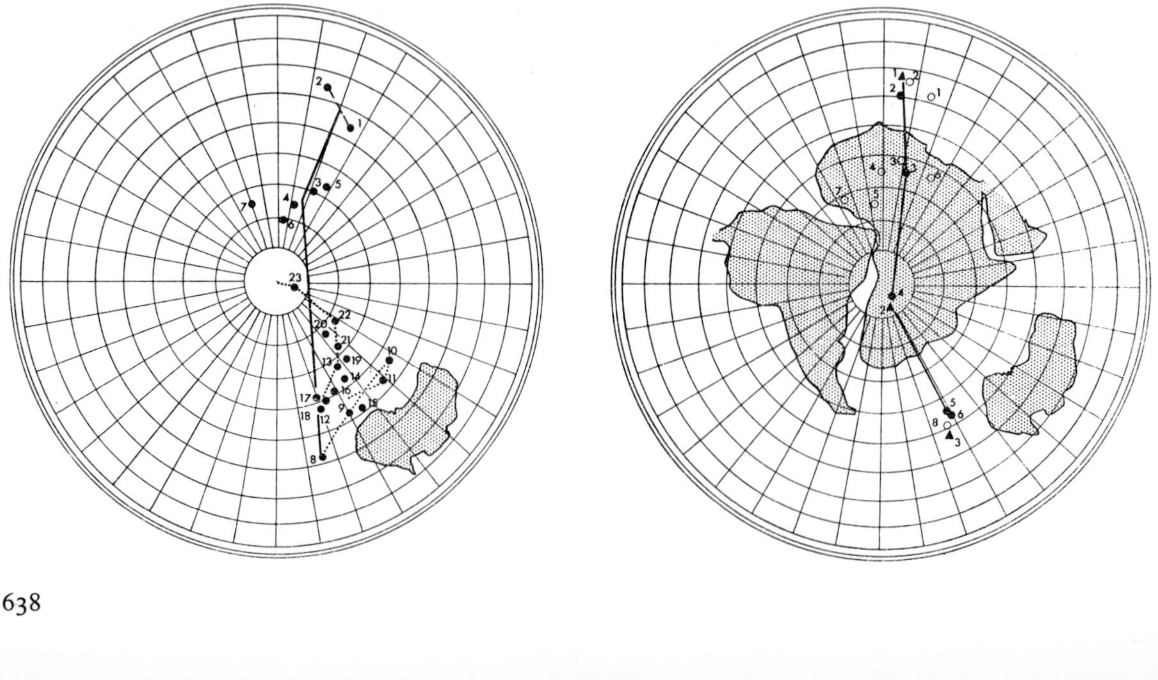

## Selected References


AHMAD, F., 1961, 'Palaeogeography of the Gondwana Period in Gondwanaland, with special reference to India and Australia, and its bearing on the theory of continental drift', *Memoirs of the Geological Survey of India*, vol. 90, pp. 142.

BULLARD, E. C., EVERETT, J. E., and SMITH, A. C., 1965, 'The fit of the continents around the Atlantic', *Philosophical Transactions of the Royal Society*, vol. 258, pp. 41–51.

CAREY, S. W., 1958, *Continental Drift: A Symposium*, Geology Department, University of Tasmania, pp. 177–355.

CREER, K. M., 1965, 'A Symposium on continental drift: III, Palaeomagnetic Data from the Gondwanic Continents', *Philosophical Transactions of the Royal Society*, vol. 258, pp. 27–40.

— 1973, 'On the arrangement of the landmasses and the configuration of the geomagnetic field during the Phanerozoic', in *Implications of Continental Drift to the Earth Sciences*, vol. I, edited by D. H. Tarling and S. K. Runcorn, Academic Press, London and New York, pp. 47–76.

DU TOIT, A. L., 1937, *Our Wandering Continents*, Oliver and Boyd, Edinburgh.

KENT, P. E., 1973, 'East African evidence of the Palaeoposition of Madagascar', in *Implications of Continental Drift to the Earth Sciences*, vol. 2, edited by D. H. Tarling and S. K. Runcorn, Academic Press, London and New York, pp. 873–5.

KING, L. C., 1973, 'An improved reconstruction of Gondwanaland', in *Implications of Continental Drift to the Earth Sciences*, vol. 2, edited by D. H. Tarling and S. K. Runcorn, Academic Press, London and New York, pp. 873–5.

— 1973, 'An improved reconstruction of Gondwanaland', in *Implications of Continental Drift to the Earth Sciences*, vol. 2, edited by D. H. Tarling and S. K. Runcorn, Academic Press, London and New York, pp. 851–63.

KRISHNAN, M. S., 1967, 'Continental Drift', *Journal of the Indian Geophysical Union*, vol. VI, pp. 1–34.

MARTIN, H., 1961, 'The hypothesis of continental drift in the light of recent advances of geological knowledge in Brazil and in South-West Africa', Alexander L. du Toit Lecture, No. 7, *Transactions of the Geological Society of South Africa*, Annexure to vol. 64, pp. 1–47.

SMITH, A. G., and HALLAM, A., 1970, 'The fit of the southern continents', *Nature*, vol. 225, pp. 139–44.

TARLING, D. H., 1972, 'Another Gondwanaland', *Nature*, vol. 238, pp. 92–3.

— 1973, 'Comments on "East African Evidence by P. E. Kent"' (see above under KENT), pp. 876–7.

TAYLOR, F. B., 1910, 'Bearing of the Tertiary mountain belt on the origin of the earth's plan', *Bulletin of the Geological Society of America*, vol. 21, pp. 179–226.

VAN DER GRACH, *et al.*, 1928, *Theory of Continental Drift: A Symposium*, The American Association of Petroleum Geologists.

WEGENER, A., 1912, 'Die Entstehung der Kontinente', *Petermann's Mitteilungen*, pp. 185–95, 253–6, 305–9.

— 1966, *The Origin of Continents and Oceans*, translated from the fourth (revised) German edition of 1929, by J. Birman, Methuen, London.

WILSON, J. T., 1963, 'Continental drift', *Scientific American*, vol. 208, pp. 86–100.

# Chapter 29

# Plate Tectonics

It is my opinion that the Earth is very noble and admirable . . . and if it had continued an immense globe of crystal, wherein nothing had ever changed, I should have esteemed it a wretched lump of no benefit to the Universe.

*Galileo* (1564–1642)

## Plates of Lithosphere

The greatest obstruction to acceptance of continental drift earlier in the century was the difficulty of understanding how continents composed of sial could possibly sail on and displace sima just as ships sail on and displace water. At that time it was thought that the oceanic crust was a thick continuation of a continental basaltic layer. In a lecture to the Geological Society of Glasgow, in 1928, Holmes suggested a possible way out of the difficulty. By comparison with the planetary system of winds he envisaged thermal convection currents within the earth's substratum (now the upper part of the mantle), with subsidiary cyclonic and anticyclonic systems induced by regions of greater and lesser radioactivity. In the subsidiary systems hot currents rose beneath the continents,

**Figure 29.1** Diagrams to illustrate a convective-current mechanism for 'engineering' continental drift and the development of new ocean basins, proposed by A. Holmes in 1928, when it was thought that the oceanic crust was a thick continuation of the continental basaltic layer (horizontal line shading).
(a) A current ascending at A spreads out laterally, extends the continental block and drags the two main parts aside, provided that the obstruction of the old ocean floor can be overcome. This is accomplished by the formation of eclogite at B and C, where sub-continental currents meet sub-oceanic currents and turn downwards. Being heavy, the eclogite is carried down, so making room for the continents to advance.
(b) The foundering masses of eclogite at B and C share in the main convective circulation and, melting at depth to form basaltic magma, the material rises in ascending currents: e.g. at A, healing the gaps in the disrupted continent and forming new ocean floors (locally with a swell of old sial left behind, such as Iceland). Other smaller current systems, set going by the buoyancy of basaltic magma, ascend beneath the continents and feed great floods of plateau basalts, or beneath the 'old' (Pacific) ocean floor to feed the outpourings responsible for the volcanic islands and seamounts. (*Arthur Holmes, Transactions of the Geological Society of Glasgow, 1928–1929, vol 18, p. 579*)

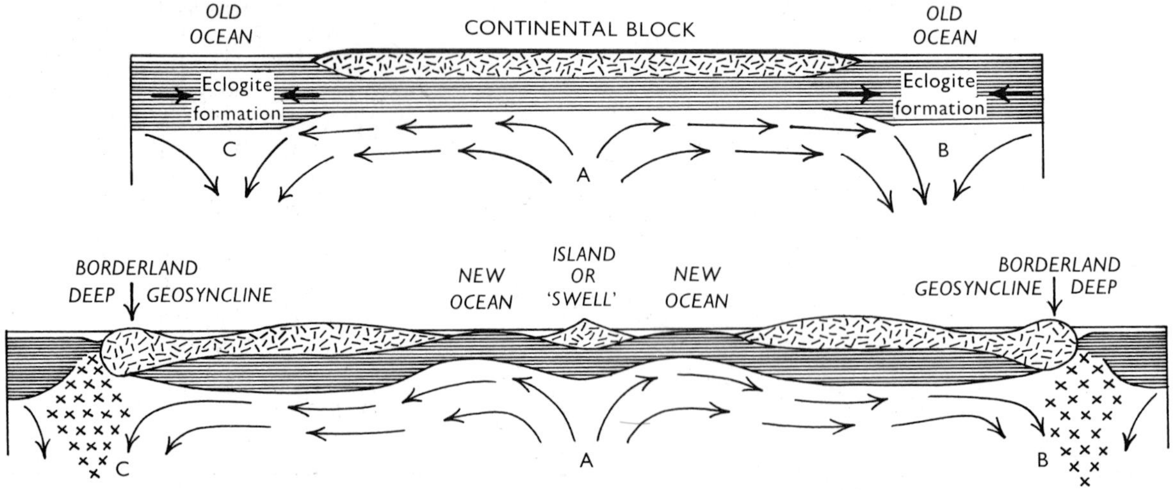

and spreading out laterally dragged the continental blocks apart (Figure 29.1). Where continental currents met oceanic currents they both turned downwards, and here Holmes envisaged the high-pressure facies eclogite as being formed from basalt or amphibolite as the currents descended. The resultant increase in density he thought to result in the subsidence of eclogite, and the speeding up of the descending currents, thus making room for the continents to advance. The ascending currents, where continents were torn apart, healed the gaps by forming new ocean floor (Figure 29.1). This concept, which made little if any impact on geological thought at the time when it was conceived, now appears as a prelude to future geophysical discoveries.

Since 1964 there has been a complete change of viewpoint relating to continental drift, arising out of the wealth of new magnetic data relating to the ocean floor, and its imaginative interpretation and confirmation, with the resulting establishment of the reality of ocean-floor spreading (Chapter 27). Drifting continents are now regarded as parts of rigid plates of lithosphere, about 100 km thick, composed of crustal rocks and upper mantle down to the low velocity zone (p. 594). Individual plates may include both continental and oceanic crust, so that the expression 'continental drift' is no longer strictly appropriate (Figure 29.2). The plates are the inert aseismic regions of the earth, bounded by narrow mobile belts, characterized by earthquakes and volcanic activity, and where continents form the leading edges of plates by orogenic belts.

The term plate was first used by Tuzo Wilson when he defined transform faults (p. 621), but it did not come into general use until 1968, after Euler's theorem had been applied to the study of movements along transform faults. According to Euler's theorem, if the earth be considered as being spherical, then a carapace-like plate of its lithosphere can be moved to another position on its surface by rotation about an axis—*the axis of rotation*—passing through the centre of the sphere. Moreover, the plate cannot move in any other way. Any point on a plate, in moving to any other position on the earth's surface, must move along a small circle at right angles to the axis of rotation. The point where the axis of rotation intersects the earth's surface being known as the *pole of rotation*.

In 1967 McKenzie and Parker outlined 'a paving stone' hypothesis, in which ocean crust was newly formed at mid-oceanic ridges and destroyed at the trenches. By reference to Euler's theorem they determined the pole of rotation for the north-west Pacific from slip directions along marginal

**Figure 29.2** The earth's surface divided into 6 major plates, according to Le Pichon (1968), and a few smaller ones, all of which move in the directions of the arrows

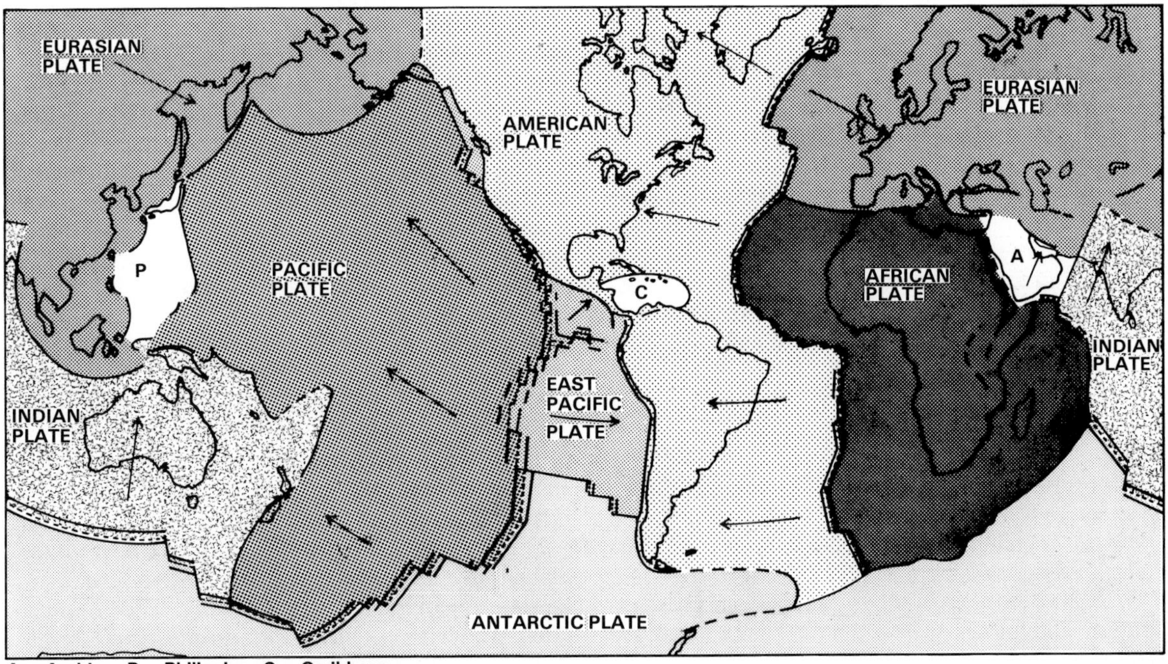

A = Arabian P = Philippine C = Caribbean

fault-planes, and located it at 50°N 85°W. If their paving stone hypothesis was correct then the North Pacific should have moved as a rigid slab, and the slip directions of all earthquakes within this slab should lie on small circles around the pole of rotation. Examination of the published data revealed that this was so, a conclusion that was confirmed by Isacks and Sykes the following year.

In 1968 Morgan outlined the hypothesis of plate tectonics. He established that transform faults, like those intersecting mid-oceanic ridges, which form parts of the boundary between two plates that are moving relative to one another, lie on small circles that are concentric about a pole of rotation (Figure 29.3). The movement of a plate is defined by the position of its pole of rotation and its angle of rotation about the rotation axis; its rate of movement varying with distance from the pole of rotation, being nil at the pole and reaching a maximum on the equator relative to the pole of rotation. Morgan divided the earth's surface into twenty plates, and determined their poles of rotation by drawing great circles at right angles to relevant groups of transform faults. If the members of a group of transform faults have a common axis of rotation then such great circles should intersect at the pole of rotation, e.g. the great circles intersection at A in Figure 29.3. In actuality the intersections have a small spread; an inevitable result of small errors that arise in mapping the ocean floors.

Le Pichon (1968) simplified the concept of plate tectonics by dividing the earth's surface into six major plates, and a few small ones. He determined the poles of rotation, about which the plates may be supposed to have rotated, by two different methods. In one he utilized the spreading rates calculated from the ocean-floor magnetic anomalies, and in the other he made use of the strike of transform faults intersecting mid-ocean ridges, as described above. The directions of movement of these plates are illustrated in Figure 29.2.

## Varieties of Plate Margins

The margins of plates can be grouped into three varieties, as follows:

(*a*) *Shear margins* where plates glide past one another along transform faults.

(*b*) *Accreting margins* where new ocean crust is formed, and plates grow. These occur at mid-oceanic ridges, and where continental crust splits

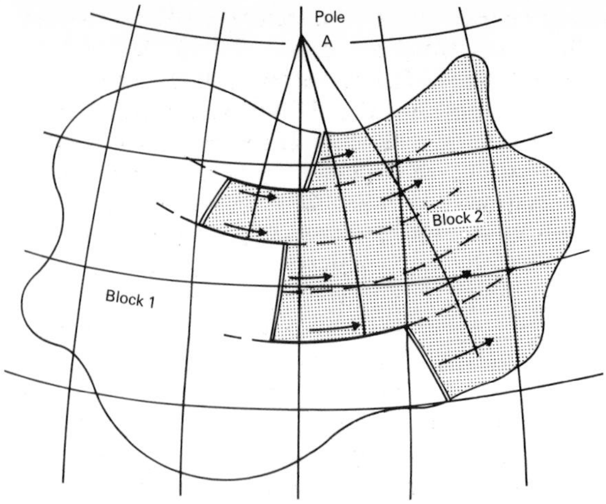


**Figure 29.3** Transform faults drawn on a sphere lie on small circles concentric about a pole of rotation (A), the position of which can be found at the intersection of great circles drawn at right angles to the relevant transform faults. The movement of Block 2, relative to Block 1 must result from rotation about A. (*After W. J. Morgan, 1968*)

apart and new oceans are born, e.g. the Red Sea (p. 646).

(*c*) *Consuming margins* where crust disappears or is narrowed. If the earth remains the same size then the amount of crust consumed must equal the amount of new crust that is formed. Examples of consumption are found at ocean trenches where oceanic crust disappears down into the mantle along Benioff zones; where continental crust underthrusts continental crust (Figure 29.4); where oceanic crust is thrust over continental crust (Figure 29.8); and where continental crust is narrowed by folding, supposedly in orogenic belts.

The structures of orogenic belts are considered in Chapter 30. Here it is convenient to consider examples of continental crust advancing under continental crust (Himalayas and Tibet) and of oceanic crust thrust over continental crust (Cyprus).

*Tibet* is the loftiest of the world's plateaus, with an average height of 4900 m. Nevertheless the southern part lay beneath the Tethys—the seaway between Laurasia and Gondwanaland—as recently as Cretaceous times. The northern part was already land during the Jurassic. During the late Cretaceous and Tertiary orogenic cycles the whole region, from the Pamirs in the west to the Yunnan plateau in the east, was upheaved by thousands of metres. It can hardly be pure coincidence that the crust underlying Tibet has double the normal continental thickness and that this doubling was brought about during the latter part of peninsular India's drift from the other side of the present equator. The northern borders of the Indian shield disappear beneath the Himalayan ranges after dipping downwards to form the floor of the deep Indo-Gangetic trough. It is natural to conjecture that what was formerly the northern prow of the advancing shield may now extend beneath Tibet, perhaps as far as the Altyn Tag (Figure 29.4).

That an area so vast as Tibet should stand at so extraordinary a height remained a geological enigma until the glacial evidence for continental drift suggested that the crust of the ancestral Indian shield had somehow migrated from Africa to Asia. Instead of staging a head-on collision, the shield was drawn downwards to continue its unfinished journey beneath the area that has become Tibet, leaving the distorted sides—Afghanistan and Baluchistan on the west, and Szechwan, northern Burma and Yunnan on the east—to accommodate themselves around the invader as best thy could. The isostatic uplift of Tibet was a tectonic consequence of this stupendous encounter.

If one continental region can underthrust another (as proposed to account for the double crustal thickness of Tibet) and since the oceanic floor can underthrust the continents at Benioff zones, then one might reasonably expect to find examples where, by some local tectonic 'accident', the oceanic floor has ridden over the adjoining continental crust instead of gliding beneath it. Alternatively, such relative movement could be described by saying that an advancing continental crust had locally 'plunged' into the upper mantle beneath an oceanic crust, instead of riding over the latter. In such an event the basic and ultrabasic rocks would be upheaved by the buoyant effect of the underlying sial and, if the sial were sufficiently thick, they would emerge as land with a thin veneer of oceanic sediments. The latter would soon be removed by erosion from the rising upwarp, and eventually there would be a growing area of land consisting of heaved up rocks which were originally part of the oceanic lithosphere. The southern part of Cyprus is one example of this kind for which there is good supporting evidence.

Cyprus has two E–W mountain ranges separated by a central plain. The northern range is part of an Alpine orogenic belt: a link, so to speak, between southern Turkey and Syria. The southern, Troodos range is a vast igneous massif of basic and ultrabasic rocks, made up of three main units, as indicated in Figure 29.5.

(*a*) The oldest unit, known as the *Sheeted intrusive complex*, consists of a swarm of N–S basic dykes invading a series of spilitic lavas, including pillow lavas. The dykes make up over 90 per cent of the whole and, being nearly vertical, their total thickness implies that during their intrusion the area was extended 130 km or more in an E–W direction.

**Figure 29.4** Section across Asia from the Verkhoyansk Mts to the Arabian Sea to illustrate a possible cause for the great height of Tibet and the Himalayas—the underthrusting of the Asiatic crust by a northern extension of the Indian crust. The current system shown in the mantle is purely hypothetical and makes no pretence to be more than suggestive. A possible source for the plateau basalts of the Deccan is indicated. Length of section about 9000 km

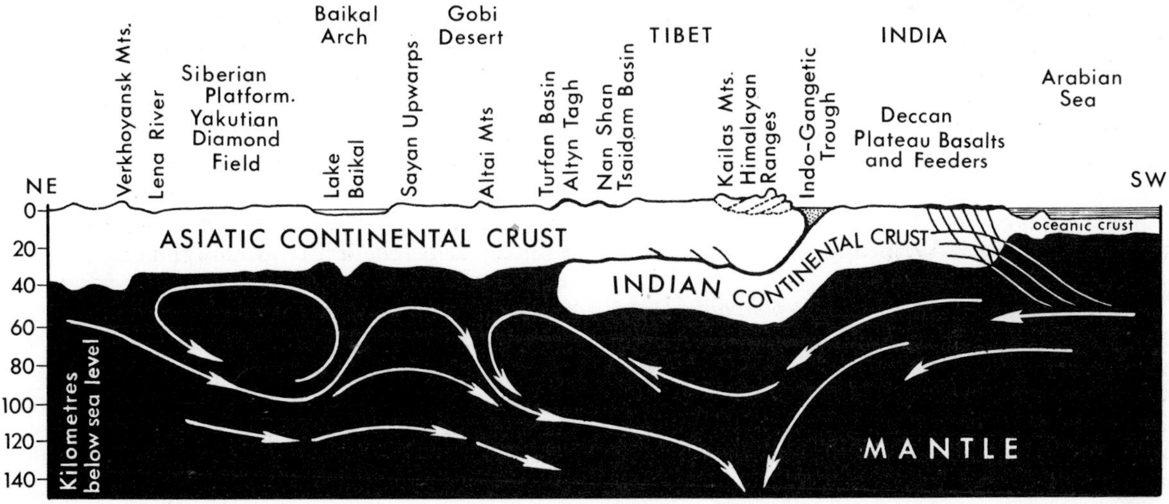

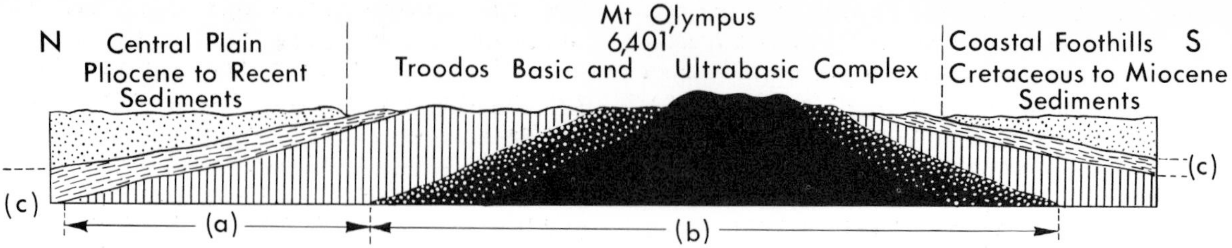


**Figure 29.5** Diagrammatic section through the Troodos Massif, Cyprus, showing (a) the Sheeted intrusive complex; (b) the Troodos plutonic complex: ultrabasic rocks, black; marginal basic rocks, black and white dots; (c) the Pillow Lava series. Length of section 60 km (*After I. G. Gass and D. Masson-Smith, 1963*)

**Figure 29.6** The crest of the Troodos Massif, Cyprus, looking south from the village of Moutoullas, which is situated where the marginal gabbro begins. The range beyond is peridotite. (*By courtesy of British Airways*)

(*b*) The central part of the massif (Figure 29.6), called the *Troodos plutonic complex*, consists of a variety of ultrabasic rocks, veined by gabbro, which becomes the dominant type towards the margins, where it is locally accompanied by granophyre veins.

(*c*) The third main unit—the *Troodos pillow lava series*—is a thick sequence of pillow lavas, with associated sills and dykes, which now form a rim surrounding (*a*), as a result of the elliptical dome-like structure of the whole massif. The marine sediments overlying the pillow lavas locally begin with early Triassic strata, and as the lavas rest unconformably on (*a*) and (*b*), a late Palaeozoic age is the latest that can be assigned to the Troodos massif as a whole.

Since the early years of the century it has been difficult to reconcile the geomorphological evidence that Cyprus is an area of present-day uplift with the geological evidence that the 2600 km$^2$ of the Troodos range are underlain by rocks of abnormally high density. Downwarping would be expected, not uplift. Later investigations only intensified the difficulty. The first gravity measurements, carried out in 1939, added to the puzzle by showing that the whole island was the site of strong positive anomalies, and that the large surplus of mass so disclosed was greatest along the Troodos range. Later it was found that the range had been uplifted at least 3000 m since the Cretaceous, while the rest of the island had risen only 900 m or less, much of it having been under the sea during the Tertiary.

In their record (1963) of their gravity survey, I. G. Gass and D. Masson-Smith show that the distribution of the positive anomalies implies that the thickness of high-density rock (such as peridotite) must be at least 11 km under Mount Olympus and at least 32 km under the NW end of the Troodos range. Corresponding to this contrast they find that the anomaly around Mount Olympus falls to about half the average for the range, and is even lower over the summit. This secondary anomaly within the major anomaly points unmistakeably to the presence of a thick layer of light rocks, such as sialic crustal rocks, beneath the slab of Troodos ultrabasic rocks. This suggests that the 'slab was once part of the oceanic lithosphere underlying an oceanic area between the African and Eurasian continents.' Developing an idea first put forward by D. W. Bishopp—that the Troodos massif might represent an immense volcanic pile that originated within the oceanic part of the Tethys (Figure 29.7) at a time when the bordering continents were much farther apart—Gass and Masson-Smith suggest that 'when the continental shields approached each other during the Alpine orogeny this slab of oceanic lithosphere was underthrust by the edge of the African shield and thereby raised to its present level in the upper part of the crust' (Figure 29.8). Similar complexes, now known as ophiolite assemblages, with similar origins, have more recently been recorded from other areas, e.g. the Appenines and California.

Overlying the upper pillow lavas of the Troodos Massif are *umbers*, that is mudstones enriched in iron, manganese and trace elements, e.g. barium, cobalt, copper, nickel, lead, vanadium, zinc and zirconium. The umbers are chemical precipitates, deposited from submarine thermal springs during the waning phase of the volcanic activity (Robertson, 1975). The high significance of these precipitates is that they are comparable with metalliferous deposits found along the axial zones of present-day mid-oceanic ridges.

## The Birth, Growth and Decline of Ocean Basins

As du Toit foresaw in 1937, new ocean basins are born when continental crust splits apart. Early stages of growth of ocean basins are represented by the Red Sea and the Gulf of Aden (p. 646).

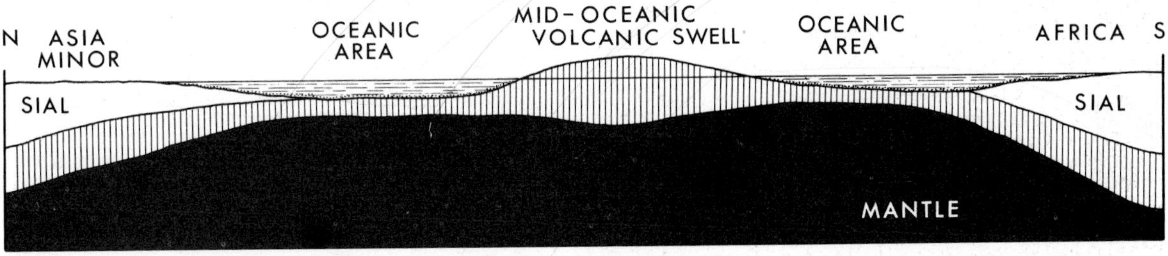


**Figure 29.7** (*above*) Diagram to illustrate an early stage in the development of the Troodos Massif: the formation of the Sheeted intrusive complex, as a volcanic swell in the oceanic area of the Tethys, which then lay between Africa and Eurasia. (*After I. G. Gass and D. Masson-Smith*)

**Figure 29.8** (*below*) Diagram to illustrate a later stage in the development of the Troodos Massif: with the approach of Africa and Eurasia the African crust has underthrust the Cyprus area, heaving up the ultrabasic and related rocks of the mantle. (*After I. G. Gass and D. Masson-Smith*)

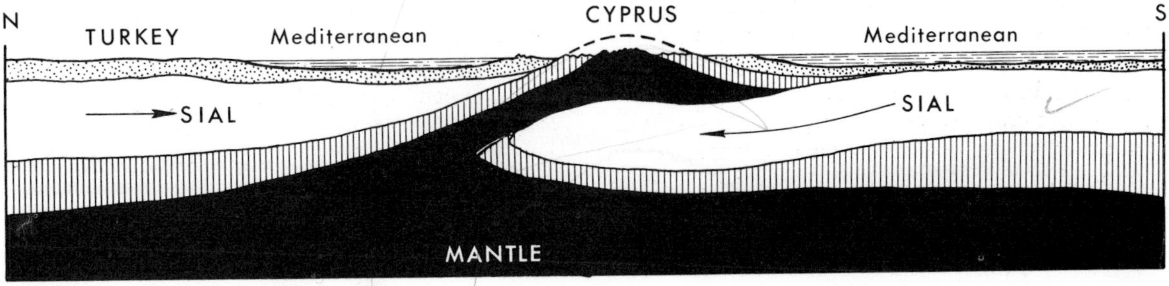

Both of these narrow seaways are floored by oceanic crust, with typical magnetic 'stripes', off-set by transform faults in the Gulf of Aden, and possibly in the Red Sea also. It is widely thought that the East African rift valleys represent a still earlier stage of continental splitting. These rift valleys, however, are floored with continental crust, and differ from mid-oceanic rifts both tectonically and in the composition and mineralogy of the associated volcanic rocks.

The various stages of growth of an ocean, like the Atlantic, can be read from the isochrons (Figure 27.16). Ten thousand million years ago, for example, the isochron 10 lay along the mid-Atlantic ridge. The opposing shores of the Atlantic were then closer together by a distance equal to twice that between isochron 10 and the mid-Atlantic ridge, and so on. The isochrons thus depict the progressive widening of oceanic floors, and moving them towards the relevant mid-oceanic ridges provides a further method of checking the reassembly of *Pangaea*.

The mid-oceanic ridges are thought to slowly change their position. For example, except on its northern side, Africa is surrounded by mid-oceanic ridges with no intervening subduction zones. The oceanic part of the African plate is therefore widening to the west, south and east. This implies that the mid-Atlantic ridge is migrating westward, and, since there is no intervening subduction zone, the American plate is consequently also moving westward.

## The Red Sea and the Gulf of Aden

*The Red Sea*, as illustrated by Figure 29.9, has an axial trough about 50 or 60 km across at its widest part, margined by shallow shelves. Gravity anomalies over the axial trough are strongly positive, and in 1958 were interpreted by Girdler as indicating the presence of a mass of basic rock beneath the sea floor. The presence of such heavy rock, approximately 2 km below the sea floor was confirmed by seismic surveys, and found to be overlain by consolidated sediment and/or lavas, surfaced with loose sediments (Drake and Girdler, 1964). Magnetic surveys, whereby magnetic anomalies were located over the central trough, further confirmed the presence of basic rock. The linear pattern of the magnetic anomalies, resembling that which characterizes the ocean floors, led to the interpretation of the basic rock beneath the Red Sea as new oceanic crust, emplaced during the separation of the continental shores of the Sea (Figure 29.10).

In 1966 Vine calculated the spreading rate across the Red Sea, from data recorded by Drake and Girdler (1964), and Allan *et al.* (1964). He found it to be 1 cm a year, away from the median line, during the last 3 or 4 million years. Subsequently a more detailed assessment by Allen and Morelli (1969) indicated a spreading rate of 1·1 cm a year away from the median line. That is, the shores of the Red Sea have separated at a rate of 2·2 cm a year.

A further seismic refraction survey by Davies and Tramonti (1970) suggests that the basaltic ocean floor may be wider than the median trough, but as yet the exact boundary between continental and oceanic crust has not been defined. No transform faults have so far been mapped, although apparent offsets of the magnetic anomalies suggest that they exist.

The *Gulf of Aden* has been found by seismic refraction investigations to be underlain by oceanic crust; a branch from the Carlsberg mid-oceanic ridge extending along its median zone. Accordingly the median zone, like the mid-Atlantic ridge, has a rough topographic surface. The median zone is intersected at high angles by scarps that are not continued through the bordering

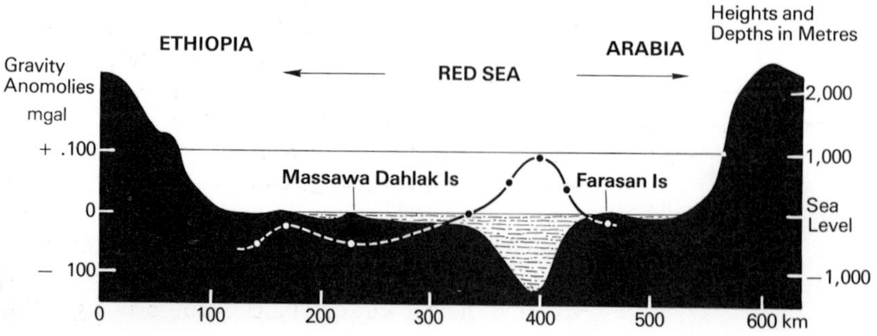


**Figure 29.9** Section across the Red Sea to show the relation between the bottom topography and the gravity anomalies (broken line with actual determinations marked by dots) (*After R. W. Girdler, 1958*)

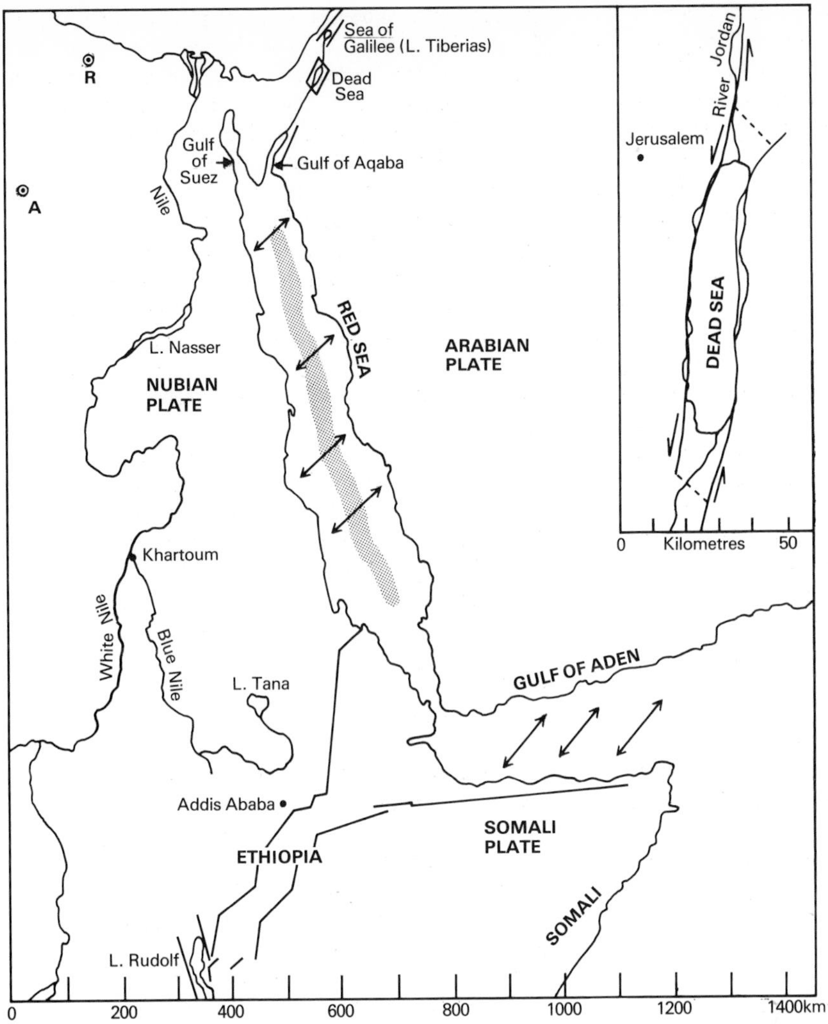


**Figure 29.10** Diagrammatic illustration of the crustal separation of Africa and Arabia resulting from the opening of the Red Sea and the Gulf of Aden in the direction of the arrows. R and A are the poles of rotation active respectively during the opening of the Red Sea and the Gulf of Aden. (*After R. W. Girdler and B. M. Darracott, 1972*). The Red Sea rift continues southward into Ethiopia and northward between Israel and Jordan; the inset map illustrates the movements associated with the formation of the Dead Sea rift. (*After A. M. Quennell, 1958*)

continental crust (Laughton 1966). Epicentres are located along the rough median zone, and extend into the axial trough of the Red Sea. Seismic solutions for two earthquakes whose epicentres were located on scarps in the Gulf of Aden, show the directions of movement to be consistent with those to be expected along transform faults. The rough median zone is characterized by 'striped' magnetic anomalies, and the spreading rate they depict, outward from the median line of the Gulf, varies from 0·9 to 1·1 cm a year. That is the Gulf of Aden is opening at the rate of 1·8 to 2·2 cm a year (Figure 29.10).

## The East African Rift Valleys

Many geophysicists consider the East African rift valleys (Figure 29.11) to represent a still earlier phase of the separation of continental crust than the Red Sea and the Gulf of Aden, and several attempts have been made to locate poles of rotation for plates thought to border the rifts. The rifts are characterized by a broad zone of negative gravity anomalies, up to 1000 km wide, on which is superposed a narrow zone of positive gravity anomalies, of the order of from 40 to 80 km wide, related to the actual rift valleys. It is important to realize, however, that no oceanic crust has been located within the rift valleys, and that geological evidence shows that they are floored with continental rocks, largely of Precambrian age. The

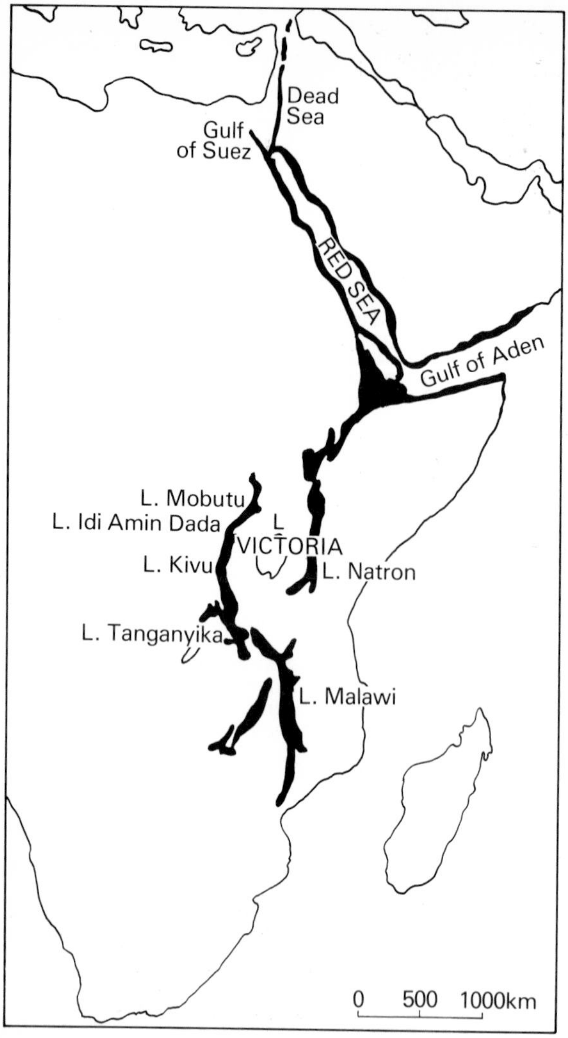


**Figure 29.11** Sketch map of the east African rift valleys (*After du Toit, 1937*)

**Figure 29.12** Normal fault between Miocene beds (on the right) and Precambrian basement rocks; exposed on the northern bank of a tributary flowing into the Semliki River near its outlet from the southern end of Lake Mobutu. The tributary has cut through one of the main boundary faults on the Zaïre side of the Rift. (*J. Lepersonne*)

positive gravity anomaly is accounted for by intrusions of basic and ultrabasic rocks.

The term *rift valley* was introduced by Gregory for the great rift valley of East Africa, which he was the first to recognize as a tectonic feature due to faulting. Gregory defined the term to mean a long strip of country let down between normal faults, or a parallel series of step faults. There is ample evidence that the dominant type of faulting associated with the East African rifts is normal. The north-west side of the Mobutu rift, for example, is a normal fault (Figure 29.12). On the south eastern side of the Mobutu rift deep bore-holes were sunk through gently folded sediments at various distances from the foot of the scarp. As the distance increased so did the depth at which the fault and crystalline basement rocks were encountered, the range being from 130 m to 1233 m. The south-eastern fault was thus proved to be a normal fault dipping inwards towards the lake at 65°. Further south, where the depressions around Ruwenzori begin to converge to form the Mobutu rift, the 'nose' of Ruwenzori (Figure 29.13) has also been shown to be bordered by normal faults both by visible geological evidence, and by the results of borings.

The floors of some exceptionally deep rift valleys have obviously subsided (Figure 29.14), but in far more examples it is clear that they have merely lagged behind the surface of the adjoining plateau during the course of the general uplift.

The African rift valleys do not constitute one long continuous trough with a curving branch to the west. Some of the individual faults can be traced for long distances, but others are shorter

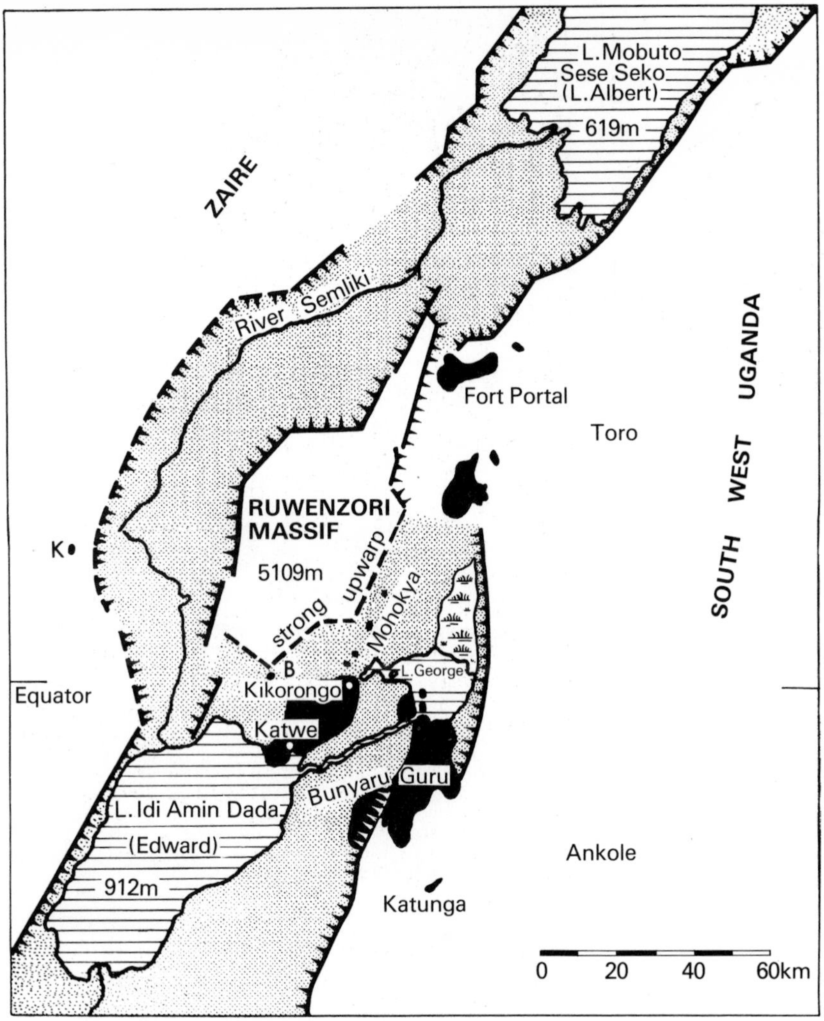


**Figure 29.13** Map of part of the Western Rift Valley, showing the massif of Ruwenzori and the adjoining volcanic areas (black). Two small areas are marked B (Bukangara) and K (Karibumba).

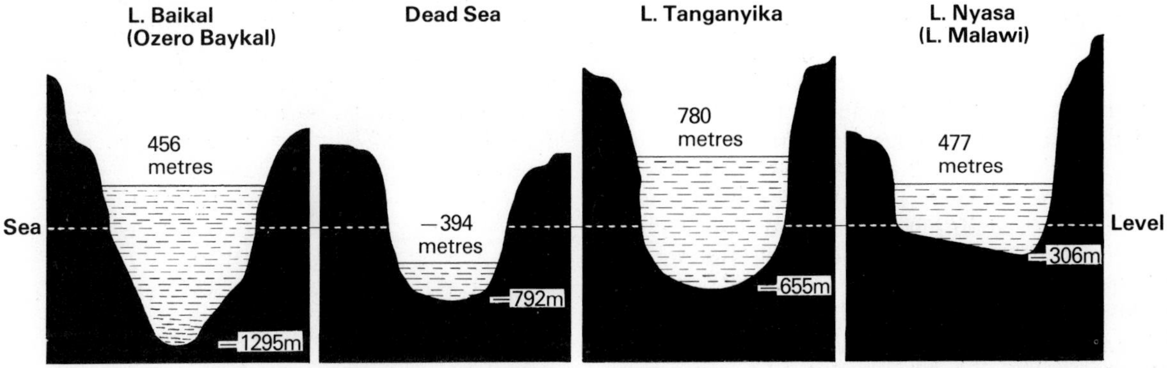


**Figure 29.14** Sections through rift-valley lakes to illustrate local depression of the floor below sea-level. Vertical scale greatly exaggerated. Each section is 80 km across.

and arranged *en echelon*. But looking at the East African rift faults as a whole, it is impossible not to recognize that they are all closely related parts of a single system of tectonic features which extends

**Figure 29.15** Lake Nakura in the Eastern or Gregory Rift, north of Lake Naivasha (see Fig. 29.28) with the stepped western walls of the Rift in the background (*Dorien Leigh Ltd*)

from the Zambezi to the Red Sea, over a total distance of nearly 3000 km. The component rift valleys are usually grouped as follows:

(*a*) The Lake Rudolf and Ethiopian section

(*b*) The Eastern or Gregory Rift (Figure 29.15), east of Lake Victoria

(*c*) The Western Rift from Lake Tanganyika, through Lake Kivu, Lake Idi Amin Dada (L. Edward) and Lake Mobutu (L. Albert) (Figure 29.16), with the uplifted massif of Ruwenzori between the last two sections (Figure 29.28)

(*d*) The Southern Rift, the Malawi (Nyasa) section and its bifurcations into the southern end of the Central Plateau.

There is a remarkable uniformity in the widths of most of the rift valleys, as indicated by the following measurements:

| | |
|---|---|
| Lake Mobutu (Albert) | 45 km |
| Lake Tanganyika (north) | 50 |
| Lake Tanganyika (south) | 40 |
| Rukwa | 40 |
| Lake Rudolf | 55 |
| Lake Natron | 30–50 |
| Ruaha (north) | 40 |
| Lake Malawi (Nyasa) | 40–60 |

Rift valleys in other continents also have similar widths, e.g. the Rhine graben, 30–45 km, and Lake Baikal south 55, north 70 km (Figures 29.1 and 29.30). These widths are all of the same order as the thickness of the continental crust, a significant relationship also brought out in the scaled-down model experiments carried out by Cloos (Figure 29.18).

Some of the oldest scarps of the African rifts date from Karroo times (Carboniferous) and are then deeply dissected by erosion, but more recent ones are steep and sharply defined. From many radiometric age determinations of lavas, tuffs and intrusions, combined with geological evidence, the early stages of modern rifting are known to have

**Figure 29.16** A 6 km stretch along the south-eastern fault scarp of the Albert (Mobutu) Rift, viewed from the air over the winding road that leads from the plateau to the Lake port of Butiaba, Uganda (*Overseas Geological Surveys*)

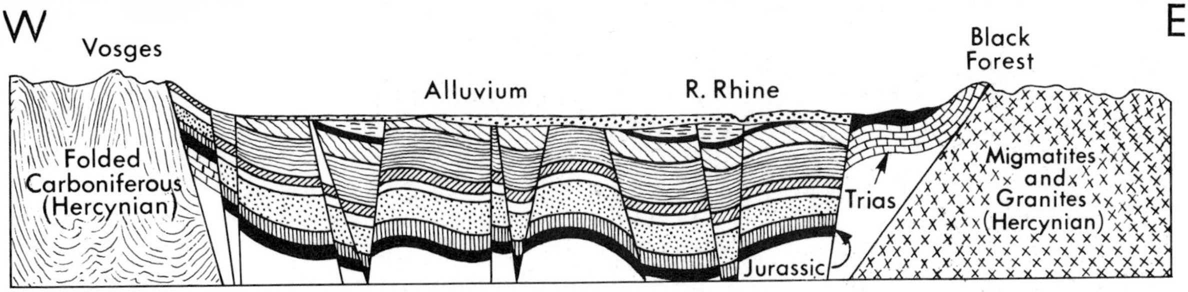


**Figure 29.17** Section across the Rhine graben, north of Mülhausen, showing the folded and faulted structure of the strata at the top of the rift-valley block as determined by numerous borings. Miocene beds appear locally beneath the alluvium, but most of the strata here shown are of Oligocene age, resting on Jurassic. Length of section about 48 km. Width of the rift valley at the surface 32 km

been taking place for 23 million years, since Lower Miocene times.

Long ago Hans Cloos (1939) attempted experimentally to reproduce the Rhine graben (Figure 29.17), with scaled-down models, by using layers of clay, just moist enough to flow and subside, while being able to fracture and thus produce joints and faults. His first apparatus was designed with the clay layers on two wooden blocks which could be pulled apart, as it were towards the Black Forest on the right, and the Vosges on the left, at rates of a few millimetres an hour. This model which, on looking back, might well be thought to have been devised to show what would happen as a result of separation of continental crust, reproduced all the structural features of the rift valleys beautifully, but two essential features of the structures as a whole were still missing—the adjoining horsts.

Cloos overcame this difficulty by simply arching up the artificial crust, thus automatically stretching the upper side. Figure 29.18 illustrates one of these experiments, in which not only the features of rift valleys were reproduced, but also the adjoining horsts and scarps. The important conclusion to be drawn from these experiments is that rift valleys are not a response to tension alone, but to uplift of the whole region, e.g. the Rhine upwarp or 'shield' as it is sometimes called, and the swell on which the East African rift valleys are located.

Figure 29.19 illustrates another of the ingenious experiments by Cloos. Here the updoming is imitated by the slow upheaval of the swelling surface of a rubber hot-water bottle coated with moist clay. The minute faults produced by uplift and extension faithfully reproduced the Y-shaped

**Figure 29.18** Experimental production of a rift valley by slow upheaval of layers of moist clay. Note that the surface width of the rift valley is of the same order as the total thickness of the clay 'crust'. (*Hans Cloos, 1939*)

pattern at each end of the Rhine graben (Figure 29.20), and of various sections of the East African rift valley.

For hundreds of millions of years the movements of the African shield have been persistently epeirogenic, giving a structural pattern of broad basins separated by irregular swells which rise towards the east to form a coalescing series of plateaus, the latter being traversed by the spectacular system of rift valleys (Figure 29.21). A significant feature of the plateau is the characteristic rise of the surface towards the edges of the rifts (Figure 29.22). Such upwarping on the western side of the rift valley is so recent that rivers are now reversed, so that they flow inwards and drain either into Lake Victoria or into the Victoria Nile. The curious shape of Lake Kyoga and its swampy margins, filled with papyrus and sudd, is a clear indication that the Kafu river and its tributaries have been ponded back towards a subsidiary depression north of Lake Victoria (Figure 29.23).

Reporting on the findings relating to the rifts by the Soviet East African Expedition in 1967–9, Logatchev *et al.* (1972) conclude that 'the hypothesis of disruption and separation of the crust, which is often used to explain the tectonics of ocean bottoms, cannot be applied easily to continental rifting'. If the rift valleys of East Africa do represent an early stage of evolution of new ocean floor, or an abortive attempt towards crustal separation, then it is clear that updoming plays a major role in crustal separation. Indeed if confirmation is required of the important part played

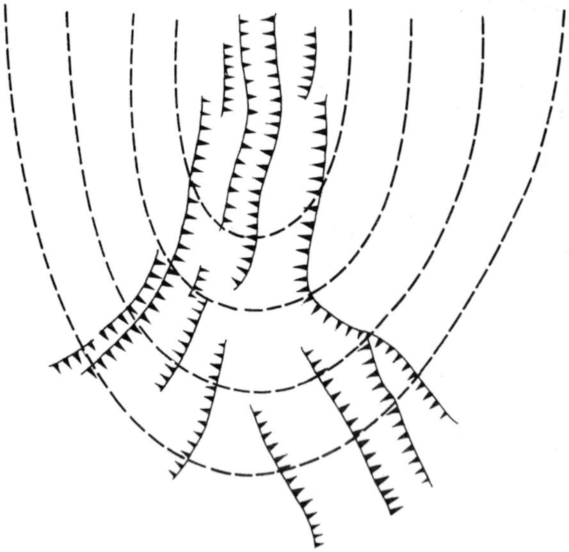

**Figure 29.19** The bifurcating pattern of rift faults developed towards the end of an elongated dome or upwarp, here produced by the swelling of an oval-shaped hot-water bottle coated with moist clay (*After Hans Cloos*)

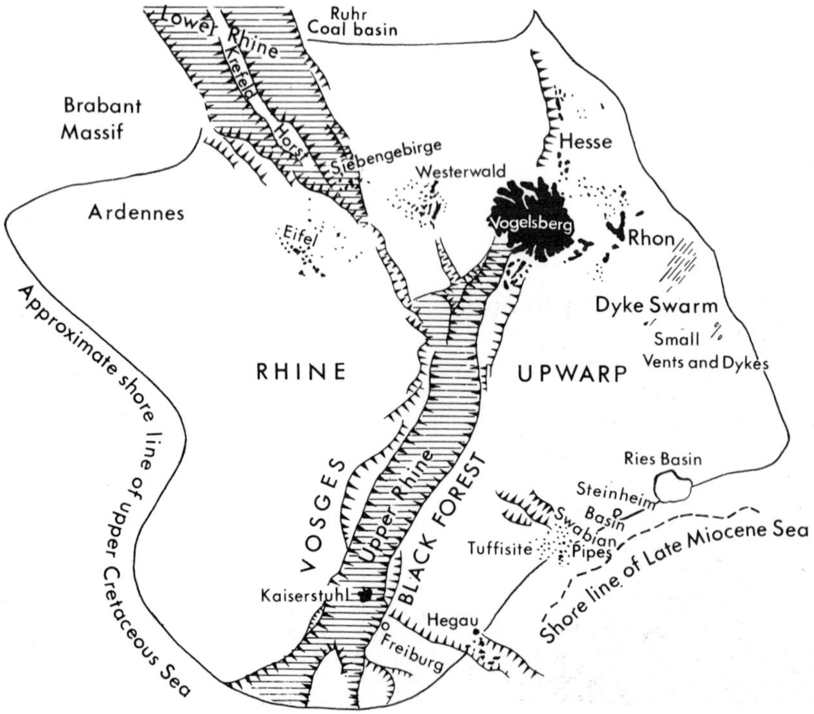


**Figure 29.20** Sketch map of the Rhine upwarp, showing the rift valley of the Upper Rhine and its northerly bifurcations into the rifts of the Lower Rhine and Hesse. Volcanic areas in black (*After Hans Cloos*)

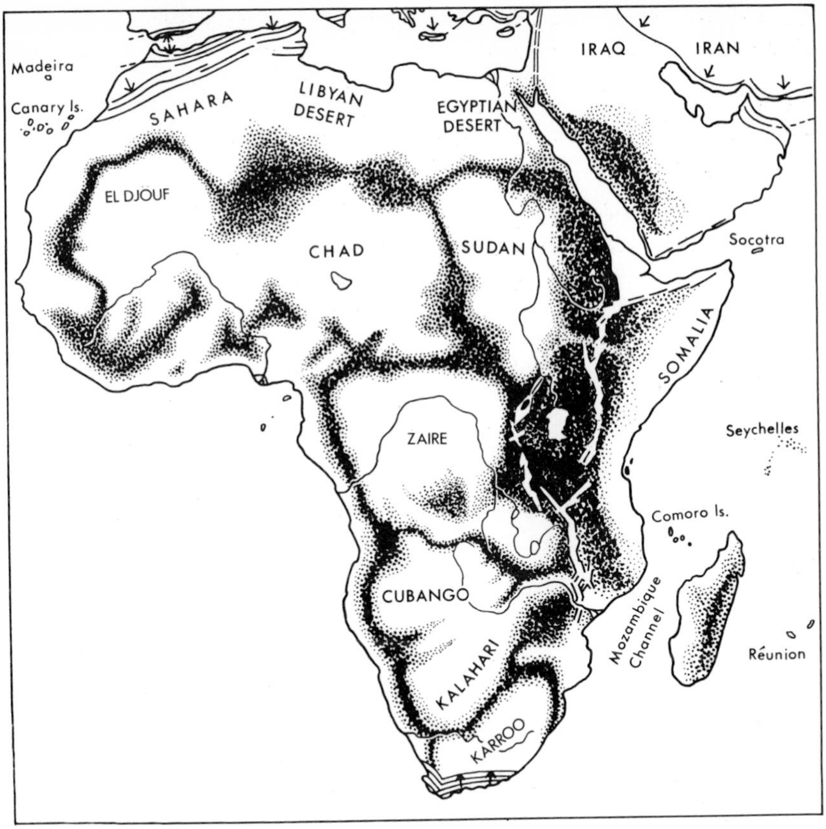


**Figure 29.21** Map showing in a generalized way the tectonic basins of Africa and the intervening swells, plateaus and rift valleys

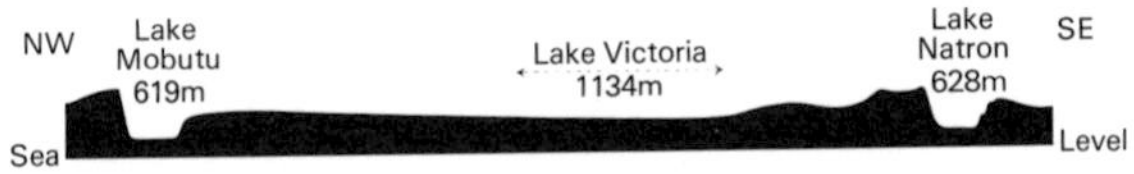


**Figure 29.22** Profile across the shallow sag of Lake Victoria and the deep troughs of the Western and Eastern Rifts. Length of section 800 km

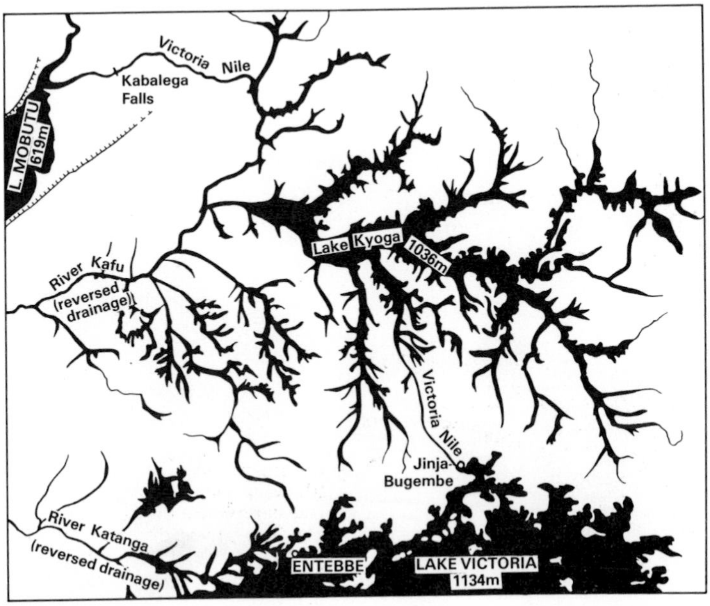


**Figure 29.23** Map of Lake Kyoga, Uganda, to illustrate the reversed drainage caused by late Pleistocene back-tilting of the plateau east of the Western Rift

by uplift associated with rift-formation, there is the astonishing feature of the Western Rift—unique in its immensity—the towering horst of Ruwenzori, which is by far the highest volcanic mountain in Africa. It rises from within a bifurcation of the rift valley (Figure 29.13) to snow-clad peaks (the 'Mountains of the Moon') up to 5119 m high, i.e. to nearly 4 km above the general level of the plateau (Figure 29.24). Towards Lake Mobutu (L. Albert) the great massif narrows to a long 'nose' flanked by fault scarps. Here recent uplift is proved by the occurrence of raised terraces of alluvium which were originally part of the valley floor of the Semliki River.

R. B. McConnell regards the African rift system as being analogous in origin to mid-oceanic rifts, crustal separation and emplacement of new ocean floor having been prevented by E–W confinement of Africa. This he suggests depended upon sea-floor spreading towards Africa, within the African plate, from the mid-Atlantic oceanic ridge on the west and the mid-Indian oceanic ridge on the east. The complete difference between the volcanic rocks of rift valleys and those of mid-oceanic ridges still has to be explained.

Lavas and pyroclasts associated with rift valleys are surprisingly rich in alkalis, sodium and potassium, even when they are highly basic or ultrabasic. In the Rhine upwarp this association is typically seen in the Kaiserstuhl, and in the Hegau volcanoes to the east where the lava includes many varieties rich in nepheline and other feldspathoids, and relatively poor in silica. Unlike the volcanic activity of mid-ocean ridges, that of rift valleys is characterized by the eruption of voluminous pyroclasts, evidencing the escape of great volumes of gas. It may be recalled that the Swabian tuff pipes are the type example of fluidization.

In the *Gregory Rift*, in East Africa, as in the Rhine upwarp, the proportion of alkalis is high relative to silica, soda being generally more abundant than potash. Silica itself rapidly decreases through rock series like soda-rhyolite—soda trachite—phonolite—nephelinite—olivine-nephelinite—melilite-nephelinite. In this series alkali feldspars first become increasingly abundant at the expense of quartz, which eventually disappears. Then feldspathoids begin to take the place of alkali feldspars until the latter also disappear. At the same time, with growing deficiency of silica,

**Figure 29.24** An exceptionally clear view of the uplifted massif of Ruwenzori, looking across Lake George 913 m from Mweya Lodge towards the summit of the range 5119 m. From the southern end to the 'nose' on the right the distance is about 80 km (*Paul H. Temple*)

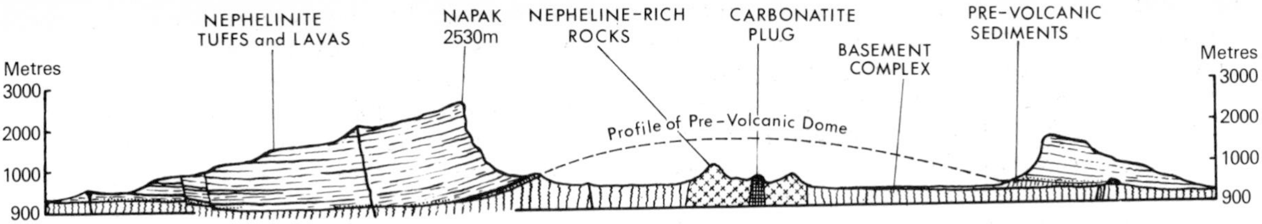


**Figure 29.25** Generalized section across the deeply eviscerated Napak volcano, souther Karamoja, Uganda, showing the core of carbonatite (black). Length of section 30 km *(After B. C. King, 1949)*

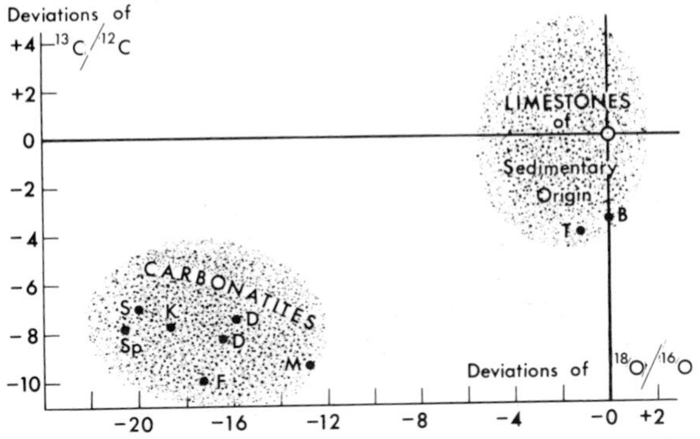


**Figure 29.26** The points representing deviations of $^{18}O/^{16}O$ from an arbitrary standard limestone (marked o), plotted against the corresponding deviations of $^{13}C/^{12}C$, fall into two well-separated fields (shaded) for carbonatites and sedimentary carbonate rocks respectively. The travertine, K, from Lake Katwe falls within the carbonatite field, while an ejected block, B, from one of the Bunyaruguru craters (Fig. 29.13) falls within the sedimentary field. Similarly, Sp from the Spitzkop alkaline complex in the Bushveld, South Africa, can clearly be distinguished as a carbonatite from the sedimentary limestone, T, from the Transvaal System, with which it was formerly identified. The other lettered carbonatites are:
D, dykes from the Pretoria Diamond Mine, South Africa
F, lava from the Fort Portal volcanic area (Fig. 29.13)
M, carbonatite lava from Mbuga crater, Katwe-Kikorongo volcanic area (Fig. 29.13)
S, carbonatite core of Sukulu volcano, E. Uganda
*(Isotopic determinations by P. Baertschi)*

pyroxenes and biotite increase, and olivine and melilite appear instead of pyroxenes and calcic feldspars. Associated with these alkali-rich rocks are others consisting of carbonates ($CaCO_3$, with variable amounts of $MgCO_3$ and $FeCO_3$). The latter are genuine igneous rocks and to distinguish them from limestone and marbles they are called *carbonatites* (Figure 29.25). As shown by the data plotted in Figure 29.26 ordinary limestones contain appreciably higher proportions of the isotopes $^{18}O$ and $^{13}C$ than carbonatites. These results are confirmed by marked differences between isotopic constitution of the strontium in carbonatites and limestones of sedimentary origin.

In the region of internal drainage associated with the Eastern Rift are glaring white salt-encrusted plains (Figure 29.27). Lakes Natron and Magadi and the plains round them contain vast reserves of soda ($Na_2CO_3$) and other chemical deposits which are exploited commercially on an

**Figure 29.27** Salt deposits of Lake Magadi in the Eastern Rift valley, Kenya (see Fig. 29.13 for locality) *(E. N. A.)*

extensive scale. The salts are derived from the soda-rich volcanic rocks of the district, partly from the direct products of eruption (which include carbonates rich in soda from the active volcano Oldoinyo Lengai in Tanzania) and from soda-rich hot springs, and partly from normal weathering and drainage of surface waters into the main depressions. The lakes are natural evaporating pans in which the soluble volcanic salts are concentrated.

The volcanic rocks of the *Western Rift* (Figure 29.28), differ dramatically from those of the Gregory Rift, in their richness in potash. The Fort Portal lavas and pyroclasts range from carbonatites with very little silica and low alkalis, to potash-rich ultrabasic lavas and lapilli composed of such minerals as leucite, potash-rich nepheline, kalsilite ($KAlSiO_4$), pyroxene, olivine and melilite. Silicate bombs and lapilli of this kind (Figure 29.29) characterize all the volcanic fields around Ruwenzori. Lava flows are rare except from the isolated volcano of Katunga in the south, and from the clustered volcanoes of Fort Portal in the north, where carbonate lavas that can be no more than a thousand years old have been recognized. In some of the ring craters of the Katwe area intrusive tongues and ejected blocks of carbonatite have been found.

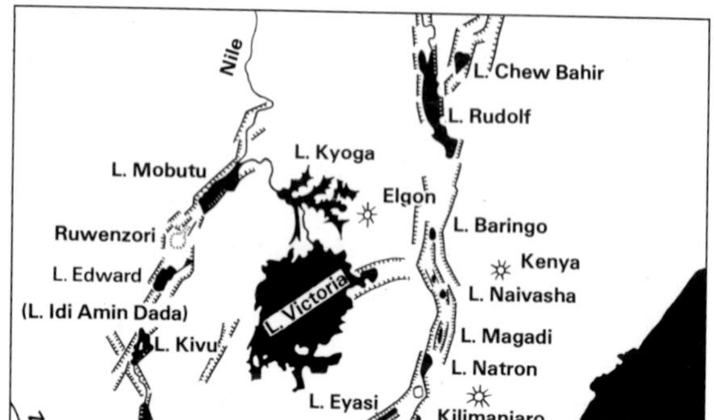


**Figure 29.28** Map of the African rift valleys from the Zambesi to the Ethiopian border. For the northern continuation into the Red Sea see Fig. 29.10. The Luangwa–Mid-Zambesi rifts are older features now largely occupied by Karroo sediments

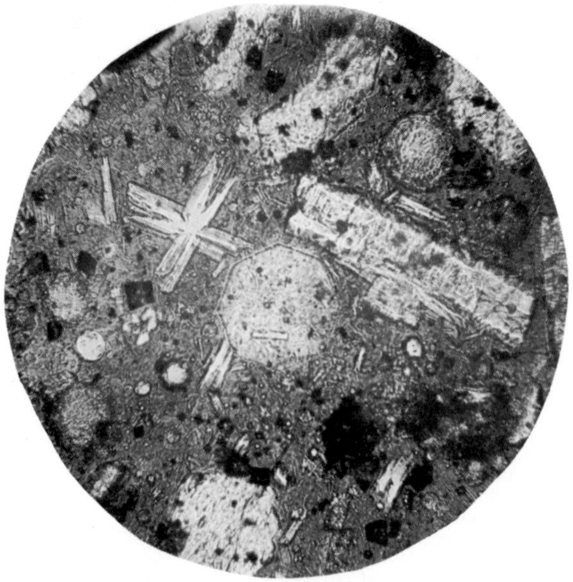

**Figure 29.29** Photomicrograph of *kalsilite-katungite* occurring as lapilli in tuffs in the Bunyaruguru volcanic field, SW Uganda (see Fig. 29.13). The minerals seen include melitite (rectangular and polygonal sections); kalsilite (forked and frayed-looking sections and one cross-like intergrowth); olivine (irregular form near top of field); and black titanium-rich grains; in a matrix of greenish-yellow glass. × 140 (*Arthur Holmes, 1945*)

It remains to be emphasized that we know nothing about the origin of the carbonatites that are so curiously involved in uplifts, rift valleys and alkaline rocks. Carbonatites seem to be distinctly more abundant in Africa than elsewhere, and moreover their occurrences seem to have been strikingly more numerous during the last 200 m.y. than at any earlier time. Only a few Precambrian carbonatite intrusions have been found (e.g. in the alkaline complexes of the Bushveld, p. 169). It may well be that most of the Precambrian examples have been rolled out of recognition during orogenic movements and are now indistinguishable from bands of marble. But it is certainly remarkable that after the dykes and sills and floods of basaltic magma that brought the Karroo régime to a close in Africa (i.e during the early Jurassic) carbonatites and their products began to be intruded on such a scale that they soon outnumbered all the pre-Karroo examples that have so far been discovered. The late Cretaceous diamond pipes seem to have originated in much the same way as the volcanic vents around Ruwenzori. Possibly water was a more abundant fluidizing agent in the production of kimberlite (p. 180) but the co-operation of carbonatite activity is fully attested by the occurrence of carbonatite dykes and the association of calcite with the hydrous minerals serpentine and chlorite. Then at intervals during the Tertiary, carbonatite intrusions appeared in greater numbers than before, accompanied and followed by carbonatite eruptions, such as still continue at the present day. And intermittently during this long period—from the time of the Karroo basalts—the great uplifts of the plateau regions of Africa have been taking place. A similar sequence of events occurred in Central Siberia.

*The Baikal Rift Valleys, Siberia* Apart from the incorporation of Caledonian structural elements, the Sayan upwarp, the Baikal Arch and the Aldan upwarp have geologically much in common with the rifted plateaus of eastern Africa (Figure 29.30). The similarity is emphasized by the rift valley system of which the deeper parts are occupied by Lake Baikal, and by the occurrence on the Aldan upwarp of a number of complexes of alkaline rocks, some of them with carbonatite cores. There are also Pleistocene and recent basaltic flows on the rifted belt and beyond it both to north and south. Near the upper reaches of the Vitim River three barely extinct volcanoes have been discovered and two of them have been named after the eminent Russian geologists Mushketov and Obruchev. Much more significant, however, was the discovery in 1954–5 of two groups of kimberlite pipes perforating the Siberian platform between the Lena and Tunguska rivers (Figure 29.30). The kimberlite shares all the essential characteristics of its African prototype, including the occurrence of diamonds and the abundance of inclusions, amongst which the presence of eclogite is es-

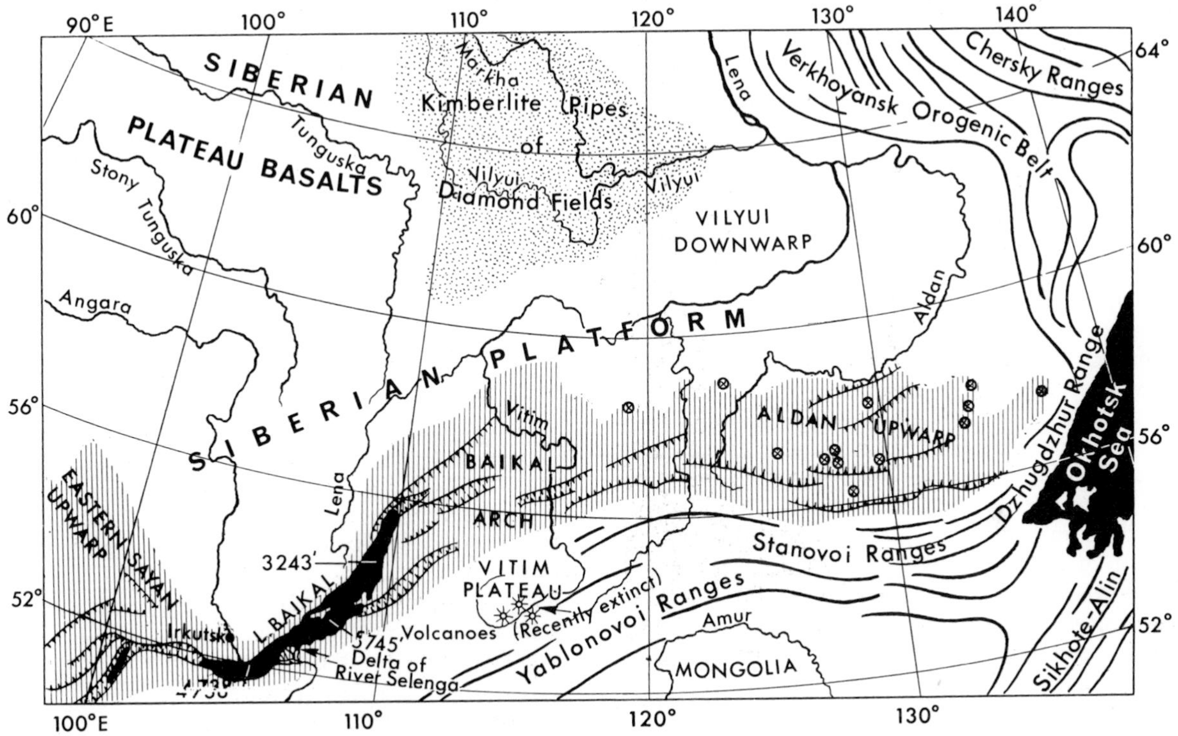


**Figure 29.30** Map of the Baikalian rift system and the Baikal Arch and its continuations (vertical ruling), showing their relationship to the Caledonian folds to the south and to the Siberian Platform to the north with its diamond fields (dotted) and vast areas of plateau basalts. The figures attached to Lake Baikal refer to the maximum depths, in feet, of water in the three depressions occupied by the Lake

pecially noteworthy. To the west and north-west of the diamond fields enormous areas of the platform are occupied by the 'Siberian traps': basaltic flows and sills that intimately resemble their Karroo counterparts in Africa. The only difference in a long catalogue of geological similarities is that the corresponding events began earlier in Siberia than in Africa. The 'traps' are mainly of Triassic age in Siberia, though locally they began in the late Carboniferous. In Africa they were mainly concentrated into the early Jurassic. In both regions, however, basaltic lavas and alkaline rocks associated with carbonatites accompanied the later vertical displacements.

## A Waning Ocean Floor: the Pacific

For a very long time the Pacific has been regarded as the oldest ocean, a relic of *Panthalassa*, the oceanic contemporary of *Pangaea*, but it now appears to have passed its prime. By analogy with the mid-Atlantic rise, the mid-Pacific rise may be supposed, at one time, to have been mid-Pacific along its length. Now, its position is asymmetric with respect to its shore-lines, and its north-eastern end vanishes at or beneath the coastline of North America (Figure 29.31). The mid-Pacific rise now extends into the Gulf of California and ends as an oceanic ridge, at its head, only appearing again as an oceanic feature, as the Gorda and Juan de Fuca ridges (Figure 29.31).

Within the *Gulf of California* seismic reflection and magnetic surveys revealed the presence of the east Pacific rise. The magnetic 'stripes' on its floor indicate that it has been widening for the last 4 million years by movements along transform faults (Figure 29.31). The peninsula of Baja California was thus progressively separated from the mainland. There has been difference of opinion as to what has happened to the east Pacific rise at the head of the Gulf of California. The view now favoured by many geophysicists is that it is displaced by the San Andreas fault system northwestward to the Juan de Fuca ridge (Figure 29.31). The oceanic trench that margins the west coast of South America, and continues northward as the mid-American trench, dies out as the mouth of the Gulf of California is approached. The current hypothesis is that a deep ocean trench, that once lay between the east Pacific rise and California, vanished when it was overrun by the American plate, being replaced by the San Andreas fault zone.

The seismic belt associated with the east Pacific rise can be traced, like the rise itself, along the Gulf of California. It continues within the continent of North America where it bulges eastward, including the Great Basin and Colorado plateau (Figure

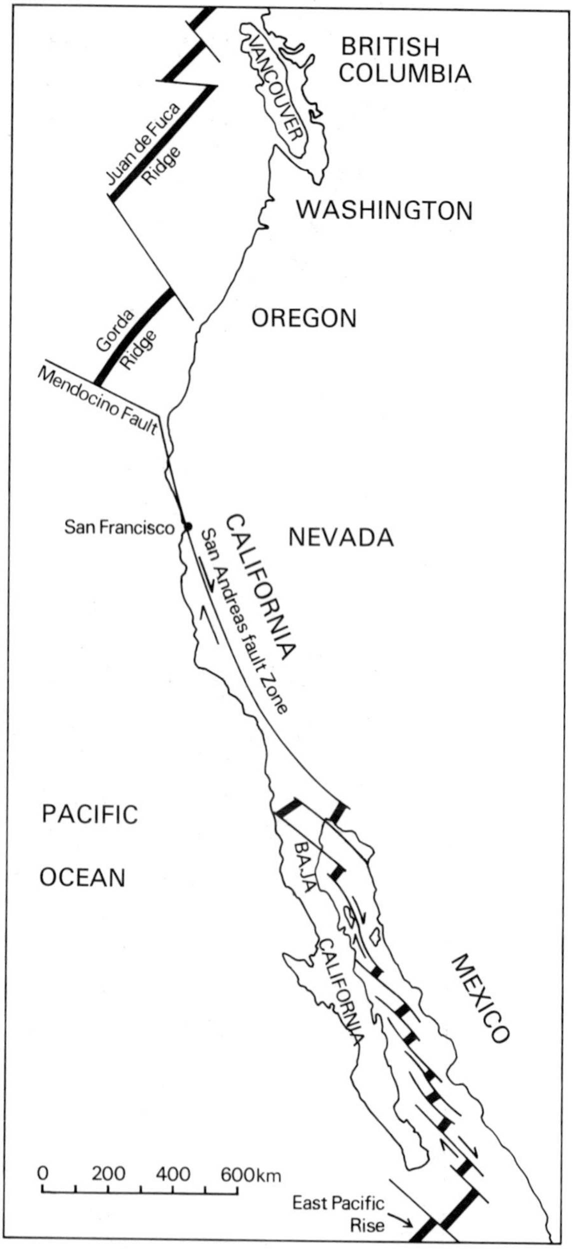


**Figure 29.31** Map illustrating the continuation of the east Pacific Rise along the Gulf of California, where it is offset by many transform faults. At the head of the Gulf the Rise is apparently displaced by the San Andreas fault-zone, and appears again as an oceanic feature in the Gorda and Juan de Fuca Ridges.

25.13). Menard (1960) suggested that the east Pacific rise still persists beneath the Colorado plateau. Others think that at least relics of the rise remain beneath the continent in the form of plumes rising from the mantle, so accounting for the high heat flow of the area.

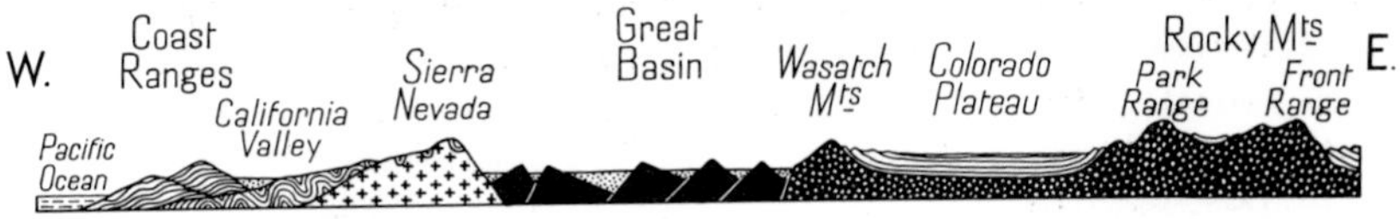


**Figure 29.32** Schematic section across the Cordillera of the western United States. Length of section about 3,000 km

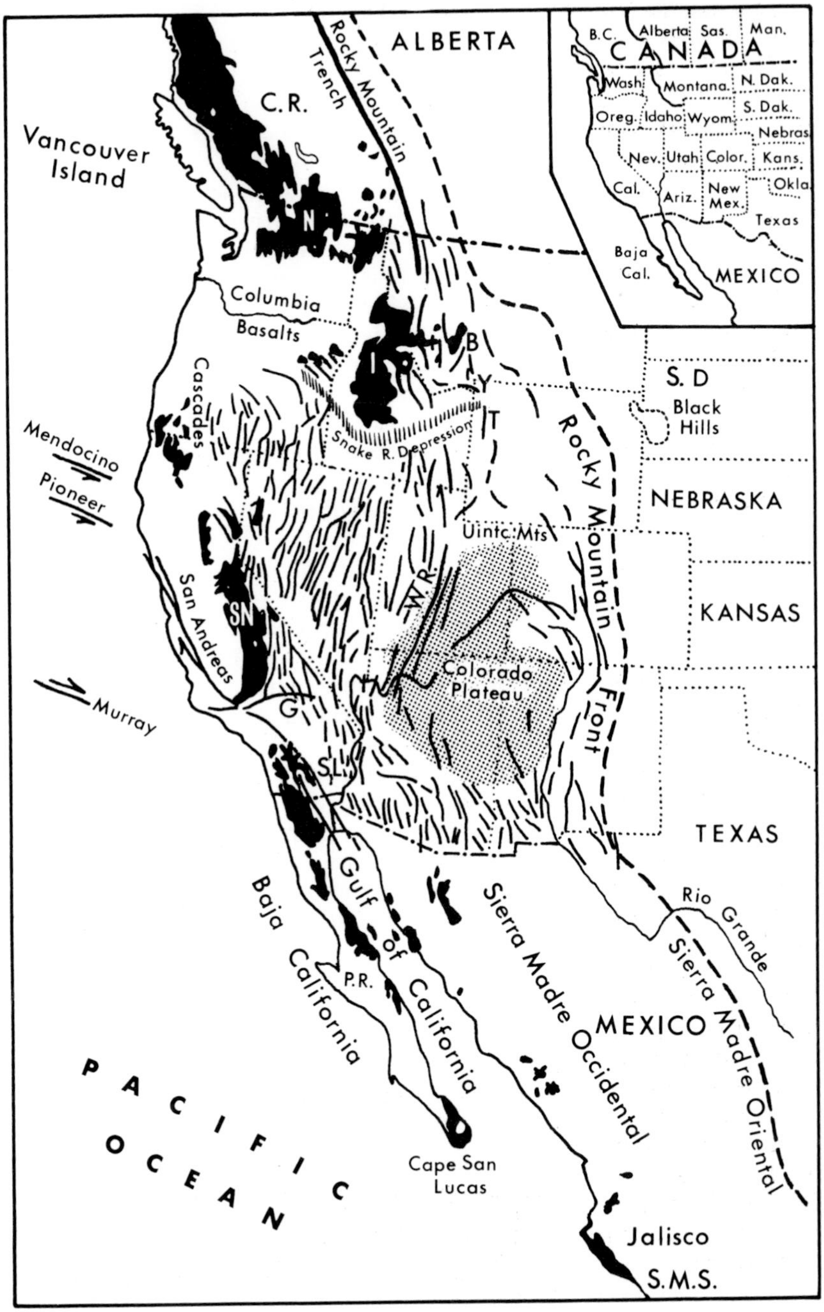


**Figure 29.33** Sketch map to indicate some of the major tectonic features of the North American Cordillera. The main outcrops of granitic batholiths (black) from north to south are C.R., Coast Range, extending NW to Alaska, where it swings round as the core of the Alaska Range (including Mt McKinley, Fig. 30.1); N, Nelson; B, Boulder; I, Idaho; S.N., Sierra Nevada; P.R., Peninsular Ranges; and in part S.M.S., Sierra Madre del Sur. The chief faults of the Basin and Range Province (between the Colorado Plateau and the Sierra Nevada) are drawn as portrayed by J. Gilluly (1963) for the U.S.A.; similar faults continue into Mexico, but have not been mapped in detail. Other localities indicated by letters are: G, Garlock fault; S.L., Salton Lake; T, Teton Range (Fig. 29.35); W.R., Wasatch Range; Y, Yellowstone Park

The Great *Cordillera* of North America includes all the ranges and plateaus west of the Great Plains, and the coastal lowlands of Mexico. The term Cordillera signifies a broad assemblage of mountain ranges belonging to orogenic belts of different ages, together with their associated plateaus and intermontane basins. The Cordillera of North America (Figures 29.32, 29.33) is composed of the *Coast Ranges* on its western side, of the *Rocky Mountains* on the east, and between is the *Great Basin*, characterized by upstanding blocks, known as the *Basin Ranges*, rising to heights of over 3000 m (Figure 29.34). Generally speaking, the Basin ranges strike in a north–south direction, but some trend north-west, whilst in the north, in Oregon, they swing round into an east–west direction. Gilluly has compared the way in which the Basin Ranges margining the Colorado plateau (Figure 29.33) with the splitting of the East African rift valley into the Western and Eastern Rifts which margin the elevated Victoria plateau (Figure 29.28).

At the close of the Cretaceous the Colorado plateau was nearly at sea-level (Figure 19.31). Since then the average uplift has been about 2500 m in the south, and about 1800 m in the north, and as the plateau is almost as extensive as the British Isles, the regional tilt is very gentle. The uplift occurred mainly in two pulses, one in late Miocene or Pliocene; the other, considerably later making possible the deep incision of the Grand Canyon.

The Basin Ranges, rising to heights of 2000 m and more, are tilted blocks, generally 30 to 40 km across, with steep scarps on one side and gentle slopes on the other (Figure 29.34). The bordering ranges of the Sierra Nevada and Wasatch Range are gigantic tilted blocks of the same type, with their high fault scarps facing inwards in each case. The older scarps have been considerably eroded into ravines and spurs, and the lower slopes are commonly aproned with screes and fans of rock-waste (Figure 22.29). Both these and the spurs are truncated by triangular facets along the lower slopes of some of the ranges, showing that the latest fault movements have been quite recent (Figure 29.35). Commonly, however, the scarps rise from pediments, rather than being mantled by screes and fans.

The point to be emphasized about the Basin and Range province is that the faults are normal, indicating extension. For the Great Basin between the Wasatch Range and the Sierra Nevada, Eardley estimates that the faults imply an extension of 50 km during the last 15 million years.

Gilluly (1972), although sceptical about the presence, or former presence, of the east Pacific rise beneath the continent, suggests that if it should be there it might account for the extension of the crust, evidenced by the innumerable normal faults, the high heat flow, and the eruption of ash flows, during the Oligocene and after, in the southern and eastern parts of the province. Another problem presented by the Basin and Range province is the thinness of its crust, which is less than the average 30 km in the Western States. On the other hand the crust of the Colorado plateau is nearly 45 km thick, more than the average. Gilluly suggested (1972) that if the eastern Pacific rise does lie beneath the Basin and Range province, then crustal material may have been dragged eastward, from beneath the Basin and Range province, and added to the crust beneath the Colorado plateau, and dragged westward to form the root of the Sierra Nevada. He concluded, however, that the rate of extension in the Basin and Range province has been about 50 km on average, and 90 km as a maximum in the 35 million years since the faulting began, whereas if the spreading rate had equalled that of the oceanic rise it should have been about 1300 km.

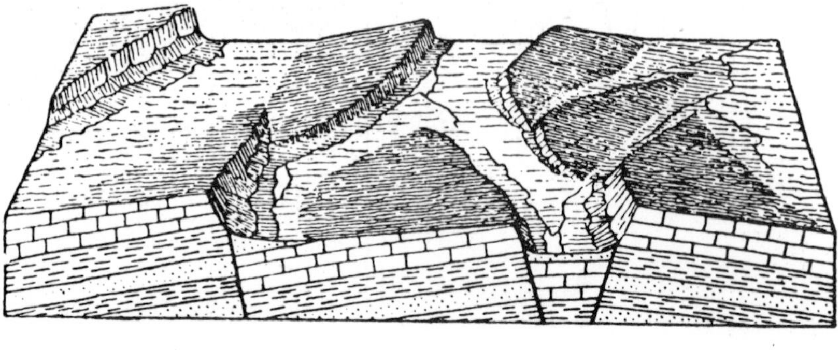

**Figure 29.34** Diagram to illustrate the fault block structure of the ranges of the Great Basin, Utah (*After W. M. Davis*)

## Vulcanism in Relation to Plate Tectonics

Nearly 800 volcanoes are known to be active or to have been in eruption during historical times (Figure 29.36). Besides these, many thousands of extinct cones and craters are so perfectly preserved that they must equally be taken into account in considering the distribution of recent vulcanism. These volcanoes are of three distinct types, both in their distribution and in the composition of the volcanic rocks that they erupt. Their distribution is related to plate boundaries, and they are:

(*a*) situated on or near, or have migrated from, mid-oceanic ridges, or (*b*) associated with the Circum-Pacific and Alpine orogenic belts of the latest orogenic cycle, or (*c*) associated with rift valleys.

It should not pass unnoticed that there are volcanoes that do not appear to conform with present plate boundaries. The Auvergne in France, and the Eifel in Germany are familiar examples of volcanic districts, now extinct, that were active only a few thousand years ago. Others not much older occur in the Hungarian Basin, enclosed by the Carpathians, and along the Mediterranean coasts of Spain and Algeria. Many sporadic groups have broken through the crust at considerable distances from the ranges of the Alpine orogenic belt. Farther east, active cones have been recorded from the borders of the Tarim depression, north of Tibet; in Mongolia and Manchuria; and north-east of the Verkhoyansk Mountains, which strike across Siberia between the Arctic Ocean and the Sea of Okhotsk. Tibet, the loftiest of the world's plateaus, appears to be free from volcanoes. It has, however, several regions of hot springs, some of them very hot.

The lavas of type (*a*) are tholeiites, that is basalts poor in potash, and thought to be derived by differential melting from the mantle. The lavas and pyroclasts of type (*b*) are, as a group, rich in silica. Here, basalts are associated with abundant andesites, dacites and rhyolites. The lavas and tuffs of type (*c*), associated with rift valleys, are characteristically very rich in alkalis. They are sometimes referred to as alkali-rich basalts, but they differ fundamentally from the basalts to which this term is usually applied, both in

**Figure 29.35** Eroded fault scarp of the Teton Range, Wyoming, south of Yellowstone Park. The highest peak 4190 m, is Grand Teton. The Precambrian rocks of the Range are separated from the depression of Jackson Hole, occupied by Palaeozoic and later sediments, by a normal fault down-throwing to the east. (*United States National Park Service*)

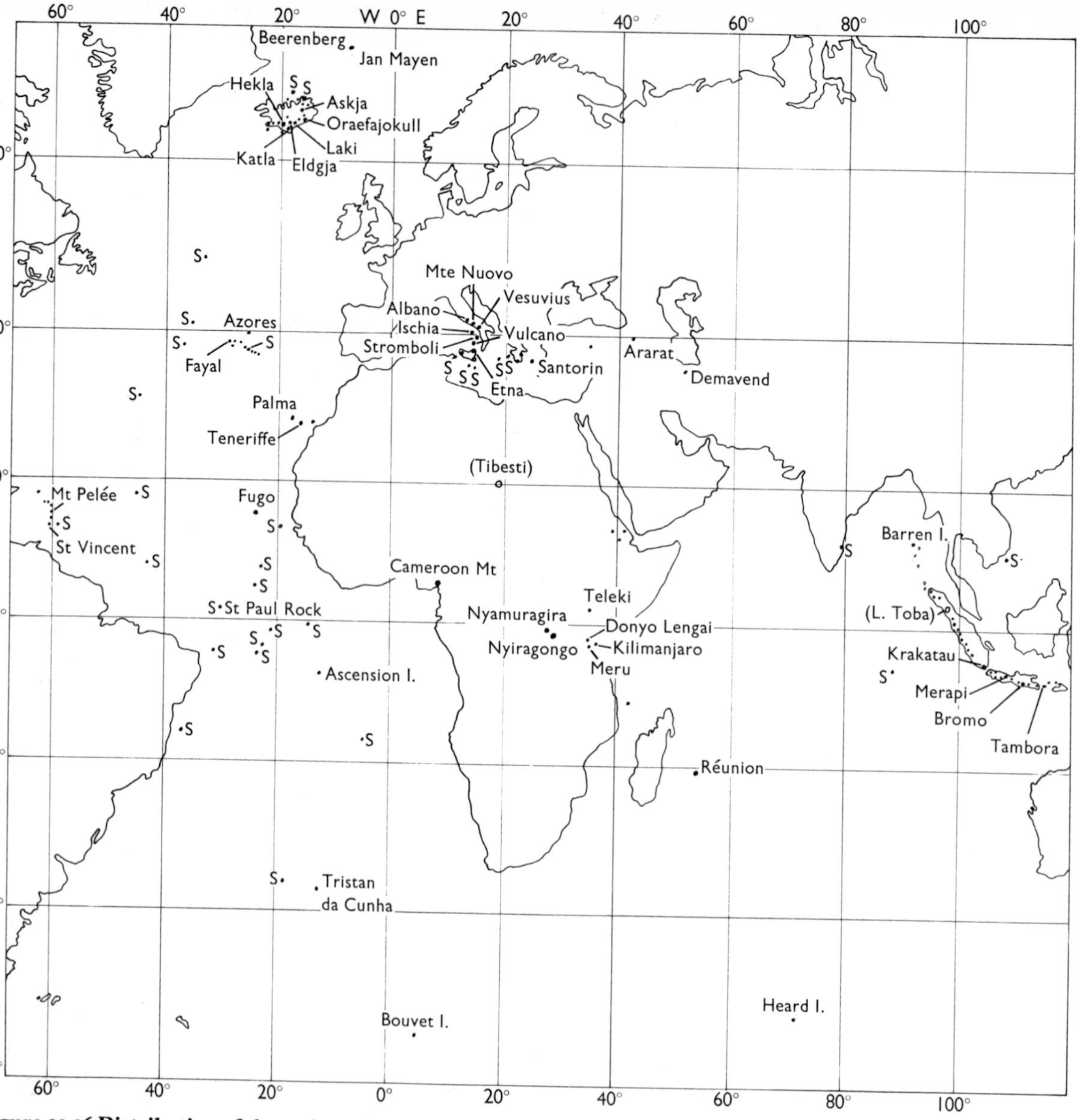


**Figure 29.36 Distribution of the active volcanoes of the world** (S=submarine eruptions)

| | |
|---|---|
| *Circum-Pacific Girdle* | |
| Alaska | 15 |
| Western U.S.A. | 3 |
| Mexico | 12 |
| Lesser Antilles | 8 |
| Guatemala | 7 |
| El Salvador | 7 |
| Nicaragua | 11 |
| Costa Rica | 6 |
| Santa Cruz (Galápagos Is.) | 1 |
| Colombia | 6 |
| Ecuador | 2 |
| Southern Antilles | 2 |
| N. Chile | 6 |
| Central Chile | 14 |
| S. Chile | 6 |
| Antarctica | 7 |
| New Zealand | 5 |
| Kermadec Is. | 3 |
| Tonga Is. | 9 |
| Samoa Is. | 4 |
| Loyalty Is. | 1 |
| New Hebrides | 7 |
| Solomon Is. | 3 |
| New Britain | 10 |
| New Guinea | 9 |
| Halmahera | 6 |
| Sulawesi (Celebes) | 6 |
| Sangihe Is. | 6 |
| Philippines | 12 |
| S. Japan–Ryukyu–Taiwan (Formosa) | 16 |
| Ogasawara Gunto group (Bonin Is.) | 20 |
| Kurile Is. | 39 |
| Hokkaido and N. Honshu (Japan | 28 |
| Kamchatka | 25 |
| Aleutian Is. | 18 |
| *Indonesian branch* | |
| Banda Sea | 5 |
| Wetar–Flores–Bali | 25 |
| Java | 17 |
| Krakatau | 1 |
| Sumatra | 11 |
| Barren Is. | 1 |

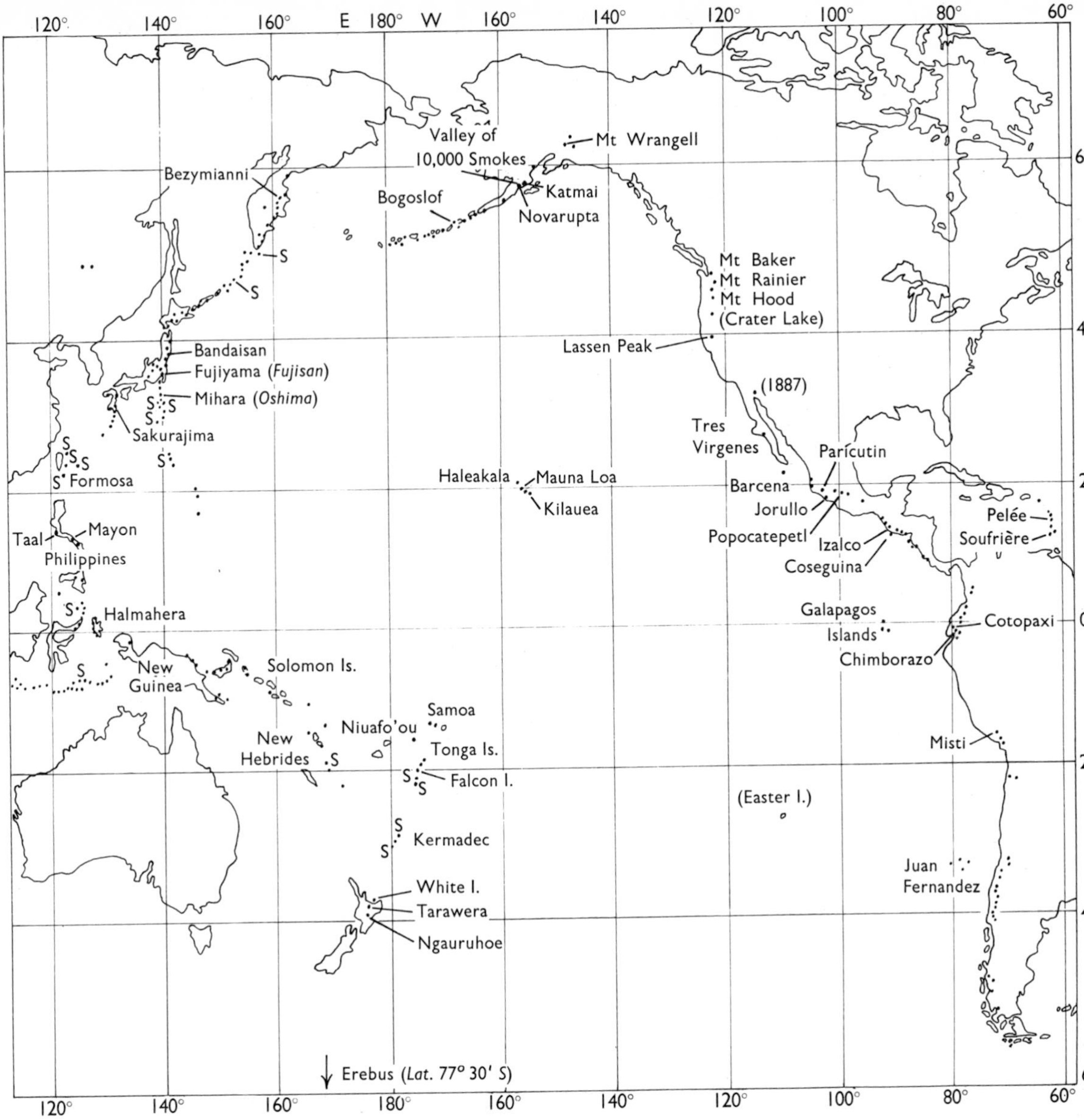


chemical composition and in their mineralogy (pp. 654–5). Holmes and Harwood (1937) long ago recorded evidence that the potash-rich rocks of the Western Rift are formed by transfusion of pre-existing rocks such as augite peridotite. Transfusion implies introduction of some constituents and removal of others, the process sometimes culminating in fusion. A simple example of transfusion described by Holmes (1936) is the transfusion of quartz to obsidian-like glass in Uganda lavas. Briefly stated, inclusions of vein quartz contain networks of veins of colourless glass. Three samples of the glass, separated from quartz were analysed microchemically by Hecht of Vienna. The results showed it to be obsidian with silica ranging from 74·51 to 79·99 per cent, and potash ranging from 6·16 to 5·53 per cent respectively. The glass is neither fused silica alone, nor a solution of silica in the material of the enclosing lava. Materials have been introduced into the quartz ($SiO_2$), chiefly alumina, potash and water, although they may not have migrated as oxides. Moreover, they have been introduced in proportions completely different from those in which they could have occurred in the enclosing lava.

The boundary between the basalts of the oceanic crust and islands, and the andesite-dacite-rhyolite volcanic rocks of the Circum-Pacific belt, is called the *Andesite Line* (see Figure 26.10), and it turns out to be essentially the boundary between the continental crust with sial and the oceanic crust without sial, and to follow the ocean trenches. The only exceptions to this generalization are due to the occurrence of a few isolated shreds and patches of sial left stranded in the Pacific floor, e.g. Macquarie Island and possibly Easter Island, and some ocean crust on the continental side of trenches. The Andesite Line around the Pacific, and its continuation around the Indonesian arc is, as Lester King puts it, 'one of the fundamental geological boundaries of the earth'. It has turned out to be the line where oceanic lithosphere is subducted along Benioff zones.

Before the advent of plate tectonics it seemed to be well established that most andesites and dacites of the Circum-Pacific belt, were hybrids, resulting from mixing and reaction between basaltic magma and sialic crust. At the present time this calc-alkaline suite is explained in terms of subducting lithosphere. Kuno showed, for example, that there is a chemical change from abundant tholeiites on the eastern side of the Japanese islands, through alumina-rich basalts to alkali-rich basalts in the Sea of Japan on the west. The basalts thus become richer in alkalis, particularly potash, as their distance above the Benioff zone increases. This observation has since been confirmed by others in relation to Benioff zones (Figure 29.37).

Experimental petrologists usually consider the andesite–dacite–rhyolite suite to be derived by partial melting either of pyrolite (a pyroxene-olivine rock) in the mantle overlying a Benioff zone, or of the tholeiite of the sinking lithosphere itself, or some derivative from this source such as amphibolite or eclogite. Ringwood (1974) has reviewed the present position in the light of experimental investigations. He considers that water, derived by dehydration of amphibolite formed from oceanic tholeiite in subduction zones, rises into the overlying mantle and causes a decrease in viscosity. As a result, he suggests that pyrolite flows upwards from a site adjacent to the subducting oceanic crust (amphibolite). Partial melting of the rising pyrolite, in the presence of high water-vapour pressure, he regards as the origin of hydrous tholeiite magma. At a later stage

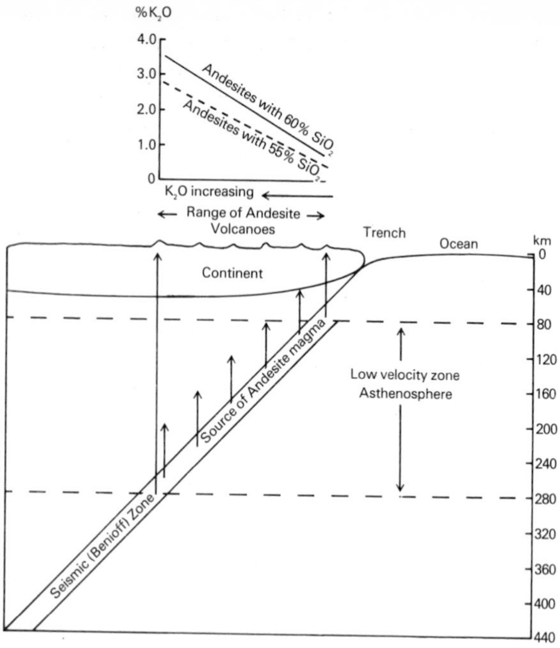


**Figure 29.37** Diagrammatic representation of subducting oceanic lithosphere illustrating a possible source of andesite magma. The graph above illustrates the increase of potash within andesites as their distance away from the trench and above the Benioff zone increases. (*Adapted from T. Hatherton and W. R. Dickinson, 1969*)

he considers that partial melting of subducting amphibolite, and at a deeper level of quartz eclogite (both higher pressure representatives of tholeiite), result in the evolution of the andesite–dacite–rhyolite suite.

Ringwood emphasizes that, residual materials, after differential melting of subducted lithosphere, can never again be the source of basaltic magma. If the present rate of sea-floor spreading has been maintained for the past 3·5 million years, he concludes that 30 to 60 per cent of the mantle is already depleted in andesite–rhyolite forming materials, and can never again be the source of basalt, such as is erupted at mid-ocean ridges, and forms new ocean floor. It therefore seems, if the andesite–dacite–rhyolite suite is derived as he suggests, that the earth is slowly becoming inert.

Gilluly has very different ideas. He suggests that the andesite–dacite–rhyolite suite is derived by partial melting of oceanic tholeiite (or amphibolite or eclogite) admixed with ocean-floor sediment such as sandstone, chert and radiolarian ooze. Gilluly argues that if the hypothesis that the American plate is overriding oceanic plates on its western side is correct, then, since no sediments older than Jurassic remain on the ocean floor, the American plate must have overridden 1500 km of oceanic crust. Furthermore, if the mid-Atlantic ridge is really migrating westward (p. 646), then the overridden part of the oceanic crust must be increased to 3000 km. Since the oceanic crust consists of a thickness of about 5 or 6 km of tholeiite, and a thickness of about 1 km of sediment, Gilluly calculates that between 7500 and 15,000 $km^3$ of tholeiite, and about 1500 to 3000 $km^3$ of sediment, for each kilometre of plate margin, have already disappeared beneath the western side of the American plate. Gilluly concludes that melting of tholeiite, or of its high pressure derivatives amphibolite or eclogite, in the subducting lithosphere, together with a little of the overlying sediment, may well be the source of the voluminous andesites and more acid rocks of the Circum-Pacific belt.

Gilluly considers the following criticisms that might be raised against his hypothesis but strongly holds the view that the missing sediment and basalt must be considered when discussing magma generation. Firstly, it is usually argued that sediment is too light to sink down subduction zones, and that rather than descend, the surface sediment on subducting plates is plastered on the landward walls of trenches. As yet, however, this is no more than hypothesis. Another argument that might be raised against Gilluly's suggestion, is based on the $^{87}Sr/^{86}Sr$ ratios of the rocks concerned. The subject has been reviewed by Faure and Powell (1972). The mean $^{87}Sr/^{86}Sr$ ratio of the andesite suite of the Marianas, New Britain, and Central America is $0{\cdot}7037 \pm 0{\cdot}0003$, which is identical with the average for basalts from oceanic islands, and the strontium ratio appears to be lower in oceanic crust. On the other hand the $^{87}Sr/^{86}Sr$ ratios of andesites from New Zealand average about 0·7055, a figure high enough to represent some contamination of tholeiite by sediment.

Besides separating regions with profound structural and petrological differences, the Andesite Line delineates another contrast of great significance. The pyroclasts, including ignimbrites, erupted from the andesite–dacite–rhyolite volcanoes on the island arc or continental side of the Andesite Line are immensely more voluminous than the lava flows, whereas on the oceanic side the proportion of pyroclasts is small. The percentage of fragmental material in the total material erupted is called the 'explosion index' by Rittmann and referred to as $E$. For the Circum-Pacific volcanoes $E$ is generally between 80 and 95, rising to 99 in Central America, a maximum which is shared by the Indonesian volcanoes. In contrast, $E$ rarely exceeds 3 in the oceanic island volcanoes of the Pacific. Evidently the volcanoes on the continental side of the Andesite Line are supplied with vast quantities of high-pressure gases from a source which—apart from superficial ground-water—is not available to the oceanic volcanoes. Since the gases are mainly $H_2O$ this is a paradoxical situation that calls for special attention.

## Sea Water and Volcanic Activity

In the early days of geology, while the interiors of the continents still remained largely unexplored, it was not unnaturally assumed that all volcanoes occur near the sea. This apparent rule suggested the idea that sea water infiltrated into the heated depths and so caused eruptions by being converted into steam. Later, however, the discovery of great spreads of Jurassic and Tertiary plateau basalts showed that continental regions far from the seas and oceans of the time had been flooded with lavas on a prodigious scale. The 'rule' was even more effectively broken when volcanoes were found to be active in Central Africa and Mongolia. As an all-embracing generalization the old idea had to be abandoned. But its application to the highly explosive eruptions of the Circum-Pacific volcanoes has been revived. These volcanoes, situated on the continental side of trenches, above downward moving oceanic lithosphere, lie above Benioff zones. If Gilluly is correct in thinking that ocean-floor sediments are subducted, they will transport ocean water in their pore spaces. As the oceanic lithosphere advances under the continents, one would expect the top layer of loose sediment saturated with sea water, to be compressed and the salt water to be squeezed out, i.e. to be returned to the ocean. There is a limit to this process, however. Loose sand with a porosity of 35 per cent can be compacted during burial until, when the pressure corresponds to an overhead load of about 2 km of water and 12 km of sediment, the porosity is reduced to about 25 per cent. With further increase of pressure the porosity remains about the same until metamorphism sets in. The reduction of porosity in the case of clay sediments may be considerably greater when the materials are dry, but not when they are saturated with water which cannot easily escape.

Volcanic emanations, apart from known contamination by air and surface waters have generally been thought to be largely of juvenile origin. Probably some of them are: notably nitrogen, carbon monoxide and carbon dioxide amongst the major constituents. But judging by their content of 'heavy' water (deuterium water, $D_2O$) samples of water from volcanic sources turn out to have the hall-marks of deep-seated connate waters. In percolating through fine pores $D_2O$ does not get through so easily as $H_2O$. Thus surface water that has been down to great depths becomes impoverished in D relative to H, i.e. the isotopic ratio D/H is appreciably reduced. Similarly, water vapour evaporated from the surface of the sea is enriched in 'light' $H_2O$ at the expense of 'heavy' $D_2O$, i.e. D/H is lower in rain and rivers than in sea water. The following values are representative:

*Isotopic ratios of D/H in various types of natural waters*

| | |
|---|---|
| Ocean water | 0·0155 |
| Rain water, Bermuda | 0·0138 |
| River water, Mississippi | 0·0146 |
| Spring water, near Parícutin, Mexico | 0·0144 |
| Fumarole water, Parícutin Volcano | 0·0147 |

For juvenile water the D/H value would be expected to be about the same as for ocean water, or perhaps a little higher. All the volcanic condensates that have been tested have markedly lower values, most of them much lower than the Parícutin sample. This unexpected result aroused new interest in an all-but-forgotten hypothesis proposed in 1903 by Svante Arrhenius (1859–1927). The mechanism suggested by this stimulating Swedish scientist was designed to show how water could soak into heated rocks, and form part of a silicate solution or a hydrous melt against a head of pressure that might have been expected to hold it back. The principle involved is that if a melt contains less water in solution than the maximum amount it can dissolve under the given conditions of temperature and pressure, it will absorb water until it is saturated. One effect of this can sometimes be seen when highly mobile basic lavas flow over a cliff into the sea (e.g. from Mauna Loa) or into a lake (e.g. from Nyamuragira into Lake Kivu) without the expected production of obscuring clouds of steam. Here we are concerned with the same phenomenon in reverse, i.e. where water is soaking down, or being carried down, to depths where lava is forming.

Arrhenius pointed out that with rise of temperature an increasing proportion of the water in buried sediments would be dissociated into its ions $H^+$ and $(OH)^-$, so that it became a strong solvent, capable of dissolving silica and silicates, and of facilitating reactions such as the transformation of clay minerals into feldspars. In certain conditions the process would go no farther than metamorphism and granitization, but in others the hydrothermal solutions would become increasingly concentrated and eventually form a water-rich silicate melt.

If the interstitial solutions or melts could not escape from the place of their formation, equilibrium would sooner or later be established and further inflow of water would cease. But if they were removed from the place of their formation, e.g. by suction into low-pressure fissures, fresh supplies would occupy the space vacated and the process would continue. Such removal would be inevitable whenever the enclosing rock was fractured, and fractures are well attested by the occurrence of intermediate and deep earthquakes. At the moment of fracture pressure suddenly becomes negligible between the walls of the newly formed crack, which thus provides a vacuum-like passageway for the escape of all mobile interstitial material from its immediate neighbourhood. At least some of the cracks are directed upwards and these constitute the 'roots' of the feeding channels of volcanoes. Invoking the analogy with glaciers and their marginal crevasses (Figure 9.4) we find part of the explanation for the fact that the active volcanoes of the Circum-Pacific and Indonesian belts appear, not on the bottom of the ocean trenches, but at a considerable distance on the continental side.

Meanwhile, it is possible to show that, however little we know of the processes concerned, most of the water emitted during volcanic eruptions has come from the sea. A. B. Ronov and his team have measured the amount of $CO_2$ now locked up in the sediments and volcanic rocks formed during the 265 m.y. between the beginning of the Devonian (400 m.y. ago) to the end of the Jurassic (135 m.y. ago). This task involved many years' work and was completed in 1959. They found (*a*) that the quantity of carbonate sediments increases when the sea transgresses over the land, and decreases when it withdraws; and (*b*) that the volume of carbonates deposited during each epoch is proportional to the volume of volcanic rocks erupted during the same time. Evidently the carbon dioxide fixed in sediments, as well as that retained by the lavas, is supplied by volcanic activity. Over the whole period of 265 m.y. the carbon dioxide traced in this way amounts to $9{\cdot}5 \times 10^{15}$ tonnes. On average the proportion of water in volcanic gases is 65 times that of carbon dioxide. Adopting this estimate for the Devonian–Jurassic interval, the water liberated from volcanoes would be $6 \times 10^{17}$ tonnes, i.e. $0{\cdot}6 \times 10^{24}$g in 265 m.y. This should be a minimum estimate, because no account has been taken of the carbon concentrated in fuels and dispersed through sediments, nearly all of which came from carbon dioxide reduced by plant life. But the total mass of the seas and oceans is only $1{\cdot}4 \times 10^{24}$g (p. 15). Even if the carbon dioxide statistics were considerably in error they would still imply that volcanoes could have provided all the water of the seas and oceans during the last few hundred million years. Since it is impossible to suppose that oceans did not exist before the Upper Palaeozoic, it follows that the ocean waters must themselves have been re-cycled, that is, used over and over again during geological time.

The same conclusion emerges from André Rivière's estimate that the average annual output of water and other volatiles from volcanoes, fumaroles and hydrothermal sources is of an order corresponding to 16 $km^3$ of water. If we assume 90 per cent of this to be actually water, i.e. $1{\cdot}4 \times 10^{16}$g, the whole volume of the seas and oceans could be supplied in $1{\cdot}4 \times 10^{24}/1{\cdot}4 \times 10^{16}$ years, i.e. in 100 million years. Obviously very little of the annual supply of 'volcanic' water can be juvenile, and any hypothesis based on the assumption that much of it *is* juvenile cannot be entertained. The supply of water that appears to be income is mostly capital in a continuous round of circulation.

## Selected References


ALLAN, T. D., and MORELLI, C., 1969, 'The Red Sea', in *The Sea, Ideas and Observations on Progress in the Study of The Seas*, edited by A. E. Maxwell., vol. 4, Pt. II, 1970.

COX, A., 1973, *Plate Tectonics and Geomagnetic Reversals*, Readings with Introductions by Allan Cox, W. H. Freeman, San Francisco.

DARRACOTT, B. W., FAIRHEAD, J. D., GIRDLER, R. W., and HALL, S. A., 1973, 'The East African Rift System', in *Implications of Continental Drift to the Earth Sciences*, vol. 2, edited by D. H. Tarling and S. K. Runcorn, Academic Press, London and New York, pp. 757–66.

DRAKE, C. E., and GIRDLER, R. W., 1964, 'A geophysical study of the Red Sea', *The Geophysical Journal of the Royal Astronomical Society*, vol. 8, pp. 473–95.

GILLULY, J., 1971, 'Plate tectonics and magmatic evolution', *Bulletin of the Geological Society of America*, vol. 82, pp. 2382–96.

HOLMES, A., 1928–9, 'Radioactivity and earth movements', *Transactions of the Geological Society of Glasgow*, vol. 18, Pt. 3, pp. 559–606.

— 1936, 'Transfusion of quartz xenoliths in alkali basic and ultrabasic lavas, South-West Uganda', *Mineralogical Magazine*, vol. 24, pp. 408–21.

HOLMES, A., and HARWOOD, H. F., 1937, *The Volcanic Area of Bufumbira'*, Pt. 2, *The Petrology of the Volcanic Field of Bufumbira, South-West Uganda, and of other parts of the Birunga Field*, Geological Survey of Uganda, Memoir No. III.

KING, B. C., 1949, *The Napak Area of Southern Karamoja, Uganda: A Study of a Dissected Late Tertiary Volcano*, Geological Survey of Uganda.

KING, B. C., LE BAS, M. J., and SUTHERLAND, D. S., 1972, 'The history of the alkaline volcanoes and intrusive complexes of Eastern Uganda and Western Kenya', *Journal of the Geological Society*, vol. 128, pp. 173–205.

LE PICHON, X., 1968, 'Sea-Floor spreading and continental drift', *Journal of Geophysical Research*, vol. 73, pp. 3661–97.

LE PICHON, X., FRANCHETEAU, J., and BONNIN, J., 1973, *Plate Tectonics*, Elsevier, Amsterdam.

LOGATCHEV, N. A., BELOUSSOV, V. V., and MILANOVSKY, E. E., 1972, 'East African rift development', *Tectonophysics*, vol. 15, pp. 71–81.

MCKENZIE, D. P., and PARKER, R. L., 1967, 'The North Pacific: an example of tectonics on a sphere', *Nature*, vol. 216, pp. 1276–80.

MOORE, D. G., 1968, 'Transform faulting and growth of the Gulf of California since the Late Pliocene', *Science*, vol. 161, pp. 1238–41.

MORGAN, W. J., 1968, 'Rises, trenches, great faults and crustal blocks', *Journal of Geophysical Research*, vol. 73, No. 6, pp. 1959–82.

RINGWOOD, A. E., 1974, 'The petrological evolution of Island Arc systems', Twenty-seventh William Smith Lecture. *Journal of the Geological Society*, vol. 130, pp. 183–204.

WYLLIE, P. J., 1971, *The Dynamic Earth*, Wiley, New York.

# Chapter 30

# Orogenic Belts: The Evolution of Fold Mountains

The question what the motive forces are is the crowning question. . . . First we must know the existing forms exactly: morphology is always the beginning. Next we must know the time sequence of the forms, in the order of their origin.

*G. F. von Weizacker* (1951)

## The Nature of Orogenic Belts

As already indicated on p. 104, belts of the earth's crust where the tectonic structures—folds, faults, overfolds and thrusts—indicate the existence of fold-mountains in past geological ages, whether or not they are now morphologically mountains, plateaus, or plains, are called orogenic belts (Greek, *oros* = mountain), a term introduced by Gilbert, a famous Canadian geologist in 1890. Most of the great continental mountain ranges of today, amongst which are the Alps and the Himalayas, came into existence during past orogenic periods from the Jurassic to the present. From

**Figure 30.1** Part of the 1000–km long Alaska Range, looking SW, showing the exceptional height reached by Mt McKinley (6187 m) which, like Mt Blanc in the Alps, occurs in the middle of the great bend of the Range, where the latter is 100 km wide. The mountains in the foreground and to the right beyond the river range from 1500 to 2500 metres and consist of Triassic and Cretaceous lavas, shales and greywackes. Tertiary sands and gravels, tightly folded, are truncated by the 600-metres plain seen in the distance. (*Bradford Washburn*)

the foothills to the axes of the main ranges structural complexities and metamorphism tend to become increasingly intense, until the climax is reached in the great masses of granitic rocks which form the median cores of the ranges. An astonishing fact is that these granitic rocks, deeply buried at the time of crystallization, now commonly form some of the highest peaks. The explanation is that whilst the higher parts of rising orogenic belts are being removed by denudation or gravity gliding, the solid granitic cores continue to rise until they are exposed first in the flanking valleys, and eventually at the summits themselves. Everest and some of its Himalayan neighbours (Figures 2.5 and 19.35), and Mt. McKinley in the Alaska Range (Figure 30.1) are celebrated examples of this culminating feature of mountain building.

Some of the structures characteristic of fold mountains have already been illustrated. These include alternations of more or less open anticlines and synclines (Figures 9.12 and 9.13); tightly packed isoclinal folds (Figures 9.18 and 23.28); recumbent folds (Figure 9.9); and thrusts (Figures 9.32 and 9.33). The crustal block towards or over which the structures splay is called the *foreland*, and the movement is characteristically directed away from an adjacent ocean. The axes of the major structures of a range of fold mountains (e.g. those that can be shown on small-scale maps) are generally roughly parallel to the trend of the ranges, but detailed mapping of smaller scale structures (depicted on large-scale maps) commonly reveals wide departures from the dominant strike. This complexity results from the fact that the structures of an orogenic belt are not produced by a single deformation, but represent the accumulated effects of a long sequence of tectonic events. In particular the main trend of orogenic belts is characteristically intersected by cross-folds which strike at high angles to the main trend, the two directions not uncommonly striking at right angles to one another. For example, in the Highlands of Scotland, the main folds of the Caledonian orogenic belt strike north-eastward, and are intersected by cross-folds striking roughly north-westward. In the Swiss Alps the main folds of the Alpine orogeny strike north-eastward, and are intersected by cross-folds striking approximately north-westward.

Movements on the main and cross-fold directions may alternate, and folds with axes parallel to either direction may become folded about axes striking in the other direction. Such folded folds may give rise to complex outcrop-patterns, which, as illustrated in Figure 30.2 may be pronged or annular, and in either case there is a rough bilateral symmetry. The direction of the strike of the older fold-axis is parallel to a line joining the tips of the prongs, whilst the second fold-axis is parallel to the axis of bilateral symmetry. Figures

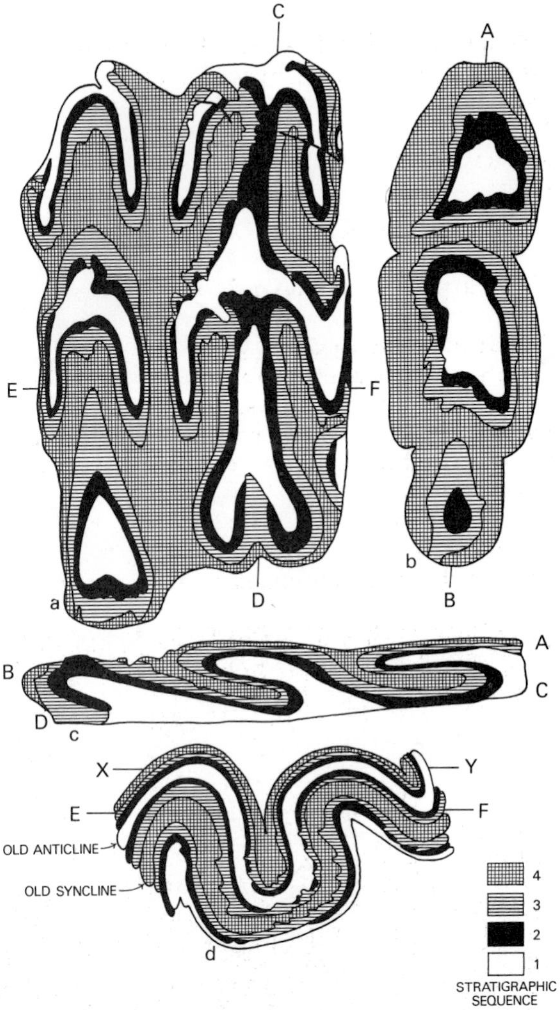


**Figure 30.2** A plasticene model of folded folds showing the resulting pronged and concentric outcrop patterns. E–F is the axial direction of the first folding, and C–D (or A–B that of the second folding.
(a) and (b) Surface outcrops after slicing the model horizontally to represent denudation, (a) at a lower level than (b)
(c) A section of the first folds at right angles to their fold axis E–F before slicing off the top of the model. A–B and C–D represent the levels at which (b) and (a) were respectively cut
(d) Section of the second folds at right angles to their fold axis A–B (or C–D), before slicing off the top of the model. X–Y and E–F represent the levels at which (b) and (a) were respectively sliced. (*D. Reynolds and A. Holmes, 1954*).

7.13 and 30.3 indicate the position and main trends of the orogenic belts of Europe and America, and Figure 30.4 illustrates the exposed parts of the Hercynian orogenic belts of Europe.

## Geosynclines

Many orogenic belts, for example the Caledonian of Britain, Scandinavia and Greenland, and the

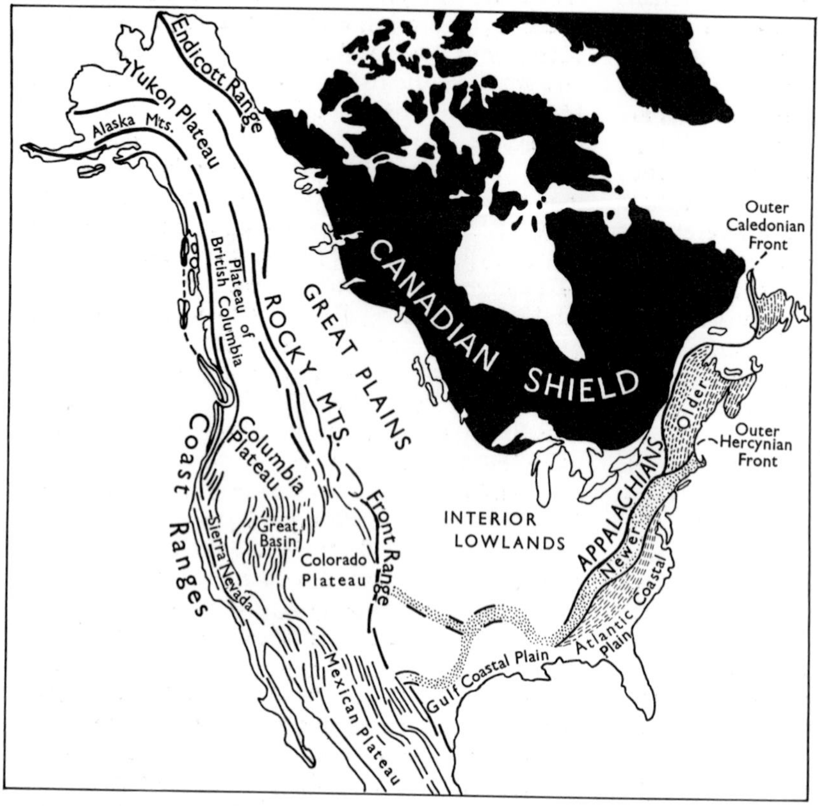


**Figure 30.3** Tectonic map of North America

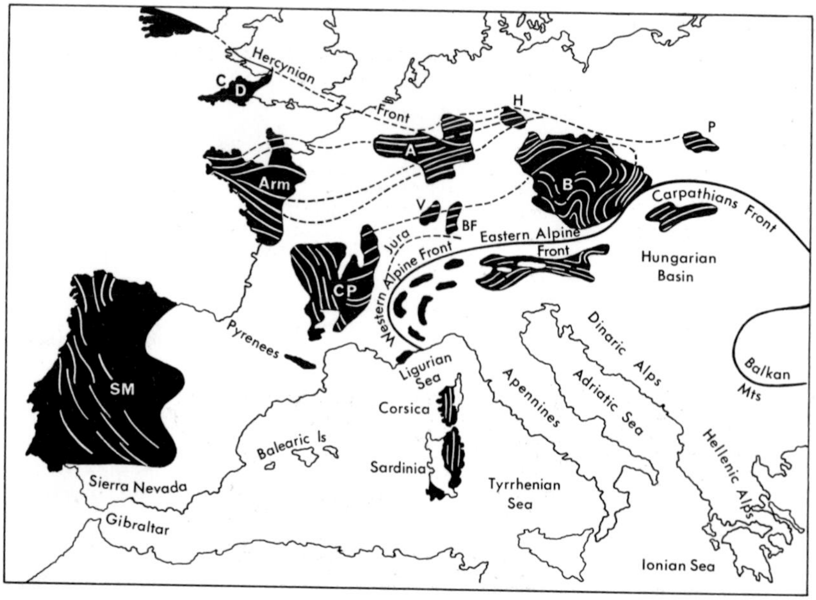


**Figure 30.4** Hercynian (Variscan) massifs of Europe, showing dominant trend lines of late Palaeozoic folding. A, Ardennes; Arm, *Armorica*; B, Bohemia, western province of Czechoslovakia; BF, Black Forest; CD, Cornwall and Devon (*Cornubia*); CP, Central Plateau of France (Massif Central); H, Harz Mts (*Hercynia*); P, Polish Massif; SM, Spanish Meseta; V, Vosges

Tertiary Alpine orogenic belt, began as elongated troughs known as geosynclines, in which the accumulated sediments were abnormally thick compared with those of the same age on the adjacent platform or shield. Not long ago it was thought that all orogenic belts began in this way, but recognition that young fold-belts, like those of the island arcs, rise in relation to subduction zones has cast doubt on the existence of geosynclines as precursors to orogeny in every example.

The geosynclinal concept was born in the Appalachians. In 1859, James Hall, as a result of his work on the stratigraphy and structure of the northern Appalachians, recorded his discovery that the folded Palaeozoic sediments of the mountain ranges are shallow-water marine types which reach a thickness of up to 12 km, and are ten to twenty times as thick as the unfolded strata of corresponding ages in the interior lowlands to the west.

The accumulation of so thick a sequence of sandstones, shales and limestones clearly implies that the underlying floor of older rocks must have subsided by a like amount. The mountains were evidently preceded by long periods of downwarping during which sedimentation more or less kept pace with depression of the crust. Such elongated belts of long-continued subsidence and sedimentation were called *geosynclinals* by Dana in 1873. Later the modified form *geosyncline* came into use.

Following a classification devised by Stille, geosynclines characterized by intermittent volcanic activity during their infilling are commonly referred to as *eugeosynclines*, whilst those with a paucity of volcanic products are distinguished as *miogeosynclines*. The Greek prefixes *eu* and *mio* are meant to indicate relatively high and low status from an igneous or mobility point of view. As exemplified by Figure 30.5 the two classes tend to occur side by side, the miogeosyncline lying close to the stable foreland, and the eugeosyncline further away.

Attempts have been made to correlate certain assemblages of sediments with each of the two classes of geosynclines. Typical sediments of the 'eu' class are dark shales and poorly sorted sandstones (graywackes), often exhibiting graded bedding, and interpreted as deposits from turbidity currents. Sediments of the 'mio' class characteristically include well-sorted sandstones, often seen as white- or pale-tinted quartzites; and limestones, some of which may be dolomitic. Two other terms should be introduced here, both of which originated in Switzerland, but are now of world-wide use, although not always in the original sense. Trümpy (1960) has given a clear description of the meaning attached to these terms in Switzerland. Briefly stated *flysch* is a marine deposit, of geosynclinal type, deposited within separate marine troughs, particularly in the Helvetian area, and consisting of alternations of sandstone and micaceous shale. It differs, however, from the older 'eu' and 'mio' geosynclinal deposits in that it commonly shows the influence of tectonic interference. Characteristically, deposition of flysch began with the early fold-movements but ended before the main orogenic phases. *Molasse*, on the other hand, ranging in age from Oligocene to Miocene, is mainly continental, having been deposited within marginal troughs or intermontane basins. It consists of soft sandstone cemented with calcareous or argillaceous material. Unlike the flysch, the molasse was deposited during and after the main tectonic movements.

The volcanic rocks interbedded with eugeosynclinal sediments are characteristically submarine lavas of the spilite suite. Spilites are basic lavas in which minerals like serpentine and chlorite pre-

**Figure 30.5** Restored section of the Cambrian and Ordovician strata and volcanic rocks, as they were before the Taconic orogenic phase of the Older Appalachians of New England; illustrating the division of the Appalachian geosyncline from the foreland on the west of the 'borderland' (now the Atlantic) on the east. Length of section 425 km (*After M. Kay, 1951*)

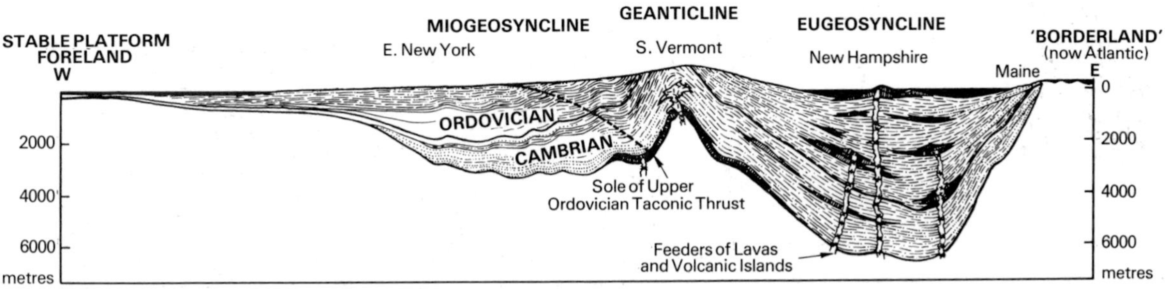

dominate instead of olivine and pyroxene, together with albite and epidote instead of calcic plagioclase. Since early days of geology these rocks have been called *spilite* (Gr. *spilos*, spot or blemish; *spilotos*, spoiled). Associated with spilites are *keratophyres*, more acid lavas and intrusives, again rich in soda.

A stumbling block over the geosyncline concept has been that modern examples have seemed to be non-existent. In 1963, however, Dietz suggested that the sediments of present-day continental shelves may be equated with miogeosynclines, the thick prisms of sediment piled up against continental shelves, and forming the continental rises, representing eugeosynclines. Comparison of Figure 30.6 with Figure 30.5 shows how this newer concept compares with the older idea of a miogeosyncline and eugeosyncline couple.

The thick prisms of sediment piled up against the continental slope, and overlying oceanic floor, are largely turbidites, which Pettijohn has equated with graywackes characteristic of eugeosynclines. Furthermore, Dietz points out that the sliding and slumping commonly shown by graywackes would be readily initiated on the continental rise by any disturbance, such as earthquakes. He accounts for the spilite-keratophyre suite, associated with eugeosynclinal sediments as a consequence of the metasomatic enrichment of basaltic lavas in soda from sea water; keratophyres resulting from their contamination with sial.

## The Uplift of Orogenic Belts

Long ago Dana realized that Hall's discovery of geosynclines, momentous as it was, failed to account for the folding and uplift of the mountains. His sense of humour led him to declare that Hall had promoted 'a theory for the origin of mountains with the origin of mountains left out'. Tertiary orogenic belts are characteristically located along continental margins that form the leading edges of plates, and they are commonly attributed to lateral compression due to the collision of plates. A commonly cited example is the association of the Andes in South America, with the Peru–Chile trench, where oceanic lithosphere plunges downward. Gansser (1973) after summarizing many facts about the Andes casts doubt on the supposition that the Andes originated as a result of a collision. Only for pre-Mesozoic times is there evidence of compression along the interface between continent and ocean, and the tectonic strike is then not parallel to the continental margin. Since the Mesozoic, the coastal belt has been characterized by block faulting, evidencing uplift and extension, possibly related to the emplacement of the enormous volume of granitic plutonic rocks. These plutons, which extend parallel to the coast are still rising. Conformable with current interpretations, however, are the facts that the epicentres of earthquakes become deeper from west to east, indicating the landward slope of the downward moving oceanic lithosphere along the Peru–Chile trench. Furthermore, the recent volcanic belt strikes parallel to the oceanic trench and lies inland, approximately 250 km east of the trench.

Dewey and Bird (1970) consider an early event of orogenesis to be the stripping of slices of oceanic crust and mantle from the downward moving plate of oceanic lithosphere, and its mechanical injection, as ophiolite, into a mélange of sediment and blue schist on the landward side of the trench, in advance of the plate with an island arc or continent at its leading edge.

The term *ophiolite* (from the Greek, *ophis*, a

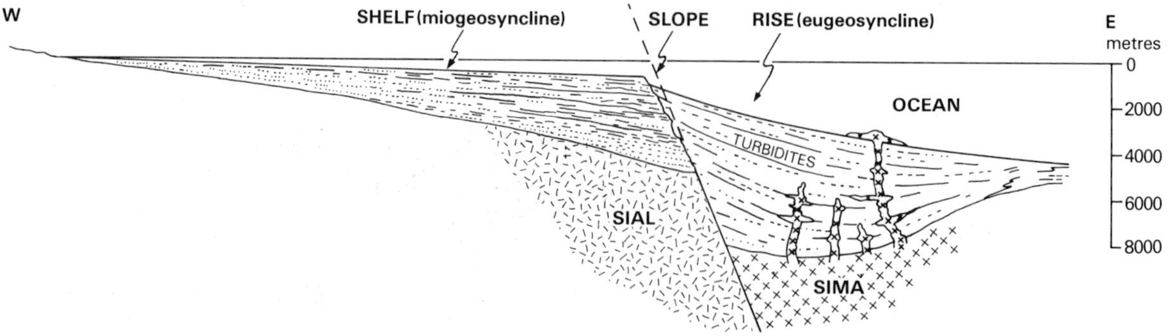


**Figure 30.6** An actualistic reinterpretation by Dietz (1963) of the miogeosyncline illustrated in Fig. 30.5. Dietz' interpretation is made to conform with conditions thought to exist at the present day off the east coast of the U.S.A.

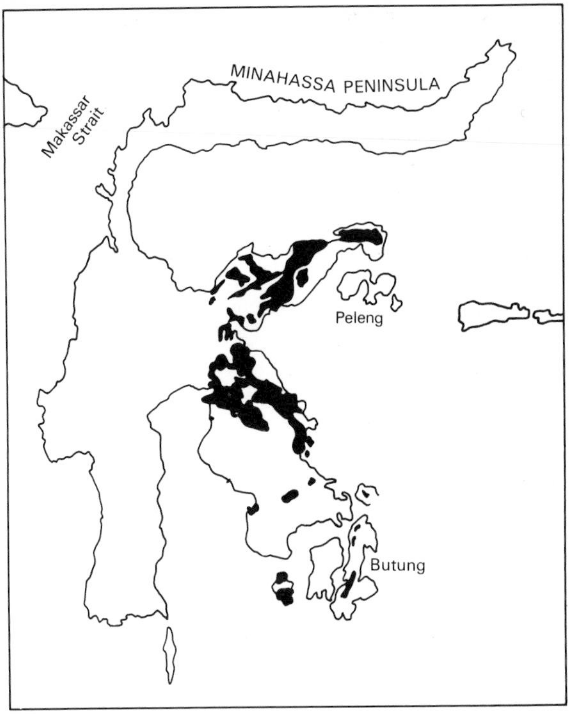


**Figure 30.7** The distribution of ophiolites (black) in the island of Sulawesi (Celebes), Indonesia (*After E. Künig, 1956*)

serpent) was introduced by Steinmann, who drew attention in 1905 to the common association of spilites, serpentinites and chert in mountain systems of Alpine type. This association, together with ultrabasic rocks which have caused no contact metamorphism, are known, for example, from Cyprus (Figures 29.7 and 29.8), Turkey, New Caledonia and Kurdistan.

A particularly instructive occurrence of ophiolites is that in the eastern part of the island of Sulawesi (Celebes) in Indonesia (Figure 30.7). The ophiolites, consisting principally of serpentinites, peridotites, gabbros, diabases and rare volcanic rocks, possibly including pillow lavas, were described by Kündig in 1956. They appear to be of Upper Cretaceous age, have no associated metamorphic aureoles, and are associated with Jurassic and Cretaceous limestones of deep sea origin, characterized by globigerina. The limestones occur in crushed slices, alternating with ophiolites in imbricate style (Figure 30.8). The whole region has been recently uplifted, for modern barrier reefs, when traced towards the north-east from Tomori Bay, continue along the coast as raised coral platforms which eventually reach a height of 500 m. It is important to notice that in Sulawesi (Celebes) the thrust slices of ophiolites are directed south-eastward towards the sea, and lie above the deep earthquake zone associated with oceanic lithosphere moving downward at the Indonesian trench (see Figure 26.14), and about 600 km away from the trench.

Another example from Indonesia showing that early thrust movements towards the trench have occurred on the island arc side is clearly displayed in a seismic profile extending across the Indonesian trench, the Island of Bali, the Java Sea, western Borneo and Sarawak (Beck, 1972). As illustrated in the sketch (Figure 30.9) the oceanic lithosphere slopes gently beneath the island arc from the trench. On the island arc side of the trench, a mass of sediments has been thrust towards the trench in imbricate style. This seismic profile clearly shows that it is the island arc margining the continental plate, that is being thrust over the downward moving lithosphere of the oceanic plate.

The same phenomenon is illustrated by Figure 30.10, depicting the results of a seismic survey of the Aleutian trench, by Grow (1972), in which the instruments were towed 100 m above the ocean floor. That the descending oceanic lithosphere is in a state of tension is indicated by the presence of

**Figure 30.8** Section across an alternation of ophiolites and deepsea Cretaceous limestone thrust, in imbricate style, towards the sea in the eastern arms of Sulawesi (Celebes), Indonesia (compare Fig. 26.14) (*After E. Künig, 1956*)

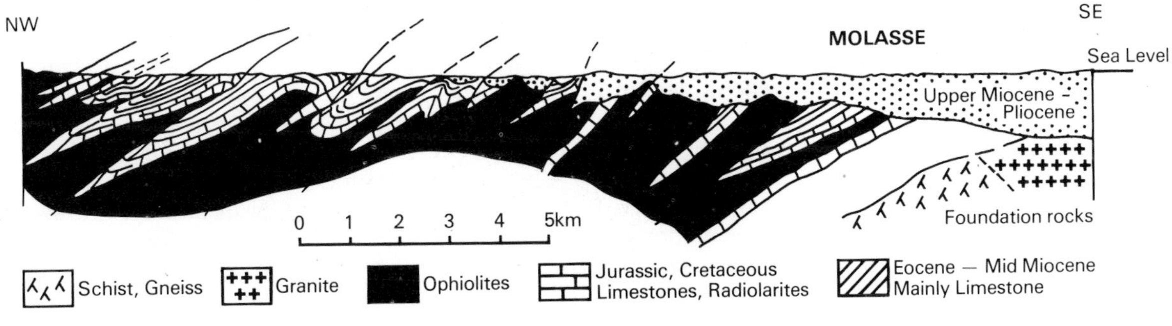

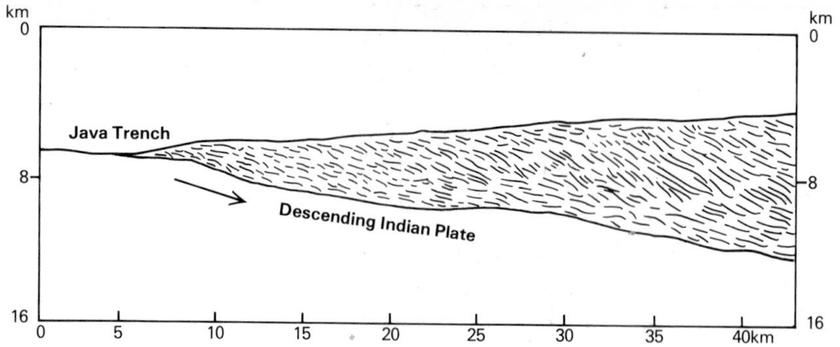


**Figure 30.9** Sketch made from a seismic profile by R. H. Beck, 1972, across the Java trench. The profile illustrates sediments thrust in imbricate style from the island side (on the right) over the subducting oceanic lithosphere.

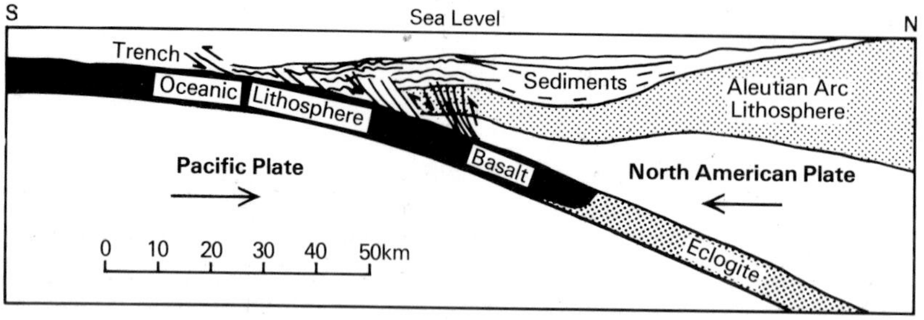


**Figure 30.10** An interpretive section across the Central Aleutian trench, with no vertical exaggeration, showing the overriding of the Pacific Plate by the North American Plate (*After J. A. Grow, 1972*)

normal tension faults. On the other hand, the Aleutian arc, margined by a wedge of sediments, is being thrust, together with the sediments, over the sinking oceanic lithosphere. Here, then, is a third example in which an island arc, on the continental side, is being thrust over oceanic lithosphere, moving downward at an oceanic trench.

The following facts appear to be opposed to lateral compression being the cause of orogenesis:

(*a*) Downward turning oceanic lithosphere, at oceanic trenches, is in a state of tension, as evidenced by the presence of normal faults, downthrowing towards the trenches.

(*b*) Such evidence as there is indicates that plates with continental lithosphere or island arcs as the leading edge are riding over plates with oceanic lithosphere as the leading edge, without actual collision

(*c*) Early thrusting when plates meet, is in the opposite direction (oceanwards) to that of the main orogenic movements, which are directed towards continents.

(*d*) Sediments deposited within trenches are not folded (Figure 24.6).

The actual orogenic belt most commonly quoted in support of an origin dependent on lateral compression, resulting from collision between plates, is that of the Swiss Alps. This is no doubt because far-travelled overfolds (nappes) can be reproduced in illustration from the writings of long ago by Heim and Argand, during what might be called the age of great syntheses. It seems therefore appropriate here to consider some of the investigations that have been made by Alpine geologists since 1925. This is the date of Argand's stupendous work, *La Tectonique de l'Asie*—now out of date, and some of the regions now better known, but still so exciting to read.

## The Discovery of Nappes

In Switzerland, the discovery that the Mesozoic and Tertiary rocks of the Helvetian zone, the High Calcareous Alps (Figure 30.11), form nappes (sheets) or recumbent folds began in 1841 with Escher von der Linth's observation, in the Canton Glarus, that Mesozoic rocks lie horizontally upon the upturned edges of beds of Tertiary Flysch, the Mesozoic rocks being in turn overlain by crystalline schists. A few years later Escher reached the conclusion that the inversion and arched form of these rocks could best be explained as a double fold, in which two recumbent folds were directed towards one another (Figure 30.12(*a*)). It fell to Albert Heim, a student of Escher's, to make the supposed double fold of Glarus famous, however, by describing it in his *Mechanismus der Gebirgsbildung* (1878), which he dedicated to Escher von der Linth. Escher himself refused to publish because,

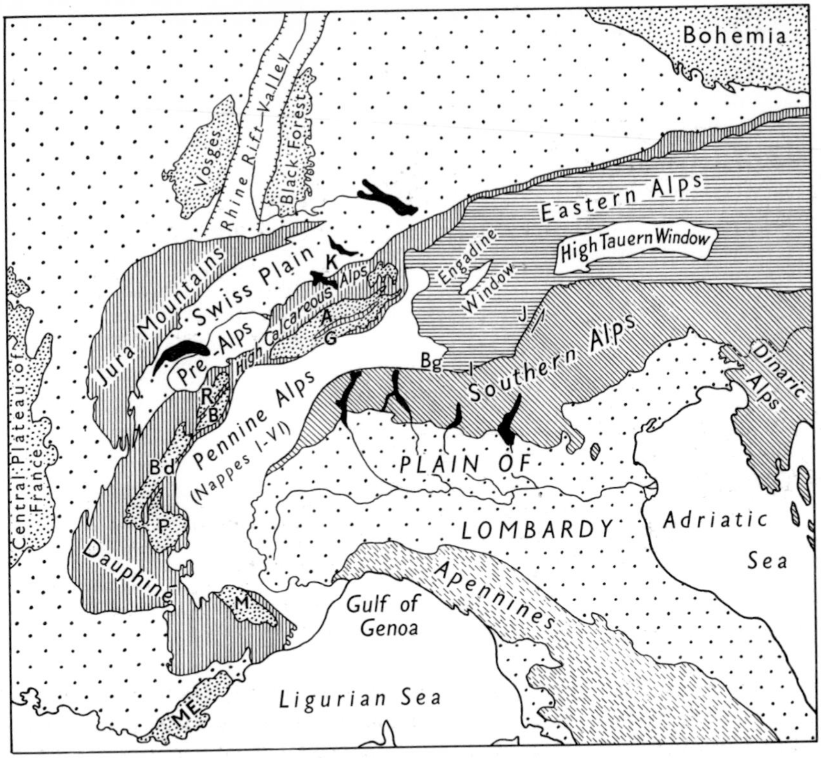


**Figure 30.11** Tectonic map of the Alps. Hercynian massifs are indicated by close dotting. A, Aar massif; B, Mt Blanc; Bd, Belledonne; Bg, Bergeller Granite; G, St Gotthard massif; I, Insubric Line; J, Judicaria Line; K, Mythen Klippes; M, Mercantour massif; ME, Maures and Esterel massifs; P, Pelvoux massif; R, Aiguilles Rouges

he said, 'No one would believe me, they would put me in an asylum'.

A major difficulty in interpreting large-scale structures like the Glarus nappe arises from the fact that, as a result of denudation, parts of the structure are missing, and much depends on reconstruction, as can be seen from Figure 30.12. Escher's interpretation of the Glarus structure as a double fold implies the smallest possible amount of lateral movement consistent with the arrangement of the strata. If it were correct, then the folds were essentially *autochthonous,* i.e. on the site where they were formed. When in 1884 Marcel Bertrand suggested that the Glarus structure could be better interpreted as a single recumbent

**Figure 30.12** (a) Glarus 'double-fold' of Escher van der Linth; and (b) its re-interpretation as a nappe by M. Bertrand. P, Palaeozoic strata, including Verrucano; Tr, Triassic; J, Jurassic; C, Cretaceous; E, Eocene Flysch. Length of section 40 km

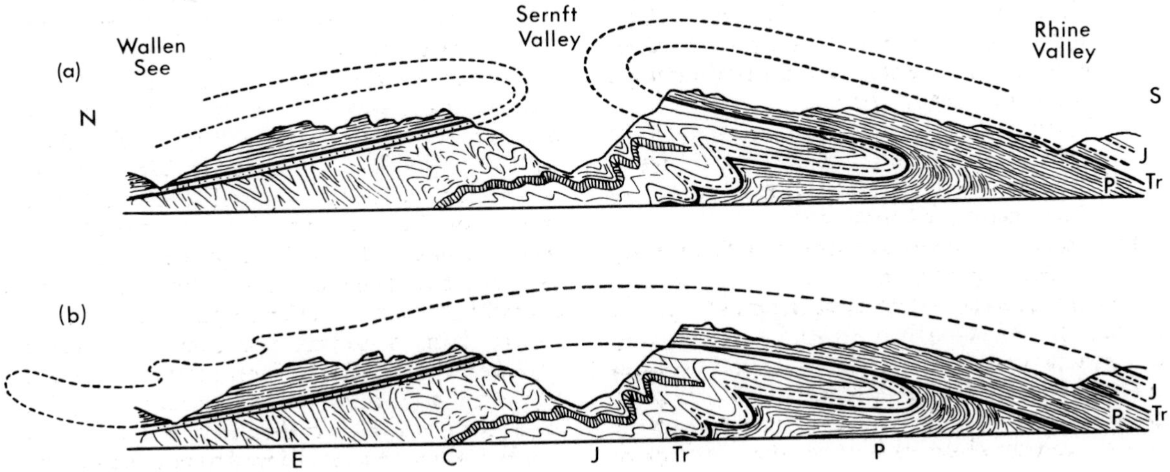

fold that had travelled northwards, the enormous lateral movement that this implies prevented the suggestion from being immediately accepted. (Figure 30.12(*b*)). The idea that rock masses could travel long distances laterally and become *allochthonous* was at that time so novel as to appear fantastic to many geologists, including Albert Heim. However, by this time several equally impossible overthrusts of gigantic proportions had been discovered in other mountain ranges. The celebrated Moine thrust (Figures 9.33 and 30.13) had been described by Calloway and Lapworth in 1883 and confirmed by the Geological Survey in 1884. By 1883 Törnebohm had recognized Caledonian overthrusts directed from the Scandinavian mountains over their foreland, the Baltic Shield; and in 1888 he showed that there had been a lateral displacement of at least 100 km. By 1896 he felt justified in raising the total to 130 km.

Obviously if progress was to be made, methods had to be devised by reference to which the direction and amount of movement of rock-masses could be established. One of the first to be used depended on the lateral variation within sedimentary strata. As early as 1838 Gressly traced the lateral variation of sedimentary rocks in the Juras. He found that the rocks varied laterally both petrographically and palaeontologically, and he introduced the term *facies* to designate the combined lithological and palaeontological characteristics of a deposit at any particular place. Furthermore, Gressly recognized that facies varied according to the palaeographic position of the deposit, and he compared this variation with that in the present-day seas, distinguishing between littoral and pelagic deposits (see Wegmann, 1963).

At the time of their deposition, sedimentary rocks have an orderly distribution relative to their appropriate shore-lines. When, therefore, in traverses across the Alps, sedimentary facies were found to change abruptly, or to occur in abnormal sequences, it became evident that there had been tectonic movement. This method was used by Albert Heim, who found the sedimentary facies exposed at the southern end of one nappe to be the continuation of the facies exposed at the northern end of the next higher nappe in the Helvetian zone. The distribution of the sedimentary facies thus indicated movement of successive layers of rocks, and the amount of movement from the south-east

**Figure 30.13** Outcrop of the Moine Thrust-plane, Highland Region, Scotland, showing Moine schist and an underlying wedge of Lewisian gneiss on white crystalline marble (Cambrian) (*Institute of Geological Sciences*)

towards the north-west, when they became piled one upon another.

As the investigation of the Helvetian nappes proceeded, a geometrical method began to play an important role. This method which was introduced by Argand, in his investigation of the Pennine nappes, is based on the recognition that the folds may have an approximately cylindroidal form, out of which the concept of a fold-axis arose. Provided the fold is cylindroidal, the axis is an imaginary line, parallel to the folded rock-layers, about which the layers may be supposed to be folded. Stated in a different way, the fold-axis may be regarded as the generatrix of the fold, for the form of a cylindroidal fold can be generated by moving its fold axis parallel to itself along the appropriate curves.

In the Swiss Alps the fold-axes of the Helvetian nappes strike approximately north-east. But they undulate, rising upwards to their highest level against the Hercynian massifs which rise like horsts, and plunging downwards between them (Figure 30.14). In other words, the nappes show tectonic *culminations* towards the Hercynian massifs, and tectonic *depressions* between them (Figure 30.14). Indeed the possibility of discovering the nappe structure largely depended on this undulation of the fold-axes, as a result of which cross-sections of the nappes became exposed to view by denudation of the culminations. Thus it is possible to look down the plunge of the fold-axes and so to recognize successively higher structural levels, even though these levels cannot be seen in a single exposure. If, for example, one could transpose oneself high into the air above the Aiguilles Rouges in Figure 30.14 it would be possible to look down in the direction of the fold axis of the Morcles nappe, and to see that the Diablerets nappe and the Wildhorn nappe form successively higher structural levels.

The nappes are not only flexed in the direction of their fold-axes (Figure 30.14) they are also flexed in the direction at right angles to their fold-axes (Figure 30.15). The nappes thus mantle the Hercynian massifs as though they were a pile of shells. In fact Lugeon used the term carapace to describe the shape of nappes that form structural culminations of this kind. It should be noticed that

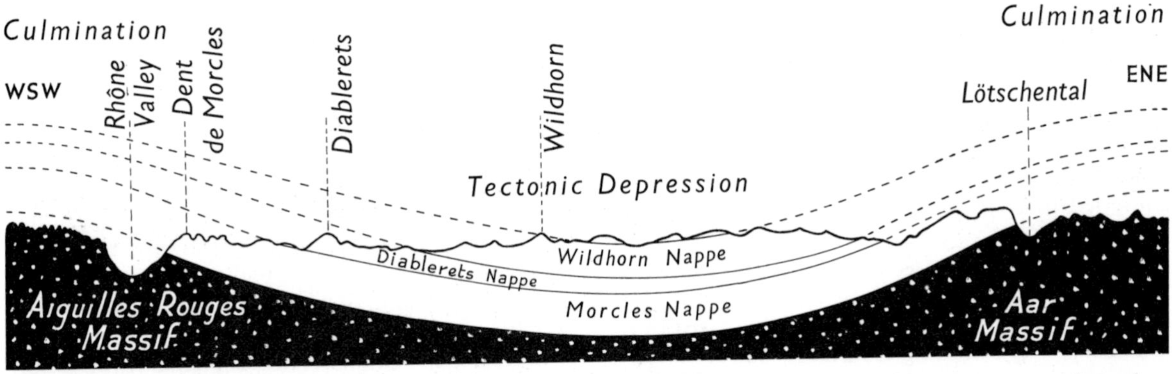


**Figure 30.14** Longitudinal section showing the nappes of the High Calcareous Alps exposed in the tectonic depression between the Aiguilles Rouges and the Aar culminations. The nappes advanced at right angles to the plane of the section in the direction away from the observer.

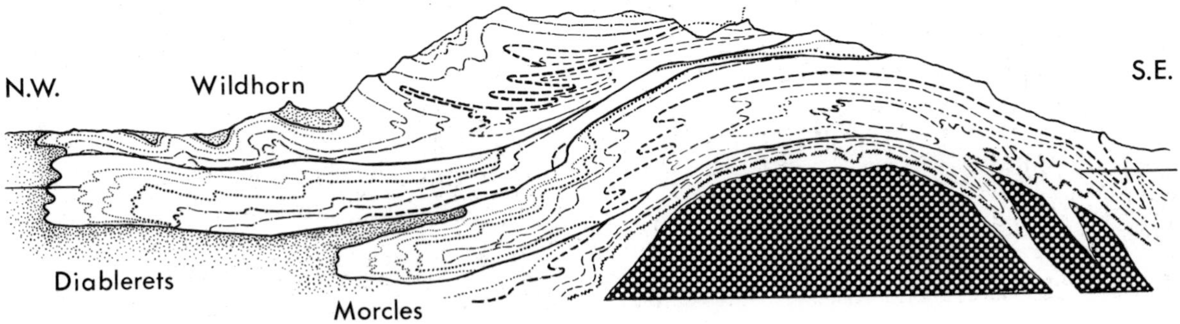


**Figure 30.15** Section of the Helvetian nappes at right angles to the fold-axis, according to the old interpretation of Lugeon and Argand. The shaded foundation rocks are at the SE end of the Aiguilles Rouges massif. Length of section 23 km. (*After* Tectonics *by Jean Gougel, translated by Hans E. Thalmann; San Francisco, W. H. Freeman and Company, 1962*)

these nappes have no lower limbs. It was wrongly supposed, at the time when Lugeon synthesized them as far-travelled thrust folds that the lower limbs had been destroyed by grinding and crushing as the nappes were driven forward. As will be described directly, it is now known that the mylonite which would inevitably be formed at the soles of great thrusts, like the Moine thrust in Scotland, does not exist.

## Movements of Foundation Rocks in the Swiss Alps

Early in the last century the Hercynian massifs, e.g. the Aar massif, were thought to have played a passive role during the Alpine orogenesis (Figure 30.11). Argand, however, recognized that their broad tumour-like forms resulted from upward-movements of the foundation rocks that had formed the floor of the Alpine geosyncline, either concomitantly with or subsequent to the tangential movements of the overlying nappes. He called such an upbulge a *pli de fond*, a foundation fold. In the Alpine geosyncline, Mesozoic and Tertiary rocks were deposited upon a foundation of older rocks that dipped steeply as a result of earlier Hercynian orogenesis. The junction between the foundation rocks, and the overlying geosynclinal sediments is therefore unconformable, since the foundation rocks had been both folded and denuded prior to the deposition of the overlying Mesozoic strata. Some of the foundation rocks had, moreover, been folded during more than one geological period, and they have to be pictured as consisting of steeply dipping bands of rocks of various kinds, including Carboniferous sediments and older crystalline schists, gneiss and granite.

Argand recognized that during the Alpine orogeny, the foundation rocks and their covering strata moved disharmonically. The foundation rocks moved upwards as steeply dipping slices (Figure 30.16), whilst the covering rocks moved tangentially. Movement was thus recognized to be three-dimensional. Argand thought, however, that the upward movements of the foundation rocks, and the tangential movements of the covering rocks were both consequences of Alpine tangential push or squeeze of Alpine age. He attributed this squeeze to collision between Africa and Europe. The Swiss Alps were thought to be squeezed as if in the jaws of a vice. This old interpretation, although more than half a century old, is still a favoured hypothesis outside Switzerland, because it accords so well with plate tectonics. Many discoveries have been made since Argand's time, however, as will gradually emerge. For example, the steeply dipping slices of crystalline foundation rocks, thought by Argand to result from Alpine squeeze, are now known to be bounded by faults dating from Hercynian times.

## The Hercynian Nappes Found to be Gravity-glide Folds without Roots

As early as 1883, Reyer recognized an alternative to compression for explaining the folding and thrusting of strata, which obviated the impossibility of transmitting forces for long distances through relative thin, and sometimes incompetent, slabs of rock by applying pressure from behind. If the slabs should have an opportunity of gliding down sloping surfaces of low frictional resistance as a result of their own weight (as in landslides), then transmission of force is no longer a necessity. Gravitation is not transmitted but acts throughout the layers of rock concerned as its does in a glacier.

**Figure 30.16** Section through the Pelvoux massif, SW France, showing the slicing of the crystalline foundation rocks (shaded) by faults, and the covering Mesozoic strata (unshaded) folded and inpinched between uprising wedges of foundation rocks (*After J. Gougel, 1948*)

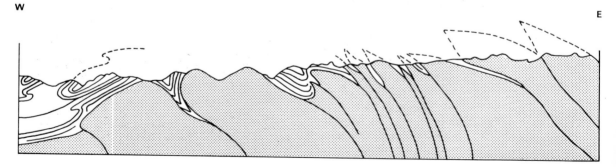

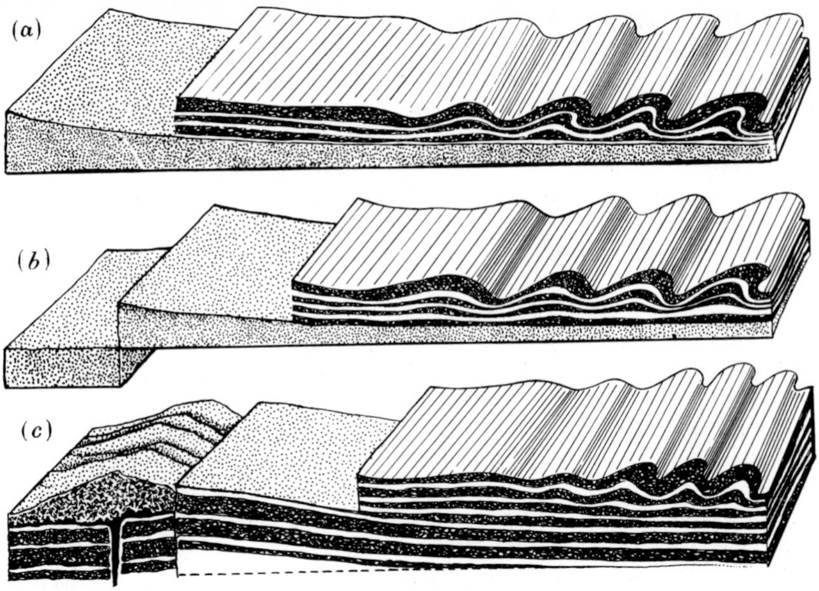


**Figure 30.17** Gravitational tectonics, as illustrated by E. Reyer in 1888 (*Theoretische Geologie*, Figs 652–654, p. 787). (a) folding of strata due to sliding down an originally inclined bedding plane, as in submarine slumping; (b) here the inclination is caused by uplift, indicated by the fault on the left; (c) here the upheaval has thrown the crest of the uplifted region into tension with consequent rifting and volcanic activity (*E. Reyer*)

With *gravity gliding*, folds, thrusts, and imbricate structures produce a thickening or piling up of the layers towards the front of the advancing mass (Figure 30.17), whereas if pressure comes from behind, the resulting structures thicken the mass at the point where the push is applied. Where gravity gliding is the solution, suitable slopes must have existed at the time when the gliding occurred, and consequently vertical movements, both up and down, become the primary necessity.

In 1893 Schardt suggested that the nappes of the Pre-Alps (Figure 30.11) resulted from gravity gliding of rock-layers which became detached (*decollé*) from their substratum, down the slopes of which they slid for great distances. Such isolated piles of nappes (layers) are known as *klippen*. Although Schart's interpretation as to the origin of the Pre-Alps is now generally accepted, when it was first proposed it was mocked at by the authorities of the time.

Two types of folds thrusts, now attributed to gravity are: (*a*) *gravity collapse* structures, and (*b*) *gravity glide* structures.

Small-scale examples of gravity collapse structures were described by Harrison and Falcon in 1934 and 1936, from south-west Iran. These are superimposed on broad anticlines and synclines of thick limestones with incompetent beds of soft marls and siltstones, etc. between and above (Figure 30.18). The diagram shows the Miocene Fars Series, deeply eroded in the synclines, overlying massive beds of Asmari Limestone (see Figure 9.16). The latter, together with the underlying marls, illustrate a remarkable variety of folds and slides, some of which show displacements of 2 or 3 km. Below the marls a thick Middle Creta-

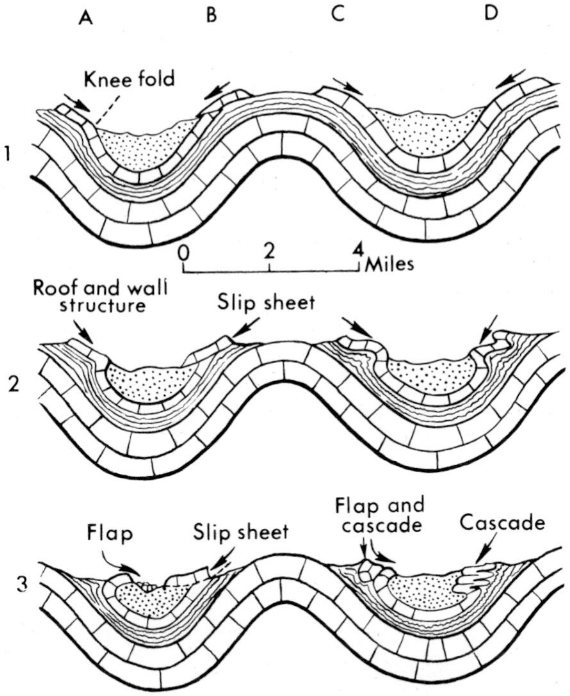


**Figure 30.18** Diagrams of gravity-collapse structures, representing stages in the development of flaps, slip-sheets and cascades of folds. No vertical exaggeration (*After J. V. Harrison and N. L. Falcon, as modified by E. S. Hills*)

ceous limestone serves as a control bed, unmoved laterally since the original major folds. It was found that as denudation had progressed, the structural sequences (Figure 30.18) include:

(A) bending from a *knee-fold* to an overturned *flap*;
(B) breaking and sliding with production of *slip-sheets*;
(C) development of a flap into a broken *cascade* of folds; and
(D) flowage into an unbroken *cascade* of folds.

From the anticlinal summits to the synclinal stream beds the present vertical relief may be as much as 2·4 km, but the surprising feature is that these secondary structures begin to show themselves in the eroded limestone cover and its incompetent support when the relief is no more than 600 m.

Cascades of folds of the kind described by Harrison and Falcon, including many on immensely greater scales, have since been recognized as highly characteristic features on the flanks of uplifts in orogenic belts. The well-known recumbent folds seen along the Axenstrasse (Lake Lucerne), illustrated by Figure 3.11, are part of a broken cascade belonging to one of the Helvetian nappes associated with the Glarus nappe. Figures 30.19 and 30.20 illustrate typical examples from the northern Pyrenees and the Montana Rockies. Nearer home the sudden folds of the Stair Hole (Figure 1.5), continued through Lulworth Cove (Figure 23.57) and the Isle of Wight (Figure 23.19) may be due to the draping of sediments over the side of an uplifted wedge of the underlying basement rocks.

Of greater importance in the present connection, because it applies to the Hercynian nappes of the Swiss Alps, is gravity gliding of sedimentary strata, sometimes on the sea floor, and sometimes as a result of oblique upward movements of crystalline foundation rocks. For about half a century the Morcle nappe was thought to be a fold-thrust, with a slice of mylonitized Hercynian granitic rock as its sole. The latter, supposedly being derived from between the Aiguilles Rouges and Mont Blanc, formed the basis for supposing that the nappe rooted between these massifs and had been squeezed up and thrust north-westward. In 1946 Schroeder recognized

**Figure 30.19** A cascade of gravity folds seen on the cliffs north of the harbour of St Jean de Luz, Bay of Biscay end of the Basses Pyrénées, France (*F. N. Ashcroft*)

**Figure 30.20** Heaven's Peak, Rocky Mts, Montana: a typical gravity fold on the flanks of a block uplift (*United States Geological Survey*)

the supposed mylonite to be sedimentary breccia, with a sandy matrix derived by erosion of granite. This necessitated the reinterpretation of the Morcle nappe. About the same time Lugeon took Marc Vuagnat, then a student, to see the Morcle nappe, and he recognized the supposed mylonite to be a feldspathic sandstone. This sandstone, obviously derived by marine erosion, was reinterpreted by Lugeon (1947) as being derived from cliffs of Hercynian granite which represented the slowly emerging Aiguilles Rouges. Under the influence of rising blocks of Hercynian crystalline rocks, which formed the foundation upon which the sediments composing the Morcle nappe were deposited, the soft sedimentary rocks of the nappe glided north-westward over a sea floor surfaced with arkose and granite blocks. Specimens of the arkose were sent to petrologists for microscopic examination and it was pronounced to be truly a sedimentary rock. The present-day elevation of the Morcle nappe, and the steep dip of the hinder part towards the south-east, Lugeon attributed to upward movement of the Hercynian foundation rocks subsequent to the emplacement of the gliding nappe.

The history of geology reveals that all hypotheses are provisional and have to be altered as factual knowledge increases. As Lugeon himself expressed it, mistakes are wiped out by the 'floorcloth of time'.

In the early part of the century, the Wildhorn nappe was also thought to result from thrusting resulting from Alpine 'squeeze'. The clue to a more actualistic interpretation of this nappe, as resulting from gravity gliding, was provided by a series of folded faults. From sedimentary evidence, and by laboriously unwinding the folded faults, Günzler-Seiffert (1952) discovered that prior to folding these faults were normal strike faults near the margin of the geosynclinal trough in which the Mesozoic and Tertiary sediments were deposited. The faults, initially of Hercynian age and intersecting the crystalline foundation rocks, delimited north-easterly striking blocks of the geosynclinal sediments, and they functioned as normal strike faults during Mesozoic and Tertiary times. During the Tertiary orogeny, as a result of oblique upward movement of steeply dipping slices of the crystalline foundation rocks, bounded by Hercynian faults, on which the Mesozoic sediments were deposited, the sediments became detached and, under the influence of gravity, slid slowly north-westward down the increasing slope of Hercynian foundation rocks. During the gravity gliding fault-bounded blocks of geosynclinal sediments remained distinct and recognizable. They can now be seen from Interlaken, on looking up the valley of the Lutechine, as successive folds, of cascade type, composed of Jurassic, Cretaceous and Tertiary sedimentary rocks. Figure 30.21

**Figure 30.21** The Wildhorn nappe originating as autochthonous sediments first glided forward as a result of the uprise of wedges of foundation rocks (1), between which portions of the gliding nappe eventually became entrapped (*pincée*) (2). Subsequently, through continued uprise of wedges of foundation rocks the Wildhorn, together with overlying nappes of sedimentary origin, continued to glide north-westward overriding the flysch (4 and 5) and eventually the molasse (8, 9 and 10) which were successively deposited within troughs farther north-west. Portions of the Wildhorn nappe entrapped between wedges of foundation rocks are not roots but inpinched portions of overlying sedimentary rocks. (*After H. Gunzler-Seiffert, 1946*)

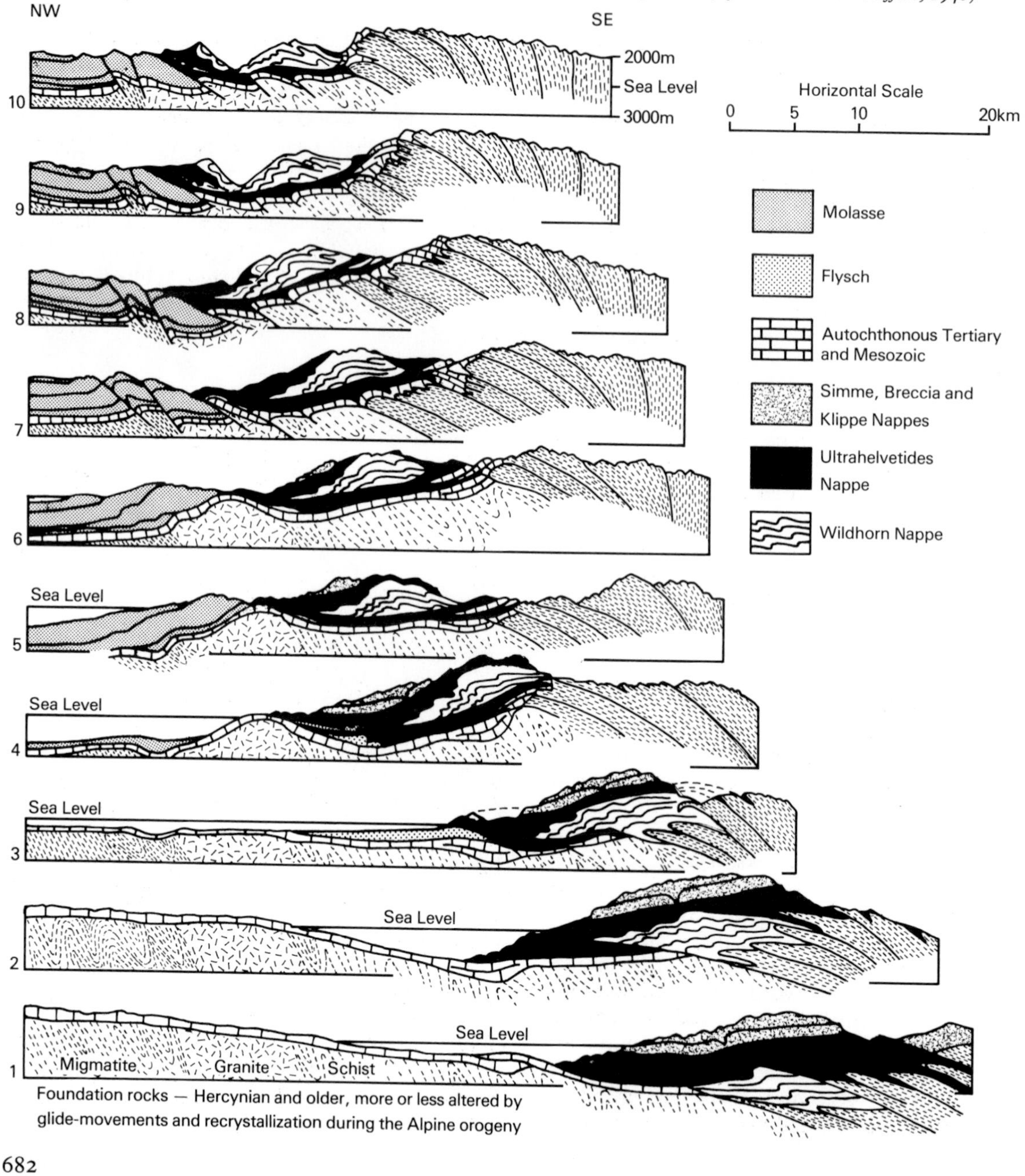

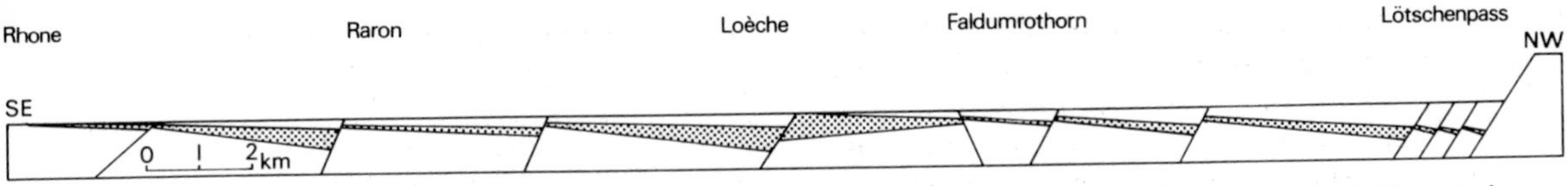


**Figure 30.22** Reconstructed section across north-easterly trending troughs of the Aar massif within which Mesozoic sediments were deposited. The troughs were separated from one another by more or less emerged ridges of Hercynian foundation gneiss which continued to move along normal Hercynian faults and progressively tilted the floors of the troughs towards the north-west (*After A. Baer, 1959*)

illustrates the gradual gliding of the Wildhorn nappe from the position of the geosynclinal trough, to its final resting place in the Helvetian area.

## The South-western End of the Aar Massif

A detailed study of the relationship between Hercynian foundation rocks and their covering of autochthonous strata has been made by Baer (1959), a student of Wegmann's. In plan the western end of the Aar massif is characterized by an interdigitated alternation of Hercynian foundation rocks, and Mesozoic and Tertiary covering rocks (Figure 30.23(*a*)). The Hercynian foundation rocks include Carboniferous and Permian strata, gneiss of pre-Hercynian age, and granite.

The crystalline Hercynian foundation rocks of the Aar massif are cut into thick slices by steeply dipping faults that strike north-eastward. Although of pre-Triassic age, the old faults continued to function as normal faults throughout Mesozoic times. Baer made this discovery through a detailed investigation of the autochthonous Mesozoic and Tertiary sedimentary rocks that overlie the crystalline foundation rocks. By reference to facies differences he established that from early Jurassic times, and possibly as early as Permian times, the sedimentary rocks were deposited in north-easterly trending troughs, that were separated from one another by more or less emerged ridges of the Hercynian foundation rocks. The successive sedimentary strata within each individual trough thicken towards the north-west, implying that the floors of the parallel marine troughs were progressively tilted in that direction. This in turn indicates slow, but continued downthrow on the south-eastern sides of the Hercynian faults, which would thus control the tilt of the upper surfaces of the blocks of crystalline foundation rocks (Figure 30.22).

During the Alpine orogeny, the sedimentary infillings of the separate marine troughs assumed a synclinal form. The limbs of some of the synclines dip steeply and are even overturned towards the north-north-west, but they were formed in a different way from normal synclines. Baer compares the steeply dipping slices of foundation rocks with gigantic dominoes, which moved differentially upwards during the Alpine orogeny, their old boundary faults now functioning in a reversed direction. As a result of these differential upward movements of the foundation rocks the lower sedimentary strata became entrapped between the 'dominoes'. French geologists call such synclinal structures *pincées*, entrapments. The higher strata remained free to move, however, and glided north-westward as nappes (Figure 30.23 a, b). Baer remarks that the only structure

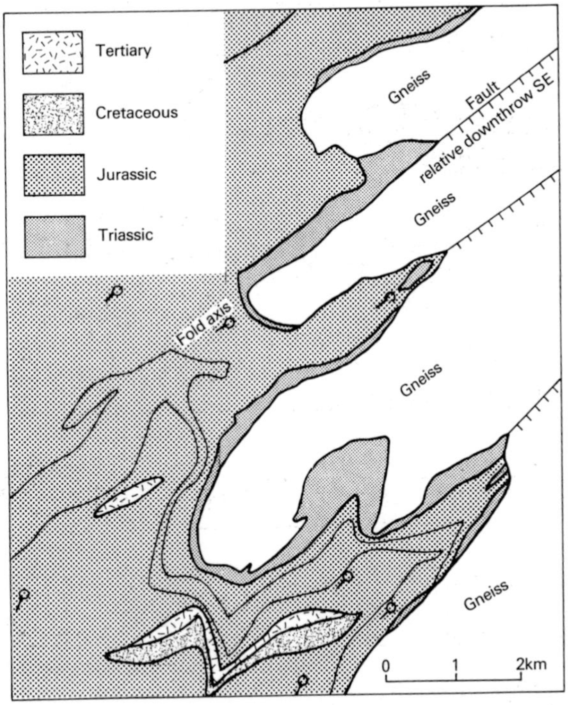


**Figure 30.23** (a) Simplified map of the south-western end of the Aar massif showing in plan the downfolding and entrapment of Triassic and Jurassic strata between steeply dipping slices of uprisen foundation gneiss (*After A. Baer, 1959*)

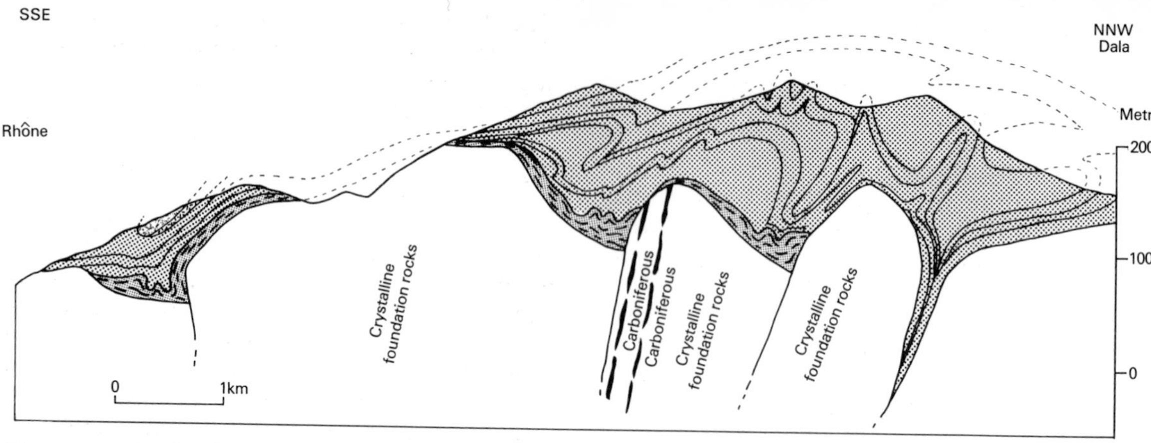


**Figure 30.23** (b) Simplified section across the Aar massif from SE (on the left) to NW (on the right) illustrating the downfolding and entrapment (*pincée*) of Triassic and Jurassic strata between uprising slices of foundation rocks (*After A. Baer, 1959*)

that might be called a 'root' is that part of a nappe that has abutted against a block of foundation rocks where it has risen upwards along an old Hercynian fault (Figure 30.23(*b*)). He concludes that compression only began during the Tertiary, and that it did little more than compress the autochthonous strata into a pre-existing framework, the relative importance of the Helvetian nappes having been previously exaggerated.

That the foundation rocks continued to move upwards after the folding of the sedimentary superstructure, is evidenced by the plunge of the fold-axes of the sedimentary strata, away from the Hercynian foundation rocks of the Aar massif at about 20° (Figure 30.23(*a*)). Furthermore Baer found that these fold axes plunge at right angles to a system of joints that intersect the whole rock-framework, foundation rocks and sedimentary covering rocks alike.

The joints which strike roughly north-westward, and dip north-eastward at 70° cut all the other structures, and are therefore relatively young. Moreover they evidence an extension of from 10 to 20 per cent in a north-easterly direction. This evidence, combined with the south-westerly plunge of the fold-axes of the sedimentary strata, indicates that after the evolution of the main structures, the crystalline foundation rocks continued to move differentially upwards, the movement now reaching a maximum about the centre of the massif, so that the Hercynian rocks now present an appearance of the kind that Argand described as a foundation fold.

The Hercynian rocks of the Aar and Mont Blanc massifs were not uncovered in Miocene times, for there are no pebbles derived from them within the younger sub-Alpine Molasse. The source of the detritus composing the Molasse was the thick cover of Helvetian sediments and younger nappes under which the Hercynian rocks were buried. This does not mean that the Alps were once much higher than they are now; upward movement and denudation have proceeded concomitantly.

By collating all the relevant evidence, Wegmann (1957) has calculated the average rate of elevation, i.e. the average rate of upward movement of the foundation rocks of the Aar massif, during the 15 million years that have elapsed since the time of deposition of the youngest Molasse, to be of the order of 5 or 6 cm a century. Elevation has not been continuous, however, nor has the rate of movement been constant. For example, the great conglomerates of the Molasse probably represent periods of rapid movement. By reference to other data, the average rate of denudation was calculated to be of the order of 5·7 cm a century. Wegmann points out that with this rate of denudation, without concomitant uprise, it would take only 3·5 million years for the Alps to be reduced to base level.

Were it not for upward movements of the foundation rocks, during Miocene, Pliocene and Pleistocene times, the Swiss mountains would not exist today. Nor is it very likely that the mountains of the Helvetian zone were ever much higher than they are now, for uplift and denudation appear to be roughly balanced. Furthermore the Aar massif is still moving upwards, for traces of recent movements have been observed in the area, and seismic records of minor earthquakes provide further evidence of such movements.

## The Jura Mountains

The Jura Mountains are composed of an assemblage of hills, standing well in front of the Alps (30.11). Their component sedimentary rocks were deposited not in a geosyncline, but on a flat tabular area. The folds rest on a swell of the Hercynian foundation rocks, which do not outcrop, but have been located in boreholes. At each end of the Juras there is one open anticline, but the number of folds increases greatly towards the middle, where there are many anticlines, some distorted and some more perfect and outlining actual hills (Figure 30.24). Unfolded tabular Jura intervene between some of the groups of folds. Here the sedimentary strata lie undisturbed over areas measuring up to 77 by 24 km. Moreover, on the convex side of the Jura arc the *folded Jura* pass into the tableland of the *tabular* Jura, where flat-lying strata overlie Hercynian foundation rocks like those that outcrop in the Vosges, Black Forest and the Massif Central.

The strata composing the Jura Mountains are largely of Jurassic age, but they range from Triassic to mid-Cretaceous, with some Tertiary, mainly Miocene, preserved within the synclines. The thickness of the strata, moreover, increases from east to west, and from north to south, and this applied to each separate horizon as well as to the whole thickness.

From photographs (Figure 30.24) one might suppose the folded Jura to be composed of the simplest of folds. Their actual complexity is revealed, however, by the many hypotheses that have been devised to explain them. Buxdorf thought the folding to result from the detachment and gliding of the covering rocks above the level of the Triassic, which includes layers of salt that could act as lubricants. Aubert first thought that wedges of the Hercynian foundation rocks had been upthrust and so flexed and folded the covering rocks. Later he suggested that the old foundation rocks had been mylonitized along the faults that intersected them, this crushing having contracted the foundation rocks and so folded the covering rocks. Glandgeau also related the folding to the movements of the Hercynian foundation

**Figure 30.24** Anticline in the Moutier Gorge, Jura Mts, Switzerland; an example of concordant morphology (*Peter Christ*)
*Left to right: A. Buxtorf, Arthur Holmes, M. Reinhard*

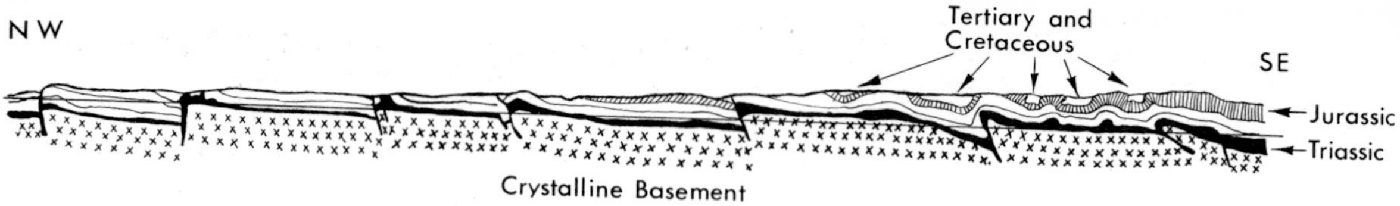


rocks, but he supposed the foundation rocks to be initially intersected by normal faults, over which the covering Mesozoic strata were flexed, and subsequently folded (Figure 30.25). Wegmann, working on a finer scale, noticed that within individual folds, the strata have moved disharmonically, and that the Juras are not only intersected by major transcurrent faults, they are also dissected by minor faults which sometimes cut only partly through a single fold. He detected in this the pattern of movements within the foundation rocks, and from the striated and channelled surfaces of faults in the Hercynian of the Vosges he recognized the importance of horizontal movements. Figure 30.26 illustrates the faulted Hercynian foundation rocks of the Jura Mountains, according to Wegmann's hypothesis. Movements along the faults have been both horizontal and sometimes vertical. In a highly diagrammatic way, Figure 30.27 indicates how the Mesozoic strata may have been folded as a result of horizontal movements within the foundation rocks, combined with detachment and slipping along the basal layer of Triassic salt.

**Figure 30.25** Section across the French Jura Mountains from the neighbourhood of Besançon to the Swiss frontier to illustrate the hypothesis that uplifts of blocks of the crystalline basement are the activating agent for the faulting and folding of the sedimentary cover. Length of section about 80 km (*After L. Glangeaud*)

There are still other hypotheses that have been proposed to explain the folds of the Jura Mountains. The object in presenting a selection of the hypotheses here is to show how impossible it would be to find a correct solution by sitting at a desk and thinking, or looking at pictures. The geologist's job is to go out into the field, like these Swiss geologists, and look.

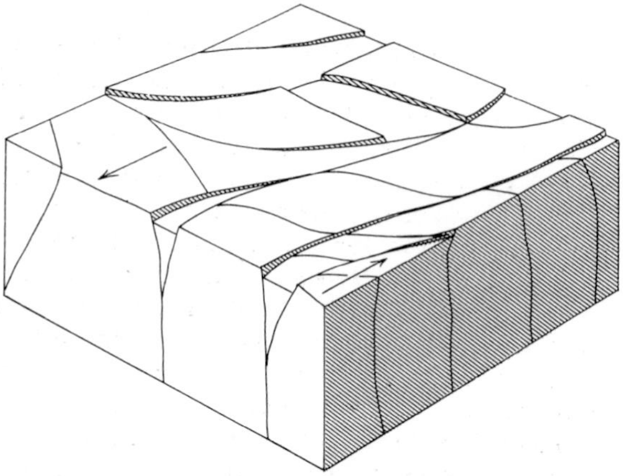

**Figure 30.26** Block diagram illustrating the possible complex movements of the Hercynian foundation rocks of the Jura, based on observations of the exposed Hercynian of the Vosges. The movements although predominantly horizontal (transcurrent), as indicated by arrows, commonly have a vertical component, indicated by stippling. (*E. Wegmann, 1961*)

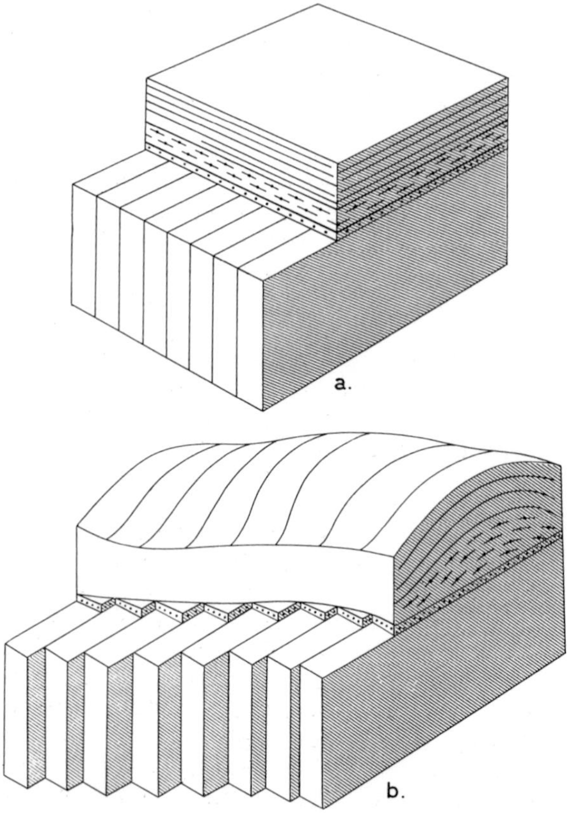


**Figure 30.27** Schematic block diagrams illustrating the detachment and folding of the sedimentary covering rocks of the Jura, resulting from unequal horizontal movements of steeply dipping slices of the foundation rocks along transcurrent faults (*E. Wegmann, 1961*)

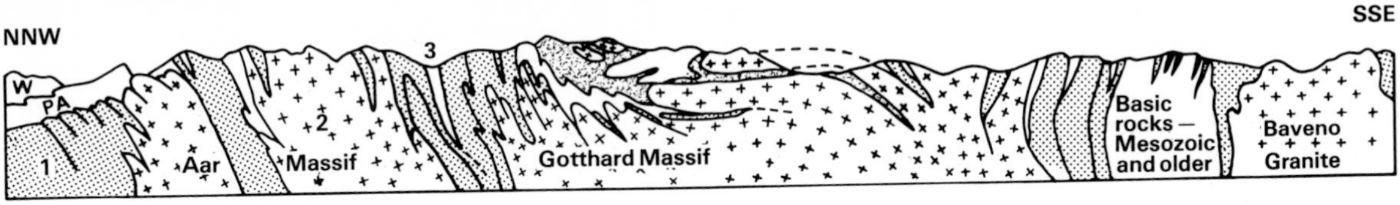


**Figure 30.28** Section across the Alps showing the tongue-like form of the foundation rocks of the Pennine Alps
PA =parauthonous and authonous strata of the Aar massif
W =The Wildhorn nappe
3 =entrapments (white) of Mesozoic rocks (stippled entrapments are the *schistes lustres*, metamorphic rocks originally of Triassic to Eocene age)
2 =Hercynian granitic rocks partly transformed to Tertiary within the tongue-like forms of foundation rocks
1 =Pre-Hercynian gneiss
(*After W. H. Nabholz, 1954*)

## The Deepest Part of the Pennine Zone

The foundation rocks in the deepest part of the Pennine zone of the Alps correspond to the steeply dipping slices of crystalline rocks, bounded by Hercynian faults, in the Aar massif, and elsewhere in the Helvetian zone. Since they have been buried under a greater superincumbent load, however, they have a more supple style. Each crystalline slice has a curved tongue-like form, to which the term nappe is again applied (Figure 30.28). As Trümpy (1960) has summarized in English, the term nappe in Switzerland includes structures of different styles and origins.

The deepest observable part of the Alpine edifice consists of the crystalline tongue-like forms of the Tessin–Simplon culmination. Each of these nappes has a core of granitic composition with a carapace of gneiss, the gneiss being composed of an inner zone of migmatite (*embrechite*), and an outer zone of gneiss (*ectinite*). Nabholz (1954) considers the mantling gneisses to be pre-Hercynian in age, and the granitic cores to represent intrusions emplaced within the former during the Hercynian orogeny. He recognized, however, that all these crystalline rocks have been metamorphosed during the Alpine orogeny. He found a great part of the post-Hercynian sediments to have become detached and travelled forwards towards the frontal region. Like the Helvetian nappes, the higher elements of the Pennine zone are no longer considered to be exaggerated recumbent folds with crushed lower limbs.

The orientation of the linear structures of these crystalline nappes of the Tessin area have been studied, both in the field and by petrographic techniques under the microscope by Wenk (1943, 1955), Hasler (1949) and Günthert (1956). These studies leave no doubt that the crystalline rocks of the Tessin nappes were both folded and metamorphosed during the Alpine orogeny, recrystallization having been synchronous with movement. Furthermore the granitic (granite, granodiorite and quartz-diorite) cores of some of the nappes have broken through their mantling rocks and, whatever the age of their initial materials, now appear as Tertiary intrusions. Relict structures of pre-Alpine age still remain within the crystalline rocks mantling the granite. They are, however, largely effaced even within the higher levels. If Lyell's suggestion of long ago were to be followed, the ages of some of the granitic rocks in the core of the Alps would be Hercynian-Alpine, and others possibly Precambrian-Alpine. In the Ticino area, samples of mica from the granitic core and the gneissic shell of the Lepontine nappe were dated by Jager and Faul, by the potassium-argon method, and found to be 18 million years old, thus confirming the petrographic evidence that the micas were completely recrystallized during the Alpine movements.

## Disharmonic Movements between Three Structural levels

It will already be apparent that in the Swiss Alps the covering geosynclinal strata, and the crystalline foundation rocks on which they rest, have moved disharmonically during the Alpine orogeny. It is interesting to note the evolution of ideas through successive generations. Maurice Lugeon was responsible for the initial synthesis and sequence of the Helvetian nappes in 1903. It remained for his student Emile Argand, however, to discover that the culminations of the nappes over the Hercynian massifs resulted from upward movements of the foundation rocks, but he thought the movement to be wholly of Alpine age. Through the efforts of many younger investigators it is now well known that the crystalline foundation rocks have been moving upwards at least since early Jurassic times, as great steeply dipping slices or dominoes, to use Baer's expression. The third generation of great Alpine tectonicians is represented by Eugene Wegmann, a student of Argand's. When he was a young man, Wegmann recognized that in order to understand orogenic movements it was necessary to study still deeper levels than those exposed in the core of the Pennine zone of the Alps. Consequently, he began his career by spending many years, approximately three in each country he visited, studying the Hercynian in France, the Caledonian in Scandinavia, the Precambrian in Finland, the Caledonian in eastern Greenland, and the Precambrian in south-western Greenland.

In Finland, under Sederholm's guidance, Wegmann saw migmatites within which tectonic movements of some of the deepest exposed levels of the sialic crust are preserved. Jointly with Kranck, he mapped the structures within the Svecofennian belt to the east of Helsinki, and found them to curve in two major arcs, concave northwards, around updomed cores of Hangö granite. The mantling rocks consist of leptites, marbles, schists, grey gneiss and metabasalts, including pillow lavas, any of which may be more or less migmatized. Like Sederholm before him, Wegmann used swarms of basalt (metamorphosed) dykes as events, representing definite time-intervals, by reference to which older and younger events could be distinguished. Within the mantles of the granitic domes are recumbent folds, overturned towards the south, intersected by a swarm of basaltic dykes. The curvature of the limbs, axial planes, and fold axes of these recumbent folds occurred after the emplacement of the dykes, for members of the latter have been flexed and pulled apart by the extension-movements imposed by the uprising granitic domes. This investigation is of great interest, because it represents an early (1931) structural map of mantled gneiss domes, such as are now well known phenomena.

It was after working in the Scorsby Sund region of eastern Greenland, where even the deepest structural levels are exposed on the high fjord walls, that Wegmann formulated his interpretation of the styles of movement in three structural zones at different depths. The deepest structural zone—the *infrastructure*—is reconstructed from crystalline rocks formed during one or more older geological eras. This is the zone of migmatites, but it is not a zone of wholesale melting, because old structures can still be deciphered and the direction of movement read. A major part of the highest structural level—the *superstructure*—is composed of a great thickness of sedimentary and volcanic rocks, representing the geosynclinal phase of the particular orogeny concerned. The superstructure may also include parts of the crystalline foundation rocks upon which the geosynclinal rocks were deposited, as in the Helvetian Alps (Figure 30.29). Between these two structural zones is the *transition* zone, the zone of detachment between the higher and lower movement levels, within which adjustment movements take place.

When a sufficiently great thickness of geosynclinal sedimentary and volcanic rocks, of the order of say 10 to 20 km, has accumulated, both by direct deposition and by possible addition of nappes gliding under the influence of gravity, the lower parts of the underlying foundation rocks become so deeply depressed beneath this load that they lie within the zone of rock flowage and migmatization. The upper irregularly curved surface of this migmatite zone Wegmann called the *migmatite front*. Even at an early stage of an orogeny, the migmatite front, as illustrated in Figure 30.29), may locally have ascended into the lower levels of the geosynclinal rocks themselves. On the other hand, as Figure 30.29 also illustrates, towards the margins of a rising orogen, parts of the crystalline foundation rocks may still lie above the migmatite front, as in the Helvetian Alps. In other words the migmatite front cross-cuts both the stratigraphy and the older structures. It may also cross-cut structures currently forming, in so far as some foundation folds lie above the migmatite front and others below.

Just as salt-domes flow upwards, by rheid flow, so in the more deeply buried part of an orogen, the rocks undergoing migmatization begin to rise upwards as major ridge- and dome-structures, over which the superstructure becomes flexed into great arches (Figure 30.30). Continued uprise of the infrastructure imparts further movements to the superstructure, and the component strata glide downwards in various ways, away from the rising ridges and domes above which they become thinned. Extension of superstructure over rising infrastructure is, however, accomplished in different ways. Beneath a superincumbent load of 5 or 6 km, Fourmarier, a famous Belgian geologist, found that extension might be expressed by the cleavage patterns. At higher structural levels extension is depicted by the patterns of fault-blocks, and by jointing, as described on p. 140.

Between the actively rising infrastructure, and

the relatively inert but stretching and downgliding superstructure, is the transition zone. The rocks of the transition zone also become stretched and thinned out by glide movements above and around the rising infrastructure. Since the rocks of the transition zone lie beneath a great superincumbent load, and in a region of elevated temperature, their glide movements, combined with recrystallization, lead to the evolution of schistosity and foliation. The transition zone is a zone of detachment between the movements of the rising infrastructure and the downgliding superstructure. It may be equated with the zone of regional metamorphism, within which rocks are transformed to *crystallophyllines*, to use the French term.

Many investigators have recognized that the flat-lying orientation of the schistosity of crystalline rocks indicates that during their formation the greatest compressive stresses acted in a vertical direction, but they have traditionally thought only in terms of superincumbent load. Indeed Grubenmann's depth zones, supplemented by Eskola's concept of metamorphic facies, are expressions of this idea. On the other hand a famous American geologist, Van Hise, in his great *Treatise on Metamorphism*, published by the United States Geological Survey in 1904, regarded crystalline schists and gneisses as expressions of rock-flowage resulting from extension under load in depth.

Wegmann recognized the flat-lying schistosity of rocks of the transition zone to be expressions of extension movements with concomitant recrystallization. Roques made a similar discovery around the dome of Agout in the Massif Central, France (see p. 185). He found that as the migmatite dome of Agout rose upwards, the crystalline schists became extended over it, so that their schistosity is more or less parallel to the surface of the dome. Here, uprise of a migmatite dome, and recrystallization and alignment of minerals giving rise to schistosity were contemporaneous. Wegmann observed that as the schists of the transition zone become extended and thinned above rising infrastructure, they become correspondingly thickened between the infrastructural domes or ridges.

In actuality, the structural forms within the superstructure and the transition zone of an orogenic belt are much more complex than the above brief statement would suggest. This arises from the fact that a variety of movement-structures may be overprinted one upon another. From the sequence of some of these overprinted structures it is commonly possible to decipher whether the rocks concerned were progressively descending and becoming more deeply buried, or whether they were ascending and progressively approaching the earth's surface.

A schematic representation of the relationship between Wegmann's three structural levels, as depicted by Haller (1955) who worked in eastern Greenland for many years, is reproduced in Figure 30.31. The forms of actual bodies of infrastructure, that have flowed upwards like salt domes by

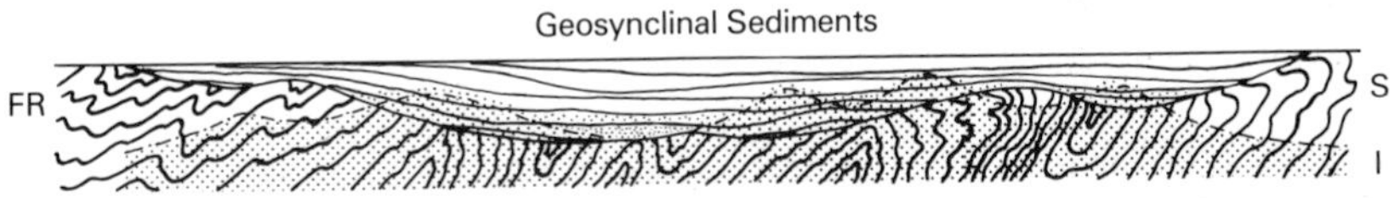


**Figure 30.29** Diagrammatic representation of geosynclinal sediments overlying foundation rocks (FR) of older periods. The hachured part cross-cutting the fold-structure represents the Infrastructure (I), the part free from hachuring, either geosynclinal sediments or folded foundation rocks representing Superstructure (S). The dashed line separating Infrastructure from Superstructure represents the migmatite front, the upward limit to which migmatization extended. The dotted part of the geosynclinal sediments immediately above the Infrastructure represents the Transition zone, the zone of regional metamorphism (*After E. Wegmann, 1935*)

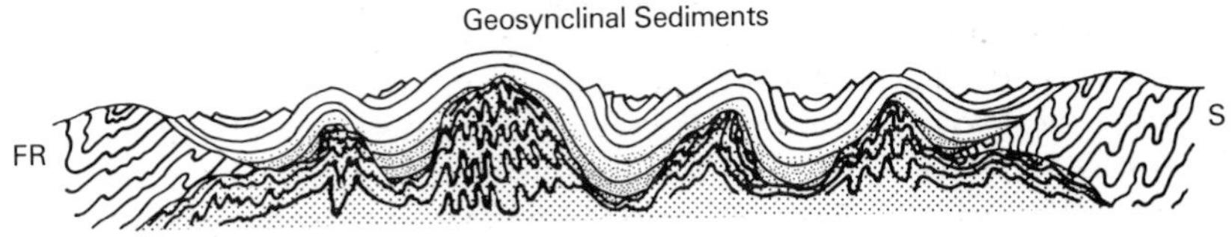


**Figure 30.30** During an orogeny the migmatized Infrastructure flows bodily upwards like salt domes, above sites where the migmatite front was initially relatively elevated. The Superstructure, like an inert blanket, becomes flexed over the rising domes (mantled gneiss domes) of Infrastructure forming major anticlines and synclines. The dotted part of the geosynclinal sediments is the zone of detachment (*décollement*) between Infrastructure and Superstructure: the zone of regional metamorphism. (*After E. Wegmann, 1935*)

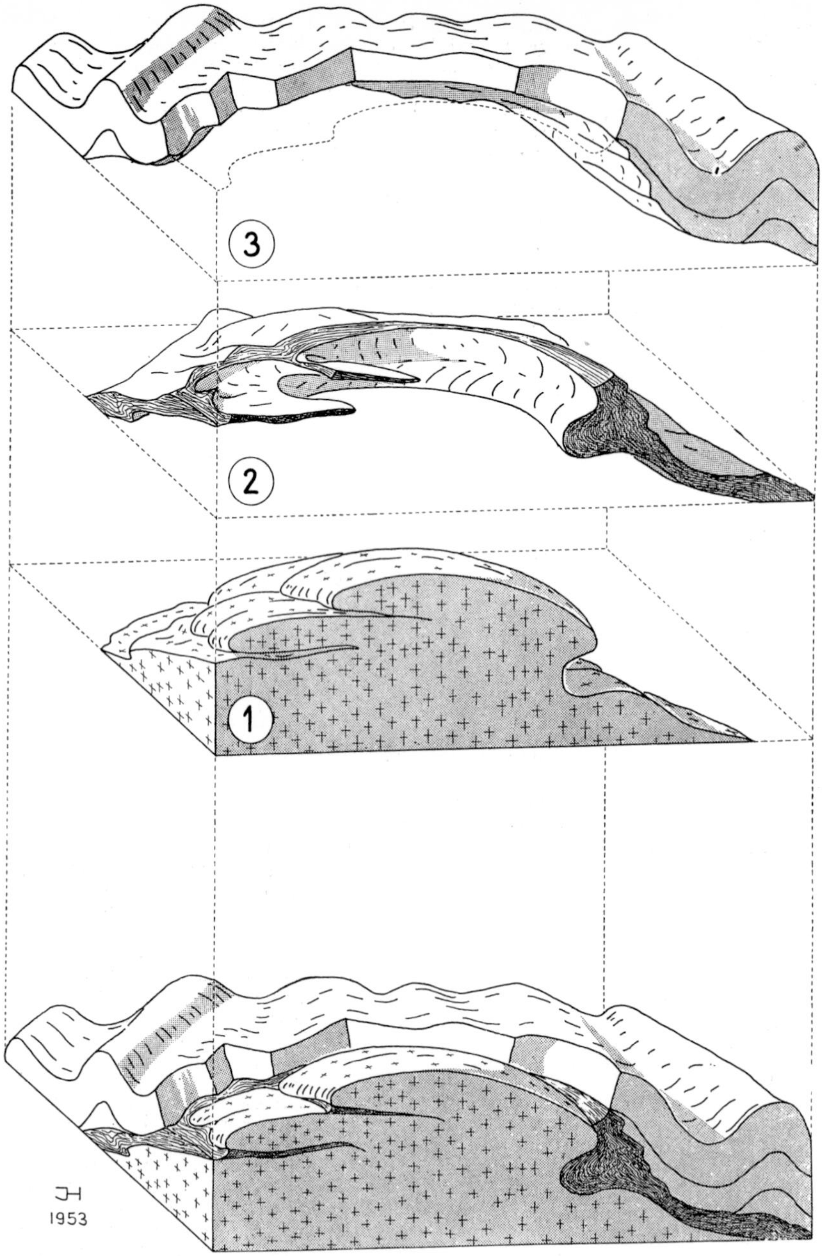


**Figure 30.31** A schematic representation of the styles of movement of each of the three structural levels. The lower figure shows their interrelationship, and the three upper figures depict each level separately.
1 Infrastructure flowing upwards with an overall dome-like form
2 Transition zone, the zone of detachment and shearing movements partly entrapped between recumbent tongues of Infrastructure
3 Superstructure, showing gravity glide-folds away from the summit of the rising Infrastructure
(*J. Haller, 1955*)

rheid flow are illustrated in Figure 30.32.

Another instructive illustration from this region is Figure 9.9 which, like Figure 30.32 (*b* and *c*), shows recumbent folding as a result of lateral flow, at levels where this mode of deformation met with less resistance than an upward continuation of the intrusion from which it spread would have done. Figure 30.32 clearly explains how it comes about that in some parts of old orogenic belts the foliation or flow-lines of the migmatites and associated rocks may gently undulate without departing far from a horizontal plane, while in other parts—those that have been more deeply denuded relative to the diapiric structures—they remain nearly vertical for long distances. In both cases the flow-lines as exposed at the surface may be roughly circular or elliptical e.g. where the surface cuts across the upper part of a dome (Figure 30.32(*a*)) or through the 'stalk' of a mushroom structure (Figure 30.32(*c*)).

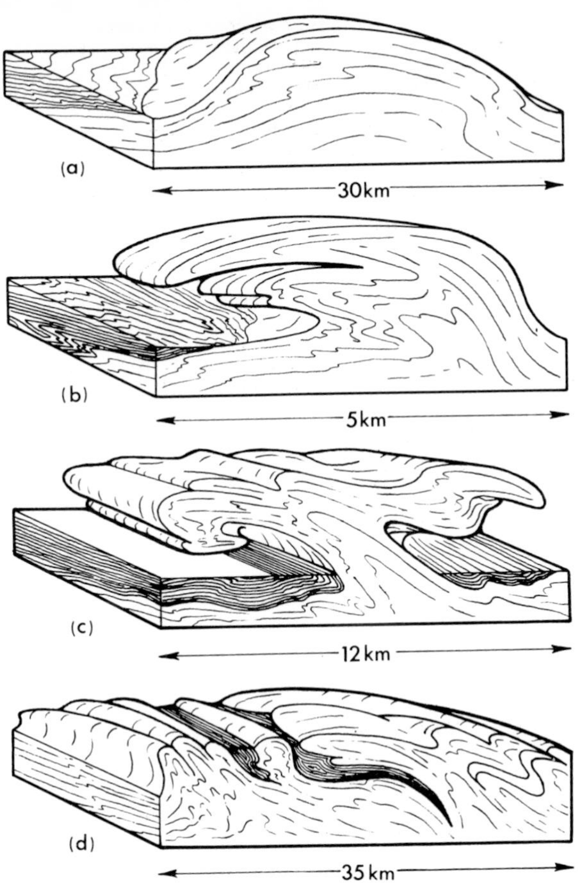

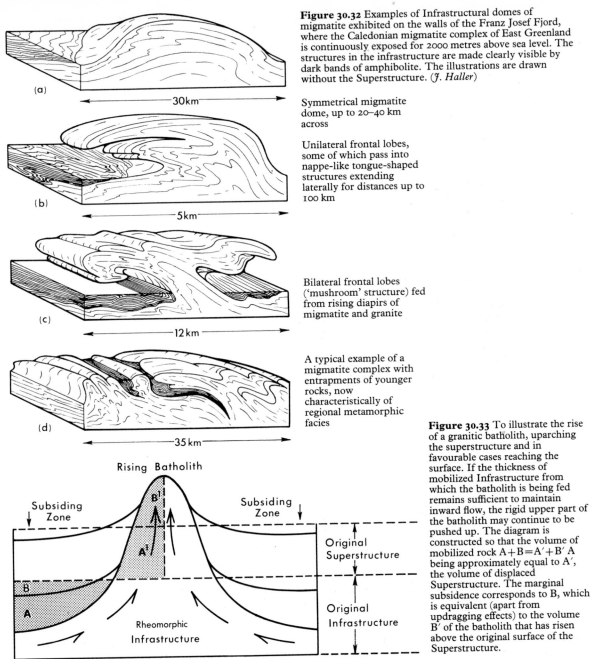

**Figure 30.32** Examples of Infrastructural domes of migmatite exhibited on the walls of the Franz Josef Fjord, where the Caledonian migmatite complex of East Greenland is continuously exposed for 2000 metres above sea level. The structures in the infrastructure are made clearly visible by dark bands of amphibolite. The illustrations are drawn without the Superstructure. (*J. Haller*)

Symmetrical migmatite dome, up to 20–40 km across

Unilateral frontal lobes, some of which pass into nappe-like tongue-shaped structures extending laterally for distances up to 100 km

Bilateral frontal lobes ('mushroom' structure) fed from rising diapirs of migmatite and granite

A typical example of a migmatite complex with entrapments of younger rocks, now characteristically of regional metamorphic facies

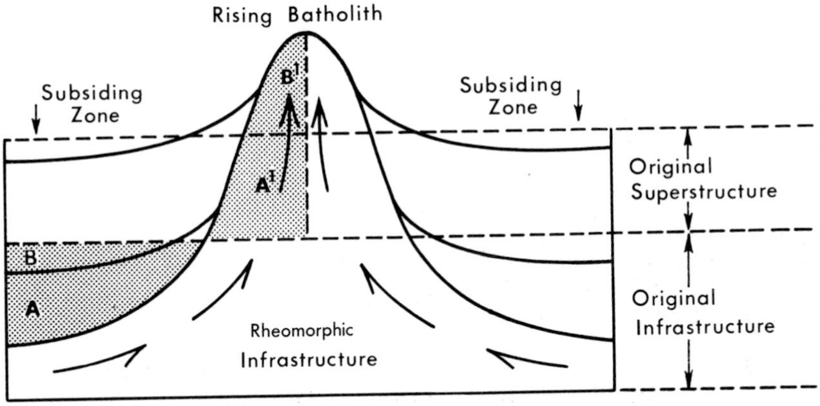


**Figure 30.33** To illustrate the rise of a granitic batholith, uparching the superstructure and in favourable cases reaching the surface. If the thickness of mobilized Infrastructure from which the batholith is being fed remains sufficient to maintain inward flow, the rigid upper part of the batholith may continue to be pushed up. The diagram is constructed so that the volume of mobilized rock A+B=A′+B′ A being approximately equal to A′, the volume of displaced Superstructure. The marginal subsidence corresponds to B, which is equivalent (apart from updragging effects) to the volume B′ of the batholith that has risen above the original surface of the Superstructure.

Figure 30.33 serves to illustrate in the simplest possible way how migmatites and granitic rocks can ascend diapirically to great heights, displacing downwards and sideways the mantling rocks through which they rise. The latter necessarily become folded and faulted as a result of (*a*) uplift and extension of the superstructure over the top of the rising infrastructure, with production of nappes and/or cascades of folds by gravity gliding; (*b*) differential up-drag on the flanks of the rising infrastructure; and (*c*) differential subsidence into the marginal or peripheral sink. For the corresponding phenomena in salt domes see Figures 10.2 and 10.3). Figure 30.34 shows a striking assemblage of diapiric domes of migmatites and granitic rocks in Rhodesia. The structures are

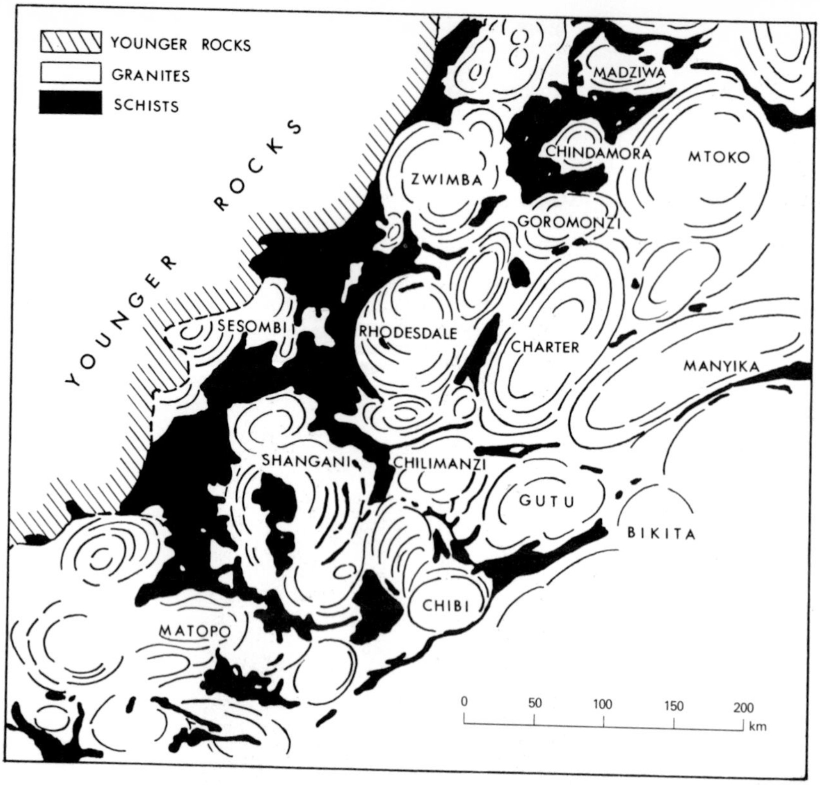


**Figure 30.34** A generalized sketch-map of diapiric domes of granite rocks and migmatites in the Shamvaiian-Bulawayan-Sebakwian orogenic belt of Rhodesia (*A. M. Macgregor*)

clearly due to inward and upward rheid flow of highly mobilized sialic rocks, with complementary downward flow of schists representing metamorphosed basic lavas and sediments. The youngest granitic rocks (pegmatites) have been dated at 2680 m.y., while the oldest probably go back beyond 3000 m.y. Similar 'gregarious batholiths' are known in other Precambrian shields which have been denuded down to what were once very deep levels in the sial.

It should be remembered that the 'flow' exhibited by these rocks is *rheid* flow, not liquid flow. Many geologists refer to rocks showing such flow structures as having been 'magma' at the time when the flowage took place, so giving the erroneous impression that the rocks originated from a liquid state, although they know full well that they were no more liquid than a glacier or a salt dome. If conviction is necessary then reference should be made to a detailed investigation by Berthelsen (1960) of a granitic dome of Precambrian age in western Greenland. The dome has an isoclinally folded mantle (Figure 30.36), but the whole structure results from three fold periods, with the refolding of earlier folds, and final diapiric uprise of one part (Figure 30.35).

Returning to the structure of the Swiss Alps, it will now be apparent that the Pennine nappes of

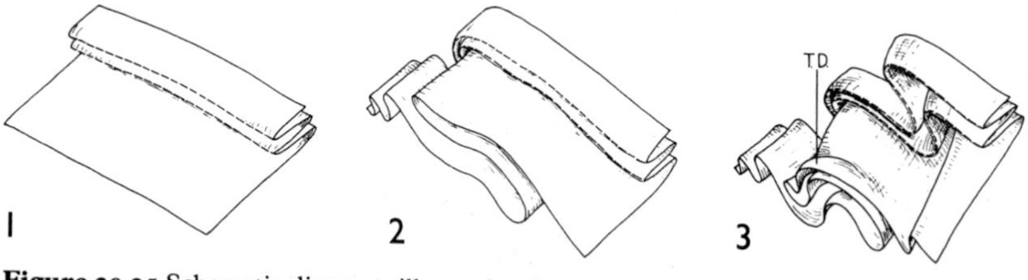


**Figure 30.35** Schematic diagrams illustrating the sequence of folding resulting in the evolution of the mantled gneiss dome of Tovqussap:
(1) Folding about north-west striking fold axes
(2) Recumbent folding of folds (1) about fold axes striking north-eastward
(3) Updoming of (2), resulting in the dome of Tovqussap (TD) and its mantle (Fig. 30.36) (*A. Berthelsen, 1960*)

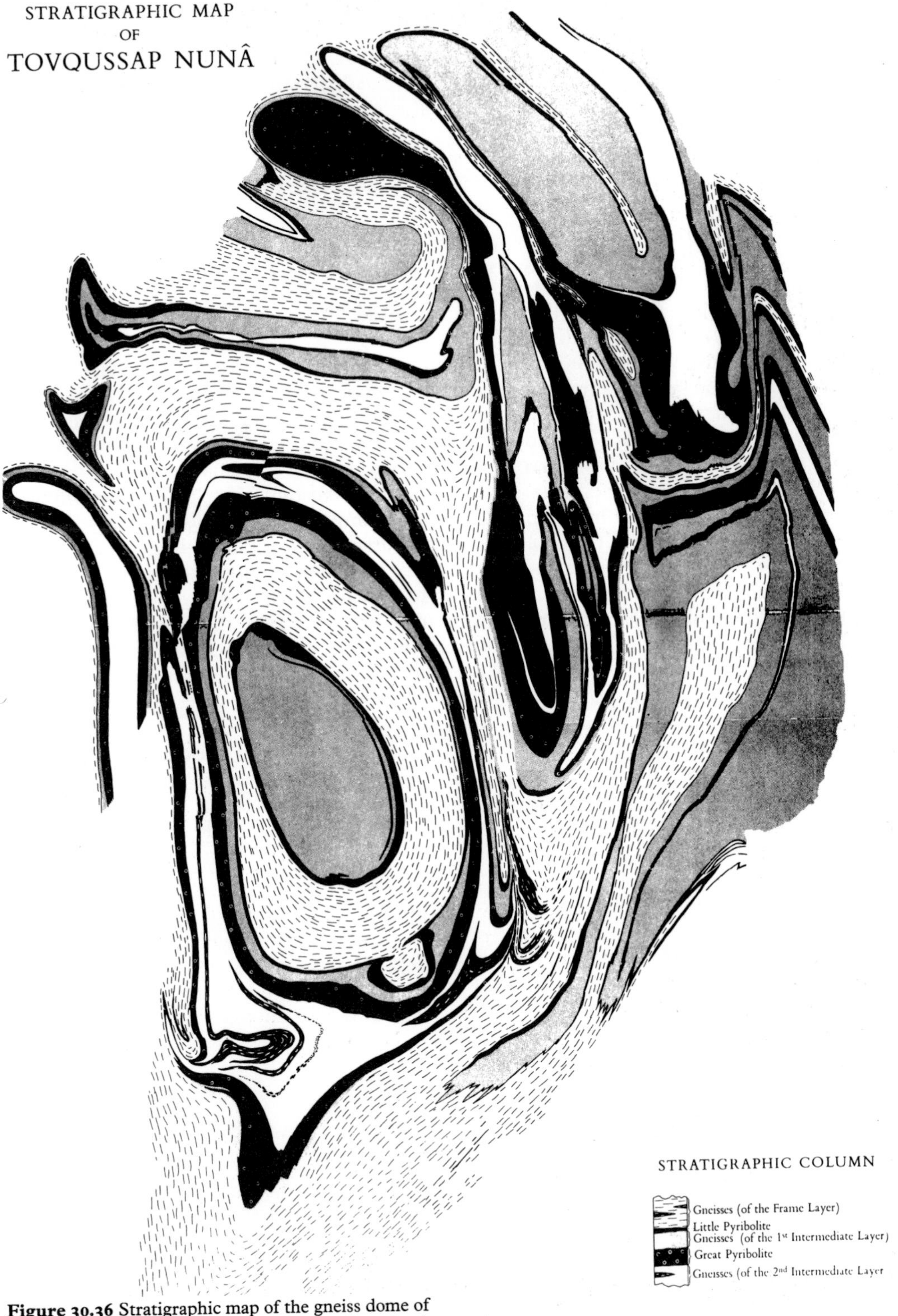


**Figure 30.36** Stratigraphic map of the gneiss dome of Tovqussap Nunâ, and its mantle, Western Greenland (*A. Berthelsen, 1960*)

the Tessin region represent the uppermost level of infrastructure comparable with that depicted in Figure 30.32(*d*). The Helvetian nappes, conjointly with the Helvetian autochthone, and their Hercynian foundation rocks, lie above the migmatite front and form parts of the superstructure. Intervening parts of the Pennine zone, characterized by thrust crystalline rocks, represent the transition zone.

The three structural levels are not unique to Greenland and Switzerland; they are equally characteristic of the Norwegian Caledonian. During the early part of the century the crystalline rocks of western Norway were thought to be Precambrian. It was found, however, that during the Caledonian orogeny they had been migmatized and had moved upwards as infrastructural domes. The overlying Eocambrian, that once rested unconformably on the Precambrian, has been moved into parallelism during the movements. Figure 30.37 illustrates the relationship between the infrastructural domes and overlying regional metamorphic rocks. The linear and arcuate inclusions of Eocambrian within the infrastructural domes of migmatized Precambrian, represent entrapments (*pincées*) like those of Figure 30.32(*d*). Where such entrapments are confined by tongue-like forms of infrastructure, as in Figure 30.32(*d*), they appear as linear inclusions on the horizontal plane of the map. When entrapments have been first infolded in this way and then refolded in the cross-fold direction, they appear as arcuate inclusions like those of the Nonshö area in Figure 30.37. For arcuate outcrops resulting from the folding of folds see Figure 30.2. The gneiss domes of the Haliburton-Bancroft area, so vividly described by Adams and Barlow in 1910, and by others since, seem to bear the hall-marks of infrastructural domes, with arcuate entrapments resulting from the two directions of folding, the main direction striking north-eastward, and the cross-folds striking approximately northwestward.

## Cross-folds

During the early part of the century what are now known as cross-folds were usually referred to as posthumous folds. The reason for this name was that the folds conformed with the direction of folding within the foundation rocks. It was therefore inferred that after 'death', posthumous folds had moved again and impressed their direction upon the overlying younger rocks. These so-called posthumous folds are now known as cross-folds. Movement on the cross-fold direction may be later than that on the main fold direction, it may be synchronous, or it may occur prior to movements on the main fold direction. Each of these possibilities gives rise to somewhat different forms. John Sutton (1960) has summarized the evidence, and illustrated the results of some of the fold sequences, with a full bibliography for northern Scotland. Very little has been written as to the cause of cross-folds, however, and

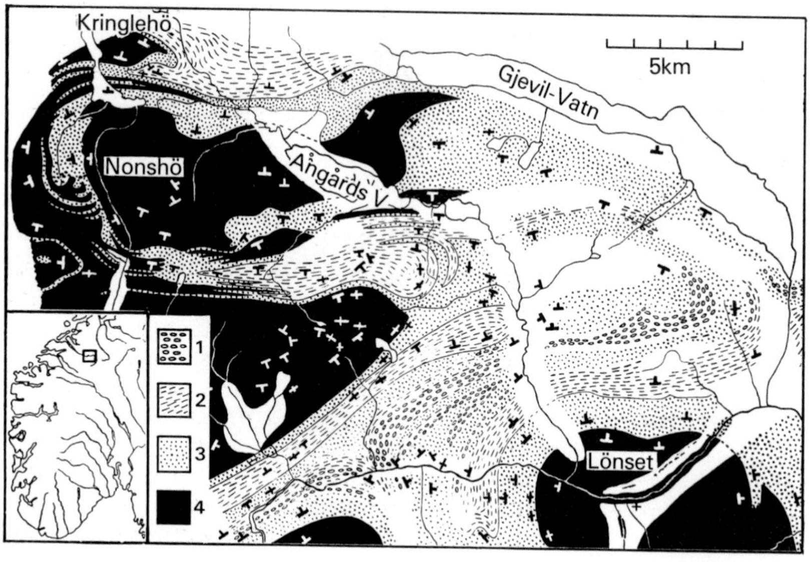


**Figure 30.37** Region to the west of Oppdal, Norway, showing 'basal gneiss' (4) representing domes of Infrastructure, mantled by Eocambrian (3), 'Trondheim schists' (2) and augen-gneisses (1). The domes of 'basal gneiss' are characterized by entrapments (*pincées*) of Eocambrian quartzite and flagstone (3), with arcuate and linear outcrops. (*After H. Holtedahl, 1952*)

it is of interest that Wegmann has illustrated a possible way in which they may be formed. Foundation rocks, whether they lie above or below the migmatite front, retain fold structures which conform with the orogenic movements to which they were subjected, and generally speaking the old direction of folding will make a high angle with that of the younger current orogeny. Whether the foundation rocks move in rigid style as in the Aar massif, or in rheid style as infrastructure, the more granitic bands will tend to rise higher than the others. Like salt, granite is a rock with relatively low specific gravity. Consequently anticlinal cross-folds are impressed on the overlying superstructural rocks, wherever the underlying foundation rocks are granitic (Figure 30.38). Conversely, in studying a map, like that of Scotland, one can sometimes observe that granite appears where cross-folds intersect the main folds.

## The Cycle of Rock Change

On p. 105 reference was made to Hutton's celebrated conclusion that he could find 'no vestige of a beginning'. It will now be abundantly clear how it comes about that no relics of the original crustal rocks have ever been detected. All such rocks, together with the earliest formations that rested upon them, have long ago been transformed into younger rocks by successive periods of orogenic movements and granitization. Since migmatites and granites of a dozen groups of different ages are known, spanning a time range of at least 3,400 million years, it is clear that the granitization process has been many times repeated.

Figure 30.39 is an attempt to summarize graphically the major cycle of rock change, together with a few representative examples of the associated minor cycles. There must be, of course, innumerable 'short circuits'. Many deposits are hardly buried before they are eroded again, and many deep-seated rocks fail to reach the surface before the region in which they occur subsides again. The diagram represents with surprisingly little modification an outline of the geological concepts that Hutton for the first time wrought into an organized science.

The main point of disagreement at the present time relates to the particular level or range of levels where fusion occurs that is responsible for the eruption of acid lavas and tuff flows. In Figure 30.39 the region of fusion is placed well up on the left side,

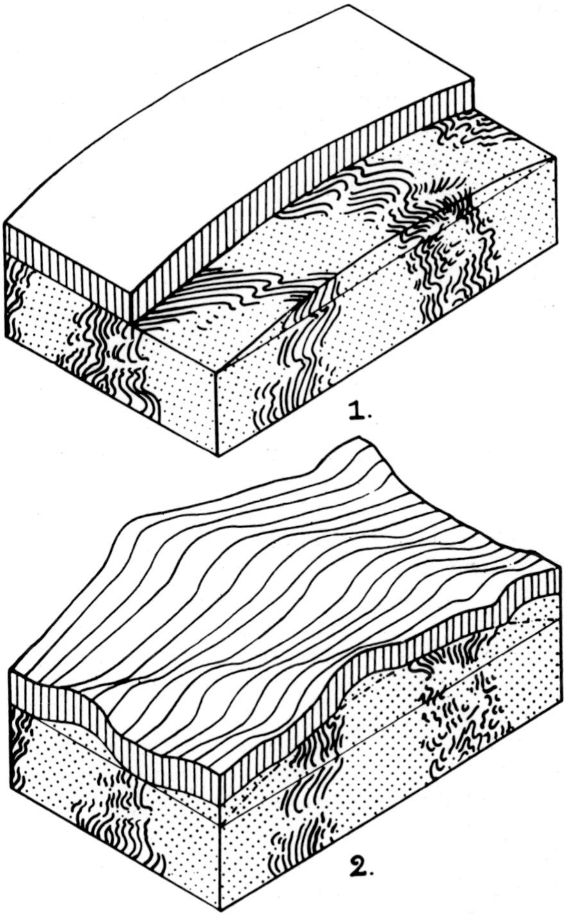


**Figure 30.38** Diagrammatic illustration of a possible origin of cross-folds:

1 Relatively young rocks overlying foundation rocks characterized by granite-rich (dotted) and granite-poor (lined) bands.
2 With the uprise of the migmatite front the granite-rich bands flow upwards, upfolding the overlying superstructure. The resultant cross-folds within the superstructure conform with the strike of the foundation rocks. (*E. Wegmann, 1935*)

consistent with the tectonic evidence, related above, concerned the uprise of granitic rocks essentially as solid bodies. Those who regard granite as having crystallized from melt would place the region of fusion of rocks of granitic composition beneath the bottom of the outermost circle.

A major problem that remains is to account for the concentration of heat, on the continental side of the 'andesite line', above *cold* oceanic crust that is moving downward into the mantle. Where volcanoes are particularly active today at such sites, the great heat is accompanied by voluminous outbursts of volcanic gases. It is therefore natural to infer that

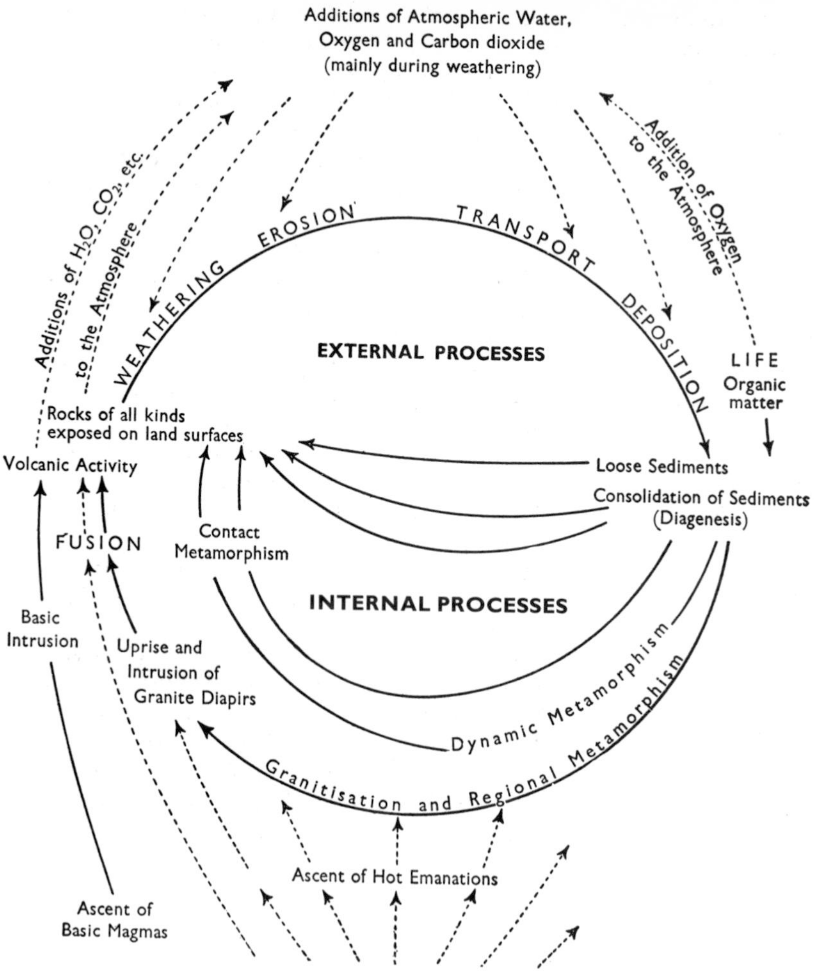


**Figure 30.39** The geological cycle of rock change

such 'hot spots' of the globe, past and present, owe at least part of their heat to the localized ascent of hot reactive gases (Figure 31.13). These gases are the emanations of Figures 30.39 and 31.13.

This kind of problem hardly arises in connection with igneous rocks of basaltic origin, since most basaltic lavas originate in the mantle. At the other extreme there are certain rocks, of which rock-salt is the most familiar example, which can go round the 'cycle' within the upper part of the crust under the influence of little or nothing more than differential pressure due to gravitation. Thick layers of salt accumulate as already described (p. 91), and after burial beneath other sediments they become the source of the astonishing intrusions known as salt plugs or domes. Some of these eventually reach the surface, where the salt is dissolved and returned to the ocean waters from which it came long millions of years beforehand.

## Selected References


BAER, A., 1959, 'L'extrémité occidentale du Massif de l'Aar. (Relations du socle avec la couverture)', *Bulletin de la Société Neuchâteloise des Sciences Naturelles*, vol. 82, pp. 1–160.

BERTHELSEN, A., 1960, 'Structural studies in the Pre-Cambrian of Western Greenland, II: Geology of Tovqussap Nunâ', *Meddelelser om Grønland*, vol. 123, No. 1.

DEWEY. J. F., and BIRD, J. M., 1970, 'Mountain belts and the new global tectonics', *Journal of Geophysical Research*, vol. 75, pp. 2625–47.
DIETZ, R. S., 1963, 'Collapsing continental rises: an actualistic concept of geosynclines and mountain building', *Journal of Geology*, vol. 71, pp. 314–33.
GANSSER, A., 1973, 'Facts and Theories on the Andes' (Twenty-sixth William Smith Lecture), *Journal of the Geological Society*, vol. 129, pp. 93–131.
GÜNTHERT, A. W., 1956–7, 'Über das Alpine Alter der Penninischen Deckengesteine des W-Tessins und der angrenzenden Simplon Region', *Geologische Rundschau*, vol. 45, pp. 194–202.
GÜNZLER-SEIFFERT, H., 1952, 'Alte Brüche im Kreide-Tertiär-Anteil der Wildhorndecke zwischen Rhone und Rhein', *Geologische Rundschau*, vol. 40, pp. 211–39.
HALLER, J., 1955, 'Der "Zentrale Metamorphe Komplex" von NE Grönland', *Meddelelser Om Grønland*, vol. 73, København.
HARRISON, J. V., and FALCON, N. L., 1934, 'Collapse structures', *Geological Magazine*, vol. 71, pp. 529–39; see also *Quarterly Journal of the Geological Society*, vol. 92, pp. 91–102, 1936.
KAY, M., 1951, 'North American geosynclines', *Geological Society of America*, Memoir 48.
KÜNIG, E., 1956, 'Geology and ophiolite problems of East-Celebes', *Verhandelingen van het Koninklijk Nederlandsch Geolofisisch-Mijnbouwkundig Genootschap, Geological Series*, vol. 16. Gedenkboek H.A. Brouwer.
NABHOLZ, M., 1954, 'Gesteinsmaterial und Gebirgsbildung im Alpenquerschnitt Aar Massif-Feengebirge, *Geologische Rundschau*, vol. 42, pp. 155–71.
SCHROEDER, J. W., 1946, 'Mylonites ou brèches sédimentaires', *Compte Rendu des Séances de la Société de Physique et d'Histoire Naturelle de Genève*, pp. 37–9.
SUTTON, J., 1960, 'Some crossfolds and related structures in Northern Scotland', *Geologie en Mijnbouw*, vol. 39, pp. 149–62.
TRÜMPY, R., 1960, 'Paleotectonic evolution of the central and western Alps', *Bulletin of the Geological Society of America*, vol. 71, pp. 843–908.
WEGMANN, C. E., 1935, 'Zur Deutung der Migmatite', *Geologische Rundschau*, vol. 26, pp. 305–50.
— 1961, 'Anatomie comparée des hypothèses sur les plissements de couverture (le Jura plissé)', *Bulletin of the Geological Institution of the University of Uppsala*, vol. 40, pp. 169–82.
— 1962–3, 'L'exposé original de la notion de faciès par A. Gressly (1814–1865)', *Sciences de la Terre*, vol. 9, Nancy, pp. 84–119.

# Chapter 31
# Some Mechanisms

It is useful to be assured that the heavings of the earth are not the work of angry deities. These phenomena have causes of their own.

*Seneca* (4 B.C.–A.D. 65)

## The Problems

Before considering some mechanisms that have been proposed to explain various aspects of global tectonics, and the forms of energy on which they depend, it is important to consider the evidence relevant to the movements to be explained.

Considering the oceanic lithosphere, which spreads away from oceanic ridges and rises, particularly where the movement is relatively slow, as across the mid-Atlantic ridge, the evidence indicates the lithosphere to be in a state of tension, at least for a distance of from 10 to 20 km on either side of the rift. This can be inferred from the presence of normal step faults which have been discovered in various ways. The faults illustrated in Figure 31.1 were inferred from the topography, that is from the presence of steep scarps revealed in seismic profiles obtained by towing a 'fish', containing the instruments, less than 100 m above the sea floor. Profiles recorded from surface ships do not show steep scarps. In other localities similar steep fault scarps have been located by dredging. Since the basaltic rocks collected from these scarps are metamorphosed to green schist towards the foot of the scarp, a transformation that can only have happened below the surface, it is apparent that the scarps lie on the upthrow sides of faults, and that the faults are normal, and indicate extension. Within the mid-Atlantic rift such fault scarps have been examined directly from submersibles (p. 595).

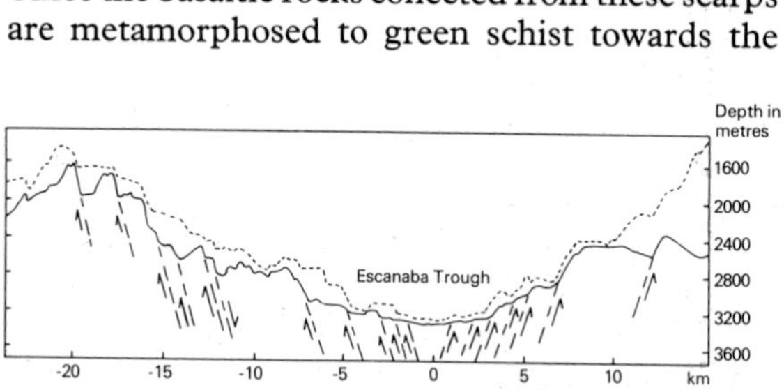


**Figure 31.1** Steep scarps revealed by seismic profiles and most probably resulting from faulting indicative of extension (*After T. M. Atwater and J. D. Mudie, 1968*)

Oceanic ridges and rises are essentially in isostatic equilibrium, having only minor positive anomalies over the median rifts, and minor negative anomalies over the flanks. The ridges and rises are compensated by anomalous, relatively low-density material within the upper mantle beneath them. The uplift of the oceanic lithosphere to form ridges and rises may be sufficient in itself to account for the observed extension. Such a mechanism is akin to that which gives rise to graben over salt domes (Figure 10.8), and to continental rift valleys.

It has been suggested that oceanic lithosphere is pushed away, on either side of the ridges, by basaltic magma intruded along the rifts, but this would imply compression, not tension. Another suggestion that has been made is that the oceanic lithosphere glides away down the flanking slopes of the ridges and rises.

When plates of lithosphere disappear at ocean trenches, they are in a state of tension where they turn over. This is evidenced by the presence of normal faults downthrowing towards the trenches. Seismic studies, however, have revealed that as the oceanic plates travel downwards they come under compression in a down-dip direction. This is of importance in connection with orogenesis which is generally attributed to compression due to the collision of plates, continental lithosphere of low density, and unable to sink, supposedly being thrust and folded. There is, however, no evidence of compression across the down-sinking oceanic plates, nor are the sediments within oceanic trenches folded and compressed; they are characteristically bedded horizontally (Figure 24.6), or tilted at a slight angle.

In 1930 Haarmann proposed an 'oscillation' theory to explain orogenesis with vertical movements of considerable amplitude. He postulated that under certain undefined 'cosmic causes' the viscous materials of the mantle were displaced laterally and gave rise to 'geotumors' (i.e. swells, where the crust was bulged upwards), separated by complementary depressions, 'geodepressions' (i.e. troughs where the crust subsided). The production of these elevated and lowered regions he called 'primary tectogenesis'. Wherever the slopes produced became sufficiently steep, strata carried up with a geotumour were enabled to glide down the flanks under their own weight, and so become folded and piled up in the adjoining depressions. These results he called 'secondary tectogenesis'. Haarmann emphasized that his theory assumes no crustal shortening, the extra length of all the folds in an orogenic belt being ascribed to extension of the original unfolded strata as they were stretched over the geotumor that rose beneath them.

Haarmann's vague appeal to 'cosmic causes' weakened his case and led to an unfavourable reception. But the dominant part assigned to gravity gliding arrested the attention of a few geologists, notably van Bemmelen (since 1933) and Gignoux (since 1937). Van Bemmelen called his development of the hypothesis the 'undation theory', the undations being wave-like alternations of crustal swellings and depressions: rise and fall being ascribed to physico-chemical changes in the mantle. In deference to lack of knowledge these are left vague, but the mantle material is supposed to differentiate into light and heavy portions under appropriate conditions of temperature and pressure. The upward movements are then due to the ascent of lighter materials, balance being preserved by the sinking of the heavier, whereby subsidence of the surface is achieved. This is, of course, a kind of convection.

A similar role played in orogenesis by the uprise of migmatite domes, with corresponding depression between them, was recorded in the last chapter as proposed by Wegmann (1932). It differs from the previous hypotheses in that it rests on observation of continental crust. Since migmatized areas have commonly been invaded by sequences of dyke swarms, prior to migmatization, Wegmann regarded these dykes as heat carriers that had helped to raise the temperature and initiate the chemical migrations that resulted in the rise of the migmatite front (Figure 31.2). A modern edition of all these ideas appears in Dewey and Bird's (1970) proposal that cordillerean-type orogens are 'thermally driven'. Dewey and Bird depict the uprise of a 'mobile core' around which migmatites and granites are evolved as a mantle

**Figure 31.2** Intersecting basic dykes emplaced with gneiss and leptite, depicting three-dimensional extension movements at a high structural level. The hammer head is 10 cm long. Grisselhamn, north-east of Uppsala (*H. G. Backlund*).

When dyke swarms precede migmatization they are curved and pulled apart as a result of the rheid flow of the migmatites. (See the many illustrations in *Sederholm, On Migmatites, Pt II. Bull. Comm. Geol. Finlande, No 77*)

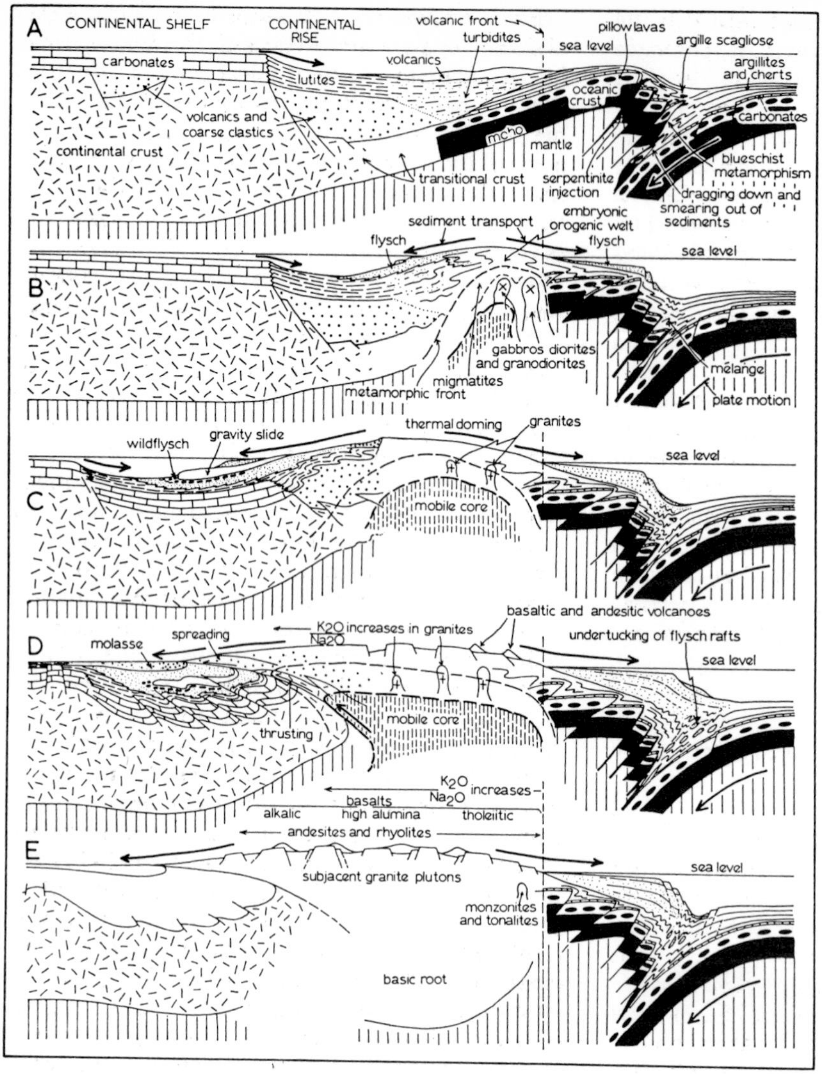


**Figure 31.3** The evolution of a cordillerean type orogenic belt showing the uprise of a 'mobile core' around which migmatites and granites are supposedly evolved (*J. F. Dewey and J. M. Bird, 1970, copyright American Geophysical Union*)

(Figure 31.3). As a consequence of this 'thermal doming' gravity gliding occurs, as in all the previous versions of this hypothesis.

Obviously no satisfactory conclusion can be reached, in the present state of knowledge, as to the fundamental causes of global tectonics. The following pages are therefore confined to a summary review of some of the earth mechanisms that have been proposed.

## The Contraction Hypothesis

In a celebrated paper of a century ago (see p. 230) Kelvin assumed the earth to have cooled from a molten state, partly because the earth is undoubtedly losing heat, and partly because volcanoes had to be accounted for. It is of course, now known, that the fact that the earth is losing heat does not necessarily imply cooling. However, at the moment, the important point is that Kelvin's treatment of his concept, or 'model' of a cooling earth provided a physical background for Élie de Beaumont's hypothesis of a *contracting earth*, devised in 1829 to account for the folding and thrusting of the rocks seen so conspicuously in fold mountain ranges. In its time-honoured form the contraction hypothesis assumed that the earth was contracting because it was supposedly cooling as a result of losing heat. At depths greater than a few hundred kilometres in this model the interior would still retain most of its original heat. The outermost zone, including what we now call the lithosphere and upper mantle, was supposed to have become relatively cold, so that the still rapidly cooling zone beneath it shrank away from it. Obliged by gravity to accommodate itself to the shrinking interior, and so to fit into a smaller space than before, it was thought that the outer zone was thrown into a state of lateral compression. The stress field was supposedly built up until the weaker parts of the outer zone gave way by faulting, thickening and crumpling, thus giving rise to mountain ranges with folded and overthrust structures.

Objections to the contraction hypothesis, which is now abandoned, are firstly that it cannot be proved that the earth is cooling; it may be in thermal equilibrium or it may actually be slightly heating up. Secondly, widespread regions of tension indicated by normal faults are left unaccounted for, and the hypothesis has become totally inadequate since the recognition of sea-floor spreading with rifting and evolution of new ocean floors.

## Physical Evidence that Gravity is Probably Decreasing

Gravity is by far the most familiar form of energy, but certainly the most mysterious. *G*, the universal constant of gravitation, is $6{\cdot}674\,(\pm 0{\cdot}004) \times 10^{-8}$ in c.g.s. units. But that is its value *now*. We have no certainty that it has not varied with time, e.g. with the stage reached by the expanding universe. If *G* has varied with time, the geological consequences will have been fundamental. If *G* has decreased during geological time the earth, like the universe, is likely to have expanded; before considering whether the earth has indeed expanded it is necessary to consider the evidence that suggests that *G* is decreasing.

In 1938 Dirac formulated the hypothesis that *G* decreases with time. Eddington recognized some very remarkable numerical relations and cosmic 'constants' (e.g. the velocity of light and the value of *G*). Dirac examined the apparent coincidences and reached the conclusion that they depended on the particular epoch of time in the history of the universe when the basic observations were made (i.e. the present century). For example, 4000 m.y. ago the galaxies were closer together than they are now, and the value *G* might be correspondingly greater.

For many years the grounds for supposing that *G* decreases with time were entirely philosophical, but there is now some physical evidence that this conclusion is correct. Preliminary results of timing the eclipses of stars by the moon provide evidence in this direction. During the moon's monthly revolution around the earth it passes in front of far distant stars, that are indeed far enough away to be essentially point sources of light. To the eye, the disappearance of one such star at the moon's leading edge, is instantaneous with its reappearance at the moon's trailing edge. If *G* is decreasing, the moon will be very slowly receding from the earth. This would imply that its orbit is increasing and that the time it takes to complete a revolution round the earth is increasing. If it were possible to measure the exact time of the eclipse of a far distant star by the moon then it would be possible to time exactly a revolution of the moon round the earth. If *G* is constant then it would be possible to predict exactly the time at which the moon would next eclipse the same star. If, on the other hand *G* is decreasing then the next eclipse would be later than the predicted time—later by such a minute amount, however, that it could only be accurately detected with some very specialized method of timing. Since 1955, atomic time has been used for this purpose, based on the extremely regular vibrations of caesium atoms. Such measurements have already shown that the moon's period is increasing at the rate of $22{\cdot}2 \pm 3{\cdot}5$ parts in $10^{11}$ a year. The moon's period is partly increasing as a result of tidal friction and this has to be subtracted from the measured increase. In calculating the related decrease in *G* certain assumptions have to be made, but Van Flandern (1975, 1976) points out that agreement between the results is sufficiently good to suggest that the value of *G* is decreasing with time.

Van Flandern (1976) describes two other astronomical experiments that are in progress, and by which it will soon be possible to measure any changes there may be in *G*. In one of these experiments pulses of laser light are beamed, through a telescope, on to one of the retro-reflectors placed on the moon by the Apollo astronauts. The reflected light is picked up by the same telescope, and the time it takes for light to travel to and from the moon is measured, so providing a measurement of the distance between earth and moon. Direct measurements of expansion of the moon's orbit can now therefore be made.

The other astronomical experiment is concerned with radar-ranging measurements of the distances of the planets Mercury and Venus from the earth to determine whether there is any change in their orbital periods. This experiment does not involve tidal friction. Laboratory experiments are also starting at the University of Virginia with the object of detecting any variation in *G*. Van Flandern writes: 'Although it is too early to be certain that the value of *G* is decreasing, the data now available favor that conclusion'.

## The Expansion Hypothesis

In 1935 Halm, a South African astronomer, in the light of views on stellar evolution current at the time, proposed the hypothesis that the earth's original density was considerably higher than it is now. He reached the conclusion that the earth's original mean density was 9·13. The mean radius was then 5430 km, instead of the present 6371 km. He followed Hilgenberg (see Figure 2.13) in thinking that terrestrial expansion has brought about splitting and gradual dispersal of the continents as they moved radially outwards during geological time. Upwellings from the mantle would fill the gaps and so increase the area of the ocean basins. Halm pointed to the Red Sea as a great fracture representing the early stage of this process, twenty years before it was recognized as such by geophysical methods. In 1940 Keindl returned to the idea that continental drift might be explained by slow expansion of an inner core consisting of *degenerate* material of high density (see p. 603). Others who have considered that the earth has expanded are Carey (1956), who found that he could reassemble the continents, as they were before dispersal, provided he assumed the Palaeozoic earth to have had a suitably shorter radius. Heezen attributed the oceanic rifts to cracking and tearing apart of the crust as a result of the earth's expansion.

It should be said at once that consensus of opinion is against expansion of the earth being an explanation of sea-floor spreading. Van Flanders has suggested, however, that it may help to explain the anomaly of a continent, such as Africa or Antarctica, that is almost surrounded by a mid-oceanic ridge, with concomitant evolution of new ocean floor, and no possibility of consumption of the additional oceanic lithosphere. The way in which ideas about continental drift were kept alive by only a small minority of geologists, is an emphatic reminder not to discard hypotheses lightly. It is therefore important to record the palaeogeographic evidence that the earth has expanded, as assembled by Egyed, in 1955, when he was Director of the Geophysical Institute at Budapest.

Egyed's elegant method for deciding between a contracting and an expanding earth was based on the variation of water-covered areas since the early Palaeozoic. Figure 31.4 shows diagrammatically how such areas might be expected to vary if the earth's volume remained unaffected by general global contraction or expansion, and if the total volume of water also remained unchanged. The oscillations superimposed on the horizontal line represent an alternation of marine transgressions and regressions, such as are well known to have occurred. The horizontal line should, however, almost certainly rise to the right, to take account of the very high probability that the volume of the seas and oceans has increased during geological time instead of remaining constant. There has been much discussion about the extent to which volcanic gases, mainly $H_2O$, may have added to the volume of the hydrosphere (see p. 667). After a careful review of all the relevant data, Kuenen reached the conclusion that since the beginning of the Cambrian the inrease has not been more than 4 per cent of the present volume.

On a contracting earth the reduction of area would involve a rise of sea-level, leading to an increase of the water-covered continental areas, as

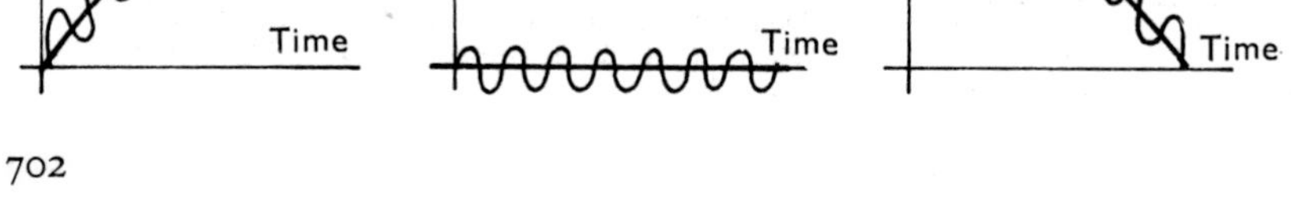

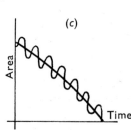


**Figure 31.4** Variations of water-covered continental areas during geological time: (a) global contraction, with cumulative increase of water-covered areas; (b) constant volume, with no cumulative change; (c) global expansion, with cumulative decrease of water-covered areas. The oscillations superimposed on the lines showing the general tendencies result from crust extensions and contractions due to other processes, such as convection. (*L. Egyed*)

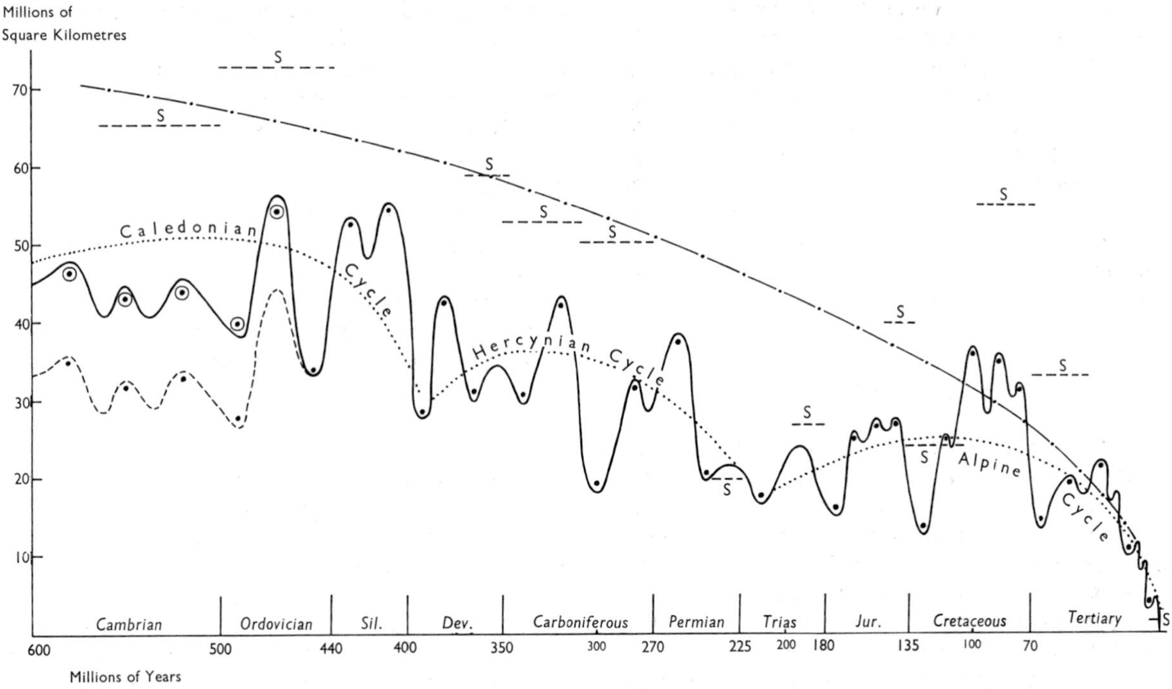


**Figure 31.5** Graphs showing estimates of sea-covered continental areas since the beginning of the Cambrian, computed by L. Egyed from palaeogeographic maps of N. M. Strahov (represented by the intervals marked -----S-----) and from those of H. and G. Termier (represented by black dots). The latter have been increased for the Cambrian and part of the Ordovician, as explained in the text, to points represented by circles enclosing black dots. It will be seen that the general tendency corresponds with that indicated theoretically in Fig. 31.4(*c*)

shown by Figure 31.4(*a*). On the other hand, with an expanding earth, progressive growth of the ocean basins would lead to emergence of the land (Figure 31.4(*c*)).

In order to judge how the actual earth has behaved, Egyed took two sets of palaeogeographical maps, one compiled by N. M. Strahov (a series of 12), the other, independently, by H. and G. Termier (a series of 34); to ensure accurate comparison he transferred the outlines to equal-area projections and then measured the continental areas that had been submerged at each of the epochs represented. In Figure 31.5 Egyed's results are plotted against the corresponding ages in years. The main curve for the Termier series has been modified towards the Cambrian end to allow for the probability, not known when the maps were prepared, that the Cambrian and Ordovician orogenic belts of the lands surrounding the Indian Ocean and the South Atlantic, including part of Antarctica, were then geosynclinal seaways (Figure 31.6). Since the rocks now exposed in these regions were formerly regarded as wholly Precambrian, they could not have been included in Egyed's measurements. To allow for this each of the five older areas of the Termier series has been increased by $12 \times 10^6$ km$^2$. The estimate based on Strahov's Cambrian map should perhaps be increased by the same amount, but this has not been done in Figure 31.5, because several of

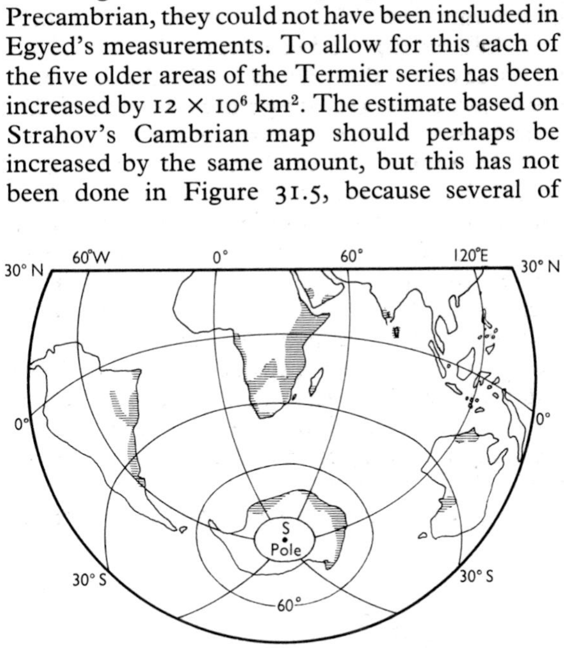


**Figure 31.6** Map showing the distribution of metamorphic rocks, granites and pegmatites formerly thought to be Precambrian but now known by their radiometric ages to have been formed during the Cambrian and Ordovician periods, when the shaded areas were probably in part geosynclinal seas

Strahov's areas (including the Cambrian) are systematically bigger than the Termiers', the reason being that they cover much longer intervals. On the whole, most of the Strahov areas are probably too big, and some of the original Termier areas too small. However this may be, 'the result is equally clear,' as Carey remarked, 'whether we use the Russian or the French palaeogeographical restorations. Further advance in stratigraphic knowledge is likely to improve the details of these curves but not to change their general form.' Both sets of curves descend with time, consistently with the hypothesis of an expanding earth.

It will be noticed, however, that the fluctuations are very considerable. These indicate that superimposed on the general expansion there have been other processes at work which were much more effective than expansion alone in bringing about changes of sea-level. One of these would be subcrustal convection, and another would be the growth and consolidation of new ocean floors. These would lead sometimes to marine invasion of the lands, a reversal of the normal trend; sometimes to an abnormally rapid marine withdrawal from the lands. Another important process, superimposed on the others during ice ages, is the alternation between piling-up of ice on the land and its restoration to the sea by melting. This is responsible for relatively rapid downward and upward swings of sea-level.

## Rate of Increase of the Earth's Radius

Considering only the effects of global expansion during the interval between some past time (e.g. the beginning of the Cambrian) and the present day, let us suppose with Egyed, for the purposes of a first approximation:

(i) that the increase, $a$, of the earth's surface area was confined to the ocean basins, $A$ being the sea-covered area of today, $361 \times 10^6$ km², and $D$ the mean depth of the water, 3·8 km; and

(ii) that the volume of water $AD$ remained constant, so that while the area increased from $(A-a)$ to $A$, the mean depth decreased from $(D+d)$ to $D$, according to the relationship

$$AD = (A-a)(D+d)$$

from which we see that $a = Ad/(D+d)$.

Figure 31.5 indicates $45 \times 10^6$ km² as the area which is now land but was submerged at the beginning of the Cambrian, according to the modified measurements of the Termiers' maps. If the general relief of the late Precambrian or early Cambrian coastal and continental shelf regions was similar to that of today (judged by the hypsometric curve, Figure 2.4), then the sea-level of that time would have been about 250 m higher than now, in the sense that the mean depth of the sea was that much greater. 250 m, thus estimated, is the value of $d$ in the above equation. Since $A$ and $D$ are known, it is easy to calculate that on this basis $a = 22 \times 10^6$ km². The corresponding area of the earth 600 m.y. ago was $(510-22) \times 10^6$ km² and its radius 6230 km. The radius has increased by 141 km to the present value of 6371 km, giving an average rate of 0·24 mm/year.

Proceeding in the same way with the data from Strahov's maps, we find the sea-level at the beginning of the Cambrian to have been about 450 m above that of today. The corresponding value of $a$ is $38 \times 10^6$ km²; the radius being 6130 km. The radius has increased by 241 km in 600 m.y., giving an average rate of 0·40 mm/year.

The rapid oscillating descent of the curves at the right-hand end of Figure 31.5 should not be interpreted to mean that the earth is now expanding at an accelerating rate. If the Tertiary and Pleistocene changes of sea-level were dealt with on the above lines, the results would suggest that during the Tertiary the average rate of increase of the radius of the earth rose to 1·7 mm/year, and that during the Pleistocene the average rate had reached 30–40 mm/year. The latter estimate varies according to the allowance made for water still to be restored to the oceans from ice-sheets. The high Tertiary estimates are based on changes of sea-level which can be almost entirely accounted for by the growth of the ocean floors.

Another method of estimating the rate of global expansion suggested by Egyed is based on the well-known fact that the length of the day is slowly increasing, or in other words, that the rate of rotation of the earth is slowly decreasing. Although they are extremely minute, the actual changes are subject to considerable and rather sudden fluctuations, which are far from being properly understood. However, on average the day has been lengthened at a rate of 2 seconds per 100,000 years. H. Spencer Jones, late Astronomer Royal, wrote 'If there were an expansion of the earth as a whole, so that its radius increased by 6 inches, the day would increase by 5 milliseconds' (i.e. by 5/1,000 of a second). Thus the lengthening of the day by 2 milliseconds in a century corresponds to a radius increase of $\frac{2}{5}$ of 6 inches, i.e. about 60 mm, or 0·6 mm/year. It is remarkable that this figure is of the same order as the estimates based on the cumulative withdrawal of the sea from the land.

Slowing down of the earth's rotation has generally been ascribed to tidal friction, especially in shallow seas such as the North Sea, where the tides act like a brake on the rotating earth. However, at best, this accounts for only part of the slowing down. Moreover, the braking effect is more or less compensated by the tidal effect exerted on the atmosphere by the sun. It was first noticed by Kelvin that the atmospheric tide caused by solar attraction speeds up the earth's rotation. Another explanation has therefore to be found for the fact that, nevertheless, the earth's rotation *is* slowing down and the day *is* growing longer. Expansion of the earth at a rate which is consistent with other evidence appears to meet the requirements.

John W. Wells, of Cornell University, discovered a method of measuring the lengthening of the day, not merely over the last few thousand years, but over periods going back to the Silurian. This places the whole subject on a firm, long-term basis. In the corals of reefs and atolls the rate of secretion of calcium carbonate is greatest during bright sunshine and falls off at night or in darkness. One would expect this rhythm to result in a waxing and waning growth-front, which would leave a record of minute ridges and furrows as seen in Figures 31.7 and 31.8. These 'growth-lines' are themselves subject to seasonal fluctuations which give rise to well-marked swellings that make it possible to recognize annual increments of growth. Wells finds that in corals now living (Figure 31.7) the growth lines can be counted like varves. The number 'hovers around 360 in the space of a year's growth', an average that strongly suggests that the growth lines represent days. A gloomy day in the rainy season might fail to give a distinct ridge and

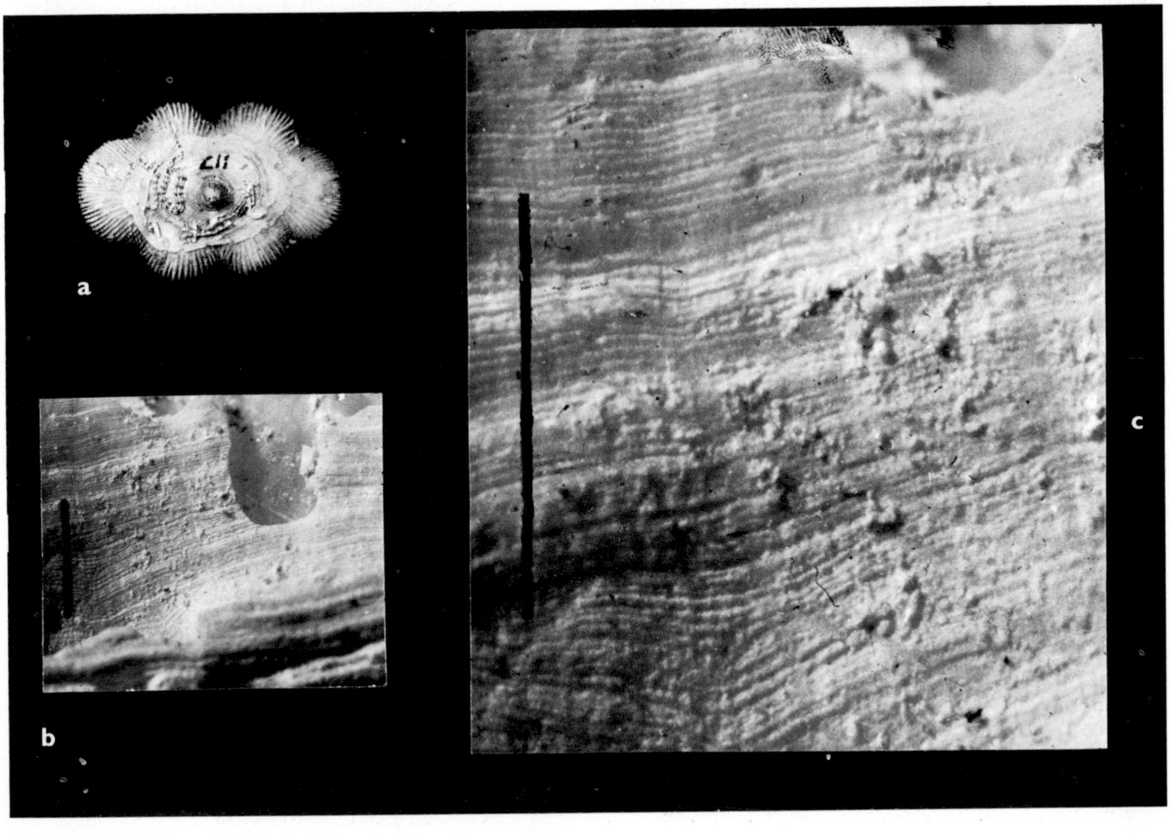


**Daily growth lines of recent and Silurian corals**
**Figure 31.7** *Manicina areolata* from the living reef (depth 34–37 m) off Dry Tortugas, Florida; (a) basal aspect of coral; (b) epithecum, showing growth lines ×17; (c) part of (b) ×60 Vertical line = 1 mm (*John W. Wells*)

**Daily growth lines of recent and Silurian corals**
**Figure 31.8** *Holophragma calceolides* from the Lower Silurian Gotland, Baltic Sea: (a) young coral with growth lines indicating an age of about 240 days, ×6; (b) part of (a) ×23 (*John W. Wells*)

so one might expect a count to be rather on the low side of 365.

Few fossil corals have retained these fine markings without corrosion or abrasion, but Wells has found well-preserved specimens of three species of corals from the Middle Devonian of New York and Ontario. In all of them the growth lines are well over 365 per annum, usually about 400, and ranging between extremes of 385 and 410. Two upper Carboniferous corals from Pennsylvania and Texas gave 390 and 385 growth lines per annum. These preliminary results, plotted on a time-scale in Figure 31.9, leave no doubt that in the past the days were shorter and that there were more of them in a year. If we take the Middle Devonian average of 400 days in a year, the average rate of lengthening of the day works out at 2·2 seconds per 100,000 years, corresponding to an average increase in the earth's radius of 0·66 mm/year. If the maximum of 410 days be adopted the radius increase would be 0·85 mm/year. For the Upper Carboniferous 390 days gives a radius increase of 0·6 mm/year. All these estimates are necessarily maxima, because some allowance must be made for tidal friction, which would be considerably greater in the past when more of the continental areas were submerged, and also for the retarding effects of the earth's internal tides. Expansion on a sufficient

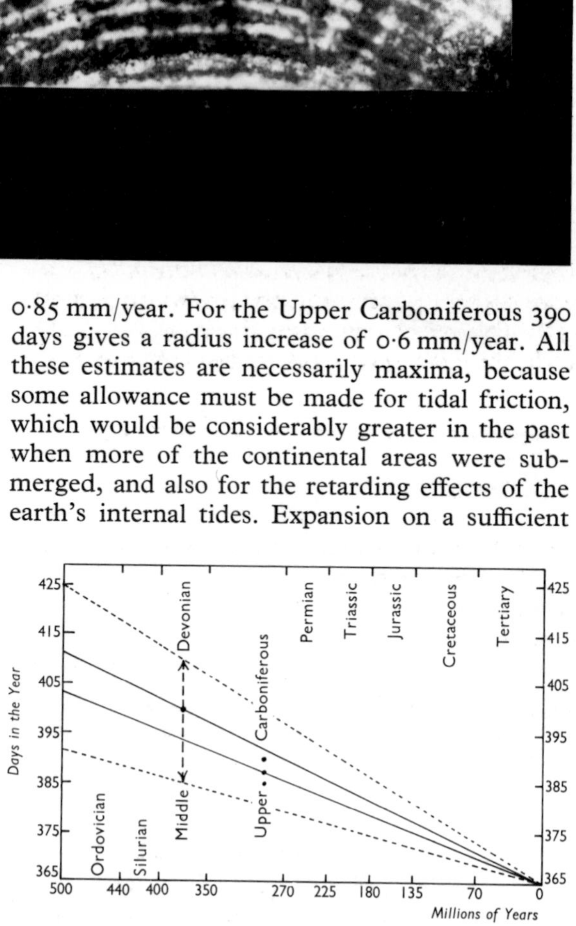


**Figure 31.9** Daily growth lines per year of Middle Devonian and Upper Carboniferous corals plotted against geological time, to show the decrease in the number of days in the year, i.e. the lengthening of the day during geological time (*From measurements by John W. Wells*)

scale to account for the whole of the growth of the Atlantic and Indian Oceans since the early Mesozoic could not have occurred by any process known to us. The time was too short. Suppose the radius has been increasing at 0·25 mm/year, or 0·25 km/m.y. Then, 200 m.y. ago the radius was 50 km shorter and the area of the earth $8 \times 10^6$ km² less extensive than today. Now consider the oceanic areas involved. The Arctic, Atlantic and Indian Oceans cover a total area of $170 \times 10^6$ km² and even if only half of this represents 'new' oceanic areas, it is still very large compared with the amount that can be ascribed to expansion. However one manipulates the figures—even doubling the rate of expansion adopted above—it remains obvious that the new ocean floors have grown very largely at the expense of the Pacific floor. This can only mean that circulation processes operating in the mantle, such as thermal convection currents, have been of greater importance than global expansion in 'engineering' the latest separation of the continents.

## Heat Flow

Penetration of the earth's crust by bore-holes and mines shows that the temperature increases with depth. The rate of temperature increase or, in other words, the *geothermal gradient* varies considerably from place to place. In assessing the downward increase of temperature we have to take into consideration not only the gradient observed near the surface, but also the thermal conductivity of the rocks through which the heat is passing. Since *thermal conductivity* is defined as the amount of heat transmitted across unit area (1 cm²) in unit time (1 s) as a result of unit temperature gradient (1°C/cm) at right angles to the area, it follows that the heat flow through a given rock varies much less from place to place than the corresponding temperature gradient taken alone. Heat flow is expressed in units of cal $\times 10^{-6}$/cm²/s. The average heat flow through continents and oceans does not differ significantly, the repective arithmetic mean values worked out by Lee and Uyeda for grids of equal area being respectively 1·41 and 1·46 in units of cal $\times 10^{-6}$/cm²/s (1 cal $= 4{\cdot}186 \times 10^7$ erg).

Apart from the heat inherited from the early days of our planet, the earth is endowed with a source of heat by radioactivity. From the facts that granites are richer in the radio-elements than granodiorites and andesites, andesites richer than basalts, and basic rocks richer than ultrabasic rocks, it is clear that these heat-generating elements are strongly concentrated in the crust, and particularly in rocks of sialic composition. The relevant data are briefly summarized in the following table

Average Abundances of the Radio-elements in grams per $10^6$ g of rock (p.p.m.)

| ROCK GROUPS | U | Th | $^{40}$K |
|---|---|---|---|
| *Sialic* rocks: mainly granites, granodiorites, migmatites, schists and sediments | 3·0 | 10·0 | 2·2 |
| *Basaltic* rocks | 0·7 | 3·0 | 1·1 |
| *Ultrabasic* rocks | 0·013 | 0·05 | 0·001 |

Heat Production from Radioactivity in Rocks

| ROCK GROUPS | In calories per year | | | | | In cal per second per cm³ |
|---|---|---|---|---|---|---|
| | per $10^6$ g of rock | | | | per $10^6$ cm³ | |
| | U | Th | $^{40}$K | *Total* | *Total* | *Total* |
| *Sialic* | 2·3 | 2·1 | 0·5 | 4·9 | 13·7 | $4{\cdot}34 \times 10^{-13}$ |
| *Basaltic* | 0·5 | 0·6 | 0·2 | 1·3 | 3·8 | $1{\cdot}20 \times 10^{-13}$ |
| *Ultrabasic* | 0·01 | 0·01 | 0·0002 | 0·02 | 0·07 | $0{\cdot}02 \times 10^{-13}$ |

It is apparent from the table on p. 707 that the sialic crust is much richer in radioactive elements than the oceanic crust, and that the mantle is poor in such elements. About three quarters of continental heat flow is of radioactive origin. The remarkable similarity between the average heat flow from continents and oceans therefore presents a problem. It has been suggested that the explanation lies in a difference between sub-oceanic and subcontinental mantle, with a concentration of the radioactive elements in the former. Since continents and oceans gradually exchange places with time, such an explanation would raise even further problems. One is tempted to think that sialic crust is opaque to some form of heat flow from the mantle, but at present there is no scientific explanation.

The maximum heat flow (average 2·5) through

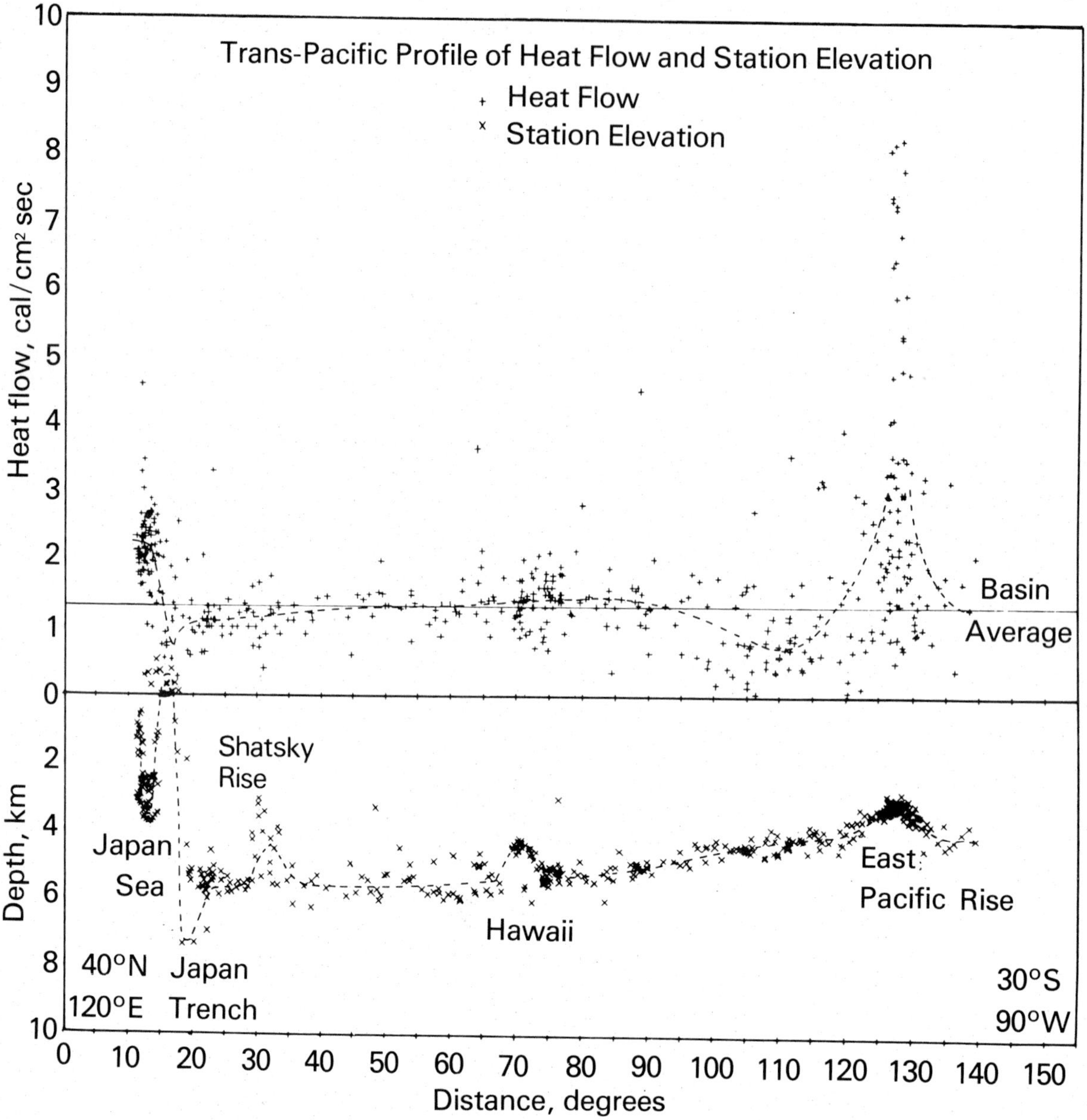


**Figure 31.10** Graphs illustrating the variation in heat flow (top) and in station elevation (bottom) across the Pacific

*(M. G. Langseth, Jr, and R. P. Von Herzen 1968. Reproduced by permission of John Wiley and Sons)*

the ocean floors is confined to narrow bands along the crests of the mid-oceanic ridges. On the flanks of the ridges, however, the heat flow is relatively low, the average being only 0·8. This gives the surprising result that the heat flow right across the ridges, including highs and low, is very little higher than that of the ocean basins, average 1·3. The variation in heat flow across the Pacific is illustrated in Figure 31.10. It will be seen that the heat flow is at its lowest at the bottom of the trench, and this applies to all trenches. On the continental side of trenches, as illustrated by Japan (Figures 31.10 and 31.11), the regions of the earth's highest heat flow occur, in close proximity to the lowest within the trenches. As Langseth and Von Herzen (1970) remark, it is quite paradoxical that the net heat flow over the mid-oceanic ridges, where oceanic lithosphere is created, is actually lower than across the deep ocean trenches and their landward sides, where cold oceanic lithosphere is disappearing into the mantle. That the heat flow to the landward side of the trenches is indeed high, apart from actual measurements, is witnessed by the volcanic activity and hydrothermal springs of the Pacific 'girdle of fire'.

**Figure 31.11** The distribution of heat flow in Japan from data by the *Earthquake Research Institute, University of Tokyo*, 1962. The figures are in units of cal × $10^{-6}$/cm²/sec. Small stars indicate active or recently active volcanoes. For additional confirmatory data see *Nature*, vol. 199 (1963) p. 364.

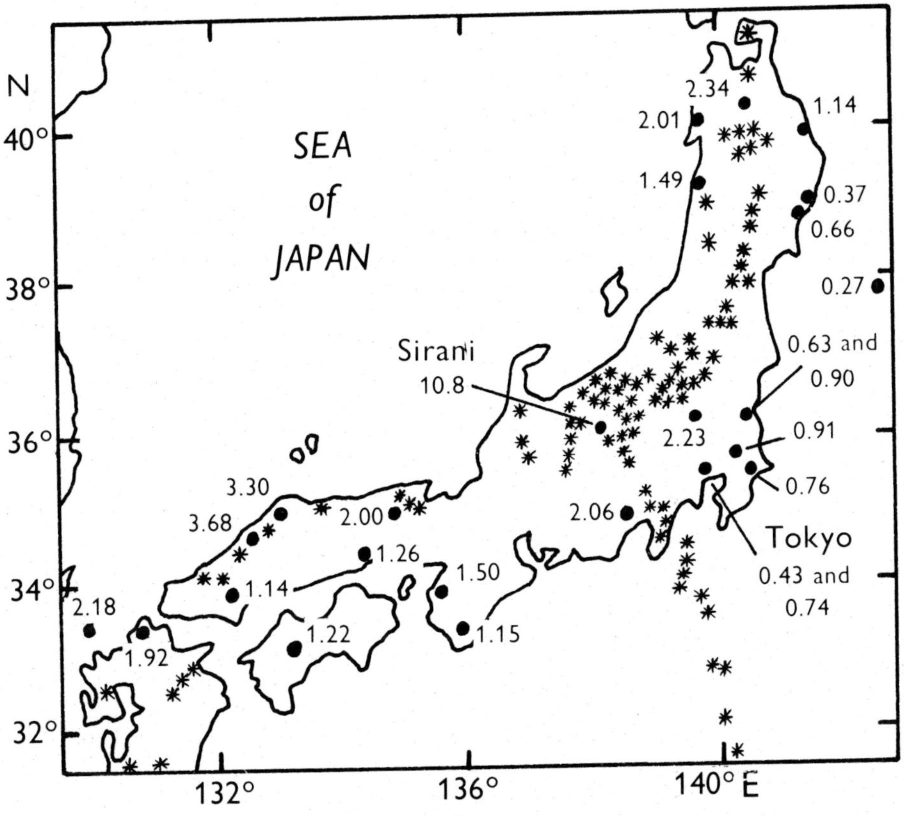


## The Thermal Convection Hypothesis

The operation of thermal convection currents within the mantle, as well as in the core, has already been tacitly assumed to be probable in previous chapters. This is the most favoured hypothesis, at the present time, for accounting for the transport of oceanic lithosphere necessitated by sea floor spreading. As a mode of propagating heat, convection was discovered in 1797 by Count Rumford. Sub-crustal convection within the earth was suggested by William Hopkins in 1839 and specific applications to geological problems were discussed by Osmond Fisher in his *Physics of the*

*Earth's Crust* of 1881. He pointed out that the frictional drag on the underside of the crust would be expected to promote mountain building along the margins of the continents, where two systems of currents approached one another and turned downwards. He also recognized that 'The existence of convection currents beneath the cooled crust of the earth at once furnishes a means of obtaining those local increments of temperature which in some form or another appear to be needful in order to explain volcanic phenomena.'

Although there seems to be no adequate alternative, the relationship between movement of plates of lithosphere and convection remains hypothetical, partly because all kinds of convection require an ample supply of energy. Ewing has expressed the position very clearly: 'Whatever the nature of the energy reservoir there is a continual drain on it for various thermal processes as well as for the mechanical deformations. Ultimately the supply of available energy must be depleted. There is no clear evidence of a decrease in the rate of the various geological processes, hence the terrestrial reservoir of available driving energy must be large compared with the total drain throughout geological history'. Like Ewing, but considering mountain building alone, van Bemmelen has clearly recognized that radioactivity and any relics of original heat that may remain are far from fulfilling the requirements. In addition to these, he writes: 'The earth apparently contains stores of energy of enormous potentiality, which are more or less gradually released during the lifetime of our planet'. Van Bemmelen suggests the liberation of inter-atomic energy as a result of changes of temperature and pressure in depths. This very broad generalization includes the more specific source of energy derived from phase changes mainly at the boundary between the outer core and the mantle (p. 603). At each phase discontinuity, the more compressed material at the greater depth has more energy of deformation—like a compressed spring—than the material above. If the overhead pressure were reduced the material would eventually respond by a certain amount of expansion, partly by springing back and partly by a creep-like flow. The work done by compression in rupturing the chemical bonds of material like peridotite, and in changing its density from 3·32 at the top of the mantle to 9·71 at the top of the core, has been calculated to be about $16 \times 10^{10}$ erg* per gram. Still more energy is locked up in material that has suffered pressure-ionization. The energy required to move an electron from an atom varies considerably from one element to another. To strip off additional electrons from an atom already ionized takes still more energy: the second more than the first, the third more than the second and so on. The following figures, in electron-volts, refer to the first electron to be removed from each of the elements of olivine, $(Mg,Fe)_2SiO_4$, the chief mineral of peridotite:

| *Element* | eV | *Element* | eV |
|---|---|---|---|
| Si | 8·15 | Mg | 7·64 |
| O | 13·62 | Fe | 7·90 |

If, for simplicity, we regard the mantle and core material as having the composition of olivine, the energy required to ionize the atoms of a single molecule is $78 \text{ eV} = 125 \times 10^{-12}$ erg. Since the number of molecules in 1 gram of olivine is about $4{\cdot}2 \times 10^{21}$, the energy required to ionize all the atoms in 1 gram of olivine is

$$4{\cdot}2 \times 10^{21} \times 125 \times 10^{-12} = 5{\cdot}25 \times 10^{11} \text{ erg}.$$

The minimum amount of energy locked up by compression and ionization in the equivalent of each gram of olivine near the top of the outer core is therefore:

| | |
|---|---|
| By compression and dissociation into atoms | $1{\cdot}6 \times 10^{11}$ erg |
| By ionization | $5{\cdot}2 \times 10^{11}$ erg |
| *Total* | $6{\cdot}8 \times 10^{11}$ erg |

*If there should be a relief of pressure*, expansion will occur and a corresponding amount of energy will be released. From the average rate of radial expansion already adopted, 0·25 mm/year, it is easily calculated that the mass of the outer core which is transformed into mantle material each year averages $37 \times 10^{16}$ g. Throughout the mantle the discontinuities and other compression effects are similarly raised by 0·25 mm. The approximate amount of energy released, on the simplifying assumptions made above, is

$$6{\cdot}8 \times 10^{11} \times 37 \times 10^{16} = 2{\cdot}5 \times 10^{29} \text{ erg/year}.$$

Some of this energy will be used in raising the temperature of the materials along the mantle/core boundary, so helping to maintain convection

* See p. 297 for the definition of the *erg*. Another unit of energy, in common use by atomic physicists, is the *electron-volt* (eV). $1 \text{ eV} = 1{\cdot}6 \times 10^{-12}$ erg. 1 kilowatt hour $= 2{\cdot}247 \times 10^{25}$ eV.

currents in the mantle. The remainder will be expended in mechanical work.

The sub-crustal currents that are interpreted as the outer horizontal parts of convection cells have speeds that are all of the same order, however they are estimated. In 1928 Harold Jeffreys showed that the viscosity of the mantle would have to be $10^{26}$ or $10^{27}$ (c.g.s. units) to prevent convectional movements. From rates of isostatic recovery and certain astronomical indications it is known that the actual viscosity is of the order $10^{22}$. Convection is therefore physically possible, given slight differences of density such as might be due to differences of temperature in the same material, or differences of material at the same temperature. Since 1928 various theoretical studies based on a wide range of geophysical data have all indicated that speeds of a few centimetres a year would suffice to exert an adequate drag on the undersurface of the crust, and that such speeds are easily reached in all the earth-models investigated. The rates of lateral movements of the lithosphere, whether by sea-floor spreading, or by present-day displacement along transform faults (San Andreas 5 cm/year, Great Alpine fault, New Zealand 2·5 cm/year) are of the right order.

Whatever may be the source of the activating energy, the general consistency of the above estimates is striking, but cannot be regarded as evidence, as the following example of another consistency clearly shows. The earth is now known to be (in highly exaggerated language) very slightly 'pear-shaped'. This was suspected from gravity anomalies and has been confirmed by the differences between (*a*) the calculated orbits and velocities of the satellites used in space research and (*b*) those actually observed. The northern hemisphere is nipped in, so to speak, while the southern hemisphere bulges out. The total radial distortion is very little, being less than 160 m, compared with 23,500 m for the polar flattening and equatorial bulge. But small as it is, this distortion implies the existence of deep-seated lateral variations in density which in turn imply convection currents. A. L. Licht has shown that a global system of convection currents consisting of three main cells (Figure 31.12) would produce and maintain the required distortion if the average speed of the currents was 3·6 cm/year.

This is a particular convection hypothesis devised for a particular purpose. But the same global system of three main cells would not fit the present-day distribution of continents and oceans,

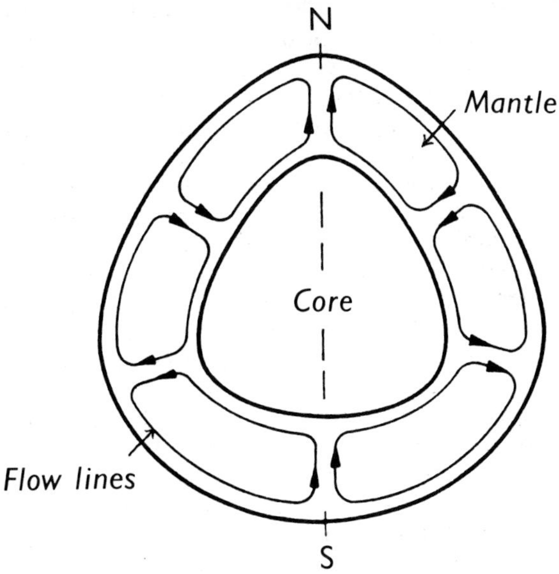


**Figure 31.12** A global system of major convection currents about a N-S axis of symmetry that would account for the indications that the earth is slightly pear-shaped (*After A. L. Licht 1960, Journal of Geophysical Research, vol. 65, p. 349*)

which requires a system of peculiar complexity. If we knew the latter it might account for the anomalies of our planet's shape as well as Licht's system, or even better.

A thermal convection current mechanism for 'engineering' the development and growth of new ocean basins is illustrated by Figure 29.1. At the time when it was proposed it was not known that the ocean floor is moving; it was thought that the sialic continental crust 'sailed' on a basaltic layer. It is now generally thought that thermal convection is confined to the upper part of the mantle, but the form of convection cells that will 'fit' the transform faults intersecting mid-oceanic ridges, and conform with the high heat flow, and volcanic activity, on the continental side of trenches has not yet been suggested. Figure 31.13, although completely hypothetical, is reproduced from the second edition of this book as Holmes' (1964) suggestion for explaining the high heat flow and volcanic activity, close to the side of a descending plate of oceanic lithosphere.

## Experimental Studies of Thermal Convection

The first systematic experiments on layers of fluids heated from below were carried out by H. Bénard

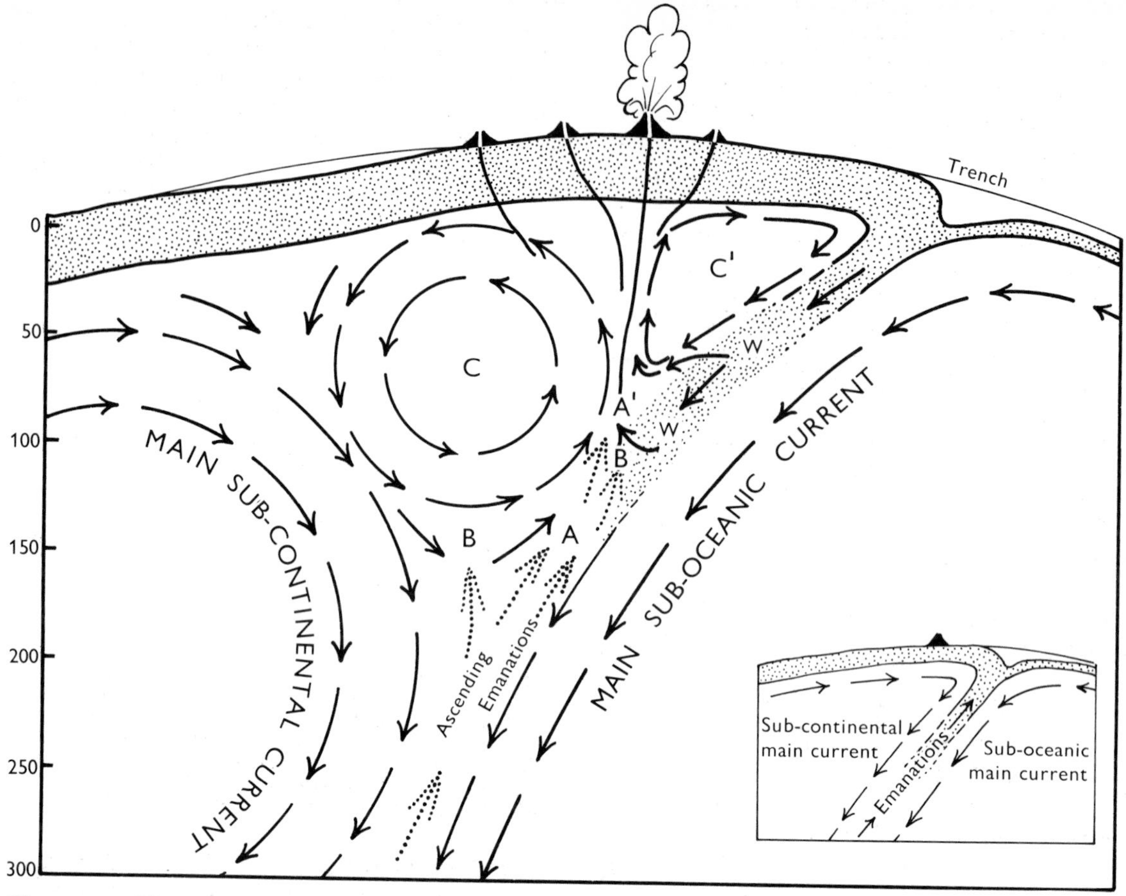


**Figure 31.13** Diagrammatic section across the Andesite Line and a typical island arc (e.g. Japan or Kamchatka, cf. Fig. 26.13) to illustrate a speculative arrangement of convection currents which might account for the origin of andesitic lavas and their associates, and for the concentration of volcanoes over the region of intermediate earthquakes.

The small inset diagram indicates the downturn of the major currents along the belt of intermediate and deep earthquakes. This dips beneath the continental crust and, being subject to continual fracturing, provides passageways through which hot emanations from great depths can ascend towards the surface. At the same time various crustal rocks, including sedimentary rocks charged with brine, are being dragged down by the main currents. These materials will be heated up: partly by strain energy liberated as heat during earthquakes; partly by radiogenic heat; but chiefly by rising emanations. Eventually the lighter and more mobile materials converge through a considerable range of depth, such as AA′, to generate lavas of the andesite–dacite–rhyolite series. Being light and highly charged with gases, these will rise in a more nearly vertical direction to feed volcanoes at the surface, and to form the radial axis of a subsidiary convection cell CC′. Most of the crustal rock of basaltic composition is likely to be transformed into eclogite and carried deeper into the mantle, before melting and ascending elsewhere. But some, in situations like B, will become basaltic magma which may either contribute to the generation of andesite or form independent basaltic volcanoes, such as commonly occur at greater distances from the trenches than the andesitic volcanoes.

in 1900. Figure 2.15 represents a section through one of the 'convection cells' into which the layer is divided when a steady state is reached, so that the heat supplied to the lower surface is carried upwards by the ascent of hotter and lighter material to the top, while an equivalent amount of cooler and heavier material sinks to the bottom. Before the circulation gets going, a certain critical temperature gradient between top and bottom must be exceeded so that the resulting density differences can overcome the resistance to movement that internal friction and other opposing factors bring into operation. When circulation is uniformly distributed, the layer is divided into cells of even size. By their interference these form a network like a honeycomb (hexagonal in ideal conditions), i.e. a pattern of polygonal cells separated by vertical walls. The breadth of the cells tends to be about three times the depth of the fluid layer, but increases with the viscosity of the latter. The stream lines of flow within each cell form a closed vortex ring, like a smoke ring (Figure 31.14). The circulation is upwards in the middle of the cell and downwards along the walls.

If convection in the earth followed this ideally simple pattern, the diameter of a cell would require to be of ocean width. Three or four major cells, like those ideally depicted in Figure 31.14, but of much more irregular form, might therefore provide a first approximation towards explaining the crustal rearrangements (continental drift and the growth of new ocean floors at the expense of the Pacific) that have occurred since the close of the Palaeozoic. But, as already indicated by Figure 31.13, subsidiary cells of much smaller size, situated between the margins of two adjoining major cells where their stream lines turn downwards, appear to be necessary to account for the circum-Pacific vulcanism.

In Bénard's later experiments a 'velocity shear' was introduced by allowing the upper part of the fluid layer to flow relatively to the bottom. This would be analogous to 'polar migration' in the earth's behaviour. The experimental effect was that each polygonal cell became drawn out into a long spindle, which might be straight or arc-shaped. The radial *axis* of Figure 31.14 thus becomes a greatly elongated radial *plane*, as in Figure 31.15, while the circuit axis is pulled out like a stretched rubber band. The circuit axis still remains closed. If it is further considered that the radial plane may be curved instead of straight, the cross-section would be banana-shaped.

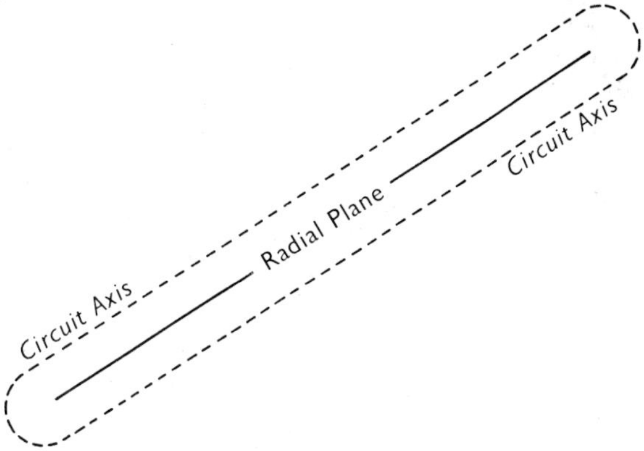


**Figure 31.15** To illustrate a convection cell extended like a stretched-out rubber band, so that the radial axis becomes a radial plane along the belt of ascending currents, while the circuit axis, though greatly elongated, remains closed. If the radial plane were curved, as it commonly would be in a globe, the pattern would be like that of an island arc and would represent in plan the cell CC′ drawn in section in Fig. 31.13.

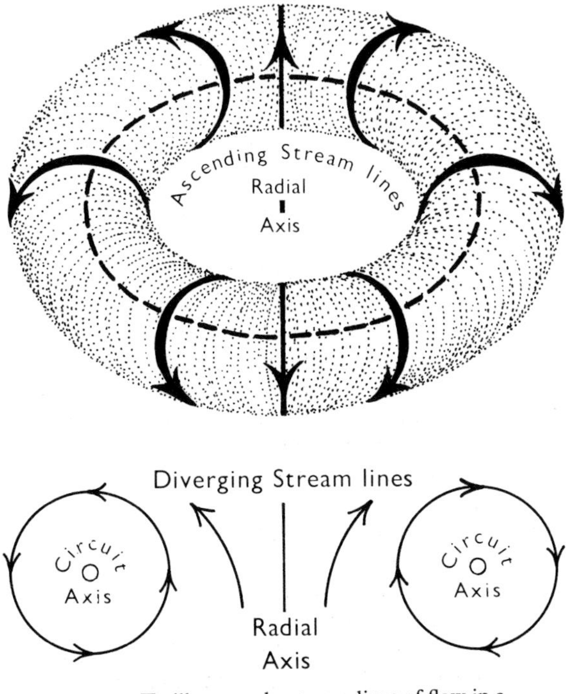


**Figure 31.14** To illustrate the stream lines of flow in a convection cell of the smoke-ring type. Assuming the cell to be parallel to the earth's surface it has a central, vertical or *radial axis*, and a closed horizontal or *circuit axis*, shown in the upper diagram by a broken line. The lower diagram is a vertical section through the cell to emphasize the fact that the two stream lines (here drawn as circles) share the same circuit axis and are part of the same convection cell.

To elucidate the effects of sub-crustal currents on the overlying crust, David Griggs, then at Harvard (1939), made a very effective series of experiments with properly scaled-down models. He used materials related to the dimensions of the models exactly as the properties of the earth's materials are related to the size of the earth. In one model, designed so that a natural process requiring a million years could be reproduced in 60 seconds, the crust was represented by a mixture of sand and heavy oil, while viscous water-glass represented the mantle. Obviously it was not practicable to produce convection currents to scale by thermal means. Instead, currents were generated mechanically by drums placed in the 'mantle' and rotated by hand at the properly scaled-down rates (Figure 31.16).

The results of Griggs' experiments (Figure 31.17) are less applicable to global tectonics as they are now known, than they were, as they were thought to be, at the time when the experiments were performed. Figure 31.17(B), however, illustrates a mechanism for bringing about oceanward

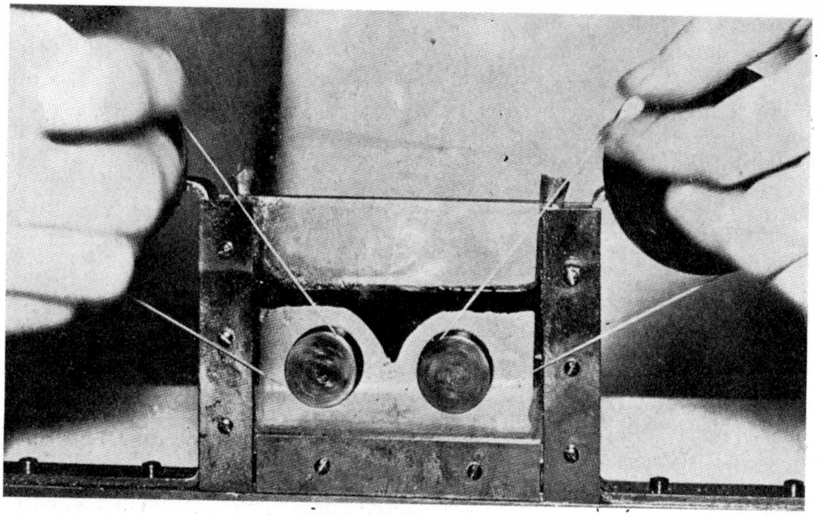

**Figure 31.16** A mechanical scale-model designed by D. Griggs to simulate the action of convection currents on the overlying crustal layers by means of rotating drums. The materials of the 'crust' (black) and the 'mantle' (light) have properties such as strength and viscosity commensurate with the small size of the model and the short time of the experiments. In the stage illustrated convergent currents have produced marked downfolding of the 'crust'. (*D. Griggs, 1939*)

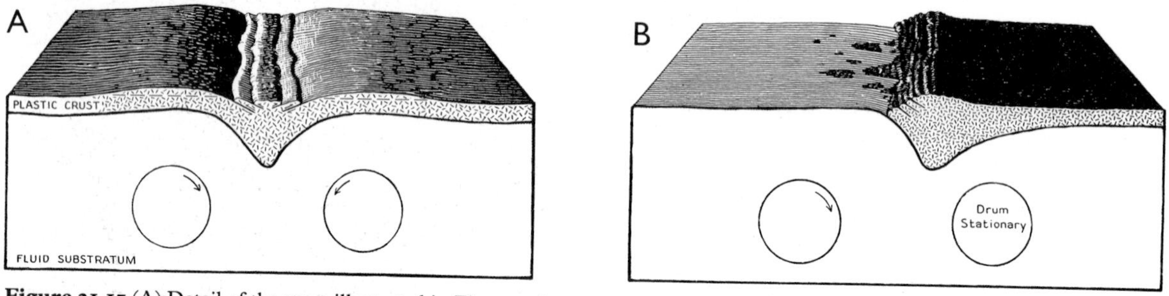


**Figure 31.17** (A) Detail of the stage illustrated in Fig. 31.16. In addition to the crustal downfold (root or tectogene) outward thrusting has developed near the surface in response to the inward drag of the crustal material imposed by the rotation of the drums.
(B) Detail of the result when only one drum was rotated, showing the capacity of a sub-crustal current to sweep the crust towards the stationary continent, which is correspondingly thickened. Outward thrusts near the surface can again be seen. (*D. Griggs, 1939*)

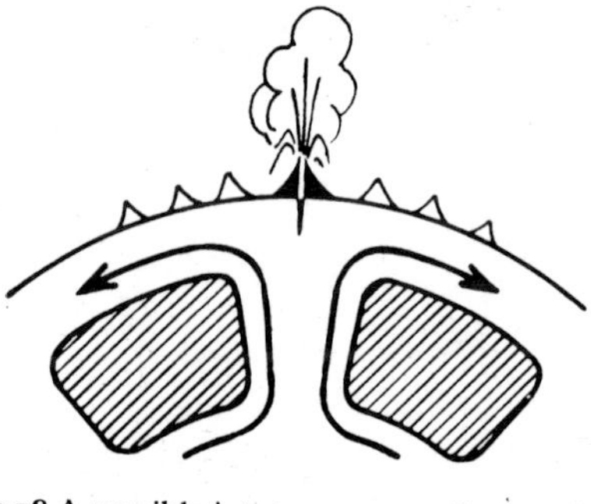

**Figure 31.18** A possible jet-stream type of convection circulating around a nearly stagnant region (shaded)

thrusting, towards a trench, when an island arc or continent forms the leading edge of a plate of lithosphere (see Figures 30.9, 30.10).

So far we have been considering what may be called *cellular convection* in a medium otherwise kept stationary. In the resulting cells the mean velocity increases with the length of the stream lines around the circuit axis, i.e. the cell tends to roll bodily around its own circuit axis, the outside moving faster than the interior, apart from frictional slowing down at the boundaries of the convecting layer. Thermal convection in a rotating fluid introduces great complexities. Here again cells are produced at first, but the convective flow tends to be increasingly confined to the outside stream lines, which thus become 'jet streams' flowing round a nearly stagnant core in which the convective motion becomes less and less towards the circuit axis (Figure 31.18). With increasing temperature differences between the hot and cold boundaries of a layer of given thickness, a major jet stream begins to develop, eventually meandering in a continuous flow of wave-like form between the warmer and the cooler sides, as can be clearly seen in Figure 31.19.

**Figure 31.19** An example of the top-surface flow pattern produced in one of R. Hide's experiments on thermal convection in rotating liquids (see text). Rotation clockwise: aluminium powder used to make the stream lines visible. The two pipes seen at one side feed the inner cylinder with water to maintain the temperature of the inner wall of the convection chamber. A 'jet stream' of four 'waves' is clearly seen, with subsidiary currents in the outer loops. The inner loops are relatively stagnant (*R. Hide, 1958*)

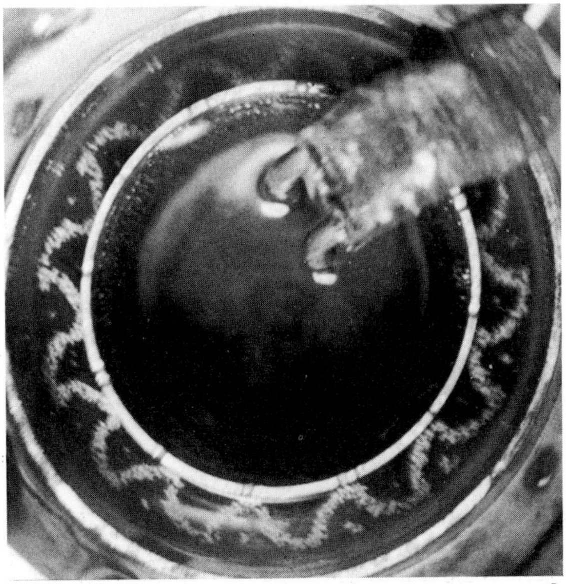

**A**

**Figure 31.20** Examples of top-surface flow patterns produced in a series of experiments with inner cylinders of different diameters. Rotation clockwise; aluminium powder used as indicator. In A the jet stream has 13 'waves'; in B there are 6 'waves'; and in C, 4 'waves'. These patterns may simulate those produced in the earth's mantle if the radius of the core at successive stages in the earth's history has been diminishing. (*R. Hide, 1958*)

The experiments that have thrown most light on *jet-stream* convection have been carried out by Hide, using three concentric cylinders mounted vertically on a horizontal turntable which could be rotated at a desired rate. The space between the inner and middle cylinders was filled with water to the required depth and used as the convecting chamber. Thermal convection was brought about by maintaining the boundaries of the convecting

**B** (*above*)

(*below*) **C**

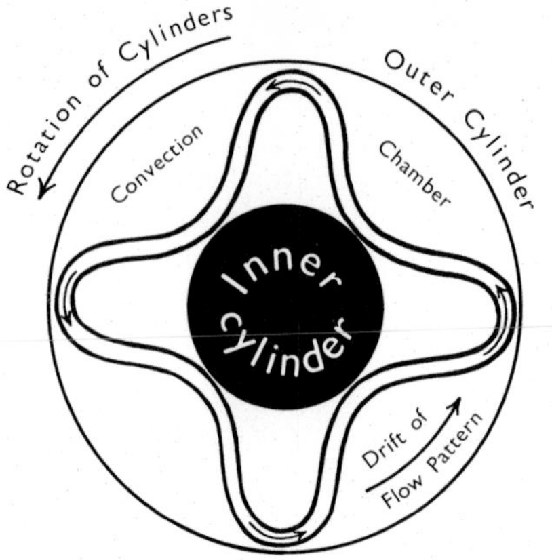


**Figure 31.21** Schematic diagram of a top-surface flow pattern, showing the directions of jet-stream flow, and the drift of the flow pattern as a whole; rotation is anti-clockwise (*After R. Hide, 1938*)

chamber at different temperatures by the controlled inflow of warm water through one of the bounding cylinders and of cold through the other. To make the convective stream lines visible, suitable dyes were introduced or, more commonly, fine aluminium powder (Figures 31.19–20). These illustrations, typical of scores of others, show the top-surface flow pattern, photographed through a 'rotoscope' coupled to the turntable, this being a device for obtaining a stationary image of a rotating system.

It is not possible to say to what extent the top-surface patterns produced in Hide's experiments correspond to those developed in the mantle: say, for example, through an equatorial section of the earth. But at least they serve to suggest possibilities differing widely from the simple cellular patterns of non-rotating layers. Hide records other details which seem likely to have far-reaching geological applications. For example, under certain conditions the wave-like pattern drifts slowly forward in the same direction as the jet stream (Figure 31.21). Behaviour of this kind in the mantle might help to account for the lateral migration of mid-oceanic ridges. With higher rates of rotation the wave-like pattern of the experiment ceases to drift progressively forward, but fluctuates or vacillates to and fro. Another possible variation during geological time is simulated by Figures 31.20, A, B, and C, which show a decreasing number of 'waves' in the convecting chamber when its width is increased by reducing the radius of the inner cylinder. If the photographs are thought of as equatorial sections through the earth, they suggest a decreasing number of 'waves' as the thickness of the mantle (the convecting chamber) increases at the expense of the core (the inner cylinder and source of energy). In this connection it may be noted in passing that mathematicians who have analysed the spherical harmonics representing the relief of the globe have reached the conclusion that mantle currents having four or five major cells or 'waves' are about equally likely at the present time. A number of minor cells would also be expected, such as CC′ in Figure 31.13, to account for the localization of volcanoes, and those implied by the rifting of continents and the formation of new ocean floor, e.g. The Red Sea and Gulf of Aden.

## Selected References

BEMMELEN, R. W. VAN, 1954, *Mountain Building*, Martinus Nijhoff, The Hague.

DEWEY, J. F., and BIRD, J. M., 1970, 'Mountain belts and the new global tectonics', *Journal of Geophysical Research*, vol. 75, pp. 2625–47.

DIRAC, P. A. M., 1938, 'A new basis of cosmology', *Proceedings of the Royal Society*, A, vol. 165, pp. 199–208.

EGYED, L., 1956, 'The change of the earth's dimensions determined from palaeogeographical data', *Geofisica Pura e Applicata*, vol. 33, pp. 42–8, Milan, Italy, (in English).

— 1957, 'A new dynamic conception of the internal constitution of the earth', *Geologische Rundschau*, vol. 46, pp. 101–21 (in English).

FLANDERN, G. T. VAN, 1975, 'A determination of the rate of change of *G*', *Monthly Notices of the Royal Astronomical Society*, vol. 170, No. 2, pp. 333–42.

— 1976, 'Is gravity getting weaker?' *Scientific American*, vol. 234, pp. 44–52.

GILBERT, C., 1956, 'Dirac's cosmology and the General Theory of Relativity', *Monthly Notices of the Royal Astronomical Society*, vol. 116, pp. 684–90; see also *Nature*, vol. 192, 1961, pp. 57 and 440.

GRIGGS, D., 1939, 'A theory of mountain building', *American Journal of Science*, vol. 237, pp. 611–50.

LANGSETH, M. G., Jr., and VON HERZEN, R. P., 1970, 'Heat flow through the floor of the world ocean', in *The Sea*, vol. 4, Part 1, Wiley-Interscience, New York.

WELLS, J. W., 1963, 'Coral growth and geochronology', *Nature*, vol. 197, pp. 948–50.

# Index

All references in this index are to pages.
Numbers in *italics* refer to pages with relevant illustrations.